FUNDAMENTALS of

BIOCHEMISTRY

AF633396

FUNDAMENTALS of
BIOCHEMISTRY

[For Medical Science, General Sciences, Veterinary Science, Home Science, Life Sciences, Agriculture, and Homoeopathy Courses]

Dr A C Deb PhD
Ex-Professor and Head, Department of Biochemistry
J. L. N. Medical College, Bhagalpur, Bihar

New Central Book Agency (P) Ltd

Co-Published by

Bijay Publications

E-mail : bijaypublications@gmail.com

Exclusively Distributed by

Bijay Pustak Bhandar (P) Ltd

90/6A M. G. Road, Bhawani Dutta Lane, Kolkata-700007

✆ : (033) 4605-7856, 8777898576, 8274916104

E-mail : bijaypustakbhandar@gmail.com

NCBA

REGD OFFICE
8/1 Chintamoni Das Lane, Kolkata 700 009, India
email: ncbakolkata@gmail.com

OVERSEAS
NCBA (UK) Ltd, 149 Park Avenue North, Northampton, NN3 2HY, UK
email: ncbauk@yahoo.com

CORPORATE OFFICE
963/8 Govind Puri, Kalkaji, New Delhi 110 019
email: delhincba@gmail.com

BIJAY PUBLICATIONS
90/6A M. G. Road, Bhawani Dutta Lane, Kolkata-7
E-mail : bijaypublications@gmail.com

BIJAY PUSTAK BHANDAR (P) LTD
90/6A M. G. Road, Bhawani Dutta Lane, Kolkata-7
E-mail : bijaypustakbhandar@gmail.com

FUNDAMENTALS OF BIOCHEMISTRY ● Dr A C Deb

© Copyright reserved by the Author

Publication, Distribution, and Promotion Rights exclusively reserved by the Publisher

All rights reserved. No part of the text in general, and the figures, diagrams, page layout, and cover design in particular, may be reproduced or transmitted in any form or by any means—electronic, mechanical, photocopying, recording, or by any information storage and retrieval system—without the prior written permission of the Publisher

First Published: 1983
Ninth Edition: 2008
Thoroughly Revised, Updated & Enlarged Tenth Edition: 2011
Reprinted: 2012, 2014, 2015, 2016, 2018, 2019, **2024**

PUBLISHER AND TYPESETTER
New Central Book Agency (P) Ltd
8/1 Chintamoni Das Lane, Kolkata 700 009

PRINTER
New Central Book Agency (P) Ltd
Web-Offset Division, Dhulagarh, Sankrail, Howrah

TECHNICAL EDITOR
Dr Mita Sen, MBBS

PROJECT SUPERVISOR
Prabir Ghosh

ISBN: 978 81 7381 144 9

Price ₹ 985.00

ॐ

Akhanda mandalakarang byaptang jena characharam
Tadpadang darshitang jena tasmai Sree Gurube namah

Consecrated to
the deep sweet memories
of my respected late father, mother and father-in-law

Dr. Sanatan Deb
Nayantara Deb
and
Dr. Nim Chand Bhadra

Contents

Preface to the ninth edition

It is really a great pleasure for the author and the publisher to be able to present the thoroughly revised and updated ninth edition of **Fundamentals of Biochemistry.**

Exhaustive research work of renowned biochemists of various countries of the world have brought remarkable advancement of Biochemistry which renders extensive benefits to teachers as well as students. The author is really glad to note that such development of Biochemistry has taken place.

Fundamentals of Biochemistry provides a bay-window to significant coverage of the principles of Biochemistry and Molecular Biology along with recent developments.

Biochemistry is essential to life sciences. Life fully depends on biochemical reactions and processes. The interrelationship of Biochemistry and Medicine in actuality is a two-way traffic. Biochemical studies have shown many aspects of health and disease. Nutrition and preventive medicine depend to a great extent on the knowledge of Biochemistry. Therefore, a sound knowledge of Biochemistry is most essential for the practice of medical and related health sciences. The present representation is sure to prove highly beneficial for students of both biochemistry and medicine, as also the practicing physicians—they would understand better the valuable principles of maintenance of health as well as the actual causes and rational treatment of different diseases.

Suggestions of readers are undoubtedly helpful for eliminating errors and updating concepts and therefore, the same are welcome.

CHANGES IN THE NINTH EDITION

The major goal of the author is to provide both medical and other students of health sciences with a textbook that not only describes the basics of Biochemistry but also reflects the most significant advances in Biochemistry that are important to medicine.

The following are the new features with updated information :

- New information on **Prions and Alzheimer's disease** in Protein chapter have been incorporated.
- A new chapter on the revolutionary ultra-new **The Human Genome Project** has been included.
- New information on Vitamins have been introduced.
- New advanced clinical information on **Atherosclerosis** and Hypertension have been included.
- New information on **High sensitive c-reactive protein** and **Hyperhomocysteinaemia** (coronary artery disease) have been introduced.
- New advanced clinical information on **Hyperlipidemia** has been included.
- Significantly updated **Normal Laboratory Values** have also been introduced.
- **The effective information on effect of whole grains in health has also been included.**

4 September 2007

A C DEB

Acknowledgements

The author is grateful to his professional colleagues and friends around the world for their ample interest, advice, and support for the improvement of this textbook. He immensely admires the dignified officials of the health and education departments of different State Governments of India as well as other countries for selecting his untiring efforts as a textbook for undergraduate and postgraduate medical students including students of related sciences.

The author heartily accepts comments from students and learners. Significant criticism and suggestions from learners of the world related to other branches of science are always highly admired. The author is quite sure that this updated and thoroughly revised edition will stimulate the learners to have updated knowledge in the vast area of Biochemistry. Further beneficial criticism and valuable suggestions will be heartily accepted and honoured. The author looks forward for receiving similar inputs in the future.

Sincere devotion and helpful cooperation have been provided by the publishers—Mr Jogendra Nath Sen and Mr Amitabha Sen, Technical Editor Smt Mita Sen (MBBS), the project supervisor Mr Prabir Ghosh, and others to successfully complete this most vital task.

The author significantly expresses his wholehearted appreciation to his wife Smt Bela Deb (MSc, BEd), his daughter-in-law Smt Debarna Deb (MBA), his brother-in-law Mr Rathindra Nath Bhadra, and sister-in-law's son Mr Sourav Das (BE), for their essential assistance throughout the preparation of this edition. It is heartening for the author to acknowledge the importance of his son Dr Pradipta Deb, MBBS (Cal), for his wholehearted untiring efforts in inserting updated information and continuous toil for highly essential modifications throughout the preparation of this entire edition.

CHAPTER 1

INTRODUCTION

Biochemistry deals with the chemistry of living organisms — both plant and animal. Protoplasm is the basis of all forms of life. Although the protoplasm of each different kind of cell in each kind of animal or plant is different in characteristic yet the chemical composition, organization and chemical processes in these many different forms of protoplasm are similar in many respects. Animal biochemistry is significantly developed by the assistance of investigations of the chemical processes of plants and microorganisms, and vice versa. The observations on various chemical reactions first made in microorganisms or plants have been finally brought to light in higher animals and the reverse is followed in the true sense.

Biochemistry can be divided broadly into four branches : (i) Medical biochemistry, (ii) Animal biochemistry, (iii) Plant biochemistry, and (iv) Biochemistry of microorganisms.

Medical biochemistry deals with the following with which the medical students are mainly concerned :

Chemistry of Tissues and Foods

Since the foods are largely derived from animals or plant tissue, the study of the chemistry of one is identical with the other. The components of tissues and foods are carbohydrates, fats, proteins, and nucleic acids. Carbohydrates provide a greater part of the energy-needs for the body. Improper utilization of glucose leads to the development of diseases like diabetes mellitus. Lipids provide a part of the energy-needs of the body. Essential fatty acids are required for maintaining normal health. These essential fatty acids are also the precursors of prostaglandins. Complex lipids occur as integral parts of nervous tissue. Proteins are necessary for growth and maintenance of positive nitrogen balance. Nucleic acids (DNA & RNA) of nucleus and cytoplasm, respectively, are involved in the regulation of gene.

Chemistry of Digestion and Absorption

The food is chiefly composed of large molecules of starch, proteins, and fats. These large molecules are digested into smaller molecules like glucose, amino acids, and fatty acids, for absorption by the organic catalysts (the enzymes) present in the digestive juice. Along with their absorption, the entry of water, minerals, vitamins, and other diffusible molecules of the food also takes place. The absence of one of the digestive enzymes leads to serious disorders.

Chemistry of Respiration

During respiration, in man, oxygen is taken into the lungs and it diffuses into the blood across the membranes. It combines with the haemoglobin of red cells and, in this form, it is carried to the tissues where it is released for the oxidation of food to liberate energy. The carbon dioxide—formed equal in volume to that of oxygen—passes from the tissues to the blood stream and is exhaled from the lungs, being reacted by chemical combination.

Chemistry of Blood

The blood carries foods to the tissues and waste products from them to the excretory organ. The hormones produced by various glands pass into the blood and through its circulation these hormones reach the specific tissues. The blood distributes heat from one part of the body to another and exerts cooling effect. The blood also distributes water and salts properly and maintains acid-base balance of the body. It contains substances that combat infection by microorganisms. The composition of blood is normally constant but it is dynamic. Even one of the substances present in it, increased or decreased, causes pathological condition.

Chemistry of Cell Membrane and Physical Chemistry

Hormones, foods, waste products, and other necessary substances required for cells pass through the cell membrane. Some substances easily pass through

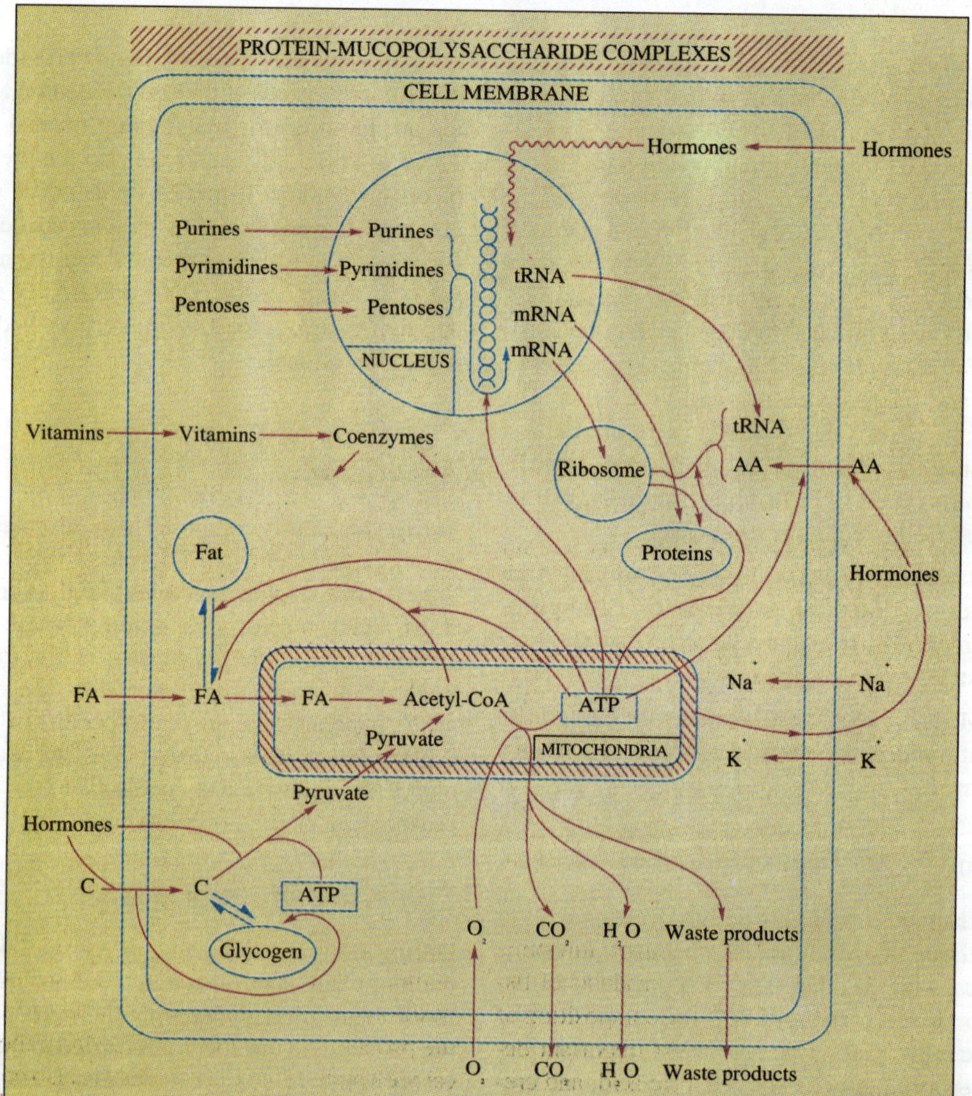

Fig. 1.1 : Schematic representation of Biochemistry in brief

the membranes but some other substances pass by other mechanisms. Normal permeability of cell membranes maintains normal physiological processes and abnormal permeability is associated with pathological conditions.

The colloidal substances which do not diffuse through membranes exhibit Donnan membrane equilibrium resulting in secretion and excretion of substances. The buffering system is most essential for maintaining pH of blood slightly alkaline for normal functioning of the body. The electrical potentials across cell membranes is of much significance. The ion exchange across the cell membrane maintains the cellular integrity. The pH of blood, if altered from the normal value, leads to complicated disorders.

Chemistry of Tissue Metabolism

The oxidation of foodstuff in tissues occurs by chemical processes with the liberation of energy and water. This is a very complicated process in biochemistry. Several diseases occur in the disorder of the metabolism of these foodstuffs.

Chemistry of Glands of Internal Secretion

The control through glands of internal secretion is barely dependent on the activities of hormones and nervous system. The hormone of one gland regulates the activity of another gland. Biological achievement is recorded by the biochemistry of the glands of internal secretion—this is one of the most significant chapters. The improvement of this chapter has influenced the effective agents of treating diseases. Overproduction of hormones also leads to serious disorders.

Chemistry of Excretion

The excretory organs—kidneys, lungs, intestine, and skin—remove decomposition products of tissues and foods in order to make the composition of the body fluids constant. The most important decomposition products are urea, uric acid, and creatinine formed from proteins; carbon dioxide and water formed from carbohydrates, fats, and proteins. The particular level of these decomposition products is compatible with health but, on excess, may lead to dysfunction and illness. The kidneys and the lungs chiefly perform the excretory functions of the body. The medical students must be quite aware of the concentration of the excretory products which may indicate certain diseases. In severe diabetes mellitus, sugar is excreted to the urine as a result of high blood sugar level and along with this sugar the kidney is to excrete more water also. Consequently, there develops both hunger and thirst. These symptoms give a hint to the disease.

Biochemical Disorders in Diseases

The knowledge of biochemical disorders in various diseases aids in the correct diagnosis and treatment of the disease. Blood glucose level increases and the presence of sugar in the urine indicates diabetes mellitus. The severity of the disease is assessed by the concentration of blood sugar and urine sugar. So the treatment is instituted by antidiabetic drugs. Jaundice is accompanied by a high level of bile pigments in blood and tissues. In kidney failure, urea, uric acid, etc. in blood are highly increased leading to uremia. The functions of kidneys, liver, pancreas, small intestine, etc. are determined by biochemical tests.

From the above accounts, it is quite evident that knowledge of biochemistry is most essential for understanding the functioning of human body in health and disease. From the time of Aristotle, students of biology and medicine have endeavoured to correlate structure and function.

Importance of Biochemistry in medicine

Biochemistry is important to physiology, and other medical subjects. Both physiology and biochemistry overlap and merge. Pathological conditions in the body are caused by deranged chemical compositions and functioning of tissues and many of the problems of pathology occur from the chemical viewpoint. The bacteriologist is also concerned with the chemical changes caused by bacteria in tissues resulting in various diseased conditions. He is to have ideas of vaccines, serum, and antitoxins. The pharmacologist must know the chemical aspects of the body since the action of drugs always involves some alterations in the biochemical events occurring in the body. The physicians too have to acquire knowledge of biochemical changes of different foodstuffs, hormones, and vitamins, etc. to diagnose a disease properly for its cure. They have also to depend on the large number of biochemical tests for treating diseases.

STRUCTURE AND FUNCTIONS OF CELL CONSTITUENTS

The cell is the basic structural and functional unit of all living organisms. Very recent studies have revealed the use of electron microscope and differential centrifugation to find out the presence of various structural units of the cell with a view to have an idea of their functions too. The structure of an animal cell is shown in Figure 1.2.

Cell Constituents

1. Cell membrane (plasma membrane or unit membrane).
2. Nucleus.
3. Mitochondria.
4. Endoplasmic reticulum.
5. Ribosomes or microsomes.
6. Centrioles.
7. Lysosomes.
8. Cytosol.

1. Cell Membrane

(i) The boundary of every cell has a thin membrane of thickness 75Å units (although the thickness varies from 70Å to 100Å). This is also known as unit membrane or plasma membrane which is formed by the ingredients of the plasma (cytoplasm).

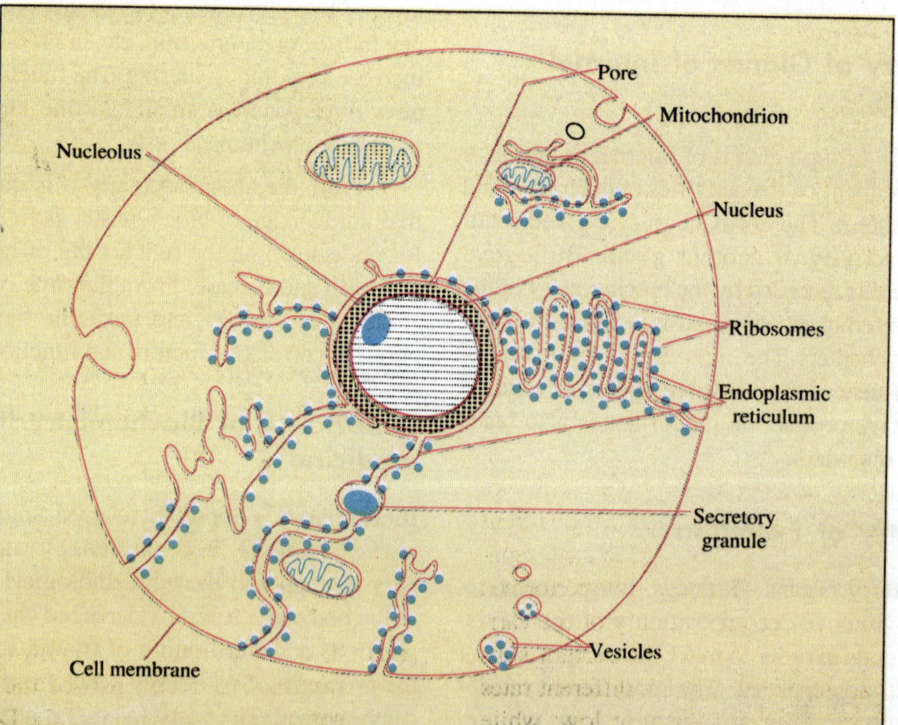

Fig. 1.2 : Structure of animal cell

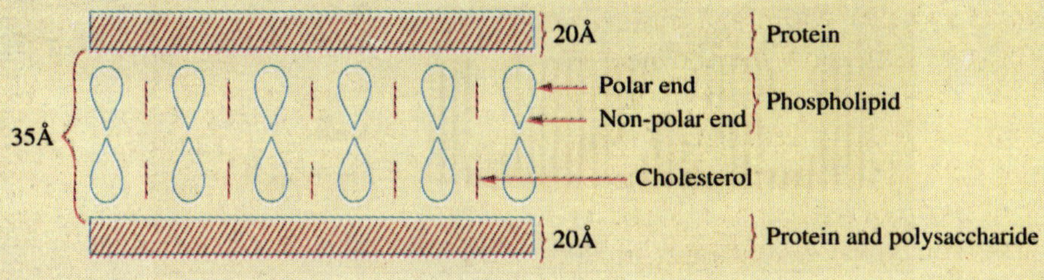

Fig. 1.3 : A cross-section of cell membrane

(ii) This membrane consists of three layers : (a) Inner layer (20Å) of proteins, (b) Middle layer (35Å) consisting of phospholipids and cholesterol, and (c) The outer layer (20Å) consisting of protein and polysaccharide. These are arranged continuously to form pores of about 8Å.

The phospholipid molecules in the light band are arranged in two rows that the phosphate containing ends (the polar or hydrophobic ends) point to the outside while the non-polar or hydrophobic ends point to the inside.

(iii) The orientation and nature of the lipid-protein components in the cell membrane determines the charge in the membrane and movement of molecules across the membrane.

(iv) The surface of the cell membrane bears pores, of internal diameter 3Å and length about 75Å, which lead from cytoplasm to the exterior.

(v) The most important function of biologic membranes is to restrict the exchange of substances among various compartments. Thus different types of body fluids are separated from each other by different membranes, eg, R.B.C. membrane separates plasma from R.B.C. This is performed by the selective permeability of the various membranes for which certain substances are passed through different rates —namely high, moderate or low; while others are prevented from passing through.

2. Nucleus

(i) Nucleus is the heaviest component of the cell and is separated from the cytoplasm by a double membrane. The outer membrane is continuous with the endoplasmic reticulum. The two membranes are separated from each other by the perinuclear cisterns (150Å) and join at intervals to form pores which allow the passage of materials from the cytoplasm to the nucleus, and vice versa.

(ii) The pores on nuclear membranes are of higher diameter than those on the cell walls and allow the passage of larger molecules such as protein, nucleic acids, etc.

(iii) The nucleolus within the nucleus contains ribonucleic acid (RNA) in granular form.

(iv) "Chromatin", the regular clumps of thread-like material, distributed throughout the nucleus, is the most important component of the nucleus. This consists of deoxyribonucleic acid (DNA).

(v) The chromatin, immediately before cell division, organises into simple small thread-like structures known as "chromosomes", which consist of nucleoprotein molecules. Most of the DNA present in the nucleus occur exclusively in chromosomes.

(vi) The gene or cistron, the unit of genetic information, forms part of the chromosomes and is a segment of the DNA molecule (about 600 bases). The genes are,

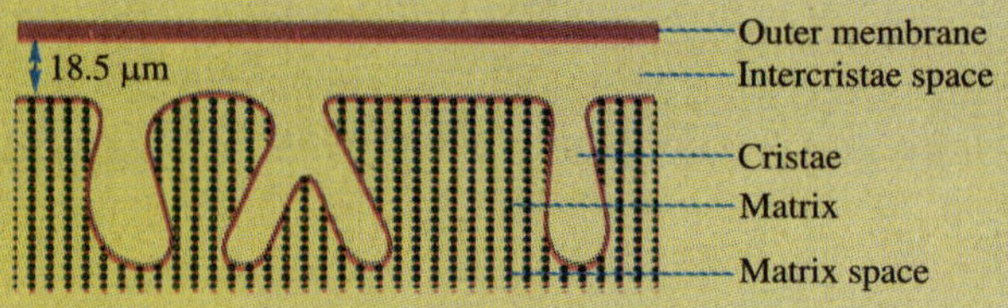

Fig. 1.4 : Structure of Mitrochondria

therefore, arranged in an orderly manner along the length of the DNA molecule in the chromosomes.

(vii) The DNA molecules form the template for the synthesis of RNA which is later on transported into the cytoplasm and the ribosomes. The RNA molecules then synthesize different types of proteins (hormones and enzymes). Since the enzymes regulate the metabolic processes of the body nucleus is, therefore, said to be governing the metabolic processes in the true sense.

(viii) In a nutshell, the nucleus contains proteins (about 15 per cent of the cell), RNA (about 30 per cent of the cell of which 20 per cent in nucleolus and 10 per cent in chromosomes), and DNA.

3. Mitochondria

(i) Mitochondria are the largest components in cytoplasm.

(ii) They contain about 35 per cent of the total protein of the liver. In addition to proteins, they also contain lipids (25 per cent) most of which (about two-thirds) are phospholipids, and a small amount of nucleic acids.

(iii) They are the "power house of the cell" and each cell may contain 50 to 2,500 mitochondria depending on the respiratory activity of the cell. The cells of skeletal muscle, kidney, and liver contain large number of mitochondria while those of heart muscle contain less.

(iv) They vary in shape (spherical, filamental or sausage shaped) and in size (0.5 to 3μ long and 0.1 to 0.6μ wide).

(v) They have two membranes – the outer membrane is smooth while the inner membrane is folded to form ridges or cristae which extend into the matrix of the mitochondrion. Two spaces – the intracristae space and the matrix space are thereby developed. The matrix space is rich in enzymes; while the intracristae space contains substances of low molecular weight in solution. The outer surface of the outer membrane and inner surface of the inner membrane possess indefinite protuberances or saucer-like structures which are rich in proteins, most of which are the enzymes in biologic oxidation namely NAD– and NADP– dependent dehydrogenases, flavin-linked enzymes, and cytochromes.

(vi) Mitochondria swell extensively in a hypotonic medium and contract again in a hypertonic medium, faster with the addition of ATP. This mechanism facilitates the exchange of metabolites.

(vii) The mitochondria contain a large number of enzyme systems known as "cyclophorase" which are involved in : (1) oxidation of pyruvic acid, acetyl–CoA through Kreb's cycle, fatty acids, amino acids; (b) electron transport and oxidative phosphorylation; (c) synthesis of fatty acids. They also contain enzymes concerned with the biosynthesis of amino acids, porphyrin, phosphatides, heme, hippuric acid, and urea. They also contain transhydrogenases, adenylate kinase, glutaminase, and the enzymes of phosphorylation of nucleoside diphosphatase and carboxylations.

4. Endoplasmic reticulum

(i) This is the complicated and organised system of membranes in the cytoplasm of the cell. This membrane is constituted of protein lipid double layers and is very well-developed in tissues with active protein synthesis.

(ii) The Golgi bodies serve as a means of producing and maintaining the internal membrane. The interior of this membrane is connected with perinuclear space and sometimes also with the extracellular space. The cisternae (enclosed spaces) of this endoplasmic reticulum play a role in the exchange of material between the cell and the extracellular fluid. This exchange of material takes place by the processes of diffusion, active transport and pinocytosis.

(iii) This and the surrounding matrix are not only meant for specific metabolic activity, but also provide the means of communication between the nucleus, mitochondria and ribosomes.

(iv) The important functions of this membrane and cytoplasm matrix are :

(a) Transport of nutrients and metabolites from in and out of the cell ; (b) Regulation of protein synthesis, (c) Glycolysis; (d) Glycogenesis; (e) Pentose phosphate pathway; (f) Fatty acid synthesis.

5. Ribosomes

(i) Ribosomes or microsomes are small granules on the outsides of the membranes of the endoplasmic reticulum.

(ii) The size of the ribosomes ranges from 15 to 20 millimicrons (10^{-6} mm); the diameter being 150Å.

(iii) They are composed of ribonucleic acid (RNA) protein complex. They also contain lipids, mainly phospholipids.

(iv) They occur separately or in a polysomal cluster. They consist mainly of particles of two sizes—one with a sedimentation constant of 50S (S=Svedberg unit) and the other with 30S. The combination of the two particles is 70S. They are the main site of protein synthesis.

(v) They also possess steroid reductase (responsible for metabolism of cholesterol and steroid hormones), phosphatases, hydrolases, hydroxylases, glucuronyl transferase, and ATPase.

(vi) The Golgi apparatus acts as the store house of various hormones and secretory enzymes which are released at the time of requirement by the process of emeiocytosis.

6. Centrioles

(i) Centrioles are two short cylindrical structures known to exist on either side of the nucleus at right angles to each other.

(ii) They are not bound by any membrane.

(iii) They help in the equal division of the chromosomes by taking them apart and they are, thereby, responsible for equal distribution of the characteristics in the offsprings.

7. Lysosomes

(i) The size (mean diameter 0.4μ) of the lysosomes is in-between that of microsomes and mitochondria. They are surrounded by a lipoprotein membrane.

(ii) Since they contain hydrolytic enzymes they are named lysosomes.

(iii) They contain digestive enzymes such as cathepsin, acid phosphatase, ribonuclease, deoxyribonuclease, catalase, collagenase, α-glucosidase, β-galactosidase, α-mannosidase, phosphoprotein phosphatase.

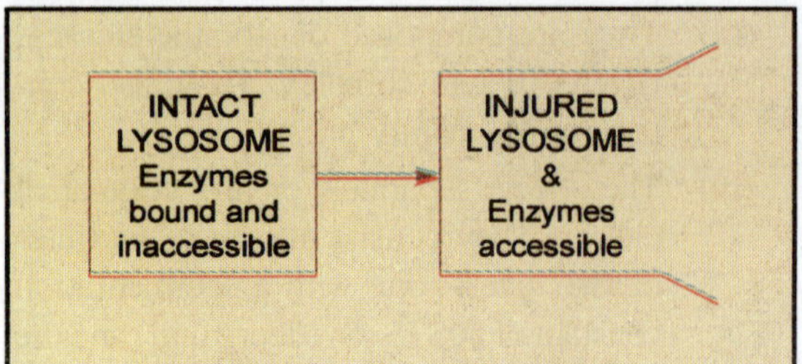

Fig. 1.5 : The Lysosome

(iv) Unless the lipoprotein membrane of the lysosomes are ruptured, the enzymes are not released into the cytoplasm. Hydrolases in lysosomes are separated from their substrates by means of lipoprotein membrane—only when this membrane is ruptured, the hydrolases enter the cytoplasm and destroy the bacteria or foreign particles when the latter enter the cells. The rupture of the lipoprotein membrane takes place under conditions of cytolysis which include fat solvents, detergents, protease, lecithinase, acid pH, and high temperature.

8. Cytosol

(i) Lardy introduced the term 'cytosol' for the cell sap. It is the unstructured "soluble" portion of the cell. It is the 'supernatant' that remains after centrifugation of mitochondria, microsomes, lysosomes, and nuclei. The supernatant is a complex mixture of proteins (35-40 per cent of the total cell), RNA, and other organic or inorganic compounds of low molecular weight.

(ii) It mainly contains the enzyme systems involved in : (a) Glycolysis, (b) Fatty acid synthesis, (c) Pentose phosphate pathway, (d) Activation of amino acids in protein synthesis, and (e) Catabolism of purines and pyrimidines.

Separation of Structural Units of the Cell

(i) The cells are disintegrated in the isotonic media by differential centrifugation to separate the structural units of the cell.

(ii) Potter-Elvehjem homogenizer or similar equipment is used to prepare the tissue homogenates with the help of 0.25M. sucrose solution at 0°C.

(iii) The homogenates are then separated by differential centrifugation below 4°C. The results of separation are given :

Table 1 : Separation of structural units of the cell

Centrifugal field	Time in minutes	Structural units
700 g.	10	Nuclei, cell membrane, and erythrocytes
5,000 g.	10	Mitochondria and Lysosomes
57,000 g.	60	Microsomes and Lysosomes
1,50,000 g.	30	Ribosomes
Unsedimented		Cytosol

Necessity of knowledge of biochemistry to all life sciences

Physiology, which deals with the body functions, matches biochemistry almost completely. Immunology utilizes many biochemical techniques and the biochemists use many immunologic approaches. Pharmacology and pharmacy mainly depend on a sound knowledge of biochemistry and physiology. Most drugs are metabolized by enzyme catalyzed reactions and the complex interactions among drugs are best understood biochemically. Poisons of toxicological affairs act on biochemical reactions or processes. Biochemical approaches are highly needed to study basic aspects of pathology such as inflammation, cell injury, and cancer. Many researchers in microbiology, zoology, and botany use biochemical approaches. In fact, biochemistry has become their common language.

Effect of biochemical research on nutrition and preventive medicine

The optimal dietary intake of a number of chemicals such as vitamins, certain amino acids, certain fatty acids, various minerals, and water are most essential for the maintenance of health. Much of the subject matter of biochemistry and nutrition is concerned with the study of various aspects of these chemicals and these two sciences have a very close relation. Dental science also has close relation with nutrition. More care is given to preventive medicine to minimise the rising costs of medical care with a view to maintain health. The cited example for this is the prevention of atherosclerosis and cancer.

Causes of diseases by various biochemical mechanisms in the body

All diseases are abnormalities of molecules, chemical reactions, or processes. The main factors responsible for causing diseases in animals and human are mentioned below and all of them affect one or more critical chemical reactions in the body.

(a) Physical agents : Extremes of temperature, sudden changes in atmospheric pressure, radiation, and electric shock.

(b) Chemical agents : Certain toxic compounds and therapeutic drugs.

(c) Biological agents : Viruses, bacteria, fungi etc.

(d) Lack of Oxygen : Loss of blood supply, poisoning of the oxidative enzymes.

(e) Immunologic reactions : Anaphylaxis, autoimmune disease.

(f) Nutritional imbalance : Deficiencies, excesses.

(g) Genetic disorders : congenital, molecular.

(h) Endocrine imbalances : Hormonal deficiencies, excesses.

Relationship between biochemistry and medicine

Biochemical studies have enlightened many aspects of health and disease and, conversely, the study of various aspects of health and disease has opened up new areas of biochemistry. For example, a knowledge of protein structure and function was essential to identify the single biochemical difference between normal and sickle cell hemoglobin. On the other hand, analysis of sickle cell hemoglobin has contributed significantly to the understanding of the structure and function of both normal hemoglobin and other proteins.

Garrod studied patients with a number of disorders such as alkaptonuria, albinism, cystinuria, pentosuria and established these conditions which were genetically determined. He designated these conditions as inborn errors of metabolism His dedication provided a strong foundation for these developments in the field of human biochemical genetics. So long medical treatment is firmly based on the knowledge of biochemistry and other basic sciences such as physiology, microbiology, and nutrition, the practice of medicine can be implemented to accommodate new knowledge.

Contribution of biochemical studies to diagnosis, prognosis, and treatment

(a) Vitamins—the complex organic molecules, are most essential to be ingested by humans to maintain health. The deficiency of a particular vitamin in the diet causes deficiency disease such as scurvy or rickets. Lack of intake of vitamins in animal and human cells are the concern of biochemists and nutritionists. Once the disease from a vitamin deficiency is established, it becomes necessary to treat it by the administration of appropriate vitamin.

(b) Many plants in Africa are deficient in one or more essential amino acids. Persons who consume these plants as dietary source suffer from malnutrition i.e. Kwashiorkor. This shows that a well-balanced diet containing sufficient amount of all the essential amino acids must be provided to maintain health.

(c) Biochemical findings have shown that diet containing fish oils rich in certain **polyunsaturated fatty acids** can reduce the plasma levels of cholesterol resulting in the low incidence of atherosclerosis. These observations have stimulated interest in the use of polyunsaturated fatty acids to reduce plasma levels of cholesterol.

(d) The defect in the conversion of phenylalanine to tyrosine causes **phenylketonuria** due to the lack of an enzyme. If this condition is left untreated, it may lead to severe mental retardation in infancy. Biochemical screening tests for diagnosing phyenylketonuria at birth is a must for effective treatment to be started immediatedly by prescribing diet low in phenylalanine.

(e) **Cystic fibrosis** is a common genetic disease of the exocrine glands and of the eccrine sweat glands. Chloride level is elevated in the sweats of the persons suffering from this disease and the victims often die at an early age from lung infections. The normal gene codes for a transmembrane protein containing 1,480 amino acids in length which functions as a chloride channel. But in case of this disease, the transmembrane protein lacks in amino acids number 508, a phenylalanine residue. It has now become possible to design drugs that can correct the abnormality in the transmembrane protein. It may also be possible to introduce the normal gene into lung cells by gene therapy.

(f) Biochemical analysis has also been done on the mechanism of the action of the bacterial toxin that causes cholera and has also provided direction as to how the clinical manifestations of this disease having copious diarrhoea and loss of salt and water are brought about.

Biochemical studies illuminate disease mechanisms and diseases inspire biochemical research

In the early 1900s, inborn errors of metabolism had stimulated the investigation of the biochemical pathways affected in these conditions. Attempts to understand the basis of the genetic disease known as **familial hypercholesterolemia**—which is caused in severe atherosclerosis at an early age—have led to dramatic progress in knowledge of cell receptors and of mechanisms of uptake of cholesterol into cells. The studies of **oncogenes** in cancer cells have turned attention to the molecular mechanisms involved in the control of normal cell growth.

Conclusion

1. Biochemistry is the science that studies various molecules occurring in living cells and organisms as well as their chemical reactions.
2. Life fully depends on biochemical reactions and biochemistry has become the basic language of all biologic sciences.
3. Biochemistry is concerned with the whole spectrum of life forms–from simple viruses and bacteria to complex human beings.
4. Biochemistry and medicine are closely related. Health depends on the proper balance of biochemical reactions occurring in the body and disease indicates abnormalities in biochemical reactions or biochemical processes.
5. Advancement of biochemical knowledge has enlightened many areas of medicine.
6. Biochemical approaches have directed the causes of disease and in designing suitable therapies.
7. The different biochemical laboratory tests help the proper diagnosis and satisfactory treatment.
8. The perfect knowledge of biochemistry is most essential for the practice of medical and related health sciences.

CHAPTER 2

BIOPHYSICS

ACIDITY AND ALKALINITY

1. The reaction of the tissues and tissue fluids of the body is nearly natural excepting gastric juice. The slight changes in reaction have a significant physiological effect. In case the reaction of the solution is made slightly more acid or more alkaline, the heart stops.

2. Acidity is considered by an excess of hydrogen ions (H^+) over hydroxyl ions (OH^-) and alkalinity by an excess of OH^- over H^+.

3. The intensity of acidity depends upon the amount of hydrogen ions in excess, and the intensity of alkalinity depends upon the excess of hydroxyl ions. If the amounts of hydrogen and hydroxyl ions are the same, the solution is neutral. Pure water is a neutral solution and it contains H^+ and OH^- in equal proportions. It is never free from H^+ and OH^-. The ions are formed by ionisation, thus :

$$H_2O \rightleftharpoons H^+ + OH^-$$

4. The hydrogen ion (H^+ or proton) with its high ratio of charge cannot exist free in aqueous solution. It becomes associated with one or more water molecules to form a hydrated hydrogen ion called hydroxynium ion (H_3O^+).

5. The dissociation of an acid should be written as :

$$H_2O + HA \rightleftharpoons H_3O^+ + A^-$$

But it is convenient to neglect the hydration and must be written as :

$$HA \rightleftharpoons H^+ + A^-$$

6. The concentration of H^+ can be increased without increasing OH^- by adding an acid which ionises to form H^+. Example :

$$HCl \rightleftharpoons H^+ + Cl^-$$

Similarly, the concentration of OH^- can be increased without increasing H^+ by adding an alkali. Example :

$$NaOH \rightleftharpoons Na^+ + OH^-$$

7. It must be noted that even in the most acid solutions there are always some OH^- and in alkaline solutions there are always some H^+.

8. Aqueous solutions always contain both H^+ and OH^- because water always ionises to some extent.

9. The ionisation of water is a reversible reaction admitting the Law of Mass Action. This clearly explains that the product of the concentrations of H^+ and OH^- maintains a constant ratio to the concentration of water in any system. This is expressed as :

$$[H^+] \times [OH^-] = K[H_2O]$$

where K = constant, [ ] = concentration.

10. If H^+ is added to this above system, the left hand side of the equation will be increased. To reestablish equilibrium, OH^- must decrease by combining with H^+ to form water until the equation is balanced. In other words, if H^+ are added to a solution, the ionisation of water will be depressed and there will be fewer OH^- then in pure water. Simi-

larly, addition of OH^- will depress the ionisation of water and reduce the concentration of H^+.

11. As a matter of fact, water is ionised to such a small extent that even if all the H^+ and OH^- are recombined to form water, the increase in the concentration of water molecules will be negligible. Considering the effect of the addition of H^+ or OH^- to a system, it can be neglected that the addition will cause a minute increase in the concentration of H_2O.

12. In pure water or a perfectly neutral solution, $[H^+] = [OH^-]$. Hence $[H^+]^2 = Kw$.

13. If $[H^+]$ is greater than 10^{-7} (e.g. 10^{-4}), the solution is acid. If $[H^+]$ is less than 10^{-7} (e.g. 10^{-11}), the solution is alkaline.

14. HCl and NaOH are practically 100% ionised. Solutions of these substances are almost completely ionised when the solutions are more diluted than N/10. The ionisation is not complete at concentration greater than this.

15. With weak acids (eg, acetic acid) 100% dissociation is attained when the solution is weaker than N/10. Therefore, N/10 acetic acid is much less acid than N/10 HCl.

16. The reactions of all biological fluids occur between $[H^+]10^{-1}$ and 10^{-10}.

INDICATORS

Definition

Certain dyes which can determine the hydrogen ions in a solution are said to be Indicators.

Characteristics

1. The dyes are weak acids (or, less commonly, weak bases).

2. They have the property of dissociating in solution to yield ions of different colours from the undissociated molecules.

3. The pH at which dissociation will occur depends upon the strength of the acid; the stronger the acid the lower the pH at which it dissociates.

4. The free acid of methyl orange, which is red, can be represented as MH. It is dissociated as below on dissolving in water :

$$MH \rightleftharpoons H^+ + M^-$$

M^- being a yellow ion, the colour of the solution will depend upon the extent of ionisation, so that in very dilute solution it will be yellow.

5. It may be orange in stronger solutions with incomplete dissociation, or red with no dissociation.

6. If acid (i.e. H^+) is added, the ionisation will be depressed and MH will be reformed and the colour will be red. But if alkali (i.e. OH^-) is added, the OH^- will combine with H^+ to form water and MH will ionise to great extent. Hence, the colour will be yellow.

Uses

1. Indicators are used to determine the titratable acidity or alkalinity.

2. When an indicator gives its intermediate colour (i.e. it is partially dissociated), it can be easily assumed that the pH of the solution is at about the middle of the range.

3. Solutions of mixed indicators which give a number of colour changes over a wide range of pH are called "universal" indicators. They are used for indicating quickly in what part of the pH range the pH of a given solution lies. For accurate determination of pH, the "pH meter" is used.

4. The following table gives a list of some common indicators which are used for the biological pH range :

Indicator	pH range	Colour change
Methyl orange	3.1-4.4	Red → Yellow
Bromophenol blue	3.0-4.6	Yellow → Blue
Congo red	3.0-5.0	Blue → Red
Topfer's reagent	2.9-4.0	Red → Yellow
Methyl red	4.3-6.3	Red → Yellow
Phenol red	6.8-8.4	Yellow → Red
Thymol blue	8.0-9.6	Yellow → Blue
Phenolphthalein	8.3-10.0	Colourless → Red

pH

Definition

pH is defined as the negative of logarithm (base 10) of the hydrogen ion concentration.

Or,

It is defined as the logarithm of the reciprocal of the hydrogen ion concentration.

i.e. $pH = -\log [H^+]$, or $pH = \log \frac{1}{[H^+]}$

All vital activities are affected by H^+ concentration. Hydrogen ion concentration must be ascertained before the pH is calculated. For strong electrolytes, $[H^+]$ may be substantially the same as the total concentration, if complete ionisation is assumed. But for weak electrolytes, $[H^+]$ must be obtained by the calculation from the ionisation constant.

Variation in pH and interpretation

A solution of pH 3 contains 10^{-3} gram H^+ per litre.

A solution of pH 5 contains 10^{-5} gram H^+ per litre.

A solution of pH 8 contains 10^{-8} gram H^+ per litre.

A solution of pH 3 has 10 times the $[H^+]$ of one of pH 4 and 100 times that of a solution of pH 5. As $[H^+]$ increases, pH decreases in such a way that for one unit increase in pH the $[H^+]$ increases 10 times. It means that the higher the pH the lower will be the acidity.

A neutral solution has pH 7. Pure water (the neutral solution) is ionised as follows :

$$[H^+] \times [OH^-] = Kw$$

or, $[H^+] \times [OH^-] = 1 \times 10^{-14}$

Putting logarithm on both sides,

$\log [H^+] + \log [OH^-] = \log 1 + \log 10^{-14}$.

or, $-\log[H^+] - \log[OH^-] = 0 + 14 \log 10$. [multiplying both sides by – sign]

or, $pH + pOH = 14$ [Since, pH = 7].

$\therefore$ $pH = pOH$ i.e. $[H^+] = [OH^-]$.

Therefore, a solution with pH less than 7 is acid and higher than 7 is alkaline. The pH range is 0 to 14 only. It should be noted that the pH scale is logarithmic, not numerical.

pH 6.5 does not represent a $[H^+]$ half-way between 6 and 7.

Actually, pH 6.5 = $[H^+]$ 3.2×10^{-7}

pH 6.0 = $[H^+]$ 10×10^{-7} = $[H^+]$ 10^{-7}

Determination of pH

For weak electrolytes with which the physiologic chemistry is concerned, this may be calculated by the law of mass action :

$$HA = [H^+] + [A^-]$$

$$\therefore \frac{[H^+][A^-]}{[HA]} = K_a$$

HA = Undissociated weak acid, K_a = Dissociation constant for the acid.

HENDERSON-HASSELBALCH EQUATION

This concerns the dissociation of weak acid in equilibrium.

Let us consider HA, a weak acid that ionises as follows :

$$HA \rightleftharpoons [H^+] + [A^-]$$

$\therefore K_a = \frac{[H^+][A^-]}{[HA]}$ [$\because$ Ka = Dissociation constant]

$$\therefore [H^+] = \frac{K_a \times [HA]}{[A^-]}$$

$\therefore \log[H^+] = \log K_a + \log \frac{[HA]}{[A^-]}$ [Taking log on both sides].

$\therefore -\log[H^+] = -\log K_a - \log \frac{[HA]}{[A^-]}$ [Changing sign on both sides]

$$\therefore\ pH = pK_a + \log \frac{[A^-]}{[HA]} \quad [\because -\log [H^+] = pH \text{ and } -\log K_a = pK_a].$$

$$\therefore\ pH = pKa + \log$$

The above 1 no. equation is known as Henderson-Hasselbalch equation.

Example

In case of blood, the ratio between $[BHCO_3]$: $[H_2CO_3]$ or $[B_2HPO_4]$: $[BH_2PO_4]$ can be found out by applying the above eqaution to maintain average pH of blood 7.4.

If ionisation constant of $H_2CO_3 = 6.1$

With the help of H.H equation

$$7.4 = 6.1 + \log \frac{[BHCO_3]}{[H_2CO_3]}$$

$$\text{Or,}\quad 1.3 = \log \frac{[BHCO_3]}{[H_2CO_3]}$$

$$\text{Or,}\quad \log 20 = \log \frac{[BHCO_3]}{[H_2CO_3]}$$

$$\text{Or,}\quad \frac{20}{1} = \frac{[BHCO_3]}{[H_2CO_3]}$$

So, $[BHCO_3] : [H_2CO_3] = 20 : 1$

If ionisation constant of $BH_2PO_4 = 6.8$

$\therefore$ With the help of Henderson-Hasselbalch equation

$$7.4 = 6.8 + \log \frac{[B_2HPO_4]}{[BH_2PO_4]}$$

$$\text{Or,}\quad 0.6 = \log \frac{[B_2HPO_4]}{[BH_2PO_4]}$$

$$\text{Or,}\quad \log 4 = \log \frac{[B_2HPO_4]}{[BH_2PO_4]}$$

$$\text{Or,}\quad \frac{4}{1} = \frac{[B_2HPO_4]}{[BH_2PO_4]}$$

So, $[B_2HPO_4] : [BH_2PO_4] = 4 : 1$

Application

This equation can be used to determine the pH of blood.

Importance

1. With the help of this equation an individual is judged of his/her normal condition when concentration of $[BHCO_3]$: $[H_2CO_3] = 20 : 1$
   Or, $[B_2HPO_4] : [BH_2PO_4] = 4 : 1$ at his/her of blood is 7.4.
2. If the above-mentioned ratios are altered the individuals are considered as suffering from acidosis or alkalosis. They are treated with medicine according to the conditions (acidosis or alkalosis) to cure them to lead a normal life.
   The table below gives the pH of some typical solutions :

Solution	pH	Solution	pH
N/10 HCl	1.04	Pure water	7.0
N/100 HCl	2.00	Milk	7.1
Gastric Juice (adult)	1.2-1.5	Tears	7.2
Gastric Juice (infants)	5.0	Blood (average)	7.4
Urine (average)	6.0	Intestinal Juice	7.8
Saliva	6.8	Pancreatic Juice	8.0
		N/100 NaOH	12.0

Effects of temperature

1. The pH of a solution may vary with temperature because ionisation is increased with the rise in temperature.
2. Pure water at 22°C has pH 7.0 but at 38°C it has pH 6.74.
3. The pH of a solution (e.g. N/10 HCl) which is completely ionised will not vary with the temperature.

CLINICAL ORIENTATION

- Enzymes need an optimum pH for best activity in the body under normal conditions, the body fluids and cell organelles maintain a specific pH for the activity of the local enzymes. Alternation in the pH leads to non-functioning of enzymes resulting in various diseases.
- Acidosis or alkalosis results in the alternation of blood pH 7.4.
- Specific tautomeric form of nucleic acid exists at pH 7.4 which helps in proper hydrogen bonding between the complementary base pairs.
- Protein with its optimum pH maintains its 3-dimensional conformation for its biological activity.
- The physiological pH 7.4 of body protein exists as anions.

  They do not cross semipermeable biological membranes resulting in unequal distribution of diffusible cations and anions on two sides of membranes.

BUFFERS

Definition

Buffers are the mixtures of weak acids and their salts of strong bases (or strong acids and their salts of weak bases).

Example

Acetic acid (CH_3COOH) + Sodium acetate (CH_3COONa).

Principles of buffers

$$HAC + NaAC \rightarrow Na^+ + H^+ + 2AC^-$$

where, HAC=Acetic acid; NaAC = Sodium acetate.

If alkali (NaOH) is added to this system, it will form salt and no free H^+ or OH^- will be available.

$$HAC + NaAC + NaOH \rightarrow 2NaAC + H_2O$$

If acid (HCl) is added to this system, it will also form salt and no free H^+ or OH^- will be available.

$$HAC + NaAC + HCl \rightarrow NaCl + 2HAC$$

In either cases there is no change in hydrogen ion concentration. The buffer acts almost as if it were "absorbing" the added free hydrogen or hydroxyl ions.

Determination of pH of buffers

The pH of buffers can be determined by the Henderson-Hasselbalch equation :

$$pH = PK_a + \log\frac{[salt]}{[acid]}$$

where, $PK_a = \log\frac{1}{K_a}$, Ka = Dissociation constant of the acid.

$$\text{and } K_a = \frac{[H^+][AC^-]}{[HAC]}$$

In case of blood, the ratio between $[BHCO_3]$: $[H_2CO_3]$ can be found out by applying the above equation to maintain average pH of blood 7.4 :

$$7.4 = 6.1 + \log\frac{[BHCO_3]}{[H_2CO_3]}$$

$$\therefore \quad \log\frac{[BHCO_3]}{[H_2CO_3]} = 1.3$$

$$\therefore \quad \log\frac{[BHCO_3]}{[H_2CO_3]} = \log 20$$

$$\therefore \quad \frac{[BHCO_3]}{[H_2CO_3]} = \frac{20}{1}$$

So, $[BHCO_3] : [H_2CO_3] = 20 : 1$.

The pH of human blood is 7.4. In normal health, it lies between 7.3 and 7.5 although CO_2 (i.e. carbonic acid) is always added. If the pH of human blood becomes 7.0 and 7.6, it alarms danger; if not fatal. The difference in pH between arterial and venous blood is rarely more than 0.04. A marked decrease in pH of blood has been observed during severe muscular exercise when the blood lactic acid content rises over 100 mg per 100 ml.

Buffer mixture

(a) Glycine and HCl.

(b) Acid potassium phthalate and HCl.

(c) Acid potassium phosphate and NaOH.

(d) Sodium bicarbonate and Sodium carbonate.

Buffer pairs in the blood

Actually in blood the buffering of carbonic acid is complicated by the presence of the red cells :

H_2CO_3	H. Hemoglobin	BH_2PO_4	→
$BHCO_3$	B. Hemoglobin	B_2HPO_4	
→ H. Protein	H. Oxyhemoglobin		
B. Protein	B. Oxyhemoglobin		

Uses of buffers

1. Buffers are used for preparing standard solutions in which it is always desired to maintain a constant pH. This is required for the colorimetric determination of the pH of unknown solutions.
2. These are used to maintain H^+ concentration which is necessary for optimal activity of enzymes.
3. These are practically important in all physiological systems.

Buffers of tissue fluids and tissues

1. The buffering system of lymph, cerebrospinal fluid etc., are similar to that of blood although the quantity is much less.
2. The chief buffering system in these fluids is $BHCO_3$ and H_2CO_3.
3. The buffering system in tissues is mainly $BHCO_3$ and H_2CO_3; protein buffers and organic acid salt.

Role of buffers in pH regulation

(a) The principal buffers of extracellular fluids :	(i) Bicarbonate buffer. (ii) Phosphate buffer. (iii) Protein buffer.
(b) The principal intracellular buffers :	(i) Phosphate buffer. (ii) Protein buffer.
(c) The principal buffer in erythrocytes :	(i) Hemoglobin buffer.

(i) Bicarbonate buffer

1. It is the main buffer in blood plasma and consists of bicarbonate (HCO_3^-) and carbonic acid (H_2CO_3).
2. The bicarbonate buffer neutralizes stronger dietary and metabolic acids (HA) converting them into weak bases (A^-) with the increase in H_2CO_3. Stronger bases (B) are also changed into weak acids (BH^+) with the rise in HCO_3^-.

$$HA + HCO_3^- \rightleftharpoons A^- + H_2CO_3$$

$$B + H_2CO_3 \rightleftharpoons BH^+ + HCO_3^-$$

3. The pH of blood is maintained 7.4 when the buffer ratio becomes 20. If the bicarbonate buffer neutralizes any acid or base, there may be the change of buffer ratio and the blood pH value. But the buffer ratio remains unchanged by the respiratory elimination of H_2CO_3 as CO_2 or the urinary elimination of HCO_3^-.
4. Since cells contain much lower amounts of HCO_3^- the importance of bicarbonate buffer inside the cell is negligible.

(ii) Phosphate buffer

1. Since the concentration of phosphate buffer in the blood plasma is about 8 per cent of that of the bicarbonate buffer, its buffering capacity is much lower than bicarbonate in the plasma.
2. The phosphate buffer consists of dibasic phosphate (HPO_4^{--}) and monobasic phosphate ($H_2PO_4^-$). Its PKa value is about 6.8. It is more effective in the pH range 5.8 to 7.8. Plasma has a $[HPO_4^{--}]$: $[H_2PO_4^-]$ is 4.

Therefore

$$pH = PK_a + \log \frac{[HPO_4^{--}]}{[H_2PO_4^-]}$$

$$\therefore\ pH = 6.8 + \log 4 = 7.4$$

3. The concentration of phosphate buffer is much higher in intra-cellular fluid than in extracellular fluids. The pH of intracellular fluids (6.0 – 6.9) is nearer to the PK_a of the phosphate buffer. Therefore, the buffering capacity of the phosphate buffer is highly elevated inside the cells and the phosphate buffer is also effective in the urine inside the renal distal tubules and collecting ducts.
4. In case the ratio of $[HPO_4^{--}] : [H_2PO_4^-]$ tends to be changed by the formation of more $H_2PO_4^-$, there occurs the renal elimination of $H_2PO_4^-$ for which the ratio ultimately remains unaltered.

(iii) Protein buffers

1. The protein buffers are very important in the plasma and the intracellular fluids but their concentration is very low in C.S.F., lymph, and interstitial fluids.
2. They exist as anions serving as conjugate bases (Pr^-) at the blood pH 7.4 and form conjugate acids (HPr) accepting H^+.
3. They have the capacity to buffer some H_2CO_3 in the blood :

$$H_2CO_3 + Pr^- \rightleftharpoons HCO_3^- + HPr$$

(iv) Hemoglobin buffers

1. They are involved in buffering CO_2 inside erythrocytes. The buffering capacity of hemoglobin depends on its oxygenation and deoxygenation. Inside the erythrocytes CO_2 combines with H_2O to form H_2CO_3 under the action of carbonic anhydrase. At the blood pH 7.4, H_2CO_3 dissociates into H^+ and HCO_3^- and needs immediate buffering. Oxyhemoglobin (HBO_2^-) on the other side loses O_2 to form deoxyhemoglobin (Hb^-) which remains undissociated (HHb) by accepting H^+ from the ionization of H_2CO_3. Thus, Hb^- buffers H_2CO_3 in erythrocytes :

$$HbO_2^- \rightleftharpoons Hb^- + O_2$$

$$Hb^- + H_2CO_3 \rightleftharpoons HHb + HCO_3^-$$

Some of the HCO_3^- diffuse out into the plasma to maintain the balance between intracellular and plasma bicarbonates. This causes influx of some Cl^- into erythrocytes along the electrical gradient produced by the HCO_3^- outflow (chloride shift).

2. $HHbo_2$, produced in lungs by oxygenation of HHb, immediately ionizes into H^+ and HbO_2^-. The released hydrogen ions (H^+) are buffered by HCO_3^- inside erythrocyte to form H_2CO_3 which is dissociated into H_2O and CO_2 by carbonic anhydrase. CO_2 diffuses out of erythrocytes and escapes in the alveolar air. Some HCO_3^- return from the plasma to erythrocytes in exchange of Cl^- and are changed to CO_2.

$$HHb + O_2 \rightleftharpoons HHbO_2 \rightleftharpoons HbO_2^- + H^+$$

$$HCO_3^- + H^+ \rightleftharpoons H_2CO_3 \rightarrow H_2O + CO_2$$

Acidosis and Alkalosis

Acidosis

1. Accumulation of acid or loss of alkali is called acidosis.
2. It occurs due to the loss or fall of $[HCO_3^-]$: $[H_2CO_3]$ of blood below 20.
3. There are two types of acidosis : (a) Metabolic; (b) Respiratory.

(a) Metabolic

(i) The concentration of plasma bicarbonate is decreased in excessive loss of bases.

(ii) It happens in renal failure, diabetic ketosis, severe diarrhoea.

(b) Respiratory

(i) The retention of CO_2 is caused in hypoventilation resulting in the rise of H_2CO_3. This lowers $[HCO_3^-]:[H_2CO_3]$ ratio.

(ii) It happens in chronic obstructive airway diseases (asthma, respiratory paralysis), prolonged anaesthesia, and unconsciousness due to any cause.

Alkalosis

1. Accumulation of alkali and loss of acid is called alkalosis.
2. There is increase in the ratio of $[HCO_3^-]:[H_2CO_3]$ of blood above 20, resulting the rise in blood pH.
3. There are two types : (a) Metabolic; (b) Respiratory.

(a) Metabolic

(i) High intake of alkaline substances like $NAHCO_3$ may elevate the plasma HCO_3^-.

(ii) It happens in severe vomiting due to any cause, indiscriminate use of antacid.

(b) Respiratory

(i) The excess CO_2 is removed from the blood due to hyperventilation and causes decrease of H_2CO_3.

(ii) It happens in high altitude (hyper-ventilation syndrome), hysteria.

Role of Lungs and Kidneys in pH regulation by means of buffers

Lungs

1. Lungs maintain the normal ratio of $[HCO_3^-]:[H_2CO_3]$ and the pH of blood by altering the rate of respiratory elimination of CO_2 from the blood. This lowers the alveolar PCO_2 and increases the diffusion of CO_2 to the alveolar air. Since the concentration H_2CO_3 is in equilibrium with that of dissolved CO_2 in the blood, hyperventilation increases the ratio of $[HCO_3^-]:[H_2CO_3]$ with the fall in the CO_2 concentration. The doubling of ventilation may raise the blood pH by 0.4.
2. Hypoventilation, on the other hand, raises the blood concentration of dissolved CO_2 and consequently lowers the buffer ratio. A fall in alveolar ventilation to one-fourth of the normal value may lower the blood pH by 0.46. The pulmonary ventilation is adjusted according to the blood pH. Hypoventilation not only retains CO_2 decrease the ratio of $[HCO_3^-]:[H_2CO_3]$ and the blood pH but also curtails the O_2 supply—an undesirable effect.
3. Lungs also have a role in the functioning of hemoglobin buffers through oxygenation and deoxygenation which has been discussed earlier.

Kidney

1. The pH of the glomerular filtrate is about 7.4. But the pH falls to about 6.9 in the proximal tubule, then to about 6–6.5 in the distal tubule and ultimately to about 4.5–4.7 in the collecting duct.
2. The urinary pH is maintained by a cooperation between the urinary buffers and the renal ion-exchange mechanism. The major urinary buffers are bicarbonate and phosphate buffers. As the filtrate proceeds along the tubules, the ratio between base member and the acid member of each urinary buffer falls progressively with a consequent fall in the urinary pH.
3. In the renal ion exchange mechanism, some urinary Na^+ are actively reabsorbed in the exchange of H^+ secreted in the tubular filtrate. 85 percent of H^+ are secreted

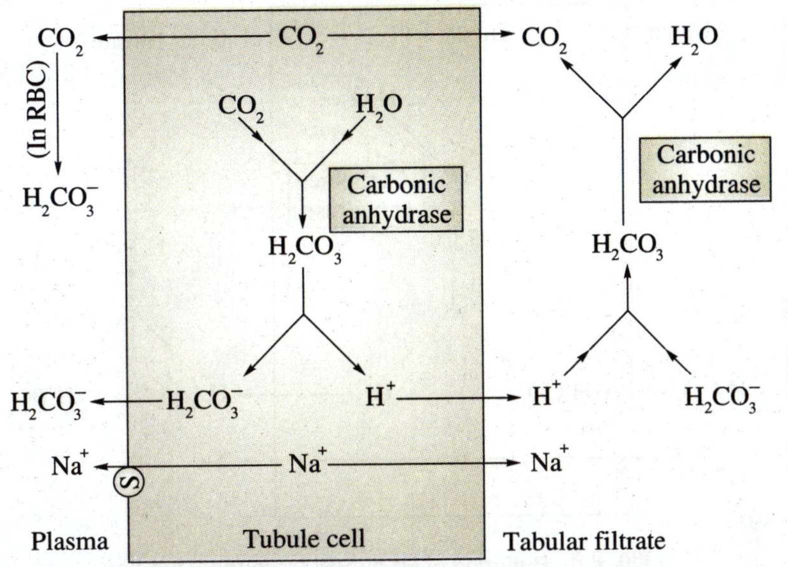

Fig. 2.1 : Buffering of H^+ by HCO_3^- in renal tubules

by the proximal tubule and 15 per cent from the distal tubules and collecting ducts. The hydrogen ions (H^+) are mainly formed from the ionization of H_2CO_3 formed from CO_2 and H_2O by carbonic anhydrase in the tubule cells.

$$H_2O + CO_2 \rightleftharpoons H_2CO_3 \rightleftharpoons H^+ + HCO_3^-$$

The bicarbonate formed is returned to blood along with the reabsorption of Na^+.

4. Most of the secreted H^+ are immediately buffered by HCO_3^- filtered from the plasma into the glomerular filtrate.
5. H_2CO_3 produced by such buffering in the proximal tubules is immediately removed through its cleavage to CO_2 and cannot play any role in lowering the pH there.
6. The buffering of the secreted H^+ is essential for continuing their secretion in the urine. This buffering is held in the tubular filtrate in several ways.

(a) Buffering by bicarbonate

(i) Normally about 3.50 mM of H^+ are secreted in the tubules per minute.

(ii) The major portion of the secreted H^+ can be buffered by HCO_3^- in the tubular filtrate to form H_2CO_3, except a small amount of free H^+ to pass into the urine.

(iii) H_2CO_3 is then immediately cleaved into H_2O and CO_2 in the proximal tubular lumen by carbonic anhydrase.

(iv) CO_2 then diffuses very readily into the proximal tubule cell and therefrom to the blood.

(v) In erythrocytes, carbonic anhydrase converts this CO_2 into H_2CO_3 which dissociates to form fresh HCO_3^-. This HCO_3^- is restored in the plasma.

(vi) When H^+ are secreted in excess due to the fall in blood pH, almost all the filtered HCO_3^- changes into H_2CO_3 and is returned to the plasma. So the urinary HCO_3^- is negligible so long as the urinary pH does not exceed 6. But whenever the blood pH tends to rise, much more HCO_3^- is filtered than the amount of H^+ secreted. So, some of the filtered HCO_3^-

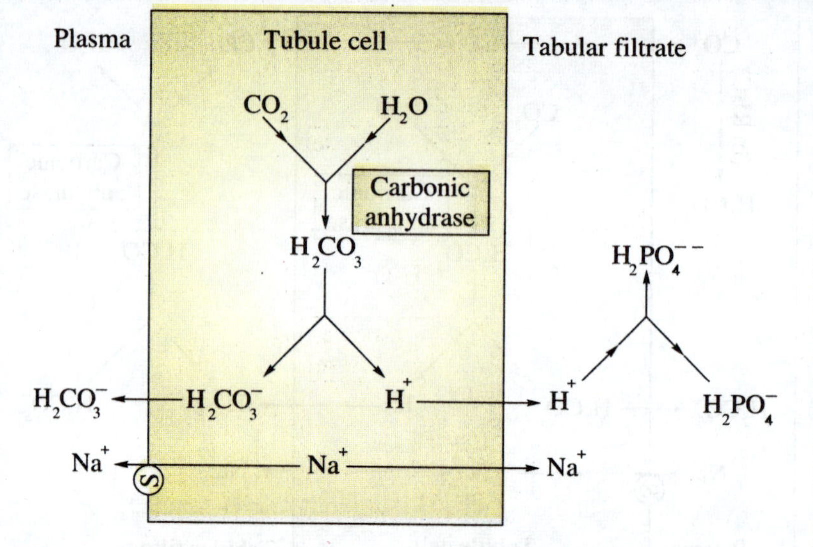

Fig. 2.2 : Buffering of H^+ by HPO_4^{--} in renal tubules

fails to get H^+ to combine and fails to return to the plasma. This causes the urinary elimination of bicarbonate.

(vii) Buffering of the secreted H^+ by the filtered HCO_3^- serves two purposes :

1. It does not allow the pH to fall below 6.9 in the proximal tubule and allows more H^+ to be eliminated by tubule cells into the urine.
2. It helps to reabsorb the filtered HCO_3^- and to restore it in the blood.

(b) Buffering by phosphate buffer

(i) In the distal tubules, some secreted H^+ is buffered by the phosphate buffer. HPO_4^{--}, filtered into the glomerular filtrate, receives the secreted H^+ to form $H_2PO_4^-$. This changes the ratio of $[HPO_4^{--}] : [H_2PO_4^-]$ from 4 of Bowman's capsule to 0.02–0.05 in the final urine. The lower this ratio, the more acidic is the urine.

(ii) $H_2PO_4^-$ is eliminated in urine carrying some Na^+ with it resulting in the urinary loss of Na^+.

(iii) The total buffering capacity of urinary phosphate is much less than that of bicarbonate.

(iv) As the urine gets more concentrated in the distal tubules and collecting ducts, the rise in tubular concentration of HPO_4^{--} enhances the buffing capacity of the phosphate buffer.

(c) Buffering by ammonia

(i) The base NH_3, synthesized and secreted by the tubule cells, can buffer some H^+ in the distal tubule.

(ii) In the tubular lumen, NH_3 combines with H^+ to form NH_4^+ which is excreted in the urine in association with Cl^- and SO_4^{--} left behind by the reabsorbed Na^+.

(iii) NH_4^+ behaves like a weak acid and does not dissociate much. Its formation lowers the tubular concentration of free H^+, enabling further secretion of H^+. The tubular

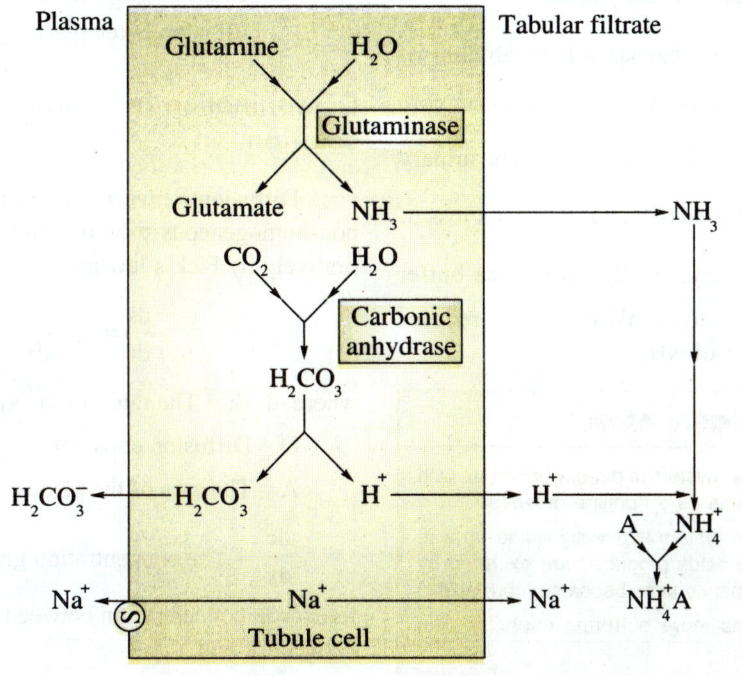

Fig. 2.3 : Buffering of H^+ by NH_3 in renal tubules

membrane is not permeable to NH_4^+ which is retained in the tubular filtrate instead of diffusing back into the tubule cells (diffusion trapping).

(iv) The secretion of NH_3 rises whenever the free H^+ concentration is high enough to lower the urinary pH below 6; the more acidic the urine, the higher is the urinary ammonia.

(v) An alkaline urine contains little or no ammonia, kidneys normally excrete about 40 mEq of NH_4^+ in 24 hours. The elimination of highly acidic urine may enhance more ammonia secretion.

Elimination of free acids

Strong conjugate bases such as lactate, acetoacetate, urate and oxalate anions accept some H^+ replacing the Na^+ reabsorbed from their salts. As a result, the free acids such as lactic acid, acetoacetic acid, uric acid are excreted. Their elimination changes the urinary pH a little only.

Renal correction of acidosis

In acidosis, the blood carries a high amount of dissolved CO_2 compared to that of HCO_3^- and the tubule cells secrete far more H^+ than the HCO_3^- filtered from glomeruli. As a result, all the filtered HCO_3^- combines with H^+ to form H_2CO_3. Thus, the urine does not contain HCO_3^- while the reabsorbed HCO_3^- is retained in blood increasing the buffer ratio. In addition, tubular secretion of NH_3 increases for buffering the H^+ left in the tubules after all the HCO_3^- has been reabsorbed.

Renal correction of alkalosis

In alkalosis, the blood carries a high amount of HCO_3^- and the glomerular filtrate contains far more HCO_3^- than H^+ secreted in the tubules. The urinary elimination of unabsorbed HCO_3^- causes a loss of HCO_3^- from blood and finally lowers the buffer ratio and the pH of blood. Alkalosis also reduces the tubular secretion of NH_3.

> **CLINICAL ASPECT**
>
> - Bicarbonate buffer system is directly linked up with respiration and acts as a frontline defence.
> - Phosphate buffer system is directly linked up with the kidneys. The acids produced are excreted by the kidneys and hence urine becomes more acidic.
> - Haemoglobin has more buffering capacity than plasma proteins.
> - Oxygenated haemoglobin is a stronger acid than deoxygenated Hb. Acidity of the medium favours delivery of oxygen and alkalinity of the medium favours oxygenation of Hb.

DIFFUSION

Definition

The spreading of solute molecules throughout the water molecules is called diffusion.

Experiment

When a lump of sugar is added to a glass of water, the sugar gradually disappears and, ultimately, a homogeneous solution is formed automatically. This is due to the energy possessed by all molecules which keeps them in a state of motion or vibration.

Characteristics

1. Diffusion in liquids is comparatively a slow process because the molecules are so crowded together that they are continually colliding.
2. In case of gas, diffusion is rapid.
3. Diffusion takes place against gravity.

Determination in a non-homogeneous solution

Diffusion between two planes X and Y in a non-homogeneous solution can be expressed quantitatively by Fick's law as

$$\frac{ds}{dt} = DA\frac{dc}{dx}$$

where, ds/dt = The rate of movement of solute.

D = Diffusion constant.

A = The area of the planes.

$\frac{dc}{dx}$ = The concentration gradient i.e. the difference in concentration between X and Y/distance between X and Y.

Physiological role

1. It helps in passive transport of substances across the cell membrane.
2. Substances move in the direction of the physical gradients which follows the rule of diffusion.

> **CLINICAL ASPECTS**
>
> - The reduction in the diffusion of oxygen in pneumonia is caused by increased thickness of inflamed alveolar membrane.
> - Diffusion causes intestinal absorption of pentoses, minerals, water-soluble vitamins and renal absorption of urea.
> - Keratinized epidermis diminishes loss of water.
> - Diffusion is involved in the whole process of respiration at various sites.
> - The polarization of membranes and membrane potentials are influenced by diffusion of ions across cell membranes.
> - Water, ions and small molecules pass easily by diffusion through plasma membrane.
> - Protein cations and Ca^{++} present in the plasma membrane cause the diffusion of cations and enhance the diffusion of anions.

OSMOSIS AND OSMOTIC PRESSURE

Definition

Osmosis is defined as the spontaneous flow of water into a solution, or, from a more dilute to a more concentrated solution when the two solutions are separated from each other by a semi-permeable membrane.

Osmosis occurs in the direction opposite to that in which diffusion occurs. The animal membranes are not completely "semi- permeable". The artificial membrane, $Cu_2Fe(CN)_6$, is the most selective. This is prepared by keeping copper crystals in the pores and dissolved in $K_4Fe(CN)_6$ solution.

Experiment

In Fig. 2.4 below, the semipermeable membrane is fitted at the mouth of the thistle funnel. The thistle funnel is provided with a sucrose solution up to a mark and inverted over water in a trough. After some time it has been found that the level of sucrose solution is increased in the thistle funnel owing to the entry of water through the semi-permeable membrane. The flow of water will be prevented when both the pressure (inside and outside) are equal. The excess pressure can be measured by a mercury manometer.

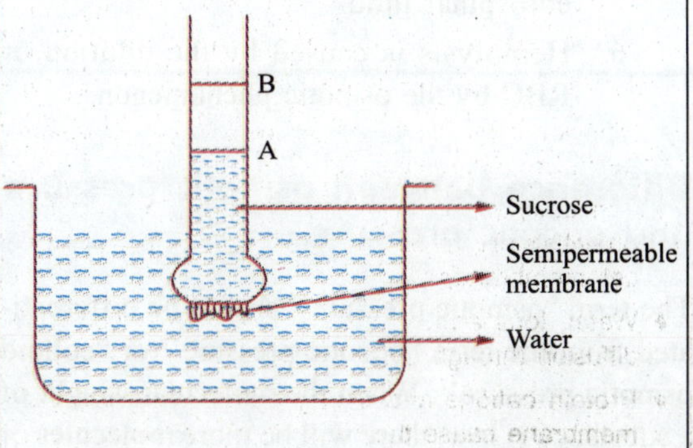

Fig. 2.4 : Osmosis and osmotic pressure

Definition

Osmotic pressure can be defined as the excess pressure which must be applied to a solution to prevent the flow of solvent of low osmotic pressure when they are separated by a perfectly semi-permeable membrane.

Points of some significance in the absorption of food

1. Osmotic pressure is only permanent if the membrane is truely semipermeable, i.e. if it stops all solute molecules and only passes the solvent molecules.
2. In case of dialysis, if the collodion or cellophane bag is filled with a solution of a dye with small molecules and placed in contact with water, water will pass into the dye, but, at the same time, water plus dye will pass out from the bag. As water molecules are smaller than the dye molecules, water will pass into the bag more quickly than water plus dye will leave it. Therefore, an osmotic pressure will be developed, but it will only be small and transient because the membrane is permeable to both water and dye. But, if the bag is filled with a dye of large molecules, the dye cannot pass out into the water but water will pass into the dye—causing a permanent pressure difference. If the bag is surrounded with a solution of NaCl of greater osmotic pressure than the dye, water will at first pass out of the bag faster but, at the same time, NaCl will pass into the bag. The final equilibrium will be attained when the osmotic pressure of the salt solution outside the bag equals the osmotic pressure of salt inside, so that the ultimate permanent osmotic pressure difference will be that of the dye alone.

Factors regulating osmotic pressure

1. Osmotic pressure depends on the number of solute molecules but not on the size of the molecules.

Example : A solution of urea (Mol. wt. 60) of 60 g. per litre has the same osmotic pressure as a solution of cane sugar (mol. wt. 342) of 342 g. per litre; because these two solutions contain the same number of molecules per litre.

2. In case of salts which ionise, it is the number of ions plus molecules which count, so that fully ionised NaCl has twice the osmotic pressure it would have as judged by the number of molecules.
3. Other substances, such as soaps, form molecular aggregates, so that their solutions have lower osmotic pressure.
4. The osmotic pressure increases with the rise in temperature.

Osmotic pressure in cells

1. If the cell is kept in a hypotonic solution, the cell wall and the vacuolar membrane both will allow water to pass into it and will set up an excess pressure in the interior of the cell causing the cytoplasm to be forced tightly against the cell wall. In normal health, this condition is known as "turgor" and the cell is said to be **turgid**.
2. If the cell is immersed in a concentrated solution (high osmotic pressure), water will pass out of the interior of the cell. The cytoplasm will then shrink and detach itself from the cell wall. This phenomenon is said to be "**plasmolysis**".

   Iso-osmotics : Solutions with the same pressure are termed iso-osmotics.

   Isotonic solution : A pair of solutions which produce no flow through a semipermeable membrane are said to be isotonic solutions.

Physiological importance

1. Absorption from gastro-intestinal tract, as also fluid interchange in various compartments of the body follow the principles of osmosis.
2. The osmotic pressure of plasma proteins regulates water to flow from the protein-free intestinal fluid into the blood vessels.
3. Living red cells, if suspended in 0.92% NaCl solution, neither gain nor lose water. Briefly speaking, intracellular fluid of red cells is isotonic with the red cell membrane in 0.92% NaCl solution.

Application of Osmosis

1. The purgative action of Epsom ($MgSO_4$ $7H_2O$) or Glauber's (Na_2SO_4 $10H_2O$) salts is an osmotic phenomenon. A strong solution of a salt in the intestine prevents absorption of water or withdrawal of water from the body causing dilution of the intestinal contents.
2. The pain caused by contact of sugar with exposed nerves of teeth is due to the osmotic withdrawal of water from the exposed area by strong sugar solution.
3. The pain experienced by the application of salt on the cut of the skin follows the same osmotic phenomenon as in No. 2.
4. Water or salts (chiefly NaCl) are excreted by the kidney to keep the blood isotonic with the cells. Hence, osmosis is of great importance in the process of urine secretion.
5. The clinical application of osmotic force is the injection of hypertonic solution of magnesium sulphate to reduce the volume of the brain or lower the pressure of cerebrospinal fluid.
6. Hemolysis is caused by the dilution of RBC by the osmotic phenomenon.

Difference between osmotic pressure and oncotic pressure

The term "osmotic pressure" should be differentiated from the term "oncotic pressure" or "colloid osmotic pressure". When the molecular weight of a substance is low, there will be more molecules of the substance per unit weight, and the osmotic pressure will be greater. Mannitol of low molecular weight increases the osmotic pressure when given intravenously.

But a substance such as human albumin or dextran having high molecular weight around 80,000 does not increase the osmotic pressure greatly. Such a substance exerts an oncotic pressure since it is confined by a membrane to which it is relatively impermeable. Hence, an infusion of albumin or dextran increases the oncotic pressure of the blood but it has almost negligible effect on the osmotic pressure of the blood. Oncotic pressure is measured in terms of pressure units (mm Hg).

Factors regulating the osmotic pressure and the volume of the extracellular water

In the maintenance of life, the body has homeostatic or regulatory mechanisms which help to maintain the osmotic pressure and the volume of the extracellular water within physiological limits. The osmotic pressure of the extracellular water is controlled by antidiuretic hormone ADH and the volume of the extracellular water is controlled by aldosterone, the adrenocortical hormone.

ADH

ADH, the antidiuretic hormone, is an octapeptide. It regulates the osmotic pressure of the extracellular water and of the cells by regulating the retention or excretion of water by the kidneys.

The posterior pituitary gland is stimulated to secrete ADH, when the osmotic pressure of the extracellular water becomes relatively greater than the osmotic pressure of the cells. ADH liberation is stopped, when the osmotic pressure of the extracellular water is relatively less than that of the cells.

The site of action of ADH is probably the distal tubules and collecting ducts of the kidneys. ADH increases the permeability of the structures to water. When this happens; an increased amount of water is reabsorbed resulting in the volume of urine decrease.

ADH secretion is not affected when the total osmotic pressure of the body changes. If urea is infused, it does not cause an antidiuretic effect even though it increases the osmotic pressure of the extracellular water. Because, it is able to penetrate the cells freely. Hence it also raises the osmotic pressures of the cells to the same degree as the extracellular water. If, for any reason, the osmotic pressure of the cells in relation to the osmotic pressure of the extracellular water changes, ADH secretion either decreases or increases. If a large amount of hypertonic glucose or sucrose is infused into the body, it raises the osmotic pressure of the extracellular water. The glucose or sucrose is unable to penetrate cells freely. Therefore, the osmotic pressure of the extracellular water rises above that of the cells. As a result, ADH secretion occurs and water excretion by the kidneys decreases.

If a person drinks plain water, the immediate effect of the imbibed water is to dilute the extracellular water. This causes the osmotic pressure of the extracellular water to become lower than that of the cells. Therefore, ADH secretion stops and water diuresis takes place.

ADH stimulation occurs in various ways, eg, it occurs as the result of fear, pain or in acute infections such as pneumonia. The mechanism by which ADH secretion occurs in these conditions is unknown.

Recently it has been found that the liver acts as an osmoreceptor to regulate the intake of water from the gastro-intestinal tract. When water is imbibed, it is absorbed into the splanchnic circulation and is brought to the liver by way of the portal vein. This causes the osmolality of the portal vein to decrease very rapidly. The hypothalamus then stimulates the kidneys to excrete water.

Aldosterone and the volume of the extracellular water

Aldosterone regulates the volume of blood and extracellular water. It happens due to the reabsorption of sodium from the kidneys. The efficiency of the heart as a pump on the volume of blood causes the secretion of aldosterone.

The site of action of aldosterone is the distal tubules of the kidneys. Sodium and water are retained by the kidneys after the secretion of aldosterone. This tends to increase renal arterial pressure and flow. As a result, the juxtaglomerular secretion of renal is inhibited and aldosterone secretion stops.

Changes in body potassium also directly influence aldosterone secretion by a mechanism which is independent of changes in body sodium or arterial pressure. Potassium depletion is associated with a decreased aldosterone secretion and potassium repletion is associated with an increased aldosterone secretion.

The third factor

Recent studies have indicated that the increased excretion of sodium is related to a decreased tubular reabsorbtion of sodium from the proximal tubules of the kidneys. A similar loss or 'escape' of sodium from the kidneys occurs when salt-retaining hormones such as aldosterone are administered over a prolonged period of time to normal human subjects. It is believed that the tubular excretion of sodium is controlled by a third factor which may be a hormone. Its site of origin and its chemical structure are both unknown.

Vant Hoff's laws of osmotic pressure

This is stated as :

1. The osmotic pressure of a solution varies directly with the concentration of the solute in the solution and is equal to the pressure the solute would exert if it would be a gas in the volume occupied by the solution, if the volume of the solute molecules relative to volume of solvent be negligible.
2. The osmotic pressure of a solution varies directly with absolute temperature in the same way as the pressure of a gas varies when its volume is kept constant.

These laws of osmotic pressure have been thoroughly verified by accurate observations. As with gases, the laws of osmotic pressure hold closely only for dilute solutions. Proper correction must be made for concentrated solution.

Electroendosmosis

If a potential develops across the ends of the capillary when a liquid is forced through it, it is not surprising to find that the liquid flows through the capillary when a potential is applied across it. The movement of a liquid through a capillary under the influence of an electric potential is termed **electroendosmosis**.

Membranes prepared from collodion, tissue membranes etc. may be considered structures containing great numbers of pores. The walls of the pores of those and most other membranes are negatively charged, and the water in the pore is positively charged. When such a membrane is placed in water between electrodes so that the liquid is separated into two compartments, water flows from the compartment containing the positive pole through the membrane into the compartment containing the negative pole. This process continues until the level of water in the negative electrode compartment is sufficiently above that in the positive electrode compartment to balance the rates of water movement through the membrane in opposite direction. Membranes with positively charged pole walls give electroendosmosis in the reverse direction because the water is negatively charged and mobile. The addition of acid "hydrogen ion (H^+)" causes negative pole walls to become less negative and positive pole walls more positive. Alkali (OH^-) has the reverse effect. Salts also tend to change the pole charge relating to the liquid, the change depending upon the selective adsorption of salt ions on the pole walls. The electroendosmotic effect is changed accordingly as the relative charges of pole walls and mobile liquid are changed.

Calculations involving osmotic pressure

Osmotic pressure is due to the difference in activity of pure solvent molecules and of solvent molecules associated with solute molecules. It is proportional to the number of and independent of the kind of dissolved particles present.

The general equation for gases applies to osmotic pressure :

$$\pi V = nRT, \text{ or, } \pi = \frac{nRT}{V}$$

where π = the osmotic pressure in atmosphere, n = the number of moles of solutes, R = the molar gas constant (0.082 litre-atmosphere/mol/Å), T = the absolute temperature, V = the volume in litres.

Since $n = \frac{g}{M}$, where g = grams of solute and M = the molecular weight of the solute, the equation may be written

$$\pi V = \frac{g}{M} RT, \text{ or, } \pi = CRT$$

where C = the molal concentration of the solution and corrects for the volume occupied by the solute molecules.

Distribution of a solute between two immiscible solvents

When a water solution of succinic acid is shaken with ether, the molecules of acid distribute themselves in such a way that the ratio of acid molecules dissolved in water to those dissolved in ether is constant, regardless of the total amount of acid dissolved.

The ratio of concentration of acid in water C_1 to concentration of acid in ether C_2 is relatively constant

$$\frac{C_1}{C_2} = K.$$

The principle of the distribution of solutes between immiscible solvents is of much importance in the body. In general, a substance that is more soluble in organic solvents than in water is also more soluble in lipids. Consequently, drugs and other molecules that are more soluble in organic solvents than in water will tend to concentrate in the tissues and fluids of the body that contain more lipid material. On the other hand, molecules that are highly soluble in water and slightly soluble in lipids will be present in greater concentration in the body fluids and tissues that contain little lipid or fat-like material. For example, when ether is used as an anaesthetic, the concentration in the brain and nervous tissue (rich in lipids) will be much greater than the concentration in the blood and tissues, which are much poorer in lipid material. Theoretically, the ether will be distributed in the body according to its distribution coefficients for the various fluids and tissues that serve as solvents for it.

CLINICAL ORIENTATION

- RBC remains well-distended in fluid in blood due to intracellular and plasma levels of electrolytes etc. Even slightest distension can cause hemolysis as in malformation of RBC, genetic disorders of enzyme deficiencies such as glucose-6-phosphate dehydrogenase, Vit. E, and selenium deficiency.
- In Kwashiorkor, hepatic cirrhosis, and nephrosis, a fall in plasma concentration of proteins minimises the colloid osmotic pressure of blood and reduces the retention of water in circulation leading to oedema.
- Polyuria results in diabetes due to a fall in water reabsorption by osmotic effect of urinary glucose. Facultative renal absorption of water also depends on the osmotic effects of the more concentrated intestinal fluid in the renal medullary tissue.

SURFACE TENSION

Definition

The force with which the surface molecules are held together is called surface tension.

Explanation

The interior molecules of a homogeneous liquid are equally attracted in all directions by surrounding molecules. They are free to move in all directions. But the molecules in the surface of the liquid are attracted downward and sideways but not upward (except for the little attraction of air

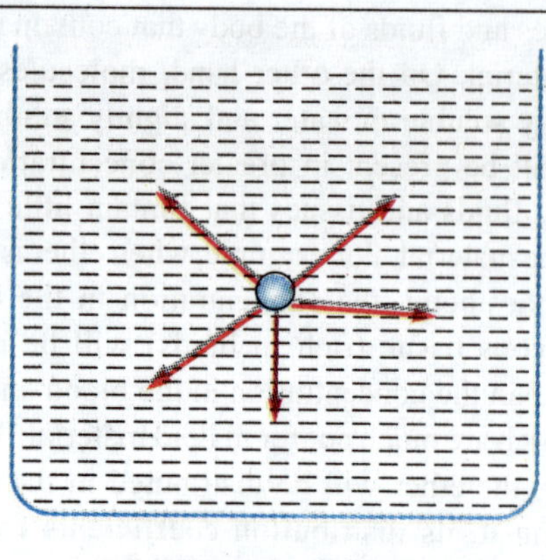

Fig. 2.5 : Surface tension

molecules). As a result, the molecules of the surface are not so free to move. They are held together and form a membrane over the surface of the liquid. Therefore, when finely powdered sulphur or other non-wetting powders are sprinkled upon water, they do not sink but are suspended on the surface.

A great part of the energy required to convert a liquid into a gas is essential to overcome surface tension and drag the molecules free from the surface of the liquid.

There is also an interfacial tension which is biochemically very important, especially in the process of adsorption. This tension lies at the boundary between immiscible liquids, eg, oil drops emulsified in water. The tension is due to unequal attraction of the film molecules as compared with the molecules in the interior of the liquid.

Surface tension × Surface area = Surface energy. A falling drop of liquid assumes a spherical form because the ratio of surface area and total free surface energy is the least.

Method of determination of surface tension

$$\gamma = \frac{1}{2} \text{hdgr}$$

where, h = height of the liquid.
d = density of the liquid.
g = acceleration due to gravity.
r = radius of the capillary tube.

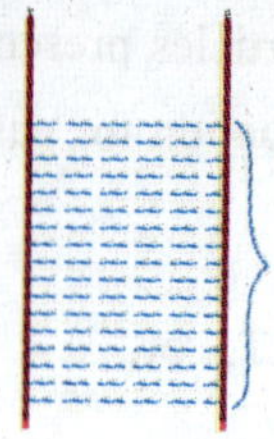

Fig. 2.6 : Capillary tube

Factors affecting surface tension

1. Temperature : Surface tension decreases with the increase in temperature.
2. Dissolved substances :
   (a) Most inorganic salts slightly raise surface tension of water although potassium permanganate lowers it.
   (b) Organic substances usually lower surface tension. Soaps and bile salts are most effective in this respect.
   (c) Alkalis increase surface tension but ammonia lowers it. Strong mineral acids also decrease surface tension.
   (d) In liquid-liquid and solid-liquid systems, dissolved substances generally lower interfacial tension.

Gibbs-Thomson Principle in relation to surface tension

1. Substances that lower the surface tension become concentrated at the interface.
2. Substances that increase surface tension tend to move away from the interface.
3. Lipids and proteins which are both effective in lowering surface tension are found concentrated in the cell wall.

Physiological importance

Surface tension is involved in the process of digestion; because bile salts reduce the surface tension of lipids and thus assist emulsification.

As a result, the surface area is increased which favours lipase activity on lipids.

CLINICAL ASPECTS

- The inward force generated by alveolar surface tension increases diffusion of fluids into alveoli from alveolar capillaries. Lung surfactant reduces such fluid exudation and prevents oedema by lowering surface tension.
- The surface tension lining the inner surface of alveoli determines the stretchability of lung. This accelerates the tendency of lungs to recoil from the thoracic wall and when unhibited keeps the lungs collapsed. This situation is found in foetal life. In postnatal life, type 2 alveolar epithelial cells in lungs secrete a surfactant called DPL (Dipalmityl lecithin) which prevents collapse of lung alveoli in expiration.

ADSORPTION

Definition

The process of taking up substances from solution on surface is called adsorption.

Characteristics

1. Adsorption is a surface phenomenon.
2. The attractive forces on the surface are limited to distances one molecule deep.
3. The extent to which adsorption takes place is dependent upon the nature of both adsorbing agent and the substances adsorbed.
4. The greater the surface of the adsorbing agent, the greater is the adsorption.
5. Charcoal becomes activated when it is heated at 700°-800°C in a closed container and adsorption takes place on the activated charcoal due to the attraction of oppositely charged ions. Salts, acids and alkalis restrict it.
6. It has got much importance in industry.

Principles governing adsorption

1. Adsorption is a reversible process.
2. It decreases with the rise in temperature.
3. This process takes place relatively quickly. Equilibrium is reached within one hour.
4. Adsorption is proportional to the surface area and it varies with the nature of the surface of the adsorbent and of the substances to be adsorbed.
5. It proceeds best from dilute solutions.
6. Narrow pores on the surface of the adsorbing agent are more effective than globular openings.
7. Heat is given off in all adsorption.
8. The molecules adsorbed on the surface are oriented and arranged in a definite manner.

Importance of adsorption

1. Many chemical reactions are speeded up by the presence of adsorptive surface. Oxygen and hydrogen are adsorbed together upon platinum black and combine rapidly at ordinary temperature to form water.
2. Surface adsorption helps to combine enzymes with substrates to give reaction products.
3. Adsorption processes taking place on the cell membranes promote many vital chemical reactions and also cause changes in surface tension and cell consistency.
4. Drugs and poisons which are adsorbed on cell surfaces exert their effects from that location. Selective adsorption may be related to specific action.
5. The process of adsorption is applied in the purification of enzymes.

Arrangement of molecules at an interface

Solute molecules accumulated at an interface tend to arrange themselves in a definite pattern if the molecules are unsymmetrical, (i.e. if they have –COOH, –OH, and $-NH_2$ groups), they have water attractions; whereas the arrangement is symmetrical within the solution.

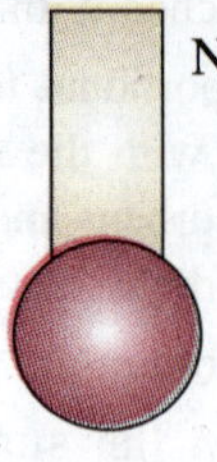

Fig. 2.7

The unsymmetrical molecules align themselves so that the polar groups are directed towards water and the non-polar groups at the other end of the molecule remain away from it (orientation).

At an oil-water interface the orientation would be especially favoured since the oil would attract the non-polar groups in addition to the water attracting the polar groups.

In the cell (which contains both water and lipids), it is likely that some cell constituents are oriented in this way. The cell membrane absorbs oriented molecules. Orientation is an important factor in adsorption and enzyme reactions. In this way, one part of a molecule can be presented to a reacting substance.

HYDROTROPY

Introduction

Moore and Parker (1901) studied the solvent action of 5% aqueous solution of bile salts on fatty acids (and their salts). The solubility of acids in water was less than 0.1%, in bile salt solution was 0.5% and the solubility was increased to 4.0% when 1% of lecithin was added. The fatty acids could be readily recovered from solutions by extraction with solvents showing that the fatty acid molecule had not been altered.

Neuberg (1916) suggested that substances having the power of making water-insoluble substances water-soluble should be called "hydrotropic substances."

Definition

It is the process by which water-insoluble substances are made water-soluble by hydrotropic substances.

Hydrotropic substances : Cholic acids, benzoic acids, hippuric acids, phenyl acetic acid, and soaps of higher fatty acids.

Water-insoluble substances : Fats, phospholipids, sterols, calcium carbonate, calcium phosphate, magnesium phosphate, and uric acid.

Biological importance

1. Substances so dissolved by hydrotropy are diffusible through membranes.
2. In the body, hydrotropic substances are found not only in bile but also in intestinal juice, extracts of intestinal mucosa, blood plasma and places where such substances are particularly useful in helping absorption and transport of insoluble substances such as cholesterol and fatty acids formed by digestion.
3. The hydrotropic substances have the power of decreasing surface tension, eg, bile salts reduce the surface tension of fats and make emulsification for lipase action.

VISCOSITY

Definition

The resistance experienced by one layer of a liquid in moving over another layer is called "viscosity".

Viscosity varies greatly. Ether and gasoline have little viscosity and are quite mobile. But honey and coal-tar have high viscosities. The unit of viscosity is the "Poise", named after Poiseuille. It is expressed as

$$\eta(\text{eta}) = \frac{\pi P r^4}{8VL} t$$

Factors affecting viscosity

1. *Temperature :* The viscosity of liquids decreases by about 2% for each degree rise in temperature.
2. *Chemical composition :* The viscosity of liquids generally depends upon the size, shape and chemical nature of their molecules. It is greater with larger than with smaller molecules; with elongated than with spherical molecules. Large amounts of dissolved solids generally increase viscosity. Small amounts of electrolytes lower the viscosity of water slightly.
3. *Colloid systems :* The viscosity of lyophilic colloid solution is generally relatively high.
4. *Suspended material :* Suspended particles cause an increase in the viscosity. The viscosity of blood is important in relation to the resistance offered to the heart in circulating the blood. The heart muscle functions best while working against a certain resistance. The viscosity of blood is due largely to the emulsoid colloid system present in plasma and to the great proportion of suspended corpuscles.

CLINICAL ASPECTS

- Blood cells increase the viscosity of blood. Therefore, higher the number of blood cells, the greater is the viscosity. So viscosity is high in polycythemia while it is low in chronic anaemia.
- Increase in blood viscosity decreases blood viscosity, especially in capillaries. This helps in the blood flow through narrow vessels.
- Blood viscosity helps in streamlining blood flow by reducing turbulence.
- Blood viscosity helps in hemodynamics by influencing resistance and turbulence. This rise in viscosity increases cardiac work load. Increase in viscosity may reduce circulation in chilled extremities.

COLLOIDS

Introduction

Graham (1861) first distinguished two types of solutions–crystalloidal which diffused through a parchment or animal membrane, and colloidal which did not. The difference is only one of the size of the solute particles and is unrelated to their chemical nature. The particles are of the size of small molecules (like sugar or urea) and they form a crystalloidal or a true solution. In such a solution, the difference in size of solute and solvent molecules is relatively small, so that the solution can be called homogeneous. But if the particles are large compared to the solvent molecules, the solution can be called heterogeneous, and if the particles do not separate when the mixture is allowed to stand, this is called a colloidal solution.

About 90% of the organic matter of living tissues is present in the colloidal state.

The colloids depend much on surface area. A physiological example of the importance of surface area may be pointed out in case of the R.B.C. which contains the oxygen-carrying pigment hemoglobin. As blood circulates through the capillaries of the lungs, oxygen diffuses the surface of the corpuscles and unite with the hemoglobin to form oxyhemoglobin. This is performed by a quick process. This is facilitated by the great surface area for taking up oxygen.

Definition

Certain substances such as proteins, polysaccharides do not diffuse through parchment or animal membrane although they form homogeneous or heterogeneous solution. These substances are called colloids.

Colloid terminology

All colloid systems are composed of two phases of matter, one of which, represented by the colloidal particles, is called the "dispersed phase". The second phase is the "dispersions medium" which is also designated the external phase. The term "colloid sol" or "simply sol" is synonymous with "colloidal solution". A hydrosol presents a colloidal system in which water is the dispersion medium while alcohol is the dispersed medium in an alcosol. A lyophobic colloid system "suspensoid" is one in which there is little attraction between the colloid particles and the dispersion medium. A hydrophobic colloid system is a lyophobic system in which the dispersion medium is water. A lyophilic colloid system "emulsoid" is one in which the colloidal particles have a high affinity for the dispersion medium and are combined with some of the medium.

A Gel is a lyophilic colloid system that is more or less rigid. Gels generally are made up of "brush heap" fibrillar structures surrounded by dispersions medium. Gels are freely permeable to non-colloid ions and the molecules. The larger aggregates of colloidal particles formed in the process of gel formation are called "miscelles".

Size of the colloidal particles

Generally the size of each particle is 10-100 nm (submicrons). Only visible under ultramicroscope.

Determination of the size of colloidal particles

1. By ultrafiltration. 2. By diffusion rate. 3. By the ultracentrifuge. 4. By light scattering. 5. By X-ray analysis. 6. By osmotic pressure.

1. By ultrafiltration

(a) Colloidal particles readily pass through ordinary filter paper and porcelain filters. A series of collodion membranes with graded pore-sizes may be prepared and used for the determination of colloidal particle size.

(b) The colloidal solution tested is filtered under pressure through the series of filters.

(c) The size of the colloidal particle is between the pore size of the first filter through which it will not pass and the next larger pore size through which it will pass.

2. By diffusion rate

(a) The rate at which particles diffuse in solution has been used for the estimation of particle size and of molecular weights.

(b) The method is applicable to particles of colloidal and subcolloidal size such as those of the sugars and amino acids.

(c) The rate of diffusion is best measured by optical methods, such as refractive index, light absorption and Tyndall effect, in which the solution is not disturbed.

(d) The diffusion coefficient represents the number of moles of solute diffusing across unit area per unit time under a concentration gradient of unity. In colloids with spherical particle, the diffusion coefficient, D, is related to the molecular weight, M, by the equation

$$D = \frac{RT}{N6\pi\eta}\sqrt[3]{\frac{4\pi\mu}{3MV}}$$

where, R = gas constant, N = Avogadro's number, η = Viscosity of the medium in poises.

V = Partial specific volume, M = Molecular weight, T = Absolute temperature.

3. By the ultracentrifuge

(a) Ultracentrifuges have been constructed in which the rotor revolves 60,000 times per minute giving centrifugal forces of the order of 500,000 times gravity.

(b) When solution of proteins are placed in glass in the rotor of the ultracentrifuge, they are subjected to a force depending upon the angular velocity of the centrifuge and the distance of the colloidal particles from the axis of rotation.

(c) The rate at which the particles move is dependent upon this force and also upon the shape, size, and density of the particles, and upon the density and viscosity of the suspending medium.

(d) In case of homogeneous substances where the particles are all alike the particles move under the centrifugal force as a sharp boundary in the medium. Since the particles absorb more light than the medium, their rate of movement may be determined by photographing the position of the particle boundary in the cells at various time intervals.

(e) Ultracentrifugation is used in the separation and purification of virus proteins mixed with tissue proteins. The virus proteins of very large molecular size are centrifuged to the bottom of cell at a centrifuge speed at which this smaller tissue proteins are left suspended in the media. Centrifugation is used in the separation of mitochondrial, microsomal, and nuclear fractions from disrupted tissue cells.

4. By light scattering

(a) The scattering by light by colloidal particles affords a valuable method for determination of particle size and it has specific application in estimating the molecular weights of proteins.

(b) When a beam of light is passed through a colloidal solution, a part of the light is transmitted and a part is scattered.

(c) Determination of molecular weight from light scattering is useful over a much wider range than determination from osmotic pressure.

(d) The lower limit is about 12,000 but for very high molecular weights light scattering is greatly superior since the intensity of scattering goes up with increasing molecular weights, osmotic pressure goes down.

5. By X-ray analysis

From the diffraction patterns of X-rays that have passed through particles of matter, it is possible to calculate particle size.

6. By osmotic pressure

This method is successful as some of the smaller molecules leak through the membrane used, while the very large molecules give a very low osmotic pressure relative to the weight concentration and a large probable error of measurement.

Electric charge of colloidal particles

The colloidal particles are electrically charged. This charge is of much importance in stabilizing colloidal solution. The particles repel each other and remain in suspension. The neutralization of the charges causes precipitation or flocculation of colloids.

Streaming potential or flow potential

When water is present in a glass capillary, the walls of the capillary are negatively charged and the water positively charged. If water is forced to flow through the capillary from left to right, a potential develops between the ends of the capillary. This potential is referred to as the streaming or flow potential.

It varies with the nature of the liquid and the capillary. The glass capillary probably becomes negative by absorbing OH^- ions from water, leav-

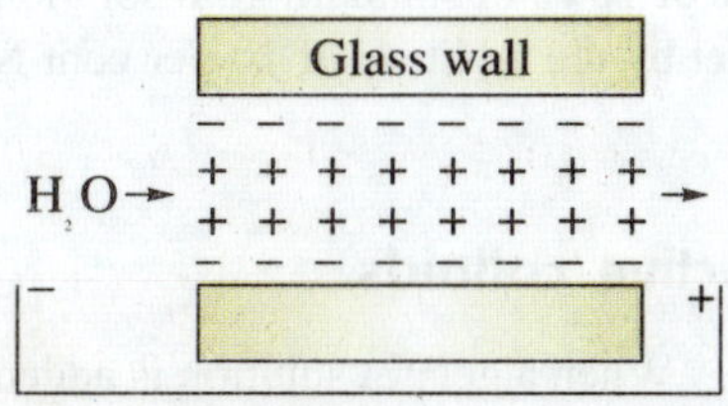

Fig. 2.8 : Streaming potential or flow potential

ing the H^+ ions to charge water molecules. This positively charged water molecules are mobile. As water flows through the capillaries, the positive charges accumulate in excess at one end leaving an excess of negative charges at the other. This causes potential differences which opposes the flow.

The flow of fluids through the membranes of tissues develops streaming potentials which contribute to the potential differences across such membranes.

Precipitation or Flocculation of colloidal particles

In order to precipitate emulsoids (hydrated colloids) by salts, sufficient salts must be added to dehydrate the emulsoid. At the same time the positive or negative ion of the salt neutralizes the charge upon the particles leading to precipitation.

The colloidal protein particles of blood serum may be precipitated by adding a large amount of $(NH_4)_2SO_4$. The precipitations of emulsoids by adding large amount of soluble salts is referred to as "salting out''.

Lyophilic colloids are precipitated by addition of colloids of the opposite charge, then neutralization of colloids occurs.

In case of lyophobic colloids they can be precipitated by addition of electrolytes. It neutralizes the charge.

This precipitation of lyophobic colloid by the addition of electrolytes prevented by the addition of lyophilic colloids. This is called protective action of lyophilic colloids on lyophobic colloids, this is expressed by gold number (least weight in mg of lyophilic colloid which can prevent the conversion of 10 cc of standard gold solution from red to violet by the addition of 10 per cent NaCl solution)

Protective colloids

1. When a gelatin solution is added to a gold sol, the particles of the emulsoids (gelatin) are absorbed by the particles of the suspensoids (gold) and the gold particles become much more resistant to precipitation.
2. Protective colloids play an important role physiologically. Some of the calcium phosphate of blood is held in colloidal suspension by the protective action of the proteins present.
3. The bile salts also act as protective colloids to keep sparingly soluble cholesterol and the calcium salt of bilirubin in colloidal suspension. Gallstones may result from the precipitation of such substances in the absence of sufficient protective colloids.
4. Protective colloids in urine may prevent bladder stone formation.

Stability of colloidal system

1. Hydration shell in case of lyophilic protein.
2. Charge of each colloidal particle in the colloidal system.
3. Brownian movement.

General properties of colloidal solution

A. Brownian Movement

1. The continuous motion of the particles is known as Brownian Movement.
2. The particles are kept in movement by continuous buffeting by the solvent molecules which are themselves always in motion.
3. The rate of Brownian Movement depends on the size of the particles, the smaller particles are more easily moved than the big ones.
4. If the particles are too large, they gradually sink under the influence of gravity.
5. Brownian Movement is quite haphazard.
6. The particles move in a straight line with sudden irregular changes of direction.

B. Osmotic pressure of colloidal solutions

1. Since the size of the colloidal particles are bigger than that of solvent molecules the number of colloidal particles are less relative to solvent molecules in a solution.
2. The osmotic pressure of a solution is directly proportional to the number but not the size of the dissolved particles. Therefore, the colloidal solutions have a low osmotic pressure.
3. The serum proteins which are present to the extent of 7% to 8% exert only an osmotic pressure of about 30 mm Hg.; whereas the crystalloids of serum (0.9% NaCl) have an osmotic pressure of about 5,200 mm Hg. and 6% sucrose has an osmotic pressure of about 3,000 mm Hg.
4. Soap solutions have small osmotic pressure because the soap molecules aggregate to form micelles of colloidal dimensions.
5. Although the osmotic pressure of colloidal solutions is very small, it has immense biological importance in providing the driving force for the passage of water and other substances through cell membranes.

C. Dialysis

1. The process of separation of crystalloids from colloids by diffusion through a membrane by osmotic force is called dialysis.
2. The most usual membranes used today for dialysis are the various grades of collodion (made from solutions of cellulose–nitrates or acetates—in solvents such as alcohol, ether or acetic acid) or cellophane parchment.
3. Dialysis is particularly needed for removing salt from proteins after precipitation by "salting out''.
4. The precipitate with a little water is placed in a colloidion bag and immersed in water.
5. The passage of salt into water is accelerated by keeping salt content of the water low, i.e. by running water.
6. Arrangement must be made for a big volume increase inside the bag in the early stages of dialysis, since it is a strongly hypertonic solution and water molecules will pass in more quickly than salt molecules pass out.
7. The pressure will only be transient if the water outside is frequently altered and the final pressure inside the bag may be very slightly more than the original.
8. Dialysis is applied in medicine in the "artificial kidney''.
9. This mechanism is inserted into the patient's circulation and urea passes out from the blood, substituting for the action of the faulty kidneys.
10. Dialysis of electrolytes can be done by passing an electric current through the solution. A cell consisting of three compartments separated by membranes is used; the colloidal solution is placed in the centre compartment and the electrodes in the outer ones. Positive ions are attached to the cathode and negative ions to the anode. This process is called electrodialysis.
11. Haemodialysis—Blood from arterial side is allowed to flow through the inner of the two concentric cellophane tubes with dialyser fluid passing through the annular space. The noncolloidal impurities in the blood are dialysed out and the purified blood is returned to venous side.
12. Peritoneal dialysis — Dialysing fluid is run into peritoneal cavity and after allowing time for exchange across the peritoneal membrane, the fluid is run off.

Types of colloidal solution

There are two types of colloidal solutions–Emulsoids and Suspensoids.

A. Suspensoids

1. The surface tension and viscosity of suspensoids are nearly the same as those of the solvent.
2. The suspensoid particles carry a definite electric charge which determines the stability of the suspensoid.
3. They are very easily precipitated if the charge is neutralised.
4. Once they are precipitated, they are not brought back into colloidal solution again.
5. Suspensoids are not hydrated, hence they are said to be hydrophobic or lyophobic colloids (water-fearing colloids).

B. Emulsoids

1. They are very stable and not easily precipitated by salts.
2. If they are precipitated, they are easily redissolved to form a colloidal solution.
3. They have a lower surface tension and much higher viscosity than the solvent.
4. The particles carry electric charges; some carry positive and negative charges simultaneously, eg, proteins.
5. The nature of the charge on a protein particle can be changed by altering the pH of the solution.
6. Practically, all the colloids of the living cell exist as emulsoids. They have a great affinity for water, hence they are called hydrophilic.
7. The emulsoid particles are molecules surrounded by shells of adsorbed water.
8. An emulsoid may be changed into a suspensoid by dehydration and then is precipitated.

An emulsoid can exist in two forms–Sol and Gel.

Sol :

1. Sol can be converted into gel by changes of temperature, hydrogen ion concentration or salt concentration.
2. The continuous phase in sol is water (or a dilute solution), and the disperse phase is a concentrated solution.
3. Addition of $CaCl_2$ to an alkaline solution of caseinogen gives a sol.

Gel :

1. Solutions stronger than about 1% form gels on cooling.
2. Gels are quasi-solids.
3. They do not assume the shape of the vessel in which they are placed.
4. Great pressure is required to squeeze water out of a gel.
5. In a gel, the concentrated solution forms the continuous phase and water the disperse phase.
6. Formaldehyde can remove water from the gel and hence it is used in the histological process of "fixing" tissues.
7. Many tissue structures are essentially gels. Cytoplasm may undergo sol $\rightleftharpoons$ gel transformation.
8. Tissues, especially the skin, are not dehydrated owing to the existence of gels.

Lyophobic colloids (Suspensoids)	Lyophilic colloids (emulsoids)
1. Mostly inorganic materials eg, gold sol.	1. Mostly organic eg, proteins, starch.
2. Well defined under microscope or ultramicroscope.	2. Invisible under microscope or ultramicroscope.
3. Easy coagulation by electrolytes.	3. Not easily coagulated by addition of substance or by cooling.
4. Undergo irreversible coagulation.	4. Coagulation is reversible.
5. Viscosity same as solvent (usually water)	5. Much more viscous than water.
6. Do not gelatinise readily.	6. Gelatinise readily.
7. Stabilize mainly by surface charge.	7. Stabilize by force of solvation.

Imbibition

1. Dried emulsoids take up (imbibe) water and swell considerably; this process is called imbibition.
2. The process of germination of seeds is the taking up of water by imbibition.
3. Heat is liberated during imbibition.
4. Most dried animal and vegetable tissues show imbibition.
5. In imbibition, water is not expelled by squeezing.

Physiological application of colloids

1. **Enzyme action :** Lyophilic colloids such as enzyme proteins are big particles and intern consists of aggregation of thousands of smaller particles. This again increases of its area, the substance it acts upon, it is absorbed (eg, enzyme catalysis change this action).
2. **Maintenance of blood pH :** Lyophilic colloids such as proteins (eg, haemoglobin) are independent of blood pH.
3. **Protective action of lyophilic colloids (such as proteins) :** Lyophilic colloids afford protection to lyophobic colloids such as Ca-bilirubinate and cholesterol in bile and $Ca_3(PO_4)_2$ in blood and milk from being precipitated by electrolytes such as Na and K and hence prevent formation of stone.
4. **Water transport and urine formation :** Colloidal osmotic pressure of plasma proteins in blood is known as oncotic pressure which is responsible for the maintenance of blood volume, hence, water transport and urine formation.
5. **Glandular secretion :** Because of taking up of water by lyophilic colloids and excreting water in the form of glandular secretion.
6. **Gibbs–Donnan membrane equilibrium :** Consider a semipermeable membrane separating solution of non-diffusible ions from another solution with diffusible ions. Non diffusible ions will enhance the diffusion of oppositely charged diffusible ions towards their side. At the same time they will reduce the diffusion of identically charged ions to their side. Only passive forces are involved in the movement of ions across semi-permeable membrane. So, oppositely charged diffusible ions are more concentrated on the side of non-diffusible ions while diffusible ions with like charges are concentrated on the opposite side. This is called Gibbs-Donnan effect.

On both sides of semi-permeable membrane, since total cation concentration equals total anion concentration on either side at equilibrium, so total charges are being contributed both by diffusible and non-diffusible ions.

SPM (Semi-permeable membrane)

Side 1	Side 2
$Na^+ (C_1)$	$Na^+ (C_2)$
$Pr^- (C_1)$	$Cl^- (C_2)$

SPM

Side 1	Side 2
$Na^+ (C_1 + x)$	$Na^+ (C_2 - x)$
$Pr^- (C_1)$	$Cl^- (C_2 - x)$
$Cl^- (x)$	

If any NaPr solution is separated by semi-permeable membrane from solution of Na^+ and Cl^- each having molar concentration of C_1 and C_2, respectively, then at equilibrium Gibbs-Donnan effect becomes

Side 1 has : $[Na^+] = (C_1 + x)$
$[Pr^-] = C_1$
$[Cl^-] = x$

Side 2 has : $[Na^+]_2 = (C_2 - x)$
$[Cl^-]_2 = (C_2 - x)$

Only diffusible ions Na^+ and Cl^- can cross the membrane.

According to Gibbs-Donnan membrane equilibrium, product of molar concentration of diffusible ions on one side equals on the other where non-diffusible ion is retained by membrane on one side :

$$[Na^+]_1\,[Cl^-]_1 = [Na^+]_2\,[Cl^-]_2$$

$$\text{Or,}\quad (C_1 + x)x = (C_2 - x)^2$$

$$x = \frac{(C_2 - x)^2}{C_1 + x}$$

$$\text{Or,}\quad C_1x + x^2 = C_2^2 - 2C_2x + x^2$$

$$\text{Or,}\quad C_1x + 2C_2x = C_2^2$$

$$\text{Or,}\quad x(C_1 + 2C_2) = C_2^2$$

$$\text{Or,}\quad x = \frac{C_2^2}{C_1 + 2C_2}$$

Application of Gibbs–Donnan effect

1. **Chloride shift :** High PCO_2 in tissues promote diffusion of CO_2 combines with H_2O to form H_2CO_3 with the help of carbonic anhydrase. H_2CO_3 dissociates into H^+ and HCO_3^-.

   The portion is buffered by deoxygenated Hb (isohydric exchange) under the influence of HHb, HCO_3^- diffuses out resulting in simultaneous inflow of Cl^- into the cell to maintain electroneutrality.

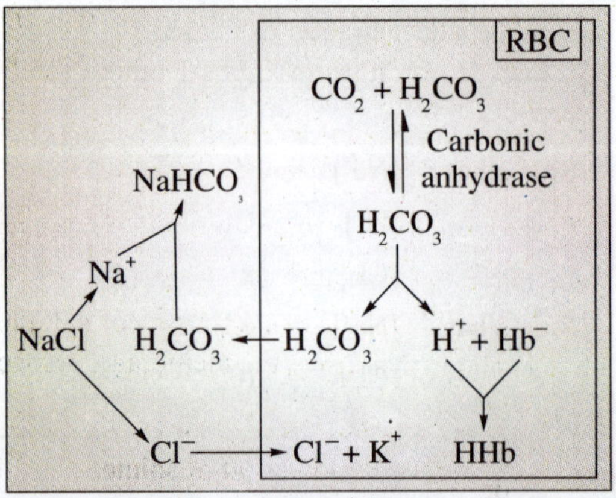

2. Gibbs-Donnan effect of non-diffusible Hb^- promotes retention of H^+ and outward diffusion of some OH^-, thus lowering pH in the erythrocyte slightly below plasma.

CLINICAL ASPECTS

- Urinary Protective Colloids prevent urinary stone formation.
- Bile salts act as protective colloids and prevent precipitation of cholesterol and calcium salts of bilirubin. In absence of bile salts, cholesterol and calcium salts of bilirubin get precipitated to form gallstones.

TRANSPORT THROUGH BIOLOGICAL CELL MEMBRANE

The living organisms can be resolved into organs, glands, tissues, cells and organelles. It is very interesting in biology to know how solutes and water get into and out of cells and organelles. Most attention is to be paid to erythrocytes and mitochondrion. The cell membrane is a complex lipoprotein structure.

Some channels are continuously open, whereas others are gated i.e. they have gates that open or close. Some are gated by alterations in membrane potential (voltage gated) whereas others are opened or closed when they bind a ligand (ligand gated). The ligand is often external (neurotransmitter or hormone) or internal (intracellular Ca^{++}, cAMP).

Other transport proteins are carriers that bind ions and other molecules and then change their configuration, moving the bound molecule from one side of the cell membrane to the other.

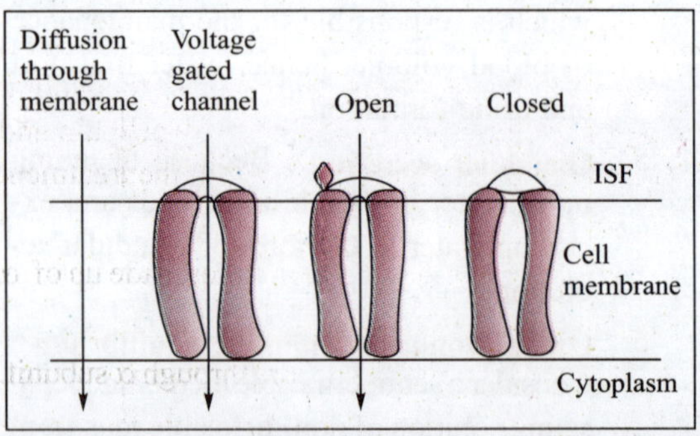

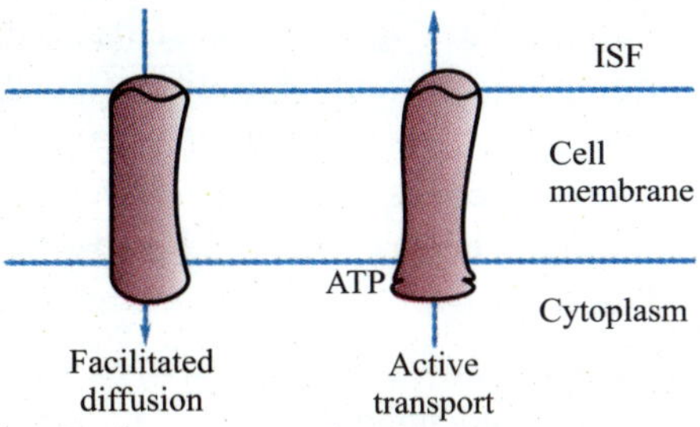

Fig. 2.9 : Transport of ions and other small molecules across cell membrane

Molecules move from areas of high concentration to areas of low concentration (down their chemical gradient).

Cations move to negatively charged areas whereas anions move to positively charged areas (down their electrical gradient), ligand gated channel.

Some of the carrier proteins are called uniports because they transport only one substance. Others are called symports because transport requires the binding of more than one substance to the transport protein and the substances are transported across the membrane together. Example : In the intestinal mucosa that is responsible for the cotransport by facilitated diffusion of Na^+ and glucose from the intestinal lumen into mucosal cells.

Other transporters are called antiports because they exchange one substance for another. Example : Na^+-K^+ ATPase.

Na^+-K^+ ATPase

It catalyzes the hydrolysis of ATP to ADP and uses the energy to extrude $3Na^+$ from the cell and take $2K^+$ into the cell for each mole of ATP hydrolyzed. The pump is said to have a coupling ratio of 3/2. Its activity is inhibited by ouabain and related to digitalis glycosides used in the treatment of heart failure.

Na^+-K^+ ATPase is a heterodimer made up of α and β subunit.

Na^+ and K^+ transport occurs through α subunit.

β subunit is a glycoprotein.

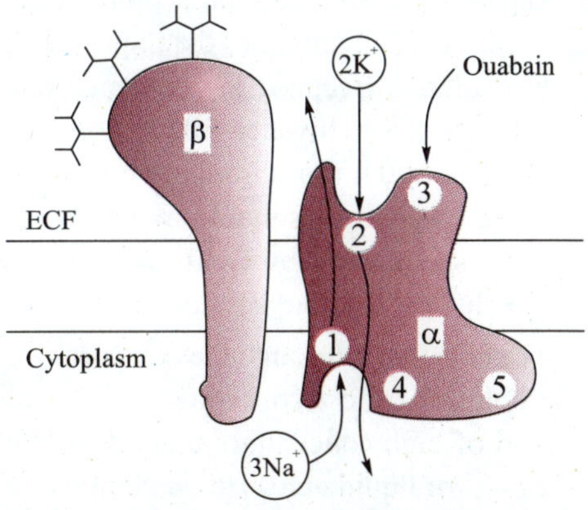

Fig. 2.10 : Na^+-K^+ ATPase. The intracellular portion of the α subunit has a Na^+ binding site (1), a phosphorylation site (4) and an ATP binding site (5). The extracellular portion has a K^+ binding site (2) and a Ouabain binding site (3)

Substances passing through the lipid bilayer of the cell membrane by simple diffusion are :

1. All lipid soluble substances.
2. Lipid soluble gases mainly CO_2, O_2 and N_2.
3. Water—though not lipid soluble—passes because of small molecular size and high kinetic energy.

Substances passing through protein channels of cell membrane by simple diffusion are :

1. Ions mainly Na^+, K^+ and Ca^{++}.
2. Water molecules.

A. Passive diffusion

1. Some solutes pass through cell membrane by simple diffusion with the concentration gradient. This can be expressed by the modification of Fick's law :

$$\frac{ds}{dt} = PA(C_0 - C_i)$$

where, P = the permeability coefficient.

A = area of membranes.

C_0 and C_i = the concentration of solution outside and inside the membrane, respectively.

$\frac{ds}{dt}$ = rate of movement of solute.

2. Lipid-soluble solutes pass more readily through cell membranes than lipid-insoluble solutes. Because the cell membrane consists of small water-filled pores of radius about 0.4 nm. through which water-soluble solute of suitable molecular size pass, surrounded by lipid areas through which lipid-soluble solutes penetrate.
3. Water diffuses through the cell pores from a solution of low concentration to a solution of high concentration and this "bulk flow" of liquid across the membrane will speed up molecules diffusing in the direction of the flow and slow down those moving in the opposite direction. This "drag" effect is a second force acting in passive diffusion.
4. The third force which may operate is an electric potential across the membrane. Many cell membranes can maintain potential difference between their inside and outside and the potential gradient acts as a driving force for passive transport across the cell. The membrane acts as a passive barrier.

B. Facilitated transfer

1. Some compounds, eg, sugar, amino acids, pass through membranes at a greater rate than expectations. This is because of the effect of a carrier.
2. The carrier in the membrane combines with the substance to be transported and in some way ferried through the membrane and released on the other side.
3. In case of enzymic reactions, there is a "saturation effect". The rate of transport of the solute increases when the carrier, enzyme, is saturated. This type is sometimes termed "catalysed diffusion".
4. Another mechanism is that the substance to be transferred is converted into another which will penetrate the membrane more easily, eg, the mitochondrial membrane is impermeable to acyl coenzyme A derivatives. The acyl group is transferred to carnitine to form acyl carnitine derivative which can pass through the membrane. The acyl coenzyme A derivative is then reformed on the other side of the membrane.

   Fatty acids can also be transferred into and out of mitochondria.

   Acetyl-CoA within the mitochondria can be transferred to oxaloacetate to yield citrate to which the mitochondrial membrane is permeable. The citrate passes out into the cytoplasm where it is split enzymically to give acetyl-CoA again.

C. Pinocytosis

1. The cell membrane forms pockets or invaginations which can draw materials on the outside towards the cell interior.
2. The vesicles extend into the cell where they are pinched off and finally release their contents into the cell by some unknown way.
3. This process occurs in the foetal and newborn animals and helps the absorption of intact protein from the gut.

D. Transport of ions

1. The membrane itself contains polar groups and is, therefore, electrically charged.
2. The transport of most ions occur more slowly than the non- electrolytes. But H^+, OH^- penetrate all cell membranes easily. The red cell is easily penetrated by Cl^- and HCO_3^-.
3. In the case of ions, especially, Na^+ and K^+, the permeability is very small. The high concentration of K^+ and low concentration of Na^+ which are often found in cells are maintained by special mechanism which involve the expenditure of energy.

E. Active transport

1. The process by which solutes can often pass through membranes against their concentration gradient requires energy. This process is termed active transport.
2. Active transport is involved in the absorption from the small intestine of glucose and galactose, amino acids and other substances important to the body.
3. An active transport device which forces Na^+ out and K^+ in has been referred to as the "Sodium Pump".
4. The mechanism requires a carrier which can exist in two forms with different affinities for Na^+ and K^+. ATPase is involved in it (see active transport of glucose).

ISOTOPES

Definition

Isotopes may be defined as atoms having the same atomic number but different atomic weights.

They are the subspecies of the same chemical element and occupy the same position in the periodic table, but have different physical properties.

The atomic constitution of three isotopes of hydrogen are illustrated :

H^1 or ordinary hydrogen consists of a nucleus containing a proton (charge +1, mass 1) around which revolves one electron (charge –1).

Heavy hydrogen contains an additional nuclear particle, a neutron (charge 0, mass 1), H^3 contains two neutrons.

There are two classes of isotopes. The "stable" isotopes which have no distinguishing characteristics other than their mass. They are obtained from natural sources by fractional procedure.

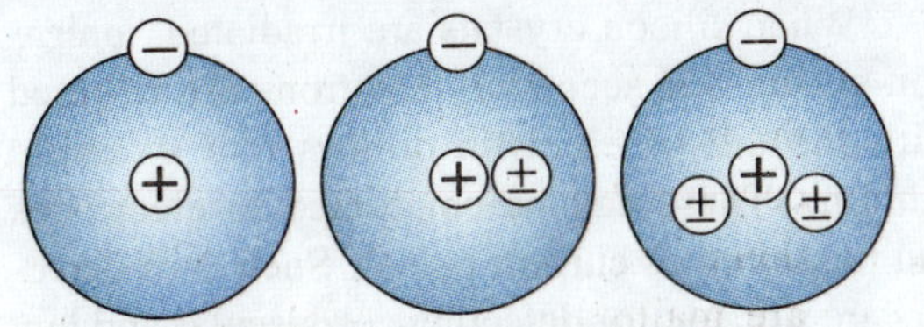

Fig. 2.11 : Isotopes of hydrogen

"Radioactive" isotopes not only differ in mass but they are also characterized by unstable nuclei. This causes them to decompose spontaneously emitting radiation in the form of waves. They occur in nature in traces. They are prepared for experimental use by bombardment in the cyclotron or the atomic pile. The concentration of radioactive isotopes is expressed in terms of intensities of radiation emitted. The unstability of these isotopes is commonly expressed as their "half life"—the time required to decrease to half its initial value.

Preparation of labelled material

It can be prepared by Grignard reaction as :

$$\text{R–CH}_2\text{–Mg–Br.} \xrightarrow[\text{then acidify}]{\text{C}^{14}\text{O}_2} \text{R–CH}_2\text{–C}^{14}\text{OOH.}$$

Chemical properties of Radioactive substances

The artificial isotopes of a chemical element have the same properties as the natural ones and also a similar metabolic activity. Quantitative dissimilarities may occur since differences in the mass numbers may result in different mass in the diffusion and dissociation constant. The more the relative atomic weight of the isotope differs from that of natural element the greater is this "isotope effect". For the same reason, the greater the number of isotopic atoms a molecule contains, the more will its metabolic turnover differ from that of ordinary molecule.

The replacement of a stable by a radioactive atom in the molecule of a chemical compound is known as "tagging" or "labelling"—resulting in isotope as a 'tracer'. Biosynthesis by living organisms and enzymatic synthesis in vitro are the least harmful methods of tagging organic substances.

Compounds of very high specific activity tagged with carrier free short-lived radioactive substances should not contain more than one radioactive atom per molecule.

The disadvantage of high specific activities is that organic compounds may be decomposed by their own radiation.

Radio substances are designated "carrier free'' when they are not contaminated by any other isotope of the same element. Most of the radioactive substances commercially available are not absolutely carrier free owing to the unavoidable adsorption phenomena occurring at the surfaces of pipettes and other glassware at the extremely low concentrations used in their preparations. Apart from this apparatus, the solvents often contain traces of the stable isotopes. For this reason, the specific activity of many carrier free radio substances falls during storage.

Half-life

The time taken for half the atoms in a quantity of a radioactive substance to disintegrate is known as the half-life.

The concept of the half-life is also applicable to the biological disappearance of a substance tagged with the radioactive substance in a biological system. This disappearance, at least initially, is often not a uniformly exponential process.

The effective half-life which is always shorter than both the biological (T_b) and physical ($T_{1/2}$) half-lifes, can be calculated from the following formulae :

$$\frac{T_{\frac{1}{2}} \times T_b}{T_{\frac{1}{2}} + T_b} = T_{eff}$$

This formula is valid only for uniform system often it suffices to estimate T_{eff} by graphic lines.

DEDUCTION OF RADIOACTIVITY

1. Ionization

The passage of high energy radiation through matter results in the formation of ions as a result of collision of electrons with atoms. So much energy is transferred to the orbital electrons that on escape from the atom, it gives rise to a slow positively charged ions and a very fast secondary electron. If the ions and electrons are not immediately separated by means of an electric field they recombine and remain undecayed. Electrons slowed by multiple collision can be captured by reactive gas molecules and give rise to negative ions. An ion strongly accelerated in an electric field may collide with a neutral gas molecule and thereby give rise to a fresh positive ion and electron. X- and γ-rays must also first give rise to free electron before they can be detected and since the probability of ionization occurring decreases rapidly with increasing energy. Such a radiation is more difficult to detect.

2. Ionization chambers

While ionization chamber measurements form the absolute basis of descimetry, the method is too slow and insensitive for detecting short-lived radioactivity.

3. Geiger counters

These are gas-filled counters operating at reduced pressure. They do not measure continuous currents like ionization chambers but register collision ionization. The primary ions in the counter gas are multiplied by applying an electric field of 800 to 2,000 V. In the range of 200 to 600 V, number of ions present is strictly proportional to the number of primary ions, and for this reason proportional counters can be used to distinguish β rays from highly ionizing particles. The life-time of a gas-filled counter is limited by the capacity of the gas to a total of 10^9 to 10^{10} collision discharges.

4. Semi-conductor detectors

When silicon crystals are irradiated, ionization occurs and secondary electrons are released with the aid of electron donors (for example lithium). These can be conducted to electrodes and measured as current pulses. Such drift detectors are suitable for detecting corpuscular and low energy X- and γ- rays at room temperature. On account of their extremely small size they can even be implanted.

5. Scintillation

This is the name given to the light flashes emitted by luminescent substances when excited by high-energy radiation. The flashes can liberate photo electrons from photosensitive substances. The photo electrons are amplified 10^7 to 10^8 times by means of a photo tube multiplier before being converted into current pulses. The pulse height depends on the energy of the original. γ-radiation and the pulses can be sorted by means of a discriminator. By using different discriminator or channels, the different radioactive substances in a mixture of isotopes can be determined either successively or simultaneously.

(a) Solid scintillators : The commonest type in use in nuclear medicine consists of single crystals of thalium activated sodium iodide. Since the decay time of fluorescence is only 0.25μs the scintillation crystals have resolving power about a thousand time greater than gas-filled counters. Since there is no dissipation of the crystals the life-time of these scintillation detectors is limited only by that of the replaceable multiplier and its semi-permanent photosensitive layer. The wavelength of the luminescent radiation is about 410 nm.

(b) Liquid scintillators : The radiation from preparations emitting β-rays, soft X-rays or γ-rays can be measured with particularly high pulse yield if they are mixed directly with a scintillator solution.

Only a few highly purified alkylbenzenes (mainly toluene and xylene) and ethers are suitable as solvents. They transmit the radiation energy to the scintillators by ionization via metastable excited states.

The first scintillator usually consists of a solution of 2, 5- diphenyloxazole in toluene. Its fluorescence has a maximum at 380 nm. Since the photocathodes of many multipliers develop optimum activity only at wavelength above 400 nm the spectrum of the primary fluorescence must be displaced to higher wavelengths by using a second fluorescent substance.

6. Autoradiography

The second scintillator converts the ultraviolet radiation from the first by fluorescence into radiation with a wavelength of about 420 nm.

If the radioactive preparation is insoluble in the solvent, other substances must be added to make it soluble.

The oldest method of detecting radioactivity is the photographic one. Both electromagnetic and corpuscular radiation cause electrons to be expelled from the halogen atoms in the silver halide grains in a gelatin emulsion. Each electron reduces a silver ion to metallic silver. The sites at which this occurs in the silver halide grains constitute "development centres'' where the developer begins the reduction of the whole grain to black metallic silver. The resulting degree of intensification is about 10^{12}.

Macroscopic autoradiography is used in nuclear medicine mainly for radiation exposure monitors; another use is the localization of radioactivity in chromatograms and in organ sections of large surface area. The film used is high-sensitivity X-ray film; for pure γ-ray sources an intensifying screen is usually necessary, though this reduces the sharpness of the image.

Methods of assay

1. Radioactive isotopes can be determined by means of Geiger-Muller counter.
2. Particles or rays from the decay of the isotope enter an ionisation chamber and interact with the gas to form ions which causes a discharge of current by positive and negative plates.
3. The current is amplified and registered automatically in a counting device which registers the total number of counts for any desired period of time.

4. Isotope concentration can then be expressed as number of counts per unit time per unit weight of the sample.

Biochemical and diagnostic importance of isotopes

Biochemical importance :

1. I^{131} in the form NaI^{131} is administered in the body to study the thyroid physiology. About 1/3 of the iodine ingested is taken up by the thyroid and 2/3 is excreted by the kidney in normal human adults.
2. Plasma volume is measured by using I^{131} labelled serum albumin and erythrocyte volume is measured by using Cr^{51} labelled erythrocytes. Total body water is measured by using I^{131} labelled iodoantipyrin.
3. Absorption of fat is studied by using I^{131}-oleic acid, absorption of iron by using Fe^{59} ferrous salts and intestinal protein loss by intravenous injection of Cr^{51} labelled protein.
4. The life span of RBC has been determined by tagging Cr^{51} to RBC.
5. Radioactive isotopes are widely applied in the study of the intermediary metabolism, almost every phase of metabolism, eg, TCA cycle, amino acid metabolism, protein biosynthesis, nucleic acid synthesis, fatty acid synthesis, biosynthesis of haem and cholesterol etc. have been studied by using compounds containing C^{14}, N^{15}, H^2, H^3, P^{32}, S^{35} etc. Mineral metabolism has also been studied by using Ca^{45}, Fe^{59}, I^{131}, Na^{24}, K^{42} and Cl^{36} etc.

Radioisotopes in Medical Science

In recent years radioactive substances are used for the following purposes :

1. Experimental.
2. Diagnostic.
3. Therapeutic.

1. Experimental purposes

(i) The research work needs radioisotopes to investigate absorption, mobilization, storage, and distribution of different substances in the body.

(ii) The substance under investigation is 'tagged' with a radioactive isotope which acts as a tracer substance. The tracer substance is detected in the body by the radiation it emits. The best method used for the detection of the radioactive isotope is Geiger-Muller counter and the less sensitive is auto-radiograph method.

(iii) The absorption, mobilization, and transport of iron have been studied by the use of radio-active iron (Fe^{59}).

(iv) Iodine clearance of blood is calculated by radio-active iodine (I^{131}).

(v) The radio-active phosphorus (P^{32}) has been used for studying the formation of antibody in the reticulo-endothelial system.

2. Diagnostic purposes

(i) The volume of fluid compartments of the body is measured by radioactive isotopes.

(ii) Thyroid function is determined by the use of radio-iodine (I^{131}).

(iii) The radioactive cobalt (Co^{60}) is used for diagnosing pernicious anemia.

(iv) The radioactive sodium (Na^{24}) has been used for circulatory studies and diagnosis of arterial diseases.

(v) The diagnosis for the localisation of brain tumours has been done by the use of radioactive di-iodofluorescein.

3. Therapeutic purposes

(i) The isotopes which are selectively taken up by certain tissues and emit mainly β-rays without causing any damage to the surrounding healthy tissues are only used for this purpose. The most important radio-isotopes for the purpose are radio-iodine (I^{131}), radiogold (Au^{198}), and radiophosphorus (P^{32}).

(ii) The radioactive (I^{131}) as a solution of NaI^{131} is administered in the body after a light breakfast. The effect of a dose is observed within 3 to 4 weeks and maximum effect within 3 to 4 months. The second dose, if required, is given at the interval of 2 to 3 months. It acts by emitting β- and γ-radiations. This is used in the treatment of hyperthyroidism and thyroid cancers. The usefulness of I^{131} depends on the capacity of the thyroid gland to capture and concentrate iodine from the blood stream. NaI^{131}, in a very high dose, is used in the treatment of congested heart failure or angina pectoris.

I^{125} is more convenient than I^{131} because its half life is 7 times greater than that of I^{131}, the half-life of I^{125} is 57.4 days, whereas the half-life of I^{131} is 8.04 days. I^{125} can be stored for more periods with losing lesser activity and this causes lesser damage to the gland because it does not emit β-radiation.

(iii) Radioactive phosphorus (P^{32}) is administered in the form of sodium phosphate, either orally or intravenously. It emits β-radiation and its half-life is 14.3 days. Its end product is S^{32}. P^{32} is distributed to all tissues of the body as it has high metabolic role. When administered, it is rapidly taken up by multiplying cells, bone marrow, spleen, and lymph nodes. It is used in the treatment of various types of cancers. Especially, it is used in polycythemia vera and chronic lymphatic leukemia.

(iv) Radioactive gold (Au^{198}) emits β- and γ-rays and its half-life is 2.69 days. Its end product is Hg^{198}. It is chemically and biologically inert. It is deposited in tissue area when injected in the body. It produces local irradiation since very little of it enters the blood. The β-rays emitted by it are completely trapped by tissues. It is used in the treatment of malignant pleural and peritoneal effusions.

(v) Radioactive cobalt (Co^{60}) is the most important isotope in clinical medicine. It emits both β- and γ-rays and its half-life is 5.3 years. It can be used as a substitute for sodium in interstitial surface. It has been found to be cheaper and dues not require any closed container.

Radioimmunoassay (RIA)

1. It is a simultaneous competition between labelled and unlabelled antigen for the fixed number of antibodies.
2. It has been seen that the more the amount of unlabelled antigen, less is the binding of labelled antigen to antibody.

ELISA (Enzymelinkedimmuno-sorbentassay)

1. Three samples A, B, C will be required :

   A → contains highly purified antibody,

   B → serum of the patient containing the hormone (acts as antigen whose concentration is to be measured)

   C → Purified antigen but conjugated with horse-raddish peroxidase.
2. A, B, C are mixed together. Some of the horse-raddish peroxidase leaves C and conjugates with complex formed by A and B.
3. The complex of A and B is isolated and washed to remove extraneous materials. Finally, the complex is stained with dye and colour is measured under spectrophotometer.
4. When the concentration of serum hormone is high, the colour will be more intense.

Applied

1. In steroid hormones.
2. AIDS.
3. Protein in the serum.

EXERCISES

1. What are colloids? Discuss some specific properties of colloids. How can these properties be related to proteins and other macromolecules in the biological system? (P.U. 69S)
2. What do you mean by permeability of membranes? Illustrate your answer with examples from biological materials. Have pH change, composition of the membrane of electrolytes any role on the permeability of the membrane? (P.U. 69A)
3. What are buffers? How do you derive Henderson-Hasselbalch Equation? (Mith. 75A)
4. Define pH. What is the pH of blood? How is it regulated? (Luck. 65A; 70A; P.U. 60S)
5. What is osmotic pressure? Discuss its role in maintenance of fluid volume. State the application of osmosis. (P.U. 63S)
6. Describe the properties of colloids. What is protective colloid? (P.U. 64A)
7. What do you mean by co-efficient of viscosity? How is viscosity of blood determined? Describe its physiological importance. (Luck. 67S)
8. Define buffer. Name the important buffer systems of the body. (P.U. 73A)
9. Discuss the following :
   (a) pH of blood is 7.4.
   (b) Osmotic pressure of blood is seven times the atmospheric pressure.
   (c) Membrane of RBC is semi-permeable.
   (P.U. 69A)
10. Write short notes on :
    (i) Surface tension. (C.U. 1981) (M.U. 73A, R.U. 79S)
    (ii) pH (C.U. 1981) R.U.66S, 69A; Mith. 74A)
    (iii) Radioactive isotope. (C.U. 1981) (R.U. 65A)
    (iv) Diffusion. (M.U. 73s, R.U. 64A, Mith, 60S)
    (v) Buffers. (P.U. 75A, Bh. U. 75A, R.U. 64S)
    (vi) Colloidal osmotic pressure. (P.U. 75A)
    (vii) Osmosis and osmotic pressure. (C.U. 1981; 1983) (R.U. 70A; Mith. 62A)
    (viii) Protective colloids. (Mith. 73A, 75A)
    (ix) Colloids. (Mith. 61A)
    (x) Hydrotropy. (Bh. U. 75S; M.U. 73A; P.U. 68A)
    (xi) Adsorption. (Bh. U. 75A)
    (xii) Brownian Movement. (Mith. 74A)
    (xiii) Donnan membrane equilibrium. (Bh. U. 74S)
    (xiv) Viscosity. (P.U. 1983)
    (xv) Gibb's adsorption equation. (C.U. 1983)
    (xvi) Coefficient of viscosity and application of viscosity. (C.U. 1983)
11. What is active transport? Discuss the mechanism by which this process operates in the body. (C.U. 1981)
12. Describe the principles of buffer action. Write the major constituents of a buffer solution. (C.U. 1983)
13. State the laws of osmotic pressure. Write any two factors controlling O.P. of blood. (C.U. 1981)
14. What is meant by protective action of colloids? How 'protective action' is expressed? (C.U. 1981)
15. What is meant by a colloidal solution? Discuss the properties of a colloidal solution having special emphasis on electrical properties of colloids. (C.U. 1981)
16. Give briefly the properties of colloidal solution. (C.U. 1983)
17. Indicate "True" or "False"
    (a) Radioactive iodine has a half-life of 15 days...... (C.U. 1981)

CHAPTER 3

CARBOHYDRATES

The carbohydrates are widely distributed both in animal and plant tissues. In plants, they are produced by photosynthesis. In animal cells, carbohydrate in the form of glucose and glycogen serves as an important source of energy for vital activities. Some carbohydrates have highly specific functions, eg, ribose in the nucleoproteins of the cells, galactose in certain lipids and the lactose of milk.

Definition

Carbohydrates may be defined as polyhydroxy alcohols with aldehydes or ketones and their derivatives.

Classification

Carbohydrates are divided into 4 major groups as :

1. Monosaccharides.
2. Disaccharides.
3. Oligosaccharides.
4. Polysaccharides.

1. *Monosaccharides :* These are often called "simple sugars'' which cannot be hydrolyzed into a simpler form. The general formula is $Cn(H_2O)n$. These can be further subdivided as :

Carbon atoms	Aldoses	Ketoses
Trioses	Glyceraldehyde or Glycerose.	Dihydroxy-acetone.
Tetroses	Erythrose.	Erythrulose.
Pentoses	Ribose. Xylose. Arabinose.	Ribulose Xylulose.
Hexoses	Glucose. Galactose. Mannose.	Fructose.
Heptoses	Glucoheptose. Galactoheptose.	Sedoheptulose.

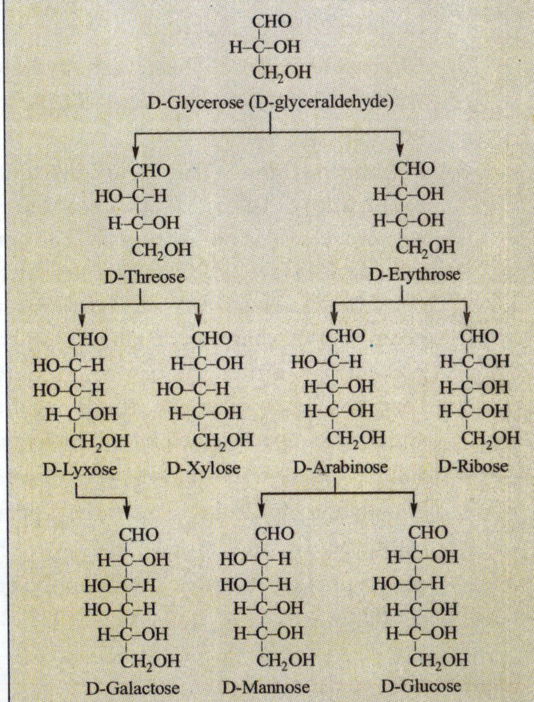

Fig. 3.1 : The structural relations of the aldoses, D series, of physiologic significance

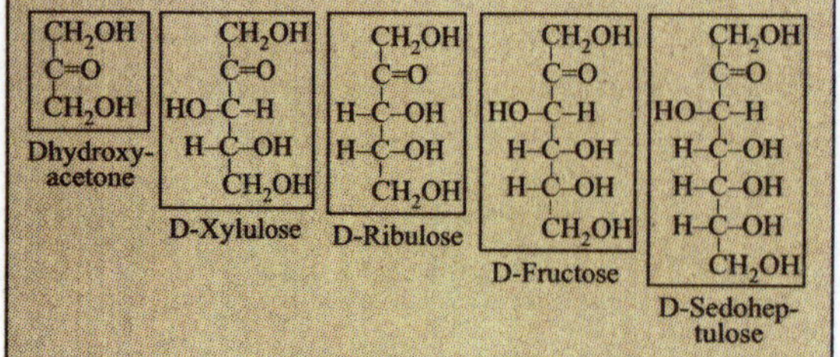

Fig. 3.2 : Examples of ketoses of physiologic significance

Simplest form of aldose — Glyceraldehyde.

Simplest form of ketose — Dihydroxyacetone.

Commonest aldose — Glucose.

Commonest ketose — Fructose.

Natural glucose — α-D-Glucose.

2. ***Disaccharides :*** These carbohydrates produce two molecules of the same or of different monosaccharides on hydrolysis. The general formula is $C_n(H_2O)_{n-1}$. Examples–Lactose, Maltose, Sucrose.
3. ***Oligosaccharides :*** These carbohydrates yield 2-10 monosaccharide units on hydrolysis, eg, Maltotriose.
4. ***Polysaccharides :*** These carbohydrates yield more than 10 molecules of monosaccharides on hydrolysis. The general formula is $(C_6H_{10}O_5)x$. These may be classified as homopolysaccharides and heteropolysaccharides depending on the presence of either the same monosaccharides or more than one simple sugar in alternating repeating sequence. Examples are :

   Homopolysaccharides : Starch, Glycogen, Cellulose, Dextrin.

   Heteropolysaccharides : Mucopolysaccharides.

Asymmetric carbon atom

A carbon atom to which four different atoms or groups of atoms are attached is said to be asymmetric carbon atom.

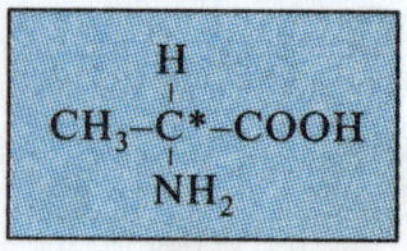

Many biochemicals contain 2 or more asymmetric C atoms.

All monosaccharides except dihydroxyacetone have asymmetric carbon atom.

Isomerism

The presence of asymmetric carbon atoms allow the formation of isomers. The compounds which have the same structural formula but differ only in spatial configuration are called Stereoisomers or Geometric isomers. Glucose with 4 asymmetric carbon atoms has 2^n (16) isomers, n is the number of asymmetric carbon atoms.

Classification

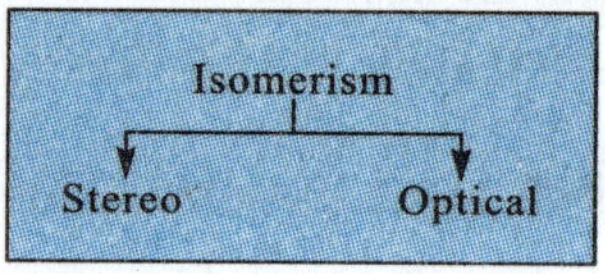

(i) D and L isomer.

(ii) Epimer.

(iii) Anomer (α and β)

(iv) Pyranose and Furanose.

(v) Aldose and Ketose.

Cis-Trans Isomerism

Cis-trans isomerism (Latin *cis* "this side", *trans* "across") occurs in compounds with double bonds. As the double bond is rigid, the atoms attached to it are not free to rotate like those attached to a single bond. The structures are :

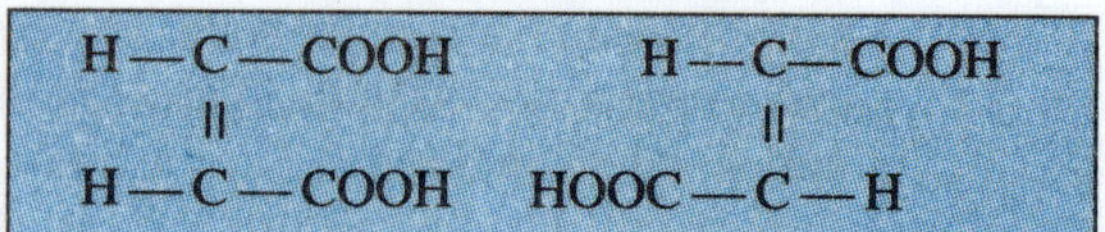

Maleic acid (Cis) Fumaric acid (trans)

These structures have different chemical and physiologic properties. Fumaric acid is physiologically active.

1. D and L

The designation of an isomer as D- or L- form is determined by its spatial configuration to the parent compound. When the OH group around the carbon atom adjacent to the terminal primary alcohol carbon (carbon atom 5 in glucose) is on the right, the sugar is a member of the D series. When it is on the left, it is a member of the L series. The majority of the monosaccharides occurring in mammalian metabolism are of the D configuration.

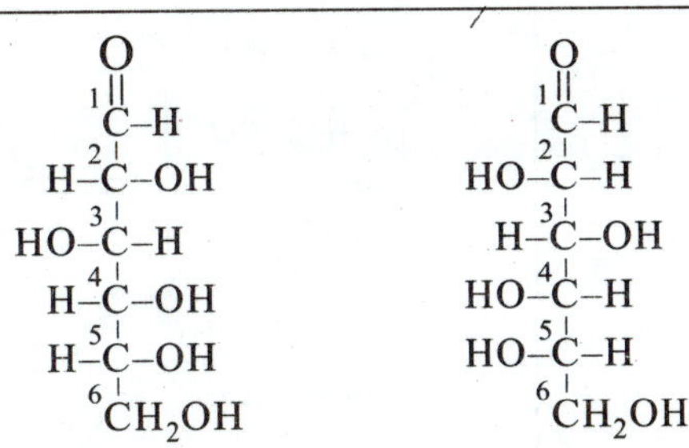

D-Glucose L-Glucose

Fig. 3.3

Optical isomer

When a beam of polarized light is passed through a solution exhibiting optical activity, it will be rotated to the right or to the left in accordance with the type of compound, i.e. the optical isomer, which is present. A compound which causes rotation of polarized light to the right is said to be dextrorotatory and a plus (+) sign is used to designate the fact. Rotation of the beam to the left (levorotatory action) is designated by a minus (–) sign. When equal amounts of dextrorotatory and levorotatory isomers are present, the resulting mixture has no optical activity since the activities of each isomer cancel the other. Such a mixture is said to be racemic or a DL mixture. Stereoisomerism and optical isomerism are independent properties. Thus a compound might be designated as D (–) or L (+). Synthetically produced compounds are necessarily racemic. The separation of optically active isomers from a racemic mixture is called resolution i.e. the racemic mixture is said to be "resolved" into the optically active components.

2. α- and β-anomers

The cyclic structure of glucose is retained in solution, but isomerism takes place about position 1. This is accomplished by optical rotation (mutarotation) by which the positions of –H and –OH groups are changed around carbon 1.

Mutarotation : Intra conversion of α & β-glucose in solution with change of optical activity, called mutarotation.

Mutarotation is explained by the α and β form of glucose requires the presence of asymmetric carbon c_1.

In mutarotation hemi acetal ring is opened and reformed with change of position of –H and –OH.

Example 1. When D-glucose is dissolved in water its specific rotation is 111°. When ammonia is passed, the value is 52.5°.

2. If it is recrystallised from boiling pyridine, the rotation is 19° and finally 52.5°.

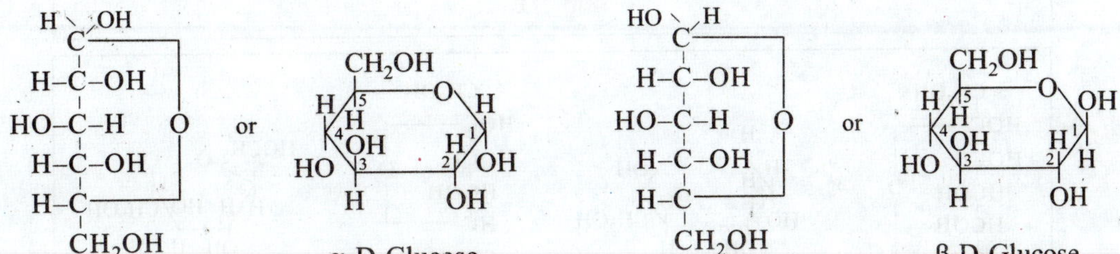

α-D-Glucose α-D-Glucose β-D-Glucose β-D-Glucose

Fig. 3.4

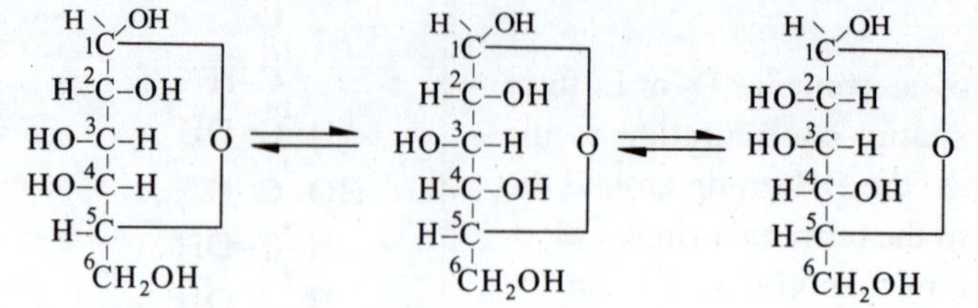

α-D-Galactose α-D-Glucose α-D-Mannose

Fig. 3.5

In mutarotation 38% $\left(\frac{1}{3}\right)$ α-D-glucose

62% $\left(\frac{2}{3}\right)$ β-D-glucose

3. Epimers

Isomers formed as a result of interchange of the –OH and –H on carbon atoms 2, 3 and 4 of glucose are known as epimers. Biologically, the most important epimers of glucose are mannose and galactose formed by epimerization at carbons 2 and 4, respectively. In the body, epimerization takes place by the enzyme epimerase.

4. Pyranose and Furanose Ring Structures

Haworth, in 1929, proposed a scheme in which all sugars forming six-membered rings are called pyranoses from their relation to pyrane and those forming five-membered rings furanoses after furane. The amylene oxide form of glucose would be called glucopyranose and the butylene oxide form glucofuranose. If a sugar is referred to by its ordinary name only, the pyranose form is to be inferred. The pyranose and furanose forms of some sugars are set out below in straight chain formulae as well as in ring formulae.

5. Aldose-ketose isomerism

Fructose has the same molecular formula as glucose but differs in its structural formula since

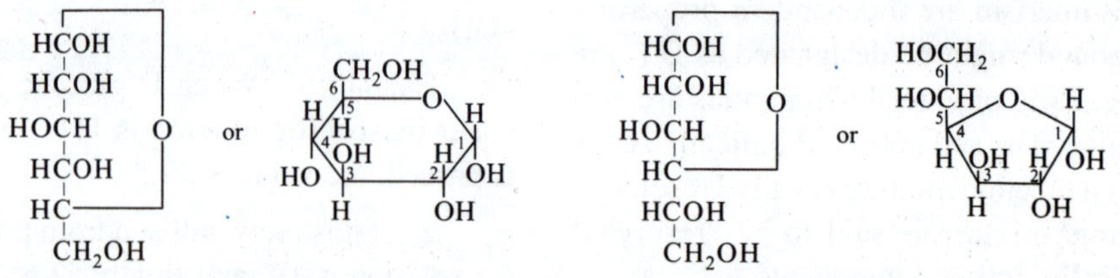

α-D-Glucopyranose α-D-Glucofuranose

Fig. 3.6

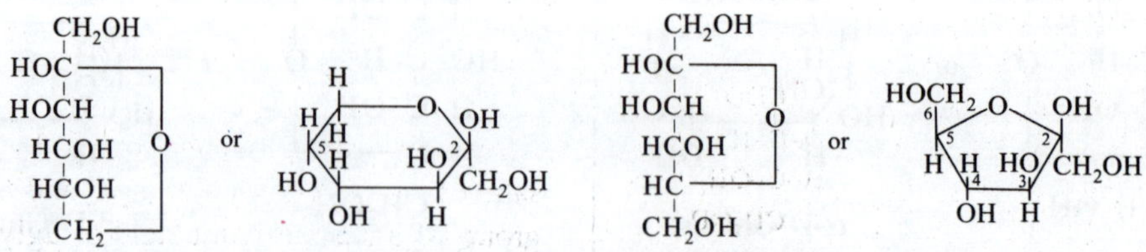

β-D-Fructopyranose β-D-Fructofuranose

Fig. 3.7

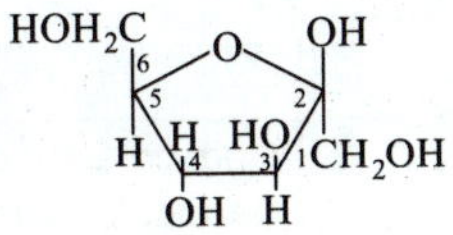

β-D-Fructose

Fig. 3.8

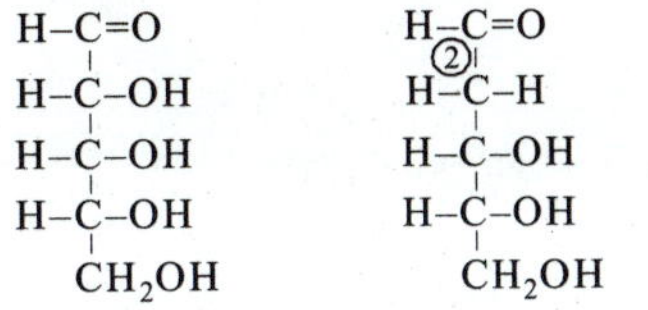

D-Ribose D-2-Deoxyribose

Fig. 3.10

carbon 2 is a part of a > Co group, which makes fructose a ketose rather than an aldose. Generally, there is a free –H on carbon 1, the sugar is an aldose, but if a $–CH_2OH$ group is substituted, the sugar is a ketose (Fig. 3.8).

THE KILIANI SYNTHESIS

A method for the synthesis of monosaccha-rides was first proposed by Kiliani. It is based upon the addition of HCN to the carbonyl group of aldehyde or ketones. The application of the Kiliani synthesis to the production of the two tetroses, erythrose and threose from glycerose is shown Fig. 3.9.

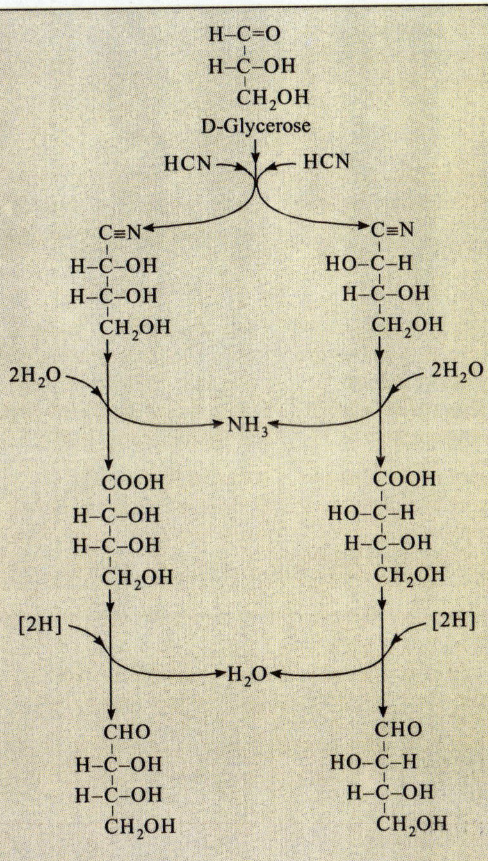

Fig. 3.9

MONOSACCHARIDES

Trioses and Tetroses

Phosphates of the two trioses, glyceraldehyde and dihydroxyacetone, are formed from fructose-1, 6-diphosphate by glycolysis. 3-phosphoglyceraldehyde and erythrose-4- phosphate are formed by the hexose monophosphate shunt.

Pentoses

Ribose and Deoxyribose are important constituents of nucleic acids and many coenzymes. The structures are given at the right. Deoxyribose is lacking in one atom of oxygen from carbon no. 2.

Heptoses

Sedoheptulose is a ketoheptose found in plants. Its phosphate is important as an intermediate in the hexose monophosphate shunt and has been identified as a product of photosynthesis.

GLYCOSIDES

Glycosides are compounds formed by the condensation reaction between a sugar and the hydroxyl group of a second compound or aglycone which may or may not be another sugar. If the carbohydrate portion is glucose, the resulting compound is a glucoside : if galactose, a galactoside, etc.

Table 1 : Pentoses of physiologic importance

Sugar	*Source*	*Importance*	*Reactions*
D-Ribose	Nucleic acids.	Structural elements of nucleic acids and coenzymes, eg, ATP, NAD, NADP, flavoproteins.	Reduce Benedict's, Fehling's, Barfoed's solution. Form osazones with phenylhydrazine.
D-Ribulose	Formed in metabolic processes.	Intermediates in hexose monophosphate shunt.	Those of keto sugars.
D-Arabinose	Plum and cherry gums.	Used in bacterial metabolism. No known physiological functions in man.	With orcinol-HCl reagent gives colours : violet, blue, red, green.
D-Xylose	Wood gums.	Same as above.	Gives red colour with phloroglucinol -HCl
D-Lyxose	Heart muscle.	A constituent of a lyxoflavin isolated from human heart muscle.	—

Hexoses

Table 2 : Hexoses of physiologic importance

Sugar	*Source*	*Importance*	*Reactions*
D-Glucose	Fruit juices, Hydrolysis of starch, canesugar, maltose and lactose.	It is carried by the blood and used by the tissues. Its presence in the urine causes glycosuria.	Reduces Benedict's, Barfoed's reagent. Gives osazones with phenyl hydrazine. Forms saccharic acid with HNO_3.
D-Fructose.	Fruit juices, Honey, Hydrolysis of canesugar and of Inulin.	It can be changed to glucose in the liver and intestine and thus used in the body.	Reduces Benedict's, Barfoed's reagents. Forms osazone with phenylhydrazine. Gives red colour with Seliwanoff's reagent.
D-Galactose.	Hydrolysis of lactose.	It can be changed to glucose in the liver and metabolized. Synthesized in the mammary gland to form lactose of milk. It is constituent of glycolipids and glycoproteins.	Reduces Benedict's, Barfoed's reagents. Forms osazone with Phenylhydrazine. Forms mucic acid with HNO_3. Not fermented by yeast.
D-Mannose.	Hydrolysis of plant mannosans and gums.	It is the constituent of prosthetic polysaccharides of albumins, globulins, mucoprotiens. It frequently occurs in glycoproteins.	Reduces Benedict's, Barfoed's reagents. Forms same osazone as glucose.

Structures

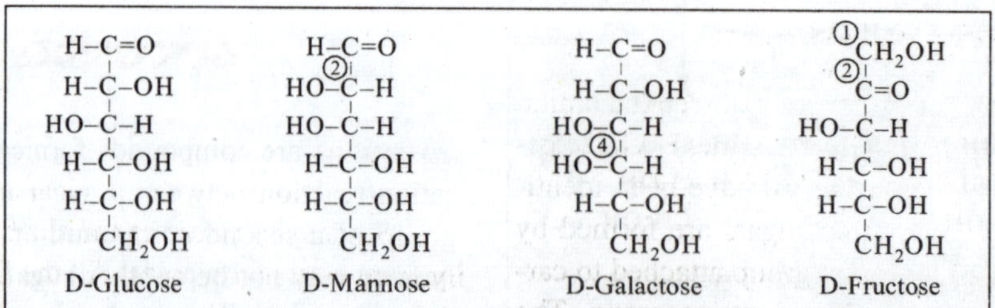

Fig. 3.11

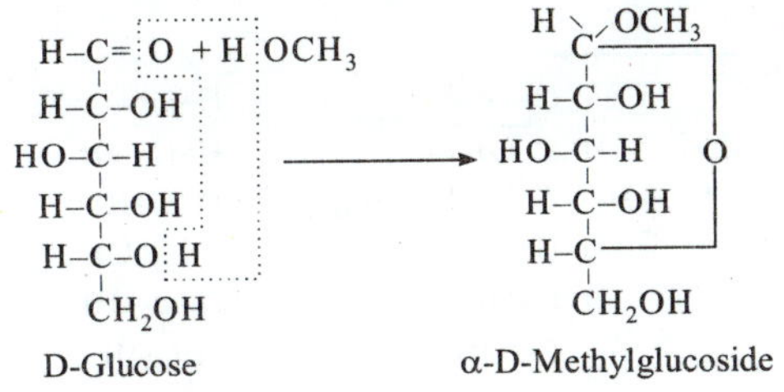

Fig. 3.12

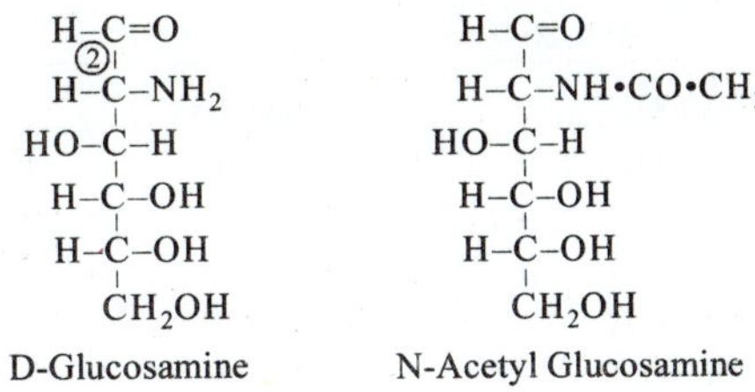

Fig. 3.13

All reducing sugars will condense with dry methyl alcohol under catalytic action of dry HCl to form glycosides. Thus, glucose forms α- and β-methyl glucosides.

The aglycone may be –OH containing substances such as methyl alcohol, glycerol, a sterol, a phenol or another sugar (as in the disaccharides).

Importance

(i) Glycosides are found in many drugs, spices and in the constituents of animal tissues.

(ii) The glycosides are important in medicine because of their action on the heart (cardiac glycosides). All the glycosides contain steroids as the aglycone component.

(iii) The glycosides include derivatives of digitalis and strophanthus such as Ouabain, which is an inhibitor of the Na^+– K^+ ATPase of cell membranes.

(iv) Other glycosides include antibiotics such as streptomycin.

AMINO SUGARS (HEXOSAMINES)

Sugars containing an amino group are called amino sugars. Examples are D-Glucosamine, D-Galactosamine, and D-Mannosamine, all have been identified in nature. These amino sugars are formed by the replacement of hydroxyl group attached to carbon atom 2 of the sugar by an amino group. The structure of an amino sugar is given in Fig. 3.13.

Importance

(i) Glucosamine is a constituent of hyaluronic acid. It is the organic constituent of lobster shell and fairly distributed in nature.

(ii) Galactosamine is a constituent of glycoproteins and of chondroitin as the N-acetyl derivative.

(iii) Mannosamine is an important constituent of mucoprotein.

(iv) Several antibiotics (erythromycin, carbomycin) contain amino sugars. Erythromycin contains a dimethylamino sugar. Carbomycin contains the first known 3-amino sugar, 3-amino-D-ribose. The amino sugars are believed to be related to the antibiotic activity of these drugs.

IMPORTANT CHEMICAL REACTIONS OF MONOSACCHARIDES

(i) Iodo Compounds

An aldose sugar, when heated with concentrated hydriodic acid (HI), loses all of its oxygen and is converted into an iodo compound (glucose to iodohexane, $C_6H_{13}I$).

(ii) Ester Formation

Sugars, by virtue of the alcohol groups, readily form esters with acids. All the free –OH groups are replaceable. The greater biochemical significance is the ester with phosphoric acid and, to a

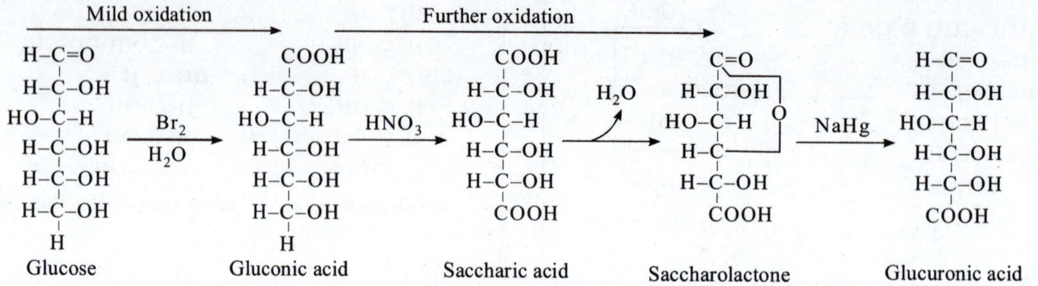

Fig. 3.14

lesser extent, with sulphuric acid. Pentose phosphates are involved in the formation of nucleic acids.

(iii) Acetylation

The acetylation with acetylchloride indicates the presence of –OH group present in the sugar. The presence of 5 OH groups of glucose results in a pentaacetate.

(iv) Oxidation

Oxidation of the aldehyde group forms "aldonic acids". If the aldehyde group remains intact and the primary alcohol group is oxidized "uronic acids" are formed. Oxidation of galactose with concentrated HNO_3 yields the dicarboxylic mucic acid. This compound crystallizes readily.

(v) Reduction

The monosaccharides are reduced to their corresponding alcohols by reducing agents such as sodium amalgam.

Thus, Glucose yields sorbitol (Protects liver).

Galactose yields dulcitol.

Mannose yields mannitol.

Fructose yields mannitol and sorbitol.

Mannitol

(a) It is neither reabsorbed nor secreted in tubules.

(b) It is used as force diuretics i.e. to increase the volume of urine.

(vi) With Strong Mineral Acids

There is a change of hydroxyl groups towards and of hydrogen away from the aldehyde end of the chain. Reaction products with acids will condense with certain organic phenols to form compounds having characteristic colours.

(vii) Heat

Gluconic acid on heating produces lactones. These are cyclic structures which resemble pyranoses and furanoses.

(viii) With Alkali

Monosaccharides react in various ways :

(a) In dilute alkali the sugar will change to the cyclic alpha and beta structures, with an equilibrium between the 2 isomeric forms.

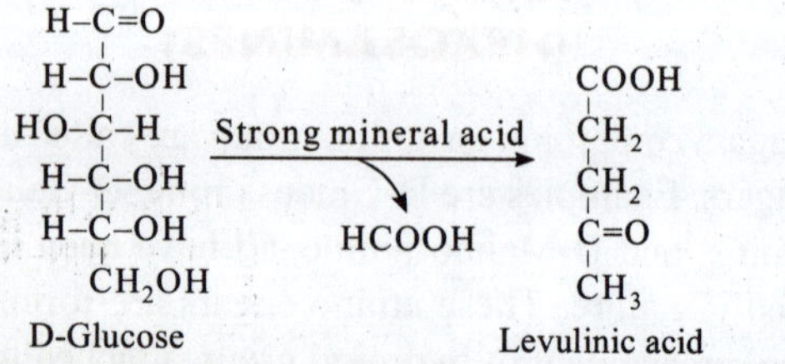

Fig. 3.15

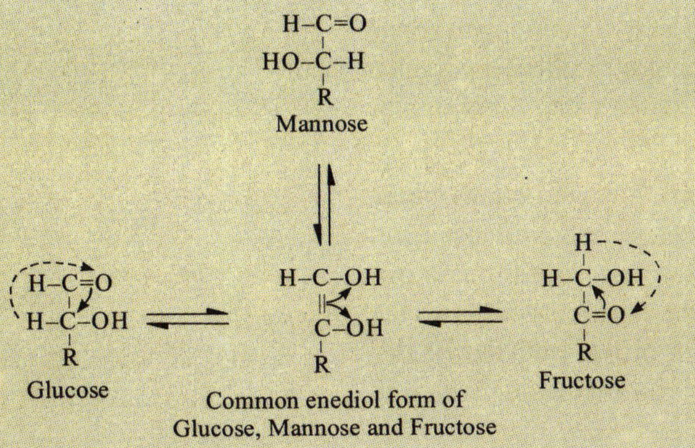

Fig. 3.16

On standing, a rearrangement will occur which produces an equilibrated mixture of glucose, fructose and mannose through the enediol form.

(b) In concentrated alkali, sugar produces a series of decomposition products. Yellow and brown pigments develop, salts may form, many double bonds between carbon atoms are formed, and carbon-to-carbon bonds may rupture.

(ix) Osazone Formation

It is nothing but the formation of crystalline derivatives of the sugars which are valuable in the identification of sugars.

These crystals are obtained by adding a mixture of phenylhydrazine hydrochloride and sodium acetate to the sugar solution and heating in a boiling water bath. The carbonyl group (i.e. aldehyde or ketone group) and the next adjacent carbon are

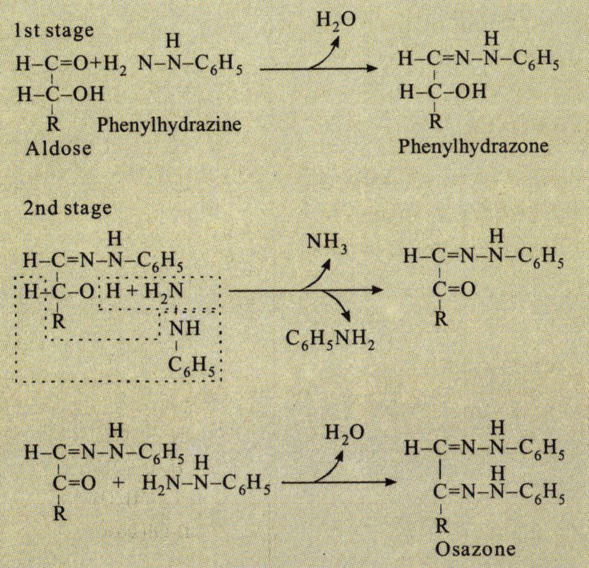

Fig. 3.17

involved in this reaction. With an aldose the reaction is shown Fig. 3.17. The hydrazone then reacts with two additional molecules of phenylhydrazine to form the osazone. The ketoses also show similar reaction.

From the comparison of their structures it may be noted that glucose, fructose and mannose form the same osazone; whereas galactose forms a different osazone because the part (carbon 4) in the structure of galactose which differs from that of glucose, fructose and mannose remains unaffected in osazone formation.

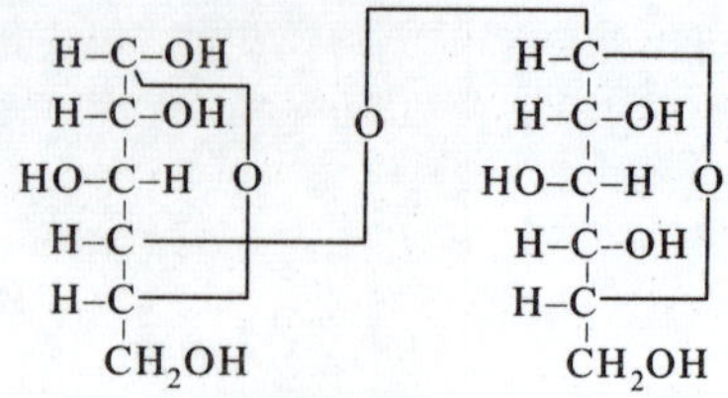

Two-α-D-Glucopyranose components
Maltose (α form)

Fig. 3.18

(x) Other Reactions

Various other reactions take place due to the presence of aldehyde or ketone groups of the sugars which are important for analytical purposes.

The *best-known* tests are reduction of metallic hydroxides together with oxidation of the sugar. The alkaline metal is kept in solution with sodium potassium tartrate (Fehling's solution) or sodium citrate (Benedict's solution). Other metallic hydroxides may be used (Bismuth, Ammoniacal silver, Tollen's test, Nylander's test).

Barfoed's test distinguishes between monosaccharides and disaccharides. The copper acetate in dilute acid is reduced in 30 seconds by monosaccharides; whereas reduction of the same takes several minutes by disaccharides.

DISACCHARIDES

The disaccharides are composed of two monosaccharide units united by a glycosidic linkage. The physiologically important disaccharides are maltose, lactose and sucrose :

Maltose = 1 mol. glucose + 1 mol. glucose.

Lactose = 1 mol. glucose + 1 mol. galactose.

Sucrose = 1 mol. glucose + 1 mol. fructose.

1. Maltose (Malt sugar)

(i) It does not occur in the body.

(ii) The sources of it are germinating cereals and malt and it is the intermediate product in the breakdown of starch by amylase in the alimentary canal.

(iii) It is hydrolysed to glucose by the enzyme maltase and the products are absorbed.

(iv) It has one free aldehyde group and hence shows mutarotation and the final rotation of the solution is +130°. It can exist in α or β forms.

(v) It can reduce Fehling's and Benedict's solutions since it is a reducing sugar but cannot reduce Barfoed's solution.

(vi) It forms an osazone with phenylhydrazine.

2. Lactose (Milk sugar)

(i) It is present in milk and formed in the lactating mammary gland. It may occur in urine during pregnancy.

(ii) It is hydrolysed to glucose and galactose by the enzyme lactase in the alimentary canal and the products are absorbed.

Lactose (β form)

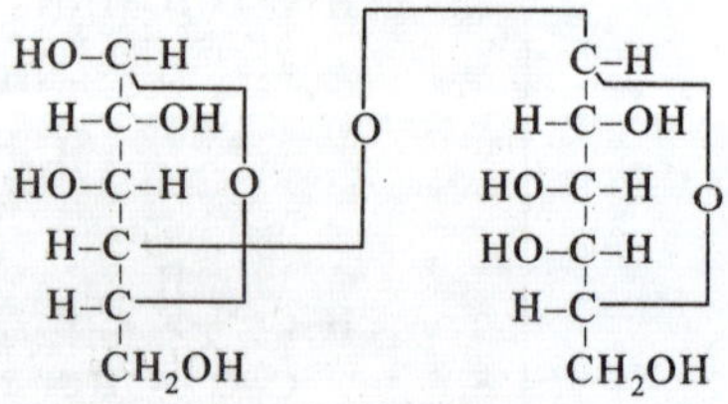

β-D-Glucopyranose component β-D-Galactopyranose component

Fig. 3.19

(iii) It has free aldehyde group and hence shows mutarotation and the final constant specific rotation of the solution is +55.2°. It can also exist in α or β forms.

(iv) Since it is a reducing sugar it can reduce Fehling's and Benedict's solutions but cannot reduce Barfoed's solution.

(v) It forms an osazone with phenylhydrazine.

3. Sucrose (Cane sugar)

(i) It does not exist in the body but occurs in cane sugar, pineapple, carrot roots, sweet potato and honey.

(ii) It is hydrolysed to glucose and fructose by the enzyme invertase (sucrase) in the alimentary canal. The products of hydrolysis are absorbed.

(iii) It has no free aldehyde or keto group because the linkage is between the aldehyde group of glucose and keto group of fructose. Hence it is a non-reducing sugar. It does not exhibit mutarotation and cannot exist in α or β forms.

(iv) Since it is a *non-reducing* sugar it does not reduce Fehling's or Benedict's solutions. It cannot reduce Barfoed's solution too.

(v) It cannot form osazone with phenylhydrazine.

(vi) The specific rotation of sucrose solution is +66.5°. During hydrolysis this rotation changes to –19.5°; since the laevorotation of fructose is greater than the dextrorotation of glucose. The product of the hydrolysis is used to be referred to as "invert sugar".

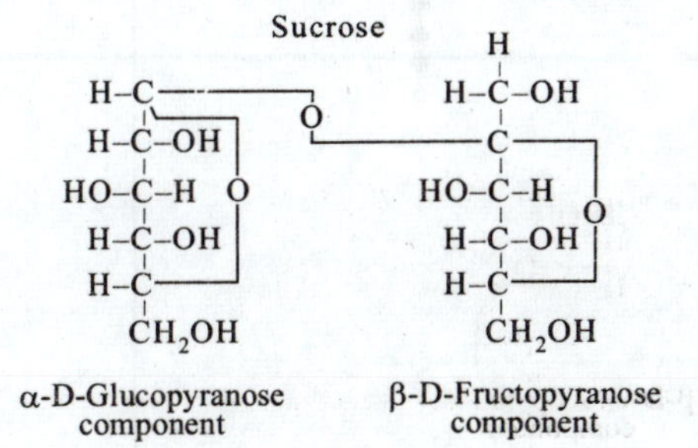

Fig. 3.20

POLYSACCHARIDES

Polysaccharides are classified as homopolysaccharides and heteropolysaccharides.

The physiological important homopolysaccharides are Cellulose, Glycogen, Starch, Dextrins, and Inulin.

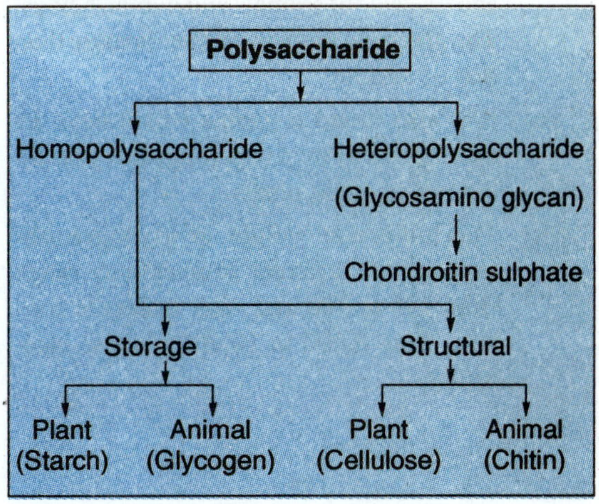

1. Cellulose

(i) It is the main constituent of the supporting tissues of plants and forms a considerable part of our vegetable food. It does not occur in the animal body.

(ii) It is made up of β-glucose molecules which are linked by 1:4 linkages.

(iii) Owing to the difference in chemical structure, cellulose is not acted upon by amylases present in the digestive juices.

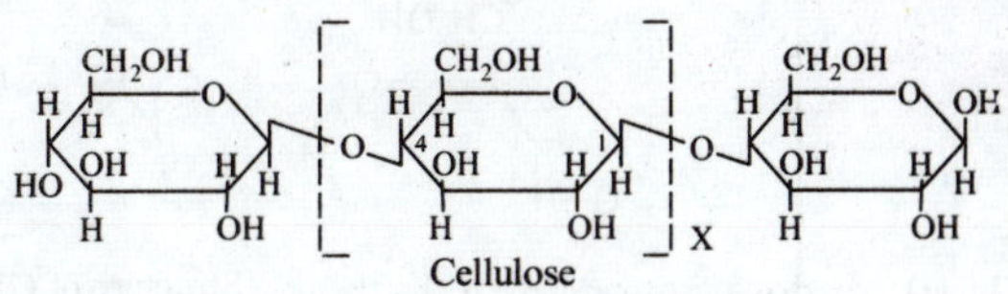

Fig. 3.21

(iv) It is of considerable human dietetic value only because it adds "bulk" to the intestinal contents, thereby stimulating peristalsis and elimination of food residues.

(v) It is insoluble in ordinary solvents and gives no colour with iodine.

(vi) Cotton and filter paper are nearly pure cellulose.

2. Glycogen (Animal starch)

(i) It is the reserve carbohydrate found in liver and muscles of animals and human beings. The glycogen content of liver is more than that of muscle.

(ii) It is also found in plants which have no chlorophyll system (eg, fungi and yeasts), but not in green plants.

(iii) The molecular weight of glycogen obtained from different sources may range from 105 to 108 and each molecule contains from 5,000 to 10,000 glucose molecules.

(iv) It has a branched structure with straight chain units of 12- 18-α-D-glucopyranose [in a 1-4 glucosidic linkage] with branching by means of a [1-6]-glucosidic bonds.

(v) It is nonreducing, readily soluble in water and gives a red colour with iodine.

3. Starch

(i) It is the stored carbohydrate of chlorophyll-containing plants. In plants, the starch is laid down in the cells in granules. The microscopic form of the granules is characteristic or the source of the starch.

(ii) It is formed by an α-glucosidic chain. Such a compound—which produces only glucose on hydrolysis—is called a *glucosan*.

(iii) It is the most important source of carbohydrate in our food and is found in cereals, potatoes, legumes and other vegetables in high concentrations.

(iv) It is a mixture of two substances – amylose and amylopectin – both are composed

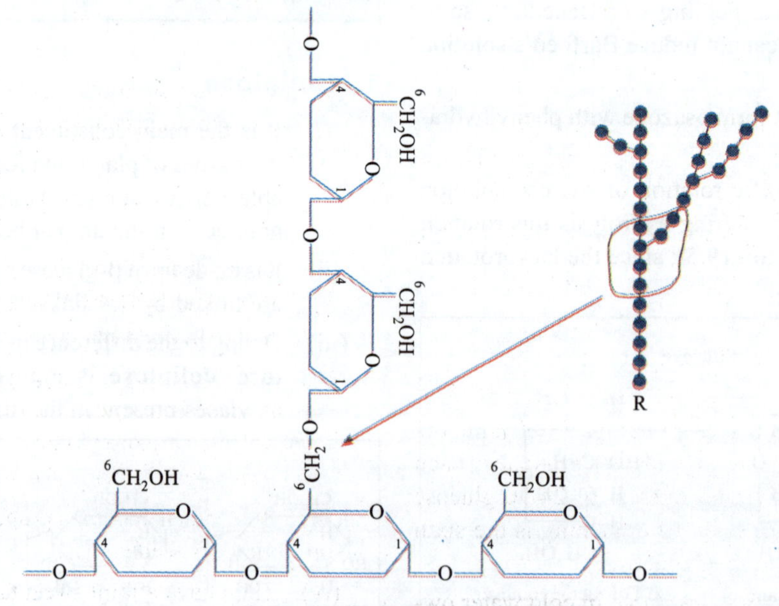

(Structure of Glycogen)

Fig. 3.22

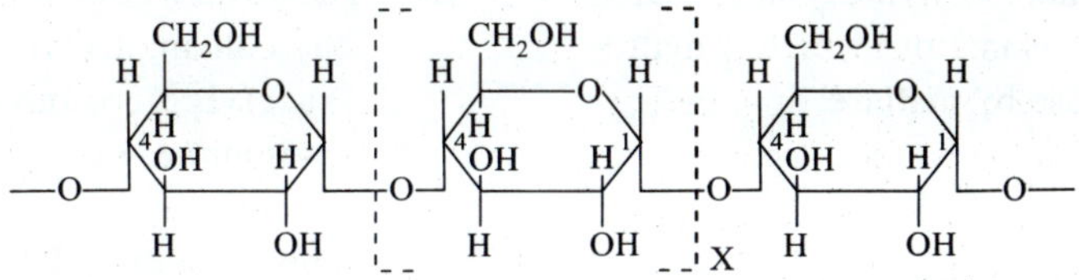

Amylose

Fig. 3.23

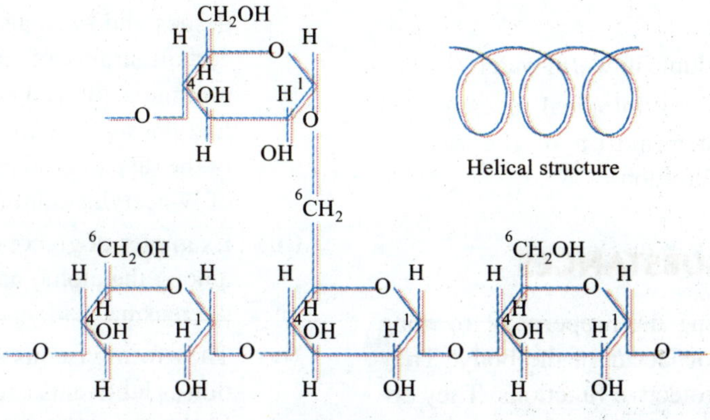

Amylopectin
(1, 6-Glucosidic linkage)

Fig. 3.24

of glucopyranose units. In the amyloses, the glucose units are joined by 1, 4 – α links to form unbranched chains which are in the form of a helix with six glucose units per turn. Their molecular weights are about 60,000 which are equivalent to about 300 – 400 glucose units and are responsible for the development of blue colour with iodine.

Amylopectins have much larger molecular weights of about 500,000 and the chains have at least 80 branches; each branch is at an interval of 24-30 glucose units. The point of branching is the sixth carbon atom of glucose.

(v) Raw starch is insoluble in cold water owing to the resistance of the outer cellulose layer of the granule. When this is ruptured by heating in water, starch is soluble. Concentrated solutions gelatinise on cooling and are used as an adhesive-starch paste.

4. Dextrins

(i) Dextrins are formed by the partial hydrolysis of starch by an enzyme (salivary amylase), dilute mineral acids or heat.

(ii) Amylodextrin, erythrodextrin and achrodextrin give blue, red-brown and no colour, respectively, with iodine, Achrodextrin being the simplest.

(iii) If they have reducing properties at all, they are very feeble.

(iv) They have a faint sweet taste.

(v) They form sticky solutions in water and are frequently used as adhesive, eg, on postage stamps.

(vi) The final product of hydrolysis of starch by an amylase is maltose which is hydrolysed to glucose by maltase.

5. Inulin

(i) It is found in tubers and roots of dahlias, and dandelions.

(ii) It is hydrolysed to fructose; hence it is a fructosan.

(iii) It does not produce any colour with iodine.

(iv) It is easily soluble in warm water.

(v) It is used in physiological investigation for the determination of the rate of glomerular filtration.

MUCOSUBSTANCES

Mucosubstances having the property of forming slimy, viscous solution occur in the body. They have structural and protective functions. They are divided into Glycoproteins and Mucopolysaccharides :

1. Glycoproteins (mucoproteins)

(i) They are protein-polysaccharide compounds occurring in the tissues, particularly in mucus secretions.

(ii) They do not contain uronic acids although they contain acetyl hexosamines such as N-acetylglucosamine and N-acetylgalactosamine. Hexoses such as mannose or galactose are also found. In addition, a methyl pentose (L-fucose) and the sialic acids commonly occur in these conjugated proteins.

The sialic acids are actually a family of compounds derived from neuraminic acid. They are widely distributed in vertebrate tissues and have also been isolated from certain strains of bacteria. N-acetylneuraminic acid is an example of sialic acid. Enzymes which are identified in the liver of the rat can accomplish the biosynthesis of N-acetylneuraminic acid.

(iii) Examples of glycoproteins are also found among the alpha$_1$ and alpha$_2$ globulins of the plasma.

(iv) They form viscous solutions which function as lubricants and protective "screens" in the body. The continuously secreted mucus of the respiratory tract is a protection against invasion by bacteria and the uterus is protected from the vaginal microbial flora by the cervical mucus. Intestinal mucus constitutes a protection for the intestinal cells against mechanical damage.

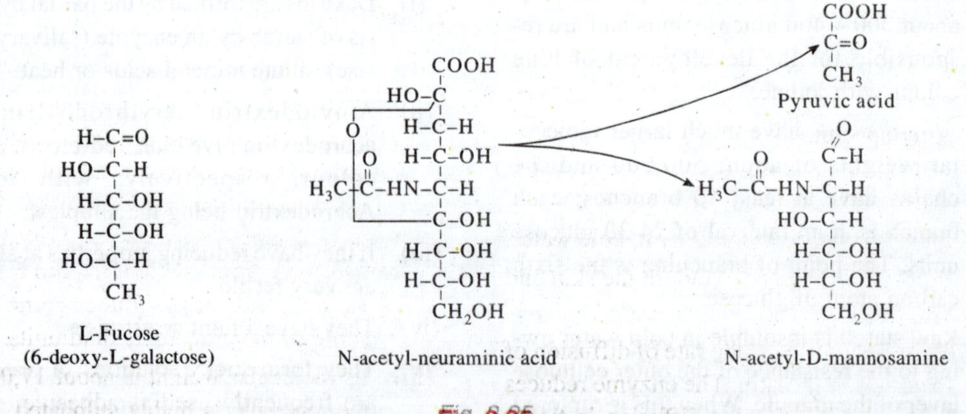

Fig. 3.25

(a) The gastric mucus glycoprotein is a high molecular-weight (2×10^6) component of the viscous coat that protects the gastric mucosa from chemical, enzymatic, and mechanical damage. The protein's resistance to pepsin digestion and its rheologic properties in solution correlates with its degree of acylation.

(b) Normally, about 20 moles of fatty acids (palmitate, stearate, and oleate) are bound in ester linkage per mole of mucus glycoprotein.

(c) In patient with cystic fibrosis, the stoichiometry of acylation is increased to 60-70 moles fatty acid/mol protein. Hyperacylation of mucus glycoprotein may be partially responsible for the extreme viscosity of the mucus produced by persons afflicted with the disease.

2. Mucopolysaccharides (Heteropolysaccharides)

The heteropolysaccharide situated in extracellular matrix is called glycosaminoglycan, such as hyaluronic acid, chondroitin sulphate, heparin, kerato sulphate.

Glycosaminoglycan + Protein = Proteoglycan.

A. Hyaluronic acid

(i) It has a molecular weight of 1 to 4 million.

(ii) It consists of alternate units of glucuronic acid and N-acetylglucosamine linked to give a thread-like structure.

(iii) It forms viscous solutions in water and is important in the body as a lubricant.

(iv) It has an enormous capacity to hold water.

(v) It occurs in synovial fluid, in the skin and vitreous humour.

(vi) Its presence affects the rate of diffusion of materials in the skin. The enzyme reduces its viscosity and helps diffusion in tissue sapce.

(vii) The presence of the enzyme hyaluronidase increase the rate of diffusion of substances through tissues containing hyaluronic acid. The richest source of the enzyme in mammals is the testis and its presence in the semen facilitates fertilization. It is secreted by some pathogenic organisms which can, therefore, invade the tissues of the host animal more easily. Hyaluronic acid in tissue acts as a cementing substance.

B. Chondroitin sulphates

(i) They occur in the ground substance of connective tissue and they are the components of cartilage, tendon and skin.

(ii) Chondroitin sulphates A and C consist of alternate units of glucuronic acid and 2-N-acetylamino galactose.

(iii) Chondroitin sulphate B contains iduronic acid in place of glucuronic acid.

(iv) In all three forms, one hydroxyl group of each amino sugar is esterified with sulphuric acid and the molecules are long and linear.

(v) They have a marked capacity to bind water and contribute to the resistance to compression of connective tissue.

(vi) Since they are polyvalent anions they may also regulate the flow and concentration of cations round the cells.

C. Heparin

(i) It occurs in most cells and is present in liver, lung and the arterial wall.

(ii) It consists of an unbranched chain of alternate units of glucosamine and glucuronic acid joined mainly by 1, 4-α-links. Sulphate groups are linked to the hydroxyl group of carbon atom 6 and the amino group of the glucosamine and to some of the glucuronic acid units.

(iii) Its molecular weight is about 17,000 and the molecule is highly sulphated.

(iv) It is used in medicine as an anticoagulant.

D. Blood group substances

(i) They consist of N-acetylglucosamine, galactosamine, galactose, fucose and sialic acid.

(ii) They are used to determine blood group.

E. Keratosulphate

(i) It consists of N-acetylglucosamine, galactose, sulphuric acid.

(ii) It is the component of cartilage and cornea.

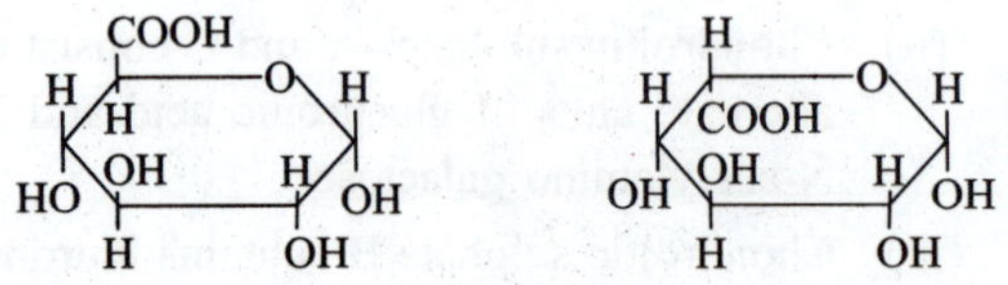

Glucuronic acid Iduronic acid

Fig. 3.26

CARBOHYDRATES IN CELL MEMBRANE

1. The mammalian cell membrane contains about 5 per cent carbohydrates present in glycoproteins and glycolipids.
2. Glycophorin, the major integral membrane glycoprotein of human erythrocytes, consists of 130 amino acid residues and spans the lipid membrane having free polypeptide proteins outside both the external and internal (cytoplasmic) surfaces. Carbohydrate chains are only attached to the N-terminal portion outside the external surface.
3. In composition of bacterial cell wall, there is presence of N-acetylglucosamine and N-acetylgalactosamine by β(1-4) linkage. Lysozyme breaks this linkage and protects the eye.

QUALITATIVE TESTS FOR IDENTIFICATION OF CARBOHYDRATES

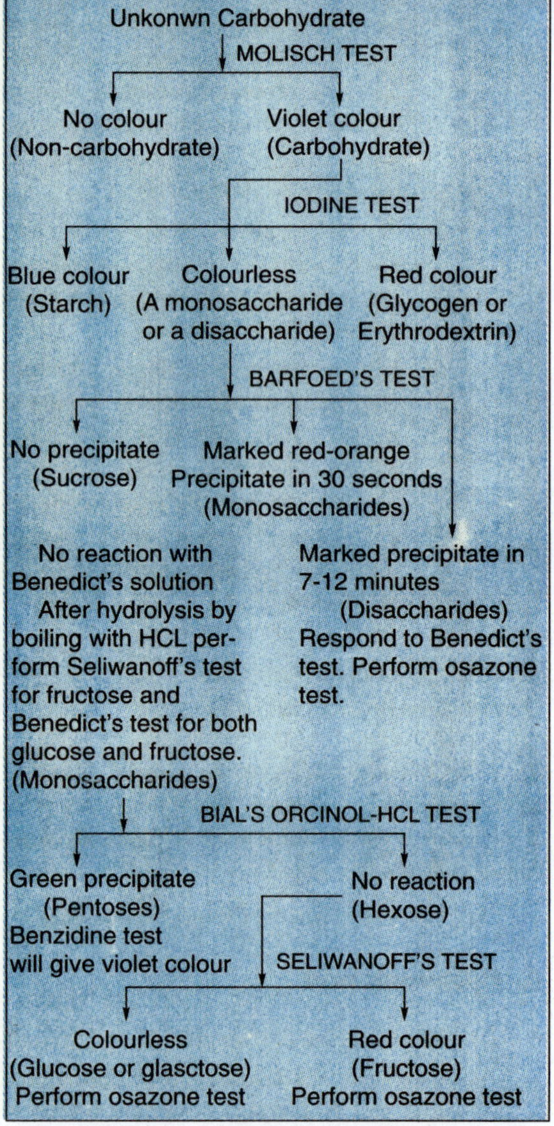

Benedict's test is preferred than the Fehling's test for the detection of glucose in urine due to the following causes :

Benedict's Qualitative solution is composed of copper sulphate, sodium citrate and sodium carbonate.

Fehling's solution is composed of copper sulphate, sodium potassium tartrate, and potassium hydroxide.

(i) Benedict's test is less likely to give weakly positive results with concentrated urine due to the action of creatinine and uric acid.

(ii) It is more sensitive to small quantities of glucose because of the weaker alkali it contains.

(iii) The strong alkali (KOH) of Fehling's solution can polymerize small quantities of glucose when vigorously boiled.

CLINICAL ORIENTATION

- The bulk of dietary carbohydrate as glucose is absorbed into the blood stream which is then converted in the liver.
- Glucose is the universal fuel of the foetus and the principal fuel of the tissues of the mammals (except ruminants).
- Other carbohydrates having highly specific functions are also formed from glucose, eg, Glycogen for storage in liver and muscles; Ribose in nucleic acids; Galactose in lactose of milk, in certain complex lipids, in combination with protein in glycoproteins and proteoglycans; Fructose abundantly present in seminal fluid is utilized by sperms for energy; Mannose in the body is available as a constituent of glycoproteins.
- Glucuronic acid formed from glucose in the liver by uronic acid pathway conjugates with toxic substances, drugs, hormones, and bilirubin and converts them into nontoxic substances which are excreted in the urine.
- Some antibodies, eg, erythromycin, carbomycin etc. contain aminosugars.
- The glycosides are available in many drugs. The cardiac glycosides are highly significant in medicine owing to their action on heart, and, hence, used in cardiac disturbances. The glycoside phloridzin displaces Na^+ (sodium ion) from the binding site of "Carrier protein" and prevents the transport of sugar molecule, across the mucosal cells of the small intestine and, thus, produces glycosuria.
- Parenterally introduced sucrose causes a flow of water from the tissues into the blood and, hence, is used by the clinicians in cerebral oedema. In case sucrose or other disaccharides not hydrolysed due to the deficiency of appropriate enzyme it causes diarrhoea.
- The antibodies, the coagulation factors, and the peptide hormones contain oligosaccharides. Insulin is used in the estimation of body water volume.
- Although cellulose in human beings not nutritionally significant due to the absence of cellulose splitting enzyme it adds bulk to the intestinal contents thereby stimulating peristalsis for the elimination of faeces to avoid constipation.
- L-Xylulose is excreted in urine of humans in "essential pentosuria" due to hereditary abnormal metabolism.
- The diseases diabetes mellitus, galactosemia, glycogen storage diseases and milk intolerance are directly associated with carbohydrates.

EXERCISE

1. Define carbohydrates, classify them and describe the properties of Monosaccharides.
   (P. U. 68A ; R.U. 67A ; Mith. 64 S)
2. Write short notes on
   (a) Disaccharides. (Bh. U. 76 A)
   (b) Starch. (Mith. 73 A)
   (c) Mucopolysaccharides.
   (Bh. U. 75A; R.U. 68A; P.U. 72S)
   (d) Lactose. (Muz. 74 A)
   (e) Glycogen. (M.U. 73A; Mith. 62 A)
   (f) Benedict's test. (P.U. 68S)
3. Mention five characteristic reactions of an aldose hexose which occurs in our body. (C.U. 1981)
4. Give the structural formula of the following :
   (a) Lactic acid (C.U. 1981)
   (b) Fructose
5. What products will be produced by acid hydrolysis of lactose? (C.U. 1981)
6. With examples write explanatory notes on optical isomerism. (C.U. 1981)
7. Explain the mutarotation of carbohydrates. Describe in brief the structure and important properties of the following : (C.U. 1982)
   (a) Glycogen.
   (b) Starch.
   (c) Cellulose.
8. Indicate 'True' or 'False' :
   In lactose molecule there is one β-1, 2 glucoside linkage......... (C.U. 1981)
   Suggest the nature and molecular formula of A from the following data : (C.U. 1981)

   (i) $A \xrightarrow[\text{+ heat}]{\text{Fehling's solution}} \text{No reduction}$

   (ii) $A \xrightarrow[\text{heat}]{\text{dil. HCl}} B + C$

(iii) $B \xrightarrow[+\text{ heat}]{\text{Fehling's solution}}$ reduction (Red ppt.)

$C \xrightarrow[+\text{ heat}]{\text{Fehling's solution}}$ reduction (Red ppt.)

(iv) Both B and C separately on reduction produces hexahydric alcohol

(v) $B \xrightarrow{\text{Strong oxidation}}$ dicarboxylic acid with six C-atom.

(vi) $C \xrightarrow{\text{Strong oxidation}}$ two dicarboxylic acids- one with two C-atoms and the other with four C-atoms.

9. Fill up the blanks with appropriate words. (C.U. 1981)

(a) Sucrose solution is — rotatory but after hydrolysis with dil. HCl it shows — rotation.

(b) Lactose solution exhibits — rotation and after treatment with dil. HCl it shows — rotation.

DIAGRAMMATIC COMPENDIUM

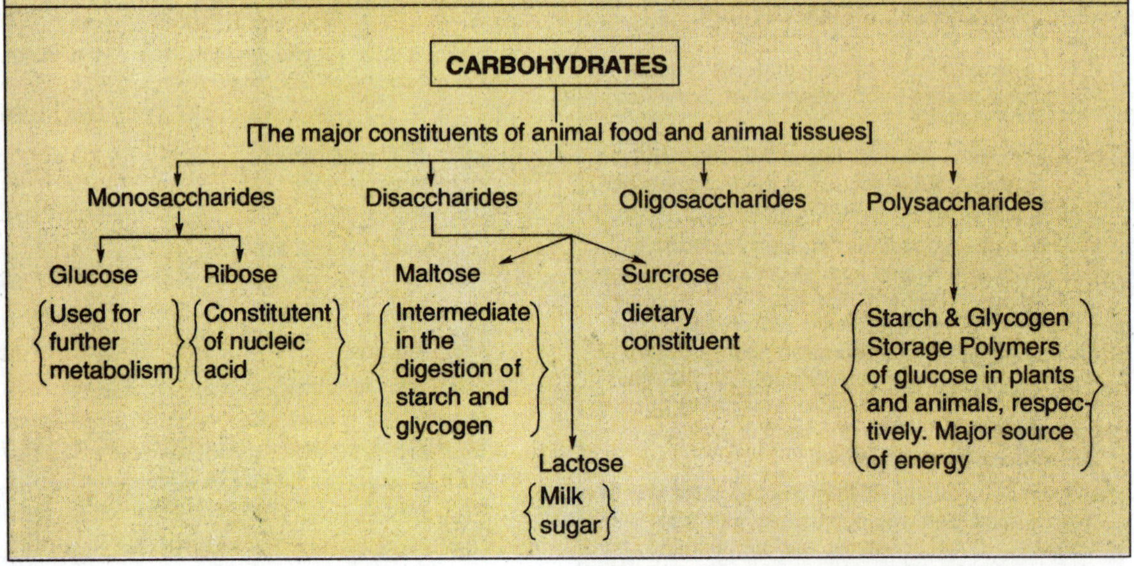

CHAPTER 4

LIPIDS

Definition

The lipids are heterogeneous group of compounds related to the fatty acids—either actual or potential, insoluble in water, soluble in other solvents such as ether, chloroform and benzene and chemically are esters of fatty acid and some alcohols.

Sources : The lipids occur widely in plant and animal kingdom.

Examples : The lipids include fats, oils, waxes and related compounds. Oils are liquids at 20°C but fats are solid at 20°C.

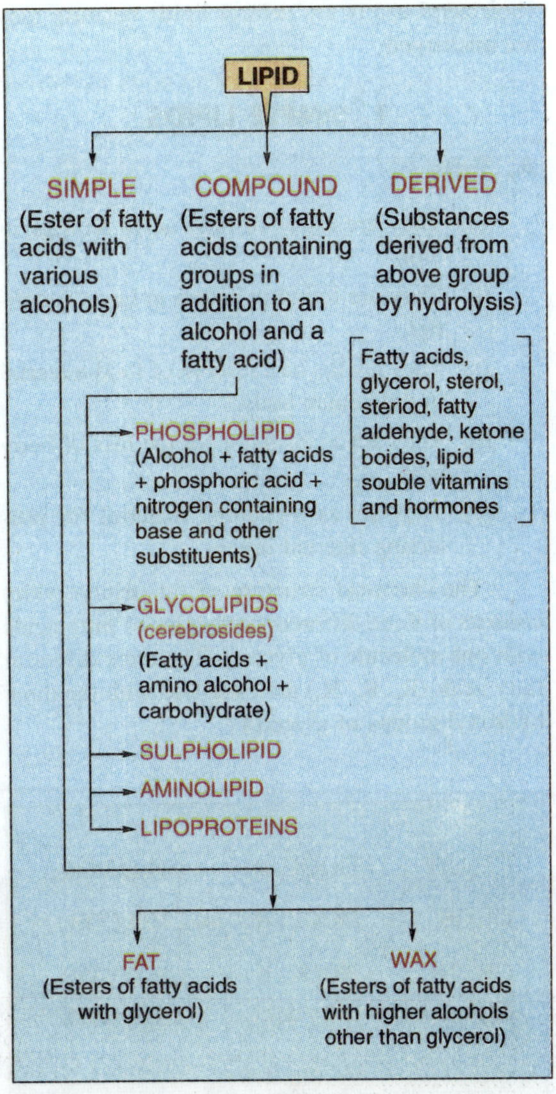

Biological Importance

1. In the body, fat serves as an efficient source of energy when stored in an adipose tissue.
2. The fat-soluble vitamins and the essential fatty acids are found with the fat of natural foods.
3. It serves as an insulating material in the subcutaneous tissues and around certain organs.
4. The phosphatides of blood platelets are involved in the production of thromboplastin activity in the early stages of blood clotting.
5. Lipoproteins (combinations of fat and protein) and glycolipids (combinations of fat and carbohydrate) are essential for maintaining cellular integrity.

6. It provides building blocks for different high molecular weight substances, eg, acetic acid and can be used for the synthesis of cholesterol and certain hormones.
7. They produce metabolites through oxidation in the tissues which are used in the interconversion of substances.

Classification

Bloor has proposed the following classification of lipids :

Glycerides (acylglycerols), cholesterol and cholesteryl esters are neutral lipids because they are uncharged.

1. SIMPLE LIPIDS

A. Fats

(a) They are esters of fatty acids with glycerol.

(b) They are found in nature in large quantities.

(c) They are the best reserve of food material in the human body.

(d) They act as insulator for the loss of body heat.

(e) They act as a padding material for protecting internal organs.

The chemical structure of fat (triglyceride) consists of three different molecules of fatty acids with one molecule of glycerol. The three different fatty acids (R_1, R_2, R_3) are esterified with the three hydroxyl groups of glycerol :

CH_2OH		$HOOC–R_1$		$CH_2O.COR_1$
$CHOH$	+	$HOOC–R_2$	$\longrightarrow$	$CHO.COR_2$
CH_2OH		$HOOC–R_3$		$CH_2O.COR_3$
Glycerol		Fatty acids		Triglyceride

Fig. 4.1

Physical Properties of Fats

(a) The fats are insoluble in water, but readily soluble in ether, chloroform, benzene, carbon tetrachloride.

(b) They are readily soluble in hot alcohol but slightly soluble in cold.

(c) They are themselves good solvents for other fats, fatty acids, etc.

(d) They are tasteless, odourless, colourless and neutral in reaction.

(e) Several neutral fats are readily crystallised, eg, beef, mutton.

(f) Their melting points are low.

(g) The specific gravity of solid fats is about 0.86. So fat people float in water more readily than thin ones.

(h) They spread uniformly over the surface of water; so the spreading effect is to lower surface tension.

Identification of Fats and Oils

(a) Hydrolysis

1. Hydrolysis of triacylglycerol takes place by lipases producing fatty acids and glycerol.
2. Phospholipases attack the ester linkage of phospholipids.

(b) Saponification

1. Boiling with an alcoholic solution of strong metallic alkali hydrolyzes triglycerides into glycerol and fatty acids —this is called saponification.
2. The products are glycerol and the alkali salts of the fatty acids which are called soaps.
3. Fats, phospholipids, glycolipids and waxes are called saponifiable lipids.
4. Steroids, polyisoprenoids and higher alcohols are grouped as unsaponifiable lipids because they cannot give rise to soap.

(c) Saponification number

1. The number of milligrams of KOH required to saponify 1 gram of fat or oil.
2. The amount of alkali needed to saponify a given quantity of fat will depend upon the number of –COOH group present. It is inversely proportional to the average molecular weight of the fatty acids in the fat i.e. the fats containing short chain fatty acids will have more –COOH groups per gram than long chain fatty acids—this will take up more alkali and, hence, will have higher saponification number.

   Example : Butter—containing a larger proportion of short chain fatty acids such as butyric and caproic acids, has relatively high saponification number 220 to 230.

(d) Acid number

1. The number of milligrams of KOH required to neutralize the free fatty acids of 1 gram of fat.
2. Significance : The acid number indicates the degree of rancidity of the given fat.

(e) Iodine number

1. This is the amount (in grams) of iodine absorbed by 100 grams of fat.
2. This is the measure of the degree of unsaturation of a fat.
3. Significance : If the fat contains higher number of unsaturated fatty acids, it becomes essential for the protection of heart disease. These unsaturated fatty acids, combined with the cholesterol, are oxidized in the liver—producing bile acids, bile salts, vit. D, gonadotrophin hormones. They prevent atherosclerosis.

(f) Acetyl number

1. The number of milligrams of KOH required to neutralize the acetic acid obtained by saponification of 1 gram of fat after it has been acetylated.
2. This is a measure of the number of hydroxy acid groups in the fat.

(g) Polenske number

1. The number of milliliters of 0.1(N) KOH required to neutralize the insoluble fatty acids from 5 grams of fat.

(h) Reichert-Miessl number

1. This is same as the Polenske number except that the soluble fatty acids are measured by titration of the distillate obtained by steam distillation of the saponification mixture.
2. Significance : It measures the amount of volatile soluble fatty acids.

(i) Halogenation

1. Chlorine, bromine and iodine atoms may be added to the double bonds of unsaturated fatty acids containing fats.

(j) Rancidity

1. Nearly all natural fats are oxidized when exposed to air, light, moisture, particularly, if warm, it develops an unpleasant odour and taste. The enzyme lipase—in the presence of moisture and warm temperature—bring about hydrolysis rapidly.
2. This happens due to the formation of peroxides at the double bonds of unsaturated fatty acids.
3. Vitamin E is an important natural antioxidant and prevents development of rancidity.

(k) Soaps

1. Soaps are metallic salts of fatty acids.
2. Soaps are formed by adding alkalis to fatty acids.
3. Soaps of unsaturated fatty acids are softer and more water soluble than those of saturated fatty acids.

4. Potassium soap of an acid is more water-soluble and softer than the sodium soap; calcium and magnesium soaps are far less soluble.

B. Waxes

1. They are esters of fatty acids with higher alcohols other than glycerol.
2. In the human body, the commonest waxes are esters of cholesterol.
3. They are mainly three types :
   (a) True waxes are esters of higher fatty acids with acetyl alcohol or other higher straight chain alcohols.
   (b) Cholesterol esters are esters of fatty acid with cholesterol.
   (c) Vitamin A and vitamin D esters are palmitic or stearic acid esters of vitamin A (Retinol) or vitamin D, respectively.

COMPOUND LIPIDS

A. Phospholipids (phosphatides)

(i) They are esters of fatty acids with glycerol containing an esterified phosphoric acid and a nitrogen base.

(ii) They are present in large amounts in nerve tissue, brain, liver, kidney, pancreas and heart.

Biological functions of phospholipids

(i) They increase the rate of fatty acid oxidation.

(ii) They act as carriers of inorganic ions across the membranes.

(iii) They help blood-clotting.

(iv) They act as prosthetic group to certain enzymes.

(v) They form the structures of membranes, matrix of cell wall, myelin sheath, microsomes and mitochondria.

Classification

It is based on the type of alcohol present in the phospholipid.

There are three types :

1. *Glycerophosphatides* — In this, glycerol is the alcohol group.

   Example :

   (i) Phosphatidyl ethanolamine (cephalin).

   (ii) Phosphatidyl choline (Lecithin).

   (iii) Phosphatidyl serine.

   (iv) Plasmalogens.

   (v) Phosphatidic acid.

2. *Phosphoinositides* — In this, inositol is the alcohol.

   Example : Phosphatidyl inositol (Lipositol).

3. *Phosphosphingosides* — In this, sphingosine is an amino alcohol.

   Example : Sphingomyelin.

The phospholipids include the following groups :

1. *Phosphatidic acid and phosphatidyl glycerols :* Phosphatidic acid is important as an intermediate in the synthesis of triacylglycerols and phospholipids.

   Cardiolipin

   (a) It is formed from phosphatidyl glycerol.

   (b) Chemically, it is diphosphatidyl glycerol.

   (c) It is found in inner membrane of mitochondria and bacterial wall.

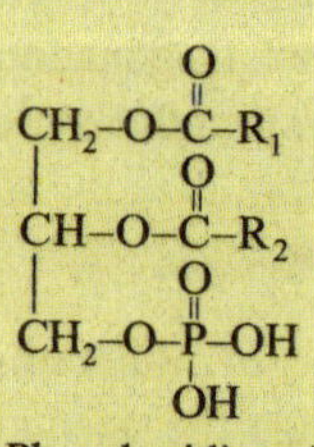

Phosphotidic acid

Fig. 4.2

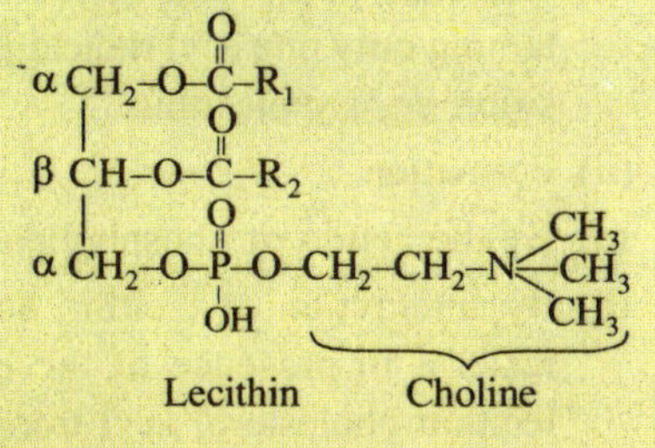

Fig. 4.3

2. *Lecithins (Phosphatidylcholine) :* The lecithins contain glycerol and fatty acids, phosphoric acid and choline (nitrogenous base). Lecithins generally contain a saturated fatty acid at α position and an unsaturated fatty acid at β position. They can exist in α or β forms.

Physical Properties

(i) Lecithins are waxy, white substances but become brown soon when exposed to air.
(ii) They are soluble in ordinary fat solvents except acetone.
(iii) They decompose when heated.
(iv) They constitute valuable agents for the emulsifications of fats and oils.

Chemical Properties of Lecithin

(i) When aqueous solution of lecithins are shaken with H_2SO_4, choline is split off, forming phosphatidic acid.
(ii) When lecithins are boiled with alkalis or mineral acids, not only choline is split off, but phosphatidic acid is further hydrolyzed to glycerophosphoric acid and 2 molecules of fatty acids :

Lecithin $\xrightarrow{H_2SO_4}$ Phosphatidic acid + choline.

Phosphatidic acid → Glycerophosphoric acid + fatty acids (2 mol)

Physiological Functions of Lecithin

(i) It facilitate the combinations with proteins to from lipoproteins of plasma and cells.
(ii) Acetylcholine formed from choline has an important role in the transmission of nervous impulses across synapses.
(iii) Choline is the most important lipotropic agent as it can prevent formation of fatty liver.
(iv) Lecithin lowers the surface tension of lung alveoli. Dipalmityl lecithin is a major constituent of "lung surfactant" which prevents the adherence of the inner surface of the alveoli of the lungs (preventing the collapse of the alveoli) by its surface tension lowering effect. The absence of this in the alveolar membrane of some premature infants causes the respiratory distress syndrome in them.
(v) It lowers the surface tension of water molecule and helps in the emulsification of fat.

Difference of Lecithin and Cephalin

Cadmium chloride compound of cephalin is soluble but cadmium chloride compound of lecithin is insoluble.

3. *Cephalins (Phosphatidyl ethanolamine) :* They always occur in the tissues in association with lecithins and are very similar in properties. The only difference is the nitrogenous base.

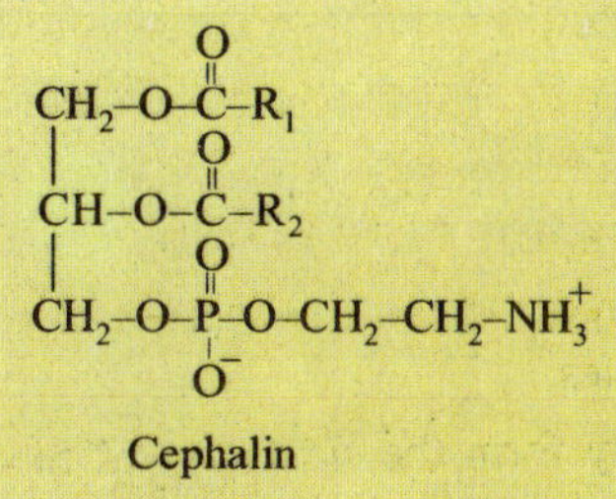

Fig. 4.4

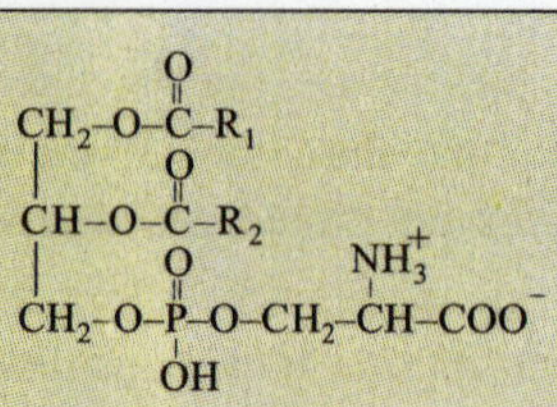

Phosphatidyl serine

Fig. 4.5

4. ***Phosphatidyl Serine***

A cephaline like phospholipid is found in tissues.

5. ***Phosphatidyl inositol (Lipositol or Phosphoinositides)***

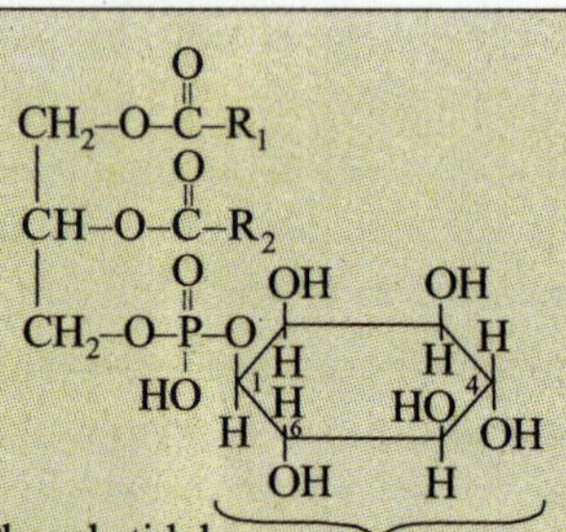

Phosphatidyl inositol — Inositol

Fig. 4.6

(i) It acts as second messenger in Ca^{++} dependent hormone action.

(ii) Some signals must provide communication between the hormone receptor on the plasma membrane and intracellular Ca^{++} reservoirs.

(iii) They are more acidic than the other phospholipids.

6. ***Lysophospholipids :***

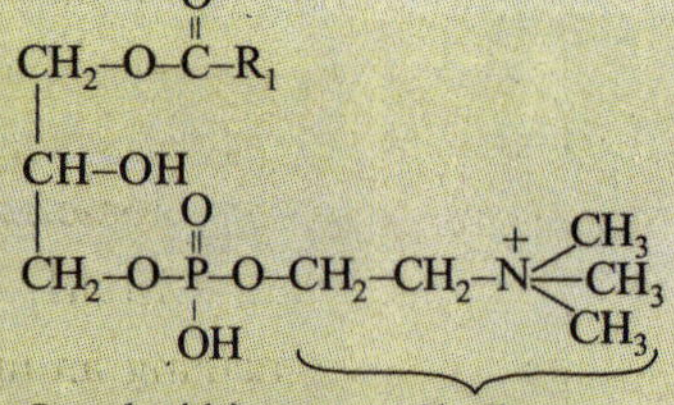

Lysolecithin — Choline

Fig. 4.7

(i) These are phosphoacylglycerols containing only one acyl radical in α position eg, Lysolecithin.

(ii) Formation :

(a) By the action of phospholipase.

(b) By interaction of lecithin and cholesterol in presence of the enzyme lecithin cholesterol acyl transferase, so lysolecithin and cholesterol ester are formed

Lecithin + cholesterol

↓ LCAT

Lysolecithin + cholesterol ester.

7. ***Plasmalogens :***

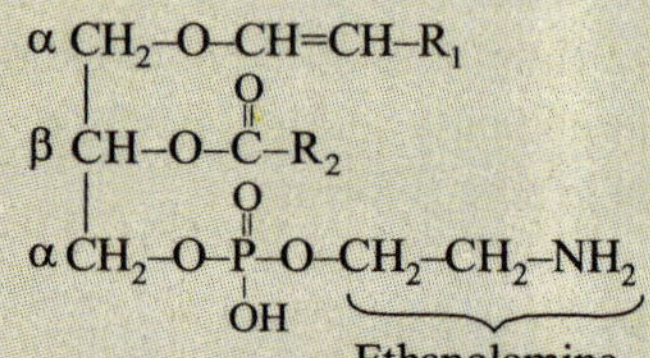

Ethanolamine

Plasmalogen (Phosphatidal ethanolamine)

Fig. 4.8

(i) These are the contents of brain and muscle.

(ii) Structurally, these resemble lecithins and cephalins but give a positive reaction when tested for aldehydes with Schiff's reagent (fuchsin-sulphurous acid) after pretreatment of the phospholipid with mercuric chloride.

(iii) They possess an ether link in a position instead of ester link. The alkyl radical is an unsaturated alcohol.

8. ***Sphingomyelins :***

(i) These are found in large quantities in brain and nerve tissue.

(ii) The concentrations of these phospholipids are increased in Niemann-Pick disease in the liver and spleen.

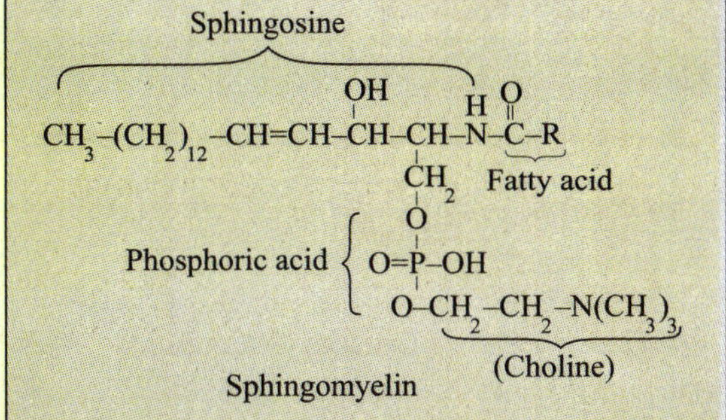

Fig. 4.9

(iii) These contain sphingosine (18 carbon) (amino alcohol) fatty acid, phosphoric acid and choline. No glycerol is present.

(iv) In sphingosine molecule $-NH_2$ group binds a fatty acid by an amide linkage to produce ceramide. When phosphate group is attached to ceramide it is called ceramide phosphate.

(v) When choline is split off from sphingomyelin, ceramide phosphate is left.

CLINICAL ASPECT

- In Niemann-Pick disease excess amount of sphingomyelin are deposited in brain, liver, spleen.
- It is a lipid storage disease (lipidoses) and hereditary. It is caused by the deficiency of enzyme sphingomyelinase.
- The clinical findings are :
  (a) Enlarged liver and spleen.
  (b) Mental retardation.
  (c) Nervous system is affected.
  (d) Anemia and leukocytosis.

Action of Phospholipase

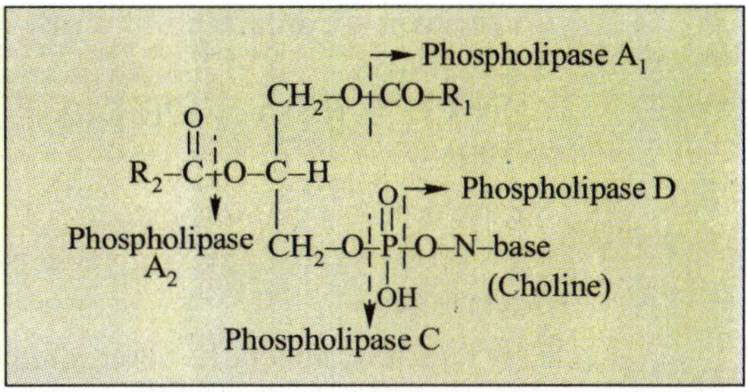

(a) Phospholipase A_1 attacks the ester bond in position 1 of phospholipid.

(b) Phospholipase A_2 attacks β position and form
Lysolecithin + one mol. fatty acid.

(c) Phospholipase B (lysophospholipase) attacks lysolecithin and hydrolyzes ester bond in α position and forms glyceryl phosphoryl choline + 1 mol. fatty acid.

(d) Phospholipase C hydrolyzes phosphate ester bond and produces α, β diacyl glycerol + phosphoryl choline.

(e) Phospholipase D-splits off choline and phosphatidic acid is formed

B. Glycolipids

These contain an amino alcohol (sphingosine or iso-sphingosine) attached with an amide linkage to a fatty acid and glycosidically to a carbohydrate moiety (sugars, amino sugar, sialic acid).

These are further classified into : (i) Cerebrosides, (ii) Gangliosides.

(i) *Cerebrosides :*

(a) Cerebrosides contain galactose, a high molecular weight fatty acid and sphingosine. Therefore, they may also be classified as sphingolipids.

(b) They are the chief constituent of myelin sheath.

(c) They may be differentiated by the type of fatty acid in the molecule. These are :

Kerasin—Containing lignoceric acid $[CH_3-(CH_2)_{22}-COOH]$.

Cerebron—Containing a hydroxylignoceric acid (cerebronic acid). $[CH_3-(CH_2)_{21}-CH(OH)-COOH]$.

Nervon—Containing an unsaturated homologue of lignoceric acid called nervonic acid. $[CH_3-(CH_2)_7-CH=CH-(CH_2)_{13}-COOH]$.

Oxynervon—Containing hydroxynervonic acid [CH_3 — $(CH_2)_7$ — CH = CH — $(CH_2)_{12}$— CH(OH) — COOH].

(d) Stearic acid is a major component of the fatty acids of rat brain cerebrosides.

(e) Cerebrosides, specially cerebronic acid, increases in Gaucher's disease and the Kerasin characterized by glucose replacing galactose.

(f) The cerebrosides are in much higher concentration in medullated than in nonmedullated nerve fibers.

CLINICAL ASPECT OF CEREBROSIDE

Gaucher's disease

- The cerebroside content of the reticuloendothelial cell (spleen) is very high.
- In cerebroside molecule, the kerasin is characterised by glucose replacing galactose.
- The disease is caused by the deficiency of enzyme glucocerebrosidase.

Symptoms

- Spleen is increased, signs of leucopenia.
- Liver is enlarged.
- Eyes show a yellow-brown wedge shaped elevation.

(ii) *Gangliosides :*

(a) These are glycolipids occurring in the brain.

(b) Gangliosides contain ceramide (sphingosine + fatty acids), glucose, galactose, N-acetylgalactosamine and sialic acid.

(c) Some gangliosides also contain dihydrosphingosine or gangliosine in place of sphingosine.

(d) Most of the gangliosides contain a glucose, two molecules of galactose, one N-acetylgalactosamine and up to three molecules of sialic acid.

Types of Ganglioside

GM_1 ganglioside :

Gal → Gal NAC → Gal → Glucose → Ceramide
↓
NANA

GM_2 ganglioside :

Gal NAC → Gal → Glucose → Ceramide
↓
NANA

GM_3 ganglioside :

Gal → Glucose → Ceramide
↓
NANA

G = Ganglioside, M = Monosialo variety.

1, 2, 3 = Position in chromatography.

GM_1 gangliosidosis : This is due to deficiency of enzyme β-galactosidase.

Tay Sach's disease (GM_2 gangliosidosis)

- This disease is characterised by increased accumulation of GM_2 ganglioside in brain and spleen.
- It is caused by deficiency of enzyme hexosaminidase A.

Symptom

- Mental retardation, blindness, muscular weakness.
- Red spot appears in the muscular region of eye in first year of life.

C. Other compound lipids :

1. Lipoproteins

(i) Triacylglycerol (45%), phospholipids (35%), cholesterol and cholesteryl esters (15%), free fatty acids (less than 5%) and also protein combine to form a hydrophilic lipoprotein complex.

(ii) Since pure fat is less dense than water, the proportion of lipid to protein in lipoproteins in plasma is by ultracentrifugation.

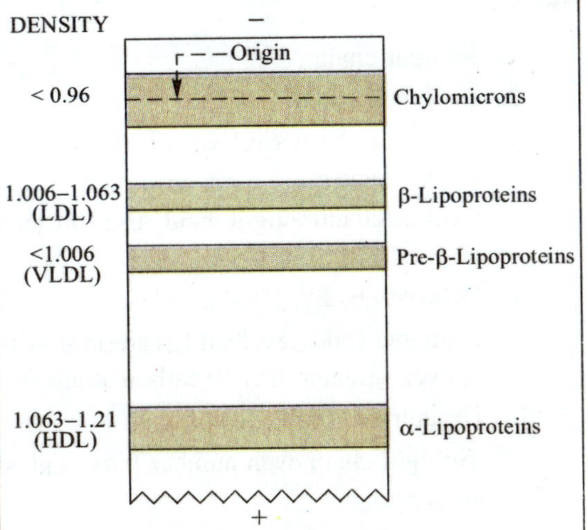

(Separation of plasma lipoproteins by paper) electrophoresis

Fig. 4.10

(iii) The density of lipoproteins increases as the protein content rises and the lipid content falls and the size of the particle becomes smaller.

(iv) Lipoproteins may be separated on the basis of their electrophoretic properties and may be identified more accurately by means of immuno-electrophoresis.

(v) Four major groups of lipoproteins have been identified which are important physiologically and in clinical diagnosis in some metabolic disorders of fat metabolism.

These are :

(a) Chylomicrons.

(b) Very low density lipoproteins (VLDL or pre-β-lipoproteins).

(c) Low density lipoproteins (LDL or β-lipoproteins).

(d) High density lipoproteins (HDL or α-lipoproteins).

(vi) Chylomicrons and VLDL : Predominant lipid is triacylglycerol (50%) and cholesterol (23%). The concentrations of these are increased in atherosclerosis and coronary thrombosis etc.

LDL : Predominant lipid is cholesterol (46%) and phospholipids (23%). Increase in atherosclerosis and coronary thrombosis, etc.

HDL : Predominant lipid is phospholipid (27%) and proteins (45%).

(vii) The protein moiety lipoprotein is known as an apoprotein which constitute nearly 60% of some HDL and 1% of chylomicrons. Many lipoproteins contain more than one type of apoprotein polypeptide.

(viii) The larger lipoproteins (such as chylomicrons and VLDL) consist of a lipid core of nonpolar triacylglycerol and cholesteryl ester surrounded by more polar phospholipid, cholesterol and apoproteins.

Importance

(i) To transport and deliver the lipids to tissues.

(ii) To maintain structural integrity of cell surface and subcellular particles like mitochondria and microsomes.

(iii) The β-lipoprotein fraction increases in severe diabetes mellitus, atherosclerosis etc. Hence determination of the relative concentrations of α-and β-lipoproteins and pre-β- lipoproteins are of diagnostic importance.

2. Aminolipids

Phosphatidyl ethanolamine and serines are aminolipids and sphingomyelins and gangliosides contain substituted amino groups.

3. Sulpholipids (Sulphatides)

(i) These have been isolated from brain and other animal tissues.

(ii) These are sulphate derivatives of the galactosyl residue in cerebrosides.

DERIVED LIPIDS

1. Fatty acids

(i) These are obtained by the hydrolysis of fats.

(ii) Fatty acids occurring in natural fats usually contain an even number of carbon atoms because they are synthesized from 2-carbon units and are straight chain derivatives.

(iii) The straight chain may be saturated (containing no double bonds) or unsaturated (containing one or more double bonds).

(iv) Carbon atoms of fatty acids are numbered from the carboxyl carbon (carbon No.1). The carbon atom adjacent to the carboxyl carbon (Carbon No. 2) is also known as the α-carbon. Carbon atom No. 3 is the β–carbon and the end methyl carbon is known as the γ-carbon.

(v) Various conventions are used for indicating the number and position of the double bonds, eg, Δ^9 indicates a double bond between carbon atoms 9 and 10 of the fatty acid.

Types

1. Straight chain.
2. Branched chain.
3. Substituted (methyl substituted – cerebronic acid)
4. Cyclic (chaulmoogric acid) used in leprosy.

1. ***Straight chain :***

(a) Saturated {odd (less than 10 carbon atoms) & even (greater than 10 carbon atoms)}.

(b) Unsaturated (odd & even).

(Straight chain even number fatty acid is common)

A. Saturated fatty acids

General formula for saturated fatty acids is $C_nH_{2n+1}COOH$. Other higher fatty acids occur in waxes. A few branched-chain fatty acids have also been isolated from both plant and animal sources.

Prostanoids include Prostaglandins (PG), and thromboxanes (TX).

General characteristics of prostanoid

(a) All are 20 carbon compounds.

Saturated Fatty Acids

Acid	Formula	Carbon atoms	Sources
Acetic	CH_3COOH	2	Product of carbohydrate fermentation by rumen organisms
Propionic	C_2H_5COOH	3	— do—
Butyric	C_3H_7COOH	4	Butter
Caproic	$C_5H_{11}COOH$	6	Product of carbohydrate fermentation by rumen organisms
Caprylic	$C_7H_{15}COOH$	8	Butter
Decanoic (Capric)	$C_9H_{19}COOH$	10	Butter
Lauric	$C_{11}H_{23}COOH$	12	Coconut oils
Myristic	$C_{13}H_{27}COOH$	14	Coconut oils
Palmitic	$C_{15}H_{31}COOH$	16	Animal and plant fats
Stearic	$C_{17}H_{35}COOH$	18	–do–
Arachidic	$C_{19}H_{39}COOH$	20	Peanut oil
Behenic	$C_{21}H_{43}COOH$	22	Seeds
Lignoceric	$C_{23}H_{47}COOH$	24	Peanut oil, crebrosides

B. Unsaturated fatty acids

General formula $C_nH_{2n-1}COOH$					
Type of acid	*Acid*	*Formula*	*Unsaturation at carbon atoms*	*Number of double bonds*	*Sources*
Mono-unsaturated	Palmitoleic	$C_{15}H_{29}COOH$	Δ^9	1	Near all fats
,,	Oleic	$C_{17}H_{33}COOH$	Δ^9	1	–do–
Poly-unsaturated	Linoleic	$C_{17}H_{31}COOH$	Δ^9, Δ^{12}	2	Animal and plant fat
,,	Linolenic	$C_{17}H_{29}COOH$	$\Delta^9, \Delta^{12}, \Delta^{15}$	3	–do–
,,	Arachidonic	$C_{19}H_{31}COOH$	$\Delta^5, \Delta^8, \Delta^{11}, \Delta^{14}$	4	Peanut oil
Eicosanoids					
Prostanoids & Leukotrienes	Timnoionic	$C_{19}H_{33}COOH$	$\Delta^5, \Delta^8, \Delta^{11}, \Delta^{14}, \Delta^{17}$	5	Fish oils, egg, cod liver oil
,,	Clupanodonic	$C_{12}H_{32}COOH$	$\Delta^7, \Delta^{10}, \Delta^{13}, \Delta^{16}, \Delta^{19}$	5	Fish oils, phospholipids in brain
,,	Cervonic	$C_{21}H_{31}COOH$	$\Delta^4, \Delta^7, \Delta^{10}, \Delta^{13}, \Delta^{16}, \Delta^{19}$	6	Fish oils, phospholipids in brain

Classification

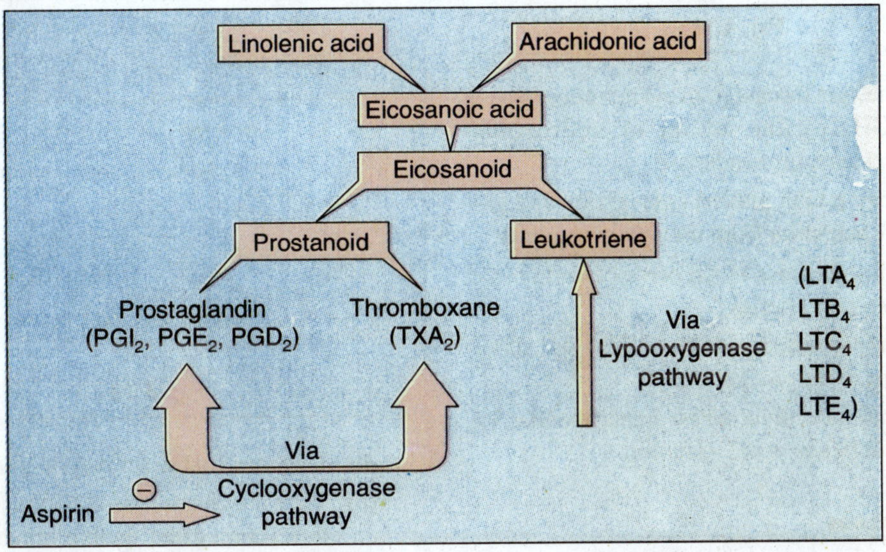

(b) Trans double bond at 13 position.

(c) –OH group at 15 position.

Prostaglandins (PG)

(a) They virtually exist in every mammalian tissue and act as local hormones.

(b) They have important physiologic and pharmacologic activities.

(c) They are synthesized in vivo by cyclization of the center of the carbon chain of 20-C polyunsaturated fatty acids (eg, arachidonic acid) to form a cyclopentane ring.

(d) Three different eicosanoic fatty acids give rise to three groups of eicosanoids characterized by the number of double bonds in the side chains, eg, PG_1, PG_2, PG_3. Variations in the substituent groups attached to the rings give rise to different types in each series of prostaglandins, as for example, "E" type of Prostaglandin has a keto group in position 9, whereas the "F" type has a hydroxyl group in this position,

Prostacyclins (PGI)

(a) They are formed in vascular endothelium and continually formed in heart. They are also formed in kidneys.

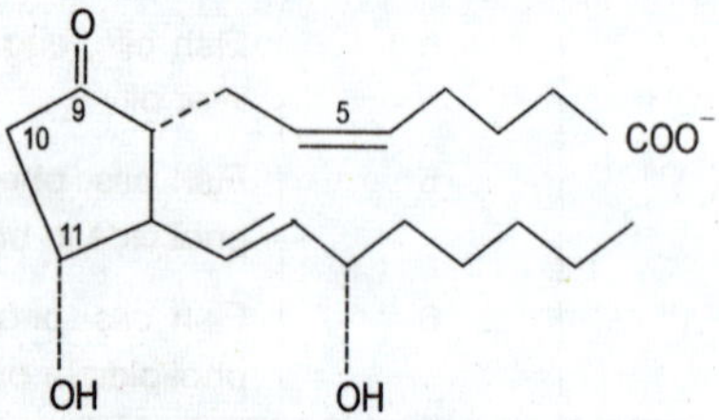

Prostaglandin E_2 (PGE_2)

Fig. 4.11

(b) They are formed from cyclic endoperoxide PGH_2 by the action of microsomal Prostacyclin synthetase.

(c) They inhibit platelet aggregation and gastric secretion from the pyloric mucosa.

(d) They decrease blood pressure and protect coronary arteries.

(e) They increase renal blood flow and stimulate renin production.

(f) They are inhibited by hyperlipemia, vit. E deficiency and radiation.

Thromboxanes

(a) They contract smooth muscles on blood vessels, GI tract, uterus, bronchioles.

(b) They are discovered in platelets, and have the cyclopentane ring interrupted with an oxygen atom (Oxane ring).

(c) The substituent groups attached to the rings being varied give rise to different types in each series of thromboxanes labelled A, B, etc.

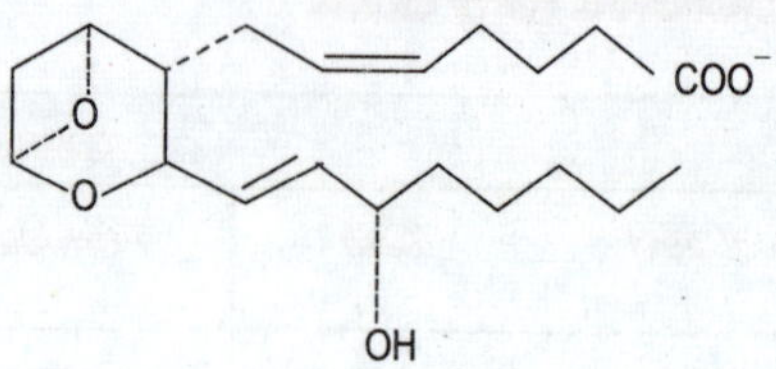

Thromboxane A_2 (TXA_2)

Fig. 4.12

(d) They produce vasoconstriction and increase blood pressure.

(e) They cause release of serotonin and calcium ion (Ca^{++}) from platelet granules.

(f) Imidazoles inhibit their synthesis.

Leukotrienes

(a) They are the third group of eicosanoid derivatives formed via the lipoxygenase pathway rather than cyclization of the fatty acid chain.

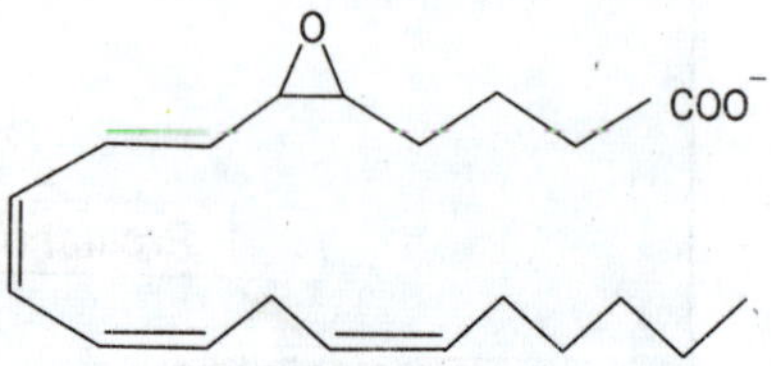

Leukotriene A_4 (LTA_4)

Fig. 4.13

(b) They are first described in leukocytes.

(c) They are characterized by the presence of three conjugated double bonds.

(d) They are stimulators of mucus secretion and are responsible for vasoconstriction of bronchial muscles.

(e) They are inhibited by prolonged use of aspirin.

The group of compounds known as prostaglandins are synthesized from arachidonic acid in the body. They have pharmacologic and biochemical activity.

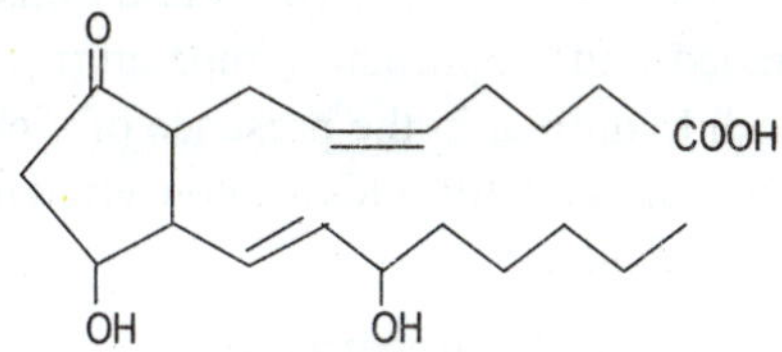

Prostaglandin E_2 (PGE_2)

Fig. 4.14

Example : Prostaglandin E_2(PGE_2)

C. Many Other Fatty Acids

(i) These have been detected in biologic material.

Example : Fish oil contain 5 and 6 unsaturated fatty acids having carbon atoms 22.

(ii) Various other structures with hydroxy groups (ricinoleic acid) or cyclic groups have been found in nature.

Example of cyclic groups is chaulmoogric acid which was used many years ago in the treatment of leprosy.

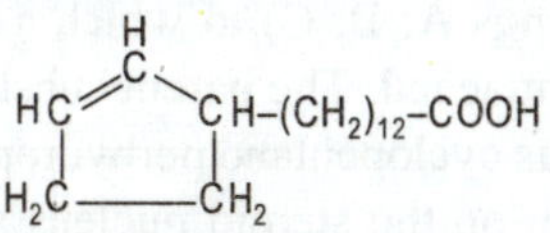

Chaulmoogric acid

Fig. 4.15

Essential Fatty Acids

Burr and Burr (1930) introduced the term "Essential Fatty Acids" (EFA) on the basis that they are essential for the growth and health of young albino rats. These polyunsaturated fatty acids which are not synthesized in the body but are taken from natural sources are called essential fatty acids. They are (mentioned above) :

Linolenic and arachidonic acids are formed from linoleic acids provided linoleic acids are available in the body in sufficient quantities.

Properties

(i) The essential fatty acids of vegetable oils have low melting points and iodine number.

(ii) They become saturated fatty acids on hydrogenation and the oils become solid fats.

Functions

1. The essential fatty acids in high concentration along with the lipids constitute the structural elements of the tissues.
2. The lipids of gonads also contain a high concentration of polyunsaturated fatty acids which suggest the importance of reproductive function.
3. They effect the prolongation of clotting time and increase the fibrinolytic activity.

Essential fatty acids	Chemical name	General formula	No. of double bonds	Structure	Sources
Linoleic	9, 2-Octadecadienoic acid	$C_nH_{2n-3}COOH$	$2(\Delta^9, \Delta^{12})$	$CH_3-(CH_2)_4(CH=CH-CH_2)_2(CH_2)_6COOH$	Corn, Peanut, Cotton seed, Soyabean oil.
Linolenic	6, 9,12-Octadecatrienoic acid	$C_nH_{2n-5}COOH$	$3(\Delta^9, \Delta^{12}, \Delta^{15})$	$CH_3-(CH_2)_4(CH=CH-CH_2)_3(CH_2)_3COOH$	Found frequently with linoleic acid but particularly in linseed oil.
Arachidonic	5,8,11,14-eicosatetraenoic acid	$C_nH_{2n-7}COOH$	$4(\Delta^5, \Delta^8, \Delta^{11}, \Delta^{14})$	$CH_3-(CH_2)_4(CH=CH-CH_2)_4(CH_2)_2COOH$	Found in small quantities with linoleic and linolenic acids but particularly in peanut oil.

4. They retard atherosclerosis being esterified and emulsified with cholesterol and are incorporated into lipoproteins for transport to the liver for further oxidation.
5. They cure skin lesions.
6. The deficiency of these acids in the diet of babies causes eczema.

ISOMERISM IN UNSATURATED FATTY ACIDS

Variations in the locations of the double bond in unsaturated fatty acid chains produce isomers. Oleic acid has 15 different positional isomers.

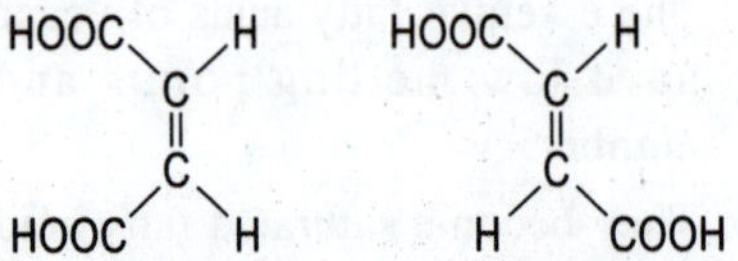

Cis form (Maleic acid) — Trans form (Fumaric acid)

Fig. 4.16

Geometric isomerism depends on the orientation of radicals around the axis of double bonds. If the radicals which are being considered are on the same side of the bond, the compound is called "cis", if on opposite side, "trans". This can be illustrated with maleic acid and fumaric acid.

There are more geometric isomers in case of acids with greater degree of unsaturation. The unsaturated long chain of fatty acids occurring in nature are nearly all in the 'cis' form and the molecules are "bent" at the position of the double bond. Thus, arachidonic acid is U-shaped.

REFINED AND HYDROGENATED OILS

Refined oil : It is prepared in the following manner :

(i) Free fatty acids are removed by alkali treatment.

(ii) Colouring matter is removed by activated carbon.

(iii) Odour is removed by superheated steam.

Hydrogenated oils : The refined oils are hydrogenated under optimum temperature and pressure with hydrogen in the presence of nickel catalyst. Unsaturated fatty acids are converted into saturated fatty acids.

$$\text{Oleic acid} \xrightarrow{\text{Hydrogenation}} \text{Stearic acid.}$$

The liquid oil becomes solid fat and the unsaturated fatty acid content decreases. Vanaspati is hydrogenated refined groundnut oil.

2. ALCOHOLS

Alcohols found in lipid molecules include glycerol, cholesterol and higher alcohols (acetyl alcohol), usually found in the waxes.

The unsaturated alcohols are important pigments. Phytyl alcohol is a constituent of chlorophyll and lycophyll ($C_{40}H_{56}O_2$); a polyunsaturated dihydroxy alcohol occurs in tomatoes as a purple pigment.

3. STEROIDS

The steroids are often found in association with fat. They have a similar cyclic nucleus resembling phenanthrene (rings A, B, C) to which a cyclopentane ring (D) is attached. The parent substance is better designated as cyclopentano perhydrophenanthrene. The position on the steroid nucleus are numbered as shown in Fig. 4.17.

Methyl side chains occur typically at positions 10 and 13 (constituting C atoms 19 and 18). A side chain at position 17 is usual (as in cholesterol). If the compound has one or more hydroxyl groups and no carbonyl or carboxyl groups, it is a sterol, and the name terminates in –OL.

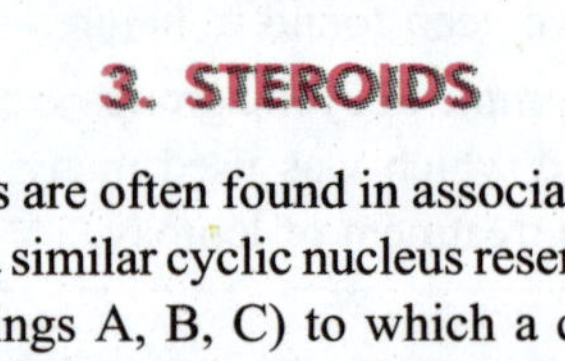

Cyclopentanoperhydrophenanthrene nucleus

Fig. 4.17

Steroids may be divided in the following manner :

Sterols—cholesterol, ergosterol, coprosterol.

Bile acids—Glycocholic acid and taurocholic acid.

Sex hormones—Testosterone, Estradiol.

Vitamin D—Vit. D_2 and D_3 .

Adrenocortical hormones—Corticosterone.

Cardiac glycosides—Stropanthin.

Saponins—Digitonin.

Cholesterol

It is widely distributed in all cells of the body. It occurs in animal fats but not in plant fats. Its structure is given below. The metabolism of cholesterol is discussed in the chapter of lipid metabolism.

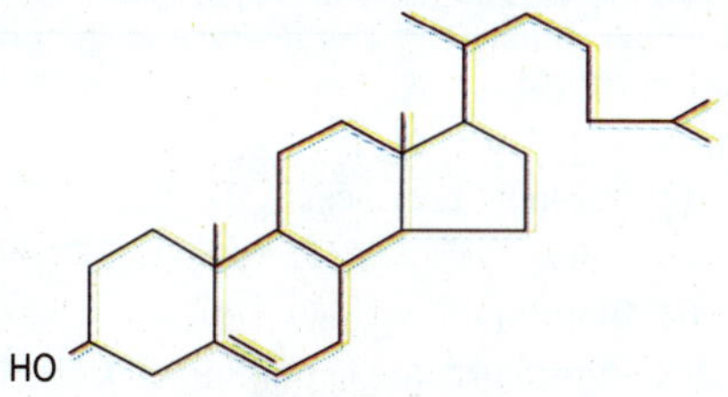

(Cholesterol)

Fig. 4.18

Erogesrol

(i) It occurs in ergot and yeast.

(ii) It is the precursor of vitamin D.

(iii) It acquires antirachitic properties with the opening of ring B when irradiated with ultraviolet light.

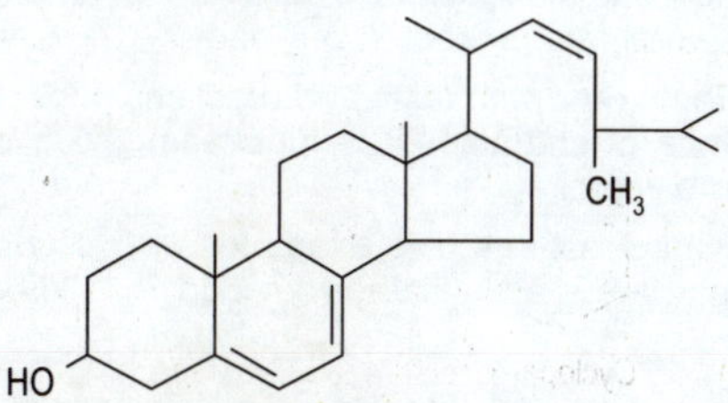

(Ergosterol)

Fig. 4.19

Coprosterol

It occurs in feces as a result of the reduction by bacteria in the intestine of the double bond between C_5 and C_6 of cholesterol.

Important tests

1. **Greese spot test :** A drop of oil placed over a piece of ordinary paper. A translucent spot is visible. This indicates the presence of fat.
2. **Emulsification test :** 2 ml water is taken in one test tube and 2 ml of diluted bile salt solution in another test tube. Add 3 drops of the given oil to each test tube and shake vigorously. Note the stability of the emulsification formed.
3. **Saponification test :** Take 10 drops of coconut oil in a test tube. Add 20 drops of 40% NaOH and 2 ml of glycerol to it. Gently boil for about 3 minutes until complete saponification occurs. If oil globules are visible, boiling must be continued. Divide the solution into 3 parts to carry the following experiments in test tube 1, 2, 3.

To test tube No. 1 add saturated solution of NaCl. Note that the soap separates out and floats to the surface (salting out process).

To test tube No. 2 add a few drops of conc. HCl. An oily layer of the fatty acids rises to the surface.

To test tube No. 3 add a few drops of $CaCl_2$ solution. The insoluble calcium soap is precipitated.

Unsaturation test : Add 10 drops of Hubble's iodine reagent to 10 ml of chloroform. The chloroform assumes a pink colour due to the free iodine. The solution is divided equally into three test tubes as (a), (b) and (c) and three types of oil are added.

Add the oil No. 1 to the test tube (a) drop by drop shaking the tube vigorously after each addition till the pink colour of the solution just disappears. The number of oil drops required are noted. The experiment is repeated by adding oil 2 and 3 to test tubes (b) and (c), respectively. The more the number of drops required to discharge the pink colour, the less is the unsaturation.

Colour Reactions to Detect Sterols

Liebermann-Burchard Reaction : A chloroform solution of a sterol when treated with acetic anhydride and sulphuric acid gives a green colour. This reaction is the basis of a colorimetric estimation of blood cholesterol.

Salkowski test : A red to purple colour appears when a chloroform solution of the sterol is treated with an equal volume of concentrated sulphuric acid.

CLINICAL ORIENTATION

- The high concentration of polyunsaturated fatty acids in the lipids of gonads are important in reproductive function.
- The essential fatty acid deficiency causes swelling of mitochondrial membrane resulting in the reduction in efficiency of oxidative phosphorylation producing increased heat.
- Docosahexenoic acid formed from dietary linolenic acids enhances the electrical response of the photoreceptors to illumination. Therefore, linolenic acid of the diet is essential for optimal vision.
- The deficiency of essential fatty acids causes skin lesions, abnormal pregnancy and lactation in adult females, fatty liver, kidney damage.
- The genetic deficiency of lecithin cholesterol acyl transferase (LCAT) causes Norum's Disease.
- Sitosterol decreases the intestinal absorption of exogenous and endogenous cholesterol and thereby lowers the blood cholesterol level.
- The deficiency of the enzyme sphingomyelinase causes the large accumulations of sphingomyelins in brain, liver and spleen of children resulting in the Niemann-Pick disease with the symptoms of enlarged abdomen, liver, spleen and mental deterioration.
- Absence of dipalmityl lecithin (DPL) in premature foetus produces respiratory distress syndrome (Hyaline-membrane disease).
- The inherited Gaucher's Disease in infancy and childhood is caused by the deficiency of the enzyme glucocerebrosidase involving the large accumulations of glucocerebrosides (usually Kerasin) in the liver, spleen, bone marrow, and brain with the manifestations of weight loss, failure in growth, and progressive mental retardation.
- The autosomal recessive Tay-Sach's Disease (GM_2 Gangliosidosis) results in the accumulation of large amounts of gangliosides in the brain and nervous tissues due to the absence of the enzyme hexosaminidase A with the association of progressive development of idiocy and blindness in infants soon after birth.

CLINICAL ORIENTATION

- The inherited disorder Metachromatic Leukodystrophy (MLD) happens on the sulfatide, formed from galactocerebroside, accumulation in various tissues owing to the deficiency of the enzyme sulfatase (Aryl sulfatase) with the symptoms of weakness, ataxia, defects in locomotion, paralysis, difficulties in speech in children before three years of age and psychiatric manifestation including progressive dementia in adults.
- Obesity and atherosclerosis are distinctly related to the concentrations of cholesterol and polyunsaturated fatty acids in the body.

EXERCISE

1. What are lipids? How are they classified? Discuss the characteristics and functions of a class of lipids having nitrogen base and phosphoric acid as constituents. (P.U. 69S)
2. What are lipids and classify them. Describe some important tests for lipids. (Mith. 65S)
3. Give the classification and characteristics of lipids. Discuss briefly the chemistry and importance of phospholipids. (Bh. U. 75A)
4. Write short notes on :
   (i) Essential fatty acids. (Mith 75A ; Bh.U. 76S; P.U. 74A)
   (ii) Sterol. (Bh.U. 76A)
   (iii) Phospholipids. (M.U. 73A; P.U. 68A; 76A)
   (iv) Saponification number. (Bh. U. 75S)
   (v) Hydrogenated fats. (Muz. 74A)
   (vi) Lecithin. (P.U. 72A ; Mith. 62S)
   (vii) Iodine number. (R.U. 71S)
   (viii) Lipoproteins. (R.U. 72S)
   (ix) Unsaturated fats. (MUz. 75A)
   (x) Saponification. (C.U. 1981)
   (xi) Rancidity. (C.U. 1981
   (xii) Cholesterol. (C.U. 1981)
5. Give the structural formula of α-glycerophosphate. (C.U.1981)
6. How can the lipids be classified ? Enumerate some neutral fats present in our body. (C.U. 1981)
7. Name, and with tests and reactions show the presence of characteristics functional group occurring in glycerol. (C.U. 1981)
8. With structures give examples of two unsaturated fatty acids essential for our body. (C.U. 1981)
9. Fill up the gaps with correct words :
   (i) Oils and fats are — of fatty acids. Oils are generally more — than fats. (C.U. 1981)
   (ii) On boiling with NaOH solution a fat will produce — and —. (C.U.1981)

DIAGRAMMATIC COMPENDIUM

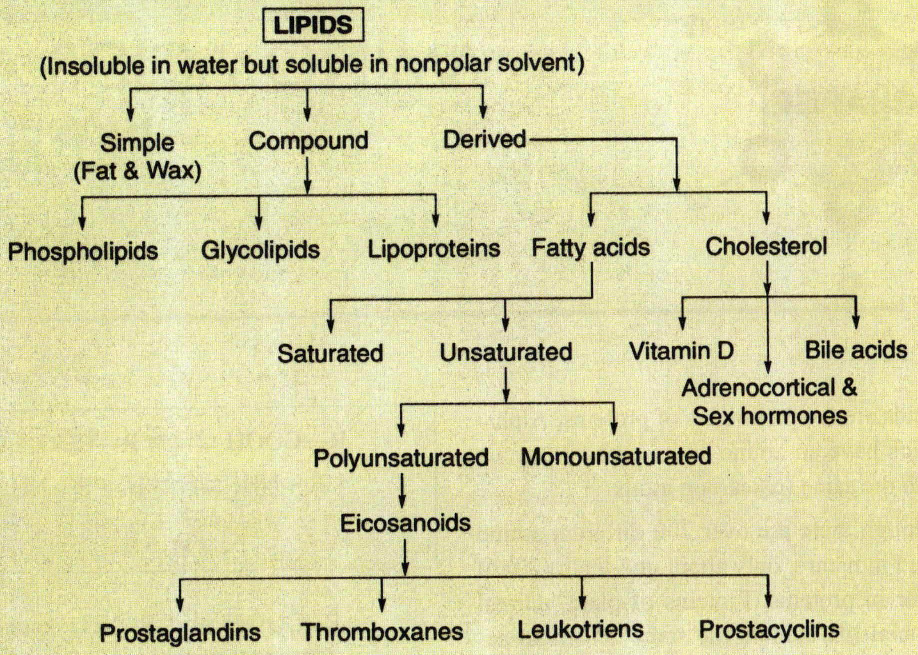

CHAPTER 5

AMINO ACIDS AND PEPTIDES

Amino acids are the monomers of proteins. Alpha-amino acids have an amino and a carboxyl group attached to the same (α) carbon atom.

Although there are over 200 different amino acids found in nature, only about one-tenth (20) of these occur in proteins. Proteins of plant, animal and microbial life contain the same 20 amino acids.

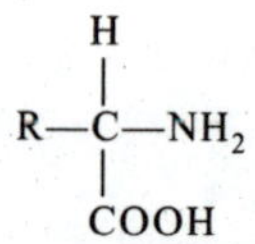

(α-amino acid)

Fig. 5.1

Complete acid-, base-, or enzyme-catalyzed hydrolysis of proteins produces the naturally occurring 20 L-α amino acids. D–amino acids are also found in some natural peptides.

Common properties of amino acids

1. Protonic equilibria of amino acids

(a) Amino acids have at least two ionizable groups i.e. $-COOH$ and $-NH_3^+$. The former dissociates more easily than the latter. In solution, two forms of these groups, one charged and one neutral, exist in protonic equilibrium with each other.

$$R{-}COOH \rightleftharpoons R{-}COO^- + H^+$$
$$R{-}NH_3^+ \rightleftharpoons H^+ + R{-}NH_2$$

Fig. 5.2

R–COOH and $R{-}NH_3^+$ represent the protonated or acid partners in these equilibria. $R{-}COO^-$ and $R{-}NH_2$ are the conjugated bases (i.e. proton acceptors) of the corresponding acids. Although both R–COOH and $R{-}NH_3^+$ are weak acids, R–COOH is a several thousand times stronger acid than is $R{-}NH_3^+$.

(b) At pH 7.4, carboxyl groups exist almost entirely as the conjugated base i.e. $R{-}COO^-$. Most amino groups exist in the form, $R{-}NH_3^+$. In blood and most tissues, amino acid structures are drawn as follows :

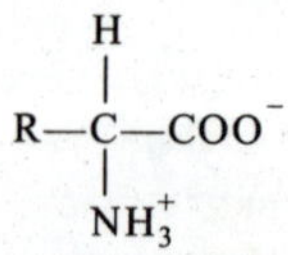

Fig. 5.3

The following structure cannot exist at any pH but is frequently used as a convenience when the chemistry of amino acids is discussed.

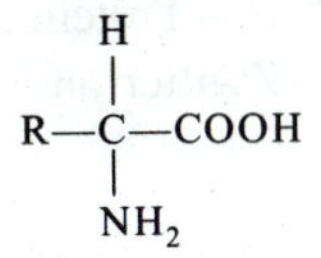

Fig. 5.4

(c) P_k of an acid is simply the negative log of the dissociation constant

$$P_k = -\log K.$$

P_k values for α–amino groups of free amino acids is about 9.8.

(d) The *isoelectric pH* (P^I) of an amino acid is that pH at which it has no net charge and hence does not move in an electric field.

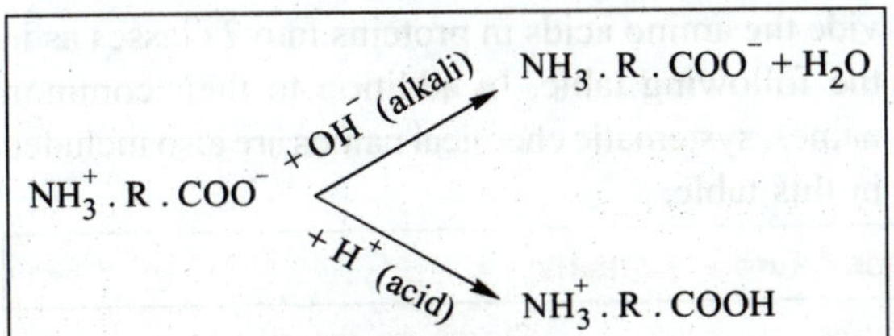

Fig. 5.5

Addition of acid or alkali depresses one type of ionisation so that the amino acid behaves as a base or an acid. The ion at the isoelectric point which carries + and – charges internally neutralised is called **"Zwitterion"**. The three types of ions are represented in Fig. 5.5.

In an acid solution, the amino acid acts as a base yielding cations. When current is allowed to pass through the solution, the amino acid migrates to the cathode or positive pool. In an alkaline solution, it behaves as an acid forming anions. In the electric field, the amino acid migrates to the anode or negative pool. On account of these opposite reactions depending on the acidity or alkalinity of the solution the amino acids are called **ampholytes**.

Since $Pk_1(RCOOH) = 2.35$ and $Pk_2(RNH_3^+) = 9.69$, the isoelectric pH (P^I) of alanine is

$$P^I = \frac{Pk_1 + Pk_2}{2} = \frac{2.35 + 9.69}{2} = 6.02$$

$$P^I \text{ of Aspartic acid} = \frac{Pk_1 + Pk_2}{2}$$

$$= \frac{2.09 + 3.86}{2} = 2.98$$

Thus P^I of lysine and arginine is 9.7 and 10.8 respectively.

The ability to perform calculations of this type is of significant value in the clinical laboratory to assess the mobility of known compounds in electric fields and to select appropriate buffers for separation of one from another.

Zwitterion

1. The ion at isoelectric point which carries + and – charges is called zwitterion.

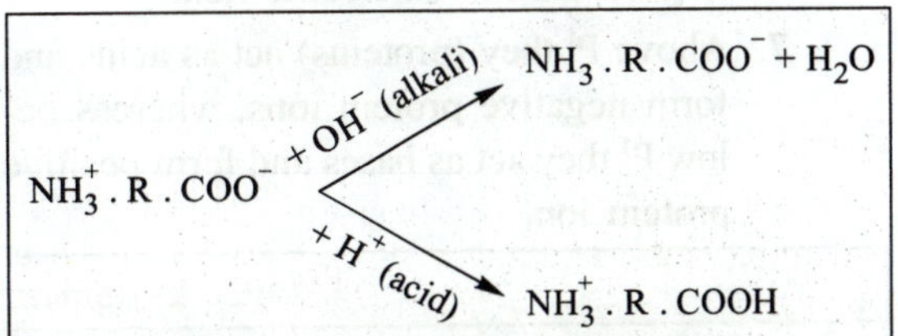

2. In acid solution zwitterion combines with H^+ ions to form base yielding cation, when the current is passed, the amino acid migrates towards cathode.
3. In alkali solution, zwitterion combines with OH^- to form acid yielding anions and migrates towards anode during the change of current.
4. Zwitterion is the ampholyte i.e. it is both a proton donor and proton acceptor.
5. Generally, the acidic and basic strength of the zwitterion are different and a solution of pure amino acid in water is not neutral.
6. Proteins like the amino acids, contain 6 acidic and basic groups exist in solution as zwitterion.

Isoelectric pH

1. The isoelectric pH of an amino acid is that pH at which it has no net charge and does not move in an electric field.

2. It can be denoted as follows :

$$P^I = \frac{Pk_1 + Pk_2}{2} \quad [\because Pk_1 = RCOO^-, \; Pk_2 = RNH_3^+]$$

3. Isoelectric pH (P^I) of
   - Aspartic acid → 2.9
   - Lysine → 9.7
   - Arginine → 10.8
4. It has a significant value in the clinical laboratory to determine the mobility of known compounds in electric field.
5. It selects appropriate buffers for separation of one from another.
6. Proteins like amino acids have isoelectric pH (P^I) at which they are least soluble and migrate least in an electric field.
7. Above P^I they (proteins) act as acids and form negative protein ions, whereas below P^I they act as bases and form positive protein ion.

$$\underset{\text{cation}}{H^+ + \text{Protein}^+} \rightleftharpoons \underset{\text{Zwitterion}}{H^+ - \text{Protein}^+} \rightleftharpoons \underset{\text{anion}}{H^+ + \text{Protein}^-}$$

Ampholyte

1. In acid solution, the amino acid (Zwitterion) acts as base yielding cation, when current → cathode.
2. In alkali solution → acid yielding anion, when current → anode.
3. On the basis of these opposite reactions depending upon the acidity and alkalinity of the solution, the amino acids are called ampholyte.

2. Structures of amino acids

For many purposes, it is convenient to subdivide the amino acids in proteins into 7 classes as in the following table. In addition to their common names, systematic chemical names are also included in this table.

TABLE : L-α-amino acids found in proteins

Group	Trivial name	Symbol	Systematic name	Structural formula
I.	With aliphatic side chains			
	Glycine	Gly (G)	Amino acetic acid	$H-CH(NH_2)-COOH$
	Alanine	Ala (A)	2-Amino propanoic acid	$CH_3-CH(NH_2)-COOH$
	Valine	Val (V)	2-Amino-3-methyl butanoic acid	$(H_3C)_2CH-CH(NH_2)-COOH$
	Leucine	Leu (L)	2-Amino-4-methyl pentanoic acid	$(H_3C)_2CH-CH_2-CH(NH_2)-COOH$
	Isoleucine	Ile (I)	2-Amino-3-methyl pentanoic acid	$CH_3-CH_2-CH(CH_3)-CH(NH_2)-COOH$
II.	With side chains containing hydroxyl (OH) groups			
	Serine	Ser (S)	2-Amino-3-hydroxy propanoic acid	$CH_2(OH)-CH(NH_2)-COOH$

Contd.

TABLE : L-α-amino acids found in proteins

Group	Trivial name	Symbol	Systematic name	Structural formula
	Threonine	Thr (T)	2-Amino-3-hydroxy butanoic acid	$CH_3-CH(OH)-CH(NH_2)-COOH$
III.	With side chains containing sulfur atoms			
	Cysteine	Cys (C)	2-Amino-3-mercapto propanoic acid	$CH_2(SH)-CH(NH_2)-COOH$
	Methionine	Met (M)	2-Amino-4-(methylthio) butanoic acid	$CH_2(S-CH_3)-CH_2-CH(NH_2)-COOH$
IV.	With side chains containing acidic group			
	Aspartic acid	Asp (D)	Amino succinic acid	$HOOC-CH_2-CH(NH_2)-COOH$
	Glutamic acid	Glu (E)	2-Amino glutaric acid	$HOOC-CH_2-CH_2-CH(NH_2)-COOH$
V.	With side chains containing basic groups			
	Arginine	Arg (R)	2-Amino-5-guanido valeric acid	$H-N(C(=NH)NH_2)-CH_2-CH_2-CH_2-CH(NH_2)-COOH$
	Lysine	Lys (K)	2, 6-Diamino hexanoic acid	$CH_2(NH_2)-CH_2-CH_2-CH_2-CH(NH_2)-COOH$
	Hydroxy lysine	Hyl.	2, 6-Diamino-5-hydroxy-hexanoic acid	$CH_2(NH_2)-CH(OH)-CH_2-CH_2-CH(NH_2)-COOH$
	Histidine	His (H)	2-Amino-1H-imidazole-4-propanoic acid	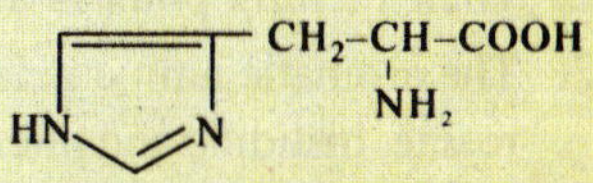
VI.	Containing aromatic rings			
	Phenyl alanine	PHe (F)	2-Amino-3-phenyle-propanoic acid	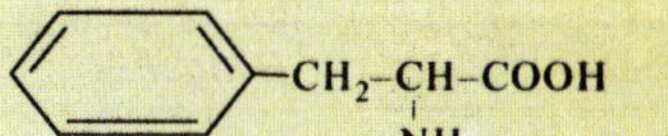
	Tyrosine	Tyr (Y)	2-Amino-3-(4-hydroxy-phenyl) propanoic acid	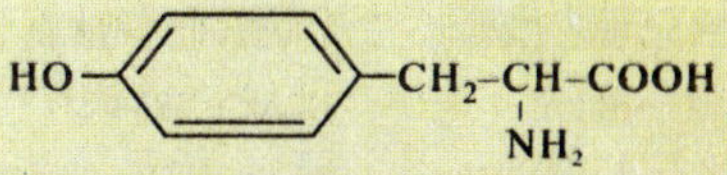
	Tryptophan	Trp (W)	2-Amino-3-(3-indolyl) propanoic acid	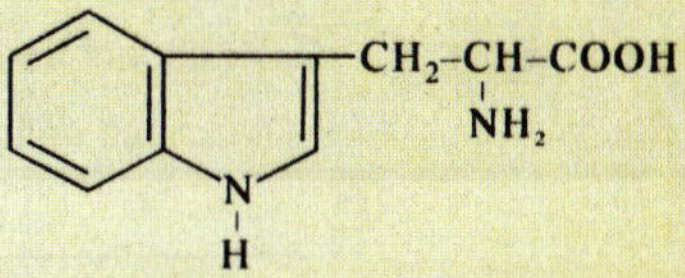
VII.	Imino acids			
	Proline	Pro (P)	2-Pyrrolidine carboxylic-acid	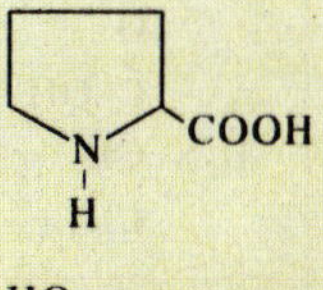
	Hydroxy proline	Hyp.	4-hydroxy-2-pyrrolidine-carboxylic acid	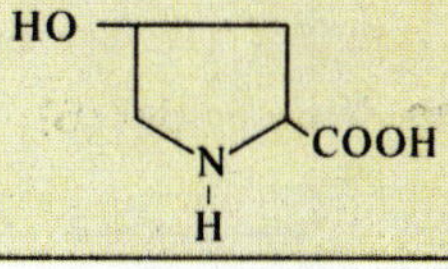

3. Optical isomers of amino acids

Except glycine, each amino acid has at least one asymmetric carbon atom and hence is optically active. Although D-amino acids occur in cells and even in polypeptides, they are not present in proteins.

Various other amino acids—Homocysteine, Homoserine, Ornithine, Citrulline, Arginosuccinic acid, Dopa, 3-monoiodotryrosine, 3, 5-Diiodotyrosine, 3, 5, 3′-triiodotyrosine, Thyroxine, β–Alanine, Taurine etc.—in free or combined states fulfil important functions in metabolic processes other than as constituents of proteins. Many additional amino acids occur in plants or in antibiotics. Over 20 D-amino acids occur naturally.

PHYSICAL PROPERTIES OF AMINO ACIDS

1. Amino acids are soluble in polar solvents such as water and ethanol but they are insoluble in nonpolar solvents such as benzene or ether.
2. Their melting point is above 200°C.
3. The aromatic amino acids tryptophan, tyrosine, histidine and phenylalanine absorb ultraviolet light.

CHEMICAL REACTIONS OF AMINO ACIDS

1. **Ninhydrin test :** Ninhydrin is a powerful oxidizing agent which causes oxidative decarboxylation of α–amino acids yielding CO_2, NH_3 and an aldehyde. The reduced ninhydrin then reacts with the liberated ammonia forming a blue complex. Proline and hydroxyproline produce a yellow rather than a purple colour with ninhydrin.
2. A variety of colour reactions specific for particular functional groups in amino acids are known which are useful in both qualitative and quantitative identification of particular amino acids. These are given below.
3. **Formation of peptide bonds :** Peptide bond formation involves removal of one mole of water between the α–amino group of one amino acid and the α–carboxyl group of a second amino acid.

PEPTIDES

Definition : A peptide consists of two or more amino acids linked by a peptide bond formed between the

Name of the test	*Reagents*	*Colour*	*Amino acid detected*
Nitroprusside	Sodium nitroprusside in dil. NH_4OH	Red	Cysteine
Hopkins Cole	Glyoxylic acid in 36N. H_2SO_4	Purple	Tryptophan
Millon	$HgNO_3$ in HNO_3, heat	Red	Tyrosine
Xanthoproteic	Boiling conc. HNO_3	Yellow	Tyrosine, Tryptophan, Phenylanine
Sakagauchi	α-Naphthol and Sodium hypochlorite	Red	Arginine
Pauly	Diazotized sulfanilic acid in alkaline solution	Red	Histidine

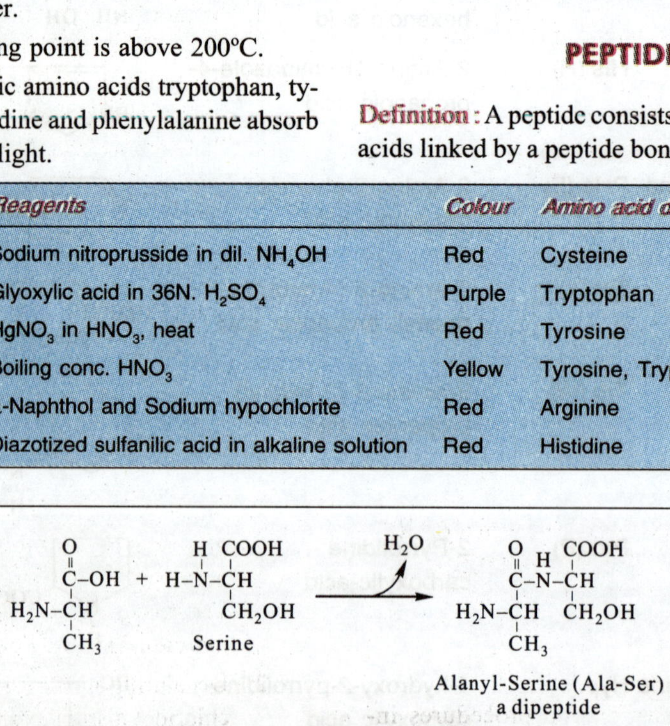

Fig. 5.6

carboxyl group of one amino acid and the amino group of another with the removal of one mole of water during a peptide bond.

Polypeptides are formed by long peptide chains containing large numbers of peptide bonds. Polymers of up to 100 amino acids are termed polypeptides and those with more than 100 are generally termed proteins.

Primary structure of peptides

The sequence of amino acids in a polypeptide is referred to as its primary structure. Chemical methods are required to determine the primary structure. Physical technics such as X-ray crystallography are employed for higher order of protein structure.

The N-terminal (amino terminal) amino acid is always shown at the left and the C-terminal (carboxyl terminal) amino acid at the right of the polypeptide chain.

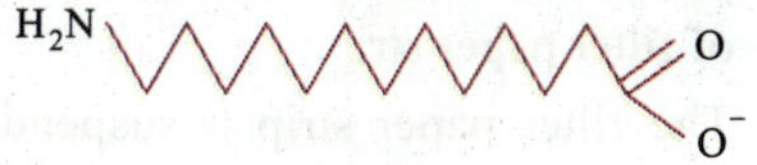

Fig. 5.7

Even small changes in the primary structure of proteins may produce remarkable physiologic effects. Substitution of a single amino acid for another in a sequence of 100 or more amino acids may abolish biologic activity with serious consequences (eg, Sickle cell disease). The biochemical basis for many inherited metabolic diseases are known owing to the introduction of new chemical and physical methods to determine protein structure.

DETERMINATION OF THE PRIMARY STRUCTURE OF PEPTIDES

A. Peptides are purified before analysis

1. The classic purification procedures include precipitation with varying salt concentrations (ammonium or sodium sulfate) or solvents (acetone or ethanol), differential centrifugation, gel filtration and electrophoresis.
2. Selective absorption and elution of proteins from the cellulose anion exchanger diethylaminoethyl (DEAE) cellulose and the cation exchanger carboxymethyl cellulose (CMC) have also been extremely successful for excessive and rapid purification of protein.

B. Determination of amino acid compostion

1. The peptide bond linking amino acid residues are first broken by hydrolysis.
2. The liberated amino acids are then separated and identified by HPLC or ion exchange chromatography.

C. Determination of sequence of polypeptide by Sanger

1. Sanger's approach was first to separate the two polypeptide chains A and B of insulin and then to convert them by specific enzymatic cleavage into smaller peptides that contained regions of overlapping sequence. Using 1 fluoro-2, 4-dinitro-benzene, he then removed and identified, one at a time, the amino terminal amino acid residues of these peptides—thus he was able to deduce an unambiguous primary structure for both the A and the B chains.

D. Primary structures are determined by the automated Edman Technique

1. Since many proteins consist of more than one polypeptide chain associated by noncovalent forces or disulfide bridges. The first step to dissociate and separate individual polypeptide chains.
2. Denaturing agents (urea, guanidine hydrochloride) disrupt hydrogen bonds and dissociate noncovalently associated polypeptides.

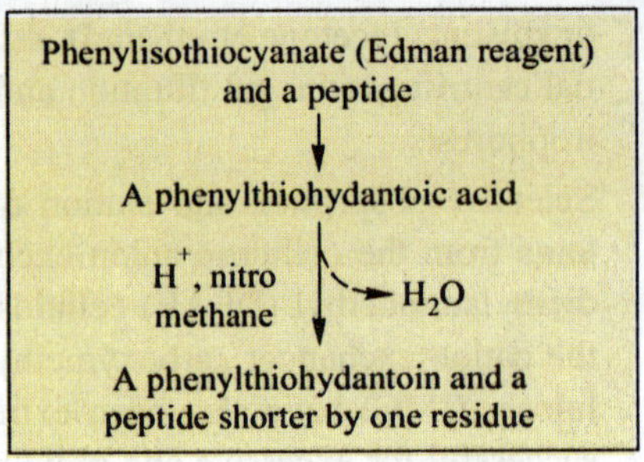

3. Oxidizing and reducing agents disrupt disulfide bridges. Polypeptides are then separated by chromatography.
4. In this automated technique, the Edman reagent is phenyl isothiocyanate. The reaction sequence releases the amino terminal amino acid as a phenylthiohydantoin derivative which is then identified by HPLC.
5. The next amino acid in sequence is then derivatized and removed, and then the process is repeated.
6. The Edman reactions take place either in a thin film on the wall of a spinning cup reaction chamber or on a solid matrix to which the carboxyl terminal of the peptide has been covalently coupled.

E. Large polypeptides are cleared before being sequenced

1. Automated sequencing instruments operate most efficiently on polypeptides of 20-60 residues long. Specific and complete cleavage at comparatively rare sites thus is desirable. The following reagents meet this requirement :
   (a) Cyanogen Bromide (CNBr)
   (b) Trypsin.
   (c) O-Iodosobenzene.
   (d) Hydroxylamine.
   (e) Protease V8.
   (f) Mild acid hydrolysis.

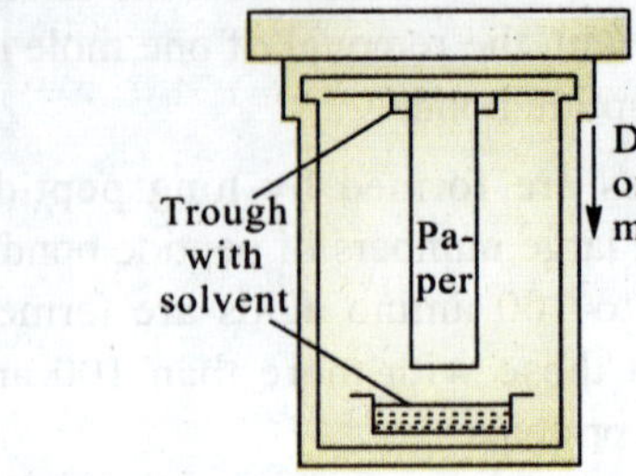

Fig. 5.8 : Descending paper chromatography

Separation and identification of the amino acids present

The free amino acids, after hydrolysis of peptides, are separated and identified by chromatography or by electrophoresis.

1. Paper chromatography

1. A small volume (about 0.005 ml containing about 0.01 mg) of amino acid is applied at a marked point 5 cm from the end of filter paper strip.
2. The filter paper strip is suspended in a sealed cylindrical jar or cabinet.
3. One end of the strip is dipped into a solvent consisting of water, an acid or base, an organic substance such as n-butyl alcohol.

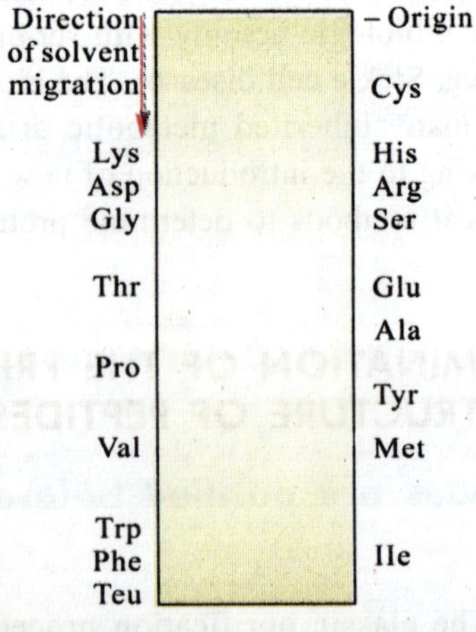

Fig. 5.9 : Identification of amino acids present in protein after descending paper chromatography

4. The solvent is placed in a trough from which the paper strip hangs (descending paper chromatography) or the strip may be suspended from the top of the jar and dipped into a trough at the bottom of the jar (ascending paper chromatography).
5. Strips are removed when the solvent migrates the marked distance and dried.
6. Then the strips are sprayed with 0.5% ninhydrin in acetone and heated for a few minutes at 80 – 100ºC.
7. Purple spots will indicate the presence of amino acids.
8. The amino acids are identified by the Rf value which is determined as follows :

$$R_f = \frac{\text{Distance travelled by an amino acid from the marked point of application}}{\text{Distance travelled by the solvent form the marked point of application}}$$

Thin layer chromatography

Thin layer chromatography may be employed in place of paper chromatography.

1. The absorbents (e.g. cellulose powder, alumina, a cellulose ion exchange resin, sephadex) are spread as a slurry on 8 × 18 inch glass plates.
2. The plates are dried and used like paper strips of paper chromatography.
3. The advantage in this is in the choice of absorbents and in the rapidity of separation. Amino acid mixtures require 3 hours for separation instead of 18 hours.

4. Lipids including sterols are rapidly and clearly separated.

2. Electrophoresis

1. Amino acid mixtures are applied at the centre of the paper or thin layers of powdered cellulose.
2. This is then moistened with buffer of an appropriate pH connected to buffer reservoirs by paper wicks and covered with a glass plate to reduce evaporative loss of buffer during passage of the current.
3. The current is applied with 2,000 – 5,000 volts for 0.5 – 2 hours. The molecules at the net positive charge at the selected pH move toward the cathode and those with a net negative charge toward the anode.
4. The electrogram (paper or thin layers of powdered cellulose) is dried and treated with ninhydrin to see the separated amino acids.
5. The amino acids are separated readily on the basis of their difference in molecular weight.
6. pH 3.5 is better for the separation of the peptides formed from the enzymic digestion of a protein.

3. Ion exchange chromatography

1. Analysis of amino acid residues after hydrolysis of a polypeptide involves ion exchange chromatography.
2. In this procedure, a short and long col-

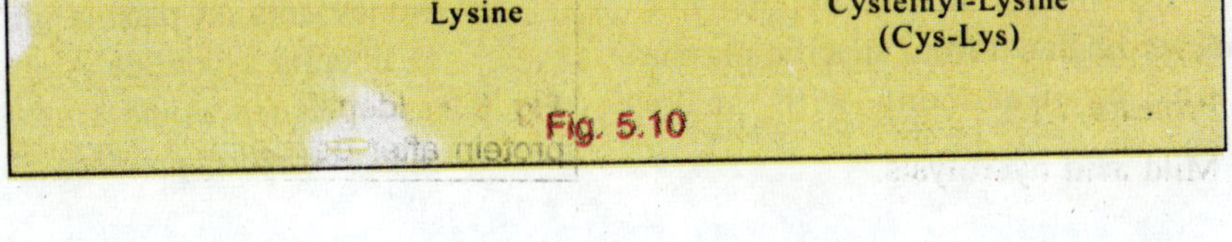

Fig. 5.10

umn that contains the Na^+ form of a sulfonated polystyrene resin.

3. When acid hydrolysate at pH_2 is applied to the columns, the amino acids bind via cation exchange with Na^+.
4. The columns are then eluted with sodium citrate under the conditions of pH and temperature.
5. Eluted material is reacted with ninhydrin reagent and colour densities are monitored in a flow through colorimeter.
6. Datas are displayed on a cathode ray tube with computer linked integration of peak areas.

SYNTHESIS OF PEPTIDES

Peptides may be synthesized by a reaction between an activated carboxyl group such as an acid chloride of one amino acid and the amino group of another, eg, cysteine acid chloride and lysine.

After the peptide bond is formed, the blocking group is removed, leaving the desired peptide.

PHYSIOLOGICALLY ACTIVE PEPTIDES

1. In some cases, they are the basis of products of protein.
2. In others, they may be hormones, antibiotics, precursors of bacterial cell walls, or even potent poisons.
3. The widely distributed tripeptide glutathione is required for the action of several enzymes including insulin. Glutathione reductase functions either in insulin degradation or in the formation of disulfide bonds.
4. Other naturally occurring important peptides are Bradykinin and Kallidin which are smooth muscle hypotensive agents are liberated from specific plasma proteins by treatment with snake venom.

Glutathione

1. It is a tripeptide consisting of glutamic acid, cysteine and glycine.
2. It is converted to disulphide form by easy dehydrogenation and involved in oxidation-reduction reaction.

Oxytocin and Vasopressin

1. These are cyclic peptide hormones consisting of 8 amino acids found in pituitary gland.
2. Oxytocin functions on uterine muscle and aids in the ejection of milk, whereas vasopressin acts in the reabsorption of water from the distal and collecting tubules of the kidney.

Carnosine

1. It is a dipeptide consisting of β–alanine and histidine.
2. It is a dipeptide of voluntary muscle.

Bradykinin

1. It is a nonapeptide consisting of 9 amino acids and the most potent pain-producing substance.
2. It mediates the production of Prostaglandin E_2 from arterial walls.
3. Kallidin I and Kallidin II consisting of 10 amino acids have relaxant effects on smooth muscles.
4. The decapeptide (10 amino acids) P_4 is present in dorsal horns of spinal cord and intestine and is involved in voluntary movement and pain relief.

Angiotensin

1. The enzyme Renin, being released from, kidney acts on plasma globulin fraction to liberate a peptide Angiotensin I consisting of ten amino acids—having slight effect on blood pressure.

2. Angiotensin II consisting of eight amino acids is formed from Angiotensin I by the splitting of two amino acids. It has prominent effect on blood pressure. It also stimulates thirst, dilatation of blood vessels of voluntary muscle and brain, increased secretion of aldosterone.
3. Angiotensin III of seven amino acids formed from Angiotensin II by the removal of aspartic acid has a role in pathology of hypertension.

Antibiotics

1. These, being elaborated by fungi, contain both D-and L-amino acids and amino acids absent in protein.
2. Tyrocidine and gramicidin S, Cyclic polypeptides contain D-phenyl alanine and ornithine which is absent in protein.

CLINICAL ORIENTATION

- The amino acids in the form of Proteins perform structural, hormonal and catalytic functions.
- The essential amino acids not synthesized in the human body support infant growth or maintain health in adults.
- Severe illness is caused by the genetic defects in the metabolism of amino acids. The rare genetic diseases eg, phenylketonuria and maple syrup urine disease etc. if not treated, result in mental retardation and early death.
- The transport defects of specific amino acids into cells result in the excretion of increased amounts of one or more amino acids in the urine causing amino acidurias.
- L-amino acids participate in nerve transmission and cell growth as well as in the biosynthesis of porphyrins, purines, pyrimidines and urea.
- Peptides are involved in endocrinology. Many principal hormones are peptides and may be given to patients in the deficiency state, eg, insulin administration to diabetic patients.
- Some peptides act in the nervous system as neurotransmitter or as neuromodulators.
- Certain antibodies are peptides eg, Valinomycin and Gramicidin A.
- The recombinant DNA technology synthesizes certain vital peptides and proteins for use in vaccines.

DIAGRAMMATIC COMPENDIUM

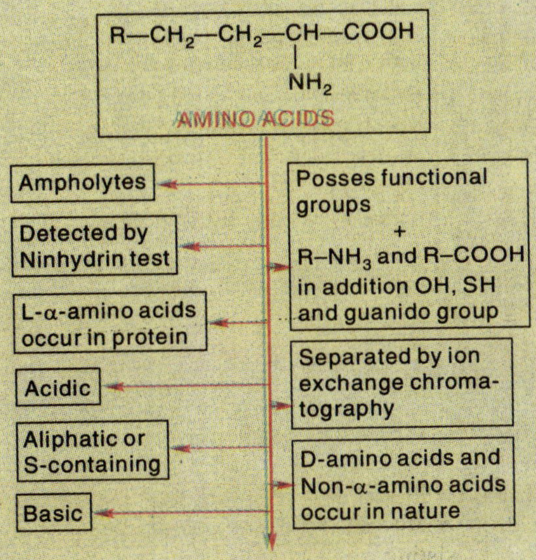

Diagram 1

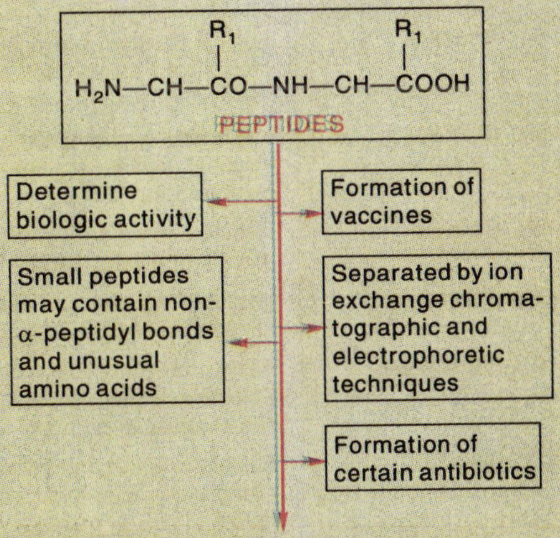

Diagram 2

CHAPTER 6

PROTEINS

Definition

Proteins may be defined as the high molecular weight mixed polymers of α–amino acids joined together with peptide linkage (–CO–N H–).

Proteins are the chief constituents of all living matter. They contain carbon, hydrogen, nitrogen and sulphur and some contain phosphorus also.

Biological importance or biochemical functions

1. Proteins are the essence of life processes.
2. They are the fundamental constituents of all protoplasm and are involved in the structure of the living cell and in its function.
3. Enzymes are made up of proteins.
4. Many of the hormones are proteins.
5. The cement substances and the reticulum which bind or hold the cells as tissues or organs are made up partly of proteins.
6. They execute their activities in the transport of oxygen and carbon dioxide by hemoglobin and special enzymes in the red cells.

Classification

1. Simple proteins

Class of Proteins	Characteristics	Examples
(i) Albumins	Soluble in water, coagulable by heat and precipitated at high salt concentrations.	Serum albumin, egg albumin, lactalbumin (Milk), leucosin (wheat), legumelin (soyabeans).
(ii) Globulins	Insoluble in water, soluble in dilute salt solutions and precipitated by half saturated salt solutions.	Serum globulin, vitellin (egg yolk), tuberin (potato), myosinogen (muscle), legumin (peas).
(iii) Glutelins	Insoluble in water but soluble in dilute acids and alkalis. Mostly found in plants.	Glutenin (wheat), oryzenin (rice).
(iv) Prolamines	Insoluble in water and absolute alcohol but soluble in 70 to 80 per cent alcohol.	Gliadin (wheat), zein (maize).
(v) Protamines	Basic proteins of low molecular weight. Soluble in water, dilute acids and alkalis. Not coagulable by heat.	Salmine (salmon sperm).
(vi) Histones	Soluble in water and insoluble in very dilute ammonium hydroxide.	Globin of hemoglobin and thymus histones.
(vii) Scleroproteins	Insoluble in water, dilute acids and alkalis.	Keratin (hair, horn, nail, hoof and feathers), collagen (bone, skin), elastin (ligament).

2. Conjugated Proteins

Class of Proteins	Characteristics	Examples
(i) Nucleoproteins	Composed of simple basic proteins (protamines or histones) with nucleic acids, found in nuclei. Soluble in water.	Nucleoprotamines and nucleohistones.
(ii) Lipoproteins	Combination of proteins with lipids, such as fatty acids, cholesterol and phospholipids etc.	Lipoproteins of egg-yolk, milk and cell membranes, lipoproteins of blood.
(iii) Glycoproteins	Combination of proteins with carbohydrate (mucopolysaccharides).	Mucin (saliva), ovomucoid (egg white), osseomucoid (bone), tendomucoid (tendon).
(iv) Phosphoproteins	Contain phosphorus radical as a prosthetic group.	Caseinogen (milk), ovovitellin (egg yolk).
(v) Metalloproteins	Contain metal ions as their prosthetic groups. The metal ions generally are Fe, Co, Mg, Mn, Zn, Cu etc.	Siderophilin (Fe), ceruloplasmin (Cu).
(vi) Chromoproteins	Contain porphyrin (with a metal ion) as their prosthetic groups.	Haemoglobin , myoglobin, catalase, peroxidase, cytochromes.
(vii) Flavoproteins	Contain riboflavin as their prosthetic groups.	Flavoproteins of liver and kidney.

3. Derived Proteins

Class of Proteins	Characteristics	Examples
A. *Primary derivatives*		
(i) Proteans	Derived in the early stage of protein hydrolysis by dilute acids, enzymes or alkalis.	Fibrin from fibrinogen.
(ii) Metaproteins	Derived in the later stage of protein hydrolysis by slightly stronger acids and alkalis.	Acid and alkali metaproteins.
(iii) Coagulated proteins	They are denatured proteins formed by the action of heat. X-rays, ultraviolet rays etc.	Cooked proteins, coagulated albumins.
B. *Secondary derivatives (Due to cleavage of peptide bonds)*		
(i) Proteoses	Formed by the action of pepsin or trypsin. Precipitated by saturated solution of ammonium sulphate, incoagulable by heat.	Albumose from albumin, globulose from globulin.
(ii) Peptones	Further stage of cleavage than the proteoses. Soluble in water, incoagulable by heat and not precipitated by saturated ammonium sulphate solutions.	
(iii) Peptides	Compounds containing two or more amino acids. They may be di-, tri-, and porypeptides.	Glycyl-alanine, leucyl-glutamic acid.

7. They function in the homostatic control of the volume of the circulating blood and that of the interstitial fluids through the plasma proteins.
8. They are involved in blood clotting through thrombin, fibrinogen and other protein factors.
9. They act as the defence against infections by means of protein antibodies.
10. They perform hereditary transmission by nucleoproteins of the cell nucleus.

Protein hydrolysing enzymes

1. Pepsin : In Gastric Juice.
2. Trypsin, Chymotrypsin and Carboxypeptidases : In Pancreatic Juice.
3. Amino-peptidases, Dipeptidases and Polypeptidases : In intestinal juice.

BONDS RELATING TO PROTEIN STRUCTURE

Protein structures are confirmed by

1. Strong bonds : Peptide and disulphide.
2. Weak bonds : Hydrogen and hydrophobic.

1. Strong bonds

A. Peptide bonds

The primary structure of proteins is derived from peptide bonds which can be established by the following evidences :

(i) Proteases produce polypeptides by hydrolyzing proteins. They can also hydrolyze peptide bonds.

(ii) The protein, insulin, is synthesized by linking amino-acids by peptide bonds.

(iii) Polypeptides, like proteins, react with biuret reagent which is suggestive of 2 or more peptide bonds.

B. Disulphide bond

1. The disulphide bond may interconnect two

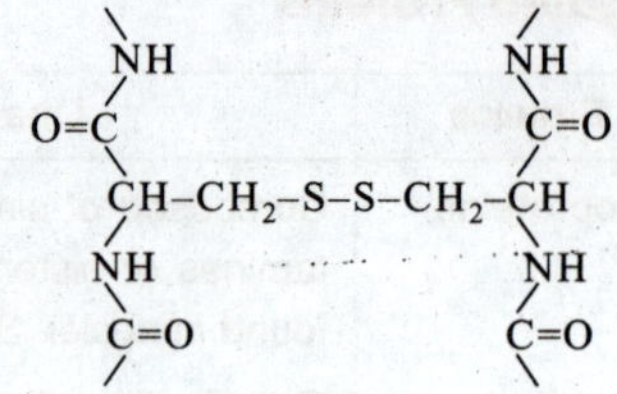

Fig. 6.1 : Two peptide chains united by a disulphide linkage

parallel chains through cysteine within each polypeptide shown in the Fig. 6.1.

2. The bond is not broken under the usual conditions of denaturation.
3. Performic acid is used to oxidize insulin to separate the protein molecule into its constituent polypeptide chains without affecting the other part of the molecule.

2. Weak bonds

A. Hydrogen bonds

1. The hydrogen bond appears from the sharing of hydrogen atoms between the nitrogen and the carbonyl oxygen of different peptide bonds as shown below by star mark.
2. Each hydrogen bond is quite weak.

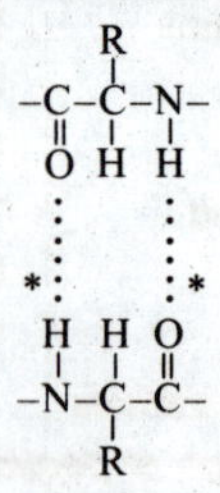

Fig. 6.2

3. During denaturation of a protein, the hydrogen and hydrophobic bonds are broken.

B. Hydrophobic bonds

1. The nonpolar side chains of neutral amino acids are closely associated with one another in proteins.

2. No true bond exists.
3. Nonetheless, they play an important role in maintaining protein structure.

STRUCTURE OF PROTEINS

The structure of protein is considered by several levels of organization on the basis of modern views. The levels of organization are:

1. Primary level of organization.
2. Secondary level of organization.
3. Tertiary level of organization.
4. Quarternary level of organization.

1. Primary structure

Peptide bond is formed by the amino acids linked by carboxyl group of one amino acid with the α-amino group of another amino acid as shown :

$$H_2N-CH_2-CO-NH-CH_2-COOH$$

Glycyl-Glycine

Thus, they form peptide bonds by several amino acids as

–Ala–Gly–gly–His–leu–

–Ala–Gly–His–Gly–leu–

The primary structure ultimately becomes as

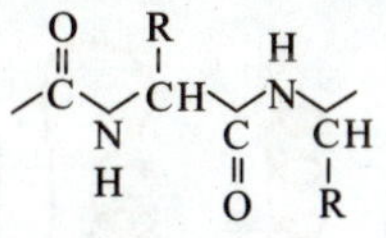

Fig. 6.3

2. Secondary structure

Globular proteins indicate a coiled structure in which peptide bonds are folded in a regular manner. The foldings are the results of linking of the carboxyl and amino groups of the peptide chains by means of hydrogen bonds and disulfide bonds. Such foldings are referred to as the secondary structure of the protein. Present evidence suggests that in many proteins, the hydrogen bonding produces a regular coiled arrangement, called α-helix.

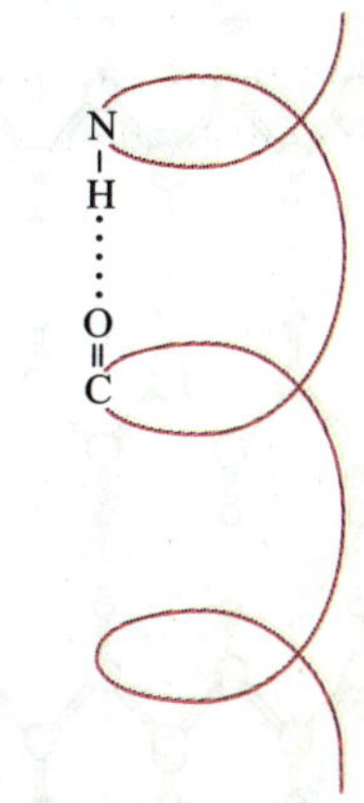

(Secondary structure)

Fig. 6.4

α-helix

1. Pauling and Corey proposed that the polypeptide chain of α-keratin is arranged as an α-helix.
2. In the structure of α-helix it has been found that R groups on the α-carbon atoms exist outward from the centre of the helix.
3. There are 3.6 amino acid residues per turn of the helix.
4. The distance travelled per turn is 0.54 nm.
5. The spacing per amino acid residue is 0.15 nm.
6. The α-helix is stabilized by hydrogen bonds.
7. Since α-helix is the lowest energy it forms spontaneously.
8. The right handed helix, when occurs in proteins, is significantly more stable than the left handed helix when the residues are L-amino acids.
9. Certain amino acids like proline tend to disrupt the α-helix.

The β-Pleated sheet

1. Pauling and Corey also proposed a second ordered structure, the β-pleated sheet (β because it was their second structure,

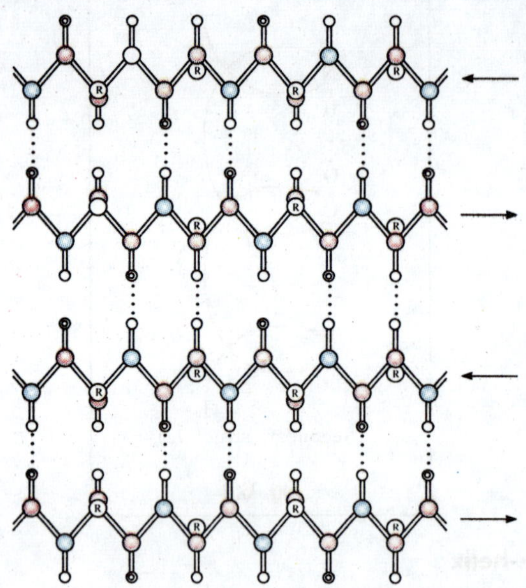

Fig. 6.5 : Antiparallel β-pleated sheet. Adjacent strands run in opposite directions. Hydrogen bonds between NH and CO group of adjacent strands stabilize the structure. The side chains (R) are above and below the plane of the sheet.
• = Carbon atoms; ◉ = Nitrogen atoms; O = Hydrogen atoms

the α-helix being the first). In the α-helix the polypeptide chain is condensed, in the β-pleated sheet it is almost fully extended (Fig. 6.5).

2. In a β-pleated sheet when the adjacent polypeptide chains run in opposite directions (N to C terminus), the structure is termed an antiparallel β-pleated sheet (Fig. 6.5) But when the chains run in the same direction, it is termed parallel.
3. Regions of β-pleated structure are present in many proteins, and both parallel and antiparallel forms occur.
4. The α-helix is stabilized by hydrogen bonding between peptide bonds 4 residues apart in a primary structural sense, stabilization of the β-pleated sheet results from formation of hydrogen bonds between peptides far removed from one another in a primary structural sense.

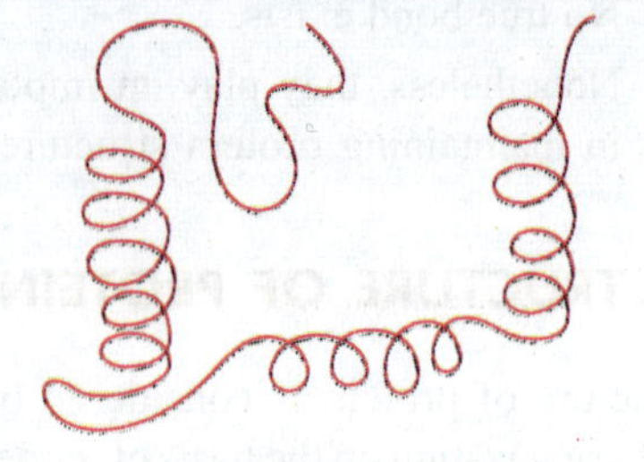

(Tertiary structure)

Fig. 6.6

3. Tertiary structure

The globular protein if completely is composed of a series of single helix, these molecules will have elongated structures with a larger axial ratio (length: breadth). Many proteins are spherical. The special structure in three dimensions is maintained by covalent or other bonds. Many globular proteins lack in disulphide bonds, yet they are stable in solution. The other bonds like hydrogen bonds, salt bonds and hydrophobic or non-polar bonds are also involved in the stability of the molecule. This has been described as the tertiary structure of protein.

4. Quarternary structure

1. When a protein consists of two or more peptide chains held together by non-cova-

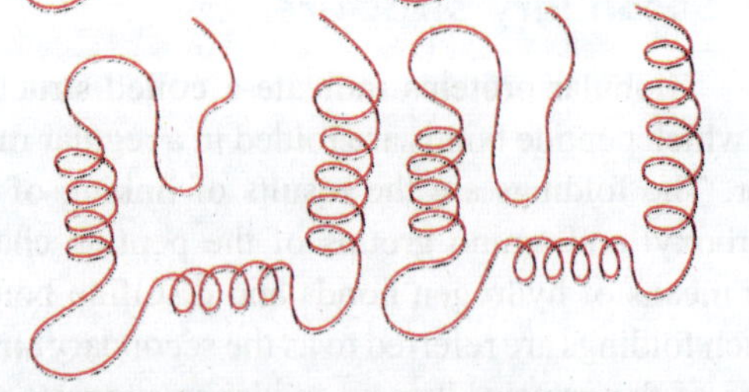

(Quarternary structure)

Fig. 6.7

lent interactions or by covalent cross-link, is referred to as quarternary structure.

2. In quarternary structure there are several monomeric units and the primary, secondary, tertiary structures may combine. The polymerization of these monomers or subunits is described as quarternary structure.
3. Glutamic dehydrogenase is tetramer and aldolase molecule is trimer.

Structures of protein related to biological functions of protein

1. Enzymes are made up of peptide by peptide linkage (–CONH–) by different amino acid sequences. This follows the primary structure of protein.
2. Many of the hormones also follow the same method forming peptides which include the primary structure of protein. If any alteration takes place in the proper sequence, there will be the abnormalities in the function.
3. The globular protein (eg, many enzymes) have folded, coiled polypeptide chains and axial ratios of less than 10 and generally not greater than 3 to 4. Fibrous proteins have axial ratio greater than 10.
4. Proteins that bind nucleotides show a nucleotide binding domain of tertiary structure.
5. Most fibrous proteins fulfil structural roles in skin, connective tissue, or fibres such as hair, silk, wool. The typical sequence of amino acid of fibrous protein follow specific secondary tertiary structures.
6. Silk fibroin contains small quantities of bulky amino acids; these bulky residues interrupt the β-sheet regions and show the flexibility essential for a successful colour.
7. Tropocollagen, the important unit of collagen, consists of 3 polypeptide chains. Each polypeptide forms a left handed helix (not an α helix) having 3 residues per turn. 3 left handed helical polypeptide then entwine to form a "right hand" triple or super, helix i.e. stabilized by H bond formed between individual polypeptide chain. Mature collagen fibres show the stable and extended helicle conformation.

Principles of determination of molecular weight of protein

The proteins have a large molecular weight and they are large macromolecules.

Molecular weight is determined by physical methods such as —

1. Analytic ultracentrifugation—It measures the rate at which a protein sediment in a gravitational field. It is an excellent physical technique for the determination of molecular weight.
2. Sucrose density gradient centrifugation—It is best applied for globular protein. Protein standards and unknowns are layered on a gradient of buffered 5-20% sucrose. Following overnight centrifugation, the tube contents are removed dropwise and analysed for protein. Mobility of unknown protein, expressed to standard of known MW, is then computed and MW is calculated.

Gel filtration—Columns of sephadex having pores of known size range are calibrated using proteins of known MW. The MW of an unknown protein is then calculated from its elution position relative to these standards.

Polyacrylamide Gel Electrophoresis—Proteins of unknowns and standards are separated by electrophoresis in 5-15% cross-linked acrylamide gels of varying porosity.

In this case, oligomeric proteins are first denatured by boiling them in presence of β-mercaptoethanol and SDS. The denatured proteins are separated on the basis of their size by SDS-PAGE on gels that contain SDS.

DENATURATION OF PROTEINS

Denaturation may be defined as the disruption of the secondary, tertiary and quarternary structure of the native protein resulting in the alterations of the physical, chemical and biological characteristics of the protein by a variety of agents.

The native proteins are said to be the proteins occurring in animal and plant tissues. They possess many characteristic properties such as solubility, viscosity, optical rotation, sedimentation rate, electrophoretic mobility etc.

For an oligomeric protein, denaturation may involve dissociation of the protomers with or without subsequent unfolding or with or without undergoing changes in protomer conformation.

Denaturing agents

1. **Physical agents :** Heat, surface action, ultraviolet light, ultrasound, high pressure etc.
2. **Chemical agents :** Acids, alkalis, heavy metal salts, urea, ethanol, guanidine detergents etc.

Urea and guanidine probably interfere with the hydrogen bonds between peptide linkages. Acids and alkalis probably attack directly the hydrogen bonds in the secondary and tertiary structure of proteins.

Physical alterations

Many proteins, especially of the globular type, can be crystallized in the native state. But denatured proteins cannot be crystallized.

Chemical alterations

The denatured protein is greatly decreased in solubility at its isoelectric point. The chemical groups are exposed to chemical reactions and more readily detected as a result of the unfolding process in denaturation. Among these are sulphydryl group of cysteine, the disulphide group of cystine and the phenolic group of tyrosine.

Biological alterations

The digestability of certain denatured proteins by proteolytic enzymes is increased. Enzymatic or hormonal activity is usually destroyed by denaturation. The antigenic or antibody functions of proteins are frequently altered.

If the denaturation is severe, the protein molecules become insoluble and precipitation results as well as the changes in the properties of the proteins are permanent and *"irreversible"*. In case of mild denaturation, there is *"reversible denaturation"* leading to the slight changes in the properties of the protein which can be restored to the native state after suitable treatment.

Significance

1. The precipitation of the native protein as a result of denaturation is used to advantage in the clinical laboratory.
2. Blood or serum samples to be analyzed for small molecules (eg, glucose, uric acid, drugs) generally are first treated with acids such as trichloroacetic acid, phosphotungstic acid or phosphomolybdic acid to precipitate most of the proteins present in the sample. This is removed by centrifugation and the protein-free supernatant liquid is then analyzed.
3. Denaturation is used to know the enzyme catalyzed reaction of an extract at the loss of the enzyme activity when boiled or acidified.

Denaturation and Renaturation of Proteins

Bovine ribonuclease of single polypeptide chain of 124 amino acid residues with small molecular weight contains four disulphide bonds. When it is treated with β-mercaptoethanol in 8 M. urea, the disulphide bonds are reduced to –SH groups as a result of the denaturation of the enzyme and the enzyme activity is also lost.

Denatured ribonuclease, when freed from urea and β-mercaptoethanol by dialysis, slowly regains

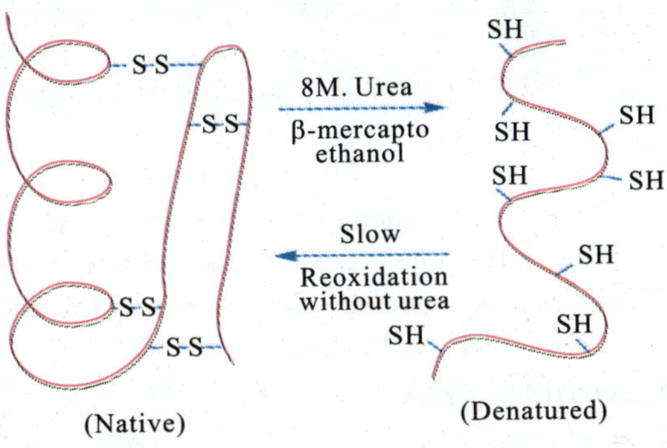

Fig. 6.8

enzymic activity as the SH groups are oxidized by oxygen of air to form S–S bonds. But if the reduced ribonuclease in 8 M. urea solution is reoxidized it loses its enzymic activity almost completely as wrong disulphide bonds are formed resulting in *'scrambled'* ribonuclease.

Similarly, when egg albumin is heated till it is coagulated, the denaturation is irreversible and the secondary and tertiary structure of the proteins are completely lost resulting in a mixture of randomly arranged polypeptide chains.

IMPORTANT TESTS OF PROTEINS

A. Colour reactions

1. *Biuret test :* To 2 ml of test solution add an equal volume of 10% NaOH and one drop of 10% $CuSO_4$ solution. A violet colour formation indicates the presence of peptide linkage.
2. *Ninhydrin test :* To 1 ml Ninhydrin solution add 1 ml protein solution and heat. Formation of violet colour indicates the presence of α-amino acids.

B. Coagulation reactions

Heat coagulation : Take the test solution up to $\frac{2}{3}$ of the test tube and heat the upper portion of the solution holding the lower part of the test tube. An opalescent appears which becomes deep on the addition of a few drops of 2% acetic acid. This indicates the presence of protein (albumin).

C. Precipitation reaction

1. **Full saturation :** Saturate 5 ml of test solution in a test tube with solid ammonium sulphate. A gelatinous precipitate appears indicating the presence of albumin.
2. **Half saturation :** Add 3 ml saturated solution of ammonium sulphate to 3 ml of test solution in a test tube and shake vigorously. A gelatinous precipitate forms indicating the presence of globulin.

D. Tests for colour reactions of amino acids

See Chapter 5 on amino acids and peptides.

ESTIMATION OF PROTEINS

Quantitative estimations of proteins of foods and other biological materials are performed by the following methods :

1. Kjeldahl method.
2. Colorimetric method.
3. Electrophoretic method.

1. Kjeldahl method

1. This method consists in digesting a known weight of the material with concentrated sulphuric acid and potassium sulphate with copper sulphate, mercuric oxide or selenium dioxide as a catalyst.
2. The amino groups of the amino acids and N present in the heterocylic rings of histidine, tryptophan, proline and hydroxyproline are converted into ammonia as well as carbon is oxidized to carbon dioxide.
3. The ammonia present as sulphate is estimated after distillation by titration with standard acid or colorimetrically using Nessler's reagent.

4. The nitrogen content of the sample is calculated and converted into *"Crude Protein"* content by multiplying by the factor 6.25.

Defects

1. The factor 6.25 is only approximate for the same proteins.
2. The nitrogen of urea, Creatinine and other N compounds (which have no nutritive value) present in animal foods, fish and milk in small amounts also forms part of the total nitrogen i.e. crude protein.

2. Colorimetric method

1. This method is entirely based on the Biuret reaction or Folin's phenol reagent.
2. This method is generally used for the estimation of concentration of protein of serum.

3. Electrophoretic method

1. This method is used for the separation and estimation of protein in serum, tissues and foodstuffs.
2. In case of tissues and foods, the proteins are extracted with suitable solvents before separating them electrophoretically.
3. The electrophoretic separation can be performed using filter paper, agar gel and starch gel.
4. The proteins, after separation, are treated with suitable dye and the intensity of the colour of the individual dye-protein complex bands are measured using a densitometer.

ALBUMIN

Chemistry

1. Albumins are simple proteins, soluble in water, coagulated by heat and precipitated by saturated salt solutions.
2. The glycine content of many albumins is low.

Sources

They are frequently found in serum, muscle, milk and egg white. They are also widely distributed in plants, especially in the seeds and fruits.

Characteristics

1. Albumin is more readily affected by nutritional factors, eg, restricted protein intake.
2. Because of its small molecular size, it is lost from the circulating plasma more easily than the other proteins by passage through capillary walls of increased permeability as in inflammation and nephrosis (albuminuria).
3. In sudden haemodilution or severe malnutrition, in the majority of the disease states, the plasma proteins are altered and the albumin is decreased.
4. It has the property of osmotic pressure and fatty acid transport.

Clinical importance

1. The plasma albumin concentration falls somewhat in the later stages of normal pregnancy owing to haemodilution and decreased synthesis.
2. The major causes for abnormal decrease in albumin (hypoalbuminemia) are outlined :
   (i) Excessive loss in urine.
   (ii) Inadequate protein supply (dietary protein. restriction, vomiting, diarrhea etc.).
   (iii) Impaired synthesis(liver dysfunction, chronic infection and severe anemia etc.).
   (iv) Sudden plasma dilution (following sudden recovery from dehydration, infantile diarrhea etc.).

3. In the prominent clinical manifestation of hypoalbuminemia edema occurs as a result of the decrease in plasma colloid osmotic pressure which favours retention of water in tissue spaces.
4. The investigation of albumin is of much importance not only in the case of edema but also of the state of liver function and of protein nutrition.
5. Pathological proteinuria may occur as a result of
   (i) Increased glomerular permeability.
   (ii) Renal tubular damage with defective reabsorption of serum albumin (nephrosis).
   (iii) Disease of the lower urinary tract.
   (iv) Abnormal protein in the plasma.

GLYCOPROTEINS

1. The polypeptide chain is attached to one or more heterosaccharide units by covalent links.
2. Some glycoproteins have many identical disaccharide units attached to the polypeptide chain.
3. In addition to the more common sugars such as glucose, mannose, N-acetyl glucosamine, and N-acetylgalactosamine, other units also occur eg, L-fucose (a methyl pentose) and sialic acids (acyl derivatives of neuraminic acid which is formed by the condensation of pyruvic acid and mannosamine).
4. The carbohydrate part is bound to the protein by a glycosidic link between N-acetylglucosamine and the amide group of asparagine; this link is more stable than a peptide bond.
5. The carbohydrate is in the form of disaccharide units. 600-700 such units are attached to the peptide chain, one per 6.4 amino acid residues.
6. They are secreted by the submaxillary glands of various animals.

Functions

1. They form viscous solutions which function as lubricants and protective *''screens''* in the body.
2. The continuously secreted mucus of the respiratory tract is a protection against invasion by bacteria and the uterus is protected from the vaginal microbial flora by the cervical mucus.
3. Intestinal mucus makes a protection for the intestinal cells against mechanical damage.

CHROMOPROTEINS

These are conjugated proteins composed of simple proteins united with a coloured prosthetic group. Many biologically important proteins belong to this group. They are :

1. Hemoglobins : These are the respiratory proteins in which the prosthetic group is the iron containing porphyrin complex heme.
2. Cytochromes : These are cellular oxidation-reduction proteins in which the prosthetic group is heme.
3. Flavoproteins : These are cellular oxidation-reduction proteins in which the prosthetic group is riboflavin.

Chromoproteins of certain animal fibres such as black wool and hair in which the prosthetic group is melanin.

Visual purple of the retina, a chromoprotein, in which the prosthetic group is a carotenoid pigment.

Catalase, a chromoprotein, in which the prosthetic group is heme. Peroxidase has a similar composition.

SCLEROPROTEINS (ALBUMINOIDS)

1. These proteins are similar to albumins and globulins.

2. They are characterized by great stability and insolubility in water and salt solutions.
3. They form the supporting structures of animals eg, collagen in cartilage and white fibres of connective tissue and in bones and teeth, elastin in the yellow or elastic fibers, keratins in horn, hair, wool, real silk and feathers.
4. The keratins differ from the collagen-like members in having high sulphur content, chiefly in the form of cystine.
5. They are readily attacked and dissolved by alkali sulphides.
6. The collagen of animal hides can be preserved in flexible condition giving the substance known as leather when treated with tannic acid, alum or various metal salts.
7. By long boiling with water collagen is partly hydrolysed giving a soluble product known as gelatin.

SILK FIBROIN

1. Fibroin, the principal protein of silk worm, fibres, consists entirely of glycines alternating with either alanine or serine. These three amino acids cover 85 per cent of the fibroin structure.
2. Silk fibres are extremely resistant to stretching and not highly flexible.
3. It also contains small quantities of *''bulky''* amino acids such as valine or tyrosine. These bulky residues interrupt the β-sheet regions and confer the flexibility essential for a successful cocoon.

COLLAGEN

1. Tropocollagen, the fundamental unit of collagen, consists of three polypeptide chains each about 1,000 residues in length. Each polypeptide forms a left-handed helix having about three residues per turn. Three left-handed helical polypeptides then entwine to form a right-hand triple helix which is stabilized by H bonds.
2. Mature Collagen fibres measuring 1.5 nm by about 300 nm reflect this stable and extended helical conformation.

PRIONS

1. *Prion diseases* are fatal neurodegenerative diseases characterized by spongiform changes, astrocytic gliomas, and neuronal loss resulting from the deposition of insoluble protein aggregates in neural cells. They include Creutzfeldt-Jakob disease in humans.
2. These diseases may manifest themselves as infactious, genetic, or sporadic disorders. Because no viral or bacterial gene encoding the pathologic prion protein could be identified. Now it is believed that prion diseases are protein conformation diseases transmitted by altering the conformation, and, hence, the physical properties, of proteins endogenous to the host.
3. Human prion-related protein, PrP, a glycoprotein encoded on the short arm of chromosome 20, normally a monomeric and rich in α helix.
4. Pathologic prion proteins serve as the templates for the conformational transformation of normal PrP, known as PrPc, into PrPsc. PrPsc is rich in β sheet with many hydrophobic aminoacyl side chains exposed to solvent. Therefore, PrPsc molecules associate strongly with one another forming insoluble protease resist aggregates. Since one pathologic prion or prion-related protein can serve as template for the conformational transformation of many times its number of PrPc molecules, prion diseases can be transmitted by the protein alone without involvement of DNA or RNA.

ALZHEIMER'S DISEASE

1. Refolding or misfolding of another protein endogenous to human brain tissue, β-amyloid, is also a prominent feature of Alzheimer's disease. The characteristic senile plaques and neurofibrillary bundles contain aggregates of the protein β-amyloid.
2. In the patients suffering from this disease, levels of β-amyloid become elevated, and this protein undergoes a conformational transformation from a soluble α helix-rich state to a state rich in β sheet and prone to self-aggregation. Apolipoprotein E has been implicated as a protentian mediator of this conformational transformation.

CLINICAL ORIENTATION

- The supply of nitrogen and sulphur is regulated by dietary protein.
- The proteins immunoglobulins act as prime defence against bacterial and viral infections.
- The contractile proteins actin and myosin aid in the movement of muscle fibre and microvilli etc.
- The enzymes are protein in nature and have high catalytic activity to speed up chemical reactions in the body for maintaining normal life.
- The hormones which are protein in nature have great effect on metabolism and reproduction.
- Certain proteins present in cell membrane bind vitamins, hormones etc. to mediate their cellular action.
- Some proteins bind specific substances to be present as storage eg, Ferritin is stored being formed by protein and iron.
- Dire consequences result in the deficiency of cofactors essential for maturation of a protein. *Ehlers-Danlos Syndrome* shows genetic defects in protein maturation and *Scurvy,* a deficiency of a cofactor essential for protein maturation.

EXERCISE

1. What are proteins? Classify them and give important tests for it. Describe briefly the biological importance of proteins. (Bihar 60S)
2. What is the primary structure of proteins? Discuss the biochemical functions of proteins. (Patna 70A; Ranchi 66S, 70A)
3. Discuss in brief the classification and biological importance of proteins. (Patna 68S)
4. What are proteins and their broad classification ? What are they composed of ? What are the special features of their structures ? Name some enzymes which can hydrolyse proteins specifically. (Patna 69S ; Lucknow 58A)
5. Describe the structure of proteins. Discuss how proteins are precipitated. (M.U. 76A)
6. Write short notes on :
   (a) Denaturation (Muz. 75S)
   (b) Basic proteins. (R.U. 70A)
   (c) Isoelectric point. (P.U. 71A ; 73A)
   (d) Glycoproteins. (R.U. 68A)
   (e) Chromoproteins. (M.U. 75A)
   (f) Conjugated proteins. (Mith. 74A)
   (g) Zwitterion. (C.U. 1981) (P.U. 75A ; M.U. 75A)
   (h) Biuret reaction and its use. (C.U. 1981)
7. Write the structure of two 'essential' amino acids. (C.U. 1981)
8. What is meant by primary structure of protein? Write an account on methods of isolation of protein. (C.U. 1982)
9. Discuss how structure of protein determines biological function. (C.U. 1983)
10. Write True or False :
    (a) Methionine is an aromatic amino acid—
    (b) The nitrogen content of protein is usually about 6.25%—
    (c) Biuret test is a test for urea — (C.U. 1983)
11. Fill up the gaps with correct words :
    (i) In special reactions of amino acids, Sakaguchi test is positive with amino acid and in reactions of protein, Niemann's test is positive with—
    (ii) On heating near 100°C a protein solution gets—; the enzyme – hydrolyses – protein. (C.U.1981)

SCHEMATIC COMPENDIUM

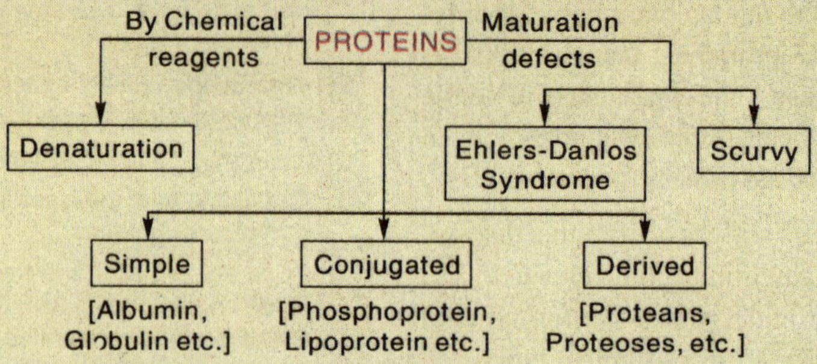

Diagram 1

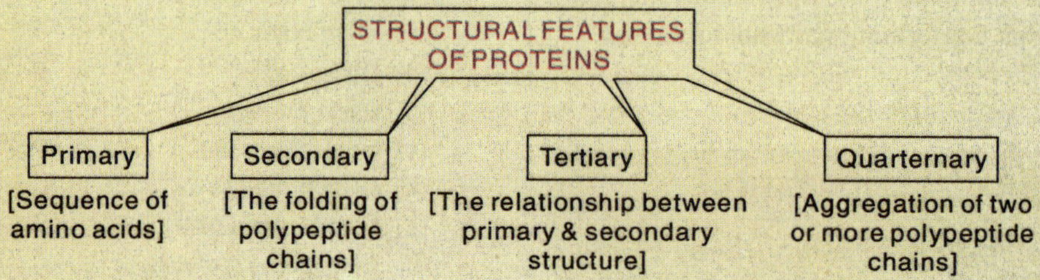

Diagram 2

CHAPTER 7

NUCLEOTIDES

Definition

These are the compounds constituted by purine or pyrimidine bases, ribose or deoxyribose sugars and phosphoric acid.

Biological importance

1. The nucleotides are important intracellular molecules of low molecular weight.
2. They play an important role in carbohydrate, fat and protein metabolism.
3. The best role of purine and pyrimidine nucleotides is to serve as the monomeric precursors of RNA and DNA.
4. The purine nucleotides also act as the high energy source ATP, cyclic AMP [cAMP] in a wide variety of tissues and organisms and as components of coenzymes FAD, NAD, NADP and of an important methyl donor, s-adenosylmethionine.
5. The pyrimidine nucleotides also act as high energy intermediates such as UDP-glucose and UDP-galactose in carbohydrate metabolism and CDP-acylglycerol in lipid synthesis.

Structure

The structures of purine and pyrimidine bases are given below. The direction of the numbering of the purine ring is different from that of the pyrimidine ring. But the number 5 carbon is the same in both.

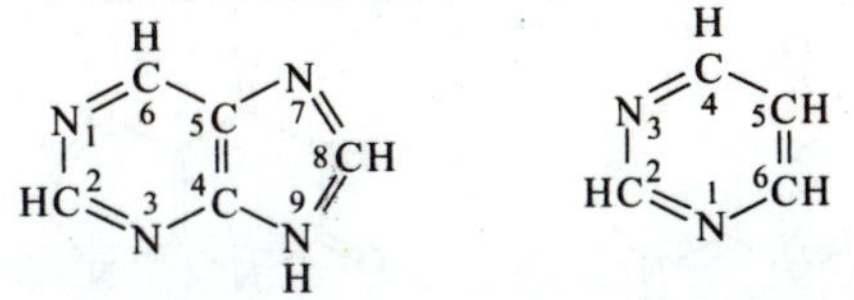

(Purine) (Pyrimidine)

Fig. 7.1

Pyrimidine bases

1. Cytosine (2-oxy-4-aminopyrimidine)
2. Thymine (2, 4-dioxy-5-methylpyrimidine)
3. Uracil (2, 4-dioxypyrimidine)

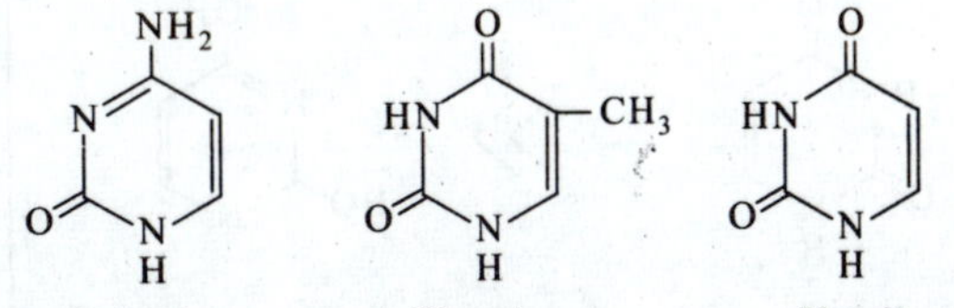

Cytosine (2-oxy-4-amino-pyrimidine) | Thymine (2, 4-dioxy-5-methylpyrimidine) | Uracil (2, 4-dioxy-pyrimidine)

Fig. 7.2

Purine bases

1. Adenine (6-aminopurine)
2. Guanine (2-amino-6-oxypurine)

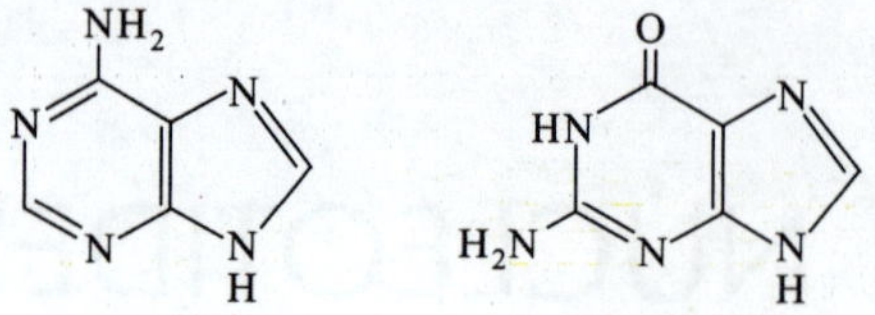

Adenine (6-aminopurine) Guanine (2-amino-6-oxypurine)

Fig. 7.3

These are the two major purines found in living organisms. Two other purine bases (Hypoxanthine and Xanthine) also occur as intermediates in the metabolism of adenine and guanine.

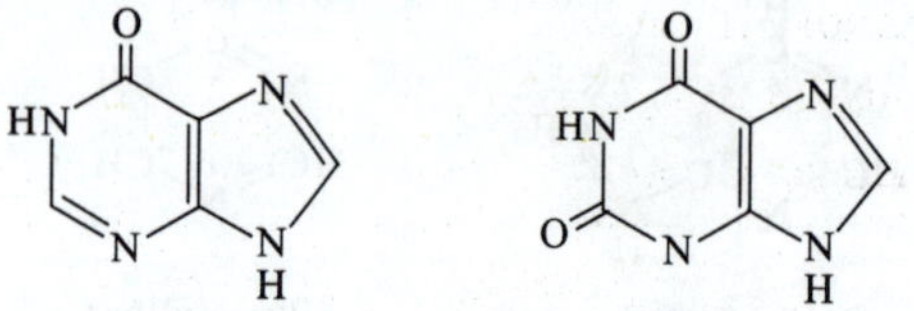

Hypoxanthine (6-oxypurine) Xanthine (2, 6-dioxypurine)

Fig. 7.4

Because of their resonant structures, they can exist in a lactim or lactam form. The lactam form is the predominant tautomer of uracil or thymine under physiologic condition.

In plants, a series of purine bases containing methyl substituents occurs. Many have pharmacologic properties.

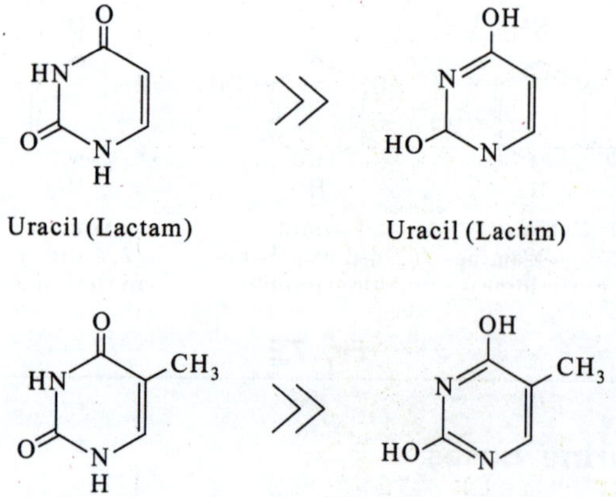

Uracil (Lactam) Uracil (Lactim)

Thymine (Lactam) Thymine (Lactim)

Fig. 7.5

Examples are :

Coffee which contains caffeine (1, 3, 7- trimethyl xanthine).

Tea which contains theophylline (1, 3- dimethylxanthine).

Cocoa which contains theobromine (3, 7- dimethylxanthine).

In *natural materials*, many minor bases occur. Some of these unusual substituted bases are found only in the nucleic acids of bacteria and viruses. 5-methylcytosine and 5-hydroxymethylcytosine are significant components of bacteria and bacteriophage, respectively.

More recently, N^6-methyladenine, N^6-dimethyladenine and N^7-methylguanine have been found in the nucleic acids of mammalian cells.

NUCLEOSIDES AND NUCLEOTIDES

The free bases occurring in nature are less in abundance than are their nucleosides and nucleotides.

A nucleoside is composed of a purine or a pyrimidine base and a ribose or a deoxyribose sugar. The composition of some nucleosides are given :

The *anti* form is necessary for the proper positioning of the complementary purine and pyrimidine bases in the double-stranded form of deoxyribonucleic acid. The structures of *syn* and *anti* configurations of adenosine are given Fig. 7.8.

Nucleotides are the nucleosides phosphorylated on one or more of the hydroxyl groups of the sugar. Thus,

Adenosine monophosphate (AMP or Adenylate) = Adenine + ribose + phosphate

2-deoxyadenosine monophosphate (dAMP or deoxyadenylate) = Adenine + 2-deoxyribose + phosphate.

Thymidylic acid (TMP) = Thymine + 2-deoxyribose + phosphate.

Uridylic acid (UMP) = Uracil + ribose + phosphate.

Name of the Nucleosides	Bases	Sugars	Attachment at the position
1. Nucleosides containing purine bases			
Adenosine	Adenine	D-ribose	9
Guanosine	Guanine	D-ribose	9
Deoxyadenosine	Adenine	2-deoxyribose	9
Deoxyguanosine	Guanine	2-deoxyribose	9
Inosine	Hypoxanthine	D-ribose	9
2. Nucleosides containing pyrimidine base			
Cytidine	Cytosine	D-ribose	1
Deoxycytidine	Cytosine	2-deoxyribose	1
Uridine	Uracil	D-ribose	1
Deoxythymidine	Thymine	2-deoxyribose	1

Structures

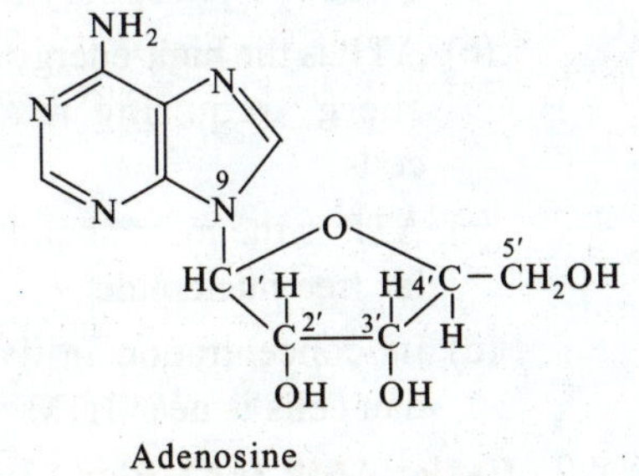

Fig. 7.6

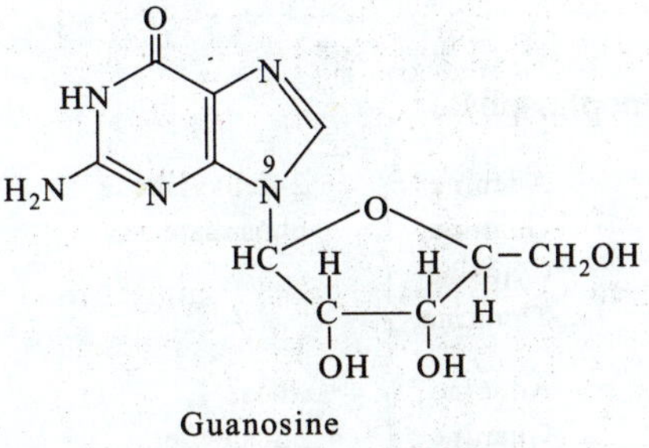

Fig. 7.7

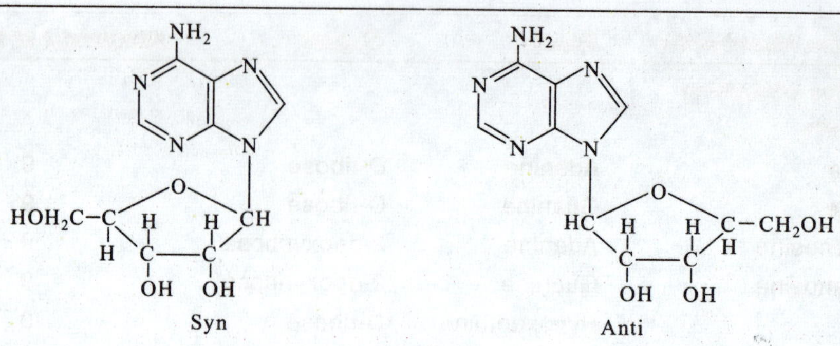

Fig. 7.8

Cytidylic acid (CMP) = Cytosine + ribose + phosphate.

or

Cytidine monophosphate.

Guanylic acid (GMP) = Guanine + ribose + phosphate.

or

Guanosine monophosphate.

Inosinic acid (IMP) = Inosine + ribose + phosphate.

or

Inosine monophosphate.

Deoxyribonucleic acid (DNA)	=	Adenine, Guanine, Cytosine, Thymine	+ 2-deoxyribose + phosphate
Ribonucleic acid (RNA)	=	Adenine, Guanine, Cytosine, Uracil	+ ribose + phosphate

Structure

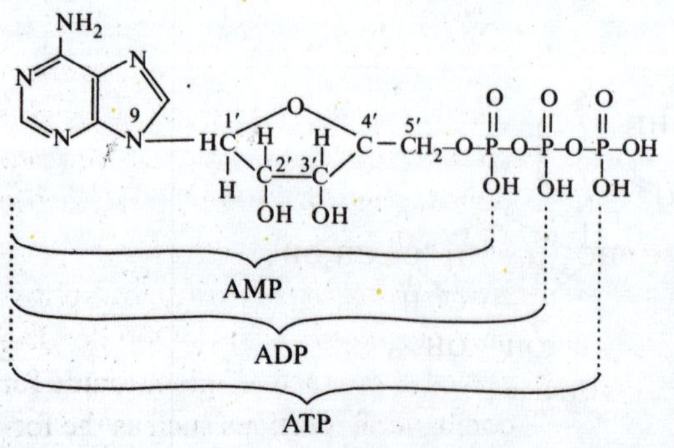

Fig. 7.9

NATURALLY OCCURRING NUCLEOTIDES

Free nucleotides which are not a part of nucleic acids are also found in tissues. Many have important roles. Some are discussed :

1. **ATP & ADP :**
   (a) These are the important compounds as regards their participation in oxidative phosphorylation.
   (b) ATP is the high-energy phosphate for energy requiring reactions in the cells.
   (c) ATP is the most abundant intracellular free nucleotide.
   (d) Its concentration in living mammalian cells is near 1mM.
2. **Cyclic AMP (cAMP or 3′, 5′-adenosine monophosphate) :**
   (a) This is present in most animal cells.

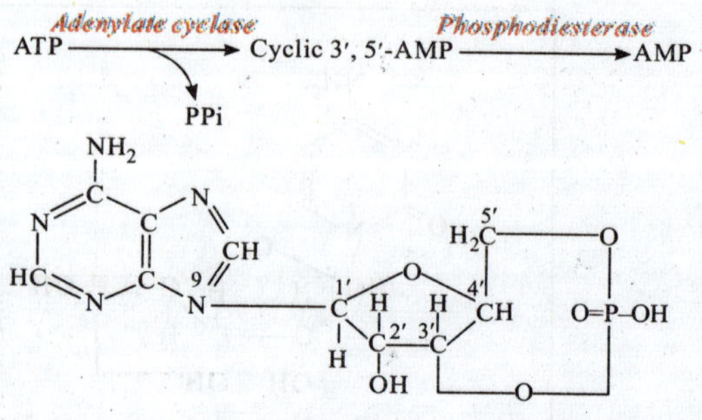

Structure of cAMP (3′, 5′- adenosine monophosphate)

Fig. 7.10

(b) It is formed from ATP by the enzyme adenylate cyclase, the activity of which is regulated practically by hormone receptors.

(c) It mediates a series of reactions.

(d) It is destroyed in tissues by its conversion to AMP by the enzyme cAMP phosphodiesterase.

(e) The intracellular cAMP concentrations are near 1 μM.

3. **S-adenosylmethionine :**

(a) It serves as a form of "active" methionine.

(b) It serves as a methyl donor in many methylation reactions.

(c) It is a source of propylamine for the synthesis of polyamines.

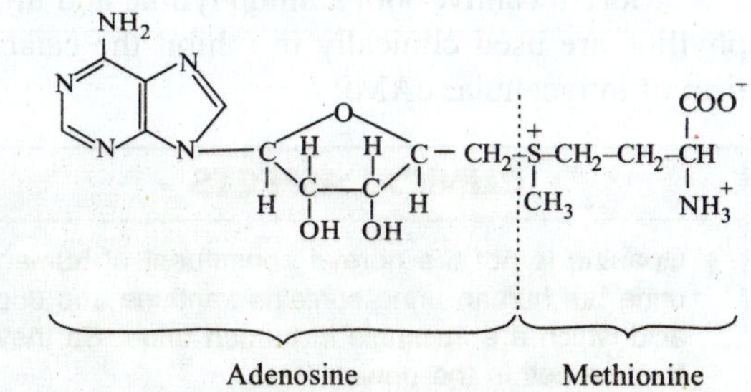

(Structure of S- adenosylmethionine)

Fig. 7.11

4. **Cyclic GMP** (cGMP or 3′, 5′-guanosine monophosphate) :

(a) It is formed from GTP by the enzyme guanylate cyclase.

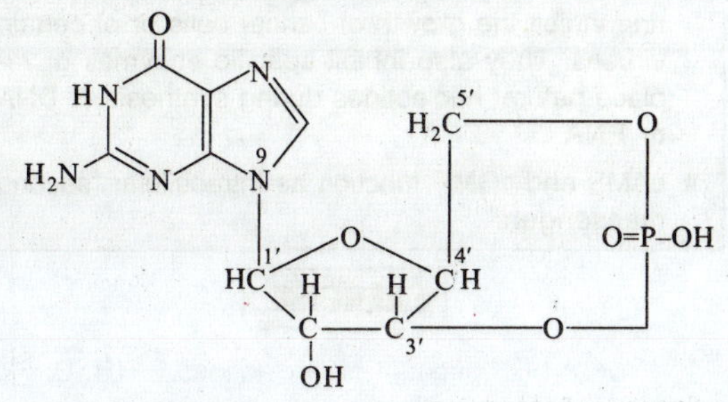
(Structure of Cyclic 3′, 5′- guanosine monophosphate)

Fig. 7.12

(b) It is catabolized by phosphodiesterase to produce 5′–monophosphate.

(c) It is an important intracellular signal of extracellular events.

(d) It acts antagonistically to cAMP.

5. **Inosine Monophosphate (IMP) :**

(a) IMP can be formed by the deamination of AMP which takes place particularly in the muscle. This IMP when reconverted to AMP results in the net production of ammonia from aspartate.

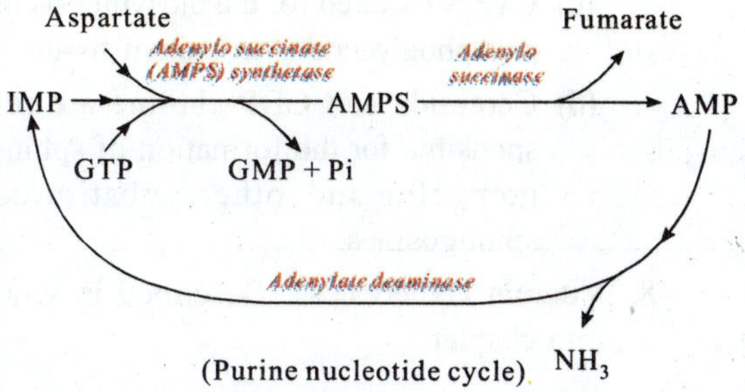

(Purine nucleotide cycle)

Fig. 7.13

(b) Inosine triphosphate (ITP) and Inosine diphosphate (IDP) play an important role in the formation of oxaloacetate from pyruvate and also in energy trapping reactions during oxidation of α-Ketoglutarate.

6. **Uridine nucleotide derivatives (UDP and UTP) :**

(a) These are the important coenzymes required in the metabolism of galactose to form lactose in the mammary gland and the polymerization of glucose to from glycogen. In these reactions the substrates are UDP Glucose and UDP Galactose.

(b) Another coenzyme uridine diphospho–glucuronic acid (UDP Glc UA) serves as the "active" glucuronide for conjugation reactions such as the formation of bilirubin glucuronide in the liver.

(c) They are involved in the epimerization of galactose to glucose and vice versa.

(d) They also participate in the formation of high-energy phosphate compounds.

(e) UTP is the precursor for the polymerization of uridine nucleotides into RNA.

7. **Cytosine derivatives (CMP, CDP & CTP) :**

(a) These derivatives are the high-energy phosphate compounds.

(b) CTP acts as the precursor for the polymerization of CMP into nucleic acids.

(c) CTP is required for the biosynthesis of phosphoglycerides in animal tissue.

(d) Ceramide and CDP-choline are responsible for the formation of sphingomyelin and other substituted sphingosines.

8. **Vitamin nucleotides :** Described in vitamin chapter.

SYNTHETIC DERIVATIVES

Synthetic nucleobases, nucleosides and nucleotides are widely used in the medical sciences and clinical medicine.

The pharmacologic view is that either the heterocyclic ring structure or the sugar moiety is changed in such a way as to induce toxic effects when they are incorporated into various cellular constituents resulting in the inhibition of enzyme activities.

6-thioguanine and 6-mercaptopurine, in which the hydroxyl groups are replaced with thiol groups at the 6 position are widely used clinically.

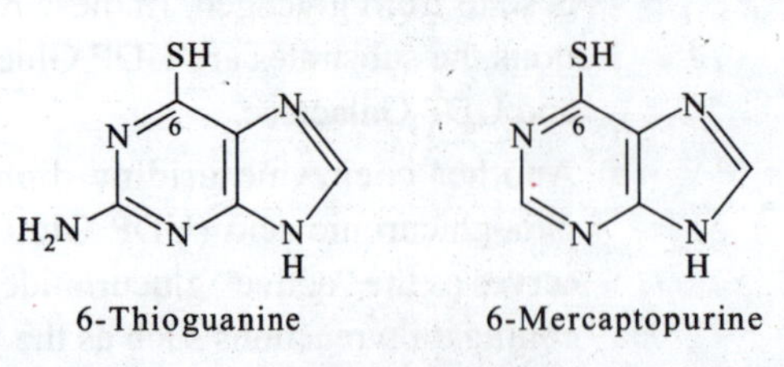

Fig. 7.14

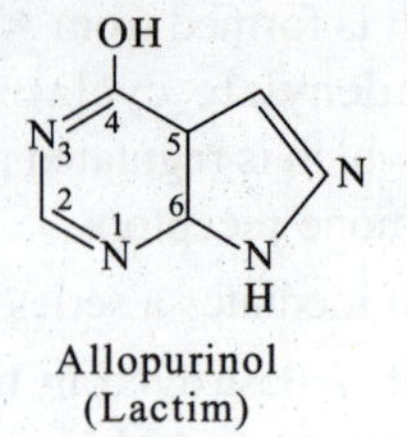

Fig. 7.15

4-hydroxypyrazole pyrimidine (allopurinol) is marketed as an inhibitor of xanthine oxidase. Hence, it is used for the treatment of hyperuricemia and gout.

Cytarabine (arabinosyl cytosine) and vidarabine (arabinosyl adenine) are used in the chemotherapy of cancer and viral infections.

Azathioprine is useful in organ transplantation.

More recently, both aminophylline and theophylline are used clinically to inhibit the catabolism of intracellular cAMP.

CLINICAL ASPECTS

- Guanine is not the normal constituent of human urine but human urine contains xanthine and uric acid which are insoluble in human urine. So they form stones in the urinary tract.
- Nucleotides form a portion of coenzyme such as FAD, NAD, NADP, coenzyme A.
- The concentration of ADP regulates the rate of mitochondrial oxidative phosphorylation.
- Nucleoside triphosphate serves as the monomer unit precursors of RNA and DNA.
- Allopurinol is used in the treatment of gout.
- The synthetic analogs of naturally occurring bases such as 5-flurouracil, 6-thioguanine, 6-mercaptopurine inhibit the growth of cancer cells or of certain viruses. They also inhibit specific enzymes or replace natural nucleotides during synthesis of DNA or RNA.
- cAMP and cGMP function as intracellular "second messengers".

EXERCISE

1. Give a brief account of nucleotides. (R. U. 72A)
2. Discuss nucleotides with clinical aspects. (Patna 87A)
3. Write short notes on : Nucleotides. (Bh.U. 89A)

DIAGRAMMATIC COMPENDIUM

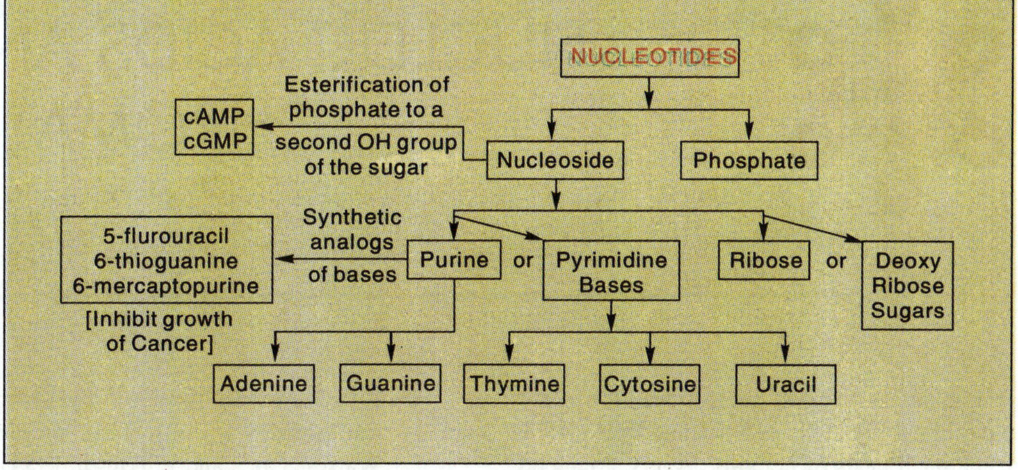

CHAPTER 8

NUCLEIC ACIDS AND CHROMATIN

Nucleic acids are the polynucleotides. There are two types of nucleic acids — DNA and RNA. DNA is the chemical basis of heredity. Avery, Macleod and McCarty in 1944 first demonstrated in a series of experiments that DNA contained the genetic information. These authors referred DNA as "transforming factor".

Chemistry of DNA

1. The four types of monomeric units of DNA are adenylate, guanylate, cytidylate and thymidylate.
2. The monomeric units form a single strand of DNA which are held in polymeric form by 3′, 5′-phosphodiester bridges.
3. The informational content of DNA is available in the sequence in which these monomers are ordered.
4. The polymer possesses a polarity; one end has a 5′-hydroxyl or phosphate terminus while the other has a 3′-phosphate or hydroxyl moiety.
5. Since the genetic information is available in the monomeric units within the polymers, there is existence of a mechanism reproducing of replicating this specific information with a high degree of fidelity.
6. In DNA molecules the concentration of Adenosine (A) nucleotides equals that of thymidine (T) nucleotides (A=T). The concentration of guanosine (G) nucleotides equals that of cytidine (C) nucleotides (G=C). This accelerated Watson, Crick and Wilkins in 1953 to propose a model of double-stranded DNA molecule.
7. The 2 strands of this double-stranded molecule are held together by hydrogen bonds between the purine and pyrimidine bases. The pairings between the purine and pyrimidine nucleotides on the opposite strands are dependent upon hydrogen bonding of A with T and G with C.
8. The 2 strands of the double helical molecule are anti-parallel i.e. one strand runs in the 5′ to 3′ direction and the other in the 3′ to 5′ direction. Since the information resides in the sequence of nucleotides on one strand, the opposite strand is considered as "antisense" i.e. the complement of the "sense" strand.
9. Three (3) hydrogen bonds hold the guanosine nucleotide to the cytosine nucleotide and A–T pair is held together by 2 hydrogen bonds. Therefore, they are represented as G ≡ C and A = T. G–C bond is stronger by 50%.
10. The B form has a pitch of 3.4 nm per turn and within the single turn 10 base pairs exist.
11. The double-stranded structure in solution can be melted by increasing temperature or decreasing salt concentration.

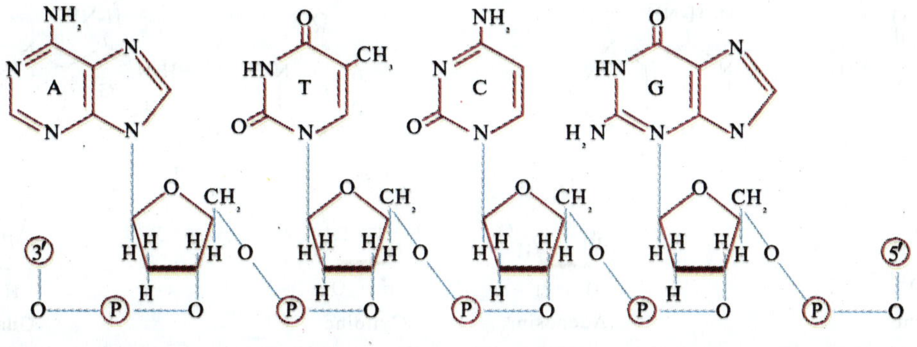

Fig. 8.1 : Segment of a structure of DNA molecule in which purine and pyrimidine bases Adenine (A), Thymine (T), Cytosine (C) and Guanine (G) are held together having 3′ at one end and 5′ at the other

12. The double-stranded DNA molecule shows the properties of a fibre and it is a viscous material in solution and loses its viscosity upon denaturation.
13. In the major and minor grooves winding along the molecule parallel to the phosphodiester backbones, specific proteins interact with DNA molecules.
14. In some organisms such as bacteria, bacteriophage and many DNA-containing animal viruses, the two ends of the DNA molecules are jointed to create a closed circle with no terminus. This does not destroy the polarity of the two molecules but it eliminates all 3′ and 5′ free hydroxyl and phosphoryl groups.
15. It contains 1,600 to 9,000 nucleotides. The molecules are long and its length is 250 times greater than its breadth. Its structure is highly complex.
16. Heat, acid and alkali denature DNA.

Chromatin

1. Chromatin is the chromosomal material extracted from nuclei of cells.
2. It consists of a long double-stranded DNA molecules and a nearly equal mass of histones as well as smaller amount of non-histone proteins and a small quantity of RNA.

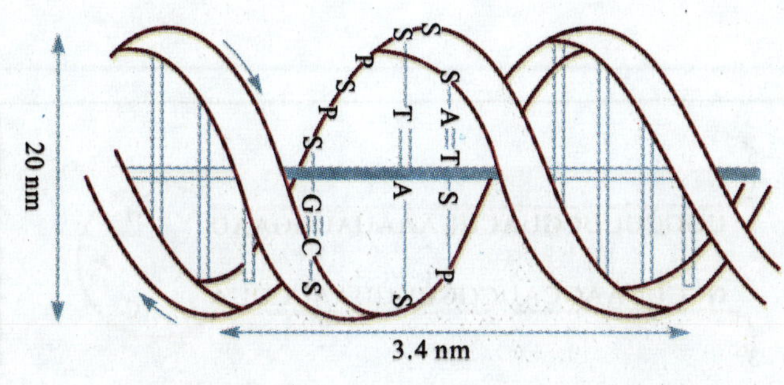

Fig. 8.2 : Double helical structure of DNA

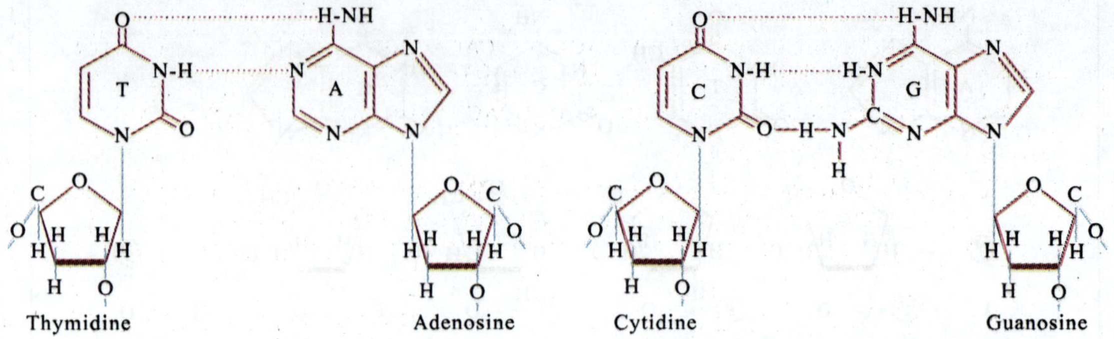

Fig. 8.3 : Base pairing between adenosine and thymidine, cytidine and guanosine

3. It contains a 10 nm repeating unit. The repeating units occur every 200 base pairs.

Chemistry of RNA

1. RNA is polymer of purine and pyrimidine ribonucleotides linked together by 3′, 5′-phosphodiester bridges.
2. The sugar moiety in RNA is ribose.
3. It contains uracil instead of thymine in addition to adenine, guanine and cytosine.
4. It exists as a single-stranded molecule rather than as a double-stranded helical molecule.
5. Since it is a single-stranded molecule, its guanine content does not necessarily equal its cytosine content and its adenine content does not necessarily equal its uracil content.
6. It can be hydrolyzed by alkali to 2′, 3′ cyclic diesters of the mononucleotides. An intermediate in this hydrolysis is the 2′, 3′, 5′-triester which cannot be formed in DNA hydrolysis by alkali because of the absence of a 2′-hydroxyl group. The alkali lability of RNA is useful diagnostically and analytically.
7. RNA molecule does not hybridize with the "antisense" strand of the DNA of its gene. Therefore, the sequence of RNA molecule (except U being replaced by T) is the same as that of the "antisense" strand of the gene. Small amounts of double-stranded RNA have been detected from mammalian organisms including humans which may probably be associated with RNA viruses.
8. It contains 60 to 6,000 nucleotides. The molecule is unbranched.

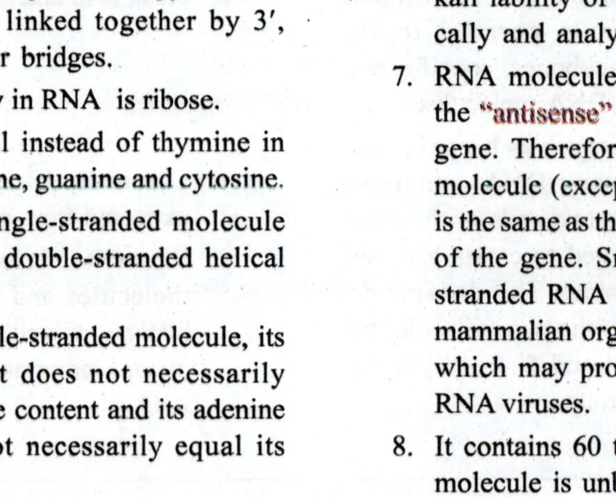

Fig. 8.4 : Hairpin structure of RNA

DNA strands

Antisense → 5′- TGGAATTGTGAGCGGATAACAATTTCACACAGGAAACAGCTATGACCATG -3′

Sense → 3′- ACCTTAACACTCGCCTATTGTTAAAGTGTGTCCTTTGTCGATACTGGTAC -5′

RNA transcript 5′- pAUUGUGAGCGGAUAACAAUUUCACACAGGAAACAGCUAUGACCAUG -3′

Fig. 8.5 : Relationship between the sequence of an RNA transcript and its gene in which the 'sense' and 'antisense' strands are shown with their polarities

Structural organization of RNA

There are 3 main classes of RNA molecules in all prokaryotic and eukaryotic organisms. These are :

1. messenger RNA (m RNA).
2. transfer RNA (tRNA) or soluble RNA.
3. ribosomal RNA (rRNA).

1. The messenger RNA (mRNA)

1. The messenger RNA is single-stranded and complementary to the sense strand of DNA.
2. The 5′ terminus of messenger RNA is "capped" by a 7-methylguanosine triphosphate which is linked to an 2′- O methyl ribonucleoside at its 5′-hydroxyl through the 3 phosphates. The protein-synthesizing machinery begins translating the mRNA into proteins at the 5′ or capped terminus. The 3′-hydroxyl terminus has attached a polymer of adenylate residues 20-250 nucleotides in length.
3. It is the most heterogeneous in size and stability.

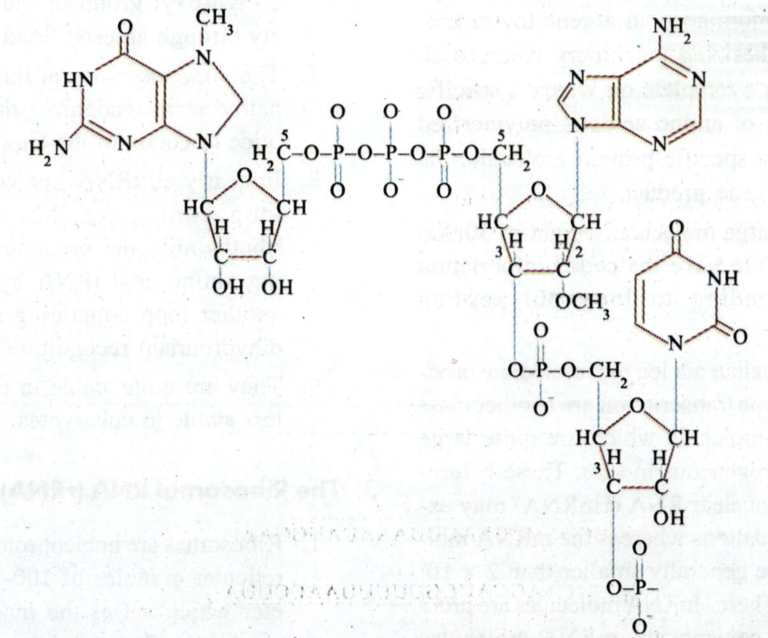

Fig. 8.6 : The expression of genetic information in DNA into the form of an mRNA transcript. This is subsequently translated by ribosomes into specific protein molecule

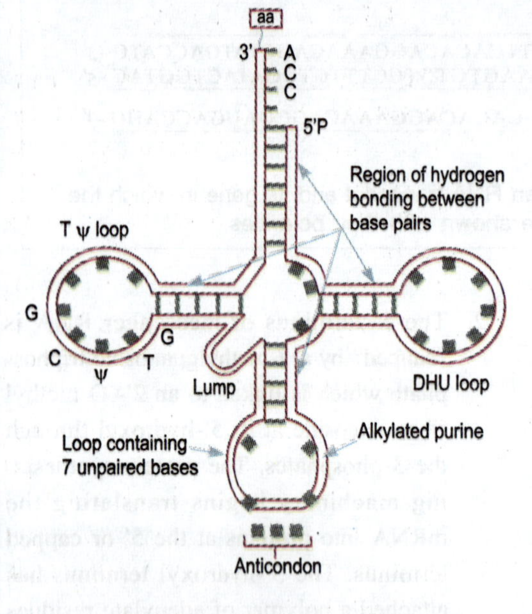

Fig. 8.7 : Structure of tRNA

4. It passes from nucleus to cytoplasm conveying information in a gene to the protein-synthesizing machinery where each serves as a template on which a specific sequence of amino acids is polymerized to form a specific protein molecule, the ultimate gene product.
5. It has a large molecular weight of 30,000 to 50,000 to have the coded information corresponding to long polypeptide chains.
6. In mammalian nuclei, the immediate products of gene transcription are another class of RNA molecule which are quite large and heterogeneous in size. These heterogeneous nuclear RNA (HnRNA) may exceed 10^7 daltons whereas the mRNA molecules are generally smaller than 2×10^6 daltons. These HnRNA molecules are processed to generate the mRNA molecules which then enter the cytoplasm to serve as templates for protein synthesis.

2. The transfer RNA (tRNA) or soluble RNA (sRNA)

1. The transfer RNA molecules amount to 10 to 20 per cent. of the total cellular RNA molecules.
2. They consist of about 75 nucleotides and have molecular weight of 25,000.
3. There are at least 20 tRNA molecules in every cell, one corresponds to each of the 20 amino acids required for protein synthesis.
4. They serve as adaptors for the translation of the information in the sequence of nucleotides of the mRNA into specific amino acids.
5. The primary structure allows extensive folding and intrastrand complementarity to generate a significant secondary structure which appears like a cloverleaf as in Fig. 8.7.
6. All tRNA molecules have a common CCA sequence at the 3′ termini. The carboxyl groups of amino acids are attached to the 3′-hydroxyl group of the adenosyl moiety through an ester bond.
7. The *anticodon loop* at the end of a base-paired stem recognizes the triplet nucleotide or codon of the template mRNA.
8. In nearly all tRNA molecules, there is a loop containing the nucleotides of ribothymine and pseudouridine for binding amino acyl tRNA by ribosome and another loop containing the minor base dihydrouracil recognition of tRNA.
9. They are quite stable in prokaryotes and less stable in eukaryotes.

3. The Ribosomal RNA (rRNA)

1. Ribosomes are nucleoprotein particles and reticular granules of 100-150Å, in diameter which act as the machinery for the synthesis of proteins from mRNA templates. They contain 80 per cent of the RNA within the cell.

2. On the ribosomes, the mRNA and tRNA molecules interact to translate into a specific protein molecule the information transcribed from the gene.
3. Ribosomal particles are very complex. The mammalian ribosome contains 2 major nucleoprotein subunits — a larger one (60 S) and a smaller one (40 S).
4. The 60 S subunit contains a 5 S ribosomal RNA, (rRNA), a 5.8 S rRNA and a 28 S rRNA.
5. There are also more than 50 specific polypeptides. The smaller (40 S) subunit contains a single 18 S rRNA and approximately 30 polypeptide chains.
6. The 5 S rRNA has its own precursor which is independently transcribed.
7. In the cytoplasm, the ribosomes remain quite stable and capable of many translations.

DNA AND GENE

1. The most important constituent of chromosomes is DNA. The double — stranded helix of DNA is tightly coiled in each chromosomes thread.
2. The unit of genetic information is the *gene* or *cistron*.
3. Gene is a segment of the DNA molecule containing about 600 base pairs. The genetic message is carried in the sequence of bases along the DNA strand.
4. In the process of transcription, the genetic message is transferred to the messenger RNA which carries it to the ribosomes.
5. The genes are arranged in orderly manner along the length of the DNA molecule in the chromosome.
6. Each gene for any particular characteristic has its counterpart in the corresponding

Comparison between DNA and RNA

DNA	RNA
1. They are formed by 4 types of monomeric units such as adenylate, guanylate, cytidylate and thymidylate.	1. They are also formed by 4 types of monomeric units such as adenylate, guanylate, cytidylate, uracidylate.
2. The sugar moiety is deoxyribose.	2. The sugar moiety is D-ribose.
3. They are present in nucleus, probably in the chromosomes.	3. They are present in cytoplasm and nucleolus.
4. They carry "genetic information" from one generation of cells to the next and undergo mutation.	4. They carry no "genetic information" and undergo no mutation.
5. They contain 1,600 to 9,000 nucleotides. The molecules are long and thread-like having a length of about 250 times greater than their breadth. Their structure is highly complex.	5. They contain 60 to 6,000 nucleotides. The molecules are unbranched. Their structure is less rigid.
6. They are concerned with protein synthesis.	6. They are also concerned with protein synthesis.
7. They have 2 strands of antiparallel double helical structure.	7. They exit as a single-stranded molecule.
8. Heat, acid and alkali denature DNA molecules.	8. The alkali hydrolysis is useful diagnostically and analytically.
9. They have 'sense' and 'antisense' strands.	9. The sequence of RNA molecule is the same as that of the 'antisense' strand of DNA.

locus on the homologous chromosome and these two genes form an allelic pair. When both loci of a pair carry genes with the same characteristics, say tallness, then the individual is said to be *homozygous* with respect to that characteristic. When one of the pair tallness and the other gene shortness, the individual is heterozygous.

Biological importance of nucleic acids

1. Nucleic acids are able to reproduce their kind or to store or express and transmit genetic information.
2. They undergo mutation.
3. In cell division, the nucleic acid chain is duplicated preserving in each daughter cell the information contained in the parent cell. So the double helix unravels and each of the two original strands then serves as a template for the synthesis of another complementary chain.
4. According to the "pairing rule" in DNA, adenine can only combine with thymine and guanine only with cytosine. The newly synthesized strand will be exactly constituted in its nucleotide sequence as was the original complementary strand of the parent strand. The result is the synthesis of two pairs of strands.
5. DNA produces a messenger RNA (mRNA) which helps in placing amino acids in the code for protein synthesis.
6. RNA functions primarily in the cytoplasm of the cell as a template in connection with the synthesis of proteins as well as in the ribosomes. The formation of RNA template is directed by nuclear DNA.
7. Ribosomal RNA (rRNA) and transfer RNA (tRNA) are also involved in protein synthesis.
8. RNA can be synthesized by RNA-polymerase which is dependent on the presence of DNA acting as a template.
9. Adenylic acid in combination with two molecules of phosphate is the biochemical unit of energy exchange in all cells which is said to be ATP.
10. Biological oxidation-reduction involves the transport of hydrogen atoms or electrons through organized systems of substances called hydrogen acceptors or electrons transport agents. The hydrogen acceptors are nucleotides such as NAD, FAD etc.

NUCLEOPROTEINS

Nucleoproteins are conjugated proteins containing nucleic acids and basic protein–like protamines or histones.

In nucleoproteins there are salt–like combination between positively charged basic proteins and negatively charged nucleic acids. Cytoplasm also contains ribonucleoprotein which is found particularly in ribosomes. All living cells contain nucleoproteins and some of the simplest systems, such as viruses, consist entirely of nucleoproteins.

Protamines

1. These basic proteins have low molecular weight and these are found in association with DNA chiefly in fish sperms.
2. They produce large quantities of arginine and less amounts of neutral amino acids on hydrolysis.
3. Their isoelectric pH is around 12.
4. The polypeptide chains of protamines are wrapped around DNA molecules in such a way that the positively charged arginine form salt linkages with the phosphates of the nucleic acids.
5. The best examples are salmin (salmon sperm) and clupeine (herring sperm).

Histones

1. They are more complex proteins than protamines.
2. They have larger molecular weights and exist in combination with DNA.

3. They produce a variety of amino acids on hydrolysis.
4. Nucleohistones mainly occur in animal sperms, birds, erythrocytes.

Deoxyribonucleoproteins

1. They are viscous material which are insoluble in 0.15 M NaCl but soluble in 1M NaCl solution.
2. They have high molecular weights of 10 million to several hundred millions.
3. They are highly elongated and thread-like in shape.

Ribonucleoproteins

1. They are present in cytoplasm of the cells and ribosomes.
2. They are composed of proteins and ribonucleic acids.
3. Ribosomes of E. coli consists entirely of ribonucleoproteins.
4. Ribosomes from mammalian and plant sources contain about 40-50 per cent RNA, the remainder being proteins.

CILINICAL ASPECTS

- The chemical basis of heredity and of genetic diseases is available in the structure of DNA.
- DNA directs the synthesis of RNA which in turn directs the synthesis of Protein.

EXERCISE

1. Write in brief the chemistry of Nucleic acids. (R.U. 70A ; M.U. 74S, P.U. 68S ; Mith. 62A, 67A)
2. What are Nucleic Acids? Describe the main points of difference between DNA and RNA. Mention the biological importance of nucleic acids. (Mith. 60A ; M.U. 72A)
3. Describe what you know of DNA and its biological role. (Bh. U. 75S ; R.U. 70S, Mith. 63S)
4. Describe the chemistry of Hereditary Unit. (P.U. 74A; Bh.U. 76S)
5. Write short notes on :
   (i) Nucleic acid. (M.U. 75A)
   (ii) DNA. (M.U. 72A ; 74A ; P.U. 68A ; R.U. 70A, 72A, Mith. 74A)
   (iii) RNA. (R.U. 68S, 76A ; Bh. U. 76S)
   (iv) Transfer RNA. (P.U. 76A; Bh. U. 76S)
   (v) Codon. (M.U. 76A)
   (vi) Messenger RNA. (Muz. 74A ; P.U. 74S)
   (vii) Adenine. (C.U. 1981)
   (viii) Adenylic acid. (C.U. 1982)
   (ix) DNA as a carrier of heredity. (C.U. 1982)
6. Give the structural formula of nucleotide. (C.U. 1981)
7. Write True or False ;
   (i) The process of protein synthesis on m-RNA linked ribosome is called transcription. (C.U. 1981)
   (ii) Uracil on methylation forms thymine. (C.U. 1983)
8. Write structure of ATP and its importance. (C.U. 1981)

DIAGRAMMATIC COMPENDIUM

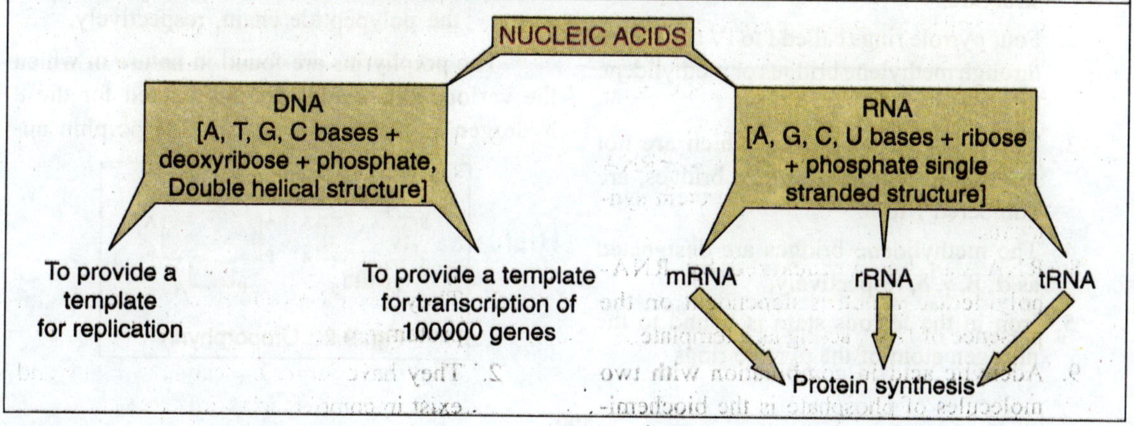

CHAPTER 9

HEMOGLOBIN, PORPHYRINS AND BILE PIGMENTS

Hemoglobin is the red colouring matter of blood which is present in the red blood cells. It is a conjugated protein consisting of heme and the protein globin. It has a molecular weight of 64,450. It can combine with oxygen and acts as the transport mechanism for oxygen within the blood. It contains 4 gram atoms of iron per mole in the ferrous (Fe^{++}) state.

Structure of Hemoglobin

The structure of Hemoglobin can be classified under two headings :

1. Structure of Heme, the prosthetic group.
2. Structure of Globin, the protein part—apoprotein.

1. Structure of Heme

1. It is an iron porphyrin. The porphyrins are cyclic compounds with "tetra pyrrole" structure.
2. Four pyrrole rings called I to IV are linked through methylene bridges or methylidene bridges.
3. The outer carbon atoms, which are not linked with the methylidene bridges, are numbered 1 to 8.
4. The methylidene bridges are designated as α, β, γ, δ, respectively.
5. Iron in the ferrous state is bound to the nitrogen atom of the pyrrole rings.

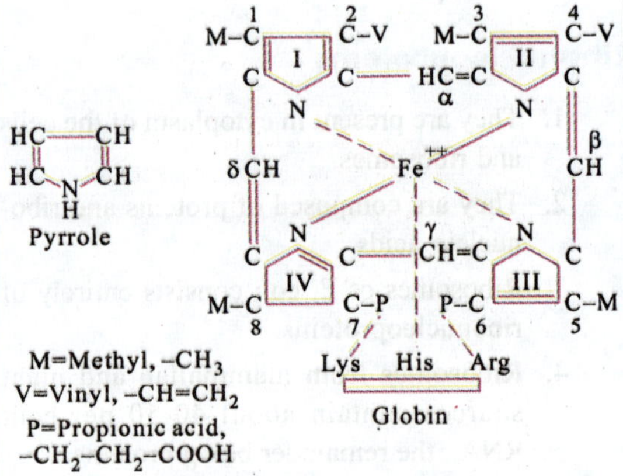

Fig. 9.1 : Structure of Heme

6. Iron is also linked internally (5th linkage) to the nitrogen of the imidazole ring of Histidine of the polypeptide chains.
7. The propionic acid of 6th and 7th position of heme of III and IV pyrroles are also linked to the amino acids Arg and Lys of the polypeptide chain, respectively.

The porphyrins are found in nature in which the various side chains are substituted for the 8 hydrogen atoms as numbered in the porphin nu-

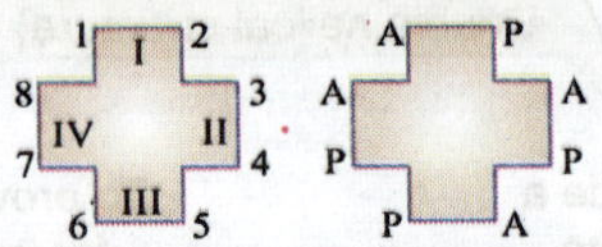

Fig. 9.2 : Uroporphyrin

cleus. The arrangement of the A and P substituents in the uroporphyrin shown here is asymmetric (in ring IV the expected order of the acetate and propionate substituents is reversed). This type of asymmetric substitution is classified as a type III porphyrin. A porphyrin with a completely symmetrical arrangement of the substituents is classified as a type I porphyrin. Only types I and III are found in nature and the type III series is more abundant.

2. Structure of Globin

1. The globin of hemoglobin is a protein which is composed of 4 parallel layers of closely packed polypeptide chains.
2. Two of the chains (α-chains) have identical amino acid composition of 141 amino acids. The two other chains may be two of the 4 polypeptide chains designated as β, γ, δ, and ε (epsilon). Each is having 146 amino acids.
3. The total number of amino acids in globin is 574.
4. α chains have Val-Leu-Ser in N terminal residues and Lys-tyr-Arg in C terminal residues.
5. β chains have Val-His-Leu in N-terminal residues and Lys-tyr-His in C-terminal residues.
6. γ chains have Gly-His-Phe. N-terminal residues and Arg-Tyr-His in C-terminal residues.
7. Hemoglobin molecule and its sub-units contain mostly hydrophobic amino acids internally and hydrophilic amino acids on their surfaces. So they form "Heme pockets".
8. In "heme pockets" α subunits are of size necessary for entry of O_2 molecule but the entry of O_2 molecule in β subunit is blocked by valine residue.

Biosynthesis of Porphyrins

Chlorophyll (magnesium-containing porphyrin), the photosynthetic pigment of plants and heme (the iron-containing porphyrin) of hemoglobin in animals are synthesized in living cells by a common pathway :

1. The starting materials are 'active succinate' (succinyl-CoA) derived from the citric acid cycle and glycine. Pyridoxal phosphate (B_6-PO_4) is necessary to activate glycine. The product of the condensation reaction is α-amino-β-ketoadipic acid which is catalyzed by the enzyme AmLev synthetase (ALA synthase).
2. α-amino-β-ketoadipic acid is rapidly decarboxylated by the same enzyme AmLev synthetase producing δ-aminolevulinic acid (AmLev). Synthesis of aminolevulinic acid occurs in the mitochondria. The anemia has been observed in the deficiency of vitamin B_6 or pantothenic acid.
3. 2 mols of AmLev condense to form porphobilinogen (the first precursor of pyrrole) which is catalyzed by the enzyme δ-aminolevulinase (AmLev dehydrase).
4. 3 mols of porphobilinogen condense first to form a tripyrrylmethane which then breaks down into a dipyrrylmethane and a monopyrrole. The dipyrryl compounds are of two types A and B. The formation of tetrapyrrole occurs by condensation of two dipyrrylmethanes. If two of the (A) components condense, a type I porphyrin results ; if one (A) and one (B) condense, a type III results.
5. The uroporphyrinogens I and III are converted to coproporphyrinogens I and III by decarboxylation being catalyzed by uroporphyrinogen decarboxylase.
6. The coproporphyrinogen III then enters the mitochondria where it is converted to protoporphyrinogen III and then to protoporphyrin III. The enzyme coproporphyrinogen oxidase catalyzes the formation of protoporphyrinogen III. The oxidation of protoporphyrinogen to protoporphyrin is catalyzed by the enzyme protoporphyrinogen oxidase. The enzyme

coproporphyrinogen oxidase is able to act on type III coproporphyrinogen only for which type I protoporphyrin has not been identified in natural materials. In mammalian liver the reaction of conversion of coproporphyrinogen to protoporphyrin requires molecular oxygen.

7. In the final step of heme synthesis ferrous ion (Fe^{++}) is incorporated into protoporphyrin III which is catalyzed by heme synthetase or ferrochelatase. The reaction takes place readily in the absence of enzymes but becomes rapid in presence of enzymes.

A summary of the steps is given :

Note :

1. The porphyrinogens are the reduced porphyrins containing 6 extra hydrogen atoms. The oxidized porphyrins cannot be used for heme or chlorophyll synthesis.
2. The porphyrinogens are readily auto-oxidized to the respective porphyrins in presence of light.

Regulation of heme biosynthesis

1. ALA synthase in the synthesis of heme is a regulated enzyme. Heme acts as a nega-

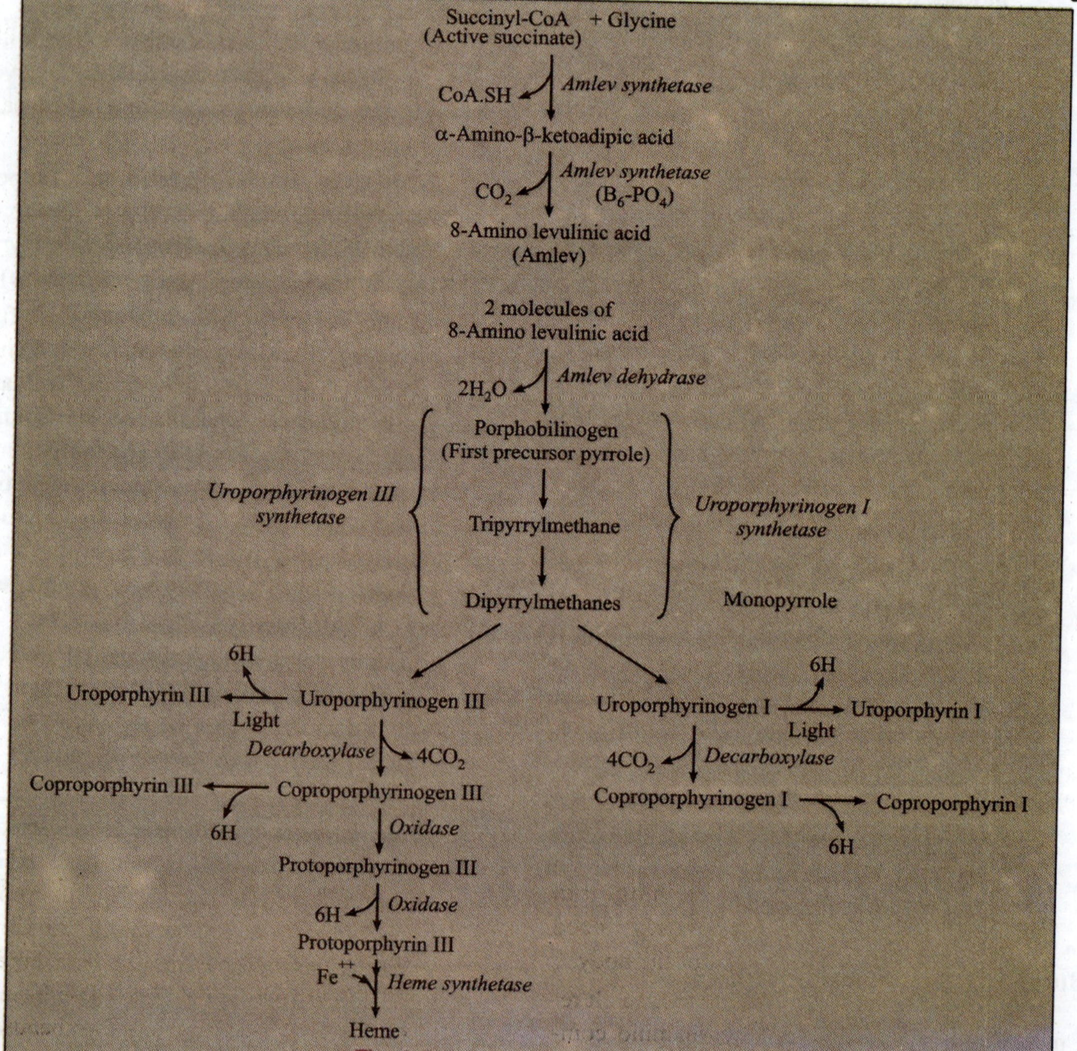

Fig. 9.3 : Biosynthesis of heme

tive regulator in the synthesis of ALA synthase. The major regulatory effect of heme appears in that the rate of synthesis of ALA synthase increases greatly in the absence of heme and is reduced in its presence. In mammalian liver, the turnover rate of ALA synthase is normally rapid (half-life is about one hour).

2. Many **Xenobiotics** on administration shows a marked increases in hepatic ALA synthase. Many of these compounds are metabolized by a system in the liver after utilizing **Cytochrome P-450**. During this metabolism, heme utilization by cytochrome P-450 is highly increased which in turn diminishes the intracellular heme concentration. This latter event effects a depression of ALA synthase with a corresponding increased rate of heme synthesis to meet the needs of the cells.

3. The induction of ALA synthase in the liver is affected by other factors. **Glucose** can prevent the induction of ALA synthase; **iron** in chelated form exerts a synergistic effect on the induction of hepatic ALA synthase, **Steroids** play a permissive role in the drug-mediated derepression of ALA synthase in vivo. **Hematin** administration in vivo can prevent the drug mediated derepression of ALA synthase.

Properties of hemoglobin

1. **Oxyhemoglobin** : It forms oxyhemoglobin in combination with oxygen. When hemoglobin is exposed to air, it takes up two atoms of oxygen for each atom of ferrous ion (Fe^{++}) present. Thus, hemoglobin will take up 4 molecules of oxygen. In low oxygen tension, oxyhemoglobin gives up O_2 readily. By this way, blood carries O_2 to different parts of the body.

2. **Formation of carbamino compound** : It reacts with CO_2 forming carbamino compounds.

$$\text{Hb-NH}_2 + CO_2 \longrightarrow \text{Hb-NH.COOH}$$

3. **Reaction with carbon monoxide** : It forms carboxy hemoglobin after reacting with carbon monoxide (CO). Carboxy hemoglobin is stable and prevents the formation of oxyhemoglobin. So inhalation of even small amounts of carbon monoxide is highly dangerous.

4. **Buffering action** : One mol of hemoglobin contains 35 histidine residues. Histidine exerts its buffering action through its basic imidazole ring. Hence, hemoglobin plays an important role in regulating the acid-base balance of blood.

5. **Formation of methemoglobin** : Methemoglobin is formed as a result of the oxidation of hemoglobin by the mild oxidizing agent, potassium ferricyanide. The ferrous ion (Fe^{++}) is oxidized to the ferric ion (Fe^{+++}). Methemoglobin cannot carry oxygen in blood. It is also formed by the action of some drugs. This is found in the blood of some individuals owing to inborn errors of metabolism. This can be reduced to hemoglobin by vitamin C which is used in the treatment of methemoglobinemia.

6. **Sulphemoglobin** : It is formed by the administration of certain drugs. It continues to remain in the blood and cannot be reconverted into hemoglobin.

7. **Cyanomethemoglobin** : It is formed by the addition of cyanide to methemoglobin. It has a bright red colour.

8. **Absorption spectra** : The different hemoglobin derivatives can be easily identified by this characteristic absorption spectra.

   (a) **Oxyhemoglobin** : Two bands—one narrow and the other wide in the green region.

   (b) **Reduced hemoglobin** : One single broad band in the green region.

   (c) **Carboxy hemoglobin** : Two bands in the green region.

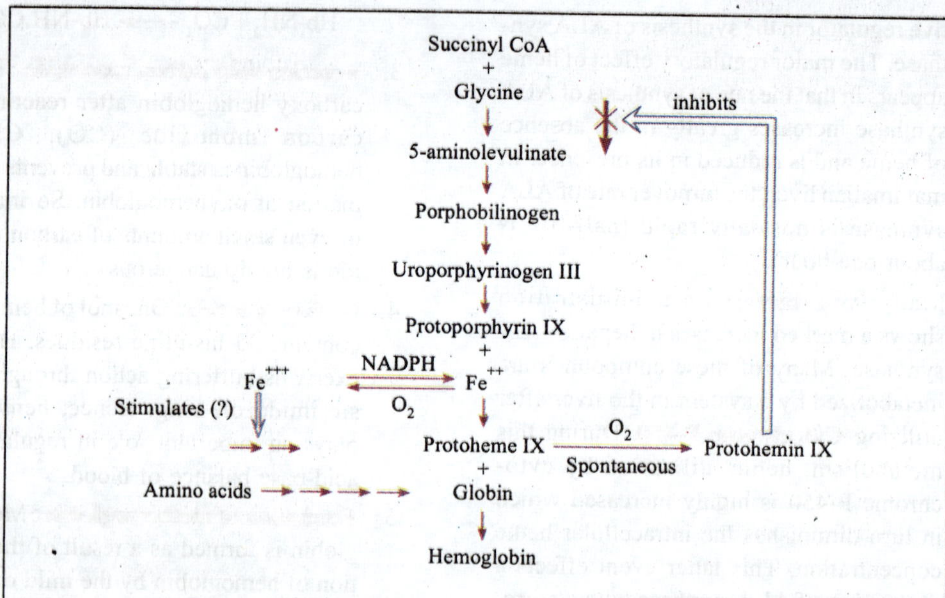

Fig. 9.4 : Regulation of hemoglobin synthesis according to Current Concepts

(d) **Methemoglobin :** Three bands – one in red and two in the green regions.

(e) **Sulphemoglobin :** Three bands similar to methemoglobin.

Biosynthesis of hemoglobin

1. The biosynthesis of hemoglobin takes place in the bone marrow in the erythroid cell during its development to erythrocyte.
2. It starts appearing at stage II (early normoblast) and the synthesis is complete when the cell reaches stage IV (late normoblast).
3. Iron in the ferrous state is incorporated into protoporphyrin to form heme.
4. The heme gets attached to the newly synthesized globin to form hemoglobin.
5. The iron of heme is coordinated to 2 imidazole nitrogen of histidine at position 38 and 87 in α-chains and 63 & 92 in β-chains.

In nature, the other metalloporphyrins which are compounds of importance in biologic processes are mentioned :

1. **Erythrocruorins :**
   (a) They are iron porphyrinoproteins occurring in blood and tissue fluids of some invertebrates.
   (b) Their function is corresponding to hemoglobin.
2. **Myoglobins :**
   (a) They are the respiratory pigments occurring in the muscle cells of vertebrates and invertebrates.
   (b) The purified one has a molecular weight of about 17,000.
   (c) They contain only 1 gram atom of iron per mole.
3. **Catalases :**
   (a) They are iron porphyrin enzymes.
   (b) They have been obtained in crystalline form.
   (c) Their molecular weight is about 225,000.
   (d) They contain 4 gram atoms of iron per mol.
   (e) In plants, their activity is minimal.

4. **Tryptophan pyrrolase :**
   (a) It is an iron porphyrin protein.
   (b) It catalyzes the oxidation of tryptophan to formyl kynurenine.
5. **Cytochromes :**
   (a) Cytochromes means the cellular pigments because these pigments are widely distributed not only in the tissues of higher animals and plants but also in yeast and bacteria.
   (b) At first, cytochromes a, b and c were identified and they had been shown to exist in oxidized and reduced forms and their fundamental role is in cellular respiration. At present, some thirty cytochromes are known to exist and according to original cytochromes they are designated as a_1, a_2, a_3 ; c_1, c_2, c_3, c_4, c_5 ; b_2, b_3, b_4 etc.
   (c) They are iron porphyrins and act as electron transfer agents in oxidation-reduction reactions.
   (d) The important example is cytochrome C which has been obtained in the purified form.
   (e) Cytochrome C has a molecular weight of about 13,000 and contains 0.43% iron.
   (f) The iron porphyrin group of cytochrome C is attached to protein more firmly than in the hemoglobin.
   (g) Cytochrome C is quite stable to heat and acids.
   (h) The reduced form of cytochrome C is not autooxidizable.
   (i) At physiological pH ferrocytochrome C does not combine with O_2 or CO as does hemoglobin.
   (j) The peptide chain of human heart cytochrome C contains 104 amino acids. Acetylglycine is the N-terminal amino acid and glutamic acid the C-terminal amino acid. The two cysteine residues are located at positions 14 and 17 in the peptide chain. The linkage of iron in heme occurs through the imidazole nitrogen of a histidine residue at position 18 in the peptide chain.
   (k) The degree of difference in primary structure among the 13 cytochrome C might be related to the degree of phytogenetic relationship between the species Eg. The cytochrome C of man as compared to that of rhesus monkey differs by only one amino acid of the 104 amino acids. Human cytochrome C differs from that of the dog in 11 amino acid residues, from that of the horse in 12.
   (l) The enzymes that catalyze the reactions of molecular oxygen are known as oxidases. Cytochrome a_3, which is found in heart muscle and other animal tissues is called cytochrome oxidase. These oxidases catalyze many reactions in addition to terminal oxidation at the electron transport chain. They can carry three general types of reactions eg, oxygen transfer, mixed function oxidation electron transfer.

Hemoglobin transports CO_2 and protons to the lungs after releasing O_2 to the tissues

1. Hemoglobin can bind CO_2 directly when oxygen is released and CO_2 reacts with the amino terminal α-amino groups of the hemoglobin forming a carbamate and releasing protons :

$$CO_2 + Hb{-}NH_3^+ \rightleftharpoons 2H^+ + Hb{-}\overset{H}{N}{-}\overset{O}{\overset{\|}{C}}{-}O^-$$

The amino terminal is converted from a positive to a negative charge favouring salt bridge formation between the α and β chains.

2. At the lungs, hemoglobin is oxygenated, being accompanied by expulsion and sub-

sequent expiration of CO_2. CO_2 is absorbed in blood and the carbonic anhydrase in erythrocytes catalyzes the formation of carbonic acid which is rapidly dissociated into bicarbonate and a proton :

$$CO_2 + H_2O \xrightleftharpoons{\text{Carbonic anhydrase}} H_2CO_3 \xrightleftharpoons{\text{Spontaneous}} HCO_3^- + H^+$$

A buffering system absorbs these excess protons to avoid the increasing acidity of blood. Hemoglobin binds two protons for every four oxygen molecules.

3. In the lungs, the process is reversed i.e. when oxygen binds to deoxygenated hemoglobin, protons are released and combines with bircarbonate forming carbonic acid which is exhaled. Thus, the binding of xygen forces the exhalation of CO_2. This reversible phenomenon is called the Bohr effect. Myoglobin does not exhibit Bohr effect.

2, 3-Biphosphoglycerate (BPG) stabilizes the T structure of hemoglobin

1. The increased accumulation of 2, 3-biphosphoglycerate is caused by an oxygen shortage in peripheral tissues. BPG is formed from 1, 3-biphosphoglycerate in the glycolytic pathway. One molecule of BPG is bound to central cavity formed by all four subunits of hemoglobin. This cavity is of sufficient size for BPG only when hemoglobin is in the T form. BPG is bound by salt bridges between its oxygen atoms and both chains as well as by Lys EF6 and His H21. Thus, BPG stabilizes the T or deoxygenated form of hemoglobin.
2. Fetal hemoglobin is more weakly bound to BPG because the H21 residue of the γ chain of HbF is Her rather than His and can not form a salt bridge with BPG. Hence, BPG has a less profound effect on he stabilization of the T form of HbF and is responsible for HbF to have a higher affinity for oxygen than does HbA.
3. The trigger for the R to T transition of hemoglobin is movement of the iron in and out of the plane of the porphyrin ring.

Deoxyhemoglobin S can form fibres that distort erythrocytes

1. After the deoxygenation of hemoglobin S the sticky patch can bind to the complementary patch on another deoxygenated HbS molecule. This binding causes polymerization of deoxyhemoglobin S forming long fibrous precipitates. These extend throughout the erythrocyte and mechanically distort it causing lysis and a good number of secondary clinical effects.
2. Deoxyhemoglobin A although contains the receptor sites for the sticky patch present on deoxygenated HbS, the binding of sticky hemoglobin S to deoxyhemoglobin A cannot extend the polymer. Because deoxyhemoglobin A does not have a sticky patch to enhance binding to another hemoglobin molecule. Therefore, the binding of deoxyhemoglobin A to the R or the T form of hemoglobin S will reject polymerization.
3. The polymer forms a twisted helical fiber whose cross section contains 14 HbS molecules. These tubular fibres distort the erythrocyte.

Varieties of human hemoglobin

Normal adult hemoglobin or hemoglobin A has a molecular weight of 64,456 and contains two pairs of peptide chains (α & β) of which α chain contains 141 and β chain contains 146 amino acids.

Fetal hemoglobin (F) is present in very small amounts.

All the normal human hemoglobins possess a common half-molecule, i.e. a pair of peptide chains (α chains); the other half consists of a pair of differ-

Some Amino acid distribution in abnormal hemoglobin

Type of abnormal hemoglobin	Replacement of the position of amino acid	Amino acid present in the normal hemoglobin at that position	Amino acid present in abnormal hemoglobin at that position
C	6	Glutamic acid	Lysine
D (Ibdan)	87	Threonine	Lysine
E	26	Glutamic acid	Lysine
M (Boston)	58	Histidine	Tyrosine
N	95	Lysine	Glutamic acid
S	6	Glutamic acid	Valine

ent types of peptide chains, one type for each hemoglobin. Hemoglobin A_2 has two δ chains and hemoglobin F has two γ chains ; both types of chains contain 136 amino acids and thus are of the same length as the β chain. Hemoglobin A is represented as $\alpha_2A\ \beta_2A$ hemoglobin A_2 as $\alpha_2A\ \delta_2A$ and hemoglobin F as $\alpha_2A\ \gamma_2A$ for describing abnormal hemoglobin. In early embryonic life, a fourth hemoglobin α_2Ae_2 exists.

Fetal hemoglobin

1. Fetal hemoglobin (F) comprises 50 to 90 per cent of the total hemoglobin in the newborn.
2. It takes up oxygen more readily at low oxygen tensions and releases carbon dioxide more readily than adult hemoglobin (A).
3. It is more resistant to denaturation by alkali and is more susceptible to conversion to methemoglobin by nitrites (contaminated water).
4. Hemoglobin F is gradually replaced by hemoglobin A during the first 6 months of extrauterine life.
5. High concentration of hemoglobin F after two years of age occur in various types of anemia, eg, sickle cell anemia and thalasse-mia.

Abnormal hemoglobins

Over one hundred different types of abnormal hemoglobins have been described. Some of these are easily differentiated by their electrophoretic mobilities and have given rise to the concept of "**molecular disease**" which explains that a defective gene (mutant) may direct the formation of a molecule similar to a normal molecule but differing from it in shape, composition and electrical charge. One amino acid of the normal hemoglobin is replaced by another amino acid, i.e. acidic amino acid is replaced by a basic or a neutral amino acid for the formation of abnormal hemoglobin. The abnormal hemoglobins are named in alphabetic order as C, D, E, F, G, H, K, L, M, N, O, P, Q, S etc.

A. **Hemoglobin C** : This occurs in the blood of some Negroes in West Africa. The abnormality is found in the β chain at position 6, the amino acid glutamic acid is replaced by Lysine. It is characterized by the mild anemia with a tendency to infarction.

B. **Hemoglobin S** : This appears among the Negroes of Africa. The abnormality occurs in β chain, glutamic acid at position 6 is replaced by valine. Sickle cell anaemia develops and the RBC becomes long and boat-shaped. The blood becomes more viscous which results in reduced blood flow.

C. **Hemoglobin F** : HbF is present in fetus and is replaced by adult hemoglobin as the child grows. It is present only in traces in normal adults, it gets hemolysed rapidly producing a severe anemia called "**Thalassemia major**".

D. **Hemoglobin M** : There are two types of HbM-HbM (Boston) and HbM I Wate which are of clinical interest. The abnormality is found in the α chain, the histidine residues in 58 and 87 position are replaced by tyrosine. Abnormal amounts of methemoglobin are found in the blood of persons affected by this condition. This methemoglobin is not reduced to hemoglobin by reducing agents.

E. **Hemoglobin D** : This occurs rarely. It exists in two forms - Dα and Dβ. The persons having HbD do not show any clinical signs and symptoms.

Technique for identification of hemoglobins

Finger print technique : Ingram developed a technique by which the peptide chains in hemoglobin could be broken down into several smaller peptide fragments by digestion with trypsin. Trypsin splits the peptides only at points where only lysine and arginine occur. A mixture of smaller peptides were obtained. He then separated this mixture using paper electrophoresis technique and paper chromatography. The peptides appeared as spots when ninhydrin was sprayed. Thus peptide maps had been prepared for different hemoglobins.

ABNORMAL HEMOGLOBIN DERIVATIVES

In various disease states, abnormal amounts of certain derivatives of hemoglobin are present in the blood which are of clinical importance. They generally result from abnormal exposure to toxic chemical agents.

These are two types :

1. If the mutation affects structural gene, it results in the replacement of single amino acid by some other amino acid.
2. If the mutation affects regulator gene i.e. the amino acid sequence is unaffected but the rate of synthesis of polypeptide chain is affected.

A. Methemoglobin

1. This is an oxidized hemoglobin (ferrous iron is oxidized to ferric iron). Oxygen is in firm combination.
2. It is incapable of functioning as an oxygen carrier. Cyanosis usually develops when the concentration of it reaches 3 grams per 100 ml of blood.
3. Methemoglobin in normal erythrocytes is 0.4 per cent of the total hemoglobin. With very high concentration, the cells may be injured, destroyed and may be liberated into the plasma along with hemoglobin and excreted in the urine.

Causes of methemoglobin

1. Drugs and other toxic agents : Nitrites, chlorates etc.
2. Other hemolytic agents.
3. Enterogenous cyanosis : Due to excessive production and absorption of nitrites from the intestine.
4. Congenital : One form is characterized by the abnormality in the molecular structure of the globin component. In another type, there is the deficiency of diaphorase enzyme system which catalyzes reduction of methemoglobin to hemoglobin.

B. Sulfhemoglobin

1. Reduced hemoglobin combines with hydrogen sulphide to form sulfhemoglobin.
2. Sulfhemoglobinemia is an unusual condition which occurs chiefly as a result of the action of nitrites and coal tar preparations in the presence of excessive amounts of sulfur.
3. It is also observed in subjects with marked constipation in the presence of nitrite-producing bacteria in the intestine.
4. Cyanosis usually occurs when the concentration of sulfhemoglobin reaches 3 to 5 grams per 100 ml of blood.

C. Carboxyhemoglobin

1. It is formed as a result of the combination of carbon monoxide with the iron in the hemoglobin molecule.
2. It has a bright, cherry-red colour.
3. It is formed by the excessive exposure to artificial illuminating gas and to automobile exhaust gases in closed or poorly ventilated rooms.

Thalassemias

Thalasa, derived from a Greek word means sea, is more common in Mediterranean countries and the α-chain thalassemia is also prevalent in South East Asia.

Causes

The impairment in the synthesis of a single kind of Hb chain results in thalassemia.

α-Chain Thalassemia

The compensatory increase in the synthesis of other chains like β-chains or γ-chain occurs due to the repression of the synthesis of α-chain :

1. Since the synthesis of α-chain is inhibited there occurs the increase on the production of β-chain which forms a large intracellular pool resulting in β_4 molecules (Hb-H). The significance of Hb-H disease is (a) Moderate degree of hemolytic anaemia (b) Red cell morphological appearance of Thalassemia (c) Variable amount of Hb-H (10–20 p.c.)
2. The repression in the synthesis of α-chains results in the great excess of γ-chains forming γ_4 molecules (Hb-Barts). The pregnant ladies suffering from this deliver still-born infants with hydrops foetalis. Eighty per cent of still-born infants of South East Asia are due to Hb-Barts.
3. Hb-Portland ($\varepsilon_2\gamma_2$) is caused by the excess production of ε-chains and γ-chains. Foetuses can survive for some time by forming increased amount of embryonic Hb, but they die shortly after delivery (Hydrops foetalis).

β-chain Thalassemia (Thalassemia Major)

1. The repression in the synthesis of β-chains results in the excess synthesis of α-chains which can combine with δ-chains producing an increase in Hb-A_2 or with γ-chains producing an increase in Hb-F. This group of thalassemia is associated with early childhood severe anaemia called Cooley's anaemia.
2. Infants suffering from this disease are characterized by "Mongoloid" features and stunted growth and are victimised by severe haemolytic anaemia.
3. The individual affected by this disease shows enlarged spleen, hypochromic microcytic anaemia (Hb may be 3 to 5 gm per cent), increased osmotic fragility rise in Hb-F (5 to 80 per cent), significant increase in Hb-A_2.
4. The blood smear of the individual suffering from this disease shows hypochromasia, polychromasia, nucleated cells +, target cell ++.
5. The bones of skull of this patient on lateral view shows hair-on-end appearance by radiological examination.

PORPHYRIAS

It is the condition in which the excretion of both coproporphyrin and uroporphyrin is increased.

Porphyrias are of two types :

1. Hereditary and familial (congenital).
2. Acquired.

1. Congenital porphyrias

A. Erythropoietic porphyria.
B. Erythropoietic protoporphyria.
C. Erythropoietic coproporphyria.

D. Hepatic porphyria.
   (a) Acute intermittent porphyria.
   (b) Porphyria variegata.
   (c) Hereditary coproporphyria.
   (d) Porphyria cutanea tarda.

A. Erythropoietic porphyria

1. Highly increased excretion of uroporphyrin I and, to a lesser extent, coproporphyrin I in both urine and feces. Both these porphyrins are present in higher concentration in the circulating erythrocytes.
2. Photosensitivity and severe skin lesions from infancy leading to severe deformities of the fingers, nose, ears and eyes.
3. Excess hair on the face and limbs.
4. Teeth and bones may be brownish or pink due to porphyrin deposition.
5. Tendency to hemolysis and defective erythropoiesis.
6. The disease is rare.

B. Erythropoietic protoporphyria

1. Increased protoporphyrin and uroporphyrin in the circulating erythrocytes, the plasma and the feces.
2. Photosensitivity is the major sign.
3. Chronic skin lesions may develop.
4. The condition is much more commonly seen.

C. Erythropoietic coproporphyria

1. Large amounts of coproporphyrin III containing increased amounts of protoporphyrins and uroporphyrins are present in the erythrocytes.
2. Swelling and itching of skin.
3. The condition is very rare.

D. Hepatic porphyria

(a) **Acute intermittent porphyria :**

(i) Excess of ALA and porphobilinogen in the urine.

(ii) Increase in serum protein-bound iodine (PBI), some degree of hypercholesterolemia and a diabetic type of glucose tolerance curve. The condition is due to a marked increase of hepatic ALA synthase.

(iii) Periodic attacks of abdominal pain which is associated with fever and leukocytosis.

(iv) There may be constipation and severe neurological and psychiatric disturbances.

(v) Freshly passed urine is often normal in colour but, on standing in sunlight, turns to red wine or black.

(vi) The disease is extremely rare before the age of 15 years and after the age of 60. It is slightly more common in females.

(b) **Porphyria variegata :**

(i) Increased excretion of porphobilinogen in the urine during acute neurological manifestations and turns to normal in remission.

(ii) Elevated urinary copro-and uroporphyrins and also fecal protoporphyrin.

(iii) Frequent uremia and electrolyte disturbances.

(iv) Increased hepatic ALA synthase.

(v) Photosensitivity and cutaneous lesions.

(c) **Hereditary coproporphyria :**

(i) Increased urinary output of porphobilinogen and ALA during acute attacks.

(ii) Marked increased excretion of coproporphyrin III, mainly in the feces, this does not always occur in the urine.

(iii) No increase in erythrocyte porphyrins.

(d) **Porphyria cutanea tarda :**

(i) Increased urinary uro- and coproporphyrins. This also occurs in feces which may also contain protoporphyrin.

(ii) Increased hepatic ALA synthase.

(iii) Frequent rise in serum iron.

(iv) Skin lesions in response to minor trauma.

(v) No pink fluorescence in teeth by ultraviolet light.

(vi) Abdominal pain with fever.

2. Acquired porphyrias

This disease is caused by

1. Severe liver diseases.
2. Ingestion of certain toxins.

Alcoholic cirrhosis often leads to the development of cutaneous hepatic porphyria. This form is common in Southern African countries. In Turkey, in 1955, the consumption of wheat contaminated with hexachlorobenzene caused an outbreak of this type of porphyria.

Important biochemical defects

1. Increased excretion of uroporphyrin in urine.
2. Fecal porphyrin level is normal.
3. ALA and PBG are not found in urine.

Effects of chemical substances on induction of porphyria

1. **Drugs :** barbiturates, sulfonamides, chloroquine and sex hormones.
2. **Chemical agents :** Hexachlorobenzene, Di- and tri-chlorophenols, phenobarbital, sulfomethane, chloropropamide, and alcohols; orally administered contraceptives.

Sickle cell anemia

1. In this case hemoglobin δ appears due to abnormality of β-chain where glutamic acid is replaced by valine at position 6.
2. The loss of glutamic acid with its two carboxyl groups alters the distribution of +ve and –ve charges on the protein surface and changes its solubility. As a result of this insolubility of Hbs, there occurs viscosity which results in reduced blood flow.
3. **Biochemical explanation of sickling :**
   (a) Though glutamic acid (polar) is replaced by valine (non polar) generates the formation of sticky patch on the outside of β-chain.
   (b) The sticky patch is present on the oxygenated Hbs (R form) and deoxygenated Hbs (T form) but never present on oxy-HbA (R form).
   (c) On the surface of deoxygenated Hbs (T form) there is complementary site which is not present in oxygenated Hbs (R form).
   (d) When HbS is deoxygenated, the sticky patch of HbS can bind with complementary site on the other deoxy HbS. This binding causes the polymerization of deoxy HbS forming a long fibrous precipitate and forming sickle cell RBC.

Effects of sickle cell erythrocyte

1. Sickle cells are more fragile and causes hemolytic anemia and jaundice.
2. Packing of these abnormal red cells into vessels, specially capillaries of organs cause vascular stasis and anoxic damage to the tissue.

Relation with other diseases

1. Persons suffering from sickle cell anemia show an increased resistance to malaria.
2. Salmonella infection is found more in this disease.

Myoglobinuria

1. Myoglobin being released from the ruptured muscle fibers appears in the urine

colouring it dark red after massive crush injury.

2. Myoglobin can be detected in plasma after myocardial infarct and assay of serum enzymes also provides a more sensitive index of myocardial injury.

Hemoglobinopathies

1. Mutation of certain residues of hemoglobin, eg, histidines E7 or F8, has serious consequences and mutation of many surface residues far removed from the heme binding site shows no clinical abnormalities.
2. Sickle cell anemia, the best exception, in which all signs and symptoms arise from mutation of a single polar residue to a nonpolar residue. Sickle cell disease being chracterized by recurring pain and early death arose independently in some locations where malaria was prevalent and is widely distributed among regions. A few patients have been cured by stem cell transplantation.
3. Drugs, cytokines, transfusion, and dialysis in neonates have improved life expectancy although adults show chronic organ dysfunction.

Glycosylated hemoglobin (HbA_{1C})

1. Hemoglobin is nonenzymatically glycosylated when blood glucose enters the erythrocytes.
2. HbA_{1C} may be separated from HbA by ion exchange chromatography or electrophoresis.
3. Normally about 5 per cent of hemoglobin glycosylated is proportionate to blood glucose concentration. The measurement of HbA_{1C} thus provides information useful for the management of diabetes mellitus.
4. The raised HbA_{1C} can guide the physician to select appropriate treatment (eg, increased insulin dosage).

CATABOLISM OF HEMOGLOBIN

The catabolism of hemoglobin can be represented under two main heads :

1. Conversion of hemoglobin to bile pigments.
2. Metabolism of bile pigments.

1. Conversion of hemoglobin to bile pigments

1. About 8 grams of hemoglobin is broken down each day in the normal adult and this amount contains 27 mg of iron. The protoporphyrin yields about 300 mg of bilirubin.
2. When hemoglobin is catabolized in the body, the protein portion globin is reutilized either as such or in the form of its constituent amino acids, and the iron enters ferritin "Pool" for reuse.
3. The porphyrin portion is broken down mainly in the reticuloendothelial cells of the liver, spleen and bone marrow.
4. First, it involves the opening of the porphyrin ring between pyrrole residue I and II and elimination of the alpha methylene carbon as carbon monoxide (CO). The iron is still present.
5. Such a green conjugated protein formed by the oxidation of hemoglobin by oxygen in presence of ascorbic acid is said to be choleglobin.

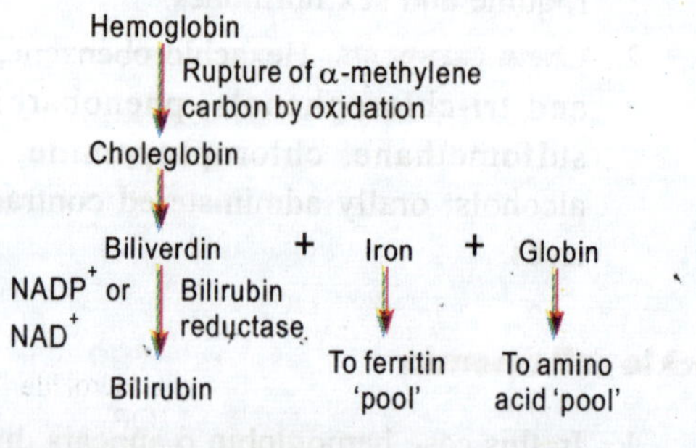

Fig. 9.5 : Conversion of hemoglobin to bile pigments

6. Then biliverdin is converted to bilirubin by bilirubin reductase in the presence of NAD^+ or NAD^+.

Normally, there are only slight traces of biliverdin in human bile but the colour is very intense. Watson (1969) states that biliverdin jaundice is largely limited to (i) biliary obstruction as a result of carcinoma. (ii) Severe parenchymal liver disease as in advanced cirrhosis of the liver.

The amounts of biliverdin in the serum are very small in obstructive jaundice caused by the presence of a stone in the common bile duct. No biliverdin occurs in the serum in hemolytic jaundice.

2. Metabolism of bile pigments

1. The normal concentration of bilirubin in human plasma is 0.3-1.1 mg/dl. One gram of hemoglobin yields 35 mg of bilirubin which is carried in loose association with albumin of the plasma by systemic circulation.
2. Then it is carried to the liver. UDP-glucose is converted to UDP-glucuronic acid by UDP-glucose dehydrogenase.
3. In liver cell, UDP glucuronic acid conjugates with bilirubin by the enzyme UDP-glucuronyl transferase. As a result, there is formation of bilirubin monoglucuronide and UDP.
4. Bilirubin monoglucuronide reacts with UDP-glucuronic acid to form bilirubin diglucuronide and UDP by the help of the above-mentioned enzyme.
5. Mono and diglucuronides of bilirubin are called conjugated bilirubins. Conjugated bilirubin can pass through glomerular filter and appear in urine (bilirubinuria). But unconjugated bilirubin cannot pass through glomerular filter.

   Glucuronyl transferase activity is inhibited by certain drugs eg,—novobiocin, pregnancy etc.
6. In the lower portions of the intestinal tract, the cecum and the colon, the bilirubin is released from the conjugate and is then reduced by enzyme systems present in the intestinal tract, mainly derived from anaerobic bacteria in the cecum.
7. Progressive hydrogenation takes place to produce a series of intermediary compounds, beginning with mesobilirubinogen which is oxidized with the loss of hydrogen to coloured compounds. The end product is colourless L-stercobilinogen (L-urobilinogen). Auto-oxidation in the presence of air produces stercobilin (L-urobilin), an orange-yellow pigment which gives the normal colour of the feces. Stercobilin is strongly levorotatory.

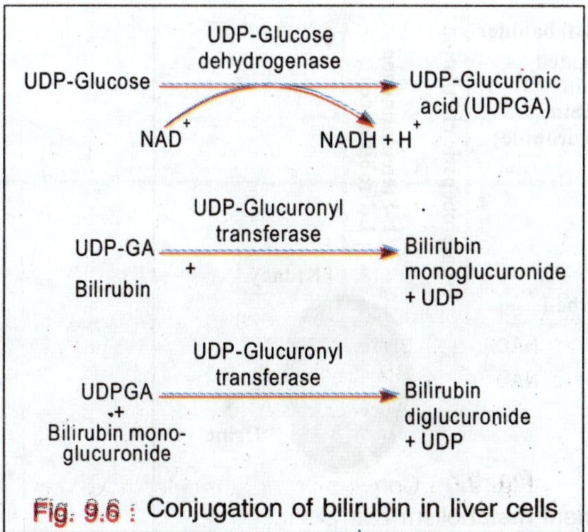

Fig. 9.6 : Conjugation of bilirubin in liver cells

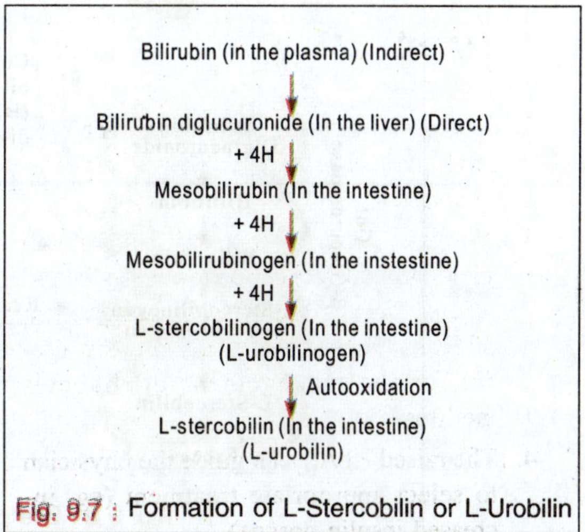

Fig. 9.7 : Formation of L-Stercobilin or L-Urobilin

8. A portion of the urobilinogen is reabsorbed from the intestine into the blood and enters into gall bladder by enterohepatic circulation.
9. Some of this is excreted by the kidney and appears in the urine in the concentration of 0-2.4 mg/day. The unabsorbed urobilinogen is excreted in the stool (40-280 mg/day). On exposure to oxygen, urobilinogen is oxidized to urobilin which gives the darkening of stools.

JAUNDICE

Definition

It is defined as yellowish discolouration of the skin and mucous membrane due to excess amount of bilirubin present in blood.

Latent Jaundice

Normal serum bilirubin level is 0.3 – 1.1 mg/dl.

Clinical jaundice is evident when serum bilirubin crosses 2.5 mg/dl.

Jaundice is latent—i.e. clinically non evident (only detected by serum analysis)—when serum bilirubin level is in between 1 – 2.5 mg/dl. It is found in mitral stenosis, myocardial infarction, cirrhosis of liver, pernicious anemia, congestive cardiac failure, etc.

Causes

1. Production of more bile pigment than the normal liver can excrete or the failure of a damaged liver to excrete the bilirubin produced in normal amounts.

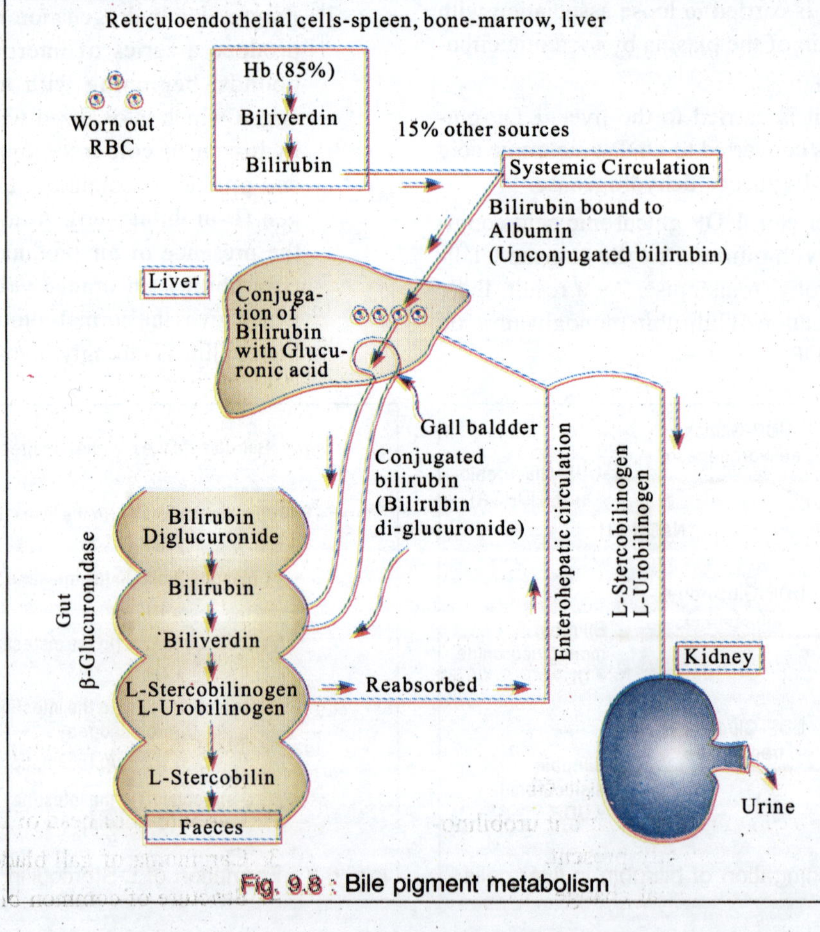

Fig. 9.8 : Bile pigment metabolism

2. Obstruction of the excretory ducts of the liver by preventing the excretion of bilirubin.

Type

A. Hemolytic Jaundice

If the concentration of bilirubin in the serum rises above normal due to more formation as a result of increased erythrocyte destruction, hemolytic jaundice results. There is lemon-yellow tinge of bulbar conjunctiva.

Examples of Haemolytic jaundice

1. Thalassaemia
2. Mismatched blood transfusion
3. Snake bite
4. Malaria
5. Rh incompatibility
6. Primaquine or sulphonamide induced (in G_6PD deficiency).

Findings in hemolytic jaundice

1. Blood :
   (a) Hemoglobin level is decreased.
   (b) Reticulocyte count is increased.
   (c) Polychromatic macrocytes and fragmented red cells are present.
   (d) Red cells are more fragile by osmotic fragility test.
   (e) Prothrombin level is not prolonged.
   (f) Coombs' test is positive.
   (g) Red colour does not appear unless methyl alcohol is added to the urine (bilirubin + diazo reagent) by van den Bergh test which indicates the presence of unconjugated bilirubin in serum.
2. Urine :
   (a) Prowine colour is produced.
   (b) Bile pigment is absent but urobilinogen is abundantly present.
   (c) No microscopical change.
   (d) Acholuric urine, i.e. freshly passed urine is of normal colour as there is no bilirubin present in urine but if the urine sample is kept for some time, it turns dark yellow due to conversion of urobilinogen to urobilin (by odixation).
3. Stool :
   (a) The colour of stool becomes dark brown.
   (b) Stercobilinogen or stercobilin is richly present.

B. Obstructive (Regurgitation) Jaundice

This condition occurs from blockage of the hepatic or common bile ducts. The bile pigment passes from blood to the liver cells as usual. However, failing to be excreted by the bile capillaries, it is absorbed into the hepatic veins and lymphatics.

The usual causes of obstruction are :

1. Blocking of the bile passage by gallstones.
2. Obstruction caused by enlarged glands, tumour of the head of the pancreas.
3. Narrowing of the bile duct as a result of surgery.

In obstructive jaundice, the bulbar conjunctiva turns greenish yellow due to oxidation of bilirubin to biliverdin (green).

Examples of obstructive jaundice

(A) Intrahepatic (medical)
   1. Viral hepatitis.
   2. Chronic active hepatitis
   3. Cirrhosis of liver
   4. Drugs.

(B) Extrahepatic (surgical)
   1. Gallstone impaction in common bile duct
   2. Carcinoma of head of pancreas
   3. Carcinoma of gall bladder
   4. Stricture of common bile duct.

Important findings in obstructive jaundice

1. **Blood**
   (a) Target cells are present.
   (b) Prothrombin time is prolonged but this becomes normal after injection of vitamin K.
   (c) Red colour is developed by van den Bergh test.
   (d) Serum alkaline phosphatase level is increased.
2. **Urine :**
   (a) Mustard oil colour is formed.
   (b) Bile pigments are abundantly found but urobilinogen is absent.
   (c) No microscopical change.
3. **Stool :** Clay-coloured stools due to the absence of stercobilinogen.

Other findings of obstructive jaundice

1. Generalised pruritus with scratch mark and shiny nails (bile acids irritate the nerve endings).
2. Sinus bradycardia — bile salts inhibit SA node
3. Gall bladder may be palpable.
4. Prolonged obstructive jaundice may produce osteomalacia, bone pain, bone fracture, night blindness due to deficiency of Vit. A or Vit. D deficiency, respectively (due to malobsorption or steatorrhoea).

   These types of bony changes in this jaundice is known as **hepatic osteodystrophy**.

C. Hepato Cellular Jaundice

This type of jaundice is caused by liver dysfunction as a result of the damage to the parenchymal cells by infection, toxins and liver poisons.

At a certain stage, the inflammation and damage to liver cells become severe leading to partial obstruction to the flow of bile. This results in the absorption of conjugated bilirubin and bile into the general circulation. Orange-yellow tinge of bulbar conjunctiva.

Examples of Hepato cellular jaundice

1. Viral hepatitis (Type A, B, C commonly)
2. Drugs — Rifampicin
3. Poison — $CuSO_4$
4. Pregnancy — Acute fatty liver of pregnancy
5. Alcoholic hepatitis
6. Weil's disease.

Important findings

1. **Blood :**
   (a) Increased levels of both conjugated and unconjugated bilirubin in serum.
   (b) Prothrombin time is prolonged but remains high after administration of vitamin K.
   (c) Increased SGPT activity.
   (d) Highly positive flocculation tests.
2. **Urine :**
   (a) Highly coloured urine due to the presence of conjugated bilirubin and urobilinogen.
   (b) Leptospira may be present.

	Urine	Fecal
Normal :	Urobilinogen 0-2.4 mg/24 hours. Bilirubin absent.	Urobilinogen 40-280 mg/24 hours.
Hepatic :	Urobilinogen increased. Bilirubin present.	Urobilinogen decreased.
Hemolytic :	Urobilinogen increased. Bilirubin absent.	Urobilinogen increased.
Obstructive :	Intermittent stone, Urobilinogen fluctuates. Complete block, Urobilinogen absent.	Urobilinogen trace to absent.

On the basis of the type of plasma bilirubin i.e. unconjugated bilirubin or conjugated bilirubin, the hyperbilirubinemia may be classified as retention hyperbilirubinemia or regurgitation hyperbilirubinemia.

Bile salts and Bile acids

(A) Bile salts — Sodium taurocholate and Sodium glycocholate,

(B) Bile acids —

1. Primary → cholic acid and chenodeoxycholic acid
2. Secondary → Deoxycholic acid and lithocholic acid.
3. Tertiary → Ursodeoxycholic acid.

Body secretions where bilirubin is found (in jaundice)

1. Found in body fluids like CSF, joint fluid, cyst.
2. Absent from true secretions like tears, saliva, pancreatic juice.

Unilateral jaundice

1. In patients with hemiplegia
2. In presence of unilateral oedema.

Differences between conjugated and uncojugated bilirubin		
	Unconjugated	Conjugated
1. Water solubility	0	+
2. Affinity for lipids	+	0
3. Renal excretion	0	+
4. van den Bergh reaction	Indirect	Direct

Sites to be examined in Jaundice patient

The patient should be observed in front of an open window :

1. Upper bulbar conjunctiva (sclera is examined by retracting the upper eyelids upwards and asking the patient to look downwards — both eyes at a time).
2. Undersurface of tongue.
3. Soft palate.
4. Palms and soles.
5. General skin surface.

van den Bergh test — This test helps in differentiation of hyperbilirubinaemia due to accumulation of conjugated bilirubin from unconjugated bilirubin.

Principle of the test — Conjugated bilirubin or soluble bilirubin + Ehrlich's diazo reagent → gets diazotised and develops azobilirubin which is coloured red.

Conjugated bilirubin develops red colouration immediately and it is water-soluble.

Unconjugated bilirubin being water-insoluble develops the colour only after addition of methyl alcohol to make it soluble.

Method — 1 ml clear serum (not hemolysed) is taken in a test tube. 0.5 ml Ehrlich's working reagent is added by side of the test tube.

Development of red coloured ring at the junction is indicative of direct positive reaction.

If no colour change takes place then 3 ml methyl alcohol is added and well-shaken. Development of red colour indicates indirect positive reaction.

If red colour develops immediately and it depends on addition of methyl alcohol the reaction is biphasic.

Interpretation of the result of van den Bergh test

- Normal serum — Negative.

Obstructive jaundice — Direct positive.

Hemolytic jaundice — Indirect positive.

Hepatocellular jaundice — Biphasic positive.

A. Crigler-Najjar Syndrome, Type I; Congenital Nonhemolytic jaundice

1. It is a rare recessive disorder of humans owing to a primary metabolic defect in the conjugation of bilirubin which is involved in the inherited absence of bi-

lirubin UDP–Glucuronyl transferase activity in hepatic tissues.

2. The disease is usually alarming within the first fifteen months of life. But a few teenagers did not develop difficulties until puberty. These children have been treated with phototherapy with some reduction in plasma bilirubin levels.
3. Phenobarbital and other drugs have no effect on the formation of bilirubin glucuronides in patients with this syndrome.
4. Serum bilirubin usually exceeds 20 mg/dL on untreatment.

B. Crigler-Najjar Syndrome, Type II

1. This rare inherited disorder is caused by a milder defect in the bilirubin conjugated system.
2. The serum bilirubin level does not exceed 20 mg/dL. All of the bilirubin accumulated is of unconjugated type. The bile in these patients contains bilirubin monoglucuronide.
3. Patients with this syndrome respond to large doses of Phenobarbital.
4. This syndrome is the homozygous state of the defect present in heterozygous form in the mild chronic hyperbilirubinemia of Gilbert.

C. Gibert's Disease

1. It is a heterozygous group of disease many of which are due to a compensated hemolysis associated with unconjugated hyperbilirubinemia.
2. There is defect in the hepatic clearance of bilirubin due to a defect in the uptake of bilirubin by the liver parenchymal cells.
3. The bilirubin UDP-Glucuronyl transferase activities in the liver are found to be reduced.

D. Toxic hyperbilirubinemia

1. Unconjugated hyperbilirubinemia results from liver dysfunction caused by chloroform, carbon tetrachloride, hepatitis virus, and cirrhosis.
2. There is a component of obstruction of the biliary tree within the liver which results in the presence of some conjugated hyperbilirubinemia.

Conjugated Bilirubin passes into urine

Since conjugated bilirubin is water-soluble it is detectable in the urine of most patients suffering from conjugated hyperbilirubinemia. They are frequently said to have choluric jaundice. The commonest causes of this condition are :

1. Chronic idiopathic jaundice (Dubin-Johnson syndrome)

(a) This syndrome happens in childhood or during adult life.

(b) This is caused by the defect in the hepatic secretion of conjugated bilirubin into the bile.

(c) The hepatocytes in the centrilobular area contain an abnormal pigment in patients with this syndrome.

2. Biliary Tree obstruction

(a) The blockage of the hepatic or common bile duct produces conjugated hyperbilirubinemia.

(b) The bile pigments pass from the blood into the liver cells but fails to be excreted. As a result, the conjugated bilirubin is absorbed into the hepatic veins and lymphatics.

(c) The term cholestatic jaundice is used to include all forms of extrahepatic obstructive jaundice that are characterized by conjugated hyperbilirubinemia.

CLINICAL ORIENTATION

- Hemoglobin provides highlights into the molecular basis of genetic diseases such as Sickle cell disease and Thalassemia.
- Cyanide and Carbon monoxide disrupt the physiologic functions of the heme, proteins, cytochrome oxidase, and hemoglobin, respectively.
- Stabilization of the quarternary structure of deoxyhemoglobin by 2, 3-biphosphoglycerate (BPG) is central to understanding the mechanism of high altitude sickness and of adaptation to high altitude.

CLINICAL ORIENTATION

- After massive crush injury, myoglobin being released from ruptured muscle fibres appears in the urine colouring it dark red.
- Common anemias result from impaired synthesis of hemoglobin.

DIAGRAMMATIC COMPENDIUM

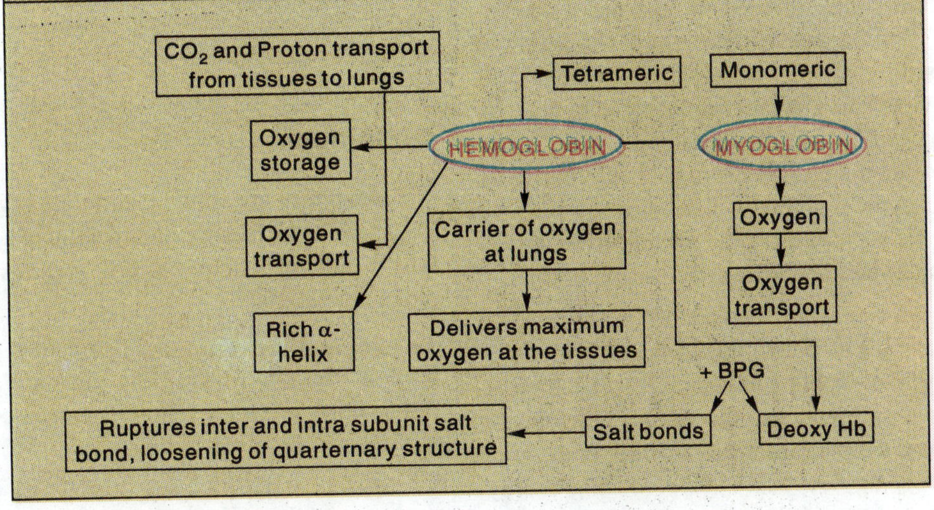

CHAPTER 10

ENZYMES AND COENZYMES

Introduction

Enzymes were first divided into two classes "unorganised" and "organised ferments". The action of "organised ferments" was thought to be due to "vital" activity. This view was encouraged by the "rediscovery" of bacteria by Pasteur and Koch in the middle of the nineteenth century. The "vital" concept of enzyme was finally discredited by Buchner in 1897. He separated active enzymes from the living cells of yeast and later of bacteria.

Enzymes promote and control the conversion of the complex carbohydrates, fats and proteins of our body into simple substances which the intestines can absorb and also the various reactions by which these simple substances are used in the body for building up new tissues or producing energy. The enzymes are not broken down or changed in the process. They are as potent at the end of the reaction as at the beginning and very small amounts can effect the conversion of large quantity of material. They are really the true catalysts.

Hydrolysis of a protein takes at least a day by the action of strong acid at 100°C. The same change takes place in a few hours in the alimentary canal at 37°C. The acid in the laboratory can cause the hydrolysis of various substances but each enzyme attacks a special class of substances or even a single substance.

Many enzymes have been purified and found to be active. Willstatter and Pollinger purified peroxidase and found it to be the most active. Even the enzymes are active, although they are not purified. The total number of molecules of a particular enzyme in a cell is very small although the number of different enzymes in a cell is very large.

Definition

Enzymes are soluble, colloidal organic catalysts formed by living cells, specific in action, protein in nature, inactive at 0°C and destroyed by moist heat at 100°C.

Intracellular Enzymes : Enzymes which are used in the cells which make them are said to be intracellular enzymes. These enzymes correspond to the old "organised ferments".

Extracellular Enzymes : Enzymes which are produced by other cells and are secreted to other parts of the body (eg, digestive juice) are called extracellular enzymes. These enzymes correspond to the old "unorganised ferments".

Zymase secretion : An extracellular enzyme which is secreted ready for action is called a zymase secretion. Example : Amylase of saliva.

Zymogen secretion : An enzyme which is secreted in inactive form and ultimately activated by an agent secreted by other cell is said to be zymogen secretion. Example : Trypsinogen (in pancreatic juice) activated by enterokinase (in intestinal mucosa) to give active trypsin, prothrombin (in blood) activated by thromboplastin (in tissues) to give active thrombin.

Zymogen secretion is probably a protective mechanism to prevent digestion of cell walls and ducts, since it is most frequently found with protein-splitting enzymes.

Substrate : The substance on which an enzyme acts is called the substrate.

Example : Maltose is the substrate on which the enzyme maltase acts to form glucose.

Names

Except the enzymes ptyalin, pepsin, trypsin and erepsin, enzymes are usually named by adding the suffix -ase to the main part of the name of the substrate on which they act.

Examples

Maltase acts on maltose.
Lactase acts on lactose.
Lipases act on lipids.
Carbohydrases act on carbohydrates.
Proteases act on proteins.
Amylases act on starch (Amylum).

But there are many substances which are acted on by some enzymes in different ways. A dipeptide can be attacked by three enzymes. These enzymes are named by their function.

Example

A dipeptide can be hydrolysed by dipeptidase into amino acids. The free amino group of the amino acid is removed by another enzyme and the free carboxyl group is also removed by another enzyme. So the names of these three enzymes acting on a dipeptide are a dipeptidase, a deaminase and a decarboxylase.

Other enzymes are named by their functions only. Examples : transferases, dehydrogenases, hydrolases, oxidases and reductases.

Many enzymes with the same function act on one substance; it is, therefore, better to specify the enzyme by its source. Example : Pancreatic amylase, bone phosphatase, liver esterase.

Some enzymes acting on the substrates are freely described by the adjectives. Example : Amylolytic, lipolytic, proteolytic.

Catalytic RNAs

1. Certain ribonucleic acids (RNAs) show highly substrate-specific catalytic activity.
2. These RNAs satisfy the classic criteria for definition as enzymes and are termed **ribozymes**.
3. Ribozymes catalyse trans-esterification and finally hydrolysis of phosphodiester bonds in RNA molecules. These reactions are enhanced by free OH gorups.
4. Ribozymes are involved in the main roles in the **intron splicing events** essential for the conversion of Pre-mRNA in mature mRNAs.

Classification

The 6 major classes of enzymes with some examples are :

1. **Oxidoreductases** : Enzymes catalyzing oxidoreductions between two substrates A and B :

$$A_{reduced} + B_{oxidized} = A_{oxidized} + B_{reduced}.$$

These enzymes can be grouped in many different ways. Three main groups can be explained in order to get the simpler expression :

Oxidases : The enzymes which use oxygen as hydrogen acceptor. Examples : tyrosinase, cytochrome oxidase, uricase.

Anaerobic dehydrogenases : The enzymes which use some other substance as hydrogen acceptor. Examples : Malate dehydrogenase, succinate dehydrogenase, lactate dehydrogenase.

Hydroperoxidases : The enzymes which use hydrogen peroxide as substrate. Examples : Peroxidase, catalase.

Aerobic dehydrogenases : The enzymes which use either oxygen or another substance as hydrogen acceptor. Examples : D- and L-amino acid oxidases, xanthine oxidase, aldehyde oxidase.

Two other groups are :

Oxygenases : The enzymes which act on single hydrogen donors with incorporation of oxygen.

Example : Tryptophan oxygenase.

Hydroxylases : The enzymes which act on paired donors with incorporation of oxygen into one donor. Examples : Steroid hydroxylases, phenylalanine 4-hydroxylase.

2. **Transferases (Transferring enzymes)** : They catalyse the transfer of some group or radical, R, from one molecule A, to another molecule, B :

$$A.R + B \rightleftharpoons A + B.R$$

Fig. 10.1

The group includes

(a) *Transphosphorylases* : Examples : Hexokinase, phosphoglucomutases, phosphoglycerate kinase :

$$ATP + \text{D-hexose} \xrightarrow{\text{Hexokinase}} ADP + \text{D-hexose-6-phosphate}$$

(b) *Transglycosidases* : Examples : Phosphorylase.

(c) *Transaminases* : Examples : Glutamate-pyruvate transaminase, aspartate amino transferase.

(d) *Transacylase* : Examples : Choline acetyl transferase, acetoacetate transacetylase, amino acid transacetylase :

$$\text{Acetyl-CoA} + \text{Choline} \xrightarrow{\text{Choline acetyl transferase}} \text{CoA} + \text{Acetylcholine}$$

Fig. 10.2

(e) *Transmethylase*

3. **Hydrolases** : Enzymes catalyzing hydrolysis of ester, ether, peptide, glycosyl, acid anhydride by the addition of water.

The group includes the extracellular digestive enzymes and many intracellular enzymes.

(a) Enzymes acting on glycosyl compounds, eg, β-galactosidase. β-D-galactoside + H_2O = An alcohol + D-galactose.

(b) Enzymes acting on peptide bonds, eg, pepsin, rennin, chymotrypsin.

(c) Esterases, eg, lipases, phosphatases, sulphatases.

(d) Amidases, eg, urease, arginase, glutaminase.

(e) Hydrolytic deaminases, eg, guanine deaminase.

Monomeric eg, Ribonuclease, trypsin carboxypeptidase.

Oligomeric eg, Acetyl CoA carboxylase, lactate dehydrogenase.

4. **Lyases** : Enzymes that catalyze removal of groups from substrates by mechanisms other than hydrolysis leaving double bonds :

$$\underset{\text{C}}{\overset{\text{X}}{|}}\!-\!\underset{\text{C}}{\overset{\text{Y}}{|}} = X - Y + C = C$$

Fig. 10.3

(a) Aldehyde-lyases, eg, aldolase. Ketose-1-phosphate = dihydroxy-acetone phosphate + an aldehyde.

(b) Carbon-oxygen lyases, eg, fumerase. L-malate = Fumerate + H_2O.

5. **Isomerases** : Enzymes catalyzing interconversion of optical, geometric, or positional isomers.

(a) Racemases and epimerases, eg, alanine racemase.

L-alanine=D-alanine.

(b) Cis-trans isomerases, eg, retinene isomerase.

All transretinene = II-cis-retinene.

(c) Enzymes catalyzing interconversion of aldoses and ketoses, eg, triose phosphate isomerase.

D-glyceraldehyde-3-phosphate = Dihydroxy-acetone phosphate.

6. **Ligases (ligare = to bind) :** Enzymes catalyzing the linking together of two compounds couple to the breaking of a pyrophosphate bond in ATP or a similar compound :

(a) Enzymes catalyzing formation of C–S bonds, eg, succinate thiokinase.

GTP+ succinate + CoA = GDP + Pi + succinyl-CoA.

(b) Enzymes catalyzing formation of C–N bonds, eg, glutamine synthetase.

ATP + L-glutamate + NH_4^+ = ADP + orthophosphate + L-glutamine.

(c) Enzymes catalyzing formation of C-C bonds, eg, Acetyl-CoA carboxylase.

ATP + acetyl-CoA + CO_2 = ADP + Pi + malonyl -CoA.

PROPERTIES OF ENZYMES

1. Enzymes are proteins. So the amino, carboxyl and sulfhydryl groups of the side chains of amino acids are available for linkage between polypeptide chains.
2. The above-mentioned groups, together with others such as the imidazole ring and the alcoholic group of serine, are also responsible for the union between substrate and enzyme. These groups are termed "active groups" and the region of the protein surface at which they are located is termed "active centre".
3. The enzyme-substrate complex theory assumes combination of enzymes and substrate and then liberation of enzyme and the reaction product. Thus,

   Enzyme + Substrate = Enzyme – Substrate Complex.

   Enzyme – Substrate Complex = Enzymes + Reaction Product(s).
4. In many cases, a series of reactions may be involved in addition to the above one. This may be indicated thus :

$$A \rightleftharpoons B \rightleftharpoons C \rightleftharpoons D \rightleftharpoons E$$

Fig. 10.4

5. Alkaline phosphatase hydrolyses a number of phosphate esters to produce orthophosphate and an alcohol such as Glucose-6-phosphate → Glucose + Orthophosphate.
6. From an energy standpoint enzyme reactions are divided into three classes :

   (a) Exergonic reaction means the system undergoes a loss of free energy, eg, lipase, catalase, urease.

   (b) There is little change in free energy and the reactions come to equilibrium; so that product molecules and the substrate molecules are present in constant amounts, eg, glycogen + inorganic phosphate = glucose-1-phosphate under the influence of phosphorylase.

   (c) Endergonic reaction means the supply of energy in order to proceed the reaction.
7. Many enzymes are conjugated proteins. The prosthetic groups (coenyzmes) are readily detached. These enzymes have special role in metaboic processes. The enzymes catalase and peroxidase have cofactors but no coenzymes.

ENZYME SPECIFICITY

Enzymes, the organic catalysts, differ from inorganic catalysts in their extraordinary specificity.

A. Reaction specificity

1. An enzyme catalyzes only a very few reactions (frequently only one).

Example : Arginase, catalase, urease only attack arginine, hydrogen peroxide and urea, respectively.

2. Most enzymes can catalyze the same type of reaction (phosphate transfer, oxidation-reduction) with several structurally related substrates.
3. Reactions with alternate substrates take place if they are present in high concentration and also dependent on their affinity to enzyme.

   Example : Histidase acts on histidine and this enzyme acts on tryptophan when it is present in high concentration and it is also dependent on due affinity of that enzyme.
4. A substrate can undergo many reactions but in a reaction specificity one enzyme can catalyse only one of the various reactions.

   Example : In the presence of Acetyl-CoA the enzyme citrate synthetase acts on or catalyse oxaloacetate to citrate. But when only oxaloacetate is present then another specific enzyme i.e. malate dehydrogenase will catalyze it and forms different products i.e malate :

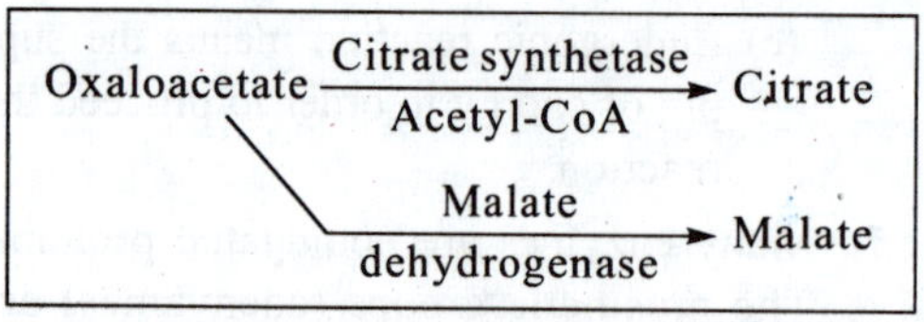

B. Optical specificity

1. Enzymes show absolute optical specificity for at least a portion of substrate molecule.

   Example : Maltase catalyzes the hydrolysis of α-glycosides but not β-glycosides, while Embden-Meyerhof and direct oxidative pathway catalyze the interconversion of D-phosphosugar but not L-phosphosugars. Exceptions, D-amino acid oxidase in kidney acts on L-amino acid.
2. The glycosidases which catalyze the hydrolysis of glycosidic bonds between sugars and alcohols are highly specific for the sugar portion but relatively non-specific for the alcohol portion or aglycone.
3. Many substrates form 3 bonds enzymes. This 3-point attachment can thus confer asymmetry on the symmetric molecule.
4. When enzyme acts on a substrate and produces one kind of isomer which will be the product.

   Example : Succinate dehydrogenase while acts on succinic acid will produce only fumaric acid and not malic acid which is an isomer of it :

$$\text{Succinic acid} \xrightarrow{\text{S.D.}} \text{Fumaric acid}$$

C. Bond specificity

1. A particular enzyme acts on particular chemical bond.

   Example : Glycosidases on glycosides, alcohol dehydrogenase on alcohol, pepsin and trypsin on peptide bond, esterases on ester linkages.

D. Group specificity

1. Particular enzyme acts on particular group.

   Example : Chymotrypsin hydrolyzes peptide bond in which the carboxyl group is contributed by the aromatic amino acids such as tyrosin, tryptophan etc. Carboxypeptidases split off carboxyl group, aminopeptidases split off amino group from the polypeptide chain.

E. Oxidoreductases function in biosynthetic processes (eg, fatty acid synthesis) tend to use NADPH as reductant but those which function in degradative processes tend to use NAD^+ as oxidant.

F. In liver, about 90% of the NADP-specific enzyme occurs extramitochondrially. But NAD-specific enzyme of mitochondria is activated by ADP.

Fe	containing	→	Cytochrome oxidase.
Cu	"	→	Cerruloplasmin, uricase.
K	"	→	Pyruvate kinase.
Mn	"	→	DNA-polymerase, Pyrrolase, Superoxide dismutase (Mitochondria)
Ni	"	→	Urease
Se	"	→	Glutathione peroxidase
Zn	"	→	Carbonic anhydrase.
Mo	"	→	Xanthine oxidase.

Preparation and Isolation of Enzymes

1. Tissues are extracted with saline or glycerol. Such extracts do not keep for any length of time.
2. These extracts are carefully dried at low temperature.
3. The dried substance is then grinded to a fine powder which keeps well.
4. Then these powders are extracted with suitable solvents.
5. More active preparations are made by various processes of purification involving dialysis to remove inorganic impurities, adsorption in suitable material or precipitation with suitable reagents. These are done at low temperature.
6. In this way much inactive materials can be removed and the enzymes are only present in tissues in very small amounts. They have the high potency. Over 100 enzymes have been obtained as crystalline proteins which are regarded as the pure enzymes.

Recognition of Enzymes

If by adding a neutral solution to some starch at 37°C we obtain sugar, we can conclude that a catalyst is present.

If the neutral solution after boiling fails to produce sugar under similar conditions, we may then conclude that the catalyst is an enzyme and not an inorganic catalyst.

Therefore, in the experiments on enzymes, it is necessary to arrange a control with the boiled enzyme solution along with the test.

THE CONDITIONS OF ACTION OF ENZYMES OR FACTORS INFLUENCING ACTION OF ENZYMES

1. Contact between Enzyme and Substrate

1. Enzyme and substrate must be in contact to form the enzyme-substrate complex.
2. The enzyme and substrate should be well mixed for efficient reaction to proceed.
3. Even the insoluble substrates must be made soluble by the help of hydrotropic substances whenever required for good mixing with enzymes.
4. The digestion of fats which are not in solution is greatly facilitated by fine emulsification of the fat giving a greater area of contact for the lipases.
5. Enzymes which are not soluble but are bound to membranes, their contact with substrate must be maintained by setting up suitable concentration gradient.

2. Concentration of Enzyme and Substrate

1. The rate of enzyme action is influenced by the concentration of substrate, the amount of enzyme and the time.
2. In case the concentration of substrate is low in comparison to the concentration of enzyme, not all the active groups of the enzyme molecule will be utilized for the formation of enzyme-substrate complex. The rate of reaction will be low.
3. In case the concentration of substrate increases, more and more active sites of enzyme molecule will be used for the formation of enzyme-substrate complex. The rate of reaction will increase.
4. If the substrate concentration is such that all the active groups of the enzyme are

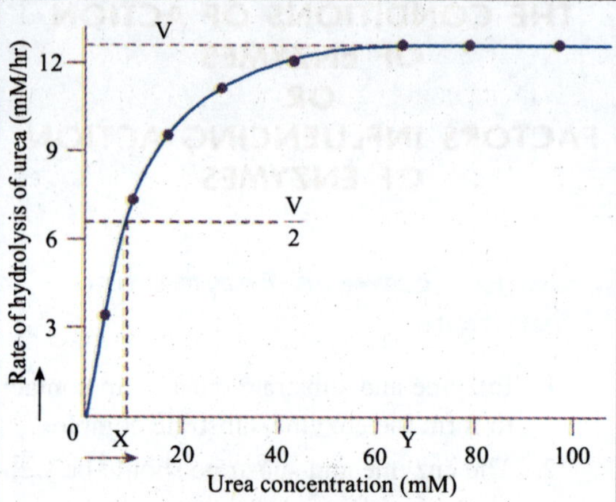

Fig. 10.5 : Michaelis curve

utilized, the rate of reaction will be proportional to the substrate concentration.

5. Above this concentration mentioned in (4), no matter how high, the rate is maximum and remains constant unless the substrate inhibits the reaction. If the substrate inhibits, the reaction rate declines.

The effect of substrate concentration on the initial rate of an enzyme reaction is shown above by a graphical representation which is known as a Michaelis curve.

It is seen that at low substrate concentration (up to Y in the Fig. 10.5) the rate increases with increasing substrate concentration. At substrate concentration above Y the rate of the reaction has a constant maximum value, V. This is called the maximum velocity of the reaction. The kinetic equation can be expressed mathematically by Michaelis-Menten equation :

$$V_i = \frac{V_{max}[S]}{K_m + [S]}$$

V_i is the initial velocity, V_{max} the maximum velocity, S the substrate concentration, K_m Michaelis constant.

The substrate concentration that produces half-maximal velocity is called Michaelis constant or K_m value :

$$V_i = \frac{V_{max}}{2} \quad \text{where} \quad K_m = [S]$$

when the enzyme concentration is constant, the rate of velocity of enzyme action increases as the concentration of substrate in a hyperbolic manner.

In the following reaction, a typical enzyme action is considered

$$E + S \underset{K_2}{\overset{K_1}{\rightleftharpoons}} ES \underset{K_4}{\overset{K_3}{\rightleftharpoons}} E + P$$

The rate of formation of ES is proportional to the concentration of uncombined enzyme i.e. [E] – [ES] and concentation of substrate [S].

$$\therefore \quad V_1 = K_1 \{[E] - [ES]\} [S] \quad \text{................ (1)}$$

Rate of dissociation :

$$V_2 = K_2 [ES] + K_3 [ES] \quad \text{........................ (2)}$$

$$\therefore \quad V_1 = V_2$$

$$\therefore \quad K_1 \{[E] - [ES]\} [S] = K_2 [ES] + K_3 [ES]$$

$$\text{or,} \quad K_1 [E][S] - K_1 [ES][S] = K_2 [ES] + K_3 [ES]$$

$$\text{or,} \quad K_2 [ES] + K_3 [ES] + K_1 [ES][S] = K_1 [E][S]$$

$$\text{or,} \quad [ES] \{(K_2 + K_3) + K_1 [S]\} = K_1 [E][S]$$

$$\text{or,} \quad [ES] = \frac{K_1[E][S]}{K_1[S] + (K_2 + K_3)}$$

At maximum velocity all the enzymes are present in combination with substrates.

$$\therefore \quad [E] = [ES]$$

$$\therefore \quad V_i = K_2 [ES]$$

$$\text{or,} \quad V_i = K_2 \frac{K_1[E][S]}{(K_2 + K_3) + K_1[S]}$$

$$\text{or,} \quad V_i = \frac{K_2[E][S]}{\frac{K_2 + K_3}{K_1} + [S]}$$

$$\because \quad V_{max} = K_2[E]$$

$$\therefore \quad V_i = \frac{V_{max}[S]}{K_m + [S]} \quad \left[\because \; K_m = \frac{K_2 + K_3}{K_1}\right]$$

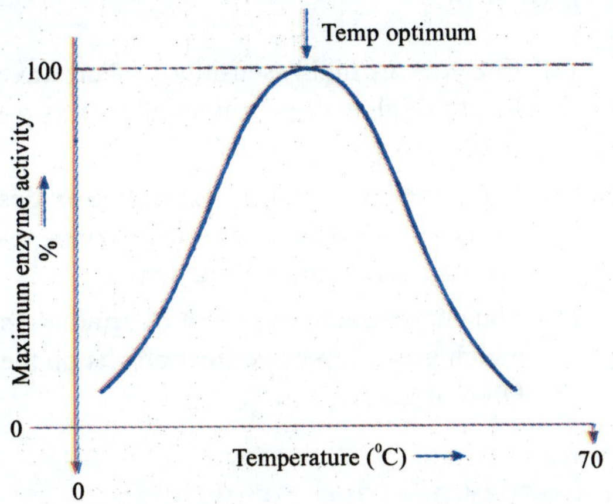

Fig. 10.6 : Effect of temperature on enzyme activity

3. Temperature

(a) The rise in temperature accelerates an enzyme reaction but at the same time causes inactivation of the enzymes due to denaturation of the protein.

(b) At a certain temperature known as the optimum temperature the activity is maximum.

(c) The optimum temperature for man is in the region of 37°C and for most animals in the region of 40°C. The optimum temperature of plant urease is 60°C. The optimum temperature for enzymes from microorganisms adapted to growth in natural hot springs is close to the boiling point of water.

(d) If the temperature is lowered, the rate of an enzyme reaction is diminished. At the temperature 0°C, most enzymes are practically inactive.

Q_{10} or temperature coefficient

1. The rate of an enzyme-catalyzed reaction increases depending over a strictly limited range of temperature owing to increased kinetic energy of the reacting molecules. The kinetic energy of the enzyme exceeds the energy barrier for breaking the weak hydrogen and hydrophobic bonds that maintain its secondary-tertiary structure.
2. The factor responsible for the increase of the rate of a biologic process for a 10°C temperature rise is the Q_{10} or temperature coefficient.
3. The rate of contraction of an excised heart approximately doubles with a 10°C rise in temperature ($Q_{10} = 2$).
4. The change in the rates of many enzyme-catalyzed reactions being accompanied by a rise or fall in body temperature constitutes an essential survival feature for life forms such as lizards which do not maintain a constant body temperature.
5. Homeothermic organisms such as humans can tolerate only a strictly limited change in body temperature and the alterations in reaction rate owing to a change in temperature are of little physiologic significance but are significant in fever or hypothermia. The enzymes in case of homeothermic organisms are able to lower the energy barriers for reactions and to elevate local reactant concentrations.

4. Hydrogen ion concentration or pH

(a) A small change in pH may inhibit enzyme activity.

(b) Pepsin works only in acid medium and is inactivated by making the medium alkaline.

(c) Trypsin rather works in alkaline solution and cannot digest protein in acid solution.

(d) The maximum activity of the enzyme is at the optimum pH. This value is generally between 5.0 and 9.0.

(e) (i) At the optimum pH, the enzyme (E^-) will react with substrate (S^+) as

$$E^- + S^+ \longrightarrow ES$$

(ii) At low pH values E will be protonated and lose its negative charge :

$$E^- + H^+ \longrightarrow E$$

(iii) At the very high pH values, S^+ will lose its positive charge :

$$S^+ \longrightarrow S + H^+$$

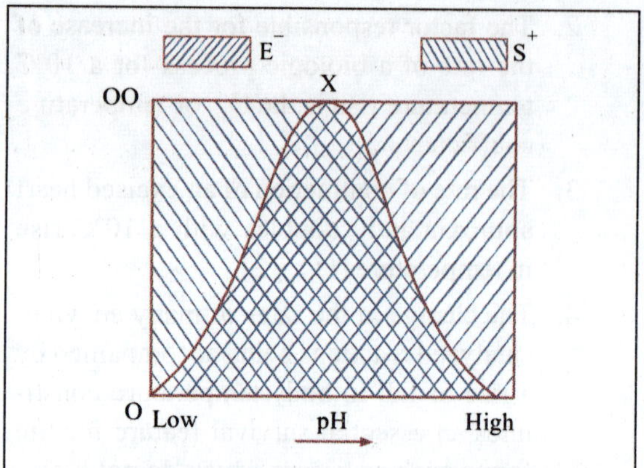

Fig. 10.7 : Effect of pH on enzyme activity

Enzymes	Optimum pH
Pepsin	1.5
Salivary amylase	6.8
Trypsin	8.0
Intestinal lipase	7.8

5. Oxidation

(a) The sulfhydryl (SH) groups of many enzymes are essential for enzyme activity.

(b) Oxidation of these (SH) groups by many oxidizing agents including the oxygen of air forms the disulfide (S-S) linkages and results in loss of enzyme activity.

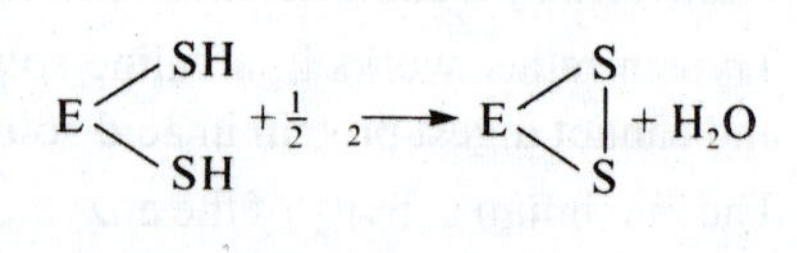

Fig. 10.8

(c) Full activity may be regained by reduced sulfhydryl compounds such as cysteine or glutathione (R-SH).

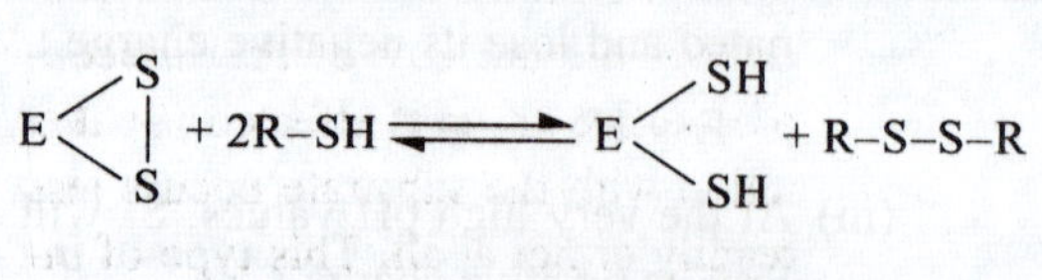

Fig. 10.9

6. Radiation

(a) Enzymes are highly sensitive to short wave length (high energy) radiation such as χ-, β- or γ- rays.

(b) High energy radiation forms peroxides which causes oxidation of the enzyme resulting in loss in enzyme activity.

(c) The enzyme activity is lost by irradiation which may support indirect effects on the DNA of genes.

7. Coenzymes and Activators

(a) Enzymes (excepting the enzymes of G.I.T.) work efficiently in presence of some other substances. These substances may be organic and inorganic known as coenzymes and activators respectively.

(b) In absence of coenzymes and activators the enzymes may be inactive or sluggish.

(c) The activators (Cl^-, Mg^{++}, Ca^{++}, Mn^{++} etc.) may take part in the formation of enzyme-substrate complex.

(d) Mn^{++} in the action of some peptidases may prevent the inactivation of the enzyme by inhibitors.

(e) Some enzymes which are activators are called kinases, e.g. enterokinase converts trypsinogen to trypsin by removing hexapeptide from trypsinogen.

8. Inhibiting agents

(a) Many enzymes are inhibited by the salts of mercury, silver, gold and salts of heavy metals or fluorides.

(b) Oxidases are generally inhibited by cyanides.

(c) Certain preservatives such as chloroform, glycerol and thymol inhibit some enzymes.

(d) Toluene has no action on enzymes but is the best preservative for enzyme solutions.

(e) Formaldehyde destroys enzymes.

(f) The inhibitors present in an enzyme solution occupy the active sites of the enzyme leaving free active sites for substrates to combine.

For example, many enzymes (known as "–SH enzymes") depend for their activity on the presence of free –SH groups. They can be inactivated by mercuric chloride which reacts with free –SH groups, thus :

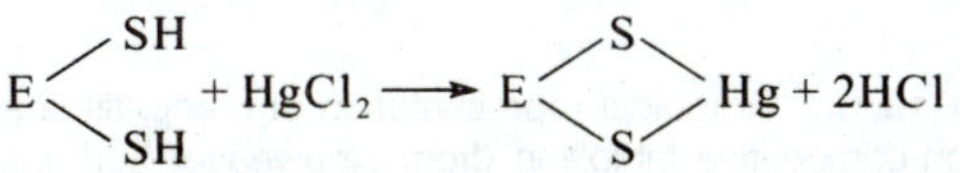

Fig. 10.10

This type of inhibition is non-specific.

(g) (i) The specific inhibitor which is structurally so similar to the substrate molecule that the inhibitor can combine with the enzyme in place of the substrate decreasing the enzyme-substrate union. The inhibitor competes with the substrate for the active group. This is known as "**Competitive inhibition**".

The followings are the examples of competitive inhibition :

(a) Succinic acid ($HOOC.CH_2.CH_2$ COOH) is converted to fumaric acid when it combines with succinate dehydrogenase; the fumaric acid is released from the enzyme complex leaving the enzyme free to unite with more succinic acid :

$$\text{Succinic acid} \xrightarrow[\text{dehydrogenase}]{\text{Succinate}} \text{Fumaric acid}$$

Fig. 10.11

Malonic acid ($HOOC.CH_2$. COOH) can also combine with the enzyme but does not undergo any change and is not readily released from the enzyme complex. So the enzyme is not available for union with succinic acid. In such a way malonic acid inhibits succinate dehydrogenase.

(b) Sulphanilamide competes P-aminobenzoic acid :

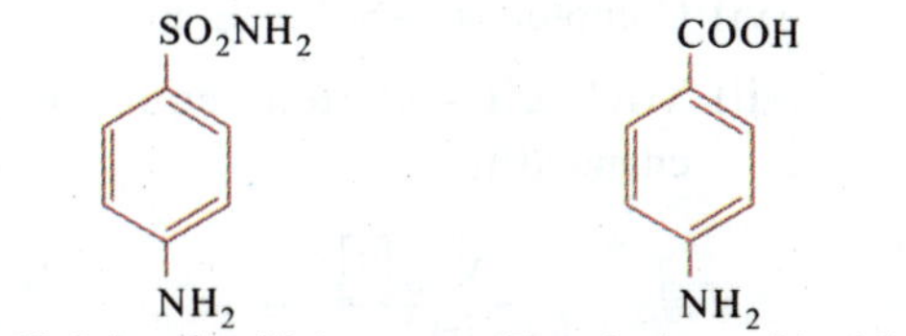

(Sulphanilamide) (P-aminobenzoic acid)

Fig. 10.12

P-aminobenzoic acid is essential for the synthesis of folic acid by the enzyme action of the enzyme in the bacteria. The union of sulphanilamide with the enzyme prevents the union of the enzyme with the P-aminobenzoic acid. Synthesis of folic acid is thus completely prevented.

(ii) Competitive inhibition is reversible.

(iii) V_{max} is same.

(iv) K_m is increased.

(v) Inhibitor cannot bind with ES complex.

(vi) Complex is EI.

(vii) Michaelis-Menten equation is changed to

$$V = \frac{V_{max}[S]}{K_m\left[1+\frac{[I]}{K_i}\right]+S}$$

(h) (i) Some inhibitors are attached not to the active group but to some other group of enzymes. Under such conditions the activity of the unoccupied active group is affected ; so that union with the substrate occurs less readily or not at all. This type of inhibition is called "**Non-competitive inhibition**".

(ii) Non competitive inhibition is reversible or irreversible.

(iii) V_{max} is lowered.

(iv) K_m is unaltered.

(v) Inhibitor can bind with ES complex.

(vi) Complex is E–S–I or E–I.

(vii) Michaelis-Menten equation is changed to

$$V = \frac{V_{max}[S]}{K_m\left[\frac{1+[I]}{K_i}\right]\{K_m+[S]\}}$$

Enzymes which show "allosteric" inhibition have two sites : (i) Isosteric site, (ii) Allosteric site.

(i) *Isosteric site :* It can bind the substrate or other molecules structurally similar to it.

(ii) *Allosteric site :* It can bind other substances.

When the allosteric site binds a molecule, the isosteric site is changed in shape so that the substrate can no longer be bound and hence the enzyme is inhibited. This is superficially similiar to non-competitive inhibition. But it is more complex because the substrate can also affect the shape of the allosteric site and for this reason allosteric inhibition is regared as a distinct process.

Allosteric inhibitors change the substrate saturation curve to the right. The presence of activators shifts the curve to the left. Aspartate carbamoylase and phosphofructokinases, the allosteric inhibitor, lower the substrate affinity to increase the K_m of the enzyme, but the V_{max} is unchanged. Acetyl-CoA Carboxylase, the allosteric inhibitor, reduces the maximum velocity but no change in K_m or substrate affinity. Feedback allosteric inhibition results when the enzyme is allosterically inhibited by the final product eg, CTP. The pyrimidine nucleotide inhibits aspartate transcarbamoylase allosterically in the synthesis of Pyrimidine. The feed-forward allosteric activation of an enzyme for a subsequent step of its metabolism is regulated by a metabolite, eg, Fructose-1, 6-diphosphate allosterically activates pyruvate kinase catalyzing subsequent step. Two allosteric enzymes catalyzing reverse reactions are influenced by an allosteric effector, eg, AMP allosterically activates phosphofructokinase and allosterically inhibits fructose diphosphatase.

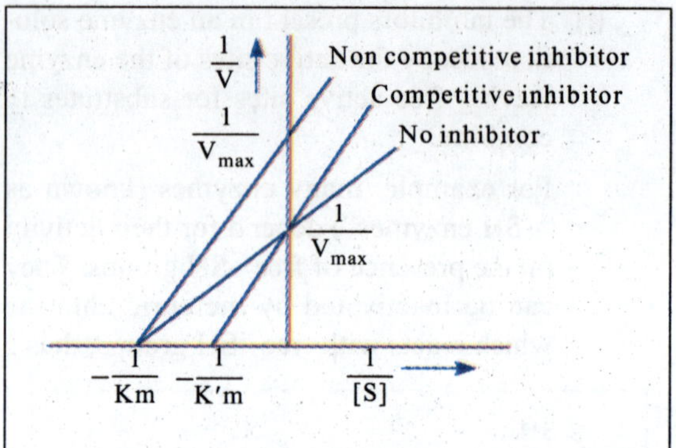

Fig. 10.13 : Graphical representation of competitive and non-competitive inhibition (from Lineweaver burk plot)

Examples of allosteric modulation

Both the allosteric site and active site are located on different subunits of oligomeric enzymes. Alternation in the enzyme substrate interaction owing to the allosteric effects of regulatory molecules other than the substrate are said to be heterotropic allosteric modulations.

The positive and negative arrangements with the substrate are exhibited by allosteric activators and inhibitors, respectively. Binding of substrate to one promoter accelerates the binding of the same to another promoter on the same enzyme molecule.

Homotropic allosteric effect results in the binding of a substrate enhancing the interaction between the allosteric enzyme and more molecules of the same substrate.

(i) Reversal of inhibition can be brought about by increasing the amount of substrate relative to inhibitor.

(ii) Inhibition may also be reversed by removal of the inhibitor by treatment with hydrogen sulphide (H_2S) :

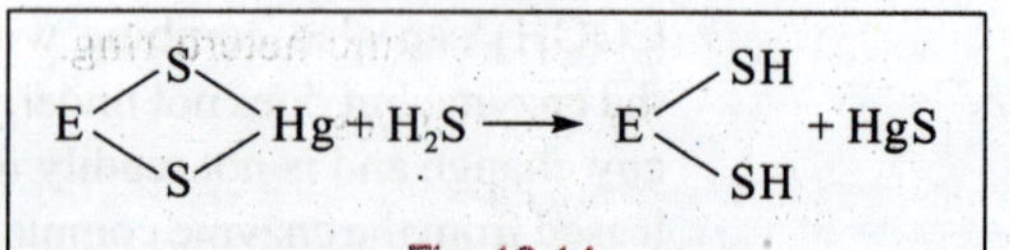

Fig. 10.14

Examples of allosteric modulation					
	Enzymes		Allosteric activator		Allosteric inhibitor
(a)	Pyruvate Carboxylase	...	Acetyl-CoA	...	ADP
(b)	Hexokinase, ICD	...	ADP	...	ATP, Glucose-6-phosphate
(c)	Glutamate Dehydrogenase	...	ADP	...	ATP, NADH
(d)	Protein Kinases	...	cAMP	...	—

9. Anti-enzymes

1. If certain enzymes are repeatedly injected into an animal, a substance will be produced in the animal's serum which can prevent the normal action of the enzymes injected. This substance is called an anti-enzyme. Examples : Antipepsin, anti-trypsin, anti-rennin, anti-urease.
2. Some people believe that the mucus membrane contains suitable antienzymes for which the alimentary canal is not digested by its own secretions.

COENZYMES

Many reactions of substrates are catalyzed by enzymes only in the presence of a specific non-protein organic molecule called the coenzyme. Coenzymes combine with the apoenzyme (the protein part) to form holoenzyme. The coenzymes are also regarded as cosubstrates.

Definition

Coenzymes are heat-stable, dialyzable non-protein organic molecules and the prosthetic groups of enzymes.

Classification

I. Based on chemical characteristics

A. Containing an aromatic hetero ring.
   1. ATP & its relatives.
   2. NAD, NADP.
   3. FMN, TPP, B_6-PO_4.

B. Containing a non-aromatic hetero ring.
   Biotin, lipoic acid.

C. No hetero ring.
   Sugar phosphate, coenzyme Q.

II. Based on functional characteristics

A. Group transferring coenzymes :
   1. ATP and its relatives.
   2. Sugar phosphates.
   3. Thiamine pyrophosphate (TPP).
   4. CoA.
   5. Pyridoxal phosphate (B_6-PO_4).
   6. Biotin.

B. Hydrogen transferring coenzymes :
   1. Nicotinamide adenine dinucleotide (NAD) and Nicotinamide adenine dinucleotide phosphate (NADP).
   2. Flavin adenine dinucleotide (FAD) and flavin mono-nucleotide (FMN).
   3. Coenzyme Q.

III. Based on nutritional characteristics

(a) Containing B vitamins :
   1. CoA.
   2. TPP.
   3. NAD & NADP.
   4. B_6-PO_4.
   5. FMN, FAD.
   6. Folic acid coenzyme.
   7. B_{12} coenzyme.
   8. Biotin.

Function

1. Their function is usually to accept atoms or groups from a substrate and to transfer them to other molecules.
2. They are less specific than are enzymes and the same coenzyme can act as such in a number of different reactions.
3. The coenzymes are also attached to the protein at a different but adjacent site so as to be in a position to accept the atoms or groups which are removed from the substrate.
4. NAD and NADP coenzymes function as hydrogen acceptors in dehydrogenation reactions.
5. The chief function of CoA is to carry acyl groups and they are used in the oxidative decarboxylation of pyruvic acid and synthesis of fatty acids and acetylation.
6. The function of TPP (co-carboxylase) is to carry 'active aldehyde' (R. $CH(OH)^-$) group.
7. The chief function of pyridoxal phosphate (B_6–PO_4) is involved in transamination reactions.
8. The chief function of tetrahydrofolic acid is expressed as a carrier of formate and it is used in the synthesis of purines and pyrimidines.

LYSOSOMES

1. Lysosomes are one type of organelles in the cytoplasm which can be regarded as a type or 'ground substance'.
2. They are similar in size to mitochondria and consist of sacs containing a solution of hydrolytic enzymes which are active at an acid pH.
3. Protein, carbohydrate or fat molecules are brought to the lysosome where they are hydrolysed, so that lysosomes constitute an intracellular digestive system.
4. They are responsible for postmortem autolysis.
5. Congenital disease such as glycogen storage disease is due to the absence of specific lysosomal enzymes.
6. The enzyme systems present in lysosomes are for hydrolytic processes, eg, phosphatases, glycosidases, glucuronidases.
7. The enzymes within the lysosomes are unable to act on substances within the cytoplasm so long the lipoprotein membrane of the lysosome remains intact. Once the membrane is ruptured, there will be the release of lysosomal enzymes.

LYSOZYME

1. Lysozyme is an enzyme present in tears, nasal mucus, sputum, tissues, gastric secretions, milk and eggwhite.
2. It catalyzes the hydrolysis of β-1, 4 linkages of N- acetylneuraminic acid in mucopolysaccharides or mucopeptides :

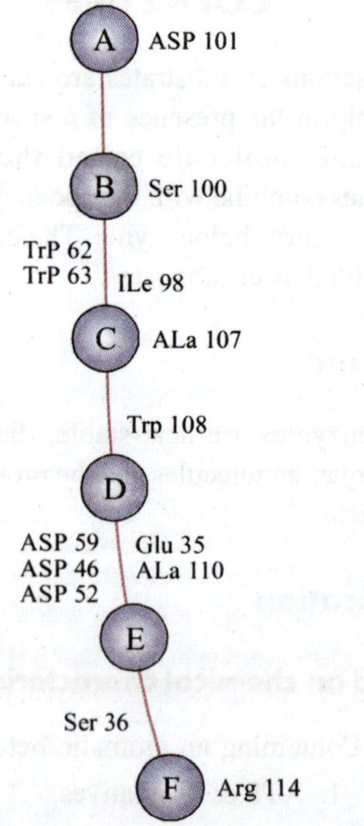

Fig. 10.15

3. It destroys the cell walls of many airborne Grampositive bacteria in tears and nasal mucus.
4. Its molecular weight is about 15,000.
5. It consists of a single polypeptide chain of 129 amino acid residues having no coenzymes or metal ion cofactors.
6. As the lysozyme moelcule may readily be unfolded and refolded, catalysis and specificity as well as 3-dimensional structure are determined solely by these residues. Recent research work states that the central feature of its mechanism is the formation of an ion pair intermediate, rather than a covalent species.
7. There are little α-helix and large regions without secondary structure.
8. It has a central catalytic site with 6 subsites which bind various substrates or inhibitors.
9. The residues responsible for bond cleavage lie between sites D and E, close to the carboxyl groups of ASP. 52 and Glu. 35.

ISOENZYMES or ISOZYMES

Definition

The enzymes that occur in a number of different forms and differ from each other chemically, immunologically and electrophoretically are called "Isoenzymes" or "Isozymes".

Occurrence

Isozymes are present in the serum and tissues of mammals, amphibians, birds, insects, plants and unicellular organisms.

Examples : Isozymes of numerous dehydrogenases, and several oxidases, transaminases, phosphatases, transphosphorylases, proteolytic enzymes, aldolases.

Characteristics

1. They catalyze the same reaction but they can be distinguished by physical methods such as electrophoresis or by immunological methods.
2. The difference between some isozymes are due to differences in the quarternary structure of the enzymes, eg, lactate dehydrogenase exists in five isozymic forms.
3. The isozymic forms of lactate dehydrogenase are tetramers, each is made up from two types of units H and M. The molecular weight of active lactate dehydrogenase is 1,30,000. Only the tetrameric molecule possesses catalytic activity. The subunits are expressed in the following 5 ways :

HHHH
HHHM
HHMM
HMMM
MMMM

4. Splitting and reconstitution of lactate dehydrogenase –I_1 or lactate dehydrogenase-I5 produces on new isozymes. Therefore, each consists of a single subunit.

But when a mixture of purified lactate dehydrogenase – I_1 and lactate dehydrogenase – I_5 is subjected to splitting and reconstitution, lactate dehydrogenase – I_2, – I_3 and – I_4 are also produced. The approximate proportions of the isozymes result if the relationships are :

Lactate dehydrogenase isozyme		Subunits
I_1	...	H H H H
I_2	...	H H H M
I_3	...	H H M M
I_4	...	H M M M
I_5	...	M M M M

Synthesis of H and M subunits are controlled by distinct genetic loci.

5. Lactate dehydrogenase (LDH) catalyzes the transfer of two electrons and one hydrogen ion from lactate to NAD :

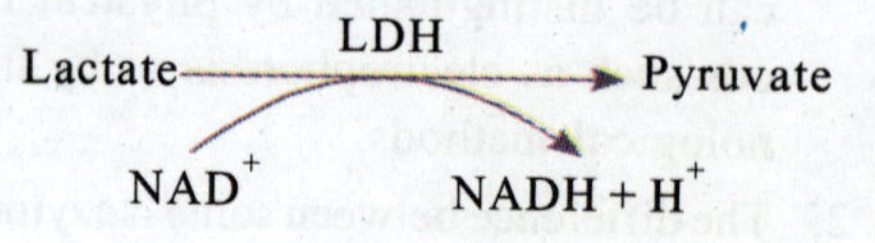

Fig. 10.16

6. Medical discovery in 1957 had shown that the relative proportions of several lactate dehydrogenase isozymes of human serum were changed significantly in some pathologic conditions.

Method of assay

1. Serum sample is subjected to electrophoresis at pH 8.6 using starch, agar medium.
2. The isozymes have different charges at this pH and migrate to 5 regions of the electrophoretogram.
3. Isozymes are then localized by means of their ability to catalyze reduction of a colourless dye to a coloured form.

Diagnostic importance

1. In certain solid tumors, there is an increase in serum LD_1 and LD_2. These isoenzymes are also present in the blood of patients with acute leukemia.
2. LD_5 is usually the predominant isozyme in the tumors.
3. Serum isozyme levels are elevated in acute leukemia.
4. Alkaline phosphatase isozymes can distinguish liver lesions from bone lesions in metastatic carcinoma.
5. The normal serum is 0 enzyme levels of LDH are :

LDH_1	20-34%
LDH_2	28-41%
LDH_3	15-25%
LDH_4	8-16%
LDH_5	6-15%

KINETIC THEORY OF REACTION

The kinetic or collision theory states that the molecules to react must collide and must possess sufficient energy to overcome the energy barrier for reaction.

If the molecules have sufficient kinetic energy to react, factors that increase the frequency of collision between molecules will increase their rate of reaction. Factors that decrease the frequency of collision or the kinetic energy will decrease the rate of reaction.

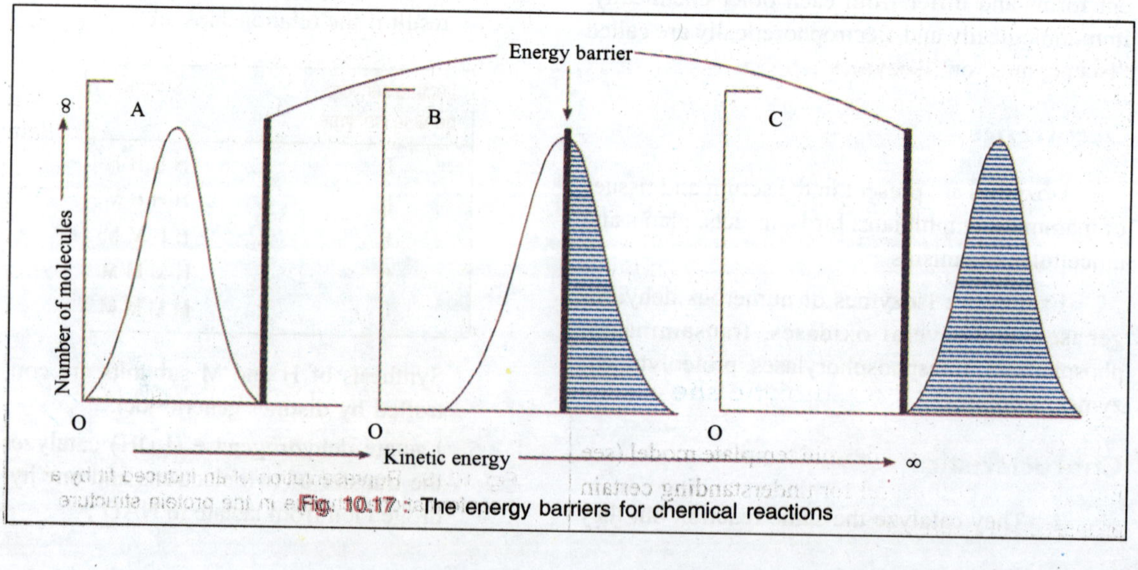

Fig. 10.17 : The energy barriers for chemical reactions

If some molecules have insufficient energy to react, increased temperature which increase their kinetic energy will increase the rate of the reaction. These concepts are illustrated diagrammatically in Fig. 10.17. In A none, in B a portion and in C all of the molecules have sufficient kinetic energy to overcome the energy barrier for reaction.

Molecules are in motion at all temperatures above absolute zero (–273°C) at which all molecular motion ceases. With the increasing temperature, the rate of diffusion increases.

In the absence of enzymic catalysis, many chemical reactions proceed slowly at the temperature of living cells. At this temperature even the molecules are in active motion and are undergoing collision. They fail to react rapidly because most of them possess insufficient kinetic energy to overcome the energy barrier for reaction. At the lower temperature, the reaction is spontaneous but slow; at the higher temperature the reaction is spontaneous and fast.

The overall energy changes in chemical reactions are independent of the path or mechanism of the reaction.

THE CATALYTIC SITE OF ENZYME

Some restricted region of the enzyme which was concerned with the process of catalysis was termed the active site. In the beginning, the biochemists were puzzled why enzymes were so large, when only a portion of their structure was involved in substrate binding and catalysis. Today, it is clear from 3-dimensional models of enzymes that a far greater portion of the protein interacts with the substrate.

Rigid model of the catalytic site

The lock and key or rigid template model (see Fig. 10.18) is still useful for understanding certain properties of enzymes :

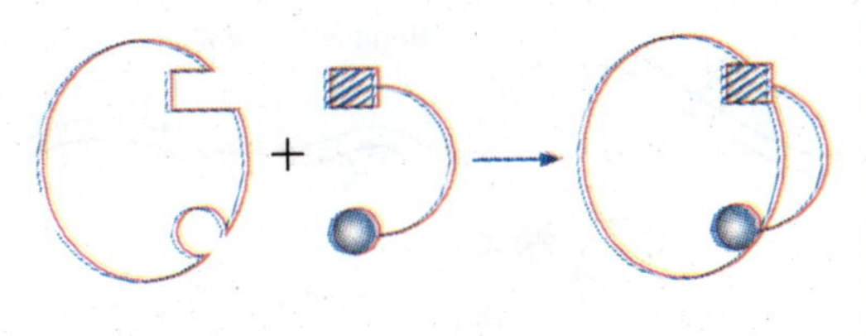

Fig. 10.18 : Representation of formation of enzyme complex

Flexible model of the catalytic site

In the Fischer model, the catalytic is presumed to be pre-shaped to fit the substrate. In the induced fit model, the substrate induces a conformational change in the enzyme. This aligns amino acid residues or other groups on the enzyme in the correct spatial orientation for substrate binding, catalysis, or both. At the same time, other amino acid residues may be buried in the interior of the molecule. Hydrophobic groups (hatched portion) and charged groups (dots) both are involved in substrate binding. A phosphoserine (–P) and the –SH of a cysteine are involved in catalysis. Other residues involved in neither process are represented by the side chains of lysine and methionine.

In the absence of substrate, the catalytic and the substrate-binding groups are several bond distances removed from one another. Approach of the substrate induces a conformational change in the enzyme. At the same time, the spatial orientations of other regions are also altered, the lysine and methionine are now closer together.

In the representation of a catalytic site shown in Fig. 10.20, several regions of a polypeptide chain

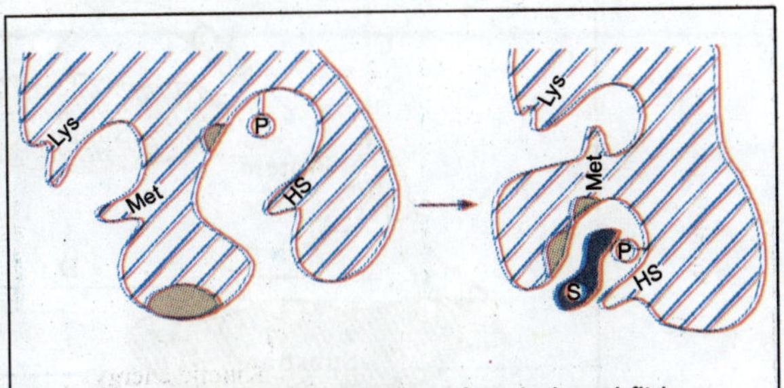

Fig. 10.19 : Representation of an induced fit by a conformational change in the protein structure

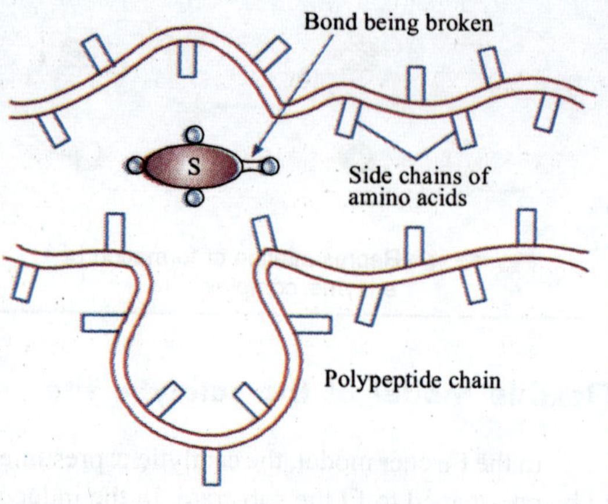

Fig. 10.20 : The catalytic site

contribute amino acid residues to the site. Three types of amino acid residues are distinguished in enzymes :

1. **Contact residue** : An amino acid residue with one bond distance (0.2 nm) of the substrate.

2. **Specificity residue** : An amino residue is involved in substrate binding as well as in catalytic process.

3. **Catalytic residue** : An amino acid residue is directly involved in covalent bond changes during enzyme action.

Substrate analogs may cause some of the conformational changes (Fig.. 10.21). On attachment of the true substrate (A), all groups (shown as closed circles) are brought into correct alignment. Attachment of a substrate analog that is too **"bulky"** (Fig. 10.21B) or too **"slim"** (Fig. 10.21C) induces incorrect alignment.

The exact sequence of events in a substrate induced conformational change remains to be established. Several possibilities stand (Fig. 10.22).

It is still difficult to decide exactly which residues constitute the catalytic site even after knowing the complete primary structure of an enzyme.

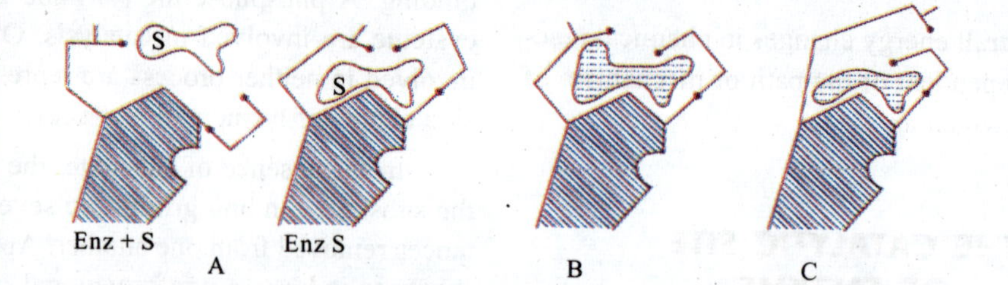

Fig. 10.21 : Representation of a conformational changes in an enzyme

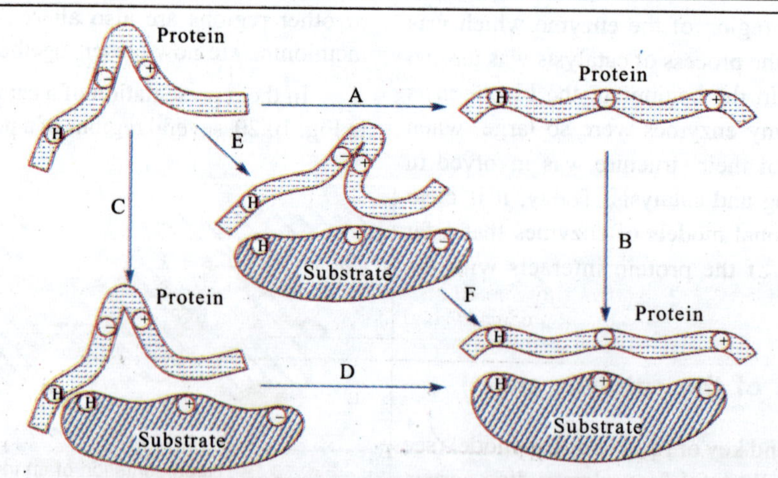

Fig. 10.22 : Alternative reaction paths for a substrate-induced conformational change

Modifiers of enzyme activity

Like all mammalian proteins, enzymes are degraded to amino acids. Although this mechanism reduces enzyme concentration and, hence, catalytic activity, still they are slow, wasteful of carbon and energy and, rather, they are like turning out a light by smashing the bulb, then inserting a new one when light is needed.

The catalytic activity of certain key enzymes can be reversibly decreased or increased by small molecules. Small molecule **modifiers** which decrease catalytic activity are termed **negative modifiers** and those which increase or stimulate activity are called **positive modifers**.

ORDERED AND RANDOM BINDING OF SUBSTRATES

Many enzymes catalyze a reaction between two or more substrates producing one or more products. But for some enzymes, all substrates must be present simultaneously for the reaction to take place. For others, the enzyme first changes one substrate and then catalyzes its reaction with a second substrate. The order in which an enzyme binds its substrates may be random or ordered (Fig. 10.23).

Many reactions which need coenzymes proceed by "ping-pong" mechanisms (Since the enzyme alternates between forms E and E' (Fig. 10.24).

ENZYMES AS GENERAL ACID OR BASE CATALYSTS

Reactions whose rates vary as regards to changes in H^+ or H_3O^+ concentration but are independent of the concentrations of other acids or bases present in the solution are said to be **specific acid or specific base catalysis**.

Reactions whose rates are responsive to all acids or bases present in solution are said to be general acid or general base catalysis. Mutarotation of glucose is one reaction subject to **general acid-base catalysis**.

Role of Metal Ions

More than 25 per cent of the enzymes contain tightly-bound metal ions for their activity. The functions of these metal ions may be studied by X-ray crystallography, nuclear magnetic resonance (NMR) and electron spin resonance (ESR).

Metalloenzymes and metal-activated enzymes : A definite quantity of functional metal ion

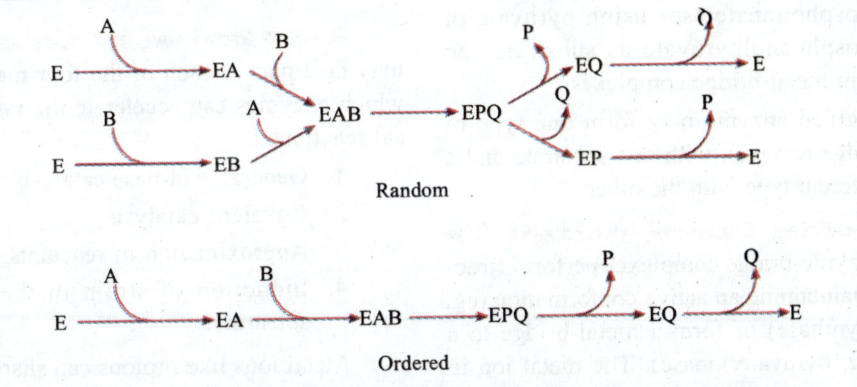

Fig. 10.23 : Random and ordered addition of substrates A and B and dissociation of products P and Q from an enzyme E

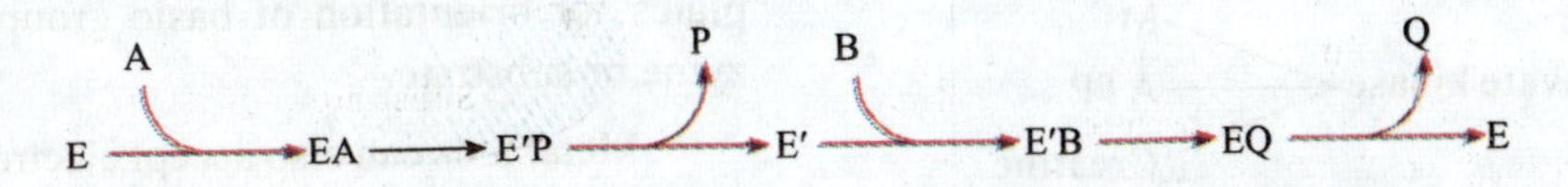

Fig. 10.24 : Generalized "Ping-Pong" mechanism for enzymic catalysis

Enz—S—M Substrate-bridge complex	M—Enz—S Enzyme-bridge complex
Enz—M—S Simple metal-bridge complex	Enz<(M | S) Cyclic metal-bridge complex

Fig. 10.25

is present in metalloenzymes and that metal ion is also retained throughout purification. Metal-activated enzymes bind metals less tightly but require added metals. The mechanism of action in both cases appears to be similar.

Ternary Enzyme-Metal-Substrate Complexes : Four schemes (shown below) are possible for ternary (3-component) complexes of the catalytic site (Enz), a metal ion (M), and substrate (S) that exhibit 1 : 1 : 1 stoichiometry.

The above four schemes are possible for metal-activated enzymes. Metalloenzymes cannot form the Enz-S-M complex, because they retain the metal throughout purification. Three stages can be stated:

1. Most kinases form substrate-bridge complexes of the type Enz-nucleotide-M.
2. Phosphotransferases using pyruvate or phosphoenolpyruvate as substrate can form metal-bridge complexes.
3. A given enzyme may form one type of bridge complex with one substrate and a different type with the other.

Enzyme-Bridge Complexes (M-Enz-S) : The metals in enzyme-bridge complexes perform structural roles maintaining an active conformation (eg, glutamine synthase) or form a metal-bridge to a substrate (eg, pyruvate kinase). The metal ion in pyruvate kinase also holds one substrate (ATP) to activate it :

Pyruvate kinase < M—ATP ; Creatine

Fig. 10.26

Substrate-Bridge Complexes (Enz-S-M) : This complex with ATP involves the displacement of water from the coordination sphere of the metal by ATP.

$$ATP^{4-} + M(H_2O)_6^{2+} \rightleftharpoons ATP\text{–}M(H_2O)_3^{2-} + 3H_2O$$

$$ATP\text{–}M(H_2O)_3^{2-} + Enz \rightleftharpoons Enz\text{–}ATP\text{–}M(H_2O)_3^{2-}$$

Fig. 10.27

Metal-Bridge Complexes (Enz<(M | S)) **:** Activation by metal ions for many peptidases is a slow process which requires many hours. The slow reaction is probably conformational re-arrangement of the binary Enz-M complex to an active conformation.

However, for metalloenzymes, the ternary metal-bridge complex is formed by combination of the substrate (S) with the binary Enz-M complex. Role of Metal Ions in Catalysis :

Enz—M + S ⇌ Enz—M—S or Enz<(M | S)

Fig. 10.28

Role of Metal Ions in Catalysis : Metal ions may take part in each of the four mechanisms by which enzymes can accelerate the rates of chemical reactions :

1. General acid-base catalysis,
2. Covalent catalysis,
3. Approximation of reactants,
4. Induction of strain in the enzyme or substrate.

Metal ions like protons can share an electron pair forming a sigma bond. They may also be considered **"super acids"** and may form pi bonds. Unlike protons, metals can serve as 3-dimensional templates for orientation of basic groups on the enzyme or substrate.

Metal ions can also accept electrons via sigma or pi bonds to activate electrophiles or nucleophiles (general acid-base catalysis). They can also activate nucleophiles or act as nucleophiles themselves by donating electrons.

The coordination sphere of a metal may bring together enzyme and substrate or form chelate-producing distortion in either the enzyme or substrate (strain). A metal ion may also "mask" a nucleophile and thus prevent side reaction.

FEEDBACK INHIBITION

The changes in the concentration of substrates, coenzymes, activators or inhibitors affect the catalytic efficiency of an enzyme. Feedback inhibition inhibits the activity of an enzyme early in the biosynthetic pathway. In the biosynthetic reaction, sequences leading from A to D is catalyzed by enzymes Enz_1 through Enz_3 :

$$A \xrightarrow{Enz_1} B \xrightarrow{Enz_2} C \xrightarrow{Enz_3} D$$

Fig. 10.29

A high concentration of D will inhibit conversion of A to B. D is specifically able to bind to and inhibit Enz_1. D thus acts as a negative allosteric effector or feedback inhibitor of Enz_1. This feedback inhibition on an early enzyme by an end product of its own biosynthesis achieves regulation of synthesis of D.

Frequ[illegible] the feedback inhibitor is the last small molecule before a macromolecule, eg, amino acid before proteins or nucleotides before nucleic acids.

Feedback inhibition is a mechanism for regulation of many bacterial and mammalian enzymes, eg, dietary cholesterol restricts the synthesis of cholesterol from acetate in mammalian tissues. The feedback regulation is not involved in feedback inhibition of an early enzyme of cholesterol biosynthesis. An early enzyme (HMG-CoA reductase) is affected, but the mechanism involves curtailment by cholesterol or a cholesterol metabolite of the expression of the gene that code for the formation of HMG-CoA reductase (i.e. enzyme repression). The catalytic activity of HMG-CoA reductase is not effected by the direct addition of cholesterol.

The kinetics of feedback inhibition may be competitive, non-competitive, mixed etc. It is the common path of regulation of a biosynthetic pathway. In cumulative feedback inhibition, the inhibitory effect of two or more end products on a single regulatory enzyme is strictly additive. Complete inhibition occurs only when two or more end products are present in excess in multivalent feedback inhibition. In cooperative feedback inhibition, a single end product present in excess inhibits the regulatory enzyme.

Control of enzyme synthesis : Induction

For a molecule to be metabolized or for an inducer to act, it first must enter the cell. In some cases, a specific transport system or permease is needed. Permeases have many properties in common with enzymes and perform functions like cytochromes in electron transport.

Escherichia coli grown on glucose will not ferment lactose due to the absence of specific permeases for a β-galactoside (Lactose) and of β-galactosidase. If lactose is added to the medium, both the permease and the β-galactosidases are induced and the culture will now ferment lactose. The inducer (lactose) is a substrate for the induced proteins, the permease and the β-galactosidase. Compounds structurally similar to the substrate [illegible] are termed **gratuitous inducers**.

Enzymes whose concentration in a cell is independent of an added inducer are termed **constitutive enzymes**.

The structural genes which specify a group of catabolic enzymes comprise an operon. All the enzymes of that operon are induced by a single inducer. This process is known as coordinate induction.

Control of enzyme synthesis : Repression and Derepression

In bacteria, the presence of the synthesized particular amino acid in the culture medium prevents new synthesis of that amino acid via repression. A small molecule such as histidine or leucine, acting as a corepressor, can ultimately block the synthesis of the enzymes involved in its own synthesis.

The genetic information coding for the biosynthesis of enzymes is again expressed after the removal or exhaustion of an essential biosynthetic intermediate from the medium. This is termed as derepression.

Location of Enzymes in the Cell	
Subcellular Fraction	Enzyme systems present for
Cytoplasm	(i) Glycolysis. (ii) Glycogenolysis. (iii) Glycogenesis. (iv) Synthesis of fatty acid.
Mitochondria	(i) Electron transport chain. (ii) Citric acid cycle. (iii) β-oxidation. (iv) Urea cycle. (v) Oxidative phosphorylation.
Microsomes	(i) Protein synthesis. (ii) Hydroxylation.
Lysosomes	(i) Hydrolytic processes, eg, cathepsin, glycosidases, glucuronidases, phosphatases.
Nuclei	(i) Synthesis of NAD. (ii) Synthesis of RNA. (iii) Synthesis of Histone.

MECHANISM OF ENZYME ACTION

1. Fisher in 1894 first proposed that enzyme is like key which fits into lock.
2. In 1913, Michaelis and Menton proposed that enzyme (E) combines with the substrate (s) to form enzyme-substrate complex (ES). This complex finally gives rise to the reaction product (s) (P) and the enzymes becomes free for further reaction :

$$E + S \longrightarrow ES$$
$$ES \longrightarrow E + S$$

Fig. 10.30

During the complex formation the substrate molecules are attached to the specific groups on the enzyme molecules. These specific groups are said to be 'active groups' and the regions on which these groups are located are said to be "active sites". The active groups are sulfhydryl (–SH) group of cysteine, phenolic group of tyrosine (as in pepsin), alcoholic group of serine (as in trypsin and chymotrypsin) and imidazole group of histidine.

3. In 1963, Koshland proposed "induced fit mechanism" for mode of enzyme action. According to this, the substrate fits the active sites effectively by inducing a configurational change in the enzyme. Thus, the complexes are formed by multiple bonding (covalent, hydrogen or electrostatic). The functional groups of the active sites are arranged in a definite spatial configuration for which d-or *l* - isomer of a substance can act as a substrate for an enzyme.

Many hormones act through allosteric Second messengers

1. Many hormones bind to their cell surface receptors and many nerve impulses and bring about changes in the rate of enzyme-catalyzed reactions inside target cells by the release or synthesis of specialized allosteric effectors called Second messengers.
2. 3′, 5′-cAMP is the best-known second messenger. This is synthesized from ATP by the enzyme adenylyl cyclase by the effect of the hormone epinephrine.
3. Calcium ion, stored inside the sarcoplamic reticulum of muscle cells, is released into the cytoplasm by the effect of depolarization of membrane resulting from a nerve impulse opening the membrane channel. Calcium ion binds to and activates enzymes which are involved in the regulation of contraction and the mobilization of stored glucose from glycogen. The glu-

cose then supplies the increased energy demands of muscle contraction.

4. Another second messengers are 3′, 5′-cGMP and polyphosphoinositols which are produced by the hydrolysis of inositol phospholipids by hormone-regulated phospholipases.

Proenzymes facilitate rapid mobilization of an activity in response to physiologic demand

1. Certain proteins are required essentially at all times.
2. Others (eg, the enzymes of blood clot formation and dissolution) are required only intermittently. When these enzymes are required, they are needed rapidly.
3. Some physiologic processes such as digestion are intermittent but fairly regular.
4. Others such as blood clot formation, dissolution, and tissue repair need to be brought "on line" in response to physiologic or pathophysiologic need.
5. The synthesis of proteases as inactive precursor proteins protect the tissue of origin (eg, the pancreas) form autodigestion which can occur in pancreatitis.
6. De novo synthesis of the required proteins is not sufficiently rapid to respond to pathophysiologic demand such as the loss of blood.
7. The secretion process is slow relative to the physiologic demand.

CLINICAL ORIENTATION

- The change in intracellular pH due to certain enzymes catalyzed reactions results in metabolic acidosis or alkalosis. The rates of enzymatic catalysis by fluctuations in temperature, fever and hypothermia disturb homeostasis by altering the rates of many enzyme catalysed reactions. These changes become the physician's advantage. The decrease in activity of all enzymes being accompanied by hypothermia is exploited to reduce overall metabolic demand during open heart surgery or transplantation surgery.

CLINICAL ORIENTATION

- Toxicology and pharmacology are the eminent factors that affect the rates of enzyme-catalyzed reactions. The metabolic poisons such a mercurials, curare and nerve gases exert their toxicity by inhibiting enzymes and ultimately abolishing essential metabolic reactions. Many therapeutically important drugs which lower the rates of key metabolic reactions by competing with the natural substrates resemble natural substrates, eg, Lovastatin and AZT drugs are used to treat hypercholesterolemia and AIDS, respectively.
- High resolution structural information obtained by X-ray crystallography facilitates the design of drugs to inhibit specific enzymes such as HMG-CoA reductase, the pacemaker enzyme of cholesterol biosynthesis.
- The recombinant DNA technology and site-directed mutagenesis may be used to modify enzyme specificity or catalytic efficiency. These techniques will finally facilitate the design and introduction into human subjects of enzymes with specific desired properties.
- The concentration of a specific enzyme in serum is increased by diseases due to an increase in its rate of release or a decrease in its rate of excretion. The increased release of enzyme is due to the following reasons :
  - (a) Necrosis of cells is due to damage to the cells of the tissues.
  - (b) Increased permeability of cell membrane without necrosis of cells due to the early stage of viral hepatitis leading to increased levels of SGPT and progressive muscular dystrophy leading to elevated levels of aldolase, SGOT, CPK.
  - (c) Increased production of enzyme within cells found in treatment of patients with protein anabolic drugs resulting in increased synthesis of liver cell transaminases and serum transaminases.
  - (d) An increase in tissue source of enzymes due to increased rate of production in cells or increased number of cells as seen in malignancies. eg, alkaline phosphatase level increased in osteoblastic bone lesions or acid phosphatase level increased in carcinoma prostate.

  The decreased serum level of enzymes is due to the following reasons :
  - (a) The decreased formation of the enzymes may be due to genetics involving hypophosphatasia with decreased ALP level in serum and Wilson's disease with decreased serum ceruloplasmin.

Contd.

CLINICAL ORIENTATION

(b) Acquired involving hepatitis with decreased serum level of pseudocholinesterase due to decreased production and decreased serum amylase in patients with chronic hepatic or pancreatic diseases or severely malnourished patients.

(c) Lack of cofactors involving decreased SGOT level in pregnancy and cirrhosis.

(d) Enzyme inhibition involving decreased level of serum pseudocholinesterase in insecticide poisoning.

- Serum amylase level is increased in acute pancreatitis, mumps, other forms of parotitis, salivary duct stone, meningoencephalitis, facial swellings, perforated peptic ulcer, intestinal obstruction.
- Increased serum lipase is observed in pancreatic disorders, perforated duodenal and peptic ulcers, intestinal obstruction and cirrhosis.
- Serum aldolase level increases in dermatomyositis, muscular dystrophies, Deuchenne type of muscular dystrophy, viral hepatitis, myocardial infarction.
- Elevated serum CPK level is found in neurogenic muscular atrophy, dermatomyositis, Deuchenne type of muscular dystrophies, hypothyroidism owing to secondary muscle disease.
- Many cancer cells exhibit abnormalities in the regulation of their enzyme complement (lack of induction or repression). Alternations of gene control are basic events in cancer cells. Certain oncogenic viruses contain a gene that codes for a tyrosine-protein kinase. When this kinase is expressed in host cells, it can phosphorylate proteins and enzymes.
- Enzyme induction is one important biochemical cause of a drug interaction, the situation in which the administration of one drug results in a significant change in the metabolism of another.
- The level of serum alkaline phosphatase (ALP) is increased in rickets, osteomalacia, hyperparathyroidism, Paget's disease, primary and secondary malignancies of bone.
- Serum acid phosphatase level is also elevated in metastasizing prostate carcinoma, Gaucher's disease, Paget's disease, Osteolytic metastasis from breast and other carcinomas, thrombocytosis, chronic granulocytic leukaemia, hemolytic anemia, pulmonary embolism.
- The concentration of serum and urinary β-glucuronidase is increased in cancer of urinary bladder. β-glucuronidase activity of vaginal fluid is useful in diagnosis of malignancies of female genital tract.

DIAGNOSTIC VALUE OF SERUM ENZYMES

Very small amounts of enzymes which are involved in the reactions in the tissues are present in blood under normal conditions. The concentrations of these enzymes are significantly increased due to their more liberation by the affected tissues to the blood stream under certain clinical conditions.

Increase in the enzyme activities in cerebrospinal fluid is not reflected in the blood.

Changes may occur in the cerebrospinal fluid enzymes in certain diseases of the central nervous system. Lactate dehydrogenase activity in cerebrospinal fluid is increased frequently in meningitis, cerebral thrombosis and hemorrhage. Glucose phosphate isomerase concentration in the cerebrospinal fluid is also elevated in malignant tumors of the brain and frequently in meningitis and cerebral thrombosis.

The determination of the activity of the following enzymes shown in the list can give valuable confirmatory or suggestive diagnostic evidence to the physicians.

COENZYME A (CoA)

Chemistry

1. It is composed of adenosine triphosphate (ATP), pantothenic acid and β-mercaptoethalamine. So it is the coenzyme form of pantothenic acid, a vitamin.
2. It is a group transferring coenzyme.
3. The reaction group is the sulfhydryl (–SH) group.
4. The acyl group is accepted by the sulfhydril group to form acetyl coenzyme A ($CH_3CoS.CoA$). The acyl coenzyme derivatives are the high energy compounds.

Function

1. Carrier of acyl groups, eg, acetyl, sccinyl, benzoyl.
2. Some of the pantothenic acid is bound to

Serum Enzymes	Concentration increased in	Concentration decreased in
1. Lipase	Acute pancreatitis, pancreatic carcinoma.	Liver disease, vitamin A deficiency, diabetes mellitus.
2. Amylase	High intestinal obstruction, acute pancreatitis, parotitis, diabetes.	Liver disease.
3. Trypsin	Acute disease of the pancreas.	—
4. Cholinesterase	Nephrotic syndrome.	Liver disease, malnutrition, acute infectious diseases.
5. Alkaline phosphatase	Rickets, Paget's disease, hyperparathyroidism, obstructive jaundice, metastatic carcinoma, osteoblastic sarcoma, kidney diseases. Alkaline phosphatase, isozymes can distinguish liver lesions from bone lesions in cases of metastatic carcinoma.	—
6. Acid phosphatase	Metastatic prostatic carcinoma.	—
7. SGOT	Myocardial infarction. Slightly elevated in acute liver diseases.	—
8. SGPT	Acute liver diseases, Slightly elevated in cardiac necrosis.	—
9. Lactate dehydrogenase	Myocardial infarction, acute hepatitis, renal tubular necrosis, In certain solid tumors LD_1 and LD_2 are increased.	—
10. Isocitrate dehydrogenase	Liver diseases.	—
11. Creatine phosphokinase (CK or CPK)	Muscular dystrophy, myocardial infarction.	—
12. Glucose-6-phosphate dehydrogenase	Myocardial infarction.	Congenital deficiency causes hemolytic anemia.
13. Ceruloplasmin (Ferroxidase activity)	Cirrhosis, bacterial infection, pregnancy.	Wilson's disease. (hepatolenticular degeneration)
14. Aldolase	Muscular dystrophy, diabetes mellitus, acute infective hepatitis, leukemia, infectious diseases, hemolytic anemia etc.	—
15. Oxytocinase	Normal pregnancy from fourth month. Increasing level shows good fetal prognosis.	Intrauterine fetal death.

protein in the form of "acyl carrier protein". This can be regarded as coenzyme A in which the adenine dinucleotide is replaced by protein. "Acyl carrier protein" chiefly functions in the synthetic processes, eg, of fatty acids and cholesterol.

3. It is required in the oxidative decarboxylation of pyruvic acid and α-ketoglutaric acid, in the breakdown and synthesis of fatty acids and in the synthesis of cholesterol which is involved in bile acids, bile salts, steroid hormones and vitamin D formation.
4. It is used for conjugation with amino compounds to form N-acetyl compounds and in the formation of hippuric acid (Benzoyl glycine).
5. It is involved in the formation of ketone bodies.
6. It is used in the formation of acetyl choline.
7. It is finally oxidized to CO_2, H_2O and ATP via citric acid cycle.

Normal Values of Serum Enzymies

Serum Enzymes	*Normal values (Values vary with procedure used)*
1. Amylase	25-125 U/L
2. Lipase	0-160 U/L
3. Lactate dehydrogenase	90-200 IU/L
4. Acid phosphatase	< 4 U/L
5. Alkaline phosphatase	Adult : 20-70 U/L
6. SGOT	8-40 Karmen units/dL
7. SGPT	0-40 Karmen units/dL
8. Aldolase	up to 7.6 U/L

LACTATE DEHYDROGENASES

It is a glycolytic enzyme involved in the conversion of lactic acid to pyruvic acid and vice versa by the help of the coenzyme NAD.

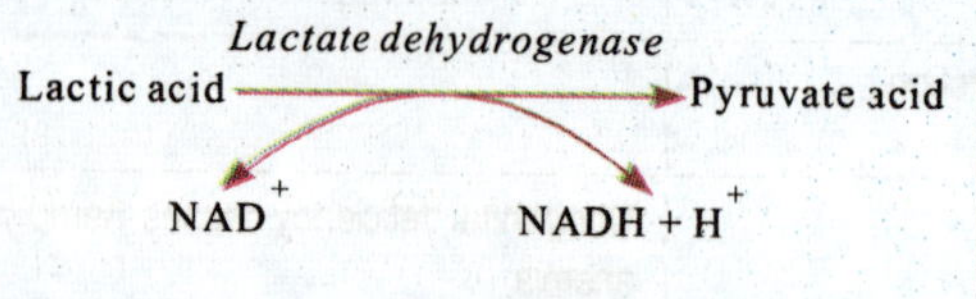

Fig. 10.30

It exists in five different isozyme forms eg, LDH_1, LDH_2, LDH_3, LDH_4 and LDH_5.

Source

Found in heart muscle as well as in many other tissues. It is liberated to the blood stream during myocardial injury.

Concentration

The normal level of serum LDH is 200-425 units/100 ml. or 90-200 I.U./liter.

Diagnostic importance

1. The rise of serum LDH is not so prompt as that of SGOT and the peak is not reached until the fourth or fifth day but the level may remain abnormally high for 6 to 12 days.
2. In case of suspected myocardial infarction, the absence or elevation of serum LDH is more significant in ruling out the disease.
3. In hepatitis, there is an insignificant rise in LDH; but a great rise in LDH is seen in untreated pernicious anemia.
4. Analysis of an overnight eight-hour urine specimen for lactate dehydrogenase activity is of great value in the diagnosis of serious diseases of kidney and bladder. Elevated findings are found, particularly in carcinoma of these organs.
5. An occasional high value is found in non-carcinomatous lesions of the kidney and bladder.

TRANSAMINASES

The enzymes which are involved in transamination reactions in the living organisms are said to be transaminases.

Examples are : Glutamate oxaloacetate transaminase (GOT).

Glutamate aspartate transaminase (GPT).

The transaminases are widely distributed in nature.

They also exist in isozyme forms.

Source

GOT is present in highest amounts in heart muscle, skeletal muscle, brain, liver, and kidney. Traces of these enzymes normally escapes into the blood stream.

Concentration

The normal level of SGOT is 8-40 Karmen units/dL.

The normal level of SGPT is 0-40 Karmen units/dL.

Diagnostic importance

1. When even a small portion of heart muscle is actually injured by occlusion of a branch of the coronary arteries an appreciable amount of GOT escapes into the blood stream and can be recognized by a sharp rise in the serum concentration.
2. In myocardial injury, a rise in SGOT concentration may occur in a few hours after onset of symptoms. The peak level is as high as 160 units within 48 hours and returns to the normal level within 3 to 5 days.
3. In myocardial infarction, the rise in SGPT is not so high as that of SGOT. But in liver disease SGPT is greatly increased.
4. In viral hepatitis, SGPT is always higher than SGOT ; but in cirrhosis and hemolytic jaundice SGOT is higher than SGPT.

CREATINE KINASE (CK or CPK)

This enzyme is involved in the formation of phosphocreatine from creatine and ATP (adenosine triphosphate). It is frequently referred to as creatine phosphokinase (CPK). It exists in isozyme forms.

Source

1. It is found in greatest amount in skeletal muscle.
2. It is present in appreciable amounts in cardiac muscle and in brain.
3. Small amounts are present in the lung, thyroid and adrenals.
4. It is apparently not present in liver and kidney and is not detectable in erythrocyte. Hence, serum levels are not affected by hemolysis.

Concentration

The normal level in serum is 12 to 99 U/I for males and 10 to 66 U/I for females.

(by Rosalki method.)

Diagnostic significance

1. There is a marked increase in serum creatine kinase levels, up to 50 times the normal, in patients suffering from the Deuchenne type of muscular dystrophy. There is a small elevation in this enzyme activity in patients suffering from certain other forms of muscular dystrophy like limb-girdle and myotonic types.
2. Enzyme levels are normal in muscular atrophy resulting from neurological disorders or associated with hyperthyroidism.
3. Increased levels are also found in muscle trauma, in polymyositis, in MC Ardle's syndrome and after severe exercise.
4. The activity of this enzyme is increased within 12 hours after the onset of an acute myocardial infarction. The level returns to normal in 3 or 4 days.
5. Increased levels of it can also occur in individuals with hypothyroidism, in patients with diabetic ketoacidosis, after pulmonary infarction, in the presence of convulsive disorders and in malignant hyperthermia.

CARBONIC ANHYDRASE

This is an enzyme which is associated with the hemoglobin in the red cells (never in the plasma). It has been isolated in highly purified form and shown to be a zinc-protein complex.

Source

1. Small amounts of carbonic anhydrase are found in muscle tissue, in the pancreas and in spermatozoa.
2. Much larger quantities occur in the parietal cells of the stomach.

Function

1. It specifically catalyzes the removal of CO_2 from H_2CO_3. The reaction is reversible :

$$CO_2 + H_2O \rightleftharpoons H_2CO_3$$

2. In the stomach, it is involved in the secretion of hydrochloric acid :

$$H_2CO_3 \rightleftharpoons H^+ + HCO_3^-$$

3. Since it occurs in the tubules of the kidney, it is also involved in hydrogen ion secretion.

EXERCISE

1. Define Enzyme. What is mechanism of their action? Discuss the factors which influence enzyme action. (R.U. 76A ; Muz. 72 March ; Mith. 67S, 74A ; R.U. 72S)
2. How are enzymes classified ? Describe the general properties of enzymes and mention their mechanism of action. (M.U. 75S)
3. Describe the mechanism of enzyme action and mention the role of coenzyme if required. (Bh. U. 76A)
4. Most of the vitamins are involved in Enzyme catalyzed reactions – Elucidate with at least two examples. (P.U. 69A)
5. Discuss the factors that regulate enzyme catalyzed reaction in the body. (P.U. 72S)
6. Short notes on :
   (a) Mechanism of enzyme action. (Bh.U. 75S ; R.U. 73A)
   (b) Specificity of enzyme action. (Mith. 62A)
   (c) Competitive inhibition of enzymes. (R.U. 66A. 70A ; Mith. 73A)
   (d) Enzyme inhibitor. (Bh.U. 75S)
   (e) Isoenzymes. (Bh.U. 76S ; P.U. 73A)
   (f) Coenzymes. (Muz. 75S ; R. U. 64A, 70S ; Mith. 61S, 68A)
   (g) Lactate dehydrogenase. (R.U. 69A)
   (h) Coenzyme A. (Bh.U. 76A ; P.U. 72A)
   (i) Allosteric enzymes. (P.U. 76A)
   (j) Carbonic anhydrase. (P.U. 74A)
   (k) Transaminases. (M.U. 76A)
   (l) Creatine kinase. (P.U. 72S)
   (m) Lysozyme. (B.U. 76A)
   (n) Lysosome. (B.U. 77S)
   (o) Apoenzyme, holoenzyme and coenzyme. (C.U. 83A)
7. Write in short the effects of the following factors on the activity of an enzyme.
   Explain each effect along with graph : (C.U. 1982)
   (a) Temperature.
   (b) Concentration of substrate.
   (c) Competitive inhibitors. (C.U. 1981)
8. Fill up the gaps with correct words :
   Activation of enzymes can be brought about by — and —. (C.U. 1981)
9. Enumerate coenzymes of nicotinic acid. Write one reaction for each of them to illustrate their actions. (C.U. 1981)

DIAGRAMMATIC COMPENDIUM

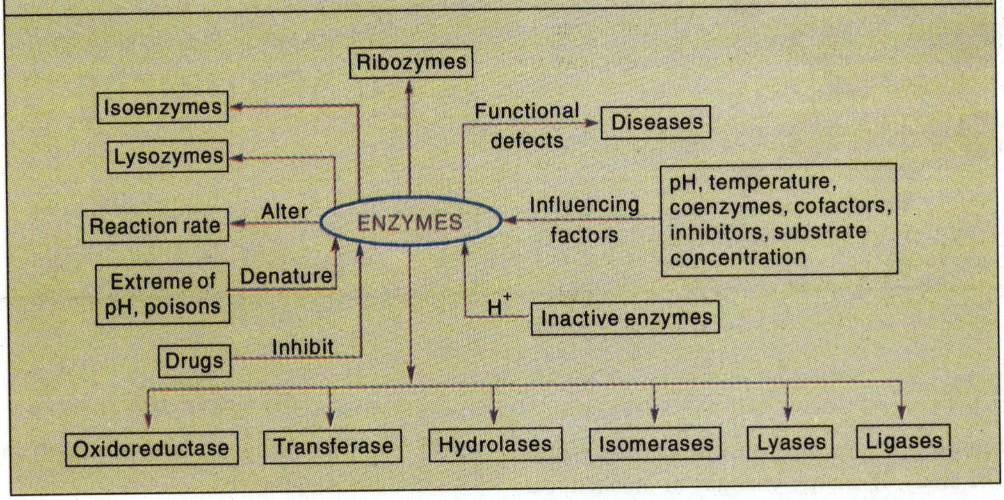

CHAPTER 11

BIOENERGETICS

Definition

Bioenergetics or biochemical thermodynamics is the study of energy changes in biochemical reactions. Non-biologic systems use heat energy to accomplish work but biologic systems are isothermic and utilise chemical energy for the living process.

Free energy and the laws of thermodyanmics

1. Free energy (ΔG) is the useful energy also known as the chemical potential.
2. The first law of thermodynamics states that "the total energy of a system plus its surroundings remains constant". This is also the laws of conservation of energy. Energy may be transferred from one part to another or may be transformed into another form of energy.
3. The second law of thermodynamics states that "the total entropy of a system must increase if a process is to occur spontaneously". Entropy represents the extent of disorder of the system and becomes maximum when it approaches true equilibrium. Under constant temperature and pressure, the relationship between the free energy change (ΔG) and the change in entropy (ΔS) is given by the following equation which combines the two laws of thermodynamics

$$\Delta G = \Delta H - T\Delta S$$

where ΔH = the change in enthalpy (heat) and T = the absolute temperature.

Under biochemical reactions ΔH is approximately equal to ΔE. So the above relationship may be expressed in the following manner :

$$\Delta G = \Delta H - T\Delta S$$

If ΔG is negative in sign, the reaction proceeds spontaneously with loss of free energy i.e. it is exergonic. On the other hand, if ΔG is positive, the reaction proceeds with the gain of energy i.e. it is endergonic. If the magnitude of ΔG is great, the system is stable. If ΔG is zero, the system is at equilibrium.

Coupling of endergonic to exergonic processes

1. The vital processes (Synthetic reactions, muscular contraction, nerve impulse conduction, and active transport) obtain energy by chemical linkage or coupling to oxidative reactions.
2. Metabolite A is converted to metabolite B with the release of free energy. It is coupled to another reaction in which free energy is required to convert metabolite C to metabolite D. Some of the energy liberated in the degradative reaction is transferred to the synthetic reaction. The exergonic reactions are termed Catabolism

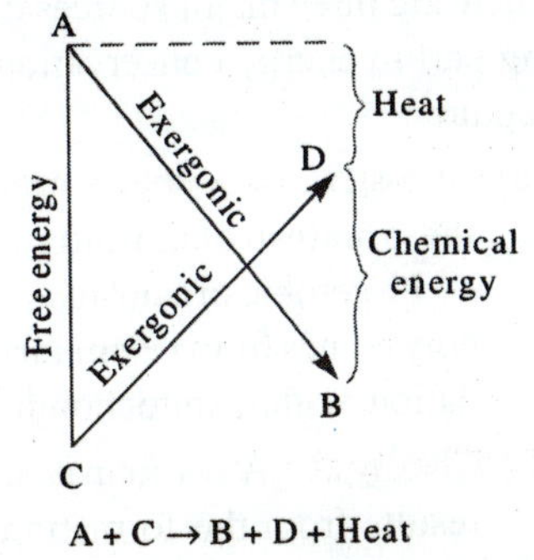

Fig. 11.1 : Coupling of exergonic to an endergonic reaction

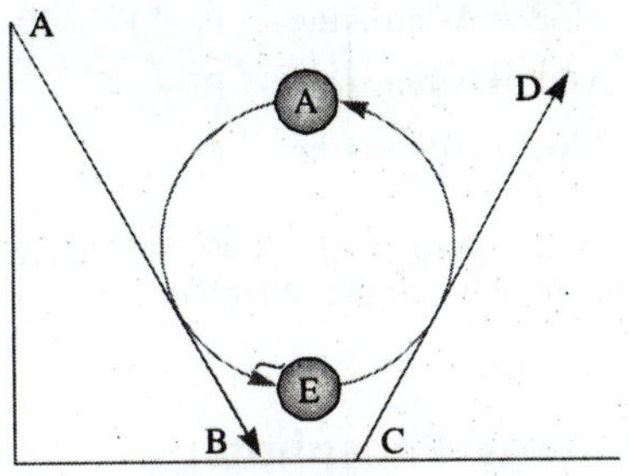

Fig. 11.3 : Transfer of free energy from an exergonic to endergonic reaction via higher energy intermediate compound

(the breakdown or oxidation of fuel molecules), whereas the synthetic reactions are termed Anabolism.

3. Reaction shown in Fig. 11.1 has to go from left to right, then the overall process must be accompained by loss of free energy as heat. One possible mechanism of coupling is shown :

$$A + C \rightarrow I \rightarrow B + D.$$

Some exergonic and endergonic reactions in biologic systems are coupled in this way. An extension of the coupling concept is provided by dehydrogenation reactions which are coupled to hydrogenations by an intermediate carrier (Fig. 11.2).

4. The alternative process of coupling from an exergonic to an endergonic process is to synthesize a compound of high energy potential in the exergonic reaction and to incorporate this new compound into the endergonic reaction, thus effecting a transference of free energy from the exergonic to the endergonic pathway (Fig. 11.3).

In the living cell, the principal high energy intermediate or carrier compound (designated ~Ⓔ) is ATP.

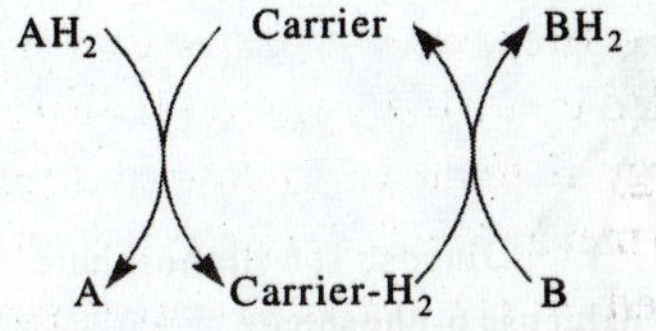

Fig. 11.2 : Coupling of hydrogenation and dehydrogenation reactions by and intermediate carrier

Role of energy Phosphates in Bioenergetics & energy capture

1. Autotrophic organisms couple their metabolism to some simple exergonic process in their surroundings eg, green plants utilize the energy of sunlight, and some autotrophic bacteria utilize the reaction $Fe^{++} \rightarrow Fe^{+++}$.
2. Heterotrophic organisms obtain free energy by coupling their metabolism to the breakdown of complex organic molecules in their environment.
3. In all these processes, ATP plays an important role in the transfer of free energy from the exergonic to the endergonic processes. ATP is a nucleotide consisting of adenine, ribose, and three phosphate groups. In its reaction in the cell, it functions as the Mg^{++} complex.
4. ATP was considered to be a means of transferring phosphate radicals in the process of phosphorylation. Lipmann introduced the concept of "high-energy phosphates" and the "high-energy phosphate bond" :

ATP = Adenosine – Ⓟ ~ Ⓟ ~ Ⓟ

ADP = Adenosine – Ⓟ ~ Ⓟ

AMP = Adenosine – Ⓟ

Fig. 11.4 : Structure of ATP, ADP, AMP showing the position and the number of high-energy bonds (~)

High-energy Phosphates

Lipmann introduced the symbol ~ Ⓟ, indicating **high-energy phosphate bond**. The term **group transfer potential** is preferred to "high-energy bond". Thus, ATP contains 2 high-energy phosphate groups and ADP contains one. The phosphate bond in AMP is of the low energy type, since it is a normal ester link (Fig. 11.4).

Role of High-energy phosphates as the "Energy Currency" of the Cell

1. ATP is the donor of high-energy phosphate and ADP can accept high-energy phosphate to form ATP. ATP/ADP cycle connects these processes which generate ~ Ⓟ to those processes that utilize ~ Ⓟ.
2. There are three major sources of ~ Ⓟ taking part in energy conservation or energy capture :
   (a) **Oxidative phosphorylation :** This is the greatest quantitative source of ~ Ⓟ aerobic organisms. The free energy comes from respiratory chain oxidation within mitochondria.
   (b) **Glycolysis :** A net formation of 2 ~ Ⓟ results from the formation of lactate from one molecule of glucose, generated in two reactions catalyzed by phosphoglycerate kinase and pyruvate kinase.
   (c) **Citric acid cycle :** One ~ Ⓟ is generated directly in the cycle at the succinyl thiokinase step.
3. Another group of compounds (Phosphagens) act as storage forms of high-energy phosphate. These include creatine phosphate in vertebrate muscle and brain, arginine phosphate in invertebrate muscle.
4. In physiologic conditions, phosphagens permit ATP concentrations to be maintained in muscle when ATP is being rap-

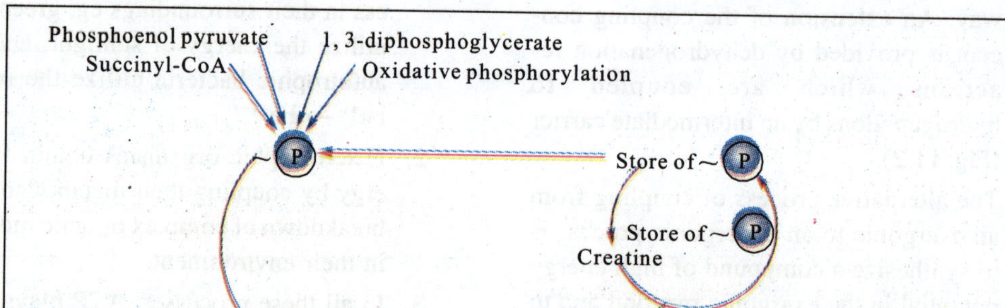

Fig. 11.5 : Role of ATP/ADP cycle in transfer of high-energy phosphate. ~ Ⓟ does not exist in a free state but is transferred in the reaction

idly used as a source of energy for muscular contraction. When ATP is abundant, its concentration can cause the reverse reaction to take place and allow the concentration of creatine phosphate to increase abundantly so as to act as a store of high-energy phosphate. When ATP acts as a phosphate donor to form those compounds of lower free energy of hydrolysis, the phosphate group is invariably converted to one of low energy eg :

$$\text{Glycerol} + \text{Adenosine} - \text{Ⓟ} \sim \text{Ⓟ} \sim \text{Ⓟ} \xrightarrow{\text{Glycerol kinase}} \text{Glycerol} - \text{Ⓟ} + \text{Adenosine} - \text{Ⓟ} \sim \text{Ⓟ}$$

Interconversion of Adenine Nucleotides

The enzyme Adenylate Kinase (myokinase) is present in most cells. It catalyzes the interconversion of ATP and AMP on the one hand and ADP on the other.

The reaction has three functions :

1. It allows high-energy phosphate in ADP to be used in the synthesis of ATP.
2. It allows AMP to be recovered by rephosphorylation to ADP
3. It allows AMP to increase in concentration when ATP becomes depleted and act as a metabolic signal to increase the rate of catabolic reactions which, in turn leads, to the generation of more ATP :

$$\underset{\text{(ATP)}}{\text{Adenosine} - \text{Ⓟ} \sim \text{Ⓟ} \sim \text{Ⓟ}} + \underset{\text{(AMP)}}{\text{Adenosine} - \text{Ⓟ}} \xrightarrow{\text{Adenylate kinase}} \underset{\text{(2ADP)}}{2\ \text{Adenosine} - \text{Ⓟ} \sim \text{Ⓟ}}$$

Metabolism of Pyrophosphate

Inorganic pyrophosphate is formed as occurs in the activation of long chain fatty acids when ATP reacts to form AMP :

$$\text{ATP} + \text{CoA.SH} + \text{R.COOH} \xrightarrow{\text{Acyl-CoA Synthetase}} \text{AMP} + \text{PPi} + \text{R.Co.} \sim \text{S CoA}$$

This reaction is accompained by loss of free energy as heat which ensures that the activation reaction will go to the right. This is further aided by the hydrolytic splitting of PPi by inorganic Pyrophosphatase. The activation via the pyrophosphate pathway results in the loss of 2 ~ Ⓟ rather than one ~ Ⓟ as occurs when ADP and Pi are formed.

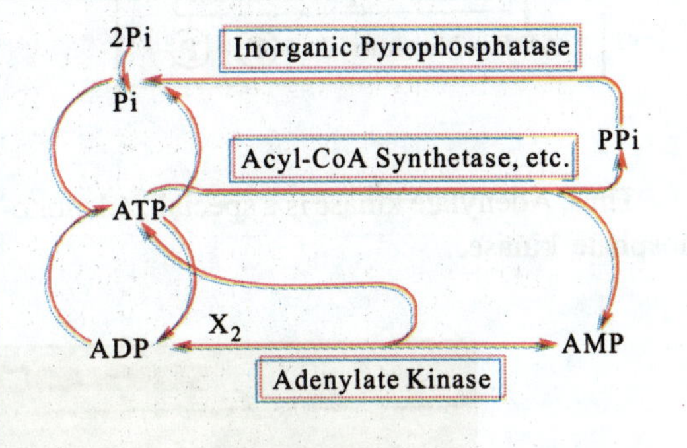

Fig. 11.6 : Phosphate cycles and interchange of adenine nucleotides

$$\text{PPi} + H_2O \xrightarrow[\text{Pyrophosphatase}]{\text{Inorganic}} 2\text{Pi}$$

Nucleoside phosphates related to ATP and ADP

Nucleoside triphosphates similar to ATP but containing an alternative base to adenine can be synthesized from their diphosphates by means of the enzyme nucleoside diphosphate kinase, eg :

ATP + UDP
↕ Nucleoeside diphosphate kinase
ADP + UTP

ATP + GDP ⟷ ADP + GTP

ATP + CDP ⟷ ADP + CTP

All these triphosphates take part in phosphorylations in the cell. Similarly, **nucleoside monophosphate kinases** catalyze the formation of nucleoside diphosphates from the corresponding monophosphates :

ATP + Nucleoside –Ⓟ
↕ Specific nucleoside monophosphate kinase
ADP + Nucleoside –Ⓟ~Ⓟ

Thus, Adenylate kinase is a specialized monophosphate kinase.

CLINICAL ASPECTS

- Energy production from food is the fundamental role of normal nutrition and metabolism.
- Death results from starvation when energy reserves are depleted and types of malnutrition are associated with energy imbalance (marasmus).
- The rate of energy release is controlled by the thyroid hormones whose malfunction is a cause of disease.
- Obesity is caused by the storage of surplus energy.

EXERCISE

1. Define Bioenergetics. State the Laws of Thermodynamics. Show the interconversion of Adenine Nucleotides. (Mith. 76S; R.U. 74A; Bh.U. 79S)
2. Discuss the role of high-energy phosphates as the "Energy Currency" of the cell. Enumerate nucleoside phosphates related to ATP and ADP. (P.U. 77A; C.U. 1982; R.U. 75S)
3. Explain the role of energy phosphates in Bioenergetics and energy capture. Describe the metabolism of pyrophosphate. (M.U. 78A; Bh. U. 79A; Muz. 76A)
4. Short notes on :
   (a) High-energy phosphates. (C.U. 1980)
   (b) Interconversion of Adenine Nucleotides. (C.U. 1984)

SCHEMATIC COMPENDIUM

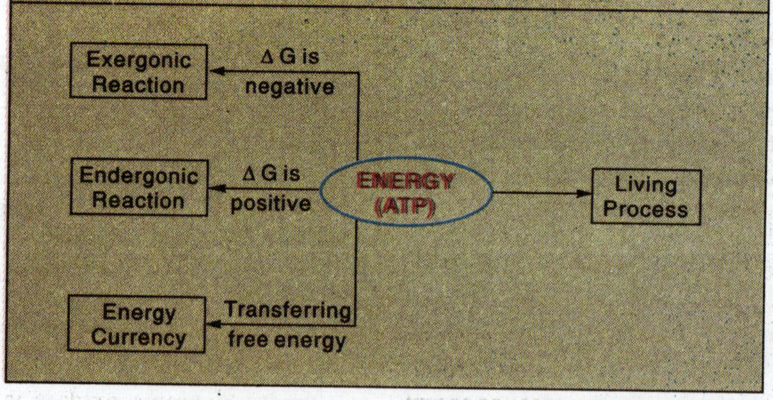

CHAPTER 12

BIOLOGIC OXIDATION

Oxidation is chemically defined as the removal of electrons and **Reduction** is the gain of electrons.

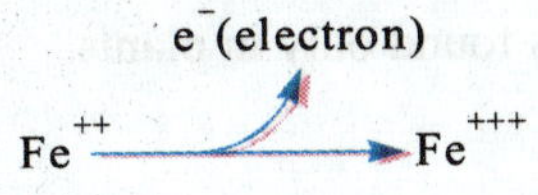

Fig. 12.1

Substrate molecules are oxidized by removal of hydrogen by dehydrogenases. The reduced dehydrogenases are then reoxidized by a group of respiratory catalysts known as cytochrome system. The substrate is thus oxidized by both processes. The reducing equivalents ultimately react with molecular oxygen in the presence of cytochrome oxidase, the last member of the cytochrome system.

Respiratory Chain : The sequence of enzymes and carriers responsible for the transport of reducing equivalents from substrates to molecular oxygen is known as respiratory chain.

The respiratory chain is localized within the mitochondria. Formation of ATP in the mitochondria is the active area of research.

Redox Potential : In oxidation and reduction reactions, the free energy exchange is proportionate to the tendency of reactants to donate or accept electrons. This is expressed as an ***oxidation-reduction or redox potential***.

Enzymes and coenzymes involved in oxidation and reduction

In 1961, the International Union of Biochemistry has designated all enzymes concerned in oxidative processes as oxidoreductases. These are classified into 5 groups :

1. Oxidases

(a) Enzymes that catalyze the removal of hydrogen from a substrate but use only oxygen as a hydrogen acceptor to form water as a reaction product (with the exception of uricase and monoamine oxidase which form H_2O_2) :

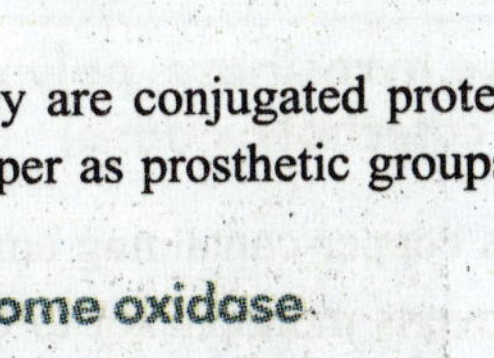

Fig. 12.2 : Oxidation of a metabolite by oxidase

(b) They are conjugated proteins containing copper as prosthetic groups.

(i) Cytochrome oxidase

(a) Cytochrome oxidase is a hemoprotein widely distributed in plants and animal tissues.

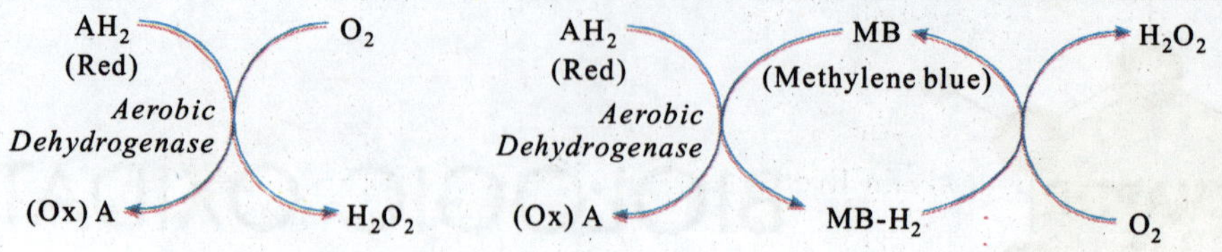

Fig. 12.3 : Oxidation of a metabolite by an aerobic dehydrogenase

(b) It is the terminal component of respiratory chain found in mitochondria.

(c) It is poisoned by cyanide and hydrogen sulfide.

(d) More recent studies show that 2 cytochromes are combined with the same protein and the complex is known as cytochrome aa_3.

(e) Cytochrome aa_3 contains 2 molecules of heme A, each having one Fe atom. 2 atoms of Cu are also present which are associated with the cytochrome oxidase activity.

(f) The terminal cytochrome aa_3 is responsible for the final combination of reducing equivalents with molecular oxygen.

(g) This enzyme system contains copper, a component of several oxidase enzymes.

(h) It has a high affinity for oxygen.

(i) It is the only one in the chain which signifies the irreversible reaction.

(j) It gives direction to the movement of reducing equivalents in the respiratory chain and to the production of ATP, to which it is coupled.

(ii) Phenolase (tyrosinase, polyphenol oxidase, catechol oxidase)

(a) It is a copper-containing enzyme.

(b) It converts monophenol to O-quinones.

(iii) Laccase

(a) It is widely distributed in plants and animals.

(b) It converts P-hydroquinones to P-quinones.

(c) It also contains copper.

(iv) Ascorbic oxidase

(a) It contains copper.

(b) It is found only in plants.

(v) Uricase

(a) It also contains copper.

(b) It catalyzes the oxidation of uric acid to allantoin.

(vi) Monoamine oxidase

(a) It is found in the mitochondria of several tissues.

(b) It oxidizes epinephrine and tyramine.

2. Aerobic dehydrogenases

(a) They catalyze the removal of hydrogen from a substrate and use either oxygen or artificial substances such a methylene blue as hydrogen acceptor.

(b) H_2O_2 is formed as a product.

(c) They are flavoprotein enzymes having FMN (flavin mononucleotide) or FAD (Flavin adenine dinucleotide) as prosthetic groups.

(d) Many of the flavoprotein enzymes contain a metal for which they are known as metalloflavoprotein enzymes.

(i) D-amino acid dehydrogenase (D-amino acid oxidase)

(a) It is an FAD-linked enzyme.

(b) It is found particularly in liver and kidney.

(c) It catalyzes the oxidative deamination of the unnatural (D-) forms of amino acids.

(ii) L-amino acid dehydrogenase (L-amino acid oxidase)

(a) It is an FMN-linked enzyme.

(b) It is found in kidney.

(c) It catalyzes the oxidative deamination of naturally occurring L-amino acids.

(iii) Xanthine dehydrogenase (Xanthine oxidase)

(a) It occurs in milk and liver.

(b) In the liver, it converts purine bases to uric acid.

(c) It contains FAD as the prosthetic group.

(d) It is highly significant in the liver and kidneys of birds which excrete uric acid as the end product of purine metabolism and also of protein and amino acid catabolism.

(e) It is a metalloflavoprotein containing nonheme iron and molybdenum.

(f) It also oxidizes all aldehydes.

(iv) Aldehyde dehydrogenase (aldehyde oxidase)

(a) It is an FAD-linked enzyme.

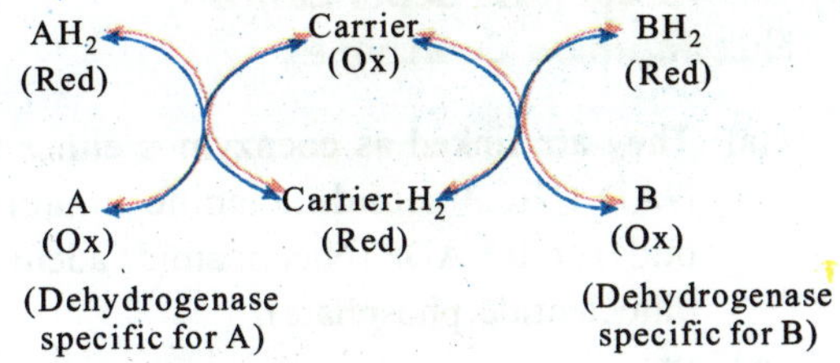

Fig. 12.4 : Oxidation of metabolite by anaerobic dehydrogenases not involving a respiratory chain

(b) It is present in pig and other mammalian liver.

(c) It is also a metalloflavoprotein containing nonheme iron and molybdenum.

(d) It oxidizes aldehydes.

(v) Glucose oxidase

(a) It is an FAD-linked enzyme.

(b) It is prepared from fungi.

(c) It is used in estimating glucose.

3. Anaerobic dehydrogenases

(a) They catalyze the removal of hydrogen from a substrate but not able to use oxygen as hydrogen aceptor.

(b) They transfer hydrogen from one substrate to another by oxidation-reduction reaction not involving a respiratory chain (shown in Fig. 12.4.).

(c) They perform oxidation of metabolite utilizing several components of a respiratory chain (shown in Fig. 12.5).

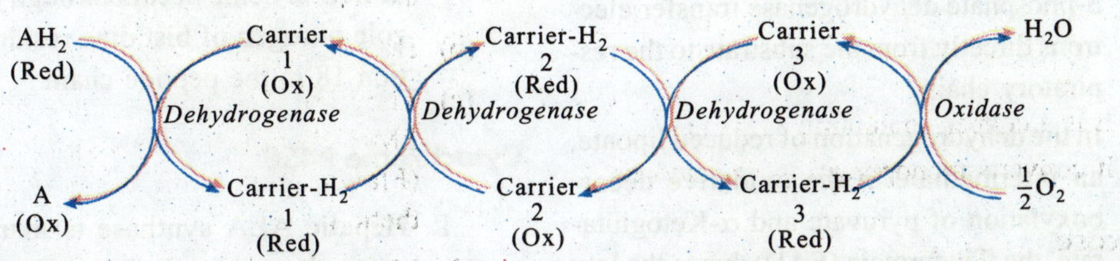

Fig. 12.5 : Oxidation of metabolite by aerobic dehydrogenases by several components of respiratory chain

(i) Dehydrogenase dependent on Nicotinamide Coenzymes

(a) They are linked as coenzymes either to NAD (Nicotinamide adenine dinucleotide) or to NADP (Nicotinamide adenine dinucleotide phosphate).

(b) The coenzymes are reduced by the particular substrate of the dehydrogenase and reoxidized by a suitable electron acceptor and synthesized from the vitamin niacin (nicotinic acid and nicotinamide).

(c) NAD-linked dehydrogenases catalyze oxidoreduction reactions in glycolysis, the citric acid cycle and in the respiratory chain of mitochondria.

(d) NADP-linked dehydrogenases are found in fatty acid and steroid synthesis in the extramitochondria. They are also found in hexose monophosphate shunt.

(e) Some nicotinamide coenzyme-dependent dehydrogenases contain zinc, particularly alcohol dehydrogenase from liver and glyceraldehyde-3-phosphate dehydrogenase from skeletal muscle. The zinc ions do not take part in the oxidation and reduction.

(ii) Dehydrogenases dependent on Riboflavin Prosthetic Groups

(a) Most of the riboflavin-linked anaerobic dehydrogenases are concerned with electron transport in the respiratory chain.

(b) Succinate dehydrogenase, acyl–CoA dehydrogenase and mitochondrial glycerol-3-phosphate dehydrogenase transfer electrons directly from the substrate to the respiratory chain.

(c) In the dehydrogenation of reduced lipoate, an intermediate in the oxidative decarboxylation of pyruvate and α-Ketoglutarate, the flavoprotein (FAD) due to the low redox potential acts as a carrier of electrons from reduced lipoate of NAD^+. The electron transferring flavoprotein in an intermediary carrier of electrons between acyl–CoA dehydrogenase and the respiratory chain.

(iii) The Cytochromes

(a) The cytochromes excepting cytochrome oxidase are anaerobic dehydrogenases. They are involved as carriers of electrons from flavoproteins to cytochrome oxidase in the respiratory chain.

(b) They are iron-containing hemoproteins in which iron becomes Fe^{+++} and Fe^{++} oxidation and reduction. The cytochromes in the respiratory chain are b, c_1, c, a and a_3.

(c) Cytochromes are also found in the endoplasmic reticulum (cytochromes P-450 and b_5) plant cells, bacteria and yeast.

Cytochrome C

1. It has a molecular wt. of 13,000.
2. The iron porphyrin group of cytochrome c is attached to protein more firmly than in the hemoglobin.
3. It is quite stable to heat and acids.
4. The reduced form of cytochrome c is not autooxidizable.
5. The peptide chain of human heart cytochrome c contains 104 amino acids. Acetylglycine is the N-terminal amino acid and glutamic acid is the c-terminal amino acid. Two cysteine residues are located at positions 14 and 17. The linkage of iron in heme occurs through the imidazole nitrogen of histidine residue at position 18 in the peptide chain.

Cytochrome P450

1. Hepatic ALA synthase is markedly increased on the administration of many drugs to humans. Most of these drugs are metabolized in the liver by a system that

utilizes a specific hemoprotein, Cytochrome P450. The utilization of heme by cytochrome P450 is highly increased during the metabolism of these drugs and in turn diminishes the intracellular heme concentration.

2. Hydroxylation reactions take place by the enzymes monooxygenases or cytochrome P450 The reaction catalyzed by a monooxygenase (cytochrome P450) is :

$$RH + O_2 + NADPH + H^+ \rightarrow R\text{–}OH + H_2O + NADP$$

[RH represents a very wide variety of xenobiotics including drugs, carcinogens, pesticides, petroleum products, and pollutants].

3. Cytochrome P450 is considered the most versatile biocatalyst. It has been shown by the use of $18O_2$ that one atom of oxygen enters R–OH and one atom enters water. This dual fate of the oxygen is responsible for naming of monooxygenases as "mixed-function oxidases". The reaction catalyzed by cytochrome P450 is represented :

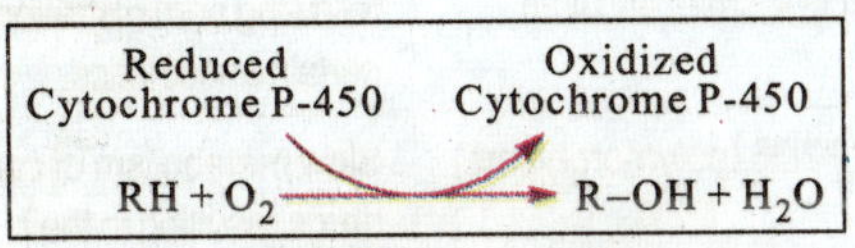

4. The major monooxygenases in the endoplasmic reticulum are cytochrome P-450s—so named because the enzyme was prepared from microsomes that had been chemically reduced and exhibited a distinct peak at 450 nm on exposure to carbon monoxide. About 50 per cent of the drugs that humans ingest are metabolized by isoforms of cytochrome P450; these enzymes also act on various carcinogens and pollutants.

5. About 150 isoforms of cytochrome P450 have been discovered so far. The abbreviated root symbol CYP denotes a cytochrome P450. CYPA1 denotes a cytochrome P450 that is a member of family 1 and subfamily A and is the first individual member of that subfamily.

6. Like hemoglobin, they are hemoproteins.

7. These are widely distributed among species. Bacteria possess cytochrome P450s, and $P450_{cam}$ (involved in the metabolism of camphor) of pseudomonas putida is the only P450 isoform.

8. They are present in large amount in liver. They are present mainly in the membranes of the smooth endoplasmic reticulum in liver and most other tissues. In the adrenal, they are found in mitochondria as well as in the endoplasmic reticulum; the various hydroxylases present in that organ play an important role in cholesterol and steroid biosynthesis. The mitochondrial cytochrome P450 system differs form the microsomal system in that it uses an NADPH-linked flavoprotein, adrenodoxin reductase, and a non-heme iron-sulphur protein, adrenodoxin.

9. There are at least six isoforms of cytochrome P450 present in the endoplasmic reticulum of human liver, each is acting on both xenobiotics and endogenous compounds. In recent years, the genes for many isoforms of P450 have been isolated and studied.

10. NADPH is involved in the reaction mechanism of cytochrome P450. The enzyme that uses NADPH to form the reduced cytochrome P450 is called **NADPH-cytochrome P450 reductase**. Electrons are transferred from NADPH to NADPH-cytochrome P450 reductase and then to cytochrome P450. This leads to the reductive activation of molecular oxygen, and one atom of oxygen is subsequently inserted into the substrate.

11. Lipids are also components of the cytochrome P450 system. The suitable lipid is phosphatidyl choline which is found in membranes of the endoplasmic reticulum.

12. Most isoforms of cytochrome P450 are inducible. The mechanism of induction involves increased transcription of mRNA for cytochrome P450. Induction of cytochrome P450 has important clinical implications, since it is a biochemical mechanism of drug interaction. The example of enzyme induction is CYP2E1 which is induced by consumption of ethanol. This P450 metabolizes certain widely used solvents and also components found in tobacco smoke, many of which are established carcinogens. If the activity of CYP2E1 is elevated by induction, this may increase the risk of carcinogenicity.

13. Certain isoforms of cytochrome P450 (eg. CYP1A1) are particularly involved in the metabolism of polycyclic aromatic hydrocarbons (PAHs) and related molecules. That is why they were formerly called aromatic hydrocarbon hydroxylases (AHH). These enzymes are important in the metabolism of (PAHs) and in carcinogenesis produced by these agents. Smokers have higher levels of this enzyme in some of their cells and tissues than do non-smokers. Some reports have mentioned that the activity of this enzyme may be elevated (induced) in the placenta of a woman who smokes, thus altering the quantities of metabolites of PAHs to which the fetus is exposed.

14. Recently, it has been shown that certain cytochrome P450s exist in polymorphic forms, some of which exhibit low catalytic activity. These observations are the important explanation for the variations in drug responses noted among many patients. One P450 exhibiting polymorphism is CYP2D6 which is involved in the metabolism of debrisoquin (an antihypertensive drug). Certain polymorphisms of CYP2D6 cause poor metabolism of these and many other drugs, so that they can accumulate in the body resulting in aweful consequences.

Table 12.1 : Some properties of cytochrome P450s

— Involved in the metabolism of many xenobiotics and steroids.
— Exhibit broad substrate specificity.
— All are hemoproteins.
— Cause introduction of one atom of oxygen into the substrate and one into water.
— The hydroxylated products are more water-soluble.
— Present in large amounts in the liver but found in most tissues.
— Found in the smooth endoplasmic reticulum or in mitochondria.
— Many are inducible.
— Molecular mass of about 55 KDa.
— Their products are carcinogenic in some cases.
— Some exhibit polymorphism.

Table 12.2 : Some important drug reactions due to polymorphic forms of ehzymes

Enzyme or protein affected	Reaction or consequence
Glucose-6-phosphate dehydrogenase	Hemolytic anemia following ingestion of drugs such as primaquine
Ca^{2+} release channel in the sarcoplasmic reticulum	Malignant hyperthermia following administration of certain anesthetics.
CYP2D6 (polymorphisms)	Slow metabolism of certain drugs resulting in their accumulation

4. Hydroperoxidases

They utilize hydrogen peroxide as a substrate. Two enzymes fall into this category (i) Peroxidase, (ii) Catalase.

(i) Peroxidase

(a) It is found in milk and leukocytes and the prosthetic group is protoheme.

(b) It catalyzes the reduction of hydrogen peroxide by the help of ascorbic acid, quinones and cytochrome C which act as electron acceptors. The reaction is complex but the overall reaction is :

$$H_2O_2 + AH_2 \xrightarrow{\text{Peroxidase}} 2H_2O + A$$

Fig. 12.6

(ii) Catalase

(a) It is hemoprotein and found in blood and liver.

(b) It uses one molecule of H_2O_2 as a substrate electron donor and another molecule as electron acceptor :

$$2H_2O_2 \xrightarrow{\text{Catalase}} 2H_2O + O_2$$

Fig. 12.7

(c) Its function is to destroy H_2O_2 formed by the action of aerobic dehydrogenases.

5. Oxygenases

They catalyze the incorporation of oxygen into a substrate molecule.

(i) Dioxygenases (Oxygen transferases, true oxygenases)

(a) They catalyze the incorporation of two atoms of oxygen (O_2) into the substrate :

$$A + O_2 \longrightarrow AO_2$$

Fig. 12.8

(b) Enzymes containing iron as a prosthetic group eg, homogentisate dioxygenase, 3-hydroxyxanthranilate dioxygenase and enzymes utilizing heme as a prosthetic group such as L-tryptophan dioxygenase (tryptophan pyrrolase) from the liver.

(ii) Mono-oxygenase (Mixed function oxidases, Hydroxylases)

(a) They catalyze the incorporation of only one atom of the oxygen molecule into a substrate. The other oxygen atom is reduced to water. A cosubstrate is necessary for this purpose :

$$A\text{–}H + O_2 + BH_2 \longrightarrow A\text{–}OH + H_2O + B$$

Fig. 12.9

(b) Many of the enzymes involved in steroid synthesis are mono-oxygenase using NADPH as a cosubstrate. They are found mainly in the endoplasmic reticulum (Microsomes) of the liver and in mitochondria and the microsomes of the adrenal glands.

(c) They are also involved in the metabolism of many drugs by hydroxylation. They are found in the microsomes of the liver together with cytochrome P450 and cytochrome b_5. The drugs metabolized by this system are benzpyrine, aminopyrine, aniline, morphine and benzphetamine. But phenobarbital induce the formation of microsomal enzymes and of cytochrome P450 :

$$\underset{(P450)}{DRUG - H + O_2 + Fe^{++} + 2H^+} \xrightarrow{\text{Hydroxylase}} \underset{(P450)}{DRUG - OH + H_2O + 2Fe^{+++}}$$

Fig. 12.10

(d) They are concerned with the synthesis or degradation of many different types of metabolites.

SUPEROXIDE METABOLISM

1. Friedovich (1975) had suggested that the toxicity of oxygen is due to its conversion to superoxide. Super-oxide is formed when reduced flavins eg, xanthine dehydrogenases are reoxidized by molecular oxygen in the respiratory chain :

$$Enz.H_2 + O_2 \longrightarrow Enz.H + O_2^- + H^+$$

Fig. 12.11

2. Superoxide can reduce oxidized cytochrome C :

$$O_2^- + Fe^{+++} \longrightarrow O_2 + Fe^{++}$$

Fig. 12.12

3. Superoxide can be removed by the specific enzyme **superoxide dismutase** in the presence of protons :

$$O_2^- + O_2^- + 2H^+ \xrightarrow{\text{Superoxide dismutase}} H_2O_2 + O_2$$

Fig. 12.13

Superoxide dismutase

(a) The cytosolic enzyme is composed of two similar subunits, each one contains Cu^{++} and Zn^{++}.

(b) The mitochondrial superoxide dismutase contains Mn^{++}.

(c) This enzyme is present in all tissues.

(d) It decomposes superoxide as soon as it is formed and thus protects the tissues against the harmful action of superoxide.

(e) The enzyme is increased particularly in the lungs on the exposure of the animals to an atmosphere of 100% oxygen.

RESPIRATORY CHAIN

The mitochondrion has been termed the **"Power House"** of the cell due to the following reasons :

1. Most of the useful energy derived from oxidation within the tissues is captured in the form of the high-energy intermediate, ATP.
2. All the useful energy formed from the oxidation of fatty acids and amino acids and virtually all the energy formed from the oxidation of carbohydrates are available in the mitochondrion.
3. For the production of this energy the mitochondrion contains the series of catalysts known as the respiratory chain which are concerned with the transport of reducing equivalents (Hydrogen and electrons) and with their final reaction with oxygen to form water.
4. Mitochondrion also contains the enzyme systems i.e. the enzymes of β-oxidation and of the citric acid cycle which are responsible for producing the reducing equivalents in the first place.

These relationships are shown in (Fig. 12.16).

Organisation of the Respiratory Chain in Mitochondria

1. The major components of the respiratory chain are arranged in order of increasing redox potential shown in (Fig 12.17).

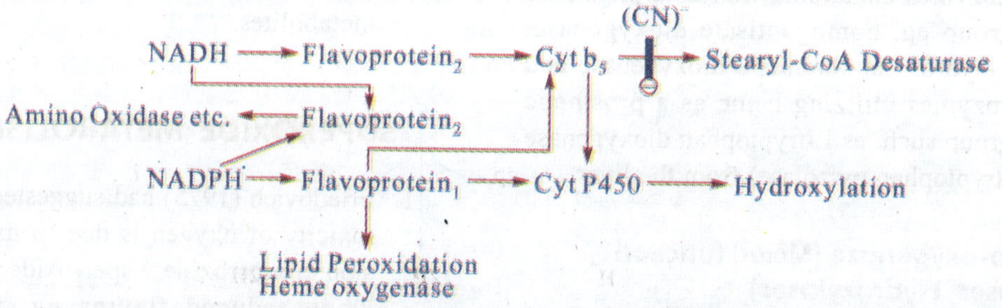

Fig. 12.14 : Electron trasport chain in microsome

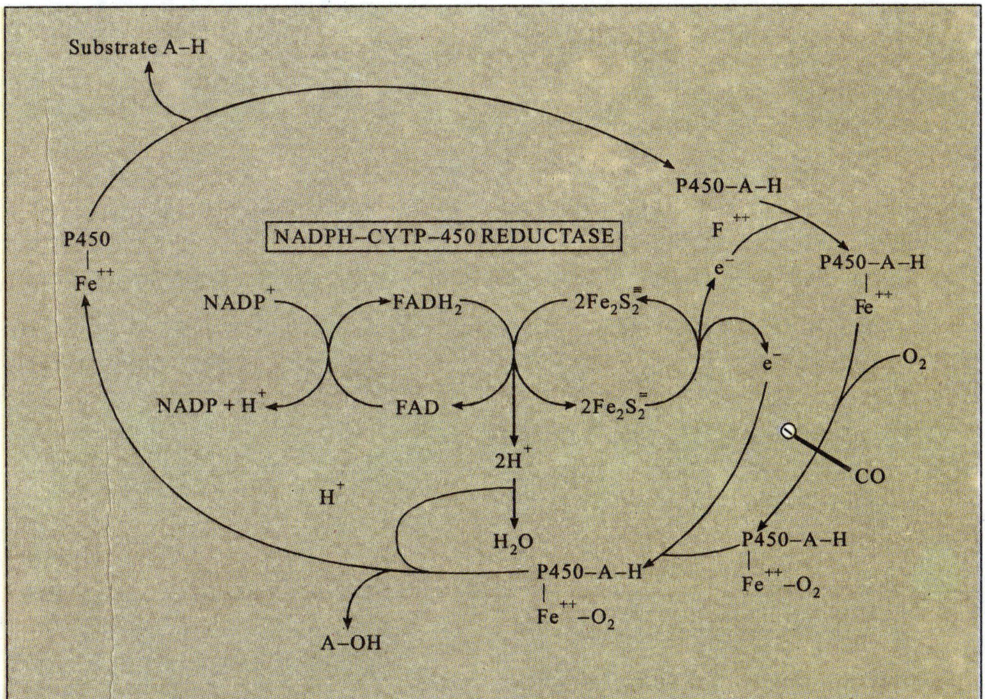

Fig. 12.15 : Cytochrome P450 hydroxylase cycle in microsomes

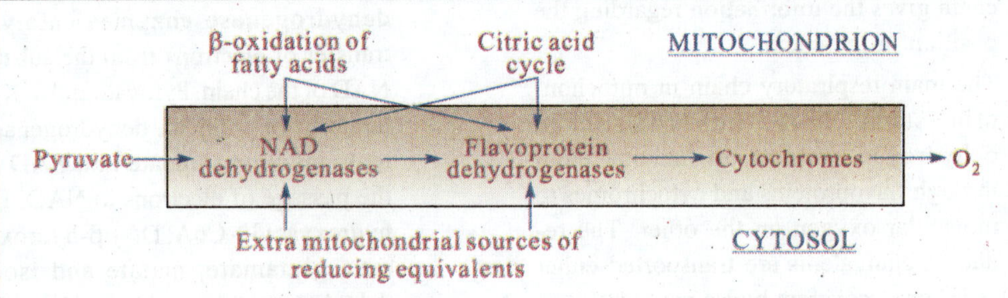

Fig. 12.16 : Relationship of electron transport in the respiratory chain

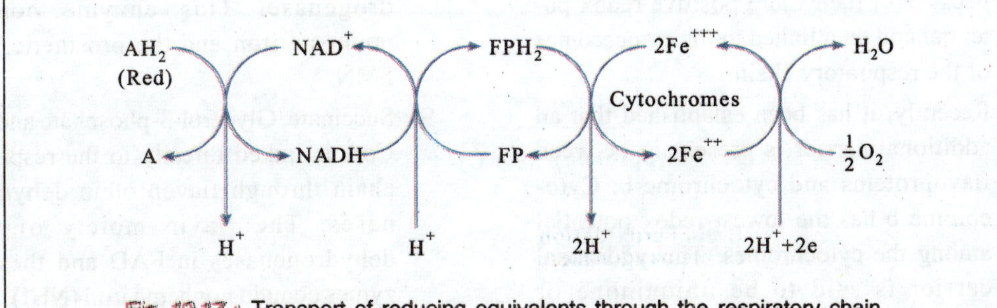

Fig. 12.17 : Transport of reducing equivalents through the respiratory chain

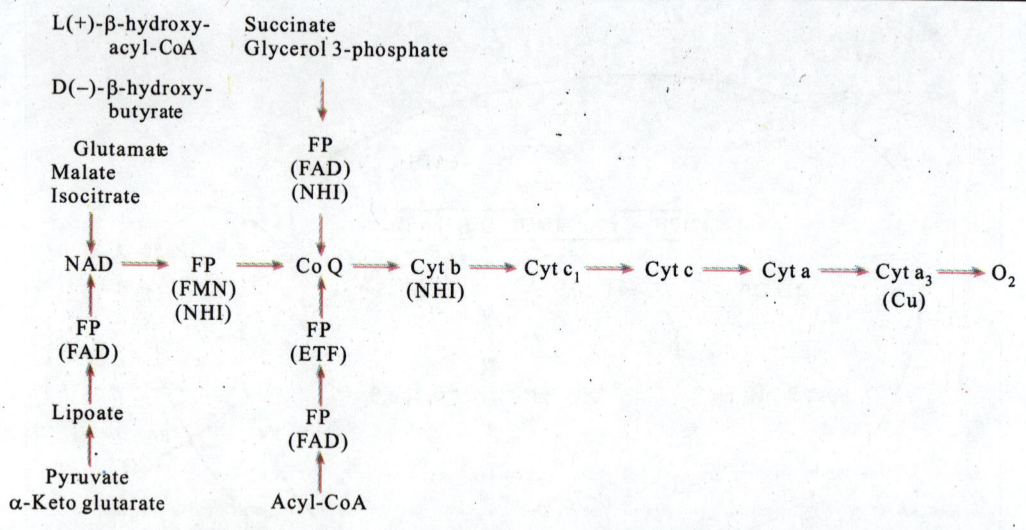

Fig. 12.18 : Components of the respiratory chain in mitochondria

2. Electron flow through the chain from the more electronegative components to the more electropositive oxygen. Thus, the redox potential of a component of the chain gives the information regarding the position in the chain.
3. The main respiratory chain in mitochondria starts from the NAD-linked dehydrogen systems on the one hand through flavoproteins and cytochromes to molecular oxygen on the other. The reducing equivalents are transported either as H^+ or as covalent hydrogen.
4. All substrates are not linked through NAD-specific dehydrogenases; some are linked directly to flavoprotein dehydrogenases because of their more positive redox potential and then linked to the cytochromes of the respiratory chain.
5. Recently, it has been established that an additional carrier is present in between flavoproteins and cytochrome b. Cytochrome b has the lowest redox potential among the cytochromes. This additional carrier is said to be ubiquinone or coenzyme Q (CoQ).
6. An additional component, nonheme iron (NHI), is associated with the flavoproteins and with cytochrome b.
7. At the electronegative end of the chain, dehydrogenase enzymes catalyze the transfer of electrons from the substrate to NAD of the chain. Pyruvate and α-Ketoglutarate have complex dehydrogenase systems involving lipoate and FAD before the passage of electrons to NAD. L(+)-β-hydroxyacyl – CoA, D(-)-β-hydroxybutyrate, glutamate, malate and isocitrate dehydrogenases couple directly with NAD of the respiratory chain.
8. The reduced NAD is oxidized by a metalloflavoprotein enzyme NADH dehydrogenase. This enzyme contains nonheme iron and the prosthetic group FMN.
9. Succinate, Glycerol-3-phosphate and acyl-CoA is linked directly to the respiratory chain through flavoprotein dehydrogenases. The flavin moiety of these dehydrogenases in FAD and these enzymes contain nonheme iron (NHI). In the dehydrogenation of acyl-CoA, an addi-

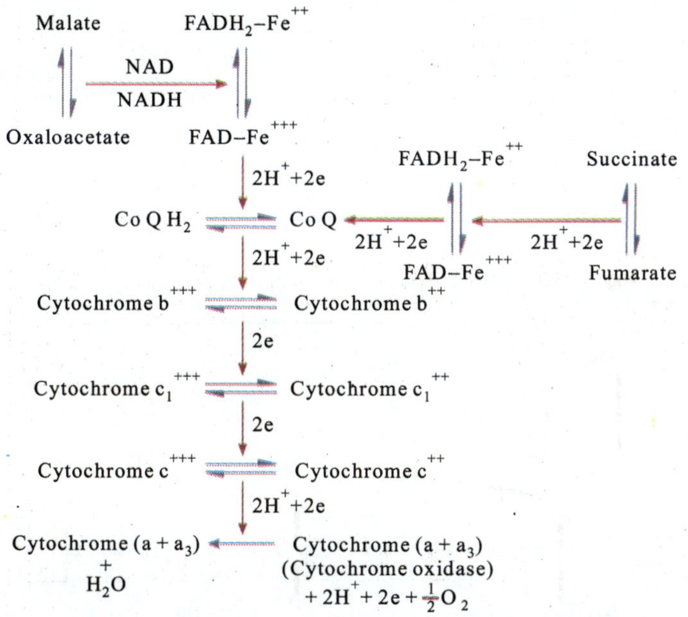

Fig. 12.19 : Electron transport oxidation of hydrogen removed from substrate to water

tional flavoprotein, electron-transporting flavoprotein (ETF), is essential for transferring of electrons to the respiratory chain.

10. The cytochromes are arranged in order of increasing redox potential. The terminal cytochrome a_3 (cytochrome oxidase) is responsible for the final combination of reducing equivalents with molecular oxygen to form H_2O. This enzyme allows the respiratory chain to function at the maximum rate until the tissue becomes virtually anoxic.

ENERGY CAPTURE IN THE RESPIRATORY CHAIN

1. When substrates are oxidized through an NAD-linked dehydrogenase, 3 mol inorganic phosphate are incorporated into 3 mol of ADP to form 3 mol of ATP per half mol of O_2 ($\frac{1}{2}O_2$) consumed, i.e. the P : O ratio = 3.
2. When a substrate is oxidized through a flavoprotein-linked dehydrogenase, only 2 mol of ATP are formed, i.e. P : O ratio = 2. These reactions are known as oxidative phosphorylation at the respiratory chain level.
3. There must be a redox potential of about 0.2 volts or free energy change of about 9 K Cal between components of the respiratory chain if that particular site is to support the coupled formation of 1 mol of ATP.

   Four sites fulfil these requirements in the respiratory chain :

   (a) between NAD and flavoprotein.

   (b) between flavoprotein and cytochrome b.

   (c) between cytochrome b and cytochrome c.

   (d) between cytochrome a and oxygen.
4. When ADP is deficient in the presence of excess substrate, 3 crossover points can be identified. These crossover points coincide with 3 of the possible sites on thermodynamic grounds. The 3 sites of phosphorylation have been designated as sites

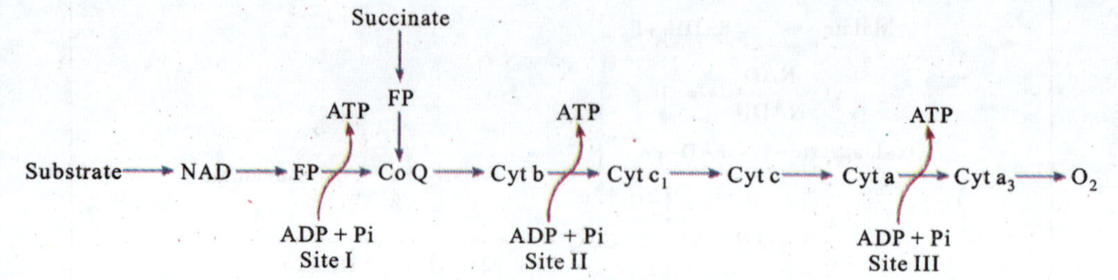

Fig. 12.20 : Sites of phosphorylation in the respiratory chain

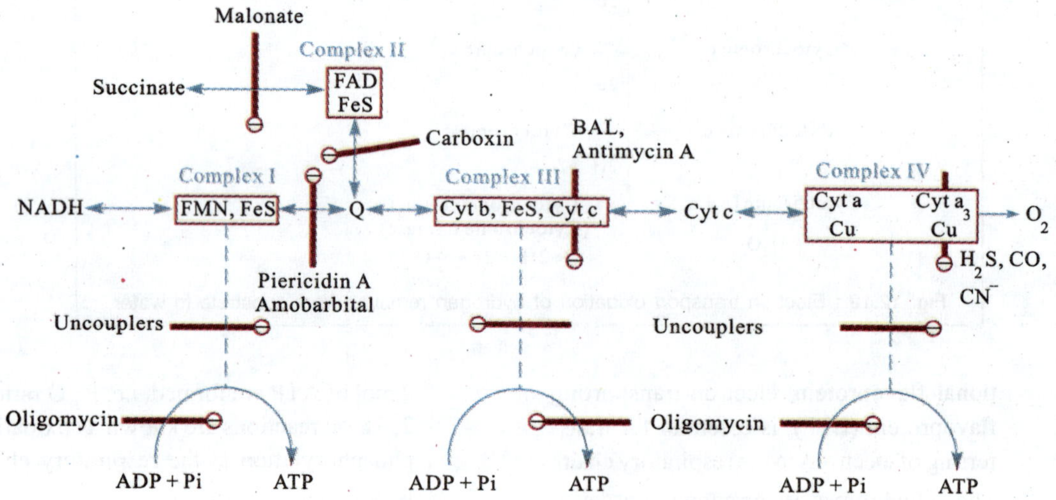

Fig. 12.21 : Sites of phosphorylation and inhibition of respiratory chain by specific drugs, chemicals, antibiotics etc.

I, II and III, respectively. So P : O ratio of succinate is only 2 as site I is bypassed by the flavoprotein-linked succinate dehydrogenase.

Inhibitors

(a) Inhibitors act at three sites of the respiratory chain to arrest respiration by blocking this chain.

(b) The first is inhibited by barbiturates such as **amobarbital**, **antibiotic piericidin A**, **fish poison rotenone** etc. These inhibitors prevent the oxidation of substrates that communicate directly with the respiratory chain via an NAD-linked dehydrogenase by blocking the transfer from FeS to Q.

(c) Antimycin A and dimercaprol inhibit the respiratory chain between cyt b and cyt c.

(d) H_2S, CO, cyanide totally block respiration because they inhibit cytochrome oxidase.

(e) Carboxin and TTFA inhibit transfer of reducing equivalents from succinate dehydrogenase to Q, whereas malonate is a competitive inhibitor of succinate dehydrogenase.

Some redox potentials in mammalian oxidation systems

System	E_0' volts
Oxygen/water	+ 0.82
Cytochrome a; Fe^{+++}/Fe^{++}	+ 0.29
Cytochrome c; Fe^{+++}/Fe^{++}	+ 0.22
CoQ; ox/red.	+ 0.10

Contd.

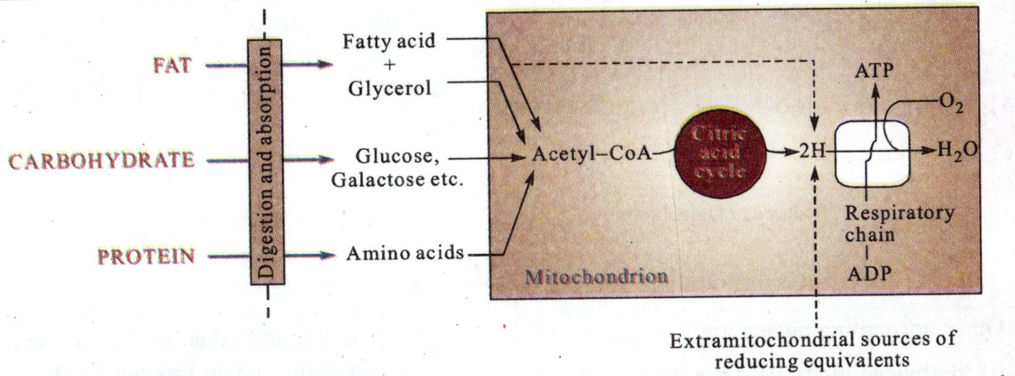

Fig. 12.22 : Roles of the respiratory chain of mitochondria in energy formation from food

Some redox potentials in mammalian oxidation systems	
System	*E_0' volts*
Cytochrome b; Fe^{+++}/Fe^{++}	+ 0.08
Flavoprotein; ox./red.	− 0.12
Fumarate/succinate	+ 0.03
NAD^+/NADH	− 0.32
Oxalo acetate/Malate	− 0.17
Lipoate; ox./red	− 0.29

Free energy change

$$\Delta G = -nF\ DE_0' \text{ coulomb joules.}$$

where, n = number of electrons transferred.

F = Faraday constant.

$\Delta E_0'$ = Difference in redox potential between two systems.

ΔG = Free energy.

This can be converted into calories by dividing by the factor 4.18.

$$\therefore \quad \Delta G = \frac{nF \times \Delta E_0'}{4.18} \text{ calories.}$$

Oxidative phosphorylation

The process by which ADP is phosphorylated by P_i to ATP in the respiratory chain is called oxidative phosphorylation.

This process is limited to the mitochondria. Various oxidations occur in other parts of the cell and liberate heat instead of energy.

Various mechanisms have been proposed for this process. Chance and Williams have proposed the following mechanism in which I and X are intermediates of unknown nature. B is the oxidized form of a carrier in the chain. AH_2 is the reduced carrier which reduces B.

$$B + I \rightleftharpoons BI.$$

$$BI + AH_2 \rightleftharpoons BH_2 \sim I + A.$$

$$BH_2 \sim I + X \rightleftharpoons X \sim I + BH_2$$

$$X \sim I + Pi \rightleftharpoons X \sim Pi + I.$$

$$X \sim Pi + ADP \rightleftharpoons ATP + X.$$

Uncoupling agent

A large number of substances uncouple oxidative phosphorylation in the respiratory chain; i.e. they prevent the formation of ATP but permit oxidation to proceed with the generation of heat. One of the uncoupling agents is 2, 4-dinitrophenol (DNP).

DNP uncouples phosphorylation by the hydrolysis of X~I or X~Pi. It has been considered to increase the ATPase activity of mitochondria :

$$X \sim I + DNP \rightleftharpoons X + DNP.I.$$

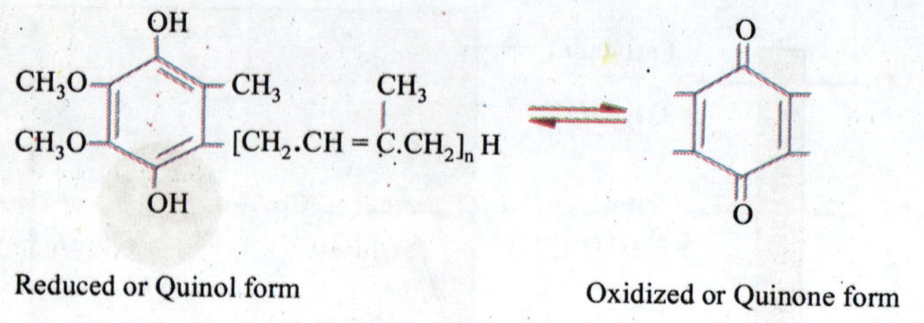

Fig. 12.23

Other uncoupling agents are :

(i) Methylene blue, (ii) Arsenite, (iii) Dicoumarol, (iv) Aureomycin, (v) Gramicidin.

Thyroxin is an uncoupling agent and causes swelling of the mitochondria.

Inhibitors

1. Oligomycin completely blocks oxidation and phosphorylation in intact mitochondria.
2. Atractyloside inhibits oxidative phosphorylation. It inhibits the transporter of ADP into the mitochondrion and of ATP out of the mitochondrion.

Coenzyme Q (Ubiquinone, CoQ)

1. It exists in mitochondria in the oxidized quinone form under aerobic conditions and in the reduced quinol form under anaerobic conditions.
2. It has a structure very similar to vitamin K and vitamin E. It is also similar to plastoquinone found in chloroplast. All of these possess polyisoprenoid side chain.
3. It is the additional carrier present in the respiratory chain linking the flavoprotein to cytochrome b.
4. It is a constituent of the mitochondrial lipids.
5. In mitochondria, there is a large stoichometric excess of CoQ compared to other members of the respiratory chain.
6. It is possible that there is more than one pool of CoQ and that some of it is not in the direct pathway of oxidation.

Role of high energy phosphates

Low-energy phosphates : These are the esterphosphates i.e. AMP found in the intermediates of glycolysis.

High-energy phosphates : The value is higher than that of ATP, eg. ATP, ADP, phosphoenolpyruvate, creatine phosphate, arginine phosphate.

High-energy compounds : Acetyl-CoA, active methionine (S-adenosylmethionine) and UDPG (Uridine diphosphate glucose).

Lipmann introduced the symbol ~ p indicating high-energy phosphate bond. ATP contains 2 high-energy phosphate groups and ADP contains one.

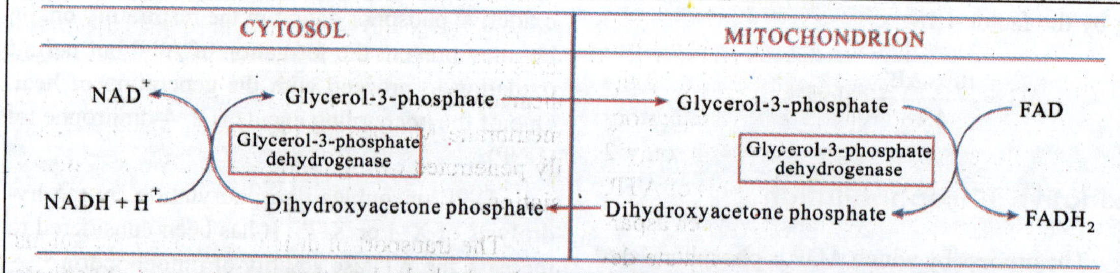

Fig. 12.24 : Glycerophosphate shuttle for transfer of reducing equivalents from the cytosol into the mitochondrion

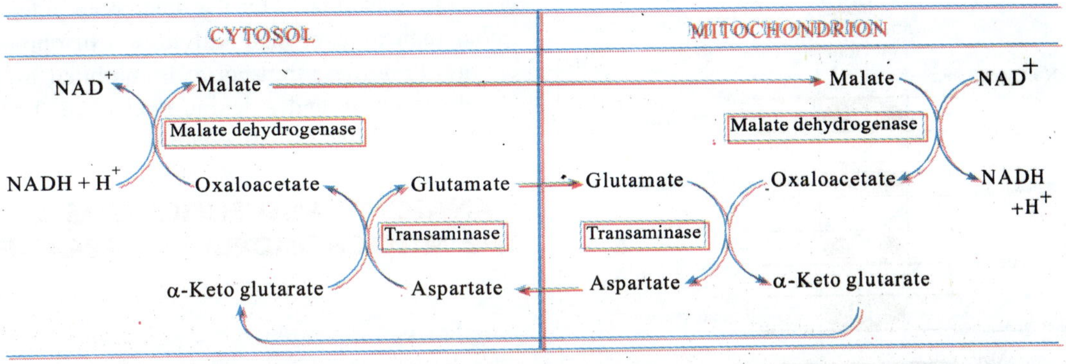

Fig. 12.25 : Malate shuttle for transfer of reducing equivalents from the cytosol into the mitochondrion

TRANSPORT OF SUBSTANCES INTO AND OUT OF MITOCHONDRIA

Oxidation of Extramitochondrial NADH

NADH is produced continuously in the cytosol by 3-phosphoglyceraldehyde dehydrogenase, an enzyme in the Embden-Meyerhof glycolysis but it cannot penetrate the mitochondrial membrane. Still it is oxidized by the respiratory chain in the mitochondria. This is possible by the transfer of reducing equivalents through the mitochondrial membrane via substrate pairs linked by suitable dehydrogenase. The substrate pairs are lactate/pyruvate, dihydroxyacetone phosphate/glycerol-3-phosphate, and malate/oxaloacetate. The specific dehydrogenase is essential to be present on both sides of the mitochondrial membrane. Lactate dehydrogenase is found only in the cytosol and glycerol-3-phosphate dehydrogenase is NAD-linked in the cytosol, whereas the enzyme is a flavoprotein enzyme found in the mitochondria. The mechanism of transfer is shown in the Fig. 12.25. Since the mitochondrial enzyme is linked to the respiratory chain via a flavoprotein rather than NAD, only 2 mols of ATP are formed instead of 3 mols of ATP. Rapid oxidation of NADH occurs only when aspartate -α- ketoglutarate transaminase and malate dehydrogenase together with glutamate, aspartate and malate are added to mitochondria. The complexity of this system is due to the impermeability of the mitochondrial membrane to oxaloacetate.

The malate "shuttle" system is also shown in (Fig. 12.25).

Energy-linked ion transport in Mitochondria

Actively respiring mitochondria maintain or accumulate cations such as K^+, Na^+, Ca^{++}, and Mg^{++}. Loss of ions from the mitochondria is due to the uncoupling with dinitrophenol. The ion uptake is not inhibited by oligomycin on the fact that the energy need not be supplied by phosphorylation of ADP.

Mitochondrial Transporter Systems

Oxygen, water, CO_2, β-hydroxybutyrate, acetoacetate and acetate are freely permeable to the inner mitochondrial membrane. Long chain fatty acids are transported into the mitochondria by the help of carnitine system. There is a special carrier for pyruvate also. Specific transporter or carrier systems are required for the transport of amino acids, dicarboxylate and tricarboxylate anions across the membrane. Monocarboxylate anions are more readily penetrated owing to the lesser degree of dissociation.

The transport of di-and tricarboxylate anions is closely linked to that of inorganic phosphate which penetrates readily as the $H_2PO_4^-$ ion in ex-

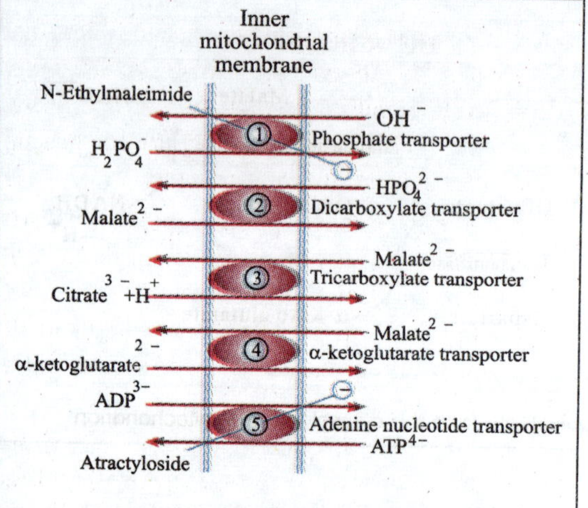

Fig. 12.26 : Transporter systems in the mitochondrial membrane

change for OH^-. The uptake of malate by the dicarboxylate transporter requires inorganic phosphate for exchange in the opposite direction. The uptake of citrate, isocitrate or cis-aconitate by the tricarboxylate transporter requires malate in exchange. α- ketoglutarate transport also requires an exchange with malate. Thus, osmotic balance is maintained by the use of exchange mechanisms. Citrate transport across the mitochondrial membrane depends not only on malate transport but on the transport of inorganic phosphate also. The adenine nucleotide transporter allows the exchange of ATP and ADP but not AMP. Na^+ can be exchanged for H^+. Active uptake of Ca^{++} by mitochondria is facilitated by the membrane potential rather than by exchange with an ion of opposite charge. Phosphate transporter is inhibited by N-Ethylmaleimide and Adenine nucleotide Transporter is inhibited by Atractyloside.

ANATOMY AND FUNCTIONS OF THE MITOCHONDRIAL MEMBRANES

Anatomy

1. The outer membrane of mitochondria is permeable to most metabolites, but the inner membrane is selectively permeable.
2. The inner membrane has a matrix and the inner membrane is thrown into folds or cristae.
3. The presence of monoamine oxidase and a few other enzymes such as acyl-CoA synthetase, phospholipase A_2, glycerophosphate acyltransferase, monoacyl glycerophosphate acyltransferase is found in the outer membrane and the outer membrane may be removed by the treatment with digitonin.
4. Adenylate kinase is found in the intermembrane space.
5. Cardiolipin is concentrated in the inner membrane where most of the lipid is phospholipid.

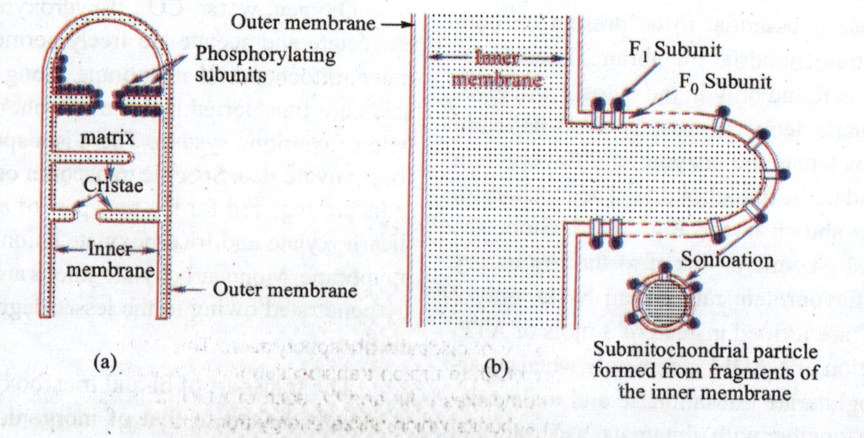

Fig. 12.27 : Structure of the mitochondrial membranes

Functions

1. The phosphorylating subunits responsible for the production of ATP are scattered over the surface of the inner membrane. These subunits consist of several proteins collectively known as an F_1 subunit which project into the matrix. These F_1 subunits also contain the ATP synthetase. These subunits are attached, probably by a stalk, to a membrane protein subunit known as Fo which extends through the membrane.
2. One ATP molecule is formed from ADP and Pi when every proton pair passes through the $F_0 - F_1$ complex. Similar phosphorylating units are found inside the plasma membrane of bacteria but outside the membrane of chloroplasts.
3. Vesicles (Submitochondrial particles) that are "inside-out" are formed on the inner mitochondrial membrane and, therefore, the phosphorylating units are located on the outside of the membrane.
4. The enzymes of the TCA cycle and β-oxidation of fatty acids are found in the matrix to associate mechanisms for transporting ions and fatty and other organic acids as well as nucleotides across the inner membrane.
5. Succinate dehydrogenase is found on the inner surface of the inner mitochondrial membrane and it transports reducing equivalents in the respiratory chain at ubiquinone (CoQ).
6. β-hydroxybutyrate dehydrogenase is also bound to the matrix side of the inner mitochondrial membrane.

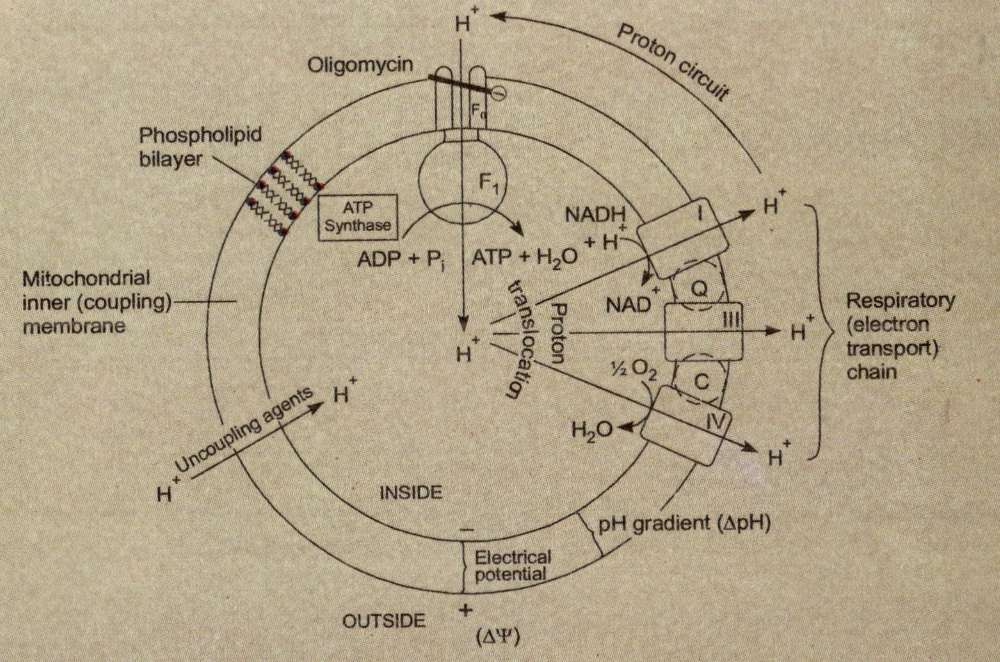

Fig. 12.28 : Principles of the chemiosmotic theory of oxidative phosphorylation. The main proton circuit is created by the coupling of oxidation in the respiratory chain to proton translocation from the inside to the outside of the membrane, given by the respiratory chain complexes I, III, and IV, each of which acts as a proton pump. Q, ubiquinone; C, cytochrome c; F_1, F_0 protein subunits which utilize energy from the proton gradient to promote phosphorylation. Uncoupling agents such as dinitrophenol allow leakage of H^+ across the membrane, thus collapsing the elecrochemical proton gradient. Oligomycin specifically blocks conduction of H^+ through F_0.

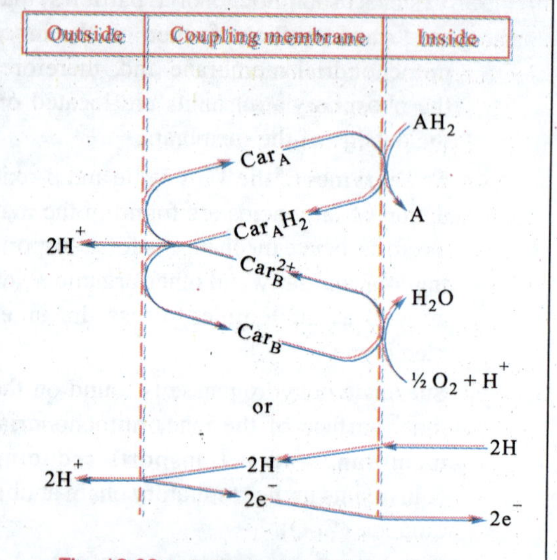

Fig. 12.29 : Oxidation/reduction (o/r) loop (Chemiosmotic theory)

THE CHEMIOSMOTIC THEORY

Mitchell explains that the primary event in oxidative phosphorylation is the translocation of protons (H^+) to the exterior of a coupling membrane driven by oxidation in the respiratory chain. The membrane is impermeable to protons which accumulate outside the membrane creating an *electrochemical potential difference across the membrane*. This electrochemical potential difference is utilized to drive a *membrane-located ATP synthetase* which, in the presence of P_i+A DP, forms ATP (Fig. 12.28).

It is suggested that the respiratory chain is folded into three oxidation/reduction (o/r) loops in the membrane, each loop functionally corresponds to site I, site II, and site III of the respiratory chain. A single loop consists of a hydrogen carrier and an electron carrier shown in Fig. 12.29. A configuration of the respiratory chain folded into three functional o/r loops is shown in Fig. 12.30.

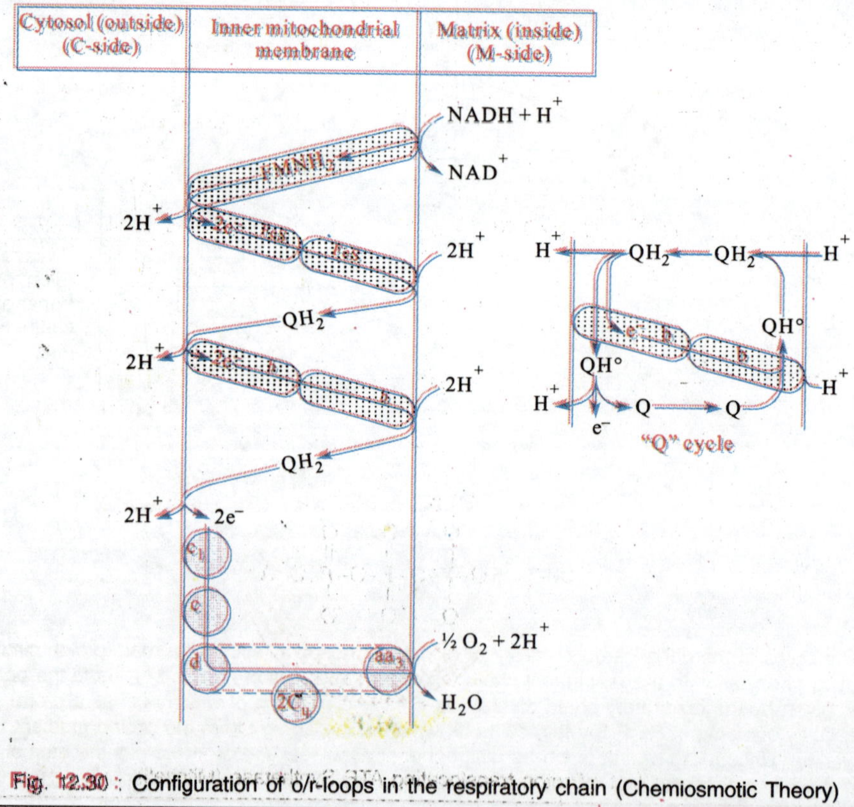

Fig. 12.30 : Configuration of o/r-loops in the respiratory chain (Chemiosmotic Theory)

Each electron pair transferred from NADH to oxygen causes six protons to be translocated from the inside to the outside of the mitochondrial membrane. NADH first donates one proton and two electrons, which, together with another proton from the internal medium, reduce FMN to $FMNH_2$. FMN is considered to extend the full width of the membrane and enables it to release two protons to the outside of the membrane and then to return two electrons to the inside surface via FeS proteins which is reduced. Each reduced FeS complex donates one electron to an ubiquinone (Q) molecule which, upon taking up a proton from inside the membrane, forms QH_2. QH_2, being lipid-soluble and a small molecule, is free to move to the outside of the membrane where it discharges a proton pair into the cytosol and donates two electrons to two molecules of the next carrier in the respiratory chain, cytochrome b. This electron carrier is thought to span the mitochondrial membrane, enabling the electrons to join another molecule of ubiquinone together with two more protons from the internal medium. The resulting QH_2 shuttles to the outer surface where two protons are liberated and two electrons passed to two molecules of cytochrome C. These electrons then pass through the remainder of the cytochrome chain, traversing the membrane to cytochrome aa_3 which lies on the inside of the membrane. At this site, two electrons combine with two H^+ from the internal medium and an oxygen atom to form water.

The phosphorylating subunits responsible for the production of ATP are scattered over the surface of the inner membrane. These consist of several proteins collectively known as an F_1 subunit which projects into the matrix and which contains the ATP synthetase (Fig. 12.28). These subunits are attached, possibly by the stalk, to a membrane protein subunit known as F_0, which probably extends through the membrane. For every proton pair passing through the $F_0 - F_1$ complex, one ATP molecule is formed from ADP and P_i.

Mitchell suggested a model which is shown in Fig. 12.31.

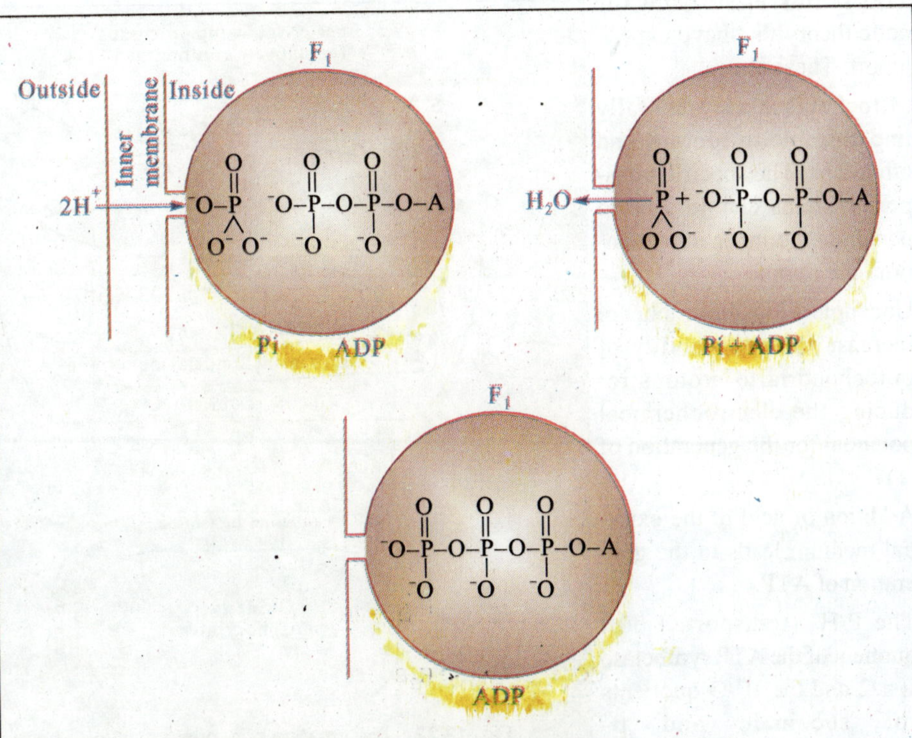

Fig. 12.31 : Proton translocating ATP Synthetase (Mitchell)

A proton pair attacks one oxygen of P_i to form H_2O and an active form of P_i which immediately combines with ADP to form ATP. Other studies have suggested that ATP synthesis is not the main energy-requiring step–rather it is the release of ATP from the active site. This may involve conformational changes in the F_1 subunit.

The ion leakage would have to be balanced by extrusion of ions against the electric gradient to prevent swelling and lysis. It was, therefore, postulated that the coupling membrane contains exchange diffusion systems for exchange of anions against OH^- ions and of cations against H^+ ions. Such system would be necessary for uptake of ionized metabolites through the membrane.

The electrochemical potential difference across the membrane would inhibit further transport of reducing equivalents through the o/r loops unless it was discharged by back translocation of protons across the membrane through the vectorial ATP synthetase system. This depends on the availability of ADP and P_i.

Several points arise from the chemiosmotic theory that have experimental support. These are :

1. Mitochondria are generally impermeable to protons and other ions. The specific transport systems enable ions to penetrate the inner mitochondrial membrane.
2. Uncouplers like dinitrophenol increase the permeability of mitochondria to protons reducing the electrochemical potential for the generation of ATP.
3. Addition of acid to the external medium leads to the generation of ATP.
4. The P/H^+ (transported out) quotient of the ATP synthetase is 1/2 and the H^+/O quotients for succinate and β-hydroxybutyrate oxidation are 4 and 6, respectively, conforming with the expected P/O ratios of 2 and 3, respectively. These ratios are compatible with the postulated existence of o/r loops in the respiratory chain.
5. Oxidative phosphorylation does not occur in soluble systems where there is no possibility of a vectorial ATP synthetase.

The Creatine phosphate shuttle facilitates transport of high energy phosphate from mitochondria :

1. An isoenzyme of creatine kinase is found in the mitochondria catalyzing the transfer of high energy phosphate to creatine from ATP. The creatine phosphate is transported into the cytosol via protein pores in the outer mitochondrial membrane.
2. Different isoenzymes of creatine kinase cause transfer of high-energy phosphate to and from the various systems that utilize it (eg, muscle contraction, glycolysis).

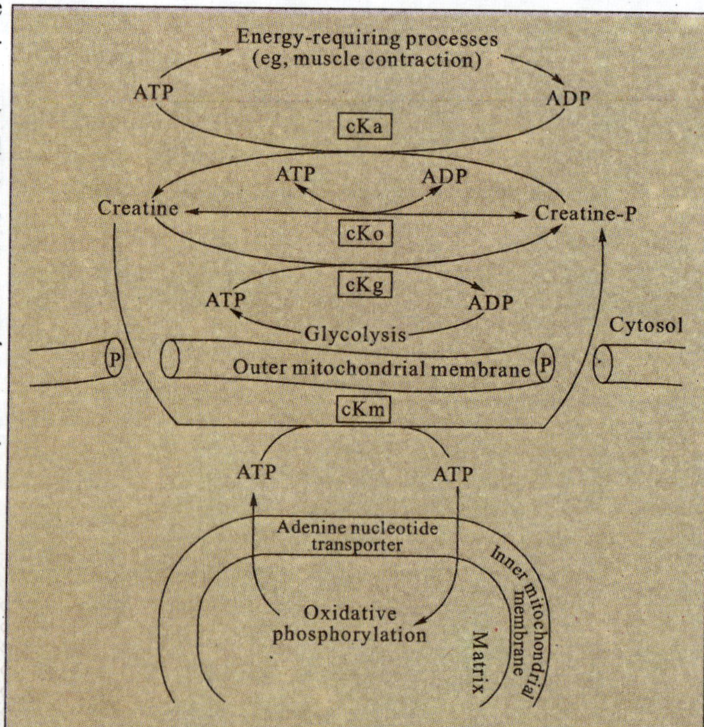

Fig. 12.32 : The creatine phosphate shuttle of heart and skeletal muscle

CLINICAL ORIENTATION

- Oxygen is principally used in respiration which is defined as the process by which cells derive energy in the form of ATP from the reaction of hydrogen with oxygen to form water. Molecular oxygen is incorporated into a variety of substrates by oxygenases.
- Many drugs, pollutants and chemical carcinogens (Xenobiotics) are metabolized by cytochrome P-450 system.
- Administration of oxygen at high pressure has proved of value, although this can result in oxygen toxicity.
- Free energy becomes available from oxidative phosphorylation in aerobic organisms.
- A number of drugs (eg, amobarbital) and poisons (eg, cyanide, carbon monoxide) inhibit oxidative phosphorylation exhibiting fatal result.
- The inherited defects of mitochondria involving components of the respiratory chain and oxidative phosphorylation exhibit myopathy, encephalopathy, lactacidosis.
- The condition of fatal infertile mitochondrial myopathy and renal dysfunction involves severe diminution or absence of oxidoreductases of the respiratory chain. MELAS is an inherited condition due to cytochrome oxidase deficiency.

EXERCISE

1. Discuss the oxidation and reduction in biological system. Name some types of enzymes involved in the process. (P.U. 69S)
2. Discuss the respiratory chain. (P.U. 72A)
3. Write short notes on :
   (a) Coenzyme Q. (R.U.70A)
   (b) Oxidative phosphorylation. (B.U. 74A)
   (c) Superoxide dismutase. (Bh. U. 73S)
4. What do you mean by biologic oxidation and oxidative phosphorylation? Describe in detail how the oxidation of a substrate takes place through the respiratory chain in mitochondria mentioning the possible sites of ATP formation with reasons.

   Mention the names of some inhibitors of respiratory chain. (C.U. 1983)

SCHEMATIC COMPENDIUM

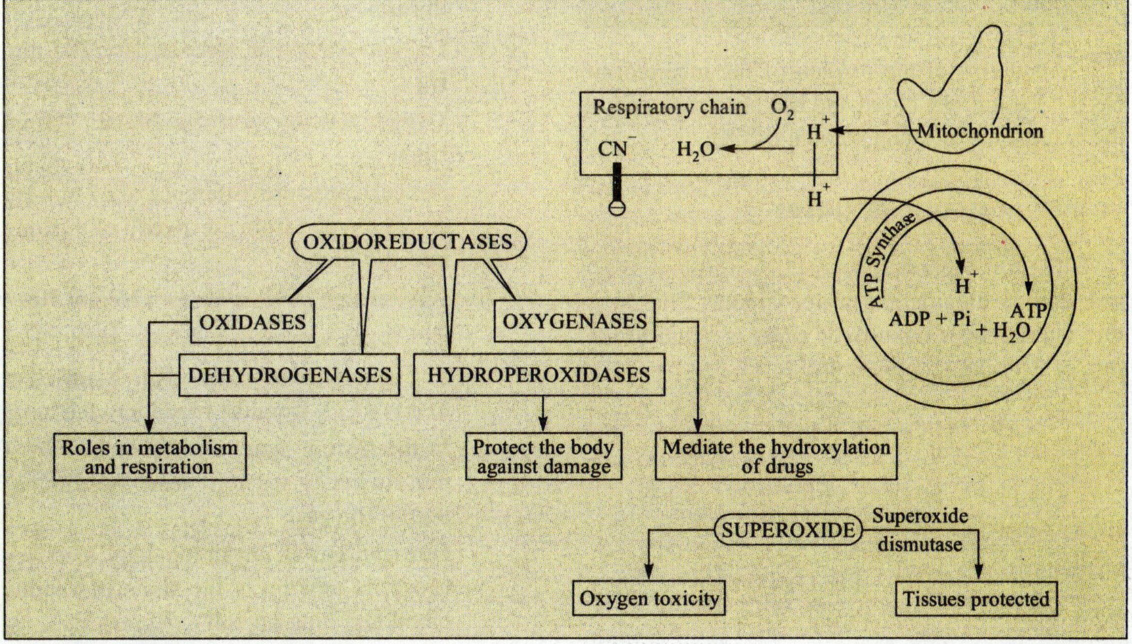

CHAPTER 13

THE CHEMISTRY OF RESPIRATION

Definition

Respiration is defined as the utilization of O_2 from inspired air by the body and elimination of CO_2 produced by the metabolites in the cells in the expired air.

Composition of Atmospheric and expired Air		
	Atmospheric air (Inspired air)	Expired air
Oxygen	20.96%	15%
Carbon dioxide	0.04%	5%
Nitrogen	79%	79%

About 1/4 of the oxygen of the inspired air has passed into the blood and has been replaced in the expired air by equal amount of CO_2.

Partial pressure of gases

The atmospheric air contains nitrogen, oxygen, carbon dioxide and water vapour. The pressure exerted by a mixture of gases is equal to the sum of their partial pressures :

$$AP - PH_2O = PN_2 + PCO_2 + PO_2$$

The partial pressure of the gases (Y) N_2, O_2 and CO_2 can be calculated as :

$$Py = \frac{(760 - PH_2O) \times \%Y}{100}$$

Diffusion of gases in the lungs

1. When the gases of the inspired air come in contact with the alveolar membrane of the lung, the exchange of gases takes place following the usual laws of diffusion. Thus, the gas passes into the blood through the membrane.
2. The gas pressure in the blood is usually expressed as gas **"tensions"**.

   Example : The oxygen tension (PO_2) is the pressure of the dry gas with which the dissolved oxygen in the blood is in equilibrium.

3. **Oxygen tension in alveolar air :** 107 mm Hg.

   Oxygen tension in venous blood : 40 mm Hg.

   Hence, a pressure difference of 67 mm Hg drives oxygen from the alveoli of the lung into the blood.

   CO_2 tension in alveolar air : 36 mm Hg.

   CO_2 tension in venous blood : 46 mm Hg.

   The difference of 10 mm Hg. is sufficient to drive CO_2 from the blood into the lung. In the resting state, a difference of 0.12 mm Hg in CO_2 tension causes the elimination of the gas.

   Nitrogen gas is physiologically inert because its tension is the same in venous blood and lung alveoli (570 mm Hg).

4. After this exchange of gases, the blood becomes arterial.

   Oxygen tension of arterial blood : 100 mm Hg.

   CO_2 tension of arterial blood : 40 mm Hg.

   Nitrogen tension of arterial blood : 570 mm Hg.

   These gases are dissolved in the blood.

5. The quantities of each of these gases which can be dissolved (calculated content in blood) is compared with the actual quantities found in the blood by the following data.

	ml./dl		
	O_2	CO_2	N_2
Calculated content in blood	0.393	2.96	1.04
Actual quantities found in arterial blood	20.0	50.0	1.70
Actual quantities found in venous blood	14.0	56.0	1.70

TRANSPORT OF OXYGEN BY THE BLOOD

Function of hemoglobin

1. The transport of oxygen from the lungs to the tissues by the blood is due to the ability of hemoglobin to combine reversibly with oxygen :

$$Hb + O_2 = HbO_2$$

(Hb=Reduced (deoxygenated) hemoglobin; HbO_2 = Oxyhemoglobin).

2. At a tension of 100 mm Hg or more, hemoglobin is completely saturated. The oxygen carrying power of the blood is absolutely a function of the hemoglobin (red cell) concentration

Dissociation of oxyhemoglobin

1. The vital relationship between the saturation of hemoglobin and the oxygen tension is shown below by the dissociation curve of oxyhemoglobin in which the per cent saturation is plotted against the oxygen tension.
2. The curve drawn with CO_2 at a tension of 40 mm Hg is considered as representative of the normal physiologic condition.
3. The hemoglobin is 95-98% saturated when the oxygen tension is 100 mm Hg in arterial blood. There is a slight effect on the saturation of hemoglobin with further increase in oxygen tension.
4. The saturation of hemoglobin declines slowly with the fall in oxygen tension and a rapid evolution of oxygen takes place at the oxygen tension of 50 mm Hg. This is the "unloading tension" of hemoglobin.
5. In the tissues, the oxygen tension is about 40 mm Hg and the oxyhemoglobin dissociates and oxygen is readily available to the cells.
6. During the passage of blood through the tissues the oxygen content of the blood falls from 20 to 15 vol%. This gives a considerable reserve of oxygenated blood in the event of inadequate oxygenation at the lung.

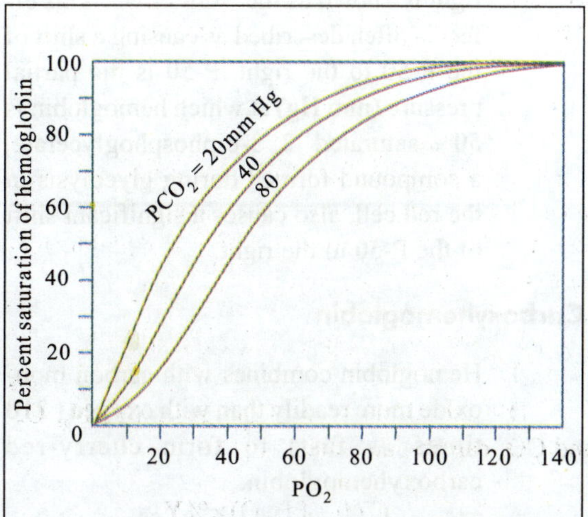

Fig. 13.1 : Dissociation of hemoglobin

Factors affecting dissociation of oxyhemoglobin

1. Temperature

(a) A rise in temperature decreases hemoglobin saturation.

(b) At 25°C, hemoglobin is 88% saturated but at 37°C, it is only 56% saturated. Therefore, hemoglobin gives up oxygen more readily while passing from high to low oxygen tension (as from lungs to tissues) in warm-blooded animals than in cold-blooded animals.

2. Electrolytes

At low oxygen tensions, oxyhemoglobin gives up oxygen more readily in the presence of electrolytes.

3. Effect of CO_2

(a) The influence of CO_2 on the shape of the dissociation curve is actually the effect of carbonic acid formation with the lowering of the pH of the environment.

(b) The increase in acidity facilitates the desiccation of oxyhemoglobin.

(c) The ability of CO_2 to shift the slope of the oxyhemoglobin dissociation curve to the right is known as the Bohr effect. This effect is often described as causing a shift of the P-50 to the right. P-50 is the partial pressure (mm Hg) at which hemoglobin is 50% saturated. 2, 3-biphosphoglycerate, a compound formed during glycolysis in the red cell, also causes a significant shift of the P-50 to the right.

Carboxyhemoglobin

1. Hemoglobin combines with carbon monoxide more readily than with oxygen (210 times as fast) to form cherry-red carboxyhemoglobin.
2. This reduces the amount of hemoglobin to carry oxygen.
3. When the carbon monoxide in the inspired air is 0.02%, headache and nausea occur.
4. In case the carbon monoxide concentration is only $\frac{1}{210}$ that of oxygen in the air (about 0.1% carbon monoxide), unconsciousness occurs in 1 hour and death in 4 hours.

Clinical signs of variation in hemoglobin saturation

1. A decrease in normal oxygenation of blood gives a characteristic bluish appearance to the skin. This is said to be cyanosis. It is characteristic of cyanide poisoning where respiration is also impaired.
2. In severe anemia, the concentration of hemoglobin is too low and cyanosis does not take place although the oxygen content of the blood is reduced.
3. In CO poisoning, the formation of cherry-red carboxyhemoglobin often produces a ruddy appearance in the lips.

TRANSPORT OF CO_2 IN BLOOD

CO_2 is carried in the cells and the plasma by the blood. It exists in three forms. The three main fractions are :

1. A small amount of carbonic acid .
2. The "carbamino-bound" CO_2 which is transported in combination with proteins (mainly hemoglobin).
3. That carried as bicarbonate in combination with cations of sodium or potassium.

The carbamino-bound CO_2 is important in the exchange of this gas because of the high rate of reaction :

$$\text{Hb–NH}_2 \xrightleftharpoons{CO_2} \text{Hb–NH–COOH}$$

The amount of CO_2 dissolved in the blood is not high, but it is important because any change in its concentration causes the following equilibrium to shift :

$$CO_2 + H_2O \rightleftharpoons H_2CO_3 \rightleftharpoons H^+ + HCO_3^-$$

The above reaction is catalyzed by the enzyme carbonic anhydrase.

Effect of CO_2 on blood pH

1. CO_2 evolved from the tissues forms carbonic acid. Most of the carbonic acid formed is promptly converted to bicarbonate as shown in the equation below (B^+ represents, principally, Na^+ or K^+).

$$H_2CO_3 \rightleftharpoons H^+ + HCO_3^- + B^+ \rightleftharpoons BHCO_3^- + H^+$$

2. At the pH of blood (7.4), a ratio of 20 : 1 must exist between the bicarbonate and carbonic acid. This ratio is calculated from the Henderson-Hasselbalch equation. Any change in H^+ activity is met by an adjustment in the reaction. Any alteration in the ratio disturbs the acid-base balance of the blood in the direction of acidemia or alkalemia.

BUFFER SYSTEMS OF BLOOD

1. Venous blood carries more CO_2 than arterial blood. Hence, the pH of venous blood is more acid than that of arterial blood by 0.01-0.03 units i.e. pH 7.40 and 7.43, respectively.
2. The blood buffers consists of the plasma proteins, hemoglobin, oxyhemoglobin, bicarbonates and inorganic phosphates.
3. When CO_2 enters the venous blood, the small decrease in pH shifts the ratio of acid to salt in all the buffer pairs. When the ratio is shifted to form more of the acid, cations become available to form additional bicarbonates. In this respect, the plasma phosphates and bicarbonates play a minor role.
4. The buffering action of the plasma proteins is important because they release sufficient cations for the carriage of about 10% of the total CO_2.
5. The phosphates in the red cells carry 25% of the total CO_2.

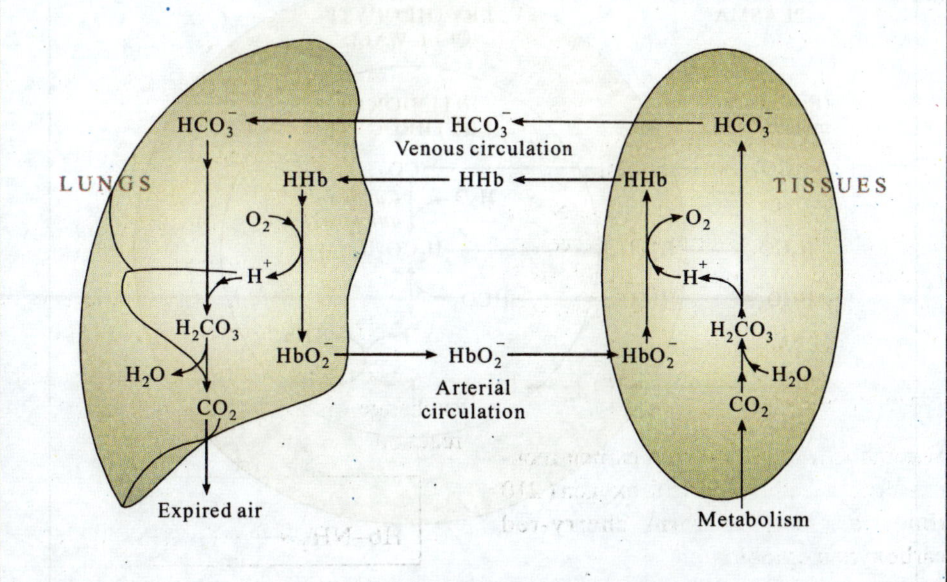

Fig. 13.2 : The buffering action of hemoglobin

6. The most important buffering role is of hemoglobin and oxyhemoglobin which carry 60% of the CO_2 of the whole blood.

Hemoglobin buffers

At the lungs the formation of oxyhemoglobin from reduced hemoglobin realeases hydrogen ions which react with bicarbonate to form carbonic acid. The low CO_2 tension in the lung shifts the equilibrium towards the production of CO_2 which is continually eliminated in the expired air :

$$H^+ + HCO_3^- \longleftrightarrow H_2CO_3 \longleftrightarrow H_2O + CO_2$$

In the tissues, the oxygen tension is reduced and hence oxyhemoglobin dissociates delivering O_2 to the cells and reduced hemoglobin is formed. CO_2 produced by metabolism enters the blood, where it is hydrated to form H_2CO_3 which ionizes to form H^+ and HCO_3^-. Reduced hemoglobin acting as an anion accepts the H^+ ions forming acid-reduced hemoglobin (HHb). Very little change in pH occurs because the newly arrived H^+ ions are buffered by formation of very weak acid.

When the blood returns to the lungs, these H^+ ions are released as a result of the formation of stronger acid (oxyhemoglobin) and the newly released H^+ ion is promptly neutralized by HCO_3^-. This reaction is necessary for the liberation of CO_2 in the lungs.

Chloride shift

1. CO_2 reacts with water to form carbonic acid (H_2CO_3), mainly inside the red cell by the enzyme carbonic anhydrase present in the red cells.
2. The carbonic acid is then buffered by the intracellular buffers (Phosphate and hemoglobin) combining with potassium.
3. Bicarbonate ion also returns to the plasma and exchanges with chloride which shifts into the cell when the tension of CO_2 increases in the blood.
4. When the CO_2 tension is reduced, chloride leaves the cell and enters the plasma.
5. Under normal conditions the red cell is impermeable to sodium or potassium. But

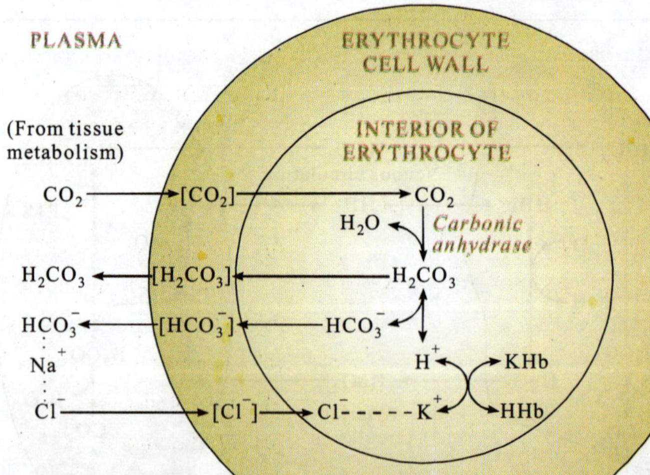

Fig. 13.3 : Chloride shift

it is permeable to hydrogen, bicarbonate and chloride ions and intracellular sources of cation (potassium) are indirectly available to the plasma by chloride (anion) exchange. This permits the carriage of additional CO_2 (as sodium bicarbonate) by plasma.

6. The CO_2 entering the blood from the tissues passes into the red cells where it forms carbonic acid by carbonic anhydrase. Some of the carbonic acid returns to the plasma. The remainder reacts with hemoglobin buffers to form bicarbonate which then returns to the plasma in exchange of chloride. The chloride is neutralized by potassium in the red cells.
7. All of these reactions are reversible. At the lung, when the blood becomes arterial, chloride shifts back into the plasma, thus liberating intracellular potassium to buffer the oxyhemoglobin and in the plasma, neutralizing the sodium which is liberated by the removal of CO_2 during respiration.

ACID-BASE BALANCE

Respiratory acidosis : This condition occurs when the accumulation of H_2CO_3 is caused in the blood under certain circumstances.

Respiratory alkalosis : This condition occurs when the rate of elimination of CO_2 is excessive following the reduction of H_2CO_3 in the blood.

Metabolic acidosis : This occurs in the deficit of bicarbonate without any change in H_2CO_3.

Metabolic alkalosis : This happens in the excess of bicarbonate.

Causes of disturbances in acid-base balance

A. Metabolic Acidosis

1. It occurs in uncontrolled diabetes with ketosis.
2. It occurs in some cases of vomiting when the fluids lost are not acid.
3. It also occurs in renal disease, poisoning by an acid salt, excessive loss of intestinal fluids (as in diarrhea or colitis) and excessive losses of electrolytes.
4. Increased respiration is an important sign of uncompensated acidosis.

B. Respiratory acidosis

1. This occurs in any disease which impairs respiration such as pneumonia, congestive failure and asthma.
2. It is also found to occur in the depression of the respiratory center (as by morphine poisoning).
3. A poorly functioning respirator also causes this condition.

C. Metabolic alkalosis

1. Causes

(a) By the ingestion of large quantities of alkali in the treatment of peptic ulcer.

(b) In the high intestinal obstruction (as in pyloric stenosis).

(c) After prolonged vomiting or after the excessive removal of gastric secretions containing hydrochloric acid.

(d) In Cushing's disease and during corticotropin or cortisone administration.

2. Symptoms

(a) Tetany is caused due to the decrease in ionized serum calcium.

(b) The respirations are slow and shallow.

(c) The urine is alkaline.

(d) Elevated blood bicarbonate.

(e) Deficiency of potassium.

D. Respiratory alkalosis

1. This is brought about by hyperventilation.
2. It may occur in patients in hepatic coma.

Role of the kidney in acid-base balance

1. The non-volatile acids including lactic and pyruvic acids, hydrochloric acid, phosphoric acid and sulphuric acid are produced by metabolic processes. About 50-150 mEq. of the inorganic acids are eliminated by the kidneys in 24 hours. These acids are partially buffered with cations, largely sodium. In the distal tubules of the kidney, some of the cations are reabsorbed and the pH of the urine falls.
2. Kidney can buffer acids and conserve fixed base in the production of ammonia from amino acids. The ammonia is utilized to neutralize acids when formed in excess.
3. In kidney disease, tubular reabsorption of sodium in exchange for hydrogen is poor and excessive retention of phosphates and sulphates. As a result, acidosis occurs.

HYPOXIA

Hypoxia means oxygen-deficiency in any condition of insufficiency of tissue oxidation processes. There are four types of hypoxia :

1. Hypoxic hypoxia. 2. Anemic hypoxia. 3. Stagnant hypoxia. 4. Histotoxic hypoxia.

1. Hypoxic hypoxia

This is characterized by normal oxygen capacity but diminished oxygen tension in the arterial blood with the result of varying degree of hemoglobin unsaturation.

(a) High altitude : Hypoxia is caused by the diminished oxygen content in the atmospheric air and consequently in the alveolar air and blood stream. This leads to respiratory alkalosis.

(b) Rapid, shallow respiration : Shallow breathing is due to insufficient oxygenation of blood. This happens in diseases.

(c) Congenital heart disease : In the congenital cardiac defect a portion of the blood may flow directly form the right to the left side of the heart without passing through the lungs. As a result, the mixture of aerated and non-aerated blood in the systemic circulation causes hypoxic hypoxia.

2. Anemic hypoxia

This is characterized by diminution in the oxygen capacity of arterial blood due to the decrease in the amount of functioning hemoglobin.

3. Stagnant hypoxia

This is due to circulatory insufficiency—the rate of blood flow through the tissues being retarded, with resulting increase in the percentage volume of oxygen removed from the blood in its passage through the capillaries.

4. Histotoxic hypoxia

In this condition the oxygen supply is normal in every respect but the degree of oxygen utilization by the tissues is diminished because the tissue cells are poisoned in such a way that they cannot use oxygen properly.

EXERCISE

1. Discuss the transport of O_2 and CO_2 in the blood. (Lucknow Uni. 71A)
2. Explain the buffer systems of the blood. (Bombay Uni. 72A)
3. Write short notes on :
   (a) Acid-base balance. (R.U. 73A)
   (b) Diffusion of gases in the lungs. (M.U. 74S)

CHAPTER 14

MUSCLE TISSUE

Chemistry

There are three types of muscle tissue in the body :

1. Striated (Voluntary) or skeletal muscle.
2. Non-striated (Involuntary) or smooth muscle.
3. Cardiac muscle.

Chemical compostion of skeletal muscle	
Water	75%
Protein	20%
Inorganic material, certain organic extractives and carbohydrate (glycogen and its derivatives)	5 %

MUSCLE STRUCTURE

1. Skeletal muscle is composed of fibrils surrounded by electrically excitable membrane, the **sarcolemma**.
2. The individual muscle fibre consists of a bundle of many myofibrils arranged in parallel. These are embedded by intramuscular fluid termed the **sarcoplasm**.
3. The fluid contains glycogen, the high energy compounds ATP and phosphocreatine, and the enzymes of glycolysis.
4. The functional unit of muscle is **sarcomere**. It exists along the axis of a fibril at distances of 2.5 µm.

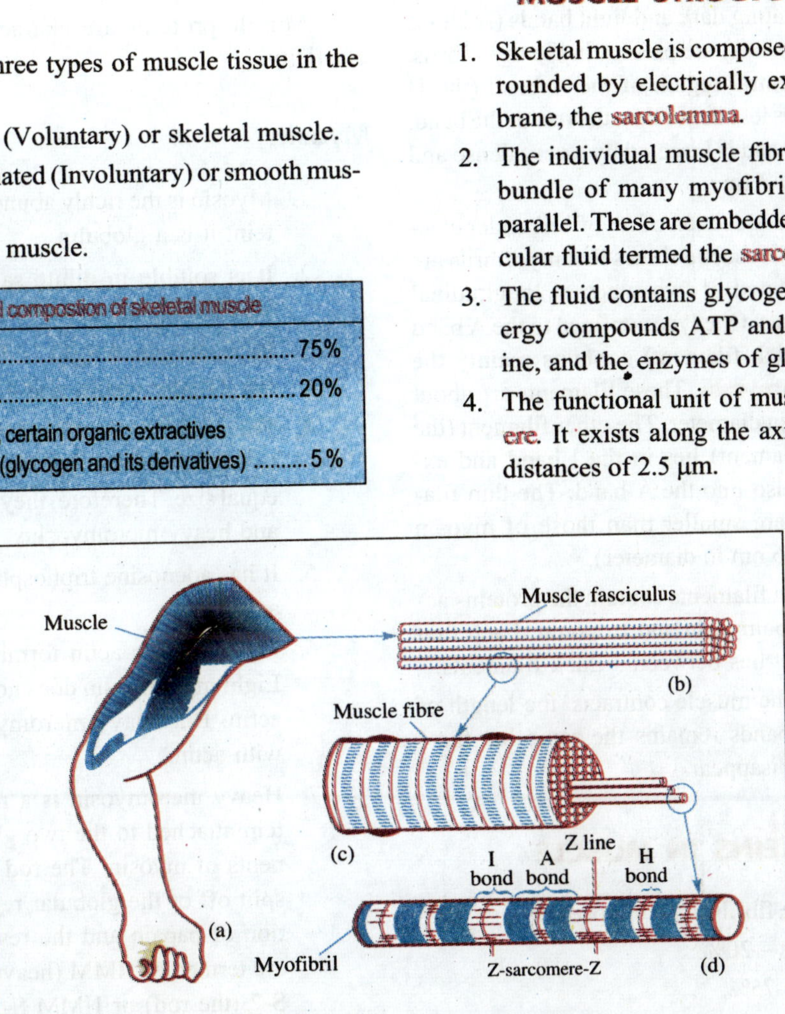

Fig. 14.1 : Structure of voluntary muscle

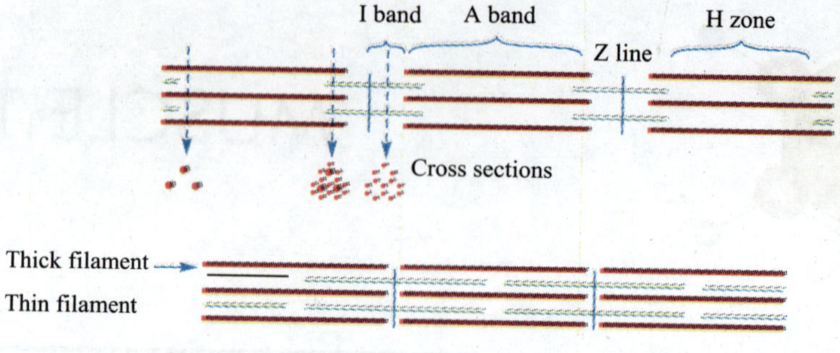

Fig. 14.2 : Arrangement of filaments in striated muscle

5. Alternating dark and light bands (A bands and I bands) are present in the myofibrils. The central region of the A band (the H zone) is less dense than the rest of the band. The I band is bisected by a very dense and narrow Z line.
6. The cross-section of myofibril under electron microscope shows that myofibrils are constructed of two types of longitudinal filaments. One type confined to the A band (the thick filament) contains mainly the protein **myosin**. These filaments are about 16 nm in diameter. The other filament (the thin filament) lies in the I band and extends also into the A band. The thin filaments are smaller than those of myosin (about 6 nm in diameter).
7. The thin filaments contain the proteins actin, tropomyosin and troponin. Each thin filament lies between 3 thick filaments.
8. When the muscle contracts, the length of the A bands remains the same but the I bands disappear.

PROTEINS IN MUSCLE

The muscle fibrils are composed of :

1. Proteins—20%.
2. Water—75%.
3. Inorganic materials, certain organic extractives and carbohydrates (glycogen and its derivatives)—5%.

Muscle proteins are characterized by their elasticity.

Myosin

1. Myosin is the richly abundant muscle protein; it is a globulin.
2. It is soluble in dilute salt solutions and insoluble in water.
3. Its molecular weight is 500,000 containing 2 major chains and 4 light chains.
4. The enzyme **trypsin** cleaves myosin into two components—**meromyosins**—of unequal size. Therefore, they are termed light and heavy meromyosins.
5. It has adenosine triphosphatase (ATPase) activity.
6. Myosin binds actin forming **actomyosin**. Light meromyosin does not combine with actin. But heavy meromyosin combines with actin.
7. Heavy meromyosin is a rod-shaped protein attached to the two globular components of myosin. The rod portion can be split off of the globular region by the action of papain and the resulting portions are termed as HMM (heavy meromyosin) S-2 (the rod) or HMM S-1 (the globular portion). Each HMM S-1 possesses an active site for ATPase activity and a binding site for actin.

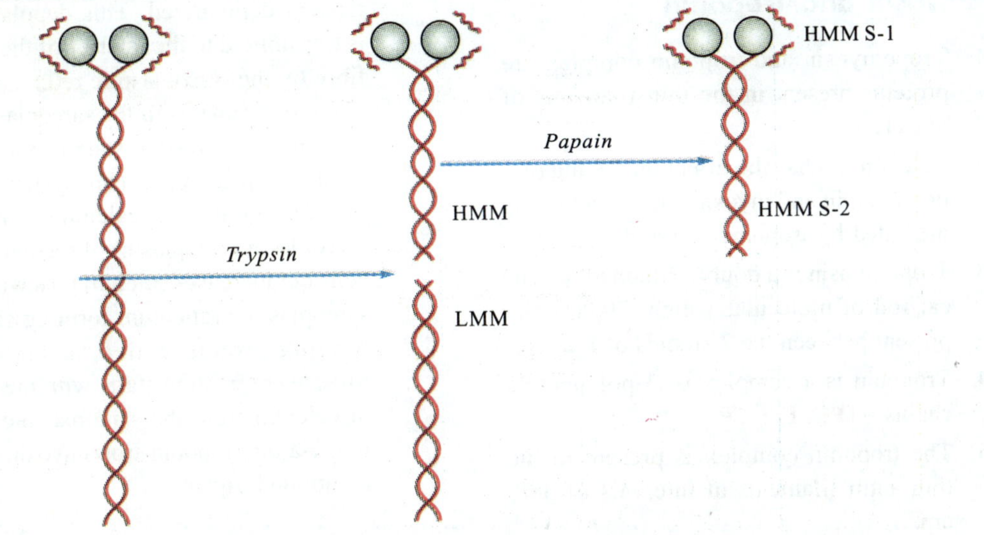

Fig. 14.3 : Enzymatic cleavage of myosin (HMM – Heavy meromyosin; LMM – Light meromyosin)

8. The 4 light chains of myosin molecule are bound to the HMM S-1 fragments. The light chains function as modulators of ATPase activity.
9. Actomyosin is formed by 3 myosin molecules with 1 actin molecule.

Actin

1. Actin is a globulin of molecular weight 60,000.
2. It is the major constituent of thin filaments in striated muscle.
3. During the preparation of actin by extraction with solution of low ionic strength it is obtained as a molecular weight of 42,000 in a globular configuration called **G-actin**.
4. G-actin polymerizes to the fibrous form F-actin in the presence of Mg^{++} and ATP is hydrolyzed to ADP and Pi is released. Hence, ATP must be added to get depolymerization of F-actin to G-actin.
5. Actomyosin is formed by actin and myosin. But ATP dissociates it into actin and myosin :

$$\text{Actin} + \text{Myosin} \underset{\text{ATP}}{\rightleftharpoons} \text{Actomyosin}$$

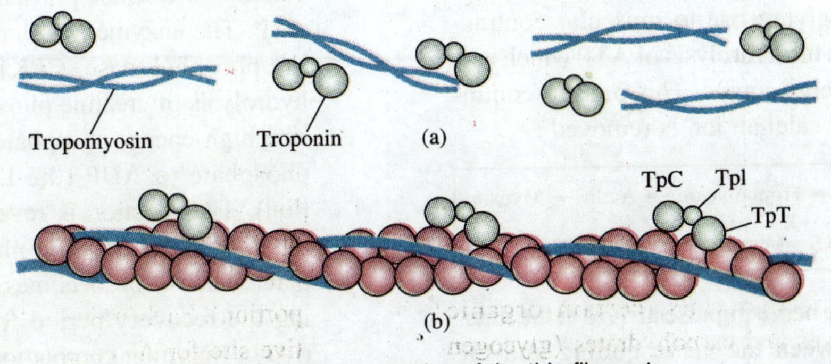

Fig. 14.4 : Representation of the thin filament

Tropomyosin and Troponin

1. Tropomyosin and troponin complex are proteins present in the thin filaments of muscle.
2. Calcium ion has the effect on the interaction of actin and myosin which, in turn, is mediated by tropomyosin and troponin.
3. Tropomyosin is a double-stranded α-helical rod of molecular weight 70,000 and present between the 2 strands of F-actin.
4. Troponin is a complex of 3 polypeptide chains—TPC, TPI, TPT.
5. The troponin complex is present in the thin actin filaments at intervals of 38.5 nm.
6. A troponin complex with tropomyosin molecule regulates the activity of 7 actin monomers.

MOLECULAR EVENTS IN MUSCLE CONTRACTION

1. In absence of calcium ions, the interaction of actin and myosin is inhibited by troponin and tropomyosin.
2. Calcium ions are released from the sarcolemma by the excitation of a motor nerve to the muscle.
3. The released calcium binds to the TPC portion of the troponin complex and is transmitted to tropomyosin and then to actin. This allows actin to interact with myosin giving rise to muscular contraction with the hydrolysis of ATP which acts as an energy source. This process continues until calcium ion is removed :

Ca^{++} → Troponin → Tropomyosin → Actin → Myosin

Fig. 14.5 : Muscle Contraction

4. When the nerve impulse arrives at the junction between the nerve ending and the muscle, the outer membrane of a muscle fibre is depolarized. This depolarization is transmitted to the interior of the muscle fibre by the **sarcoplasmic reticulum**. Calcium ions remains in the sarcoplasmic reticulum in the resting state of the muscle by the help of ATP. This system lowers the concentration of calciums ions in the sarcoplasm (cytoplasm) of resting muscle cells but increases calcium ions within the sarcoplasmic reticulum forming calcium-binding-protein called **calsequestrin**. Muscle contraction starts with the release of calcium from the sarcoplasmic reticulum via the troponin-tropomyosin system mentioned above.

MUSCLE PHOSPHAGENS

1. ATP is the immediate source of energy for muscular contraction. But the amount of it in muscle is very small. It is just enough to cause contraction for a fraction of a second.
2. In vertebrate muscle, the source of high-energy phosphate is phosphocreatine which can transfer higher phosphate groups than ATP, and can donate a high-energy phosphate group to ADP to reform ATP.

   Phosphocreatine is the vertebrate phosphagen; whereas ***phosphoarginine*** is the invertebrate phosphagen.
3. In the ***resting state,*** mammalian muscle contains 4-6 times phosphocreatine than ATP. The enzyme ***creatine kinase*** (creatine phosphokinase, CPK) catalyzes the hydrolysis of creatine phosphate transferring high-energy phosphate from creatine phosphate to ADP (the Lohmann reaction). The reaction is reversible and the resynthesis of creatine phosphate takes place when ATP becomes available during the recovery period following a period of muscular contraction. The transfer of phosphate from ATP to creatine to form

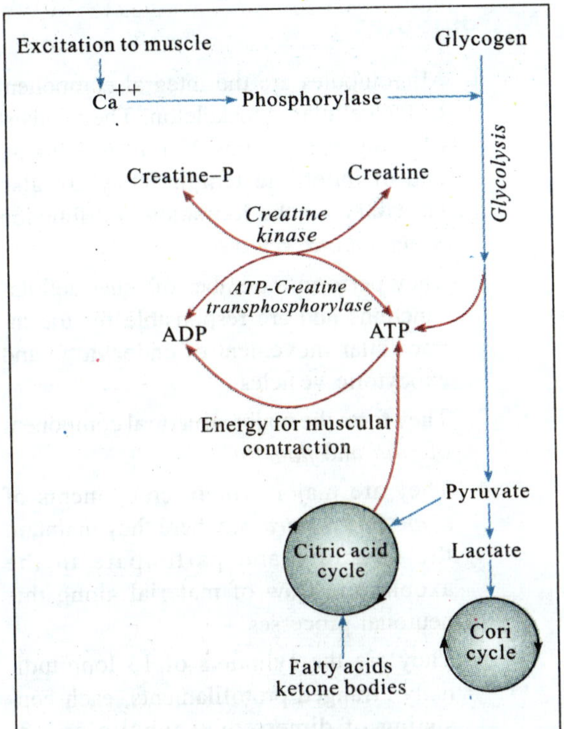

Fig. 14.6 : Muscle contraction

creatine phosphate is catalyzed by the enzyme ATP-creatine transphosphorylase (shown in Fig. 14.6).

4. ATP is also available in the muscle from ADP by the enzyme myokinase (adenylate kinase) which catalyzes the transfer of a high-energy phosphate from ADP; AMP is also formed :

$$2ADP \xrightarrow{\text{Myokinase}} AMP + ATP$$

5. Iodoacetate can block resynthesis of ATP and phosphocreatine by preventing glycolysis.

6. ATP is also formed in the muscle from glycogen under anaerobic conditions in which glycogen is converted to pyruvate or lactate by glycolysis. The initial breakdown of glycogen takes place by the enzyme phosphorylase which is activated by the calcium ions. Lactate thus formed is reconverted to glucose in liver by Cori cycle and re-enters the muscle.

   ATP is also formed from the oxidation of other fuels such as free fatty acids, ketone bodies and glucose. Muscle contraction generally takes place under aerobic conditions.

7. Fat is the ultimate source of energy for long periods of muscular exertion as in the case of migratory birds, whose total carbohydrate stores are quite inadequate to act as the sole source of energy and whose fat stores become depleted during migration. The flight muscles of birds (red meat) are particularly well-developed for aerobic oxidation of pyruvate, free fatty acids and ketone bodies. These muscles also have a well-developed vasculature to increase oxygenation and a high content of enzymes of the respiratory chain as well as of cytochromes and myoglobin.

8. Heart muscle is also similar to the flight muscles of birds. It can utilize free fatty acids, ketone bodies and even lactate which are oxidized under aerobic conditions.

INORGANIC CONSTITUENTS OF MUSCLE

1. The cations (potassium, sodium, magnesium and calcium) of muscle are the same as in extracellular fluids excepting potassium which predominates in muscle.
2. The anions are phosphate, chloride and small amounts of sulphates.
3. A good amount of potassium is incorporated into the tissue during deposition of glycogen in the muscle and synthesis of protein. Muscle weakness is a sign of potassium deficiency.
4. Muscle calcium and magnesium act as activators or inhibitors of intramuscular enzyme systems.

CELL MOTILITY AND THE CYTOSKELETON

Nonmuscle cells perform mechanical work, morphogenesis, cleavage, endocytosis, exocytosis and intracellular transport. These cellular functions are carried out by an extensive intracellular network of filamentous structures constituting the cytoskeleton. All eukaryotic cells contain three types of filamentous structure :

1. Actin filaments (7-9.5 nm in diameter)
2. Microtubules (25 nm in diameter)
3. Intermediate filaments (10-12 nm in diameter).

Actin filaments

1. The G-actin protein of nonmuscle cells has a molecular weight of about 43,000 and contains N-methylhistidyl residues.
2. This actin spontaneously polymerizes in the presence of magnesium and potassium chloride to form the double helical *F-actin filaments* like those in muscle.
3. There are two types of actin in nonmuscle cells : β-actin and γ-actin. Both types can coexist in the same cell and even copolymerize in the same filament.
4. Actin forms *microfilaments* of 7-9.5 nm in the cellular cytoplasm. The bundles of filaments are prominent just underlying the plasma membrane of resting cells and are there referred to as *stress fibres*. These stress fibers disappear as cell motility increases.
5. Actin microfilaments are associated with other muscle-like proteins in nonmuscle cells. α-Actinin is present at the plasma membrane sites to which microfilaments are attached.
6. Myosin is found in association with actin microfilaments at the bases of microvilli.
7. Tropomyosin promotes the formation of bundles of actin stress fibres. α-Actinin promotes the attachment of actin microfilaments to membranes, substratum and other cell organelles.

Microtubules

1. Microtubules are the integral component of the cellular cytoskeleton. They consist of cytoplasmic tubes 25 nm in diameter and of indefinite length. They are also necessary for the formation and function of the *mitotic spindle*.
2. They perform a number of other cellular functions and are responsible for the intracellular movement of endocytotic and exocytotic vesicles.
3. They form the major structural component of *cilia* and *flagella*.
4. They are major protein components of *axons* and *dendrites*, where they maintain the structure and participate in the axoplasmic flow of material along this neuronal processes.
5. They are the cylinders of 13 longitudinally arranged protofilaments, each consisting of dimers of α-tubulin and β-tubulin which are closely related protein molecules.
6. Many important alkaloids can present microtubule assembly.
7. The movement of chromosomes during anaphase of mitosis is dependent upon microtubules.

Intermediate filaments

1. They are distinct from microfilaments and microtubules. There are five major classes of these filaments.
2. Each filament consists of distinct subunits —biochemically and immunologically.

COLLAGEN

1. Collagen is the major macromolecule of connective tissues and it is the most common protein in the animal kingdom.
2. In mammalian tissues, there are *five distinct types* of collagen and they exist as a family of molecules sharing many properties.

3. The most important property of collagen molecules is their *triple helix*, a coiled coil of three polypeptide subunits. Each polypeptide subunit is twisted into a left-handed helix of 3 residues per turn. Three of these left-handed helices are then wound to a right-handed superhelix to form a stiff rodlike molecule 1.4 nm in diameter and about 300 nm long. These triple helical molecules are associated into fibrils. There is a gap between the end of one triple helix and the beginning of the next where there is the deposition of hydroxyapatite crystals in bone formation.
4. The collagen fibres range from 10 to 100 nm in diameter and are visible by microscopy as banded structures in the extracellular matrix of connective tissues.
5. Glycine is the only amino acid to exist in the limited space available down the central core of the triple helical molecule.
6. In mammalian collagen, about 100 of the X positions are proline and 100 of the γ positions are 4-hydroxyproline. They also contain 3-hydroxyproline in some X positions and 5-*hydroxylysine in γ positions*.

Synthesis of collagen

1. Collagen is an extracellular protein but it is synthesized as an intracellular precursor molecule before becoming a mature collagen fibril.
2. *Preprocollagen*, a precursor of collagen, contains a leader or signal sequence of about 100 amino acids at its amino-terminus. This preprocollagen is generated by ribosomes attached to the endoplasmic reticulum.
3. When the signal sequence penetrates into the vesicular space of the endoplasmic reticulum, the leader sequence is cleaved off and the amino-terminal end of procollagen continues to protrude into the endoplasmic reticular space. At this site, prolyl 4-*hydroxylase* and *lysyl hydroxylase* acts on proline or lysine residues respectively in the γ position of the (Gly-X-γ)n Peptide. A prolyl 3-hydroxylase in the X position immediately preceeding a 4-hydroxyproline in the γ position.
4. The procollagen molecule contains at its aminoterminus a 20,000-MW peptide and at its carboxyl-terminus a 30-35 thousand-MW peptide. Neither of these is present in mature collagen.
5. After formation of triple helix, further hydroxylation of prolyl and lysyl residues cannot take place.
6. After this intercellular processing, the glycosylated procollagen molecule reaches the outside of the cell by way of the Golgi complex. Extracellular *procollagen aminoprotease* and *procollagen carboxyprotease* remove the amino-terminal and carboxyl-terminal propeptides, respectively. The newly formed collagen molecules have about 1,000 amino acids and spontaneously assemble into *collagen fibrils* that are indistinguishable from the mature fibrils found in tissues. These fibrils do not have the tensile strength of mature collagen fibrils until they are *cross-linked by a series of covalent bonds*.

Inherited defect of collagen

Type VI Ehlers-Danlos Syndrome :

1. This is an inherited deficiency of lysyl hydroxylase.
2. This disease is characterized by frequent abnormalities of the eye, severe scoliosis (abnormal vertebral column curvature) and hyperextensibility of the skin and joints.

Type V Ehlers-Danlos Syndrome :

1. This is due to the deficiency of lysyl oxidase activity. The absence of this enzyme prevents normal cross-linking of collagen. Lysyl oxidase is a copper-requiring enzymes.

2. There is severe arteriovascular and skeletal change.

Type VII Ehlers-Danlos Syndrome :

1. This is caused due to the non-serving of procollagen as a substrate for the procollagen aminoprotease.
2. The patients have hip dislocations, increased skin elasticity, and short stature.

Role of nitric oxide (NO) in health and disease

EDRF (Endothelium-derived relaxing factor) is the gas nitric oxide (NO).

Formation

1. Nitric oxide (NO) is formed by the action of the enzyme NO synthase, which is cytosolic. The endothelial and neuronal forms of NO synthase are activated by Ca^{2+}. The substrate is arginine, and the products are citrulline and NO :

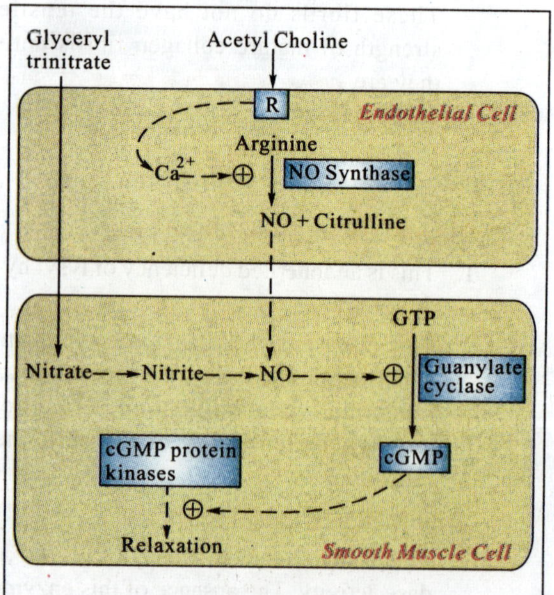

Fig. 14.7 : Formation of Nitric oxide (NO) in an endothelial cell, interaction of an agonist (eg, Acetyicholine) with a receptor (R) leads to intracellular release of Ca^{2+}

$$\text{Arginine} \xrightarrow{\text{NO Synthase}} \text{Citrulline} + \text{NO}$$

2. NO synthase is a complex enzyme containing five redox cofactors : NADPH, FAD, FMN, heme, and tetrahydrobiopterin.
3. NO can also be formed from nitrite, derived from vasodilators such as glyceryl trintrate during their metabolism.
4. NO has a very short half-life (about 3-4 seconds) in tissues because it reacts with oxygen and superoxide. The reaction with superoxide produces peroxynitrite ($ONOO^-$) which decomposes to form the highly reactive OH^- radical.

Relaxative action on smooth muscle of blood vessels

1. Acetylcholine is a vasodilator and causes relaxation of the smooth muscle of blood vessels. It does not act directly on smooth muscle.
2. Acetylcholine initially interacts with the endothelial cells of small blood vessels via receptors. The receptors are coupled to the phosphoinositide cycle leading to the intracellular release of Ca^{2+} through the action of inositol triphosphate. The elevation of Ca^{2+} leads to the liberation of endothelium-derived relaxing factor (EDRF) which diffuses into the adjacent smooth muscle.
3. EDRF then reacts with the heme moiety of a soluble guanylyl cyclase with the elevation of intracellular levels of cGMP which stimulates the activities of certain cGMP-dependent protein kinases that phosphorylate specific muscle proteins causing relaxation.
4. The coronary artery vasodilator nitroglycerin, widely used to relieve angina pectoris, acts to increase intracellular release of EDRF and thus of cGMP.

Important biologic functions

1. Nitric Oxide (NO) is inhibited by hemoglobin and other heme proteins which bind it strongly.
2. Chemical inhibitors of NO synthase can markedly decrease formation of NO. The use of such inhibitors to animals and humans leads to vasoconstriction and a high elevation of blood pressure. This indicates that NO is important in the maintenance of blood pressure in vivo.
3. It has the important cardiovascular effect that it acts as an inhibitor of platelet aggregation by increasing synthesis of cGMP.

Table 14.1 : Some physiologic and pathologic role of nitric oxide (NO)

- Vasodilator and regulator of blood pressure.
- Neurotransmitter in the brain and peripheral autonomic nervous system.
- Involved in penile erection.
- Role in long-term potentiation.
- Role in neurotoxicity.
- Role in relaxation of skeletal muscle.
- Inhibits adhesion, activation and aggregation of platelets.
- Low level causes pylorospasm in infantile hypertrophic pyloric stenosis.
- Constitutes part of a primitive immune system.

CLINICAL ORIENTATION

- Duchenne-type muscular dystrophy is a genetic disease associated with muscle weakness in both the pelvic and shoulder girdle and enlargement of the calf muscles.
- The malignant hyperthermia, a serious complication during certain types of anesthesia, is due to abnormal accumulation of Ca^{++} in the cytoplasm of skeletal muscle cells causing rigor and the generation of excessive heat.
- The increase in Ca^{++} is due to mutation in the structural gene for the Ca^{++} release channel of the sarcoplasmic reticulum. This condition relates to serious financial implications for the Swine industry.
- Cardiomyopathies are genetically determined and shown by recent work that mutations in the β-Cardiac myosin heavy chain gene are one cause of familial hypertrophic cardiomyopathy.
- Many widely used Vasodilators such as glyceryl trinitrate is used in the treatment of angina pectoris.
- The endothelium-derived relaxing factor, synthesized by endothelial cells, is the gas nitric oxide (NO) which is a neurotransmitter.

EXERCISE

1. Discuss the chemistry of muscle. (M.U. 74S)
2. Discuss muscle contraction. (P.U. 72A)
3. Describe the structure of muscle. (R.U. 71A)
4. Write short notes on :
   (a) Actomyosin. (M.U. 73S)
   (b) Tropomyosin. (Bh. U. 72S)

SCHEMATIC COMPENDIUM

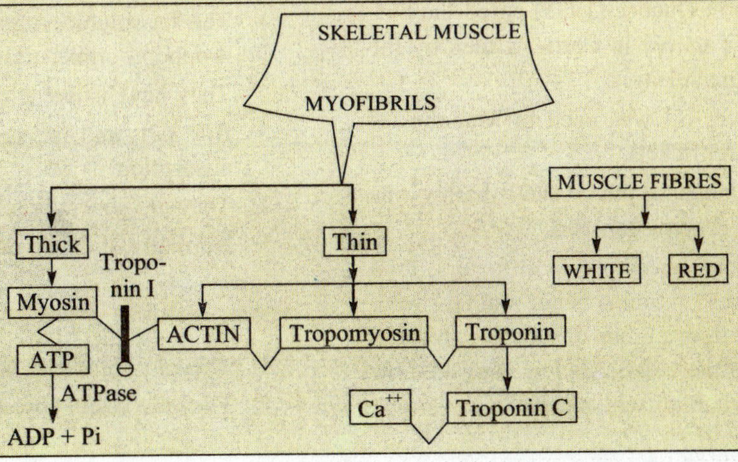

CHAPTER 15

VITAMINS

History

1. Lunin (1881) and Pakelharing (1905) showed in an experiment that the cessation of growth and death of animals had been prevented on the addition of small amounts of milk to their diet.
2. Eijkman (1906) concluded that the dietary deficiency of a substance which was present in rice polishings caused the disease beri-beri.
3. Hopkins (1906-1912) by experiments gave more importance to the "growth-promoting" substance in milk. These experiments had been confirmed by Osborne and Mendel and by McCollum and Davies (1913).
4. Hippocrates (C. 400 B.C.) recommended ox liver and honey for nightblindness.
5. Indians in Quebec (1935) cured the outbreak of scurvy in Cartier's men by the use of green leaves.
6. Cod liver oil was used in Manchester (1770) for rickets.
7. In the Japanese navy (1860) barley was used to prevent beri-beri.
8. The names accessory food factors, food hormones, vitamins (vital amines) were derived from a group of substance responsible for these deficiencies. They are now called vitamin (without the "e").

Definition

Vitamins are the accessary food factors or essential compounds which are organic in nature, occurring in natural foods, synthesized in body, required in minute amounts for normal growth, reproduction etc.

Classification

1. Fat-soluble vitamins : Vitamins A, D, E and K.
2. Water-soluble vitamins : Vitamins C and B-complex. (B_1, B_2, B_6, niacin, pantothenic acid, biotin, lipoic acid, folic acid, B_{12} , inositol)

Common properties of vitamins

A. Fat-soluble vitamins

1. The fat-soluble vitamins are apolar, hydrophobic molecules.
2. They have isoprene derivatives.
3. Bile salts and fats are essential for their absorption.
4. They are generally stored in liver.
5. Normally, they are not excreted in urine.

B. Water-soluble vitamins

1. They are soluble in water.
2. They are easily absorbed.

3. They are not stored in the body (except B_{12}).
4. They have a threshold for urinary excretion.

FAT-SOLUBLE VITAMINS

VITAMIN A

Chemistry

1. Vitamin A_1 aldehyde (retinal) is derived from β-carotene by cleavage at the midpoint of polyene chain connecting the β-ionone rings.
2. The biosynthesis of vitamin A is likely a dioxygenase reaction in which molecular oxygen reacts with the two central carbon atoms of β-carotene followed by the cleavage of the central double bond of β-carotene to produce 2 moles of vitamin A_1 aldehyde (retinal). The aldehyde is then reduced by retinene reductase depending on NADH to yield vitamin A alcohol (retinol). The reaction is shown below.
3. The potency of vitamin A_2 (3-dehydroretinol) is 40% that of vitamin A_1 and it differs from A_1 structurally by the presence of an additional double bond between carbon 3 and 4 of the β-ionone ring.
4. Retinol exists, as an ester with higher fatty acids, in the liver, kidney, lung and fat depots.
5. The plant pigments, carotenes or carotenoids, synthesized by all plants except parasites and saprophytes are said to be provitamin A. In the human diet, vitamin A is derived from the preformed vitamin (retinol) and the provitamin carotenoids, β-carotene has the highest vitamin A activity and is the most abundant in human diets.

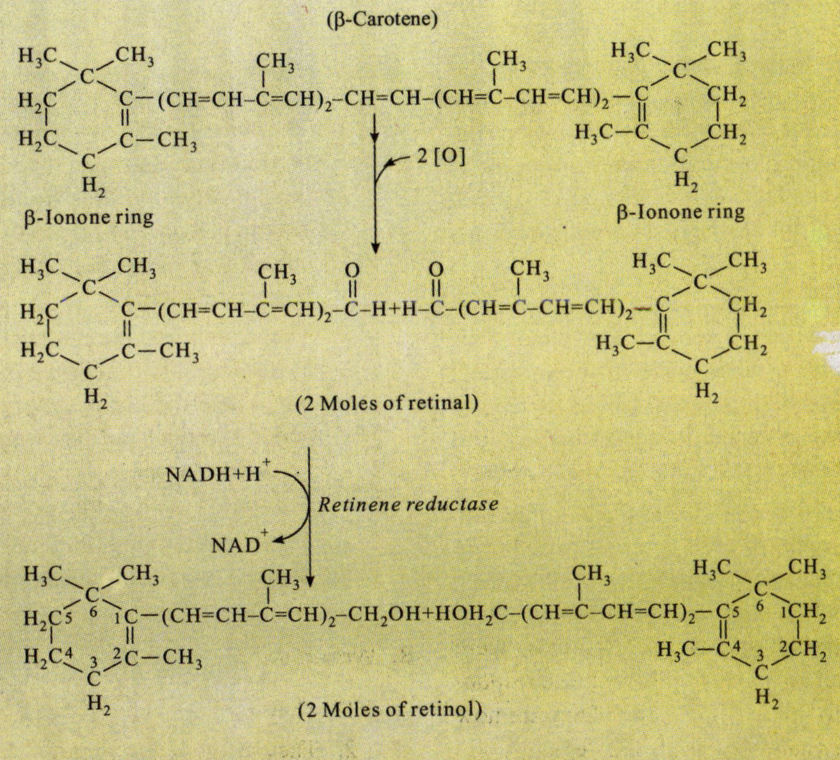

Fig. 15.1 : Vitamin A biosynthesis

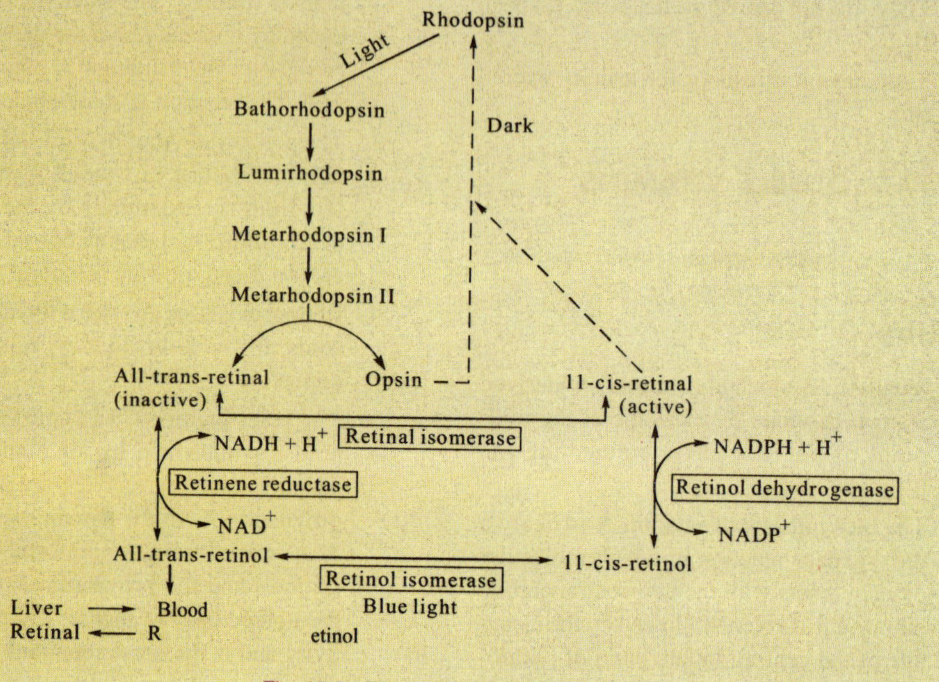

Fig. 15.2 : Rhodopsin — vitamin A cycle

β-carotene is an antioxidant and effective at low oxygen concentration but vitamin E which is effective at higher oxygen concentration. The antioxidant property of these two lipid-soluble vitamins may well account for their possible anticancer activity.

6. Carotenes are only active as a vitamin precursor when it is converted to retinol. This conversion takes place in the intestinal wall in rats, pigs, goats, rabbits, sheep and chickens although liver also participates; but in man, it takes place in the liver only.

   Retinol is transported to the blood as retinol-binding protein (retinol attached to α_1-globulin). Whereas, the carotenes are transported with the lipoproteins.

7. A small fraction of retinal is oxidized to retinoic acid :

$$\text{Retinal} \xrightarrow[\text{FAD}]{\text{NAD}} \text{Retinoic acid}$$

Absorption

1. The dietary intake of vitamin A in the form of vitamin A esters are hydrolyzed in the lumen of the intestine by the enzyme lipase in presence of bile salts and fats. Tocopherol prevents the oxidative destruction of the vitamin.
2. The vitamin and the carotene are taken up by the intestinal mucosa where the vitamin is esterified and carotene is first converted to retinal and then to retinol by retinol dehydrogenase which is then esterified mostly with palmitic acid.
3. These esters—together with some unchanged retinal and carotenoids which are not precursors of vitamin A—are absorbed and enter the intestinal lymphatics and eventually the circulation, in chylomicrons.
4. In the blood, the vitamin esters are attached to β-lipoproteins and are then taken

up by the liver (Kupffer cells) which contains almost all the body store.

5. The vitamin is then released as retinol-binding protein for use elsewhere, eg, in the retinal rods.
6. Since retinol dehydrogenase is present in the liver serum, retinal can be converted to retinol.
7. The impaired absorption of dietary fat (eg, obstructive jaundice, chronic pancreatitis) leads to the impaired absorption of vitamin A in the absence of pancreatic enzymes (impaired hydrolysis of vitamin A esters) and in celiac disease or other conditions.
8. In subjects with liver damage, the capacity for storage and formation of vitamin A is impaired and its concentration in the blood is also decreased.

Storage

1. 95 per cent of vitamin A in the form of esters is stored in the liver. The hepatic storage increases with age. The quantity stored in the liver varies in different species but the storage capacity in man is relatively large.
2. A small amount of it is also present in other tissues, eg, lactating breast, adrenals, lung, intestine.

Sources

1. Plant sources : All pigmented (particularly yellow) vegetables and fruits (eg, sweet potatoes, carrots, pumpkins, papayas, tomatoes, apricots and peaches) and the leafy green vegetables which supply provitamin A (carotene) in the diet. Cereals also contain carotene.
2. Animal sources : Preformed vitamin A (retinol) is supplied by foods of animal origin; they are liver, milk, butter, eggs, kidney, the fat of muscle, meats, and fish liver oil which is very rich in the vitamin.

Daily requirements

Adults	5,000 I.U.
During pregnancy and lactation	6,000-8,000 I.U.
Children	2,000-3,000 I.U.
Infants	1,500 I.U.
1 I.U. = 0.3 mg of retinol = 0.6 μg of β-carotene	

Normal concentration in blood

0.15 to 0.60 μg/dL

Physiological functions

1. Role in vision : Wald and Morton gave more emphasis on the specific role of vitamin A in physiological mechanism of vision.

   Retina contains two types of receptor cells :

   (a) Cones—which are specialized for colour and detail vision in bright light.

   (b) Rods—which are specialized for visual activity in dim light (night vision).

   Photolysis of rhodopsin closes sodium channels in rod membrane through either a release of Ca^{++} into cytosol or a hydrolytic fall in the cytosolic cyclic GMP level. This results in lowering the permeability (membrane) to Na^{+} causing hyperpolarization of rod membrane and changes its conformation through 5° and initiates the visual signal from the rod. The basic factors stimulating the rods are :

   (i) Cis-trans isomerization

   (ii) Conformational changes in the rhodopsin due to cGMP.

   Visual activity of rod cells is dependent on their content of photosensitive pigment called "rhodopsin" which is a conjugated protein. It contains opsin as its apoprotein and retinene as its prosthetic group. Retinene or retinal present in rhodopsin is 11-cis retinal. The aldehyde group of 11-cis retinal is bound to $-NH_2$ group of lysine of opsin. Rhodopsin has a light absorbing property due to polyene group of 11-cis retinal.

When the light falls on rhodopsin it splits into opsin and all trans retinal in a series of event like bathorhodopsin → lumirhodopsin → metarhodopsin 1 → metarhodopsin II. All trans retinal is inactive in synthesis of rhodopsin, it has to be converted to 11-cis retinal in following ways :

(a) All trans retinal may be isomerized to its 11-cis isomer in presence of blue light but in the eye this isomerization is not significant.

(b) The all trans retinol from blood can be converted to all trans retinal by retinene reductase by making use of NAD^+ and all trans retinal then can be isomerized to its cis-isomer.

(c) All trans retinol from blood can be first isomerized to 11-cis-retinol. All 11-cis-retinol then can be converted to 11-cis retinal by retinol dehydrogenase in presence of coenzyme $NADP^+$.

Thus the visual process involves continual removal of the active retinol from blood into retina.

2. Vitamin A helps in maintaining the integrity of epithelial tissues as epithelial layers of skin, respiratory mucosa, oesophagus and genitourinary tract.
3. It helps in the preservation of the structural integrity and the normal permeability of the cell membrane as well as the membrane of the sub-cellular organelles such as mitochondria, lysosome etc.
4. It accelerates the normal formation of bones and teeth.
5. Retinoic acid is found to be more involved in glycoprotein synthesis probably forming its phosphate which acts as carrier of oligosaccharides in glycoprotein synthesis. Thus, vitamin A is involved in the development and maintenance of ground substances in collagen tissues.
6. Vitamin A is involved in the synthesis of chondroitin sulphate.
7. When retinol is taken up into cellular retinol binding protein, it is transported around the cell and binds to nuclear protein and involves in gene expression. Thus, vitamin A behaves similar to steroid hormone.

Retinol, Retinal & Retinoic acid

1. Retinol and Retinal are interconverted in the presence of NAD or NADP-requiring dehydrogenases or reductases present in many tissues.
2. Retinoic acid once formed from retinal cannot be converted back to retinal or to retinol.
3. Retinoic acid can support growth and differentiation but cannot replace retinal in its role in vision or retinol in its support of the reproductive system.

Retinol acts as a steroid hormone

Retinol being taken up into CRBP is transported around the cell and binds to nuclear proteins where it is involved in the control of the expression of certain genes. Thus, vitamin A behaves similarly to steroid hormones.

Retinal is a component of Rhodopsin

Rhodopsin occurring in the rod cells of the retina are responsible for vision in poor light. II-Cisretinal, an isomer of all-trans-retinal, is specially bound to the visual protein opsin to form rhodopsin. Rhodopsin on exposure to light dissociates into all-trans-retinal and opsin. This reaction is induced by a calcium ion channel in the membrane of the rod cell.

Retinoic acid participates in Glycoprotein synthesis

Retinoyl phosphate functions as a carrier of oligosaccharides across the lipid layer of the cell by way of an enzymatic trans- cis-isomerization. Thus, retinoic acid is involved in glycoprotein synthesis, since a deficiency of vitamin A results in the accumulation of abnormally low molecular weight oligosaccharide-lipid intermediates of glycoprotein synthesis.

Deficiency of Vitamin A

1. **Xerophthalmia** : Thickening and loss of transparency of the bulbar conjunctiva with yellowish pigmentation. The keratinization of ocular tissue progresses to blindness in the late result of vitamin A deficiency.
2. **Follicular conjunctivitis**.
3. **Keratomalacia** : Softening of the cornea, in advanced cases with ulceration and necrosis. Defective vision due to keratinization of epithelium of cornea.
4. **Night blindness (nyctalopia)** : This is one of the earliest signals of vitamin A deficiency which is impairment of dark adaptation. Some retinal liberated by rhodopsin is not available for resynthesis and must be replaced by synthesising fresh 11-cis-retinal from retinol. For this all transretinol is collected from blood and dehydrogenated and isomerized. So, continual supply of retinol is essential for normal visual function. Retinol deficiency depresses the resynthesis of rhodopsin and interferes with the function of rods resulting in night blindness.
5. **Follicular hyperkeratosis** of the skin.
6. **Keratinizing metaplasia** of epithelium of nose, respiratory mucosa, oesophagus and genitourinary tract.
7. Formation of renal calculi (**Urolithiasis**).

Hypervitaminosis

1. Acute symptoms are developed on the ingestion of very large amounts of vitamin A. These include drowsiness, sluggishness, severe headache, vomiting and peeling of the skin about the mouth and elsewhere. In infants and young children, there may be a sudden rise of intracranial pressure.
2. Continued intake of excessive amounts, especially in children, produces roughening of the skin, irritability, coarsening and falling of the hair, anorexia, loss of weight, headache, hyperesthesia, occasionally anemia, leukopenia.
3. Serum level of vitamin A, lipid and acid phosphatase are elevated.

   All the symptoms vanish on withdrawal of vitamin A.

Coenzyme activities

Vitamin A has no coenzyme activities.

Determination of vitamin A

Colorimetric determination utilizes the Carr-Price reaction, in which a blue colour is obtained when a solution of antimony trichloride in chloroform is added to the vitamin-containing mixture. This is used to determine the vitamin A content of blood plasma.

Effects

Vitamin A in excess intake causes toxicity.

Management

(a) On diagnosis in the deficiency of vitamin A a single large dose of 60 mg retinol as palmitate or acetate (200,000 I.U.) should be given orally. The oral dose should be repeated the next day and again prior to discharge.

(b) 55 mg retinol palmitate should be given by intramuscular injection if there is vomiting or severe diarrhoea in addition to the deficiency of vitamin A.

(c) Antibiotics are of value in case of secondary bacterial infection.

(d) Local treatment of the eye will be required only if there is the presence of disorganisation and in this case the services of an ophthalmic surgeon is essential.

Toxicity

(a) Acute overdose leads to nausea and headache, increased intracranial pressure and skin desquamation.

(b) Chronic high doses can cause liver damage and hyperostosis.

THE VITAMINS D

The vitamins D are a group of compounds. All are sterols occurring chiefly in animal organisms. Provitamins D possess the property of curing or preventing rickets when subjected to long- wave ultraviolet light (about 265 nm).

Vitamin D is a steroid hormone. It is represented by a group of steroids that occur mainly in animals

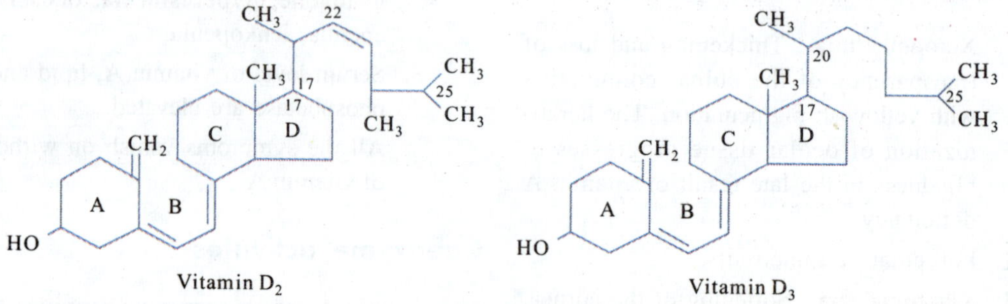

Fig. 15.3 : Structure of vitamin D_2 and vitamin D_3

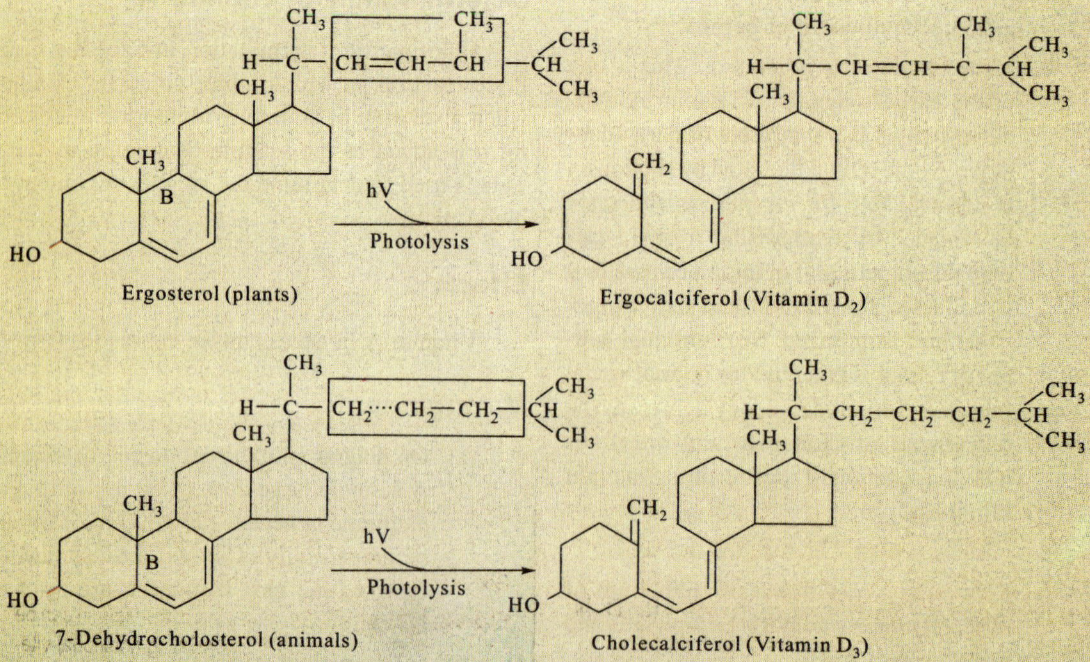

Fig. 15.4 : Formation of Vitamins D_2 & D_3

but also in plants and yeasts. They give rise to a hormone known as calcitriol by various metabolic changes. Calcitriol has a role in calcium and phosphate metabolism.

Chemistry

1. Ergosterol occurs in plants and 7-dehydrocholesterol in animals. Ergosterol is unsaturated and contains an extra methyl group in its side chain than 7-dehydrocholesterol. Ultraviolet irradiation from sunlight cleaves B ring of both compounds. Ergocalciferol (Vit. D_2) is formed in plants and cholecalciferol (Vit D_3) is formed in exposed skin. Both vitamins are of equal potency giving rise to D_2 calcitriol and D_3 calcitriol.
2. The most important D vitamins are D_2 (activated ergosterol) or ergocalciferol or viosterol and D_3 (activated 7-dehydrocholesterol, cholecalciferol) on the basis of nutrition.
3. Provitamin D_2 (ergosterol) occurs in the plant kingdom; whereas vitamin D_3 occurs in fish liver oil in nature.
4. The structure of vitamin D_2 is the same as that of D_3 except that the side chain on position 17 is that of cholesterol.
5. Provitamin D_3 are synthesized in the body of man and other mammals. This is then activated in the skin by sunlight or ultraviolet rays and carried to various organs in the body for utilization or storage in the liver.
6. The D vitamins are more stable than vitamin A and are not lost by ordinary cooking and preserving processes.

Formation of Calcitriol

The biologically active form of vit. D is calcitriol (synthesized in liver end kidney) :

Calcitriol (1, 25 dihydroxy cholecalciferol)

Synthesis of calcitriol

7-Dehydrocholesterol (Provitamin) —[Skin in presence of sunlight or UV]→ Vit. D_3

Vit. D_3 —[O_2, NADPH, Mg^{++}, Cyt P450; Liver 25 Hydroxylase]→ 25 (OH) D_3

25 (OH) D_3 —[Kidney mitochondria; NADPH, O_2, Mg^{++}; 1-α hydroxylase]→ 1, 25 $(OH)_2$ D_3 (Calcitriol)

Vit. D has intracellular receptor

Receptor carries calcitriol in target cells such as intestinal mucosa, kidney etc.

↓

Calcitriol receptor binds to chromatin in the nucleus

↓

Stimulating transcription of mRNA that codes for calcium binding protein (eg, osteocalcin)

Absorption and Storage

1. The D vitamins are readily absorbed in the small intestines.
2. Since they are fat-soluble their absorption is enhanced by factors which favour fat absorption such as sufficient quantity of bile salts.
3. After absorption, the vitamin is stored largely in the liver, kidneys, intestines, adrenals and bones.
4. A small amount is excreted in the bile but is partly reabsorbed in the intestine. No amount is eliminated in the urine.

Sources

In the active form, vitamin D is not well distributed in nature.

Rich sources	: The liver and viscera of fish and the liver of animals which feed on fish.
Good sources	: Eggs and butter.
Poor sources	: Milk. The amount of the vitamin can be increased by providing additional vitamin D in the cow's diet.

Cheapest source : The cheapest source is sunlight which forms D_3 from 7-dehydrocholesterol in the skin.

Daily requirements

Infants and children	400 I.U.
Adults	200 I.U.
Women during pregnancy and lactation	400 I.U.

1 I.U. = 0.025 μgm. of cholecalciferol

Normal concentration in blood

In adults	700-3,100 I.U./1.
In children	860-2,100 I.U./1.

Physiological functions

1. **Intestinal absorption of calcium and phosphate :** Vitamin D binds to the chromatin of target tissue and expresses the genes for calcium binding protein as well as Ca^{++}–ATP ase in intestinal cells. This increases the Ca^{++} absorption by actively transporting Ca^{++} across the plasma membrane against electrochemical gradient.
2. **Mineralization of bones :** The synthesis of Ca^{++}–binding protein (osteocalcin) and alkaline phosphatase increase Ca^{++} and phosphate ions in the bone and enhances the mineralization of bone.
3. Vitamin D stimulates the transcription of mRNA for calcium-binding protein.
4. It increases the excretion of phosphate by kidney and decreases the concentration of serum phosphate.
5. It decreases the pH in the lower intestinal tract.
6. It increases the citrate level of blood, bone, kidney and heart tissues and also the excretion of citric acid.
7. It stimulates the activity of phytase which catalyzes the hydrolysis of phytic acid in intestine.
8. It helps in the development and growth of bone.

Deficiency of Vitamin D

In absence of it, instead of growth occurring normally, the osteoblast proliferation does not take place in a regular manner and there is no vascularization and mineralization of bones. As a result, irregularity is at the zone of calcification. The cartilage cells do not degenerate and ends of the long bones become bulky and soft.

Bow legs and knock knees occur due to deficiency of it. The ankles, knees, wrists, elbows are swollen due to swelling of epiphyseal cartilages. The ribs give beaded appearance and chest gives pigeon breast appearance.

Vitamin D deficiency produces rickets in children and osteomalacia in adult.

1. There are two types of rickets :
   (a) Type I : Deficiency of 1α-hydroxylase and as a result no conversion of $25(OH)D_3$ to $1,25(OH)_2D_3$.
   (b) Type II : Autosomal recessive disorder where there is a single amino acid change in one of the zinc fingers of the DNA binding sites for receptors.
2. **Osteomalacia :** In fully grown bones in adults, there is a type of defect in mineralization of osteoid tissue termed osteomalacia. Vitamin D deficiency occurs due to absorption defect and less synthesis defect.
3. **Renal osteodystrophy :** When renal parenchyma is lost, it is unable to form calcitriol and calcium absorption is impaired.

Hypocalcemia leads to increase in PTH (parathyroid hormone) which acts on bone to increase Ca^{++}. This condition is known as Renal osteodystrophy.

Causes of deficiency of vitamin D

1. Dietary insufficiency or insufficient exposure to sunlight.
2. Gastrointestinal disorder.
3. Chronic obstructive jaundice.

4. Prolonged treatment with anticonvulsant drugs.

Hypervitaminosis D

Extremely large amounts (500 to 1,000 times the normal requirements) cause hypervitaminosis D :

1. The early symptoms are anorexia, thirst, constipation and polyuria, followed later by nausea, vomiting and diarrhoea. Hyperphosphatemia also occurs.
2. Increased urinary excretion of calcium and phosphate may lead to urinary lithiasis, and the hypercalcemia and hyperphosphatemia may lead to metastatic calcification. The kidney, arteries, muscles and gastric mucosa are mainly involved. The development of renal failure leads to death.

Coenzyme activities—No coenzyme activities.

William syndrome :

1. Infants with hypercalcemia, hypervitaminosis D, aortic stenosis and mental retardation (congenital).
2. Hyperparathyroidism.
3. Over dosage of vitamin D.

Management

According to **WHO** the recommended oral daily intake in case of deficiency of vitamin D is as follows :

For infants and children up to 5 years of age and for pregnant or lactating women	10 μg or 400 I.U.
For older children and adults	2.5 μg or 100 I.U.

Toxicity

(a) Large oral doses cause hypercalcemia with the symptoms of nausea, vomiting, constipation, drowsiness and signs of renal failure; metastatic calcification in the arteries, kidneys and other tissues.

(b) Renal damage may occur before clinical sings of toxicity and patients on large doses of vitamin D should have their serum calcium level checked regularly at an interval of three months and if serum calcium level is found to be above 10.5 mg/100 ml, it is an early indication of overdosage.

VITAMIN E (TOCOPHEROLS)

Chemistry

1. Vitamin E refers to a group of compounds known as tocopherols.
2. Four tocopherols namely α-, β-, γ- and δ- have been isolated. β- and γ-tocopherols have two methyl groups in the aromatic nucleus whereas α-tocopherol has three methyl groups in the aromatic nucleus and δ- has only one methyl group.
3. Tocopherols are soluble in fat solvents and destroyed in an alkaline medium. The vitamin activity is destroyed by oxidation.
4. The tocopherols are largely methyl derivatives of the parent compound tocol; they are yellow oily substances.
5. Some tocopherols are derivatives of tocotrienol which contains three double bonds in the aliphatic side chain and is, therefore, terpenoid in structure.

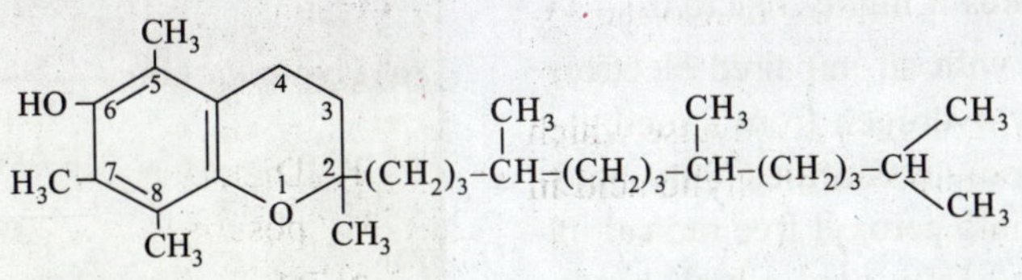

Fig. 15.5 : α-Tocopherol

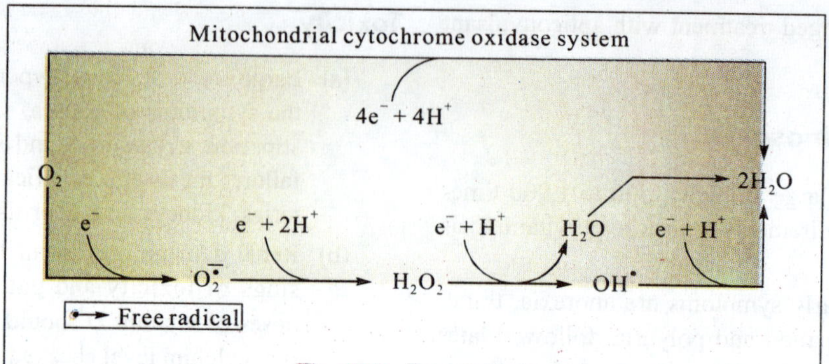

Fig. 15.6 : Free radical

Absorption and Storage

1. It is absorbed in the intestine in the presence of bile salts.
2. It is stored in the liver (mitochondria; microsomes) and fatty tissues.
3. It is present in high concentration in the adrenals, the pituitary, the uterus and the testes.

Sources

Good sources : Eggs, meat, liver, fish, chicken, oatmeal, corn oil and cotton seed oil.

Daily requirements

Adults : 25-30 mg.

Normal concentration in blood

10 mg/I.

Free Radical

1. A free radical is an atom or molecule that has one or more impaired electrons. Its tendency to acquire an electron from other substances makes it highly reactive.
2. A free radical with an impaired electron may take away hydrogen from a methylene group of polyunsaturated fatty acid and convert it into peroxyl free radical of polyunsaturated fatty acid which binds with O_2 to give a fatty acid peroxyl radical. This then changes into fatty acid hydroperoxide by acccepting hydrogen from methylene group of another polyunsaturated fatty acid, converting it into another free radical of fatty acid. Thus, the free radical peroxidates the unsaturated fatty acids of the cell membranes, mitochondrial membranes etc.
3. Free radicals like OH^-, superoxide anion O_2^- are formed during the action of some oxidoreductases such as microsomal NADPH oxidase.
4. When oxygen is reduced to water by cytochrome oxidase, 4 electrons are required. Electrons can be gained one at a time by univalent reduction. The individual molecules in univalent reaction are highly reactive and potentially damaging the tissues.
5. Other sources : Xanthine oxidase generates superoxide. Cyclooxygenase and lipooxygenase produce hydroxyl and peroxyl radicals. Superoxide may also be produced during the metabolism of xenobiotics by cytochrome P450 etc. :

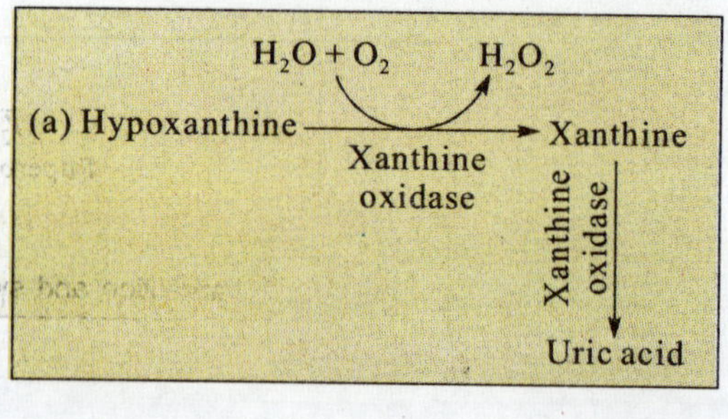

$$\text{(b) } 2O_2 + NADPH \xrightarrow[\text{oxidase}]{\text{NADPH}} 2O_2^- + NADP + H^+$$

$$\text{(c) } O_2^- + O_2^- + 2H^+ \xrightarrow[\text{dismutase}]{\text{Superoxide}} H_2O_2 + O_2$$

6. Superoxide anion is produced :
   (i) When Fe^{++} is oxidised to Fe^{+++}.
   (ii) By the action of drugs eg, primaquine.
   (iii) By peroxidation or antioxidation of membrane lipids.

Physiological function

1. Removal of free radicals
   (a) Vitamin E appears to be the first line of defense against peroxidation of polyunsaturated fatty acids contained in cellular and subcellular membrane phospholipids.
   (b) The phospholipids of mitochondria, endoplasmic reticulum, plasma membranes possess affinities for α-tocopherol and the vitamin E appears to concentrate at these sites.
   (c) The tocopherol acts as antioxidant breaking free radical chain reactions as a result of their ability to transfer a phenolic hydrogen to a peroxyl free radical of a peroxidized polyunsaturated fatty acid.
   (d) The phenoxy free radical formed may react with vitamin C to regenerate tocopherol or it reacts with a further peroxyl free radical so that the chromane ring and the side chain are oxidized to the non-free radical product :

$$ROO^\bullet + TOCOH \rightarrow ROOH + TOCO^\bullet$$

$$ROO^\bullet + TOCO^\bullet \rightarrow ROOH + \text{Non free product.}$$

$ROO^\bullet \rightarrow$ Peroxyl radicals

$TOCOH \rightarrow$ Tocopherol.

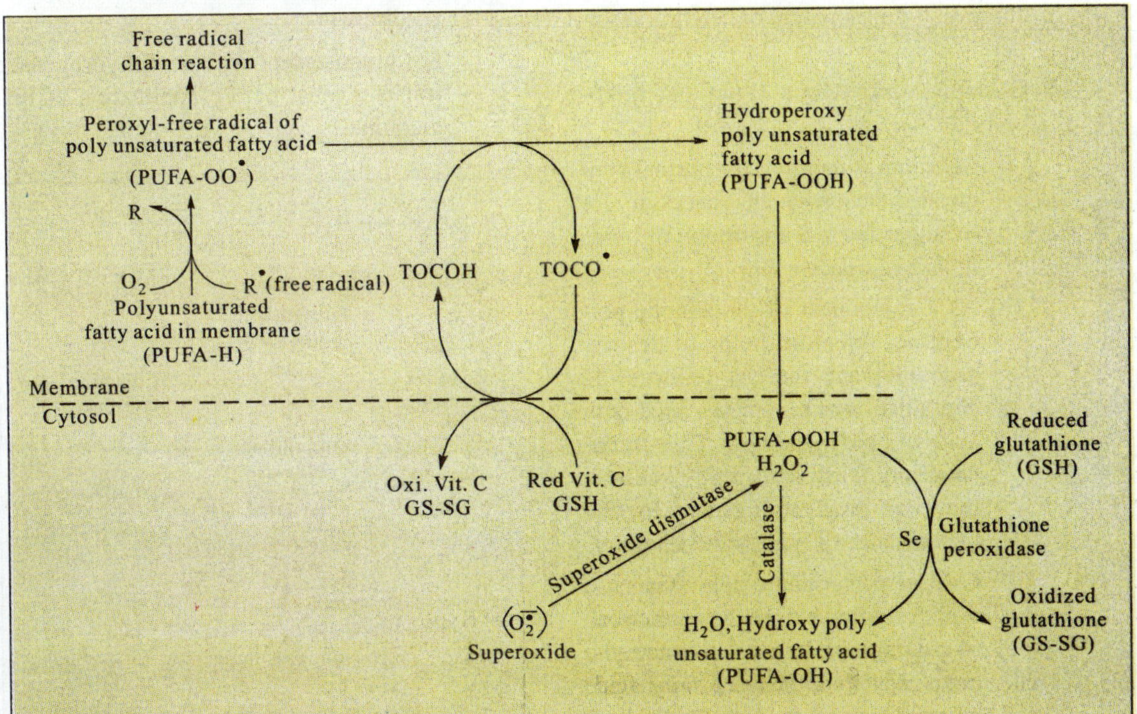

Fig. 15.7 : Interaction and synergism between antioxidant system

(e) This oxidation product is conjugated with glucuronic acid and excreted in the bile. If it reacts in this manner, tocopherol is not recycled after carrying out its function but must be replaced totally to continue its biological role in the cell.

(f) Antioxidant action of tocopherol is effective at high oxygen concentration.

2. Antioxidant action of vitamin E alongwith other factors prevent the peroxidative effect of O_3, H_2O_2 and NO_2 on respiratory membrane and prevent their damage.
3. Vitamin E prevents the peroxidative changes in membranes of mitochondria and helps in maintaining the smooth translocation of phosphate ions into mitochondria. So the oxidative phosphorylation is enhanced.
4. It prevents the hepatic necrosis produced by the lack of sulphur containing amino acids in dietary proteins.
5. It prevents development of cerebral disorder.
6. Selenium and vitamin E act synergistically :

(i) Selenium is required for normal pancreatic function which is necessary for the digestion and absorption of lipids including vitamin E.

(ii) As a component of glutathione peroxidase, selenium helps to destroy peroxides and thereby reduces the peroxidation of polyunsaturated fatty acids of lipid membrane. This diminished peroxidation greatly reduces the vitamin E requirement for the maintenance of membrane integrity.

(iii) Glucose-6-phosphate dehydrogenase test is required before the application of primaquine because primaquine has a capacity to produce superoxide anion (O_2^-) and to damage the cell membrane of RBC or decrease the life span of erythrocyte by increasing the rate of oxidation of hemoglobin to methemoglobin.

In the HMP shunt, when glucose-6-phosphate is converted to 6-phosphoglucono-lactone in presence of glucose-6-phosphate dehydrogenase, there is a formation of $NADPH + H^+$.

In the erythrocyte, NADPH is required for the reduction of oxidized glutathione to reduced glutathione catalyzed by glutathione reductase. Reduced glutathione removes H_2O_2 from the erythrocyte (H_2O_2 is formed by the combination of O_2^- and H^+) in a reaction catalyzed by glutathione peroxidase (containing selenium).

In absence of glucose-6-phosphate dehydrogenase, NADPH is not formed and reduction of oxidized glutathione to reduced glutathione does not occur, so removal of H_2O_2 from the erythrocyte becomes impossible. As a result, accumulation of H_2O_2 may decrease the lifespan of erythrocyte.

So, glucose-6-phosphate dehydrogenase test is a must before application of primaquine.

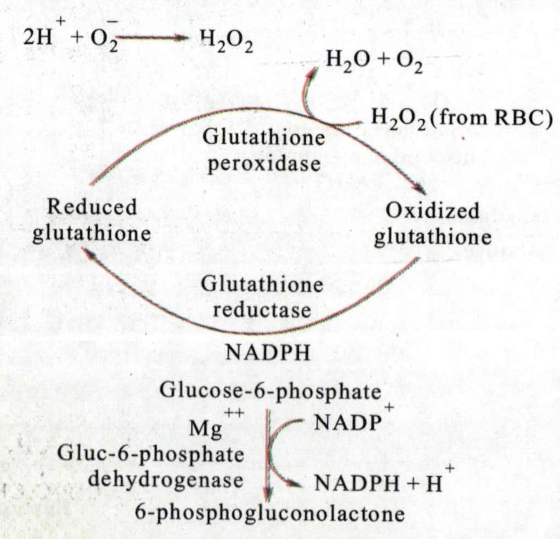

Fig. 15.8 : Role of glucose-6-phosphate dehydrogenase

Deficiency manifestations

1. Increased susceptibility of erythrocytes to hemolysis by hydrogen peroxide and decrease in the erythrocyte lifespan in human adult males.
2. Anemia, edema and skin changes are observed in infants when fed unsaturated oils.
3. Decreased erythrocyte lifespan, hemolysis, creatinuria and ceriod deposition in malabsorption syndromes in children.
4. Smaller testis and permanent sterility in male rats.
5. Resorption of foetus in female rats.
6. Increased oxygen consumption by skeletal muscle.
7. Hepatic necrosis.
8. Anemia occurs in pregnant and lactating women and in newborn infants due to the deficiency of vitamin E.
9. Its deficiency causes neurologic disorder.
10. Vitamin E deficiency causes muscular dystrophy leading to the increased oxidation of polyunsaturated fatty acids in the muscle with a consequent rise in O_2 consumption and peroxide production which causes an increase in hydrolase activity producing muscular dystrophy and exhibiting increased respiration.
11. Production of nocturnal muscle cramp.
12. Its deficiency causes fibrocytic breast disease and atherosclerosis.

Coenzyme activities : No coenzyme activities.

Disorder

Vitamin E is destroyed by commercial cooking and food processing including deep-freezing.

Toxicity

Few adverse effects of vitamin E have been reported even up to 80 times the recommended intake.

VITAMIN K

Dam (1935) named it vitamin K since it is a koagulation-vitamin. Dam, Karrer and co-workers (1939) isolated pure vitamin K_1. In the same year, Doisy and co-workers isolated pure vitamin K_2.

Chemistry

1. The several substances (natural and synthetic) with vitamin K (antihemorrhagic) activity are naphthoquinones.
2. Two naturally occurring vitamins K have been identified. Vitamin K_1 was isolated

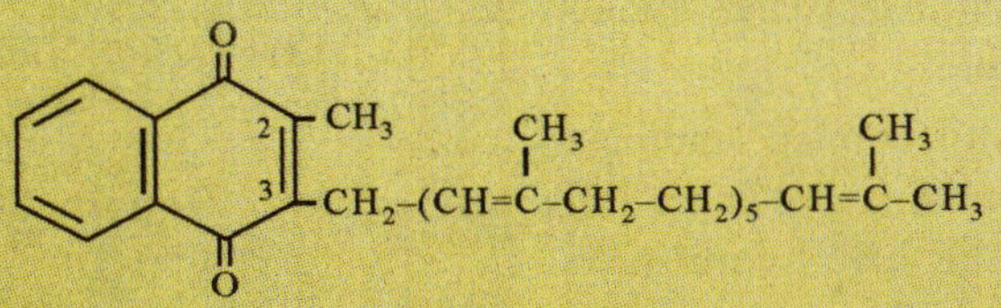

Fig. 15.9 : Vitamin K_1 (2-methyl-3-phytyl-1, 4-naphthaquinone)

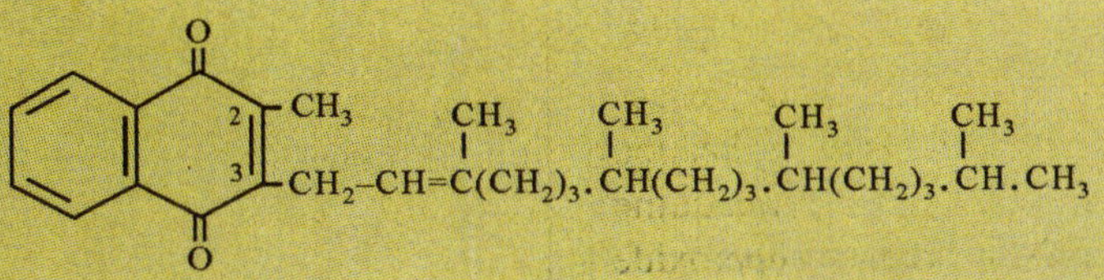

Fig. 15.10 : Vitamin K_2 (2-methyl-3-difarnesyl-1, 4-naphthaquinone)

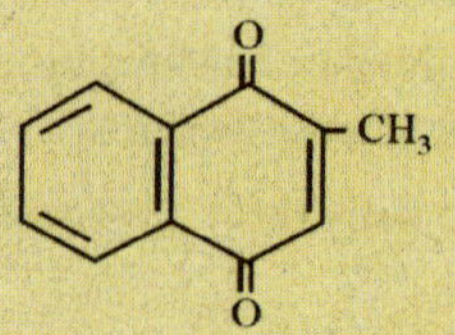

Fig. 15.11 : Menadione (2-methyl-1, 4-naphthaquinone)

originally from alfalfa and vitamin K_2 was originally isolated from putrid fish meal. Menadione is the most important of the synthetic vitamins K.

3. The natural vitamins K are "fat-soluble" vitamins. The synthetic vitamins K are lacking the long hydrocarbon chain and hence are soluble in water to some extent.
4. Vitamin K_1 possesses a phytyl radical on position 3 which occurs in plants and vitamin K_2 (menaquinone, farnoquinone) possesses a difarnesyl radical which occurs in bacteria.

Absorption and Storage

1. Being fat-soluble, its absorption is enhanced by sufficient amount of bile salts mainly in the jejunum by way of lymphatics.
2. Absorption is diminished by large amounts of liquid petroleum.
3. Its absorption is interrupted leading to hemorrhage in jaundice and other liver diseases when the bile secretion is scanty.
4. It is present in blood stream in significant amounts.
5. Liver stores appreciable amounts.
6. All tissues contain small amounts of vitamin K.
7. The feces contain large amounts of it which is produced by the bacterial flora, eg, E. coli.

Sources

Vitamin K_1		
Best sources	:	Green leafy vegetables, eg, alfalfa, spinach, cabbage, kale etc.
Good sources	:	Cauliflower, soyabean, wheat germ etc.
Fair sources	:	Carrots and potatoes.
Poor sources	:	Milk, meat and fish.

Vitamin K_1 is produced by most bacteria present in the human intestine if it is not supplied in the diet.

Daily requirements

The average diet contains adequate amounts of vitamin K_1 and K_2 being synthesized by bacteria in the intestine. So the vitamin K deficiency has not been reported in healthy individuals except in newborn infants fed on mother's milk when mother's diet has low vitamin K content.

Physiological functions

1. Vitamin K catalyzes the synthesis of prothrombin by the liver after transcription from information carried on messenger RNA.
2. It reduces the prothrombin time.
3. It regulates the synthesis of plasma clotting factors (factors VII, IX and X).

   Vit K is first converted to its hydroquinone form in liver microsomes by dehydrogenase using NADPH. K-hydroquinone acts

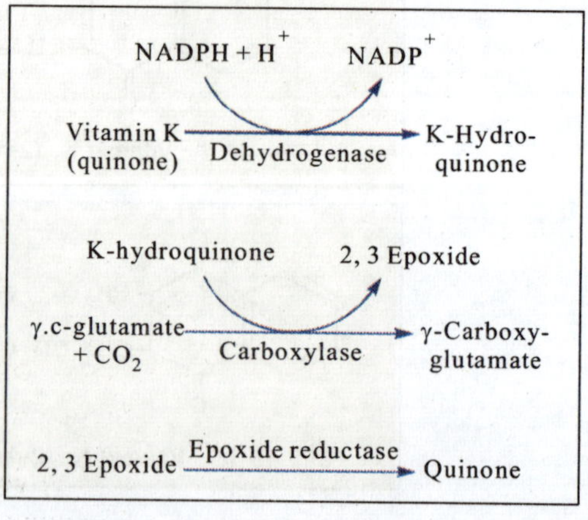

as a coenzyme for carboxylase and converts γc-glutamate to γ-carboxyglutamate with the help of CO_2. Hydroquinone may change to 2, 3 epoxide which is reduced back to quinone by microsomal epoxide reductase. Dicoumarol is found to inhibit epoxide reductase.

γ-carboxyglutamate residues now provide calcium binding sites in N-terminal portion. This brings together activated factor and accelerin close to the phospholipid membrane of platelets. This enhances blood coagulation.

4. Vitamin K_1 is an essential component of the phosphorylation processes involved in photosynthesis in green plants. It is also involved in oxidative phosphorylation in animal tissues.
5. Vitamin K is involved in the maintenance of normal levels of blood clotting factors II, VII, IX and X, all of which are synthesized in the liver initially as inactive precursor proteins.
6. Vitamin K acts as cofactor of the carboxylase that forms γ-carboxyglutamate residues in precursor proteins.
7. Vitamin K is used therapeutically as an antidote to poisoning by dicoumarol-type drugs. The quinone forms of vitamin K will bypass the inhibited epoxide reductase and provide a potential source of the active hydroquinone form of vitamin K.
8. Vitamin K acts as a cofactor of carboxylase.
9. Vitamin K is required for absorption of fat.

Deficiency manifestation

1. The deficiency of Vit. K leads to a lowering of prothrombin level and increased clotting time of blood. This may lead to hemorrhagic conditions.
2. Vitamin K deficiency causes hemorrhagic disease of the newborn.
3. Vitamin K deficiency is caused by fat malabsorption which may be associated with pancreatic dysfunction, biliary disease, atrophy of the intestinal mucosa or any cause of steatorrhea.
4. Sterilization of the large intestine by antibiotics can result in deficiency when dietary intake is limited.
5. Short circuiting of the bowel as a result of surgery may also cause deficiency which may not respond even to large doses of vitamin K. Water-soluble form of vitamin K (vitamin K_3) is alone useful in such cases.

Hypervitaminosis K

The parenteral administration of too large doses of vitamin K (eg, 30 mg/day for 3 days) to infants has been shown to produce hyperbilirubinemia in some cases.

Coenzyme activities—No coenzyme activities.

Management

(a) Vitamin K_1 is given routinely to newborn babies to prevent hemorrhagic disease of the newborn.

(b) Since dietary vitamin K is not absorbed in obstructive jaundice it is very important to administer the vitamin before biliary surgery.

(c) Warfarin and related anticoagulants act by antagonising vitamin K.

WATER-SOLUBLE VITAMINS

Characteristics

1. They are polar molecules.
2. They have diverse chemical nature, i.e. no similarity of chemical structure between them.
3. They are not stored in the body (except vitamin B_{12}).

4. Act as a co-enzyme of different enzyme catalyze reaction.

   Except Vit. C which itself acts as an enzyme.

5. Synthesized from plant tissue (such as cereals, green leafy vegetables, meat, milk etc).

   Except cobalamine which is stored in liver and not synthesized in plant tissue.

ASCORBIC ACID (VITAMIN C)

Introduction

1. The isolation of vitamin C was carried out by Zilva during 1917-1927. He obtained the highly potent substance and noted its reducing properties.
2. In 1928, Szent-Gyorgy isolated an acid with strong reducing properties from cabbages, adrenal glands and oranges. He called it Hexuronic acid.
3. In 1931, Waugh and King isolated vitamin C in a crystalline form from lemon juice.
4. In 1933, vitamin C was named ascorbic acid owing to its antiscorbutic properties.

Chemistry

1. Haworth and co-workers in 1933 established the chemical structure of ascorbic acid.
2. The synthesis of vitamin C was reported in 1933 by Haworth and co-workers in England and Reichstein and co-workers in Switzerland.
3. The chemical structures of L-ascorbic acid and L-dehydro ascorbic acid are given below.
4. The structure of L-ascorbic acid shows that it is a derivative of hexose called L-gulose.

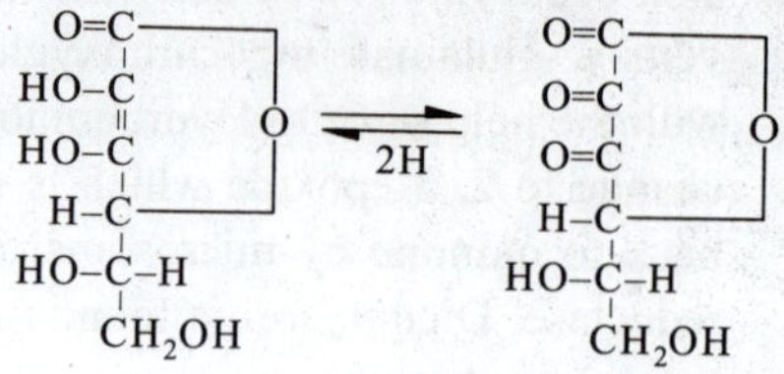

Fig. 15.12 : Structure of ascorbic acid

Properties

1. Ascorbic acid is a white crystalline water-soluble substance with sour taste.
2. It is chemically an enediol-lactone which is oxidized to dehydroascorbic acid (ascorbone). Both these forms are biologically active.
3. D-ascorbic acid does not possess any antiscorbutic activities.
4. The oxidation of ascorbic acid is catalyzed by copper and silver ions and the oxidation is faster at higher temperatures, eg, during cooking of foods.
5. It is a powerful reducing agent which can reduce Fehling's solution in the cold.
6. It can reduce 2, 6-dichlorophenolindophenol to the colourless leuco base.
7. It is easily destroyed by cooking.
8. It is stable below pH 6.8 at room temperature but readily oxidized in an alkaline medium.

Absorption and Storage

1. Ascorbic acid is readily absorbed from the small intestine, peritoneum and subcutaneous tissues.
2. It passes through the portal vein to the general circulation and to all tissues.
3. It is supplied to the fetus from the maternal circulation passing the placental barrier readily. The placenta is also able to concentrate this vitamin.

4. It is not stored in any particular organ and is distributed throughout the body.
5. Each organ or tissue has an optimal saturation level of ascorbic acid. Excessive intake of ascorbic acid does not increase the saturation level but the excess is excreted in the urine.

Sources

Richest source :	Amla.
Good sources :	Citrus fruits, tomatoes, green peppers, raw cabbage, guava, cauliflower, adrenal cortex.
Fair sources :	Grapes, apple, banana, jackfruit, milk, liver, green leafy vegetables, particularly salad greens, fresh potatoes, fresh meat, fresh raw fish, papaya.

Ascorbic acid is not synthesized in the body of the human beings. Dietary sources can only provide this vitamin. All artificial food and sterilised milk is devoid of vitamin C.

Daily requirement

Infants	35 mg
Children	40 mg
Adults	45 mg
Pregnant women	60 mg
Lactating women	80 mg

Normal concentration in blood plasma

0.4-1.5 mg/100 ml of blood plasma.

In erythrocytes $1\frac{1}{2}$ times of that of plasma, and in white blood cells and platelets 20-40 times of that of plasma.

Physiological functions

1. Since it is very sensitive to reversible oxidation (ascorbic acid $\rightleftharpoons$ dehydroascorbic acid) it is involved in the oxidation-reduction reactions of the cell.
2. It is involved in the conversion of folic acid to folinic acid (citrovorum factor).
3. It is involved in the hydroxylation of steroids in the adrenal cortex.
4. It is required in the metabolism of tyrosine and phenylalanine and also in tryptophan.
5. It is required for the absorption of iron and incorporation of plasma iron in ferritin.
6. It is involved in the formation of norepinephrine.
7. It is essential for the normal regulation of the colloidal condition of intercellular substances including the fibrils and collagen of connective tissue, osteoid tissue, dentine, the intercellular 'cement substance' of the capillaries.
8. It is concerned in the hydroxylation of proline and hydroxyproline which is an important constituent of collagen.
9. It has an inhibitory effect on the hyaluronidase-hyaluronic acid system.
10. High doses of vitamin C prevent the common cold or reduce the duration of its symptoms.

Deficiency manifestations

Severe ascorbic acid deficiency produces scurvy. The signs of this deficiency are entirely confined to supporting tissues of mesenchymal origin (bone, dentine, cartilage, and connective tissue).

Scurvy is characterized by failure in the formation and maintenance of intercellular materials which causes typical symptoms such as

1. Internal haemorrhages.
2. Loosening of the teeth.
3. Poor healing of wounds.
4. Swelling of long bones.
5. Easy fracturability of bones.
6. Swelling, sponginess, tenderness and bleeding of gums.
7. Anemia.
8. Susceptibility to infections.
9. General weakness.

Management

(a) A dose of 250 mg by mouth 3 times daily may saturate the tissues quickly.

(b) Sometimes folic acids are given if the patient is anemic iron.

(c) No patient dies of scurvy with adequate treatment and recovery is usually rapid and complete.

(d) Old or solitary people who do not eat fruit and vegetables should be advised to take 50 mg ascorbic acid tablets daily.

(e) The requirement of vitamin C is increased in case of trauma, surgery and burns, infections, smoking and certain drugs—adrenocortical steroids, aspirin, indomethacin and tetracycline. So patients affected by these require more than the recommended intake.

Toxicity

(a) The most important side-effect of large doses is to exacerbate hemochromatosis or other iron storage disease which is not a problem for the great majority of people.

(b) Oxalate is a metabolite of ascorbic acid and urinary oxalate is likely to increase with large intakes of vitamin C. It may be related to urinary tract stones. Hence, it is wise to avoid excessive vitamin C intake.

VITAMINS OF THE B-COMPLEX

1. Thiamine (vitamin B_1, antiberiberi substance, antineuritic vitamin, aneurine).
2. Riboflavin (vitamin B_2, lactoflavin).
3. Niacin (Pellagra Preventive factor, nicotinic acid).
4. Pyridoxine (vitamin B_6, rat antidermatitis factor).
5. Pantothenic acid (filtrate factor, chick antidermatitis factor).
6. Lipoic acid (thioctic acid, protogen, acetate replacement factor).
7. Biotin (vitamin H, anti-egg white injury factor).
8. Folic acid group (vitamin M, streptococcus lactis R (SLR) factor, Vitamin Bc, fermentation residue factor, pteroylglutamic acid, liver lactobacillus casei factor).

VITAMINS OF THE B-COMPLEX

9. Inositol (mouse anti-alopecia factor).
10. Vitamin B_{12} (cyanocobalamin, cobamide, antipernicious anemia factor, extrinsic factor of castle).

THIAMINE (VITAMIN B_1)

Introduction

1. In 1885, Takaki prevented the occurrence of beriberi in the Japanese navy by altering the diet.
2. In 1897, Eijkmann produced polyneuritis in fowls by feeding them a diet consisting of washed polished rice and showed that the birds recovered after the administration of extracts of rice polishings.
3. In 1926, Jansen and Donath isolated vitamin B_1 in crystalline form from rice polishings.
4. In 1931, Windaus and co-workers isolated vitamin B_1 in crystalline form from yeast and established its empirical formula.
5. In 1934, Williams and co-workers and in 1933 and 1935 Peters and co-workers made improvements in the methods of isolation of vitamin B_1.

Chemistry

1. Williams and co-workers in 1936 established the chemical structure of thiamine. The chemical structure is given (Fig. 15.13) :

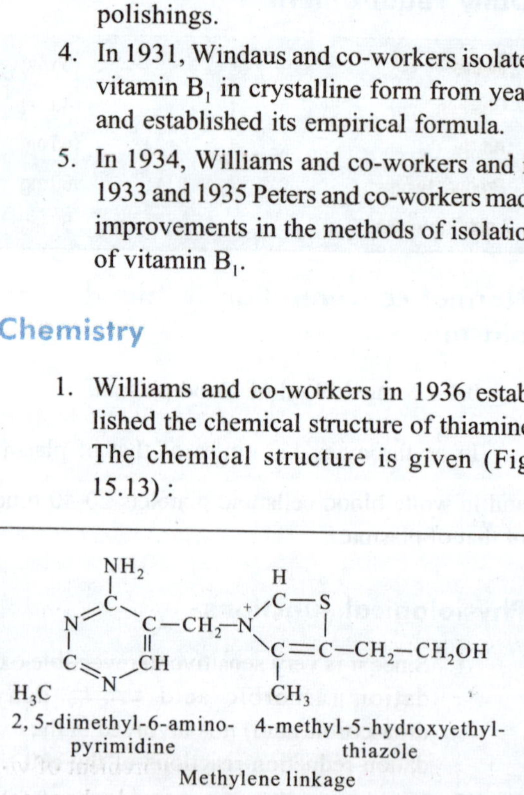

Fig. 15.13 : Structure of thiamine

2. Vitamin B_1 contains a pyrimidine and thiazole ring and they are linked by methylene bridge.
3. One compound : 2, 5 dimethyl 6 amino-pyrimidine.

   Another compound : 4 methyl 5 hydroxy ethylthiazole.

Properties

1. Thiamine is readily soluble in water.
2. It is stable in acid medium.
3. It is destroyed when autoclaved at 120°C for 30 minutes.
4. It is destroyed even at room temperature in an alkaline medium.
5. When it is dissolved in sodium bisulphite solution at pH 4.8 to 5.0 it is cleaved into pyrimidine half and thiazole half.
6. When oxidized with potassium ferricyanide in alkaline solution, it is converted into thiochrome which has a strong fluorescence in ultraviolet light.

Absorption and Storage

Free thiamine is readily absorbed from the small intestine but not TPP. Excess thiamine administered is not stored in the tissues. A part of the excess thiamine is excreted in urine and some of it is destroyed.

Sources

Rich sources :	Rice polishings, wheat germ and yeast.
Good sources :	Cereals, pulses, nuts, oilseeds.
Fair sources :	Meat, fish, eggs, milk, vegetable and fruits.

Thiamine is practically present in all plants and animal tissues commonly used as food. The thiamine contained in these foods is destroyed with improper cooking.

Daily requirement

It is difficult to fix a single requirement of vitamin B_1. The requirement is increased when metabolism is elevated as in fever, hyperthyroidism, increased muscular activity, pregnancy and lactation. Fat and protein reduce while carbohydrate increases the daily requirement of the vitamin. Some of the thiamine is synthesized by the bacteria in the intestine. Deficiencies of the vitamin occur not only by poor dietary intake but also in persons suffering from organic diseases.

Infants	0.3-0.5 mg
Children	0.7-1.2 mg
Adult (males)	1.2-1.5 mg
Adult (females)	1.0-1.1 mg
Pregnant women	1.3-1.5 mg
Lactating women	1.3-1.5 mg

[Recommended daily dietary allowance (Revised 1974)].

Normal concentration of vitamin B_1 in the blood plasma

About 1 μg/100 ml. of blood plasma in free form.

Physiological role

1. Thiamine is essential for growth.
2. It is essential for maintaining the nerves in normal condition.

Coenzyme activities

Thiamine is converted to its active form TPP (Thiamine pyrophosphate or diphosphate) by ATP-dependent thiamine pyrophosphokinase which is present in brain and liver.

Thiamine pyrophosphate is a coenzyme in enzyme reactions in which an activated aldehyde unit is transfered.

There are two types of reactions :

1. An oxidative decarboxylation of α-keto acids (eg, α-ketoglutarate, pyruvate and the α-keto analogs of leucine, isoleucine and valine).
2. Transketolase reactions (in pentose phosphate pathway)

   All of these reactions are inhibited in thiamine deficiency.

In each case, TPP provides a reactive carbon on the thiazole that forms a carbanion

↓

Which is then free to add to the carboxyl group of, for instance, pyruvate

↓

The addition compound then decarboxylates, eliminating CO_2

↓

This reaction occurs in a multienzyme complex known as pyruvate dehydrogenase complex.

Metabolic role as a coenzyme

1. TPP is involved in oxidative decarboxylation of certain important intermediates in carbohydrate metabolism eg, pyruvic acid and α-ketoglutaric acid, so it is refered to as co-carboxylase.
2. Acts as a coenzyme to the enzyme pyruvate dehydrogenase complex which converts pyruvic acid to acetyl-CoA (oxidative decarboxylation) :

$$\text{Pyruvate} \xrightarrow[\text{nase complex TPP}]{\text{Pyruvate dehydroge-}} \text{Acetyl-CoA}$$

3. Acts as a coenzyme to α-ketoglutarate dehydrogenase complex and converts α-ketoglutarate to succinyl-CoA (oxidative decarboxylation) :

$$\alpha\text{-ketoglutarate} \xrightarrow[\text{dehydrogenase complex TPP}]{\alpha\text{-ketoglutarate}} \text{Succinyl-CoA}$$

4. TPP acts as a coenzyme with the enzyme transketolase in transketolation reaction in HMP shunt pathway :

$$\text{Ribose-5-Ⓟ} + \text{Xylulose-5-Ⓟ} \xrightarrow[\text{TPP}]{\text{Transketolase}} \text{Sedoheptulose-7-Ⓟ} + \text{Glyceraldehyde-3-Ⓟ}$$

5. Vitamin B_1 is required in tryptophan metabolism for the activity of enzyme tryptophan pyrrolase.
6. Acts as a co-enzyme for mitochondrial branched chain α-ketoacid decarboxylase which catalyzes oxidative decarboxylations of branched chain α-ketoacids formed in the catabolism of valine, leucine, isoleucine.

Deficiency manifestation

1. **Beriberi** : It is a nutritional disorder i.e. thiamine deficiency. Three forms of disease are :

(a) **Wet beriberi**

(i) Oedema is the most prominent feature and develops rapidly in the legs, face etc.

(ii) Anorexia and dyspepsia are present.

(iii) The calf muscles are tense, slightly swollen and tender on pressure (pins and needles).

(iv) The pulse is fast.

(v) The apex beat of heart is displaced outwards.

(vi) Urine volume is diminished but there is no albuminuria.

(b) **Dry beriberi**

(i) The essential feature is polyneuropathy.

(ii) The subject needs stick to stand and walk and finally becomes bedridden.

(iii) The concentrations of pyruvic acids and lactic acids are increased.

(iv) Transketolase activity of red cells is decreased.

(v) The muscles become wasted and weak and it is difficult to walk.

(c) **Infantile beriberi**

(i) This occurs in breast-fed infants between 2nd to 5th months.

(ii) In acute form—Cardiac failure develops, the mother notices that infant is restless, cries a lot, is passing less urine than normal. The infant may be cyanosed with dyspnoea and tachycardia and may die within 24 to 48 hours.

(iii) In chronic form—the main symptoms are due to gastrointestinal disturbances. There is constipation and vomiting. The child is fretful and sleeps poorly.

2. Wernicke's encephalopathy is a condition associated with thiamine deficiency. It is frequently found in chronic alcoholics consuming little other food.
3. In the deficiency of vitamin B_1, there is impaired conversion of pyruvate to acetyl-CoA. In case of high carbohydrate diet, there is increased plasma concentration of pyruvate and lactate which may cause life-threatening **lactic acidosis**.

Management

(a) Treatment must be started as soon as the diagnosis is made in the deficiency of thiamine. Complete rest is most essential and 50 mg thiamine should be given intramuscularly for 3 days. Thereafter 10 mg 3 times a day should be continued by mouth until convalescence is established.

(b) Wernicke's encephalopathy should be treated with 50 mg thiamine hydrochloride by slow intravenous injection followed by 50 mg intramuscularly daily for a week. Confusion, disorientation and ophthalmoplegia should respond within 2 to 3 days. The memory disorder takes longer to improve.

Prevention

The prevention of Beriberi and Wernicke's encephalopathy in Western countries is closely related to the control of alcoholism. Bread and other staple cereal foods are prescribed with thiamine in most industrial countries. In a chronic alcoholic, vitamin B complex tablets can at least prevent the complications of thiamine deficiency. Beriberi is much less common in Asia.

Toxic effects

If the vitamin is taken in excessive amounts, the excess vitamin is promptly excreted in the urine. As a result, there is no evidence of toxicity.

RIBOFLAVIN (VITAMIN B_2)

Introduction

1. As early as 1879 the water-soluble, yellow-green, fluorescent pigment in milk whey was noted. This substance was not isolated in pure form until 1932.
2. It was first isolated from milk, hence the earlier name lactoflavin.
3. It was shown to be a constituent of oxidative tissue-enzyme system and an essential growth factor for laboratory animals.

Chemistry

1. Kuhn, Karrer and co-workers (1934, 1935) accomplished the synthesis of riboflavin.
2. It contains an isoalloxazine nucleus. The chemical structure of it is given (Fig. 15.14).
3. It is water-soluble, heat-stable and sensitive to light.
4. The solution of vitamin B_2, when exposed to ultraviolet light, emits a strong greenish yellow fluorescence.
5. On oxidation with ultraviolet rays or visible light, it undergoes irreversible decomposition.

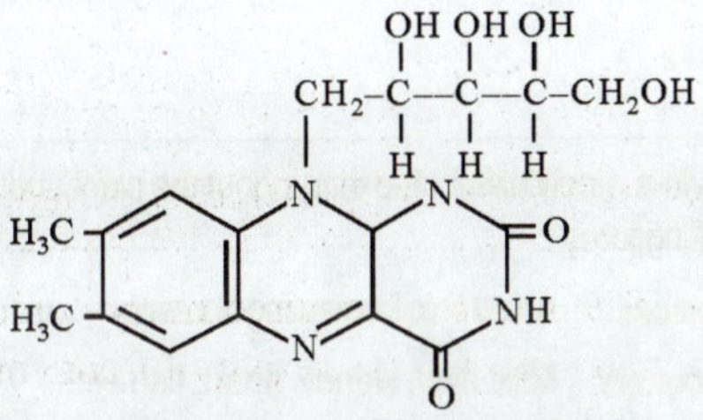

6, 7-dimethyl-9-D-ribitolisoalloxazine

Fig. 15.14 : Structure of riboflavin

6. It is rapidly destroyed when the solution is exposed to bright light.
7. Reducing agents such as stannous chloride and hydrosulphite convert it into a colourless compound without any fluorescence.
8. Riboflavin in alkaline solution when exposed to ultra-violet light is converted into lumiflavin which has a greenish yellow fluorescence in ultraviolet light.

Absorption and storage

The vitamin is phosphorylated in the intestinal mucosa during absorption. It is absorbed from the small intestine through the portal vein and is passed to all tissues being stored in the body. The major part is excreted in urine and a small part is metabolized in the body.

Normal concentration in blood plasma

2.5 to 4.0 μg/100 ml. of blood plasma.

Daily requirements

Recommended daily dietary allowances (Revised 1976)	
Infants	0.4-0.6 mg
Children	0.8-1.2 mg
Adults (males)	1.5-1.8 mg
Adults (females)	1.1-1.4 mg
Pregnant women	1.4-1.7 mg
Lactating women	1.6-1.9 mg

Sources

Riboflavin is widely distributed throughout the plant and animal kingdoms

Rich sources :	Anaerobic fermenting bacteria.
Good sources:	Milk, liver, kidney, heart, fish etc.
Fair sources :	Cereals, roots, germinating wheat and barley.

Physiological functions

1. Riboflavin is involved in the regulatory functions of some hormones connected with carbohydrate metabolism.
2. The free riboflavin present in retina is converted by light to a compound involved in stimulation of the optic nerve.

Coenzymic activities

Biological active form

1. FMN — Flavin mononucleotide (flavin ribityl PO_4).
2. FAD — Flavin adenine dinucleotide (flavin ribityl-P-P ribose adenine).

The acidic properties given by phosphoric acid group influence their capacity for combining with proteins apo-enzyme forming flava protein (holoenzyme).

Thus

FP (holoenzyme) = FMN/FAD + Protein

(co-enzyme) (apo-enzyme)

FP may also unite with metals like Fe and Mo, thus forming metallo flavoproteins.

Riboflavin is converted into FMN by flavokinase during absorption in intestine.

Free riboflavin undergoes phosphorylation, a prerequisite for absorption (thiamine).

Metabolic role

(a) It acts as a coenzyme for enzyme catalysed oxidation-reduction reaction.

(b) It is the component of aerobic dehydrogenase.

(c) This is one of the component of respiratory chain for the transpart of electron :

FMN	FAD
1. Warburg's yellow enzyme (cyt aa_3)	1. Xanthine oxidase (Xanthine → Uric acid)
2. Cyt. c reductase	2. D-amino acid oxidase

Contd.

FMN	FAD
3. L-amino acid oxidase	3. Aldehyde oxidase
	4. Fumarate dehydrogenase (Succinate → fumarate)
	5. Glycine oxidase
	6. Acetyl-CoA dehydrogenase

Deficiency manifestations

1. Lips—Redness and shiny appearance of lips.
2. Cheilosis—Lesions at the mucocutaneous junction at the angles of mouth leading to painful fissures.
3. Tongue—Painful glossitis, the tongue assuming a red purple (magenta) colour.
4. Seborrheic dermatitis—Scaly, greasy, desquamation chiefly about the ears, nose and nasolabial folds.
5. Eyes—May lead to corneal vascularization and inflammation with cloudiness of cornea, watering, burning of eyes, photophobia and cataract.

Antagonist

1. Dichlororiboflavin.
2. Isoriboflavin.

Management

In the deficiency of riboflavin, the therapeutic dose of riboflavin is 5 mg 3 times a day by mouth. It gives the patient's urine a green fluorescence. Other B complex vitamins should also be given.

Toxic effects

No toxic effect.

NIACIN (NICOTINIC ACID) (Vit. B_3)

Introduction

1. During 1922-1928 Goldberger and coworkers treated the disease pellagra in human beings and, in 1928, they showed that boiled yeast extract can cure pellagra. Hence, Goldberger named niacin as the Pellagra-Preventive (P-P) factor.
2. In 1937, three groups of workers independently reported that nicotinic acid was effective in curing Pellagra in human beings.
3. Niacin is not strictly a vitamin since it can be synthesized from the essential amino acid tryptophan. Most of the niacin in cereals is biologically unavailable which is discounted.

Chemistry

1. Niacin is pyridine 3-carboxylic acid. It occurs in tissues as niacinamide (nicotinamide). The chemical structures of both are given (Fig. 15.15).
2. It is soluble in water.
3. It is stable to heat and not destroyed by autoclaving at 120°C for 20 minutes in acid or alkaline medium.
4. Nicotinamide, when heated in a strong alkaline or acid solution, is converted into nicotinic acid.
5. Nicotinamide exists in human and animal tissues as coenzyme 1 (DPN, now called NAD) and coenzyme 11 (TPN, now called NADP).

Absorption and Storage

1. Nicotinic acid and nicotinamide are absorbed from the intestine through the portal vein into the general circulation.

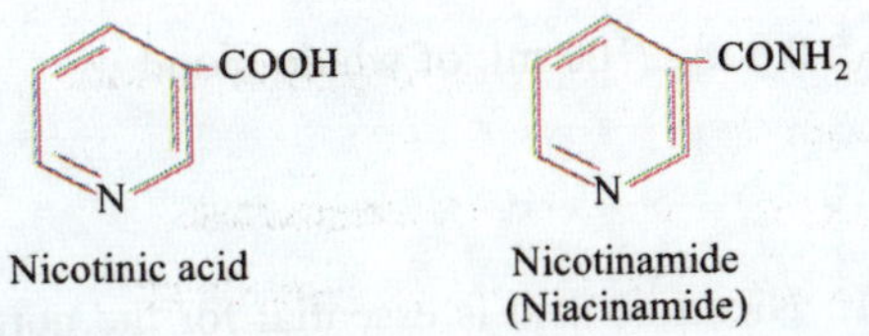

Fig. 15.15 : Structure of niacin

2. Excess nicotinic acid is not stored in the body.
3. The majority of the excess nicotinic acid is excreted in urine in the form of N-methylnicotinamide, 6-pyridone of N-methylnicotinamide, N-methylnicotinic acid and the glycine conjugates of these methyl derivatives. Methylation takes place in the liver. Methionine is the principal source of these methyl groups.
4. Niacin in the form of niacytin present in maize is not absorbed unless the food is prepared with alkali (tortilla).

Sources

Richest sources : Yeast, rice polishing.

Good sources : Meat, liver and poultry.

Fair sources : Milk, eggs, tomatoes, leafy green vegetables.

Poor sources : Most fruits and vegetables.

The amino acid tryptophan present in the dietary proteins is converted into niacin in the body. 60 mg of tryptophan produce 1 mg of niacin. So tryptophan present in the foodstuff also provides additional niacin.

Daily requirements

Recommended daily dietary allowances (Revised 1974)	
Infants	5-8 mg
Children	9-16 mg
Adult (males)	16-20 mg
Adult (females)	12-16 mg
Pregnant women	14-18 mg
Lactating women	16-20 mg

Normal concentration of niacin in the whole blood

0.5-0.8 mg./100 ml. of whole blood.

Physiological functions

1. Nicotinic acid is essential for the normal functioning of the skin, intestinal tract and the nervous system.
2. Lack of vitamin B_2 and vitamin B_6 causes deficiency of niacin due to the nonproduction from tryptophan.
3. Nicotinic acid (but not nicotinamide) is used therapeutically for lowering plasma cholesterol. This happens so due to the inhibition of the flux of FFA from adipose tissue which leads to less formation of the cholesterol bearing lipoproteins, VLDL and LDL.

Coenzyme activities

(a) Nicotinic acid principally occurs as nicotinamide or niacinamide. This niacinamide is a component of two coenzymes NAD and NADP. This reduced form of NAD is dihydronicotinamide adenine dinucleotide (NADH) and that of NADP is dihydronicotinamide adenine dinucleotide phosphate (NADPH).

(b) These coenzymes play an important role in metabolism by acting as hydrogen and electron transfer agents by means of reversible oxidation and reduction. Hence, the great importance of niacin in human nutrition as well as in the requirements of many other organisms including bacteria and yeast is stated.

The mechanism of transfer of hydrogen from a metabolite to oxidized NAD causing the oxidation of the metabolite and the formation of reduced NAD is shown in Fig. 15.16.

(c) NAD and NADP take part in many enzyme reactions involving dehydrogenases.

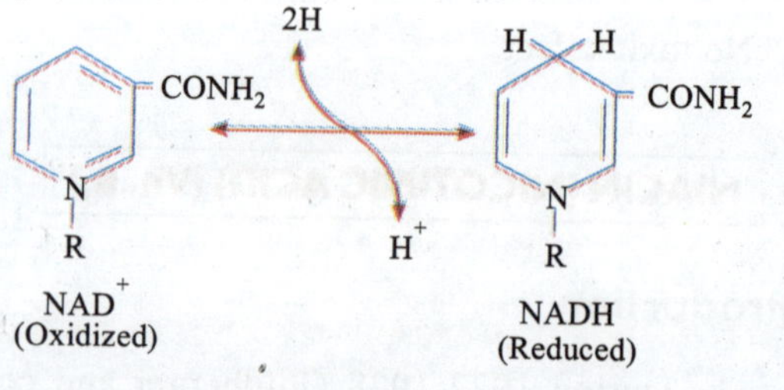

Fig. 15.16 : Formation of NADH

Some of these include oxidation of alcohol to aldehyde, of glucose to gluconic acid, of malic acid to oxaloacetic acid, of lactic acid to pyruvic acid, of glycerophosphate to phosphoglyceraldehyde, of glucose-6-phosphate to 6-phosphogluconate, of pyruvic acid to acetyl-CoA and of α-ketoglutarate to succinyl-CoA etc.

(d) The reduced NADP(NADPH) is also involved in enzyme reaction in fatty acid synthesis, synthesis of cholesterol and of steroid hormones and also in the formation of tetrahydrofolate (H_4 folate).

Deficiency manifestations

The deficiency of niacin causes the disease pellagra. The clinical features of the disease include three D'S-dermatitis (lesion of skin of face, neck, knees, breasts, thick and scaly skin), diarrhoea, dementia (headache, depression, anxiety, insomnia and forgetfulness).

Management

(a) In the deficiency of niacin, nicotinamide is given in a dose of 100 mg every 6 hours by mouth. The vitamin is well-absorbed but can be given parenterally. The response is very quick. The erythema of the skin diminishes and the diarrhoea ceases within 24 hours. Improvement is also found in the patient's behaviour and mental attitude.

(b) The deficient disease is developed due to a low intake of protein including tryptophan. Deficiencies of other B vitamins (riboflavin and vitamin B_6) are likely. Nicotinamide treatment should, therefore, be supplemented with a nutritious diet, high in protein.

(c) Vitamin B complex tablets should be given and iron, folic acid and vitamin B_{12} may be necessary in addition in some cases.

(d) Alcohol should be forbidden.

Prevention

Pellagra can be prevented by enrichment of maize meal and bread with niacin.

Toxic Effects

In large doses, nicotinic acid, causes burning sensation that may ultimately alarm the patient.

Intake of niacin in excess of 500 mg/d can cause liver damage.

PYRIDOXINE (VITAMIN B_6)

Introduction

1. In 1938, pyridoxine was isolated in a pure form by five different groups of workers.
2. In 1939, the vitamin was synthesized independently by two groups of workers in Germany and U.S.A., respectively.

Chemistry

1. Since pyridoxine occurs in nature it is a mixture of three compounds (pyridoxine, pyridoxal and pyridoxamine). The structure of these three are given (Fig. 15.17) :

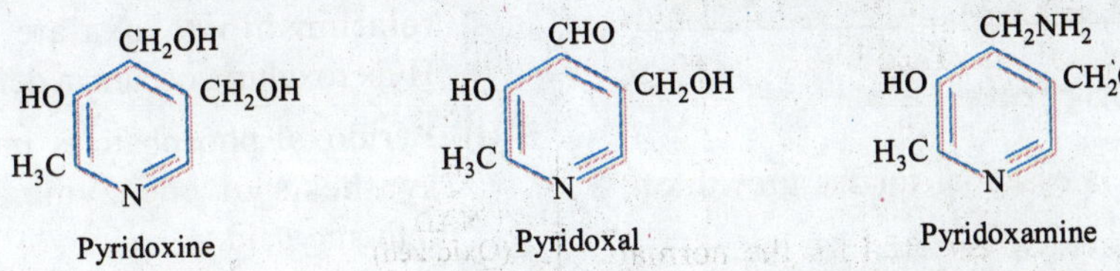

Fig. 15.17 : Structure of Pyridoxine, Pyridoxal, Pyridoxamine

The more active derivatives are pyridoxal and pyridoxamine phosphates. Pyridoxine is 3-hydroxy-4, 5-dihydroxymethyl-2-methyl pyridine. Pyridoxal contains an aldehyde instead of hydroxyl methyl in No. 4 and pyridoxamine contains a primary amine side chain in No.4 of pyridine nucleus. These three compounds are interchangeable.

2. Pyridoxine is readily soluble in water.
3. When it is in alkaline solution, it is slowly destroyed by exposure to light.
4. It reacts with phenol reagent or 2:6 dichloroquinone chlorimide producing coloured compounds.

Absorption and Storage

It is readily absorbed from the small intestine. The excess amount, if ingested, is not stored in the body but is excreted in urine.

Sources

Rich sources	:	Yeast, rice polishings, certain seeds such as a wheat and corn.
Good sources	:	Milk, meat, leafy vegetables, liver.
Fair sources	:	Fish, fruits, vegetables.

Intestinal bacteria can also synthesize this vitamin.

Daily requirements

Recommended daily dietary allowances (Revised 1974)	
Infants	0.3 mg
Children	0.6-1.2 mg
Adult (males)	1.6-2.0 mg
Adult (females)	1.6-2.0 mg
Pregnant women	2.5 mg
Lactating women	2.5 mg

Physiological functions

1. Pyridoxine is essential for the growth of infants.
2. Muscle phosphorylase contains 70-80% of total body vitamin B_6.

Coenzyme activities

(a) Pyridoxal phosphate, the active derivative of pyridoxine, functions as *codecarboxylase* in the decarboxylation of tyrosin, arginine, glutamic acid and certain other amino acids.

(b) The *deaminases* (*dehydrases*) for serine and threonine are also catalyzed by pyridoxal phosphate acting as coenzyme.

(c) Pyridoxal phosphate acts as *cotransaminase* in the transamination reactions.

(d) Pyridoxal phosphate acts as a coenzyme for *kynureninase* in the synthesis of niacin from tryptophan.

(e) Muscle phosphorylase also contains pyridoxal phosphate as coenzyme.

(f) Pyridoxal phosphate acts as coenzyme in the *transulfuration* reaction in the transfer of sulphur from methionine to serine to form cysteine.

(g) Pyridoxal phosphate is also involved in the process of absorption of amino acids from the intestine.

(h) Pyridoxal phosphate is also involved in the *desulphuration* of cysteine and homocysteine.

(i) Pyridoxal phosphate is required for the synthesis of α-aminolevulinic acid which is an important intermediate in the synthesis of porphyrin and heme nuclei.

(j) This pyridoxal phosphate especially applies to brain metabolism because it is necessary for the formation of serotonin, γ-aminobutyric acid and the catecholamines.

(k) Pyridoxal phosphate has also important relationship to oxalate metabolism. Hyperoxaluria occurs in deficiency states.

(l) Pyridoxal phosphate is involved in the synthesis of coenzyme A from pantothenic acid.

(m) Pyridoxal phosphate is concerned with immune response.

Deficiency symptoms

Fortunately, deficiency of vitamin B_6 is rare because of its easy availability in most foodstuffs :

1. Deficiency gives rise to irritability and depression. In some subjects there are lymphopenia and peripheral neuropathy.
2. Deficiency of this vitamin occurs in infants on inadequate milk formulas. The major symptoms are convulsions due to depletion of brain γ-amino-butyric acid content.
3. The drugs isonicotinic acid hydrazide and hydralazine act as B_6 antagonist causing deficiency symptoms including hypochromic anemia and peripheral neuropathy.
4. In the deficiency states, there are inborn errors of metabolism including cystathioninuria, familial xanthurenic aciduria and some pyridoxal-responsive anemias.
5. The deficiency of this vitamin is found in nursing infants whose mothers are deficient of the vitamin due to long term use of oral contraceptives.
6. Alcoholics may also be deficient due to metabolism of ethanol to acetaldehyde which stimulates hydrolysis of the phosphate of the coenzyme.
7. Isoniazid, the antituberculosis drug, can produce vitamin B_6 deficiency by forming a hydrazone with pyridoxal.

Management

Vitamin B_6 deficiency can occur in women taking oral contraceptives and the mild depression as a result of this may be relieved by a small dose of pyridoxine.

Toxicity

Doses of vitamin B_6 (200 mg per day or more) taken for some weeks cause a sensory polyneuropathy.

PANTOTHENIC ACID (Vit. B_5)

Introduction

1. In 1938, Williams and co-workers isolated this vitamin in a pure form as its calcium salt.
2. In 1940, its synthesis was accomplished by several groups of workers.

Chemistry

1. Pantothenic acid consists of β-alanine and pantoic acid joined through a peptide bond. The chemical structure is given below (Fig. 15.18) :

$$\underbrace{HOCH_2.C(CH_3)_2.CH(OH).CO}_{\text{Pantoic acid}}\ .\underbrace{NH\ CH_2CH_2COOH}_{\beta\text{-alanine}}$$

Fig. 15.18 : Structure of pantothenic acid

2. It is highly soluble in water.
3. It is stable to autoclaving at 120°C for 30 minutes in neutral solution but is destroyed in acid or alkaline medium.
4. It is not obtained in crystalline form but its sodium, potassium or calcium salts crystallise readily.
5. Pantothenic acid exists in the tissues in the active form as coenzyme A which contains pantothenic acid, β-mercaptoethylamine, adenine, ribose and phosphoric acid.

Absorption and Storage

Pantothenic acid and its salts are readily absorbed from the small intestine through the portal vein into the general circulation.

If ingested in excess of the requirements, it is not stored in the body, but is excreted in urine or metabolised by the tissues.

Sources

Richest source	:	Jelly.
Rich sources	:	Yeast, liver, rice polishings and wheat germ.
Good sources	:	Milk, meat, eggs and leafy vegetables.
Fair sources	:	Fruits and other vegetables.

Daily requirements

Infants	1-2 mg
Children	4-5 mg
Adults (males & females)	5-10 mg
Pregnant & Lactating women	10-15 mg

Physiological functions

Pantothenic acid is essential for the growth of infants and children.

Coenzyme activities

(a) Pantothenic acid as a constituent of coenzyme A is required for several fundamental reactions in metabolism.

(b) Coenzyme A combines with acetate to form **"active acetate" (acetylcoenzyme A)** which is directly utilized by combination with oxaloacetic acid to form citric acid which initiates the citric acid cycle.

(c) Acetyl-CoA derived from carbohydrates, fats or many of the amino acids undergoes further metabolism through the **"common metabolic pathway"**.

(d) In the form of active acetate, acetic acid also combines with choline to form acetylcholine or with the sulfonamide drugs which are acetylated prior to excretion.

(e) The decarboxylated product of a-ketoglutarate in the citric acid cycle is a coenzyme A derivative called **"active succinate" (succinyl-CoA)**. Succinyl-CoA and glycine are involved in the first step leading to the biosynthesis of heme. So anemia occurs in deficiency of this vitamin.

(f) In lipid metabolism, coenzyme A has got significant role. In the first step of oxidation of fatty acids, the fatty acids are to be activated by coenzyme A catalyzed by the enzyme thiokinase. In each turn of the β-oxidation cycle, one molecule of acetyl-CoA is released. This acetyl-CoA directly enters the citric acid cycle for degradation to carbon dioxide and water or two molecules of acetyl-CoA condense to form ketone bodies.

(g) Coenzyme A in the form of acetyl-CoA is also required for the synthesis of cholesterol and thus of the steroid hormones.

(h) A significant amount of the cellular pantothenic acid is protein-bound. This form is contained in a compound known as acyl carrier protein, a coenzyme required in the biosynthesis of fatty acids.

(i) Coenzyme A is also involved in the metabolism of propionate and of branch chain fatty acids.

Deficiency symptoms

1. Deficiency of this vitamin in humans results in nausea, vomiting, certain gastrointestinal disorders, irritability, inadequate growth, anemia, fatty liver, failure in gaining weights.
2. Pantothenate deficiency causes burning foot syndrome in prisoners of war and is associated with reduced capacity for acetylation.

Toxic effects

No ill-effects are still reported.

FOLIC ACID GROUP

Introduction

1. In 1934, Wills showed that tropical macrocytic anemia in human beings was cured by a vitamin present in autolysed yeast extract.
2. In 1947, Pifiner et al isolated folic acid in a crystalline form from liver.

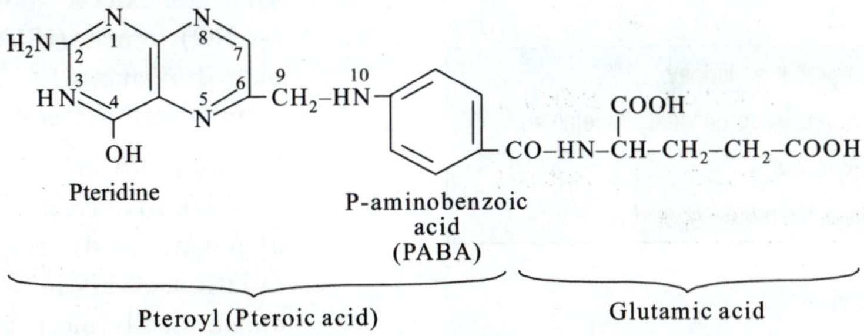

Fig. 15.19 : Structure of folic acid

Chemistry

1. Folic acid (folacin, pteroylglutamic acid) is a compound made up of the Pteridine nucleus, P-aminobenzoic acid and glutamic acid. There are at least three nutritionally important and chemically related compounds which occur in natural products belong to the folic acid group. These three compounds only differ in the number of glutamic acid residues attached to the Pteridine-aminobenzoic acid complex. The chemical structure of folic acid is given (Fig. 15.19). Folic acid is synonymous with vitamin Bc.
2. Folinic acid (leucovorin, folinic acid-SF [synthetic factor] is the reduced form of folic acid with a formyl group on position 5 (N^5-formyltetrahydrofolic acid).
3. Folic acid is soluble in water.
4. It is stable to heat at neutral pH.
5. Its activity is not lost if it is heated at 120°C for 30 minutes at neutral pH.
6. Riboflavin accelerates the photo-oxidation of folic acid.

Absorption and transport

1. Absorption of folic acid takes place along the whole length of the mucosa of the small intestine.
2. Monoglutamates are produced from polyglutamates which is ingested within the intestinal mucosa and dihydrofolates are further reduced to tetrahydrofolates by *folic acid reductases*.
3. The tetrahydrofolates are then converted to methyltetrahydrofolate which enters the portal blood to be transported to the liver.
4. The vitamin then appears in the systemic circulation to supply the tissue.
5. The vitamin is transported to the plasma as methyltetrahydrofolate bound to protein. The folate level of plasma obtained from umbilical cord blood is about 3 times that of the maternal plasma.

Excretion

1. 20% of ingested folate that remains unabsorbed is excreted in the feces.
2. 2-5 μg of folic acid is excreted in the urine daily. This may be increased after an oral dose of folate if the tissues are saturated.
3. Some folates are also excreted in saliva, sweat and bile.

Folic acid in tissues

1. Tissue folate is about 70 mg in the whole body.
2. About one-third (5-15 μg/g) is in the liver.
3. Folate is incorporated into the erythrocytes during erythropoiesis.

Normal level of folic acid in serum

3-25 μg/1 of serum in healthy subjects.

Sources

Richest sources	:	Yeast, liver, kidney.
Good sources	:	Meat, fish, green leafy vegetables.
Fair sources	:	Milk, fruit.

Intestinal bacteria also synthesize folic acid.

Daily requirements

Recommended dietary daily allowances (Revised 1974)	
Infants	50 μg
Children	100-300 μg
Adult (males)	400 μg
Adult (females)	400 μg
Pregnant women	800 μg
Lactating women	600 μg

Physiological functions

Coenzyme activities

1. Folic acid as a coenzyme is involved in the transfer and utilization of the single carbon (C-1) moiety.

   Before functioning as a C-1 carrier, folic acid is first reduced to 7, 8-dihydrofolic acid (H_2 folate) and then to the tetrahydro compound (H_4 folate) catalyzed by *folic acid reductase* which uses NADPH as hydrogen donor. The reactions are given below in Fig. 15.21 :

The "one carbon" moiety may be formyl (–CHO), formate (H^+COO^-), methyl (CH_3^-) or hydroxymethyl (–CH_2OH). These are metabolically interchangeable.

Folinic acid (formyl tetrahydrofolate, f^5–H^4 folate) is involved in the formylation of glutamic acid during the metabolic degradation of histidine; otherwise, it is metabolically inert. But f^{10}–H_4 folate is metabolically active. However, f^5 can be converted to f^{5-10} by formyl tetrahydrofolic acid isomerase as :

$f^{5\text{-}10}$–H_4 folate is also converted to N^5-methyltetrahydrofolic acid by an NAD-dependent reductase and this methyl group is then transferred to deoxyadenosyl –B_{12} (cobamide coenzyme) to form methyl –B_{12} which is an important donor of methyl group in the formation of methionine from homocysteine.

The other sources of the one-carbon moiety are the methyl groups of methionine, choline (by way of betaine) and thymine. These methyl groups are oxidized to –CH_2OH groups and carried as such on f^{5-10}–H_4 folate. The beta carbon of serine contributes hydroxymethyl group to the formation of a single carbon moiety.

The single (formyl) carbon present on the tetrahydrofolic acids is utilized in the following ways :

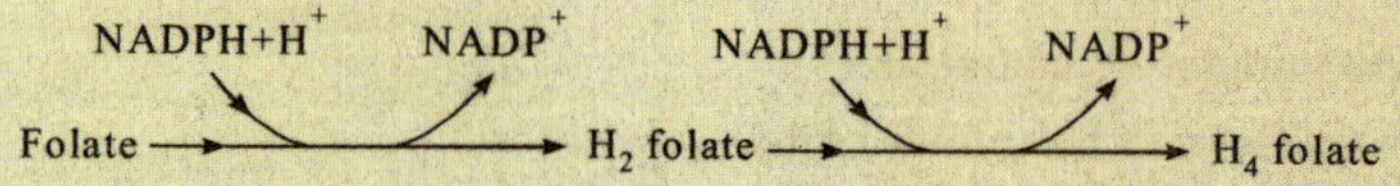

Fig. 15.20

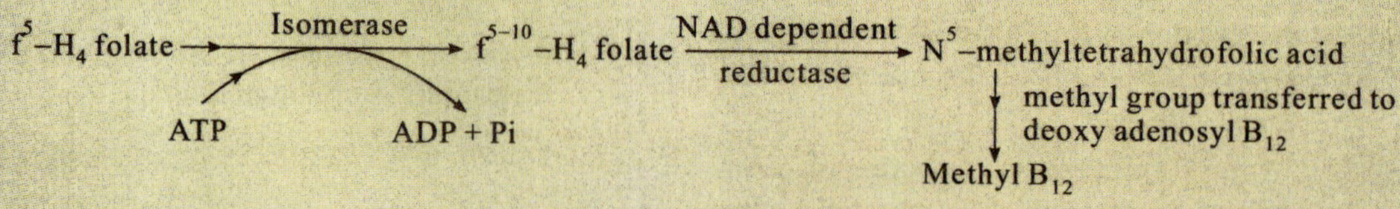

Fig. 15.21

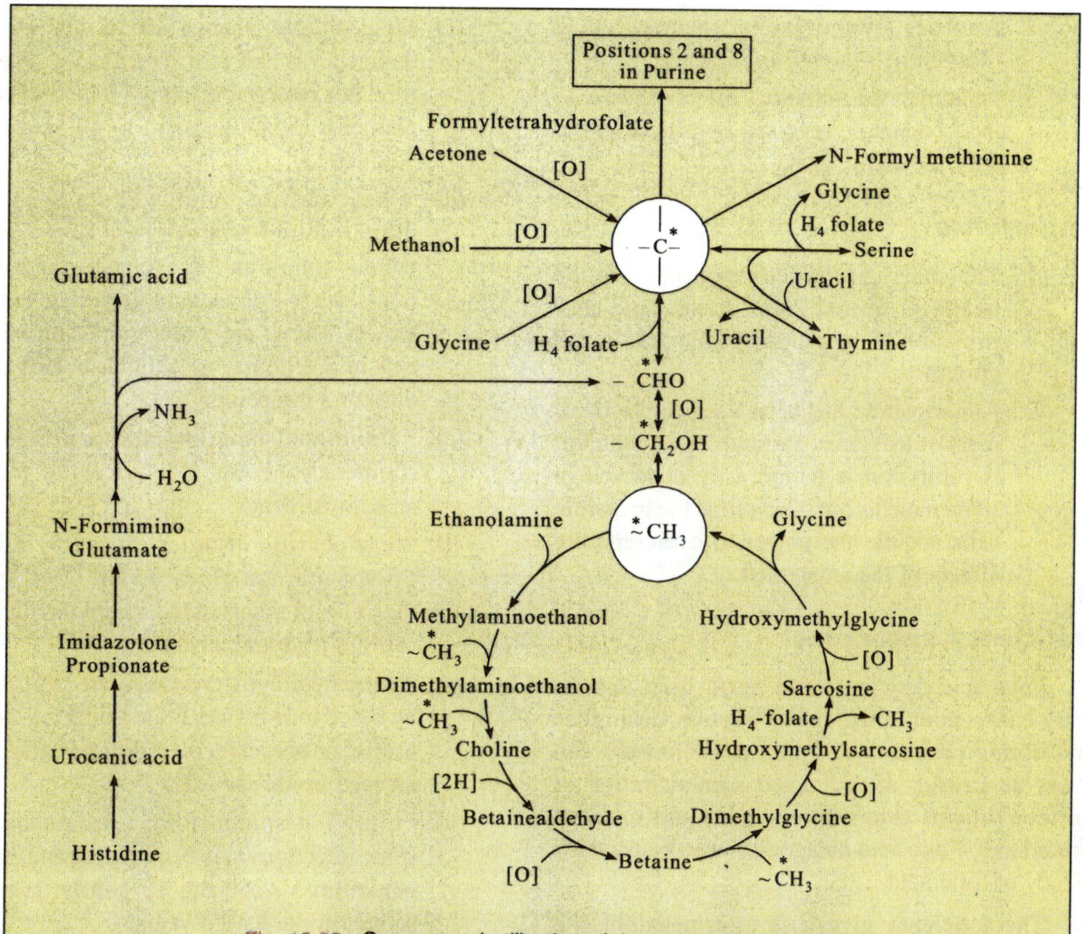

Fig. 15.22 : Sources and utilization of the one carbon moiety

(i) As a source of carbons 2 and 8 in the purine nucleus.

(ii) As a source of the formyl group on N-formyl methionine –tRNA which initiates synthesis of peptide chains on ribosomes in microorganisms.

(iii) As a source of the formyl carbon in the formation of the beta carbon of serine in conversion of glycine to serine.

(iv) In the synthesis of methyl groups for methylation of homocysteine to form methionine or methylation of uracil to form thymine and for the synthesis of choline by the way of methyl groups from methionine.

2. Since the folic acid coenzymes take part in the synthesis of purines and thymine (the methylated pyrimidine of DNA) they are fundamentally involved in growth and reproduction of cells.

3. Folic acid coenzymes are not only confined to the hematopoietic system but are also generalized throughout the body.

4. In the metabolism of histidine, there is a folic acid-dependent step at the point where formiminoglutamic acid (Figlu) is converted to glutamic acid.

Folic Acid Antagonists

1. The competitive inhibitor *aminopterin* (4-amino folic acid) is the most potent folic acid inhibitor.

2. Another antagonist is *amethopterine* (4-amino-10-methyl folic acid).
3. In animals the inhibitory effect of *aminopterin* cannot be reversed by folic acid but only by folinic acid.

Importance

1. They have got clinical application in the treatment of malignant disease and confirmation of the action of folic acid in cell growth.
2. Aminopterine has been used in the treatment of leukemia, particularly in children. A remission is temporarily observed in some patients but after a time the leukemic cells acquire the power to overcome the effects of the antagonist.

Deficiency symptoms

Folic acid deficiency can result from low dietary intake, in intestinal malabsorption syndromes and during pregnancy. A similar deficiency can occur as a result of prolonged administration of anticonvulsant drugs (phenytoin sodium and primidone). Folic acid antagonists also cause deficiency of folic acid.

The deficiency gives rise to a megaloblastic anemia. The nuclei of the neutrophil polymorphonuclear leukocytes contain more than the normal number of lobes. Other deficiency manifestations include retardation of growth, weakness, infertility, inadequate lactation in females and increased output of formiminoglutamic acid (FIGLU) in the urine after histidine loading.

Management

(a) In folic acid deficiency, treatment with a daily dose of 5 mg of folic acid by mouth is sufficient; 5 mg once a week is always adequate for maintenance therapy.

(b) Folic acid must never be given other than with vitamin B_{12} in Addisonian pernicious anemia or other vitamin B_{12} deficiency anemias because of the risk of aggravating or precipitating neurological features of vitamin B_{12} depletion.

(c) Megaloblastic change due to vitamin B_{12} deficiency is very rare in pregnancy. Therefore, it is reasonable to give folic acid supplements (350 μg daily) to all pregnant women.

(d) When a drug methotrexate inhibits dihydrofolate reductase, it is better to employ folinic acid to overcome the metabolic block. Folinic acid may be given as tablets, 15 mg daily orally or as an injection intravenously or intramuscularly at a dose of 3 mg per ml.

(e) Folinic acid mouthwashes are used to counteract the oral side-effects of folate antagonist drugs.

(f) Megaloblastic disorder caused by other cytotoxic drugs which inhibit DNA synthesis is not reversed by either vitamin B_{12} or folic acid administration.

(g) Severe hemolytic crises require treatment by blood transfusion. Folic acid 5 mg daily orally is prescribed to support the increased erythropoiesis.

(h) Following splenectomy, resistance to some infections may be impaired and daily penicillin V, 250 mg 12 hourly, is prescribed for at least 5 years.

Toxic effects

Renal injury has been observed in animals receiving large doses (50 mg./kg. body wt.) of folic acid intravenously.

Excess intake of folic acid may hasten the development of the nerve damage found in B_{12} deficiency.

VITAMIN B_{12}

Introduction

1. In 1926, Minot and Murphy discovered that liver can cure pernicious anemia.
2. In 1948, Smith and Parker in Great Britain and Rickes et al in U.S.A. independently isolated Vitamin B_{12}.

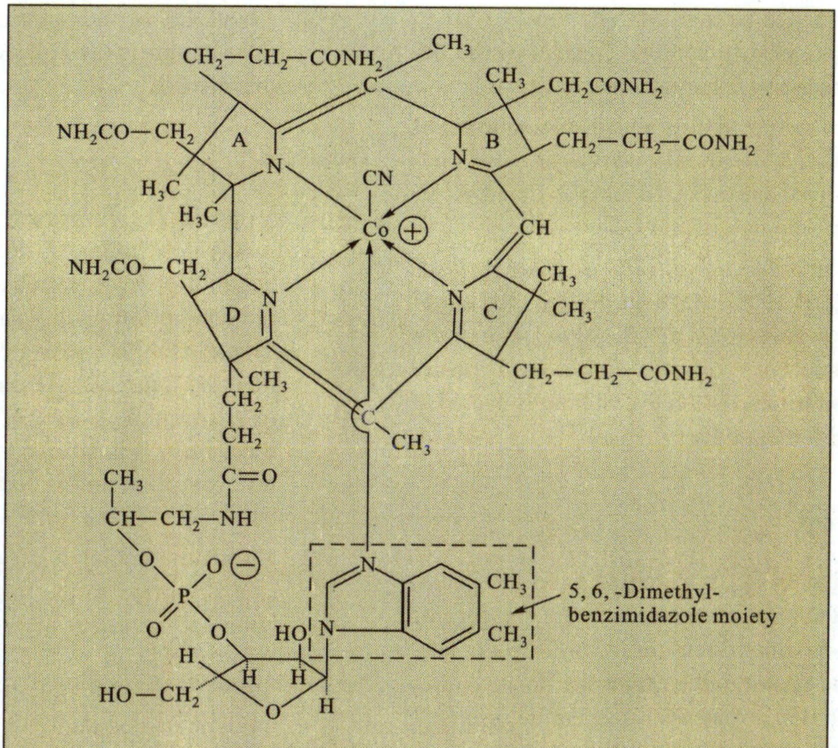

Fig. 15.23 : Structure of vitamin B_{12}

3. It is identical with extrinsic factor present in liver which cures pernicious anemia. It was found effective in curing pernicious anemia when administered intramuscularly in small quantities (5 to 10 µg.).
4. Although vitamin B_{12} is synthesized exclusively by microorganisms, for practical purposes it is found only in foods of animal origin but no plant sources of this vitamin. This indicates that the strict vegetarians are at risk of developing B_{12} deficiency. The small amounts of the vitamin formed by bacteria on the surface of fruits may be adequate to meet requirements. The preparations of vitamin B_{12} made by bacterial fermentation are available.

Chemistry

1. The structure of vitamin B_{12} is given below. The central portion of the molecule consists of 4 reduced and substituted pyrrole rings surrounding a single cobalt atom. The central structure is referred to as a *"Corrin"* ring system. Two of the Pyrrole rings (rings A and D) are joined directly rather than through a single methylidyne carbon. Below the corrin ring system there is a 5, 6-dimethylbenzimidazole riboside that is connected at one end to the central cobalt atom and at the other end from the ribose moiety through phosphate and aminopropanol to a side chain on ring D of the tetrapyrrole nucleus. Addition of cyanide forms *"Cyanocobalamin"* which is identical with the originally isolated vitamin B_{12} and the removal of cyanide group results in the formation of the compound *"Cobalamin"*. Cyanide group, if substituted by hydroxy group, forms *"hydroxocobalamin"*, by a nitro group forms *"nitro-cobalamin"* by a methyl group forms *"methylcobalamin"*. It contains cobalt 4 to 5 per cent.
2. It has molecular weight of 1,355 and an empirical formula $C_{63}H_{88}N_{14}O_{14}PCO$.

3. Aqueous solution of vitamin B_{12}, when exposured to sunlight, results in the destruction of the vitamin.
4. If ascorbic acid is added to a solution of vitamin B_{12}, the B_{12} activity is slowly destroyed due to the reducing action of ascorbic acid on vitamin B_{12}.
5. Crystalline vitamin B_{12} is stable to heating at 100°C for long periods and aqueous solutions at pH 4.0–7.0 can be autoclaved with very little loss. Destruction is rapid when the vitamin is heated at pH 9.0 or above.

Absorption

1. Vitamin B_{12} is absorbed from the ileum. Its absorption depends on the presence of hydrochloric acid and a constituent of normal gastric juice called the *intrinsic factor* (IF), a mucopolysaccharide, secreted by the parietal cells of the gastric mucosa.
2. Free vitamin B_{12} (cobalamin) is bound to intrinsic factor in the proportion of 2 mol of cobalamin to 1 mol of IF. Intrinsic factor is found in the cardia and fundus of the stomach. The combination of B_{12} with intrinsic factor results in the formation of a complex which is resistant to intestinal digestion. Vitamin B_{12} also binds to the proteins of gastric juice, bile and saliva.
3. It is currently believed that intrinsic factor possesses 2 receptor sites, one for B_{12} and the other for ideal intestinal microvilli. The former receptor site requires a neutral pH and the presence of calcium ions (Ca^{++}). The latter receptor site is readily saturated which limits the absorption of B_{12} to 1.5 μg after a single dose. The intrinsic factor is released within the intestine liberating B_{12} to pass into the intestinal mucosal cell. Intrinsic factor protects vitamin B_{12}-peptide against bacterial attack, and in the peptide form the vitamin is absorbed.
4. About 1% of the large concentration of the vitamin is also absorbed by passive mechanism.

Transport

1. Vitamin B_{12} is present in plasma as methylcobalamin, 5′-deoxyadenosylcobalamin or hydroxycobalamin bound to proteins, *transcorrin I* and *transcorrin II*. Transcorrin I is a strong binder of cobalamin and transcorrin II is a weaker binder of cobalamin. 1-10% of B_{12} carried on transcorrin II is attached immediately after absorption and is readily released to be excreted in the urine.
2. Cobalamin bound to plasma protein fraction is carried to the tissues where it is bound to a variety of protein receptors. Any excess is stored in the liver as 5′-deoxyadenosylcobalamin. The total amount of cobalamin in the bodies of adults is 2.5 mg of which about 1.5 mg is in the liver.

Excretion

1. B_{12} is excreted mainly by the way of the bile and by this pathway about 40 μg. pass into the jejunum daily. By an interohepatic circulation, most of this is reabsorbed in the ileum using the intrinsic factor mechanism.
2. Small amounts of the vitamin also enter the intestine from the gastric, pancreatic and intestinal secretions.
3. The unabsorbed vitamin leaves the body in the feces. This amount—together with that produced in the colon by bacterial synthesis—is about 3-6 μg daily.
4. Cobalamin unbound to protein is excreted in the urine. This amounts to 0.25 μg/day.

Normal concentration in serum

200 to 850 μg/dL.

Sources

Vitamin B_{12} is present only in the foods of animal origin. It is not present in foods of vegetable origin. Bacterial synthesis of cobalamin occurs in the human colon but it is not absorbed. The only source of cobalamin in nature is via synthesis by microorganisms in soil, water and the animal intestine.

Richest sources	Liver and Kidney.
Good sources	Meat, fish, eggs.
Fair sources	Milk, cheese.

Daily requirements

Recommended daily dietary allowances (Revised 1974)	
Infants	0.3 μg
Children	1-2 μg
Adult (males)	3.0 μg
Adult (females)	3.0 μg
Pregnant women	4.0 μg
Lactating women	4.0 μg

Functions

1. Vitamin B_{12} along with folic acid is required for the development of red blood cells beyond megaloblast stage.
2. It stimulates the appetite and general health of the subject.
3. It cures the neurological symptoms of pernicious anemia.
4. Coenzyme activities :
   (a) Vitamin B_{12}, as the coenzyme, is involved with tetrahydrofolate in the synthesis of labile methyl groups which can be transferred to homocysteine to form methionine. This is not sufficient to supply the body with its total methionine requirement and the amino acid is, therefore, a member of the essential amino acid group.
   (b) The vitamin is also a coenzyme for the mutase which converts methylmalonyl-CoA into succinyl-CoA. Hence, methylmalonate accumulates in the urine in deficiency states.
   (c) The vitamin is also involved in the maintenance of sulfhydryl groups in the reduced state. The deficiency of the vitamin produces the reduction of glutathione in the blood.
   (d) Vitamin B_{12} is also involved in the conversion of ribonucleotides to deoxyribonucleotide.
   (e) It is also required in the dehydrase reactions in the conversion of ethylene glycol to acetaldehyde and glycerol to β-hydroxypropionaldehyde.
   (f) As coenzyme, it is required for the isomerisation of glutamic acid to threo-β-methyl aspartate.
   (g) It is involved in the biosynthesis of proteins.

Deficiency Symptoms

In man, deficiency of vitamin B_{12} may result from poor dietary intake which occurs in the tropics, in strict vegetarians (vegans) and occasionally in the elderly. It may result from a deficiency of intrinsic factor or from interference with the function of intrinsic factor and from intestinal malabsorption and also from the prolonged use of anticonvulsant drugs and of paraminosalicylic acid which is used in the treatment of tuberculosis.

In deficiency states, the symptoms are :

1. Megaloblastic macrocytic anemia (Pernicious anemia).
2. Mucosal atrophy and inflammation of tongue (glossitis) and mouth.
3. Severe disease of the nervous system both central and peripheral.
4. Psychiatric symptoms are not uncommon.
5. There may also be amblyopia.
6. A severe form of acidosis in children due to the excessive production of methylmalonate due to congenital defect in the biosynthesis of 5′-deoxyadenosyl cobalamin.

Management

(a) In the deficiency of vitamin B_{12}, hydroxocobalamin is given parenterally in a dosage of 1,000 μg twice during the first week, then 100 μg weekly for a further 6 doses. Within 48 hours of the first injection the bone marrow shows a striking change from a megaloblastic to a normoblastic state.

(b) In some patients the rapid regeneration of the blood depletes the iron reserves of the body and recovery is stopped. To prevent this, ferrous sulphate 200 mg daily should be given soon after the commencement of the treatment. A combined deficiency of vitamin B_{12} and iron is found by the presence of macrocytosis and hypochromia.

Maintenance treatment

The patient must have regular parenteral doses of hydroxocobalamin indefinitely (1,000 μg i.m. every 3 months) for life. Blood counts should be performed annually.

Toxic Effects

No ill-effects on excessive ingestion of the vitamin.

BIOTIN

Introduction

1. Boas (1927) observed that when raw egg white was given as the main source of protein in the diets of rats, they developed symptoms of dermatitis, retarded growth, loss of hair and loss of muscular control. All these symptoms were prevented by egg yolk. The factor was called *'anti-egg white injury factor'*.
2. In 1931, Gyorgy named this factor 'vitamin H'.
3. In 1939, vitamin H was isolated by Gyorgy, Kuhn and Lederer.
4. In 1942, Melville et al isolated vitamin H from milk and named this vitamin *'Biotin'*.
5. Biotin is widely distributed in many foods as biocytin which is released on proteolysis. It is synthesized by intestinal flora in excess of requirements. Dificiency is unknown except a very small number of people who eat abnormally large amounts of uncooked egg white which contains *avidin* that binds *biotin* and renders it unavailable for absorption.

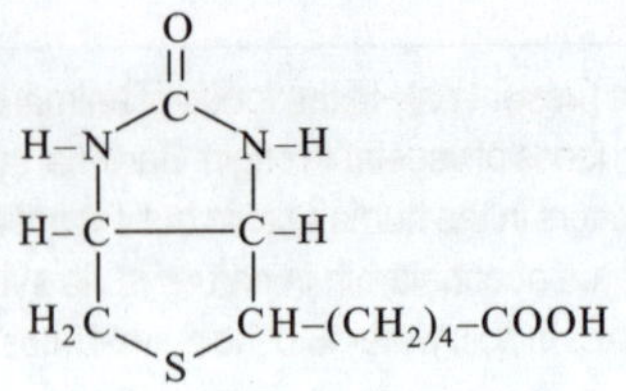

Fig. 15.24 : Structure of Biotin

Chemistry

1. The chemical structure of biotin in the free state is given (Fig. 15.24).
2. The carboxyl group of biotin combines with the terminal nitrogen of lysine residue of enzyme protein forming *biocytin* (Lysine-biotin conjugate).
3. It is highly soluble in hot water.
4. It forms salts with alkali hydroxides.
5. It is stable to autoclaving at 120°C for 30 minutes at neutral pH.

Absorption and Storage

Biotin is readily absorbed from the small intestine through the portal vein into the general circulation.

Excess of the requirements is not stored in the body but is mostly excreted in the urine.

Sources

Biotin is widely distributed in natural foods. Intestinal bacteria also supply biotin in large proportion.

Rich sources	: Egg yolk, liver, kidney.
Good sources	: Yeast, milk, tomatoes.
Fair sources	: Fruits, vegetables, human milk.

Daily requirements

Since intestinal bacteria and diets supply biotin in adequate amounts the deficiency of this vitamin in human beings is rare.

Infants	10-15 μg
Children	20-40 μg
Adults	50-60 μg

Normal concentration in blood

12-24 μg/dl in adults.

14-55 μg/dl in infants.

Functions

Coenzyme activities

Biotin is connected with the carboxylation reactions (CO_2 "fixation"). The biotin coenzyme-apoenzyme complex attaches CO_2 which is afterwards transferred to other substances. The following reactions are accomplished by this mechanism :

1. Biotin acts as coenzyme along with acetyl-CoA carboxylase to convert acetyl-CoA to malonyl-CoA which is an important step for fatty acid synthesis in extramitochondrial pathway.
2. The conversion of pyruvate to oxaloacetate is also a biotin-dependent carboxylation reaction catalyzed by pyruvate carboxylase.
3. Biotin acts as a coenzyme in the conversion of propionic acid to succinic acid.
4. Biotin is involved in the fixation of CO_2 for the formation of carbon 6 in purine synthesis.
5. Succinic acid dehydrogenase, succinic acid decarboxylase, aspartic acid deaminase, serine deaminase and threonine deaminase are influenced by biotin.
6. β-methylcrotonyl-CoA-carboxylase converts β-methylcrotonyl-CoA to β-methylglutacronyl-CoA in presence of biotin.

Deficiency symptoms

The deficiency of Biotin may result from the destruction of intestinal bacteria by sulfonamide drugs or from the adequate intake of raw egg white which contains the protein avidin (interfere absorption of biotin) in association with diarrhoea in case of infants.

The deficiency develops nausea, anorexia, anemia, muscular pain, dermatitis of extremities.

LIPOIC ACID

Introduction

Lipoic acid was first detected in lactic acid bacteria. It had replaced growth-stimulating effect of acetate. Hence, it was designated as "acetate replacement factor". This vitamin was also required for the nutrition of the protozoon Tetrahymena gelei for which it was termed as "protogen".

Chemistry

1. It is a sulfur-containing fatty acid [6, 8-dithiooctanoic acid (thioctic acid)]. The chemical structure is given (Fig. 15.25).
2. It is soluble in water.

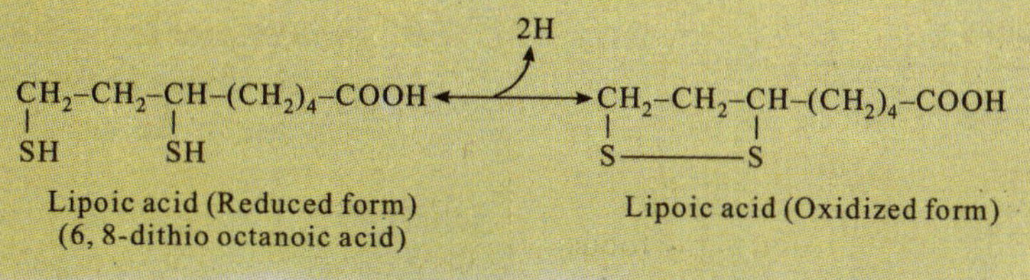

Fig. 15.25 : Structure of lipoic acid

Sources

It is widely distributed in natural foods.

Daily requirements

Not yet established.

Functions

Lipoic acid acts as a coenzyme in the oxidative decarboxylation of pyruvic acid and α-ketoglutaric acid.

Deficiency symptoms

No deficiency symptoms have so far been established for this vitamin.

INOSITOL

Introduction

1. Both plant and animal tissues contain inositol.
2. For a good length of period the hexaphosphoric ester of inositol (phytic acid) present in cereals has been known.
3. Seven optically inactive and two optically active forms of inositol are present in nature of which only one inactive form has got nutritional value.
4. In 1940, Woolley et al reported that inositol can prevent the development of alopecia in mice.

Chemistry

1. Inositol (meso-inositol or myo-inositol) is hexahydroxycyclohexane. The structure is given below :

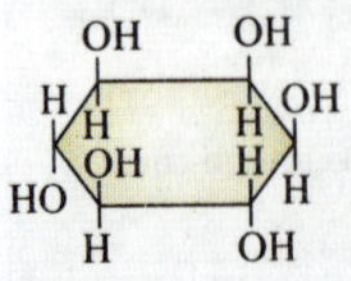

Fig. 15.26 : Strucure of inositol

2. Meso-inositol is biologically most effective.
3. It is highly soluble in water.
4. It is a crystalline compound and has a sweet taste.
5. It is stable to heat in neutral acid and alkaline medium.

Absorption and Storage

Inositol is absorbed readily from the small intestine through the portal vein and passes to the tissues through the general circulation.

Excess of the requirement of this vitamin is not stored in the body but is excreted in the urine and feces or metabolised.

Existence of inositol

1. The greater part of inositol exists in bound form; although a small amount exists in the free state in muscle and other tissues.
2. It exists in the form of phospholipids in animal tissues; whereas it exists in the form of phospholipids, phytic acid and phytin in plant tissues.
3. Brain and liver contain phospholipids containing inositol. Mitochondria and microsomes also contain large amount of inositol containing phosphatides.
4. Cereals oilseeds and nuts contain large amounts of phytic acid and phytin (calcium or magnesium salt of phytic acid). Phytic acid (inositol-hexaphosphoric acid) is hydrolyzed by an enzyme phytase present in plants into inositol and phosphoric acid. Unless phytic acid is hydrolyzed it is not absorbed from the intestines. Only a small percentage of the phytic acid present in the diet is hydrolyzed and absorbed due to the low concentration of phytase in the intestinal juice.

Sources

Yeast, meat, fruits, milk, nut, vegetables and grains contain inositol.

Daily requirements

Not yet reported.

Functions

1. Along with cholin, inositol exerts a lipotropic effect.
2. Large concentration of inositol in heart muscle increases the rate of contraction.
3. Labelled inositol gives rise to labelled glucose in the body.
4. It is oxidized to glucuronic acid in the liver by oxygenase.
5. It stimulates the growth of yeast and fungi.
6. It occurs in brain tissues as phosphatidyl inositol (lipositol).
7. It causes an increase in nerve chronaxia in the rat.
8. In animals, it increases the peristalsis of the small intestines.

Deficiency symptoms

Deficiency symptoms in mice include retarded growth, failure of lactation, alopecia (loss of hair over the body), spectacled eye (a condition due to loss of hair around the eye) and those in chicks include encephalomalacia.

CHOLINE

Introduction

1. In 1934, Best and Huntsman discovered that choline deficiency produces fatty liver in rats.
2. Choline is very essential for life processes. It is an important metabolite although it cannot be classified as a vitamin. Furthermore, the choline requirement is more than most substances and the deficiency symptoms are suggestive of vitamin deficiency diseases. Hence, it is considered as vitamin.

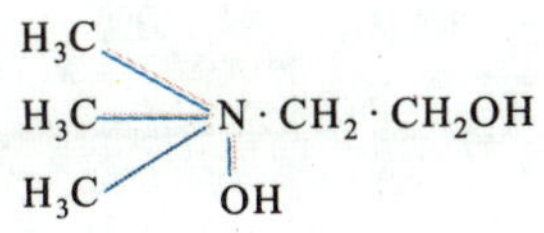

Fig. 15.27 : Strucure of choline

Chemistry

1. Choline is hydroxyethyl trimethylammonium hydroxide. The chemical structure is given (Fig. 15.27) .
2. It is a strong base and a colourless crystalline compound.
3. It is soluble in water.
4. Dilute solutions of choline (less than 4 per cent) are stable to heat.

Absorption and Storage

Choline in the free state present in food is easily absorbed from the small intestine through the portal vein into the general circulation.

Excess ingestion of choline than the requirements is not stored in the tissues. Part of it is converted into phospholipids but a greater part is metabolised in the liver into trimethylamine which is excreted in the urine.

Sources

Rich sources	:	Liver, egg yolk.
Good sources	:	Rice polishing, meat, cereals.
Fair sources	:	Milk, fruits, vegetables.

Requirements

Choline requirements of human beings are not yet established. However, the diets commonly consumed contain large amounts of choline.

Normal level of free choline in plasma in male adults

About 4.4 mg/ml.

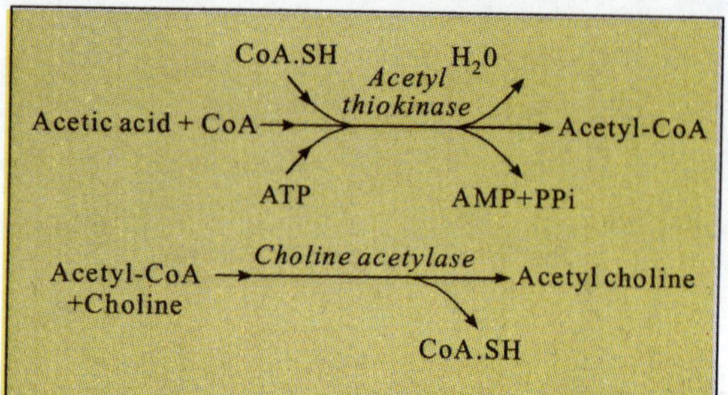

Fig. 15.28 : Formation of Acetyl Choline

Functions

1. Acetyl choline is a chemical mediator of parasympathetic and other types of activity in the nervous system. It is formed as follows :

   Acetyl cholinesterase is an enzyme present in many tissues which can hydrolyze acetylcholine to choline and acetic acid. Red blood cells can synthesize acetyl choline. Both choline acetylase and acetyl cholinesterase are present in red cells. Choline acetylase is also present in brain, skeletal muscle, spleen and placental tissue. When choline acetylase becomes inactive due to drug inhibition or lack of substrate, the red cells lose their selective permeability and undergo hemolysis.
2. Choline is a constituent of phospholipids which are essential constituents of all cells in the body.
3. Choline takes part in transmethylation reactions in the formation of methionine from homocystine :

$$\text{Choline} \xrightarrow{\text{Oxidation}} \text{Betaine}$$

$$\text{Betaine} + \text{homocystine} \longrightarrow \text{Methionine} + \text{dimethylglycine}$$

4. It prevents accumulation of fat in the liver of many animals.
5. It is essential for growth of many animals.
6. Its deficiency reduces egg formation in the hen.

Deficiency symptoms

1. Deficiency of choline causes fatty liver in rats.
2. In the young growing rat there is hemorrhagic degeneration of the kidneys and hemorrhage into the eyeballs and other organs and ultimately leads to cirrhosis.
3. In chicks, its deficiency causes slipped tendon disease in which there is a defect at the tibiotarsal joint of the bird.

Metabolism of choline

A. Biosynthesis :

1. Serine is decarboxylated to ethanol amine in presence of pyridoxal phosphate.
2. Ethanol amine is progressively methylated to choline by the incorporation of one-carbon fragment into a methyl group of methionine.

Serine → Ethanol amine → choline.

B. Catabolism :

1. Choline is first oxidized to betaine aldehyde and then to betaine. Choline acts as a methyl donor only after oxidation to betaine.
2. After loss of methyl group in the use of homocystine to methionine, betaine is converted to dimethylglycine. Dimethylglycine produces N-hydroxy-methyl sarcosine after oxidation.
3. Sarcosine is then formed by transferring hydroxymethyl to tetrahydrofolic acid.
4. Sarcosine is then converted to glycine by oxidation.
5. Glycine is converted to serine by the addition of a hydroxymethyl group derived from formylated tetrahydrofolic acid derivatives.

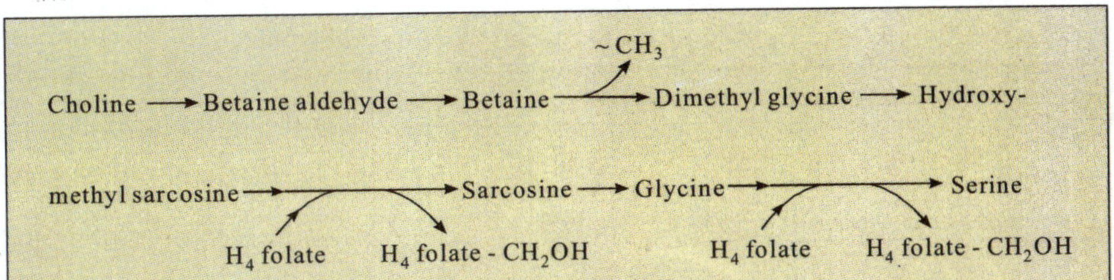

Fig. 15.29 : Catabolism of Choline

CLINICAL ORIENTATION

- Fat-soluble vitamin deficiencies are caused by the dietary inadequacy or malabsorption of the fat soluble vitamins.
- Vitamin A deficiency causes night-blindness and Xerophthalmia; Vitamin D deficiency leads to rickets in young children and osteomalacia in adults; Vitamin E deficiency, although rare, causes neurologic disorders and anemia of the newborn; Vitamin K deficiency, although very rare in adults, leads to bleeding and hemorrhage of the newborn.
- Since the body is able to store surplus lipid-soluble vitamins, Toxicity can result from excessive intake of these vitamins.
- Vitamin A and Provitamin A (β-Carotene) as well as Vitamin E can prevent cancer.
- Single vitamin of the B-complex deficiency is rare, since poor diets are associated with multiple deficiency states.
- The water-soluble vitamins :
  (a) Thiamin deficiency causes beriberi.

CLINICAL ORIENTATION

(b) Riboflavin deficiency causes cheilosis, glossitis, seborrhea and photophobia.

(c) Niacin deficiency causes Pellagra.

(d) Pyridoxine deficiency causes Peripheral neuritis.

(e) Cobalamin deficiency causes Megaloblastic anemia, methylmalonic aciduria, pernicious anemia.

(f) Folic acid deficiency causes megaloblastic anemia.

(g) Ascorbic acid deficiency causes Scurvy.

- Vitamin deficiencies can be avoided by the consumption of varieties of food in large quantities to be relieved of the various undesired complications and ailments.
- Excesses of water-soluble vitamins are easily excreted in urine and, so, rarely accumulated in toxic concentration in the body. These vitamins must be provided regularly along with the food since their storage is limited.

Daily requirements of vitamins

	Fat requirements of vitamins				Water-soluble vitamins	
	A	D	E	K	B_1	B_2
Adults :	5,000 I.U.	200 I.U.	25–30 mg	1–2 mg	1.2-1.5 mg	1.5–1.8 mg
Women during pregnancy & lactation :	6,000–8,000 I.U.	400 I.U.	—	—	1.3-1.5 mg	1.6–1.9 mg
Infants :	1,500 I.U.	400 I.U.	—	—	0.3-0.5 mg	0.4–0.6 mg
Children :	2,000–3,000 I.U.	400 I.U.	—	—	0.7-1.2 mg	0.8–1.2 mg
	One I.U. = 0.3 μg of retinol = 0.6 μg of β-carotene.	1 I.U.= 0.025 μg of cholecalciferol				

	Water-soluble vitamins						
	Niacin	Pyridoxine	Pantothenic acid	Folic acid	B_{12}	C	Biotin
Adults :	16–20 mg	1.6–2.0 mg	5–10 mg	400 μg	3.0 μg	45 mg	50–60 μg
Women during Pregnancy & lactation :	16-20 mg	2.5 mg	10–15 mg	600–800 μg	4.0 μg	60–80 mg	50–60 μg
Infants :	5–8 mg	0.3 mg	1–2 mg	50 μg	0.3 μg	35 mg	10–15 μg
Children :	9–16 mg	0.6–1.2 mg	4–5 mg	100–300 μg	1–2 μg	40 mg	20–40 μg

RICKETS

Causes

Vitamin D deficiency causes the disease rickets in children.

Symptoms

1. It is characterised by bone deformities.
2. The earliest bony lesion is often craniotabes-small round unossified areas in the membranous bones of the skull.
3. Another early sign is the 'beading' at the costochondral junctions of the ribs known as *'Rickety rosary'*.
4. Later features are *'Bossing'* of the frontal and parietal bones and delayed closure of the anterior fontanelle and deformities of the chest such as undue prominence of the sternum known as *'Pigeon chest'*.
5. The bones are soft due to the non-deposition of calcium salts and hence they are easily bent by the weight of the body.
6. In case the rickets continue for two to three years, there is serious bone deformities such as bow legs, deformities of the spine and pelvis. There is extension and widening of the epiphysis at the growing points.
7. Renal ricket is caused by defective transport of phosphate by the renal tubules.

Prevention

It can be prevented by the administration of vitamin D and calcium. Vitamin D in ordinary dose cannot cure the renal rickets and hence renal rickets are sometimes referred to as Vitamin D-resistant rickets.

OSTEOMALACIA

Causes

Vitamin D deficiency causes the disease osteomalacia in adults. It generally occurs in pregnant women of low income groups who consume a diet devoid of vitamin D and calcium and those who observe *'purda'* also. It is prevalent in India, Pakistan, Bangladesh, China and Middle-East countries.

Symptoms

1. The changes in the blood and bone are similar to those in rickets.
2. Progressive decalcification of bones—bones become soft.
3. Bone deformities occur in pelvis, legs, ribs, sacrum and lower lumbar vertebrae.
4. Owing to the deformity of the pelvis, normal delivery of the baby becomes difficult.

Prevention

This disease can be prevented by the administration of vitamin D and calcium.

SCURVY

Causes

Severe deficiency of ascorbic acid leads to the disease scurvy.

Symptoms

A. Scurvy in adults

1. The first symptoms are weakness, easy fatigue and listlessness. These are followed quickly by pain in bones, joints and muscles of the extremities, shortness in breath.
2. Hemorrhages in muscle occur particularly in calf, thigh and forearm. Hemorrhages may occur in joints with swelling and pain.
3. As the deficiency advances, the gums become swollen, blue red and spongy. The gums may be infected by bacteria. There is loss of teeth.

B. Infantile scurvy

1. The prescorbutic infants become anorexic and listless for a few days.
2. With the beginning of the disease, the infant lies with legs drawn up on the abdomen.
3. The infant cries when touched especially when its legs or arms are moved or lifted.
4. Extreme tender swellings may be felt at the end of the long bones.
5. The sternum may sink slightly inward.
6. Purpura occurs in the skin.
7. If treatment is delayed, dyspnea, cyanosis, convulsions and death may occur.

Prevention

1. The disease can be prevented by the administration of high dose of ascorbic acid.
2. Citrous fruits are to be taken regularly.

BERIBERI

Causes

The disease beriberi is caused by the deficiency of thiamine (vitamin B_1).

Types

There are three types of beriberi—wet beriberi, dry beriberi and infantile beriberi.

Symtoms

A. Wet beriberi

1. The important feature of wet beriberi is edema.
2. It develops rapidly and not only the legs but also the face, trunk, serous cavities are involved.
3. The calf muscles are slightly swollen and tender on pressure.
4. Palpitation and breathlessness appear.
5. The diastolic blood pressure is low while the systolic is high.
6. The pulse is fast.
7. The heart becomes weak and death occurs as a result of heart failure.

B. Dry beriberi

1. The muscles are progressively wasted and weak and difficult to walk.
2. The affected individual takes the help of sticks to stand and walk and, ultimately, he becomes bedridden.
3. The patient dies if untreated.
4. There is cerebral disorder.

C. Infantile beriberi

1. Those infants will suffer from the deficiency of the vitamin whose mother's breast milk contains low thiamine.

2. The infants face restlessness and sleeplessness.
3. Anorexia, vomiting and breathlessness develop.
4. Most of the symptoms are due to cardiac dilatation and failure.
5. Sudden death occurs if not treated with thiamine immediately.

Prevention

The disease can be prevented by the administration of thiamine by intramuscular injection.

PELLAGRA

Causes

1. The deficiency of niacin causes the disease pellagra.
2. It occurs whenever corn (maize) is the main diet and the corn protein is deficient in some of the essential amino acids, notably tryptophan and lysine. Hence, niacin is not formed in the body.
3. It is also caused by acute or chronic infection.

Characteristics

Pellagra is characterised by three Ds—dermatitis, diarrhoea and dementia.

Symptoms

A. Dermatitis

1. The bright red erythema resembling sunburn occurs over the exposed parts of the body.
2. The commonest sites are the back of the fingers and hands, the forearms, dorsum of the feet and ankles and the neck.
3. At the onset, the skin becomes red and slightly swollen.
4. Secondary infection is always present.
5. The dermatitis is precipitated by exposure to sunlight.

B. Diarrhoea

1. In most cases, nausea and vomiting are found.
2. The diarrhoea ranges from few to several loose stools a day with blood and mucosa.

C. Dementia

1. Dementia more frequently occurs in chronic cases.
2. Irritability, changes in disposition, depression, inability to concentrate are found in milder mental disturbances.
3. In mild cases, poor memory is common.
4. In chronic cases, spasticity, ataxia, the involvement of the bladder and rectal sphincters are seen.

Prevention

1. Niacinamide in doses of 15-25 mg three times a day are to be prescribed.
2. Adequate quantities of meat, eggs, milk and vitamin B-complex are given.
3. Any accompanying infection should be treated with the proper antibiotic.

PERNICIOUS ANEMIA

Causes

The lack of 'intrinsic factor' in the stomach resulting in the failure of absorption of vitamin B_{12} causes the disease pernicious anemia.

Symptoms

1. The RBC count is low—1.5 to 2.5 million per cubic millimetre.
2. The average diameter of the cell is above normal.
3. Excessive destruction of the abnormal circulating red cells causes the increase in the serum bilirubin.
4. The hemoglobin content is as low as 8 to 9 per cent.

5. The nucleated red cells of the bone marrow is highly increased.
6. The cells of Stage I are peculiar and are called megaloblasts.
7. The cells of the stomach responsible for acid and enzyme secretions are atrophied. So the gastric secretions are devoid of acid, pepsin and intrinsic factor (IF).
8. The tongue is sored and inflammed.
9. Numbness and tingling occur in fingers and toes.
10. Signs of involvement of the spinal cord (vitamin B_{12} neuropathy) are observed.
11. In advanced cases, demyelination of the white fibres of the spinal cord occurs.

Prevention

It can be prevented by the administration of vitamin B_{12}.

EXERCISE

1. Give the chemistry, functions and deficiency manifestations of vitamin A. (M.U. 75S)
2. Mention the sources of vitamin A. Describe the function and effects of deficiency of vitamin A. What is the daily requirement? (M.U. 73A)
3. Describe the role of vitamin A in our food requirements. What are the manifestations of deficient intake of this vitamin? (P.U. 64A)
4. Describe briefly the chemical nature, action in the body, dietary sources and nutritional importance of vitamin A. (P.U. 65S)
5. What do you know about vitamin D? (Luck. 69S)
6. Name the fat-soluble vitamins, their occurrence, daily adult requirements and their importance to the body. (Punjab 60A)
7. Give a brief account of the chemistry, sources, daily requirement and deficiency states of thiamine. Describe the mechanism of its action. (P.U. 71A)
8. Describe briefly the chemical nature, mode of action, dietary sources and nutritional importance of thiamine. (R.U. 67A)
9. Describe the chemistry, sources, daily requirement and metabolic functions of (a) thiamine, (b) niacin. Mention the disease caused by their deficiency in diet. (R.U. 76A)
10. Name the vitamins with coenzyme activities. Give an account of the sources, properties, effects of deficiency and functions of pyridoxine. (M.U. 72A)
11. Discuss the role of Riboflavin in the body. (PUN. 69A)
12. What is riboflavin? Discuss its metabolic role in the body and the deficiency symptoms. (P.U. 72S)
13. What are vitamins? Describe the functions and properties of any two vitamins of B-complex group. (Muz. 74A)
14. Mention the sources of vitamin B_{12} and the effect of its deficiency. Describe the metabolic functions of vitamin B_{12}. (M.U. 74S)
15. Describe the sources, requirements, functions and deficiency manifestations of vitamin C. (M.U. 73S, P.U. 68A)
16. Name the vitamins with coenzyme activities. Discuss the biological importance of flavoproteins. (P.U. 73A)
17. Write short notes on :
    (i) Vitamin A. (Muz. 75A)
    (ii) Deficiency symptoms of vitamin D. (M.U. 74A)
    (iii) Vitamin D. (R.U.64S)
    (iv) Rickets. (Muz. 74A; Mith. 67A)
    (v) Vitamin K. (Pun. 66A)
    (vi) Tocopherol. (M.U. 75A; Bh. U. 75A)
    (vii) Thiamine (R.U. 74A; M.U. 73A; Mith. 71A)
    (viii) Thiamine deficiency. (M.U. 72A)
    (ix) Coenzyme activity of thiamine. (Bh. U. 76A)
    (x) Functions of riboflavin. (Mith. 65A)
    (xi) Nicotinic acid. (Bh. U. 75S; P.U. 68S; R.U. 69A)
    (xii) Pyridoxine. (M.U. 76A; Mith. 67S; P.U. 70A)
    (xiii) Folic acid. (M.U. 76A; Mith. 67S; P.U. 70A)
    (xiv) Vitamin C. (P.U. 68S; R.U. 64A; Mith. 74A)
    (xv) Cyanocobalamin. (P.U. 72A)
    (xvi) Scurvy. (Bh. U. 74; R.U. 70S)
    (xvii) Choline. (R.U. 74A)
    (xviii) Pellagra. (P.U. 74A)
    (xix) Beriberi. (P.U. 72S)
    (xx) Pernicious anemia. (M.U. 76A)
    (xxi) Osteomalacia. (R.U. 74A)
    (xxii) Ascorbic acid. (Bh. U. 72S)
    (xxiii) Manifestation of Niacin deficiency. (C.U. 1982)
18. Discuss the importance of ascorbic acid as an essential food factor. (C.U. 1981)

SCHEMATIC COMPENDIUM

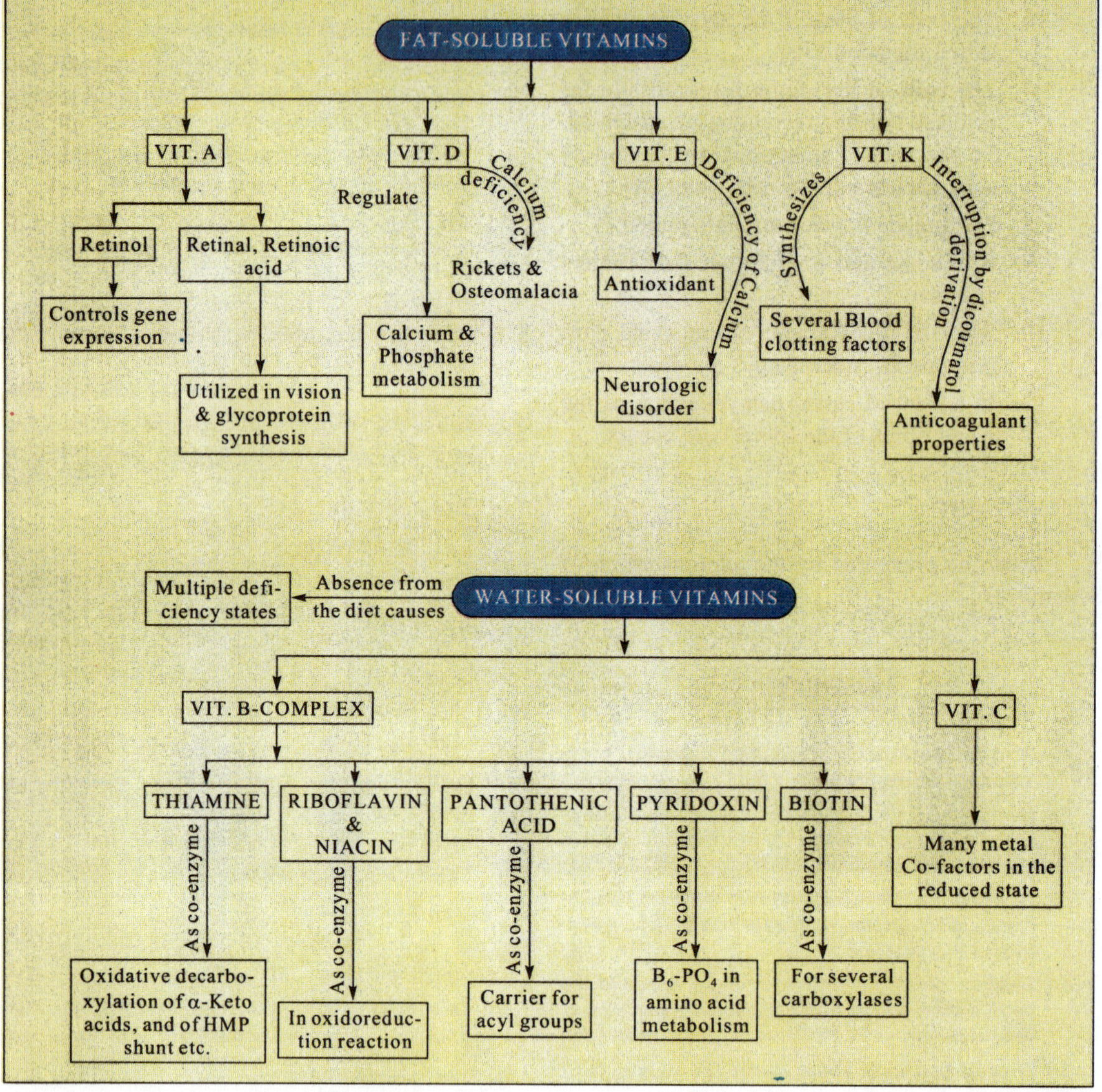

CHAPTER 16

DIGESTION AND ABSORPTION FROM THE GASTROINTESTINAL TRACT

Unless the foodstuffs taken are broken down into smaller molecules, they are not absorbed from the digestive tract. The disintegration of foodstuffs into assimilable form constitutes the process of digestion.

The digestion in the alimentary canal takes place by the hydrolytic enzymes with very small expenditure of energy. During digestion, the vitamins and the minerals are made more assimilable. This is true for the fat-soluble vitamins which are absorbed only when fat digestion proceeds normally.

Advantages of digestion

1. The compositon of the body proteins, fats and polysaccharides is not the same as that of proteins, fats and polysaccharides of food. Only the units of these complex substances are the same in the body and the blood. The food protein must be broken down into amino acids from which the tissue protein can be formed. Similarly, the polysaccharides and fats are formed in the body.
2. Disaccharides, which are absorbed easily, are not used in the body as such and are excreted as foreign substances. But after digestion to hexoses, they are utilized in the body.
3. Digestion prevents the undesirable effects of the introduction of the foreign proteins into the blood converting them into amino acids which have no harmful effect.
4. It prevents the too rapid absorption of food and thus enables the blood to distribute the absorbed substances without undesirable effects.
5. The power of digestion increases the variation in our diet and our enjoyment.

DIGESTION IN THE MOUTH

Constituents of saliva

1. Saliva is colourless, slightly acid (pH 6.4-7.1) rather than alkaline, viscous fluid secreted mainly by the parotid, submaxillary and sublingual salivary glands. The submaxillary secretion contains most of the glycoproteins (mucin) but the parotid contains none.
2. The amount of secretion from each gland varies with the stimulus and nature of the food. The amount is stated to be 1,000-1,500 ml each day.
3. A saliva of 0.6% solids will contain 0.4% organic and 0.2% inorganic material.
4. The chief organic substances are glycoprotein (mucin), salivary amylase (ptyalin), small amounts of albumin and globulin, urea and uric acid, traces of thiocyanic acid.

5. The inorganic substances are Cl^-, Na^+, K^+, Ca^{++}, Mg^{++}, HCO_3^- and HPO_4^-. Chloride ion is an important activator of amylase. Calcium ion helps to stabilise the amylase.

Salivary digestion

1. The salivary amylase or ptyalin brings about the hydrolysis of starch and glycogen to maltose at the optimum pH 6.8. Chloride ion activates the enzyme.
2. The enzyme can act on the food for a short time (5 to 6 minutes) which is of little significance.
3. Salivary amylase is readily inactivated at pH 4.0 or less. So the digestive action on food in the mouth is soon ceased in the acid environment of the stomach.
4. In many animals, salivary amylase is entirely absent.

Functions of saliva

1. Saliva moistens dry food and facilitates swallowing by the lubricating action of the glycoprotein.
2. The digestion of starch begins in the mouth.
3. Saliva contains buffering substances eg, bicarbonate, phosphate and mucin.
4. It keeps the mouth at a neutral pH and thus protects the teeth from decalcification and also keeps the mouth and teeth clean.

DIGESTION IN THE STOMACH

Gastric secretion

1. Nervous or reflex mechanisms cause the initiation of gastric secretion.
2. The continued gastric secretion is regulated by the hormone gastrin (gastric secretion). The chemical stimulant is produced by the gastric glands and absorbed into the blood, which carries back to the stomach where it stimulates gastric secretion.
3. Histamine, the decarboxylated product of the amino acid histidine, also stimulates the secretion of gastric juice.

Constituents of gastric juice

1. Normal gastric juice is a thin, light coloured fluid which is strongly acidic.
2. Hydrochloric acid secreted by the oxyntic cells (parietal cells) is 0.55 per cent which is equivalent to pH 0.9. The high acidity is neutralised by the high water content of saliva, the glycoprotein of saliva, the proteins of the food functioning as weak acid and the mucus of the stomach.
3. The gastric juice contains 97-99% water and about 0.55% solids of which about 0.4 per cent is organic.
4. The organic substances are the enzyme pepsin, glycoprotein, traces of lipase, rennin in infants.
5. The inorganic substances are mainly Cl^-, K^+, Na^+, traces of Ca^{++}, Mg^{++}, phosphate and sulfate.
6. On ordinary diet, the amount of secretion is 2-3 litres daily by an adult.
7. The intrinsic factor (HCl and mucoproteins), present in the gastric juice, helps in the absorption of vitamin B_{12}. The mucopolysaccharide present in the intrinsic factor contains fucose, hexosamine and neuraminic acid.

Gastric digestion

An ordinary meal is evacuated from the stomach in about four hours. The time may be more if the food is not properly masticated or if there is an excess of fat or gastric secretion is disturbed by emotion.

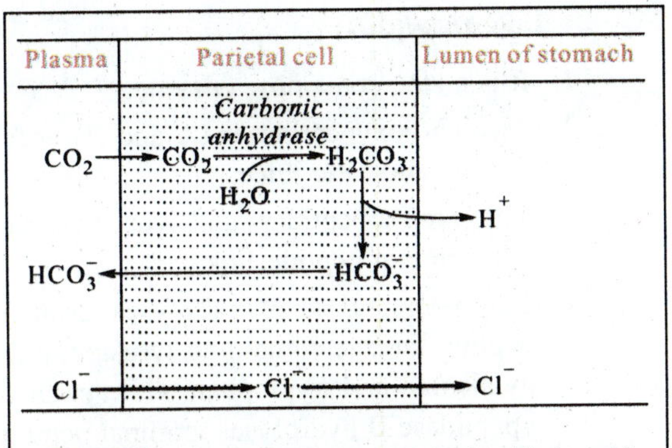

Fig. 16.1 : Production of gastric hydrochloric acid

A. Hydrochloric acid

Hydrochloric acid is formed by parietal cells and the process is similar to that of the 'chloride shift'.

Carbonic anhydrase catalyzes the formation of H_2CO_3 from H_2O and CO_2 as shown in Fig. 16.1.

Production of no hydrochloric acid leads to the condition achlorhydria.

Functions of hydrochloric acid

1. It maintains the optimal pH (1.2–1.5) for digestion of proteins by pepsin.
2. It converts inactive pepsinogen into active pepsin.
3. It stimulates duodenum to liberate secretin.
4. It helps the absorption of iron by converting ferric hydroxide of the food into ferric form which is changed to the ferrous form by reduction.
5. It denatures food proteins making them more readily digestible.
6. It has a germicidal effect on microorganisms and hence prevents the growth of microorganisms in the stomach.

B. Pepsin

1. It is secreted in the inactive form pepsinogens which is activated to pepsin by the action of HCl and a small amount of pepsin can cause the activation of the remaining pepsinogen.
2. It converts native protein into proteoses and peptones.
3. It has a molecular weight of 32,700.
4. It contains large amount of acidic amino acids.

C. Rennin (Chymosin, Rennet)

1. This enzyme occurs in the stomach of infants and is absent from the stomach of adults.
2. It causes the coagulation of milk and prevents the rapid passage of milk from the stomach.
3. It changes the casein of milk to paracasein in presence of calcium at pH 5.0. Paracasein is then acted on by pepsin :

$$\text{Casein} \xrightarrow[Ca^{++}]{\text{Rennin}} \text{Calcium paracaseinate} + \text{Peptides}$$

Fig. 16.2

D. Lipase

1. The gastric lipase has mild lipolytic action which is of no significance.
2. It can act only at the pH 5 to 7. Hence it is not so active at pH 1.3–1.5.

PANCREATIC DIGESTION

Pancreatic secretion

1. The secretion of pancreatic juice is controlled by both nervous and hormonal means.
2. Hydrochloric acid, fats, proteins, carbohydrates and partially digested foodstuffs entering the duodenum and the upper jejunum stimulate the secretion of hormone secretin and pancreozymin.

   Secretin stimulates the pancreas to produce thin, watery fluid, high in bicarbonate but low in enzyme content. Pancreozymin stimulates the pancreas to produce a viscous fluid low in bicarbonate but high in enzyme content.

Constituents of pancreatic juice

1. It is a clear alkaline fluid with a pH about 8.0.
2. It contains 1.8% of solids (including HCO_3^-) of which 0.6% is organic substances.
3. The organic substances include proteins, the enzymes trypsinogen, chymotrypsinogen, carboxypeptidase, lipase, amylase, maltase, phospholipase, ribonuclease, deoxyribonuclease, cholesteryl ester hydrolase and collagenase.
4. The inorganic substances are Na^+, Cl^-, HCO_3^- with small amounts of K^+, Ca^{++} and HPO_4^-.
5. The amount of juice secreted each day is 600-800 ml.

Pancreatic digestion

A. Trypsin

1. It is secreted in the inactive form trypsinogen which is converted into trypsin by the enzyme enterokinase secreted by the duodenal mucosa.
2. It attacks the native protein, proteoses and peptones to produce polypeptides.
3. It attacks peptide linkages containing arginine or lysine residue.

B. Chymotrypsin

1. It is also secreted in the inactive form chymotrypsinogen which is converted into chymotrypsin by the action of trypsin.
2. It also attacks the native protein, proteoses and peptones to produce polypeptides.
3. It attacks peptide linkages containing tyrosine and phenylalanine residues.

C. Carboxypeptidase

1. It is a zinc containing enzyme.
2. Two carboxypeptidases (A and B) occur in the pancreatic juice.
3. They are exopeptidase and hydrolyze only the terminal peptide linkage.
4. Carboxypeptidase A hydrolyses terminal peptide linkage containing tyrosine, phenylalanine and tryptophan while carboxypeptidase B hydrolyses terminal peptide linkages containing lysine and arginine.

D. Amylase

1. Pancreatic amylase is similar in action to salivary amylase.
2. It is an α-amylase and an endoamylase.
3. It hydrolyses starch and glycogen into maltose and a mixture of branched of (1 : 6) oligosaccharides into glucose.
4. In pancreatitis, serum and urine amylase level is increased showing diagnostic importance.

E. Lipase

1. Pancreatic lipase is most important in the digestion of fats. It hydrolyses fats into diglyceride, monoglyceride, glycerol and fatty acids.
2. It is specific for the hydrolysis of primary ester linkages which occurs in position 1 and 3 of triglyceride.
3. The hydrolysis of fat by lipase is increased when the fat is emulsified by bile salts. This is due to the larger surface exposed owing to the reduction of surface tension to the action of lipase.

F. Aminopeptidase and Dipeptidase

The aminopeptidase attacks the terminal peptide bond at the free amino end of the chain.

G. Phospholipses

These enzymes hydrolyse phospholipids (lecithin and cephalin).

H. Cholesteryl ester hydrolase (Cholesterol esterase)

This enzyme hydrolyses the esterification of free cholesterol with fatty acids.

I. Collagenase

This enzyme hydrolyses collagen present in meat and fish.

J. Ribonuclease (RNAase) and Deoxyribonuclease (DNAase)

1. These enzymes are specific for the hydrolysis of RNA and DNA, respectively.
2. These are endonucleases. Both the enzymes are capable of cleaving internal phosphodiester bonds to produce a 3′-hydroxyl and a 5′-phosphoryl or a 5′-hydroxyl and a 3′-phosphoryl terminus.
3. Some are capable of hydrolyzing both strands of a double stranded molecule whereas others can only cleave single strand of nucleic acids.
4. Some nucleases are exonucleases. These are capable of hydrolyzing a nucleotide when it is present at a terminus of a molecule.

DIGESTION IN THE INTESTINE

Intestinal secretion

Hydrochloric acid, fats, proteins, carbohydrates and partially digested foodstuffs entering the duodenum and the upper jejunum stimulates the secretion of *enterocrinin* which induces the flow of intestinal juice.

Constituents of intestinal juice

1. It is an alkaline fluid of pH about 7.8.
2. The amount of the juice secreted each day by an adult is 2 to 3 litres.
3. It contains about 1.5 per cent solids of which nearly two-thirds is inorganic and the remainder is organic substances.
4. Half the inorganic part is $NaHCO_3$ and half NaCl.
5. The organic substances are mainly enzymes and proteins. The enzymes include aminopeptidase and dipeptidase, lactase, maltase, sucrase, lipase, phosphatases, nucleases, nucleotidases, nucleosidases, phospholipases, enterokinase.

Intestinal digestion

1. Disaccharidases (lactase, maltase, sucrase) : They hydrolyse the corresponding dissaccharides into monosaccharides for absorption :

$$\text{Lactose} \xrightarrow{\text{Lactase}} \text{Glucose} + \text{Glactose}$$

$$\text{Maltose} \xrightarrow{\text{Maltase}} \text{Glucose} + \text{Glucose}$$

$$\text{Sucrose} \xrightarrow{\text{Sucrase}} \text{Glucose} + \text{Fructose}$$

2. Phosphatases : They remove phosphate from certain organic phosphates such as hexophosphate, glycerophosphate and the nucleotides derived from the food.
3. Nucleases : They cause the hydrolysis of nucleic acids to mononucleotides.
4. Nucleotidases : They hydrolyse nucleotides to nucleosides removing phosphate.
5. Nucleosidases : They hydrolyse the nucleosides into purine or pyrimidine base and pentose sugars (ribose or deoxyribose).
6. Phospholipases : They hydrolyse phospholipids into glycerol, fatty acids, phosphoric acid and bases such as choline.
7. Enterokinase : This is secreted by the duodenum. It converts inactive trypsinogen to active trypsin.
8. Lipase : This converts triglycerides and diglycerides into monoglycerides and fatty acids.

THE BILE

1. Bile is absolutely necessary for the digestion of fat.
2. Liver plays an important role in digestion by producing bile. The gall-bladder, attached to the hepatic duct, stores a certain amount of bile produced by the liver between meals. The composition of the bile in the gall-bladder is modified by addition of mucin and other substances and by removal of water, bicarbonate and chloride by reabsorption by the bladder mucosa. During digestion, the gall-bladder contracts by the stimulation of the hormone cholecystokinin which is produced by the small intestine and release bile rapidly to the small intestine by the way of common bile duct. The pancreatic secretions mix with the bile.

Properties of gall-bladder bile

1. Gall-bladder bile may be golden-yellow, brownish-yellow or olive-green in colour depending on the proportions of the bile pigments.
2. It is a viscid fluid.
3. It has a bitter taste and characteristic smell.
4. The inorganic material is mainly Na^+, K^+, Ca^{++}, Cl^-, HCO_3^-.

Daily formation

About 300 to 1,200 ml of bile formed daily in adult human beings.

Composition of bile

The composition of hepatic bile differs from that of gall-bladder bile which is shown in the following table :

Composition of human hepatic and gall-bladder bile

Constituents			Hepatic bile in per cent	Gall-bladder bile in per cent
Water	...	...	97.00	85.92
Solids	...	...	2.52	14.08
Bile acids	...	...	1.93	9.14
Mucin and pigments	...	...	0.53	2.98
Cholesterol	...	...	0.06	0.26
Fatty acids and fat	...	...	0.14	0.32
Inorganic salts	...	...	0.84	0.65
Specific gravity	...	...	1.01	1.04
pH	...	...	7.1-7.3	6.9-7.7

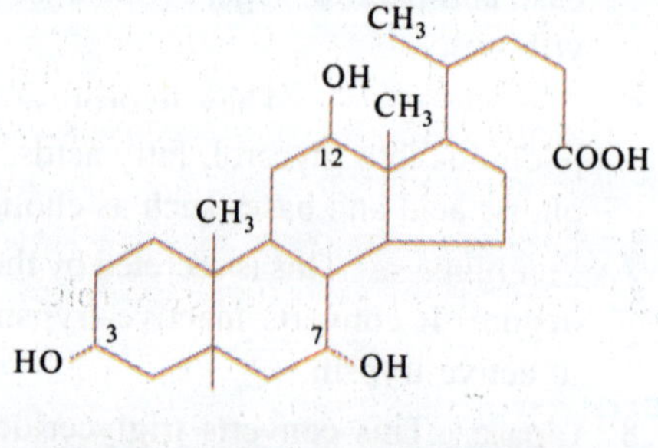

Fig. 16.3 : Structure of cholic acid (3, 7, 12-trihydroxy cholanic acid)

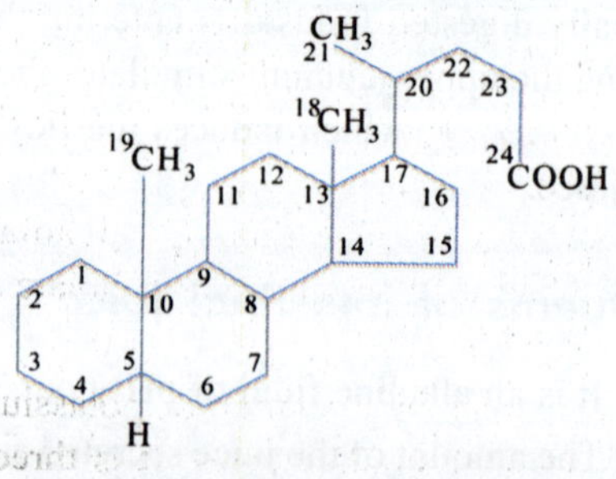

Fig. 16.4 : Structure of cholanic acid

Bile acids

1. Bile acids are synthesized in the liver from cholesterol. The synthesis of cholic acid from cholesterol is given :

   Cholesterol → 7-hydroxy cholesterol → 3, 7-dihydroxy cholestane → 3, 7, 12-trihydroxycholestane → 3, 7, 12-trihydroxy cholestanoyl-CoA → cholyl-CoA → cholic acid.
2. The bile acids are derived from the parent acid called cholanic acid. The structure of cholic acid and cholanic acid are given (Figs. 16.3, 16.4).
3. The different bile acids are hydroxy derivatives of cholanic acid which are given below :

Cholic acid	3, 7, 12-trihydroxycholanic acid.
Deoxycholic acid	3, 12-dihydroxycholanic acid.
Chenodeoxycholic acid	3, 7-dihydroxycholanic acid.
Lithocholic acid	3-Hydroxycholanic acid.

4. Many substances such as fatty acids, phenols, higher alcohols, camphor, naphthalene, combine with deoxycholic acid in various molecular proportions. The resulting compounds are called choleic acids. These choleic acids are water soluble. By such combinations insoluble fatty acids, cholesterol, fat-soluble vitamins, drugs are rendered soluble and diffusible and thus capable of being absorbed.

Bile salts

1. Cholic acid + Glycine→Glycocholic acid.
2. Cholic acid + Taurine → Taurocholic acid.

   Sodium or potassium glycocholate and sodium or potassium taurocholate are the two bile salts.
3. In human bile, sodium or potassium glycocholate predominates. It is three times as much as sodium or potassium taurocholate.

Functions

1. Bile salts act as emulsifying agents and emulsify fats increasing surface area and fats miscible with water. This helps to hydrolyse fats by pancreatic lipase.
2. They activate pancreatic lipase and cholesterol esterase.
3. They combine with free fatty acids and monoglycerides to form minute particles called miscelles and help in their absorption in the intestines.
4. They stimulate intestinal peristalsis.
5. They stimulate bile production in the liver. They follow enterohepatic circulation and cause continuous secretion of bile by the liver.
6. Cholesterol remains soluble in gall-bladder bile by bile salts.
7. They aid in absorption of fatty acids, cholesterol, carotene, and the fat-soluble vitamins D and K by forming complexes more soluble in water (hydrotropic action).

Clinical significance

1. Bile salts in the blood are increased greatly in clinical obstructive jaundice.
2. After prolonged obstruction, the concentraton of bile salts in the blood may diminish due to diminshed synthesis of these substances as a result of progressive hepatocellular damage.
3. In the absence of bile salts, gall-stones are formed.

Bile pigments

1. Biliverdin and Bilirubin are the bile pigments formed from the breakdown of hemoglobin.
2. They are excreted in the bile.
3. They have no physiological role.

Enterohepatic circulation

A portion of the bile acids in the intestine undergoes changes by the activity of the intestinal bacteria. The deconjugation and 7α-hydroxylation produce the secondary bile acids, deoxycholic acid from cholic acid and lithocholic acid from chenodeoxycholic acid. The conjugated and unconjugated bile salts are absorbed almost in the ileum. As fecal bile acids are present as the products of bacterial metabolism, it is assumed that metabolism within the intestinal lumen with reabsorption by passive diffusion is a component of the enterohepatic circulation. This mechanism helps to return 90% of the bile acids secreted into the intestine to the liver each day. But lithocholic acid is not reabsorbed to any significant extent due to its insolubility.

500 mg of bile salts per day are not absorbed and is eliminated in the feces. The enterohepatic circulation of bile salts is so efficient that a small amount of bile acids is cycled through the intestine 6-10 times a day with the loss of a small amount in the feces.

Functions of bile

1. Bile salts help to lower the surface tension of water and thus emulsify fats in the intestine and dissolve fatty acids and water-insoluble soaps. The presence of bile in the intestine helps the digestion and absorption of fats and the absorption of fat-soluble vitamins A, D, E and K.
2. Bile salts are activators of lipase.
3. Choleic acid formed by deoxycholic acid assists the absorption of many important insoluble compounds.
4. Bile salts are reabsorbed from the intestine and pass back to the liver where they stimulate further secretion of the bile (cholagogue action).
5. Bile is an important source of alkali which helps to neutralize the acid chyme from the stomach.
6. Bile is an important channel for the excretion of some substances like bile pigments, many drugs, toxins, and various inorganic substances such as copper, zinc and mercury.
7. Fat digestion is impaired in the absence of bile. The fat then covers the other food particles and prevents enzymes from attacking them. These undigested food particles ultimately leads to putrefaction in the large intestine.
8. Bile salts keep cholesterol in solution in gall-bladder bile. In the absence of bile salts, cholesterol becomes precipitated. This results in the formation of gall-stones.

GALL-STONES

In the gall-bladder, cholesterol is solubilized by being held in micelles together with conjugated bile salts and phospholipids. The solubility depends on the ratio of cholesterol to bile salts plus phospholipids. The secretion of phospholipids into the bile depends on the availability of bile salts. If bile salts content is decreased, the phospholipid content is also diminished and, hence, the solubility of cholesterol is decreased causing crystallization. These crystals grow to form stones.

Gall-stones are formed due to defects in the enterohepatic circulation and with the diseases of the terminal ileum as well as in patients with cirrhosis. In these cases, there is reduction in the bile salt pool.

Infection of bile causes the deconjugation of bile acids with a decrease in their solubility. This also results in the production of a phospholipase which converts lecithin to lysolecithin. This decreases the stability of the micelles holding cholesterol in solution. Infection can give rise to calcium-bilirubinate stones which were frequent in Japan.

Chenodeoxycholic acid decreases the rate of secretion of cholesterol into the bile. Bile then becomes unsaturated with respect to cholesterol and thus the cholesterol stone can be redissolved. Unfortunately, bacterial action in the intestine converts chenodeoxycholic acids to lithocholic acid which is very hepatotoxic in Rhesus monkeys producing proliferation of bile ducts.

Classification

Gall-stones are mainly of three types :

1. Cholesterol stones

(a) These stones may be single or multiple.
(b) They may be white or yellowish.
(c) They may be mulberry-shaped.
(d) They are not radio-opaque.

2. Pigment stones

(a) These stones are formed by bile pigments, organic material and calcium.
(b) They are small multiple stones.
(c) They are hard, dark green or black.
(d) They are rarely radio-opaque.

3. Mixed Gall-stones

(a) These are composed of a mixture of cholesterol, bile pigments, protein and calcium.
(b) They are faceted dark brown stones with a hard shell and soft centre.
(c) They may be radio-opaque.
(d) They are the commonest forms of gall-stones.

RENAL CALCULI

This incidence is more in North India due to the hot climates prevalent for more than six months in the year and the large amounts of tea (rich in oxalates) consumed by them. In addition, the diet consumed by the people are rich in calcium, phosphates, and purines. The urine volume of these people is small and they excrete more uric acid, calcium and phosphates.

Conditions of Calculi formation

1. Low urine volume due to warm climates or low intake of water.
2. Consumption of diets rich in calcium, phosphates, oxalates and purines.
3. Conditions accelerating increased excretion of uric acid, oxalic acid, calcium, and phosphates in urine.
4. Urine infection and stagnation.

Classification

Renal Calculi can be classified into the following four groups :

1. Calculi due to calcium oxalate and phosphates : High concentration of calcium oxalates and phosphates in urine forms stones which are hard, white, and radio-opaque.
2. Urate stones : This is due to hyperuricemia and gout. These stones are usually small and yellowish brown in colour.
3. Cystine stones : These are rare but only happens in cystinuria.
4. Xanthine stones : These are also rare but only happens in xanthinuria.

INTESTINAL PUTREFACTION AND FERMENTATION

1. The residue of the food after absorption in the small intestine passes into the large intestine. Water is also absorbed there and the residue becomes solid. During this period, the activity of the bacteria takes place. The bacteria produce various gases, such as CO_2, methane, hydrogen, nitrogen and H_2S as well as acetic acid, lactic acid and butyric acid by fermentation and putrefaction. Lecthin is decomposed to choline and neurine.
2. Intestinal bacteria cause the decarboxylation of many amino acids into their amines such as

   Histidine → Histamine
   Tyrosine → Tyramine
   Arginine → Agmatine
   Lysine → Cadaverine

   Many of these amines are powerful vasopressor substances.

3. After reduction, biliverdin and bilirubin produce urobilinogen and stercobilin and cholesterol to coprosterol, sulphur to H_2S.
4. By a series of reactions, tryptophan forms indole and nethylindole (skatole) which are responsible for the odour of the feces.
5. Ethyl and methyl mercaptan and H_2S are formed from cysteine by a series of transformations.
6. Intestinal bacteria putrefy nitrogenous substances to form ammonia which is absorbed into the portal circulation and is removed from the blood by the liver under normal conditions. In the liver diseases, the concentration of ammonia in the peripheral blood rises to toxic levels. Neomycin by its anti-bacterial action reduces the quantity of ammonia transported from the large intestine to the blood.
7. The large intestine is the site for the reduction of biliverdin and bilirubin to urobilinogen and stercobilinogen, cholesterol to coprosterol, cystine to cysteine, sulphur to H_2S.

Importance of intestinal bacteria

1. In man, intestinal bacteria synthesize certain vitamins, particularly vitamin K and possibly certain members of the B-complex.
2. In herbivora, the intestinal bacteria are essential for the digestion of cellulose and make it available for absorption. They also synthesize essential amino acids and vitamins.

AUTOLYSIS

After death, the intracellular enzymes digest the tissues when kept under sterile conditions. This self-digestion of tissues are called autolysis.

The group of intracellular proteases present in all mammalian tissues is called **cathepsin**. Cathepsins occur in the lysosomes. There are four cathepsins which are analogous to pepsin, trypsin, aminopeptidase and carboxypeptidase.

Autolysis takes place in the living animal under pathological conditions, eg, wasting in starvation and fevers. Atrophy is a process of autolysis in the living animal.

DIGESTION AND ABSORPTION FROM THE GASTROINTESTINAL TRACT

Digestion and absorption of carbohydrate

The carbohydrate diet mainly consists of polysaccharides (starch and glycogen) and disaccharides (sucrose and milk lactose). It also contains indigestible cellulose, hemicelluloses and pentosans etc.

Digestion

Mouth : Salivary amylase (ptyalin) starts the digestion of cooked starch in the mouth. But very little digestion takes places in the mouth since the food remains in the mouth for a very short period of time.

Stomach : Since the food gets mixed with the gastric juice the action of amylase ceases due to high acidity. Some of the sucrose present in the food get hydrolysed by the action of HCl in the stomach.

Small intestine : The pancreatic amylase in the small intestine converts starch and glycogen into a mixture of maltose and isomaltose :

$$\text{Starch and glycogen} \xrightarrow{\text{Pancreatic amylase}} \text{Maltose} + \text{Isomaltose}$$

Then maltose and isomaltose along with sucrose, lactose present in the diet are digested by the different disaccharidases present in the intestinal mucosa into their corresponding monosaccharides as shown :

$$\text{Maltose} \xrightarrow{\text{Maltase}} \text{Glucose} + \text{Glucose}$$

$$\text{Isomaltose} \xrightarrow{\text{Isomaltase}} \text{Glucose} + \text{Glucose}$$

$$\text{Sucrose} \xrightarrow{\text{Sucrase}} \text{Glucose} + \text{Fructose}$$

$$\text{Lactose} \xrightarrow{\text{Lactase}} \text{Glucose} + \text{Galactose}$$

Cellulose is not digested in human G.I. tract due to the absence of cellulase.

Absorption

The comparative rates of absorption of monosaccharides taking glucose as 100 may be indicated as follows : galactose (110), glucose (100), fructose (43), mannose (39), xylose (15) and arabinose (9). Galactose and glucose are absorbed at a faster rate than fructose. Pentoses are slowly absorbed. This is due to the fact that glucose and galactose are actively transported while fructose, mannose and pentoses are absorbed by simple diffusion.

The monosaccharides are absorbed into the mucosal cells of small intestine and pass into circulation via portal vein. A very small amount may be absorbed by the lymph. The microvilli (brush border) lining the mucosa cells greatly help the absorption by increasing the surface area. The rate of absorption of monosaccharides is independent of blood sugar concentration.

Glucose and galactose for absorption follow the active transport against a concentration gradient; because they have the same chemical characteristics which are necessary for active transport mechanism. The chemical characteristics are :

1. The OH on carbon 2 should have the same configuration as in glucose.
2. A pyranose ring should be present.
3. A methyl or a substituted methyl group should be present on carbon 5.

Active transport of glucose

1. A mobile carrier which binds both glucose and Na^+ at separate sites and which transports them through the plasma membrane of the intestinal cell is required.
2. Both the glucose and Na^+ are released into the cytosol, allowing the carrier to return for more *"cargo"*.
3. The Na^+ is transported down its concentration gradient and at the same time causes the carrier to transport glucose against its concentration gradient.
4. The free energy required for this active transport is obtained from the hydrolysis of ATP linked to sodium pump which expels Na^+ from the cell.

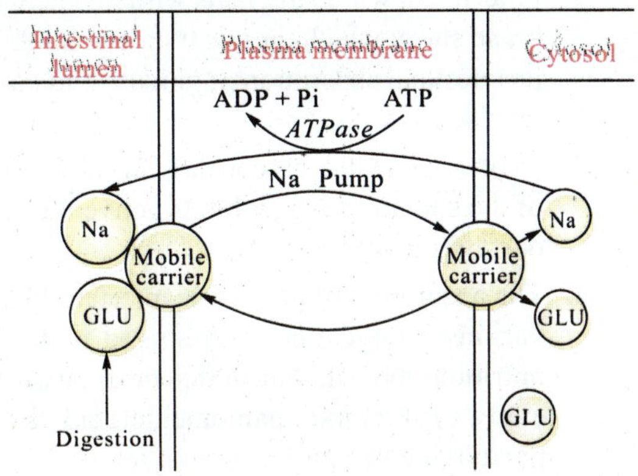

Fig. 16.5 : Active transport of glucose (GLU) across the intestinal mucosa

Since polysaccharides and oligosaccharides are quickly hydrolysed the absorptive mechanism for glucose-galactose and fructose are quickly saturated. But the hydrolysis of lactose proceeds at half the rate for sucrose. This slower rate of hydrolysis of lactose shows that the digestion of lactose does not lead to saturation of the transport mechanism for glucose and galactose.

Factors controlling active transport of glucose

1. Active transport of sugar is depressed by agents (cyanide, malonate, fluoroacetate) that inhibit respiration and by dinitrophenol which uncouples oxidative phosphorylation.

2. Phlorihizin which interacts with the membrane site at which the sugar enters, inhibits intestinal absorption of glucose and galactose.
3. Ouabain (cardiac glycoside), an inhibitor of the sodium pump, inhibits active transport of gucose.
4. The absorption of glucose is interfered by various intoxications, prolonged under-nutrition and vitamin deficiencies, especially of thiamine, pantothenic acid and pyridoxine.
5. Absorption may be decreased in the presence of abnormalities (structural or functional) of the mucous membrane, as in inflammation (enteritis), edema and celiac disease.
6. The absorption is retarded in hypothyroidism and accelerated in hyperthyroidism.
7. Diminished absorption in adrenal cortical insufficiency is dependent upon the decreased concentration of sodium in the body fluid.

Disacchariduria : The intestine normally is virtually impermeable to disaccharides. If it is absorbed, it is not metabolised. An increase in the excretion of intact disaccharides are found in some patients with disaccharidase deficiencies. The patients with intestinal damage also excrete more disaccharides. This condition is said to be disacchariduria.

DIGESTION AND ABSORPTION OF FATS

Digestion

Stomach : Lipase present in the stomach is unable to hydrolyze fats owing to the high acidity of the gastric contents. Therefore, the major part of the ingested fat is digested in the small intestine.

Small intestine : The ingested fat reaching the duodenum is mixed with the bile and pancreatic juice which contains lipase. The bile salts emulsify the fat before the action of lipase. The emulsification is also brought about by monoglycerides, phospholipid and lysolecithin.

The secreted inactive pancreatic lipase is activated by bile and Ca^{++}. The surface area of the emulsified fat becomes increased for which the rate of reaction of lipase is increased.

Pancreatic lipase hydrolyzes 1- and 3-positions of the triglycerides leaving a mixture of 2-monoglycerides, 1, 2- and 2, 3-diglycerides as well as the soaps of the free fatty acids.

The pancreatic juice also contains phospholipase and cholesterol-esterase which hydrolyze phospholipid and esterified cholesterol.

Intestinal juice also contains a lipase whose action is not of much importance as most of the fat is hydrolyzed by the pancreatic lipase.

Absorption

Several theories have been proposed for the mechanism of absorption of fats after digestion. The important theories are : A. Lipolytic hypothesis. B. Partition theory. C. More recent theory.

A. Lipolytic hypothesis

1. According to this theory, fat is completely hydrolyzed to fatty acids and glycerol which are absorbed.
2. The fatty acids combining with bile salts form a miscible complex which is absorbed into the intestinal mucosa.
3. The fatty acids are then separated from bile acids and converted into triglycerides by combining with glycerol.

$$\text{Triglyceride} \xrightarrow{\text{Pancreatic lipase}} \text{1,2-diglyceride} \xrightarrow{\text{Pancreatic lipase}} \text{2-monoglyceride}$$

Fig. 16.6 : The action of Pancreatic lipase

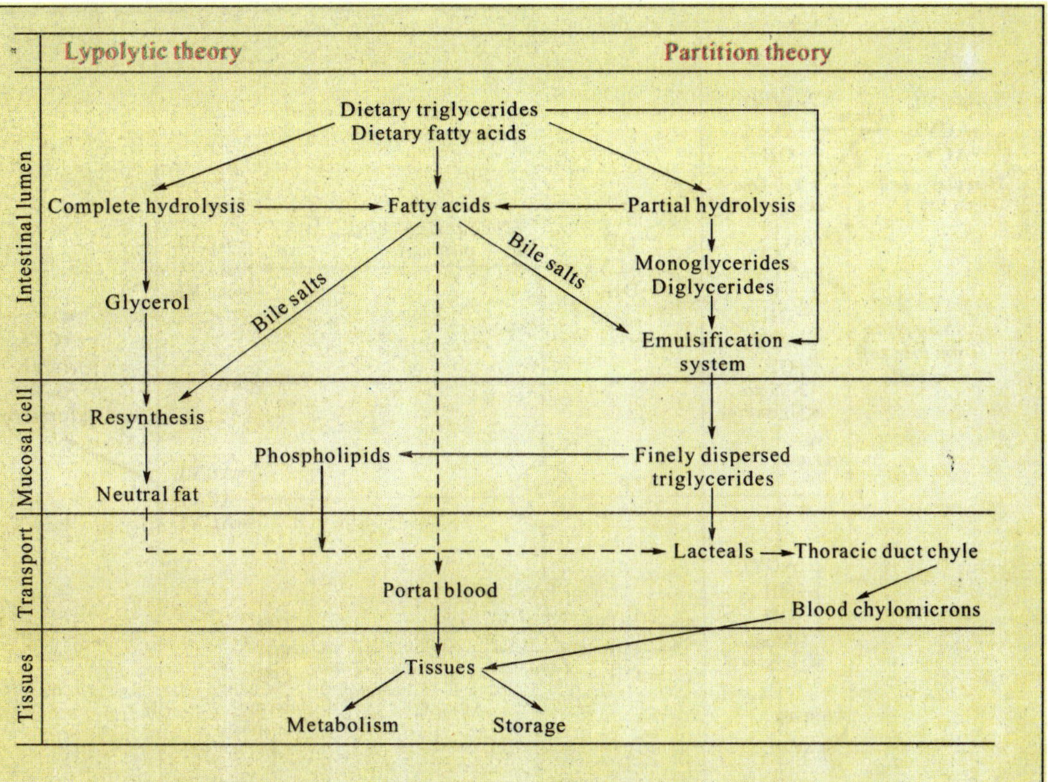

Fig. 16.7 : Old concept of fat absorption containing lipolytic theory and partition theory

4. The triglycerides are passed to the lacteals. They then enter the lymphatics and reach the systemic circulation via thoracic duct.

B. Partition theory

1. According to this theory, 30 per cent of the triglycerides are hydrolyzed to fatty acids and glycerol while 70 per cent remain unhydrolyzed.
2. The unhydrolyzed triglycerides are emulsified by monoglycerides and diglycerides in combination with bile salts to form minute particles known as "micelles" of size about 0.1 to 0.5μ.
3. The resulting mixture is absorbed into the intestines, passed on to the lacteals and then to the lymphatics. The mixture then reaches the systemic circulation via thoracic duct.
4. The free fatty acids are absorbed as bile salt-fatty acid complex into the intestinal mucosa. The fatty acids are absorbed into the portal blood to reach the liver.

C. Recent theory

1. The removal of the ester group of 2-monoacylglycerol requires isomerization to a primary ester linkage. This is a slow process. As a result, monoacylglycerols are the major end products of fat digestion and less than one-fourth of the ingested fat is completely broken down to glycerol and fatty acids.
2. Within the intestinal wall, 2-monoacylglycerols are converted to triacylglycerols and 1-monoacylglycerols are further hydrolyzed to form free glycerol and fatty acids.
3. The fatty acids are then activated by **thiokinase** in presence of ATP and

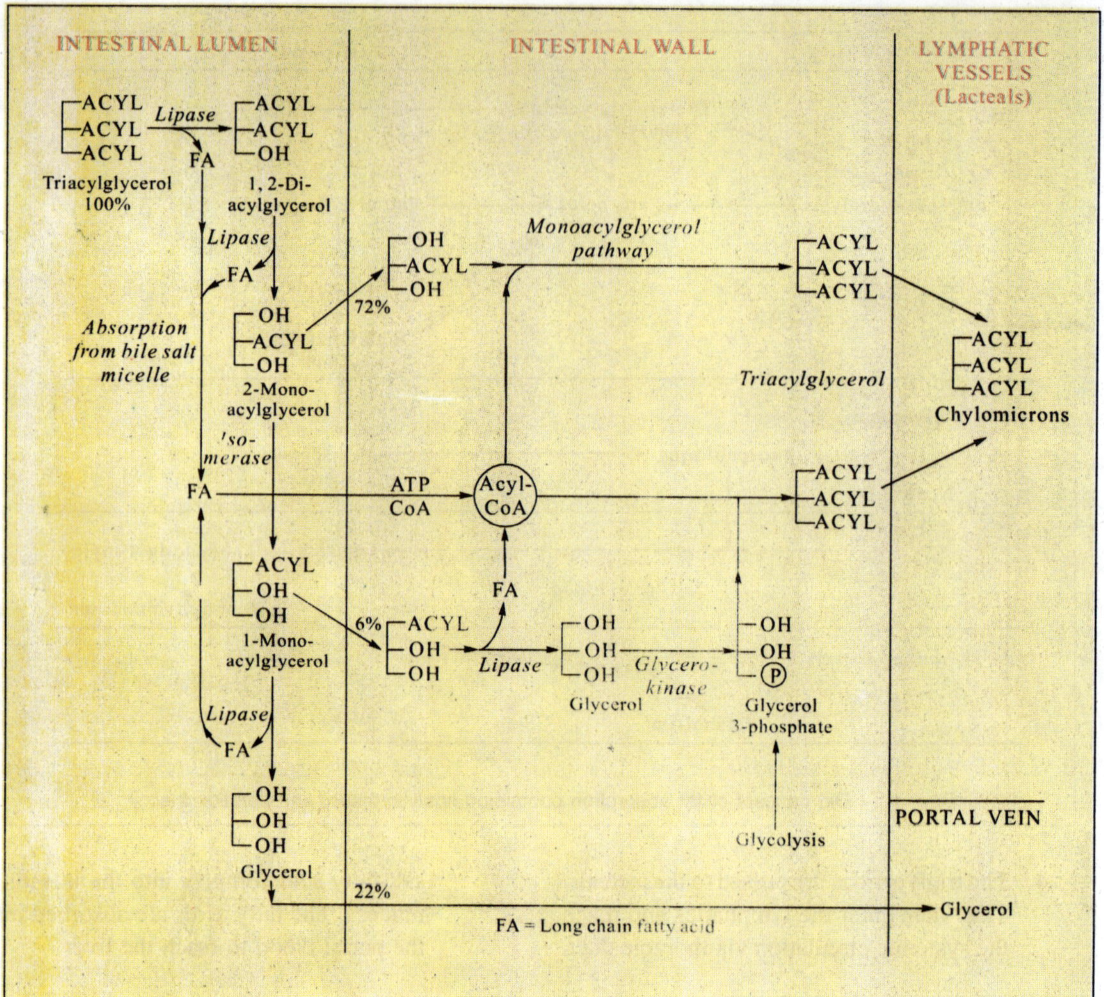

Fig. 16.8 : Digestion and absorption of fat

coenzyme A for the resynthesis of triacylglycerols.

4. The free glycerol in the intestinal lumen is about 22 per cent of total amount of triacylglycerol originally present. This passes directly to the portal vein.
5. The glycerol within the intestinal wall is activated by glycerokinase in presence of ATP to form glycerol-3-phosphate for the synthesis of triacylglycerol followed by the combination with acyl-CoA present in the intestinal wall.
6. All long chain fatty acids present in the intestinal wall are reincorporated into triacylglycerols which are transported to the lymphatic vessels of the abdominal region (the so-called lacteals) for distribution to the rest of the body.
7. The great majority of absorbed fat appears in the form of chylomicrons which appear first at the lymphatic vessels of the abdominal region and later in the systemic blood. The chylomicrons contain triacylglycerol, free and esterified cholesterol, phospholipid and 0.5 per cent protein.

All of the factors relating to digestion and absorption of fat are mentioned in Fig. 16.8.

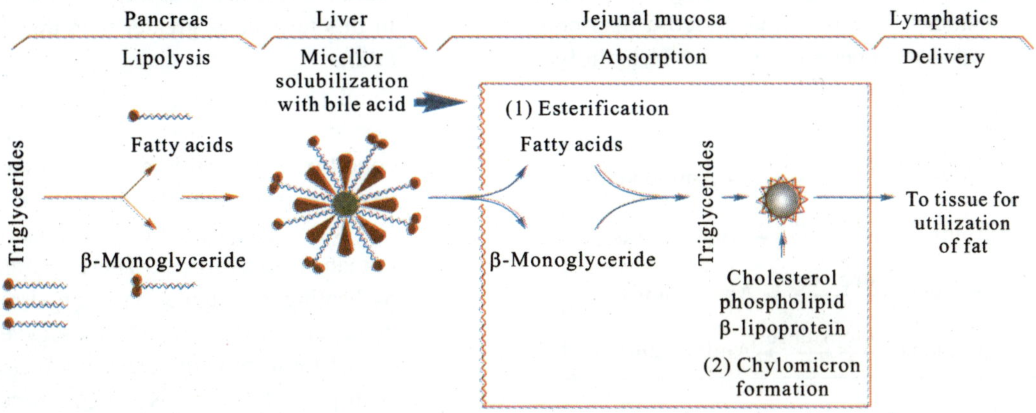

Schemic representation of lipid digestion and absorption. Dietary lipid in the form of long chain triglycerides (LCT). The overall process can be divided into (1) a digestive phase that includes both lipolysis and micelle formation requiring pancreatic lipase and conjugated bile acids, respectively, in duodenum; (2) an absorptive phase for mucosal uptake and reesterification; (3) postabsorptive phase that includes chylomicron formation and exit from the intestinal epithelial cell via lymphatics

Absorption of phospholipids : Phospholipids are split by phospholipases and their acyl chains are incorporated into chylomicrons, choline, the hydrophilic component, may be transported directly to the liver via the hepatic portal vein.

Absorption of cholestrol : It is absorbed into the lymphatics and recovered mainly as cholesteryl esters.

Chyluria : In this abnormality, the patient excretes milky urine because of the presence of an abnormal connection between the urinary tract and the lymphatic drainage system of the intestine, a so-called "chylous fistula".

DIGESTION AND ABSORPTION OF PROTEIN

Digestion : The proteolytic enzymes secreted in gastric juice, pancreatic juice and also present in the intestinal mucosa cause the hydrolysis of protein in the gastrointestinal tract.

Stomach : Pepsin, the endopeptidase, is present in gastric juice and hydrolyzes the peptide bonds in the interior of the protein molecule. Pepsin hydrolyzes the dietary protein into a mixture of polypeptides :

$$\text{Dietary protein} \xrightarrow{\text{Pepsin}} \text{Polypeptides}$$

Renin has a strong clotting action on milk. This is very important in the digestion of milk proteins in infants. The pH of the gastric juice becomes low in achlorhydria, achylia gastrica (both pepsin and HCl absent) and in pernicious anemia. Then dietary protein will not be digested in the stomach.

Small intestine : The polypeptides formed in the stomach are digested in the intestine by trypsin, chymotrypsin and carboxypeptidases secreted in pancreatic juice and aminopeptidases present in the intestinal mucosa. Trypsin hydrolyzes peptide linkages containing arginine or lysine and chymotrypsin hydrolyzes peptide linkages containing tyrosine or phenyialanine :

$$\text{Polypeptides} \xrightarrow{\text{Trypsin \& Chymotrypsin}} \text{Peptides + Amino acids}$$

Fig. 16.9

Carboxypeptidase A hydrolyzes the end group of peptides containing aromatic or aliphatic amino acid and releases free amino acids. Carboxypeptidase B hydrolyzes peptides containing arginine and lysine residues. The intestinal mucosa also contains tripeptidase, dipeptidase etc., which hydrolyze tri- and dipeptides :

$$\text{Peptides} \xrightarrow{\text{Carboxypeptidase}} \text{Amino acids}$$

$$\text{Peptides} \xrightarrow{\text{Aminopeptidase}} \text{Amino acids}$$

$$\text{Dipeptides} \xrightarrow{\text{Dipeptidase}} \text{Amino acids}$$

$$\text{Tripeptides} \xrightarrow{\text{Tripeptidase}} \text{Amino acids}$$

The final products of digestion of proteins are amino acids which are absorbed.

Absorption

1. Three different active processes are involved in the transport of amino acids. One process involves cystine and the basic amino acids, another the amino acids proline and hydroxyproline and the third the neutral (L–) amino acids.
2. D-amino acids are absorbed by simple diffusion. But the neutral (L–) amino acids require a carrier system in the absorption. Na^+ is also required. This is similar to that of active transport of glucose. Vitamin B_6 (pyridoxal phosphate) is also involved in the process. The amino acid associates with the carrier and Na^+ in the microvilli and the complex travels to the inner side of the membrane where it dissociates, releasing the amino acid Na^+ into the cytosol. The carrier returns back and functions repeatedly. Na^+ is then actively transported out of the cell.
3. If one amino acid is fed in excess, it retards the absorption of another. This is similar to those made with respect to reabsorption of amino acids by the renal tubules.
4. Sometimes the whole protein is absorbed into the blood. A protein is antigenic and accounts for food allergies. In the young mammal, the permeability of the mucosa, in this respect, is greater than that in the adult.
5. Food proteins are generally readily digested (90 to 97 per cent) under normal conditions, very little escapes in the faeces. The insoluble fibrous protein, keratin, is not hydrolyzed by enzymes of the human digestive tract. These are altered by heating to coagulation and hydrolyzed by superheated steam. The biological values of these proteins are not affected by such procedures. Cooked egg albumin is digested more readily than raw. The nutritional value of cereal proteins is lowered by overheating or toasting.

DIGESTION AND ABSORPTION OF NUCLEOPROTEINS AND NUCLEIC ACIDS

Digestion

Stomach : The nucleoproteins of many foods are cleaved by the acidity of the gastric juice into proteins (histones, protamines etc.) and nucleic acids :

$$\text{Nucleoproteins} \xrightarrow{\text{HCl}} \text{Nucleic acids + Proteins}$$

The proteins are hydrolyzed by proteases present in the digestive juices to amino acids in the same way as the dietary proteins.

Small intestine : Nucleic acids are hydrolyzed by three groups of enzymes :

1. Nucleases.
2. Nucleotidases.
3. Nucleosidases.

Nucleases are of two types : 1. Ribonucleases. 2. Deoxyribonucleases. Ribonucleases hydrolyze RNA and deoxyribonucleases hydrolyze DNA. The end products are nucleotides.

Nucleotidases hydrolyze nucleotides into nucleosides and phosphoric acid.

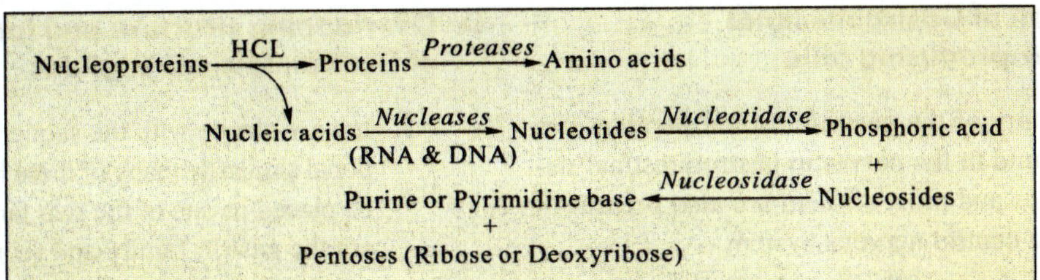

Fig. 16.10 : Digestion of nucleoproteins

Nucleosidases hydrolyze nucleosides into purine or pyrimidine base and the pentose sugars (ribose or deoxyribose).

All of the above mentioned reactions are given in Fig. 16.10.

Absorption

1. The purine or pyrimidine base and the pentoses are absorbed in the small intestine.
2. Certain amounts of nucleosides and pentose phosphates may also be absorbed. Most of the tissues contain nucleosidases which hydrolyze nucleosides absorbed from the intestines.

GASTROINTESTINAL HORMONE

Introduction

1. Many hormones are liberated by the gastrointestinal tract—more than any other single organ.
2. The gastrointestinal hormones assist in digestive processes of the foodstuffs moving the digested products into the extracellular space through the intestinal mucosal cells, moving those products to distant cells through the circulation, and expelling waste products.

Biomedical Importance

1. Excessive production of several of the gastrointestinal hormones cause disease syndrome. The physicians cannot diagnose accurately unless they are aware of these syndromes.
2. These hormones have also close link to neuropeptides.

Acronyms Used in this Chapter	
BLI	Bombesinlike immunoreactivity.
CCK	Cholecystokinin.
GIP	Gastric inhibitory polypeptide.
GRP	Gastrin-releasing peptide.
PP	Pancreatic polypeptide.
VIP	Vasoactive intestinal polypeptide.

Features of Gastrointestinal hormones

A. Diversity of actions

1. The classic definition of hormone is satisfied by many of the gastrointestinal peptides. Examples include gastrin, secretin, GIP, CCK, motilin, PP and enteroglucagon.
2. Other gastrointestinal peptides have *paracrine actions* or to act in a neurocrine fashion. Examples include VIP, somatostatin, substance P, BLI, Enkephalin, and neurotensin. Although these substances are found in high concentration in neurons, they are either not found in the circulation under normal conditions or have such short plasma half-lives that they would not be effective.

B. Location of Gastrointestinal Peptide-producing cells

1. Many of the gastrointestinal peptides are found in the nerves in gastrointestinal tissues and most of them are also present in the central nervous system.
2. Synthesis of the peptides by central nervous system tissue is difficult to prove, but new techniques of molecular biology should establish whether genes coding for these substances are active.
3. The function of these peptides in the central and peripheral nervous system is under investigation.
4. The distribution of the gastrointestinal hormones is mentioned in Table 16.2 (next page).

C. Precursors and Multiple forms

1. Secretin only exists in a single form.
2. The presence of multiple forms of gastrointestinal peptides in gastrointestinal tissues and in the circulation impedes the definition of the number and nature of these molecules.

D. Overlapping structure and function of Gastrointestinal peptides

1. In accordance with the sequence and functional similarity many of these hormones can be placed in one of the two families. These are the gastrin family and secretin family.
2. The gastrin family consists of gastrin and CCK.
3. The secretin family includes secretin, glucagon, GIP, VIP, and glicentin (which has glucagonlike immunoreactivity but is a distinct peptide).
4. The neurocrine peptides neurotensin, bombesinlike peptides, susbtance P, and somatostatin bear no structural similarity to any other gastrointestinal peptide. They have very short plasma half-lives and may play no physiologic role in plasma.

E. Mechanism of Action

1. Pancreatic acinar cells have got six different classes of receptors. These are for (a) muscarinic cholinergic agents. (b) The gastrin CCK family. (c) Bombesin and related peptides. (d) The physalaemin-sub-

Table 16.1. Gastrointestinal hormones

+ E = Endocrine, N = Neurocrine, P = Paracrine, ( ) = Suggested but not proved, + = yes, – = No.

Hormones	Mechanism of action +			Major Action
	E	N	P	
Gastrin	+	(+)	–	Gastric acid and pepsin secretion
CCK	(+)	(+)	–	Pancreatic amylase secretion.
Secretin	+	–	–	Pancreatic bicarbonate secretion.
GIP	+	–	–	Enhances glucose-mediated insulin release, inhibits gastric acid secretion.
VIP	–	+	(–)	Smooth muscle relaxation, stimulates pancreatic bicarbonate secretion.
Motilin	(+)	–	–	Initiates interdigestive intestinal motility.
Somatostatin	–	+	(+)	Numerous inhibitory effects.
PP	(+)	–	(+)	Inhibits pancreatic bicarbonate and protein secretion.
Enkephalins	–	+	(+)	Opiatelike actions.
Substance P	–	+	(+)	Physiologic action uncertain.
BLI	–	+	(+)	Stimulates release of gastrin and CCK.
Neurotensin	–	+	(+)	Physiologic actions unknown.
Enteroglucagon	(+)	(+)	(+)	Physiologic actions unknown.

Table 16.2. Distribution of gastrointestinal hormones
EC = Enterochromaffin cell. ... = yet to be named..

Hormones	Endocrine cell +	Localization	Localized in gut Nerves
Gastrin	G	Gastric antrum, duodenum	(?)
CCK	I	Duodenum, Jejunum	Yes
Secretin	S	Duodenum, Jejunum	No
GIP	K	Small bowel	No
VIP	D_1	Pancreas	Yes
Motilin	EC_2	Small bowel	No
Substance P	EC_1	Entire gastrointestinal tract	Yes
Neurotensin	N	Ileum	(?)
Somatostatin	D	Stomach, duodenum, pancreas	Yes
Enkephalins	....	Stomach, duodenum, gall-bladder	Yes
BLI	P	Stomach, duodenum	Yes
PP	D_2F	Pancreas	No
Enteroglucagon	A	Pancreas	No
	L	Small Intestine	

stance P family. (e) Secretin and VIP. (f) Cholera toxin.

2. The peptide-receptor complexes activate two distinct intracellular mechanisms. One involves the mobilization of intracellular calcium stores, and the other involves the activation of adenylate cyclase and the generation of cAMP.
3. The mechanisms do not cross-over, i.e., gastrin does not change cAMP levels, nor does secretin affect the intracellular Ca^{++} level. These two systems converge at some point.
4. The peptides causing Ca^{++} mobilization in the pancreatic acinar cells also affect the metabolism of phosphatidylinositol and enhance its conversion to diacylglycerol and various inositol phosphates. These effects are associated with depolarization of the acinar cell which may be involved in amylase secretion.
5. The mechanism of the actions of cAMP, Ca^{++}, and phospholipids on amylase secretion is similar in many respects.

SECRETIN FAMILY

A. Secretin

1. Secretin is constituted of 27 amino acids.
2. It is synthesized and released by the S cells in the duodenum and proximal jejunum in response to duodenal acidification.
3. It contains 4 arginine residues and one histidine and is basic.
4. Its structure is identical to that of glucagon in 14 of 27 amino acids.
5. It is structurally quite similar to that of GIP and VIP.
6. Its entire structure is required for stimulation of bicarbonate and water secretion from the pancreas.
7. It causes lipolysis.

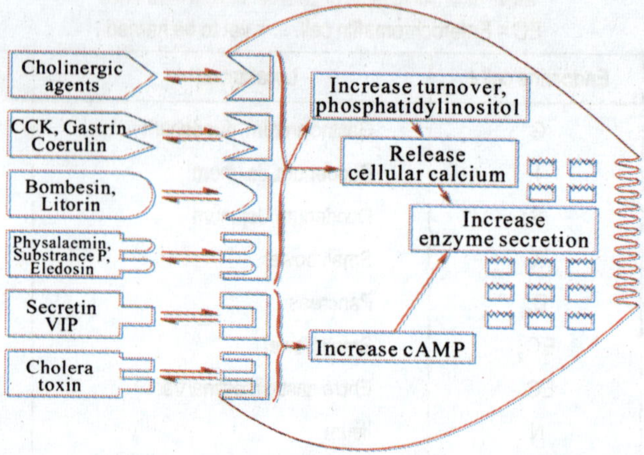

Fig. 16.11 : Mechanism of action of secretagogues on enzyme secretion from pancreatic acinar cells

8. It is administered to assess the pancreatic exocrine function.

B. Gastric Inhibitory Polypeptide (GIP)

1. GIP consists of 43 amino acids and is released from the duodenal and jejunal mucosa in response to glucose.
2. It stimulates insulin release.
3. It is the physiologic B cell stimulating hormone of G.I.T.
4. Hyperglycemia appears to be required for its action.

C. Vasoactive Intestinal Polypeptide (VIP)

1. VIP is basic 28-amino acid peptide whose physiologic function has not been defined.
2. It exists in the nerves of the submucosal plexus, the myenteric plexus, and blood vessels.
3. It causes gut motility, sphincter relaxation, and blood flow.
4. Highly concentrated VIP stimulates secretion by the pancreas and small intestine.
5. Tumors that produce this peptide, VIP-oma, cause a syndrome of watery diarrhea, hypokalemia and achlorhydria.

GASTRIN–CHOLECYSTOKININ FAMILY

A. Gastrin

1. Gastrin is produced by G cells of the antral gastric mucosa and to a small extent in the duodenal mucosa.
2. It shows more heterogeneity in size than any other gastrointestinal hormone.
3. Each of its form has sulphated and nonsulphated form.
4. The carboxy-terminal 14 amino acids of G_{34}, G_{17} and G_{14} are identical. G_{34} is more abundant in the circulation than G_{17} because its plasma half-life (15 minutes) is 5 to 7 times that of G_{17}.
5. G_{17} is the main stimulus for gastric acid secretion which is under negative feedback control, since acidification of the antral region of the stomach decreases gastrin release.
6. Gastrin also stimulates the secretion of pepsin and intrinsic factor from gastric mucosa.
7. The carboxyl end of gastrin is responsible for biologic activity.

8. The carboxyl-terminal pentapeptide has the full range of physiologic action of C_{17}.
9. Pentagastrin is a synthetic pentapeptide and is used clinically.
10. Increased gastrin release is caused by vagal stimulation (eg, caused by insulin-induced hypoglycemia) by acetyl-choline and by food intake particularly protein or amino acids, of the amino acids glycine is the most potent.
11. It stimulates secretin release and can delay delivery of gastric contents into the duodenum by reducing the rate of gastric emptying.
12. Zollinger-Ellison syndrome is characterized by gastrin-secreting tumors, gastrinomas, which cause excessive gastric acid production and intractable peptic ulcer disease.
13. Pentagastrin causes calcitionin release and is used as a provocative test in the diagnosis of medullary thyroid cancer.

B. Cholecystokinin (CCK)

1. Cholecystokinin is present in I cells in the mucosa of duodenum and proximal jejunum.
2. It is constituted of 39 amino acids.
3. It exists in at least 5 molecular forms of which CCK 8 is the most abundant and potent form in the circulation.
4. A 7-amino acid C-terminal fragment with a sulphated tyrosine is required for maximal CCK activity (stimulation of pancreatic enzyme secretion and contraction of gall-bladder).
5. It is released in response to peptides, amino acids, long-chain fatty acids, calcium and acid.
6. It effectively stimulates the release of both insulin and glucagon from pancreas.
7. Its physiologic actions are to cause gall-bladder contraction and stimulate pancreatic enzyme secretion.
8. It performs many of the actions of gastrin and secretin on water, bicarbonate and acid changes.
9. It represents as one of the "Gut factor" influencing insulin during oral administration of glucose.

OTHER GASTROINTESTINAL PEPTIDES

A. Substance P

1. Substance P is found in both the gut and the brain.
2. This peptide contains 11 amino acids.
3. The 5 carboxy-terminal amino acids of this peptide are required for its action which causes stimulation of smooth muscle contraction in the intestine.

B. Bombesin

1. Bombesin is found in frog skin, but a similar peptide, often called gastrin-releasing peptide (GRP), has been isolated from endocrine cells in the gut, gut neurons, and brain.
2. The amino acids at positions 5-14 of bombesin are identical to those at positions 18-27 of gastrin-releasing peptide except at one residue.
3. It stimulates gastric and pancreatic scretion and increases mobility of the gall-bladder and intestine.
4. It has a growth-promoting autocrine action in small cell carcinoma of the lung.

C. Motilin

1. This peptide contains 22 amino acids and is found in the intestinal mucosa.
2. It stimulates acid and pepsin secretion by gastric mucosa.
3. It is a stimulator of intestinal smooth muscle contraction.

D. Somatostatin

1. This peptide is formed in gastric D cells.
2. It inhibits (by paracrine action) the release of gastrin, secretin, CCK, motilin, and GIP.

E. Glucagon

1. This is made in the gastic mucosa A cells and is thought to contribute to the metabolic action of pancreatic glucagon.
2. Other peptides with glucagon like immunoreactivity (GLI) have been isolated from the L cells of the ileum and colon.
3. The major component of GLI is a large 100 amino acid peptide called glicentin which contains the exact sequence of pancreatic glucagon.
4. Glicentin may mimic the actions of glucagon.

F. Neurotensin (a 13- amino acid peptide)

Met-and Leu-enkephalins, and serotonin are found in intestinal cells and may be active in these tissues.

Some 40 peptides have been found in neural tissues. It is highly expected that more gastrointestinal peptide will be discovered.

Enterocrinin

1. It is released from the duodenal mucosa.
2. This polypeptide stimulates the secretion of intestinal juice by the small intestine.

Enterogastrone

1. Fat in the diet influences the secretion of this polypeptide by the duodenal mucosa.
2. It inhibits gastric secretion and gastric motility.
3. It inhibits the secretion of HCl and pepsin.

Hepatocrinin

1. This polypeptide is liberated from the duodenum.
2. It stimulates liver to secrete more bile.

CLINICAL ORIENTATION

- Ulceration is caused by excessive gastric HCl and achlorhydria develops in diminished secretion of HCl.
- The defects of bile secretion lead to gall-stones or defective lipid digestion.
- In cystic fibrosis, excrine pancreatic insufficiency causes steatorrhea.
- Malabsorption of nutrients due to a variety of defects leads to nutritional deficiency, eg,
  (a) Malabsorption of Vit B_{12} and Folic acid causes anemia.
  (b) Defects in calcium and magnesium lead to tetany.
  (c) Vitamin D malabsorption leads to rickets and Osteomalacia.
  (d) Lactase deficiency produces milk intolerance.
  (f) Neutral amino acid malabsorption leads to Hartnup diseases.
  (g) Disaccharidase deficiency causes disacchariduria.

EXERCISE

1. How are foodstuffs digested in the Gastrointestinal Tract? (P.U. 69A; R.U. 74A)
2. Discuss the mechanism of absorption of the foodstuffs after digestion. (P.U. 62A; 64S)
3. Discuss the gastrointestinal hormones in the body. (P.U. 74A, 75S)
4. Write short notes on :
   (a) HCl formation. (P.U. 68A)
   (b) Gastro-intestineal Hormones. (R.U. 70A)
   (c) CCKPZ. (B.U. 74A)
   (d) Intrinsic factor. (Bh. U. 72S)
   (e) Pancreatic Juice. (R.U. 73A)
   (f) Secretin family. (Bh. U. 79A)
   (g) Gastrin. (R.U. 81A)

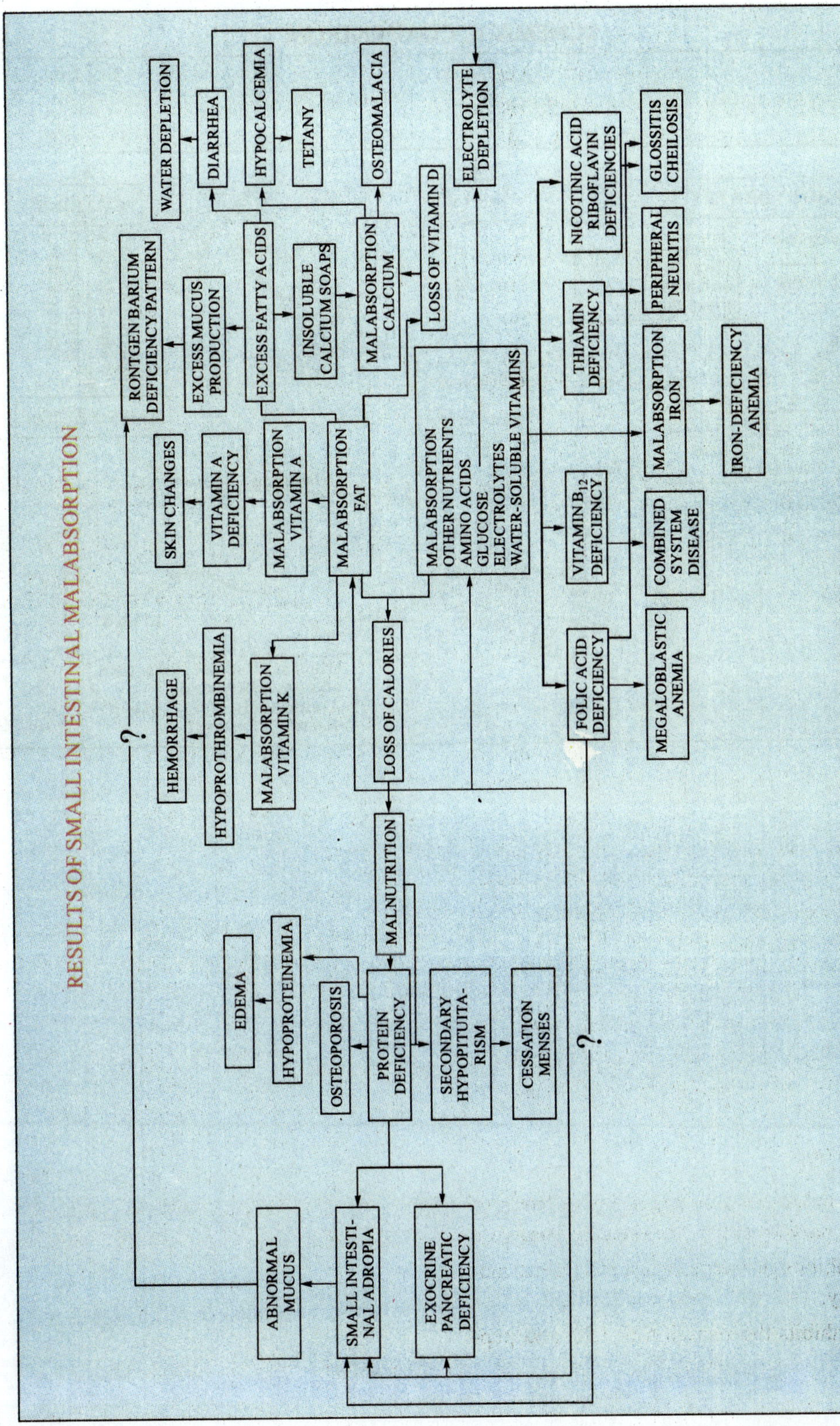

Fig. 16.12 : Gastrointestinal Malabsorption Syndromes

SCHEMATIC COMPENDIUM

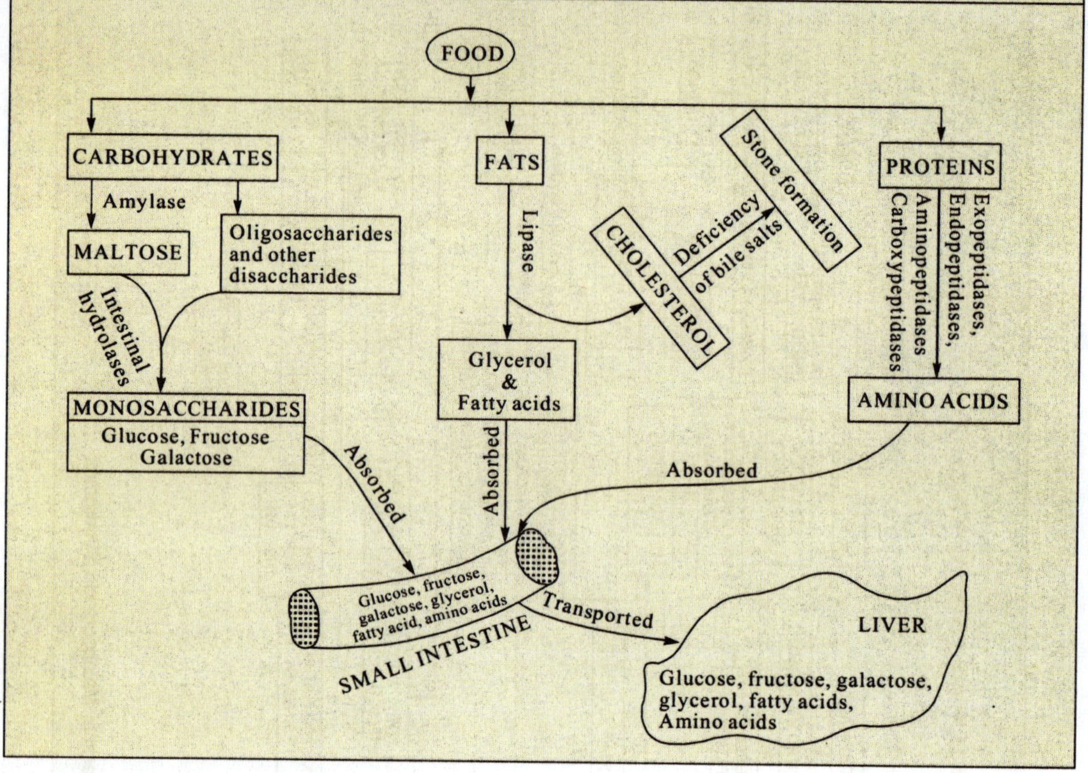

CHAPTER 17

METABOLISM OF CARBOHYDRATE

Fate of absorbed carbohydrate

Carbohydrate, after absorption into the portal blood, passes to the systemic circulation through the liver. Certain hepatic mechanisms are involved in the withdrawal of carbohydrates from the blood. The hepatic mechanisms are :

1. Uptake of hexoses like fructose and galactose for conversion to glucose by the liver cells.
2. Conversion of glucose to glycogen for storage in the liver.
3. Utilization of glucose for energy production by oxidation.
4. Utilization of glucose for the synthesis of other compounds such as fatty acids and certain amino acids.

Others which oppose these above mechanisms lead to the release of glucose by the liver to the blood by the following means :

1. Formation of blood sugar from hexoses other than glucose by the liver.
2. Conversion of liver glycogen to blood glucose.
3. Synthesis of blood glucose by the liver and kidney from non-carbohydrate sources.

The amount of glucose reaching the systemic circulation is the resultant of the operations of these two groups of opposing processes. After that the blood glucose is utilized by the extrahepatic tissues.

Utilization of glucose

A. Storage

1. In the absence of urgent physiological demands, excess glucose may be deposited as glycogen in the liver, muscles and other tissues (Glycogenesis).
2. Since the amount of glucose stored in the liver is limited, excess of this upper limit is converted to fatty acids and stored as triglycerides in the fat depots.

B. Oxidation

1. Glucose is completely oxidized in all tissues to CO_2 and H_2O for physiological demands for energy.
2. Sometimes, under special circumstances in muscle, there is only the partial degradation of glucose (Glycolysis) forming lactic acid notably by liver.

C. Conversion to fat

Glucose is converted to fatty acids when glycogen storage is exceeded. The conversion of glucose to fatty acids is irreversible but the transformation of glucose to glycerol is reversible.

D. Conversion to other carbohydrates and other substances

1. Mannose, glucosamine and galactosamine form parts of the mucopolysaccharides and glycoproteins.
2. Glucuronic acid is involved in the mucopolysaccharides and in 'detoxication' reactions.
3. Galactose is required as a component of the glycolipids and lactose.
4. Ribose and deoxyribose are required for the synthesis of the nucleic acids.

E. Conversion to amino acids

Certain amino acids are also formed from glucose. These amino acids are said to be "glucogenic".

Metabolism : Metabolism consists of anabolism (synthesis) and catabolism (breakdown).

INTERMEDIARY METABOLISM OF CARBOHYDRATE

In the mammalian organism, the metabolism of carbohydrate is subdivided as :

1. *Glycogenesis* : The synthesis of glycogen from glucose in the liver and muscle.
2. *Glycogenolysis* : The breakdown of glycogen to glucose in the liver and pyruvate and lactate in the muscle.
3. *Glycolysis* : The breakdown of glucose and glycogen to pyruvate and lactate under anaerobic conditions.
4. *Citric acid cycle* : The oxidation of pyruvate to CO_2 and H_2O under aerobic conditions.
5. *The hexose monophosphate shunt* (direct oxidative pathway; pentose phosphate cycle; phosphogluconate oxidative pathway) : The oxidation of glucose by an alternative pathway to glycolysis and citric acid cycle.
6. *Gluconeogenesis* : Formation of glucose and glycogen from noncarbohydrate sources. Citric acid cycle and glycolysis are involved in this process.

GLYCOGENESIS

1. Glucose is first phosphorylated to glucose-6-phosphate by the enzyme hexokinase and glucokinase in presence of the coenzyme ATP and the activator Mg^{++}. ATP is converted to ADP. Hexokinase has high affinity for glucose. Glucokinase is to remove glucose from the blood following a meal.
2. Glucose-6-phosphate is then converted to glucose-1-phosphate. The reaction is catalyzed by the enzyme phosphoglucomutase with Mg^{++} and the reaction is re-

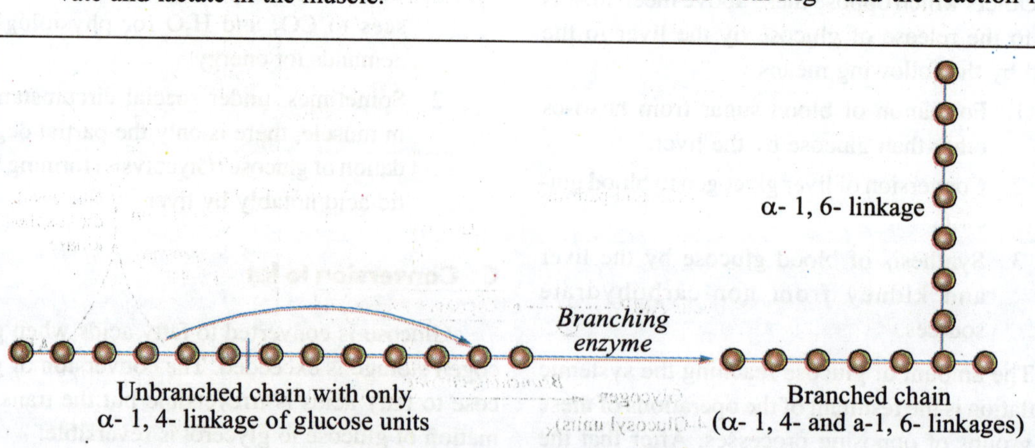

Fig. 17.1 : Action of branching enzyme

versible. Glucose-1, 6-diphosphate is formed as an intermediate due to the phosphorylation of the enzyme.

3. Glucose-1-phosphate reacts with uridine triphosphate (UTP) to form the active nucleotide *(uridine diphosphate glucose (UDPG)*. The reaction is catalyzed by the enzyme *UDPG pyrophosphorylase* with the release of inorganic pyrophosphate.
4. The C_1 of the activated glucose of UDPG forms a glycosidic bond with the C_4 of the terminal glucose residue of glycogen liberating UDP by the enzyme glycogen synthetase (glucosyl transferase). The glycogen primer is formed on a protein primer known as glycogenin which is a protein.
5. By the successive 1, 4 linkages by glucose units to the preexisting glycogen chain, the branches of the glycogen 'tree' become elongated.

When the chain has been lengthened to between 6 and 11 glucose residues, a second enzyme, the branching enzyme (amylo 1, 4 → 1, 6 transglucosidase), acts on the glycogen. This enzyme transfers a part of the -1, 4-chain (minimum length of 6 glucose residues) to a neighbouring chain to form α-1, 6–linkage which establishes a branch point in the molecule.

Storage of Carbohydrate in postabsorptive normal adult (70 kg.)		
Liver glycogen	4.0 % =	72 grams
Muscle glycogen	40.7% =	245 gms.
Extra cellular glucose	0.1 % =	10 gms.
Total		327 gms
Liver weight	1800 gms.	
Muscle mass	35 kg.	
Total Volume	10 L	

Regulation of glycogenesis

Glycogen synthase (glycogen synthetase) is the key enzyme which regulates the process of glycogenesis. This enzyme exists in "active" as well as "inactive" forms which are interconvertible. Glycogen also exerts an inhibition on its own formation and insulin also stimulates glycogen synthesis in muscle by promoting dephosphorylation and activation of inactive glycogen synthase. The in-

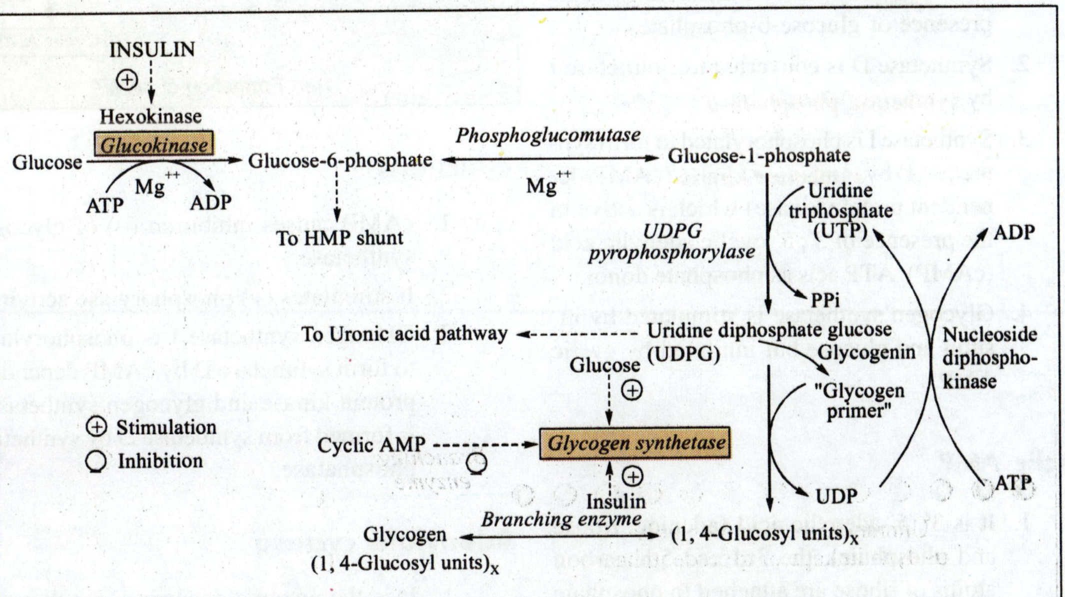

Fig. 17.2 : Pathway of glycogenesis in the liver and muscle

hibition of glycogenesis enhances net glycogenolysis. Hexokinase stimulates the oxidation of glucose in the muscle whereas glucokinase stimulates the oxidation of glucose in the liver.

Glycogen

1. It is a branched polysaccharide consisting of α-D-glucose units. These glucose units are connected to one another by glucosidic linkages between the first and fourth carbon atoms except at branch points.
2. The molecular weight varies from 1 million to 5 million or more.
3. In case some of the glucose chains terminate in the interior, a larger molecule is possible.

Glycogen synthetase

1. In muscle (and possibly liver), it is present in two interconvertible forms–**synthetase D** (dependent) whose activity is dependent on the presence of glucose-6-phosphate and **synthetase I** (independent) whose K_m value for UDPG decreases in the presence of glucose-6-phosphate.
2. Synthetase D is converted to synthetase I by **synthetase phosphatase**.
3. Synthetase I is phosphorylated to form synthetase D by **synthetase kinase** (cAMP-dependent protein kinase) which is active in the presence of 3′, 5′-cyclic adenylic acid (cAMP). ATP acts as phosphate donor.
4. Glycogen synthetase is stimulated by insulin and glucose but inhibited by cyclic AMP.

Cyclic AMP

1. It is 3′, 5′-adenylic acid (adenine, ribose and phosphate), the 3rd and 5th carbon atoms of ribose are attached to phosphate group. The structure is given (Fig. 17.3).

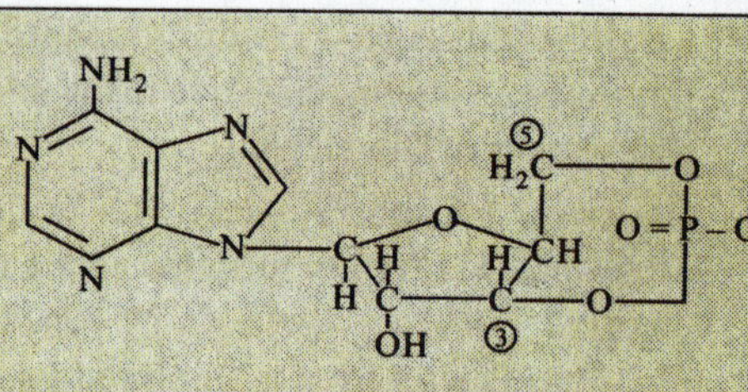

Fig. 17.3 : Structure of cAMP

2. It is the intracellular intermediate compound through which many hormones act.
3. **Adenylate cyclase** of cell membranes being activated by the hormones epinephrine, norepinephrine and glucagon form cAMP from ATP.
4. cAMP is destroyed by ***phosphodiesterase***. The normally low level of it is maintained by the activity of phosphodiesterase. In liver, insulin increases the activity of phosphodiesterase lowering the concentration of cAMP.

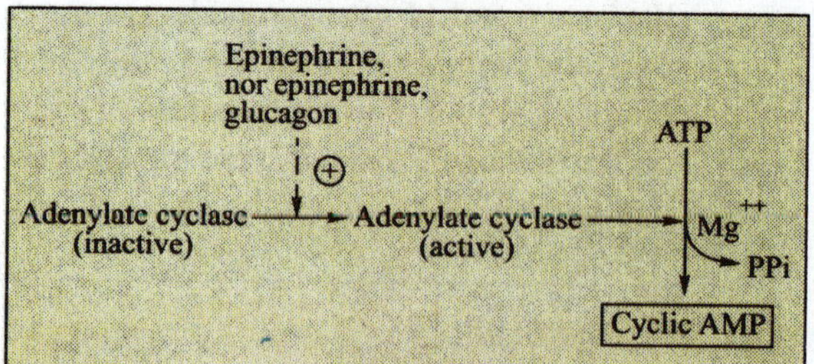

Fig. 17.4 : Formation of cAMP

Functions

1. cAMP causes inhibition (–) of glycogen synthetase.
2. It stimulates (+) phosphorylase activity.
3. Glycogen synthetase I is phosphorylated to form synthetase D by cAMP dependent protein kinase and glycogen synthetase I is formed from synthetase D by synthetase phosphatase.

Adenylate cyclase

1. It is the enzyme occurring in cell membrane.

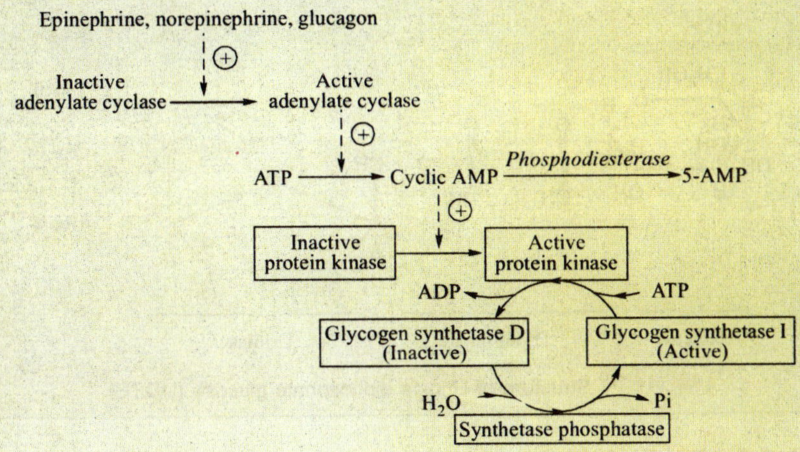

Fig. 17.5 : Control of glycogen synthetase in muscle

2. It is activated by epinephrine, norepinephrine and glucagon and the activated enzyme forms cAMP from ATP.
3. Thyroid hormones may increase the synthesis of adenylate cyclase.

Cyclic GMP

1. Cyclic GMP is 3′, 5′-guanylic acid (guanine, ribose and phosphate), the 3rd and 5th carbon atoms of ribose are attached to phosphate.
2. Prolactin causes the elevation of cGMP in mouse mammary gland which may be mediated by $PGF_{2\alpha}$ and PGF_2 and also raises the level of cGMP in the rat uterus.
3. PGE_5 at high concentration stimulates cGMP accumulation.

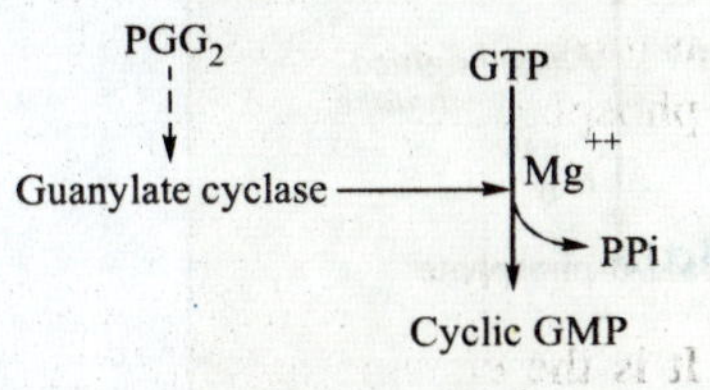

Fig. 17.6 : Formation of cyclic GMP

4. PG endoperoxides also stimulate the production of cGMP in platelets and recent evidence indicates that PGG_2 activates that soluble guanylate cyclase from platelets.

UDPG

1. UDPG has been isolated from animal tissues and from yeast.
2. Uridine diphosphate glucose is an intermediate product of glycogenesis formed from glucose-1-phosphate by the enzyme UDPG pyrophosphorylase in presence of the coenzyme UTP.
3. It is essential for the synthesis of lactose in the mammary gland thus :

$$\text{UDPG} + \text{galactose-1-}Ⓟ \xrightarrow[\text{Uridyl transferase}]{\text{Galactose-1-}Ⓟ} \text{UDP-galactose} + \text{glucose-1-}Ⓟ$$

$$\text{UDP-galactose} + \text{glucose} \xrightarrow[\text{synthetase}]{\text{Lactose}} \text{Lactose} + \text{UDP}$$

4. It can be enzymatically oxidized to UDP-glucuronic acid by UDPG- dehydrogenase and is required for the production of ascorbic acid (vitamin C) by the uronic acid pathway.

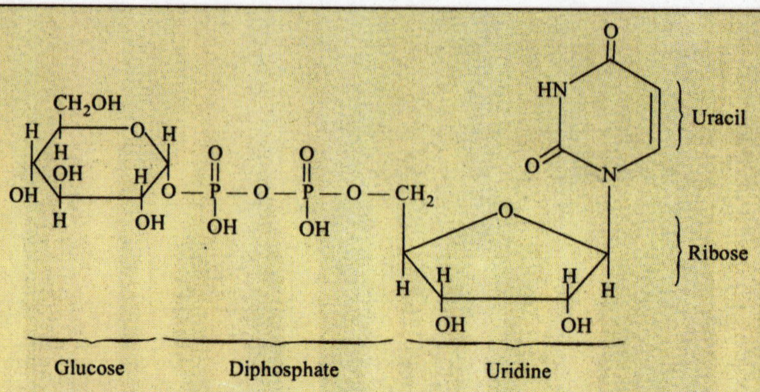

Fig. 17.7 : Structure of Uridine diphosphate glucose (UDPG)

GLYCOGENOLYSIS

1. Glycogen breakdown in the liver and muscle is initiated by the enzyme phosphorylase. This enzyme is specific for the phosphorylytic breaking of the α-1, 4-linkages of glycogen to produce glucose-1-phosphate. The removal of -1, 4-glucosyl residues continues until about 4 glucose residues remain on either side of α-1, 6-branch. Cyclic AMP stimulates phosphorylase activity.

   The debranching enzyme (amylo-1, 6-glucosidase) causes the hydrolytic splitting of -1, 6-linkages.

   By the combined action of both these enzymes, glycogen is converted to glucose-1-phosphate.

2. The action of phosphoglucomutase is reversible and glucose-6-phosphate is formed from glucose-1-phosphate.

3. In liver and kidney (but not in muscle), there exists a specific enzyme, glucose-6-phosphatase, which removes phosphate from glucose-6-phosphate, enabling the free glucose to diffuse from the cell into the extracellular spaces including the blood. The overall breakdown is shown (Fig. 17.8).

Regulation of glycogenolysis

1. The hormones catecholamines (epinephrine, norepinephrine), glucagon and thyroid cause the increase in cAMP level in cells. This cAMP activates protein kinase

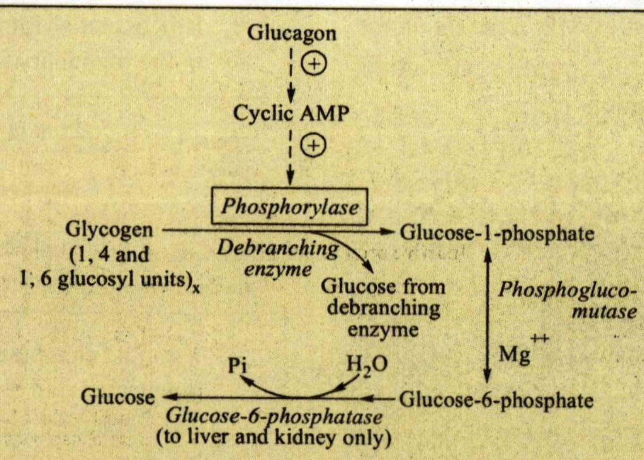

Fig. 17.8 : Pathway of glycogenolysis in the liver

which stimulates the key enzyme phosphorylase for glycogenolysis.

2. Immediately after the onset of muscle contraction, glycogenolysis is highly increased in muscle by the rapid activation of phosphorylase due to the activation of phosphorylase kinase by Ca^{++}.
3. **Calmodulin** (A Ca^{++} dependent regulatory protein) causes further activation of phosphorylase kinase for glycogenolysis.

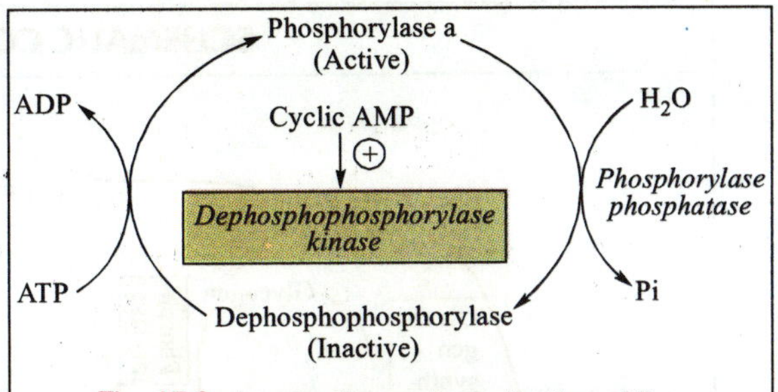

Fig. 17.9 : Inactivation and reactivation of liver phosphorylase

Phosphorylase

1. In liver, phosphorylase exists in both active and inactive form. The active phosphorylase (phosphorylase a or phosphophosphorylase) can be inactivated by phosphorylase phosphatase to **dephosphophosphorylase**. Reactivation takes place by the enzyme phosphorylase b kinase or **dephosphophosphorylase kinase** in presence of ATP.
2. In muscle, phosphorylase is present in two forms : **phosphorylase a** (active in the absence of 5′-AMP) and phosphorylase b (active only in the presence of 5′-AMP).

   Phosphorylase a is physiologically active. It is a tetramer containing 4 mols of pyridoxal phosphate. It is hydrolytically converted to phosphorylase b, a dimer, by phosphorylase phosphatase. Phosphorylase b contains 2 mols of pyridoxal phosphate.

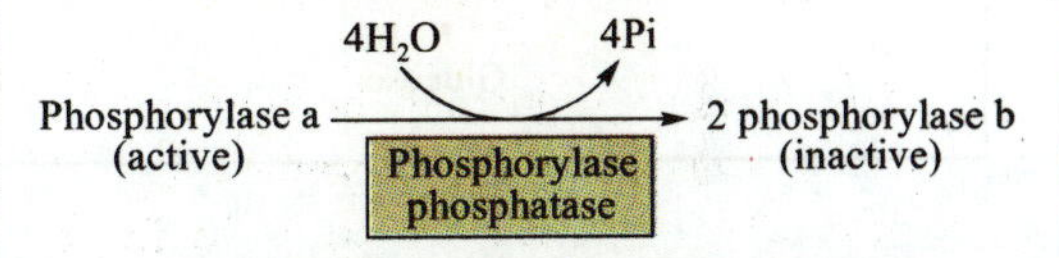

   Phosphorylase b recondenses to phosphorylase a by **phosphorylase kinase b**. Conversion of phosphorylase b to phosphorylase a signifies the mechanism for increasing glycogenolysis :

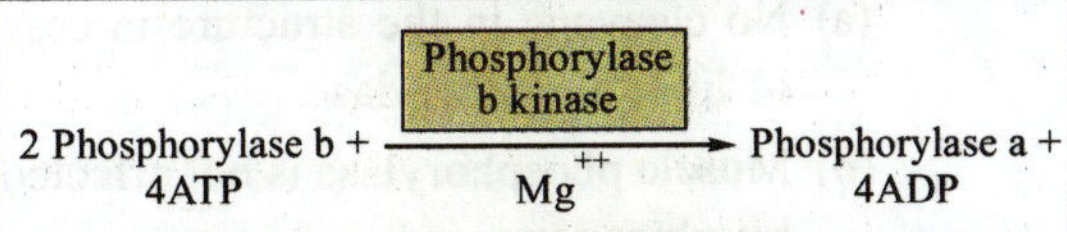

3. Inactive muscle phosphorylase b kinase is converted to active phosphorylase by ki-

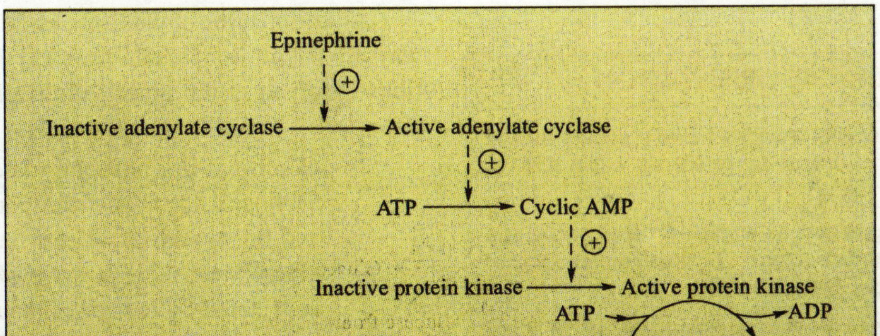

Fig. 17.10 : Control of phosphorylase in muscle

SCHEMATIC COMPENDIUM

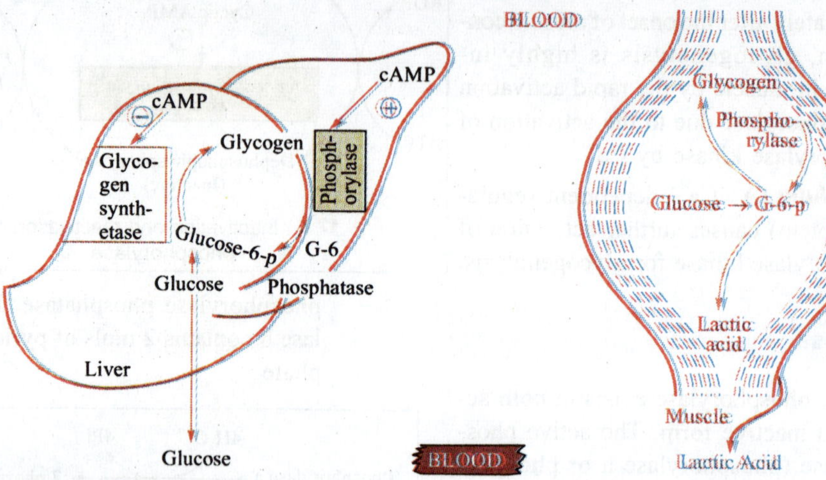

nase by the activation of active protein kinase which is being stimulated by cyclic AMP. Epinephrine is involved in cyclic AMP formation through adenylate cyclase.

4. Difference in muscle phosphorylase and liver phosphorylase :
   (a) No cleavage in the structure in case of liver phosphorylase.
   (b) Muscle phosphorylase is not affected by glucagon.
   (c) Active phosphorylase inhibits liver synthetase phosphatase.

CLINICAL ORIENTATION

- Muscle glycogen provides hexose sugars for the generation of energy by glycolysis within this tissue.
- Liver glycogen acts as storage for emergency and supplies hexose sugars for the maintenance of blood glucose between meals.
- Liver glycogen is totally depleted after 12-18 hours of fasting, whereas muscle glycogen is exhausted markedly only after prolonged vigorous exercise.
- The inherited disorders result in glycogen storage diseases due to insufficient mobilization of glycogen as well as ample deposition of the glycogen leading to muscular weakness or even death.

GLYCOLYSIS

When a muscle contracts under anaerobic condition, pyruvate and lactate become the principal end product. But if it contracts under aerobic condition, pyruvate and lactate disappear and these are further oxidized to CO_2 and H_2O. Hence, carbohydrate metabolism is divided into two phases – anaerobic and aerobic. When oxygen is in short supply, NADH is reoxidized by being coupled to the reduction of pyruvate to lactate.

Sequence of reactions in glycolysis

All the enzyme of the Embden-Meyerhof pathway are found in cytosol. They catalyze the reactions involved in the glycolysis of glucose to lactate :

1. Glucose is first phosphorylated to glucose-6-phosphate by the enzyme hexokinase and by an additional enzyme in the liver, glucokinase. The activity of glucokinase is affected by nutritional state. This reaction is accompanied by ATP and Mg^{++}. This is an irreversible reaction.

   Hexokinase has a higher affinity for glucose than glucokinase. Its functions is to supply glucose to the tissues even in low blood glucose concentration. It can

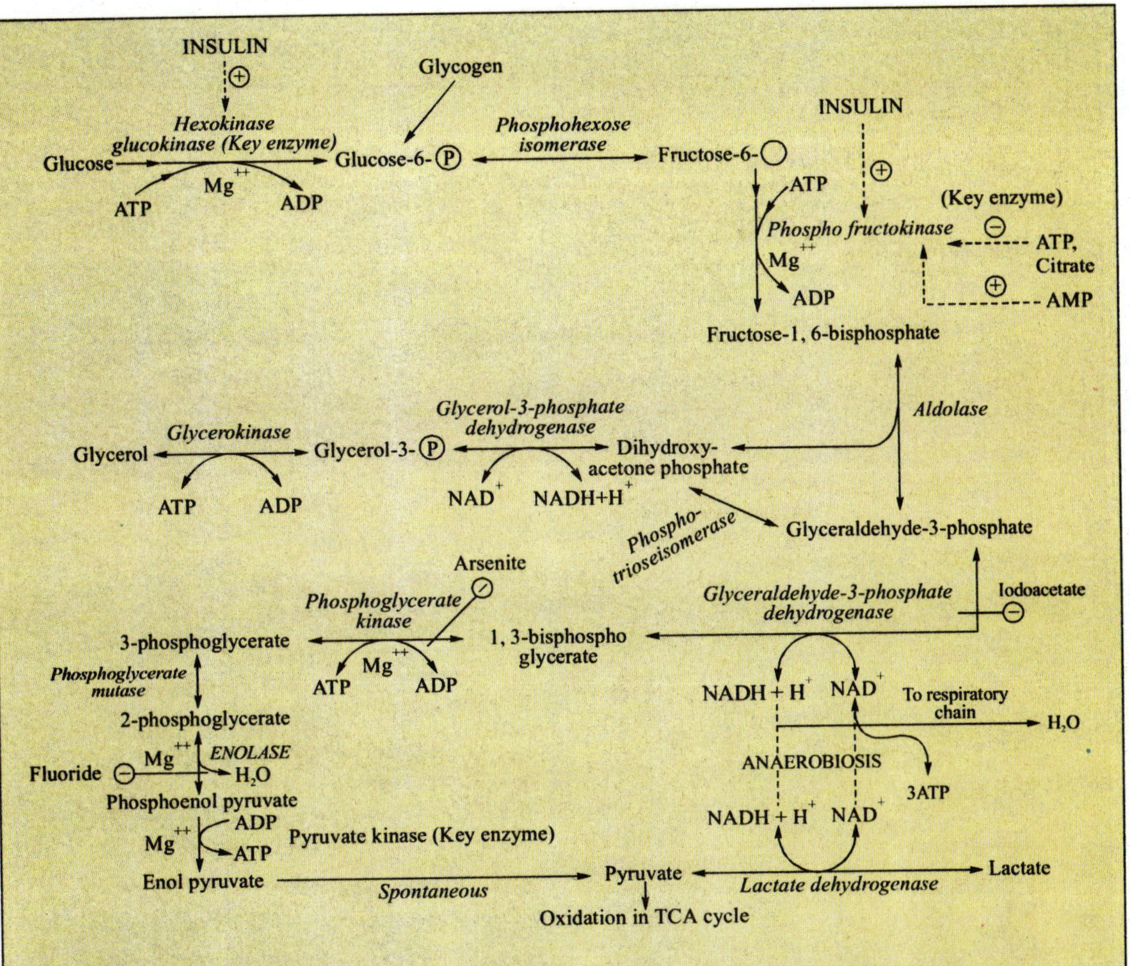

Fig. 17.11 : Flow chart of Embden-Meyerhof pathway of glycolysis

catalyze the phosphorylation of other hexoses but at a slower rate than glucose. But the function of glucokinase is to remove glucose from the blood following a meal. Both the enzymes are stimulated by insulin.

2. Glucose-6-phosphate is an important compound in the metabolic pathways (glycolysis, gluconeogenesis, hexose-monophosphate shunt, glycogenesis, glycogenolysis). It is converted to fructose-6-phosphate by phosphohexose isomerase.
3. Fructose-6-phosphate is phosphorylated with ATP by phosphofructokinase to form fructose-1, 6-bisphosphate. This is also an irreversible reaction. The enzyme is also stimulated by insulin.
4. The enzyme aldolase splits fructose-1, 6-bisphosphate into two triosephosphates : glyceraldehyde-3-phosphate and dihydroxy acetone phosphate.

   Both the triosephosphates are interconverted by phosphotrioseisomerase.

   Dihydroxy acetone phosphate is also formed from glycerol which is phosphorylated to glycerol-3-phosphate, then to dihydroxy acetone phosphate.
5. Glyceraldehyde-3-phosphate is oxidized to 1, 3- bisphosphoglycerate by glyceraldehyde-3- phosphate dehydrogenase with the help of the coenzyme NAD. Energy is released in the course of the reaction.

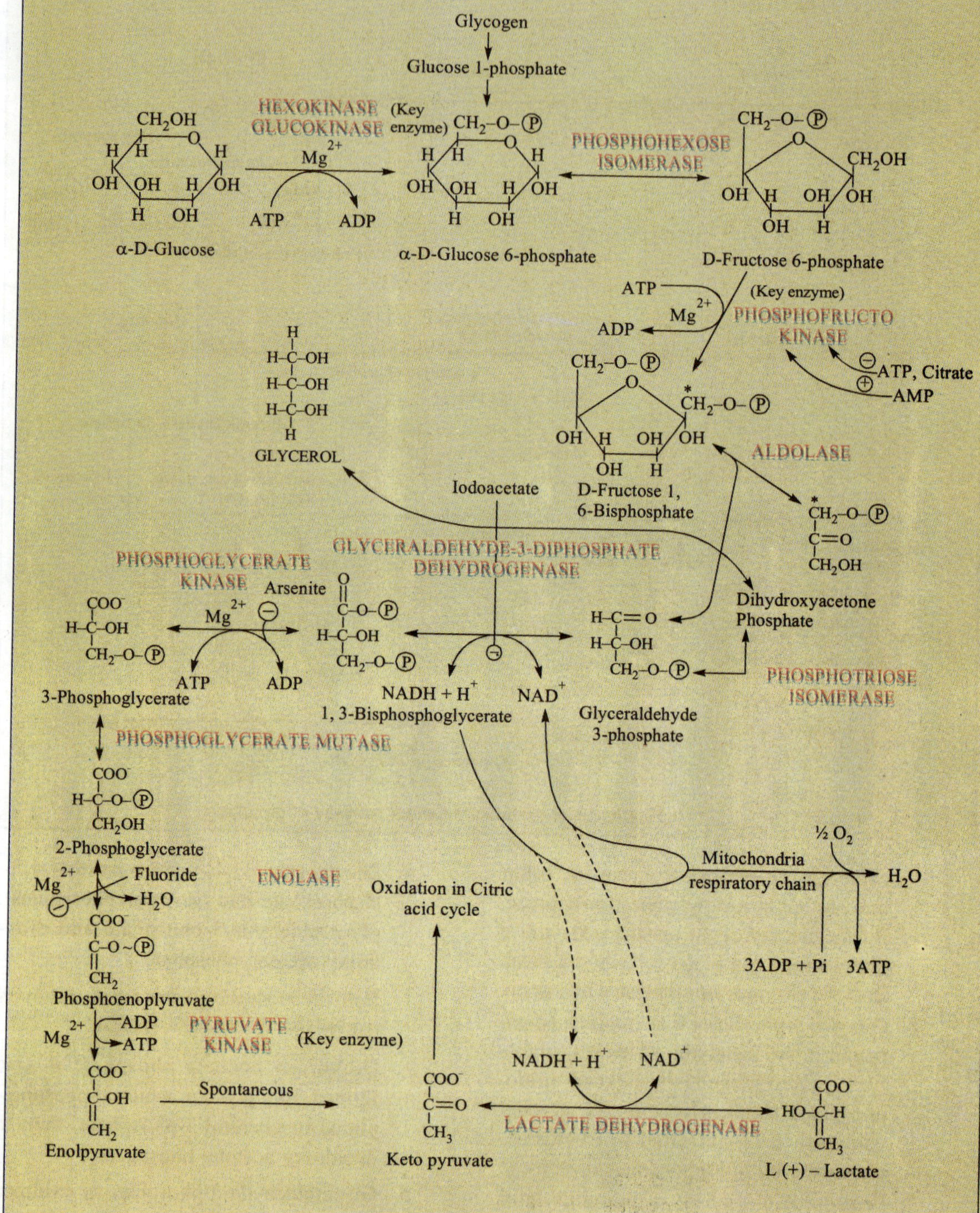

Fig. 17.12 : Reaction chart of Embden-Meyerhof pathway of glycolysis. "Bisphosphophate" indicates that the phosphate groups are separated whereas diphosphate indicates that they are joined

6. 3-phosphoglycerate is formed from 1, 3-bisphosphoglycerate by phosphoglycerate kinase with the production of ATP from ADP.

   Since two molecules of triosephosphates are formed from one molecule of glucose, 2 molecules of ATP are generated in this stage. The reaction is reversible.

7. 3-phosphoglycerate is converted to 2-phosphoglycerate by phosphoglycerate mutase.

8. 2-phosphoglycerate is catalyzed by enolase to produce phosphoenolpyruvate in presence of Mg^{++}.

9. The high-energy phosphate of phosphoenolpyruvate is transferred to ADP by pyruvate kinase to form ATP.

10. Enolpyruvate formed is converted spontaneously to the keto form of pyruvate. This is an irreversible step.

11. If anaerobic conditions prevail, the reoxidation of NADH through the respiratory chain to oxygen is prevented. Pyruvate then is reduced to lactate by the NADH. The reaction is catalyzed by lactate dehydrogenase.

Glycolysis in erythrocytes even under aerobic condition forms lactate because of the absence of enzymatic machinery for the aerobic oxidation of pyruvate.

Total number of ATP formed by glycolysis under anaerobic conditions up to pyruvic acid :

From one molecule of glucose, two molecules of glyceraldehyde-3-phosphate are formed. After that, in the reactions of glycolysis, each product is 2 molecules.

Reactions	ATP formed
1. Glyceraldehyde-3-phosphate → 1, 3-bisphosphoglycerate	6
2. 1, 3-bisphosphoglycerate → 3-phosphoglycerate	2
3. Phosphoenolpyruvate → Enol pyruvate	2
Total	10

Reactions	ATP formed
ATP consumed in :	
1. Glucose → glucose-6-phosphate	1
2. Fructose-6-phosphate → Fructose-1, 6-bisphosphate	1
Net ATP Synthesized	8

But in anaerobic conditions, the total number of ATP will be only 2 up to lactic acid.

Because the reduced NAD (NADH) in glyceraldehyde-3-phosphate dehydrogenase is utilized in lactate dehydrogenase, NADH is not oxidized in mitochondria.

The total calories in this case is 2 × 7,600= 15,200 calories.

Regulation of glycolysis

1. The enzymes involved in the utilization of glucose are activated on the more availability of glucose whereas the enzymes responsible for gluconeogenesis are inhibited at the same time. The activity of the key enzymes glucokinase, phosphofructokinase-1 and pyruvate kinase is increased by glucose.

2. The hormone insulin stimulates the synthesis of the key enzymes responsible for glycolysis and antagonizes the effects of glucocorticoids and glucagon-stimulated cAMP in enhancing the key enzymes responsible for gluconeogenesis.

3. The hormones epinephrine and glucagon increase cAMP level to activate cAMP-dependent protein kinase which phosphorylates and inactivates the key enzyme pyruvate kinase and thus, inhibit glycolysis.

4. Phosphofructokinase-1 is involved in "feedback" control. This enzyme is activated by AMP and inhibited by ATP and citrate. When ATP is utilized in energy requiring processes, the concentration of AMP is highly increased.

5. In hypoxia, the concentration of ATP in the cell is decreased with the increase in the concentration of AMP which explains clearly the inner significance of increase of glycolysis in absence of oxygen.

Effect of hormones in glycolysis

1. Insulin stimulates hexokinase and glucokinase which catalyze the conversion of glucose to glucose-6-phosphate.
2. Insulin stimulates phosphofructokinase which catalyzes the conversion of fructose-6-phosphate to fructose-1, 6-bisphosphate.
3. Glucagon stimulates liver glucose-6-phosphatase which is involved in the conversion of glucose-6-phosphate to glucose and also fructose-1, 6-bisphosphatase involved in the conversion of fructose-1, 6-bisphosphate to fructose-6 phosphate.

Inhibitors

1. Iodoacetate is the inhibitor of glyceraldehyde-3-phosphate dehydrogenase involved in the conversion of glyceraldehyde-3-phosphate to 1, 3-bisphosphoglycerate.
2. Arsenite inhibits the synthesis of ATP by accomplishing uncoupling of oxidation and phosphorylation in the conversion of 1, 3-bisphosphoglycerate to 3-phosphoglycerate.
3. Fluoride inhibits enolase involved in the conversion of 2-phosphoglycerate to phosphoenolpyruvate.

Reaction chart (Fig 17.12) of the pathway of glycolysis Ⓟ PO_3^{2-}; Pi, $HOPO_3^{2-}$; (–), inhibition carbon atoms 1-3 of fructose bisphosphate form dihydroxyacetone phosphate, whereas carbons 4-6 form glyceraldehyde-3-phosphate. The term bis, as in bisphosphate, indicates that the phosphate groups are separated, whereas diphosphate, as in adenosine diphosphate, indicates that they are joined.

Difference between Hexokinase and Glucokinase	
Hexokinase	**Glucokinase**
1. K_m is low, hence high affinity for glucose	1. K_m is high, hence low affinity for glucose
2. Non-specific, can phosphorylate any of the hexoses	2. Specific, can phosphorylate glucose only
3. Available in all tissues	3. Available only in liver
4. Not effected by insulin	4. Stimulated by glucose and insulin
5. Allosterically inhibited by glucose	5. Not inhibited by Glucose-6-Ⓟ

CLINICAL ORIENTATION

- Glycolysis is not only the main pathway for glucose metabolism leading to the formation of acetyl-coA and oxidation in the Citric Acid Cycle, but it also provides the principal pathway for the metabolism of fructose and galactose derived from the diet.
- Glycolysis produces ATP in the absence of oxygen, because this allows skeletal muscle to perform efficiently when aerobic oxidation becomes insufficient and it allows tissue with significant glycolytic ability.
- Heart muscle adapted for aerobic performance has poor glycolytic ability and very poor survival under conditions of ischemia.
- The deficient activity of Pyruvate kinase, the enzyme of glycolysis, produces the disease hemolytic anemias.
- The deficient activity of the glycolytic enzyme phosphofructokinase in skeletal muscle results in fatigue.
- Glycolysis proceeds at a much higher rate than citric acid cycle in fast-growing cancer cell, Thus, more pyruvate is formed than can be metabolized. As a result, excessive production of lactate creates acid local environment in the tumor, a situation of implications for certain types of cancer therapy.
- The deficiency of Pyruvate dehydrogenase results in lactacidosis.

SCHEMATIC COMPENDIUM

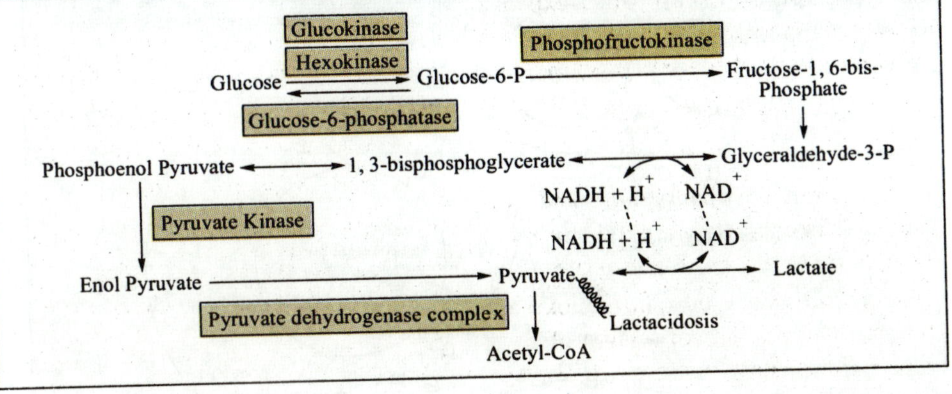

OXIDATION OF PYRUVATE TO ACETYL-CoA

1. Pyruvate oxidatively decarboxylated to acetyl-CoA ("active acetate") before entering the citric acid cycle.
2. The reaction is catalyzed by the multienzyme complex consisting of several different enzymes. This complex is known as pyruvate dehydrogenase complex.
3. Pyruvate is decarboxylated in the presence of thiamine pyrophosphate (TPP) to a hydroxymethyl derivative which reacts with oxidized lipoate to from S-acetyl lipoate being catalyzed by the enzyme pyruvate dehydrogenase.
4. S-acetyl lipoate reacts with coenzyme A to form acetyl-CoA and reduced lipoate in presence of dihydrolipoyl transacetylase.
5. The reduced lipoate is reoxidized by FAD in presence of dihydrolipoyl dehydrogenase.
6. Finally, the reduced FAD is oxidized by NAD^+. The reduced NAD ($NADH + H^+$) enters the respiratory chain producing 3 ATP.
7. The pyruvate dehydrogenase complex consists of about 29 mols of pyruvate dehydrogenase and 8 mols of dihydorlipoyl dehydrogenase distributed around 1 mol of transacetylase.

Regulation

1. The increased pyruvic acid from carbohydrate diet inhibits pyruvate dehydroge-

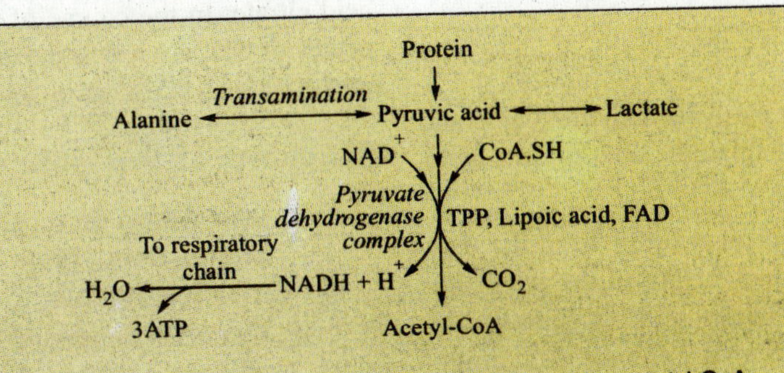

Fig. 17.13 : Flow chart of Oxidation of pyruvic acid to acetyl-CoA

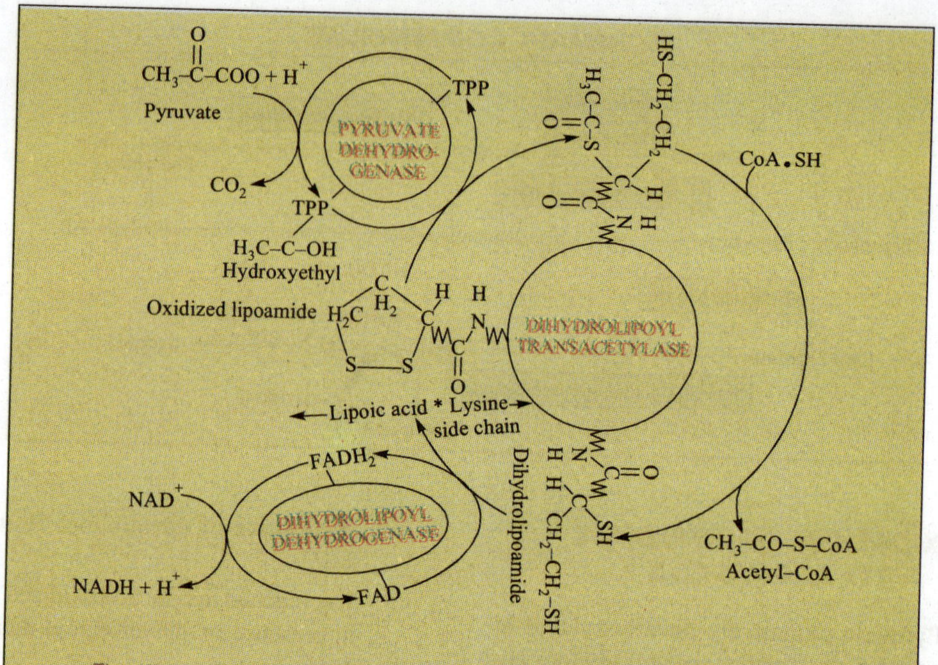

Fig. 17.14 : Reaction of oxidative decarboxylation of pyruvate to Acetyl-CoA

nase kinase for which active pyruvate dehydrogenase is formed. This causes the rapid breakdown of pyruvic acid to form acetyl-CoA.

2. Actyl-CoA and NADH formed by enhanced β-oxidation during starvation and diabetes mellitus activate pyruvate dehydrogenase kinase decreasing the "active" form of pyruvate dehydrogenase. Hence, less pyruvic acid is catabolized and glycolysis is also inhibited.

Inhibitors

Arsenite inhibits pyruvate dehydrogenase and dietary deficiency of thiamine also allows pyruvate to accumulate.

Chronic alcoholics also suffer from the deficiency of thiamine which results in the accumulation of pyruvic acid. Lactic acidosis is caused by the inherited deficiency of pyruvate dehydrogenase.

THE CITRIC ACID CYCLE (KREBS OR TRICARBOXYLIC ACID CYCLE) (IN MITOCHONDRIA) (IN AEROBIC CONDITIONS)

This cycle consists of a series of reactions in mitochondria which catabolizes the oxidation of acetyl-CoA to CO_2 and H_2O in aerobic condition. The acetyl-CoA combines with a 4-carbon dicarboxylic acid, oxaloacetate to form 6-carbon tricarboxylic acid citrate. In the course of the reactions 2 molecules of CO_2 are lost and oxaloacetate is regenerated. Oxaloacetate plays an important catalytic role.

Reactions of the citric acid cycle

1. Acetyl-CoA first combines with oxaloacetate to form citrate being catalyzed by citrate synthetase.
2. Citrate is converted to isocitrate by the enzyme aconitase (aconitate hydratase) which contains Fe^{++}. This conversion takes

place in two steps–dehydration to cis-aconitate and rehydration to isocitrate. The reaction is inhibited by **fluoroacetate**.

3. Isocitrate undergoes dehydrogenation in the presence of **isocitrate dehydrogenase** to form oxalosuccinate.
4. The decarboxylation of oxalosuccinate produces α-ketoglutarate by the catalytic action of **isocitrate dehydrogenase** in presence of Mn^{++}.
5. α-ketoglutarate undergoes oxidative decarboxylation in similar manner as that occurred in pyruvate. The reaction is catalyzed by **α-ketoglutarate dehydrogenase complex** which requires coenzymes TPP, lipoate, NAD^+, FAD^+ and CoA forming succinyl-CoA.
6. Succinyl-CoA is then converted to succinate by the enzyme **succinate thiokinase** (succinyl-CoA synthetase). The reaction requires GDP or IDP which is converted to GTP or ITP by the presence of inorganic phosphate.
7. Succinate is further metabolized undergoing dehydrogenation catalyzed by **succinate dehydrogenase**. The reaction requires FAD. Fumarate is formed.
8. Fumarate, under the influence of fumarase, is converted to malate by the addition of water.
9. Malate is converted to oxaloacetate by **malate dehydrogenase** which requires NAD^+.

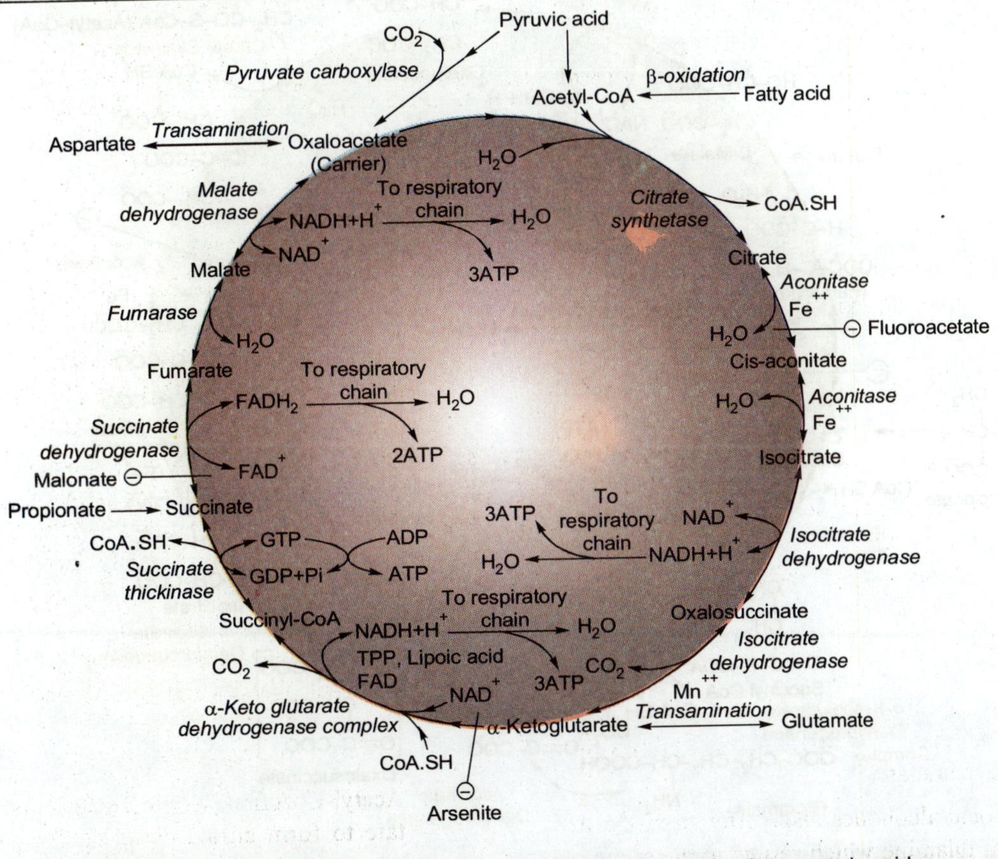

Fig. 17.15 : Flow chart of Citric Acid Cycle in mitochondria starting from pyruvic acid.
*According to the Committee of Editors of Biochemical Recommendations (1975), the Standard Biochemical Convention, the ending-ate, e.g. citrate, denotes mixture of free citric acid and the ionized form in which the cations are not specified

Energetics of the citric acid cycle :

Reaction	No. of ATP formed
Isocitrate → Oxalosuccinate	3
α-ketoglutarate → Succinyl-CoA	3
Succinyl-CoA → Succinate	1
Succinate → Fumarate	2
Malate → Oxaloacetate	3
Total	12

Total number of ATP in the complete oxidation of one molecule of glucose :

One molecule of glucose forms 2 molecules of pyruvic acid by glycolysis.	
Number of ATP formed in glycolysis	8
Number of ATP formed in the oxidation of pyruvate to acetyl-CoA	3×2 = 6
Number of ATP formed in the citric acid cycle	12×2 = 24
Total	38

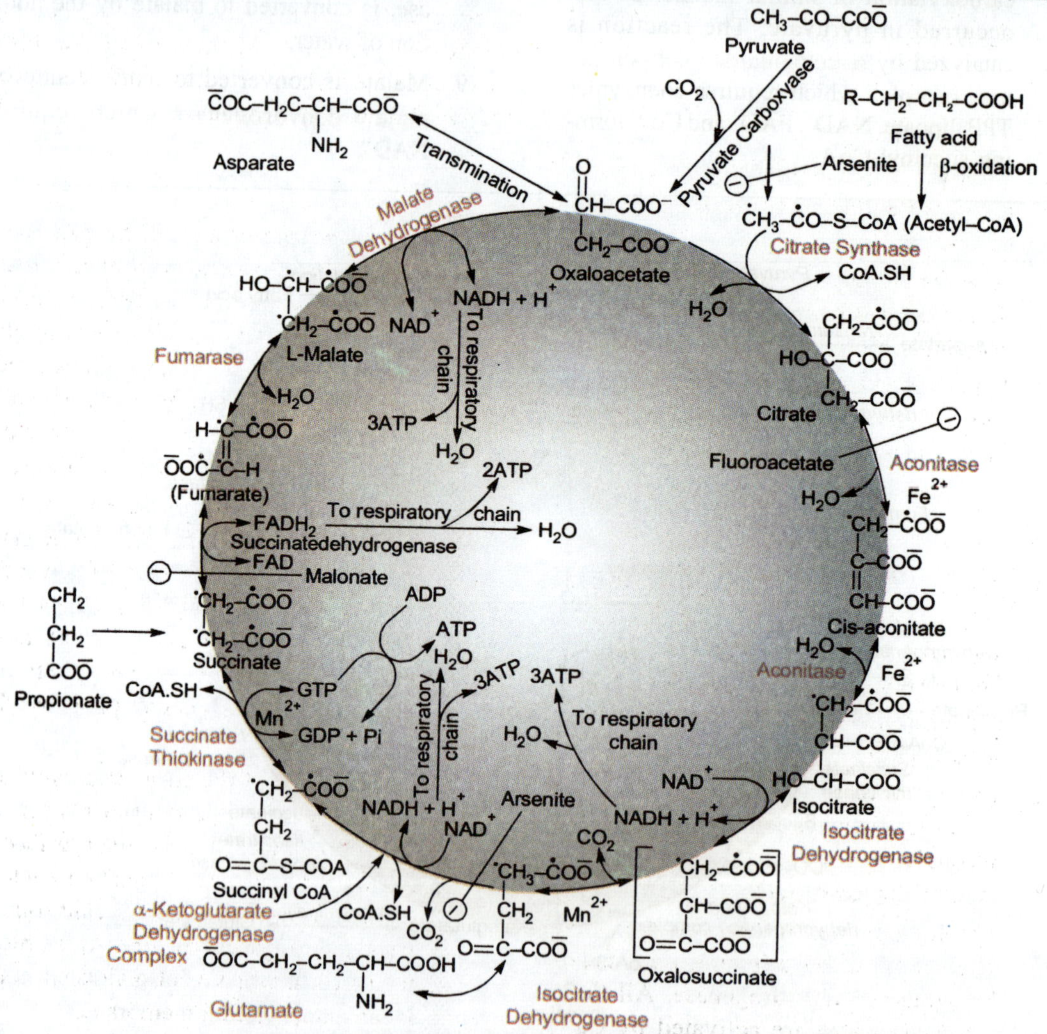

Fig. 17.16 : Reaction chart of Citric Acid Cycle in mitochondria from pyruvate. According to the Committee of Editors of Biochemical Journals Recommendations (1975), the Standard Biochemical Convention, the ending -ate, e.g., citrate, denotes mixture of free citric acid and the ionised form in which the cations are not specified. The sites of inhibition (–) by fluoroacetate, malonate and arsenite are indicated

Number of molecules of CO_2 and H_2O formed in the oxidation of pyruvic acid

Pyruvate → acetyl-CoA	1 molecule of CO_2
Oxalosuccinate → α-ketoglutarate	1 molecule of CO_2
α-ketoglutarate → Succinyl-CoA	1 molecule of CO_2
Total	3 molecules of CO_2
The oxidation of reduced NAD & FAD by repiratory chain produces	$5H_2O$
Utilized in TCA cycle	$3H_2O$
Net	$2H_2O$

Inhibitors

1. Fluoroacetate inhibits the enzyme aconitase and prevents the conversion of citrate to isocitrate.
2. Arsenite inhibits α-ketoglutarate dehydrogenase and causes α-ketoglutarate to accumulate.
3. Malonate or oxaloacetate inhibits succinate dehydrogenase competitively resulting in succinate accumulation.

Regulation of the Citric Acid Cycle

1. The main function of TCA cycle is to provide energy through the respiratory chain and oxidative phosphorylation. The activity of this cycle fully depends on NAD^+ and the availability of ADP for the formation of ATP.
2. The rate of respiration and the activity of the citric acid cycle are determined by the adequate O_2 and the rate of doing work through the use of ATP.
3. The key enzymes of this cycle are citrate synthase, isocitrate dehydrogenase, and α-ketoglutarate dehydrogenase. All these dehydrogenases are activated by Ca^{++}, which increases in concentration during muscular contraction and secretion in case of increased energy demand. These enzymes are responsive to the energy status as expressed by the [ATP] / [ADP] and [NADH] / [NAD^+] ratios.
4. Citrate synthase is allosterically inhibited by ATP and long chain fatty acyl-CoA. Mitochondrial NAD-dependent isocitrate dehydrogenase is allosterically activated by ADP and is inhibited by ATP and NADH.

α-ketoglutarate dehydrogenase complex is under the same condition of pyruvate dehydrogenase.

Succinate dehydrogenase is inhibited by oxaloacetate and the availability of oxaloacetate is controlled by malate dehydrogenase which depends on [NADH] / [NAD^+] ratio.

Significance of the Citric Acid Cycle

1. The major significance of the citric acid cycle is that it acts as the common metabolic pathway for the oxidation of carbohydrate, lipids and proteins because glucose, fatty acids and many amino acids are metabolized to acetyl-CoA which is finally oxidized in the citric acid cycle.
2. The reducing equivalents in the forms of hydrogen or of electrons are formed by the activity of specific dehydrogenases during the oxidation of acetyl-CoA in the cycle. These reducing equivalents then enter the respiratory chain, where large amounts of high-energy phosphate are generated by the oxidative phosphorylation.
3. The enzymes of the citric acid cycle are located in the mitochondrial matrix, either free or attached to the inner surface of the inner mitochondrial membrane which facilitates the transfer of reducing equivalents to the adjacent enzymes of the respiratory chain which is also situated in the inner mitochondrial membrane.
4. The citric acid cycle is amphibolic (dual) in nature which is the source for anabolic processes such as fatty acid and amino acid synthesis and gluconeogenesis.

CLINICAL ORIENTATION

- Citric acid cycle is the common pathway for the oxidation of carbohydrate, lipids and protein, because glucose, fatty acids and many amino acids are all metabolized to form acetyl-CoA which finally forms ATP.
- This cycle plays a major role in gluconeogenesis, transamination, deamination and lipogenesis.
- The disadvantage becomes prominent when a good number of hepatic cells are damaged or replaced by connective tissue as in **acute hepatitis** and **cirrhosis**.
- The intermediates of citric acid cycle help in the synthesis of heme, fatty acids, cholesterol, steroid and non-essential amino acids.

GLUCONEOGENESIS or NEOGLUCOGENESIS

1. The formation of glucose from non- carbohydrate substances such as lactic acid, amino acids and glycerol is called **gluconeogenesis** or **neoglucogenesis**.
2. When the carbohydrate is insufficient in the diet, gluconeogenesis meets the needs of the body for glucose.
3. A continued supply of glucose is necessary as a source of energy and glucose is the only fuel which supplies energy to skeletal muscle under anaerobic conditions.
4. There is always a certain basal requirement for glucose even when fat is supplied to the caloric requirement of the organism.
5. In mammals, the liver and kidney are the principal organs responsible for gluconeogenesis. Gluconeogenesis is essentially a reversal of glycolysis. Therefore, the glycolytic activity of liver and kidney is low when there is active gluconeogenesis.

Metabolic pathways in gluconeogenesis

1. The metabolic pathways in connection with gluconeogenesis are the modifications or the Embden-Meyerhof pathways and the citric acid cycle.
2. They are concerned with the conversion of glucogenic amino acids, lactate, glycerol, propionate (in ruminants) to glucoses or glycogen.
3. The energy barriers obstruct a simple reversal of glycolysis (i) between pyruvate and phosphoenolpyruvate, (ii) between fructose-1, 6-bisphosphate and fructose 6-phosphate, (iii) between glucose-6-phosphate and glucose, (iv) between glucose-1-phosphate and glycogen. These barriers are overcome by the following reactions :

   (i) The enzyme **pyruvate carboxylase** present in mitochondria converts pyruvate to oxaloacetate in presence of ATP, biotin and CO_2. A second enzyme phosphoenolpyruvate carboxykinase present in the extramitochondrial part of the cell converts oxaloacetate to phosphoenolpyruvate in presence of GTP. Lactate with the help of these two enzymes and lactate dehydrogenase is converted to phosphoenolpyruvate.

In Mitochondria :

$$\text{Pyruvate} + \text{ATP} + CO_2 \xrightarrow[\text{Biotin}]{\text{Pyruvate carboxylase}} \text{Oxaloacetate} + \text{ADP} + \text{Pi}$$

Fig. 17.17

In the extramitochondrial part of the cell :

$$\text{Oxaloacetate} \xrightarrow[\text{Phosphoenolpyruvate carboxykinase}]{\text{GTP} \rightarrow \text{GDP}} \text{Phosphoenol-pyruvate} + CO_2$$

Fig. 17.18

But, oxaloacetate does not diffuse readily from mitochondria. Alternative means are applied to convert oxaloacetate to malate which is readily diffused from mitochondria. Malate is then converted to oxaloacetate in the extramitochondrial portion of the cell.

(ii) The conversion of fructose-1, 6-bisphosphate to fructose-6-phosphate catalyzed by another enzyme fructose-1, 6-bisphosphatase. This enzyme is present in liver, kidney and striated muscle but absent from adipose tissue, heart muscle and smooth muscle.

(iii) Glucose-6-phosphate is converted to glucose by glucose-6-phosphatase which is present in intestine, liver and kidney but absent from muscle and adipose tissue.

(iv) The conversion of glucose-1-phosphate to glycogen is through UDPG and glycogen synthetase.

Glucogenic amino acids after transamination or deamination form either pyruvic acid or members of the citric acid cycle. Therefore, the glucogenic amino acids and lactate can be converted to glucose or glycogen. *Propionate in ruminants* enters the main glucogenic pathway via the citric acid cycle being converted to succinyl-CoA.

Conversion of propionate to succinyl-CoA

1. Propionate is first activated by thiokinase with ATP and CoA to form propionyl-CoA.
2. Propionyl-CoA undergoes CO_2 fixation reaction to form D-methylmalonyl-CoA, catalyzed by propionyl-CoA carboxylase and biotin is required as a coenzyme.
3. D-methylmalonyl-CoA is converted to L-methylmalonyl-CoA by methylmalonyl-CoA racemase.
4. L-methylmalonyl-CoA is isomerised to succinyl-CoA by methylmalonyl-CoA isomerase which requires vitamin B_{12} as a coenzyme.

Conversion of glycerol

1. Glycerol is first converted to glycerol-3-phosphate by glycerokinase with ATP in liver and kidney.
2. Glycerol-3-phosphate is oxidized to dihydroxyacetone phosphate by glycerol-3-phosphate dehydrogenase in presence of NAD^+. Dihydroxyacetone phosphate is then converted to glucose.

Conversion of lactate to glucose

1. Lactic acid is the major end product in muscle in anaerobic glycolysis. Muscle tissue is incapable of resynthesizing glucose from lactate. The conversion takes place entirely in the liver.
2. Muscle lactate is transported to the liver by the blood. In the liver, it is converted to glucose and glycogen by the enzymes concerned in gluconeogenesis.
3. Liver glycogen is converted to glucose which is carried back to muscle by blood.

This conversion of muscle lactate to glucose in liver and its reentry into muscle is called "Cori Cycle".

Conversion of amino acids to glucose

1. The glucogenic amino acids are converted to the intermediates of citric acid cycle either by transamination or deamination which is given in Fig. 17.19.
2. These intermediates are converted to malate and finally converted to glucose by the enzymes involved in gluconeogenesis.

Conversion of fatty acids to glucose

Fatty acids are metabolized to acetyl-CoA by β-oxidation. Acetyl-CoA enters the citric acid cycle and then converted to malate. Malate is diffused from the mitochondria to the extramitochondrial portion of the cell where it is finally converted

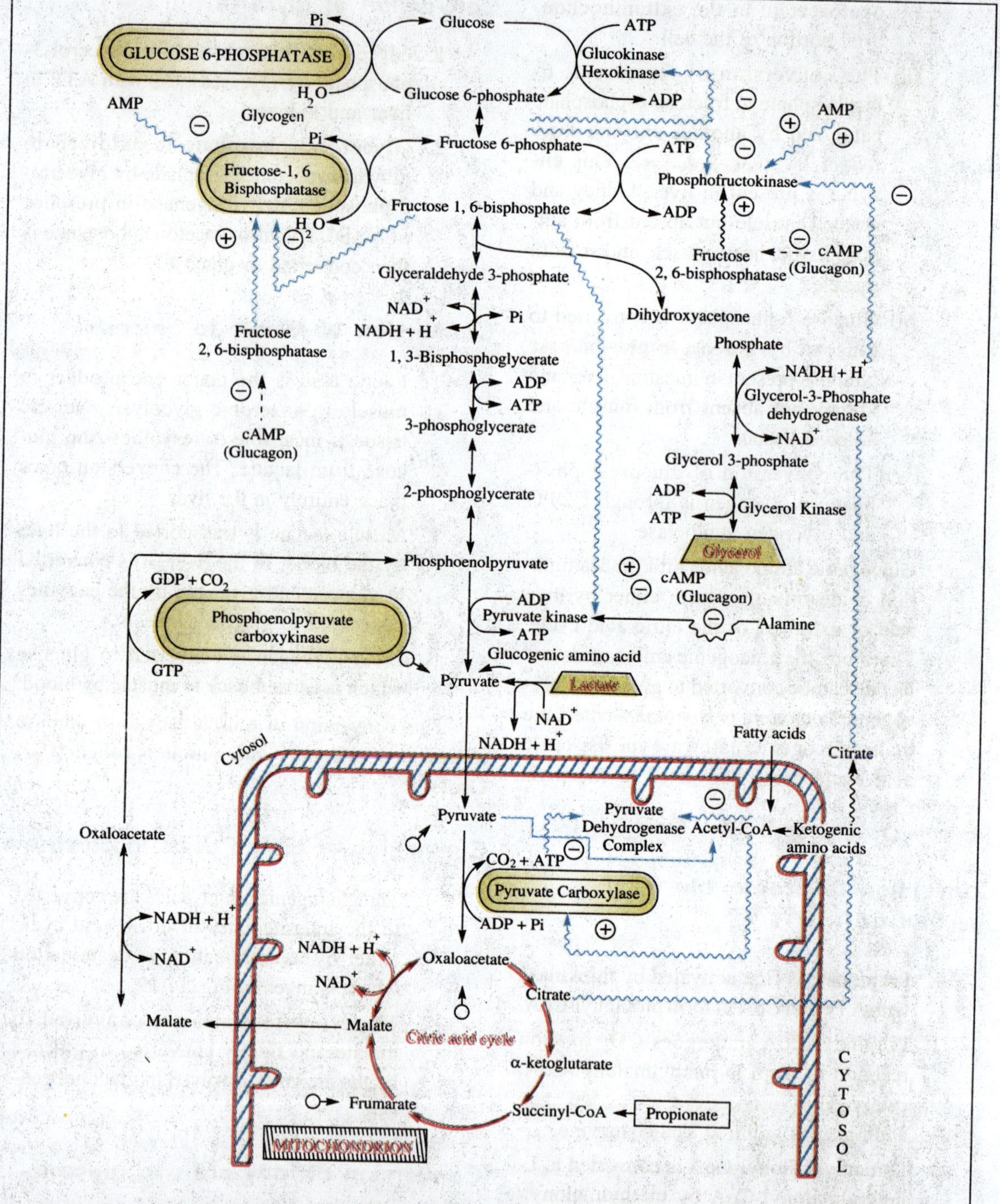

Fig. 17.19 : Major pathways and regulation of gluconeogenesis and glycolysis in the liver. Entry points of glucogenic amino acids after transamination are indicated by ○➝. The key gluconeogenic enzymes are shown thus ◎. The ATP required for gluconeogenesis is supplied by the oxidation of long chain fatty acids. Propionate is of quantitative importance only in remnants ⇝, allosteric effect ....., Covalent modification by reversible phosphorylation. High concentrations of alanine as a "gluconeogenic signal" by inhibiting glycolysis at the pyruvate kinase step

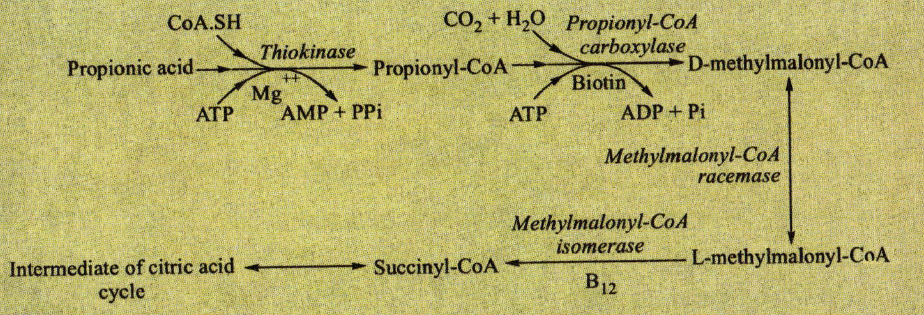

Fig. 17.20 : Propionic acid to succinyl-CoA

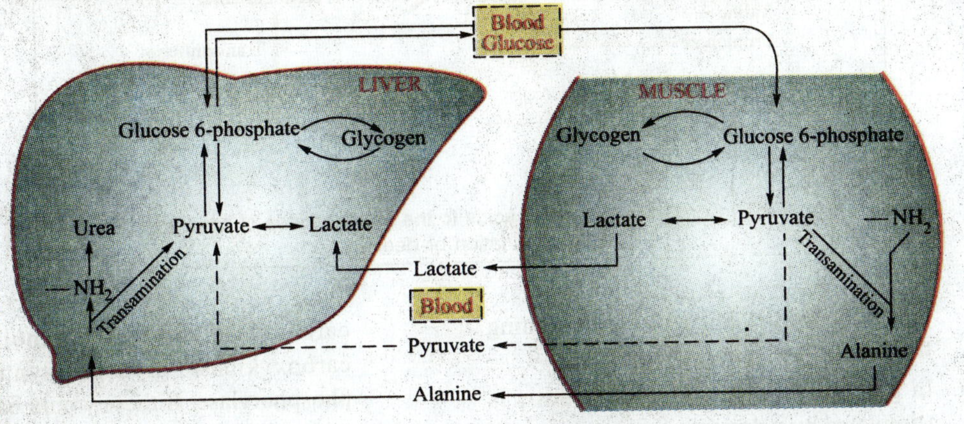

Fig. 17.21 : Lactic acid (cori) Cycle and Glucose-Alanine Cycle

to glucose by the enzymes involved in gluconeogenesis.

Acetyl-CoA is not permeable to pass from the mitochondria to the cytosol through the mitochondrial membrane. But citrate is permeable through mitochondrial membrane to pass to the cytosol where it is splitted to acetyl-CoA and oxaloacetate.

Role of fructose 2, 6-bisphosphate in the regulation of glycolysis and gluconeogenesis in liver

1. Fructose 2, 6-bisphosphate is the most potent positive allosteric effector of phosphofructokinase-I and inhibitor of fructose-1, 6-biphosphatase in liver.
2. It relieves inhibition of phosphofructokinase-I by ATP and increases affinity for fructose 6-phosphate.
3. It inhibits fructose-1, 6-bisphosphatase by increasing the K_m for fructose 1,6-biphosphate.
4. It is formed by phosphorylation of fructose 6-phosphate by phosphofructokinase-2 which is also responsible for its breakdown since it contains fructose-2, 6-bisphosphatase activity. This bifunctional enzyme is under the allosteric control of fructose 6-phosphate.
5. When glucose is short, glucagon stimulates the production of cAMP which inactivates phosphofructokinase-2 and activates fructose-2, 6-bisphosphatase by phosphorylation.
6. In the superfluity of glucose, the concentration of fructose 2, 6-bisphosphate increases stimulating glycolysis by activat-

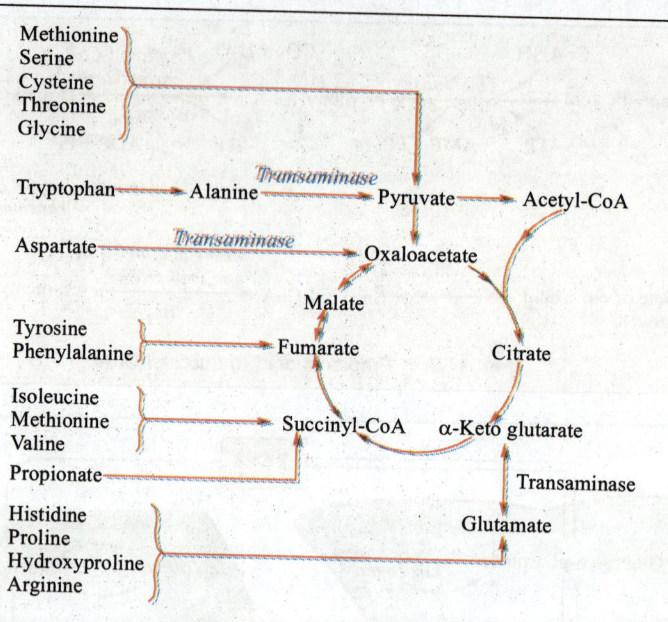

Fig. 17.22 : Conversion of amino acids to the intermediates of citric acid cycle by transamination or deamination

ing phosphofructokinase-I and inhibiting fructose-1, 6-bisphosphatase.

7. In glucose shortage, gluconeogenesis is stimulated by a decrease in the concentration of fructose 2, 6-bisphosphate which deactivates phosphofructokinase-I and deinhibits fructose-1, 6-bisphosphatase. This mechanism also shows that glucagon stimulation of glycogenolysis in liver results in glucose release rather than glycolysis.
8. Recently, it has been indicated that glucose 1, 6-bisphosphate plays a similar role in some extrahepatic tissues.

Substrate (Futile) Cycle

(a) The control points in glycolysis and glycogen metabolism involve a cycle of phosphorylation and dephosphorylation catalyzed by the enzymes like glucokinase and glucose-6-phosphatase; phosphofructokinase-1 and fructose-1, 6-bisphosphatase; pyruvate kinase, pyruvate carboxylase and phosphoenolpyruvate carboxykinase; glycogen synthase and phosphorylase. If these would be allowed to cycle unchecked, ultimately there would be hydrolysis of ATP. But this cannot happen due to the different control mechanisms which can signify that one limb of the cycle is inhibited as the other is stimulated as per the requirements of the tissue and of the body.

(b) In the phosphofructokinase and fructose-1, 6-bisphosphatase, the allosteric modifier, fructose 2, 6-bisphosphate, causes a large change in the net effect of metabolites in either direction of the cycle. This fine effect of metabolic control occurs only at the expense of some loss of ATP.

Regulation of gluconeogenesis

1. The key enzymes of gluconeogenesis are pyruvate carboxylase, phosphoenolpyruvate carboxykinase, fructose-1, 6-bisphosphatase and glucose-6-phosphatase.

2. The hormones glucagon and glucocorticoids which are secreted during starvation stimulate glucose-6-phosphatase to enhance gluconeogenesis.
3. High carbohydrate diets increase the insulin/glucagon ratio and thus reduce the activities of the key enzymes of gluconeogenesis to minimize gluconeogenesis.
4. ATP and citrate are the activators of fructose-1, 6-bisphosphatase and, hence, gluconeogenesis is increased. But the high level of AMP in liver cells inhibits fructose-1, 6-bisphosphatase and thus reduces gluconeogenesis.
5. In glucose shortage, the more secreted glucagon stimulates gluconeogenesis by decreasing the concentration of fructose-2, 6-bisphosphate which in turn inhibits phosphofructokinase-1 and activates the enzyme fructose-1, 6-bisphosphatase.
6. During starvation the excessive liberated **glucagon** stimulates the enzyme phosphoenolpyruvate carboxykinase and thus increases gluconeogenesis.
7. The enzyme pyruvate carboxylase is allosterically activated by **acetyl-CoA**. During starvation increased fatty acid oxidation provides more **acetyl-CoA** which promotes gluconeogenesis.
8. **Increased ADP** allosterically inhibits pyruvate carboxylase and thus reduces gluconeogenesis.
9. The hormones **glucagon, epinephrine,** and glucocorticoids stimulate the synthesis of pyruvate carboxylase and thus enhance gluconeogenesis. But the hormone **insulin** depresses the enzyme pyruvate carboxylase and thus reduces gluconeogenesis.

CLINICAL ORIENTATION

- The need for glucose by the body is met by gluconeogenesis when carbohydrate supply is insufficient from the diet. The supply of glucose as a source of energy should be continuous for the following purposes :
  - (a) As a source of energy for the nervous system and the erythrocytes. Failure of gluconeogenesis becomes fatal. Brain dysfunction leads to coma and death when blood glucose concentration reaches below a critical level.
  - (b) As a source of glyceride-glycerol in adipose tissue.
  - (c) Required for maintaining the level of intermediates of the citric acid cycle in many tissues.
  - (d) Acts as a precursor of milk sugar (lactose) in the mammary gland and is taken up actively by fetus.

Contd.

SCHEMATIC COMPENDIUM

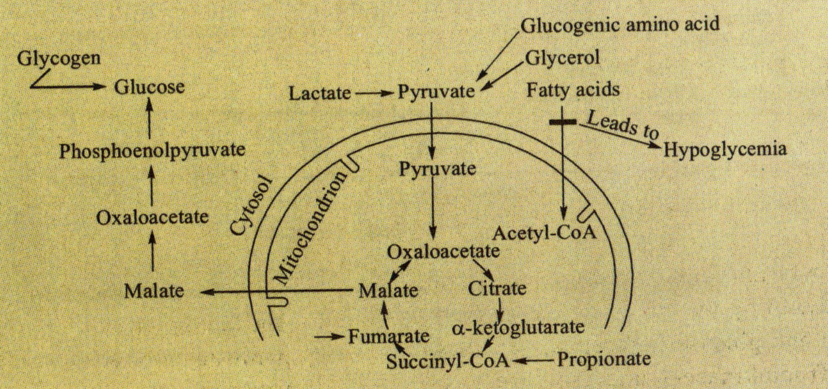

Gluconeogenesis

CLINICAL ORIENTATION

(e) Serves as the principal fuel for the supply of energy to skeletal muscle under anaerobic conditions.

- Gluconeogenic mechanisms are utilized to clear the products of the metabolism of other tissues from the blood, eg, lactate, produced by muscle and erythrocytes; glycerol, produced by adipose tissue.
- Propionate is the principal glucogenic fatty acid. It is produced in the digestion of carbohydrates by ruminants and acts as a significant substrate for gluconeogenesis in these species.

HEXOSE MONOPHOSPHATE SHUNT (HMP SHUNT) OR PENTOSE PHOSPHATE PATHWAY

1. This is an alternate aerobic pathway for the oxidation of glucose in the liver, lactating mammary gland and adipose tissue in addition to the Embden-Meyerhof pathway for glycolysis.
2. The enzymes of this pathway are present in the extramitochondrial portion of the cell. This pathway is active in liver, adipose tissue, adrenal cortex, thyroid, erythrocytes, testis and lactating mammary gland.
3. In this pathway, 3 molecules of glucose-6-phosphate yield 3 molecules of CO_2 and 3 molecules of 5 carbon residues (pentose sugars). The latter are converted ultimately to 2 molecules of glucose-6-phosphate and one molecule of glyceraldehyde-3- phosphate.
4. NADP, instead of NAD, is used as a hydrogen acceptor in this pathway.

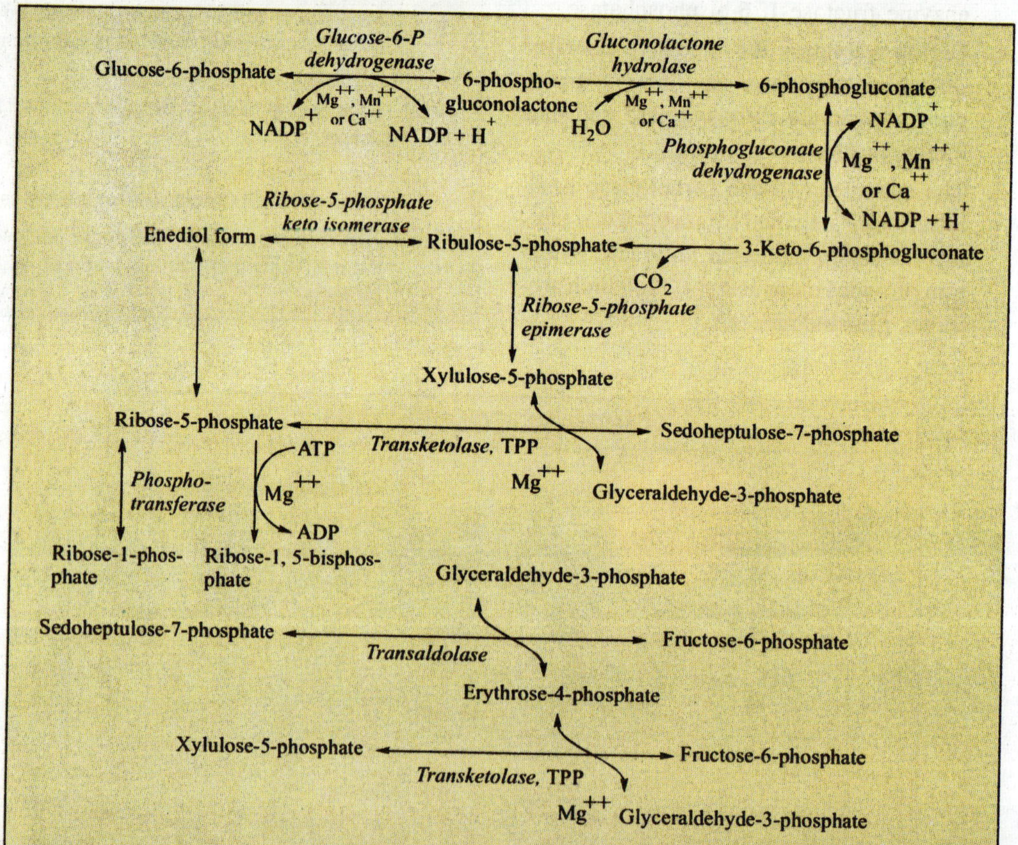

Fig. 17.23 : Flow Chart of the Hexose Monophosphate Shunt or Pentose Phosphate Pathway

Metabolic reactions

The reaction takes place in two phases :

(i) Glucose-6-phosphate by dehydrogenation and decarboxylation gives rise to ribulose-5-phosphate.

(ii) Ribulose-5-phosphate is converted back to glucose-6-phosphate by transketolase and transaldolase.

1. Glucose-6-phosphate is dehydrogenated to 6-phosphogluconate via 6-phosphogluconolactone by glucose-6-phosphate dehydrogenase in presence of NADP and the cofactors Mg^{++}, Mn^{++} or Ca^{++}. 6-phosphogluconolactone is acted by gluconolactone hydrolase in presence of Mg^{++}, Mn^{++} or Ca^{++}. Genetically, the deficiency of glucose-6-phosphate dehydrogenase in erythrocytes is associated with a tendency to hemolysis by primaquine and sulfonamide.
2. 6-phosphogluconate is oxidized by 6-phosphogluconate dehydrogenase in the presence of the coenzyme NADP and cofactors Mg^{++}, Mn^{++} or Ca^{++} to 3-keto 6-phosphogluconate which is decarboxylated to form ribulose-5-phosphate.
3. Ribulose-5-phosphate is acted on by ribulose-5-phosphate epimerase which changes the configuration about carbon 3 forming xylulose-5-phosphate and also by the enzyme ribose-5-phosphate ketoisomerase which converts ribulose-5-phosphate to ribose-5-phosphate.
4. Transketolase with the help of TPP and Mg^{++} transfers carbons 1 and 2 of xylulose-5-phosphate to the ribose-5-phosphate forming sedoheptulose-7-phosphate and glyceraldehyde-3-phosphate.
5. Transaldolase allows the transfer of a 3-carbon moiety from sedoheptulose-7-phosphate to the glyceraldehyde-3-phosphate to form fructose-6-phosphate and erythrose-4-phosphate.
6. Transketolase with the help of TPP and Mg^{++} transfers carbon 1 and 2 from xylulose-5-phosphate to erythrose-4-phosphate producing fructose-6-phosphate and glyceraldehyde-3-phosphate.

The overall reaction of the hexose monophosphate shunt is shown :

$$3\text{Glucose-6-}\textcircled{P} + 6NADP^+ \longrightarrow 3CO_2 + 6NADPH + 6H^+ + \text{Glyceraldehyde-3-}\textcircled{P} + 2\text{Fructose-6-}\textcircled{P}$$

Regulation of HMP shunt

1. The first reaction of this pathway catalyzed by glucose-6-phosphate dehydrogenase is the "rate limiting" step. This is mainly regulated by the cytoplasmic levels of $NADP^+$ and NADPH.
2. The high carbohydrate diets enhance the activities of both dehydrogenases and the rate of the pathway. But reverse occurs in case of starvation and diabetes mellitus.
3. The increased fatty acid synthesis and steroid synthesis reoxidize NADPH to $NADP^+$ for which cytoplasmic ratio of $NADP^+$/NADPH is increased enabling to enhance the shunt pathway.
4. The hormone insulin stimulates the synthesis of both the dehydrogenases and thus enhances the activity of the pathway.
5. Thyroid hormones also stimulate the activity of glucose-6-phosphate dehydrogenase and thus enhances the shunt pathway.

Metabolic significance of the HMP shunt

1. CO_2 is the characteristic product in this pathway which is not formed in Embden-Meyerhof pathway. This product is utilized for the synthesis of fatty acid and purine bases etc.

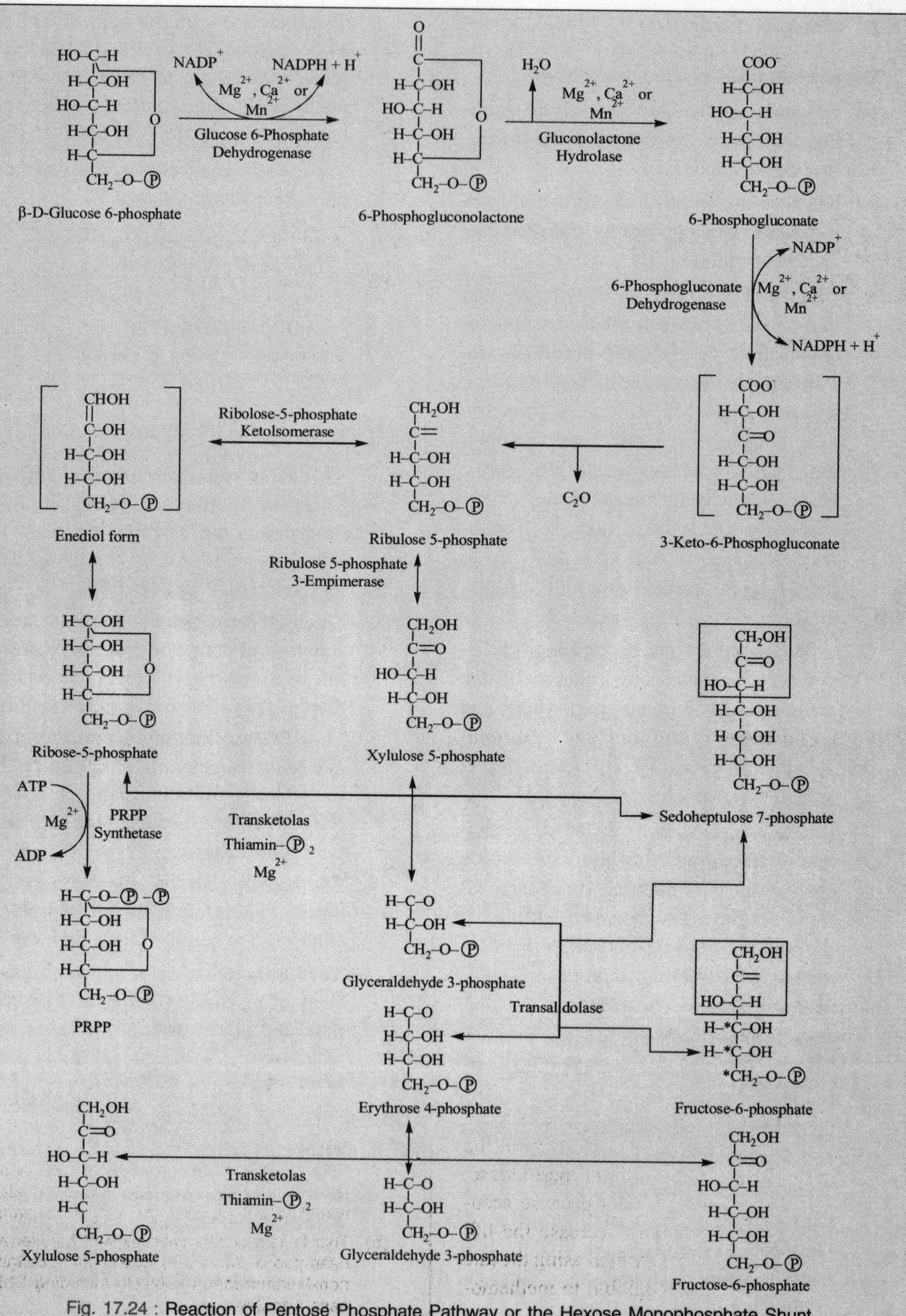

Fig. 17.24 : Reaction of Pentose Phosphate Pathway or the Hexose Monophosphate Shunt

2. The reduced NADP (NADPH) formed in this pathway is used in the synthesis of fatty acids, cholesterol or steroids and in the synthesis of amino acids via glutamate dehydrogenase outside the mitochondria. The reduced NADP is also required in uronic acid pathway for the synthesis of ascorbic acid and xylulose-5-phosphate. Ascorbic acid is not synthesized in man and other primates including guineapig. The system which utilizes reduced NADP stimulates an active degradation of glucose via this shunt pathway. The reduced NADP functions in the operation of the shunt pathway in red blood cells and it has a direct correlation between the glucose-6-phosphate dehydrogenase and the fragility of red cells (susceptibility to hemolysis) when the cells are subjected to certain drugs such as primaquine and sulfonamide.
3. The pentose sugars produced in this pathway are utilized for the synthesis of nucleic acids and nucleotides.

   Skeletal muscle is incapable of synthesizing ribose. This is probably accomplished by a reversal of the shunt pathway utilizing fructose-6-Ⓟ, glyceraldehyde-3-Ⓟ and the enzymes transketolase and transaldolase.
4. The products fructose-6-Ⓟ and glyceraldehyde-3-Ⓟ are utilized in Embden-Meyerhof pathway for glycolysis.
5. This pathway in the erythrocytes provides reduced NADP (NADPH + H^+) for the reduction of oxidized glutathione (G-S-S-G) to the reduced glutathione by the enzyme glutathione reductase. The reduced glutathione then removes H_2O_2 from the erythrocytes by glutathione peroxidase. This reaction is important because accumulation of H_2O_2 may decrease the life span of erythrocytes by increasing the rate of oxidation of hemoglobin to methemoglobin.

CLINICAL ORIENTATION

- **Hemolytic anemia due to glucose-6-phosphate dehydrogenase deficiency :**
  - (a) Hemolytic anemia is caused by the deficiency of the enzyme glucose-6-phosphate dehydrogenase with the impairment of generation of NADPH which is manifested as red cells hemolysis when the susceptible individuals are subjected to antimalarial drugs like primaquine and aspirin etc.
  - (b) The disease is generally transmitted by a sex-linked gene of intermediate dominance.
  - (c) This defect is more abundant in Negroes and also found in non-Negro races, e.g., Italians. Other races like Indians, Chinese, Malayan and thais are also affected by this defect. In India, the genetic disorder is also frequent in Parsis, Punjabis, Sindhis and Kutchiluhans.
  - (d) The severity of hemolysis depending on the dose of antimalarial drugs shows anemia, fall in blood hemoglobin, may be jaundice and blackenning of urine.
  - (e) The oxidant drugs which may cause hemolysis are primaquine, phenacetin, sulphonamide group of drugs, furadantin, furoxone, sulphoxone, phenylhydrazine etc.
  - (f) This defect confers some protection against P falciparum infection in the prevalence of genetic disorder and it may reduce the severity of malarial infections in young children and infants.
  - (g) This disorder shows a close linkage with colour blindness.
- **Wernicke-Korsakoff syndrome**
  - (a) A genetically "variant" form of transketolase. The "transketolation" reaction is affected since the enzyme transketolase cannot bind TPP.
  - (b) The patient suffering from this disorder shows neuropsychiatric symptoms having lesions and hemorrhages near the III cerebral ventricles, depression, deranged mental functions, mental confusions and loss of memory.
  - (c) There maybe paralysis of eye-movements.
- **Favism :**
  - (a) In glucose-6-phosphate dehydrogenase deficient persons with acute hemolytic anemia and hemoglobinuria become sensitive to the "fava beans" either on ingestion with uncooked or lightly cooked beans or on inhalation of pollens from the blossom of the plant.
  - (b) This favism occurs typically with Mediterranean people and it may also occur in certain non-Mediterranean subjects including Chinese and Jews.

URONIC ACID PATHWAY

1. This pathway is the alternative oxidative pathway for glucose. Glucose is converted to glucuronic acid, ascorbic acid and pentoses.
2. In man and other primates as well as in guineapigs, ascorbic acid cannot be synthesized and gulonic acid is ultimately converted to L-xylulose.

Metabolic reactions

1. Glucose-6-phosphate is converted to glucose-1-phosphate by phosphoglucomutase. The product then reacts with UTP to form UDPG by the help of the enzyme UDPG pyrophosphorylase.
2. UDPG is oxidized to uridine diphosphoglucuronic acid by UDPG dehydrogenase in presence of NAD^+. UDP-glucuronic acid is the "active" form of glucuronic acid for reactions involving incorporation of glucuronic acid into chondroitin sulfate.
3. Glucuronic acid is reduced to L-gulonic acid in the presence of reduced NADP. L-gulonic acid is the precursor of ascorbic acid.
4. In man and other primates, as well as in guineapigs, L-gulonic acid is oxidized to 3-keto-L-gulonic acid which is then decarboxylated to L-xylulose. L-xylulose is converted to D-xylulose in presence of reduced NADP. D-xylulose, being converted to D-xylulose-5-phosphate by ATP, is further metabolized in the HMP shunt.
5. In the rare hereditary disease called "essential pentosuria", large quantities of L-xylulose appear in the urine due to the

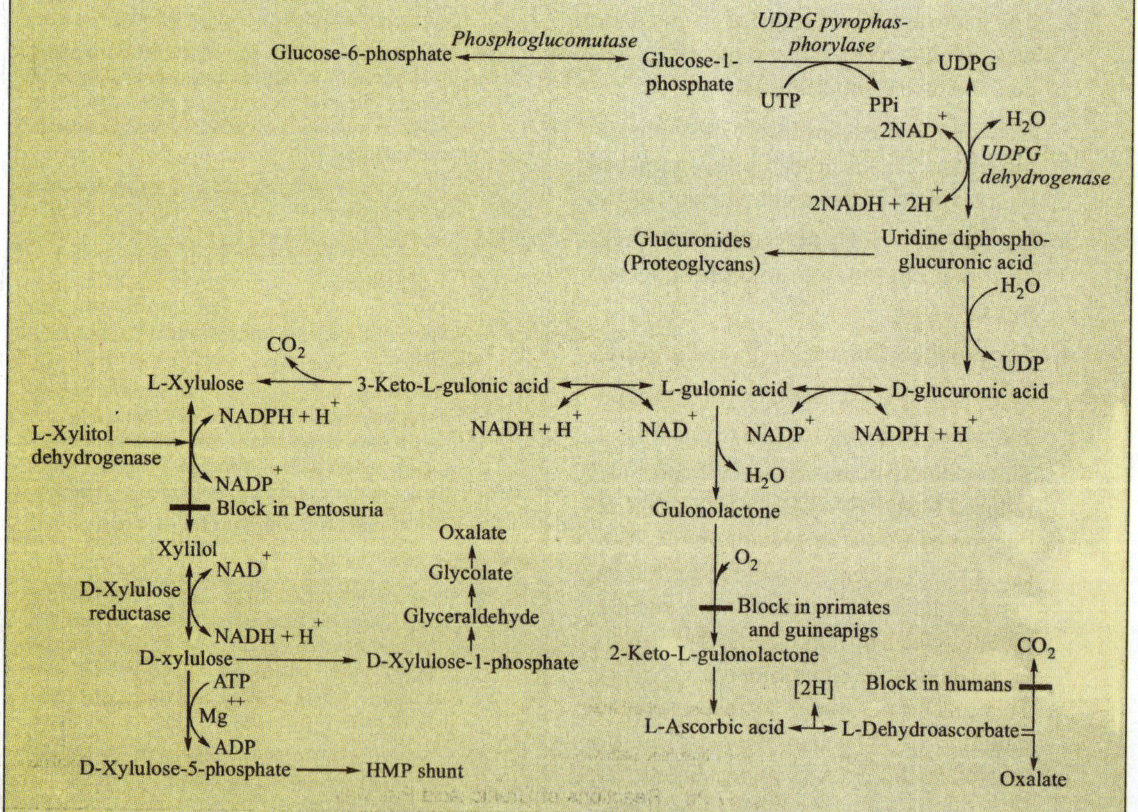

Fig. 17.25 : Flow chart of Uronic Acid Pathway

absence of the enzyme L-xylitol dehydrogenase which can cause reduction of L-xylulosè to xylitol.

Importance of glucuronic acid

1. D-glucuronic acid combines with benzoic acid to form benzoyl glucuronide.
2. Phenol combines with D-glucuronic acid to form phenylglucuronide.
3. Bilirubin is conjugated with D-glucuronic acid to form mono and di-glucuronides.
4. The drugs and other xenobiotics are first hydroxylated by monooxygenase cytochrome P-450 system and then conjugated with D-glucuronic acid.

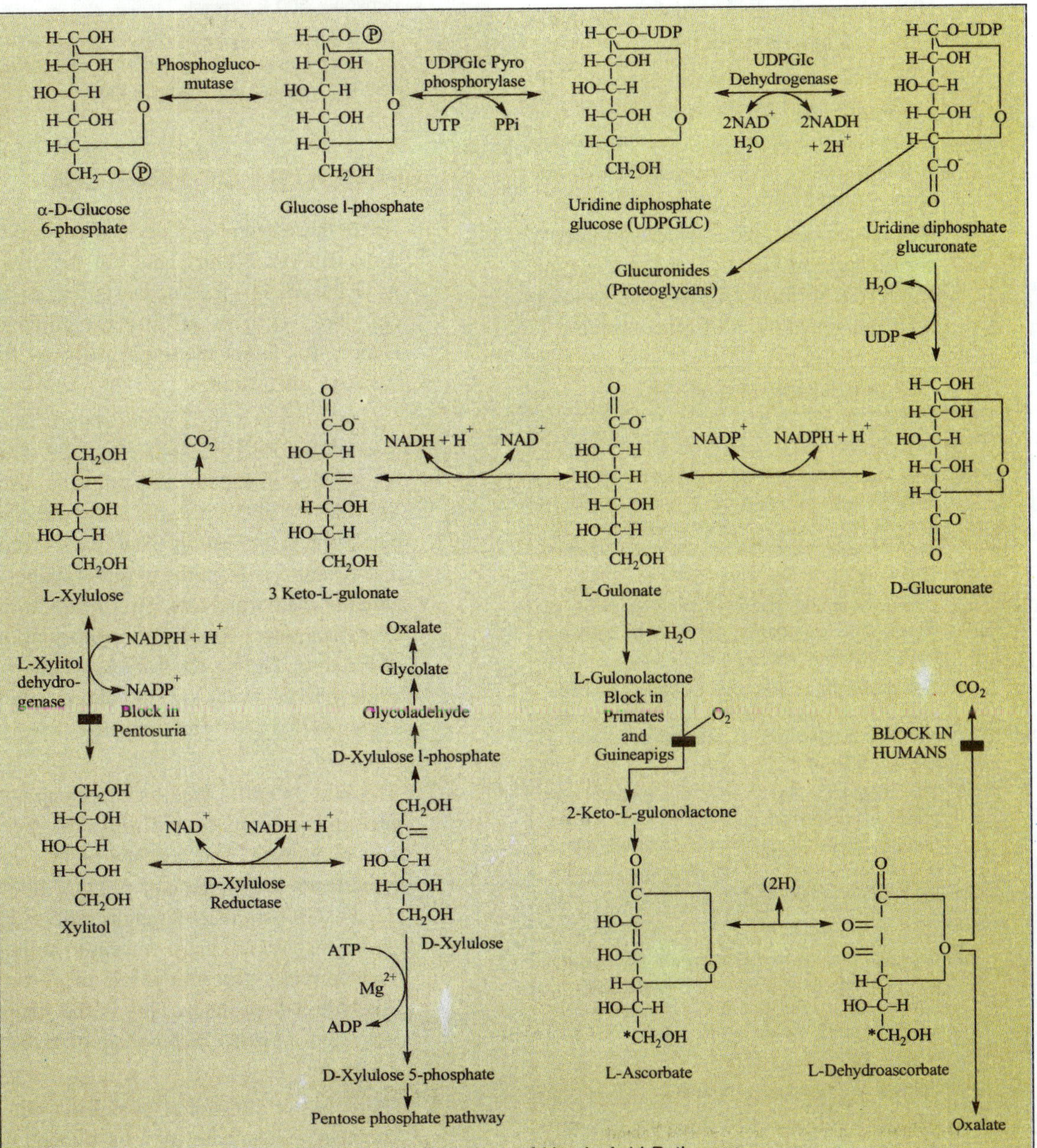

Fig. 17.26 : Reactions of Uronic Acid Pathway

5. Certain antibiotics like chloramphenicol is also conjugated with D-glucuronic acid.
6. Certain hormones like thyroid hormones and derivatives of certain steroid hormones are detoxicated by D-glucuronic acid.
7. D-glucuronic acid is incorporated to hyaluronic acid, chondroitin sulphate and heparin.
8. UDP-glucuronic acid, which is the active form, acts as a donor in liver.
9. UDP-glucuronic acid is changed to UDP-L-iduronic acid by the help of UDP-glucuronic acid-5-epimerase. L-iduronic acid is incorporated in forming dermatan sulphate (chondroitin sulphate B) in skin.
10. UDP-glucuronic acid is decarboxylated in cornea and cartilage by a specific enzyme and NAD^+ to form UDP-xylose which is used in the synthesis of mucoprotein.

CLINICAL ORIENTATION

- **Essential Pentosuria**
  - (a) This is an inherited disorder caused by the deficiency of the enzyme L-xylitol dehydrogenase for which L-xylulose cannot be converted to xylitol resulting in the accumulation of L-xylulose which is excreted in urine.
  - (b) L-arabitol is also excreted in urine in some cases. This pentose sugar may be derived from L-xylulose by reduction.
  - (c) The presence of Pentose in urine becomes harmless to health. This is found almost exclusively in Jews and causes no decrease in life expectancy.
  - (d) The subjects suffering from this disorder excrete an increased quantity of xylulose on administration of glucuronic acid; while the normal subjects do not respond like this after the administration of glucuronic acid.
- **Oxalosis**
  - (a) Oxalosis occurs by the parental administration of xylitol having calcium oxalate deposition in brain and kidneys. This oxalate is formed from D-xylulose as shown :

    D-xylulose → D-xylulose-1- P → Glyceraldehyde → Glycolate → Oxalate.

    The other sources of oxalates are ascorbic acid (Vit. C) and glycine.

CLINICAL ORIENTATION

- **Effect of drugs :**
  - (a) The conversion of glucose to glucuronic acid, L-gulonic acid and ascorbic acid is markedly increased by the administration of barbital or chlorobutanol to rats. The synthesis of L-ascorbic acid is also significantly increased by the administration of barbiturates, aminopyrine, and antipyrine.
  - (b) The drugs aminopyrine and amidopyrine also cause an increase in the excretion of L-xylulose in essential pentosuric subjects.

METABOLISM OF FRUCTOSE

1. Hexokinase can phosphorylate fructose to form fructose-6-phosphate but the affinity of the enzyme for fructose is very small compared with its affinity for glucose. Hence, this is not the major pathway for fructose utilization.
2. Fructokinase present in liver converts fructose to fructose-1-phosphate by ATP. This takes place in kidney and intestine also. The activity of this enzyme is not affected by fasting or by insulin. That is why fructose disappears from the blood of diabetic patients at a normal rate. This is why fructose disappears for the phosphorylation of fructose. The K_m for fructose of the enzyme in liver is very low, indicating a very high affinity of the enzyme for its substrate.
3. Aldolase B splits fructose-1-phosphate into glyceraldehyde and dihydroxyacetone phosphate. The absence of this enzyme leads to a hereditary fructose intolerance. Glyceraldehyde again enters the glycolysis. Triosekinase present in human liver converts glyceraldehyde to glyceraldehyde-3-phosphate. This is the major pathway for further metabolism of glyceraldehyde.
4. There is the additional possibility that fructose-1-phosphate may be phosphorylated to fructose-1, 6-bisphosphate by

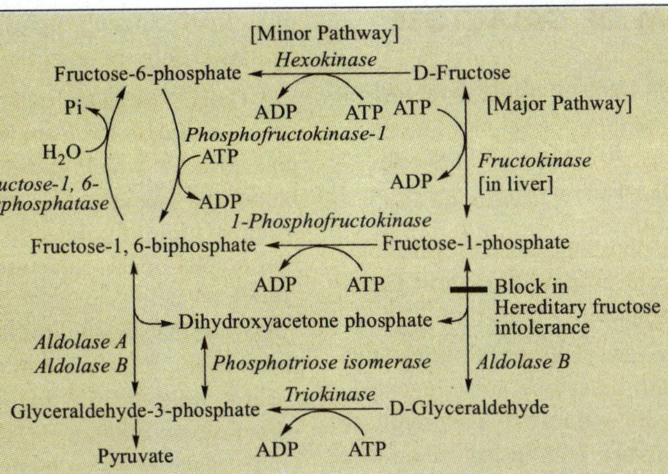

Fig. 17.27 : Metabolism of fructose

the catalyzing effect of 1-phosphofructokinase which has been found in muscle and in liver.

5. **Fructosuria** is a very rare condition in which fructose is found in the urine. This is a congenital defect–**essential fructosuria**–due to defect in its metabolism. This condition is harmless and is due to a deficiency of fructokinase.

The overall metabolism is given (Fig. 17.27).

CLINICAL ORIENTATION

- Fructose is abundantly present in seminal fluid. It is the principal energy source of spermatozoa. It is formed from glucose in the seminal vesicle in the following manner :

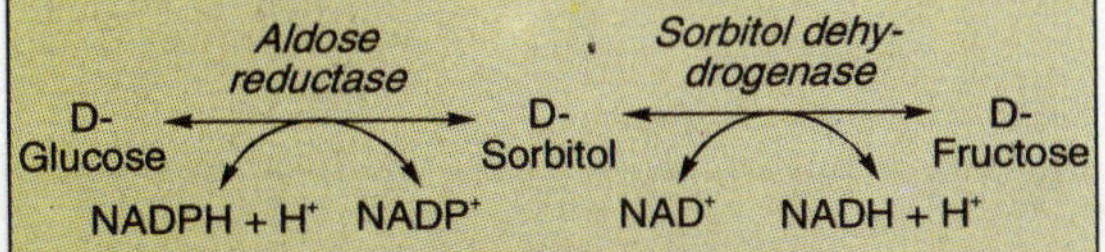

- Lactate dehydrogenase (LDH) is present in the mitochondria of the sperm. Lactate formed by fructolysis is completely oxidized to CO_2 and H_2O due to the presence of lactate dehydrogenase without shuttle system.
- In diabetic individuals, sorbitol level is highly increased in the lens of the eye since the oxidation of sorbitol to fructose is decreased. The increased level of sorbitol in these cells increases the osmotic pressure which is responsible for the development of cataracts of lens of the eye and diabetic neuropathies.
- Fructose is metabolized more rapidly than glucose by the liver since it by-passes the step in glucose metabolism catalyzed by phosphofructokinase-1. This allows fructose to enhance fatty acid synthesis, increased esterification and VLDL secretion which increases serum triglyceride level.
- Fructose, by parenteral use for nutrition, can cause depletion of adenine nucleotides in liver and liver cell necrosis.
- Sorbitol, by intravenous administration, is converted to fructose which escapes absorption from the gut and is fermented in the colon by intestinal bacteria to produce acetate and hydrogen. The abdominal pain is caused by sugar-free" sweeteness containing sorbitol (**"sorbitol intolerance"**).
- **Hereditary fructose intolerance** is an inherited disorder which is caused due to the deficiency of the enzyme **Aldolase B**. After the administration of fructose in the patients suffering from this disorder, the developing symptoms are :
  (a) Excessive and prolonged rise of fructose in blood.
  (b) Hypoglycemia accompanied by nausea and vomiting, profuse sweating.

After cessation of symptoms there may be albuminuria, aminoaciduria, and decreased serum inorganic phosphate. The cause of hypoglycemia may be due to excessive insulin secretion and inhibition of phosphoglucomutase by fructose-1-phosphate.

METABOLISM OF GALACTOSE

Galactose is available by the hydrolysis of lactose of milk.

Metabolic reactions

1. Galactose is phosphorylated by galactokinase in presence of ATP to form galactose-1-phosphate.
2. Galactose -1-phosphate reacts with UDPG to form UDP galactose and glucose-1-phosphate by galactose-1-phosphate uridyl transferase. UDP galactose may be converted to UDP glucose by uridine diphosphate galactose epimerase. UDP glucose can form glycogen.
3. UDP galactose condenses with glucose to form lactose in the mammary gland by lactose synthetase.

Galactose is required in the body not only in the formation of milk but also as a constituent of glycolipids, mucoproteins. Galactokinase shows increased activity upon the feeding of galactose.

Galactosemia, an inherited metabolic disease, occurs owing to the inability to metabolize dietary galactose for which galactose accumulates in the blood and passes to the urine. There is also high accumulation of galactose-1-phosphate in the red blood cells of the galactosemic individual, which shows no deficit of galactokinase. Recently, it is suggested that an inherited lack of galactose-1 phosphate uridyl transferase in the liver and red blood cells is responsible for glactosemia. Still, in galactosemic individual, there is the formation of UDP galactose by epimerization. In case it is untreated it may lead to cataracts, mental retardation and death. The condition may be treated by eliminating milk from the diet. The biochemical findings are increased blood galactose, hypoglycemia, decreased blood inorganic phosphate, galactosuria, albuminuria, aminoaciduria.

The overall metabolism is schemed in Fig. 17.28.

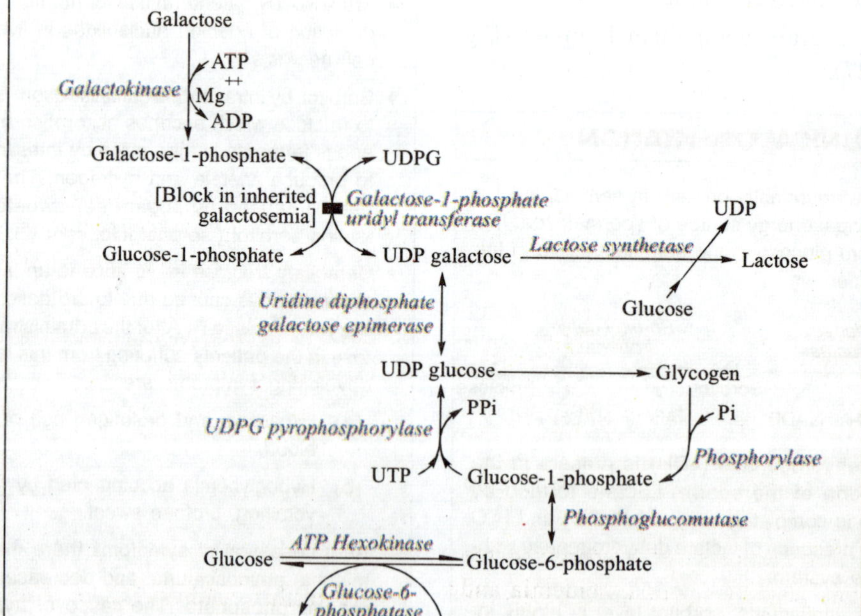

Fig. 17.28 : Metabolism of galactose

SCHEMATIC COMPENDIUM

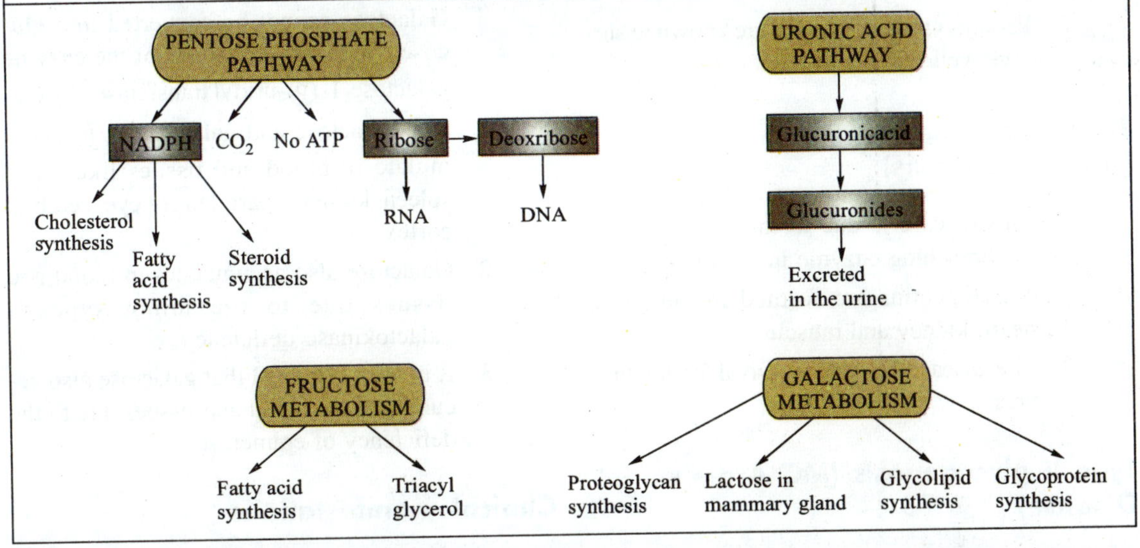

GLYCOGEN STORAGE DISEASES

The term "glycogen storage disease" is a genetic one in which abnormal quantities of glycogen are deposited in the liver, kidney, heart and muscle. This is due to the enzyme deficiency. Seven main types have been recognized :

Glycogenosis : It is a group of inherited disorder or genetic disorder and characterised by deposition of abnormal glycogen in tisue.

Type I glycogenosis (von Gierke's Disease)

1. The disease is due to the deficiency of glucose-6-phosphatase for which glycogen cannot be broken down to liberate glucose and glucose-6-phosphate promotes glycogen synthesis.
2. Children with this disease tend to develop hypoglycemia.
3. They use fat mostly as an energy source and this leads to lipemia, acidemia and ketosis.
4. There is fatty infiltration of the liver.
5. The hypoglycemia inhibits insulin secretion which, in turn, also inhibits protein synthesis and growth is ceased.
6. Hypoglycemia stimulates epinephrine production which causes the breakdown of muscle glycogen forming lactate. This lactate competes with urate for excretion by the kidney. So blood urate level is increased.

Type II glycogenosis (Pompe's Disease)

1. The disease is due to the deficiency of a lysosomal enzyme, acid maltase.
2. There is excessive amount of glycogen in all tissues.
3. The heart is enlarged and there is extreme muscle weakness.
4. Death occurs usually before the ninth month of life.

Type III glycogenosis (Limit Dextrinosis)

1. In this condition, there is an accumulation of limit dextrin in liver and muscles. This substance is produced by the action of phosphorylase on glycogen.

2. Debrancher enzyme system is deficient in this disease.
3. Patients with this disease are known to survive well into adult life.

Type IV glycogenosis (Amylopectinosis)

1. This disease is due to the deficiency of the branching enzyme in the liver.
2. Amylopectins are formed in the liver, heart, kidney and muscle.
3. The disease is fatal, survival being four years.

Type V glycogenosis (MC Ardle's Disease)

1. In this disease, there is a defect of muscle phosphorylase.
2. Cramp occurs in muscle after moderate exercise and recovery is attained during rest.
3. The exercised muscles can metabolize fructose.

Type VI glycogenosis

1. There is always a deficiency of liver phosphorylase.
2. Liver glycogen concentration is usually increased.

Type VII glycogenosis

1. This disease is due to the deficiency of phosphofructokinase.
2. There is marked accumulation of glucose-6-phosphate and fructose-6-phosphate and a marked diminution of fructose-1, 6-bisphosphate.

GALACTOSEMIA

This is an inherited disorder in which there is high concentration of galactose in the blood. The incidence occurs in 1 in 18,000 births.

Causes

1. Galactose cannot be converted into glucose due to the deficiency of the enzyme galactose-1-Ⓟ-uridyl transferase. As a result, galactose and galactose-1-Ⓟ accumulate in blood and tissues like liver, spleen, kidney, heart, lens of eye, cerebral cortex.
2. Galactose also accumulates in blood and tissues due to the minor type of galactokinase deficiency.
3. It is rarely claimed that galactose also accumulates in blood and tissues due to the deficiency of epimerase.

Clinical manifestations

1. Infants become lethargic, fail to thrive, hypoglycemic due to nonconversion of galactose to glucose, may vomit and may suffer from jaundice.
2. Infants after two to three months may suffer from cirrhosis of liver, mental retardation, development of cataracts.

Biochemical findings

1. Increased blood galactose level.
2. Decreased blood sugar level.
3. Decreased inorganic phosphate level.
4. Galactosuria.
5. Albuminuria.
6. Aminoaciduria.

HEREDITARY FRUCTOSE INTOLERANCE

1. This inherited disorder occurs due to the deficiency of the enzyme Aldolase B for which fructose-1 phosphate cannot be further metabolized.
2. Fructose administration in these patients leads to :

   (a) Elevated level of fructose and fructose-1-Ⓟ in blood.

(b) Hypoglycemia accompanied by nausea, vomiting, and profused sweating. This may be due to excessive insulin secretion.

(c) Albuminuria, aminoaciduria.

Mellituria

This is a condition in which sugars are excreted in urine.

LACTOSURIA

Lactosuria occurs in women during the period of lactation. Small to moderate amounts of lactose may be found in the urine of most of the pregnant women, the amount increases as pregnancy advances. It appears more frequently in the afternoon.

MALTOSURIA

This is a rare condition of no known clinical significance.

FRUCTOSURIA

Fructose may appear in the urine under the following circumstances :

1. The patients with hepatic insufficiency excrete fructose in urine when large quantities of fructose are ingested.
2. Essential fructosuria is a rare congenital disorder in which there is the deficiency of fructokinase and characterized by inability to utilize fructose completely. Insulin has no influence upon this condition.

There are two varieties of this condition. In one, the metabolism of other carbohydrates is undisturbed and there are no clinical symptoms. In another form, the rise in blood fructose is accompanied by a sharp drop in the blood glucose concentration with severe symptoms of hypoglycemia. This is due to the deficiency of aldolase B in the liver resulting in accumulation of fructose-1-phosphate which blocks important pathways of fructose utilization.

PENTOSURIA

Pentose may appear in the urine under the following circumstances :

1. After the ingestion of large quantities of fruits, there is alimentary pentosuria occurring in normal individuals. It has no clinical significance except the wrong idea for glycosuria.
2. Essential pentosuria is due to "inborn errors of metabolism". This occurs owing to the lack of the enzyme L-xylitol dehydrogenase which causes the reduction of L-xylulose to xylitol in the liver. As a result, L-xylulose is excreted in urine. It occurs practically in males of Jewish subjects. The utilization of other carbohydrates is unimpaired.

ENDOCRINE (HORMONAL) INFLUENCES ON CARBOHYDRATE METABOLISM

The supply of glucose to the blood by the liver and its utilization in the tissues is regulated and integrated with one another and also by the metabolism of proteins and lipids. Endocrine organs play a very important role in the metabolism of carbohydrate which is discussed :

A. Insulin

1. Causes glucose transport across cell membrane by increasing the permeability of the cells.
2. Increases glycogen formation in liver and muscle by stimulating hexokinase and glycogen synthetase.
3. Accelerates the conversion of glucose to fat.
4. Stimulates protein synthesis.
5. Inhibits ketogenesis.
6. Inhibits gluconeogenesis.
7. Increases glycolysis by stimulating hexokinase and phosphofructokinase in liver and muscle.

8. Has effect on muscle ion balance.
9. Causes initial phosphorylation (hexokinase reaction) of glucose for further oxidation.

The **sulfonylureas**–eg, tolbutamides–are the stimulators of insulin release.

Insulin is destroyed by **insulinase** which separates the A and B chain of it. **Peptidases** also play a part in the destruction of insulin.

B. Adrenocortical hormones

The adrenalectomized animal exhibits the following changes in carbohydrate metabolism :

1. Decrease in liver glycogen.
2. Less marked decrease in muscle glycogen.
3. Hypoglycemia.
4. Decreased intestinal absorption of glucose.

The changes are prevented by glucocorticoids. The glucocorticoid hormones release gluconeogenic precursors, such as amino acids, from muscle. The hormones also release FFA from adipose tissue, resulting in inhibition of hepatic glycolysis. The increase in FFA in the liver results in an increase of acetyl-CoA which activates pyruvate carboxylase and so stimulates gluconeogenesis.

C. Anterior pituitary factors

Growth hormone produces the following changes in carbohydrate metabolism :

1. Hyperglycemia and aggravation of diabetes in depancreatized animals.
2. Inhibition of insulin action with decreased utilization of carbohydrate and lowering of the respiratory quotient.
3. Diminution of the breakdown of muscle protein.
4. Increase in the mobilization of fat in the form of FFA from adipose tissue with the increase in lipase activity. The increase in FFA, and its use as the source of energy results in the excessive production of ketone bodies.
5. ACTH enhances the release of free fatty acids from adipose tissue and inhibits glucose utilization. It causes increase in blood sugar level by stimulating the secretion of adrenal cortex hormones.

D. Epinephrine

1. Causes an increase in blood sugar and in blood lactate due to the acceleration of **glycogenolysis** in the liver and muscles.
2. Diminishes the uptake of glucose by tissue cells, thus interfering with its utilization.
3. Causes diminution in the amount of insulin released from the pancreas.

E. Thyroid hormone

1. Thyroxine accelerates hepatic glycogenolysis with consequent rise in blood sugar. This is due to increased sensitivity to epinephrine.
2. Thyroxine also increases the rate of absorption of hexoses from the intestine and accelerates **gluconeogenesis**.
3. It has diabetogenic action.

F. Glucagon

1. Promotes glycogenolysis in liver stimulating phosphorylase via cyclic AMP.
2. Causes breakdown of glycogen to lactic acid in muscle. The lactic acid is again converted to glucose by **"Cori Cycle"**.
3. Stimulates gluconeogenesis in liver.

G. Sex hormones

Estrogens stimulate increased insulin secretion and thus reduce blood sugar level. **Testosterone** also decreases blood sugar level.

REGULATION OF BLOOD GLUCOSE CONCENTRATION

In the resting, postabsorptive state (overnight fast), the true glucose present in the venous blood is 70-115 mg/100 ml, the value is higher, i. e. 80-120 mg/100 ml including nonsugar reducing substances. After the ingestion of a carbohydrate meal, it may rise to 115-160 mg/100 ml. The normal blood glucose level of sheep is 40 mg/100 ml. and of cattle 60 mg/100 ml. These lower values are due to the fact that ruminants ferment virtually all dietary carbohydrate to lower fatty acids which replace glucose as the main metabolic fuel. The condition in which the blood sugar is raised above the normal range is called **hyperglycemia** and when it goes below the normal level, it is called **hypoglycemia**. The concentration of glucose in the blood depends on two general factors : (i) the rate of its entrance into the blood, (ii) the rate of its removal from the blood stream.

Sources of blood glucose

A. From carbohydrates of the diet

1. Most carbohydrates in the diet after digestion form glucose, galactose or fructose which are absorbed into the portal vein.
2. In the liver, galactose and fructose are converted to glucose.

B. From various glucogenic compounds

1. In the skeletal muscle, glucose is oxidized to lactic acid which is transported by blood to the **liver** and **kidney** where glucose is reformed and undergoes oxidation in the tissues via the circulation. This process is known as the **"Cori Cycle"** or **"Lactic Acid Cycle"**.
2. In the adipose tissue, the synthesis of triacylglycerol takes place from glycerol which is derived initially from blood glucose. Acylglycerols of adipose tissue continually undergo hydrolysis to form free glycerol which diffuses out of the tissue into the blood. This is converted back to glucose in the liver and kidney by gluconeogenesis. Hence, a continuous cycle exists in which glucose is transported to adipose tissue from the liver and kidney and glycerol is returned to be synthesized into glucose by the liver and kidney.

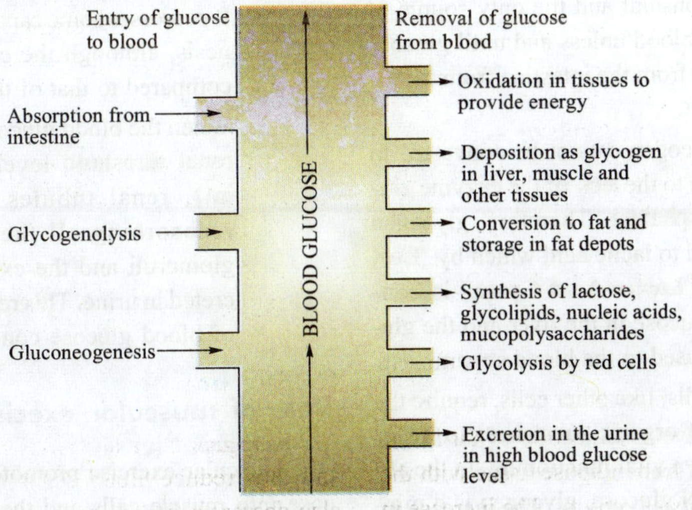

Fig. 17.29 : Entrance and removal of blood glucose

3. During starvation, the amino acid, alanine, is transported from muscle to liver with an effect of a net transfer of amino nitrogen from muscle to liver and of free energy from liver to muscle. The energy required for the synthesis of glucose from pyruvate in the liver is derived from fatty acid oxidation.

C. From liver glycogen by glycogenolysis

Glucose is formed in the liver from glycogen by phosphorylase and debranching enzyme and also by glucose-6-phosphatase by the effect of the hormone epinephrine and glucagon.

Regulation of the blood glucose

The stable blood glucose level is maintained by the role of liver, skeletal muscle, kidney, muscular exercise and hormones.

Role of liver

1. Liver is the pivot of carbohydrate metabolism of the whole body. The presence of glucose-6-phosphatase in the liver converts glucose-6-phosphate to glucose which diffuses into the blood stream to form the constant and the only source of glucose of blood unless and until glucose is available from the intestine from carbohydrate diet.
2. Muscle glycogen cannot be converted to glucose due to the lack of the enzyme glucose-6-phosphatase. Therefore, glycogen is converted to lactic acid which by "Cori Cycle" or "Lactic Acid Cycle" is converted to glucose in the liver and the glucose is diffused to the blood stream.
3. The liver cells, like other cells, require the oxidation of organic substances to maintain their own vital functioning. In the absence of fuel glucose, glycogen is diminished and the oxidation of fat occurs forming keto acids. Some of the keto acids are utilized for cellular energy. But if the concentration of keto acids is increased, the keto acids diffuse into the blood stream and accumulate producing ketosis.
4. When the glycogen reservoir diminishes, the amino acids of the body proteins are utilized by the liver for gluconeogenesis.

Role of skeletal muscle

1. Extrahepatic tissues are relatively impermeable to glucose and, therefore, insulin is required for the uptake of glucose to these cells.
2. Increased blood glucose promotes glycogenesis and oxidation of glucose in muscles. Muscle glycogen does not serve directly as a source of glucose during hypoglycemia. But glucose is supplied to the blood form muscle glycogen by "Cori Cycle" or "Lactic Acid Cycle".

Role of kidney

1. There is considerable evidence that the kidney is able to form glucose from a number of carbohydrate intermediates and also from amino acids.
2. It possesses some capacity for gluconeogenesis, although the capacity is minor as compared to that of the liver.
3. When the blood glucose level exceeds the renal threshold level (160-180 mg/100 ml), renal tubules are incapable of reabsorbing all the filtered sugar in glomeruli and the excess glucose is excreted in urine. This results in the decrease of blood glucose concentration.

Role of muscular execise

Muscular exercise promotes the entry of glucose into muscle cells and the glucose is utilized by the muscle. Thus, it lowers blood glucose level.

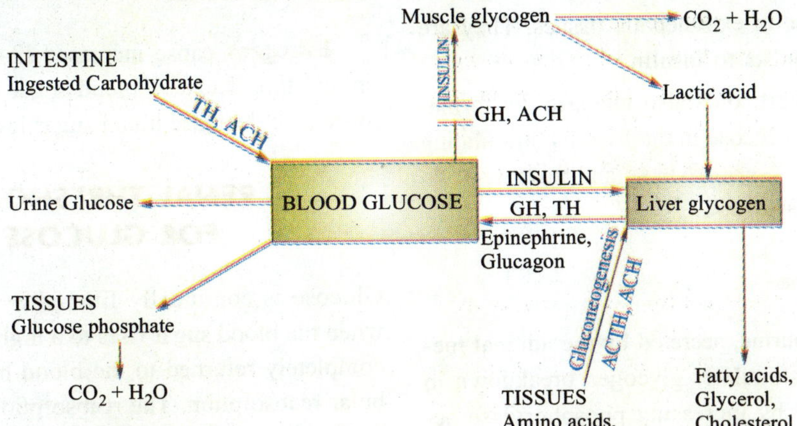

TH = Thyroid Hormone
ACH = Adreno Cortical Hormone
GH = Growth Hormone
ACTH = Adreno Corticotrophic Hormone

Fig. 17.30 : Factors controlling blood glucose concentration

Role of hormones

A. Insulin

1. It is produced by the β-cells of the islets of Langerhans in the pancreas and is liberated into the blood by the direct response to hyperglycemia. Insulin is released by amino acids, free fatty acids, ketone bodies, glucagon, secretin and tolbutamide. Epinephrine and norepinephrine block the release of insulin.
2. It increases the rate of uptake of glucose to tissues.
3. It promotes glycogenesis by stimulating hexokinase and glycogen synthetase and oxidation of glucose by stimulating phosphofructokinase.
4. It decreases hepatic glycogenolysis and gluconeogenesis.
5. It stimulates lipogenesis and protein synthesis.
6. It inhibits ketogenesis.

B. Anterior pituitary hormones

1. The anterior pituitary gland secretes hormones that elevate the blood sugar level by antagonizing the action of insulin. They are growth hormone, ACTH and other "diabetogenic" principles.
2. Growth hormone secretion is stimulated by hypoglycemia. It decreases glucose uptake in certain tissues, eg, muscle.
3. Growth hormone mobilizes free fatty acids from adipose tissue which themselves inhibit glucose utilization.
4. It produces hyperglycemia which stimulates secretion of insulin causing exhaustion of β-cells.
5. ACTH enhances the release of free fatty acids from adipose tissue and inhibits glucose utilization. It also increases blood glucose level by stimulating the secretion of adrenal cortex hormones.

C. Adrenal cortex hormones

1. Glucocorticoids (11-oxysteroids) are important in carbohydrate metabolism. They lead to gluconeogenesis by the increased protein catabolism in the tissues, increased hepatic uptake of amino acids and increased activity of transaminases in the liver.

2. Glucocorticoids inhibit the utilization of glucose in extrahepatic tissues. They are antagonists to insulin.
3. Glucocorticoids also increase the formation of glucose in the liver by stimulating glucose-6-phosphatase and fructose-1, 6-bisphosphatase.

D. Epinephrine

1. Epinephrine, secreted by the adrenal medulla, stimulates glycogen breakdown in muscle by increasing phosphorylase activity.
2. In muscle, glycogen is converted to lactic acid instead of glucose due to the lack of glucose-6-phosphatase. This lactic acid is converted to glucose in the liver by "Cori Cycle" and diffuses into the blood.
3. It diminishes the release of insulin from pancreas.

E. Glucagon

1. Glucagon is produced by the alpha cells of the islets of Langerhans of the pancreas, being stimulated by hypoglycemia. It causes glycogenolysis by activating phosphorylase in the liver.
2. It stimulates glucose-6-phosphatase in the liver to form glucose from glucose-6-phosphate.
3. It enhances gluconeogenesis from amino acids and lactate.

F. Thyroid hormone

1. Thyroxine has diabetogenic action.
2. It increases the rate of absorption of hexoses and accelerates gluconeogenesis.
3. It stimulates hepatic glycogenolysis with consequent rise in blood sugar. This is due to increased sensitivity to epinephrine.

G. Sex hormones

Estrogens cause increased liberation of insulin and thus decrease blood sugar level. Testosterones also decrease blood sugar level.

RENAL THRESHOLD FOR GLUCOSE

Glucose is continually filtered by the glomeruli when the blood sugar rises to a high level. But it is completely returned to the blood by the renal tubular reabsorption. The reabsorption is influenced by phosphorylation by enzymes. The capacity of the renal tubules to reabsorb glucose is limited to the rate of about 350 mg/minute. If the filtrate contains more glucose than can be reabsorbed, the excess passes into the urine to produce glycosuria.

Glycosuria occurs in the individuals when the venous blood sugar exceeds 160-180 mg/100 ml. This level of the venous blood sugar is said to be the renal threshold for glucose.

The maximal rate of reabsorption of glucose by the tubule (TmG– the tubular maximum for glucose) is constant. Hence, it is a more accurate measurement than the renal threshold which varies with the change of glomerular filtration rate.

DIABETES MELLITUS

1. Primary idiopathic or essential diabetes

Diabetes mellitus means the hyperglycemia due to lack of insulin. This also takes place due to the over-production of other hormones, eg, glucagon, hormones of the anterior pituitary, adrenal and thyroid which are antagonists to insulin or due to increased production of insulinase which inactivates insulin. In India, males are more prone to diabetis mellitus than females.

Classification

Primary diabetes is classified into the following subgroups, depending on the degree of severity of the disease :

(a) Potential diabetic (prediabetic) : The individual with normal glucose tolerance test but with a family history of the disease.

(b) Latent diabetic (suspected diabetic) : The individual with a normal glucose tolerance test but has a diabetic type of glucose tolerance curve after cortisone administration, pregnancy or severe infection.

(c) Asymptomatic diabetic (chemical diabetic) : The individual with a diabetic glucose tolerance curve but without signs of diabetes.

(d) Clinical diabetic (overt diabetic) :

(i) Juvenile-onset type,

(ii) Adult or maturity onset type.

Juvenile-onset type

(i) This condition develops during the first 40 years of life in normal or less than normal body weight.

(ii) The symptoms become severe and if insulin treatment is delayed, ketoacidosis and diabetic coma develop.

(iii) Insulin secretion is very low.

(iv) The β-cells of pancreas are exhausted and the cells become atrophied, resulting in diabetes mellitus.

Adult or Maturity-onset type

(i) This condition appears after 40 years of age who are obese.

(ii) This is a mild type of diabetes which can be controlled by diet and oral hypoglycemic drugs.

(iii) Insulin levels are lower than normal because the β-cells secrete less amount of insulin.

2. Secondary diabetes

Damage to pancreas : In chronic pancreatitis and pancreatic carcinoma, the pancreatic destruction results in absolute insulin deficiency.

Presence of insulin antagonists : Excess growth hormone secretion (Acromegaly) or excess glucocorticoid secretion (Cushing's syndrome) act as antagonists to insulin.

Inhibition of insulin secretion : The secretion of insulin by β-cells is inhibited by the excess secretion of epinephrine and thyroxine resulting in the breakdown of liver glycogen.

A diabetes-like condition can be produced in experimental animals by injecting alloxan which is a pyrimidine derivative containing 400×O groups. Alloxan destroys the β-cells of pancreas for which insulin production is inhibited. But the α-cells remain unaffected and the glucagon production remains intact. Dehydroascorbic acid, if injected in large doses, also destroys β-cells of pancreas and produces alloxan-type diabetes.

If growth hormone is repeatedly injected into experimental animals, the β-cells of the pancreas are stimulated to secrete insulin. The cells become hypertrophied and insulin secretion finally stops resulting in permanent diabetes mellitus in the animals.

Diabetic coma

1. In case of diabetes, if the insulin is not taken and the diet is not controlled, diabetes becomes severe leading to diabetic coma.
2. In diabetic coma, there is ketonemia and ketonuria.
3. The patient becomes semi-conscious or fully unconscious with dehydration.
4. The patient should be treated with soluble insulin (50 units) immediately followed by normal saline containing 10 per cent glucose intravenously.
5. The treatment should be controlled by frequent estimation of glucose in blood and urine every 30 minutes to make sure that blood glucose does not fall below normal level.

Treatment of diabetes mellitus

1. Administration of oral antidiabetic drugs like diabenese–which are not polypeptides–in moderate diabetes and insulin in severe diabetes.
2. High protein and low carbohydrate and fat diet are advisable. High protein meal serves as a prolonged source of carbohydrate without rapid hyperglycemic effect and it has protective effect on liver.
3. Exercise lowers the blood sugar level stimulating the β-cells of pancreas to liberate more insulin to act on blood sugar for glycogenesis and glycolysis.
4. The patient suffering from diabetes mellitus should have normal psychological affairs. The psychological stress and strain stimulate the alpha cells of pancreas and adrenal medulla causing the liberation of glucagon and epinephrine, respectively, which have glycogenolytic effect resulting in increased blood sugar (hyperglycemia).
5. A diabetic patient requiring more than 200 units/day of insulin is regarded as insulin resistant which is due to antigenic effect of insulin.

Complications of diabetes mellitus

1. Preventable complications are due to infections, injury, neglect in treatment, unsuitable diet or alcohol.
2. Delayed complications occur in all diabetics after some years since the onset of the disease. These are due to :
   (a) Hyperglycemia leading to cataract of the lens of the eyes.
   (b) Atherosclerosis involves coronary and peripheral arteries.
   (c) Hypertension, particularly systolic hypertension, is more.
   (d) Development of ulcers in the foot becomes a very difficult problem.
   (e) Severity of nephropathy results in Proteinuria.

DIABETIC KETOACIDOSIS

1. Insulin deficit may result from excess diet, irregular intake of insulin, vomiting and diarrhoea from any cause, or infection.
2. Ketoacidosis may be induced by myocardial infarction, severe urinary tract infection, acute cholecystitis, acute pancreatitis, hyperthyroidism. This condition is aggravated by increased activity of other hormones such as epinephrine, corticosteroid, growth hormone and glucagon.
3. Failure of adequate insulin action leads to hepatic over-production of glucose and peripheral under-utilization of glucose resulting in hyperglycemia, increased fatty acid catabolism, decreased fatty acid synthesis, decreased protein synthesis and increased protein catabolism. The large amounts of FFA released from adipose tissue are taken up by liver. The more acetyl-CoA formed from fatty acid oxidation, produces ketone bodies causing Ketonemia and Ketonuria.

Clinical manifestations

1. In case of the avoidance of insulin dose by the patient, the patient shows weakness, excessive thirst, fatigue, nausea, vomiting, abdominal pain.
2. In severe cases, the patient becomes unconscious. The symptoms on examination shows tachycardia, dehydration, fruity odour in breath. In more severe cases, the patient develops severe shock and, if untreated, results in coma.

Management

1. Ketosis by insulin administration.
2. Electrolyte imbalance by proper fluid therapy.
3. Infection by antibiotics or anti-bacterial drugs.
4. Treatment of any other precipitating factors.

Comparision between Hypoglycemic coma and Diabetic coma		
	Hypoglycemic coma	*Diabetic coma*
Cause	No food, overdose of insulin.	Normal food intake and no insulin.
Onset	Sudden and in good health.	The patient suffers from ill-health for several days.
Symptoms	Sweating, dizziness, tremor, mental confusion, semiconsciousness, headache, seeing double images.	Dehydration, abdominal pain and vomiting.
Signs	Moist skin, full pulse, normal or raised blood pressure, normal breathing.	Dry skin, weak pulse, low blood pressure, air hunger.
Urine	Ketone bodies absent, glucose absent.	Ketone bodies and glucose present.
Blood	Hypoglycemia, normal blood bicarbonate. Sugar level comes down to 30 mg/100 ml.	Hyperglycemia, reduced blood bicarbonate. Sugar level reaches to 550 mg/100 ml.

CHANGES IN CARBOHYDRATE, FAT AND PROTEIN METABOLISM IN DIABETES MELLITUS

Carbohydrate metabolism

1. Diminished entry and oxidation of glucose in muscle and other tissues.
2. Decreased glycogen formation in liver and muscle.
3. Decreased synthesis of fat and increased breakdown of glycogen in liver.

Fat metabolism

1. Fats are mobilized from adipose tissue and fatty acids are oxidized producing large amounts of acetyl-CoA. Due to lack of carbohydrate metabolism, acetyl-CoA is not oxidized via TCA cycle. Hence, two molecules of acetyl-CoA condense to form ketone bodies.
2. The ketone bodies in large amounts in blood produce ketosis which leads to coma in severe diabetes mellitus.
3. Administration of insulin corrects the condition.

Protein metabolism

1. Since the energy is not available by oxidation of carbohydrate and fat, the breakdown of tissue protein occurs leading to negative energy balance. The synthesis of tissue protein from dietary proteins requires insulin. Due to lack of insulin, tissue protein is not formed from dietary protein. The concentration of amino acids in blood and liver increases. The amino acids are deaminated in liver and glucose is produced from ketoacids by gluconeogenesis. The urea excreted more in urine. The body will be in negative nitrogen balance.
2. Administration of insulin corrects the condition. Insulin promotes oxidation of glucose to meet the energy needs and also the synthesis of tissue protein from dietary protein. Insulin also suppresses gluconeogenesis.

GLYCOSURIA

The condition in which abnormal quantities of glucose are excreted in urine is called glycosuria.

Normal urine contains traces of glucose which cannot be detected by Benedict's test. Beyond the renal threshold value (160-180 mg/100 ml), the tubules cannot reabsorb glucose which escapes reabsorption and is excreted in urine. Some individuals have low renal threshold value. Glycosuria occurs in two conditions : (A) In normal blood glucose level, and (B) In hyperglycemia.

A. Glycosuria in normal blood glucose level

1. **Alimentary glycosuria :** Some individuals excrete glucose in urine after the intake of large amounts of sugar or carbohydrate-rich meal in spite of their normal blood glucose level. They have normal renal theshold value for glucose but their blood glucose level shoots up (200 – 220 mg/100 ml) for a short period. Hence, a **transitory glycosuria** occurs.
2. **Emotional glycosuria :** Glycosuria occurs during periods of excessive nervous strain and emotional excitement such as intense fear, anger and severe anxiety due to increased secretion of epinephrine. This condition is termed **"emotional"** or **"psychic" glycosuria.**

   Emotional glycosuria has been observed in college students sitting for the final examination, worried athletes and candidates for competitive examinations.

   Glycosuria results in the use of ether and chloroform as anaesthetics due to **hypersecretion of epinephrine.**
3. **Glycosuria in pregnancy and lactation :** Glycosuria occurs in normal pregnant women, particularly in the later months, due to temporary reduction in maximum tubular reabsorption capacity of glucose and partly due to decreased glucose tolerance caused by temporary hypertrophy of the pituitary gland.

   Lactose is also excreted in urine during late pregnancy and lactation which should not be mistaken for glucose.
4. **Renal glycosuria (Hereditary glycosuria) :** Some individuals have low renal threshold value for glucose which may be below 150 mg/100 ml. This condition is harmless and is known as **"diabetes innocens"** or **"benign glycosuria"**. These individuals have an impaired tubular reabsorption for glucose; it is a hereditary defect.

   The following are the standards for the diagnosis of true renal glycosuria :

   (i) Normal blood sugar level in fasting.

   (ii) Glucose is present in every sample of urine, either in fasting state or after a meal.

   (iii) Carbohydrate utilization is normal.

   (iv) No disturbance of fat metabolism. Ketosis may develop when the patient fasts than when he overeats.

   (v) Moderate doses of insulin have little or no effect upon the glycosuria.

Practical danger in making the diagnosis of renal glycosuria lies in the confusion of this condition with mild diabetes mellitus. No metabolic disturbance occurs in subjects with renal glycosuria as long as the carbohydrate intake is adequate to compensate for the amount lost in the urine.

Artificially, a severe renal glycosuria can be produced by poisoning the kidney with phlorrhigin which prevents reabsorption of glucose from the tubules. As a result, there is a continuous loss of glucose and the blood sugar level becomes abnormally low. This hypoglycemia causes the secretion of epinephrine and glucagon which accelerate liver glycogenolysis. Due to the exhaustion of liver glycogen, glucose is formed from the tissue protein to maintain the blood sugar level. The fasting urine, on analysis, has shown that the ratio glucose excreted/N excreted (the **G/N** or **D/N** ratio) is constant.

B. Glycosuria in hyperglycemia

Glycosuria occurs due to increased blood glucose level in diabetes mellitus. The blood glucose level in this condition become very high and glucose is excreted in urine.

GLUCOSE TOLERANCE TEST (GTT)

It is the test to measure the degree and duration of hyperglycemia after giving a known amount of glucose.

The ability of body to utilize glucose is done

by measuring its glucose tolerance. It has a diagnostic value.

Type

1. Oral glucose tolerance.
2. Intravenous glucose tolerance.
3. Cortisone stress glucose tolerance test.
4. Multiple dose glucose tolerance test.

Utility

1. To detect the cause of abnormality of carbohydrate metabolism.
2. To detect cause of asymptomatic glycosuria.

Procedure

1. Patient should take normal meal or carbohydrate diet 3 or 4 days before the test.
2. Test is usually done after a overnight fasting. (It may be done 5 hours after fasting)
3. Fasting blood and urine samples are collected.
4. Patient should take 75 grams of glucose dissolved in 200 ml of distilled water.
5. Blood is taken every 30 minutes up to $2\frac{1}{2}$ hour (5 samples are taken).
6. Urine is collected after taking glucose at the end of 1 hour and at the end of 2 hours (3 samples).
7. Blood sugar is estimated by blood sample and urine is examined in the presence of sugar; there is a graphical representation wherein time (in hours) is along base line and blood glucose or sugar (mg/dl) along vertical line.

Different types of curves

A. Normal response
B. Abnormal response

A. Normal response

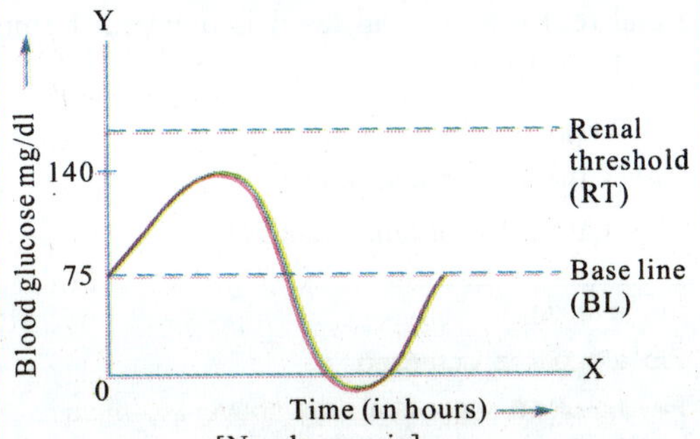

[No glycosuria]

(a) Initial rise of blood sugar is normal and it never crosses renal threshhold.
(b) It returns to normal within 1 hours usually, and almost always within 2 hours.

B. Abnormal response

(i) Diminished glucose tolerance.
(ii) Increased glucose tolerance.
(iii) Lag curve.
(iv) Decreased renal threshold.
(v) Increased renal threshold.

(i) Diminished glucose tolerance

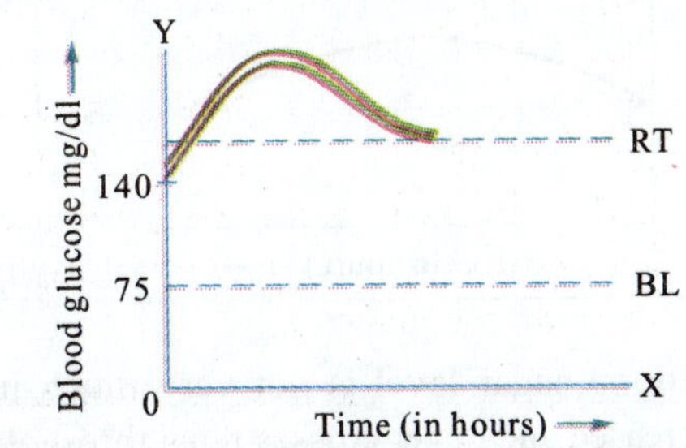

(a)

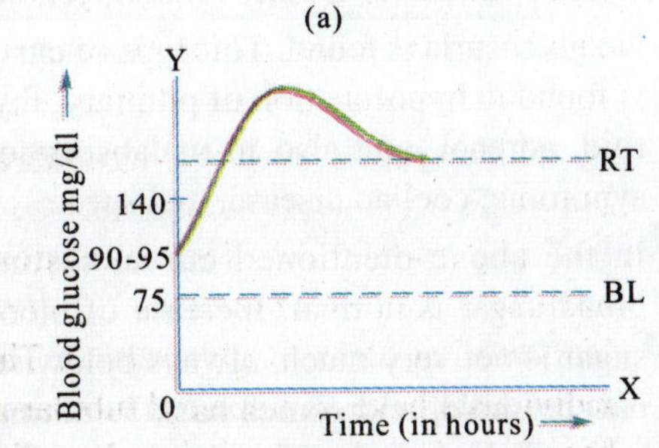

(b)

In this condition blood sugar is increased more than in normal patient, fasting blood sugar may be high and return to normal level is delayed. Urine shows glycosuria in most sample.

Two types of curves :

(a) Curve for true diabetes.

(b) Curve for potential diabetes.

In curve (a)

- Blood sugar is increased.
- Fasting blood sugar is always above 140 mg/dl.
- Sugar become high to 1 hour after taking glucose and return is delayed and glycosuria is present in all samples.

In curve (b)

- Fasting blood sugar is slightly increased than normal but always below 140 mg/dl.
- Increased up to 1 hour after taking glucose and return is delayed.
- 2 to 5% of these patients develop diabetes within 1 to 5 years.

(ii) Increased glucose tolerance

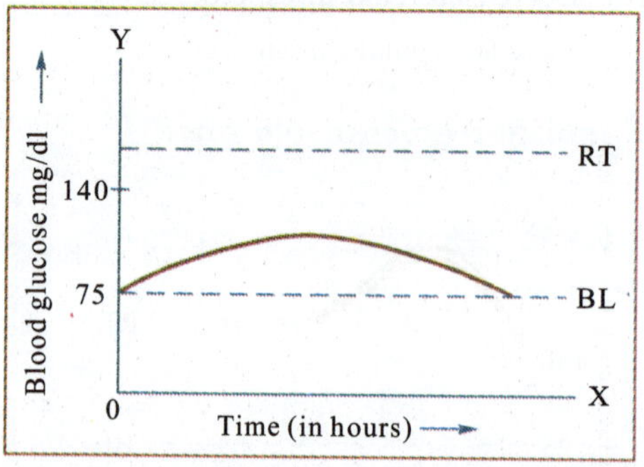

(a) Blood sugar level is not very much increased and never crosses renal threshold.

(b) No glycosuria is found. This type of curve is found in hypofunction of pituitary, thyroid, adrenal, and also in malabsorption syndrome, coeliac disease, and sprue.

(c) In the above-mentioned curve, fasting blood sugar is normal. Increase of blood sugar is not very much, always below renal threshold, and comes back to normal within half an hour. This curve is called flat type curve.

(iii) Lag curve

(a) Some apparently normal people are found to have tolerance curve in which return of blood sugar is normal but with certain exaggeration of blood sugar level. The return to normal value is sharp and rapid.

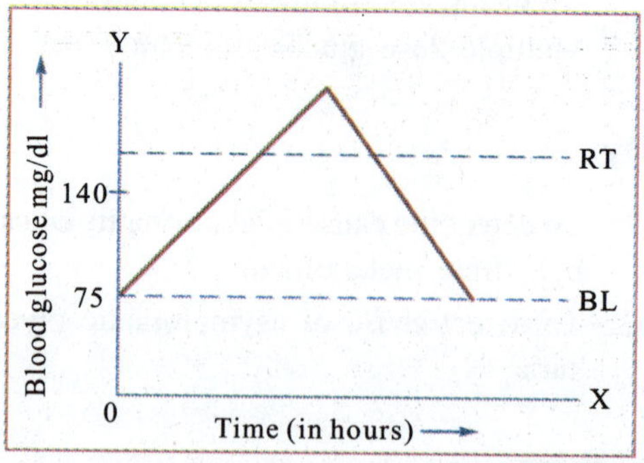

(b) So peak rise is always above renal threshold and sugar is absent in fasting sample.

Cause

Delayed insulin mechanism and there is increased absorption of glucose from intestine after rapid emptying of stomach. This is a harmless condition, these people do not develop diabetes later on.

This condition is found in :

1. Hyperthyroidism.
2. Pregnancy.
3. Gastro-enterostomy.

(iv) Decreased renal threshold

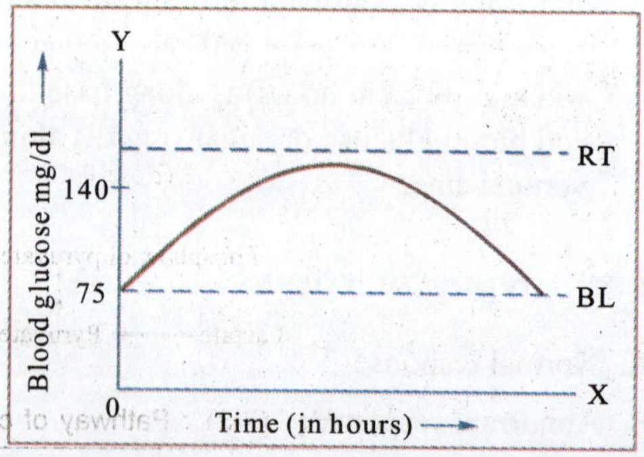

(a) In some patient, blood sugar level is not very high. But sugar is present in urine. Blood sugar level does not exceed renal threshold but still there is glycosuria. So renal threshold is decreased. This is called renal glycosuria and it is accidently detected.

(b) It is found in late stage of pregnancy and Wilson's disease.

(c) There is no defect in carbohydrate metabolism but defect in reabsorption of glucose by renal tubules $\frac{1}{2}$ to 1 hour after taking glucose.

(v) Increased renal threshold

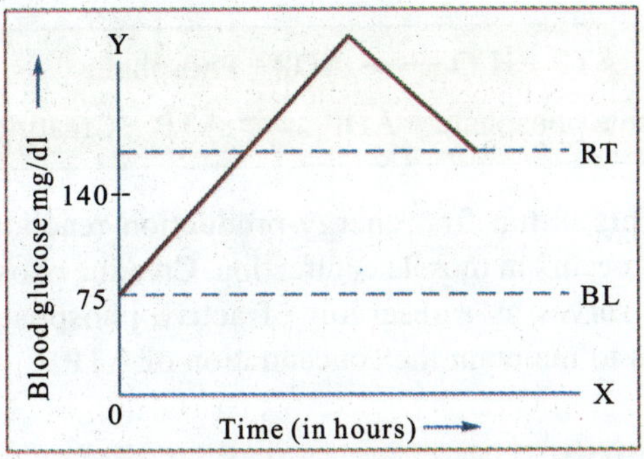

(a) Blood sugar is always high. Fasting sugar is also high.

(b) It is increased $\frac{1}{2}$ to 1 hour after taking glucose although blood sugar level is above 200 mg/dl. But there is no glycosuria.

(c) Urine examination of this group of patients does not give true picture of diabetes and this is a true diabetic curve without glycosuria. It is found in elderly diabetic patient.

CONVERSION OF CARBOHYDRATES INTO FAT

1. Glucose and the intermediates such as pyruvate and acetyl-CoA are converted into fat in the body by anabolic processes.
2. The rate of conversion is high in relation to the diet containing high proportion of carbohydrate. It is depressed in restricted caloric intake, on a high-fat diet or in the deficiency of insulin as in diabetes mellitus. Increased free fatty acids inhibit lipogenesis.
3. Hexose monophosphate shunt (HMP shunt) is closely associated with the lipogenesis because this shunt supplies reduced NADP (NADPH) for the synthesis of fat.
4. At present, it is recognized that acetyl-CoA carboxylase step is the rate-limiting reaction in the pathway of lipogenesis. Long chain acyl-CoA molecules produced from lipolysis inhibit acetyl- CoA carboxylase and thus inhibits the synthesis of new fatty acids.
5. Microsomes catalyze the esterification of acyl-CoA with glycerol-3-phosphate to form triacylglycerols and phospholipids.

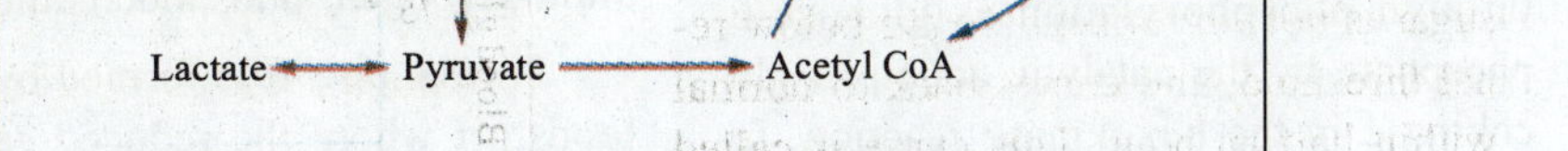

Fig. 17.31 : Pathway of conversion of carbohydrates into fat

6. Acyl-CoA also inhibits the transport of citrate from mitochondria into the cytosol and thus changes the availability of citrate for lipogenesis.
7. Insulin stimulates lipogenesis by several mechanisms :
   (i) It increases the transport of glucose into the cell and thus increases the availability of pyruvate and glycerol-3-phosphate.
   (ii) It can convert the inactive pyruvate dehydrogenase into the active form.
   (iii) It depresses cyclic AMP (cAMP) and thus inhibits lipolysis and thereby reduces the concentration of long chain acyl-CoA, which is an inhibitor of lipogenesis.
8. Oxidation of fatty acids, due to lack of insulin or increased levels of free fatty acids, causes the increased lipolysis of triacylglycerols and thus increases the concentrations of acetyl-CoA and NADH in mitochondria inhibiting pyruvate dehydrogenase. As a result, the formation of acetyl-CoA from carbohydrate via pyruvate is blocked.

   The pathway of conversion of carbohydrate into fat is shown (Fig. 17.31).

HEXOKINASE

1. The molecular weight of the enzyme hexokinase is 96,600.
2. This enzyme exists in plant and animal tissues.
3. Liver contains two hexokinase enzymes–glucokinase, probably formed in the non-parenchymal cells, and hexokinase, formed in the parenchymal cells.
4. Glucose enters the metabolic stream through phosphorylation as glucose-6-phosphate by the catalytic action of glucokinase in the hexokinase reaction. Insulin regulates the activity of glucokinase and glucose phosphorylation. Liver hexokinase appears to be unaffected by insulin.
5. Hexokinase has a high affinity for glucose.
6. This enzyme catalyzes the phosphorylation of glucose, fructose, mannose and glucosamines etc.

ATP

It is adenosine triphosphate or high energy phosphate bond.

Discovery : Lohmann isolated ATP from muscle. The phosphate groups are consecutively removed from ATP to ADP and AMP. Lohmann reaction is shown :

$$\text{ATP} + H_2O \longrightarrow \text{ADP} + \text{Phosphate}$$
$$\text{Creatine phosphate} + \text{ADP} \rightleftharpoons \text{ATP} + \text{Creatine}$$

This is the first energy-production reaction which occurs in muscle contraction. Creatine phosphate serves as a reservoir of active phosphate groups to maintain the concentration of ATP.

Function

1. It acts as a basic unit of biological energy.
2. It acts as a phosphate transfer reaction.
3. It helps photosynthesis.
4. It is involved in the activation of fatty acids and amino acids.
5. It causes phosphorylation of glucose, galactose and fructose.
6. It acts as a coenzyme in the energy metabolism.

Effects of shock : Shock causes the rapid decrease of high energy phosphates in the shocked tissue. Severe shock causes collapse and anoxia. Electric shock and cyanide poisoning decrease brain glycogen, phosphocreatine and ATP.

Formation : It is formed by the catabolism of foodstuff where the reduced equivalents are oxidized in the respiratory chain.

GLYCOGEN

Chemistry

1. It is a polysaccharide consisting of many glucose units by 1, 4 and 1, 6 linkages.
2. It forms a colloidal solution.
3. Its structure is just like a tree having numerous branches. Each branching is at an interval of 12-18 glucose units.
4. Its molecular weight is very high—of the order of several millions.

Tissue glycogen

Glycogen is the chief carbohydrate of tissues just as glucose of blood and other body fluids. The following are the contents of glycogen in the tissues.

Liver–5% (average); 10-15 per cent on carbohydrate diet.

Skeletal muscle–0.5%.

Skin and brain–0.1% or less.

Liver glycogen is reversibly convertible to blood glucose and maintain the blood sugar level in the impairment of intestinal absorption of carbohydrate. The energy is liberated when glycogen is broken down completely to CO_2 and H_2O. The formation of fat from carbohydrate is from the breakdown products of glycogen or glucose.

Glycogen is formed from glucose by glycogenesis; and glycogen is broken down to glucose in the liver by glycogenolysis.

Factors affecting tissue glycogen

1. *Dietary state :* The glycogen content of tissues is markedly altered by changes in the dietary condition.
2. *Exercise :* Heavy exercise greatly accelerates the utilization of carbohydrate in the tissues and ultimately leads to drastic reduction of muscle and liver glycogen.
3. *Insulin :* It is required for both the oxidation of glucose and its conversion to glycogen. It also accelerates the activity of hexokinase and glycogen synthetase responsible for glycogen formation.
4. *Epinephrine and glucagon :* Both these hormones cause glycogenolysis by stimulating phosphorylase and produce glucose in the liver and lactic acid in the muscle.

Glycogen storage disease : Already discussed in carbohydrate metabolism.

ACETYL-CoA

Acetyl-COA can be regarded as a focal point in metabolism. It is a common high energy substance produced by the oxidation of fatty acids, amino acids and pyruvate from carbohydrate. It is utilized

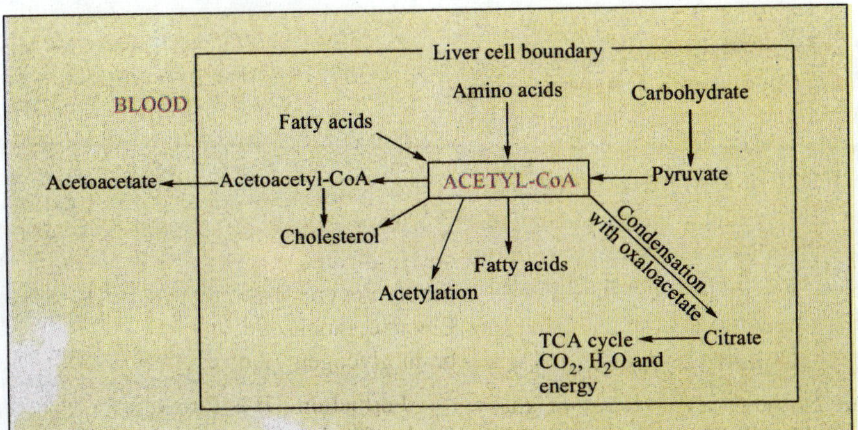

Fig. 17.32 : Formation and utilization of acetyl-CoA

in many ways and referred to as the **"Building stone"** of fatty acid synthesis.

Formation : Acetyl-CoA is formed from coenzyme A (CoA). The components of CoA are adenine-phosphoribose-phosphate-phosphate-pantothenic acid-mercapto-ethanolamine. The 'SH' is the functional group, combining with acyl substances, having the carboxyl group (R COOH) to form thioesters (R COS-CoA). In respect of this linkage, the factor is designated as CoA.SH. This linkage is a high energy bond and is formed at the expense of ATP.

Utilization : It is utilized in cells in various ways :

1. In the complete oxidation of glucose by the TCA cycle.
2. In the fatty acid synthesis.
3. In the acetylation of choline (in nerve) and of aromatic amines.
4. In the formation of acetoacetyl-CoA in ketone body formation.
5. In the synthesis of cholesterol, steroid hormones, bile acids etc.

CO-CARBOXYLASE

Thiamine pyrophosphate (TPP) is said to be the co-carboxylase. This is the coenzyme formed from thiamine. It is a group transferring coenzyme.

Function

1. TPP as coenzyme is involved in the conversion of pyruvate to acetyl-CoA along with pyruvate dehydrogenase complex. Pyruvate is decarboxylated in the presence of TPP to a hydroxyethyl derivative of the thiazole ring of enzyme bound thiamine diphosphate which reacts with oxidized lipoate to form S-acetyl lipoate.
2. Similar reaction occurs in the conversion of α-ketoglutarate to succinyl-CoA where TPP acts as coenzyme.
3. TPP also acts as coenzyme in the conversion of pentose phosphates to sedoheptulose-7-phosphate and glyceraldehyde-3-phosphate catalyzed by **transketolase** in the HMP shunt.

EXERCISES

1. Draw a flow diagram showing the reactions of Kreb's cycle. Discuss its importance. (M. U. 76A)
2. What is Krebs' cycle? Why is it so called? What is the alternate name of the cycle? Justify the name. State very briefly the steps of the Krebs' cycle. (R. U. 76A)
3. What do you mean by intermediary metabolism? Explain why glucose-6-phosphate is regarded as the key intermediate metabolic product of carbohydrate metabolism. (P. U. 75S)
4. Discuss the various steps of TCA cycle and point out where energy, water and CO_2 are formed. How many molecules of ATP are produced in the cycle? (P. U. 71S; Muz. 75S)
5. Describe schematically the citric acid cycle. (Mith. 66S, 71S; R. U. 70S; P.U. 68S)
6. Describe the pentose phosphate pathway of carbohydrate metabolism and the significance of it. (Bh. U. 75A)
7. Give an account of the citric acid cycle and explain why it is called a common metabolic pathway. (Muz. 74A; Mith. 62A)
8. Discuss the oxidation of glucose in the tissues up to the stage of pyruvic acid with particular reference to the energy-rich phosphate bonds. (Mith. 60S)
9. Describe the pathway of conversion of carbohydrates into fat. (Mith. 67S)
10. Describe briefly the mechanism by which the body derives energy from glucose. (Bh. U. 75A)
11. Discuss the formation of glucose from noncarbohydrate substances. (P. U. 73A; R. U. 76S; Bh. U. 71S)
12. Describe the hormonal influence of carbohydrate metabolism. (P. U. 76A)
13. What is the normal blood sugar level? How is it maintained at a constant level? (P. U. 65A, 69A; Luck. 67S)
14. Describe the mechanism of regulation of blood sugar level. (Mith. 75A; P. U. 73A)
15. Name the reducing sugars that may be found in pathological urine. Describe a method for the estimation of the concentration of reducing sugar in urine. (R. U. 70S)
16. Discuss the glucose tolerance test. (R. U. 73A)

17. Write notes on :
   (i) Synthesis of lactose. (Bh. U. 74S; Muz. 74A)
   (ii) Glycogen. (M. U. 73A; Mith. 62A)
   (iii) Neoglucogenesis. (M. U. 74S; R. U. 71S; Mith. 62A, 65A)
   (iv) Glycogenesis. (Muz. 75S; Mith. 65S, 75S)
   (v) Glycogenolysis. (Bh. U. 74S; R.U. 70A; Mith. 62S, 64A)
   (vi) Benedict's test. (Mith. 73A, Bh. U. 76S; P. U. 68S)
   (vii) Glucose tolerance test. (Mith. 62A, 66S; P. U. 72S)
   (viii) Glycosuria. (R. U. 65S; Luck. 64A)
   (ix) Renal glycosuria. (Mith. 75A; P.U. 68A)
   (x) ATP. (P. U. 75S; M. U. 73A; R. U. 71A, 73A; Mith. 71A, 75A)
   (xi) High energy phosphate bond. (Mith. 62A, 64S)
   (xii) UDPG. (R. U. 73S; P.U. 72S)
   (xiii) Hexokinase. (R. U. 73S)
   (xiv) Co-carboxylase. (Mith. 60S)
   (xv) Conversion of glyceraldehyde-3-phosphate to 1, 3-diphosphoglyceric acid. (R. U. 72S)
   (xvi) Fructosuria. (Bh. U. 71A)
   (xvii) Galactosemia. (M. U. 72S)
   (xviii) Glycogen storage disease. (R. U. 67S; Muz. 70A)
   (xix) Diabetes mellitus. (P. U. 68A; R. U. 66S)
18. Write in short the process of neoglucogenesis. Mention the specific enzymes necessary for the above process. (C.U. 1981)
19. Describe with diagram the metabolic pathways of glycogenesis and glycogenolysis in muscle indicating the hormonal influences on them.

   Describe how you will perform and interprete an oral glucose tolerance test. (C.U. 1983)
20. Describe with diagram how acetyl-CoA is formed from pyruvate and long chained fatty acids. Enumerate the fate of acetyl-CoA in the body.

   What will happen to acetyl-CoA if there is a lack of insulin in the body? (C.U. 1985)

CHAPTER 18

METABOLISM OF LIPIDS

The lipids of metabolic significance include triacylglycerol (triglycerides, neutral fat), phospholipids, steroids together with long chain fatty acids (free fatty acids), glycerol and ketone bodies.

Advantages of lipids over carbohydrate or protein

1. The caloric value of lipid is over twice as great (9.3 Kcal/g) and it is associated with less water in storage.

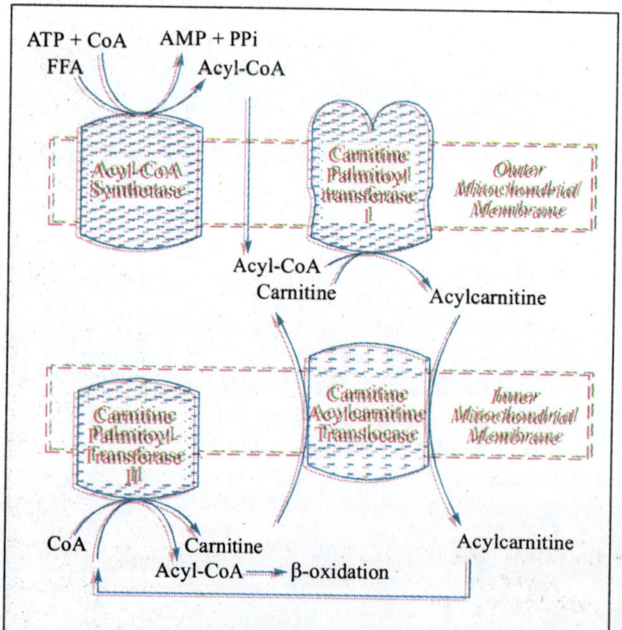

Fig. 18.1 : Role of carnitine in the transport of long chain fatty acids

2. Triacylglycerol is the most concentrated form in which potential energy can be stored.
3. Fatty acids on oxidation provide more metabolic water than other metabolic fuels. This is helpful to mammals accompanying dry environments.
4. Less amount of lipid in the diet can supply adequate amount of certain polyunsaturated fatty acids (the essential fatty acids) and it is essential also for fat-soluble vitamins. Lipid is also necessary for their efficient absorption from the gastrointestinal tract.

OXIDATION OF TRIACYLGLYCEROL

1. Triacylglycerol is first hydrolyzed to fatty acids and glycerol mostly in adipose tissue.
2. The free fatty acids are released into the plasma where they combine with serum albumin.
3. Long chain fatty acids are oxidized in liver, heart, kidney, muscle, lung, testis, brain and adipose tissue.
4. Glycerol is utilized by liver, kidney, intestine and lactating mammary gland where the activating enzyme **glycerokinase** is present abundantly.

Oxidation of fatty acids

Fatty acids are oxidized by β-, α- and ω-oxidation. Quantitatively, β-oxidation is the most important pathway. The term β-oxidation means the oxidation takes place in the β-carbon of the fatty acid with the removal of 2 carbon atoms at a time from the carboxyl end of the molecule.

The fatty acids containing even number and odd number of carbon atoms as well as the unsaturated fatty acids are oxidized by β-oxidation.

β-oxidation of fatty acids

1. Several enzymes collectively known as "fatty acid oxidase" are found in mitochondrial matrix adjacent to the respiratory chain (which is found in the inner membrane).
2. Long chain fatty acids are first activated to "active fatty acid" or acyl-CoA in the cytosol by the enzyme acyl-CoA synthetase in the presence of ATP, coenzyme A and Mg^{++}. But activation of lower fatty acids occurs within the mitochondria. Thiokinases are found both inside and outside the mitochondria.
3. Long chain acyl-CoA does not penetrate mitochondria without the presence of carnitine. The enzyme carnitine palmitoyl transferase 1 associated with the mitochondrial membranes allows long chain acyl groups to penetrate the mitochondria. The mechanism is shown (Fig. 18.1). Another enzyme, carnitine acyl carnitine translocase, present in mitochondria, catalyzes the transfer of short chain acyl groups between CoA and carnitine.

$$\text{Acyl carnitine} + \text{CoA} \xleftrightarrow[\text{transferase}]{\text{Carnitine palmitoyl}} \text{Acyl-CoA} + \text{Carnitine}$$

Acyl carnitine combines with CoA to form acyl-CoA and carnitine which is released by the enzyme carnitine palmitoyl transferase 11.

4. Acyl-CoA is then converted to α, β-unsaturated acyl-CoA by the enzyme acyl-CoA dehydrogenase in presence of the coenzyme flavoprotein which contains FAD as prosthetic groups. The reduced coenzyme is reoxidized by the respiratory chain.

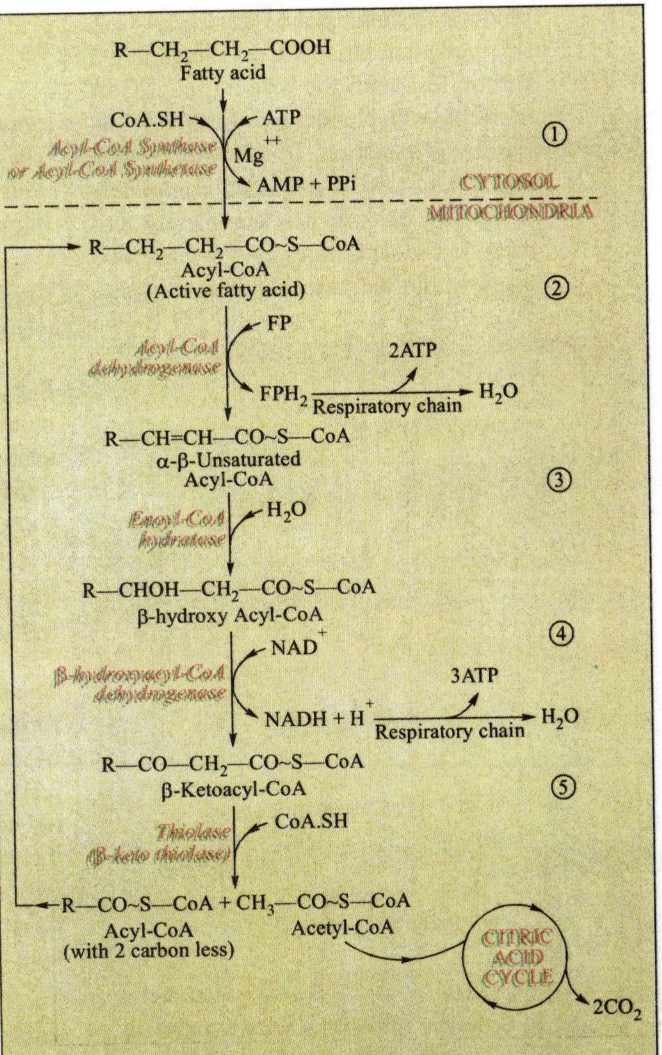

Fig. 18.2 : β-oxidation of fatty acids

5. Water is added to saturate the double bond and form β-hydroxy acyl-CoA, catalyzed by the enzyme **enoyl-CoA hydratase** (**crotonase**).
6. The β-hydroxy acyl-CoA undergoes dehydrogenation on the β-carbon forming β-keto acyl-CoA. The reaction is catalyzed by **β- hydroxy acyl-CoA dehydrogenase** in presence of NAD. The reduced coenzyme (NADH) is reoxidized by the respiratory chain.
7. Finally, β-keto acyl-CoA is split at the β-position by **thiolase** (**β-ketothiolase**) with the addition of one molecule of CoA. The products of this reaction are acetyl-CoA (C_2-units) and an acyl-CoA containing 2 carbon less than the original acyl-CoA molecule which undergoes further-oxidation by β-oxidation. In this way, a long chain fatty acid is degraded completely to acetyl-CoA which is completely oxidized to CO_2 and water via the citric acid cycle within the mitochondria.

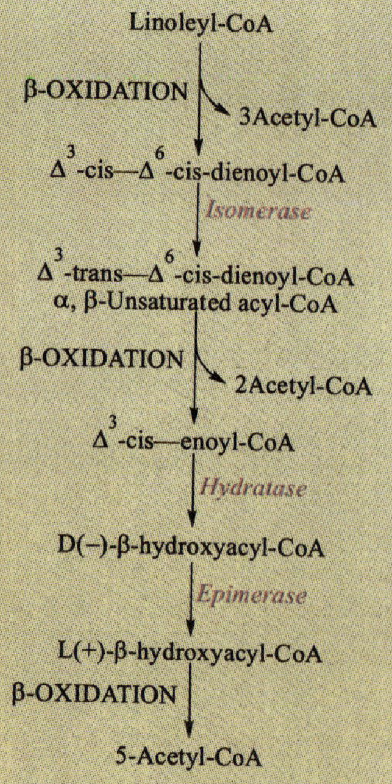

Fig. 18.3 : Sequence of reactions in the oxidation of unsaturated fatty acids, eg, linoleic acid

Fatty acids with an odd number of carbon atoms are oxidized by the pathway of **β-oxidation** until **propionyl-CoA** is formed. This compound is converted to succinyl-CoA which is a constituent of the citric acid cycle.

Unsaturated fatty acids are also oxidized following β-oxidation. The reactions are shown in Fig. 18.3.

α- and ω-oxidation of fatty acids

α-oxidation takes place by the removal of one carbon at a time from the carboxyl end of the molecule. It does not require CoA intermediates and **does not lead to the generation of high-energy phosphates.**

ω-oxidation takes place by **hydroxylase** enzymes involving cytochrome P-450 in microsomes. The $-CH_3$ group is converted to a$-CH_2OH$ group which is ultimately oxidized to –COOH group forming a **dicarboxylic acid.**

Oxidation of fatty acids with an odd number of carbon atoms

Fatty acids with odd number of carbon atoms are oxidized by the pathway of β-oxidation producing acetyl-CoA unitl a three carbon (Propinyl-CoA) remains. Propionyl-CoA is converted to succinyl-CoA, a constituent of citric acid cycle.

Peroxisomes oxidize very long chain fatty acids

The modified form of β-oxidation in Peroxisomes forms acetyl-CoA and H_2O_2 which is broken down by catalase. This first dehydrogenation is not directly linked to the generation of ATP, but the initial activation by a very long chain acyl-CoA synthetase facilitates the oxidation of very long chain fatty acids (eg, C_{20}, C_{22}). These enzymes are induced by high fat diets and in some species by hypolipidemic drugs such as clofibrate.

The enzymes in peroxisomes do not attack shorter chain fatty acids and the β-oxidation sequence ends at octanoyl-CoA. Octanoyl and acetyl groups are subsequently removed from the peroxisomes in the form of octanoyl and acetylcarnitine and both are further oxidized in mitochondria.

Another role of peroxisomal β-oxidation is to shorten the side chain of cholesterol in bile acid formation.

Peroxisomes do not contain carnitine palmitoyltransferase.

Energetics of fatty acid oxidation

Palmitic acid (16 carbons) undergoes 7 times β-oxidation and produces 8 molecules of acetyl-CoA. Each time, β-oxidation produces 5 ATP :

∴ Total number of ATP formed trough β-oxidation (7×5) = 35

Total number of ATP formed on oxidation of acetyl-CoA through citric acid cycle (8×12) = 96

Total = 131

2 ATP utilized for initial activation of fatty acid − 2

Net total yield = ATP 129

129×7.6 = 980 Kcal/mol.

The caloric value of palmitic acid is 2,340 Kcal/mol.

∴ The process captures as high-energy phosphate as

$$\frac{980 \times 100}{2{,}340} = 41\%$$

CLINICAL ORIENTATION

- Gluconeogenesis is dependent upon fatty acid oxidation. Any impairment in fatty acid oxidation leads to hypoglycemia.
- The impairment of fatty acid oxidation occurs in various states of Carnitine deficiency or Carnitine palmitoyl transferase or inhibition of fatty oxidation by poisons eg, hypoglycin.
- Jamaican vomiting sickness is caused by eating the unripe fruit of the akee tree that contains a toxin, hypoglycin, which inactivates medium-and short-chain acyl-CoA dehydrogenase inhibiting β-oxidation of fatty acid causing hypoglycemia with excretion of medium- and short-chain fatty acids.
- Dicarboxylic aciduria is characterized by the excretion of $C_6 - C_{10}$ ω-dicarboxylic acids and by non-Ketotic hypoglycemia. This is caused by the deficiency of mitochondrial medium – chain acyl-CoA dehydrogenase.
- Refsum's disease, a rare neurologic disorder, is caused by accumulation of phytanic acid, formed from phytol found in plant foodstuffs. Methyl group on carbon 3 of phytanic acid blocks β-oxidation. Persons with this disease have an inherited defect in α-oxidation that allows accumulation of phytanic acid.
- Zellweger's (Cerebrohepatorenal) syndrome occurs in individuals having inherited absence of peroxisomes in all tissues. They accumulate $C_{26} - C_{38}$ polyenoic acids in brain tissue due to inability to oxidize long chain fatty acids in peroxisomes.

SCHEMATIC COMPENDIUM

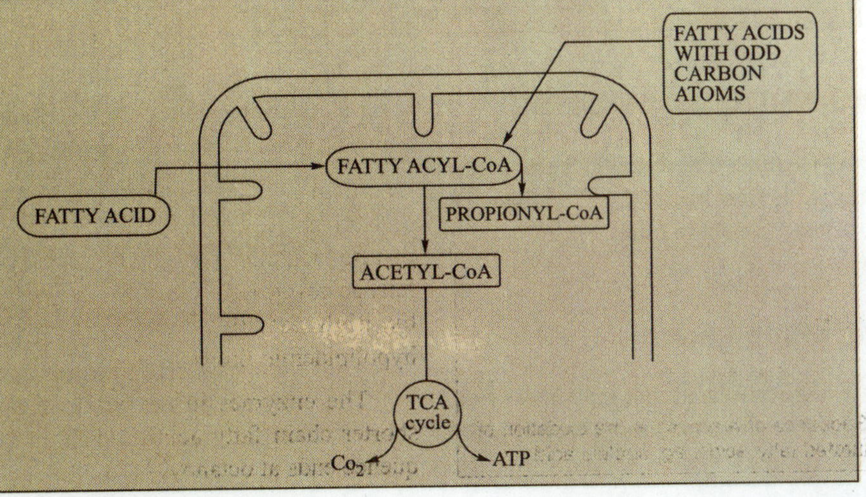

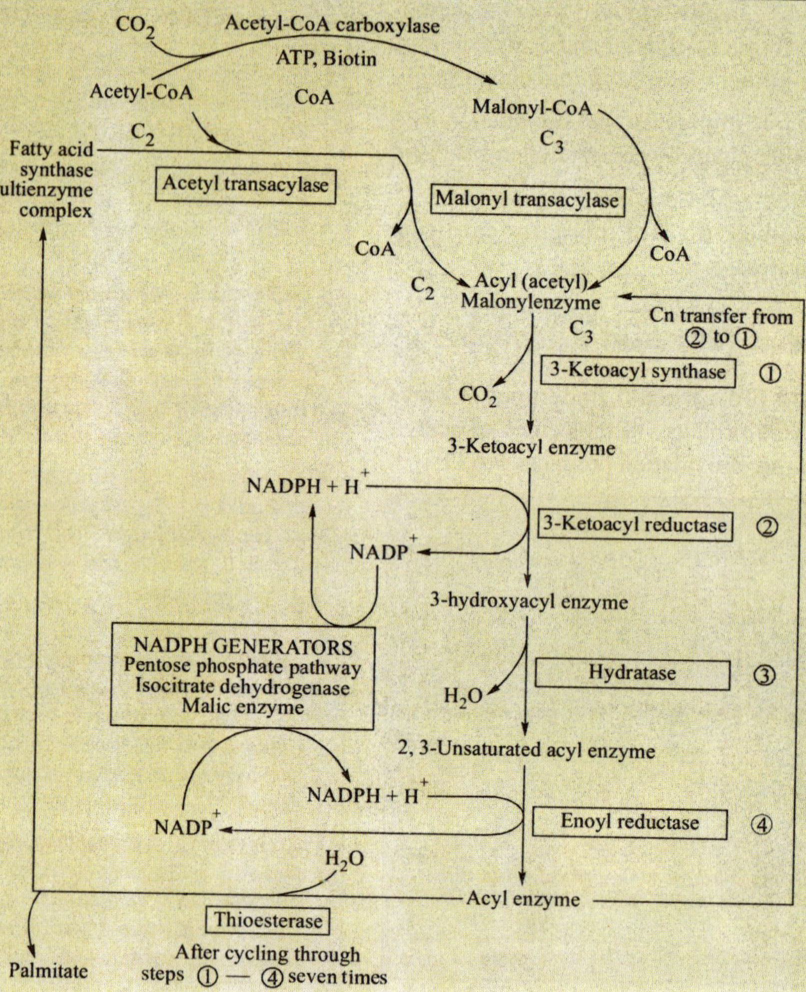

Fig. 18.4 : Flow chart of Extramitochondrial system for long chain fatty acid synthesis (The main pathway)

BIOSYNTHESIS OF LIPIDS

The process of synthesis of triacylglycerol (neutral fat) and fatty acids from acetyl-CoA in adipose tissue and to a lesser extent in other tissues is said to be lipogenesis.

Synthesis of fatty acids

Formerly, it was considered that fatty acid synthesis was the reversal of its oxidation. Now it is clear that the mitochondrial system for fatty acid synthesis is responsible only for elongation of existing fatty acids of moderate chain length. The synthesis of palmitate from acetyl-CoA is complete only by the highly active extramitochondrial system. Three different systems are mentioned :

A. Mitochondrial system

1. Mitochondria catalyze the incorporation of acetyl-CoA into long chain fatty acids under anaerobic conditions.
2. The enzymes are mostly the same as those

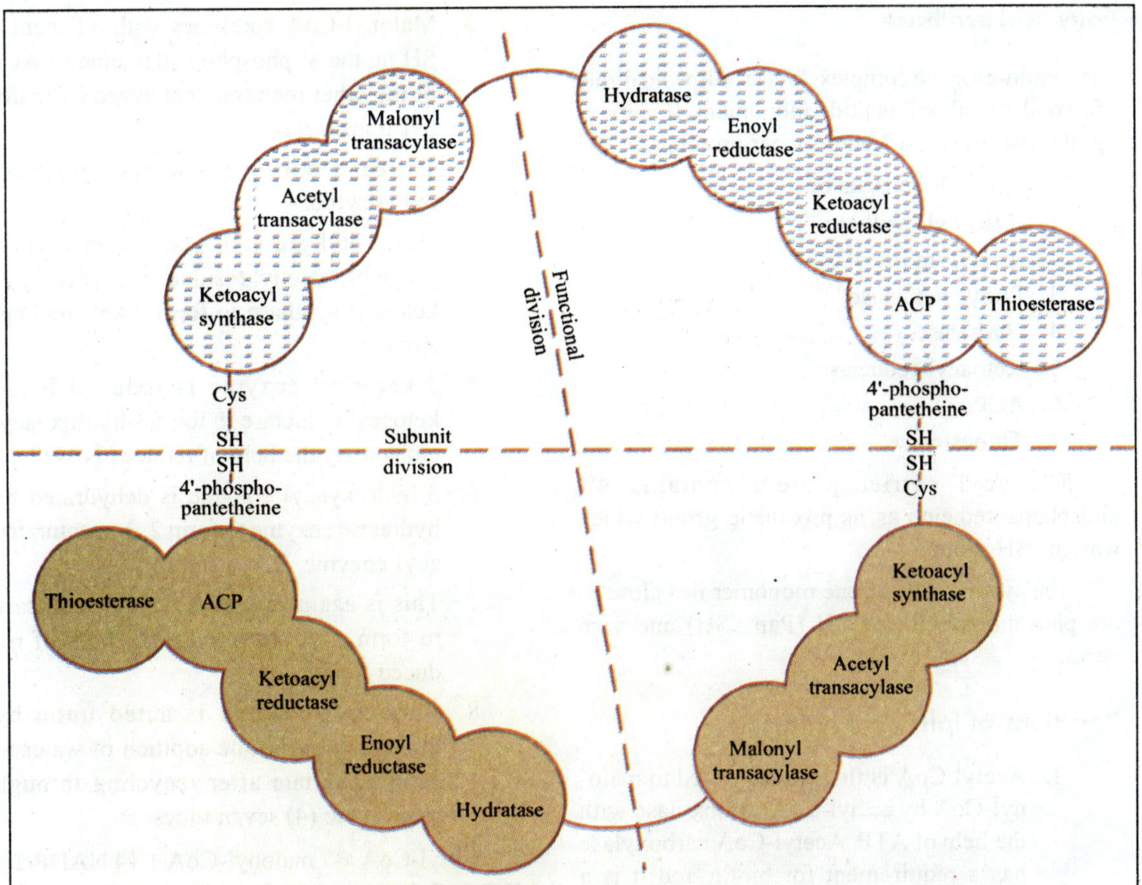

Fig. 18.5 : Fatty acid synthase multienzyme complex. The complex is a dimer of two identical polypeptide monomer 1 and 2, each containing seven enzymes and acyl carrier protein (ACP)

involved in β-oxidation excepting α, β-unsaturated acyl-CoA reductase which converts α, β-unsaturated acyl-CoA to a saturated compound, requiring NADPH + H^+

3. Pyridoxal phosphate ($B_6 - PO_4$) as a coenzyme is required for the enzyme condensing acetyl-CoA with acyl-CoA; thus, thiolase is not used in this synthetic pathway.
4. Since this system takes place under anaerobic condition the physiologic significance of this pathway is uncertain.

B. Microsomal system for chain elongation

1. This is the main pathway for the elongation of existing fatty acid molecules.
2. Acyl-CoA compounds are converted to higher fatty acids by means of malonyl-CoA along with NADPH + H^+. The acyl groups include the saturated series from $C_{10} - C_{16}$ as well as some unsaturated C_{18} fatty acids.
3. Fasting prevents chain elongation.

C. Extramitochondrial system for De NOVO synthesis (The main pathway)

De novo synthesis of palmitic acid from acetyl-CoA is carried out in the cytoplasm by fatty acid synthase (multi enzyme complex). This mainly takes place in liver, kidney, lactating mammary gland, brain etc.

Fatty acid synthase

It is a multi-enzyme complex. It is an ellipsoid dimer of two identical polypeptide monomers. It is made up of 6 enzymes and 1 acyl carrier protein (ACP).

Each subunit consists of :

1. Ketoacyl synthase.
2. Acetyl transacylase.
3. Enoyl reductase.
4. Hydratase.
5. Ketoacyl reductase.
6. ACP.
7. Thioesterase.

The acyl carrier protein contains 4′-phosphopantetheine as its prosthetic group which was an –SH group.

The cystine –SH of one monomer lies close to the phosphopantetheine SH (Pan –SH) and vice versa.

Reactions of fatty acid synthesis

1. Acetyl-CoA is first carboxylated to malonyl-CoA by acetyl-CoA carboxylase with the help of ATP. Acetyl-CoA carboxylase has a requirement for biotin and it is a multienzyme protein.

   This reaction takes place in two steps :

   (a) Carboxylation of biotin.

   (b) Transfer of the carboxyl to acetyl-CoA to form malonyl CoA.

   Acetyl-CoA carboxylase is a rate limiting enzyme, it is an allosteric enzyme with citrate as its activator and palmityl-CoA as its inhibitor.

   Once malonyl CoA is synthesized, rest of fatty acid synthesis reactions take place on fatty acid synthetase.

2. Acetyl-CoA combines with Cysteine–SH group catalyzed by acetyl transacylase :

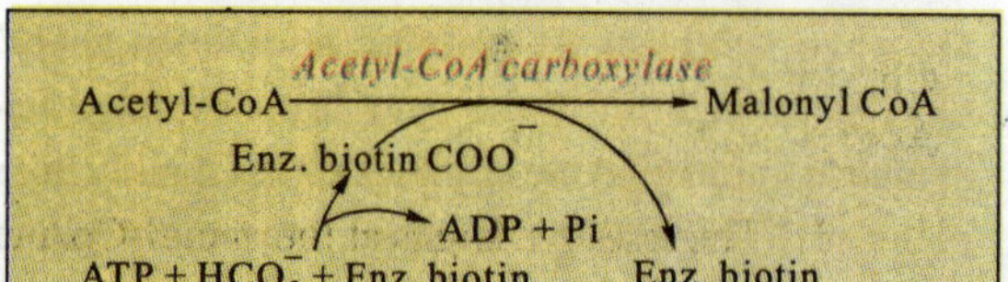

3. Malonyl-CoA combines with adjacent –SH on the 4′-phosphopantetheine of ACP of the other monomer catalyzed by malonyl transacylase.

   There is a formation of acyl (acetyl) malonyl enzyme.

4. The acetyl group attacks the methylene group of malonyl residue, catalyzed by 3-ketoacyl synthase to form 3-ketoacyl enzyme.
5. 3-ketoacyl enzyme is reduced by 3-ketoacyl reductase to form 3-hydroxyacyl enzyme by the help of reduced NADP.
6. 3-hydroxyacyl enzyme is dehydrated by hydratase enzyme to form 2, 3-unsaturated acyl enzyme.
7. This is again reacted by enoyl reductase to form acyl enzyme by the help of reduced NADP.
8. This acyl enzyme is acted upon by thioesterase with the addition of water to form palmitate after recycling through steps (1) to (4) seven times.

Acetyl-CoA + 7 malonyl-CoA + 14 NADPH + 14H$^+$ → Palmitate + 7CO_2 + 6H_2O + 8CoA.SH + 14NADP$^+$.

Recently it is suggested that butyryl-CoA is the primer molecule in mammary gland.

Sources of acetyl-CoA

1. Acetyl-CoA is mainly found in mitochondria which cannot pass out. It forms citrate by reacting with oxaloacetate. Citrate is transported out by a transporter protein in exchange of malate. By the presence of enzyme citrate lyase in cytosol, citrate is catalyzed to form Acetyl-CoA and oxaloacetate.
2. Carnitine palmitoyl transferase-1 combines acetyl-CoA with carnitine to form acyl carnitine which enters into matrix and is cleaved by carnitine palmitoyl transferase 11 to form acetyl-CoA.

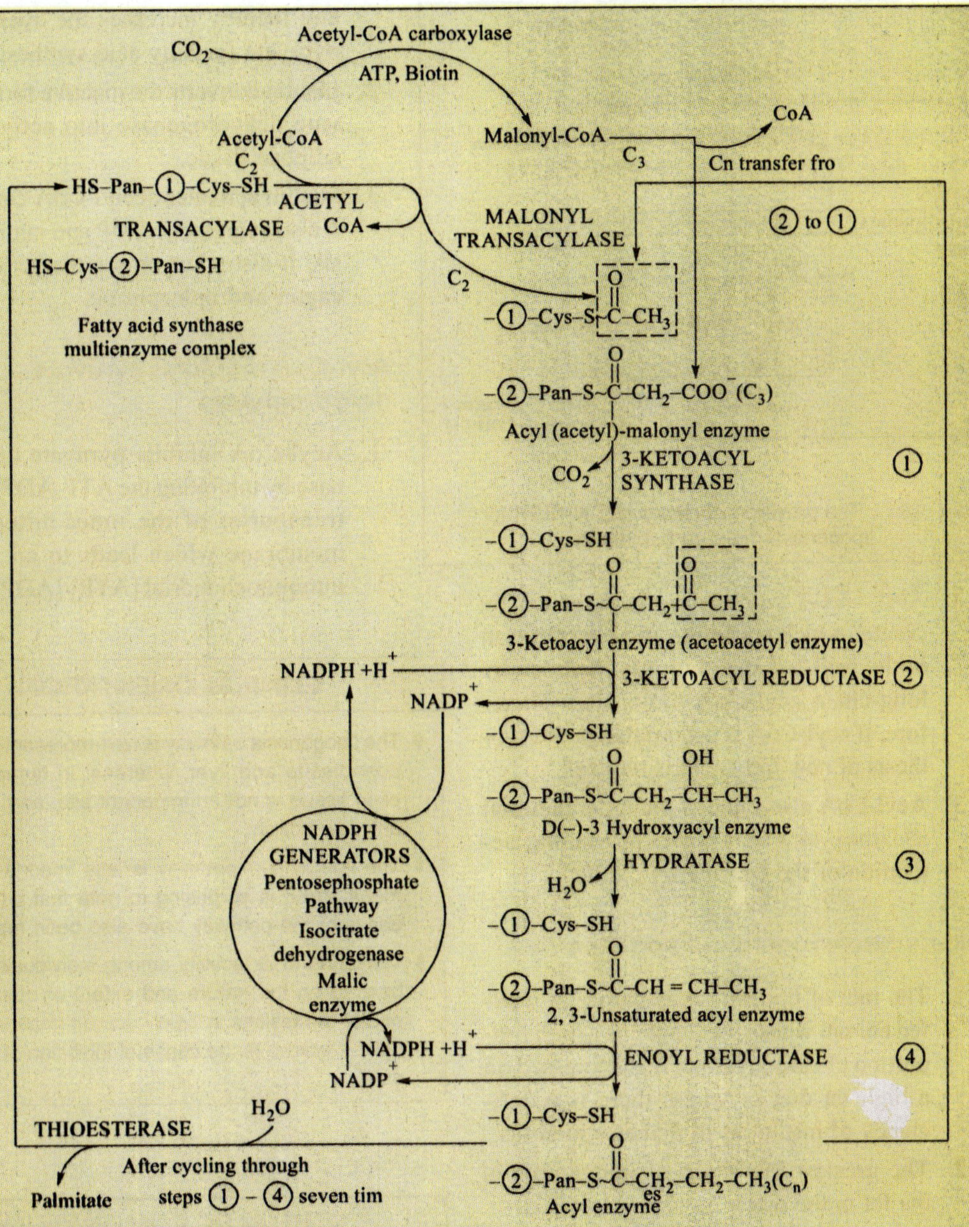

Fig. 18.6 : Reaction of Biosynthesis of Palmitate (The long chain fatty acid)

Sources of NADPH

1. Hexose monophosphate pathway is the main source of NADPH.
2. By the presence of malic enzyme malate is oxidatively decarboxylated to pyruvate and NADPH is produced.
3. The extra mitochondrial isocitrate dehydrogenase (probably not a substantial source).

Citrate and Acyl-CoA regulate Acetyl-CoA carboxylase

1. The rate limiting is at the acetyl-CoA carboxylase step.

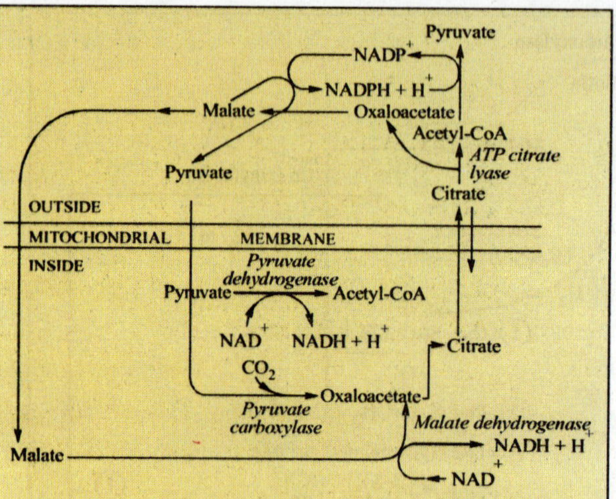

Fig. 18.7 : The provision of citrate and NADPH for lipogenesis from carbohydrate

2. Acetyl-CoA carboxylase is activated by citrate, which increases in concentration in the well-fed state but it is inhibited by long-chain acyl-CoA molecules. Therefore, if acyl-CoA is accumulated, the synthesis of new fatty acid is reduced.
3. Acyl-CoA also inhibits the mitochondrial tricarboxylate transporter preventing activation of the enzyme by citrate.

Nutritional state regulates lipogenesis

1. The rate of lipogenesis is higher in well-fed animal whose diet contains a high proportion of carbohydrate. It is depressed on a high fat diet or where there is a deficiency of insulin as in diabetes mellitus.
2. The greatest inhibition of lipogenesis is the fat in the diet.
3. Lipogenesis is higher where sucrose is fed instead of glucose. Because, fructose bypasses the phosphofructokinase control point in glycolysis and is flushed by lipogenic pathway.

Hamones regulate lipogenesis

1. Insulin stimulates lipogenesis by increasing the transport of glucose into the cell and thereby increases the availability of pyruvate for fatty acid synthesis.
2. Insulin converts the inactive form of pyruvate dehydrogenase into active form in adipose tissue.
3. Insulin activates acetyl-CoA carboxylase. It also depresses cAMP and inhibits lipolysis. It also antagonises the action of glucagon and epinephrine.

Acyl-CoA regulates pyruvate dehydrogenase

1. Acyl-CoA inhibits pyruvate dehydrogenase by inhibiting the ATP-ADP exchange transporter of the inner mitochondrial membrane which leads to an increased intramitochondrial [ATP]/[ADP].

CLINICAL ORIENTATION

- The lipogenesis pathway is well-represented in adipose tissue and liver, whereas, in humans, adipose tissue is not an important site, and liver has only low activity.
- The lipogenesis pathway is less important in humans and it is surprising to note that critical diseases of the pathway have also been reported.
- Variations in its activity among individuals have a bearing on the nature and extent of obesity, and one of the lesions in type I, insulin-dependent diabetes mellitus, is the cause of inhibition of lipogenesis.

SCHEMATIC COMPENDIUM

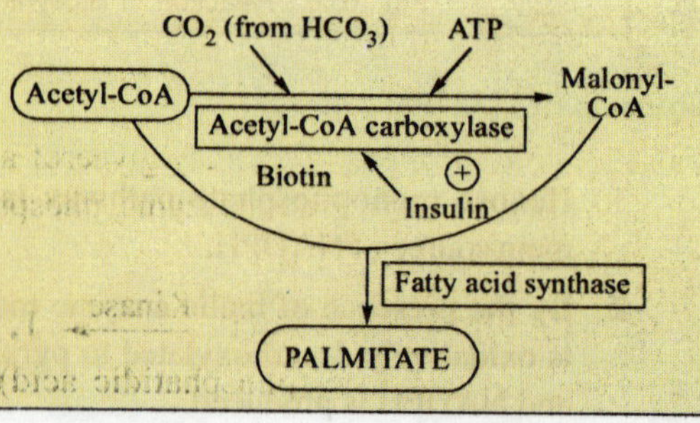

BIOSYNTHESIS OF ACYLGLYCEROLS (Glycerides)

1. Fatty acids are activated to acyl-CoA by the enzyme acyl-CoA synthase using ATP and CoA.
2. Two molecules of acyl-CoA combine with glycerol-3-phosphate to form 1, 2-diacylglycerol phosphate by glycerol-3-phosphate acyl transferase.

   In intestinal mucosa, a monoacylglycerol pathway exists by which monoacylglycerol is converted to 1, 2-diacylglycerol by monoacylglycerol acyl transferase.
3. Another molecule of acyl-CoA is esterified with the diacylglycerol to form a triacylglycerol by diacylglycerol acyl transferase.

   The overall reaction is given :

Fatty acid + ATP + CoA $\xrightarrow{\text{Acyl-CoA synthase}}$ Acyl-CoA (Fatty acid CoA)

Glycerol + ATP $\xrightarrow{\text{Glycerokinase}}$ Glycerol-3-phosphate + ADP

2Acyl-CoA + glycerol-3-phosphate $\xrightarrow{\text{Glycerophosphate acyl transferase}}$ 1, 2-diacylglycerol + phosphate

1, 2-diacylglycerol + phosphate $\xrightarrow{\text{Phosphatase}}$ 1, 2-diacylglycerol + phosphoric acid

1, 2-diacylglycerol + acyl-CoA $\xrightarrow{\text{Diacylglycerol acyl transferase}}$ Triacylglycerol (Triglyceride)

BIOSYNTHESIS OF PHOSPHOLIPIDS

1. Formation of phosphatidic acid

By the reaction between 1, 2-diacylglycerol and ATP in the presence of a kinase enzyme, phosphatidic acid is formed :

1, 2-diacylglycerol + ATP $\xrightarrow{\text{Kinase}}$ 1, 2-diacylglycerol phosphate (phosphatidic acid) + ADP

2. Formation of phosphatidyl choline (Lecithin)

Choline is activated in two stages; first by ATP and then by CTP to convert it to CDP-choline which reacts with 1, 2- diacylglycerol to form phosphatidyl choline.

Choline + ATP → choline monophosphate

Choline monophosphate + CTP →
CDP-choline + PPi
(Active choline)

CDP-choline + 1, 2-diacylglycerol → phosphatidyl choline + CMP.

3. Formation of phosphatidyl ethanolamine (Cephalin)

Ethanolamine is activated to 'active ethanolamine' in two stages. It first reacts with ATP and then to CTP to form CDP-ethanolamine which reacts with 1, 2-diacylglycerol to form cephalin.

Ethanolamine + ATP → Ethanolamine monophosphate + ADP.

Ethanolamine monophosphate + CTP →
CDP-ethanolamine + PPi.
(Active ethanolamine)

CDP-ethanolamine + 1, 2-diacylglycerol → phosphatidyl ethanolamine + CMP.

The enzyme phosphoethanolamine-acylglycerol transferase, responsible for the formation of phosphatidyl ethanolamine, is not present in liver.

4. Formation of phosphatidyl serine

This can be formed directly from phosphatidyl ethanolamine by serine.

Phosphatidyl ethanolamine + serine → phosphatidyl serine.

5. Phosphatidyl inositol

CDP-diacylglycerol + inositol $\xrightarrow{\text{Transferase}}$ phosphatidyl inositol + CMP

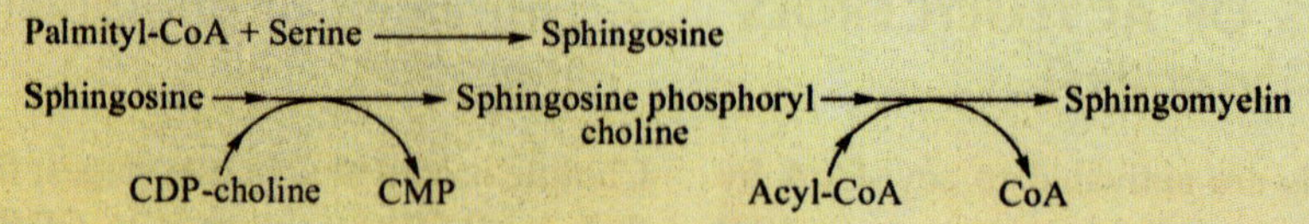

Fig. 18.8

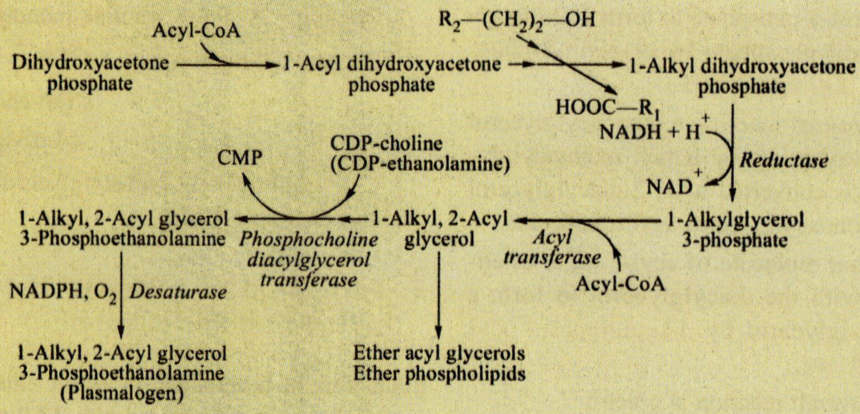

Fig. 18.9

6. Formation of sphingomyelin

Fig. 18.8.

7. Formation of glycerol ether phospholipids and plasmalogens

Fig. 18.9.

8. Formation of cardiolipin

Cardiolipin, a phospholipid, is present in mitochondria. It is formed from phosphatidyl glycerol (Fig. 18.10).

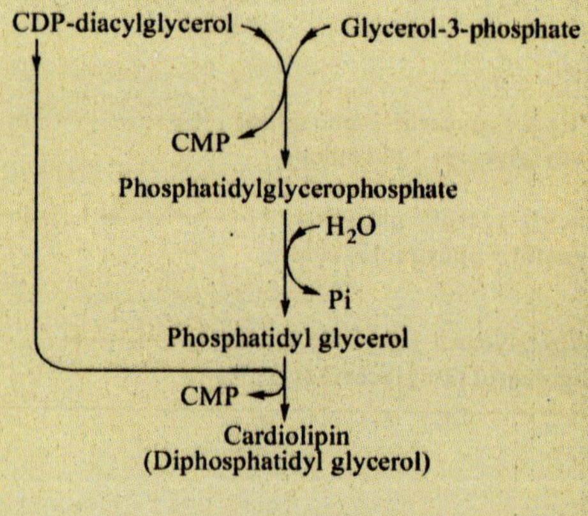

Fig. 18.10 : Formation of cardiolipin

Degradation and turnover of phospholipids

Although phospholipids are actively degraded, each portion of the molecule turns over at a different rate. The presence of enzyme allows partial degradation followed by resynthesis.

Phospholipase A_2 catalyzes the hydrolysis of the ester bond in position 2 of glycerophospholipids to form lysophospholipids which, in turn, may be reacylated by acyl-CoA in the presence of an acyltransferase. Phospholipase A_1 attacks the ester bond in position 1 of phospholipids. Phospholipase C attacks the ester bond in position 3 releasing 1, 2-diacylglycerol and a phosphoryl base. Phospolipase D is a plant enzyme which hydrolyzes the nitrogenous base from phospholipid.

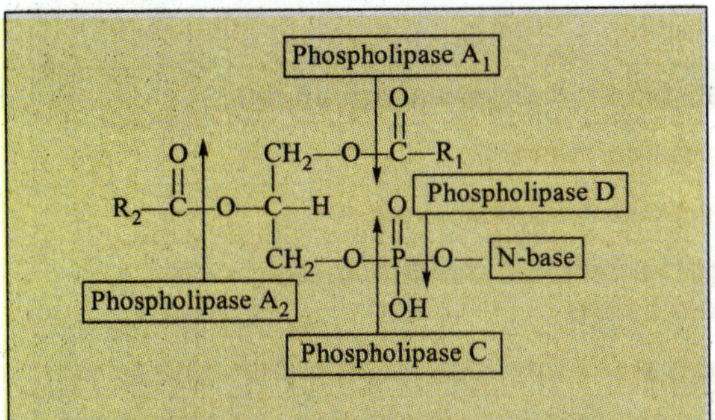

Fig. 18.11

$$\text{Stearyl-CoA} + \text{Enzyme} \longrightarrow \text{Stearyl-Enz.} + \text{CoA}$$

$$\text{Stearyl-Enz.} + O_2 + \text{NADPH} + \text{H}^+ \xrightarrow{\text{Hydroxylase cyt } b_5} \text{Hydroxystearyl-Enz.} + H_2O + \text{NADP}^+$$

$$\text{Hydroxystearyl-Enz.} \xrightarrow{\text{Hydratase}} \text{Oleyl- Enz.} + H_2O$$

$$\text{Oleyl-Enz.} + \text{CoA} \longrightarrow \text{Oleyl-CoA} + \text{Enz.}$$

Fig. 18.12 : Microsomal desaturase system

Physiological functions of phospholipids

1. Certain enzymes appear to require tightly bound phospholipids for their action.
2. These are involved in blood coagulation.
3. These help lipid absorption in intestine.
4. These facilitate lipid transport between tissues.
5. These assist in ion transport and secretion of hormones by cells.
6. These help the oxidation of fatty acids.

Biosynthesis of mono-unsaturated fatty acids

Liver is the main organ responsible for the interconversion with the saturated fatty acids. An enzyme system catalyzes the conversion of stearyl-CoA to oleonyl-CoA. Oxygen, NADPH or NADH is necessary for the reaction. The sequence of reaction is given in Fig. 18.12. It is specific for introducing a double bond in the Δ^9 position of saturated fatty acids, eg, palmitic and stearic acid,

Biosynthesis of polyunsaturated fatty acids

Linoleic and **linolenic acid** must be supplied in the diet to perform the synthesis of the other members of the ω-6 and ω-3 series of polyunsaturated fatty acids. Linoleate may be converted to **arachidonate**. The pathway is first by dehydrogenation of the CoA ester through γ-linoleate followed by the addition of a 2-carbon unit to give **eicosatrienoate** (homo γ-linoleate). The latter forms arachidonate by a further dehydrogenation which is shown in Fig. 18.13.

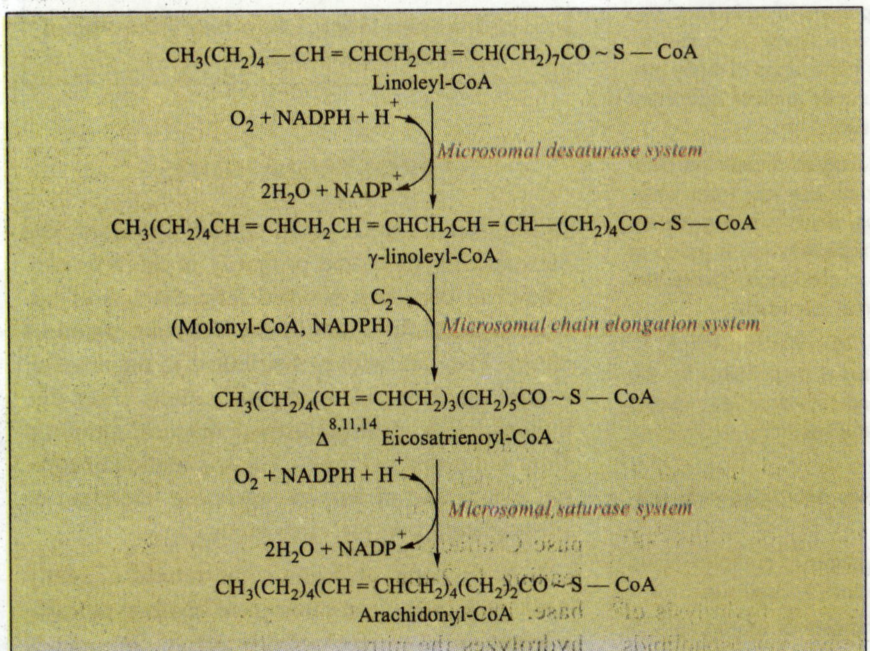

Fig. 18.13 : Conversion of linoleate to arachidonate

In the **fasting state** and in the absence of **insulin**, the desaturation and chain elongation system is greatly diminished.

Disease	Enzyme Deficiency	Clinical Symptoms
Tay Sach's disease	Hexosaminidase A	Mental retardation, muscular weakness, blindness
Fibry's disease	α-galactosidase	Skin rash, kidney failure
Krabbe's disease	β-galactosidase	Metanl retardation, myelin almost absent
Gaucher's disease	β-glucosidase	Enlarged liver and spleen, erosion of long bones, mental retardation in infants
Farber's disease	Ceramidase	Dermatitis, skeletal deformation, mental retardation, fatal in early life
Niemann-Pick disease	Sphingomyelinase	Enlarged liver and spleen, mental retardation, fatal in early life
Ceramide lactoside lipidosis	Ceramide lactosidase	Progressing brain damage, liver and spleen enlargement
Fucosidosis	a-fucosidase	Cerebral degeneration, thick skin, muscle spasticity
Tay Sach's variant or Sandhoff's disease	Hexosaminidase A and B	Same as Tay Sach's disease but progressing more rapidly
Generalized gangliosidosis	G_{M1}-β-galactosidase	Mental retardation, liver enlargement, skin deformation
Metachromatic leukodystrophy	Arylsulphatase A	Mental retardation and psychologic disturbances in adults, demyelination

CLINICAL ASPECTS

- Lung Surfactant is a secretion composed mainly of lipid with some proteins and carbohydrates that prevent the alveoli from collapsing. Dipalmitoyl phosphatidylcholine synthesized shortly before parturition in full term infants is present in surfactant. The respiratory distress syndrome develops in the deficiency of lung surfactant in the lungs of many preterm newborns. The natural or artificial surfactant may be therapeutically used.
- Both phospholidips (particularly ethanolamine, plasmalogen) and sphingolipids are lost from white matter in multiple sclerosis, which is a demyelinating disease. The sphingolipidoses are a group of inherited diseases found in childhood. These diseases are part of lysosomal disorders.
- The combined deficiency of arylsulfatases A, B, and C and steroid sulfatase is responsible for the accumulation of sulfogalactosylceramide, steroid sulfate, and proteoglycans resulting in multiple sulfatase deficiency.
- The lipid storage diseases exhibit several features :
  (a) There is an accumulation of complex lipids having common structure – ceramide.
  (b) The high rate of synthesis of stored lipid in humans.

CLINICAL ASPECTS

(c) The enzymatic defect in these diseases is a deficiency due to gene mutation of specific lysosomal hydrolytic enzyme necessary to breakdown the lipid.

(d) The extent to which the activity of the affected enzyme is decreased.

PROSTAGLANDINS

Prostaglandins are a group of naturally occurring substances synthesized primarily in the prostrate. They had been first isolated from extracts of human seminal fluid and of the vesicular gland of sheep. They are widely distributed in mammalian tissues, eg, lung, kidney, thyroid, spleen, brain, iris, endometrium, gastrointestinal mucosa, amniotic fluid and menstrual fluids. The stimulation of adrenal glands and of nerves causes the liberation of prostaglandins in the circulation.

Arachidonic acid and some related C_{20} fatty acids with methylene-interrupted bonds synthesize a group of pharmacologically active compounds known as prostaglandins.

Chemistry

1. Prostaglandins have a common structure based on prostanoic acid which contains 20 carbon atoms.
2. They are separated into four groups A, B, E and F depending on the variations in the double bonds and in the hydroxyl and ketone groups.
3. The carbon chains are bonded at the middle of the chain by a 5-membered ring.
4. A, B and E have an oxo-grouping at position 9, whereas F has a hydroxyl group in this position.
5. A has a double bond between positions 10 and 11, whereas B has a double bond between positions 8 and 12. E and F do not have a double bond in the ring but possess a hydroxyl group at position 11.
6. All active prostaglandins have at least one double bond between positions 13 and 14. Some have two double bonds, the second being between positions 5 and 6 and some prostaglandins have three double bonds, the additional bond being between positions 17 and 18.
7. The common single double bond has the trans-configuration, whereas the other double bonds have the cis-configuration.
8. All prostaglandins have a hydroxyl group at position 15 and some have another hydroxyl group at position 19.
9. When the 5-carbon ring has two hydroxyl groups, their positions give rise to the possibility of a or b isomers.
10. The six primary prostaglandins, which occur in most cells, can be converted to the 8 secondary prostaglandins that have been identified in natural materials.
11. The synthesis of prostaglandins requires the consumption of 2 molecules of O_2 and 2 molecules of reduced glutathione. The synthesis is catalyzed by the prostaglandin synthetase complex. Aspirin inhibits its synthesis.
12. Fourteen prostaglandins have been isolated from male reproductive tract. These are PGA_1, PGA_2, PGB_1, PGB_2, PGE_1, PGE_2, PGE_3, $PGF_1\alpha$, $PGF_2\alpha$, $PGF_3\alpha$, 19–OH PGA_1, 19–OH PGA_2, 19–OH PGB_1, 19–OH PGB_2.

[The number in subscript indicates the number of double bonds and the Greek letter the isomeric form.]

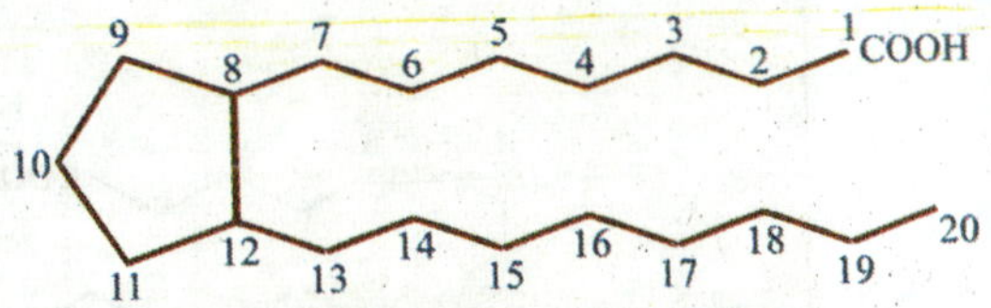

Fig. 18.14 : Prostanoic acid

Functions

1. Prostaglandins exhibit hormone-like activity. They are the most potent biologically active substance. As little as 1 ng/ml causes contraction of smooth muscle in animal.
2. They increase cAMP in platelets, thyroid, corpus luteum, fetal bone, adenohypophysis and lung; but lower cAMP in adipose tissue.
3. Although they are synthesized from the "essential fatty acids" they do not relieve symptoms of essential fatty acid deficiency, because they are too rapidly metabolized.
4. They promote the secretion of epinephrine and cause significant increase in the activity of phosphophosphorylase in liver. They produce a direct inhibitory effect on glycogen synthetase in liver.
5. They inhibit the formation of cAMP which is necessary for the activation of hormone-sensitive lipase and thus decrease lipolysis in adipose tissue. Free fatty acids stimulate more prostaglandin secretion which ultimately inhibits lypolysis.

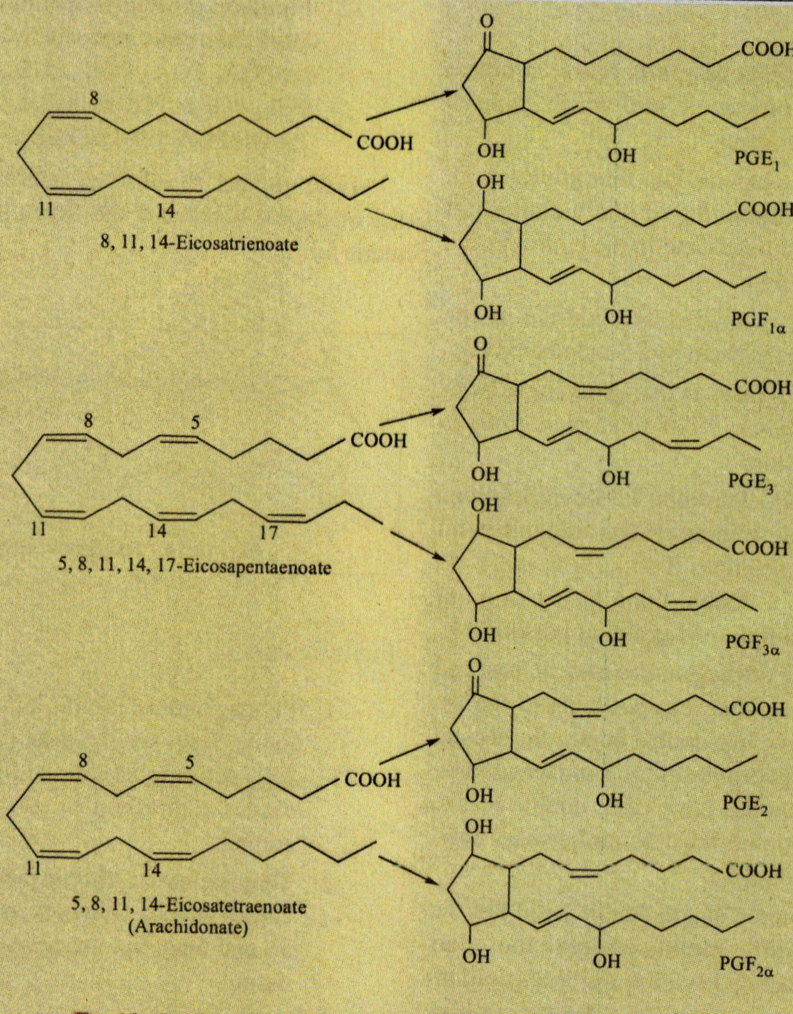

Fig. 18.15 : The six primary prostaglandins and their biosynthetic origin (PG, Prostaglandins)

6. They cause vasodilatation, decreased peripheral resistance, decreased blood pressure, enhanced capillary permeability.
7. They inhibit gastric secretion, increase intestinal motility and cause loose motion.
8. They are involved in spermatogenesis, sperm maturation and transport.
9. They are involved in the regulation of hypothalamic releasing factor release.
10. Infusion of PGE_2 into the renal artery of dogs causes an increase in urinary volume and excretion of Na^+, K^+, Cl^-.

Clinical importance

1. Prostaglandins exert stimulatory effect on contractions of the human uterus. PGE_1, PGE_2, $PGE_2\alpha$ are given intravenously to induce labor. PGE_2 and $PGF_2\alpha$ are also effective orally to induce labor.
2. They are also used as contraceptives to prevent conception.
3. PGA_1 infused intravenously acts as a vasodilator and lowers the blood pressure.
4. PGE_1 by inhalation produces improve-

ment in asthmatic patients and becomes an inhibitor of gastric acidity when administered intravenously.

5. PGA_1, PGE_1, PGE_2 and $PGE_1\alpha$ act as vasoconstrictor on the blood vessels of the nasal mucosa and PGE_1 is an effective nasal decongestant.
6. Amniotic fluid during labor contains high concentration of $PGF_2\alpha$ which causes myometrial contractions. $PGF_2\alpha$ is present in maternal venous blood immediately before uterine contractions in normal spontaneous labor. The placenta is the major source of prostaglandins found in the amniotic fluid and maternal circulations.
7. They are used to control inflammation.

Side-effects

1. PGE_2 causes uterine smooth muscle to contract when induction of labor is desired and also causes gastrointestinal smooth muscle to contract which leads to cramping and diarrhoea.
2. PGE_2 irritates the muscosa lining of the throat causing pain and coughing.

Metabolism

Prostaglandins are quickly metabolized by the enzyme **15-hydroxyprostaglandin dehydrogenase** which is present in most mammalian tissues. This enzyme is blocked by the introduction of a methyl group at the C_{15} position. Then only the half-life of a prostaglandin is prolonged.

Inhibitors

Aspirin, indomethacin, phenanzine, tranylcypromine.

Accumulation of Triacylglycerol in Liver

The factors that take part in role for the accumulation of triacylglycerol are :

1. **The raised level of plasma-free fatty acids formed** from the mobilization of fat from

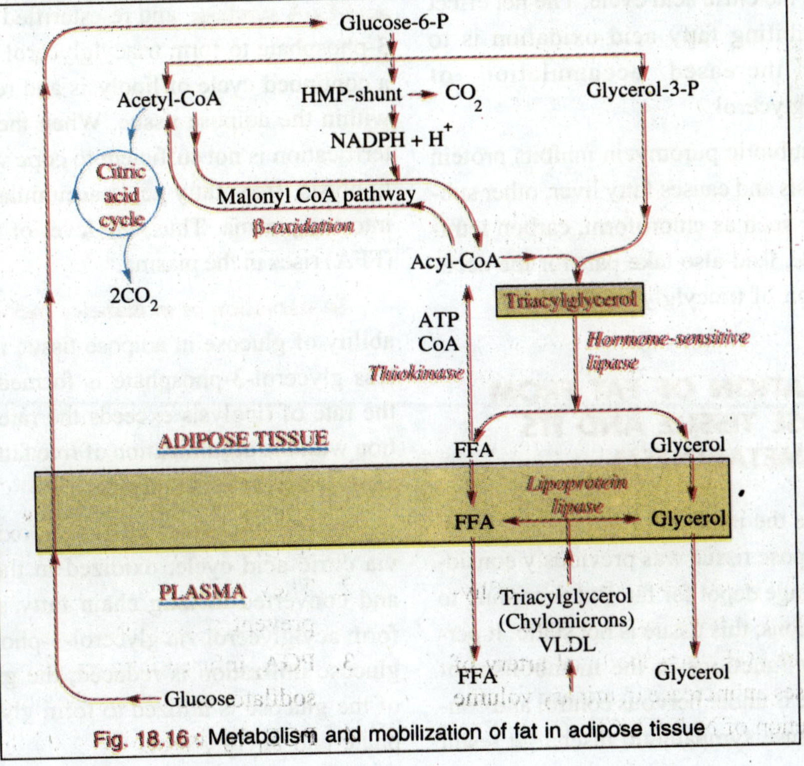

Fig. 18.16 : Metabolism and mobilization of fat in adipose tissue

adipose tissue or hydrolysis of lipoprotein by lipoprotein lipase in extrahepatic tissue occurs.

2. **In starvation and by feeding of high fat diets**, the triacylglycerol present in liver is increased.
3. **In uncontrolled diabetes mellitus** fatty appearance and enlargement of liver occurs.
4. **Due to metabolic block in the production of plasma lipoprotein** triacylglycerol accumulates.
5. **Alcoholism** leads to triacylglycerol in liver causing cirrhosis. When ethanol is oxidized to acetaldehyde by alcohol dehydrogenase there is production of NADH. NADH competes with reducing equivalents from the other substrate for the respiratory chain inhibiting their oxidation. The increased [NADH] / [NAD^+] ratio causes a shift to the left in the equilibrium, malate $\rightleftharpoons$ oxaloacetate which reduce the citric acid cycle. The net effect of inhibiting fatty acid oxidation is to cause increased accumulation of triacylglycerol.
6. The antibiotic puromycin inhibits protein synthesis and causes fatty liver, other substances such as chloroform, carbon tetrachloride, lead also take part for the accumulation of triacylglycerol in liver.

MOBILIZATION OF FAT FROM ADIPOSE TISSUE AND ITS METABOLISM

Adipose tissue is the main store of triacylglycerol in the body. Adipose tissue was previously considered an inert storage depot for fat. But according to recent investigations, this tissue is not static. It performs important functions in the metabolism of lipids. This tissue is under nervous control and conditions which cause sympathetic discharge result in the liberation of unesterified fatty acids and thus a loss of fat. Mobilization of fat is inhibited by denervation. Different types of adipose tissue (eg, white and brown) exhibit differences in metabolism.

Both the glycolytic sequence of reactions and the pentose-shunt pathway are operative in adipose tissue. **Triacylglycerol** is synthesized from acyl-CoA and glycerol-3-phosphate in adipose tissue. Since the activity of **glycerokinase** is low in adipose tissue, glycerol cannot be utilized to any great extent in the esterification of acyl-CoA to form triacylglycerol. For the provision of glycerol-3-phosphate, adipose tissue is dependent on a supply of glucose.

Triacylglycerol is hydrolyzed by a **hormone-sensitive lipase** to form free fatty acids and glycerol. But the tissue cannot utilize glycerol readily. Therefore, glycerol diffuses out into the plasma and it is utilized by liver and kidney where active glycerokinase is present. The free fatty acids can be resynthesized in adipose tissue to acyl-CoA by **acyl-CoA synthase** and re-esterified with glycerol-3-phosphate to form triacylglycerol. Thus, there is a continued cycle of lipolysis and re-esterification within the adipose tissue. When the rate of re-esterification is not sufficient to cope with the rate of lipolysis, free fatty acids accumulate and diffuse into the plasma. Thus, the level of free fatty acid (FFA) rises in the plasma.

In starvation or in diabetes mellitus, the availability of glucose in adipose tissue is reduced and less glycerol-3-phosphate is formed. As a result, the rate of lipolysis exceeds the rate of esterification with the accumulation of free fatty acids which are released into the plasma.

In **adipose tissue**, glucose is oxidized to CO_2 via citric acid cycle, oxidized in the HMP shunt and converted to long chain fatty acids and can form acylglycerol via glycerol-3-phosphate. When glucose utilization is reduced, the greater portion of the glucose is utilized to form glycerol-3-phosphate and acylglycerol.

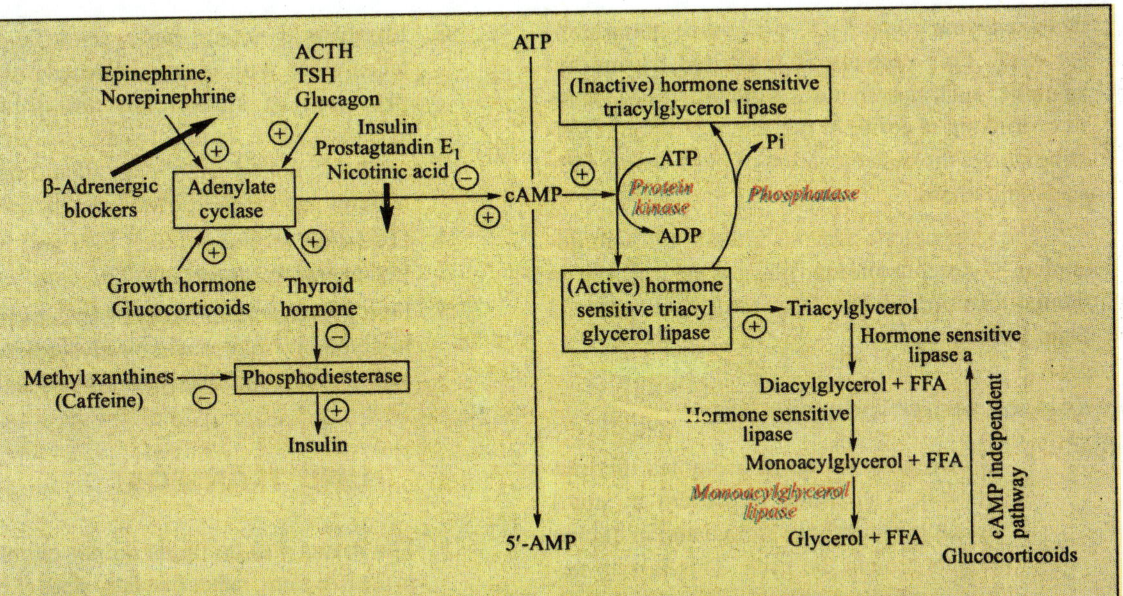

Fig. 18.17 : Influence of hormone on adipose tissue lipolysis

Influence of hormones on adipose tissue

A. Insulin

1. It inhibits the release of free fatty acids from adipose tissue, enhances lipogenesis and the synthesis of acylglycerol and increases the oxidation of glucose to CO_2 via HMP-shunt.
2. It inhibits the activity of hormone-sensitive lipase and reduces the release of free fatty acids as well as glycerol.
3. Insulin, nicotinic acid and prostaglandin E_1 inhibit the synthesis of cAMP depressing adenylate cyclase or stimulating phosphodiesterase. Prostaglandin E_1 in low concentration causes the release of catecholamines resulting in the increase in free fatty acid mobilization.

B. ACTH, MSH, TSH, growth hormone, vasopressin, epinephrine, norepinephrine and glucagon

1. These hormones accelerate the release of free fatty acids from adipose tissue by increasing the rate of lipolysis of the triacylglycerol stores.
2. Many of these activate the hormone-sensitive lipase and increase glucose utilization.
3. Glucocorticoids and thyroid hormones do not increase lipolysis but the presence of these hormones in the lipolytic processes is essential. Thyroid hormone inhibits phosphodiesterase activity. The lipolytic effect of growth hormone in presence of glucocorticoids is slow.
4. These hormones stimulate adenylate cyclase for the formation of cAMP from ATP and this cAMP stimulates the protein kinase which converts inactive hormone-sensitive triacylglycerol lipase into the active lipase. In addition to the hormone-sensitive triacylglycerol lipase, the adipose tissue contains diacylglycerol and monoacylglycerol lipases which are more active than the hormone-sensitive triacylglycerol lipase and catalyze the rate limiting step in lipolysis.

Lipolysis is controlled by the amount of cAMP present in the tissue. cAMP is degraded to 5′-AMP by the enzyme cyclic 3′, 5′- nucleotide phosphodiesterase. This enzyme is inhibited by methyl xanthines such as caffeine and theophylline. Therefore, drinking of coffee or the administration of caffeine causes the high elevation of plasma-free fatty acids in humans.

The sympathetic nervous system, through liberation of norepinephrine, plays a central role in the mobilization of free fatty acids by exerting a tonic influence.

Role of brown adipose tissue

1. Brown adipose tissue promotes thermogenesis. The tissue is present in small amounts in newborn animal and in human infants around the neck. This is very active in newborn animals. It contains high content of mitochondria, cytochromes and a well-developed blood supply.
2. Its oxygen consumption is very high with the conversion of glucose and fatty acids to CO_2.
3. Lipolysis is active; but re-esterification takes place with glycerol because of the presence of significant amounts of glycerokinase in this tissue. Norepinephrine liberated from sympathetic nerve endings increases lipolysis in this tissue.
4. Oxidation produces much heat and little free energy is tapped in ATP.
5. Free glycerol from lipolysis is converted to glycerol-3-phosphate by glycerokinase and is oxidized directly in the tissue.

LIPID TRANSPORT

1. The major plasma lipids do not circulate in the free form, rather free fatty acids (FFA) are bound to albumin, whereas cholesterol, triglycerides and phospholipids are transported in the form of lipoprotein complexes. The density of these lipoproteins is inversely proportional to their lipid content.

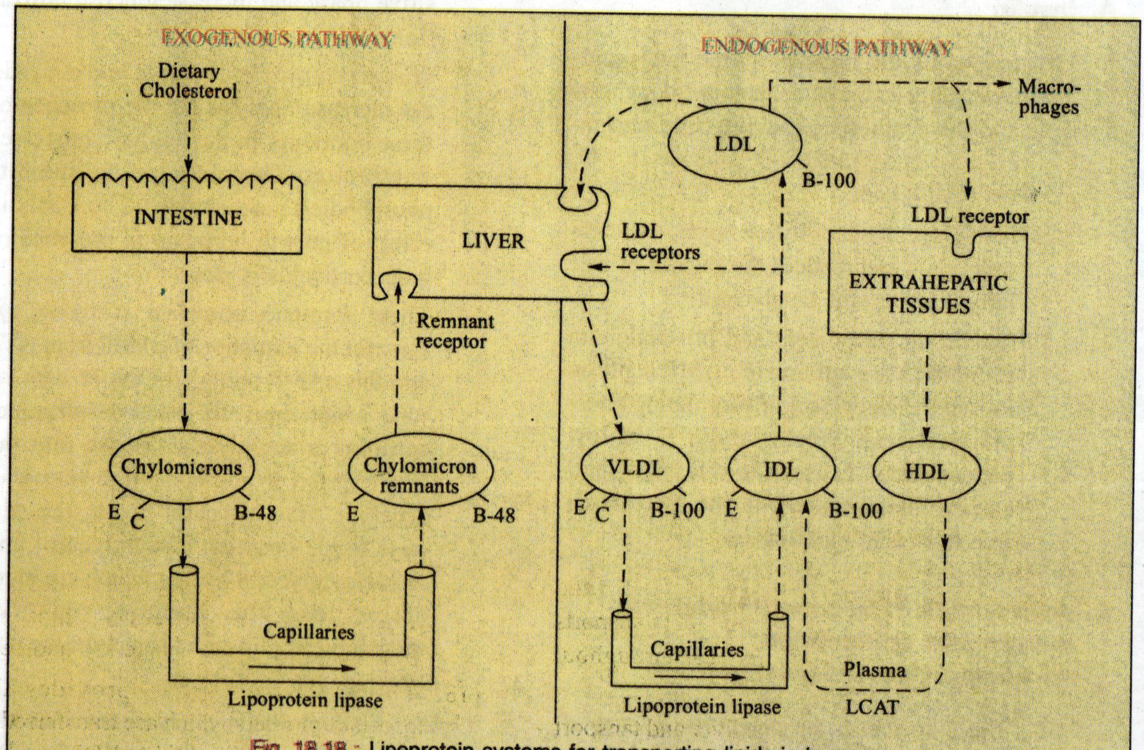

Fig. 18.18 : Lipoprotein systems for transporting lipids in humans

2. The proteins of the lipoproteins are called **apoproteins** and the major apoproteins are called apo E, apo C, apo B-48, and apo B-100. Apo B-48 having a low molecular weight is the characteristic of the exogenous system, whereas apo B-100 having a high molecular weight is the characteristic of the endogenous system.
3. **Chylomicrons**, formed in the intestinal mucosa during the absorption of the products of fat digestion, are very large lipoprotein complexes that enter the circulation through the lymphatic ducts. The plasma shows the milky appearance (**lipemia**) after meals due to the presence of many of these particles in the blood. These chylomicrons are cleared from the circulation by the action of **lipoprotein lipase** which is present on the surface of the endothelium of the capillaries. The triglycerides of these chylomicrons are broken down to glycerol and free fatty acid which then enter adipose cells and are reesterified by the catalytic action of the lipoprotein lipase. Lipoprotein lipase, with the help of heparin as a cofactor, also removes triglycerides from circulating **very low density lipoproteins** (**VLDL**). Apoliprotein C-11 activates lipoprotein lipase.
4. Chylomicrons remain in the circulation as cholesterol-rich lipoproteins called **chylomicron remnants** which are carried to the liver where they bind to chylomicron remnant and LDL receptors and are degraded in lysosomes. The chylomicrons and their remnants form a transport system for ingested exogenous lipids.
5. The endogenous system made up of **VLDL, intermediate density lipoproteins** (**IDL**), **low density lipoproteins** (**LDL**) and **high density lipoproteins** (**HDL**) transports triglycerides and cholesterol throughout the body.
6. VLDL are formed in the liver and tansport triglycerides to extrahepatic tissues.
7. **IDL** is formed when triglyceride is largely removed by the action of lipoprotein lipase. The IDL give up phospholipids and pick up cholesteryl esters formed from cholesterol in the HDL by the action of the plasma enzyme **lecithin cholesterol acyltransferase** (**LCAT**). The liver takes up some IDL and the remaining IDL then lose more triglyceride and protein in the sinusoids of the liver and become LDL. During this conversion, they lose apo E but apo B-100 remains intact.
8. **Cholesterol** is provided to the tissues by LDL. This cholesterol is an essential constituent of cell membranes and is used by the gland cells to form steroid hormones.
9. **LDL** are taken up by receptor-mediated endocytosis in coated pits in the liver and most extrahepatic tissues. The receptors recognize the apo B-100 component of the LDL and bind apo E but not apo B-48. The human LDL receptor is a complex molecule made up of a cysteine-rich region of 292 amino acid residue that binds LDL.
10. **Cholesterol** in the cells inhibits intracellular synthesis of cholesterol by inhibiting HMG-CoA reductase, stimulates esterification of any excess cholesterol that is released and inhibits the synthesis of new LDL receptors.
11. **Macrophages** specially take up LDL which are modified by oxidation and antioxidants like vitamin E and, as a result, the progress of atherosclerosis becomes slow. The LDL receptor on macrophages and related cells is called **Scavenger receptor**.
12. **Cholesterol** leaves as well as enters cells. The cholesterol leaving the cells is absorbed in HDL, lipoproteins that are synthesized in the liver and the intestine. Some of the HDL contain apo E and bind to LDL receptors on other cells transporting cholesterol from one cell to another.
13. Through LCAT, **HDL** provides the cholesteryl esters which are transferred to IDL and then back again to LDL.

14. Apo E is synthesized by astrocytes in the brain and by cells in the spleen, lung, ovary, kidney, and liver. Its concentration is highly increased in injured nerves where it plays an important role in nerve regeneration by adjusting the cholesterol concentration during the repair process.

METABOLISM OF THE PLASMA LIPOPROTEINS

Five groups of lipoproteins present in plasma exhibit important role in the transport and metabolism of lipids. These are :

1. Chylomicrons : Derived from intestinal absorption of triacylglycerol.
2. Very low density lipoproteins (VLDL or pre-β-lipoproteins) : Mainly derived from the liver for the export of triacylglycerol and also formed from dietary lipids.
3. Low density lipoproteins (LDL or β-lipoproteins) : These represent a final stage in the catabolism of VLDL and chylomicrons.
4. High density lipoproteins (HDL or α-lipoproteins) : These are involved in VLDL, chylomicrons and cholesterol metabolism.
5. Free fatty acids : These are not classified with the other plasma lipoproteins as their structure is different. These consist of long chain fatty acids attached to serum albumin.

Apolipoproteins (Apoproteins)

The lipoproteins contain one or more proteins or polypeptides known as apoprotein. The apoproteins in the lipoproteins are present as :

HDL	: A-I and A-II, C-I, C-II and C-III, arginine rich apoprotein.
LDL	: Apoprotein B.
VLDL	: Apoprotein B, C-I, C-II, C-III, arginine rich apoprotein.
Chylomicrons	: Apoprotein B, C-I, C-II, C-III.

Some lipoproteins are also glycoproteins.

Electrophoretic separation of lipoprotein and their normal concentration in human plasma

1. α-Lipoproteins : These occupy the α-globulin region after electrophoresis. These contain 45% proteins, 8% triacylglycerol, 20% cholesterol, 27% phospholipids.
   Normal concentration : 300 mg%.
2. β-Lipoproteins : These occupy the β-globulin region after electrophoresis. They contain 46% cholesterol, 23% phospholipids, 10% triacylglycerol. Their concentration increases in atherosclerosis and coronary thrombosis.
   Normal concentration : 300 mg%.
3. Pre-β-lipoproteins : These occupy the region in between α- and β-lipoproteins. These contain low protein, but contain 50% triacylglycerol, fair amount of cholesterol and phospholipids. Their concentration is also increased in atherosclerosis and coronary thrombosis etc.
   Normal concentration : 150 mg%.

Formation of chylomicrons and VLDL

1. Both chylomicrons and VLDL are found in chyle formed by the lymphatic system draining the intestine. VLDL is the vehicle of transport of triacylglycerol from the liver to the extrahepatic tissues.
2. Apoprotein B which is essential for their formation is synthesized by ribosomes in the rough endoplasmic reticulum and is incorporated into lipoproteins in the smooth endoplasmic reticulum which is the main site of synthesis of triacylglycerol, phospholipids and cholesterol. Carbohydrate is added to the lipoproteins found in the Golgi apparatus.
3. Both are released from the intestine or hepatic cell by reverse pinocytosis.
4. Chylomicrons pass into the spaces between the intestinal cells making their way

into the lacteals draining the intestine. VLDL are secreted by hepatic parenchymal cells into the space of Dissc and then into the hepatic sinusoids.

Catabolism of chylomicrons and VLDL

A. Role of lipoprotein lipase clearing factor lipase

1. It is present in the walls of blood capillaries and also found in the extracts of heart, lung, spleen, lactating mammary gland.
2. Its concentration in the normal blood is less. It is released from the tissues into the circulation following injection of heparin.
3. Phospholipids and apoprotein C-11 are required as cofactors for its activity.
4. It hydrolyzes triacylglycerol to monoacylglycerol through diacylglycerol. The monoacylglycerol is finally hydrolyzed by monoacylglycerol hydrolase.

B. Role of liver

Chylomicron remnants (about half the diameter of parent chylomicrons) are taken by the liver in vivo and by the perfused liver, in which the cholesteryl esters are hydrolyzed and the triacylglycerol, fatty acids are metabolized.

Metabolism of LDL

1. It is formed from VLDL and chylomicrons.
2. It is removed from the circulation by the liver. The half-time of disappearance of apoprotein B in LDL from the circulation is about $2\frac{1}{2}$ days.
3. Fibroblasts and lymphocytes may degrade LDL in extrahepatic tissues.

Metabolism of HDL

1. It is synthesized and secreted from liver and intestine.
2. Nascent HDL from intestine does not contain apoprotein C but only apoprotein A. Nascent HDL formed by the liver contains apoprotein and free cholesterol. These lipoproteins are similar to the particles found in the plasma of patients with a deficiency of the plasma enzyme lecithin : Cholesterol acyl transferase (LCAT) and in the plasma of patients with obstructive jaundice.
3. The liver and the intestine are the final sites of degradation of HDL apoproteins.

Plasma lipoprotein abnormalities

1. In cases of abnormal hyperlipemia, the concentration of serum VLDL is increased and the concentration of serum HDL may be decreased, increased or normal.
2. In cases of hyperlipemia with marked hypercholesterolemia, serum LDL is increased.
3. Serum LDL has been found to be increased in diabetes mellitus, hypothyroidism, obstructive jaundice, the nephrotic syndrome and in glycogen storage diseases.
4. The concentration of serum LDL and of total serum cholesterol is significantly increased in atherosclerosis.
5. Studies of serum lipoproteins are still of limited clinical value in myocardial infarction, cerebral thrombosis etc.
6. LDL and also VLDL are possible risk factors in studies related to increased susceptibility to ischemic heart disease.

Role of liver in lipid metabolism

1. The liver has active enzyme systems for synthesizing triacylglycerols, phospholipids, cholesterol, plasma lipoproteins and for converting fatty acids to ketone bodies.
2. The fatty acids used in the synthesis of liver triacylglycerol are derived from two sources : (a) from acetyl–CoA derived

from carbohydrate, (b) uptake of free fatty acids from the circulation.

3. It is the site for the synthesis of bile acids from cholesterol.
4. It is the major site for the oxidation of fatty acids.
5. Feeding of diets high in carbohydrate containing sucrose or fructose, high levels of circulating free fatty acids, ingestion of ethanol and the presence of high level of insulin enhance the synthesis of triacylglycerol and the secretion of VLDL by the liver.
6. It has the enzyme systems for lengthening and shortening of fatty acids and for saturating and desaturating fatty acids.
7. This organ is chiefly concerned in removal of phospholipids, cholesterol and lipoproteins from the plasma.
8. Hepatic glucokinase increases with the availability of carbohydrate in the diet. This increases glucose incorporation into the liver and hence glycolysis and fatty acid synthesis. The fatty acids are carried to the adipose tissue as triacylglycerol in VLDL.

Starvation diminishes glucokinase and leads to diminished fatty acid synthesis in the liver. The hypoglycemia stimulates growth hormone production which, in turn, stimulates lipolysis. The free fatty acids so liberated from adipose tissue can influence carbohydrate metabolism in the liver where they are broken down with the formation of acetyl-CoA. The lipolytic effect of growth hormone is also necessary for its protein anabolic action.

CLINICAL ORIENTATION

- Abnormalities of lipoprotein metabolism occur at the sites of production or utilization of lipoproteins causing hyper or hypolipoproteinemias. The most common is diabetes mellitus in which insulin deficiency causes excessive mobilization of FFA and underutilization of chylomicrons and VLDL leading to hypertriacylglycerolemia. Other pathologic conditions are due to inherited defects in the synthesis

CLINICAL ORIENTATION

of apoprotein portion of the lipoprotein. Some of these defects cause hypercholesterolemia and premature atherosclerosis. Excessive fat deposit causes obesity.

- Hyperlipoproteinemias are of two types : (a) Primary, (b) Secondary.

(a) Primary hyperlipoproteinemias

(i) Type I : Familial lipoprotein lipase deficiency is a rare disorder characterized by hypertriglyceridemia and hyperchylomicronemia. The clinical feature is present in early childhood characterized by eruptive Xanthomas and recurrent abdominal pain.

(ii) Type II : Familial hypercholesterolemia (FHC), a common disorder involving no enzyme deficiency is characterized by hyper β-lipoproteinemia (LDL), increased total cholesterol, increased total triglyceride. The clinical feature is characterized by increased incidence of atherosclerosis, Premature cardiovascular disease, Xanthomas tendinous, corneal arcus.

(iii) Type III : Familial Dys-Beta Lipoproteinemia is an inherited disease characterized by increase in β-lipoproteins (LDL) and increase in VLDL, hypercholesterolemia and hyperglyceridemia. The metabolic defects include increased level of apo-E, increased synthesis of apo-B, conversion of normal VLDL to β-VLDL. The clinical features exhibit premature cardiovascular diseases and atherosclerosis. The treatment includes weight reduction, low carbohydrate diet, and fat containing unsaturated fatty acids.

(iv) Type IV : Familial hyper- triglyceridemia is an inherited disease characterized by hyper VLDL, increased synthesis of TG, increased cholesterol level, subnormal HDL. The metabolic defects include increased synthesis of TG, Glucose intolerance found in early adulthood. The above-mentioned lipoprotein characteristic is found in coronary heart disease (CHD), maturity onset diabetes mellitus, obesity, chronic alcoholism. The advisable treatment is weight reduction, decrease of carbohydrate intake, low cholesterol diet.

(v) Type V : Combined Hyperlipidemia is an inherited disease characterized by increased chylomicrons and VLDL, hypertriglyceridemia, hypercholesterolemia. The clinical features exhibit obesity and diabetes mellitus, glucose intolerance, xanthomas which are found in adult life. The treatment involves weight reduction, less carbohydrate and fat intake.

Contd.

CLINICAL ORIENTATION

Wolman's Disease is caused by the deficiency of cholesteryl ester hydrolase in lysosomes and it exhibits increased levels of cholesterol in blood (Hyper-cholesterolemia).

- Hypolipoproteinemias exhibit the following disorders :
  - (i) A beta lipoproteinemia is an inherited disorder characterized by decreased plasma cholesterol due to absence of LDL, presence of lipids in low concentration, nonformation of chylomicrons and VLDL. Other clinical features are abnormal RBC, malabsorption of fats. The other form of the disorder is known as Bassenkornzweig Syndrome.
  - (ii) Tangier's Disease is caused by the deficiency of HDL. The metabolic defects of this disease is reduction in Apo-A–I and Apo-A–11, accumulation of cholesteryl esters in different tissues. The clinical features exhibit hyperplastic orange yellow tonsils and adenoids, decreased plasma cholesterol level associated with elevated TG level.

Fatty liver

Lipid (mainly as triacylglycerol) can accumulate in the liver for the following reasons causing fatty liver :

1. The increased levels of plasma free fatty acids resulting from mobilization of fat from adipose tissue.
2. The hydrolysis of lipoprotein or chylomicron triacylglycerol by lipoprotein lipase in extrahepatic tissues.
3. Increasing amounts of free fatty acids are taken up by the liver and esterified.
4. The production of plasma lipoprotein does not keep pace with the influx of free fatty acids allowing triacylglycerol to accumulate.
5. During starvation and the feeding of high-fat diets, the quantity of triacylglycerol present in the liver is significantly increased.
6. Uncontrolled diabetes mellitus and toxaemia of pregnancy cause fatty appearance and enlargement of the liver.
7. The metabolic block in the synthesis of lipoproteins from lipid and apoprotein.
8. The deficiency of lipotropic factor causes triacylglycerol to accumulate even though only a normal rate of fatty acid synthesis and uptake of free fatty acids take place.
9. Carbon tetrachloride, chloroform, phosphorus, lead, arsenic and ethionine (α-amino-γ-ethylmercaptobutyric acid) cause fatty liver. These substances inhibit hepatic protein synthesis. Orotic acid blocks apo-B synthesis.
10. Alcoholism also leads to fat accumulation in the liver, hyperlipidemia and ultimately cirrhosis.
11. Protein deficiency, essential fatty acid and vitamin deficiencies (eg, vitamin E, pyridoxine, pantothenic acid) cause fatty liver. The deficiency of essential fatty acids depresses the synthesis of phospholipids and, therefore, cholesterol is involved in esterification causing fatty livers.

Lipotropic factor

The substances that prevent the accumulation of fat in the liver are known as lipotropic factor. The phenomenon is said to be lipotropism.

Choline, methionine and betaine and β-propiothetin act as lipotropic agents in curing fatty livers due to choline deficiency. Diets poor in protein (containing methionine) or lecithin (containing choline) tend to favour the production of fatty liver. Choline is synthesized using labile methyl groups donated by methionine in the process of transmethylation. Vitamin B_{12} and folic acid which are important in hematopoiesis are also able to produce lipotropic effect. Vitamin B_{12} is concerned in the biosynthesis of labile methyl groups and folic acid in transmethylation reactions. Inositol exerts a limited lipotropic effect in fat free diets. Casein and certain other proteins possess lipotropic activity.

KETOSIS

The fatty acids undergo excessive oxidation in the liver under certain metabolic conditions producing large quantities of keto acids—acetoacetic acid and β-hydroxybutyric acid, which pass into the blood by diffusion. Acetoacetic acid then undergoes spontaneous decarboxylation to produce acetone. These three substances—acetoacetate, β-hydroxybutyrate and acetone—are collectively known as the ketone bodies (acetone bodies or ketones).

Normally, the blood of mammals contains ketone bodies not exceeding 1 mg/100 ml. The concentration is little higher than this in ruminants. Daily excretion of ketone bodies of normal person is less than 1 mg. Higher than normal quantities in the blood or urine constitute ketonemia (hyperketonemia) or ketonuria, respectively. The condition in which there is a high concentration of ketone bodies in tissues and blood is called ketosis.

Acetoacetic acid and β-hydroxybutyric acid are moderately strong acids and are buffered in blood or tissues. Their excretion in large quantities admits some loss of buffer cation (in spite of ammonia production by the kidney) which depletes the alkali reserve causing ketoacidosis.

The process of formation of ketone bodies is termed ketogenesis and the process of breakdown of ketone bodies taking place in peripheral tissues is called ketolysis.

Ketogenesis in the liver

Ketosis generally occurs in severe diabetes mellitus, prolonged starvation, glycogen storage diseases, toxemia of pregnancy, infective hepatic disease and continued fever. Experimentally, it occurs in the oral administration of fatty acids, high fat diet, low carbohydrate diet, pancreatectomy, administration of growth hormone or ACTH. Under these conditions, there is diminished carbohydrate utilization and increased fat mobilization.

In ruminants, the rumen converts butyric acid formed from fermentation to β-hydroxybutyrate which enters the blood stream. The ruminant lactating mammary gland also produces ketone bodies. But these ketone bodies do not cause ketosis in these species.

Ketone bodies are formed in the liver but utilized in the extrahepatic tissue. Enzymes responsible for ketone body formation are associated mainly in the mitochondria.

Acetyl-CoA (C_2 units) formed in β-oxidation of fatty acids is the basic unit for the formation of ketone bodies. Two molecules of acetyl-CoA condense to form acetoacetyl-CoA by a reversal of thiolase reaction. Two pathways have been proposed for the formation of acetoacetate from acetoacetyl-CoA :

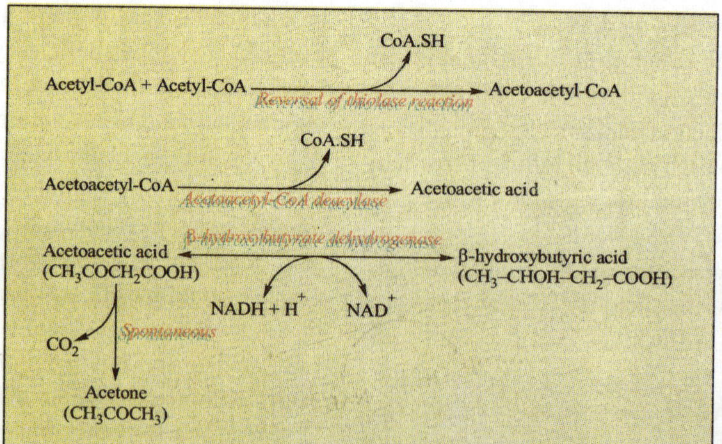

Fig. 18.19 : Minor pathway of ketogenesis in liver

1. First pathway (Minor pathway)

The first pathway is by simple deacylation catalyzed by the enzyme acetoacetyl-CoA deacylase which is shown in Fig. 18.19.

2. The second pathway (Major route)

It involves the condensation of acetoacetyl-CoA with another molecule of acetyl-CoA to form β-hydroxy-β-methyl glutaryl-CoA by HMG-CoA synthetase. HMG-CoA is splitted into acetoacetic acid and acetyl-CoA by HMG-CoA lyase present in mitochondria. From acetoacetic acid, acetone and β-hydroxybutyrate are formed which are shown in Fig. 18.20.

Regulation of Ketogenesis

1. Ketosis does not occur in vivo unless the concentration of circulating free fatty acids increases by the lipolysis of triacylglycerol in adipose tissue. Free fatty acids are the precursors of ketone bodies in the liver. The factors regulating mobilization of free fatty acids from adipose tissue are important in controlling ketogenesis.
2. The activity of carnitine palmitoyl transferase 1 in the outer mitochondrial membrane regulates the entry of long chain acyl groups into mitochondria prior to β-oxidation. Fatty acid oxidation is depressed

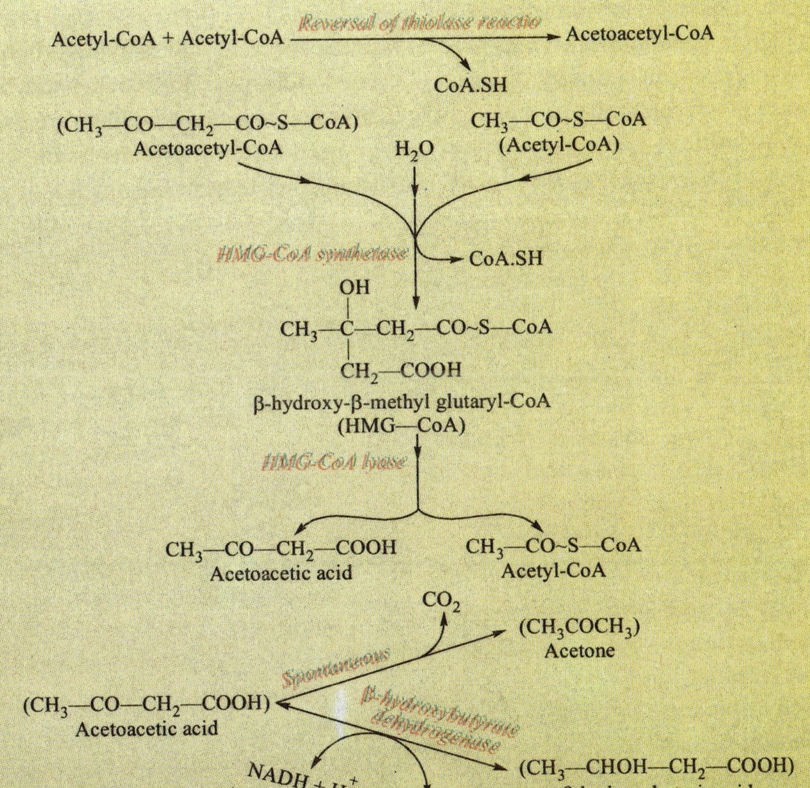

Fig. 18.20 : Major pathway for ketogenesis in liver

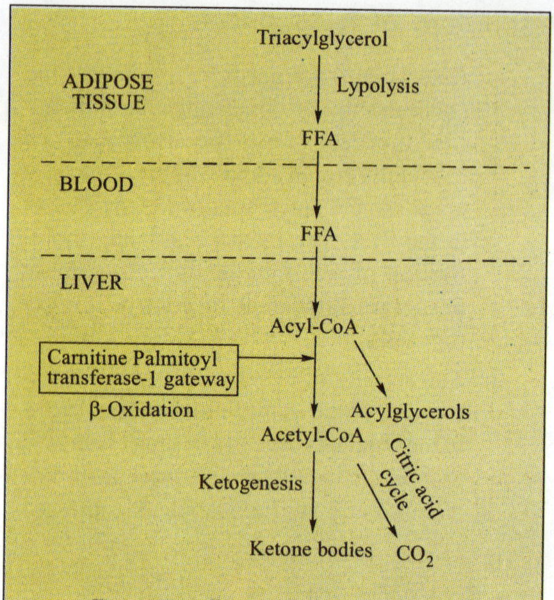

Fig. 18.21 : Regulation of ketogenesis

in fed state since the enzyme activity is low; but high in starvation when fatty acid oxidation increases. In fed state, there is the increase in concentration of malonyl-CoA which inhibits this enzyme and thereby switch off β-oxidation. In fed state, free fatty acids enter the liver cells in low concentrations and all are esterified to acylglycerols and transported out of the liver in VLDL. The concentration of free fatty acids increases with the onset of starvation, acetyl-CoA carboxylase is inhibited directly by acyl-CoA, and malonyl-CoA concentration is decreased releasing the inhibition of carnitine palmitoyl transferase-1 and as a result, more acyl-CoA is β-oxidized. There is the decrease in the [Insulin]/[Glucagon] ratio. Thus, β- oxidation from free fatty acids is controlled by the carnitine palmitoyl transferase-1 gateway into the mitochondria.

3. When the concentration of serum free fatty acids is raised, more free fatty acid is converted to ketone bodies and less is oxidized via the citric acid cycle to CO_2. The complete oxidation of one mol of palmitate produces 129 mol of ATP via β-oxidation and CO_2 production in the citric acid cycle, whereas only 33 mol of ATP is produced when acetoacetate is the end product and only 21 mol when 3-hydroxybutyrate is the end product.

Metabolism of ketone bodies or utilization of ketone bodies in the extrahepatic tissues

Acetoacetic acid and β-hydroxybutyric acids are carried from liver to extrahepatic tissues mainly kidney and muscle where they are oxidized for energy production after conversion to acetyl-CoA. The enzyme responsible for the activation of ac-

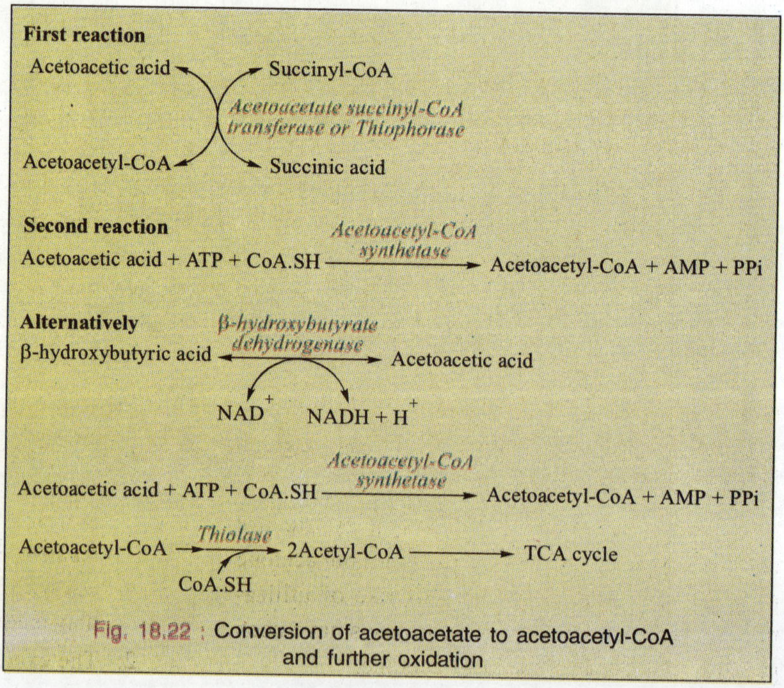

Fig. 18.22 : Conversion of acetoacetate to acetoacetyl-CoA and further oxidation

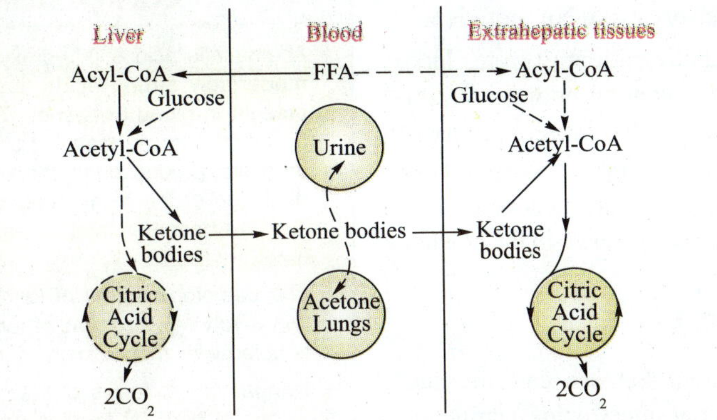

Fig. 18.23 : Formation, utilization and excretion of ketone bodies (The main pathway is indicated by the heavier arrows)

etoacetate to acetoacetyl-CoA is absent from liver for which liver cannot utilize these acids. Two reactions take place in extrahepatic tissues for the activation of acetoacetate to acetoacetyl-CoA.

In the first reaction, succinyl-CoA reacts with acetoacetic acid in presence of the enzyme acetoacetate-succinyl-CoA transferase (Thiophorase) to form acetoacetyl-CoA and succinate.

In the second reaction, acetoacetate is activated by ATP in presence of CoA catalyzed by acetoacetyl-CoA synthetase to form acetoacetyl-CoA.

Alternatively, β-hydroxybutyric acid is converted directly by β-hydroxy butyrate dehydrogenase in extrahepatic tissues to form acetoacetic acid which is then converted to acetoacetyl-CoA by any of these above two reactions.

The acetoacetyl-CoA formed by these reactions is split to acetyl-CoA by thiolase and oxidized in the citric acid cycle. The reactions are given in Fig. 18.22.

Acetoacetate and β-hydroxybutyrate are readily oxidized by extrahepatic tissues but acetone is oxidized with difficulty and its rate of utilization is also very slow. Several pathways have been proposed for the utilization of acetone.

1. First pathway

Acetone is converted to acetoacetate by reversal of decarboxylation.

2. Second pathway

Acetone is converted to propanediol which can form 1 carbon (formate) unit and 2 carbon (acetate) units.

Most of the evidence suggests that ketonemia is due to increased production of ketone bodies by the liver rather than to a deficiency in their utilization by extrahepatic tissues. In moderate ketonemia, the loss of ketone bodies through the urine is only a few per cent of the total ketone body production and utilization. The formation, utilization and excretion of ketone bodies are shown in Fig. 18.23.

Affects of ketosis

1. Both acetoacetate and β-hydroxybutyrate are moderately strong acids. They neutralize bicarbonates resulting depletion of alkali of the body and produce metabolic acidosis. In case of severe ketosis, death may ensure from acidosis.
2. The excretion of ketone body in the urine

involves the loss of Na^+ in particular, leading to total electrolyte and Na^+ deficiency.

3. The severe diabetic patient excretes large quantities of both ketone bodies and glucose in the urine with a large quantity of water and thus developing dehydration. In diabetic acidosis, there is severe alteration in cation-anion balance in the plasma.

Prevention of ketosis

1. In case of diabetic ketosis, carbohydrate diet, intramuscular injection of insulin and antiketogenic substance (aspartic acid) which may provide oxaloacetate by transamination should be administered.
2. In case of prolonged starvation ketosis, carbohydrate diet and antiketogenic substance (aspartic acid) which may provide oxaloacetate by transamination should be given.
3. The electrolytes and the fluids of the body must be restored by intravenous injection of isotonic solution of sodium salts such as NaCl, $NaHCO_3$ or sodium lactate. Potassium salts are desired to be added.

Test for ketone bodies in the urine

Rothera's test

1. 5 ml of the urine is saturated with solid ammonium sulphate by shaking it vigorously.
2. 2 drops of freshly prepared 5 per cent solution of sodium nitroprusside and 1 ml of ammonium hydroxide are added.
3. Allowed to stand for a while.

A permanganate colour which appears just above the layer of the undissolved ammonium sulphate indicates the presence of ketone bodies.

In normal individuals, the ketone bodies are excreted less in quantities in the urine. This negligible amount does not respond to Rothera's test. Hence, the ketone bodies in the normal urine are not detected by this test:

CLINICAL ASPECTS

- Acetoacetic acid & β-hydroxybutyric acid are both moderately strong acids and are buffered when present in blood or tissues. Their continual excretion directs some loss of buffer cation (in spite of ammonia production by the kidney) resulting in the depletion of the alkali reserve causing ketoacidosis. This becomes fatal in case of uncontrolled diabetes mellitus.
- The pathologic forms of ketosis occurs in starvation involving depletion of available carbohydrate coupled with mobilization of free fatty acids.
- Nonpathologic forms of ketosis occur under conditions of high fat feeding and after severe exercise in the postabsorptive state.
- Much losses of carnitine occurs in case of patients with organic aciduria. This deficiency shows episodic periods of hypoglycermia due to reduced gluconeogenesis resulting from impaired fatty acid oxidation and ketogenesis leading to a lipid accumulation with muscular weakness. The oral supplementation of carnitine is the treatment. The symptoms are similar to Reye's Syndrome (encephalopathy with fatty degeneration of the viscera) in which carnitine is sufficient.
- The deficiency of carnitine palmitoyl transferase-1 affects only the liver resulting in reduced fatty acid oxidation and ketogenesis with hypoglycemia, whereas the deficiency of carnitine palmitoyl transferase-11 mainly affects skeletal muscle. The hypoglycemic sulfonylureas (glyburide and tolbutamide) reduce fatty acid oxidation by inhibiting carnitine palmitoyl transferase. The deficiency of long chain β-hydroxyacyl-CoA dehydrogenase is the cause of acute fatty liver in pregnancy.
- Jamaican vomiting sickness occurs from the eating of unripe fruit of the akee tree which contains a toxin, hypoglycin that inhibits β-oxidation causing hypoglycemia.
- Dicarboxylic aciduria is caused by a lack of mitochondrial medium chain acyl-CoA dehydrogenase which impairs β-oxidation but increases ω-oxidation forming dicarboxylic acids which are excreted.
- Refsum's disease is a rare neurologic disorder caused by accumulation of phytanic acid which contains a methyl group on carbon 3 that blocks β-oxidation. Normally, methyl group is removed by α-oxidation but persons with Refsum's disease have an inherited defect in α- oxidation that allows accumulation of phytanic acid.
- Zellweger's (Cerebrohepatorenal) syndrome occurs in the inherited absence of peroxisomes in all tissues for which long chain fatty acids are not oxidized. As a result, there is the impaired bile acid and ether lipid synthesis.

SCHEMATIC COMPENDIUM

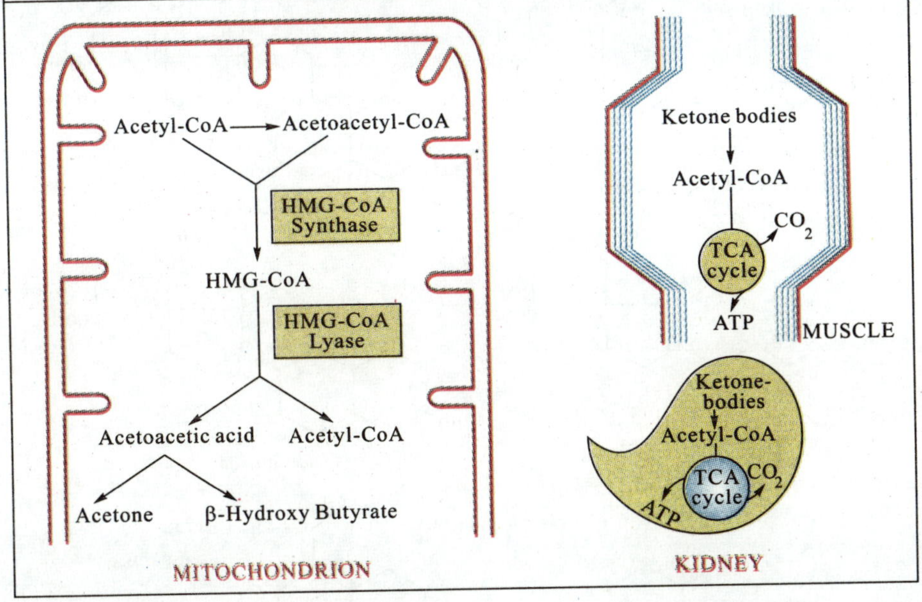

CHOLESTEROL METABOLISM

Occurence

Human body contains large quantities of cholesterol which are found in brain and nervous tissues. Other tissues such as liver, kidney, spleen and skin also contain fairly good amounts of cholesterol. The total amount of cholesterol is about 140 grams in the body of a man weighing 70 kg. The greater part of the cholesterol of the body is synthesized (about 1 gram per day) whereas about 0.3 gram per day are provided by the average diet. It is not synthesized in plants. Dietary cholesterol is obtained only from animal sources like meat, liver, brain and egg yolk (a particularly rich source).

Chemistry

1. Cholesterol is a white, waxy, solid found associated with fats but chemically different from them.
2. It has a parent nucleus which is said to be **cyclopentanoperhydrophenanthrene nucleus.**
3. It has a hydroxyl group at C_3, an unsaturated bond at C_5–C_6, two methyl groups at C_{10} and C_{13} and 8 carbon paraffin side-chain attached to C_{17}.
4. It is an alcohol.
5. It occurs free and combined with fatty acids by ester linkage at the hydroxyl group.
6. Cholesterol in ester form is often referred to as 'bound' cholesterol esters. These are normally rich in linoleic acid.
7. Steroids closely related to cholesterol include 7-dehydrocholesterol which occurs

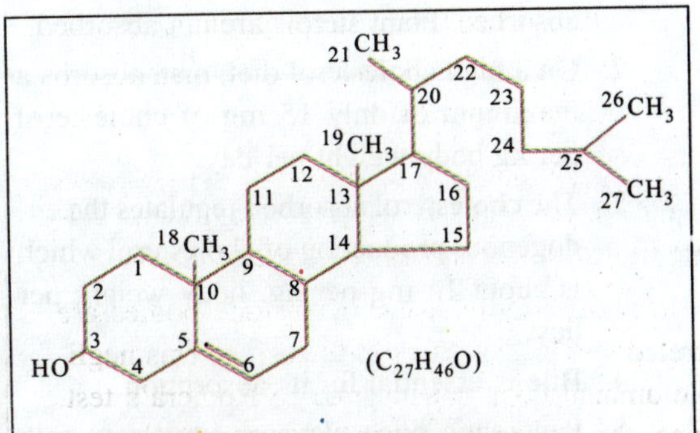

Fig. 18.24 : Structure of cholesterol

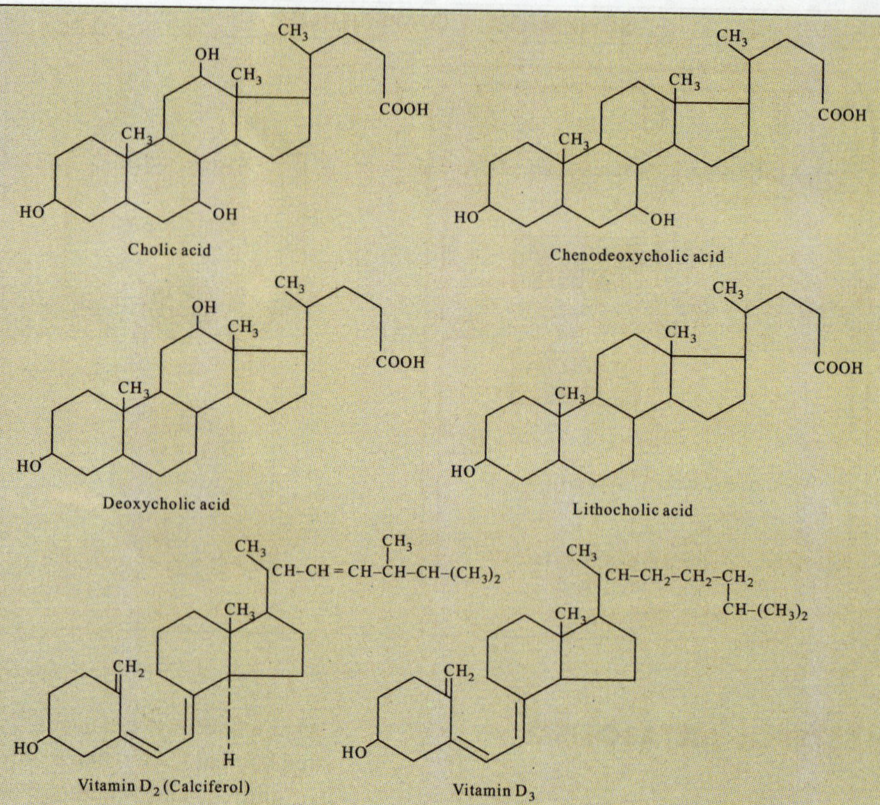

Fig. 18.25 : Structural relation of cholesterol to vitamin D and bile acids

in skin and can be converted by ultraviolet radiation to vitamin D_3, dihydrocholesterol, bile acids, hormones of the adrenal cortex and of the sex glands.

Absorption

1. A great part of the ingested cholesterol is absorbed. Plant sterols are not absorbed.
2. On a high cholesterol diet, man absorbs a maximum of only 15 mg of cholesterol per kg body weight per day.
3. The cholesterol absorbed regulates the endogenous production of cholesterol which is about 14 mg per kg. body weight per day.
4. Bile is essential for its absorption.
5. Pancreatic juice plays an important role because **cholesterol esterase** present in it hydrolyzes the esters of cholesterol present in the diet. After absorption it is again esterified in the cell and then released for transport via the lymphatic to the thoracic duct.

Normal concentration of cholesterol in blood

140-220 mg per 100 ml of blood. It increases with ages and also during pregnancy.

Physiological importance of cholesterol in the body

1. It is the essential constituent of cells.
2. It aids in the permeability of the cells.
3. It controls the red cells from being easily hemolyzed.

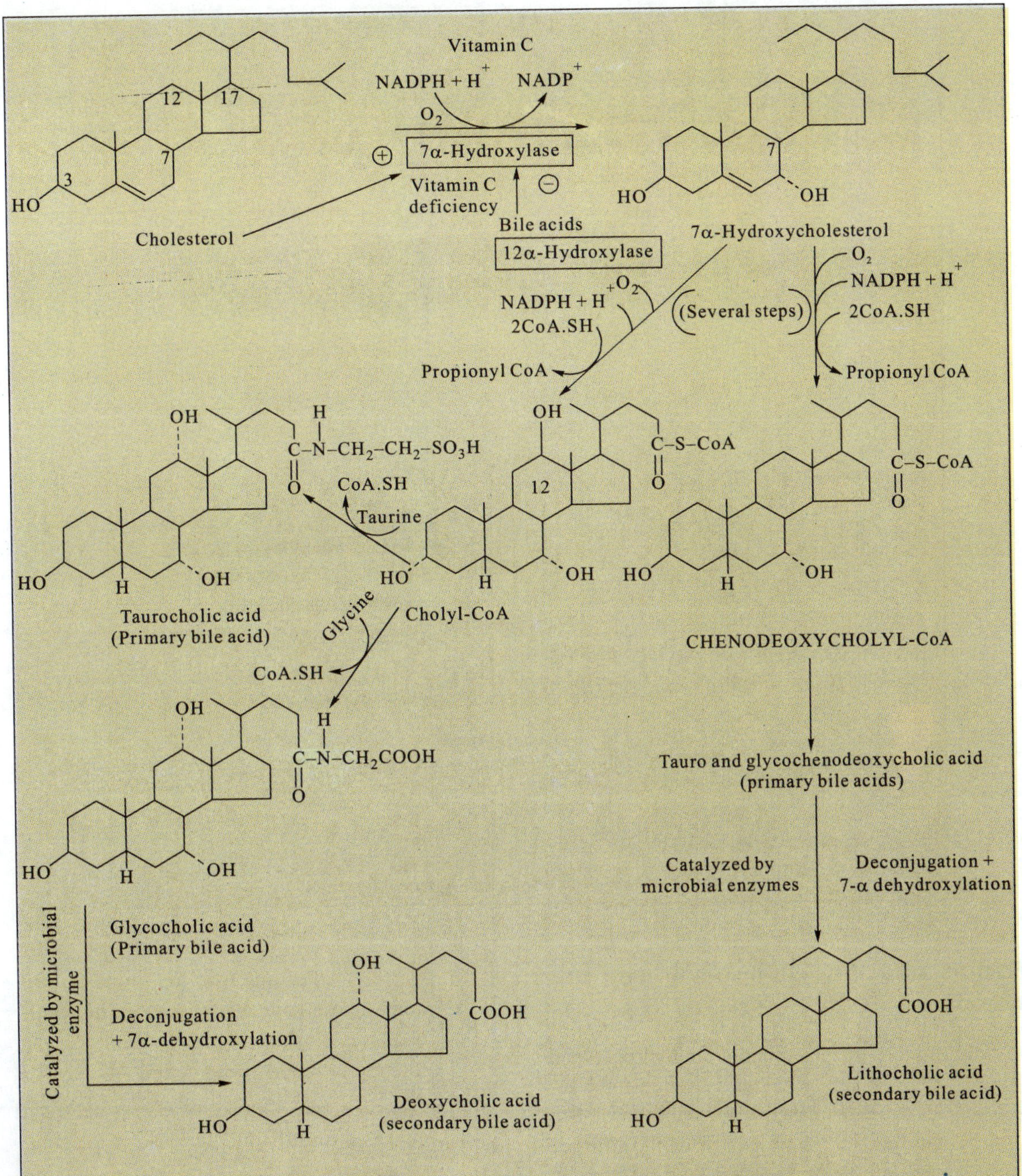

Fig. 18.26 : Reaction of formation of bile acids from cholesterol

4. It functions as the defensive action.
5. It transports fat to liver in the form of cholesterol ester for oxidation.
6. It assists the formation of bile acids and bile salts, 7-dehydrocholesterol and vitamin D_3, corticosteroid hormones, androgens (male sex hormones), estrogens and progesterone (female sex hormones).
7. It helps the granulation of cell division.
8. It acts as an antagonist to phospholipid.

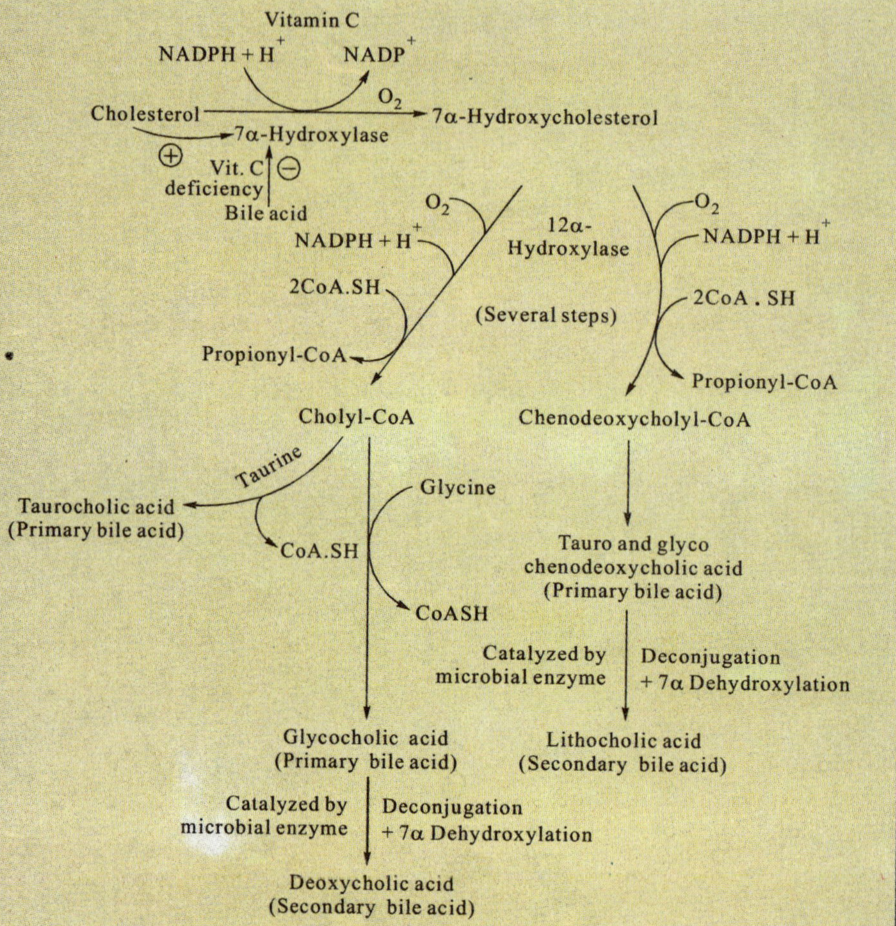

Fig. 18.27 : Flow chart of bile acids from cholesterol

Factors affecting cholesterol level in blood

1. **Dietary fat :** Fats (butter fat, hydrogenated fat) containing higher saturated fatty acids cause increased serum cholesterol. But fats (corn oil, sunflower seed oil, safflower seed oil, fish oil, cotton seed oil) rich in polyunsaturated fatty acids cause marked reduction in serum cholesterol level.
2. **Dietary cholesterol :** It tends to increase the serum cholesterol level.
3. **Dietary carbohydrates :** Consumption of excessive amounts of sucrose causes an increase in serum cholesterol level.
4. **Heredity :** Persons who are prone to become obese have a high level. The level becomes slightly higher in persons belonging to blood group A and AB.
5. **Caloric intake :** Intake of excess calories causes a significant increase in plasma cholesterol level.
6. **Proteins :** Increase in protein intake does not change the plasma cholesterol level but low protein intake causes reduction in plasma cholesterol level.
7. **Vitamin B-complex :** Nicotinic acids in large amounts causes lowering of plasma cholesterol level; whereas pyridoxine de-

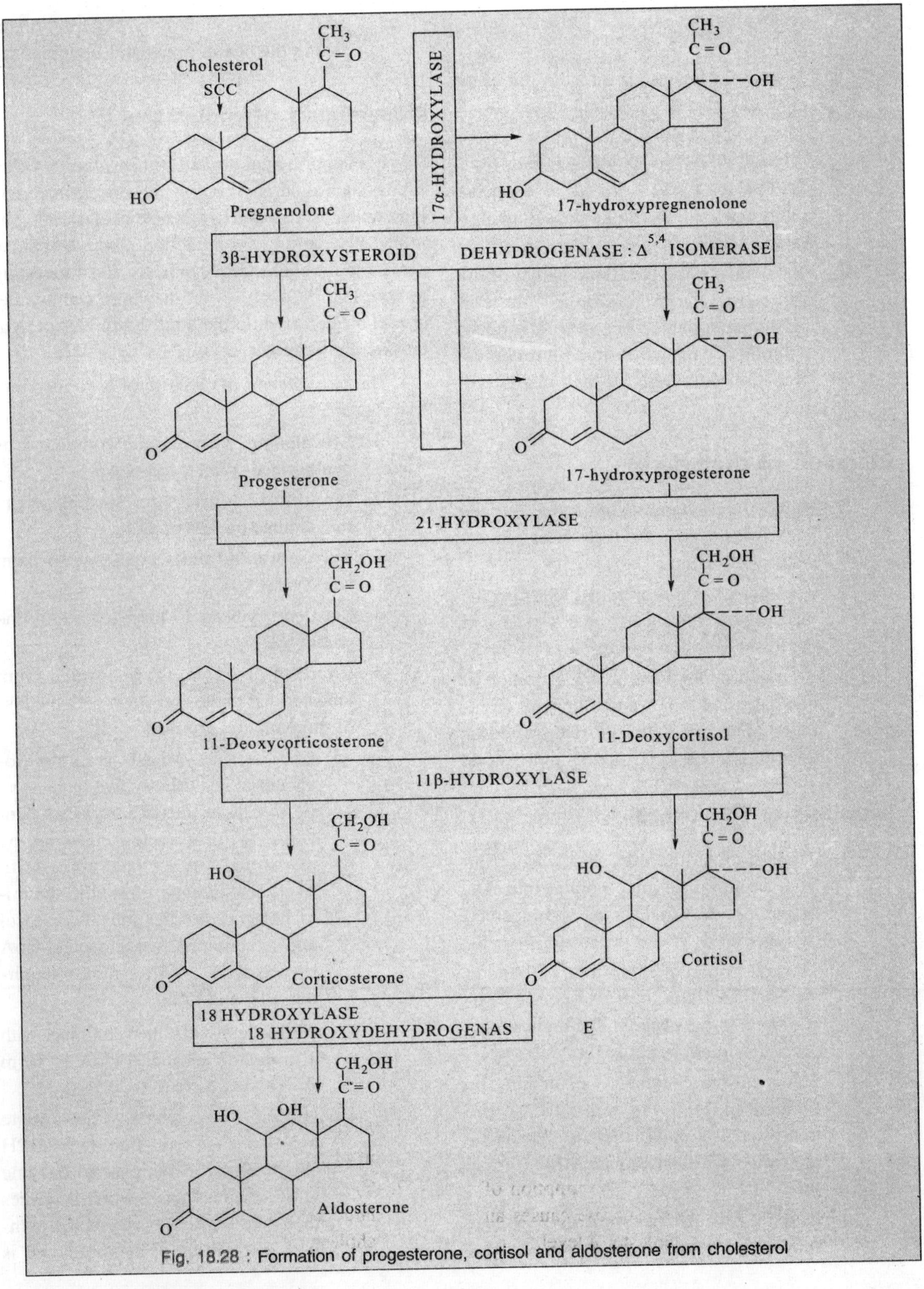

Fig. 18.28 : Formation of progesterone, cortisol and aldosterone from cholesterol

ficiency causes increased blood cholesterol level.

8. **Minerals :** Magnesium salts do not bring about any change. The conversion of acetate to cholesterol is depressed by iron salts and increased by manganese salt.
9. **Physical exercise :** Physical exercise brings about a lowering in the serum cholesterol level.
10. **Fibre :** Increasing the fibre content of the diet, causes the excretion of cholesterol and bile acids in the feces and brings about a significant reduction in serum cholesterol level on a high fat-high cholesterol diet.

Excretion of cholesterol

1. Half of the cholesterol eliminated from the body is excreted in the feces after conversion to bile salts.
2. **Coprostanol** is the principal sterol in the feces which is formed from cholesterol in the lower intestine by the **bacterial flora**.
3. A large portion of the biliary excretion of bile salts undergoes enterohepatic circulation. The bile salts not reabsorbed or their derivatives are excreted in the feces.

Transport of cholesterol

1. Cholesterol in the diet after absorption from the intestine is incorporated into chylomicrons and VLDL being accompanied by other lipids.
2. The greater part of cholesterol is found in the esterified form. It is transported as lipoprotein in the plasma. The highest proportion is found in the LDL.
3. Some plasma cholesteryl ester may be formed in HDL as a result of the transesterification reaction in plasma by lecithin : Cholesterol acyltransferase (LCAT).
4. Patients with parenchymal liver disease show a decrease of lecithin : Cholesterol acyl transferase activity and abnormalities in the serum lipids and lipoproteins.

Biosynthesis of cholesterol

Liver is the principal organ for its synthesis. Other tissues such as adrenal cortex, intestine, skin, ovary, kidney, testis also can synthesize cholesterol. The **microsomal** and **cytosol** fraction of the cell is responsible for cholesterol synthesis. It is interesting to note that **the brain of the** newborn can synthesize cholesterol while the adult brain cannot synthesize cholesterol.

The biosynthesis of cholesterol is divided into five stages :

1. Mevalonate, a six carbon compound, is synthesized from acetyl- CoA.
2. Isoprenoid units are formed from mevalonate by loss of CO_2.
3. Six isoprenoid units condense to form squalene ($C_{30}H_{50}$).
4. Squalene cyclizes to form lanosterol, the parent steroid.
5. Cholesterol ($C_{27}H_{46}O$) is formed from lanosterol after several steps with the loss of three methyl groups :
   (a) Cholesterol synthesis is extramitochondrial and follows two pathways : (i) Two molecules of acetyl-CoA condense to form acetoacetyl-CoA by **cytosolic thiolase enzyme**. (ii) In liver, acetoacetate formed in the mitochondria diffuses into the cytosol and is activated to acetoacetyl-CoA by acetoacetyl-CoA synthase requiring ATP and CoA.

   Acetoacetyl-CoA then combines with a molecule of acetyl-CoA to form **HMG-CoA** by **HMG-CoA synthase**.

   (b) HMG-CoA is converted to **mevalonate** in a two stage reduction by NADPH catalyzed by a microsomal enzyme HMG-CoA reductase which catalyzes then **rate-limiting steps** in the pathway of cholesterol synthesis and is

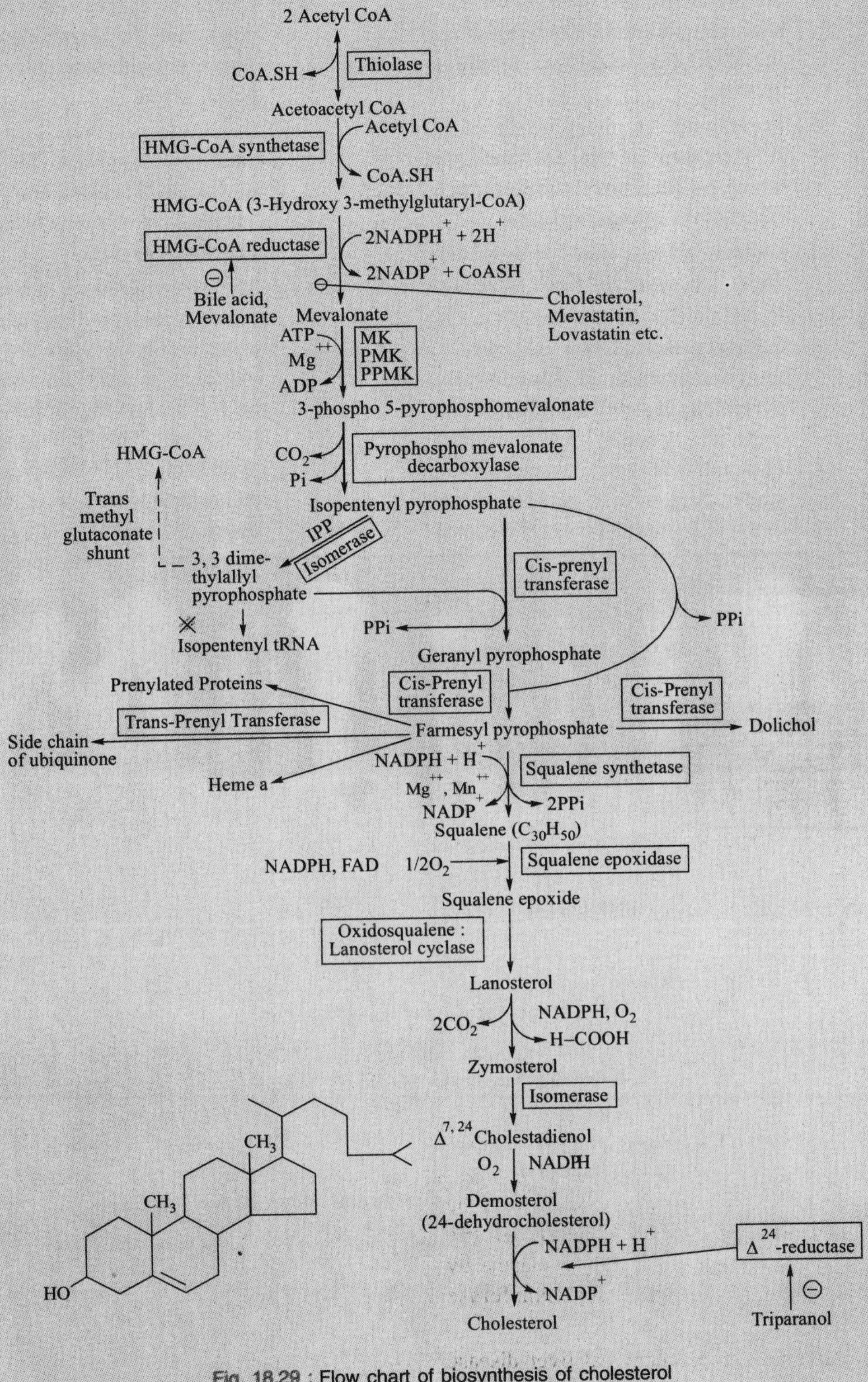

Fig. 18.29 : Flow chart of biosynthesis of cholesterol

the site of action of the most effective class of cholesterol lowering drugs, the HMG-CoA reductase inhibitors (Statins).

(c) Mevalonate is phosphorylated by ATP to form several intermediates which by decarboxylation produces **isopentenyl pyrophosphate**.

(d) **Farnesyl pyrophosphate** is formed by the condensation of three molecules of isopentenyl pyrophosphate.

(e) **Geranyl pyrophosphate** is formed by the condensation of dimethylallyl pyrophosphate with another molecule of isopentenyl pyrophosphate. This, with a further condensation with isopentenyl pyrophosphate, forms **farnesyl pyrophosphate**. Two molecules of farnesyl pyrophosphate condense to form **squalene** by squalene synthetase with the help of NADPH + H^+.

(f) Squalene is then converted to **squalene epoxide** by squalene epoxidase. Squalene epoxide is converted to lanosterol by oxidosqualene : lanosterol cyclase.

(g) The methyl group is oxidized to CO_2 from 14-desmethyl lanosterol and two more methyl groups are likely removed to produce zymosterol which on isomerisation forms $\Delta^{7,24}$-Cholestadienol. This $\Delta^{7,24}$-cholestadienol by the help of NADPH and O_2 yields demosterol which, on reduction, produces **cholesterol**.

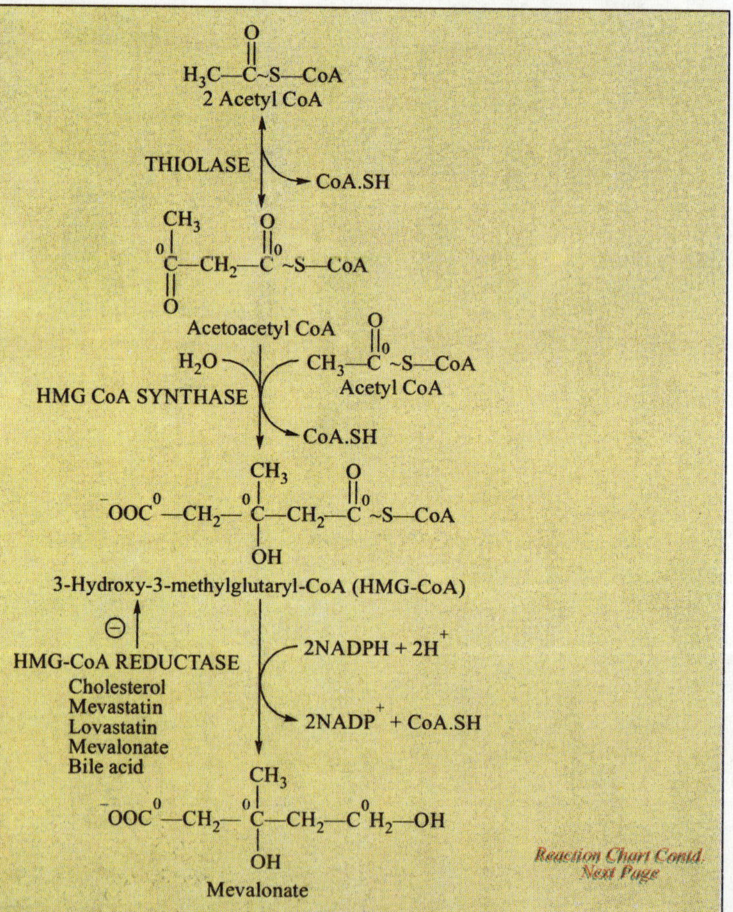

Fig. 18.30 : Reaction of biosynthesis of cholesterol

Contd.

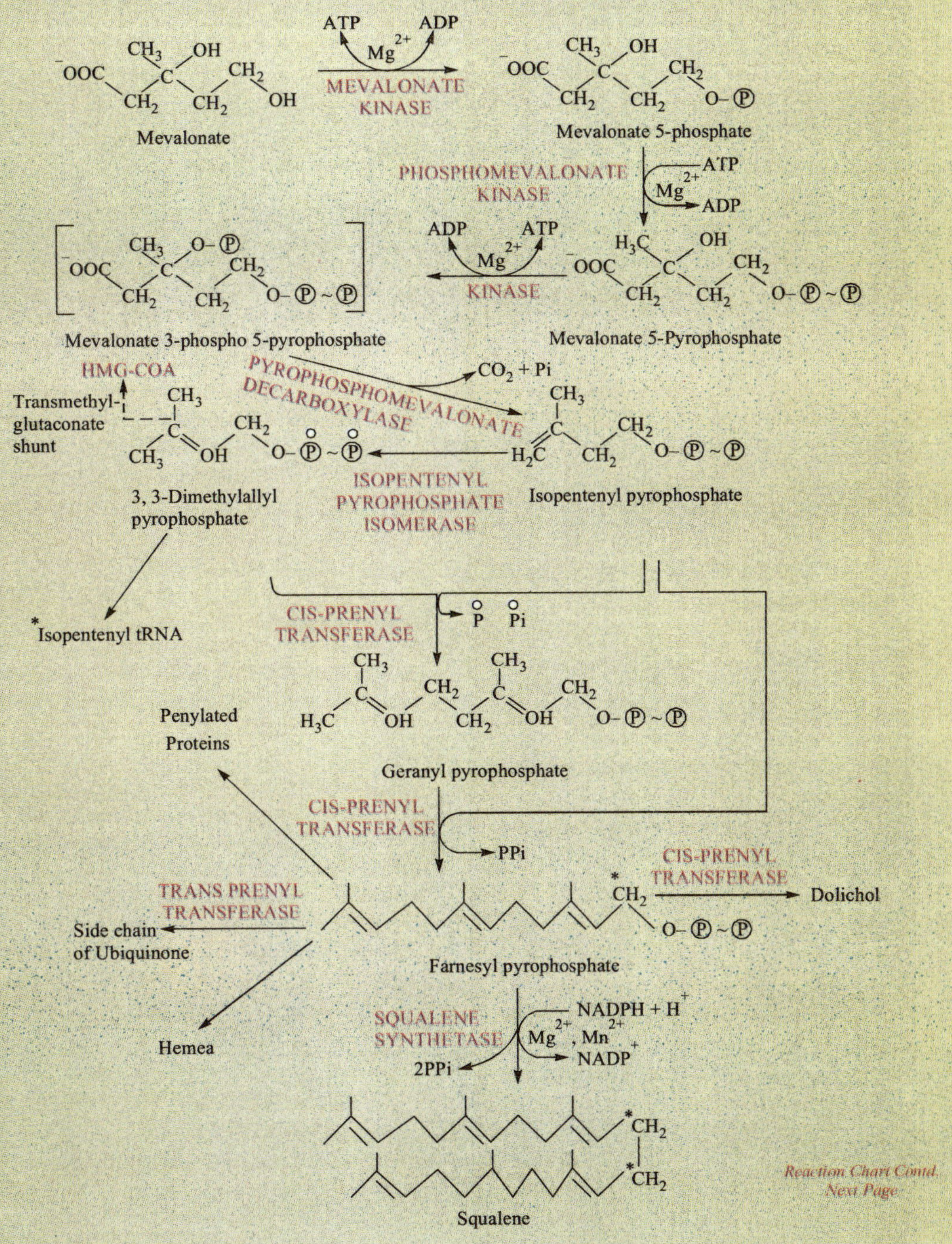

Fig. 18.30 : Reaction of biosynthesis of cholesterol

Contd.

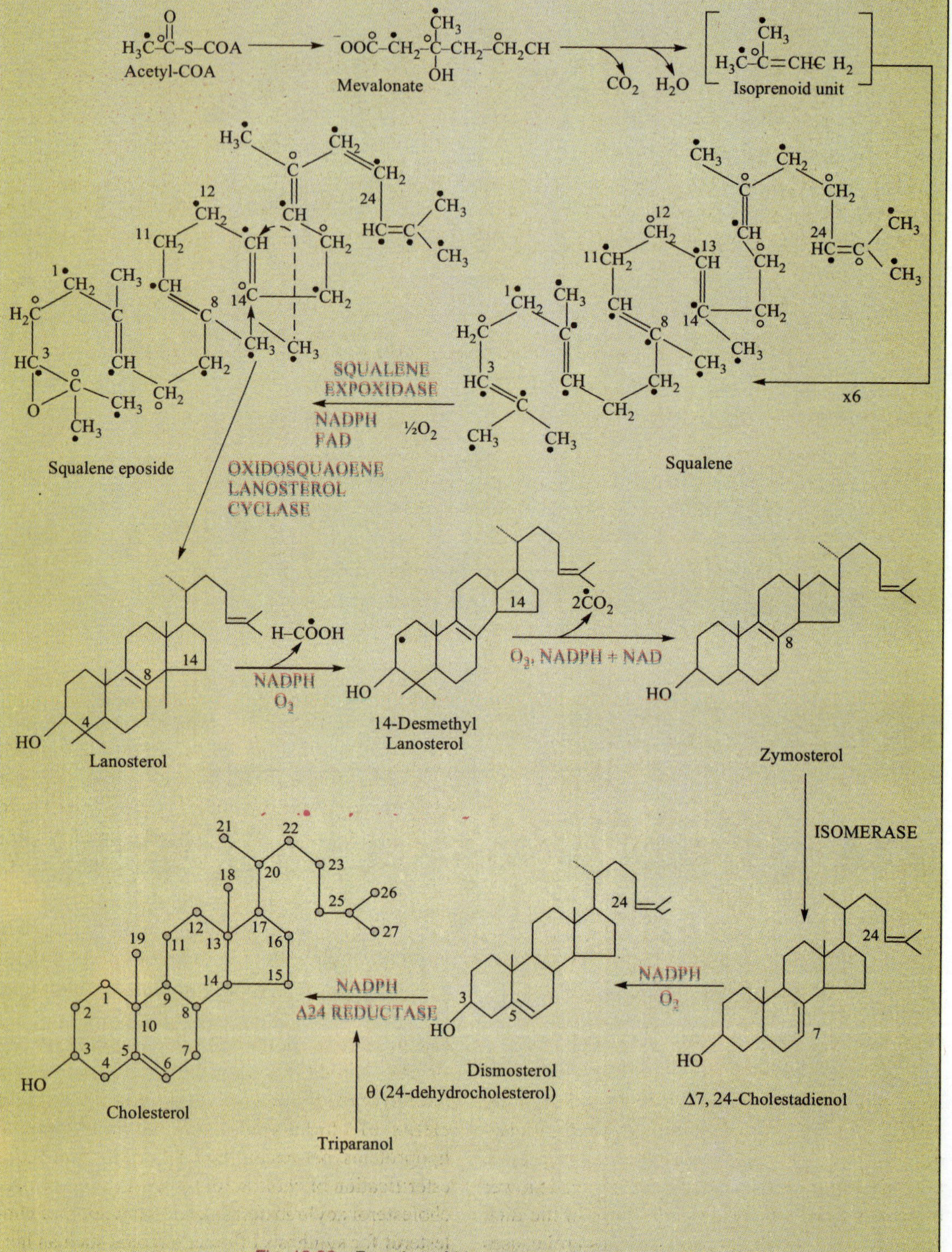

Fig. 18.30.: Reaction of biosynthesis of cholesterol

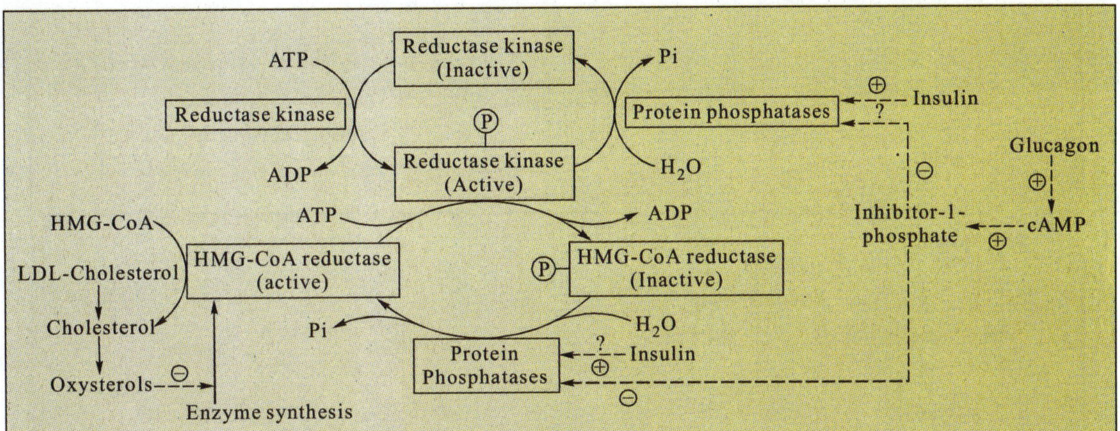

Fig. 18.30a : Mechanisms in the regulation of cholesterol synthesis by HMG-CoA reductase

(h) The intermediates from squalene to cholesterol are attached to a special carrier protein known as the squalene and sterol carrier protein. It seems likely that it is in the form of cholesterol sterol carrier protein that cholesterol is converted to steroid hormones and bile acids and participates in the formation of membranes and of lipoproteins.

(i) Farnesyl pyrophosphate is the branch point for the synthesis of the other polyisoprenoids, dolichol and ubiquinone.

The activity of HMG-CoA reductase is decreased during fasting. The activity is not reduced in diabetes mellitus. But the activity of HMG-CoA reductase is inhibited by cholesterol feeding.

Administration of insulin or thyroid hormone increases the activity of HMG-CoA reductase; whereas the administration of glucagon and glucocorticoids reduces the activity of the enzyme.

More recent experiments have shown that cholesterol synthesis is inhibited by cAMP, indicating that one or more reactions in the synthetic pathway may be controlled by a cAMP-dependent protein kinase. Plasma cholesterol in humans is made lower by reducing the amount of cholesterol in the diet. An increase of 100 mg in dietary cholesterol causes a rise of 5 mg cholesterol per 100 ml of serum.

Hypercholesterolemia has been observed in uncontrolled diabetes mellitus, impairment of liver, obstructive jaundice, glomerulonephritis, hypothyroidism and nephrosis.

Hypocholesterolemia has been found in anemia, hyperthyroidism, hepatic diseases, infection, carcinoma and acute pancreatitis.

Polyunsaturated fatty acids can lower cholesterol level. Because, the cholesteryl esters of polyunsaturated fatty acids are more rapidly metabolized by the liver and other tissues which may enhance their rate of turnover and excretion. Recently, it has been found that saturated fatty acids cause higher rates of secretion of VLDL by the perfused liver than do unsaturated free fatty acids. Cholesterol level is also decreased on the minimum intake of animal fat.

The cholesterol level in tissues is increased due to uptake of cholesterol-containing lipoproteins by receptors, eg, the LDL receptor, uptake of free cholesterol from cholesterol-rich lipoproteins to the cell membrane, cholesterol synthesis, and hydrolysis of cholesteryl esters bythe enzyme cholesteryl ester hydrolase. The level is also decreased due to efflux of cholesterol from the membrane to lipoproteins, particularly to HDL_3 or nascent HDL, esterification of cholesterol by ACAT (acyl- CoA : cholesterol acyltransferase), and utilization of cholesterol for synthesis of other steroids such as hormones or bile acids in the liver.

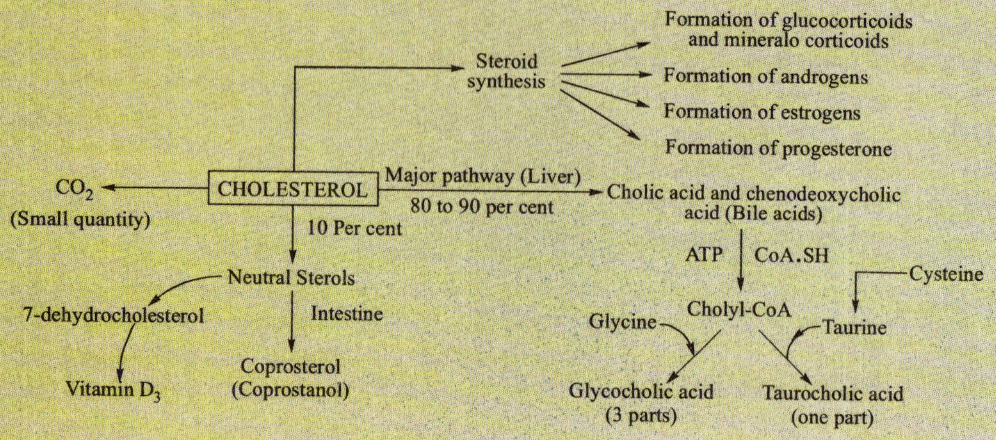

Fig. 18.31 : Fate of cholesterol in the body

Role of HMG-CoA Reductase

1. Marked decrease in the activity of HMG-CoA reductase has been observed in fasting animals which shows the reduced synthesis of cholesterol during fasting.
2. HMG-CoA reductase in liver is inhibited by mevalonate and by cholesterol.
3. Cholesterol is considered to act by repression of transcription of the HMG-CoA reductasegene, cholesterol synthesis is also inhibited by LDL-cholesterol taken up via LDL receptors. A diurnal variation takes place in both cholesterol synthesis and reductase activity.
4. HMG-CoA reductase activity is increased by the administration of insulin or thyroid hormone, whereas its activity is decreased by glucagon or glucocorticoids.
5. HMG-CoA reductase exists in both active and inactive forms.

ATHEROSCLEROSIS

Atherosclerosis is defined by the variable combination of focal accumulation of lipids, complex carbohydrates, blood and constituents, fibrous tissue and calcium deposits combined with its changes of the media (as per WHO). So it is a patehy, nodular type of arteriosclerosis.

The three major clinical forms arise from narrowing of the coronary arteries, the cerebral arteries and the femoral artery and its branches, but other arteries can be affected, such as the renal or mesenteric.

Coronary atherosclerosis is almost invariably associated with aortic atherosclerosis.

Types of lesion

It is customary to separate the lesions of atherosclerosis into fatty streaks, plaques and complicated lesions.

Fatty streaks are short, thin, slightly raised yellow lines running longitudinally along the internal surface of arteries and consist of an intracellular accumulation of lipids within the intima.

Plaques are the lesions of atherosclerosis. They are raised, focal, circumscribed lesions up to 1 cm in diameter, consisting of various amounts of fibrous tissue and lipid. The lipid accumulates mostly in extracellular amorphous masses; plaques in which this process is prominent are called soft or atheromatous plaques. In others, fibrous tissue is prominent and lipid is widely scattered or localized to the deeper portions of the lesion; these are called hard or fibrous plaques.

Four other processes may complicate the lesions : (i) the endothelium may be lost so that the surface ulcerates and the fatty contents may be exposed to the blood stream. (ii) Fibrin is commonly deposited and thrombosis occurs on the plaque surface. (iii) Free blood can be found in plaque. (iv) Calcification may occur.

Conventional risk factors for CAD (Coronary Artery Disease)

1. Dyslipidemia — elevated LDL – C and low HDL – C
2. Smoking
3. Diabetes
4. Hypertension
5. Obesity
6. Sedentary lifestyle
7. Positive family history
8. Advancing Age
9. Postmenopausal state in women.

New and emerging risk factors for CAD

1. Hyperhomocysteinaemia
2. Small dense LDL phenotype
3. Inflammation and Infectious agents
4. Hyperfibrinogenaemia
5. Elevated LP (a). [Lipoprotein (a)]

Development with ageing

A few small lesions are seen in most adolescents and they increase in number and size throughout life. It does not usually lead to clinical disease until middle age.

Thrombogenesis

An arterial thrombus almost always forms at the site of an atheromatous plaque, these are present in most of us from an early age. Small thrombi form frequently in the circulation. These microthrombi are then attached to the arterial wall where they are rapidly lysed. A failure of the fibrolytic mechanism could in this way allow a thrombus to grow. Hence one of the primary faults responsible for atherosclerosis might lie in an increased tendency to thrombosis or in an efficient fibrinolytic mechanism.

Treatment

1. Diabetes and hypertension need to be controlled.
2. Tobacco habit should be forbidden as it aggravates ischaemia of arteries.
3. Cold exposure should be avoided and suitable woollen clothing to be worn.
4. Protection against trauma and treatment of sepsis are mandatory.

Affect of drugs on Cholesterol

DRUGS USED CLINICALLY

Drug	Dosage	Effect
Lovastatin	20 mg once or twice daily	Lowers cholesterol significantly
Niacin	½ to 1.0 gm thrice daily	Lowers cholesterol and triglyceride
Gemfibrozil	30 mg capsule. Two capsules twice daily before meals	Lowers triglyceride mainly and lowers LDL also
Colestipol	10 gms thrice daily	Lowers cholesterol and triglyceride
Guggul	One capsule thrice daily	Lowers cholesterol and triglyceride
Garlic and Onion extracts	To be used in dietary preparations	Lowers cholesterol and triglyceride

5. Graduated exercise should be undertaken to improve blood flow.
6. Surgical treatment may be helpful.

Affect of drugs on cholesterol

1. Sitosterol is hypocholesterolemic agent which blocks the esterification of cholesterol in the gastrointestinal tract and thus reduces cholesterol absorption.
2. Drugs like choloxin and neomycin cause the increased fecal excretion of cholesterol and bile acids.
3. Clofibrate acts by inhibiting the secretion of VLDL by the liver or by inhibiting hepatic cholesterol synthesis.
4. The hypocholesterolemic drugs include nicotinic acid and estrogens.
5. Significant reductions of plasma cholesterol can be effected by the use of cholestyramine resin.
6. The fungal inhibitors of HMG-CoA reductase, mevastatin and lovastatin, reduce LDL cholesterol levels by up-regulation of the LDL receptors.
7. Probucol causes to increase LDL catabolism, but its antioxidant properties are more important in preventing accumulation of oxidized LDL in arterial walls. Oxidized LDL is a prime cause of atherosclerosis.

Test for cholesterol

1. Salkowski's test : A little cholesterol is dissolved in 2 ml of chloroform. An equal volume of conc. H_2SO_4 added to it. Shaked gently, upper layer of chloroform turns red and the sulphuric acid layer assumes a yellow colour with green fluorescence.
2. Libermann-Burchard reaction : A crystal of cholesterol is dissolved in 2 ml of chloroform in a dry test tube. 10 drops of acetic anhydride and 2 drops conc. H_2SO_4 are added. Mixed well. A red rose colour develops which quickly changes through blue to green.

CLINICAL ORIENTATION

- Cholesterol is a major constituent of gall-stones.
- Patients with arterial disease have any one of the following abnormalities :
  - (a) Elevated concentration of VLDL with normal concentration of LDL.
  - (b) Elevated LDL with normal VLDL.
  - (c) Elevation of both lipoprotein fractions. There is also an inverse relationship between HDL concentration and coronary heart disease.
- Atherosclerosis is characterized by the deposition of cholesterol and cholesteryl ester of lipoproteins containing apo-B-100 in the connective tissues of the arterial walls.
- Diseases (eg, diabetes mellitus, lipid nephrosis, hypothyroidism) in which prolonged elevated levels of VLDL, LDL occur in the blood are accompanied by premature or severe atherosclerosis.
- The intake of polyunsaturated and monounsaturated fatty acids along with the diet is most beneficial. Naturally occurring oils that contain a high proportion of polyunsaturated fatty acids include sunflower, cottonseed, soyabean oil, and olive oil contain a high concentration of monounsaturated fatty acids. On the other hand, butter fat, beef fat, and palm oil contain a high proportion of saturated fatty acids. Sucrose and fructose have a greater effect in raising blood lipids, particularly triacylglycerol. Diets rich in palmitate inhibit the conversion of cholesterol to bile acids. The reason of it is that the distribution of cholesterol is shifted from plasma into the tissues because of increased catabolic rate of LDL due to up-regulation of the LDL receptor by poly- and monounsaturated fatty acids. Saturated fatty acids cause the formation of smaller VLDL particles that contain relatively more cholesterol and they are utilized by extrahepatic tissues at a slower rate, tendencies that may be regarded as atherogenic.
- The coronary heart disease is caused by high blood pressure, smoking, abdominal obesity, lack of exercise, and drinking soft water. Increased plasma free fatty acids also lead to increased VLDL secretion by the liver involving extra triacylglycerol and cholesterol output into the circulation. The factors causing higher levels of free fatty acids include emotional stress, nicotine from cigarette smoking, coffee drinking etc. Premenopausal women are protected from these deleterious factors because of higher concentration of HDL they have. Moderate alcohol consumptions have a lower incidence of coronary heart disease because of elevation of HDL concentrations. But red wine is particularly beneficial because of its content of antioxidants.

SCHEMATIC COMPENDIUM

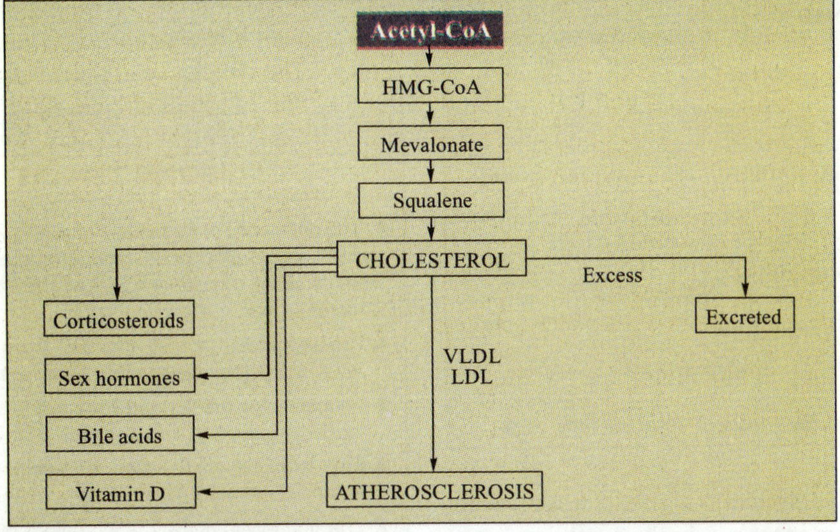

ESSENTIAL FATTY ACIDS (EFA)

In 1928, Evans and Burr observed that rats fed on a purified nonlipid diet exhibited a reduced growth rate and a reproductive deficiency. The deficiency symptom was cured by the addition of essential fatty acids. Further symptoms due to EFA were scaly skin, necrosis of the tail and lesions in the urinary system. The essential fatty acids are linoleic acid, linolenic acid and arachidonic acid.

Chemistry

Linoleic acid and linolenic acid contain 18 carbon atoms each and arachidonic acid contains 20 carbon atoms. Linoleic acid, linolenic acid and arachidonic acid have 2, 3 and 4 double bonds, respectively, in their structure.

Their structures are :

$CH_3(CH_2)_4.\ CH = CH.CH_2.CH = CH\ (CH_2)_7.COOH$

Linoleic acid (18 : 2)

$CH_3.CH_2.CH = CH.CH_2CH = CH.CH_2CH = CH(CH_2)_7.COOH$

Linolenic acid (18 : 3)

$CH_3.(CH_2)_4(CH{=}CH.CH_2)_4(CH_2)_2.COOH$

Arachidonic acid (20 : 4)

Fig. 18.32 : Structure of Essential fatty acids

Chemical properties

1. The essential fatty acids of vegetable oils have low melting point and low iodine number.
2. These essential fatty acids become saturated on hydrogenation and the oils are converted to solid fats.

Sources

1. Linoleic acid occurs in high concentrations in various edible vegetable oils eg, corn, cotton seed, peanut, safflower.
2. Arachidonic acid occurs in animal fats, although in small amounts.

Synthesis

Lenoleic acid cannot be synthesized by animals and, therefore, must be supplied preformed in the diet. Arachidonic acid can be formed from linoleic acid in the animal body.

The pathway is first by dehydrogenation of the CoA ester through γ-linolenate followed by the addition of a 2 carbon unit (as acetyl-CoA in the mitochondrial system for chain elongation or as malonyl-CoA in the microsomal system, which is

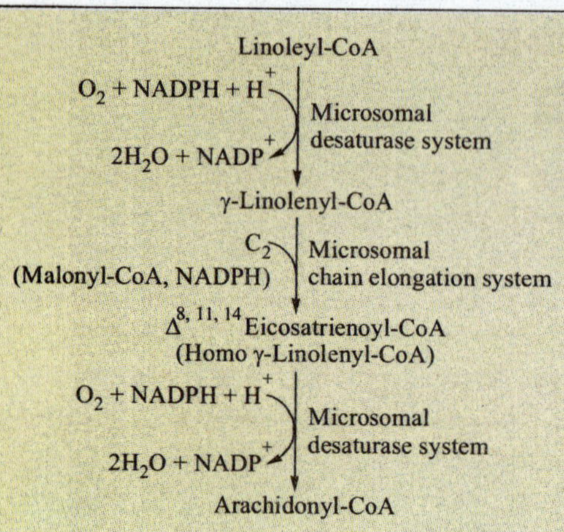

Fig. 18.33 : Synthesis of essential fatty acids

the more active system) to give eicosatrienoate (homo γ-linolenate) which further forms arachidonate by dehydrogenation. Therefore, the nutritional requirement of arachidonate is compensated by the linoleate in the diet. The synthesis is shown in Fig. 18.33.

Functions

1. Essential fatty acids found in the structural lipids of the cell are concerned with the structural integrity of the mitochondrial membrane.
2. These occur in high concentration in the reproductive organs.
3. These effect the prolongation of clotting time and increase the fibrinolytic activity.
4. These are esterified and emulsified with cholesterol and thus retard atherosclerosis.
5. These are incorporated into lipoproteins for transport to the liver for further oxidation.
6. These are present in phospholipids, mainly in the position 2.
7. The fatty livers due to deficiency of the essential fatty acids are cured only by the reintroduction of these substances into the diet.
8. The deficiency of these essential fatty acids causes skin lesion and impairment of lipid transfer.
9. The deficiency of these substances in the diet of babies also causes eczema.

CLINICAL ASPECTS

- The deficiency of Essential Fatty Acid (EFA) causes swelling of mitochondrial membrane and reduction in efficiency of oxidative phosphorylation. This may increase heat production in EFA deficiency.
- Prostaglandins are synthezised from arachidonic acid by cyclooxygenase enzyme system.
- Leukotrienes are formed from arachidonic acid in leucocytes by the lipoxygenase pathway.
- Fats with polyunsaturated fatty acids can reduce serum cholesterol level.
- The fibrinolytic activity is increased by fats rich in EFA.
- Docosahexonoic acid, formed from dietary linolenic acid, is the most abundant polyenoic fatty acid present in retinal photoreceptor membranes. It enhances the electrical response of the photoreceptors to illumination. Hence, linolenic acid is essential for optimal vision.
- Abnormal metabolism of essential fatty acids connected with dietary insufficiency has been found in cystic fibrosis, acrodermatitis enteropathica, hepatorenal syndrome, multisystem neuronal degeneration, Crohn's disease, cirrhosis and alcoholism, and Reye's syndrome. Increased levels of very long chain polyenoic acids have been found in the brains of patients with Zellweger's syndrome.

EICOSANOIDS

Formation of Eicosanoids

1. Eicosanoids formed from arachidonate and some other C_{20} fatty acids with methylene-interrupted bonds, physiologically and pharmacologically active compounds known as prostaglandins (PG), thromboxanes (TX), and leukotrienes (LT). Physiologically, they are considered to act as local hormones.
2. Arachidonic acid, usually derived from the 2-position of phospholipids in the plasma membrane by phospholipase A_2 activity,

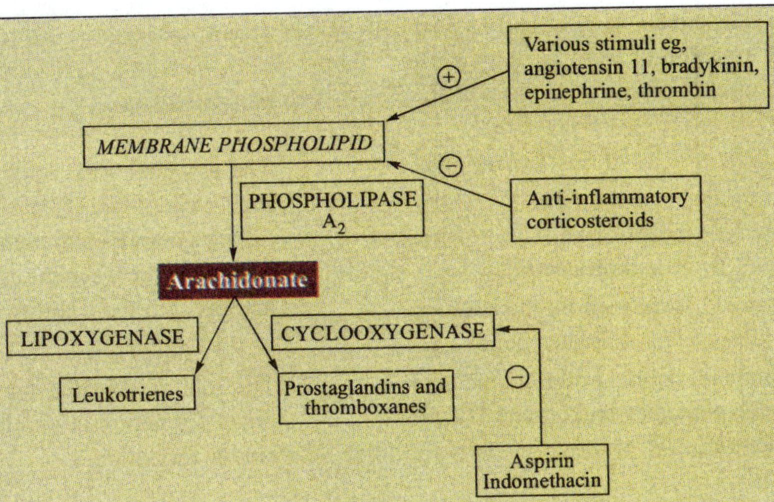

Fig. 18.34 : Conversion of arachidonic acid to prostaglandins and thromboxanes via the cyclooxygenase pathway and to leukotrienes via the lipoxygenase pathway. The figure indicates why steroids, which inhibit the total eicosanoid production, are better anti-inflammatory agents than aspirin-like drugs, which only inhibit the cyclooxygenase pathway. Anti-inflammatory steroids are thought to inhibit phospholipase A_2 through induction of an inhibitory protein named lipocortin

is the substrate for the synthesis of PG_2, TX_2, LT_4 compounds. These two pathways are known as the cyclooxygenase and lipoxygenase pathways, respectively.

3. Three groups of eicosanoids (each comprising PG, TX and LT) are synthesized from C_{20} eicosanoic acids derived from essential fatty acids linoleic acid and linolenic acid, or directly from arachidonic acid and eicosapentaenoic acid in the diet.

Prostanoid synthesis by Cyclooxygenase Pathway

1. Two molecules of O_2 catalyzed by **Prostaglandin endoperoxide synthase** which possesses two separate enzyme activities—Cyclooxygenase and Peroxidase—are responsible for Prostanoid synthesis.
2. The product of the cyclooxygenase pathway, an endoperoxide (PGH), is converted to Prostaglandins D, E and F as well as to the thromboxane (TXA_2) and Prostacylin (PGI_2).
3. Each cell type produces one type of Prostanoid.
4. **Aspirin** and **indomethacin** inhibit Cyclooxygenase.

Cyclooxygenase is a "Suicide Enzyme"

1. Prostaglandin formation is partly stopped by a remarkable property of Cyclooxygenase—that of self-catalyzed destruction (a "suicide enzyme")
2. The inactivation of Prostaglandin is rapid due to the presence of the enzyme 15-hydroxyprostaglandin dehydrogenase in most mammalian tissues.
3. The half-life of Prostaglandin in the body can be prolonged by blocking the action of this enzyme with **indomethacin** or **sulfasalazine**.

Leukotrienes are formed by the Lipoxygenase pathway

1. The **leukotrienes**, a family of conjugated trienes, are formed from eicosanoic acids in leukocytes, mastocytoma cells, platelets and macrophages by the lipoxygenase pathway in response to both immunologic and nonimmunologic stimuli.

2. Oxygen is inserted into 5, 12 and 15 positions of arachidonic acid giving rise to hydroperoxide (HPETE) by three different **lipoxygenases** (dioxygenases).
3. Only **5-lipoxygenase** forms leukotrienes.
4. The first formed is leukotriene A_4, which, in turn, is metabolized to either leukotriene B_4 or leukotriene C_4.
5. Leukotriene C_4 is formed by the addition of glutathione via a thioether bond.
6. The subsequent removal of glutamic acid and glycine generates leukotriene D_4 and Leukotriene E_4.

Correlation between Essential Fatty Acid Activity and Production of Prostaglandins

1. The essential fatty acids do not exert all their physiologic effects through Prostaglandin synthesis although there is marked correlation between the essential fatty acid activity and their ability to be converted to prostaglandins.
2. The role of essential fatty acid in membrane formation is unrelated to prostaglandin formation.

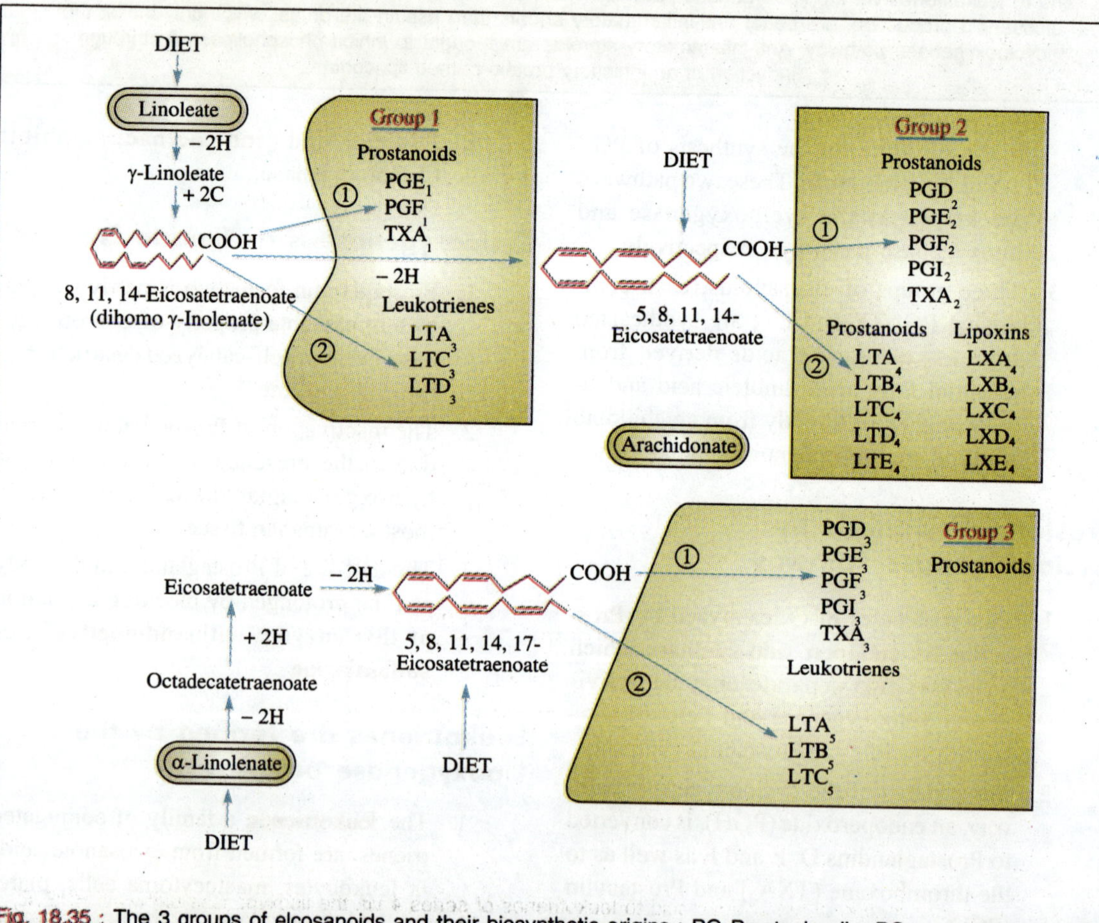

Fig. 18.35 : The 3 groups of eicosanoids and their biosynthetic origins : PG Prostaglandin: PG , prostacyclin; Tx, thromboxane, LT, leukotriene, (1) Cyclooxygenase pathway (2) lipoxygenase pathway. The subscripts denote the total number of double bonds in the molecule and the series to which the compound belongs

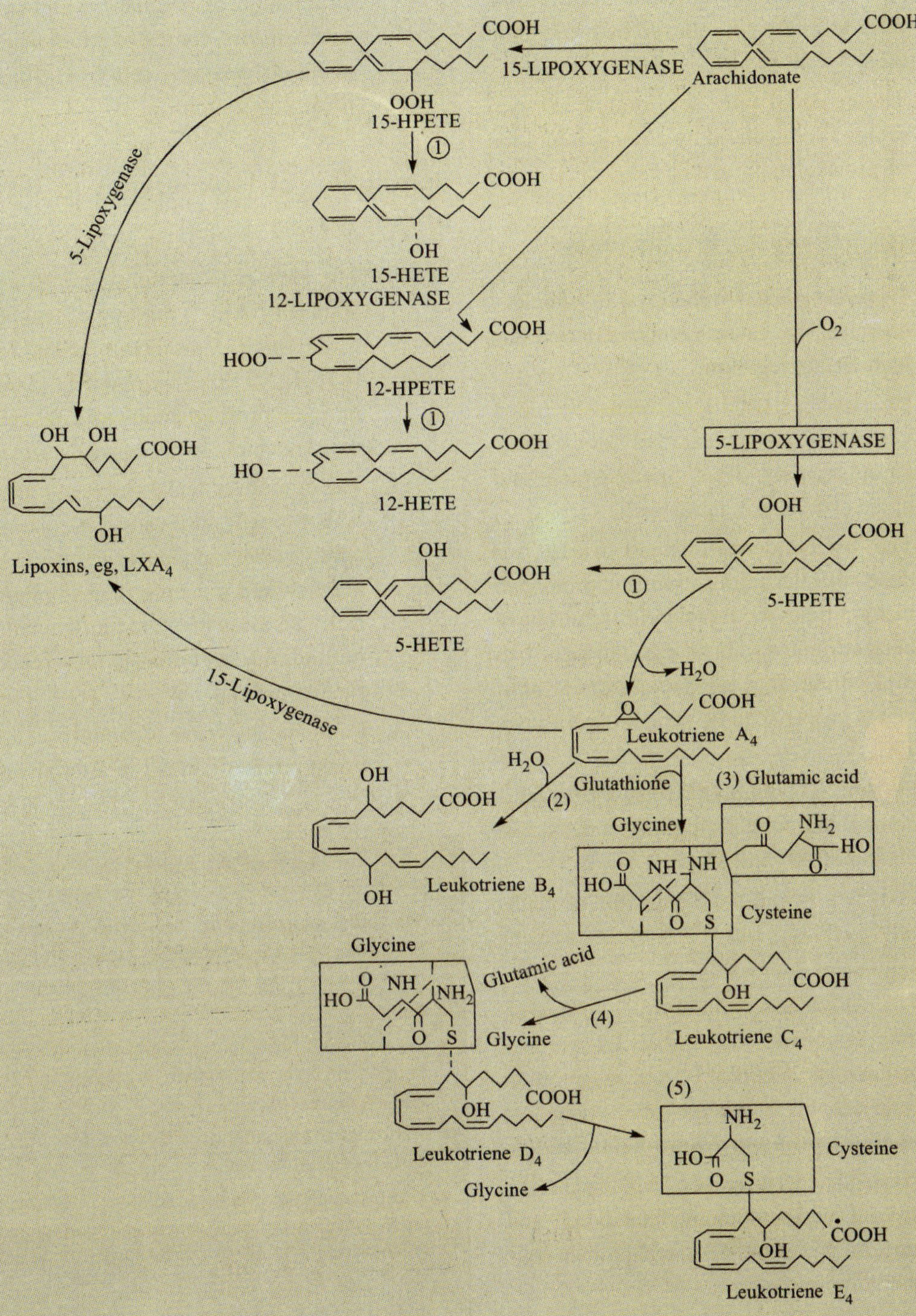

Fig. 18.36 : Conversion of arachidonic acid to leukotrienes of series 4 via the lipoxygenase pathway. HPETE, hydroperoxy eicosatetraenoate, HETE hydroxyeicosatetraenoate. Some similar conversions occur in series 3 and 5 leukotrienes (1) Peroxidase (2) Leukotriene A_4 epoxide hydrolase (3) Glutathione S-transferase (4) γ-glutamyltransferase (5) Cysteinyl-glycine dipeptidase

3. The symptoms of deficiency of essential fatty acids are not relieved by prostaglandins.
4. The essential fatty acid deficiency syndrome is not caused by chronic inhibition of prostaglandin synthesis.

Biological activity of Prostanoids

1. **Thromboxanes**, synthesized in platelets, on releasing, cause vasoconstriction and platelet aggregation.
2. **Prostacyclins (PGI_2)**, produced by blood vessel walls, are potent inhibitors of platelet aggregation. Thus, thromboxanes and prostacyclins are antagonistic.
3. The low incidence of heart disease, reduced platelet aggregation, and prolonged clotting times in Greenland Eskimos have been attributed to their high intake of fish oils containing eicosapantaenoic acid which gives rise to PG_3 and thromboxanes TX_3.
4. PG_3 and TX_3 inhibit the release of arachidonic acid from phospholipids and the formation of PG_2 and TX_2.
5. PGI_3 is a potent antiaggregator of platelets as PGI_2, but TXA_3 is a weaker aggregator than TXA_2.
6. The plasma concentrations of cholesterol, triacylglycerol, LDL and VLDL are low in Eskimos, whereas HDL concentration is raised—all factors are involved against atherosclerosis and myocardial infarction.
7. Prostaglandins increase cAMP in platelets, thyroid, corpus luteum, fetal bone and lung, but lower cAMP in renal tubule cells and adipose tissue.
8. 1ng/ml of prostaglandins causes contraction of smooth muscle in animals. When therapeutically used, it causes prevention of conception, induction of labor at term, termination of pregnancy, prevention of gastric ulcers, control of inflammation and of blood pressure, relief of asthma and nasal congestion.

Regulation of leukotrienes in disease processes

1. The slow-reacting substance of anaphylaxis (SRSA) is a mixture of leukotrienes C_4, D_4 and E_4. This mixture of leukotrienes is 10-1,000 times more potent than histamine or prostaglandins as constrictor of the bronchial airway musculature.
2. The mixture of leukotrienes together with leukotriene B_4 also cause vascular permeability and attraction and activation of leukocytes and are important regulators in many diseases involving inflammatory or immediate hypersensitivity reactions, such as asthma.
3. Leukotrienes are vasoactive and 5-lipoxygenase is found in arterial walls.

CLINICAL ASPECTS

- The prostaglandins and thromboxanes are local hormones that are synthesized rapidly on requirement and act near the site of synthesis.
- **Aspirin** inhibits prostaglandin synthesis.
- Leukotrienes have muscle contractant properties suggesting their importance in **allergic reactions** and **inflammation**.
- A mixture of leukotrienes is the slow-reacting substance of anaphylaxis.
- The prostaglandins are modulators of adenyl cyclase activity in controlling platelet aggregation and in inhibiting the effect of antidiuretic hormone in the kidney.
- It is possible to influence the type of eicosanoids synthesized by applying the variable proportions of the different polyunsaturated fatty acids in the diet indicating the influence over the course of disease by dietary means.

SCHEMATIC COMPENDIUM

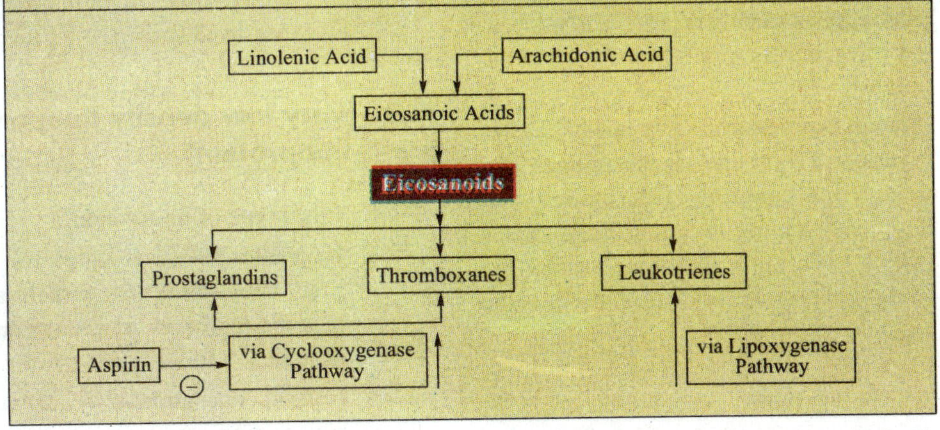

PHOSPHORYLATION

1. All cells of the body, especially muscle cells, contain ATP which is composed of adenine-ribose-phosphoric acid-phosphoric acid-phosphoric acid.
2. ATP is one of the intermediate phosphate donors of the body. It furnishes the phosphate for the phosphorylation of glucose to produce glucose-6-phosphate and ATP is converted to ADP.
3. The liberation of phosphate from ATP is an **energy-producing reaction**. But the energy is utilized for chemical transfers, not as heat. Since, phosphorylation is going on continuously in the cell and the quantity of ATP is small, that ATP is reformed from ADP and inorganic phosphate.
4. In the cell, a number of organic phosphates are known as high-energy phosphate donors that serve for the reformation of ATP. Among these are 1, 3-diphosphoglyceric acid, phosphopyruvic acid and these are reformed during the breakdown of glycogen to lactic acid.
5. In muscle, another organic phosphate is phosphocreatine (phosphagens). This compound is the emergency store of phosphorylation energy which serves to regenerate ATP during muscle contraction. During glycogenolysis or glycolysis there is accumulation of ATP, creatine is again built-up to creatine phosphate.

LIPIDOSIS, XANTHOMATOSIS

According to Fredrickson, **lipidosis** means the abnormal concentrations of lipoproteins in blood or of specific lipids in tissues.

The specific lipids are the sphingolipids (Tay-Sach's disease, Gaucher's diesease, Fabry's disease, metachromatic leukodystrophy, Niemann-Pick disease) or other lipids (Refsum's disease, Wolman's disease etc.). These disorders are discussed in chapter on inborn error in metabolism.

Each of these disorders involves the nervous system. These are hereditary enzyme deficiency disorders.

Xanthomatosis means lipid accumulation in tissues in association with large "**foam cells**". The lipid is mostly cholesterol. Xanthomata develop in **hyper-lipoproteinemias**, diabetes mellitus, biliary cirrhosis, hypothyroidism, glycogen storage disease etc.

Carnitine

(i) Carnitine is a protein. Lysine is the precursor of carnitine. The biosynthesis of carnitine requires two essential amino acids, lysine and methionine. Its biosynthesis utilizes α-ketoglutarate as cosubstrate and requires ascorbate, Fe^{++}, and molecular oxygen. It is widely distributed in nature.

(ii) Mammalian mitochondria can synthesize a variety of short-chain acylcarnitines. Human urine contains large amounts of propionylcarnitine in certain disease states which leads to the excessive urinary excretion of specific acylcarnitines in some organic acidemias.

(iii) **Carnitine acetyltransferase** (CAT) is widely distributed in nature, occurring in most mammalian tissues including brain. It is a freely reversible enzyme. Its molecular weight varies from 51,000 to 75,000. Although CAT has little function in microsomes, high activities have been reported for microsomes from brown adipose tissue in the fetal stage. Its optimum activities are usually for propionyl-CoA and lower activities with acyl chain lengths of C_8 or C_{10}.

Competitive inhibition of acetyl-CoA with acetylcarnitine indicates that the acetyl groups of ether occupy a common acetyl-binding site. CAT is also inhibited by **chloro-and bromo-acetyl derivatives and palmityl-CoA**. It is also inhibited by **iodoacetamide** and **N-ethyl maleimide**.

(iv) The possible established functions for carnitine in metabolism can be summarized as :

(a) β-oxidation of long chain fatty acids in mitochondria.

(b) Elimination of selective acyl residues/detoxification.

(c) Translocation of acetyl units into mitochondria (yeast).

(d) Modulation of the CoA.SH/Acetyl-CoA ratio (may vary depending on mitochondrial source and the levels of CAT and the carnitine/acylcarnitine translocase).

(e) Reservoir of activated acetyl units (some insect flight muscles, possible heart and sperm).

VLDL (Very low density lipoportein or pre-β-lipoprotein)

1. It is a type of lipoprotein.
2. It contains 50% triacylglycerol, phospholipid, cholesterol, low protein and occupies the region in between α and β-lipoprotein.
3. Normal concentration — 150 mg%.
4. It is found in chyle and acts as a vehicle for the transport of triacylglycerol from the liver to extrahepatic tissue.
5. Apoprotein B is essential for its formation.
6. VLDL is secreted by hepatic parenchymal cells into the space of Dissc and then into hepatic sinusoids.
7. Catabolism of VLDL occurs by lipoprotein lipase with phospholipid acting as cofactor and LDL also formed from VLDL.
8. In atherosclerosis and coronary thrombosis, VLDL concentration is increased.

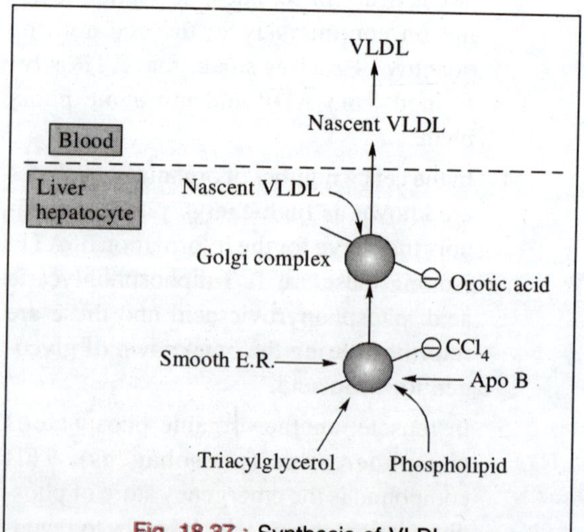

Fig. 18.37 : Synthesis of VLDL

EXERCISE

1. Give an account of the metabolism of fat with particular reference to β-oxidation. *(Mith. 64A)*
2. What is lipogenesis? Describe the cytoplasmic system of fatty acid synthesis. *(Bh. U. 75A)*
3. Describe the oxidation of fatty acids in the body. *(P.U. 68A ; Mith. 60A)*
4. How are fatty acids synthesized and degraded in the body? *(R.U. 74A)*
5. Describe the fate of lipids in our body. What are essential fatty acids? *(P.U. 66S)*
6. Describe the complete oxidation of oleic acid in the body. State how many molecules of ATP are formed during the process. *(P.U. 72A)*
   [Ans. Describe like unsaturated fatty acid oxidation (Linoleic acid)]
7. Describe briefly the chemistry of cholesterol and its physiological importance. What is the normal blood cholesterol level? *(Muz. 7SS; R.U. 64A; Pun. 70A)*
8. Describe the occurrence, properties, tests and physiological importance of cholesterol and show its structural relation to vitamin D or bile acids. *(Mith. 61A)*
9. Describe metabolism of cholesterol with reference to atherosclerosis. Discuss the role of lipoprotein in it. *(P.U. 76A)*
10. Name the ketone bodies. Describe briefly the formation and metabolism of ketone bodies. Describe the test usually done in clinical laboratory to detect the presence of ketone bodies in urine. *(B.H.U. 85A)*
11. What are ketone bodies? Describe how ketosis is produced in the body. *(Bh. U. 74S; Mith 71A, 75S; P.U. 64S)*
12. What is ketosis? How are the ketone bodies produced in the body? How does body try to combat ketosis? *(MUz. 75S)*
13. **Short notes :**
    (i) Essential fatty acids. *(Mith 75S; Bh. U. 76S; P.U. 74A)*
    (ii) Cholesterol. *(M.U. 73S, 74S; R.U. 68A, 70A)*
    (iii) Lipoprotein. *(R.U. 67A;Mith. 65S)*
    (iv) Phospholipids. *(M.U. 73A; P.U. 68A, 76A)*
    (v) Phosphorylation. *(Bh. U. 75S)*
    (vi) Ketosis. *(M.U. 73A; P.U. 68A, 76A)*
    (vii) Ketone bodies. *(Bh. U. 76A; P.U. 71A)*
    (viii) Ketogenesis. *(Mith. 64A)*
    (ix) Rothera's test. *(P.U. 68S)*
    (x) Lipotropic factors. *(P.U. 71S; M.U. 73A)*
    (xi) Unsaturated fatty acids. *(Muz. 75A)*

CHAPTER 19

REGULATION OF CARBOHYDRATE AND LIPID METABOLISM

Regulation of carbohydrate metabolism at the cellular and enzymatic level

1. The changes in the metabolism fully depend on the changes in the availability of substrates. The concentration of glucose, fatty acids and amino acids in blood influences their rate and pattern of metabolism in many tissues.
2. Alterations in the concentrations of glucose, fatty acids and amino acids in the blood owing to the changes in the dietary availability may alter the rate of secretion of hormones that influence the pattern of metabolism in metabolic pathways.
3. Three types of mechanisms are responsible for regulating the activity of enzymes concerned in carbohydrate metabolism :
   (i) Changes in the rate of enzyme synthesis.
   (ii) Conversion of an inactive to an active enzyme.
   (iii) Allosteric effects.

Regulation of Glycolysis, Gluconeogenesis and Hexose Monophosphate Shunt

1. **Glucokinase** catalyzes the conversion of glucose to glucose-6-phosphate. In the same extramitochondrial region **glucose-6-phosphatase** is also found which **catalyses** the same interconversion in the reverse direction on the supply of sufficient carbohydrate, glucokinase activity is increased whereas glucose-6-phosphatase activity is decreased. In starvation, glucokinase activity falls as compared to glucose-6-phosphatase activity.
2. Under the availability of glucose the enzymes utilizing glucose are all activated but the enzymes producing glucose by gluconeogenesis are all depressed. The secretion of **insulin** controls the activity of the enzymes responsible for glycolysis as well as gluconeogenesis.
3. Both dehydrogenases of the HMP shunt are adaptive enzymes since their activity is increased in the well-fed animal as well as when insulin is given to a diabetic animal. Their activity is low in diabetes or fasting. Similar behaviour has been found in "**Malic enzyme**" and **ATP-citrate lyase**. This indicates that these two enzymes are involved in lipogenesis rather than gluconeogenesis.
4. The activity of pyruvate dehydrogenase is decreased since it is regulated by phosphorylation involving an ATP-specific kinase and its activity is increased by dephosphorylation by a phosphatase. Thus, pyruvate dehydrogenase is inhibited during fatty acid oxidation. Its activity is in-

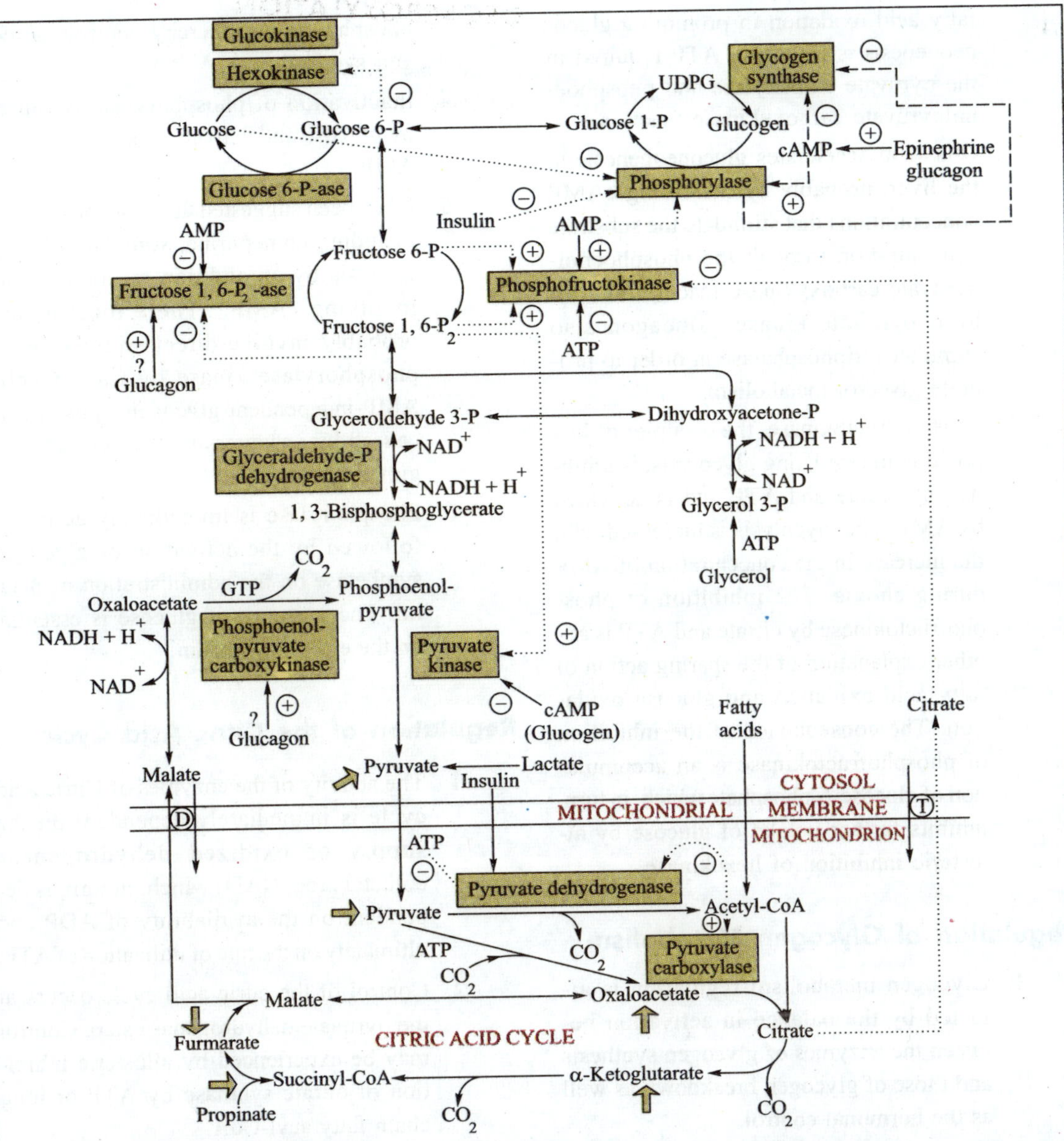

Fig. 19.1 : Key enzymes in the control of glycolysis, gluconeogenesis and glycogen metabolism in liver along with the necessary hormones

creased after administration of insulin and decreased in starvation.

5. The allosteric control of the activity of an enzyme is also available in carbohydrate metabolism. In gluconeogenesis, the synthesis of oxaloacetate from pyruvate by the enzyme pyruvate carboxylase requires the presence of acetyl-CoA as an allosteric activator. The activation of pyruvate carboxylase and the inhibition of pyruvate dehydrogenase by acetyl-CoA formed from the oxidation of fatty acids helps to explain the sparing action of fatty acid oxidation. On the oxidation of pyruvate and the stimulation of gluconeogenesis in the liver (Fig. 19.1), the main role of

fatty acid oxidation in promoting gluconeogenesis is to supply ATP required in the pyruvate carboxylase and phosphoenolpyruvate carboxykinase reactions.

6. **Glucagon** accelerates gluconeogenesis in the liver, probably by increasing cAMP concentrations that stimulate the substrate concentration through the phosphoenolpyruvate carboxykinase reaction and inhibit pyruvate kinase. Glucagon also stimulates triphosphatase in order to promote glycerol metabolism.
7. **Phosphofructokinase**, the occupier of key position in regulating glycolysis, is inhibited by citrate and ATP and is activated by AMP. The glycolysis is increased with the increase in the concentration of AMP during anoxia. The inhibition of phosphofructokinase by citrate and ATP is another explanation of the sparing action of fatty acid oxidation and glucose oxidation. The consequence of the inhibition of phosphofructokinase is an accumulation of glucose-6-phosphate which, in turn, inhibits further uptake of glucose by allosteric inhibition of hexokinase.

Regulation of Glycogen Metabolism

1. Glycogen metabolism regulation is affected by the balance in activation between the enzymes of glycogen synthesis and those of glycogen breakdown as well as the hormonal control.
2. **Cyclic AMP-dependent protein kinase** activates phosphorylase b kinase and inactivates glycogen synthetase. Thus, inhibition of glycogenolysis promotes glycogenesis and inhibition of glycogenesis enhances glycogenolysis.
3. Glycogen metabolism in the liver is controlled by the concentration of phosphorylase a. This enzyme not only controls the rate-limiting step in glycogenolysis but also inhibits the activity of synthetase, phosphatase and thereby controls glycogen synthesis.
4. Inactivation of phosphorylase is caused by glucose and activation is caused by 5′-AMP.
5. It has been suggested that catecholamines, including epinephrine, stimulate glycogenolysis by an addition mechanism not involving cAMP. These mechanisms probably involve direct stimulation of phosphorylase kinase by Ca^{++}. Cyclic AMP-independent glycogenolysis is also caused by vasopressin, oxytocin and angiotensin II.
6. Phosphorylase is immediately activated followed by the activation of glycogen synthetase on the administration of insulin. The presence of glucose is essential on the effects of insulin.

Regulation of the Citric Acid Cycle

1. The activity of the enzymes of Citric acid cycle is immediately dependent on the supply of oxidized dehydrogenase cofactors (eg, NAD) which, in turn, is dependent on the availability of ADP and ultimately on the rate of utilization of ATP.
2. Control of the citric acid cycle occurs at the pyruvate dehydrogenase step. Control may be experienced by allosteric inhibition of citrate synthase by ATP or long chain fatty acyl-CoA.
3. Oxaloacetate inhibits succinate dehydrogenase and the availability of oxaloacetate, as controlled by malate dehydrogenase, depends on the ratios of the concentrations of NADH and NAD^+.
4. The increase in the ratio of the concentration of ATP and ADP is considered to raise the ratio of the concentration of GTP and GDP at the succinate thiokinase step, thereby increasing the concentration of succinyl-CoA.

REGULATION OF LIPID METABOLISM

Regulation of Fatty Acid Synthesis (Lipogenesis)

1. Lipogenesis is concerned with the conversion of glucose and intermediate such as pyruvate, lactate and acetyl-CoA to fat.
2. The rate of lipogenesis is high in case of a diet containing a high proportion of carbohydrate. The rate is decreased on a high-fat diet or in the deficiency of insulin as in diabetes mellitus. All these conditions are related to increased concentration of plasma-free fatty acids. There is an inverse relationship between hepatic lipogenesis and the serum-free fatty acids concentration. Lipogenesis is greatly inhibited over the high range of free fatty acids. Lipogenesis is depressed by the fat diet in the liver and a little carbohydrate is converted to fat when the fat diet is more than 10%.
3. Lipogenesis is higher when sucrose is fed instead of glucose. It is also blocked in fasting due to the lack of NADPH generation from the HMP shunt pathway.
4. Acetyl-CoA carboxylase is competitively inhibited with the activator citrate by the long chain acyl-CoA molecules. Therefore, if acyl-CoA is accumulated, it will automatically reduce the synthesis of new fatty acids.
5. Acyl-CoA may also inhibit the mitochondrial transport of citrate into the cytosol. Free fatty acids also are inversely related to the proportion of active to inactive pyruvate dehydrogenase. Acyl-CoA may also inhibit pyruvate dehydrogenase by inhibiting the ATP-ADP exchange transporter of the inner mitochondrial membrane. As a result, the supply of acetyl-CoA from carbohydrate is blocked.
6. Insulin increases the transport of glucose into the cell making increased availability of pyruvate for fatty acid synthesis and glycerol 3-phosphate for esterification of the fatty acids. It also converts the inactive pyruvate dehydrogenase and acetyl-CoA carboxylase to the active form. It also depresses the level of intracellular cAMP, inhibits lipolysis and thereby reduces the concentration of long chain acyl-CoA which is an inhibitor of lipogenesis.

Regulation of Ketogenesis

1. In adipose tissue, very high concentrations of plasma-free fatty acids (FFA) are available as a result of lipolysis of triacylglycerol. In fed as well as in fasting conditions, 30% or more of the free fatty acids pass to the liver. After they are activated to acyl-CoA they are either esterified mainly to triacylglycerol and phospholipid or they undergo β-oxidation to form acetyl-CoA. The acetyl-CoA is either oxidized in the citric acid cycle or used to form ketone bodies (Fig. 19.2).
2. The esterification, which acts as an antiketogenic factor, depends on the availability of precursors in the liver to supply sufficient glycerol-3-phosphate. In the liver, antiketogenic effects of glycerol and dihydroxyacetone are not correlated with the levels of glycerol 3-phosphate.

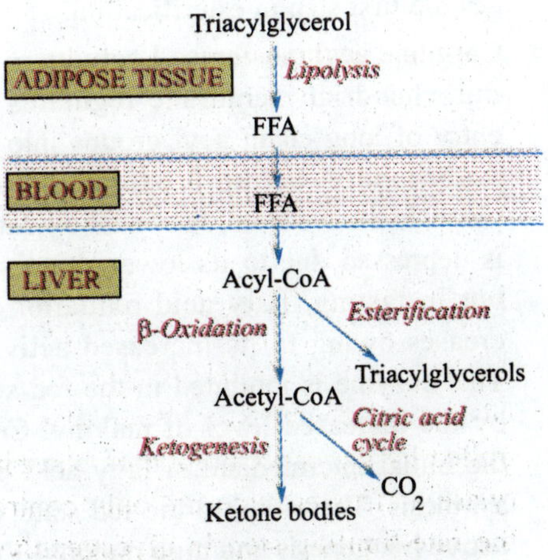

Fig. 19.2 : Regulation of ketogenesis

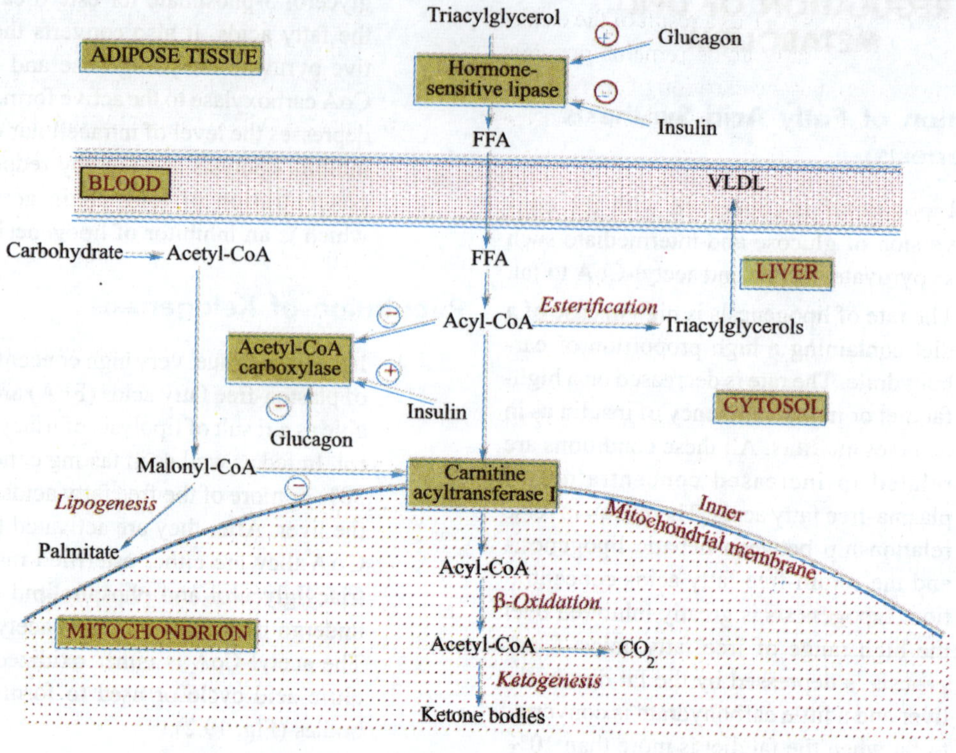

Fig. 19.3 : Regulation of long chain fatty acid oxidation in the liver

3. Phosphatidate phosphohydrolase appears to increase in activity in livers in which extra triacylglycerol synthesis takes place. Insulin increases the activity of glycerol phosphate acyltransferase which catalyses the first step in esterification.

4. Carnitine acyltransferase I activity in the mitochondrial membrane regulates the entry of long chain acyl groups into the mitochondria before β-oxidation takes place. In the fed state, fatty acid oxidation is depressed due to its lowered activity; but in fasting, fatty acid oxidation increases owing to its increased activity. This enzyme is inhibited in the fed state by the increased level of malonyl-CoA, the initial intermediate in fatty acid biosynthesis. Therefore, in the fed state, the active lipogenesis and high malonyl-CoA inhibit carnitine acyltransferase I.

5. Low level of free fatty acids entering the liver cell are nearly all esterified to triacylglycerols and transported out of the liver in VLDL. The concentration of free fatty acids increases with the onset of starvation and acetyl-CoA carboxylase is inhibited and malonyl-CoA decreases. Therefore, the inhibition of carnitine acyltransferase I is released with the permission of more acyl-CoA to be oxidized. The ratio of the concentrations of insulin and glucagon reinforces these events in starvation causing increased lipolysis in adipose tissue and inhibition of pyruvate kinase and acetyl-CoA carboxylase in the liver.

6. More free fatty acid is converted to ketone bodies and less is oxidized to CO_2 via TCA cycle with the increased concentration of serum-free fatty acids. The total

free energy as ATP as a result of the oxidation of free fatty acids remains constant on the fact that the partition of acetyl-CoA between the ketogenic pathway and the pathway of oxidation to CO_2 is regulated in such a particular manner, on complete oxidation in the Citric acid cycle. One mol of palmitate yields 129 mols of ATP, whereas this one mol of palmitate produces only 33 mols of ATP when acetoacetate is the end product. Therefore, ketogenesis is regarded as a mechanism in which the liver can oxidize large quantities of fatty acids without increasing its total energy expenditure.

7. Reduced level of oxaloacetate within the mitochondria can cause impairment of the TCA cycle to metabolize acetyl-CoA. Krebs has suggested that a fall in the concentration of oxaloacetate owing to an enhanced gluconeogenesis may be the cause of the severe forms of ketosis found in diabetes and the ketosis of cattle. It has been assumed that citrate synthase is inhibited either by long chain acyl-CoA or by increased level of ATP. It has also been shown that pyruvate carboxylase—which converts pyruvate to oxaloacetate—is activated by acetyl-CoA. Therefore, sufficient amounts of acetyl-CoA signifies the presence of significant amounts of oxaloacetate to initiate the condensing reaction of TCA cycle.

SUMMARY

1. Free fatty acids (FFA) are the main substrates for ketone body formation in the liver. Therefore, all factors, metabolic or endocrine, influence ketogenesis releasing free fatty acids from adipose tissue.
2. Free fatty acids, after entering the liver, are established by the balance between their esterification and oxidation. This balance is governed by carnitine acyltransferase I, whose activity is raised indirectly by the concentration of free fatty acids and the hormonal state of the liver.
3. The regulation of more ketone bodies formation and less CO_2 formation with the increased oxidation of fatty acid is performed in such a manner that the total ATP production remains constant. Ketone bodies, without being oxidized significantly by the liver, diffuse into the circulation and, hence, are oxidized by extrahepatic tissues to other fuels.

ECONOMICS OF CARBOHYDRATE AND LIPID METABOLISM IN THE WHOLE BODY

1. Glucose is converted to fat under optimal nutritional intake. Except glycerol, fat (as fatty acid) cannot produce glucose due to the irreversibility of the oxidative decarboxylation of pyruvate to acetyl-CoA. Certain tissues are more dependent on the continued supply of glucose than others. A minimum supply of glucose is necessary in extrahepatic tissues to maintain the integrity of TCA cycle. Glucose is the main source of glycerol-3-phosphate in tissues devoid of glycerokinase. Large quantities of glucose are also required for fetal nutrition and the synthesis of milk. Certain mechanisms are also involved to safeguard essential supplies of glucose during shortage by allowing other substrates to spare its oxidation.
2. Ketone bodies and free fatty acids spare the oxidation of glucose in muscle by impairing its entry into the cell, its phosphorylation of glucose-6-phosphate, the phosphofructokinase reaction, and the oxidative decarboxylation of pyruvate. Free fatty acids and ketone bodies oxidation increases the concentration of intracellular citrate which inhibits phosphofructokinase. Olson has justified that, under carbohydrate shortage, available fu-

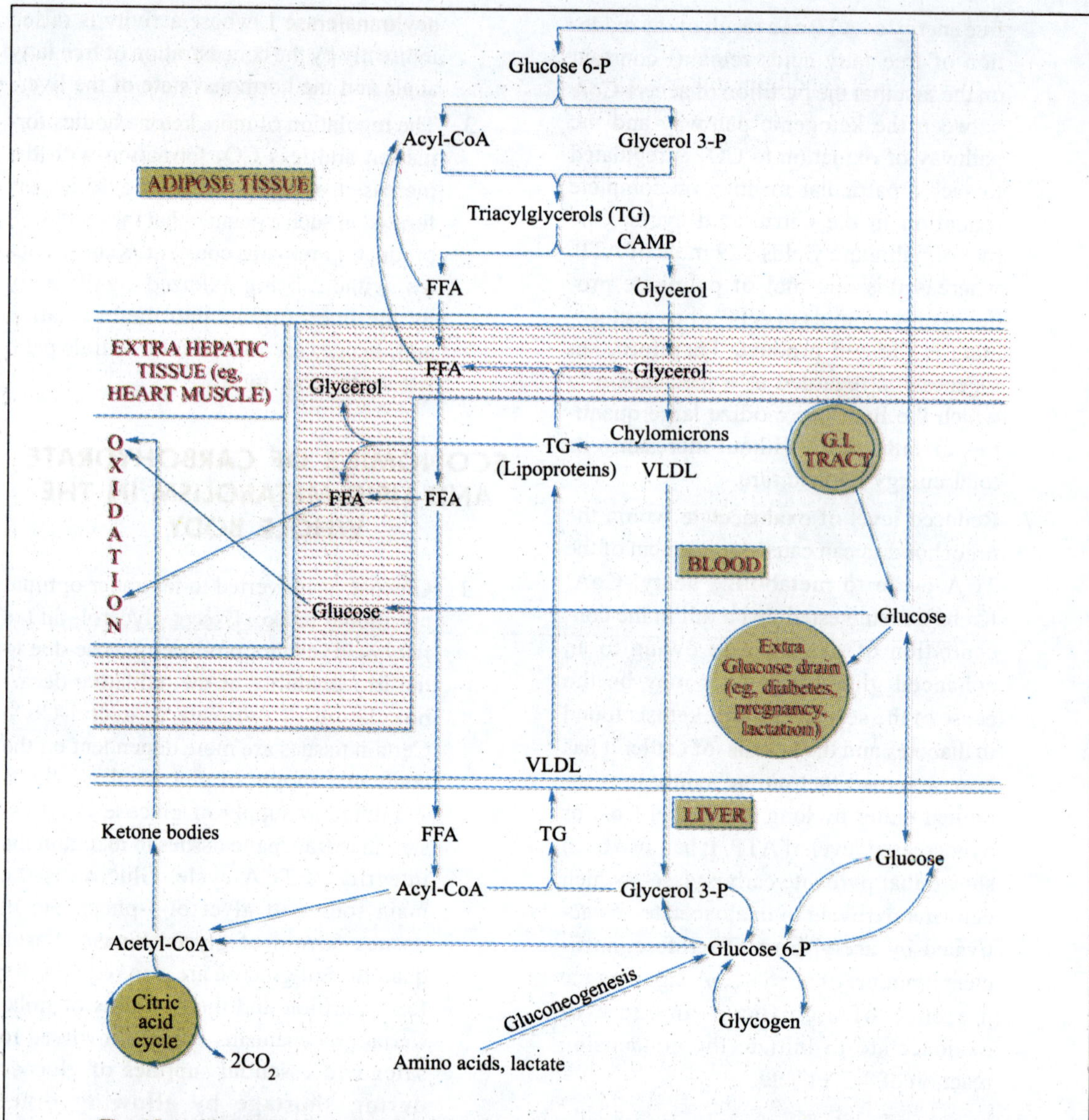

Fig. 19.4 : Metabolic interrelationship between adipose tissue, the liver and extrahepatic tissues

els are oxidized in the following order of preference : (i) ketone bodies (and other short chain fatty acids, eg, acetate) (ii) free fatty acids, (iii) glucose (shown in Fig. 19.4.).

3. Under certain conditions, fat mobilization can be reduced by the administration of noncarbohydrate calorigenic substrates. If glucose oxidation is spared by the free fatty acids and ketone bodies, more glucose will be available causing a reduction in output of free fatty acids from adipose tissue and the free fatty acid level of plasma will fall. The combination of the effects of the free fatty acids in sparing glucose utilization in muscle and heart and the effect of the spared glucose in inhibiting free fatty acid mobilization in adipose tissue has been called the "**glucose-fatty acid cycle**".

Starvation

1. Under fasting condition, glucose concentration becomes less, glycogen is drawn upon to maintain the blood glucose level. Blood insulin level decreases and glucagon level increases. Since the glucose utilization in the adipose tissues decreases and the inhibitory effect of insulin on lipolysis becomes less, fat is mobilized as free fatty acids and glycerol. The free fatty acids are esterified in the liver and the remainders are oxidized. In the liver and kidney, glycerol joins the carbohydrate pool after being converted to glycerol-3- P . The endogenous glucose production cannot keep pace with its utilization and oxidation. Therefore, the liver glycogen stores become depleted and blood glucose tends to fall. Then fat is mobilized at the increased rate but in several hours the plasma-free fatty acids and the blood glucose stabilize at the fasting level. Hence, increased fatty acid oxidation takes place —resulting in the formation of ketone bodies.
2. The adipose tissues provide carbohydrate in the form of glycerol together with that provided by gluconeogenesis from protein.
3. In prolonged starvation in humans, gluconeogenesis from protein is diminished owing to the diminished release of amino acids, particularly alanine, from muscle. The brain then utilizes ketone bodies in place of glucose.
4. Free fatty acids are under most conditions mobilized in excess of oxidative requirements since a large amount is esterified even during fasting. When the carbohydrate supplies are sufficient, most of the influx is esterified and ultimately retransported from the liver as VLDL to be utilized by other tissues. In the case of increased influx of free fatty acids, ketogenesis is available which allows the liver to retransport much of the influx of free fatty acids in a form which is readily utilized by extrahepatic tissues under all nutritional conditions.
5. The carbohydrate cycle involves the release of glycerol from adipose tissue and its conversion to glucose in the liver. The lipid cycle involves the release of free fatty acids by adipose tissue, its transport to and esterification in the liver and retransport of VLDL back to adipose tissue. These two interrelated cycles are involved during the disturbances in carbohydrate or lipid metabolism where they interact in adipose tissue and in the liver.

EXERCISE

1. Discuss in detail the regulation of carbohydrate metabolism in the body. (P.U.86A)
2. How is the metabolism of lipid regulated in the body? (B.H.U. 88A)
3. Discuss the economics of carbohydrate and lipid metabolism in the whole body. (M.U. 89S)

CHAPTER 20

PROTEIN METABOLISM

DYNAMIC STATE OF PROTEIN

1. All body proteins (plasma proteins, hemoglobin and intracellular proteins) continually undergo degradation and synthesis. More than half of the protein of the liver and intestinal mucosa are broken down and resynthesized in ten days; the rate is slower in muscle and erythrocytes.
2. Active resynthesis occurs even in starvation, and active breakdown in nitrogen equilibrium.
3. Degradation of protein in one tissue is accompanied by synthesis in others.
4. Antibody protein (γ-globulins) induced by immunization also undergo continual breakdown and synthesis. The half-life is about two weeks.
5. The balance between the rates of synthesis and degradation of its constituents is maintained in adult organism. But in a growing organism, the rate of synthesis of its constituents exceeds the rate of breakdown and the rate of catabolism is greater than the rate of anabolism in wasting diseases and starvation.

METABOLIC POOL

1. The mixture of endogenous and exogenous materials constitutes a reservoir or "metabolic pool" of the compound.
2. This pool is contributed by substances derived from catabolism of protein in the tissues by the stimulation of the excessive amounts of thyroid hormones and 11-oxygenated adrenocortical hormones and substances absorbed from the intestine. Protein anabolism is stimulated by growth hormone insulin and testosterone.
3. The metabolism of protein is integrated with that of carbohydrate and fat through pyruvate, acetate, oxaloacetate and α-ketoglutarate etc. and the members of one pool are metabolically equilibrated with other pools, eg, alanine is deaminated to pyruvic acid and hence the carbon skeleton of this amino acid is involved in the pyruvate pool which is directly connected to the carbohydrate pool.

INTERMEDIARY METABOLISM

Metabolism is generally interpreted to mean intermediary metabolism. It includes all chemical processes within cells and tissues which are concerned with their building up and breaking down and in their functional operation.

The building up and breaking down of protoplasm are concerned with protein metabolism. Various substances of lipids, carbohydrates and inorganic metals are also involved.

The functional operation of tissues, such as muscular contraction, the propagation of nerve

Fig. 20.1 : Intermediary metabolism of proteins

impulses, the secretory work of various glands, the selective absorption of intestine and the excretory processes of the kidney and other organs require a number of chemical reactions to provide necessary energy and specific chemical substances.

Metabolism does not only mean the chemistry of tissue formation and breakdown but also the chemistry of the processes in tissues necessary for the formation of various specific compounds required for the operation and regulation of the metabolic machine.

INTERMEDIARY METABOLISM OF PROTEINS

1. The amino acids abstracted from the liver are not utilized for repair or special synthesis but are broken down and the ammonia thus formed is converted into urea which is excreted.
2. The amino acids abstracted from the tissues are used for the repair of the tissues. The unwanted amino acids are either used up by the tissue, or converted into urea in the liver.
3. Amino acids provide the nitrogen for the synthesis of many nitrogenous compounds of the body including bile acids, creatine, purine, pyrimidines, epinephrine, thyroxine, amino sugars, enzymes and the bases of the phospholipids.
4. Fates of amino acids result in :
   (i) Catabolism in the liver,
   (ii) Formation of tissue protein,
   (iii) Formation of other nitrogenous substances.
5. The essential and excess metabolism of proteins are usually differentiated by the term "endogenous" and "exogenous" metabolism. Endogenous metabolism is involved in the formation of tissue and the other nitrogenous substances and also the normal breakdown of tissue proteins. Exogenous metabolism is the metabolism of all proteins ingested over and above the essential requirements.
6. In normal health, the nitrogen ingested equals to that excreted in urine, feces and sweats.
7. Fates of sulphur containing amino acids result in :
   (i) Catabolism in the liver producing sulphuric acid which is excreted.
   (ii) Synthesis of tissue protein including that of hair.
   (iii) Synthesis of other substances including taurine, glutathione, insulin, ethereal sulphate, sulphated mucopolysaccharide and lipids.

In general, it can be summarized as shown in Fig. 20.1.

CATABOLISM OF AMINO ACID NITROGEN, SYNTHESIS OF UREA

In mammalian tissues, the alpha amino groups of amino acids derived from the diet or from tissue protein breakdown are ultimately excreted in urine as urea. Several enzymes are involved in the biosynthesis of urea which involves four processes :

1. Transamination.
2. Oxidative deamination.
3. Ammonia formation and transport.
4. Reactions of the urea cycle.

In human and other ureotelic organisms, urea is the end product of amino acid nitrogen metabolism. But in uricotelic organisms (eg, reptiles and birds), uric acid is the end product and in ammonotelic organisms (eg, bony fish), ammonia is the end product of amino acid nitrogen metabolism.

1. Transamination

1. Transamination involves interconversion of a pair of α-amino acids and a pair of α-keto acids and is catalyzed by transaminases or aminotransferases.
2. Pyridoxal phosphate (B_6–PO_4) is the coenzyme essential for the transaminase activity.
3. Alanine-pyruvate transaminase (alanine transaminase) and glutamate-α-ketoglutarate transaminase (glutamate transaminase) present in most mammalian tissues, catalyze the transfer of amino groups from most amino acids to form alanine (from pyruvate) or glutamate (from a-ketoglutarate) shown in Fig. 20.2.
4. Transamination is a reversible process. This reversibility allows transaminases to function in amino acid catabolism and biosynthesis.

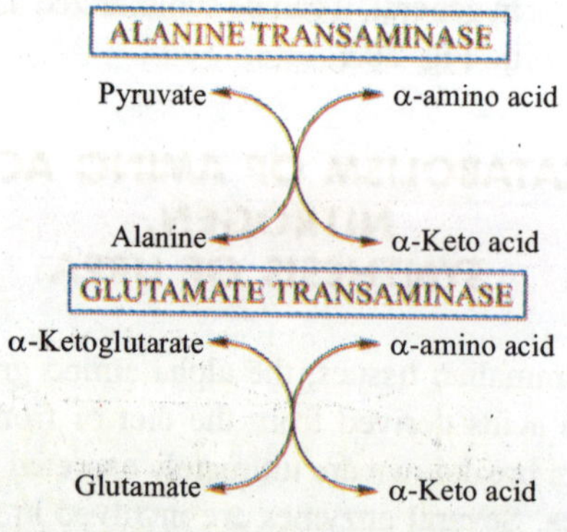

Fig. 20.2 : Alanine and glutamate transaminases

5. Each transaminase is specific for the specified pair of amino acid and keto acid as one pair of substrates.
6. Most amino acids are substrates for transmination except lysine, threonine, proline and hydroxyproline. Not only δ-amino acid, δ-amino group of ornithine undergo transamination.

CLINICAL ORIENTATION

Aspartate transaminase (SGOT) and Alanine transaminase (SGPT) are clinically important.

SGOT

- This enzyme is cytoplasmic and mitochondrial.
- Its concentration in the blood is not increased in coronary insufficiency and angina pectoris.
- Its concentration is very high in myocardium as well as in liver cells.
- Elevated level is also found in muscular dystrophies, but no raised level in muscular disease of nervous origin.
- Raised concentration is also observed in acute pancreatitis, leukaemias and in hemolytic anemia.
- Increased level is found in therapy with erythromycin and in normal persons after prolonged heavy exercise.

SGPT

- This enzyme is available mainly in liver and entirely cytoplasmic.
- Very high level of this enzyme is observed in acute viral hepatitis.
- Existence of limited value in differential diagnosis of jaundice.
- Increased level is also found in extra-hepatic biliary obstruction and also in hepatic malignancy.

2. Oxidative Deamination

1. Oxidative conversion of many amino acids to their corresponding α-keto acids occurs in mammalian liver and kidney tissues.
2. Both L- and D-amino acid oxidase activities occur in mammalian liver and kidney tissue and are also widely distributed in other animals and microorganisms.

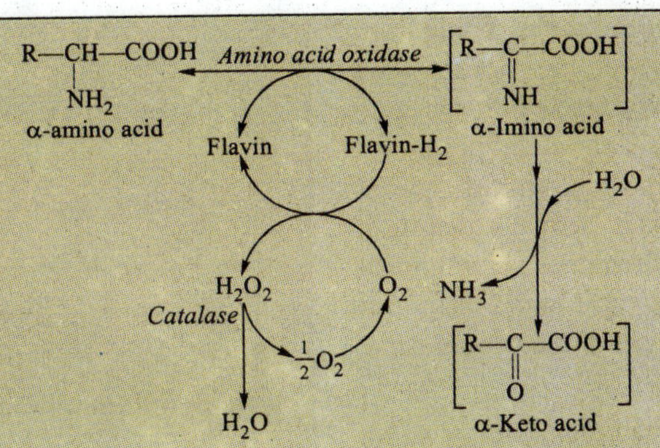

Fig. 20.3 : Oxidative deamination catalyzed by L-amino acid oxidase

3. The amino acid oxidases are auto-oxidizable flavoproteins i.e. the reduced FMN or FAD is reoxidized directly by molecular oxygen forming H_2O_2 without the help of cytochromes or other electron carriers shown in Fig. 20.3. The toxic product (H_2O_2) is then split to O_2 and H_2O by catalase which occurs widely in tissues, especially liver. If catalase is absent, the α-keto acid is decarboxylated by H_2O_2 forming a carboxylic acid with one carbon atom less.

   In the amino acid oxidase reactions shown in Fig. 20.3, the amino acid is first dehydrogenated by the flavoprotein of the oxidase forming an α-imino acid which spontaneously decomposes to the corresponding a-keto acids by the addition of water.

4. Mammalian L-amino acid oxidase, an FMN-flavoprotein, is restricted to kidney and liver tissue and its activity is quite slow. Mammalian D-amino acid oxidase, an FAD-flavoprotein, occurs in the liver and kidney tissue of most mammals.

Significant Remarks

1. If the enzyme catalase is absent genetically, α-keto acid produced by oxidative deamination is decarboxylated by H_2O_2 forming a carboxylic acid with one carbon atom less.
2. The activity of mammalian L-amino acid oxidase is restricted to liver and kidney only and the activity of this enzyme is quite slow in these tissues.
3. The enzyme L-amino acid oxidase has no effect on glycine.
4. This enzyme does not have a major role in the catabolism of mammalian amino acid and formation of ammonia.

3. Ammonia Formation and Transport

1. The amino groups of most amino acids are ultimately transferred to α-ketoglutarate by transamination. Release of this nitrogen as ammonia is catalyzed by L-glutamate dehydrogenase which is widely distributed in mammalian tissues. Certain hormones also influence glutamate dehydrogenase activity.
2. Glutamate dehydrogenase uses NAD^+ or $NADP^+$ as cosubstrate. The reaction is reversible,

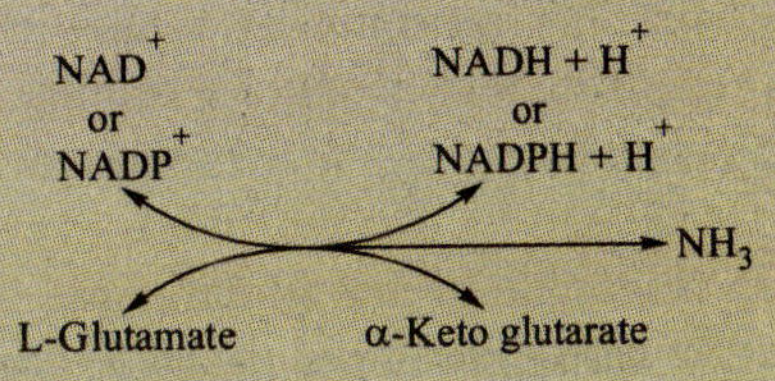

Fig. 20.4 : L-Glutamate dehydrogenation reaction

3. Intestinal bacteria produce ammonia from dietary protein as well as from the urea present in fluids secreted into the gastrointestinal tract. This ammonia is absorbed from the intestine into the portal venous blood. Under normal conditions, the liver promptly removes the ammonia from the portal blood. Minute quantities of ammonia are toxic to the central nervous system.

   The symptoms of ammonia intoxication include slurring of speech, blurring of vision, a peculiar flapping tremor and in severe cases coma and death. These symptoms occur when brain ammonia levels are increased.

   Ammonia is produced in the kidney from intracellular amino acid, glutamine, catalyzed by renal glutaminase. Ammonia production by the kidney is highly increased in metabolic acidosis and depressed in alkalosis.

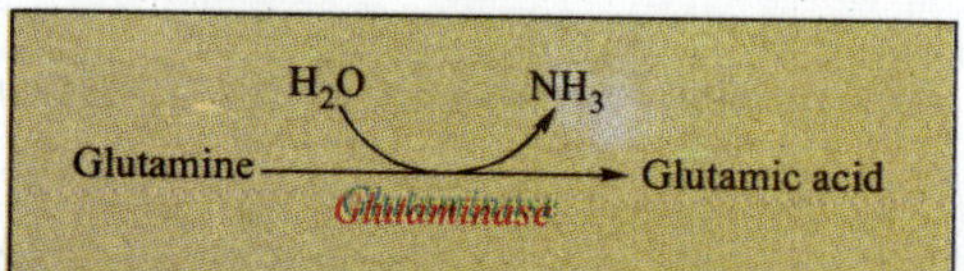

Fig. 20.5 : Ammonia production

4. Ammonia is excreted as ammonium salts during metabolic acidosis but the majority is excreted as urea. Ammonia is present only in traces in blood (10-20 μg/100 ml) because it is rapidly removed from the circulation by the liver and converted to glutamine or urea.
5. Formation of glutamine is catalyzed by glutamine synthetase, a mitochondrial enzyme present in renal tissues in highest concentration. synthesis of glutamine is accompanied by the hydrolysis of ATP to ADP and Pi. The reaction is irreversible. Asparaginase and glutaminase are employed as antitumor agents because certain tumors require glutamine and asparagine.

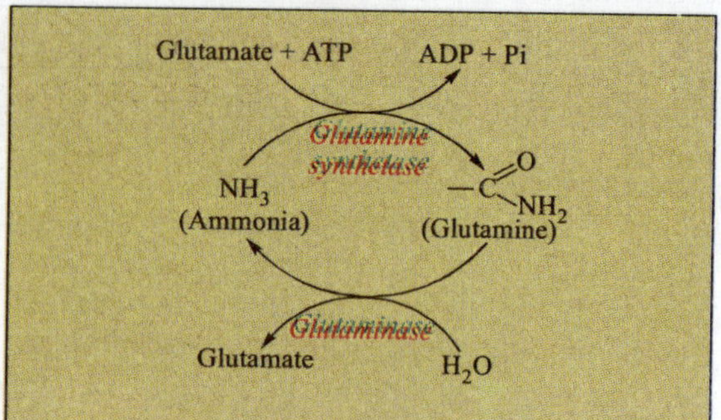

Fig. 20.6 : Formation and transport of ammonia

6. In brain, the major mechanism for removal of ammonia is glutamine formation and in the liver, the most important pathway is urea formation. Brain tissue can form urea but this does not play a significant role in ammonia removal.

UREA CYCLE

Reactions of the Urea Cycle

1. **Carbamoyl phosphate** is formed by the condensation of one mol of phosphate (derived from ATP) being catalyzed by the enzyme **carbamoyl phosphate synthetase** which is present in liver mitochondria of all ureotelic organisms including humans. In addition to magnesium ion (Mg^{++}), N-acetyl-glutamate (a dicarboxylic acid) is required. Probably, the presence of N-acetyl-glutamate brings about a marked change in the structure of carbamoyl phosphate synthetase which exposes certain sulfahydryl groups and affects the affinity of the enzyme for ATP.

   In **bacteria**, glutamine in place of ammonia serves as a substrate for carbamoyl phosphate synthesis.
2. Carbamoyl moiety is transferred to ornithine to form citrulline being catalyzed by ornithine transcarbamoylase of liver mitochondria.
3. Arginosuccinic acid is formed by the com-

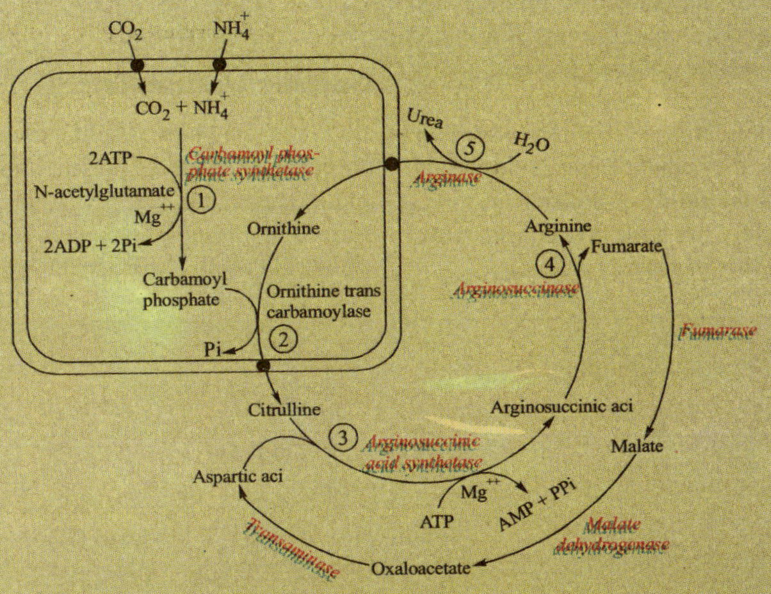

Fig. 20.7 : Flow chart of Biosynthesis of urea or ornithine-urea cycle. Reactions 1 and 2 take place in the matrix of liver mitochondria and reaction 3, 4, 5, in liver cytosol. Specific carriers (•) present in the inner membranes of liver mitochondria for the transverse of NH_4^+, CO_2, ornithine, citrulline

bination of citrulline and aspartic acid in presence of arginosuccinic acid synthetase and ATP.

4. **Arginosuccinic acid** is cleaved to arginine and fumaric acid by **arginosuccinase** which is present in mammalian liver and kidney. The fumarate formed is converted to oxaloacetate via the fumarase and malate dehydrogenase reactions and then transaminated to regenerate aspartate.
5. The hydrolytic cleavage of arginine is catalyzed by **arginase** which is present in the livers of all ureotelic organisms forming ornithine and urea.

   Smaller quantities of arginase also occur in renal tissue, brain, mammary gland, testicular tissue and skin. Ornithine and lysine are the competitive inhibitors of arginine.

The cycle for the overall reactions is given in Fig. 20.7.

The biosynthesis of urea occurs mainly in the **liver**. 1 mol of urea is synthesized from 1 mol of ammonia, 1 mol of carbon dioxide, 3 mols of ATP (2 of which are converted to ADP and Pi and 1 to AMP + PPi), 5 enzymes catalyzing the reactions and 6 amino acids involved in the reaction.

One amino acid, N-acetyl-glutamate, serves as an **enzyme activator**. The remaining 5 amino acids aspartate, arginine, ornithine, citrulline and arginosuccinic acid – all function as carriers of atoms which, ultimately, become urea. Aspartate and arginine occur in protein while ornithine, citrulline, arginosuccinate do not.

Urea formation is partly a cyclical process. Ornithine used in reaction (2) is regenerated in reaction (5).

An active man consuming about 300g of carbohydrates, 100g of fat, 100g of protein daily excrete about 16.5g of nitrogen. 95 per cent is eliminated by the kidneys and 5 per cent in the stool. The major pathway of nitrogen excretion in humans is as urea which is synthesized in the liver.

Blood level : Normal concentration of urea in the blood is 15 to 45 mg/100 ml.

Excretion : The daily output of urea through urine is 20 to 30 grams. A less quantity is excreted in the sweat. The quantity of urea excreted is proportional to the total protein metabolism. The excretion of urea is decreased in certain liver diseases. In severe acidosis, the output of urea is decreased. In nephritis, when the ability of the kidneys to excrete urea is severely impaired, the concentration of urea in the blood is increased (uremia).

TEST OF UREA IN URINE

Specific Urease Test

Principle : Urea is split into ammonia and CO_2 by the enzyme urease present in plants at the optimum temperature and pH. Ammonium carbonate formed as a result of ammonia and carbon dioxide shows the alkali characteristic by the indicator.

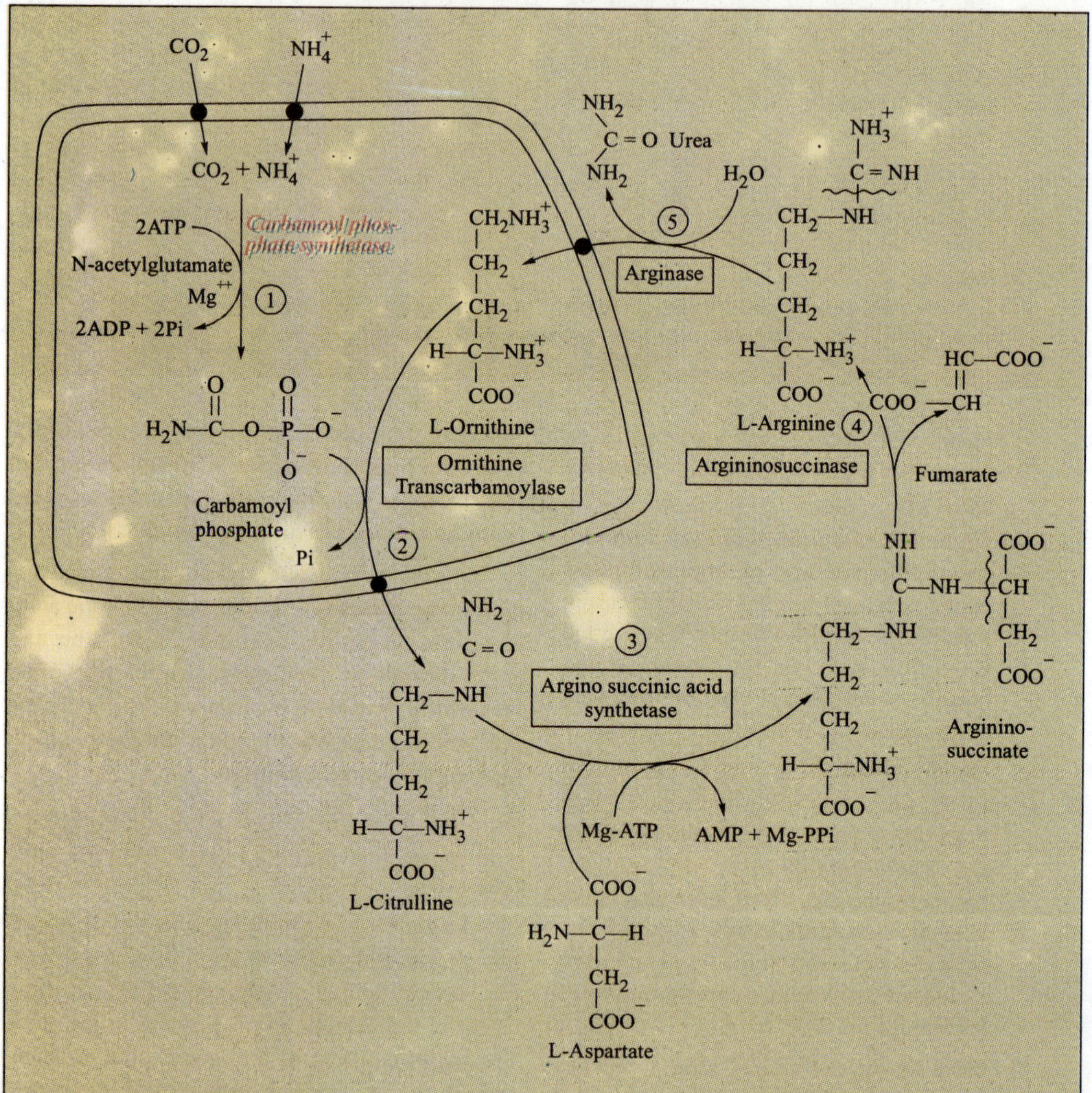

Fig. 20.8 : Reaction of biosynthesis of urea or ornithine urea cycle. Reactions 1 and 2 take place in the matrix of liver mitochondria and reactions 3, 4, 5, in liver cytosol. Specific carriers (•) present in the inner mitochondria for the transverse of NH_4^+, CO_2, ornithine and citrulline

Method

1. 2 ml of urine in one test tube and 2 ml of water in another test tube.
2. A drop of phenol red is added to each of the test tubes.
3. 2 per cent Na_2CO_3 added drop by drop till the deep pink colour develops.
4. 1 or 2 drops of acetic acid added to each test tube till the colour is just yellow.
5. A pinch of soyabean powder added to each and the test tubes are rotated between the palms or the temperature is raised to 60°C, the optimum temperature for the enzyme urease.
6. The development of pink colour in the test tube containing urine indicates the presence of urea.

CLINICAL ORIENTATION

- Increased and decreased blood urea levels occur in a number of diseases. The increased level is associated with kidney diseases mainly.

In increased levels :

- Pre-renal disease exhibits the following conditions:
  - (a) Ulcerative colitis with severe chloride loss.
  - (b) Prolonged and severe diarrhoea.
  - (c) Severe vomiting in intestinal vomiting.
  - (d) Salt and water depletion.
  - (e) Hematemesis.
  - (f) Hypoadrenalism.
  - (g) Hemorrhage and shock.
- Renal disease shows the following conditions :
  - (a) Maligant nephrosclerosis (hardening of kidney).
  - (b) Chronic pyelonephritis (inflammation of kidney)
  - (c) Mercurial poisoning.
  - (d) Renal tuberculosis.
- Conditions of Post-renal diseases :
  - (a) Enlarged prostate.
  - (b) Stones in urinary tract.
  - (c) Stricture of the urethra.
  - (d) Tumors of the bladder.

CLINICAL ORIENTATION

In decreased levels :

Decreased blood urea levels are rare and may be found in the following conditions :

(a) Severe liver damage.

(b) Pregnancy.

METABOLIC DISORDERS IN THE UREA CYCLE

All disorders of urea synthesis cause ammonia intoxication. Clinical symptoms in urea cycle disorders are vomiting in infancy, avoidance of high protein foods, irritability, lethargy and mental retardation. Significant improvement has been observed on a low-protein diet and much of the brain damage is prevented. The daily food intake should be given in frequent small feedings to avoid sudden increases in blood ammonia levels. Uses of antibiotics to decrease the absorption of ammonia formed by bacterial decomposition in the intestine.

Hyperammonemia Type I

1. This is a familial disorder.
2. This occurs due to the deficiency of the enzyme carbamoyl phosphate synthetase.
3. It produces the symptoms of ammonia toxicity.

Hyperammonemia Type II

1. Patients suffer from a deficiency of ornithine transcarbamoylase.
2. The clinical finding is an elevation of glutamine in the blood, cerebrospinal fluid and urine.
3. There is enhanced synthesis of glutamine by the glutamine synthetase reaction for which tissue level of ammonia is increased.
4. The mothers show hyperammonemia and intolerance to high protein diet.

Citrullinemia

1. Complete absence of arginosuccinate synthetase activity produces this condition.
2. Large quantities of citrulline are excreted in the urine and plasma and CSF citrulline levels are elevated.
3. Citrulline and arginosuccinate act as alternative carriers of waste nitrogen, since they contain nitrogen used for urea synthesis.
4. Arginine feeding accelerates citrulline excretion in these patients.
5. Benzoate feeding diverts ammonium nitrogen to hippuric acid via glycine.

Arginosuccinic aciduria

1. Arginosuccinic aciduria is caused due to the absence of arginosuccinase activity. This enzyme is also absent from brain, liver, kidney and erythrocytes of patients with this disease.
2. The disease is characterized by the increased level of arginosuccinic acid in blood, CSF, and urine.
3. It is frequently associated with the occurrence of tufted hair.
4. The disease always appears by age 2 and generally terminates fatally early in life.
5. It needs measurement of erythrocyte arginosuccinase activity for confirmative diagnosis.
6. Arginine and benzoate feedings accelerate nitrogen excretion in these patients.

Hyperargininemia

1. Low erythrocyte level of arginase causes this disease.
2. The urinary amino acid pattern resembles that of lysine cystinuria.
3. A low protein diet results in the lowering of plasma ammonia and abolishes urinary lysine cystinuria.

DECARBOXYLATION

Decarboxylation reaction is catalyzed by the enzyme decarboxylase which requires Pyridoxal phosphate (B_6–PO_4) as coenzyme with the removal of CO_2 resulting in the formation of an amine.

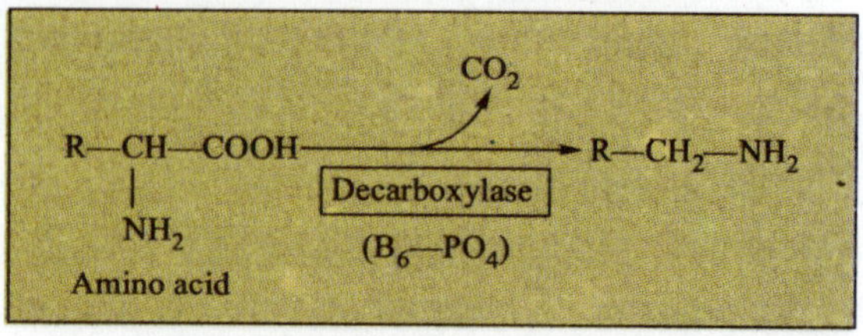

Decarboxylase enzyme is available in liver, kidney and brain and also from microorganisms of intestinal tract. The macroorganisms cause putrefaction in intestines producing amines.

1. Tyramine : (see chem. eqn. A—next page)

1. Tyramine is formed from tyrosine by tyrosine decarboxylation causing elevation of blood pressure.
2. This occurs in the gut by bacterial action.
3. The reaction takes place also in the kidney being favoured by oxygen deficiency.
4. The tissue deaminates tyrosine in presence of sufficient oxygen.

2. Tryptamine : (see chem. eqn. B—next page)

1. Tryptamine is formed from the amino acid tryptophan by the enzyme tryptophan decarboxylase present in mammalian kidney, liver and bacteria of gut involving (B_6–PO_4) as a coenzyme.
2. Derivative of 5–OH tryptamine is serotonin a tissue hormone.
3. It also increases blood pressure.

3. Histamine

1. The enzyme histidine decarboxylase, in

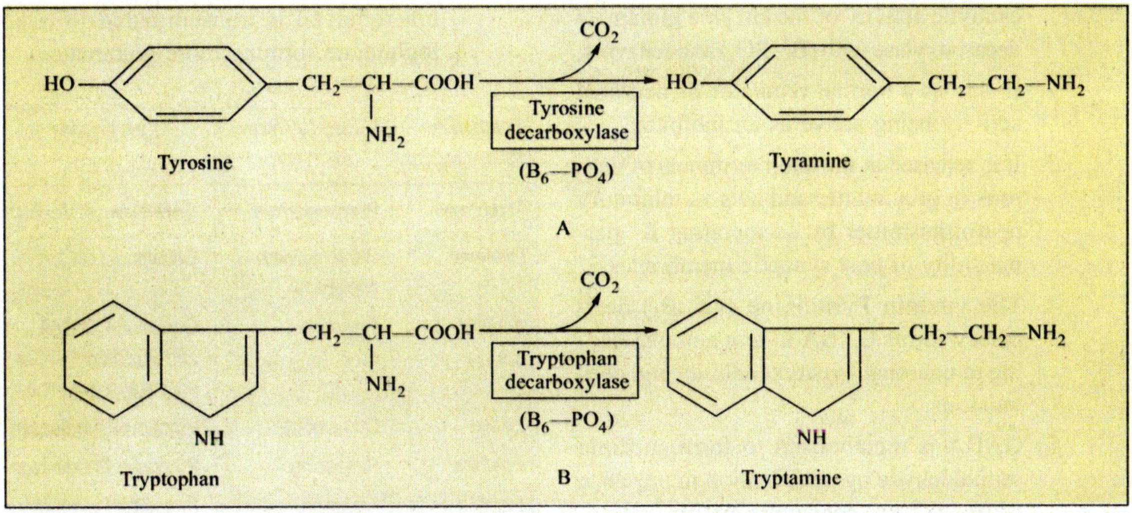

presence of the coenzyme (B_6–PO_4), catalyze the conversion of histidine to histamine.

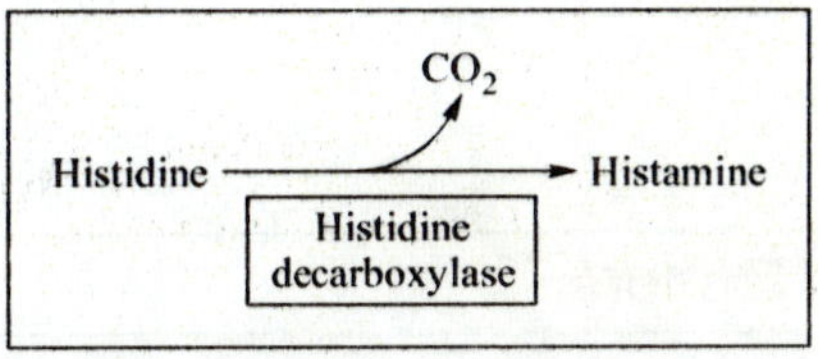

2. This is produced by gastric mucosa cells and histaminergic neurons of the central nervous system.
3. This is produced in the gut by bacterial decarboxylation of histidine.
4. Basophils, another principal source of histamine, in the circulating cells.
5. In the hypothalamus, it acts as a neurotransmitter.
6. It serves as an anaphylactic and inflammatory agent in response to antigen.
7. Effects of released histamine are mediated through H_1 and H_2 receptors.
8. It is formed in injured tissues. The liberation of histamine excessively is related to traumatic shock.
9. Its elevated level depresses blood pressure and large doses cause extreme vascular collapse.
10. The increased plasma level of histamine is found in patients with antigen-induced bronchial asthma.
11. The anaphylactic reactions of histamine is mixed by promethazine and mepyramine which block H_1 receptors and cimetidine, the blocker of H_2 receptors, reduce the gastric acidity in peptic ulcer patients.
12. In the kidneys, the enzyme histaminase oxidatively deaminates histamine to β-imidazole acetaldehyde which is further oxidized in the liver by aldehyde oxidase to form β-imidazole acetic acid which is excreted in urine—being conjugated with ribose.

4. GABA (γ-aminobutyric acid)

1. GABA is formed principally in CN system in grey matter from glutamate by the

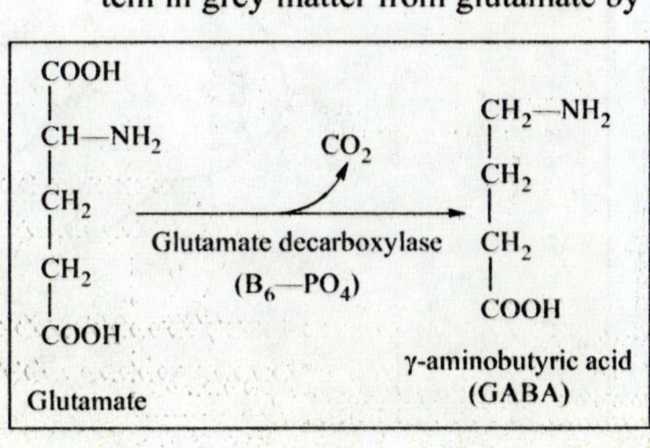

catalytic activity of the enzyme glutamate decarboxylase with (B_6–PO_4) as coenzyme.

2. It acts as a normal regulator of neuronal activity being active as an inhibitor.
3. It is released at the axon terminals of neurons in grey matter and acts as inhibitory neurotransmitter by accelerating K^+ permeability of post synaptic membranes.
4. The vitamin Pyridoxine (vit. B_6) deficiency forms GABA in less amount leading to neuronal hyperexcitability and convulsions.
5. GABA is metabolised to form succinic semialdehyde by deamination in presence of (B_6–PO_4) as coenzyme and the ammonia removed is transaminated to α-ketoglutarate forming more glutamate.

Some other amino acids	Decarboxylation Products	Importance
Threonine	Propanolamine	Constituent of Vit. B_{12}
Cysteine	β-mercaptoethanolamine	Constituent of CoA
Aspartic acid	β-alanine	Constituent of CoA and constituent of Carnosine and Anserine
Lysine	Cadaverine	Product of Putrefaction
Ornithine	Putrescine	Product of Putrefaction
Arginine	Agmatine	Product of Putrefaction

SCHEMATIC COMPENDIUM

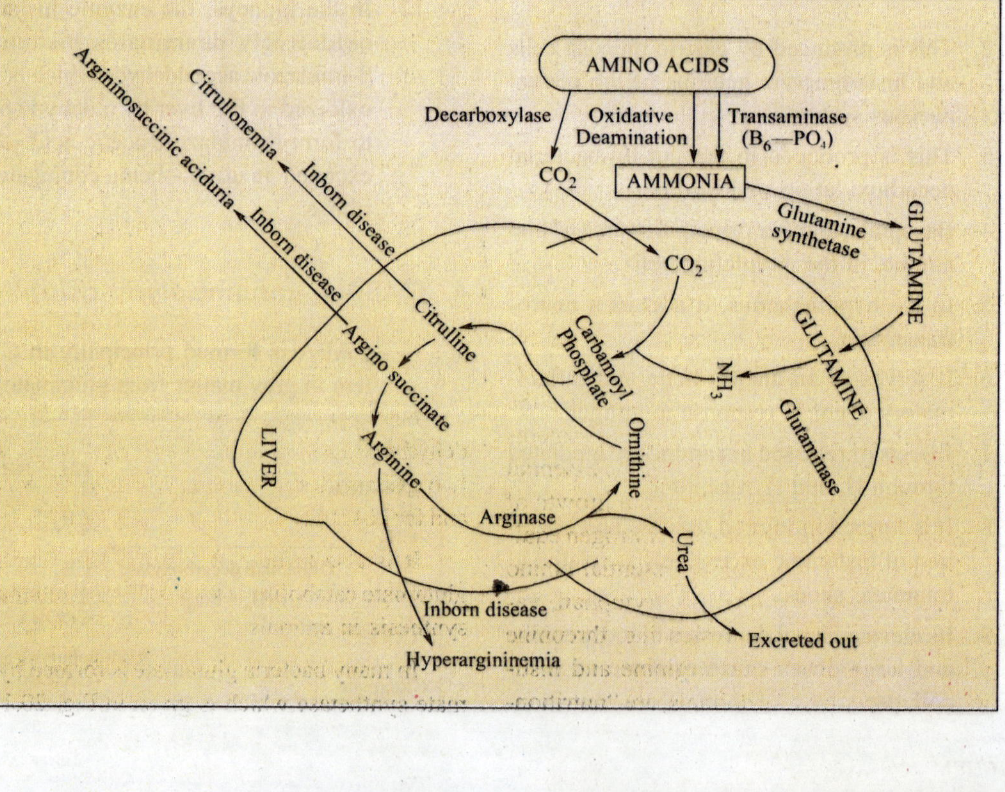

BIOSYNTHESIS OF AMINO ACIDS NUTRITIONALLY ESSENTIAL AND NONESSENTIAL AMINO ACIDS

Plants and bacteria can form all 22 amino acids from amphibolic intermediates. But humans and other animals cannot synthesize some of these. Therefore, these are supplied by the diet and are termed nutritionally essential amino acids. The remainders are synthesized in the body. Therefore, these are termed nutritionally nonessential amino acids.

According to nutritional scientists, the nutritionally essential amino acids are termed "essential" or "indispensable" amino acids and the nutritionally nonessential amino acids are termed "nonessential" or "dispensable" amino acids.

Nutritionally nonessential amino acids are more important to the cell than the nutritionally essential ones.

The essential amino acids are methionine, tryptophan, valine, leucine, isoleucine, phenylalanine, threonine, lysine, histidine.

Essential Amino Acids

Histidine, isoleucine, leucine, lysine, methionine, phenylalanine, threonine, tryptophan, and valine.

An "essential" or "indispensable" amino acid is defined as one which cannot be synthesized by the organism from substances ordinarily present in the diet at a rate commensurate with certain physiological requirements.

Certain of these essential amino acids are replaced by the corresponding α-keto acids or α-hydroxy acids. Ten amino acids are required for the optimal growth of animals found in the experiments on white rats. But in case of humans, nine essential amino acids are required for the optimal growth of the young and for the maintenance of nitrogen equilibrium in the adult. These nine essential amino acids are : Histidine, methionine, trytophan, valine, phenylalanine, leucine, isoleucine, threonine and lysine. Two amino acids, arginine and histidine, which are required for animals, are "nutritionally semi-essential" for humans because they may be synthesized in tissues at rates inadequate to support growth of children.

Certain nonessential amino acids in the diet serve as the sparing action of certain essential amino acids, eg, tyrosine spares phenylalanine and cystine spares methionine. In phenyl ketonuric individuals, who are unable to convert phenylalanine into tyrosine, the latter becomes an essential amino acid.

If a single essential amino acid is omitted from the group and fed separately several hours later, the nutritional effectiveness of the entire group is impaired.

The omission of an essential amino acid from the diet results in the negative nitrogen balance or decrease of growth.

Nutritionally nonessential amino acids formed from amphibolic intermediates

Alanine : Alanine is formed from pyruvate by transamination in presence of the coenzyme pyridoxal phosphate (B_6–PO_4) in all forms of life.

Glutamic acid : In all forms of life, glutamic acid is formed from α-ketoglutarate by glutamate dehydrogenase.

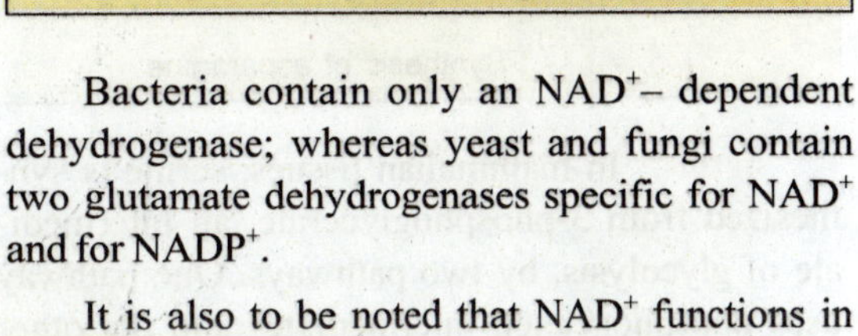

Fig. 20.9 : Synthesis of glutamic acid

Bacteria contain only an NAD^+– dependent dehydrogenase; whereas yeast and fungi contain two glutamate dehydrogenases specific for NAD^+ and for $NADP^+$.

It is also to be noted that NAD^+ functions in glutamate catabolism and $NADP^+$ in glutamate biosynthesis in animals.

In many bacteria glutamate is formed by glutamate synthetase which is given in Fig. 20.10.

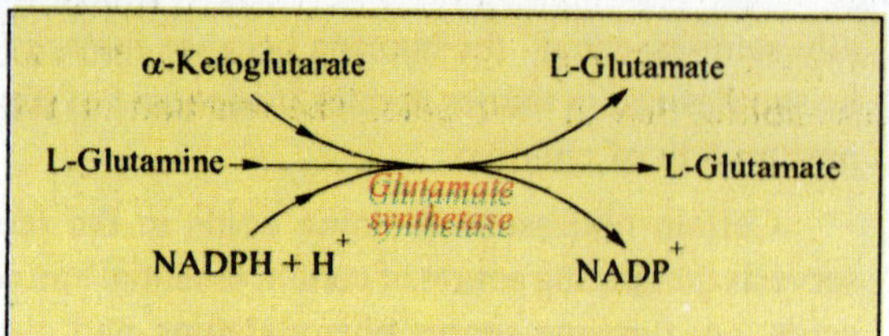

Fig. 20.10 : Formation of glutamate

Aspartic acid : Aspartic acid is formed by transamination of oxaloacetate.

Glutamine : In plants, animals and bacteria, glutamine is synthesized from glutamate by glutamine synthetase. NH_4^+ aminates glutamate requiring ATP and Mg^{++} which is shown in Fig. 20.11

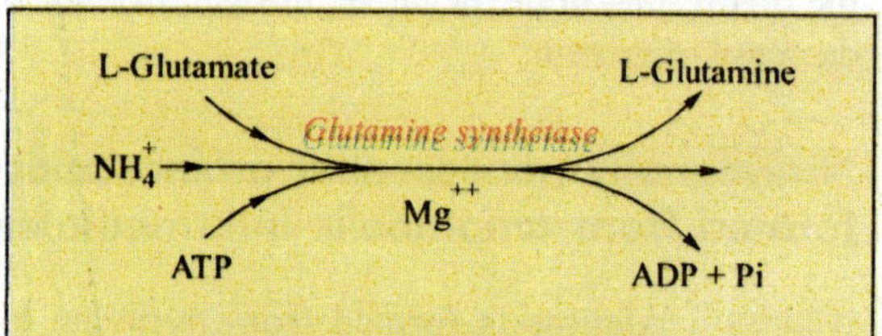

Fig. 20.11 : Synthesis of glutamine

Asparagine : Asparagine is synthesized from aspartate by asparagine synthetase. ATP and Mg^{++} are also required in this reaction. ATP is hydrolyzed to AMP + PPi.R-NH_2 aminates aspartate which is shown in Fig. 20.12.

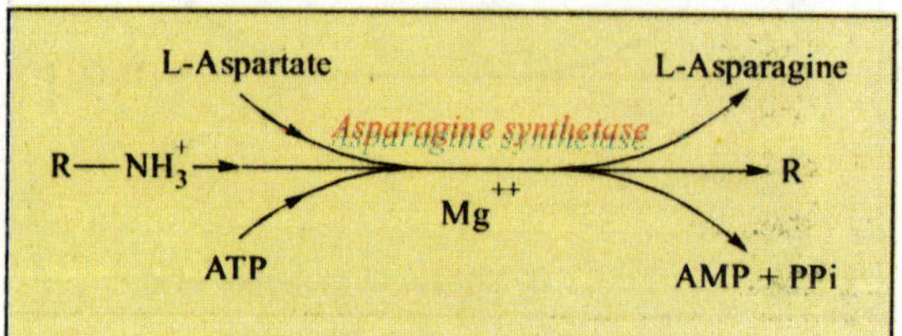

Fig. 20.12 : Synthesis of asparagine

Serine : In mammalian tissues, serine is synthesized from 3-phosphoglycerate, an intermediate of glycolysis, by two pathways. One pathway uses phosphorylated intermediates and the other uses nonphosphorylated intermediates. Majority of the serine is synthesized by the pathway via phosphorylated intermediates. Plants and microorganisms follow this pathway.

Synthesis via phosphorylated intermediates

3-phosphoglycerate is oxidized to phosphohydroxypyruvate which by transamination is converted to phosphoserine. Finally, phosphoserine is converted to serine by phosphatase.

Synthesis via nonphosphorylated intermediates

3-phosphoglycerate is dephosphorylated to glycerate by a phosphatase. Glycerate is oxidized to hydroxypyruvate which is finally transaminated to form serine. The reactions are shown in Fig. 20.13.

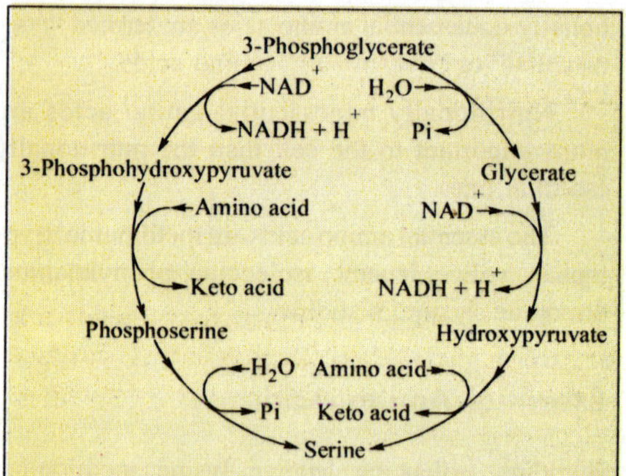

Fig. 20.13 : Synthesis of nonphosphorylated intermediates

Glycine : Glycine is synthesized from serine as well as choline as shown in Fig. 20.14.

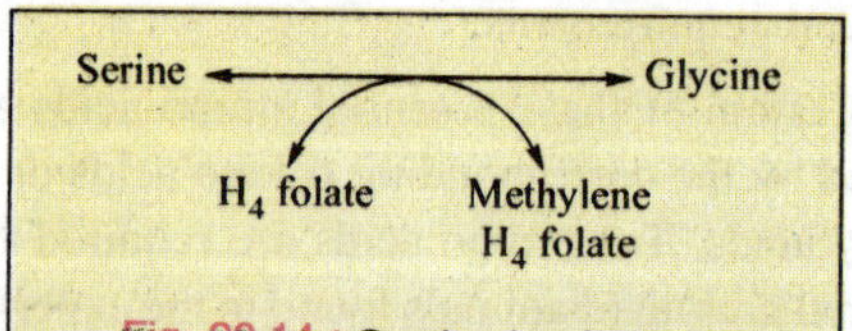

Fig. 20.14 : Synthesis of glycine

Nutritionally nonessential amino acids formed from other nutritionally nonessential amino acids

Proline : Proline is synthesized from glutamate by reversal of reactions for proline catabolism.

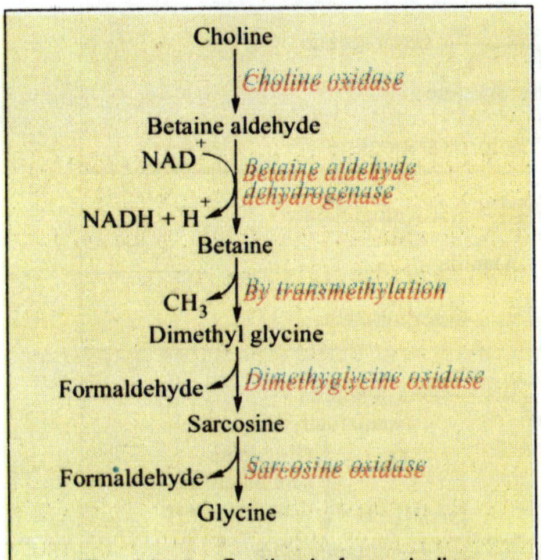

Fig. 20.15 : Synthesis from choline

Hydroxyproline : Since proline serves as a precusor of hydroxyproline, this is also synthesized from glutamate.

Nutritionally nonessential amino acids formed from nutritionally essential amino acids

Cysteine : Cysteine is formed from methionine (essential amino acid). Methionine is first converted to homocysteine which is converted to cysteine in conjugation with serine :

$$\text{Homocysteine} + \text{Serine} \xrightarrow{H_2O} \text{Cystathionine} \xrightarrow{H_2O} \text{Homoserine} + \text{Cysteine}$$

Tyrosine : The conversion of phenylalanine (an essential amino acid) to tyrosine is catalyzed by *phenylalanine hydroxylase complex*, a mixed function oxygenase present in mammalian liver but absent from other tissues. One atom of molecular oxygen is incorporated into the para position of phenylalanine and the other atom is reduced forming water (shown in Fig. 20.16). The reducing power, supplied ultimately by NADPH, is immediately provided as tetra-hydrobiopterine, a pteridine resembling that in folic acid. The reaction is not reversible.

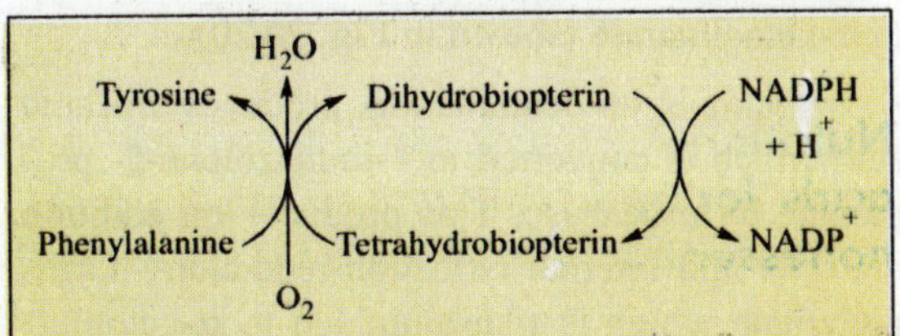

Fig. 20.16 : Synthesis from choline

Nutritionally essential amino acids

Nutritionally essential amino acids are synthesized by bacteria but the synthesis does not take place in mammalian tissues; hence the synthesis is not discussed here.

CATABOLISM OF AMINO ACIDS

List of amino acids converted to carbohydrate and fat or both :

Glycogen ["Glycogenic" L-amino acids]	Fat ["Ketogenic" L-amino acids] From Acetyl-CoA	Both Glycogen and Fat ["Glycogenic" and "Ketogenic" L-amino acids]
1. Alanine	1. Leucine	1. Isoleucine
2. Arginine		2. Lysine
3. Aspartate		3. Phenylalanine
4. Cystine		4. Tyrosine
5. Glutamate		5. Tryptophan
6. Glycine		
7. Histidine		
8. Hydroxyproline		
9. Proline		
10. Methionine		
11. Serine		
12. Threonine		
13. Valine		

L-amino acids are catabolized to amphibolic intermediates.

Amino acids forming oxaloacetate

Asparagine and aspartate are converted to oxaloacetate by the successive actions of asparaginase and a transaminase (shown in Fig. 20.17).

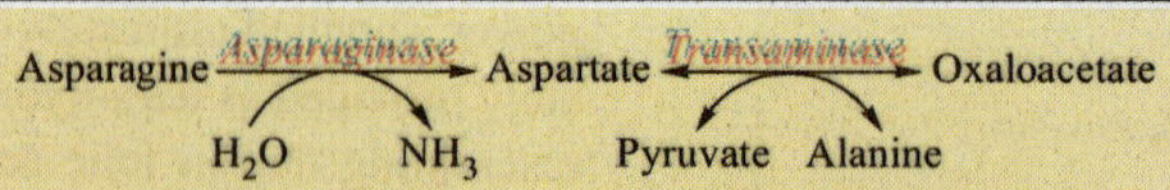

Fig. 20.17 : Conversion of asparagine to oxaloacetate

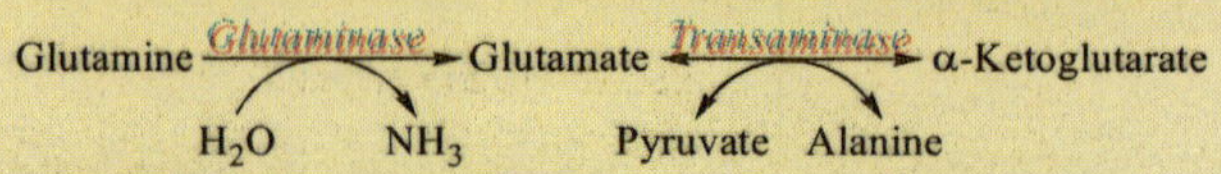

Fig. 20.18 : Conversion of glutamine to α-ketoglutarate

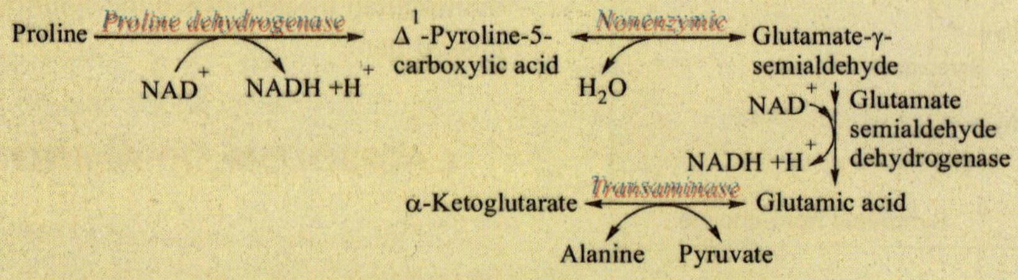

Fig. 20.19 : Conversion of proline to α-ketoglutarate

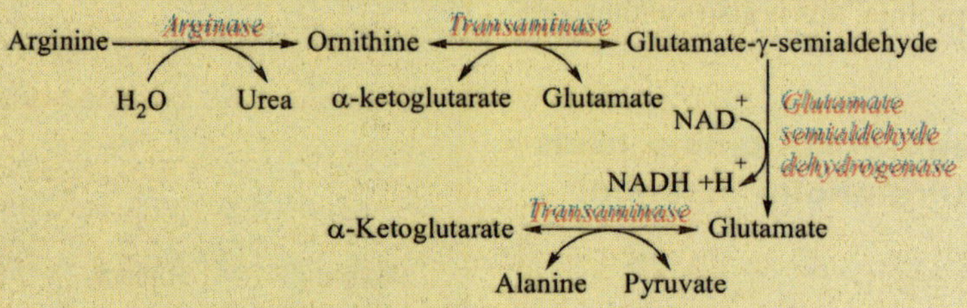

Fig. 20.20 : Conversion of arginine to α-ketoglutarate

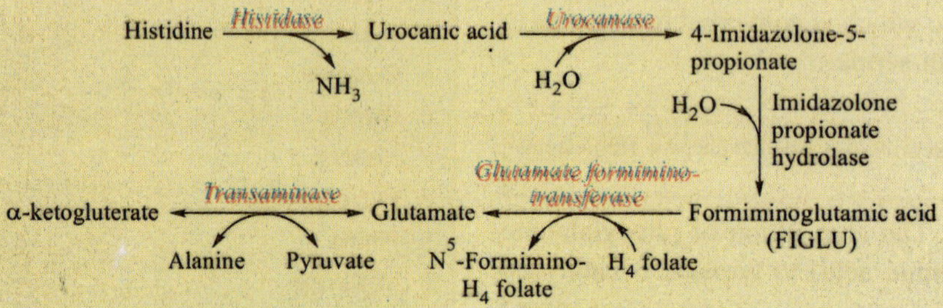

Fig. 20.21 : Conversion of histidine to α-ketoglutarate

Amino acids forming α-ketoglutarate

Glutamine and glutamate are catabolized like that of asparagine and aspartate but with the formation of α-ketoglutarate (shown in Fig. 20.18.).

Proline is oxidized to a form of dehydroproline which, on addition of water, forms glutamate γ-semialdehyde. This is then oxidized to glutamate and transaminated to α-ketoglutarate (shown in Fig. 20.19).

Arginine and histidine are both converted to α-ketoglutarate. Arginine is converted to ornithine by arginase with the removal of urea. Ornithine, by transamination forms glutamate γ-semialdehyde which is oxidized to glutamate and transaminated to α-ketoglutarate (shown in Fig. 20.20).

Histidine, on deamination, produces urocanic acid which is converted to 4-imidazolone-5- propionate by urocanase. This product—on addition of water and internal oxidation-reduction—forms glutamate which is transaminated to α-ketoglutarate (shown in Fig. 20.21).

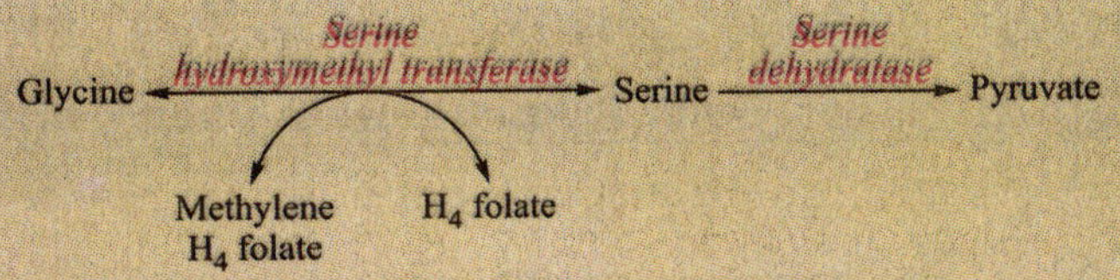

Fig. 20.22 : Conversion of glycine to pyruvate

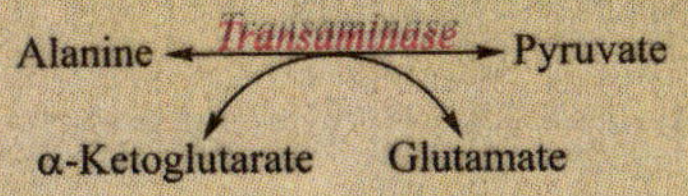

Fig. 20.23 : Conversion of alanine to pyruvate

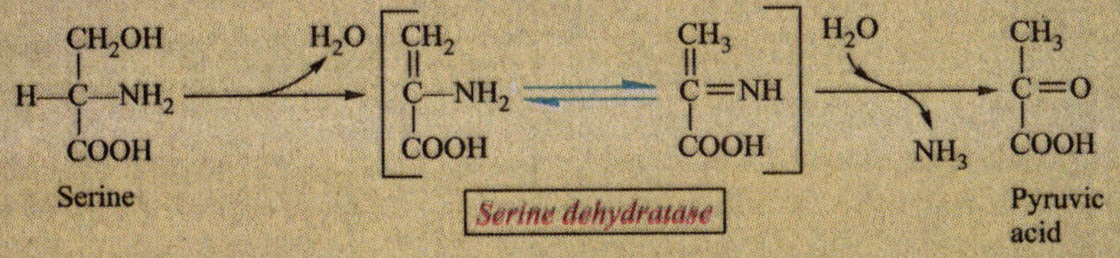

Fig. 20.24 : Conversion of serine to pyruvate

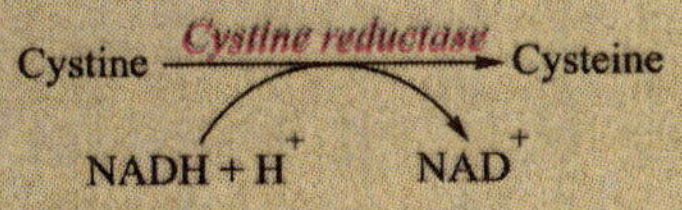

Fig. 20.25 : Cysteine formation

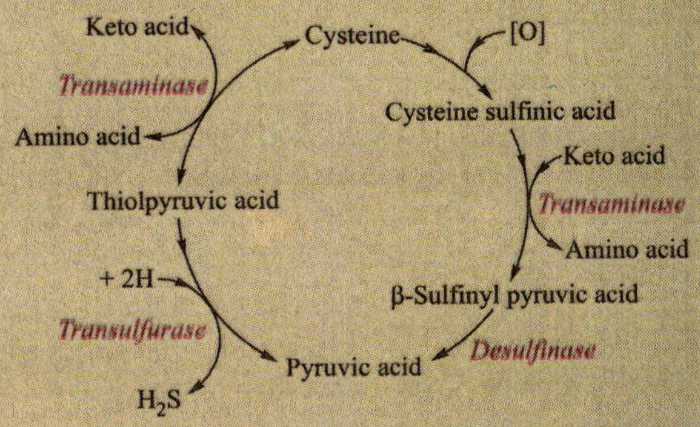

Fig. 20.26 : Conversion of cysteine to pyruvate

Amino acids forming pyruvate

Glycine is converted to serine by serine **hydroxymethyltransferase**. Serine then forms pyruvate by serine dehydratase (shown in Fig. 20.22).

Alanine forms pyruvate by transamination—Fig. 20.23.

Serine is converted to pyruvate by serine dehydratase, a pyridoxal phosphate protein. Addition and loss of water as well as loss of ammonia are involved in this reaction. The reaction is shown in Fig. 20.24.

This conversion of serine to pyruvate is prominent in the liver tissue of rats and guinea-pigs because **serine dehydratase** is rich in this tissue of these animals.

But in humans and many other vertebrates, serine is degraded to glycine by serine hydroxymethyltransferase. The further catabolism follows the glycine cleavage system.

Cystine is converted to cysteine by an NADH-dependent cystine reductase as shown in Fig. 20.25.

Cysteine is converted to pyruvate by

1. Transamination and loss of H_2S.
2. By oxidation of the sulfhydryl group forming cysteine sulfinic acid, transamination and by desulfination (shown in Fig. 20.26).

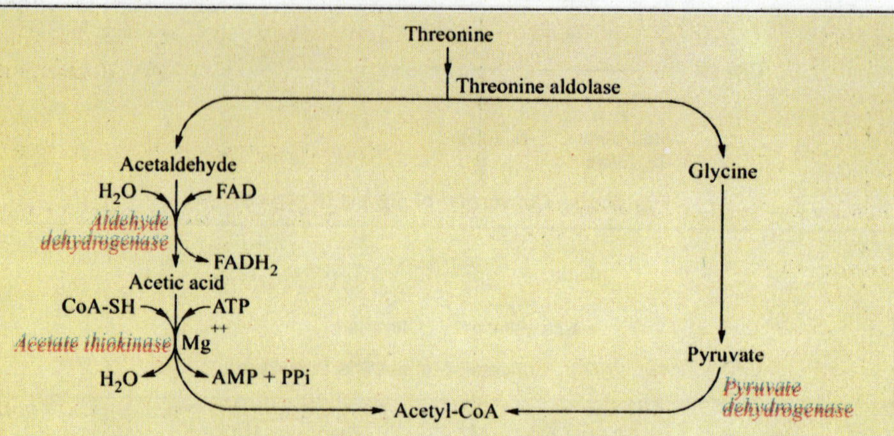

Fig. 20.27 : Conversion of threonine to acetyl-CoA

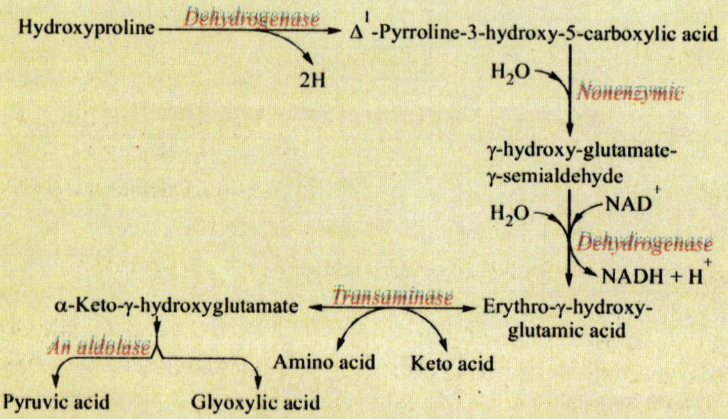

Fig. 20.28 : Conversion of hydroxyproline to pyruvate

Threonine aldolase cleaves threonine to acetaldehyde and glycine. Glycine is catabolized to pyruvate as discussed before. Both pyruvate and acetaldehyde then form acetyl-CoA (shown in Fig. 20.27).

Hydroxyproline is converted to pyruvate and glyoxylate. The conversion is indicated in Fig. 20.28.

AMINO-ACIDS FORMING ACETYL-CoA

Amino acids forming pyruvate are convertible to acetyl-CoA. In addition to this, 5 amino acids form acetyl-CoA directly, without first forming pyruvate.

PHENYLALANINE

Phenylalanine (an essential amino acid) is converted to tyrosine by phenylalanine hydroxylase; tetrahydropteridine, NADPH and O_2 are required. The reaction is not reversible.

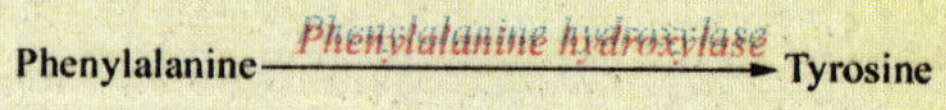

Fig. 20.29 : Tyrosine formation

The catabolism of phenylalanine and tyrosine may be discussed under the following heads :

1. Formation of fumarate and acetoacetate.
2. Formation of melanin.
3. Conversion to epinephrine.
4. Conversion to thyroxine.

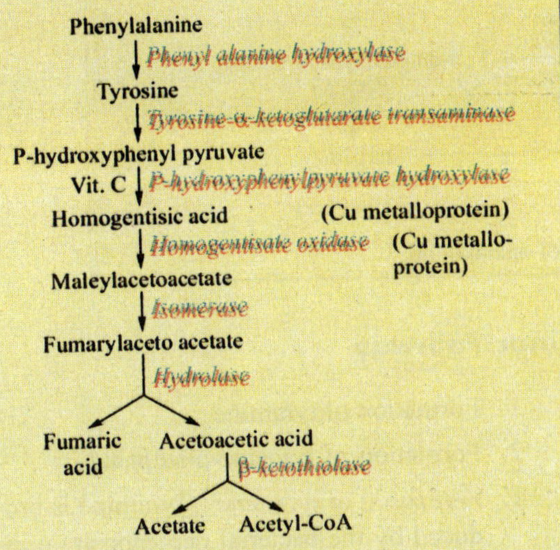

Fig. 20.30 : Formation of fumarate and acetoacetate

Major pathway

1. Formation of fumarate and acetoacetate

1. Tyrosine is transaminated to P-hydroxyphenylpyruvate by tyrosine-α-ketoglutarate transaminase, an enzyme of mammalian liver tissue, as shown in Fig. 20.30.
2. P-hydroxyphenylpyruvate hydroxylase, a copper metalloprotein with properties similar to those of tyrosinase, converts P-hydroxyphenylpyruvate to homogentisic acid. Ascorbic acid acts as a cofactor in this reaction.
3. By oxidative reaction catalyzed by homogentisate oxidase, an iron metallo-protein of mammalian liver, maleylacetoacetate, is formed by the rupture of the benzene ring of homogentisate. The reaction is inhibited by a chelating agent that binds iron.
4. Maleylacetoacetate is converted to fumarylacetoacetate by maleylacetoacetate cistrans isomerase, present in mammalian liver.
5. Fumarylacetoacetate on hydrolysis by fumarylacetoacetate hydrolase forms fumarate and acetoacetate. Acetoacetate can be converted to acetyl-CoA and acetate by β-ketothiolase.

2. Formation of melanin

1. Melanin, the black pigment present in the skin, hair and retina of the eye, is formed from phenylalanine and tyrosine in the specialised cells (melanoblasts) present in the skin. The pathway of melanin formation is given (Fig. 20.31).
2. Tyrosine is oxidized to dihydroxyphenylalanine (DOPA) catalyzed by tyrosinase in presence of ascorbic acid as cofactor.
3. DOPA is converted into dopaquinone which is further converted into 5:6 dihydroxyindole-2-carboxylic acid. This is oxidized to dihydroxyindole which polymerises spontaneously to melanin.

3. Conversion to Epinephrine

1. Phenylalanine is converted into tyrosine which is further converted into 3:4-dihydroxyphenylalanine (DOPA) by tyrosine hydroxylase with tetrahydro-pterdine as cofactor.
2. DOPA is decarboxylated to dopamine by a decarboxylase which is present in many tissues including adrenal medulla with pyridoxal phosphate as cofactor.

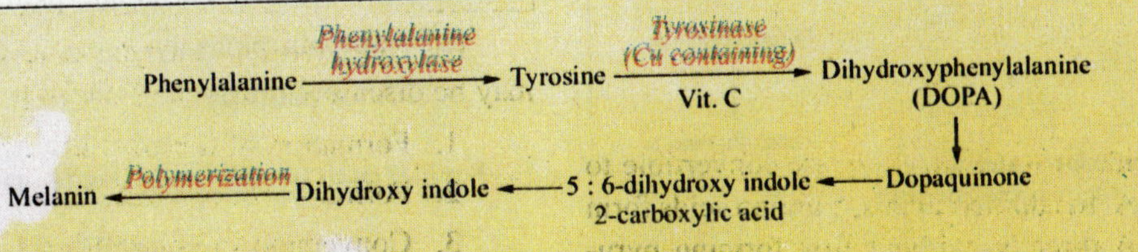

Fig. 20.31 : Formation of melanin

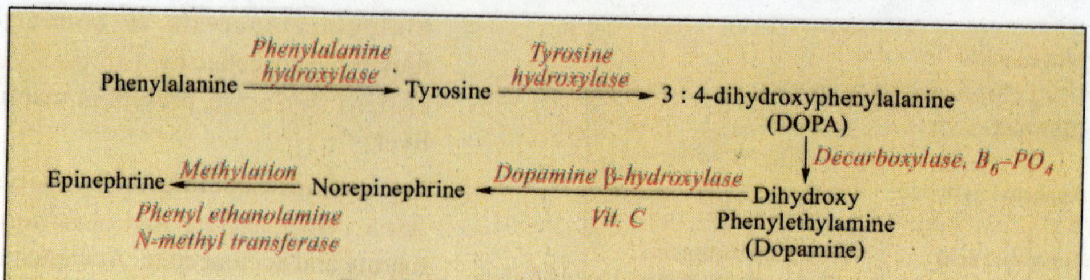

Fig. 20.32 : Formation of epinephrine

3. The hydroxylation of dopamine is carried out by **dopamine β-hydroxylase** to form norepinephrine in the presence of ascorbic acid and molecular oxygen.
4. The methylation from methionine converts norepinephrine to epinephrine.

The sequence of reactions are given in Fig. 20.32

4. Conversion to thyroxine

The synthesis of thyroxine takes place in the thyroid gland :

1. Phenylalanine is converted to tyrosine which, on iodination, forms monoiodotyrosine. This, on iodination, forms diiodotyrosine.
2. Coupling of 2 mols of diiodotyrosine yields thyroxine.
3. Coupling of 1 mol of monoiodotyrosine and 1 mol of diiodotyrosine produces triiodothyronine.

The steps are shown :

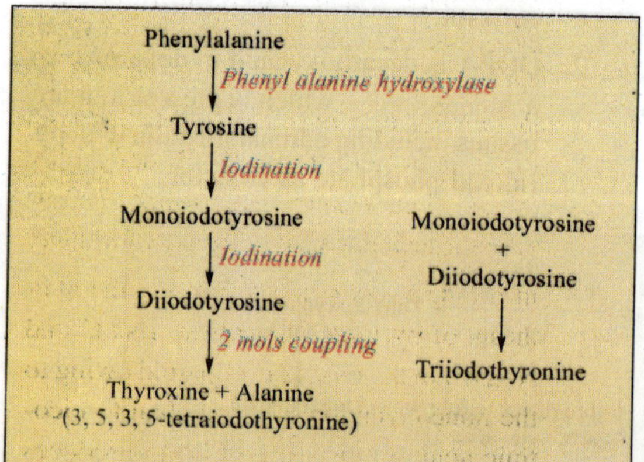

Fig. 20.33 : Formation of thyroxine

Minor Pathway

1. Formation of tyramine.
2. Formation of tyrosine-o-sulphate.

1. **Formation of tyramine :** Tyramine is produced by the bacterial decomposition involving decarboxylation of tyrosine by **decarboxylase** present in bacteria.

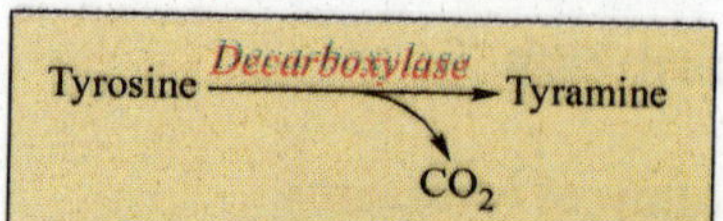

Fig. 20.34 : Formation of tyramine

2. **Formation of tyrosine-o-sulphate :** This is present in fibrinogen. Two peptides are liberated during the conversion of fibrinogen into fibrin. One of the peptides contain tyrosine-o-sulphate.

TRYPTOPHAN

The carbon atoms of the side chain and of the aromatic ring of tryptophan may be completely degraded to amphibolic intermediates. This proceeds via **kynurenine-anthranilate pathway.** This pathway is important for degradation of tryptophan as well as **niacin** formation from tryptophan.

The catabolism of tryptophan may be discussed under the following pathways :

1. Kynurenine-anthranilate pathway.
2. Serotonin pathway.
3. Minor pathways.

1. Kynurenine-anthranilate pathway

1. **Tryptophan oxygenase (Tryptophan pyrrolase)** cleaves the indole ring of tryp-

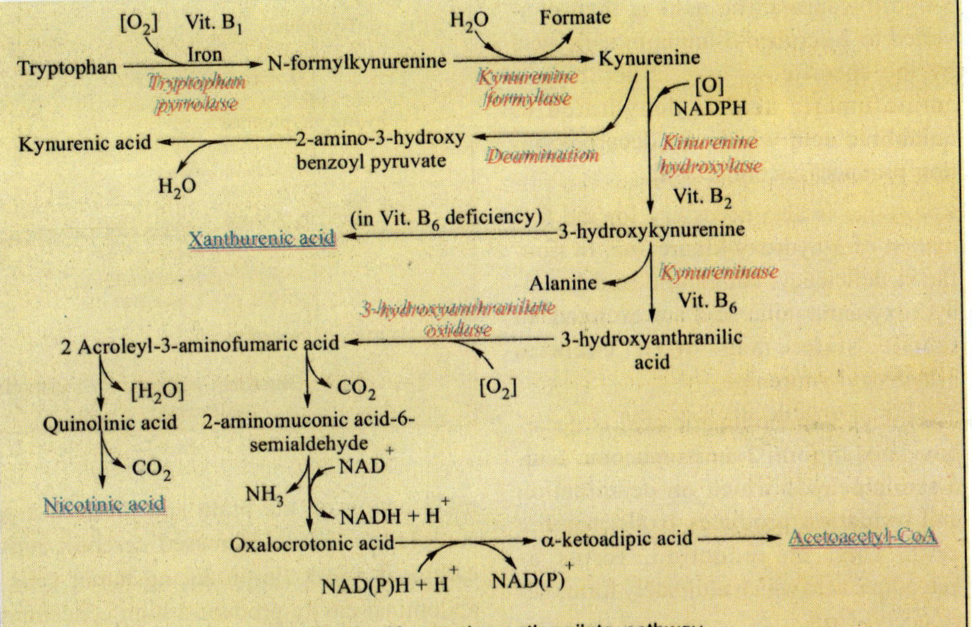

Fig. 20.35 : Kynurenine-anthranilate pathway

tophan with the incorporation of 2 atoms of molecular oxygen forming N-formylkynurenine. The oxygenase enzyme is an iron porphyrin metalloprotein which is present in the liver of mammals, amphibians, birds and insects. Four forms of hepatic tryptophan pyrrolase have been described : the active holoenzyme, the apoenzyme, third form is combined with hematin and the fourth form requires prolonged incubation. The chief inducing agents of tryptophan pyrrolase is adrenal corticosteroids, induction is blocked by puromycin. Tryptophan stabilizes the enzymes towards proteolytic degradation.

2. Kynurenine formylase of mammalian liver catalyzes the hydrolytic removal of the formyl group of N-formyl-kynurenine producing kynurenine.
3. Kynurenine on deamination produces 2-amino-3-hydroxybenzoyl pyruvate which loses water and then undergoes spontaneous ring closure forming kynurenic acid. This is not formed in the main pathway of tryptophan breakdown.
4. Kynurenine is hydroxylated by kynurenine hydroxylase with molecular oxygen in presence of NADPH to 3-hydroxy-kynurenine.
5. 3-hydroxyanthranilic acid is formed from 3-hydroxy kynurenin by the enzyme kynureninase which requires vitamin B_6 (pyridoxal phosphate) as coenzyme. In the deficiency of vitamin B_6, kynurenine derivatives reach the extrahepatic tissues where they are converted to xanthurenic acid. This is found in the urine of human, monkeys and rats when there is dietary deficiency of vitamin B_6. Excess tryptophan feeding can induce excretion of xanthurenic acid if vitamin B_6 deficiency exists. The kidney can produce xanthurenic acid derivatives from kynurenine.

   In the deficiency of vitamin B_6, the synthesis of pyridine nucleotides (NAD and NADP) in the tissues is impaired owing to the nonconversion of tryptophan to nicotinic acid. 60 mg of tryptophan produces 1 mg of nicotinic acid.

6. 3-hydroxyanthranilic acid is then converted to 2-acroleyl-3-aminofumaric acid by the specific oxidase. 2-acroleyl-3-aminofumaric acid is dehydrated to quinolinic acid which, on decarboxylation produces nicotinic acid.
7. Riboflavin is also necessary for the formation of 3-hydroxy-kinurenine. In riboflavin deficiency, anthranilic acid and 5-hydroxy-anthranilic acid are excreted. In febrile state, man often excretes 3-hydroxykynurenine.
8. 2-acroleyl-3-aminofumaric acid is decarboxylated to form 2-aminomuconic acid-6-semialdehyde which on deamination and oxidation produces oxalocrotonic acid. This, on reduction, forms a-ketoadipic acid which ultimately forms acetoacetyl-CoA.

The sequence of reactions is shown (Fig. 20.35).

In the deficiency of riboflavin

Kynurenine ⟶ Anthranilic acid ⟶ 5-hydroxy-anthranilic acid

2. Serotonin pathway

Tryptophan on hydroxylation in the liver forms 5-hydroxytryptophan which, on decarboxylation, produces 5-hydroxytryptamine (serotonin), a stimulant of the central nervous system and also a vasoconstrictor. Serotonin is stored in platelets. It also occurs in intestinal mucosa where it promotes peristalsis. Serotonin is broken down by oxidative deamination to 5-hydroxyindoleacetic acid which is excreted in the urine.

The sequence of reaction is given in Fig. 20.36.

Trytophan
↓ Hydroxylation (Liver)
5-hydroxytryptophan
↓ Decarboxylation
5-hydroxytryptamine (Serotonin) ⟶ (Acetylation / Methylation) ⟶ Melatonin
↓ Oxidative deamination/ Mono amino oxidase
5-hydroxyindoleacetic acid

Fig. 20.36 : Serotonin pathway for tryptophan catabolism

Serotonin on acetylation and methylation produces N-acetyl-5-methoxytryptamine (the hormone melatonin).

Serotonin deficiency produces depressant effect. Drugs like Ipronizide which inhibit the enzyme mono-amine oxidase (MAO), prolong serotonin action on the brain and produce a psychic stimulation due to increased cerebral activity in carcinoid, serotonin-producing tumor cells of the abdominal cavity produce a high concentration of serotonin N-acetylserotonin glucuronide; and 5-hydroxyindoleaceturate are present in the urine of patients with carcinoid. The enhanced conversion of tryptophan to serotonin reduces niacin synthesis producing symptoms of pellagra in patients with carcinoid.

1. Ten forms of human oculocutaneous albinism are differentiated on the basis of their clinical, biochemical, ultrastructural and genetic characteristics exhibiting decreased pigmentation of the skin and eye. A few are mentioned :

A. Tyrosine hydroxylase-negative albinos :
   (i) They lack all visual pigment.
   (ii) Hair bulbs from these patients fail to convert added tyrosine to pigment.
   (iii) The melanocytes contain unpigmented melanosomes.

B. Tyrosine hydroxylase-positive albinos :
   (i) They have some visible pigment and white-yellow to light tan hair.
   (ii) Their hair bulb melanocytes may contain highly pigmented melanosomes, which convert tyrosine to black melanin in vitro.

C. Ocular albinism :

(i) This occurs as an autosomal recessive and as an X-linked trait.

(ii) The melanocytes of X-linked and heterozygous ocular albinos contain macro-melanosomes.

(iii) The retinas of females heterozygous for X-linked ocular albinism exhibit a mosaic pattern of pigment distribution due to random X-chromosome inactivation.

2. This disease is involved in the defects of absorption of neutral amino acids.

3. Tryptophan metabolism is diverted to serotonin after the administration of the drug isoniazid and tryptophan absorption is also impaired in hartnup disease.

3. Minor pathways

Minor pathways include :

1. Formation of kynurenic acid.
2. Formation of tryptamine.
3. Formation of indoleacetic acid.

1. **Formation of kynurenic acid** : Mentioned in the kynurenic-anthranilate pathway (Fig. 20.35)

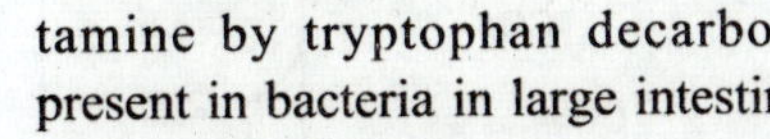

Fig. 20.37 : Formation of tryptamine

2. **Formation of tryptamine** : Small quantities of tryptophan are converted into tryptamine by tryptophan decarboxylase present in bacteria in large intestine.

3. **Formation of indoleacetic acid** : Oxidative deamination converts tryptophan to indolepyruvic acid which, in turn, is oxidized to indoleacetic acid. Indoleacetic acid is further converted to indole, skatole and indoxyl by intestinal bacteria. The reactions are shown in Fig. 20.38.

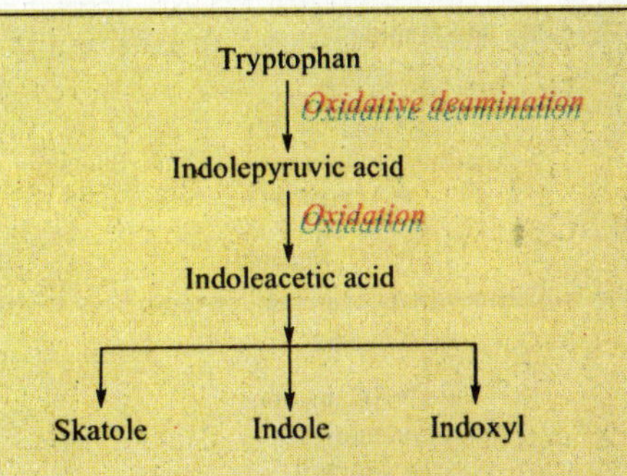

Fig. 20.38 : Conversion of tryptophan (minor pathways)

AMINO ACIDS FORMING SUCCINYL-CoA

Methionine

1. Methionine is first activated by ATP forming S-adenosylmethionine ("active methionine").

2. After removal of the methyl group, the active methionine is converted into S-adenosylhomocysteine. This is hydrolyzed to produce homocysteine and adenosine.

3. Homocysteine then condenses with a molecule of serine forming cystathionine which produces homoserine and cysteine on hydrolytic cleavage.

4. Homoserine is then converted to α-ketobutyric acid by the enzyme homoserine deaminase.

5. The oxidative decarboxylation converts a-ketobutyrate to propionyl-CoA.

6. Propionyl-CoA is converted to D-methyl malonyl-CoA by propionyl CoA carboxylase in presence of ATP and biotin as Co-enzyme.

7. **Methylmalonyl-CoA recemase** converts D-methyl malonyl-CoA to L-methylmalonyl-CoA which in turn, is converted into succinyl-CoA by the specific isomerase with B_{12} as coenzyme.

The reactions are given (Fig. 20.39).

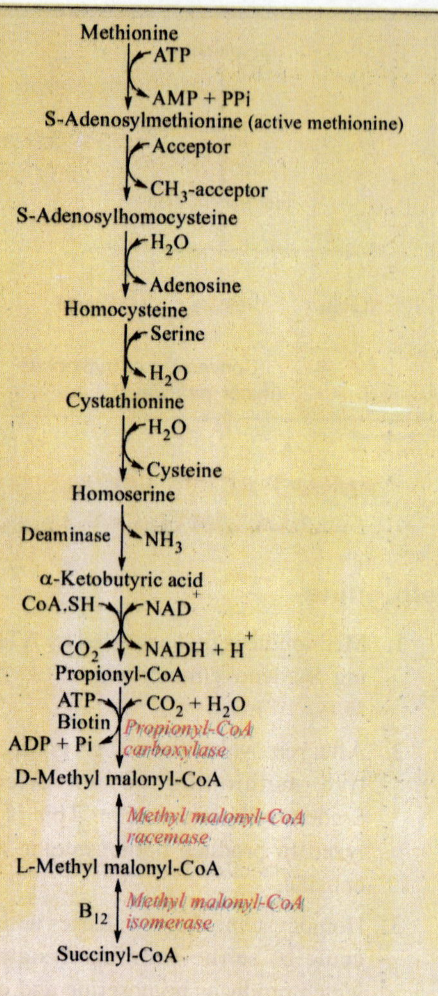

Fig. 20.39 : Catabolism of methionine

Transmethylation

1. The transfer of CH_3 group from a compound of the body to a suitable acceptor which has no CH_3 group.

   Such a reaction is said to be transmethylation reaction.

2. The CH_3 donor is biologically labile $-CH_3$ group. The most important compounds with biologically labile methyl groups are
   (a) Active methionine containing $-S-CH_3$ group.
   (b) Choline containing $N^+ CH_3CH_3CH_3$.
   (c) Betaine, an oxidative derivative of choline.

Leucine

1. Leucine is transaminated to produce α-ketoisocaproic acid, which, on decarboxylation, yields isovaleryl-CoA. This on dehydrogenation produces β-methylcrotonyl-CoA.
2. Carboxylation converts β-methylcrotonyl-CoA into β-methylglutaconyl-CoA which in turn, is converted to β-hydroxy-β-methyl-glutaryl-CoA on hydration. This product is cleaved into acetoacetate plus acetyl-CoA.

The reactions are given below (Fig. 20.40).

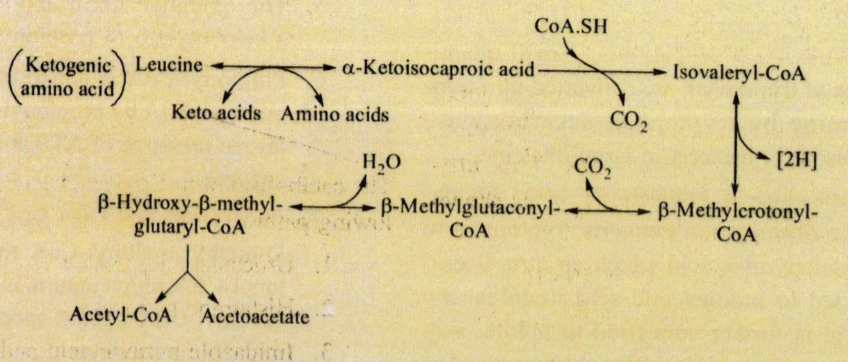

Fig. 20.40 : Catabolism of leucine

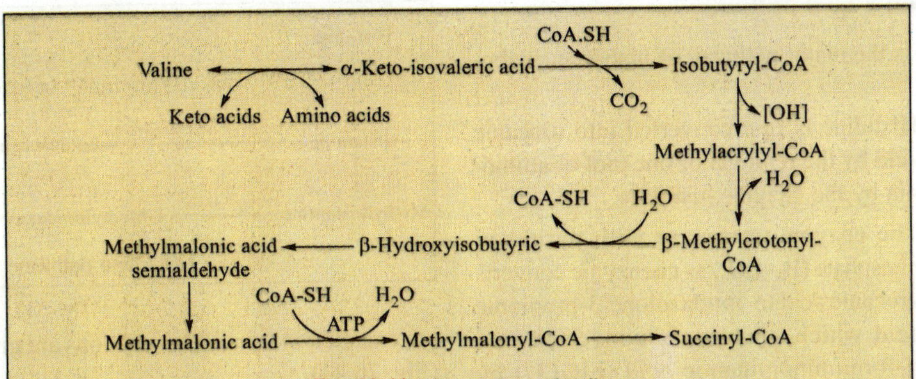

Fig. 20.41 : Catabolism of valine

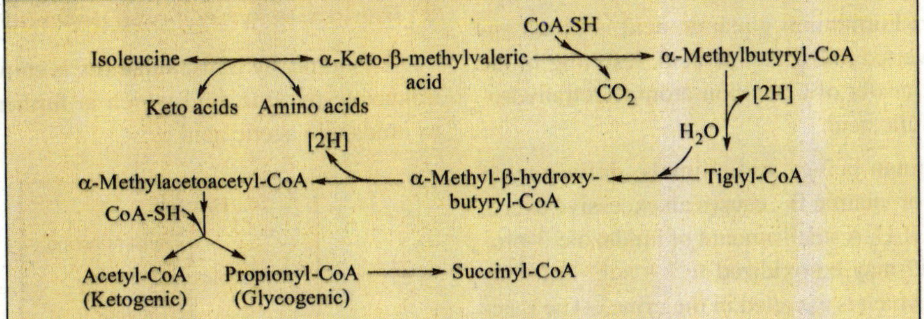

Fig. 20.42 : Catabolism of isoleucine

Valine

1. Valine, on transamination, produces α-keto-isovaleric acid which, on decarboxylation yields isobutyryl-CoA. This, on dehydrogenation, produces methyl-acrylyl-CoA.
2. Methylacrylyl-CoA is hydrated to produce β-hydroxyisobutyryl-CoA which, on hydrolysis, gives β-hydroxyisobutyric acid. This is converted into methylmalonic acid semialdehyde.
3. The semialdehyde is oxidized to yield methylmalonic acid which, on acylation, yields methylmalonyl-CoA. This ultimately produces succinyl-CoA as shown in leucine catabolism.

The reactions are given in Fig 20.41.

Isoleucine

1. Isoleucine, on transamination, produces α-keto-β-methylvaleric acid which, on decarboxylation, yields α-methylbutyryl-CoA. This, on dehydrogenation, produces tiglyl-CoA.
2. Tiglyl-CoA is hydrated to produce a-methyl-β-hydroxybutyryl-CoA which, on dehydrogenation, forms α-methylacetoacetyl-CoA. Thiolytic cleavage forms acetyl-CoA plus propionyl-CoA. Propionyl-CoA is converted to succinyl-CoA as mentioned in leucine catabolism.

Catabolism of Histidine

The catabolism of histidine proceeds under the following pathways :

1. Urocanic acid pathway.
2. Histamine pathway.
3. Imidazole pyruvic acid pathway.
4. Minor pathways.

1. Urocanic acid pathway

This is the major pathway of histidine catabolism :

1. Histidine is first converted into urocanic acid by the removal of one mol of ammonia by the enzyme histidase.
2. The enzyme urocanase with pyridoxal phosphate (B_6–PO_4) as coenzyme converts urocanic acid to imidazolone-3-propionic acid which, in turn, is converted into α-formiminoglutamic acid (FIGLU) by hydrolase.
3. α-Formimino glutamic acid is then converted into glutamic acid, resulting in the transfer of C_1 carbon atom to tetrahydrofolic acid.

In human beings and animals, deficency of folic acid or vitamin B_{12} causes an excessive excretion of FIGLU. A small amount of imidazole-3-propionic acid may be oxidized to hydantin-3-propionic acid which is excreted in the urine. The reactions are stated below (Fig. 20.43).

2. Histamine Pathway

Histidine is decarboxylated by histidine decarboxylase to form histamine. This is formed at a significant rate in fetal and regenerating liver as well as in healing skin wounds.

Histamine is detoxified as its methyl or acetyl derivatives or to oxidize to imidazole acetate which

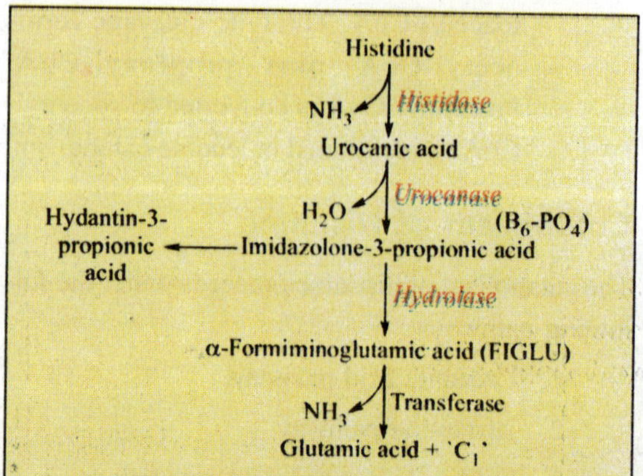

Fig. 20.43 : Urocanic acid pathway

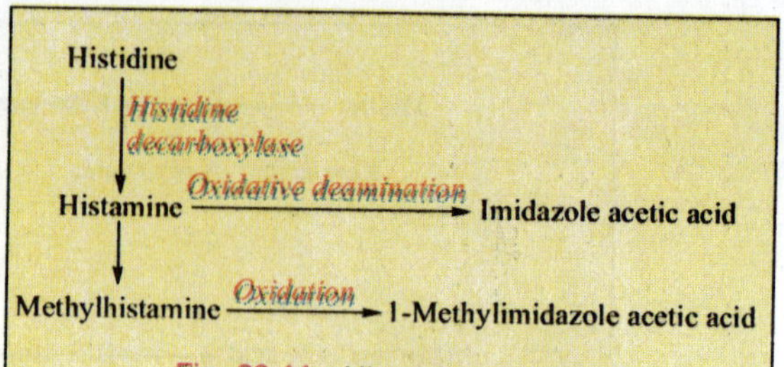

Fig. 20.44 : Histamine pathway

are excreted in urine. The reactions are given above (Fig. 20.44).

3. Imidazole Pyruvic acid pathway

Histidine, by transamination, is converted into imidazole pyruvic acid which is further oxidized to imidazole acetic acid as :

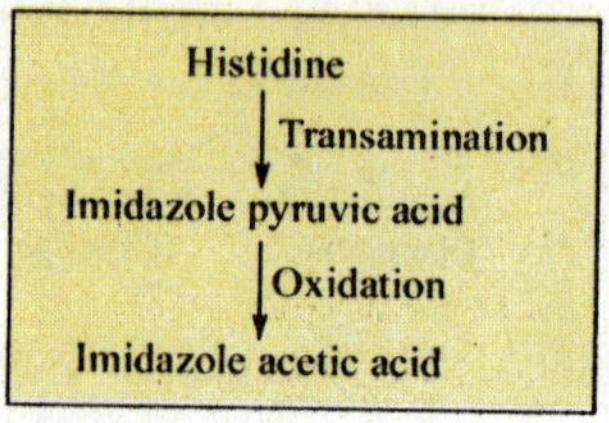

4. Minor pathways

(a) Histidine is converted into carnosine and anserine by minor pathways. These compounds are found in muscle. Carnosine is a peptide of histidine and β-alanine; whereas aniserine is a peptide of methylhistidine and β-alanine.

2. Ergothionine, the betaine of 2-mercaptohistidine, is present in high concentration in human erythrocytes. It is also found in liver and brain.

Metabolism of Glycine

Glycine is non-essential, but it is an important amino acid since it produces significant biological compounds in the body.

1. Deamination

The enzyme glycine oxidase (a flavo-protein

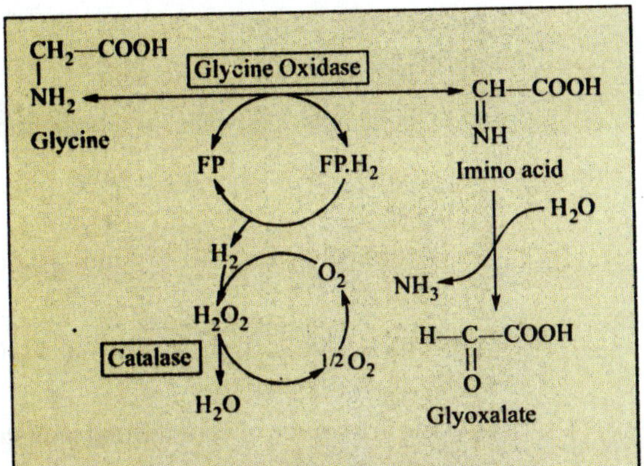

Fig. 20.45 : Glyoxalate formation

enzyme) available in liver and kidney deaminates glycine to produce glyoxalate (see Fig. 20.45).

Glyoxalate is further converted to either oxalic acid or formic acid which enters "One carbon Pool".

2. Formation of CO_2, NH_3 and N^5, N^{10}-Methylene FH_4

Glycine is catabolized by the enzyme glycine synthase complex, found as aggregate in mitochondria, to CO_2, NH_3 and N^5, N^{10}- methylene FH_4. This reversible reaction occurs in liver of most vertebrates including humans (Fig. 20.46).

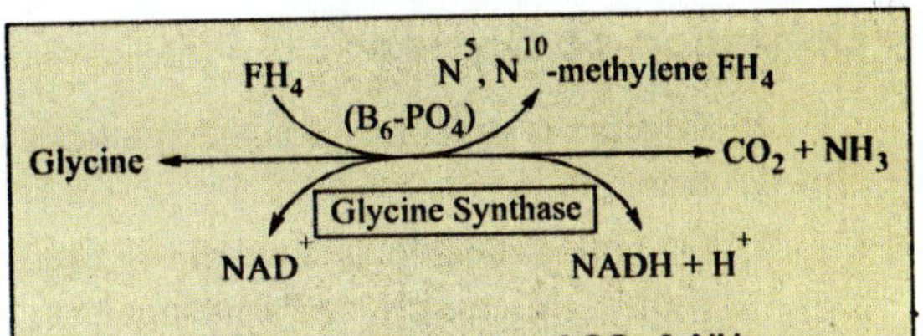

Fig. 20.46 : Formation of CO_2 & NH_3

3. Formation of Pyruvic acid through serine

Glycine is first converted to serine which is converted into pyruvic acid by non-oxidative deamination. So glycine is a glycogenic amino acid (Fig. 20.47).

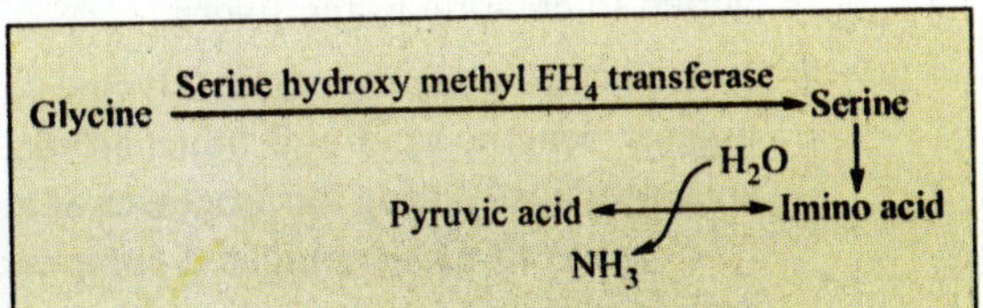

Fig. 20.47 : Pyruvic acid formation

4. Formation of amino acetone

Glycine is oxidized to amino acetone by the enzyme amino-acetone synthetase in presence of acetyl-CoA. The amino-acetone is further metabolized to lactic acid and pyruvic acid via methyl glyoxal (Fig. 20.48).

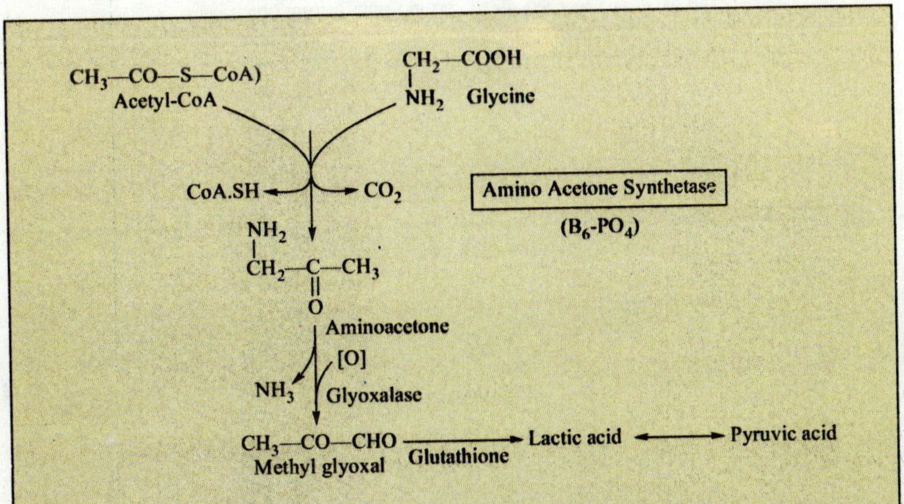

Fig. 20.48 : Formation of aminoacetone

Utilization of Glycine

1. Glycine is necessary for the synthesis of heme.
2. Glycine molecule as a whole is used to form C_4, C_5, and N_7 of Purine nucleus.
3. It is required in the formation of glutathione, a tripeptide, consisting of other amino acids, cysteine and glutamic acid in addition to Glycine.
4. The enzyme transamidinase present in kidney catalyzes the conversion of glycine and arginine to Glycocyamine (Guanido acetic acid) which is further converted to creatine in the liver.
5. It acts as a conjugating agent. Glycine conjugates benzoic acid and other aromatic acid, produced during the processes of metabolism to form a conjugated compound before excreted in urine.

   (a) See Fig. 20.49.

   (b) Cholesterol → Cholic acid

   Cholic acid + Glycine → Glycocholic acid

   ↓

   Excreted in Bile ← Sodium glycocholate

6. Glycine is converted to pyruvic acid. Hence, it is glycogenic.
7. Glycine is a source of formate and oxalate.

Inherited disorder of Glycine metabolism

1. Glycinuria – See in inborn errors of Protein metabolism.
2. Primary hyperoxaluria in inborn errors of Protein metabolism (Fig. 20.50).

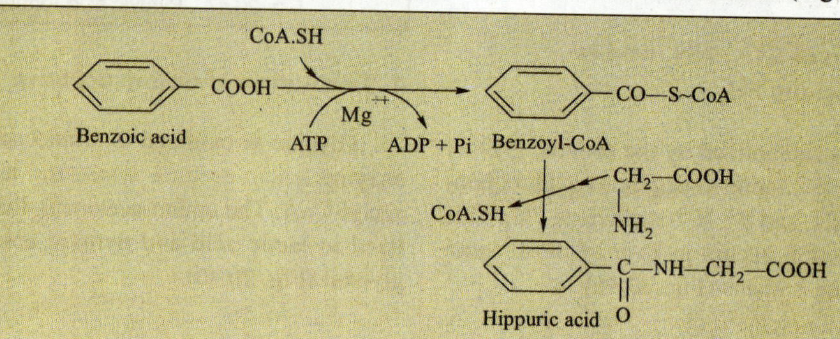

Fig. 20.49 : Hippuric acid formation

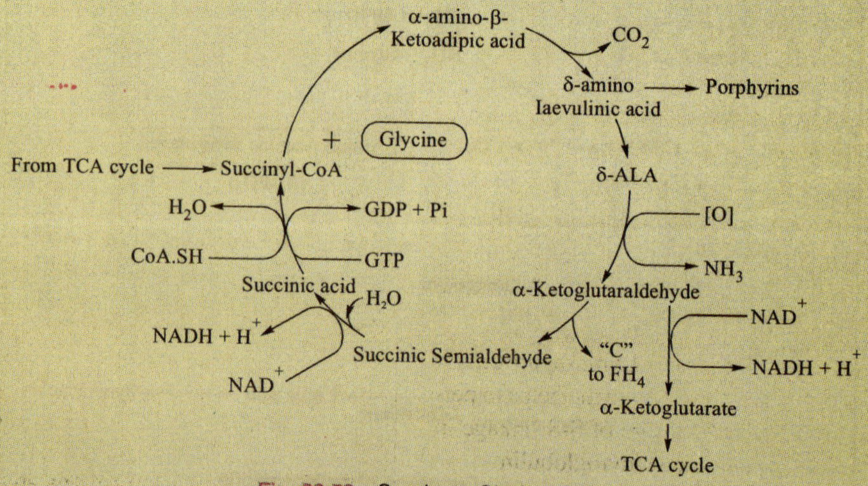

Fig. 20.50 : Succinate-Glycine Cycle

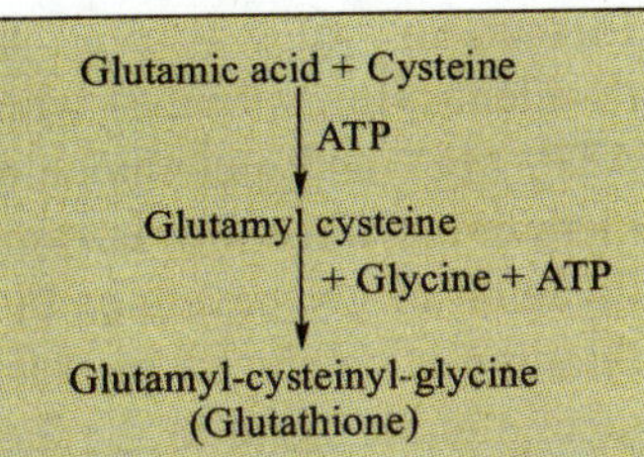

Fig. 20.51 : Synthesis of glutathione

GLUTATHIONE

Formation

Glutathione is a tripeptide of glycine, cysteine and glutamic acid. Its synthesis is shown in Fig. 20.51.

Functions

1. It acts as a coenzyme for the enzyme glyoxalase which converts methyl glyoxal to lactic acid.
2. It acts as a coenzyme for formaldehyde dehydrogenase which converts formaldehyde to formic acid.
3. It acts as coenzyme for maleyl acetoacetate isomerase.
4. It is involved as a coenzyme in the γ-glutamylacyl transferase for amino acid transport in some cells.
5. As a coenzyme, it participates in the formation of prostaglandin, PGE_2 from arachidonic acid.
6. It takes part as a coenzyme in the prevention of peroxidation of poly-unsaturated fatty acids in tissues.
7. Insulin is catabolized and degraded by the liver enzyme glutathione-insulin transhydrogenase with the help of glutathione as coenzyme.
8. The enzyme glutathione transhydrogenase with glutathione as coenzyme helps to perform reductive cleavage of S-S linkage in the glycoprotein and thyroglobulin.
9. Glutathione protects many –SH groups containing enzymes from the oxidation of their – SH groups, eg, Glyceraldehyde–3–Ⓟ dehydrogenase enzyme.
10. The enzyme glyoxylase requiring glutathione as coenzyme converts methyl glyoxal to lactic acid.
11. Oxidized glutathione is harmful for RBC membrane and lens proteins. Oxidized glutathione is reduced back to glutathione by glutathione reductase.
12. Reduced glutathione is required for conversion of dopaquinone to glutathione.
13. Glutathione is required as a coenzyme for liver enzyme for the activation of methionine to form S-adenosylmethionine.

CREATINE AND CREATININE

Chemistry

Creatinine is the anhydride of creatine and a product of endogenous metabolism. Both creatine and creatinine are readily interconverted in solution. Acid favours the formation of creatinine. The structures are shown in Fig. 20.52.

Occurrence

Creatine is widely distributed in animal tissues. It is present in muscle, brain and blood as phosphocreatine and also in the free state. Skeletal muscle contains about 0.5 per cent creatine and heart muscle about half that amount. 98 per cent of the total creatine in the body is in the muscles. Creatinine is formed largely in the muscle by the irreversible and nonenzymic removal of water from creatine phosphate. Traces of creatine are also normally

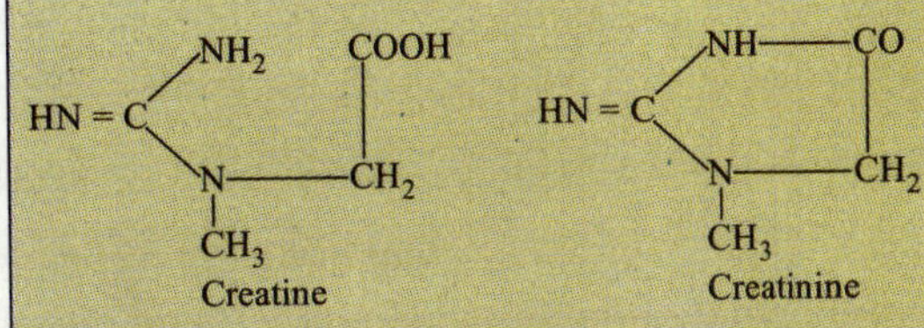

Fig. 20.52 : Structure of creatine and creatinine

present in urine. Creatinine formation is a preliminary step required for the excretion of most of the creatine.

Biosynthesis of creatine and creatinine

1. Three amino acids—Glycine, arginine, and methionine—are directly involved.
2. The first reaction is that of transamidination from arginine to glycine to form guanidoacetic acid (Glycocyamine). This reaction takes place in the kidney but not in the liver or in heart muscle. Recently, evidence has shown that nephrectomized rats can still synthesize creatine. The interpretation is that there is the existence of an extrarenal site or sites of transamidination in this animal.
3. The synthesis of creatine is completed in the liver by the methylation of glycocyamine. "Active" methionine is the methyl donor. Other methyl donors–betaine or choline after oxidation to betaine—serve indirectly by producing methionine through the methylation of homocysteine. The methylation of glycocyamine is not reversible. Creatine or creatinine cannot methylate homocysteine to methionine. In the methylation of creatine, ATP and oxygen are required.

   The enzymatic mechanisms for the methylation of glycocyamine involves first the formation of active methionine (S-adenosylmethionine) which requires ATP, Mg^{++} and glutathione as well as a methionine activating enzyme. The methylation of glycocyamine by active methionine is catalyzed by guanidoacetate methylferase found in the liver of mammals. Glutathione or other reducing substances are required for the optimal activity of the enzyme. There is no evidence for the requirement of metal ions or other cofactors.

   It has also been found that the pancreas can synthesize glycocyamine. Therefore, the pancreas may play an important role in the synthesis of creatine within the body of mammals.

   Dietary creatine or high blood creatine has no effect on the rate of synthesis of creatine in the liver. The rate of creatine biosynthesis is dependent on kidney transamidinase activity. Hyperthyroidism is associated with reduced kidney transamidinase actvity. The effect of hyperthyroidism on kidney transamidinase is mediated by the increased levels of blood creatine.
4. Creatinine is the anhydride of creatine and is formed by the non-enzymatic means in muscle.

The overall reaction is shown in Fig. 20.53.

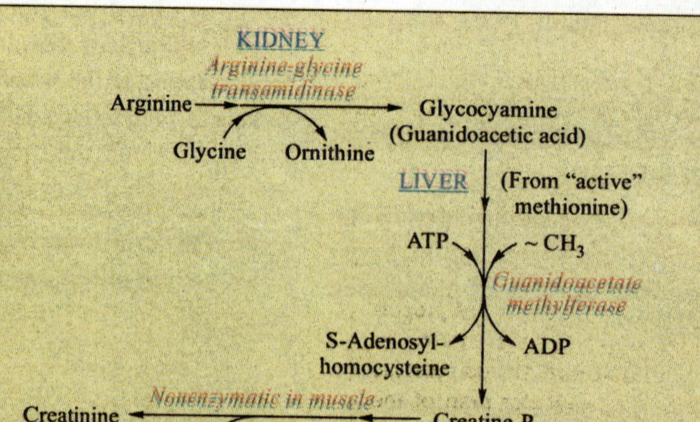

Fig. 20.53 : Flow chart of biosynthesis of creatine and creatinine

Absorption

It is completely reabsorbed in the renal tubules at lower concentrations.

Normal level of creatine and creatinine in the blood

Creatine : 2.5-7.3 mg/100 ml
Creatinine : 0.6-1.2 mg/100 ml

Daily excretion in urine

Creatine :	
Prepubertal children	4.2 mg/Kg
During pregnancy and occasionally in normal non-pregnant women	0-100 mg
Normal males on a balanced diet	Nil
Normal subject maintained on a very low carbohydrate intake and during fasting period (due to excessive catabolism of muscle tissue)	Up to 100 mg
Creatinine :	
Normal adult males & females	0-200 mg/24h
Normal children	7-11mg/24h

Creatinine coefficient

This term is used to indicate the number of milligrams of creatinine (plus creatine) nitrogen excreted per kilogram of body weight in 24 hours.

The creatinine coefficient is an index of the amount of active muscle tissue in the body.

Normal values of creatinine coefficient range from 20 to 26 mg creatinine for men and 14 to 22 mg creatinine for women.

CLINICAL IMPORTANCE

- The amount of creatinine excreted is not proportional to the size of the subject but to his muscular development. A man with well-developed muscles excrete more than a less muscular man of the same weight. Ordinary muscular exertion does not affect the daily output of creatinine, which is approximately the same on a day of exercise as on a day of rest.

CLINICAL IMPORTANCE

- Excessive creatinuria may occur in starvation, febrile and wasting diseases, diabetes mellitus, in certain myopathies (myotonia congenita, various forms of myositis, congenital muscle dystrophies, congenital muscular hypertrophy and secondary muscle atrophy).
- Excessive creatinuria has also been observed in hyperthyroidism, during the menstrual period and in pregnancy. It has been observed after fractures (to 500 mg daily). It disappears gradually during the period of healing of the fracture.
- Testosterone diminishes creatinuria in castrates by increasing the storage but not the synthesis of creatine. The administration of methyltestosterone causes the increased excretion of creatine and creatinine due to increased synthesis of creatine.
- In myasthenia gravis, the administration of large doses of glycine results in the increased excretion of creatine. Such increase in creatinuria also occurs in the majority of normal subjects after ingestion of glycine.

Creatine tolerance

1. When 1 or 2 gms of creatine are ingested by a normal adult, about 80 per cent is retained by normal men and about 70 per cent by normal nonpregnant women. About 20 and 30 per cent, respectively, are excreted in the urine during the subsequent 24 hours.
2. Diminished tolerance (i.e. excessive excretion) is present in mild cases of above mentioned disorders.
3. Creatine tolerance is decreased in the majority of patients with hyperthyroidism and increased in hypothyroidism. This procedure carries much values in the diagnosis of hyperthyroidism or hypothyroidism in children.

Creatinuria

Excretion of creatine in urine is said to be creatinuria. Creatine excretion occurs in the following conditions :

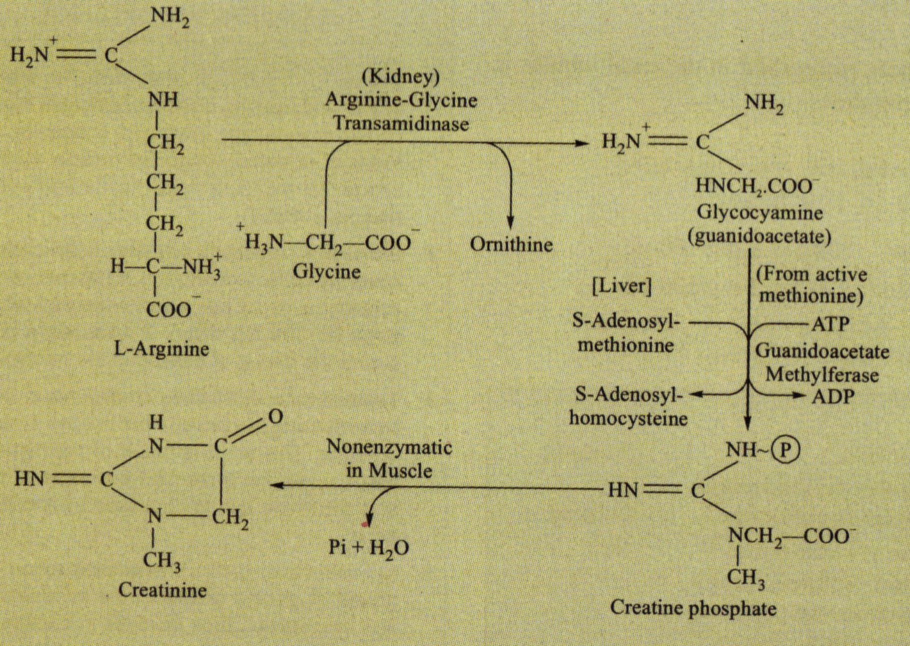

Fig. 20.54 : Reaction of biosynthesis of creatine and creatinine

1. The inability to convert creatine to creatinine in children.
2. In pregnancy, and maximum after parturition in adult females.
3. In febrile conditions.
4. In thyrotoxicosis, probably due to associated myopathies.
5. In muscular dystrophies, myositis, and myasthenia gravis.
6. In diabetes mellitus with lack of carbohydrate diet.
7. In starvation.
8. In wasting diseases, eg, in malignancies.

FATE OF AMINO ACIDS IN THE BODY

1. The amino acids are required for the synthesis of protein.
2. These are utilized for the synthesis of methionine which is methyl donor. This methyl donor is required for the synthesis of choline, betaine, creatine and epinephrine. Methionine can also be converted into cystine when this amino acid is inadequately supplied in the diet.
3. Phenylalanine can be converted into tyrosine and tyrosine is the precursor of epinephrine, thyroxine and the pigment melanin.
4. Glycine, arginine and methionine all participate in the synthesis of creatine.
5. Glutamine, formed from glutamate, is the chief source of ammonia formation in the kidney. Ammonia helps in the neutralization of acid to be excreted.
6. Glycine acts as a coupling agent to convert benzoic and salicylic acids into hippuric acid and salicyluric acid, respectively, which are readily excreted by the kidney.
7. Histidine can be decarboxylated to produce histamine.
8. Tryptophan can be converted into nicotinic acid in presence of vitamin B_6 and

B$_2$ and into serotonin which on acetylation and methylation produces the hormone melatonin.

9. Glutamate, aspartate and the ammonia formed from amino acids by deamination are utilized for purine and pyrimidine synthesis.
10. Glutamate and aspartate are involved in transamination reaction to form keto acids which are converted into glucose by gluconeogenesis.
11. Alanine, on deamination, forms pyruvic acid which is an important intermediate compound in the metabolism of glucose. The fate of alanine is the same as that of glucose. Hence, alanine can be regarded as a glucose-former.
12. In case the diet contains more proteins than is required to replace the catabolized protein, the excess of amino acids is used for energy production. The presence of an oxidative deaminating enzyme system in the liver (and kidney) can convert amino acids into keto acids liberating ammonia in the free state. The ammonia is used for the synthesis of urea.
13. Some of the amino acids are metabolized to glucose derivatives while others form acetoacetate.

CLINICAL ORIENTATION

- The deamination of the α-amino nitrogen of amino acids results in the formation of ammonia which is toxic to all animals. Human tissues detoxify it by converting it to glutamine for transport to the liver. Deamination of glutamine in the liver releases ammonia which is efficiently converted to urea essential for health.
- In cirrhosis of the liver or severe hepatitis, ammonia accumulates in the blood generating clinical signs and symptoms. The most suitable treatment is essential on the basis of the knowledge of biochemistry of urea systhesis for infants born with the deficiency of any one of five enzymes of the urea cycle.

CLINICAL ORIENTATION

- The deficiency state is observed on feeding certain grains deficient in tryptophan and lysine and not supplemented by protein sources.
- A child on starchy diet and poor protein diet suffers from Kwashiorkor and Marasmus on deficient caloric intake and specific amino acids.
- Many of the genetic disorders, if left untreated, result in irreversible brain damage and early mortality. Recombinant DNA technology may finally provide ways to correct genetic defects by "gene therapy".
- Any mutation that significantly lowers the catalytic activity of argininosuccinase will cause the metabolic disorder termed as argininosuccinic acidemia. These are the distinct molecular diseases at a molecular level.
- Histamine, formed by the decarboxylation of histidine, plays a central role in many allergic reactions.
- The neurotransmitters derived from amino acids include GABA from glutamate; Serotonin from tryptophan; dopamine, epinephrine and norepinephrine from phenylalanine. The metabolism of these neurotransmitters are affected by many drugs used to treat neurologic and psychiatric conditions.

NITROGEN BALANCE

Nitrogen balance is defined as the quantitative difference between the nitrogen intake and the nitrogen output, both expressed in gms N/day. Intake means the nitrogen of the food and output means the excretion of nitrogen as urine, feces, milk, sweat, vomiting, menstrual fluid and loss of hair.

Positive nitrogen balance exists when intake exceeds output. This condition occurs whenever new tissue is synthesized such as during growth of the young and in pregnancy.

In negative nitrogen balance, the output exceeds the intake. The condition occurs in inadequate intake of protein (fasting, diseases of the gastrointestinal tract), in accelerated catabolism of tissue protein (fevers, infections, wasting diseases and trauma) and in the absence of "essential" amino acid from the diet.

SCHEMATIC COMPENDIUM

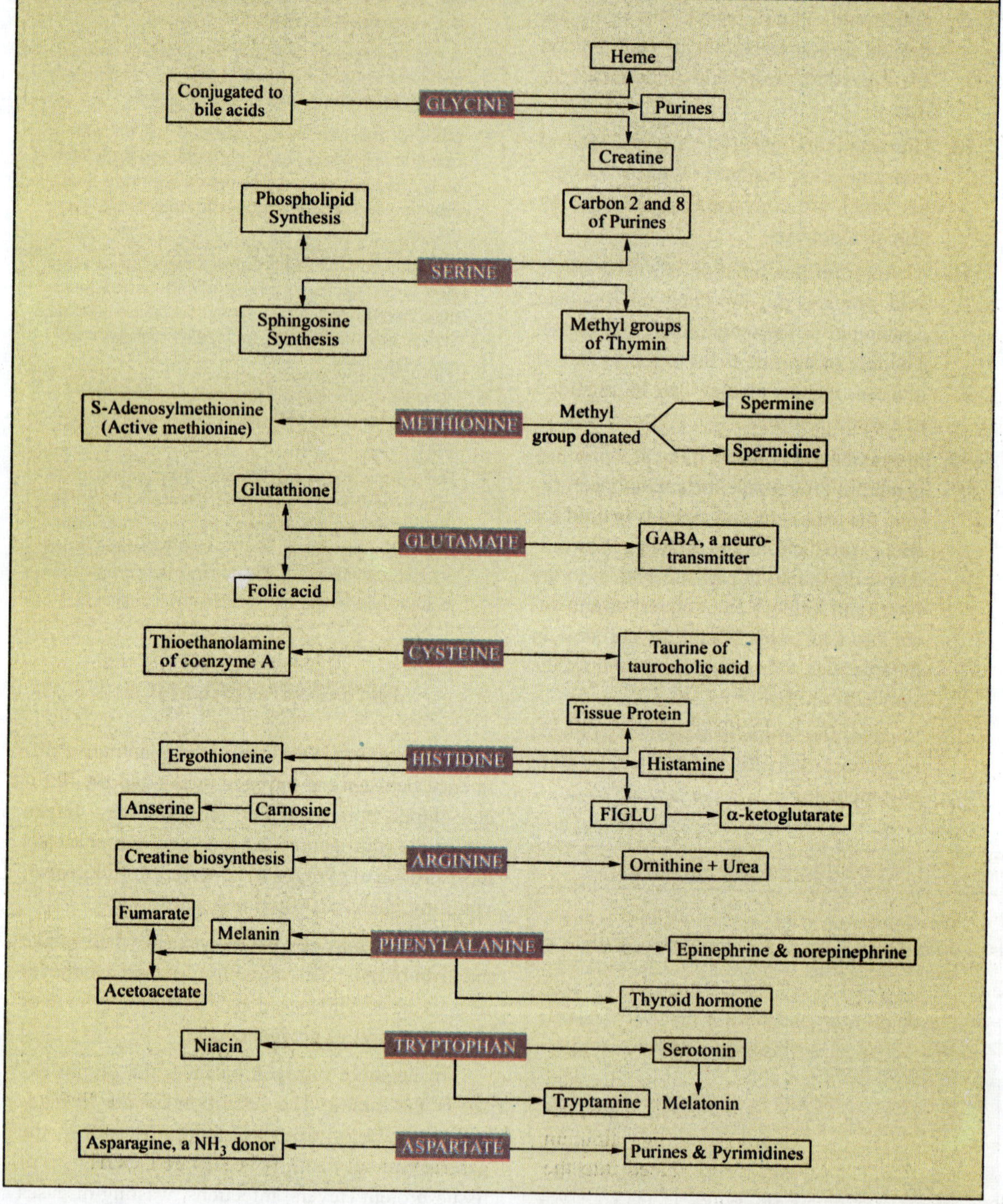

It is concluded from the above discussion that nitrogen equilibrium [intake = output] is maintained only in the adult organisms and only in the absence of the abnormal conditions.

An "adequate diet" contains all requirements for minerals and vitamins, the protein of the diet is of high "biological value" and is administered at a sufficiently high level; other caloric needs are met satisfactorily by the carbohydrate and fat of the diet. A minimum amount of dietary carbohydrate (5 grams/100 calories) is required for the maintenance of nitrogen equilibrium, independently of the adequacy of the energy provision from fat and protein. This is referred to as the "protein-sparing action" of carbohydrate. Under these circumstances, the nitrogen output equals the intake. An increase or decrease in the intake is followed by a corresponding adjustment in the output, so that nitrogen equilibrium is established at a new level. There is no marked tendency for storage of surplus nitrogen.

PROTEINURIA

The glomerular membrane permits only a very small amount of the normal plasma proteins. The glomerular filtrate normally contains 10 to 25 mg protein per 100 ml. This is almost completely reabsorbed by the tubular epithelial cells. 150 mg is the upper limit of normal daily protein excretion in the urine. Hence, it is difficult to detect albumin in the urine normally.

In nephrotic syndrome, urinary albumin : Globulin ratios are very high. In chronic glomerulonephritis, the ratio is lower. In acute nephritis, the ratio is low. Low values are found in amyloid disease of the kidneys.

Certain foreign proteins entering the plasma are eliminated by the kidney, whereas normal plasma protein is retained in the blood. Thus, hemoglobin–when free in the plasma, egg albumin and other foreign proteins–if introduced into the blood-stream, appears in the urine.

Two major mechanisms operate to produce abnormal proteinuria :

1. Humoral proteinuria : Plasma contains abnormal proteins of small molecular size which pass readily through normal glomeruli.
2. Renal proteinuria : The glomerular membranes are so injured that they become more permeable to the normal plasma proteins or tubular epithelial cell damage results in inadequate reabsorption of proteins from the glomerular filtrate.
3. Abnormal proteinuria occurs when the quantities of proteins reaching the tubules exceed their reabsorptive capacity.

The pathogenesis of proteinuria may be summarized as :

1. In the majority of instances, glomerular abnormalities—structural or functional, primary or secondary—play a fundamental role.
2. In many cases, associated tubular involvement is the primary factor.
3. A few cases are due to abnormal proteins in the plasma.
4. The urinary proteins originate in the lower urinary tract.

Proteinuria is classified into three classes :

1. Transitory proteinuria : It never indicates permanent renal diesease. It may occur in fever, after intense physical exertion, and after minor emotional stimuli.
2. Continuous proteinuria : This is always pathological and is indicative of pyelonephritis.
3. Orthostatic proteinuria : This is caused by erect posture in lumbar lordosis and is common in adolescence.

METHIONINE

Methionine is an essential and a sulphur containing amino acid.

Structure : $CH_3 . S. CH_2 . CH_2 . CH(NH_2) . COOH$

Concentration in blood : 0.85 mg/100 ml.

Daily requirements

Infants	: 85 mg/kg body weight
Adult male	: 1.5 mg/kg body weight
Adult female	: 4.7 mg/kg body weight

Physiological functions

1. Methionine, as the sole source of sulphur-containing amino acids, is capable of maintaining nitrogen equilibrium.
2. Active methionine is the main methyl donor of the body.
3. Choline can be formed from ethanolamine if enough methionine is available in the body.
4. Methionine, after donating methyl group, is converted into homocysteine from which cysteine is formed.
5. It donates methyl group for the synthesis of creatine from guanidoacetic acid.
6. It is involved in the detoxification of nicotinic acid for the formation of the excretory product trigonelline.
7. Methyl group is donated by methionine for the formation of epinephrine from norepinephrine.
8. It acts as a lipotropic factor.

HYPERHOMOCYSTEINAEMIA

By the year 2015, it is predicted that in India cardiovascular disease will replace infectious diseases as the major killer. At present 40 million patients suffer from coronary artery disease in India. Homocysteine plays a role in being an easily reversible risk factor for atherosclerosis.

1. Metabolic pathways of Homocysteine (see chart below).
2. Factors influencing Homocysteine metabolism :
   (a) Advance age, male, menopause, enzyme deficiencies.
   (b) Folic acid, Vit. B_6, Vit. B_{12} deficiencies.
   (c) Renal failure, rheumatoid arthritis, SLE, hypothyroidism, diabetes mellitus etc.

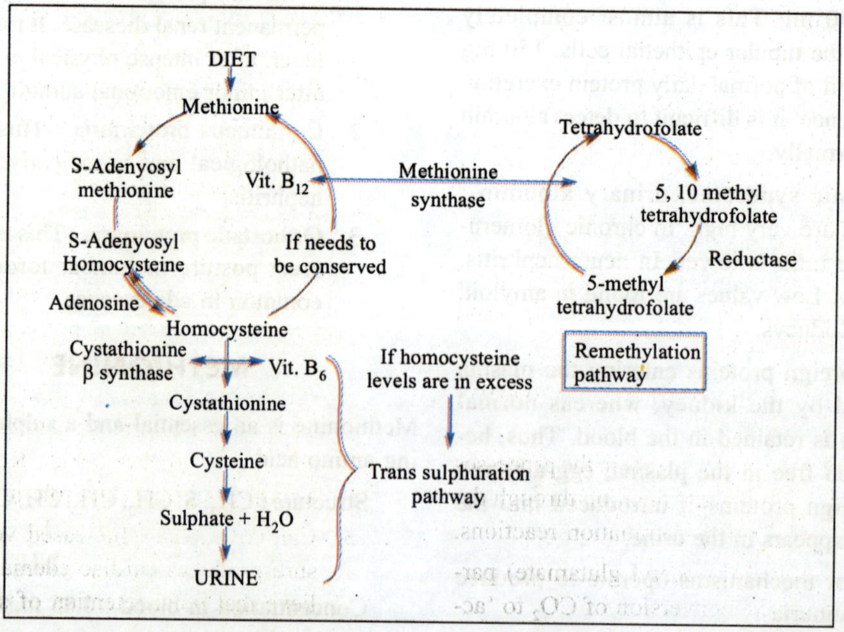

(d) Medications — Metformin, methotrexate, therophylline, anticonvulsants, niacin etc.

3. Kang and coworkers have classfied hyperhomocysteinaemia as :

   Moderate risk → 15-30 mmol/L

   Intermediate risk → 30-100 mmol/L

   Severe → > 100 mmol/L

4. So, homocysteine is a new-emerging risk factor for hypertension, stroke, peripheral vascular disease and coronary artery disease.

5. Multivitamin supplementation containing folic acid, vit. B_{12}, vit B_6 could go a long way in decreasing homocysteine levels and thereby offering potential benefits *vis-a-vis* mortality and morbidity.

ACIDIC AMINO ACIDS

The acidic amino acids are : 1. Aspartic acid (α-amino-succinic acid). 2. Glutamic acid (α-amino glutaric acid).

Structure

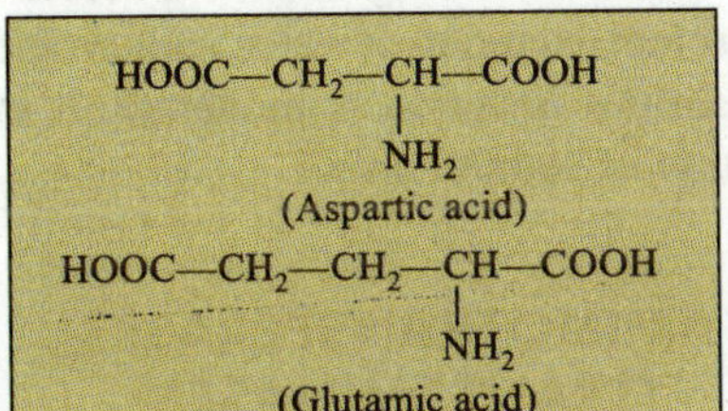

(Aspartic acid) (Glutamic acid)

Biological importance

1. These amino acids are nonessential and readily synthesized in the body.
2. These are metabolically the most reactive of the amino acids.
3. These are the active agents in the processes of deamination and amination through the participation in transamination reactions.
4. Glutamic acid (as acetyl glutamate) participates in the conversion of CO_2 to 'active CO_2' and 'active CO_2' to carbamoyl phosphate which, in combination with ornithine, forms citrulline.
5. Aspartic acid participates in the conversion of citrulline to arginine.
6. Decarboxylation of aspartic acid leads to the formation of β-alanine which is a constituent of pantothenic acid.
7. Both the amino acids are glucogenic only.
8. Asparagine formed from aspartic acid plays an important role in the metabolism of plants.
9. Glutamic acid is a component of glutathione and folic acid.
10. Glutamic acid is converted into glutamine in the kidney which is the source of urinary ammonia. Glutamine also serves as a reservoir of amino nitrogen in the body.

Metabolism

See (Fig. 20.55 next page).

EDEMA

Edema is recognized by the increase in the volume of interstitial fluid. This is of three types: Nephrotic type, nephritic type and cardiac edema.

1. Nephrotic type : This type is produced by the lowering of serum protein concentration due to severe albuminuria or to low protein intake. If the serum colloidal osmotic pressure drops below the normal, there is accumulation of interstitial fluid.
2. Nephritic type : This type is formed by damage to the capillary endothelium allowing the passage of protein into the interstitial fluid. The lowered osmotic pressure allows greater filtration and less reabsorption. This mechanism is applied in acute glomerulonephritis. Renal factors also play an important role in it.
3. Cardiac edema : Increased venous pressure produces cardiac edema. Many believe that renal retention of sodium pro-

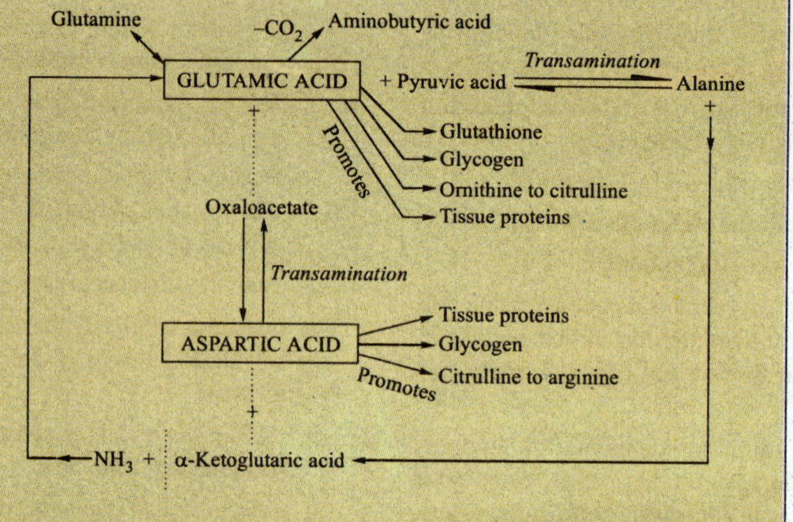

Fig. 20.55 : Metabolism of acidic amino acids

duced by forward failure of heart leads to retention of water, to increased venous pressure and then to edema.

Clinical manifestations of edema

1. Facial edema distinguishes nephritic or nephrotic edema from cardiac edema. The whole face becomes swollen. Patients with uremia often have such facial edema.
2. General dehydration of uremia usually minimizes the facial edema.
3. Edema of the brain has also been reported in **acute glomerulonephritis**.
4. Edema produced by '**forward failure**' diminishes cardiac output and thereby decreases renal circulation. The lowering of glomerular filtration produces a greater reabsorption of sodium which, in turn, causes a greater reabsorption of water.

Management of edema

Diuresis is essentially an increase in urine output sufficient to produce a significant negative balance of water; and negative balance of water signifies negative balance of sodium. In most edema, if glomerular filtration is not greatly diminished, diuresis can be established by sodium intake. The water intake need not be limited. Normal or even higher than normal ingestion of water may enhance the urine output. Such management is beneficial in cardiac, nephrotic and cirrhotic edema.

Two new powerful diuretics—Ethacrynic acid and Furosemide—have recently been introduced. **Ethacrynic** acid inhibits the reabsorption of sodium in proximal and distal segments. **Furosemide** produces a prompt excretion of water, sodium and chloride. **Potassium** injection is moderate. It inhibits reabsorption of sodium also.

AMINOTRANSFERASES [TRANSAMINASES]

Aminotransferases catalyze the transfer of amino group from an amino acid to a keto acid—a new amino and keto acid are formed in this process.

Clinically, the two most important aminotransferases are : Glutamic oxaloacetic transaminase (GOT), now designated aspartate aminotransferase and glutamic pyruvic transaminase (GPT), now designated alanine aminotransferase. These enzymes are widely distributed in tissues and are normally present also in the blood serum in concentrations up to 19 U/I or, up to 40 Karmen units/dL.

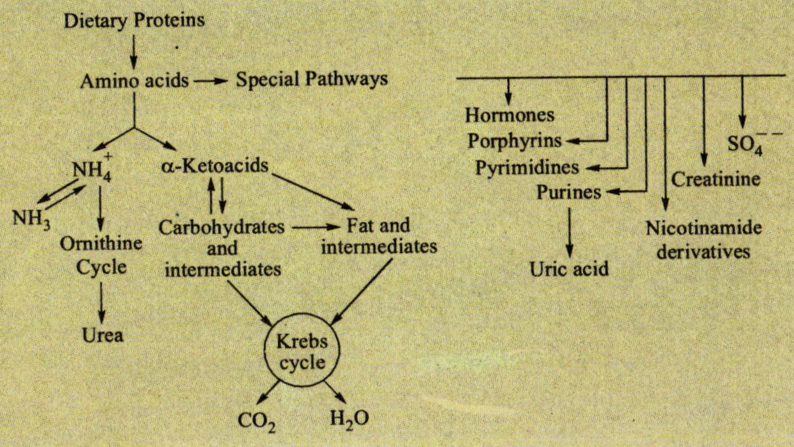

Fig. 20.56 : General Pathways of Protein and Amino acid metabolism

Aminotransferase activity is increased in certain diseases involving tissues like liver and myocardium owing to the liberation of abnormally large amounts of it from the damaged tissues.

Serum amino transferase activity is increased in inflammatory, degenerative, and neoplastic lesions of the liver. The extent of increase depends on the severity of the disease. The highest values have been obtained in **acute hepatic necrosis**. Lower values (<250U/I) are obtained in uncomplicated portal cirrhosis, biliary cirrhosis and extrahepatic biliary obstruction as well as in hepatic malignancy (primary or metastatic).

Serum GOT activity is increased in acute heart failure; the values are usually below 100 U/I, but occasionally as high as 1,500 U/I. Slight or moderate increases may occur after intracardiac operations. **In acute myocardial infarction**, GOT activity is significantly increased but no increase in GPT. The determination of GOT is diagnostically important for the following circumstances :

1. Differentiation between acute myocardial infarction and coronary insufficiency. No increase occurs in **coronary insufficiency**.
2. Diagnosis of acute myocardial infarction when the ECG changes are not definitive or are difficult to interprete because of previous infarction.
3. Diagnosis of extension of the original infarction or of recurring acute infarction.

Serum GOT may increase in muscular dystrophies, myositis and gangrene. An increase may occur in patients with acute pancreatitis, leukemia and toxemia of pregnancy. No increase in serum GOT occurs in muscular diseases of nervous origin. Increases occur in patients with acute hemolytic anemia and in normal person after prolonged severe exercise. **Erythromycin** can cause an increase in serum GOT.

Isoenzymes : GOT exists in two isomeric forms which can be demonstrated by electrophoresis. One is derived from cytoplasm and the other from mitochondria. If the isoenzyme from mitochondria appears in serum, cell death occurs. This indicates a severe lesion and affects the prognosis.

G PROTEIN

1. Adenylyl cyclase regulation is mediated by two GTP-dependent proteins designated as Gs (stimulatory) and Gi (inhibitory). Each is composed of three subunits α, β, γ adenylyl cyclase present in the inner surface of the plasma membrane and catalyzes the formation of cAMP from ATP in the presence of magnesium.
2. There are more than 20α subunits, 4β subunits, 6γ subunits, some forms of αi stimulate K^+ channels.

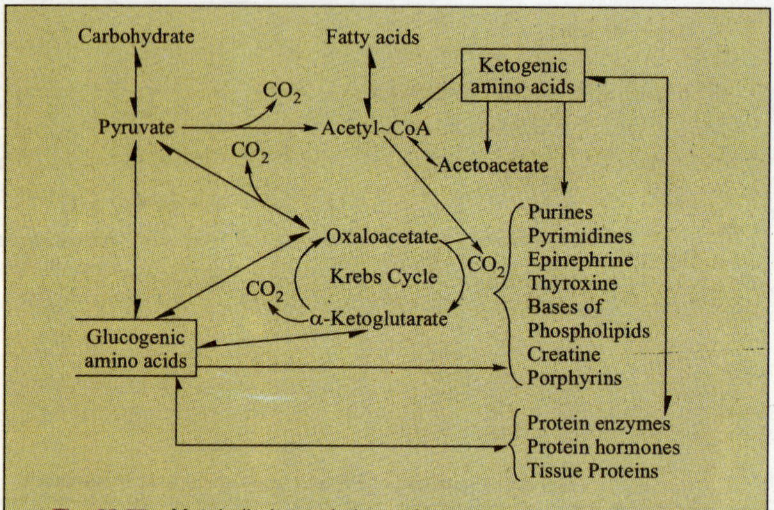

Fig. 20.57 : Metabolic interrelations of proteins with other foodstuffs

3. The hormones ACTH, ADH, glucagon, TSH, FSH etc. stimulate adenylyl cyclase and the hormones somatostatin, angiotensin 11 etc. inhibit adenylyl cyclase.
4. The families of mammalian G proteins are classified into four subfamilies (Gs, Gi, Gq, G_{12}). Members of the Gq family activate the phospholipase C group of enzymes.
5. G proteins are involved in many important biologic processes in addition to hormone action.

HIGH-SENSITIVE C-REACTIVE PROTEIN

1. C-reactive protein derives from the fact that it reacts with capsular polysaccharide of streptococcus pneumoniae, it is 1,35,000 daltone non-immunoglobulin protein, it is attached with phosphatidylcholine of cell membrane Ca^{++}.
2. High concentration of high sensitive (hs) c-reactive protein is found in following conditions :
   (a) Coronary artery disease — Angina, myocardial infarction.
   (b) Endocarditis; (c) Obesity; (d) Sepsis;
   (e) Smoking; (f) Vasculitis;
   (g) Connective tissue disorder
3. hs-c-reactive protein is nowadays a very important marker to assess cardiovascular patients in modern clinical practice :

Range of hs-CRP

< 1mg/L	—	Low risk for developing coronary artery disease (CHO)
1-3 mg/L	—	Average risk
> 3 mg/L	—	High risk

4.

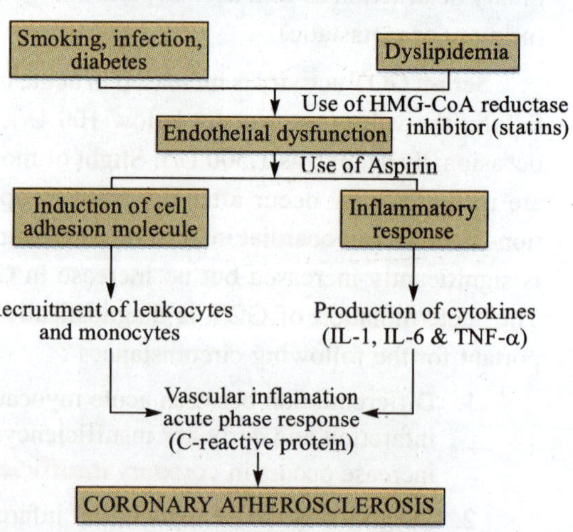

Representation of pathogenesis of inflammations and acute phase reactants, specifically c-reactive protein in coronary atherosclerosis

5. Therefore, increased hs CRP levels in patients with high risk cardiovascular diseases without documented CAD (coronary artery disease) warrant treatment with statin even if LDL-cholesterol level are within target range.

   Patients with documented CAD and high hs CRP levels should be followed and their risk factors should be managed aggressively.

6. The use of hs CRP levels to screen for CAD disease risk in entire adult population is not recommended. Further studies will clarify the utility and lost effectiveness of these measurements.

EXERCISE

1. What do you mean by intermediary metabolism? Give an account of intermediary metabolism of proteins. *(R.U. 70A)*
2. What is Urea? Describe its formation and excretion in human body. *(Bh. U. 74A; R.U. 70A)*
3. Explain the chemical reactions involved in formation of Urea. *(R.U. 73S)*
4. What is Urea? Describe the synthesis of urea in the body. *(Muz. 75S; P.U. 68A)*
5. List the excretory products of nitrogen metabolism. How are they formed in the body? *(R.U. 65A, 68S)*
6. What are the sources and main functions of amino acids in our body? Describe the process involved in the breakdown of amino acids. *(Mith. 67A)*
7. Name the essential amino acids and describe the specific metabolic role of any two of them. *(Pun. 68S, 70A)*
8. Describe the fate of amino acids in the body. *(Mith. 74A; P.U. 72S)*
9. Describe the metabolic fate of phenylalanine and tyrosine in the body. *(Mith. 65A; 75A)*
10. Name one amino acid which takes part in synthesis of many hormones in the body. Briefly outline the steps of synthesis of any one of these hormones. *(P.U. 72A)*
11. Write notes on :
    (a) Essential amino acids. *(P.U. 75S; Muz. 75A; Bh. U. 74A)*
    (b) Acidic amino acids. *(P.U. 70A)*
    (c) Methionine. *(R.U. 64S, 70S)*
    (d) Tryptophan. *(Muz. 74A; Pun. 69A)*
    (e) Aminotransferase. *(Mith. 75A)*
    (f) Edema. *(R.U. 72A)*
    (g) Nitrogen balance. *(Muz. 74A)*
    (h) Nitrogen equilibrium. *(M.U. 73S)*

CHAPTER 21

OVERVIEW OF INTERMEDIARY METABOLISM

Introduction

1. The dietary constituents after digestion and absorption follows intermediary metabolism.
2. The metabolic pathways fall into three categories :
   (i) **Anabolic Pathways** involve the synthesis of compounds and constitute the body's structure and machinery. Protein synthesis is one of them. The free energy required for these pathways are available from catabolic pathways.
   (ii) **Catabolic Pathways** involve oxidative processes and release free energy (usually in the form of high-energy phosphate) or reducing equivalents eg, biologic oxidation.
   (iii) **Amphibolic Pathways** are the links between the anabolic and catabolic pathways, eg, TCA cycle.

Biomedical Importance

1. The concept of metabolism in the normal animals provides a sound understanding of many diseases.
2. The variations and adaptation in metabolism due to periods of starvation, exercise, pregnancy, and lactation are included in normal metabolism.
3. **Abnormal metabolism** results from nutritional deficiency, enzyme deficiency, and excessive secretion of hormones.
4. The example of disease caused by abnormal metabolism is **diabetes mellitus**.

THE BASIC METABOLIC PATHWAYS

1. The basic pattern of metabolism in the tissues is set by the diet.
2. **Humans** require to process the absorbed products of digestion of dietary carbohy-

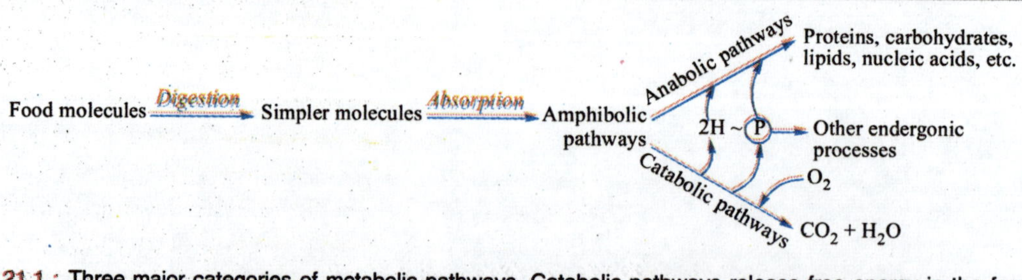

Fig. 21.1 : Three major categories of metabolic pathways. Catabolic pathways release free energy in the form of reducing equivalents (2H) or high-energy phosphate (~p) to power the anabolic pathways. Amphibolic pathways act as links between the other two categories of pathways

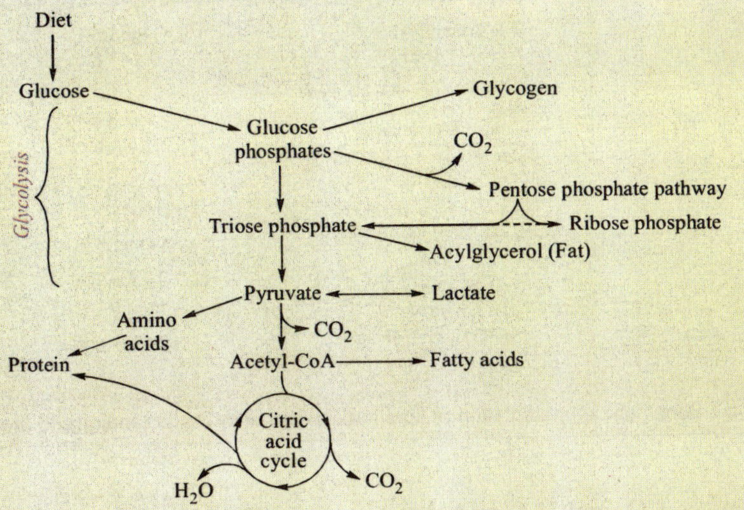

Fig. 21.2 : Schematic representation of carbohydrate metabolism showing the major end products

drate, lipid, and protein. These are mainly glucose, triacylglycerol, and amino acids, respectively.

3. In ruminants, cellulose in the diet is digested by symbiotic micro-organisms to lower fatty acids (acetic, propionic, butyric) which are utilized as major substrates in the tissue metabolism of these animals.

Carbohydrate metabolism

1. Pyruvate and lactate are formed in the mammalian cells as a result of the oxidation of glucose by glycolysis.
2. Glycolysis occurs in the cytoplasm of cells in absence of oxygen producing lactate only.
3. Under aerobic condition, pyruvate is metabolized to accetyl-CoA which enters the citric acid cycle for complete oxidation to CO_2 and H_2O.
4. Glucose also takes part in other metabolic process as follows:
   (i) It is converted to glycogen as a storage, particularly, in liver and skeletal muscle.
   (ii) The HMP shunt or the pentose phosphate pathways arising from intermediates of glycolysis is a source of reducing equivalents (2H) for biosynthesis of fatty acids, cholesterol, etc. and it is a source of ribose which is important for nucleic acid formations.
   (iii) Triose phosphate of glycolysis is a source of glycerol of fat.
   (iv) Pyruvate and the intermediates of citric acid cycle form amino acids and acetyl-CoA is the building block for long-chain fatty acids and cholesterol, the precusor of all steroid hormones in the body.

Lipid metabolism

1. The long–chain fatty acids are synthesized form acetyl-CoA derived from carbohydrate or from dietary lipid.
2. In the tissues, fatty acids are oxidized to acetyl-CoA or esterified to acylglycerol to form fat which is the main caloric reserve of the body.
3. Acetyl–CoA formed by β-oxidation has the following significant roles in the body:

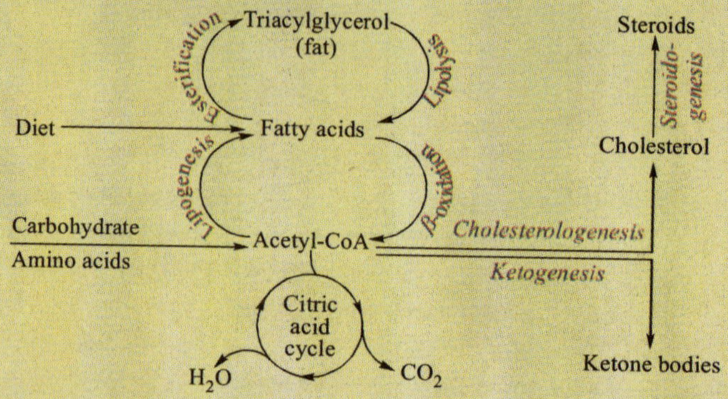

Fig. 21.3 : Schematic representation of lipid metabolism showing the major end products

(i) It liberates CO_2 and H_2O and also yields high energy. Therefore, during the oxidation of fatty acids by β-oxidation for their complete oxidation, more energy is formed.

(ii) It is a source of cholesterol biosynthesis.

(iii) In the liver, it forms ketone bodies which are alternative water-soluble tissue fuels. These fuels become important sources of energy under certain conditions (eg, starvation).

Amino acid metabolism

1. Amino acids are required for protein synthesis.

2. The essential amino acids must be supplied in the diet since these are not synthesized by the tissues.

3. Diet can supply the non-essential amino acids which are also formed from the intermediates of citric acid cycle by transamination.

4. Excess amino nitrogen as a result of deamination of amino acids is removed as urea and the carbon skeletons that remain after transamination give the following products :

   (i) Carbon dioxide and water via the citric acid cycle.

   (ii) Glucose (by gluconeogenesis).

   (iii) Ketone bodies.

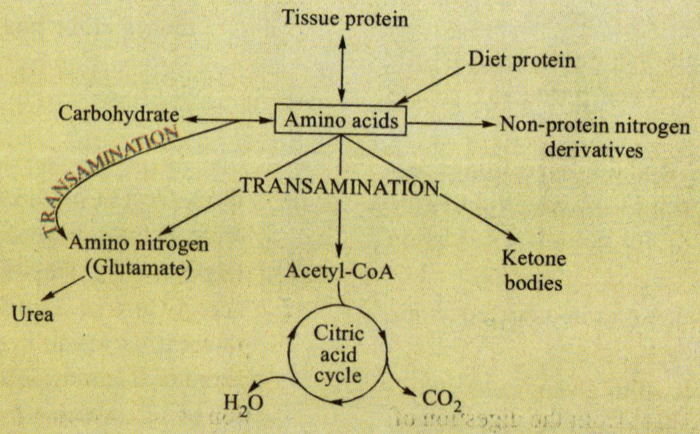

Fig. 21.4 : Schematic representation of amino acid metabolism showing the major end products

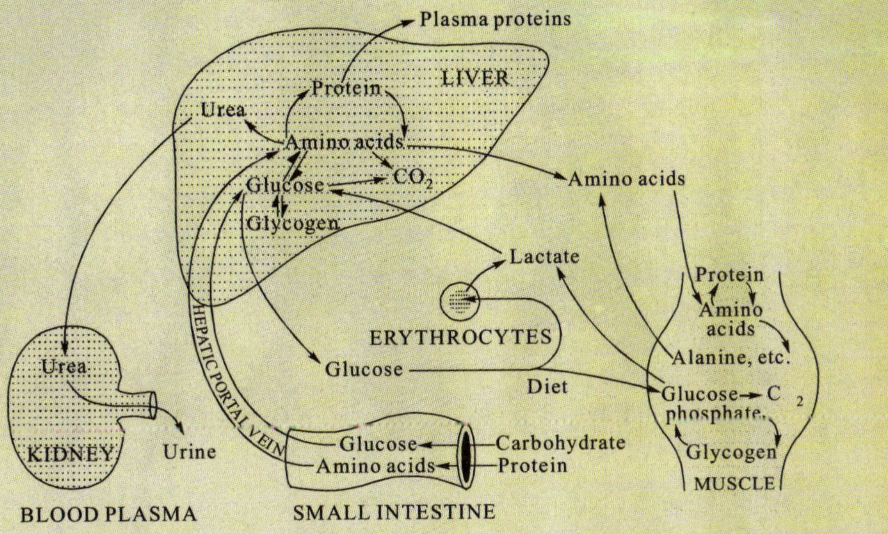

Fig. 21.5 : Transport and fate of major carbohydrate and amino acid substrates and metabolites. Little free glucose in muscle is rapidly phosphorylated upon entry

5. The amino acids are also the precursors of many other important compounds, eg, purines, pyrimidines, and hormones such as epinephrine and thyroxine.

THE LOCATION OF METABOLIC PATHWAYS

The metabolic pathways can be studied at many levels of organization. These levels are divided into two major groups:

1. **At the tissue and organ level :** The nature of the substrates entering and metabolites leaving tissues and organs is defined, and their overall fate is discussed.
2. **At the subcellular level :** Each cell organelle (eg, the mitochondrion) or compartment (eg, the cytosol) carries out specific biochemical roles showing metabolic pathways.

Intermediary Metabolism at the Tissue and Organ Level

1. Amino acids formed from the digestion of dietary protein and glucose formed from the digestion of carbohydrate follow the common path of absorption via the hepatic portal vein. Both these metabolites and other-soluble products of digestion are initially directed to the liver (Fig. 21.5).
2. The liver primarily regulates the blood concentration of particularly glucose and amino acids.
3. The liver takes up excess glucose and converts it to glycogen (glycogenesis) or to fat (lipogenesis). During meals, it causes glycogenolysis to supply glucose in the blood or in company with the kidney, it converts non-carbohydrate metabolites such as lactate, glycerol and amino acids to glucose (gluconeogenesis).
4. Sufficient concentration of blood glucose is maintained by certain tissues in which it is fuel, eg, brain and erythrocytes.
5. The liver also synthesizes the major plasma proteins (eg, albumin) and deaminates amino acids with the formation of urea which is transported via blood to the kidney and excreted.

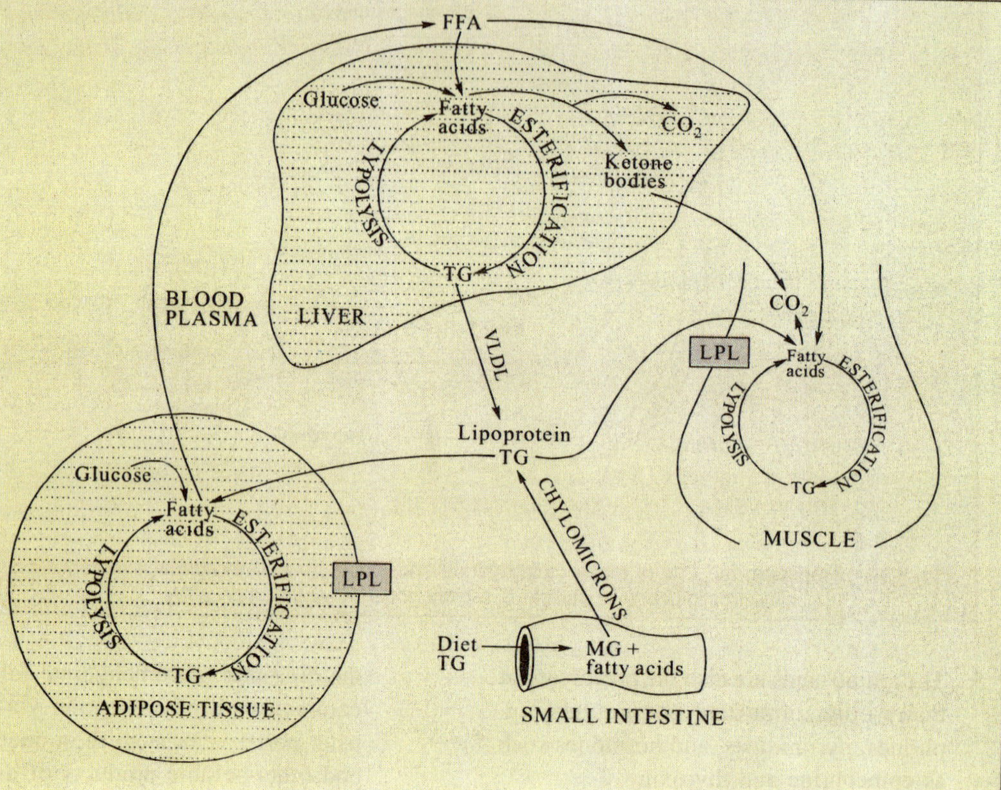

Fig. 21.6 : Transport and fate of major lipid substrates and metabolites
FFA = Free fatty acids, LPL = Lipoprotein lipase, MG = Monoacyl glycerol, TG = Triacyl glycerol. VLDL = Very low density lipoprotein

6. Skeletal muscle uses glucose as a fuel forming lactate and carbon dioxide. It stores glycogen for its use in muscular contraction and synthesizes muscle protein from plasma amino acids. During dietary shortage of protein, it can supply plasma amino acids from its stores of protein.
7. Lipids being digested form monoacylglycerols and fatty acids. These are recombined in the intestinal cells with protein and secreted into the lymphatic system and then into the circulation as a lipoprotein known as chylomicron.
8. Chylomicron triacylglycerol is not taken up by the liver and it is metabolized by extrahepatic tissue having the enzyme lipoprotein lipase which hydrolyzes the triacylglycerol releasing fatty acids that are incorporated into tissue lipids or oxidized as fuel.
9. The long-chain fatty acid is synthesized (lipogenesis) from carbohydrate mainly in adipose tissue and the liver.
10. Adipose tissue triacylglycerol is hydrolyzed (lipolysis) to fatty acids which are released in the circulation.
11. Free fatty acids are taken up by most tissues (but not brain or erythrocytes) and esterified to acylglycerols or oxidized to carbon dioxide.
12. The liver performs two additional important functions :

    (i) Surplus triacylglycerol from both lipogenesis and free fatty acids is secreted into the circulation as very low density lipoprotein (VLDL). This triacylglycerol faces the fate similar to that of chylomicrons.

(ii) Partial oxidation of free fatty acids leads to ketone body formation (ketogenesis). These ketone bodies are transported to extrahepatic tissues where they become the major fuel source.

Intermediary Metabolism at the Subcelluar Level

1. **Mitochondrion** acts as the focus and crossroad of carbohydrate, lipid, and amino acid metabolism. Particularly, it is the **house** of the enzyme of citric acid cycle, of the respiratory chain and ATP synthase, of β-oxidation of fatty acids, and ketone body production. It is also the collecting chamber for the carbon skeletons of amino acids after transamination and for utilizing these skeletons for the synthesis of nonessential amino acids.
2. In the **cytosol**; glycolysis, HMP shunt or the pentose phosphate pathway and fatty acid synthesis all take place. Substances such as lactate and pyruvate that are formed in cytosol must enter the mitochondrion and form oxaloacetate before conversion to glucose in the process of gluconeogenesis.
3. The membranes of the endoplasmic reticulum contain the enzyme system for acylglycerol synthesis and the ribosomes are responsible for **protein synthesis**.
4. The transport of metabolites of varying size, charge and solubility through the membranes separating organelles involves complex mechanisms.

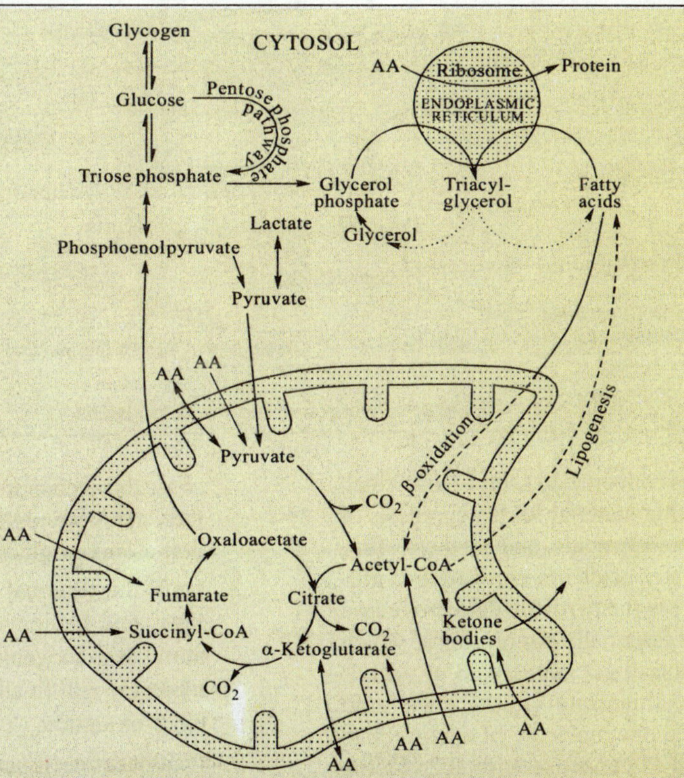

Fig. 21.7 : Intracellular location and integration of major metabolic pathways in a liver parenchymal cell
AA → Metabolism of one or more essential amino acids
AA ↔ Metabolism of one or more nonessential amino acids

CHAPTER 22

INTERRELATIONSHIP IN THE METABOLISM OF PROTEIN FAT AND CARBOHYDRATE

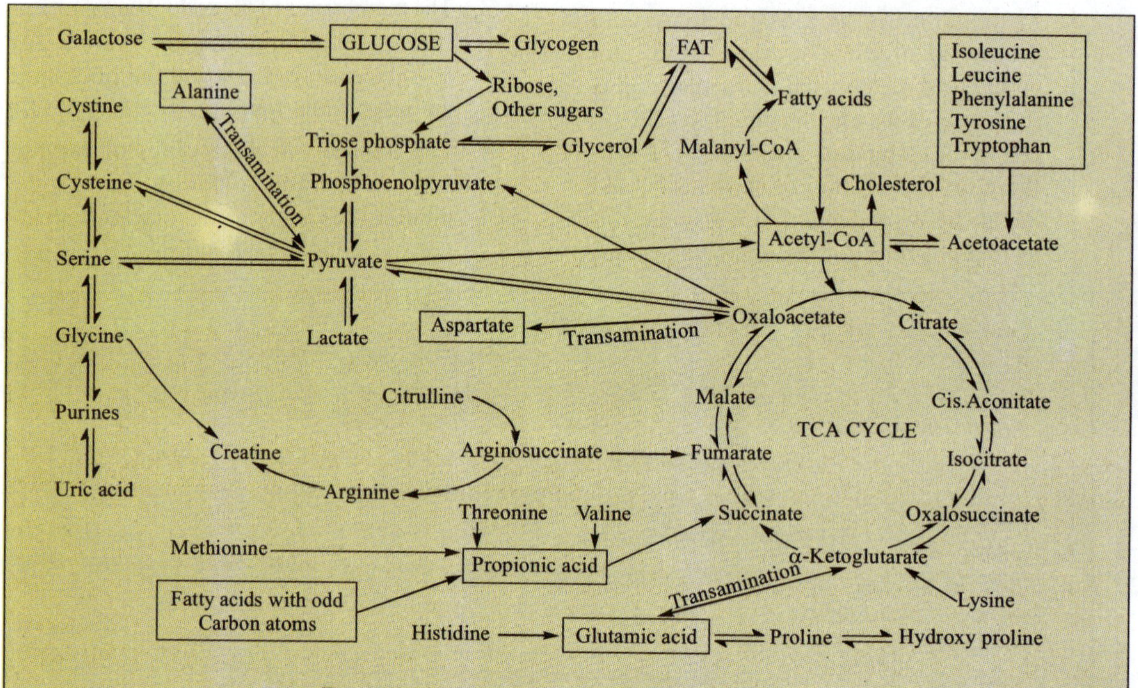

Fig. 22.1 : Inter conversion of the major foodstuffs

1. Pyruvate is converted to acetyl-CoA which is the starting material for the synthesis of long-chain fatty acids, and by the reverse process, fatty acids are converted to glucose whereas the pyruvate dehydrogenase reaction is essentially nonreversible which prevents the direct conversion of acetyl-CoA to oxaloacetate via the citric acid cycle, since one molecule of oxaloacetate is required to condense with acetyl-CoA and only one molecule of oxaloacetate is regenerated. On the basis of similar reasons, there cannot be a net conversion of fatty acids having an even number of carbon atoms to glucose or glycogen.
2. Only the terminal 3-carbon portion of a fatty acid having odd number of carbon atoms is glucogenic, as this portion of the molecule will finally form propionyl-CoA by β-oxidation.
3. Oxaloacetate is an intermediate both in the citric acid cycle and in the pathway of gluconeogenesis.

4. The glycerol moiety of triacylglycerol can form glucose after activation of glycerol 3-phosphate.
5. Many of the carbon skeletons of the non-essential amino acids can be produced from carbohydrate through the citric acid cycle and transamination. Glucogenic amino acids can also produce carbon skeletons that are precursors or members of the citric acid cycle by the reversal of these processes. They are readily converted to glucose and glycogen by gluconeogenic pathways.
6. The ketogenic amino acids are converted to acetoacetate which is metabolized as ketone bodies forming acetyl-CoA in extra hepatic tissues.
7. It is not possible for a net conversion of fatty acids to carbohydrate and also for a net conversion of fatty acids to glucogenic amino acids. However, it is possible for the conversion of the carbon skeletons of glucogenic amino acids to fatty acids either by formation of pyruvate and acetyl-CoA or by reversal of nonmitochondrial reactions of the citric acid cycle from α-ketoglutarate to citrate followed by the action of ATP-citrate lyase to give acetyl-CoA. However, under starvation, a net breakdown of protein and amino acids is accompanied by a net breakdown of fat. The net conversion of amino acids to fat is not a significant process except in animals receiving a high-protein diet.

Necessity of Glucose for the brain and erythrocytes

1. Gluconeogenesis is very important because certain tissues and cells including the central nervous system and the erythrocytes are fully dependent on a continual supply of glucose.
2. A minimal supply of glucose is also necessary for extrahepatic tissues to maintain oxaloacetate concentrations and the integrity of the citric acid cycle.
3. Glucose is the main source of glycerol 3-phosphate in tissues devoid of glycerol kinase such as adipose tissue.
4. A minimal amount of glucose is a must under all conditions.
5. Large quantities of glucose are essential for fetal nutrition and synthesis of lactose in milk.
6. Certain mechanisms are also involved in supplying glucose in times of shortage by introducing other substrates to spare its oxidation.

Necessity of continual supply of fuel during starvation

1. During starvation, glucose availability from food becomes less and liver glycogen is involved in maintaining the blood glucose. The level of insulin in the blood decreases, and glucagon increases.
2. Since the glucose utilization diminishes in adipose tissue and the inhibitory effect of insulin on lipolysis becomes less, fat becomes mobilized as free fatty acids and glycerol. The free fatty acids are oxidized or esterified in the tissues. Glycerol is activated to glycerol 3-phosphate in the liver and kidney.
3. In fasting state, endogenous glucose production (from amino acid and glycerol) does not keep pace with its utilization and oxidation. Therefore, the stores of liver glycogen become depleted and blood glucose level falls.
4. Fat is mobilized rapidly and in several hours the plasma free fatty acids and blood glucose stabilize at the fasting level. At this point, there is the increased oxidation of fatty acids forming ketone bodies. In case of demands of more glucose, carbohydrate in the form of glycerol in adipose tissue performs an important function, for it is only this source of carbohydrate together with that provided by gluconeogenesis from protein.
5. In prolonged starvation in humans, gluconeogenesis from protein is diminished. This adapts the brain to replace half of the glucose oxidized with ketone bodies.

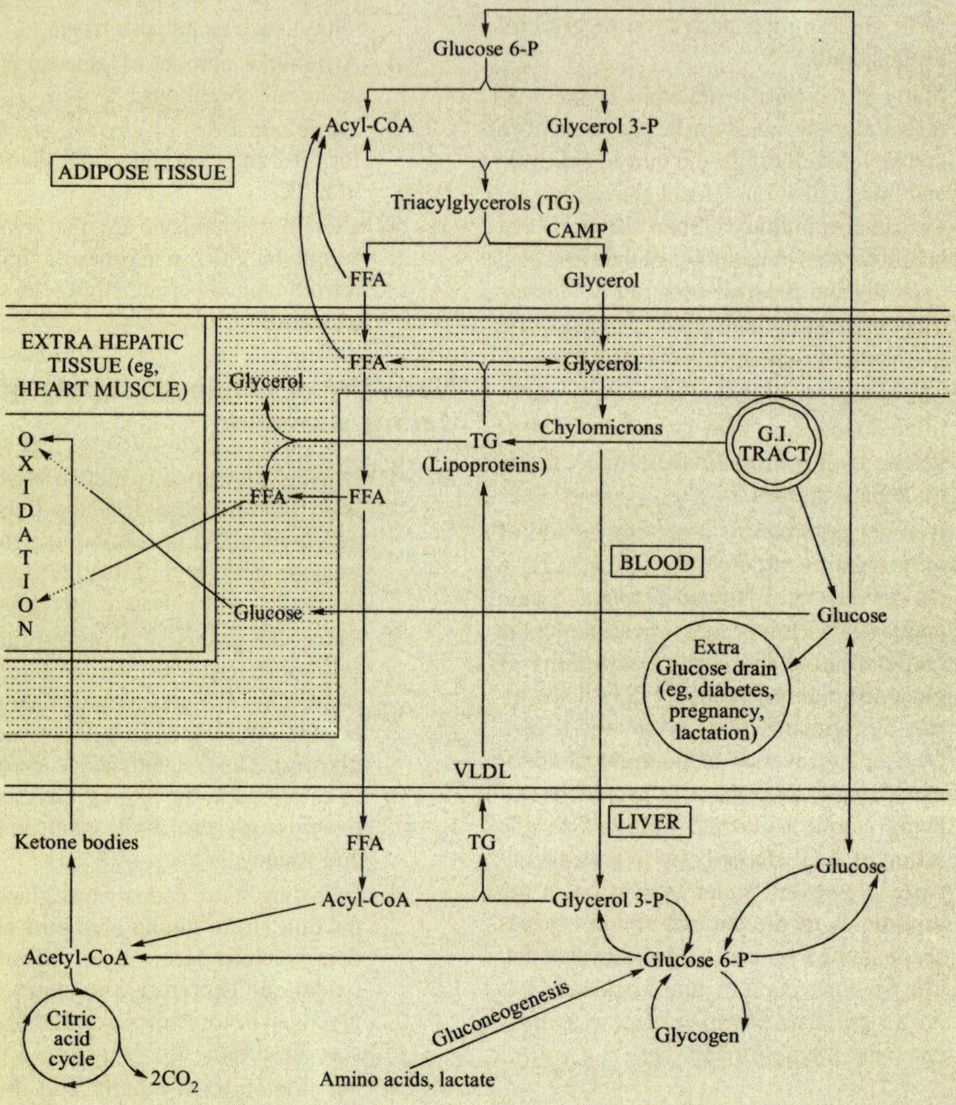

Fig. 22.2 : Metabolic interrelationships between adipose tissue, the liver, and extrahepatic tissues. (LPL, lipoprotein lipase : FFA, free fatty acids; VLDL, very low density lipoporteins.)

Table 22.1 : Summary of the major regulators of metabolic pathways

Pathway	Major Regulatory Enzymes	Activator	Inhibitor	Effector Hormone	Remarks
Citric acid cycle	Citrate synthase		ATP, long-chain acyl-CoA		Regulated mainly by the need for ATP and therefore by the supply of NAD^+
Glycolysis	Phosphofructo-kinase	AMP, fructose 2,6-bisphosphate in liver, fructose 1, 6-bisphosphate in muscle	Citrate (fatty acids, ketone bodies), ATP, cAMP	Glucagon ↓	Induced by insulin
Pyruvate oxidation	Pyruvate dehydro-genase	CoA, NAD, ADP, pyruvate	Acetyl-CoA, NADH, ATP (fatty acids, ketone bodies)	Insulin ↑ (in adipose tissue)	Also important in regulating the citric acid cycle
Gluconeogenesis	Pyruvate carboxylase	Acetyl-CoA	ADP	Glucagon?	Induced by glucocorticoids, glucagon, cAMP, Repressed by insulin
	Phosphoenol-pyruvate carboxykinase	cAMP ?			
	Fructose-1, 6-bisphosphatase	cAMP	AMP, fructose 2, 6-bisphosphate in liver, fructose 1, 6-bisphosphate in muscle	Glucagon	
Glycogenesis	Glycogen synthase		Phosphorylase (in liver), cAMP, Ca^{2+} (in muscle)	Insulin ↑ Glucagon (liver) ↓ Epinephrine ↓	Induced by insulin
Glycogenolysis	Phosphorylase	cAMP, Ca^{2+} (muscle)		Insulin ↓ Glucagon (liver) ↑ Epinephrine ↑	
Pentose phosphate pathway	Glucose-6-phosphate dehydrogenase	$NADP^+$	NADPH		Induced by insulin
Lipogenesis	Acetyl-CoA carboxylase	Citrate	Long-chain acyl-CoA, cAMP	Insulin ↑ Glucagon (liver) ↓	Induced by insulin
Cholesterol synthesis	HMG-CoA reductase		Cholesterol, cAMP, mevalonate, bile acids	Insulin ↑ Glucagon (liver) ↓	Inhibited by certain drugs, eg, lovastatin

Table 22.2 : Summary of the major and unique features of metabolism of the principal organs

Organ	Major Function	Major Pathways	Main Substrates	Major Products	Specialist Enzymes
Liver	Service for the other organs and tissues	Most represented, including gluconeogenesis; β-oxidation; ketogenesis; lipoprotein formation; urea, uric acid, and bile acid formation; cholesterol synthesis; lipogenesis[1]	Free fatty acids, glucose (well fed), lactate, glycerol, fructose, amino acids (Ethanol)	Glucose, VLDL (triacylglycerol) HDL, ketone bodies, urea, uric acid, bile acids, plasma proteins (Acetate)	Glucokinase, glucose-6-phosphatase, glycerol kinase, phosphoenolpyruvate carboxykinase, fructokinase, arginase, HMG-CoA synthase and lyase, 7α-hydroxylase (Alcohol dehydrogenase)
Brain	Coordination of the nervous system	Glycolysis, amino acid metabolism	Glucose, amino acid, ketone bodies (in starvation) Polyunsaturated fatty acids in neonate	Lactate	
Heart	Pumping of blood	Aerobic pathways, eg, β-oxidation and citric acid cycle	Free fatty acids, lactate, ketone bodies, VLDL and chylomicron triacylglycerol, some glucose		Lipoprotein lipase. Respiratory chain well developed
Adipose tissue	Storage and breakdown of triacylglycerol	Esterification of fatty acids and lipolysis; lipogenesis[1]	Glucose, lipoprotein triacylglycerol	Free fatty acids, glycerol	Lipoprotein lipase, hormone sensitive lipase
Muscle Fast twitch Slow twitch	Rapid movement Sustained movement	Glycolysis Aerobic pathways, eg, β-oxidation and citric acid cycle	Glucose Ketone bodies, triacylglycerol in VLDL and chylomicrons, free fatty acids	Lactate	Lipoprotein lipase. Respiratory chain well developed
Kidney	Excretion and gluconeogenesis	Gluconeogenesis	Free fatty acids, lactate, glycerol	Glucose	Glycerol kinase, phosphoenolpyruvate carboxykinase
Erythrocytes	Transport of O_2	Glycolysis, pentose phosphate pathway. No mitochondria and therefore no β-oxidation or citric acid cycle	Glucose	Lactate	(Hemoglobin)

[1]In many species but not very active in humans.

CHAPTER 23

METABOLISM OF PURINE AND PYRIMIDINE NUCLEOTIDES

Sources of the various atoms of the purine base

1. Glycine is utilized to form the carbon positions 4 and 5 and its α-nitrogen forms the nitrogen in position 7.
2. The amino nitrogen of aspartic acid provides the nitrogen at position 1.
3. The N atoms at positions 3 and 9 are derived from the amide nitrogen glutamine.
4. The carbon atom at position 6 is derived from CO_2.
5. The carbons in positions 2 and 8 are supplied from a one carbon (C-I) compound given by the tetrahydrofolate carrier.

This is represented below (Fig. 23.1) :

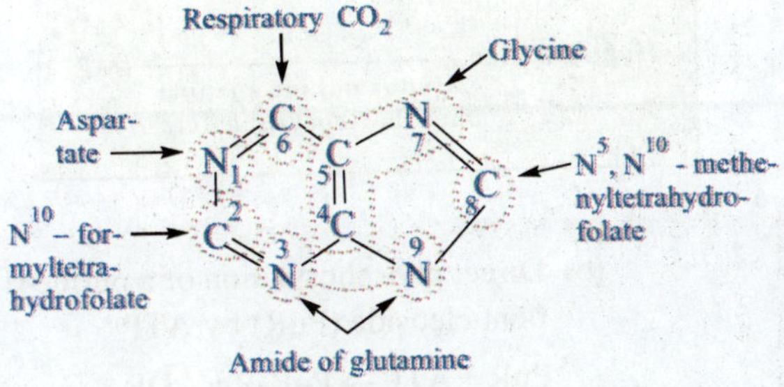

Fig. 23.1 : Sources of nitrogen and carbon atoms of the purine ring

Biosynthesis of purine nucleotides [DE NOVO]

1. Ribose-5-phosphate is converted into 1-pyrophosphoribosyl-5-phosphate (PP ribose P) by PP ribose P synthetase with ATP and Mg^{++}.
2. PP ribose P then reacts with glutamine by the enzyme **phosphoribosyl pyrophosphate glutamyl amidotransferase** to form 5-phosphoribosylamine by the displacement of pyrophosphate and the formation of glutamate.
3. 5-Phosphoribosylamine reacts with glycine to produce glycinamide ribosylphosphate by **glycinamide kinosynthetase** in presence of ATP.
4. The N_7 of glycinamide ribosylphosphate is formylated by the enzyme **glycinamide ribosylphosphate formyltransferase** to transfer the C_1 moiety.
5. Amidation from glutamine occurs at the C_4 of the formylglycinamide ribosyl-5-phosphate by formalglycinamide ribosyl-5-phos-phate synthetase requiring ATP. The amide N becomes position 3 in the purine.
6. Imidazole ring is closed by aminoimidazole ribosyl-5-phosphate synthetase requiring ATP.
7. Respiratory CO_2 is utilized requiring biotin to form aminoimidazole carboxylate

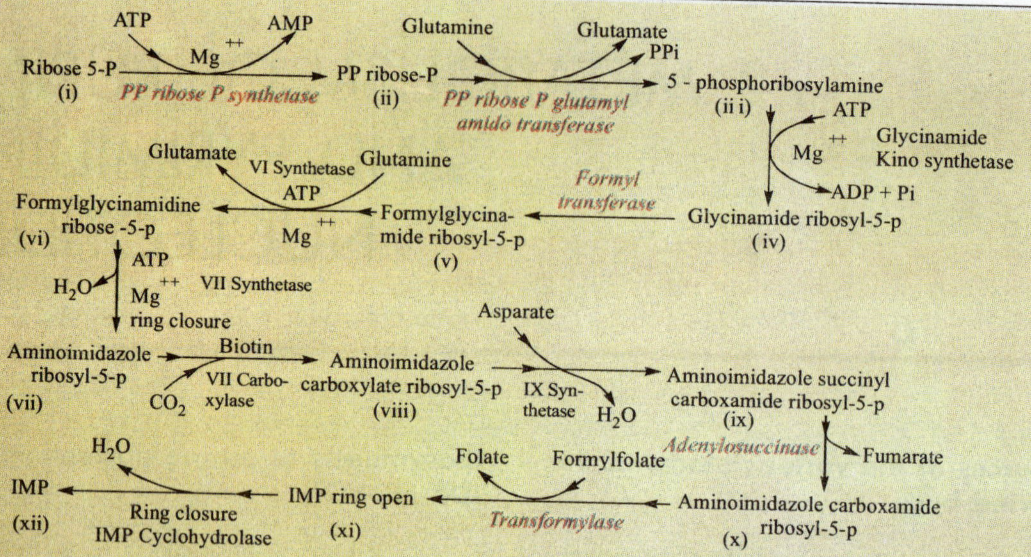

Fig. 23.2 : The pathway of de novo purine biosynthesis

ribosyl-5-phosphate by aminoimidazole ribosyl-5-phosphate carboxylase.

8. Addition of Aspartate by IX synthetase forms aminoimidazole succinyl carboxamide ribosyl-5-phosphate which is converted into aminoimidazole carboxamide ribosyl phosphate by adenylosuccinase.
9. Transformylase converts the above products into nucleoside phosphate and then the ring is closed.
10. The major determinant of the overall rate of de novo purine nucleotide biosynthesis is the concentration of PRPP.

The rate of PRPP synthesis depends both on the availability of ribose 5-phosphate and on the activity of PRPP synthetase.

SALVAGE REACTION

1. In salvage reaction purines, purine ribonucleosides and purine deoxyribonucleosides are converted to mononucleotides. There is requirement of far less energy than de novo synthesis.
2. The more important mechanisms are :

   (a) Phosphoribosylation of a free purine (Pu) by PRPP, forming a purine 5′-mononucleotide (Pu-RP)

   $$\text{Pu} + \text{PP-RP} \rightarrow \text{Pu-RP} + \text{PPi}$$

   This reaction is catalyzed by adenine phosphoribosyltransferase and hypoxanthine guanine phosphoribosyltransferase.

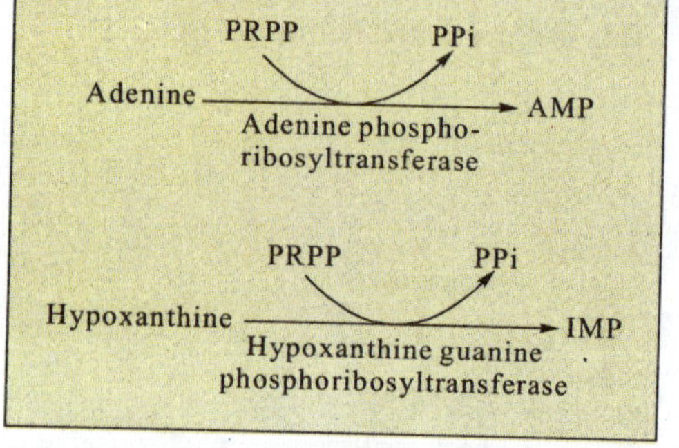

   (b) Direct phosphorylation of a purine ribonucleoside (PuR) by ATP

   $$\text{PuR} + \text{ATP} \rightarrow \text{PuR-P} + \text{ADP}$$

   Adenosine kinase catalyzes phosphorylation of adenosine to AMP or of deoxyadenosine to dAMP.

Hepatic Purine Nucleotide Biosynthesis is Stringently Regulated

1. Mammalian liver is the major site of purine nucleotide biosynthesis. Liver provides purines and their nucleosides for salvage reaction.
2. Human brain has a low level of PRPP amidotransferase and hence depends on exogenous purines.
3. Erythrocytes and polymorphonuclear leukocytes cannot synthesize 5-phosphoribosylamine and so utilize exogenous purines to form nucleotides.

AMP and GMP Feedback regulate their formation from IMP

1. Two mechanisms regulate conversion of IMP to GMP and AMP.
2. AMP feedback regulates adenylo succinate synthetase and GMP feedback inhibits IMP dehydrogenase.
3. For the conversion of IMP to AMP requires GTP and conversion of xanthinylate to GMP requires ATP.
4. Cross regulation between the pathways of IMP metabolism thus serves to decrease synthesis of one purine nucleotide when there is a deficiency of the other nucleotide.

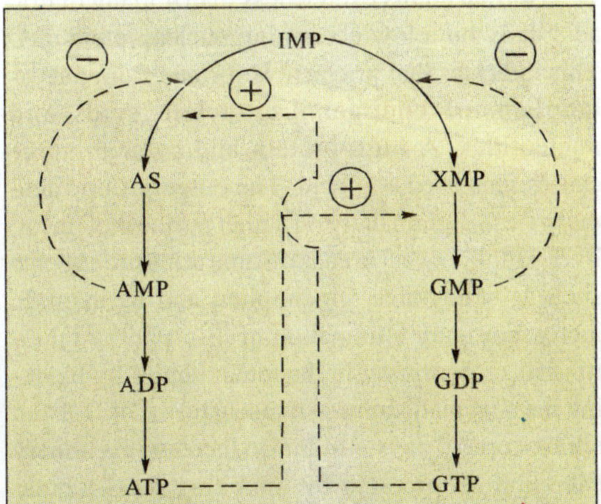

Fig. 23.3 : Regulation of the interconversion of IMP to adenosine nucleotide and guanosine nucleotides. Solid lines represent chemical flow and broken lines indicate both positive and negative feedback regulation.

Catabolism of purines

Uric acid is the chief end-product of purine catabolism in man and the higher apes. Other mammals degrade uric acid to allantoin by means of the enzyme, uricase, which is lacking in primates.

Almost all tissues contain enzymes capable of breaking nucleoprotein down to nucleoside which can be oxidized to uric acid. Uric acid is always excreted even on a purine-free diet or in starvation. Urinary uric acid is both endogenous and exogenous in origin.

Organisms that form uric acid as the major nitrogenous waste product are said to be uricotelic. Birds, amphibius and reptiles do not possess uricase activity. These animals excrete uric acid and guanine as the end-products of purine metabolism and nitrogen(protein) metabolism. In man and most of the mammals, urea is the main product of nitrogen metabolism. Hence, they are ureotelic.

In animals other than mammals, uric acid is further degraded to urea and glyoxylic acid.

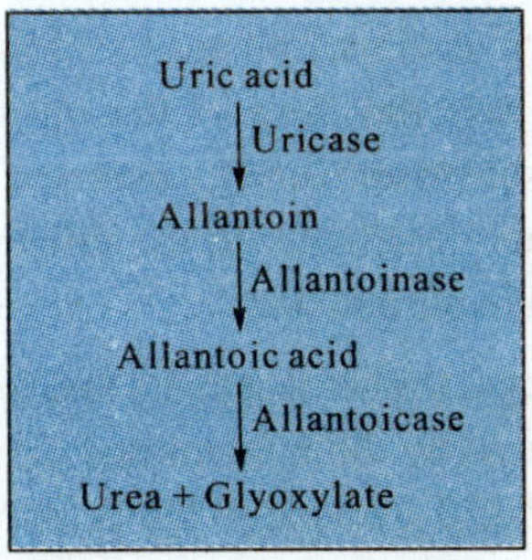

Formation of uric acid

1. Adenine (6-aminopurine) is deaminated by adenylate deaminase to form inosinic acid. Adenylate deaminase is quite abundant in skeletal muscle. Adenosine can also be deaminated to form inosine.
2. Both inosinic acid and inosine give rise to free hypoxanthine, which may be

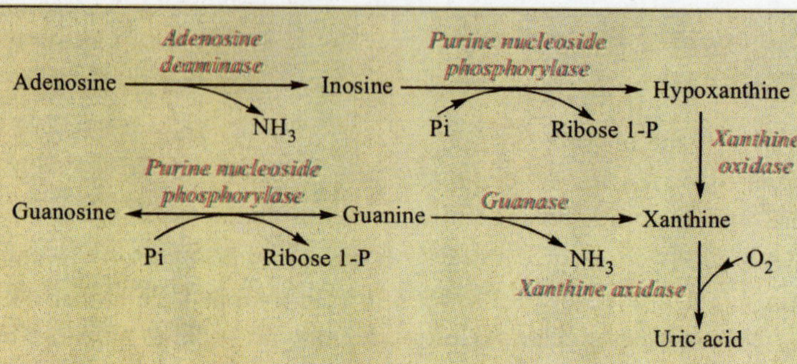

Fig. 23.4 : Flow chart of formation of uric acid

reutilized for nucleic acid synthesis but is most frequently oxidized to xanthine by the enzyme xanthine oxidase present in greatest amount in liver, small intestine and kidney.

3. Xanthine oxidase further oxidizes xanthine to uric acid (2, 6, 8-trioxypurine).
4. Free guanine (2-amino-6-oxypurine) is deaminated to form xanthine directly by the enzyme guanase, which is very active in most tissues. The liberated xanthine is then converted to uric acid by xanthine oxidase.

The pathway for the formation of uric acid is as follows (Fig. 23.4) :

Some uric acid may be produced from nucleic acid by the bacterial flora of the intestinal tract, when it is absorbed and directly excreted. This pathway is a minor contributor to the urinary uric acid of persons on a normal diet.

From recent studies it appears that sodium urate is freely filtered by the mammalian glomerulus. It is reabsorbed and secreted in the proximal tubules and the loop of Henle and partially reabsorbed in the distal convoluted tubules. The net excretion of total uric acid in normal men is 600-700 mg in 24 hours. Aspirin in high doses competitively inhibits urate excretion as well as reabsorption. Allopurinol competitively inhibits Xanthine oxidase for which uric acid cannot be formed.

Uric acid is mainly excreted in urine, to a lesser extent in digestive fluid, and in small amounts in sweat and saliva. A portion of the uric acid is destroyed by bacterial action in the intestine. This intestinal uricolysis gives rise to urea and ammonia, which are absorbed and excreted by the kidneys. Under conditions of normal production and removal, the body contains a 'readily miscible uric acid pool'.

The normal uric acid content of serum is 2.5 to 7.0 mg/100 ml for adult males and 1.5 to 6.0 mg/100 ml for premenopausal females. One-third of it is loosely bound to plasma proteins, mostly albumin, but some is bound to $\alpha_1 - \alpha_2$ globulins. Supersaturation of uric acid causes the disease gout which is much more common in males. Only about 5 per cent of gouty patients are females and most of them are menopausal.

Normal adults excrete less than 450 mg of uric acid daily on a low containing nucleoprotein diet. This indicates that uric acid is formed from the catabolism of endogenous nucleic acids and nucleotides. A high protein and caloric intake causes increased uric acid. The output of uric acid may rise to 1 gram daily on a high purine diet (meat, liver, kidney, sweet breads). Some uricosuric agents such as salicylates, cinchophen and carinamide increase urinary elimination of uric acid by inhibiting its reabsorption in the renal tubules by blocking the enzymatic transport mechanism. ACTH and adrenocortical oxysteroids also increase the urinary excretion of uric acid by inhibiting renal tubular reabosrption.

Hyperuricemia is due to overproduction, decreased destruction and decreased renal excretion.

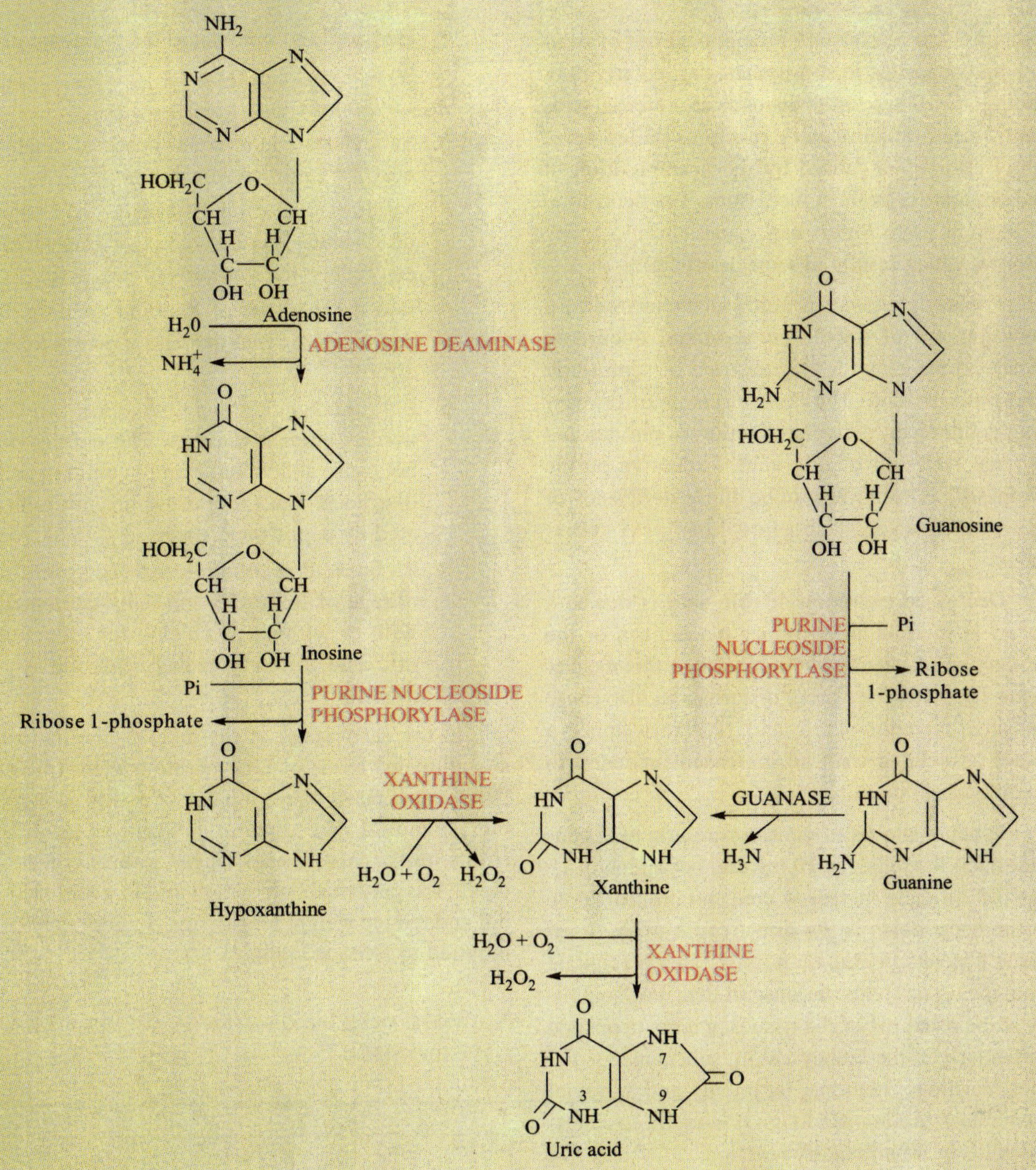

Fig. 23.5 : Reactions of formation of uric acid

Increased values are observed in all forms of nephritis with nitrogen retention. Values as high as 10 mg /100 ml are frequently observed. Serum uric acid is also increased in **eclampsia**. In **chronic leukemia**, blood uric acid level is increased. Uric acid level in blood may also shoot up in **sickle cell anemia**, **thalassemia**, **hemolytic anemia** and **macroglobulinemia**.

GOUT

Sodium urate crystals are precipitated out of solution and deposited in soft tissues, particularly in or about joints. These urate deposits are referred to as tophi. Acute inflammatory reactions called acute gouty arthritis is caused by the accumulation of sodium urate crystals in the tissues. The deposition of sodium urate tophi can cause chronic gouty arthritis which results in joint destruction.

Primary gout is an arthritis characterized by a derangement of purine metabolism, occurring mostly in males, with the elevation of serum uric acid concentration. The hyperuricemia of primary gout is due to excessive production of purines and to renal retention of uric acid. Excessive purine synthesis has been found to be due to deficiency of hypoxanthine–guanine phosphoribosyl transferase.

Decreased excretion of uric acid is due to a defect in the tubular secretory machanism or the production of a metabolic inhibitor of this mechanism. The renal circulatory disturbances also cause defective excretion of uric acid. Hyperuricemia is a factor associated with an increased tendency to myocardial infarction.

In acute attacks of gout, serum uric acid concentration is as high as 10 mg and occasionally 15 mg/100 ml. The increased concentration of urate in the body fluids is the direct cause of the local tissue deposits in chronic gout. Urates precipitate from the tissue fluids, because of their low solubility at the usual pH of the tissues and are deposited particularly in the kidneys, skin, subcutaneous tissues, cartilage, tendons, ligaments and synovial membranes. Acting as foreign bodies, these deposits incite an inflammatory reaction.

The 'urate pool', i.e., the amount of urate capable of mixing promptly with intravenously injected urate, is increased in subject with gout. 'Secondary gout' is applied to a condition with similar clinical signs with various disorders accompanied occasionally by elevation of serum uric acids. The secondary gout is accompanied by an increased rate of metabolic turnover of nucleic acids.

Treatment of Gout

1. Bed rest and colchicine for the acute attack.
2. Low purine diet.
3. Salicylates and/or small doses of colchicine for the intercritical period.
4. Recently, the drug allopurinol has been discovered. This is an effective drug which can reduce the synthesis of uric acid.

   It is the inhibitor of Xanthine oxidase. It can produce renal damage showing toxic effects.
5. Endomethacin, Phenylbutazone can be used as alternative drugs. These drugs do not cause abdominal cramp, vomiting and diarrhoea. But these drugs are not tolerated by peptic ulcer cases.
6. Probenacid or Binemid can be given in dosage of 0.5 to 2.0 gm daily combined with or without colchicine for reducing uric acid. It rarely causes nephrotic syndrome.

In the early stage of the disease, the patient should avoid foods of high purine content (meat, fish, beans, peas, sweet breads, liver and kidney) but he should take diets of low purine content (fruits, milk, cheese, eggs, sugar, sweets, vegetables). The patient should be advised to avoid alcohol and reduce weight for obesity. Surgery may be required to remove tophi.

Sources of various atoms of pyrimidines

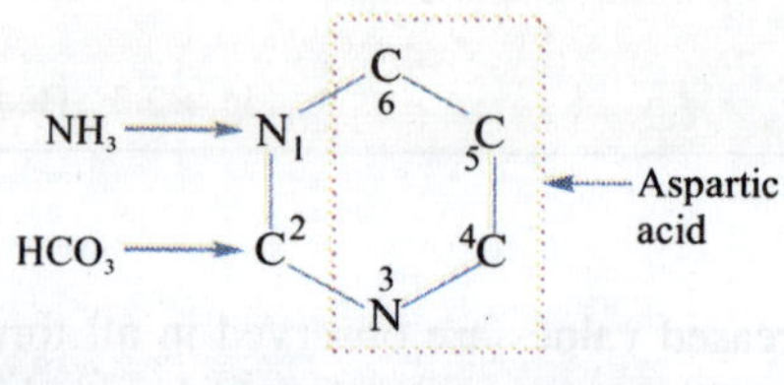

Fig. 23.6 : Sources of nitrogen and carbon atom of pyrimidine ring

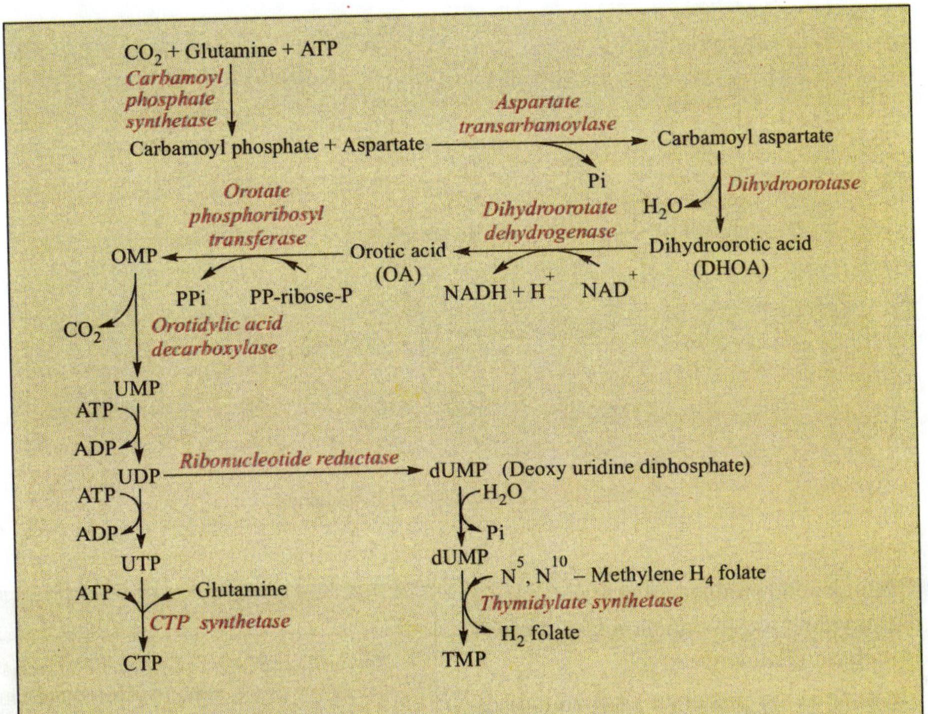

Fig. 23.7 : The biosynthetic pathway for pyrimidine nucleotides

Biosynthesis of Pyrimidines

1. The synthesis of pyrimidine ring starts with the formation of carbamoyl phosphate from glutamine, ATP and CO_2 being catalyzed by carbamoyl phosphate synthetase, present in the cytosol of the cell.
2. Carbamoyl phosphate then reacts with aspartate by aspartate transcarbamoylase to form carbamoyl aspartate which, with the loss of water, is converted into dihydroorotic acid (DHOA) by the enzyme dihydroorotase.
3. Dihydroorotic acid on dehydrogenation by dihydroorotate dehydrogenase utilizing NAD as coenzyme is converted into orotic acid (OA) which is by the action of orotate phosphoribosyl transferase, converted into orotidine monophosphate (OMP). This on subsequent decarboxylation by orotidylic acid decarboxylase forms uridine monophosphate (UMP).
4. By further phosphorylation, UMP is converted into UDP and then to UTP.
5. UTP is converted into CTP in the reaction catalyzed by CTP synthetase utilizing ATP and glutamine.
6. The enzyme Ribonucleotide reductase converts UDP into deoxyuridine diphosphate (dUDP) which is converted into dUMP. This is by the action of thymidylate synthetase with N^5, N^{10}-methylene H_4 folate, converted into thymidine monophosphate (TMP).

The overall reaction is given in Fig. 23.7.

Catabolism of Pyrimidines

1. Liver is the main site for the catabolism of pyrimidines.
2. CO_2 is released from the pyrimidine nucleus representing a major pathway for the catabolism of uracil, cytosine, and thymine.

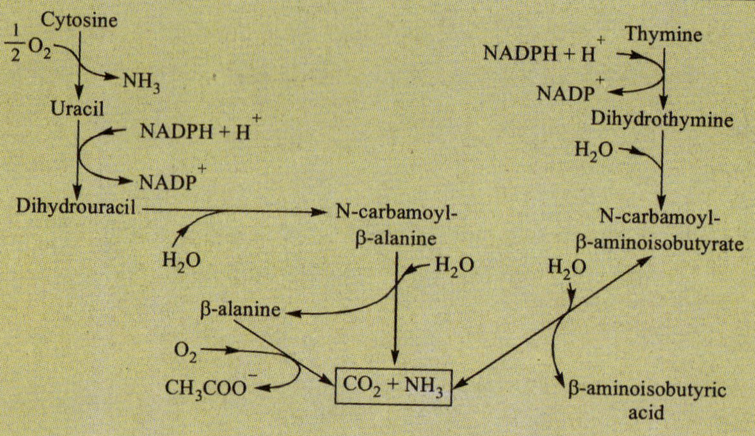

Fig. 23.8 : Catabolism of pyrimidines

3. The major end products of cytosine, uracil and thymine are β-alanine and β-amino-isobutyric acid, respectively.
4. Thymine is the precursor of β-aminoiso-butyric acid in humans and in animals. β-aminoisobutyric acid is excreted more in leukemia. This is due to increased destruction of cells and their DNA.
5. The β-aminoisobyutyric acid is converted into methylmalonic semialdehyde and then to propionate which turns to succinate.

The overall reactions for degradation is noted (Fig. 23.8).

CLINICAL ORIENTATION

- The dietary purine and pyrimidine bases are not incorporated into tissue nucleic acids of humans. Rather, humans biosynthesize the purines and pyrimidines of tissue nucleic acids, ATP, and NAD^+. CoA, etc. from amphibolic intermediates.

CLINICAL ORIENTATION

- The injected purine or pyrimidine analogs including anticancer drugs may be incorporated into DNA.
- The abnormal human diseases in purine metabolism include gout, Lesch-Nyhan syndrome, adenosine deaminase deficiency and purine nucleoside phosphorylase deficiency. These diseases should be treated properly for eliminating the sufferings.
- The diseases of pyrimidine biosynthesis, although rare, include orotic aciduria.
- The clinically singnificant disorders of pyrimidine catabolism are few, since the products of pyrimidine catabolism are highly soluble (CO_2, HN_3, and β-aminoisobutyrate).

EXERCISE

1. Discuss the catabolism of purines and pyrimidines. (R. U. 68A)
2. Write notes on :
   (a) Uric acid (P.U. 64A; 73S)
   (b) Gout. (M.U.72A; R.U. 69S)
3. Discuss the biosynthesis of purines and pyrimidines. (P.U. 68A)

SCHEMATIC COMPENDIUM

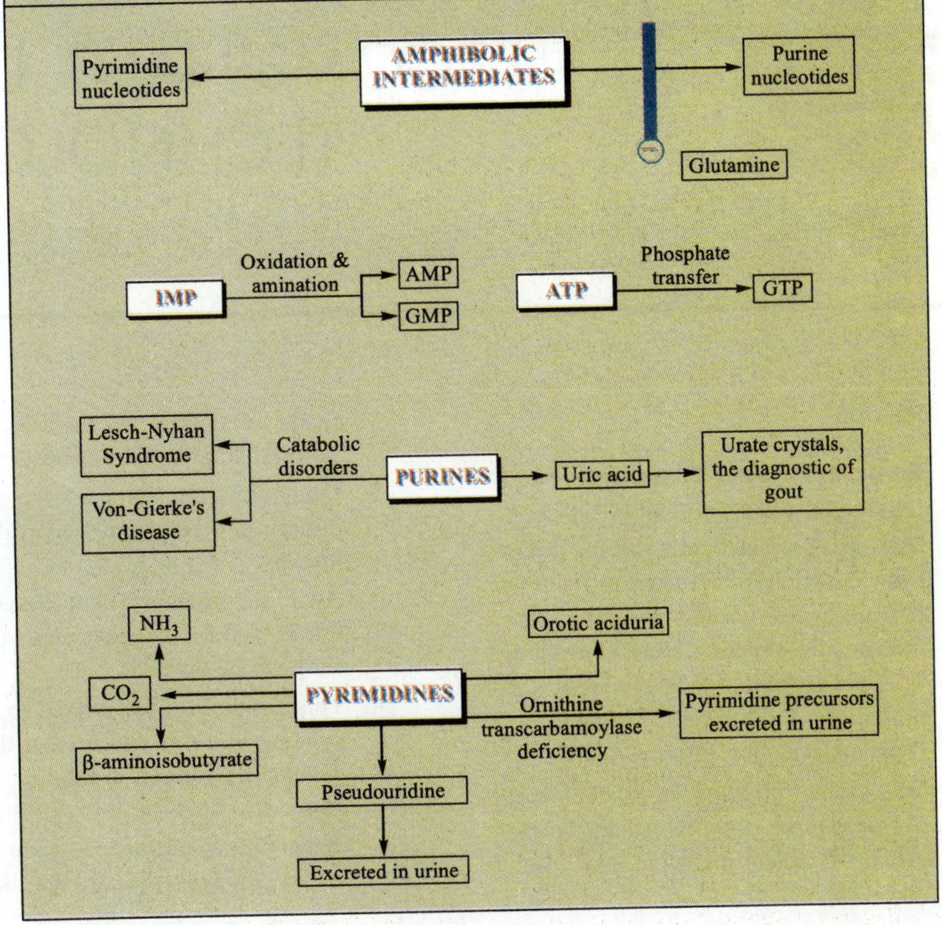

CHAPTER 24

RECOMBINANT DNA TECHNOLOGY

Introduction

1. Recombinant DNA technology is better referred to as genetic engineering.
2. Much has been learned about the diseases from the study of affected proteins, but this mechanism cannot be applied where the specific genetic defect is unknown. This new technology overcoming these limitations will approach directly to the DNA molecule for information.
3. This chapter presents the basic concepts of recombinant DNA technology, its application to clinical medicine, and a glossary.

Biomedical Importance

1. It is helpful to give clear idea regarding the molecular basis of a number of diseases (eg, familial hypercholesterolemia, sickle cell disease, the thalassemia, cystic fibrosis, Huntington's chorea).
2. Using this technology, a large quantity of human proteins can be produced for therapy.
3. By its aid proteins for vaccines (eg, hepatitis B) and for diagnostic tests (eg, AIDS test) can be obtained.
4. This technology is utilized to diagnose existing diseases and predict the risk of developing a given disease.
5. Gene therapy for sickle cell disease, the thalassemias, adenosine deaminase deficiency, and other diseases may be devised.

CONCEPTS USED IN RECOMBINANT DNA TECHNOLOGY

Isolation and manipulation of DNA is the object of recombinant DNA research. This requires several techniques and reagents.

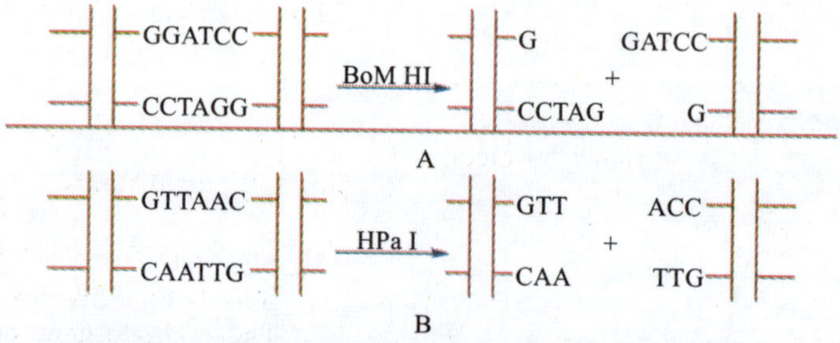

Fig. 24.1 : Results of restriction endonuclease digestion

Restriction Enzymes

1. Some endonucleases that cut DNA at specific DNA sequences within the molecule are a key tool in recombinant DNA research. These enzymes were originally said to be restriction enzymes. More than 200 defensive enzymes protect the host bacterial DNA from foreign organism (primarily infective phages). They are only present in cells that also have a companion enzyme that methylates the host DNA giving it an unsuitable substrate for digestion by the restriction enzyme.
2. The restriction enzymes are named vide the bacterium from which they are isolated (eg, Eco RI from Escherichia Coli, Bam HI from bacillus amyloliquefaciens).
3. Each enzyme recognizes and cleaves a specific double-stranded DNA sequence. These DNA cut result in blunt ends or overlapping (sticky) ends (Bam HI) (Fig. 24.1), depending on the mechanism used by the enzyme. Sticky ends are particularly useful in constructing hybrid or chimeric DNA molecules.
4. In case the nucleotides are distributed randomly in a given DNA molecule, one can easily calculate how frequently a given enzyme could cut a length of DNA.
5. For each position in the DNA molecule there are 4 possibilities (A,C, G or T); therefore, a restriction enzyme that recognizes a 4-bp sequence will cut, on average, once every (4^4), whereas another enzyme that recognizes a 6-bp sequence will cut once every (4^6).
6. When DNA is digested with a given enzyme, the ends of all the fragments will have the same DNA sequence. The fragments produced can be isolated by electrophoresis.

Preparation of chimeric DNA molecules

1. Sticky ends of a vector may reconnect with themselves with no gain of DNA. These ends of fragments can also anneal so that tandem heterogeneous inserts form. These end sites may not be available or in a convenient position.
2. To overcome the above problems, an enzyme that generates blunt ends is used and new ends are added using the enzyme terminal transferase.
3. If poly d(G) is added to the 3′ ends of the vector and poly d(C) is added to the 3′ ends of the foreign DNA, the two molecules can only anneal to each other and thus overcome the above problem. This procedure, called homopolymer tailing, also generates an Sma I restriction site.
4. Sometimes, synthetic oligonucleotide linkers with a convenient restriction enzyme sequence are ligated to the blunt-ended DNA. This technique has the advantage of joining together any pairs of ends. The disadvantanges are that there is no control over the orientation of insertion of the number of molecules annealed together.

MOLECULAR BASIS OF INHERITED DISEASES

A. Genetic Diseases

The human genome contains DNA with about 3 billion (10^9) base pairs that encode 50,000 to 100,000 genes located on 23 pairs of chromosomes. The molecular defects of inherited diseases are based on mutation in DNA causing the incorporation of an incorrect amino acid into a protein. Genetic diseases are divided into five principal categories.

1. Single gene defects

(a) Single gene disorders result from a mutation at a single site on a chromosome. These single gene defects indicate new significance to the old genetic terms "recessive" and "dominant".

(b) Recessive pattern shows the deficiencies in enzymes. In a heterzygote for a recessive disorder the one normal gene often provides sufficient enzyme to prevent clinical sysmptoms.

(c) The heterozygote with one normal and one defective gene coding for structural proteins exhibit disease symptoms.

2. Multifectoral disorders

(a) These are determined by many genes and are influenced by environmental factors, eg, diabetes mellitus, hypertensive and manic depression.

(b) These occur more frequently in the immediate family of an affected person.

(c) Twins that are genetically identical show a greater similarity than twins that have half of their genes in common.

3. Chromosome disorders

(a) These are caused by the loss, gain or abnormal arrangement of one or more of the 46 chromosomes composing the human diploid genome (two copies of each of the 22 autosomal chromosomes plus two sex chromosomes).

(b) Chromosomal abnormalities are common.

(c) The most common autosomal defect is trisomy 21 (three copies of chromosome 21) which is responsible for about 95 percent of cases of Down syndrome.

4. Somatic cell gene defects

Somatic cell gene mutations are important in tumorigenesis.

5. Mitochondrial mutations

(a) Circular molecules of DNA (mt DNA) are present in mitochondria. Each codes for 13 of 100 polypeptides required for the process of oxidative phosphorylation, as well as for 2 species of rRNAs and 22 kinds of tRNA.

(b) mt DNA is maternally inherited because mitochondria from the sperm cell do not enter the fertilized egg.

(c) mt DNA has a high rate of mutation and mitochondrial defects cause a number of human diseases.

B. Cleavage of DNA into Fragments

The process of DNA analysis has been discovered by a special group of bacterial enzymes called restriction endonucleases (restriction enzymes) that cleave double-stranded DNA into smaller fragments. Restriction enzymes are used experimentally to obtain the defined DNA segments called *restriction fragments*.

1. Specificity of restriction endonucleases

These enzymes recognize specific nucleotide sequences (generally 4 or 6 base pairs). These sequences exhibit two-fold rotational symmetry which means the nucleotide sequence on the "top" strand read 5′ → 3′ is identical to that of the "bottom" strand also read in the 5′ → 3′ direction.

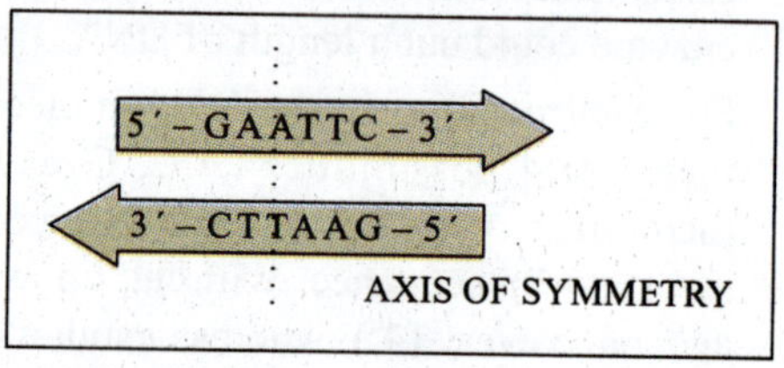

2. Nomenclature

(a) A restriction enzyme is named according to the organism from which it was isolated.

(b) The first letter of the name is from the genus of the bacteria.

(c) The next two letters are from the name of the species.

(d) An additional subscript letter indicates the type or strain.

(e) The final number appears to indicate the

order in which the enzyme was discovered in the particular organism.

Example : Hae 111 is the third restriction endonuclease isolated from the bacterium Haemophilus aegyptius.

3. "Sticky" and "blunt" ends

(a) Restriction enzymes cleave DNA to produce a 3′-hydroxyl group on one end and a 5′-phosphate group on the other.

(b) Some restriction endonucleases, such as Taq 1, produce "sticky" or cohesive ends, ie, the resulting DNA fragments have single stranded sequences that are complementary to each other.

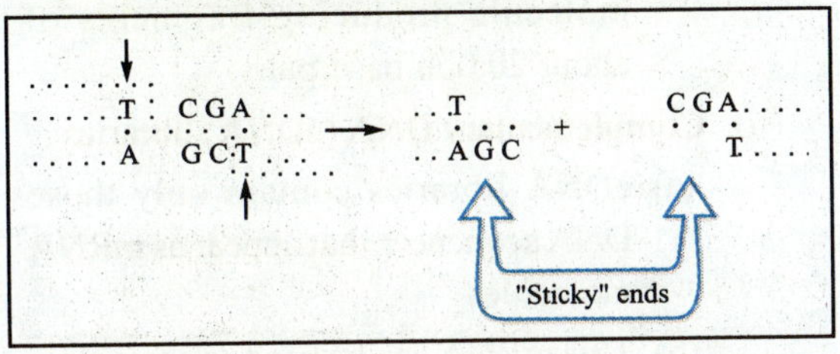

(c) Other restriction nucleases, such as Hae 111, cleave in the middle of their recognition sequence and produce fragments that have "blunt" ends that do not form hydrogen bonds with each other.

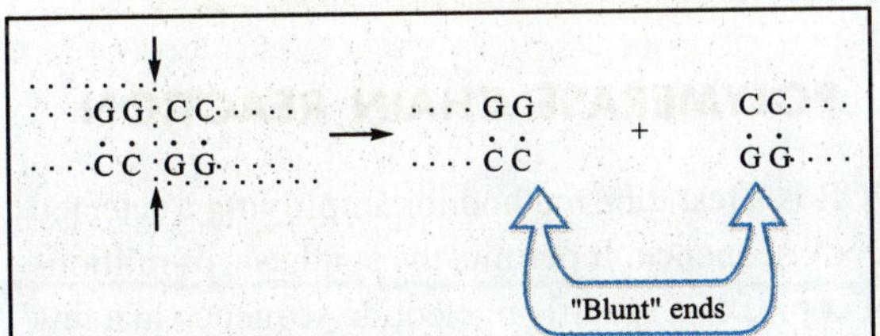

(d) Using the enzyme DNA ligase, sticky ends of a DNA fragment can be covalently joined with other DNA fragments that have sticky ends produced by cleavage with the some restriction endonulcease. The hybrid combination of two fragments is called a **recombinant DNA molecule**.

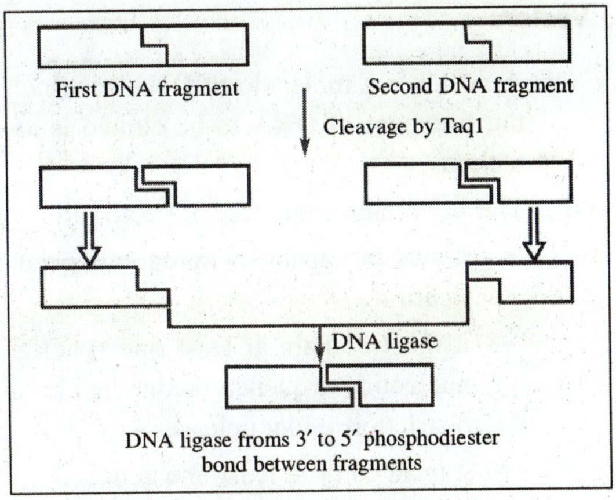

4. Restriction sites

(a) A DNA sequence recognized by a restriction enzyme is called a ***restriction site***.

(b) Since these sites occur at random, restriction endonucleases cleave DNA into fragments of different sizes.

(c) An enzyme that recognizes a specific four base-pair sequence produces many cuts in the DNA molecule. But an enzyme requiring a sequence of six base pairs produces fewer cuts and hence, longer pieces.

C. DNA Cloning

If a foreign DNA molecule is introduced into a replicating cell, it permits the **amplification** (production of many copies) of the DNA. In some cases, a single DNA fragment to be cloned can be isolated and purified. To clone a nucleotide sequence, the total cellular DNA is first cleaved with a specific restriction enzyme forming thousands of fragments. These "insert" DNA fragments are joined to DNA cloning vectors to form hybrid molecules. Each hybrid recombinant DNA molecule conveys its insert DNA into a single host cell. As the host cell multiplies, it forms a clone in which every bacterium carries copies of the some insert DNA fragment, hence the name "**cloning**". The cloned DNA is released from its vector by cleavage (using the suitable restriction endonuclease) and is isolated.

1. Vectors

(a) A vector is a molecule of DNA to which the fragment of DNA to be cloned is attached.

(b) The important properties of vector are :

— it must be capable of autonomous replication.

— it must contain at least one specific nucleotide sequence recognized by a restriction endonuclease.

— it must carry at least one gene.

— the used vectors are commonly plasmids and bacterial and animal viruses.

(i) **Prokaryotic plasmids :**

(a) Prokaryotic organisms contain single, large, circular chromosomes.

(b) Most species of bacteria also contain small circular, extrachromosomal DNA molecule called ***plasmids***.

(c) Plasmids may carry genes that convey antibiotic resistance to the host bacterium and may transfer the genetic information from one bacterium to another.

(d) Plasmids can be isolated from bacterial cells and their circular DNA can be cleaved at specific sites by restriction endonucleases and the foreign DNA can be inserted into the circle.

(e) The hybrid plasmid can be reintroduced into a bacterium and large numbers of copies of the plasmid containing the foreign DNA are produced.

(ii) **Other Vectors :**

(a) The development of improved vectors can accommodate large DNA segments showing improved molecular genetic research.

(b) Yeast plasmids, and bacterial and mammalian viruses are widely used as cloning vectors.

2. DNA libraries

A DNA library is a collection of cloned restriction fragments of the DNA of an organism. Two kinds of libraries are : (i) Genomic libraries; (ii) cDNA libraries.

(i) **Genomic DNA libraries**

(a) A genomic library is the collection of fragments of double-stranded DNA obtained by digestion of the total DNA with a restriction endonuclease.

(b) The amplified DNA fragments represent the entire genome of the organism and are called a ***genomic library***.

(c) Cleavage occurs only at a fraction of the restriction sites on any one DNA molecule producing fragments of about 20,000 base pairs.

(ii) **Complementary DNA (cDNA) libraries :**

(a) cDNA libraries contain only those DNA sequences that appear as mRNA molecules.

(b) This mRNA can be used as a template to make a cDNA using the enzyme reverse transcriptase.

(c) It can be used as a probe to locate the gene that coded for the original mRNA in mixtures containing many unrelated DNA fragments.

POLYMERASE CHAIN REACTION

PCR is a test tube method for amplifying a selected DNA sequence. It permits the synthesis of millions of copies of a specific nucleotide sequence in a few hours. The method can be used from any source — bacterial, viral, plant, or animal.

A. Steps of a PCR

PCR uses DNA polymerase to amplify repetitively targeted portions of DNA. The amplified DNA sequence can then be analyzed by southern hybridization.

1. Primer construction

(a) It is essential to know the nucleotide sequence of short segments on each side of the target DNA.

(b) The nucleotide sequences of the flanking regions are used to construct two single-stranded oligonucleotides, usually 20 to 35 nucleotides long.

(c) These synthetic oligonucleotides function as primer in the PCR.

2. Denature the DNA

The DNA to be amplified is heated to separate the double-stranded target DNA into single strands.

3. Annealing of primers to single-stranded DNA

The separated strands are cooled and allowed to anneal to the two primers.

4. Chain extension

(a) DNA polymerase and deoxyribonucleoside triphosphates are added to the mixture to initiate the synthesis of two new chains complementary to the original DNA chains.

(b) DNA polymerase adds nucleotides to the 3′- hydorxyl end of the primer, and strand growth extends.

(c) After one cycle of replication, the reaction mixture is heated again to denature the DNA strands.

(d) Each DNA strand binds a complementary primer and the cycle of chain extension is repeated.

(e) Typically 20 to 30 cycles are run during DNA amplification.

(f) Each newly synthesized polynucleotide can act as a template for the successive cycles.

B. Advantages of PCR

PCR is so sensitive that DNA sequences present in an individual cell can be amplified. The isolation and amplification of a specific DNA sequence by PCR is faster and less technically difficult than traditional cloning methods using recombinant DNA techniques.

C. Applications

1. The synthesis of the mutant DNA is allowed by PCR in sufficient quantities without cloning the altered DNA.
2. The long latency period of viruses such as HIV is difficult to detect at the early stage of infection by the help of conventional methods. PCR gives a rapid and sensitive method for detecting viral DNA sequences.
3. The isolation of DNA from a single human hair is quite ample to determine whether the sample comes from a specific individual.

PRENATAL DIAGNOSIS

It is required to analyse DNA of family members of the affected individual to set up a diagnostic tool for a genetic disease. The markers (RFLPs) that are tightly linked to the disease is needed to be identified first. The physicians are required to have a basic understanding of the application of DNA technology to diagnosis of genetic diseases in the interest of medical and legal reasons. In case of families with a history of severe genetic diseases, it is essential to determine the presence of the disorder in the developing fetus.

A. Methods available

1. The fetus should be visualized by ultrasound if the genetic abnormality is caused in gross anatomic defects.
2. The chemical composition of the amniotic fluid can also give diagnostic clues.
3. Molecular analysis of fetal DNA can provide the most detailed genetic picture permitting reproductive information.

B. Sources of DNA

DNA is obtained from white blood cells, amniotic fluid or chorionic villi. The time for DNA analysis has been shortened by the development of PCR.

C. Diagnosis of sickle cell diseases using recombinant DNA technology

1. In the past, prenatal diagnosis of hemoglobinopathies was involved in the determination of the amount and kinds of hemoglobin synthesized in red cells obtained from fetal blood.
2. Sickle cell anemia is a genetic disease caused by a point mutation. At present, diagnostic techniques for analyzing fetal DNA has become valuable since they provide safe, early detection of genetic diseases.

D. Phenylketonuria

1. The phenylalanine hydroxylase (PAH) gene has been located on chromosome 12. The presence of abnormal phenylalanine hydroxylase genes has been detected by the use of DNA polymorphism as markers to distinguish between normal and mutant genes.
2. If an unborn fetus is affected, DNA-based screening is useful not only in determining but also in detecting carriers of the PKU gene. PKU is inherited as an autosomal recessive trait. It is also important to identify heterozygotes for future family planning.

Probes

1. Probes are generally pieces of DNA or RNA labelled with a p32-containing nucleotide.
2. The probe must know a complementary sequence to be effective.
3. A cDNA synthesized from a specific mRNA can be used to screen either a cDNA library for a longer cDNA or a genomic library for a complementary sequence in the coding region of a gene.
4. cDNA probes are used to detect **DNA fragments** on southern blot transfers and to detect and quantitate RNA on Northern blot transfers.

Blotting and Hybridization Techniques

1. Observation of a specific DNA or RNA fragment in many "Contaminating" molecules requires a number of techniques which are collectively termed **blot transfer**.
2. Sometimes, if a specific base is altered and a restriction site is changed, these procedures can detect a point mutation.
3. The Northern and Western blot transfer techniques are used to size and quantitate specific RNA and protein molecules, respectively.
4. Colony or plaque hybridization is the method by which specific clones are identified and purified. Bacteria are grown on colonies on an agar plate and overlaid with a nitrocellulose filter paper.
5. Perfect matches hybridize readily and withstand high temperatures in the hybridization and washing reactions. These complexes also form in the presence of low salt concentrations.
6. Less than perfect matches do not tolerate these **stringent conditions**.
7. Gene families can be detected by varying the stringency of the hybridization and washing steps.

DNA Sequencing

1. The segments of specific DNA molecules can be analyzed for their nucleotide se-

quence. This method depends upon having a large number of identical DNA molecules.

2. This requirement can be satisfied by cloning the fragment of interest by the above techniques.
3. Enzymatic method uses specific deoxynucleotide analogs that terminate DNA strand synthesis at specific nucleotides as the strand is synthesized on purified template nucleic acid.

Manual and Automatic techniques for the determination of the sequence of DNA

1. The segments of specific DNA molecules obtained by recombinant DNA technology can be analyzed for their nucleotide sequence. The manual, enzymatic method (Sanger's) employs specific dideoxynucleotides that terminate DNA strand synthesis at specific nucleotides.
2. Polyacrylamide gel electrophoresis can be used to separate the fragments according to size. Each of the fragments produces an image (band) on an x-ray film.
3. Another manual method, that of Maxam and Gilbert employs chemical methods to cleave the DNA molecules where they contain the specific nucleotides.
4. The most commonly procedure is the four different fluorescent labels, each representing one nucleotide. Each emits a specific signal by a laser beam and this can be recorded by a computer.

SOME PRACTICAL APPLICATIONS ON RECOMBINANT DNA TECHNOLOGY

Gene Mapping

1. Specific genes to distinct chromosomes are localized by this technique and thus to define a map of the human genome. This is already producing useful information in the definition of human disease.
2. Somatic cell hybridization and in situ hybridization are two techniques used to accomplish this.
3. In situ hybridization, the simpler and more direct procedure, a radioactive probe is added to a metaphase spread of chromosomes on a glass slide. The exact area of hybridization is localized by layering photographic emulsion over the slide and, after exposure lining up the grains with some histologic identification of the chromosome. Some of the human genes are localized by this technique.
4. Genes that code for proteins with similar functions can be located on separate chromosomes.
5. Genes that form part of a family can also be on separate chromosomes (Growth hormone and prolactin).
6. The genes involved in many hereditary disorders known to be due to specific protein deficiencies, including X chromosome-linked conditions, are really located at specific sites.

Protein Production

1. This technology has two prominent merits :
   (i) It can supply large amounts of materials that could not be obtained by conventional purification methods.
   (ii) It can provide human material (eg, insulin, growth hormone).
2. Although the primary aim is to supply products, generally proteins, for treatment (Insulin) and diagnosis (AIDS test) of human and other animal disease and for disease prevention (hepatitis B vaccine), there are other real and potential commercial applications, especially in agriculture.

RECOMBINANT DNA TECHNOLOGY IN MOLECULAR ANALYSIS OF DISEASE

Normal gene variations

1. **Polymorphisms** occur once in every 500 nucleotides, or about 10^7 times per genome.
2. There are deletions and insertions of DNA as well as single base substitutions.
3. In healthy people, these alterations occur in noncoding regions of DNA or at site that cause no change in function of the encoded protein.
4. The polymorphism of DNA structure can be associated with certain diseases.

Gene variations causing Disease

1. Most genetic diseases were due to point mutations that resulted in an impaired protein.
2. β-globin gene is located in a cluster on chromosome 11. Defective production of β-globin results in a variety of diseases and is due to many different lesions in and around the β-globin gene.

Point Mutations

1. **Sickle cell disease** is caused by mutation of a single base out of the 3×10^9 in the genome.
2. The altered codon specifies a different amino acid (valine rather than glutamic acid) and this causes a structural abnormality of the β-globin molecule. β-thalassemia is the result of these mutations.

Deletions, Insertions and Rearrangements of DNA

1. Pieces of DNA can move from one place to another within a genome on the study of bacteria, viruses, yeasts, and fruit flies.
2. Disease is caused by the deletion of critical piece of DNA, the rearrangement of DNA within a gene, or the insertion of a piece of DNA within a coding of regulatory region can also cause changes in gene expression.
3. Deletions in the alpha-globin cluster, located on chromosome 16, cause alpha thalassemia.
4. Deletions or insertions of DNA larger than 50bp can often be detected by the southern blotting procedure.

Pedigree Analysis

1. **Incubation** of DNA from normal (AA), heterozygous (AS), and homozygous (SS) individuals results in three different patterns on southern blot transfer.
2. **Pedigree** analysis has been applied to a number of genetic diseases and is most useful in those caused by deletions and insertions or the rarer instances in which a restriction endonuclease cleavage site is affected.

Restriction Fragment Length Polymorphism (RFLP)

1. Inherited differences in the pattern of restriction is known as RFLP.
2. This results from single base changes (eg, sickle cell disease) or from deletions or insertions of DNA into a restriction fragment (eg, the thalassemia) and are proving to be a useful diagnostic measure.
3. PFLPs may disrupt the function of the gene or may have no biological significance.
4. RFLPs are inherited and they segregate in a Mendelian fashion.
5. These can be used to establish linkage groups, which in turn, by the process of chromosome walking, will define the disease locus.
6. In chromosome walking, a fragment representing one end of a long piece of DNA

is used to isolate another that overlaps but extends the first.

7. 20 p.c. of the defined RFLPs are on the X chromosome.
8. X-linked disorders such as **Duchenne type muscular dystrophy** will be defined using RFLPs.
9. The defect that causes polycystic kidney disease is linked to the alpha globin locus on chromosome 16.

Gene Therapy

1. Bone marrow precusor cells are being investigated because they presumably will resettle in the marrow and replicate there. The introduced gene would begin to direct the expression of its protein product and this would correct the deficiency in the host cell.
2. Some percentage of genes injected into fertilized mouse ovum will be incorporated into the genome and found in both somatic and germ cells. These trangenic animals are useful for analysis of tissue—specific effect on gene expression.
3. The trangenic approach has recently been used to correct a genetic deficiency in mice.
4. Fertilized ova obtained from mice with genetic hypogonadism were injected with DNA containing the coding sequence for the gonadotropin—releasing hormone (GnRH) precursor protein. This gene was expressed and regulated normally in the hypothalamus of a certain number of the resultant mice, and these animals were in all respects normal. Their offspring also showed no evidence of GnRH deficiency.

Glossary

cDNA : A single-stranded DNA molecule which is complementary to an mRNA molecule and is synthesized from it by the action of reverse transcriptase.

Chimeric molecule : A molecule (eg, DNA, RNA, protein) containing sequences derived from two different species.

Clone : A large number of cells or molecules which are identical with a single parental cell or molecule.

Exon : The sequence of a gene that is represented as mRNA.

Endonuclease : An enzyme which cleaves internal bonds in DNA or RNA.

Cosmid : A plasmid into which the DNA sequences from bacteriophage lambda that are necessary for the packaging of DNA have been inserted.

Exonuclease : An enzyme which cleaves nucleotides from either the 3′ or 5′ ends of DNA or RNA.

Hybridization : The specific reassociation of complementary strands of nucleic acids (DNA with DNA, DNA with RNA, or RNA with RNA).

Insert : An additional length of base pairs in DNA.

Library : A collection of cloned fragments that represents the entire genome. Libraries may be either genomic DNA or cDNA.

Ligation : The enzyme-catalyzed joining in phosphodiester linkage of two stretches of DNA or RNA into one. The respective enzymes are DNA and RNA ligases.

Plasmid : A small, extrachromosal, circular molecule of DNA that replicates independently of the host DNA.

Probe : A molecule used to detect the presence of a specific fragment of DNA or RNA in common probes are cDNA molecules.

Recombinant DNA : The altered DNA that results from the insertion of a sequence of deoxynucleotides not previously present into an existing molecule of DNA by enzymatic or by chemical means.

Restriction enzyme : An endodeoxynuclease that causes cleavage of both strands of DNA at highly specific sites dictated by the base sequence.

Reverse transcription : RNA-directed synthesis of DNA catalyzed by reverse transcriptase.

Signal : The end product observed when a specific sequence of DNA or RNA is detected by autoradiography or some other method.

Sticky-ended DNA : Complementary single strands of DNA that protrude form opposite ends of DNA duplex.

Tandem : It is described as the multiple copies of the same sequence (eg, DNA) that lie adjacent to one another.

Tanscription : DNA-directed synthesis of RNA.

Transgenic : It describes the introduction of new DNA into germ cells by its injection into the nucleus of the ovum.

Translation : Synthesis of protein using mRNA as template.

Vector : A plasmid or bacteriophage into which foreign DNA can be introduced for the purposes of cloning.

Northern blot : A method for transferring RNA from an agarose gel to a nitrocellulose filter on which the RNA can be detected by a suitable probe.

Southern blot : A method for transferring DNA from an agarise gel to a nitrocellulose filter on which the DNA can be detected by a suitable probe.

Western blot : A method for transferring protein to nitrocellulose filter on which the protein can be detected by a suitable probe (eg, an antibody).

CLINICAL ASPECTS

- Recombinant DNA technology offers a knowledge to understand the molecular basis of a number of diseases (eg, sickle cell disease, the thalassemias, familial hypercholesterolemia, cystic fibrosis, muscular dystrophy).
- Using this technology, human proteins can be produced abundantly for therapy (eg, insulin, growth hormone, plasminogen activator).
- Proteins for vaccines (eg, hepatitis B) and for diagnostic tests (eg, AIDS test) can be obtained.
- It is used to diagnose existing diseases and predict the risk of developing a given disease.
- Advancement has been made in forensic medicine by the use of special techniques.
- Gene therapy has been applied for sickle cell diease, the thalassemias, adenosine deaminase deficiency, and other diseases.
- The fundamental concept of this technique is very helpful in the diagnosis of diseases to have the satisfactory healthy life.

EXERCISE

1. Explain briefly Recombinant DNA Technology. (B.H.U. '92A)
2. Discuss the importance of Recombinant DNA technology in diseases. (Delhi University '93A)
3. Write notes on :
   (a) Cloning (P.U. '91A)
   (b) Libraries (Punjab University '90S)
   (c) Gene Mapping (B.H.U. '91A)
   (e) RFLP (B. U. '92A)
4. Throw a light on clinical aspects of Recombinant DNA Technology. (B. H. U. '92S)

CHAPTER 25

METABOLISM OF NUCLEIC ACID

THE NATURE OF DNA

1. DNA is a very long polymer of purine and pyrimidine mononucleotide monomers bound one to the other by phosphodiester bridges.
2. It exists as a double-stranded molecule in nature. The strands are held together by Vander Waals' forces and the purine and pyrimidine bases of the two strands are held together by hydrogen bonding.
3. The two strands extend in opposite direction, i.e., each is antiparallel.
4. Each purine or pyrimidine base in one strand is related to pyrimidine or purine base of the other strand.

   i.e., Adenine (A) is paired with Thymine (T) and Guanine (G) is paired with Cytosine (C).
5. Genetic information is contained in the sequence of mononucleotide of the DNA molecule. For each gene in the DNA molecule, there is a "sense" strand and its complementary "antisense" strand.
6. Watson and Crick double-stranded model of DNA strongly suggests the replication of DNA molecule in a semiconservative manner. When each strand of the double-stranded DNA molecule separates during replication, each can then serve as a template on which a new complementary strand can be synthesized.
7. Each of the 2 newly formed double-stranded DNA molecule contains one strand (but complementary) from the parent double-stranded DNA molecule. The two newly formed double-stranded DNA molecule can then be sorted between 2

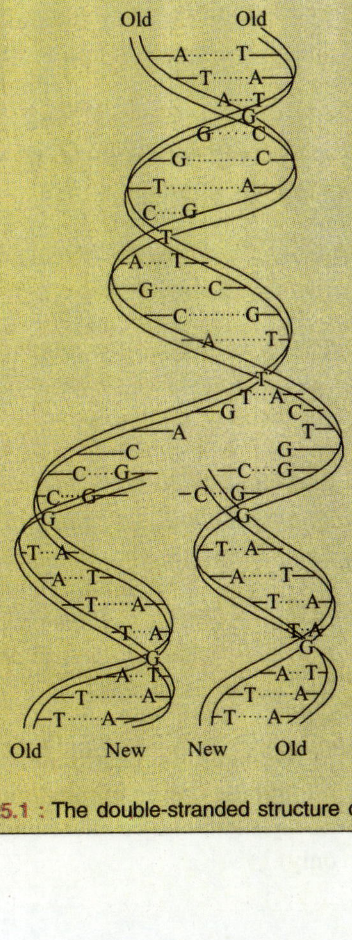

Fig. 25.1 : The double-stranded structure of DNA

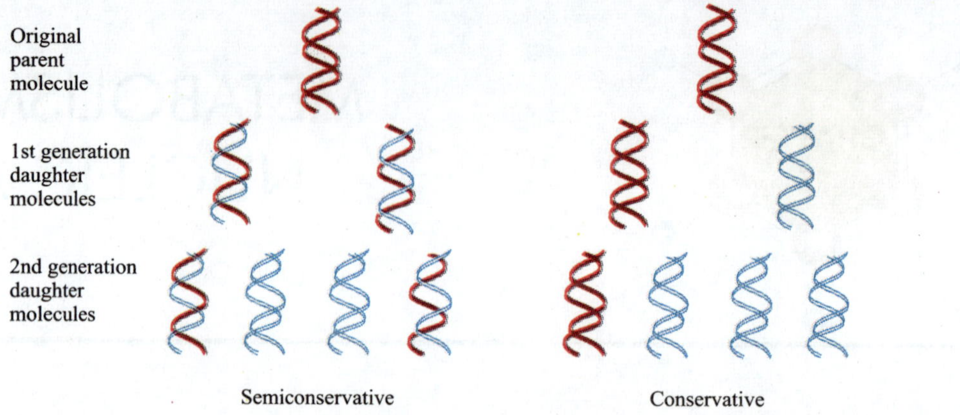

Fig. 25.2 : The distributions of Parental DNA strands during semiconservative and conservative replication. The parental strands are solid and the newly synthesized strands are open

daughter cells. Each daughter cell will contain DNA molecules with information identical to that which the parent possesses.

The double-stranded structure of DNA and the template function of each old strand on which a new complementary strand synthesized and also the semiconservative and conservative replication are given in Fig. 25.2.

Histones are the most abundant chromatin proteins

1. The histones are the small family of closely related basic proteins. HI histones are the least tightly bound to chromatin and are easily removed with a salt solution, after which chromatin becomes soluble.
2. The core nucleosomes contain four classes of histones : H2A, H2B, H3, and H4. The slightly lysine — rich histones are H2A and H2B, whereas the arginine — rich histones are H3 and H4.
3. Four core histones are subject to five types of covalent modifications : acetylation, methylation, phosphorylation, ADP-ribosylation, and covalent linkage (H2A only).
4. The histones interact with each other in very specific ways. H3 and H4 form a tetramer, while H2A and H2B form dimers. Two of these histone tetramers associate to form the histone octamer.

Some regions of chromatin are 'active' and others are 'inactive'

1. Chromatin containing active genes differs in several ways from that of nonactive regions. The nucleosome structure of active chromatin is absent in highly active regions. DNA in active chromatin contains large regions (about 100,000 bases long) that are sensitive to digestion by a nuclease such as DNase 1.
2. DNase 1 will digest DNA not protected by protein into deoxynucleotides. The sensitivity to DNase 1 of chromatin regions reflects only a potential for transcription.
3. There exists shorter stretches of 100-300 nucleotides that exhibit an even greater sensitivity to DNase 1 within the large regions of active chromatin. In many cases, if a gene is capable of being transcribed, it must have a DNase hypersensitive site in the chromatin.
4. Transcriptionally inactive chromatin is densely packed during interphase and is

referred to as heterochromatin. Transcriptionally active chromatin stains less densely and is referred to as euchromatin. Generally, euchromatin is replicated earlier in the mammalian cell cycle than is heterochromatin.

5. Heterochromatins are of two types — constitutive heterochromatin and facultative heterochromatin. Constitutive heterochromatin is always condensed and inactive. It is found in the regions near the chromosomal centromere and at chromosomal ends (telomeres). Facultative heterochromatin is at times condensed but at other times it is actively transcribed and appears as euchromatin.

DNA is organized into chromosomes

1. At metaphase, mammalian chromosomes possess a two-fold symmetry with identical sister chromatids connected at a centromere.
2. The centromere is an A—T region of about 130bp. It binds several proteins with high affinity. This complex, called the kinetochore, provides the anchor for the mitotic spindle.
3. The ends of each chromosome contain structure called telomeres. These consist of short, repeat TG-rich sequences. Human telomeres have a number of repeats of the sequence 5′-TTAGGG-3′, which can extend for several kilobases.
4. Telomerase is the enzyme responsible for telomere synthesis and maintains the length of the telomere. Since telomere shortening is associated with malignant transformation (and aging), telomerase has become an attractive target for cancer chemotheraphy.
5. Each of the 23 chromatids in the human haploid genome contains averagely 1.3×10^8 nucleotides in one double-stranded DNA molecule. The length of each DNA molecule must be compressed about 8000-fold to generate the structure of a condensed metaphase chromosome.

Chromosomal recombination is one way of rearranging genetic material

1. Genetic information can be exchanged between similar or homologous chromosomes. The exchange or recombination event occurs during meiosis in mammalian cells and requires homologous chromosomes.
2. A process of crossing over results in an equal and reciprocal exchange of genetic information between homologous chromosomes. If the homologous chromosomes possess different alleles of the same genes, the cross-over may produce remarkable and heritable genetic linkage differences.
3. In rare cases, if the alignment of recombination event may result in an unequal exchange of information.

DNA METABOLISM

DNA Synthesis

1. The initiation of DNA synthesis requires priming by a short length of RNA (10-200 nucleotides long).
2. This priming process involves the nucleophilic attack by the 3′-hydroxyl group of the RNA primer on the α-phosphate of the deoxynucleoside triphosphate with splitting off pyrophosphate.
3. The 3′-hydroxyl group of the recently attached deoxyribonucleoside monophosphate is then free to carry out a nucleophilic attack on the next entering deoxyribonucleoside triphosphate on its α-phosphate moiety with the splitting off pyrophosphate.
4. The selection of proper deoxyribonucleotide is dependent upon proper pairing

with the other strand (template) of DNA molecule.

5. The template dictates in which dNTP is complementary and holds this by hydrogen bonding.
6. The polymerization of deoxyribonucleotides takes place by such a process in a discontinuous phase of about 100 nucleotides in length.
7. This newly synthesized DNA strand attached to RNA primer is called okazaki fragments.
8. When many okazaki fragments are generated, the replication complex begins to remove the RNA primers by DNA polymerase I and the gaps left by their removal are filled up by proper base paired deoxynucleotide. The enzyme DNA ligase seal the fragments of newly synthesized DNA.

Replication

1. Due to unwinding of double helix of DNA, each strand acts as a template for the formation of a new strand. This process is called *replication*.
2. Types of replication :

   (a) Conservative replication : The parental strand is never completely separated. So, after one round of replica-

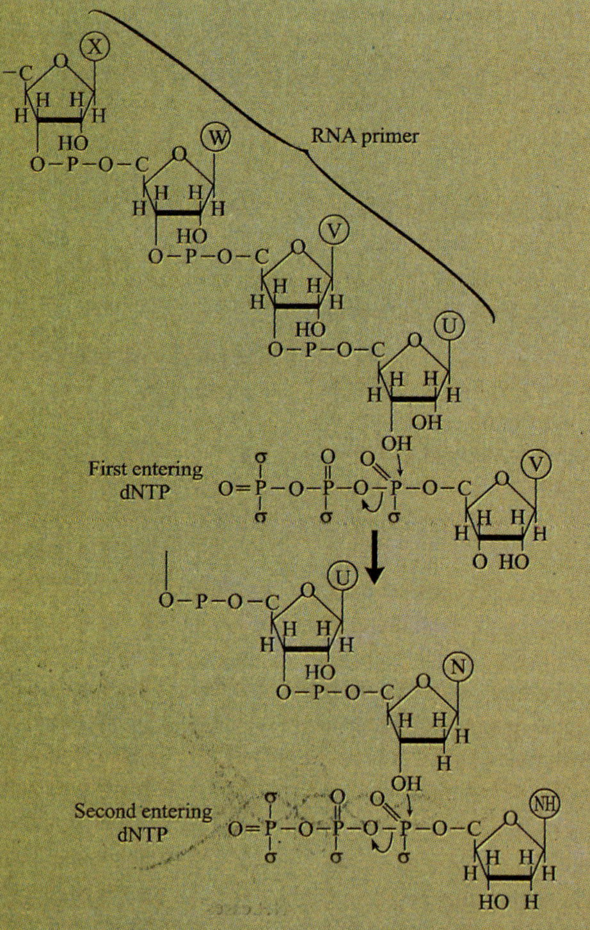

Fig. 25.3 : The initiation of DNA synthesis upon a primer of RNA

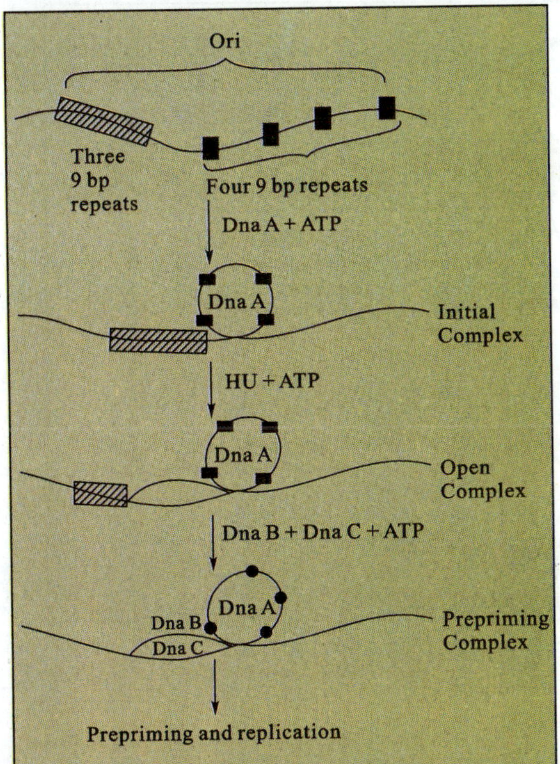

Fig. 25.4 : Initiation of DNA replication

tion, one daughter duplex contains only parental strands and the other only daughter strands.

(b) **Semiconservative replication** : The process of unwinding of the double helical daughter molecules, each of which is composed of a parental strand and a newly synthesized strand formed from the complementary strand, called *semiconservative replication*.

3. Process of Replication :

(A) **Initiation of DNA replication** :

(i) Replication begins at a specific initiation point and this is a unique sequence of bases called *Ori*.

(ii) In *Ori* there are two series of short repeats such as three repeats of a 13-base pair sequence and four repeats of a 9-base pair sequence.

(iii) In the initiation process about 20 Dna A protein molecule each with a bound ATP, bind at the four repeats of 9-base pair sequence, DNA is wrapped around the complex (initial).

(iv) With the help of ATP and histone like protein HU, the three 13-base pair repeats are denatured to give open complex.

(v) With the help of Dna C protein ATP Dna B protein binds to the open complex to form prepriming complex. As a result, unwinding of DNA occurs and priming replication starts.

(B) **Elongation** : In the elongation process of replication two operations occur such as leading strand synthesis and lagging strand synthesis.

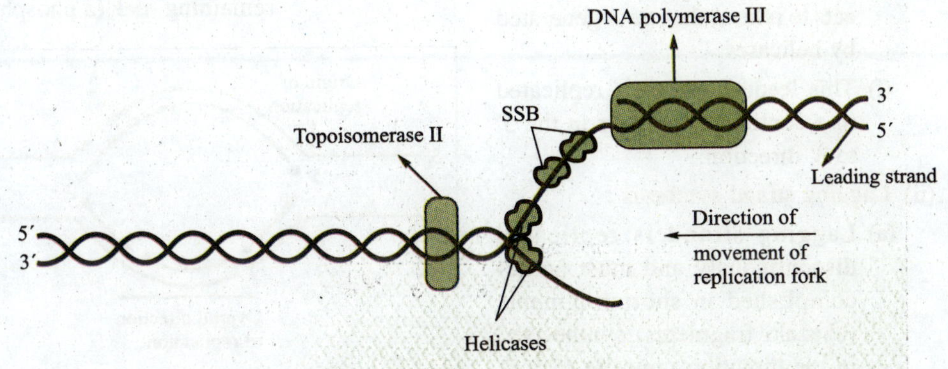

Fig. 25.5 : Leading strand synthesis of elongation of replication

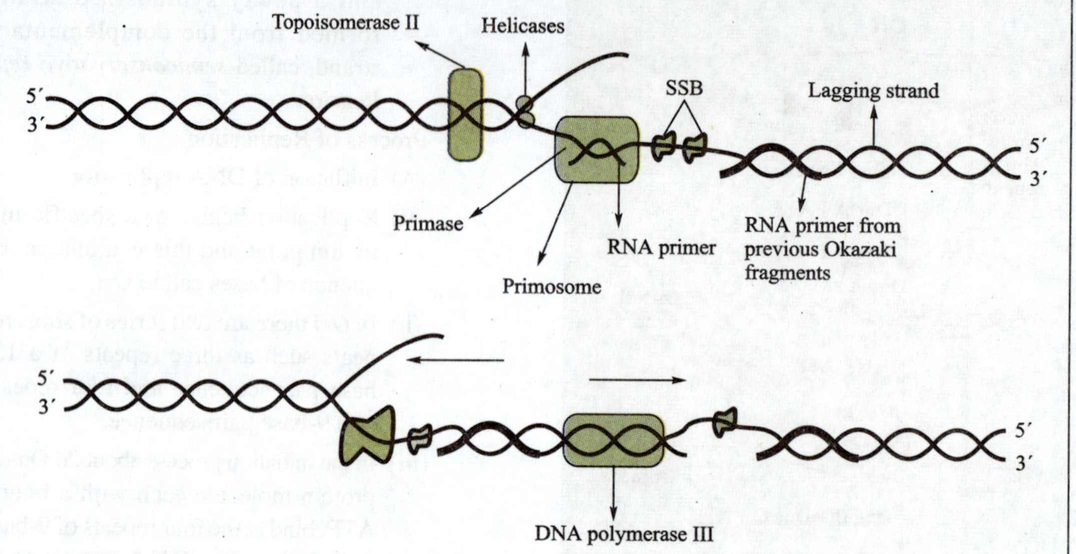

Fig. 25.6 : Lagging strand synthesis of replication

(i) *Leading strand synthesis :*

(a) Leading strand synthesis begins with the synthesis of RNA primer (10 to 60 nucleotides) by Dna G protein at the replication origin.

(b) Deoxyribonucleotides are then added to this primer by DNA polymerase III (it keeps pace with the replication fork).

(c) SSB molecules help to stabilize the separated strand.

(d) Helicases separate the two DNA strands at the fork.

(e) Topoisomerase II (DNA gyrase) acts to relieve the stress generated by helicases.

(f) This leading strand is replicated in a continuous manner in the 5′ to 3′ direction.

(ii) **Lagging strand synthesis :**

(a) Lagging strand is replicated discontinously and must be accomplished in short fragments (okazaki fragments) synthesized in the direction opposite to fork movement.

(b) **Synthesis of okazaki fragments :** The multiprotein primosome complex travels in the same direction as the replication fork.

At intervals, Dna G protein synthesizes an RNA primer for a new okazaki fragment. The synthesis proceeds in the direction opposite to fork movement.

Each primer is extended by DNA polymerase III.

(c) When the new okazaki fragment is complete, the RNA primer is removed by DNA polymerase I. The remaining **nick** (a phosphodiester

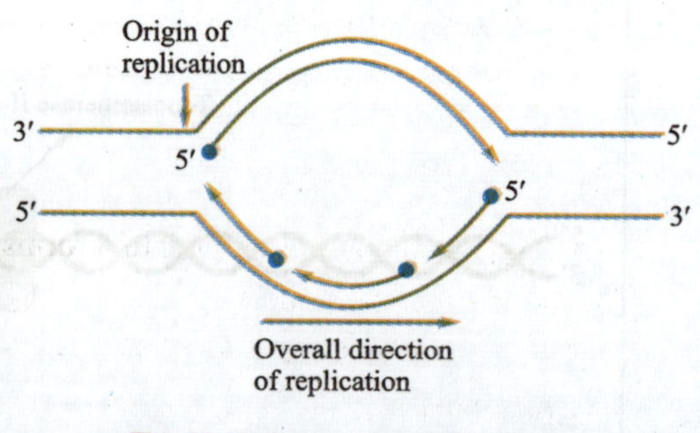

Fig. 25.7 : Termination of replication

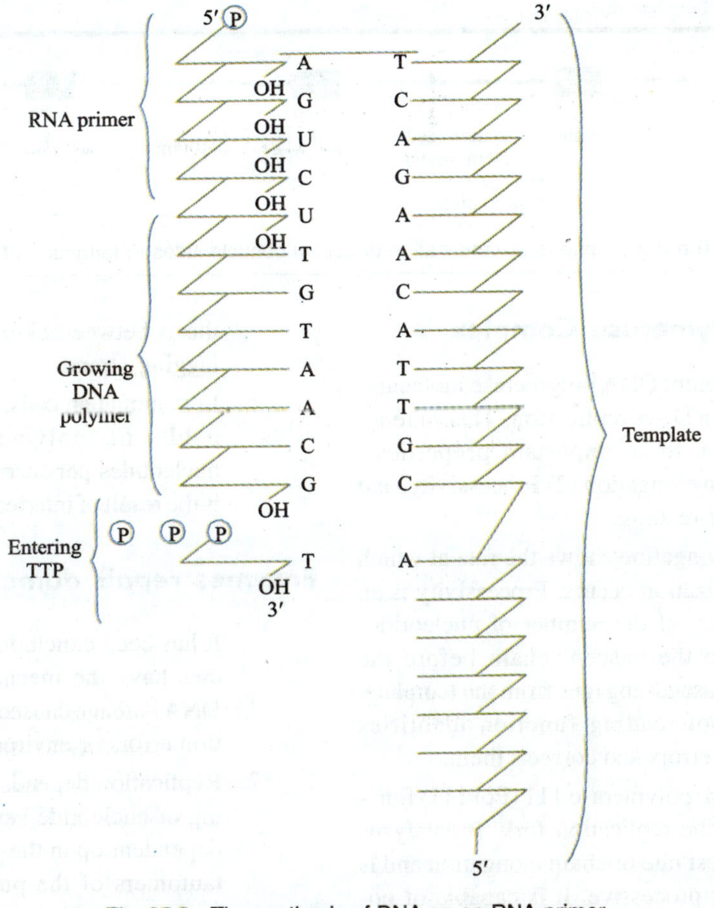

Fig. 25.8 : The synthesis of DNA on an RNA primer

bond broken to leave a free 3′ OH and 5′ phosphate) is sealed by DNA ligase.

(C) **Termination** : Very little is known about this process, it is assumed that DNA topoisomerase IV appears to be necessary for final separation of the two completed circular DNA molecules.

Degradation of DNA

DNA damage may be classified into 4 forms :

1. Single base alteration.
2. Two base alteration.
3. Chain breaks.
4. Cross-linkage.

(i) ***Single base alteration*** : The one base damage includes the hydration of the cytosine residue by ultraviolet irradiation.

(ii) ***Two base alteration*** : The two base damage includes Thymine-thymine dimer formation via a cyclobutane moiety.

(iii) ***Chain breaks*** : Chain breaks may be created by irradiation such as X-ray exposure.

(iv) ***Cross-linkage*** : Cross-linkage agents which link bases of opposite strands also induce 2 base alterations. Cross-links can also occur between the DNA molecule and histones.

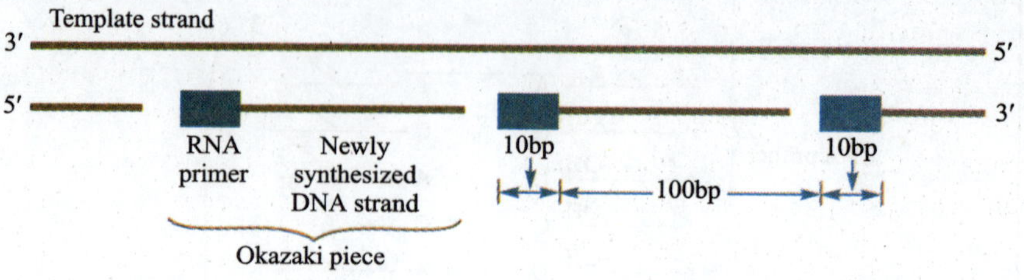

Fig. 25.9 : The discontinuous polymerization of deoxyribonucleotides in formation of Okazaki pieces

The DNA Polymerase Complex

1. The different DNA polymerase molecules engage in DNA replication. These are involved in three important properties : (1) Chain elongation (2) Processivity, and (3) Proof-reading.
2. Chain elongation shows the rate at which polymerization occurs. Processivity is an expression of the number of nucleotides added to the nascent chain before the polymerase disengages from the template. The proof-reading function identifies copying errors and corrects them.
3. In E. Coli, polymerase 111 (Pol 111) functions at the replication fork. It catalyzes the highest rate of chain elongation and is the most processive. It is capable of polymerizing 0.5 Mb of DNA during one cycle on the leading strand. It is the product of Dna E gene in E. Coli.
4. Polymerase 11 (Pol 11) is mostly concerned with proof-reading and DNA repair.
5. Polymerase 1 (Pol 1) completes chain synthesis between okazaki fragments on the lagging strand.
6. In mammalian cells, the polymerase is capable of polymerizing about 100 nucleotides per second. This reduced rate is the result of interference by nucleosomes.

Enzymes repair damaged DNA

1. It has been concluded that surviving species have the mechanisms for repairing DNA damage caused as a result of replication errors or environmental insults.
2. Replication depends on the specific pairing of nucleotide bases. Proper pairing is dependent upon the presence of favoured tautomers of the purine and pyrimidine nucleotides. The proper base pairing can be assured by monitoring the base pairing twice. Such double monitoring occurs in both bacterial and mammalian systems. This double monitoring does not produce errors of mispairing due to the presence of the unfavoured tautomers.

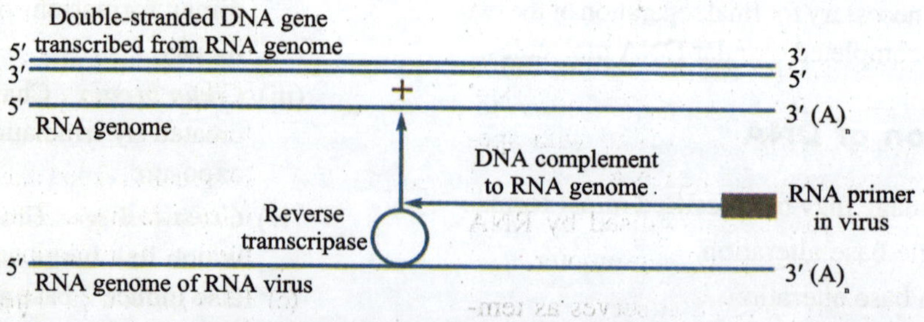

Fig. 25.10 : The generation of a double-stranded DNA molecule from an RNA template by the action of reverse transcriptase

3. Replication errors lead to the accumulation of mutations.
4. DNA damages are replaced by mismatch repair, base excision repair, nucleotide excision repair, and double strand break repair. The defective in one strand can be returned to its original form by relying on the complementary information stored in the unaffected strand.

Some repair enzymes are multifunctional

1. Some repair enzymes are found as components of the large TF11H complex that plays central role in gene transcription. Another component of TF11H is involved in cell cycle regulation. Some repair ezymes are involved in gene rearrangement that normally occur.
2. In patients with ataxia-telangiectasia, an autosomal recessive disease in humans, there exists an increased sensitivity to damage by x-ray. Patients with Fanconi's anemia have defective repair of cross-linking damage.
3. All these clinical syndromes are associated with increased frequency of cancer.

RNA METABOLISM

RNA Synthesis or Transcription

1. Transcription is the process by which the synthesis of RNA molecule is initiated and terminated.
2. Stages of transcription
   (a) Formation of transcription complex : The enzyme RNA polymerase (DNA dependent) needs to bind with specific sequences on a DNA.

   The sequences recognised by RNA polymerase are called promoter.

   The DNA strand that serves as template for RNA synthesis is called *template strand* (–strand). The DNA strand complementary to the template is called the *nontemplate strand* (+ strand). This nontemplate strand is called *coding strand*.

   RNA polymerase (holoenzyme) consists of core enzyme and sigma (σ) factor.

   Sigma factor recognises the promoter sequences.

   RNA polymerase attaches to promoter region.

   RNA polymerase melts the helical structure and separates 2 strands of DNA locally.

   RNA polymerase initiates RNA synthesis. The site at which the first nucleotide is incorporated is called the *start point*.

   (b) Initiation : Core enzyme starts transcription at the separated DNA strand of an initiation complex.

   The subunit of RNA polymerase has two specific binding sites for the binding of nucleotide triphosphate (NTP).

   The first incoming NTP binds to RNA polymerase at the start point of initiation site and H bonds to the complementary base on the DNA within complex. This site binds only purine nucleotide triphosphate either A or G. The binding is with 3′ end of the NTP leaving 5′ end to be free.

   The second incoming NTP binds to the elongation site on the polymerase. This dinucleotide tetraphosphate has either as the 5′ terminal nucleotide. After this phosphodiester bond formation σ factor is released.

   First base is then dissociated from the initiation site and indicates the completion of initiation.

   (c) Elongation : The core enzyme polymerase moves in 3′-5′ direction of the coding strand and it adds suc-

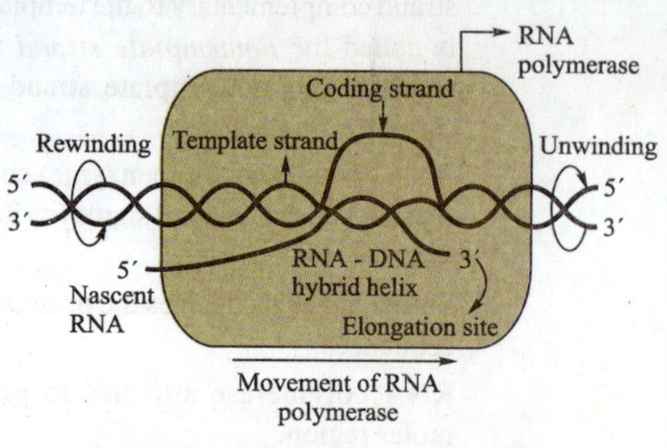

Fig. 25.11 : Model of transcription bubble

cessive NTPs at the 3′-OH end of the ribonucleotide chain already laid down in 5′-3′ direction.

The incoming NTP forms a phosphodiester bond with 3′-OH end of the preceeding ribonucleotide.

The DNA helix recloses after RNA polymerase transcribes through it and growing RNA chain dissociates from the DNA.

(d) Termination : Termination of the synthesis of the RNA molecule is signaled by a sequence in the template strand of the DNA molecule, a signal that is recognized by a termination protein, Rho (ρ) factor. At the specific termination sites the new RNA chain is released.

After the termination, core enzyme separates from the DNA template and the core enzyme again binds with σ factor and takes part in the formation of new RNA molecule.

After the end of termination, Rho (ρ) factor dissociates RNA and this factor is again recycled.

The overall process is given in Fig. 25.12.

RNA Polymerase

1. It is the key enzyme for transcription.
2. A single type of RNA polymerase is responsible for the synthesis of mRNA, rRNA, tRNA etc.
3. Different enzymes are required to synthesize different types of RNA. They are called RNA polymerase I, RNA polymerase II, III, etc.
4. Subunits composition of RNA polymerase of E. Coli.

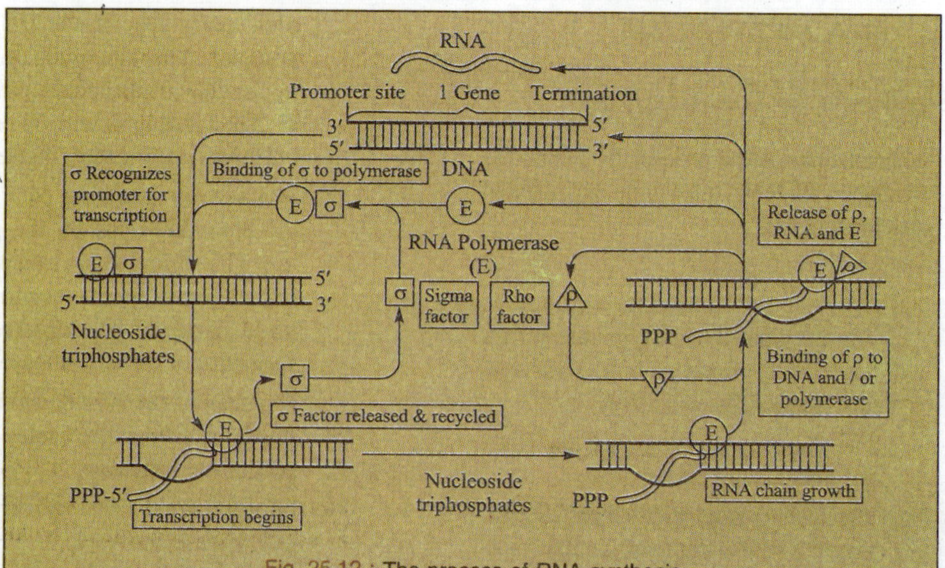

Fig. 25.12 : The process of RNA synthesis

Sub unit	No.	Location	Function
α	2	Core enzyme	Promoter binding
β	1	Core enzyme	Nucleotide binding
β′	1	Core enzyme	Template binding
σ	1	Sigma factor	Initiation

5. The holoenzyme $\alpha_2\beta\beta'\sigma$ can be separated into two components, the core enzyme ($\alpha_2\beta\beta'$) and the sigma factor.
6. The holoenzyme requires template of double stranded DNA, four ribonucleotide triphosphates (GTP, ATP, UTP, CTP) and Mg^{++} or Mn^{++}.
7. Holoenzyme initiates transcription.
8. Core enzyme has the ability to synthesize RNA on a DNA template.

SIGMA (σ) FACTORS

Transcription initiation requires specificity proteins (known as σ factors). These sigma factors bind reversibly to the catalytically active core RNA polymerase.

The core RNA polymerase interacts with sigma (σ) factors to form a holoenzyme. After the release of σ factor the core enzyme elongates an RNA chain.

The majority of cellular transcription requires the primary σ factor. In E. coli the primary σ factor is referred to as σ^{70} and in B. subtilis σ^{43}.

In recent years, the genes encoding bacterial and phage σ factors have been isolated and their amino acid sequences have been deduced from the corresponding DNA sequences. This shows that many σ factors have similar amino acid sequences and form a homologous protein family. σ factors thus may use a variety of protein structures to perform related functions in bacterial cells.

Biochemical activities of σ factors

The biochemical activities of bacterial σ factors include

(a) Core binding

(b) Activation of promoter recognition

(c) DNA melting

(d) Inhibition of nonspecific transcription.

All σ factors have at least the first two of these activities.

Different σ factors interact with the same region of the core RNA polymerase complex. Therefore, many σ factors may share a common protein structure that allows core binding.

Promoter recognition can occur only when holoenzyme is formed. It seems most likely that σ factors themselves make sequence-specific contacts with promotion. Bacterial σ factors do not bind lightly to DNA in the absence of core RNA polymerase. As with promoter recognition, σ factors do not show a DNA melting activity in the absence of the core RNA polymerase subunits.

Lastly, the E. coli σ^{70} factors have been shown to reduce the affinity of the core subunits for non-promoter containing DNA promoter by about 10^4. It has been argued that this activity is important in allowing promoter recognition against large background of non-specific DNA. In B. subtilis, a separate protein component, δ^{21}, inhibits core-mediated transcription of non-promoter containing DNA. The δ^{21} protein binds to the B. subtilis core enzyme in the presence and absence of σ factors and reduces the interaction of RNA polymerase with the DNA template. Interestingly, the ability to inhibit nonspecific transcription has also been attributed to a specificity factor from yeast mitochondria that may be functionally similar to a σ factor.

General Features of σ factors

(a) σ factors form a heterogeneous group of proteins.

(b) E.coli σ^{70} and B. subtilis σ^{43} have approximately 50% amino acid identity allowing for a 245-amino acid gap introduced into the smaller σ^{43} factor.

(c) All the related σ proteins are acidic with an excess of negative charge at pH 7.0. E. coli σ^{70} factor is one of the most acidic proteins in E. coli.

(d) Most bacterial σ factors are single polypeptides.

Ribozymes

1. The RNAs which show highly substrate-specific catalytic activity and meet all the classic criteria for definition as enzymes are termed ribozymes.
2. They catalyze trans-esterification and ultimately hydrolysis of phosphodiester bonds in RNA molecules. These reactions are facilitated by free –OH groups eg, on guanosyl residues.
3. They play key roles in the intron splicing events essential for the conversion of Pre-mRNAs to mature mRNAs.
4. Recently, a ribosomal RNA component had been noted to hydrolyze an aminoacyl ester and thus to play a central role in peptide bond function. These observations, made in organelles from plants, yeast, viruses, and higher eukaryotic cells, show that RNA can act as an enzyme.

Inhibitors of DNA and RNA synthesis

Many antibiotics and nucleotide analogs inhibit the synthesis of DNA and RNA.

Inhibitor	Inhibition
Actinomycin D	RNA and DNA chain elongation.
Mitomycin (reduced)	DNA replication.
Anthramycin	DNA and RNA synthesis.
Rifampicin(Refampin)	RNA initiation
5-hydroxyuridine	RNA and DNA synthesis.

Catabolism of RNA

1. Ribonucleases specifically hydrolyze ribonucleic acids.
2. Endonucleases cleave internal phosphodiester bonds to produce a 3′-hydroxyl and a 5′-phospharyl or a 5′-hydroxyl and a 3′-phospharyl terminus.
3. Exonucleases are capable of hydrolyzing a nucleotide only when it is present at a terminus of a molecule.

RNA molecules are often processed before they become functional

1. All eukaryotic RNA primary transcripts undergo processing before their function whether it be as mRNA or rRNA or tRNA. This processing occurs primarily within the nucleus. The processing includes nucleolytic cleavage to smaller molecules and coupled nucleolytic and ligation reactions.
2. In mammalian cells, 50-75 per cent of the nuclear RNA does not contribute to the cytoplasmic mRNA. This nuclear RNA loss is significantly greater than the loss of intervening sequences alone.

The coding portions (Exons) of most Eukaryotic genes are interrupted by introns

1. The exons of many genes are long sequences of DNA that do not supply the genetic information and are finally translated into the amino acid sequence of a protein molecule. These sequences interrupt the coding region of structural genes. These intervening sequences (introns) exist in most eukaryotes.
2. The intron RNA sequences are cleaved out of the transcript and the exons of the transcript are spliced together in the nucleus before the resulting mRNA molecule appears in the cytoplasm for translation. Exons are the means of supplying genetic information.

Introns are removed and exons are spliced together

1. After removal of the introns, the exons are ligated to form the mRNA molecule which is transported to the cytoplasm.

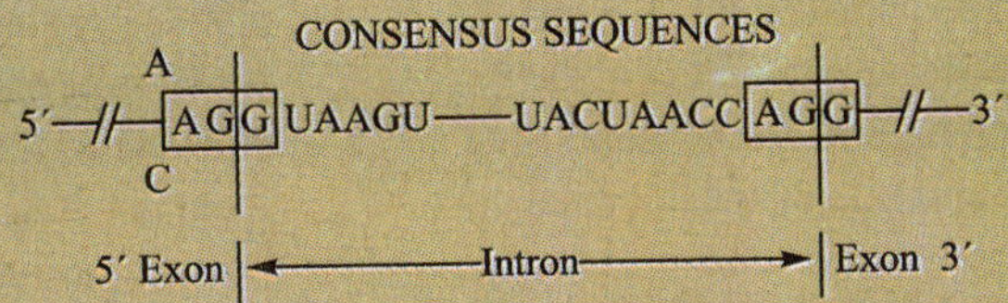

Fig. 25.13 : Consensus sequences at splice junctions. The 5′ (donor or left) and 3′ (acceptor or right) sequences are shown. The yeast consensus sequence (UACUAAC) for the branch site. In mammalian cells, the consensus sequence is PyNPyPyPuAPy, where Py is a pyridine, Pu is a purine, N is any nucleotide. The branch site is located 20-40 nucleotides upstream from the 3′ site

2. Of the four different splicing reaction mechanisms, the one most frequently used in eukaryotic cells is described here. There are reasonably conserved sequences at each of the two exon-intron (splice) junctions and at the branch site which is located 20-40 nucleotides upstream from the 3′ splice site. The spliceosome, a special structure, is involved in converting the primary transcript into mRNA.
3. Spliceosomes consist of the primary transcript, five small nuclear RNAs (U1, U2, U5, and U4/U6) and more than 50 proteins. Collectively, these form a small nucleoprotein (snRNP) complex, sometimes called a snurp. Snurps are used to position the RNA segments for the necessary splicing reactions. The splicing reaction starts with a cut at the junction of the 5′ exon. This is performed by a nucleophilic attack by an adenylyl residue in the branch point sequence located just upstream from the 3′ end of this intron.
4. The free 5′ terminal then forms a loop which is linked by an unusual 5′ — 2′ phosphodiester bond to the reactive A in the PyNPyPyPuAPy branch site sequence (Fig. 25.13). The adenylyl residue is typically loated 28-37 nucleotides upstream from the 3′ end of the intron being removed. The branch site identifies the 3′ splice site. A second cut is made at the junction of the intron with the 3′ exon. In this second transesterification reaction, the 3′ hydroxyl of the upstream exon attacks the 5′ phosphate at the downstream exon-intron boundary and the structure containing the intron is released and hydrolyzed. The 5′ and 3′ exons are ligated to form a continuous sequence.
5. U1 binds first to the 5 exon-intron boundary by base pairing U2 then binds, by base pairing, to the branch site. A U5/U4/U6 complex next joins the spliceosome. An ATP-dependent protein-mediated unwinding results in the disruption of the base paired U4/U6 complex with the release of U4. U6 is then able to interact first with U2, then with U1. These interactions serve to approximate the 5′ splice site, the branch point with its reactive A, and then 3′ splice site. This is enchanced by U5. The two ends are cleaved by the U2/U6 complex. RNA serves as a catalytic agent. This sequence is then repeated in genes containing multiple introns. The introns are not necessarily removed in sequence — 1, then 2, then 3, etc.
6. The hn RNA molecules are the primary transcripts plus their early processed products which are transported to the cytoplasm as mature mRNA molecules after the addition of Caps and Poly (A) tails and removal of the portion corresponding to the introns.

Ribosomal RNAs are processed from a large precursor

1. In mammalian cells, three rRNA molecules are transcribed to act as large precursor molecule. The precursor is then processed

in the nucleolus to provide the RNA component for the ribosome subunits of the cytoplasm. The rRNA genes are present in the nucleoli and hundreds of these genes are present in every cell. These large number of genes ultimately synthesize many copies of each type of rRNA to form the 10^7 ribosomes required for each cell replication. A single mRNA molecule is copied into 10^5 protein molecules, providing a large amplification, the rRNAs are end products.

2. Each of units being transcribed by the rRNA genes encodes (5′ to 3′) an 18s, a 5.8s and a 28s ribosomal RNA. The primary transcript is a 45s molecule which is highly methylated in the nucleolus. In the 45s precursor, the 28s segment contains 65 ribose-methyl groups and 5 base-methyl groups. The 45s precursor is nucleolytically processed.
3. During the processing of rRNA further methylation occurs. In the nucleoli, the larger 60s subunit is formed by the assembling of 28s chains with 50 ribosomal proteins. The 5.8s rRNA molecule is also formed from the 45s precursor RNA in the nucleolus. The 18s rRNA molecule is associated with 30 ribosome proteins to form the ribosomal subunits (40s).

Messenger RNA (mRNA) is modified at the 5′ and 3′ ends

1. Mammalian mRNA molecules contain a 7-methylguanosine cap structure at their 5 terminal, and most have a poly (A) tail at the 3′ terminal. The cap structure is added to the 5′ end of the newly transcribed mRNA precursor in the nucleus prior to transport of the mRNA molecule to the cytoplasm. The 5′ cap of the RNA transcript is necessary for translation initiation and protection of the 5′ end of mRNA from attack by 5′ → 3′ exonucleases.
2. Poly (A) tails are added to the 3′ end of mRNA molecules in the post transcriptional processing step. The mRNA is first cleaved about 20 nucleotides downstream from an AAUAA recognition sequence. Poly (A) polymerase adds a short poly (A) tail which is extended to 200 A residues. This enzyme protects the 3′ end of mRNA from 3′ → 5′ exonuclease attack. In mammalian cells, cytoplasmic processes can add and remove adenylate residues from the poly (A) tails.
3. The extra nucleotides occur in noncoding regions both 5′ and 3′ to the coding region, the longest noncoding sequences are usually at the 3′ end.

Transfer RNA (tRNA) is extensively processed and modified

1. The tRNA molecules act as adaptor molecules for the translation of mRNA into protein sequences. The tRNAs are reduced in size by a specific class of ribonucleases. The genes of some tRNA molecules contain a single intron 10—40 nucleotides long. These introns are transcribed. The processing of the precursor transcripts of many tRNA molecules must include removal of the introns and proper splicing of the anticodon region to generate an active adaptor molecule for protein synthesis.
2. The further modification of the tRNA molecules includes nucleotide alkylations and the attachment of the characteristic C·C·A terminal at the 3′ end of the molecule by the enzyme nucleotidyl transferase. The 3′OH of the A ribose is the point of attachment for the specific amino acid that is to enter the polymerization reaction of protein synthesis. The methylation of mammalian tRNA precursors occurs in the nucleus, but the cleavage and attachment of C·C·A are cytoplasmic functions. The attachment of amino acids to the C·C·A residues require the enzymes within the cytoplasm of mammalian cells.

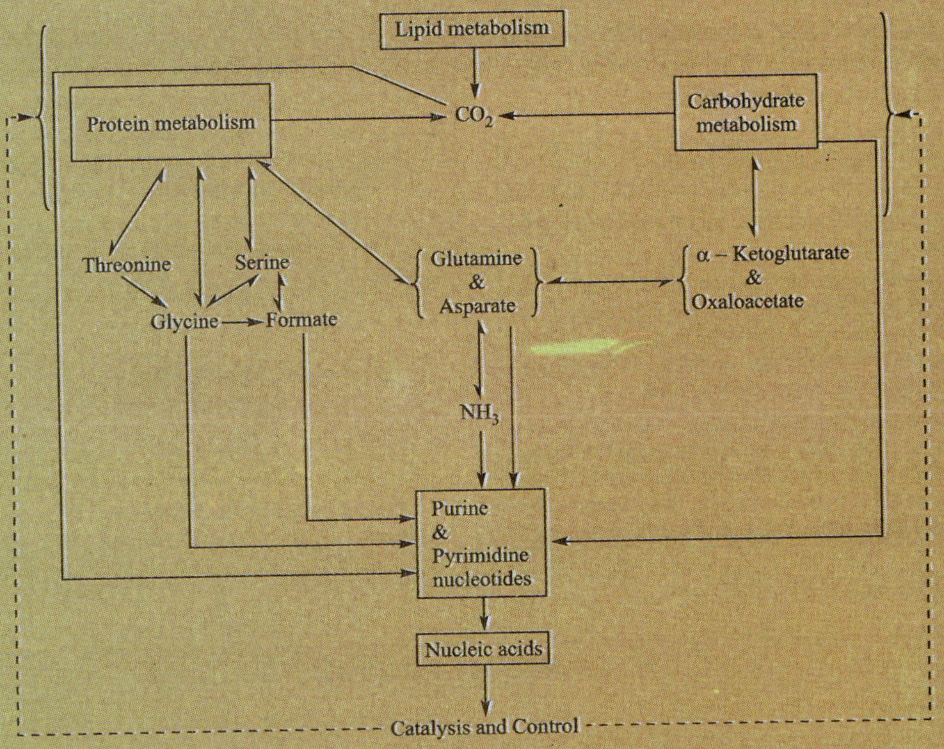

Fig. 25.14 : Metabolic interrelations of nucleic acids with other foodstuffs

Biological Importance of Nucleic Acids

1. Nucleic acids are able to reproduce their own kind or to store, express and transmit genetic information.
2. They undergo mutation.
3. In cell division, the nucleic acid chain is duplicated preserving in each daughter cell the information contained in the parent cell. So the double helix unravels and each of the two original strands then serves as a template for the synthesis of another complementary chain.
4. DNA produces a messenger RNA which helps in placing amino acids in the code for protein synthesis.
5. RNA functions primarily in the cytoplasm of the cell as a template in connection with the synthesis of proteins as well as in the ribosomes. The formation of RNA template is directed by nuclear DNA.
6. Ribosomal RNA and tRNA are also involved in protein synthesis.
7. RNA can be synthesized by RNA polymerase which is dependent on the presence of DNA acting as a template.
8. Adenylic acid in combination with two molecules of phosphoric acid is the biochemical unit of energy exchange in all cells. It is called ATP.
9. Biological oxidation-reduction involves the transport of hydrogen atoms or electrons through organized systems of substances called hydrogen acceptors or electron transport agents. These hydrogen acceptors are nucleotides such as NAD, FAD etc.
10. According to the "pairing rule" in DNA, adenine can combine only with thymine and guanine only with cytosine. The newly synthesized strand will be exactly

constituted in its nucleotide sequences as was the original complementary strand of the parent strand. The result is the synthesis of two pairs of strands.

CLINICAL ASPECTS

- The heredity and genetic diseases are found in the structure of DNA.
- The knowledge of the basic information pathway (i.e. DNA directs the synthesis of RNA which in turn directs protein synthesis) is used to define normal cellular physiology and the pathophysiology of disease at the molecular level.
- Mutations are due to a change in the base sequence of DNA. These are formed from the faulty replication or repair of DNA and occur within a frequency of about one in every 10^6 cell divisions. An abnormal gene product is the result of mutations.

CLINICAL ASPECTS

- Viruses, chemicals, ultraviolet light, and ionizing radiation increase the rate of mutation.
- Mutations often attack somatic cells of an organism and are passed on to successive generations of cells within an organism.
- A number of diseases and most cancers are due to this horizontal transmission of induced mutations.

EXERCISE

1. Discuss the biological importance of nucleic acids. (M.U. 72A; R.U. 68A)
2. Describe in short the metabolism of nucleic acid in man. (R.U. 71A)

SCHEMATIC COMPENDIUM

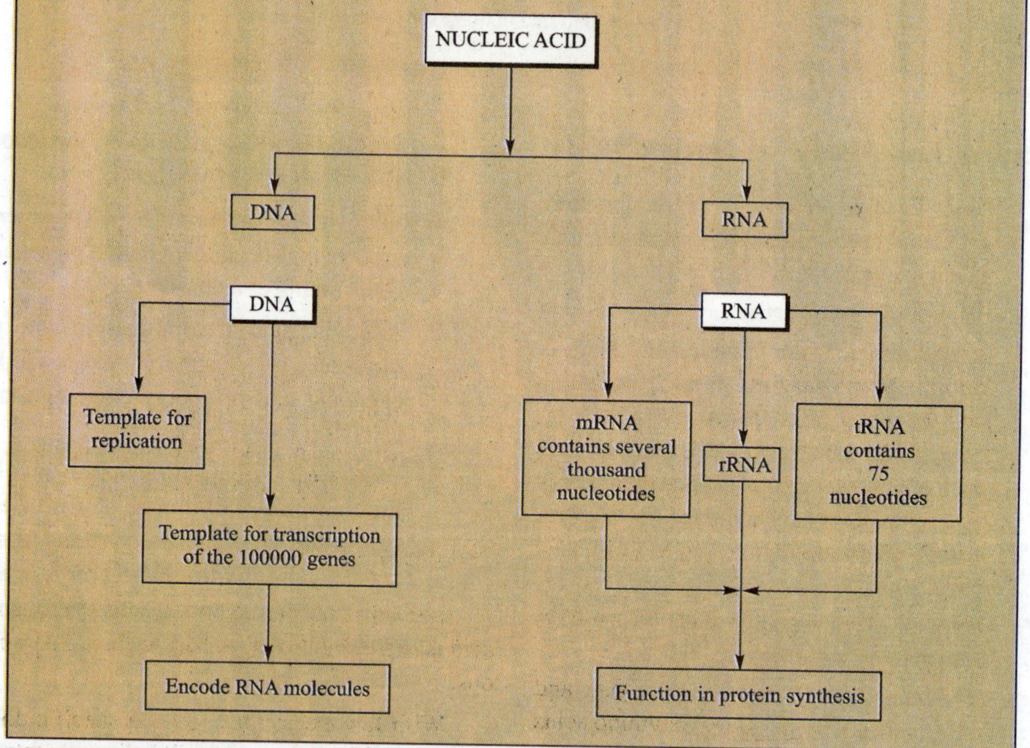

CHAPTER 26

PROTEIN SYNTHESIS AND THE GENETIC CODE

THE GENETIC CODE

The genetic code is composed of words formed by a sequence of nucleotide bases and a sequence of amino acids. Each word in the code is composed of three nucleotide bases. The genetic words are called codons.

Codons

1. The codons are usually presented in the messenger RNA.
2. The four nucleotide bases [Adenine (A), Guanine (G), Cytosine (C), and Uracil (U)] are used to form the three-base codons. Therefore, 64 different combination of bases, taken three at a time (4^3), are formed.
3. The nucleotide sequences are always written from the 5′-end to the 3′-end.
   (a) **Translation of codons :** Sixty one of 64 codons code for the 20 common amino acids. For example, the codon 5′-CAU-3′ codes for histidine, whereas 5′AUG-3′ codes for methionine.
   (b) **Termination (stop or nonsense) codons**
      (i) Three codons (UAG, UGA, and UAA) do not code for amino acids and hence, termed as *termination codons*.
      (ii) If one of these codons appears in an mRNA sequence, it signals the complete synthesis of the peptide chain coded for by that mRNA.

Characteristics of Genetic Code

1. Degeneracy

(a) Three codons (UAA, UAG, UGA) do not code for specific amino acids, these have been termed nonsense codons. At least 2 of these nonsense codons are utilized in the cell as termination signals, they specify where the polymerization of amino acids into a protein molecule is to stop. The remaining 61codons code for 20 amino acids. So there must be degeneracy in the genetic code.

(b) Every amino acid except methionine is represented by several codons.

(c) Codons that represent the same amino acid are called synonyms.

2. Unambiguity

For any specific codon, only a single amino acid is indicated. So the genetic code is unambiguous.

With few exceptions, given a specific codon, only a specific amino acid will be incorporated although, given a specific amino acid, more than one codon may call for it.

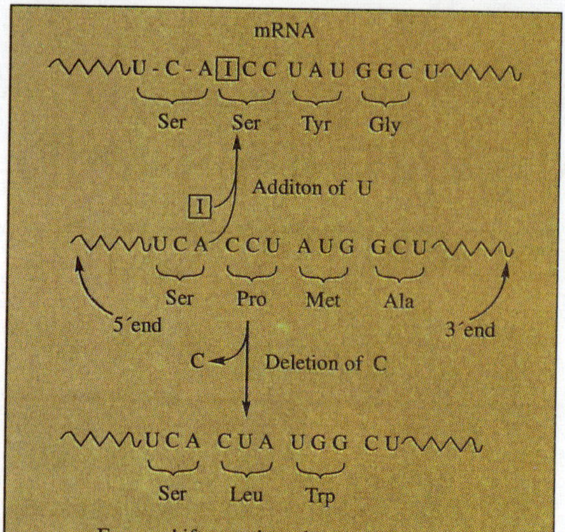

Frame–shift mutations due to addition or deletion of a base

3. Non-overlapping

All codons are independent sets of 3 bases. There is no overlapping.

4. Universality

(a) In all the living organisms the genetic code is the same.

(b) The codon AUA is read as Met and UGA codes for Trp in mammalian mitochondria, codons AGA and AGG are read as stop or chain terminator codons. So mitochondria require only 22 tRNA molecules to read their genetic code, whereas cytoplasmic translation system passes a few complement of 31 tRNA species. These exceptions represent that the genetic code is universal.

5. Commalessness

The genetic code is read from a fixed starting point as a continuous sequence of bases taken three at a time. If one or two nucleotides are added to or deleted from the interior of a message sequence, a frame-shift mutation occurs and the reading frame is changed. It is to be noted that three nucleotides, if are added, a new amino acid is added to the peptide, or if three nucleotides are deleted, an amino acid is lost. In such cases, the reading frame is not affected.

Wobble

The wobble hypothesis describes the mechanism by which tRNAs can recognize more than one codon for a specific amino acid. In this hypothesis, the base at the 5′-end of the anticodon is not defined as the other two bases. The movement of the first base allows nontraditional base pairing with the 3′-base of the codon. This movement is called **wobble**.

Nucleoid : A Prokaryotic organism contains a single, double- stranded, supercoiled, circular chromosome. Each chromosome is associated with histone-like proteins and RNA that can condense the DNA to form a nucleoid.

Plasmids :

(i) In addition to nucleoid, most species of bacteria also contain small, circular, extrachromosomal DNA molecules called plasmids.

(ii) Plasmid DNA carries genetic information and undergoes replication that may or may not be related to chromosomal division.

(iii) Plasmid may carry genes that cause antibiotic resistance to the host bacterium and may facilitate the transfer of genetic information from one bacterium to another.

(iv) Plasmids are used as vectors in recombinant DNA technology.

Semiconservative replication : Each strand of the DNA as a result of the separation of the two strands of the DNA double helix can act as a template for the replication of a new complementary strand producing two daughter molecules each of which contains two DNA strands with an antiparallel orientation. This process is called semiconservative replication.

Mutation

An alteration in the sequence of bases in a gene due to change, removal or insertion of one or more bases in a gene resulting in an altered gene product is called mutation.

(a) Single base changes (point mutations) may be **transitions** or **transversions**.

In transitions, a given pyrimidine is changed to the other pyrimidine or a given purine is changed to other purine.

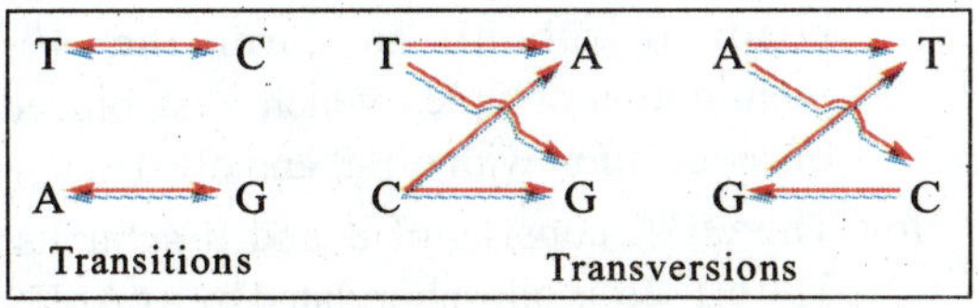

Transversions are changed from a purine to either of the 2 pyrimidines or the change of a pyrimidine into either of the 2 purines.

(b) **Missense mutation** : The codon containing the changed base may code for a different amino acid, eg, if the serine codon UCA is given a different first base to become, say, CCA, it codes for a different amino acid, in this case, proline. The substitution of an incorrect amino acid is called a **missense** mutation.

(c) **Nonsense mutation** : The codon containing the changed base may become a termination codon, eg, if the serine codon UCA is given a different second base to become, say, UAA, the new codon causes termination of translation at that point. The formation of a termination codon at an inappropriate place is called a **nonsense** mutation.

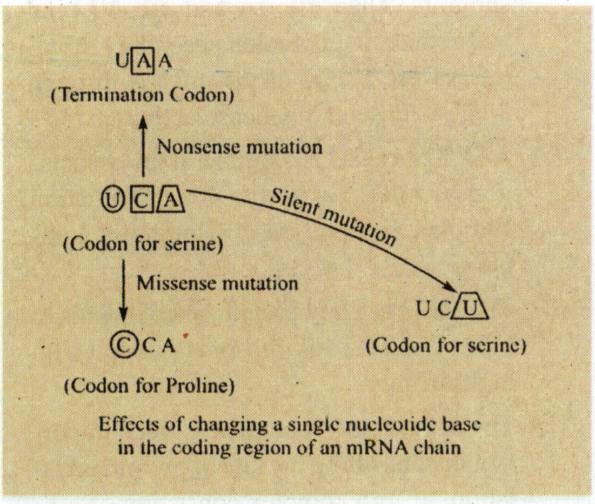

Effects of changing a single nucleotide base in the coding region of an mRNA chain

Genetic information flows from DNA to RNA to protein

1. The genetic information within the nucleotide sequence of DNA is transcribed in the nucleus into the specific nucleotide sequence of an RNA molecule. A number of different classes of RNA combine to direct the synthesis of proteins.
2. In eukaryotic cells, the **heterogeneous nuclear RNA (hn RNA)** is much larger than the mature mRNA. The large hn RNA contains coding regions (**exons**) that will form the mature mRNA and long intervening sequence (**introns**) that separate the exons. After the removal of introns by the processing of hnRNA within the nucleus, the exons are spliced to form mature mRNA which is translated into protein being transported to the cytoplasm.
3. The cell possesses the machinery to translate information accurately and efficiently from the nucleotide sequence of an mRNA into the sequence of amino acids of the corresponding specific protein.

The nucleotide sequence of an mRNA molecule consists of a series of codons

1. **Transfer RNA** (**tRNA**) molecules are the adaptor molecules that translate the codons into the amino acid sequence of a protein.
2. The ribosome is the cellular machine to assemble the protein molecule. Ribosomes can assemble to translate a single mRNA molecule and form a **polyribosome**.
3. The **rough endoplasmic reticulum** is a compartment of membrane—attached polyribosomes that provides for the synthesis of membrane proteins. Polyribosomal structures also exist free in the cytoplasm where they synthesize proteins that remain within the cell.

4. Twenty different amino acids are required for the synthesis of proteins and there must be at least 20 distinct codons that make up the genetic code. In mRNA, there are only four different nucleotides and each codon must consist of more than a single purine or pyrimidine nucleotide.

Protein Biosynthesis

Protein biosynthesis can be explained by initiation, elongation and termination.

The mRNA nucleotide sequence is translated into the sequence of amino acids of the specified protein. The translation of the mRNA starts near its 5′-terminal with the formation of the corresponding amino terminal of the protein molecule. In eukaryotic organisms, the process of transcription is a nuclear one; mRNA translation occurs in the cytoplasm. The simultaneous transcription and translation in eukaryotic organisms make the processing necessary to generate mature mRNA from the primary transcript –hnRNA.

Initiation

The process of initiation needs tRNA, rRNA, mRNA, and at least 10 eukaryotic initiation factors (eIFs), some of which have multiple (3 to 8) subunits. GTP, ATP and amino acids are also involved. This process can be divided into four steps :

1. Dissociation of the ribosome into its 40s and 60s subunits.
2. Formation of the 40s preinitiation complex.
3. Formation of the 40s initiation complex.
4. Formation of the 80s initiation complex.

1. Dissociation of the ribosome :

(a) The binding of two eukaryotic initiation factors eIF-3 and eIF-IA to the 40s subunit favours dissociation of the 80s ribosome into its 40s and 60s subunits and prevents reassociation.

(b) Reassociation is also prevented by the binding of eIF-3A to the 60s subunits.

2. Formation of the 40s preinitiation complex :

(a) GTP binds eIF-2 and this binary complex then binds to met- tRNA, a tRNA is involved in binding to the initiation codon AUG.

(b) This above complex binds to the 40s ribosomal subunit to form the 40s preinitiation complex which is stabilized by association with eIF-3 and eIF-IA.

(c) The eIF-2 consists of α and β subunits. The eIF-2α is phosphorylated by a cAMP-independent protein kinase which is activated when a cell is under stress and also when the energy expenditure is deleterious. Amino acid and glucose starvation, virus infection, serum deprivation, and heat shock are included in such conditions.

(d) The phosphorylated eIF-2α inactivates and binds tightly to the GTP-GDP recycling protein eIF-2β. This prevents formation of the 40s preinitiation complex and blocks protein synthesis.

3. Formation of the 40s initiation complex :

(a) The methyl-guanosyl triphosphate cap helps the binding of mRNA to the 40s preinitiation complex.

(b) A cap binding protein complex, eIF-4F, binds to the cap through one of its subunits. Then eIF-4A and eIF-4B bind and reduce the complex secondary structure of the 5′-end of the mRNA through their ATPase and helicase activities.

(c) ATP hydrolysis is required in the association of mRNA with the 40s preinitiation complex to form the 40s initiation complex.

(d) The eIF-3 is a key protein since it binds to mRNA and the 40s ribosomal subunit with high affinity.

(e) The eIF-4F (4F) complex is very important in controlling the rate of protein translation. 4F is a complex consisting of eIF-

4E (4E), which binds to the m^7G cap structure at the 5′-end of the mRNA, eIF-4A, which is an RNA-dependent ATPase that helps unwind the RNA, and p220. 4E is important since cap-binding is a rate-limiting step in translation. Overexpression of 4E causes malignant transformation of cells.

(f) The phosphorylation of 4E (on serine 53) and activation of the protein are caused by insulin and mitogenic growth factors.

(g) A recently discovered family of proteins 4E-BP1 (and the closely related protein PHAS-1) and 4E-BP2 bind to and inactivate 4E. These proteins are again phosphorylated by insulin and growth factors via the stimulation of serine protein kinase resulting in the dissociation of the binding protein from 4E, the assembly of 4E and other components of the 4F complex in the mRNA cap, and initiation of protein synthesis.

4. Formation of the 80s initiation complex

(a) The hydrolysis of the GTP bound to eIF-2 by eIF-5 is caused by the binding of the 60s ribosomal subunit to the 40s initiation complex. This reaction promotes the release of the initiation factors bound to the 40s initiation complex (these factors are then recycled) and the rapid association of the 40s and 60s subunits to form the 80s ribosome.

(b) Ultimately, the met-tRNA on the P site of the ribosome becomes ready for the elongation cycle to start.

Elongation

Elongation is a cyclic process consisting of several steps catalyzed by elongation factors (eEF). The steps are

1. Binding of aminoacyl-tRNA to the A site.
2. Peptide formation.
3. Translocation.

1. Binding of Aminoacyl-tRNA to the A site

(a) The aminoacyl-tRNA binds to the A site of the complete 80s ribosome requiring proper codon recognition.

(b) The complex formed by the elongation factor eEF-1α with GTP and the aminoacyl-tRNA enters the A site with the release of eEF-1α-GDP and phosphate. The eEF-1α-GDP then recycles to eEF-1α-GTP by the help of other soluble protein factors and GTP.

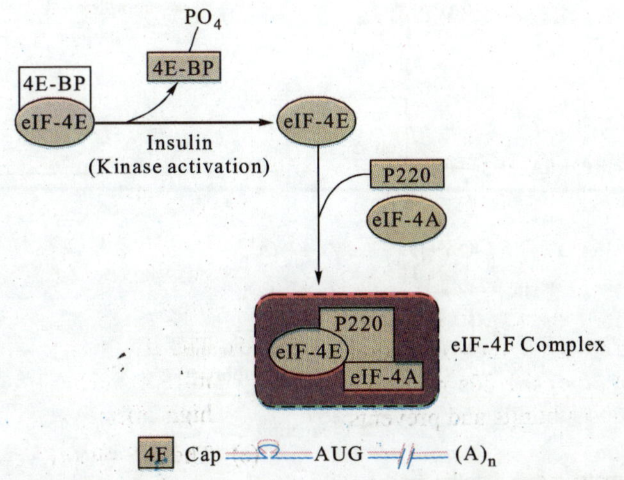

Fig. 26.1 : Activation of eIF-4E by insulin and formation of the cap binding eIF-4F complex

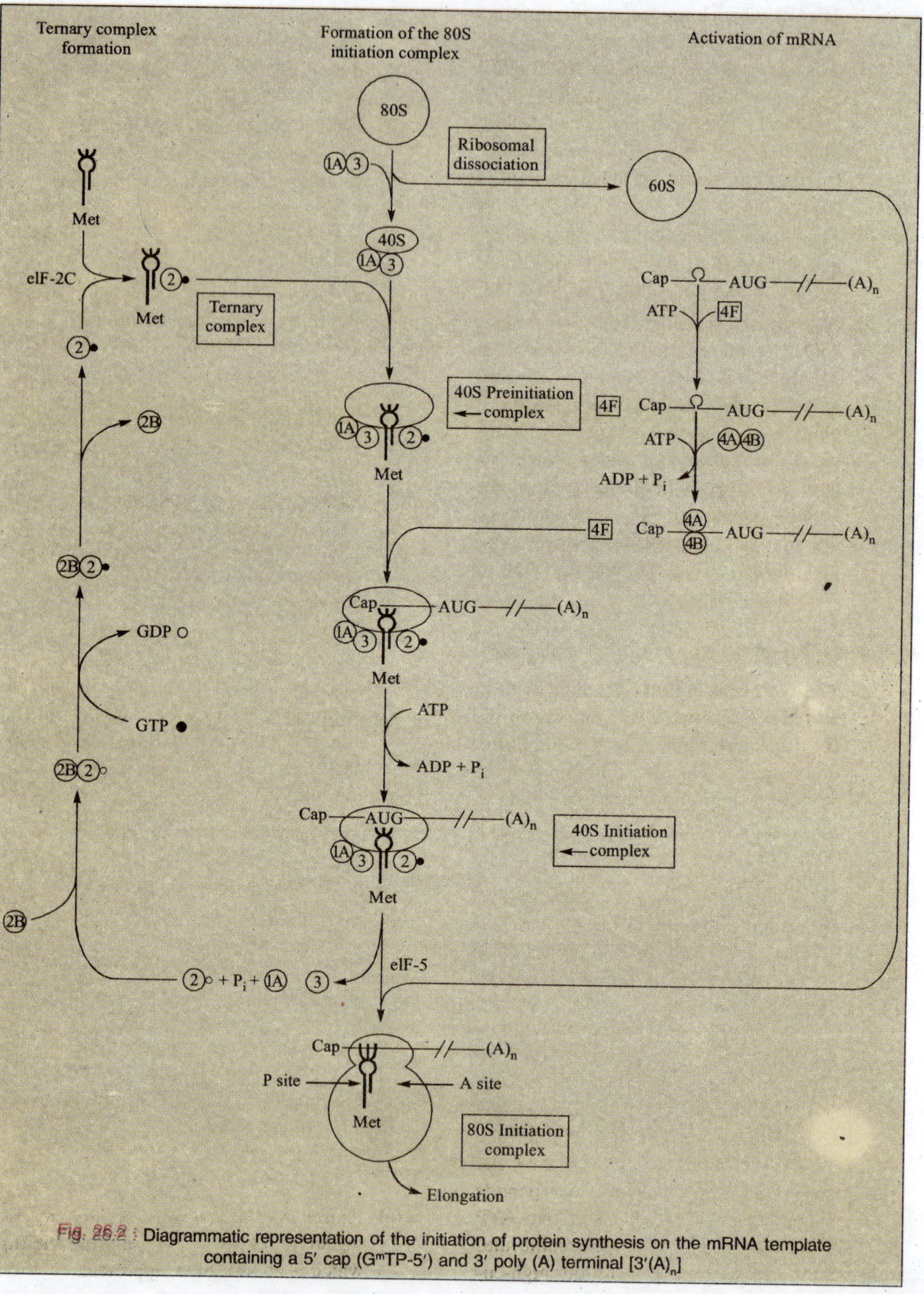

Fig. 26.2 : Diagrammatic representation of the initiation of protein synthesis on the mRNA template containing a 5′ cap (G^mTP-5′) and 3′ poly (A) terminal [3′$(A)_n$]

2. Peptide bond formation :

(a) The α-amino group of the new aminoacyl-tRNA in the A site has a nucleophilic attack on the esterified carboxyl group of the peptidyl-tRNA occupying the P site being catalyzed by a protein component peptidyltransferase of the 60s ribosomal subunit. This enzymic activity is performed by some RNA component of the ribosome, perhaps the 23s rRNA. This shows the direct role for RNA in protein synthesis.

(b) The amino acid on the aminoacyl-tRNA is already activated for which no further energy source is required for this reaction.

3. Translocation :

(a) The discharged tRNA is quickly dissociated from the P site on the removal of the peptidyl moiety from the tRNA in the P site.

(b) The elongation factor 2 (eEF-2) and GTP are responsible for the translocation of the newly formed peptidyl-tRNA at the A site into the empty P site. GTP is hydrolyzed to GDP and phosphate during the translocation.

(c) The translocation of the newly formed peptidyl-tRNA frees the A site for another cycle of aminoacyl-tRNA codon recognition and elongation.

(d) ATP is hydrolyzed to an AMP which is equivalent to the hydrolysis of two ATPs to two ADPs and phosphates in the charging of the tRNA molecule with the aminoacyl moiety. GTP is also hydrolyzed to GDP during the entry of the aminoacyl-tRNA into the A site. The hydrolysis of GTP to GDP and phosphate is caused by the translocation of the newly formed peptidyl-tRNA in the A site into the P site by eEF-2. The formation of one peptide bond causes the hydrolysis of two ATP molecules to ADP and two GTP molecules to GDP or the hydrolysis of four high-energy phosphate bonds. This process occurs rapidly.

(e) As many as six amino acids are incorporated per second in case of an eukaryotic ribosome and eighteen amino acids are incorporated per second in case of prokaryotic ribosome. Thus, the process of peptide synthesis is enhanced with great speed and accuracy until a termination codon is reached.

Termination

(a) The nonsense or the terminating codon of mRNA (UUA, UAG, UGA) appears in the A site after many cycles of elongation in the process of polymerization of the specific amino acids into a protein molecule.

(b) Normally there is no tRNA with an anticodon to recognize such a termination signal.

(c) The existance of a termination signal in the A site is recognized by the releasing factors (eRF) which are proteins that hydrolyze the peptidyl-tRNA bond when a nonsense codon occupies the A site.

(d) The releasing factor enhances the hydrolysis of the bond between the peptide and the tRNA occupying the P site in conjugation with GTP and the peptidyl transferase. This hydrolysis caused by the addition of a water molecule releases the protein and the tRNA from the P site.

(e) The 80s ribosome dissociates into its 40s and 60s subunits upon hydrolysis and release. These subunits are recycled.

(f) The mRNA is then released from the ribosome which dissociates into its component 40s and 60s subunits and another cycle can be repeated.

The machinery of Protein synthesis can respond to environmental threats

(a) Ionized iron (Fe^{++}) is prevented from reaching toxic levels within cells by ferritin which is an iron binding protein.

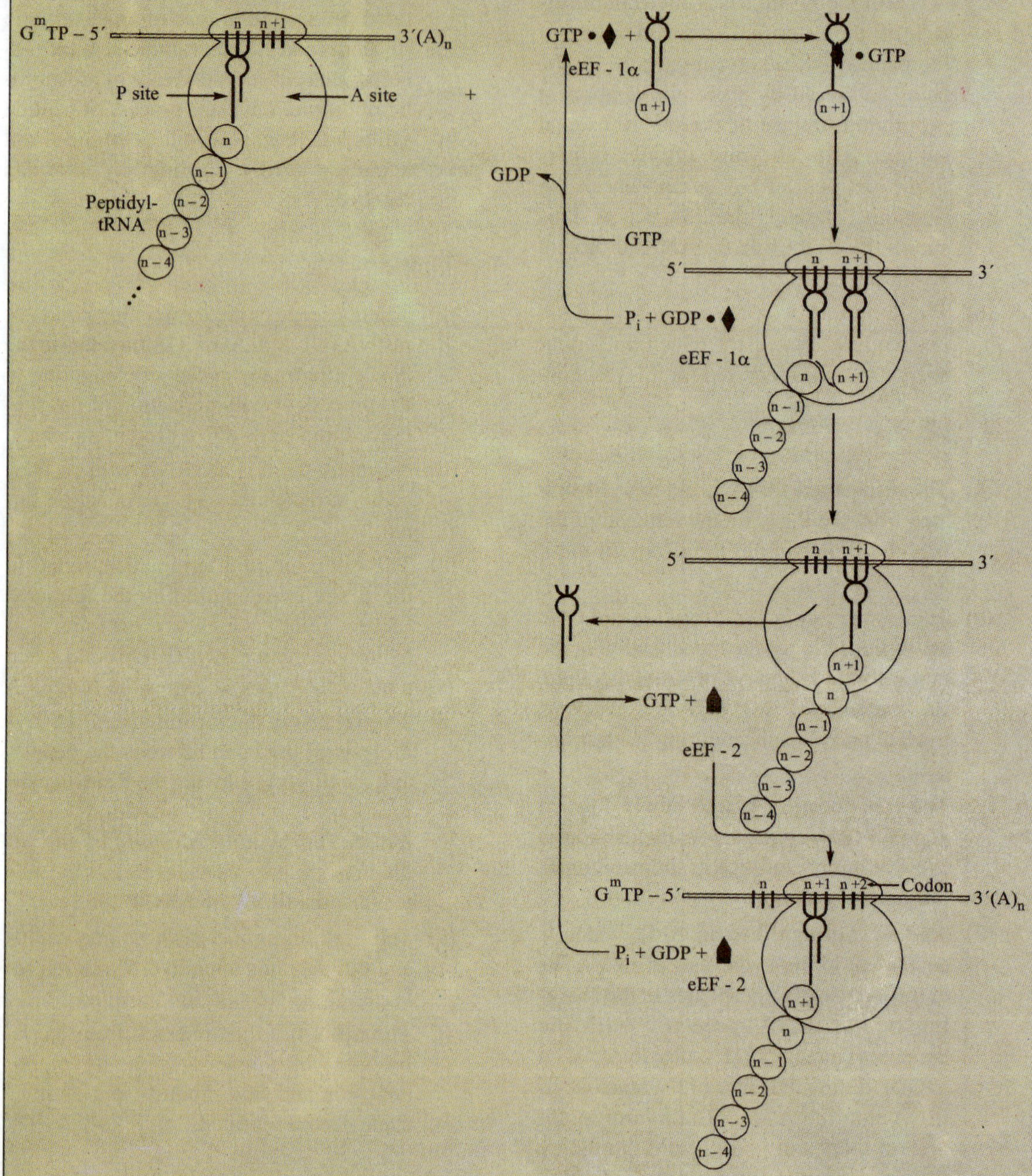

Fig. 26.3 : Diagrammatic representation of the peptide elongation process of protein synthesis. The small circles labelled n – 1, n, n + 1 etc. represent the amino acid residues of the newly formed protein molecule

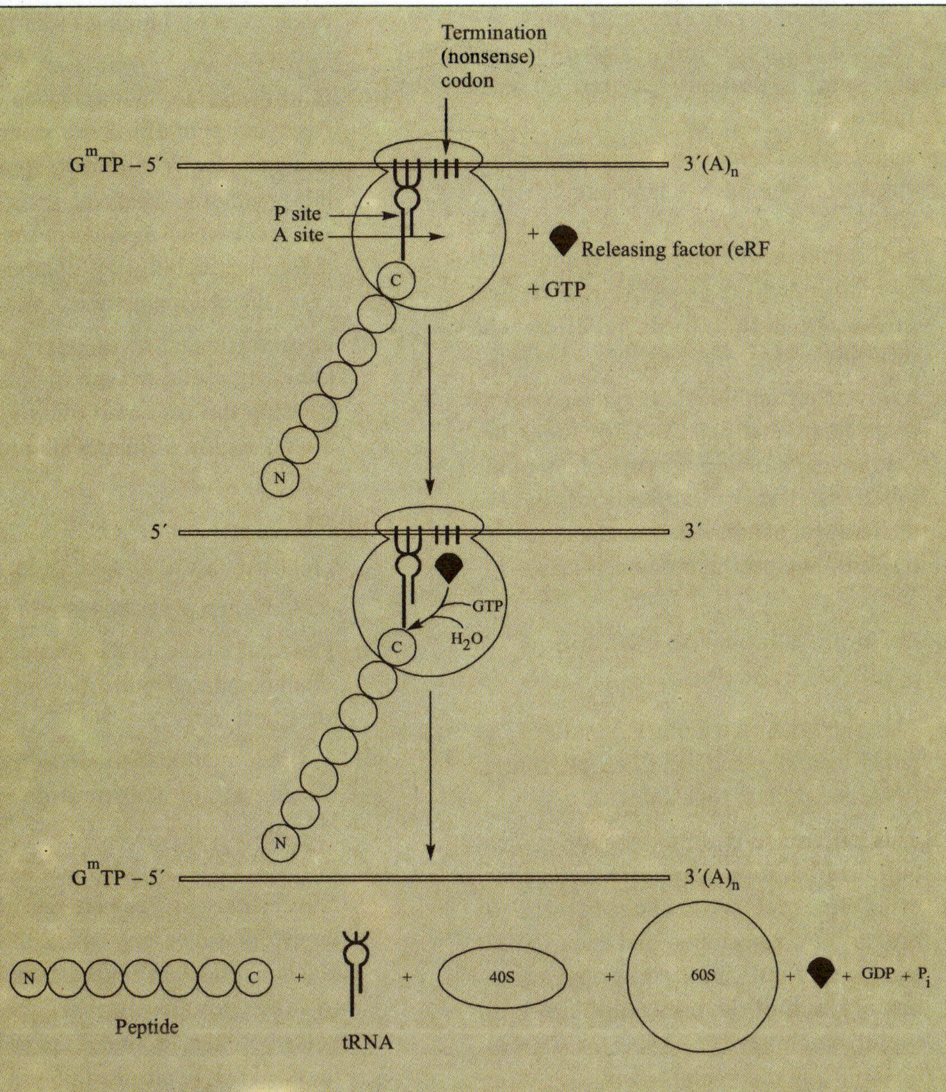

Fig. 26.4 : Diagrammatic representation of the termination process of protein synthesis. The peptidyl-tRNA and amino acyl-tRNA sites are indicated as P site and A site respectively

(b) Ferritin synthesis is stimulated by elemental iron by causing the release of a cytoplasmic protein that binds to a specific region in the 5′ nontranslated region of ferritin mRNA. The disruption of this protein –mRNA activates ferritin mRNA and results in its translation. This mechanism causes rapid control of the synthesis of a protein that resists Fe^{++}, a toxic molecule.

(c) Some viral mRNAs are tanslated more efficiently than those of the host cell. Other viruses inhibit host cell protein synthesis by preventing the association of mRNA with the 40s ribosome.

The activity of many proteins is affected by posttranslational processing

(a) The long polycistronic proteins are synthesized from one long mRNA molecule by some animal viruses and hepatitis A virus. These protein molecules are cleaved at specific sites to give the several specific proteins required for viral function.

(b) In animal cells, many proteins are synthesized from the mRNA template as a precursor molecule which is then modified to form the active protein. The prototype is insulin which is synthesized as a **prohormone** that folds to allow the disulfide bridges to form.

(c) Many other peptides are synthesized as proproteins that need modifications before having biologic activity. Many of the posttranslational modifications involve the removal of amino terminal amino acid residues by specific aminopeptidases.

The role of antibiotics in the inhibition of protein synthesis in bacteria

(a) The bacterial ribosome is smaller (70s rather than 80s) and has a complement of RNA and protein molecules.

This difference is important for clinical purposes because many effective antibiotics interact with the proteins of prokaryotic ribosomes and thus inhibit protein synthesis. This causes growth arrest or death of the bacterium. The most useful antibiotics eg, **tetracyclines**, **erythromycin** and **chloramphenicol**, do not interact with components of eukaryotic ribosomal particles and hence are not toxic to eukaryotes.

(b) Other antibiotics eg, **puromycin** inhibit protein synthesis on all ribosomes. Puromycin is structurally similar to tyrosinyl-tRNA. This drug is incorporated via the A site of the ribosome into the carboxy-terminal position of a peptide but causes the premature release of the polypeptide. Puromycin effectively inhibits protein synthesis in both prokaryotes and eukaryotes. **Cycloheximide** inhibits peptidyltransferase in the 60s ribosomal subunit in eukaryotes by binding to an rRNA component.

(c) **Diphtheria toxin** catalyzes the ADP-ribosylation of eEF-2 in mammalian cells and specifically inhibits mammalian protein synthesis. Many animals (eg, mice) are resistant to diphtheria toxin. This is due to the inability of diphtheria toxin to cross the cell membrane.

(d) **Puromycin** and **Cycloheximide** are not clinically useful but are important in elucidating the role of protein synthesis in the regulation of metabolic processes.

Polysomes

(a) Many ribosomes on the same mRNA molecule form a polyribosome or polysome.

(b) The mass of the mRNA molecule is quite small compared to the mass of even a single ribosome.

(c) A single mammalian ribosome can synthesize about 100 peptide bonds per minute.

(d) Polyribosomes actively synthesizing protein can exist as free particles in the cellular cytoplasm or may be attached to endoplasmic reticulum. This attachment is due to the rough appearance of the polyribosomes. Some of the proteins synthesized by the attached polyribosmes are packaged by the golgi apparatus.

(e) The polyribosomal particles free in the cytosol are responsible for the synthesis of proteins required for intracellular functions.

The energy requirements for the formation of one peptide bond

1. 2 ATP (charging of amino acids).
2. 2 GTP (one for transferring amino acyl tRNA and another for transferring tRNA from A site to P site).

REGULATION OF GENE EXPRESSION

A single ribosome is capable of translating about 400 codons in 10 seconds into a protein with a molecular weight of 40,000.

Mammalian cells possess only about 1,000 times more genetic information than does the bacterium E. Coli. Much of this additional genetic information is probably involved in the regulation of gene expression.

There are only two types of gene regulation: *positive regulation* and *negative regulation*. When the expression of genetic information is quantitatively increased by the presence of a specific regulatory element, regulation is said to be *positive*; whereas when the expression of genetic information is diminished by the presence of a specific regulatory element, regulation is said to be *negative*. The element or molecule mediating the negative regulation is said to be a negative regulator, that mediating positive regulation is a positive regulator. A *double negative* has the effect of acting as a positive. Thus, an effector that inhibits the function of a negative regulator will appear to bring about a positive regulation. In many regulated systems that appear to be induced, they are depressed at the molecular level.

In biologic organisms, there are three types of temporal responses to a regulatory signal. These three responses are shown in the Fig. 26.5 as rate of gene expression in temporal response to an inducing signal.

A *Type A response* is characterized by an increased rate of gene expression that is dependent upon the continued presence of the inducing signal. When the inducing signal is removed, the rate of gene expression is diminished to its basal level, but the rate repeatedly increases in response to the reappearance of the specific signal. This type of response is commonly observed in many higher organisms after exposures to inducers such as *steroid hormones*.

A *Type B response* exhibits an increased rate of gene expression that is transient even in the continued presence of the regulatory signal. After the regulatory signal has terminated and the cell has

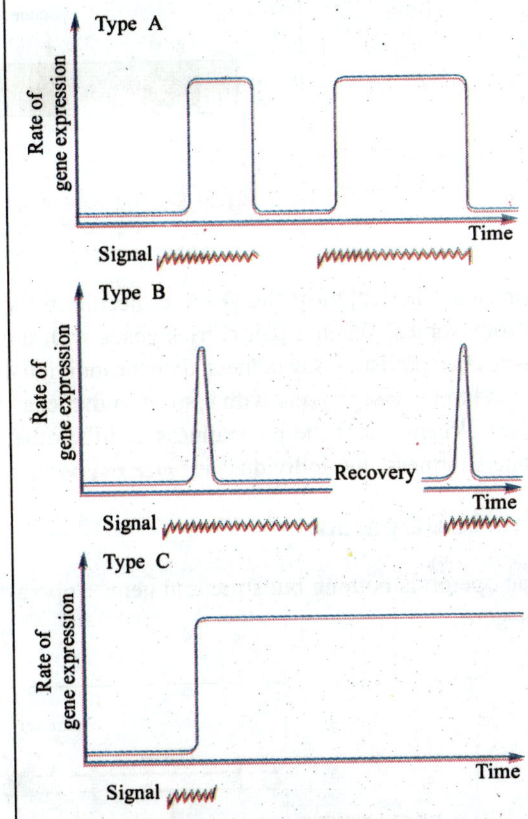

Fig. 26.5 : Diagrammatic representation of the responses of the rate of expression of gene to specific regulatory signals such as a hormone

been allowed to recover, a second transient response to a subsequent regulatory signal may be observed. This type of response may commonly occur during development of an organism when only the transient appearance of a specific gene product is required although the signal persists.

The *Type C response* exhibits an increased rate of gene expression that persists indefinitely even after the termination of the signal. The signal acts as a trigger in this pattern. Once the gene expression is initiated in the cell, it cannot be terminated even in the daughter cells; it is, therefore, an irreversible and inherited alteration.

The *gene* or *cistron* is the unit of genetic information. It is the smallest segment of the DNA molecule containing about 600 base pairs. The genetic message is carried in the sequence of bases along the DNA strand. These are arranged in orderly man-

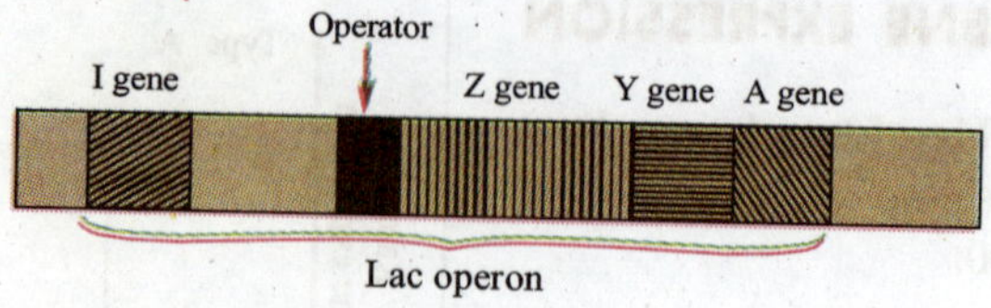

Fig. 26.6 : The positional relationship of the structural and regulatory genes of the lac operon

ner along the length of the DNA molecule in the chromosomes. When a pair carries genes with the same characteristics, say tallness, then the individual is said to be ***homozygous*** with respect to that character. When one of the pair tallness and the other gene shortness, the individual is ***heterozygous***.

The Lac Operon

Lac operon is nothing but structural gene + operator gene.

A lactose analog which is capable of inducing the lac operon while not itself serving as a substrate for β-galactosidase is called a **gratuitous inducer**.

The expression of some gene is constitutive i.e., they are expressed at a reasonably high rate in the absence of any specific regulatory signal.

The regulator gene controls the activity of the operator gene. The regulator gene induces the synthesis of protein macromolecules (probably RNA protein) called ***repressors***.

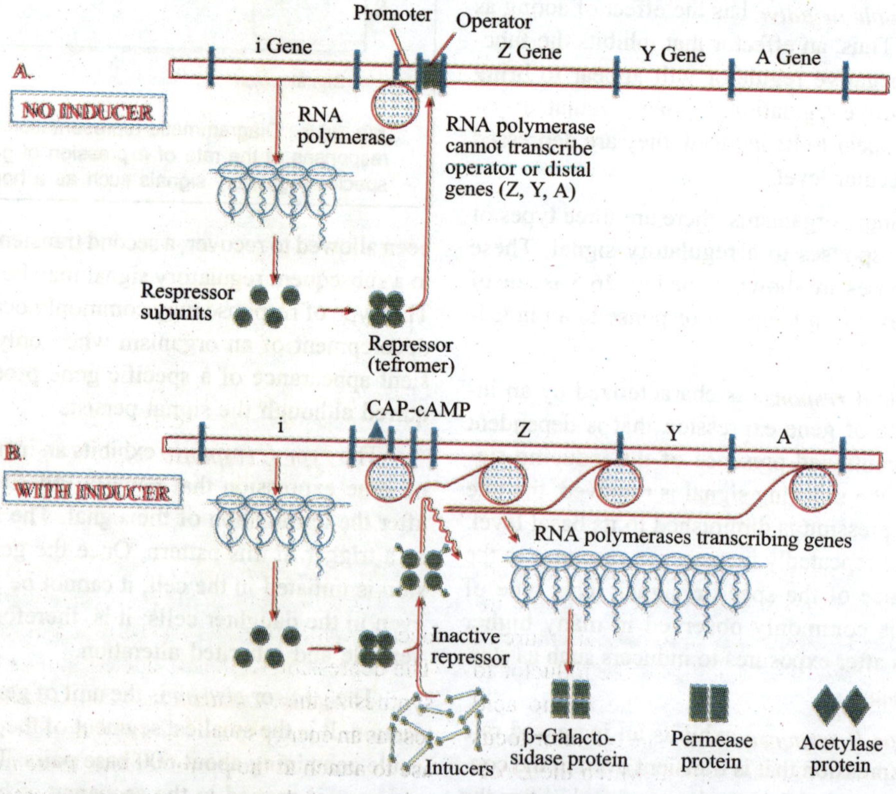

Fig. 26.7 : The mechanism of repression and derepression of the lactose operon

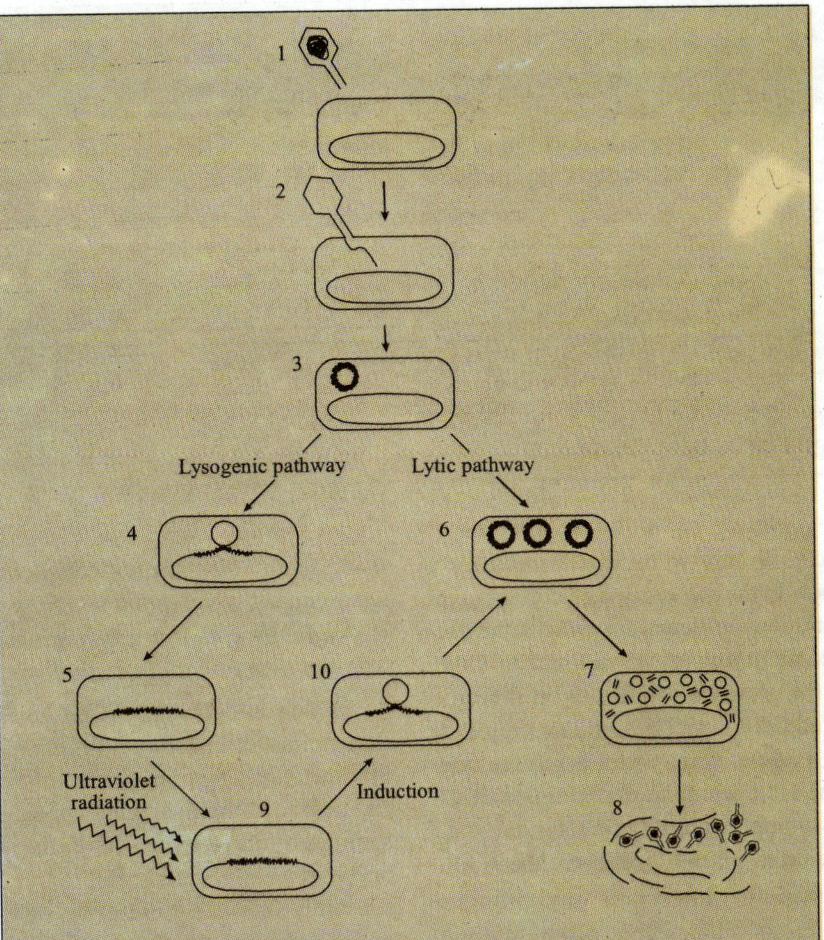

Fig. 26.8 : Infection of the bacterium E. Coli by the lamda virus

The operon becomes active because the repressor system is itself inactivated. The phenomenon is said to be Derepression.

The operator locus is a region of double-stranded DNA of 27 base pairs long with a 2-fold rotational symmetry in a region that is 21 base pairs long. The minimum effective size of an operator for lac repressor binding is 17 base pairs. The binding occurs mostly in the major groove without interrupting the base-paired double helical nature of the operator DNA. The binding of the inductor to the repressor molecule involves the amino acid residues in positions 74 and 75. The operator locus lies between the Promoter site, at which the *DNA-dependent RNA polymerase* attaches to commence transcription, and the beginning of the Z gene. When attached to the operator locus the repressor molecule prevents the transcription of the operator locus and of the distal structural genes, Z, Y, A. Thus, the repressor molecule is a *negative regulator* and in its presence the expression of the Z, Y, and A genes is prevented. Normally 20-40 repressor molecules are present and one or two operator loci per cell are also found to be present.

The translation of the polycistronic mRNA can occur even before the transcription is completed. The depression of the lac operon allows the cell to synthesize the enzymes necessary to catabolize lactose as an energy source. Only for the RNA polymerase to attach at the promoter site, there must be the presence of the catabolite gene activation protein (CAP) to which cAMP is attached. The bacterium

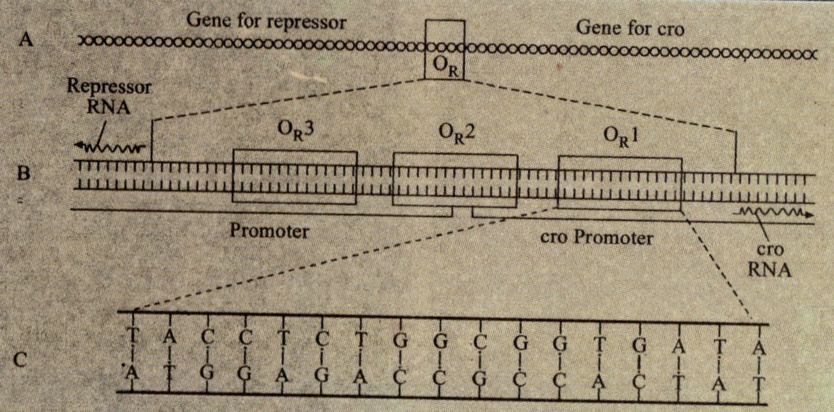

Fig. 26.9 : Right operator (OR) is shown in increasing detail in this series of drawing

accumulates cAMP only when it is starved for a source of carbon. In the presence of glucose or glycerol in concentrations sufficient for growth, the bacteria will lack sufficient cAMP to bind to CAP. Therefore, in the presence of glucose on glycerol, cAMP-saturated CAP is lacking, so that the *DNA-dependent RNA polymerase* cannot begin the transcription of the lac operon. In the presence of the CAP- cAMP complex on the promoter site transcription then takes place. Therefore, the CAP–cAMP regulator is acting as a *positive regulator*.

When the i gene is mutated so that its product, the lac repressor, is not capable of binding to operator DNA, the organism will exhibit constitutive expression of the lac operon. An organism with an i gene mutation that prevents the binding of an inducer to the repressor will remain repressed even in the presence of the inducer molecule, because the inducer cannot bind to the repressor on the operator locus in order to depress the operon.

Bacteriophage Lambda (λ)

When a lamda infects a sensitive E. Coli, it injects its 45,000 base-pair, double stranded, linear DNA molecule into the cell (Fig. 26.8). Depending upon the nutritional state of the cell, the lambda DNA will either integrate into the host genome (Lysogenic Pathway) and remain dormant until activated, or it will commence replicating until it has made about 100 copies of complete, protein-packaged virus at which point it effects lysis of its host (Lytic Pathway). The newly generated virus articles can then infect other sensitive host.

When integrated into the host genome in its dormant state, lambda will remain in such a state until activated by exposure of its lysogenic bacterial host to DNA-damaging agents. In response to such a noxious stimulus, the dormant bacteriophage becomes *induced* and begins to transcribe and subsequently translate those genes of its own genome which are necessary for its excision from the host chromosome, its DNA replication, and its protein cost and lysis enzymes. This event acts like a trigger of type C responses.

The switching event in lambda is centred around an eighty base pair region in its double-stranded DNA molecule referred to as the *right operator* (*OR*) Fig. 26.9A. The *right operator* is flanked on its left side by the structural gene for another regulatory protein called cro. When lambda is in its prophage state, the *repressor gene* is the only lambda gene that is expressed. When the bacteriophage is undergoing lytic growth, the repressor gene is not expressed, but the cro gene, as well as many other genes in lambda, is expressed. That is, when *the repressor gene is on, the cro gene is off*, and when *the cro gene is on, the repressor gene is off*. These two genes regulate each other's expression.

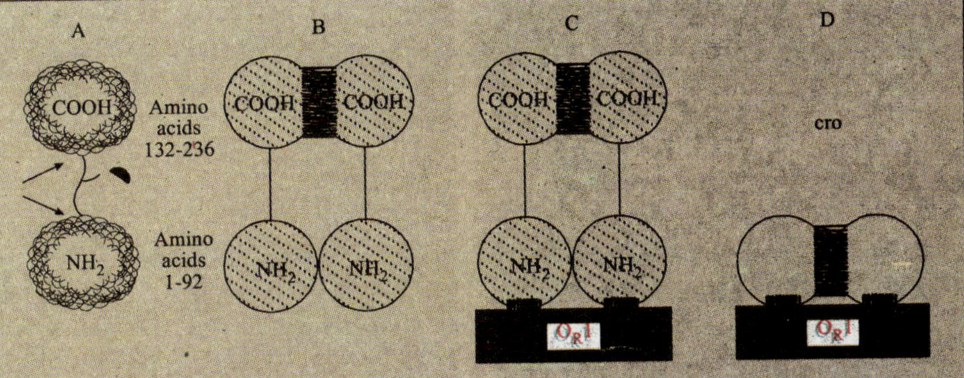

Fig. 26.10 : Lambda repressor protein is a chain of 236 amino acids. The chain folds itself into a dumbbell shape with two substructures

The operator region can be subdivided into three discrete sites, each consists of 17 base pairs of similar but not identical DNA sequence (Fig. 26.9B). Each of these 3 subregions, (OR^1, OR^2, and OR^3) can bind either repressor or cro proteins in the major groove of the DNA double helix. The DNA region between the cro and repressor genes also contain two promoter sequences that direct the binding of RNA polymerase in a specified orientation where it commences transcribing the adjacent genes. One promoter directs RNA polymerase to begin transcription in the ***rightward direction*** and, thus, to transcribe cro and other distal genes, while the other promoter directs the transcription of the ***repressor*** gene in the ***leftward direction*** (Fig. 26.9B).

The product of the repressor gene, the 236 amino acid repressor protein, exists as a 2-domain molecule in which the amino-terminal domain binds to operator DNA and the carboxy-terminal domain promotes the association of one repressor protein with another to form a dimer. A dimer of repressor molecules binds to operator DNA much more tightly than does the monomeric form (Fig. 26.10–A to C).

The product of the cro gene, the 66-amino acid cro protein has a single domain but also binds the operator DNA more tightly as a dimer (Fig. 26.10D).

In a lysogenic bacterium i.e., a bacterium containing a lambda prophage, the lambda repressor dimer binds preferentially to OR^1 but in doing so by a cooperative interaction, enhances the binding of another repressor dimer to OR^2 (Fig 26.11). The affinity of repressor of OR^3 is the least of the three operator subregions. The binding of repressor to OR^1 has two major effects. The occupation of OR1 by repressor blocks the binding of RNA polymerase to the ***rightward Promoter*** and thereby prevents the expression of the cro gene. Secondly, repressor dimer bound to OR^1 enhances the binding of repressor dimer OR^2. The binding of repressor to OR^2 has the important effect of enhancing the binding of RNA polymerase to the leftward promoter that overlaps OR^2 and thereby enhances the transcription and subsequent expression of the repressor gene. Thus, the lambda repressor is both a negative regulator by preventing transcription of the cro gene, and a positive regulator, by enhancing the transcription of its own gene, the repressor gene. This dual effect of ***repressor*** is responsible for the stable state of the dormant lambda bacteriophage.

When a DNA damaging signal, such as ultraviolet light, strikes the lysogenic host bacterium, fragments of signal-stranded DNA are generated that activate a specific protease coded by a bacterial gene and referred to as **rec A**. The activated rec A protease hydrolyzes the portion of the repressor protein that connects the amino-terminal and carboxy-terminal domains of that molecule. Such cleavage of the repressor domains causes the repressor dimers to dissociate, which in turn causes a dissociation of the repressor molecules from OR^2

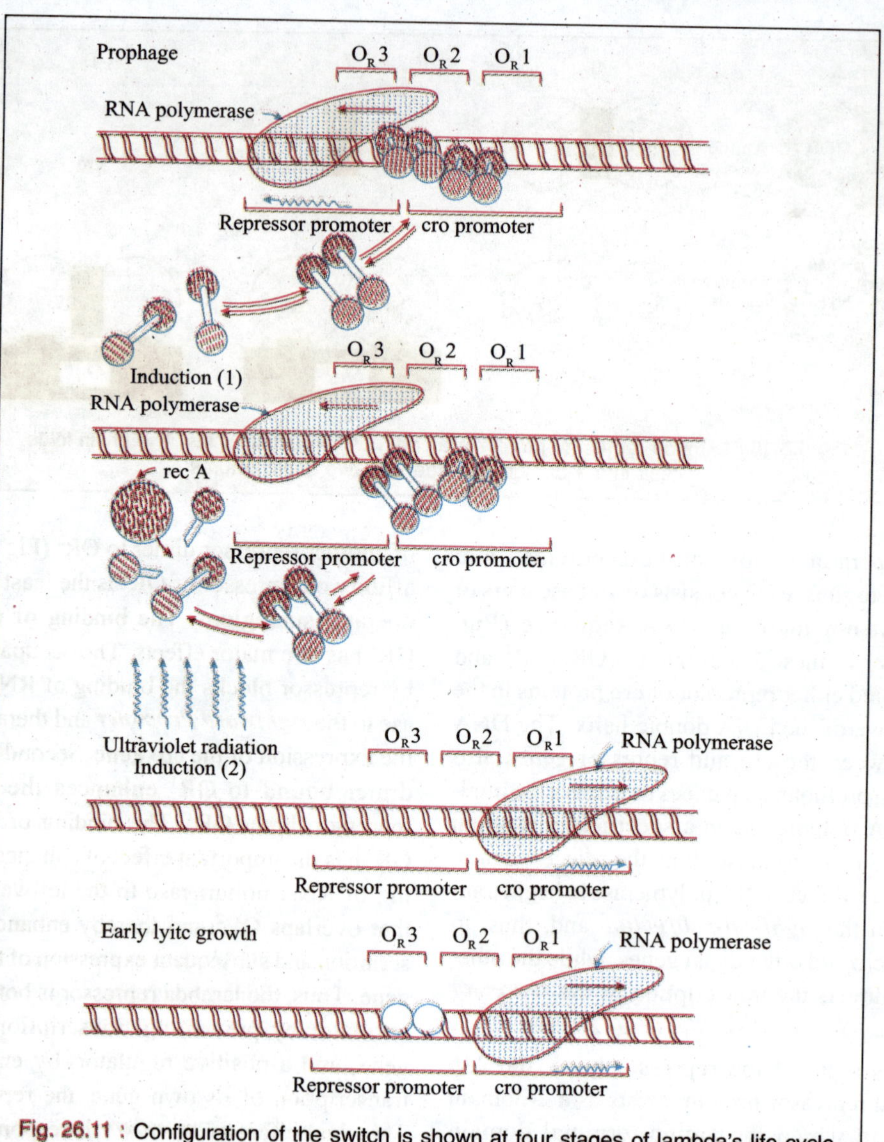

Fig. 26.11 : Configuration of the switch is shown at four stages of lambda's life cycle

and eventually from OR^1. The effects of removal of repressor from OR^1 and OR^2 are speculated. RNA polymerase immediately has access to the rightward promoter and begins transcribing the cro gene, and the enhancement effect of the repressor at OR^2 on leftward transcription is lost.

The cro protein translated from the newly transcribed cro gene also binds to the operator region as dimers, but its order of preference is the opposite of that of repressor. That is, cro binds most lightly to OR^3, but there is no cooperative effect of cro at OR^3 on the binding of Cro to OR^2. At increasingly higher concentrations of Cro, the protein will bind to OR^2 and eventually to OR^1.

The occupancy of OR^3 Cro immediately turns off the transcription from the leftward promoter and hence, prevents any *further expression of the repressor gene*. Therefore, the switch is completely effected. The Cro gene is now expressed and the repressor gene is fully turned off. When Cro repressor concentration becomes very high, it will eventually occupy OR^1 and in doing so turn down

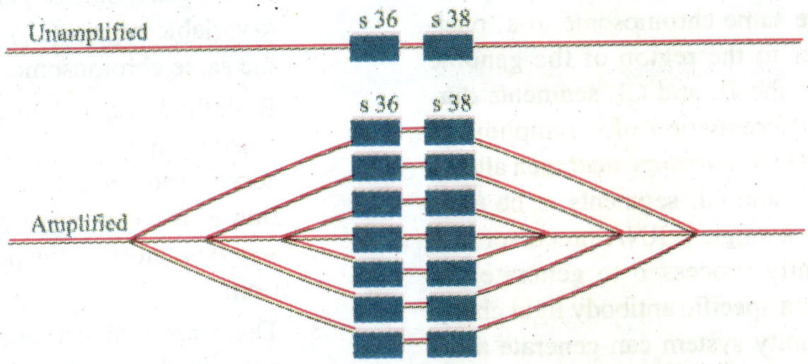

Fig. 26.12 : Amplification of chorion protein genes S36 and S38

the expression of its own gene, a process that is necessary to effect the final stages of the lytic cycle.

Gene Amplification During Development

During early development of metazoans, there is an abrupt increase in the need for specific genetic molecules such as ribosomal RNAs and messenger RNA molecules for proteins that make up such organs as the egg-shell. One way to increase the rate at which such molecules can be formed is to increase the number of genes available for transcription of these specific molecules. Among the repetitive DNA sequences are thousands of copies of ribosomal RNA genes and tRNA genes. These genes preexist repetitively in the genomic material of the gametes and, thus, are transmitted in high copy number from generation to generation. In some specific organisms such as the fruit fly (Drosophila), there occurs during oogenesis an amplification of a few pre-existing genes, such as these for the chorion (eggshell) proteins, S36 and S38. Subsequently, these amplified genes, presumably generated by a process of repeated initiations during DNA synthesis provide multiple sites for gene transcription.

In recent years, it has been possible to promote the amplification of specific genetic regions in cultured mammalian cells. In some cases, a several thousand-fold increase in the copy number of specific genes can be achieved over a period of time involving increasing doses of selective drugs. In fact, it has been demonstrated in patients receiving ***methotrexate*** for treatment of cancer that malignant cells can develop drug resistance by increasing the number of genes for ***dihydrofolate reductase***, the target of methotrexate.

Immunoglobulin Gene Rearrangement

1. The coding segments responsible for the generation of specific protein molecules are frequently not contiguous in the mammalian genome. Immunoglobulin molecules consist of two types of polypeptide chains, light(L) and heavy(H) chains. The L and H chains are each divided into N-terminal variable (V) and carboxyterminal constant (C) regions. The V regions are responsible for the recognition of antigens (foreign molecules) and the constant regions for effector functions that determine how the antibody molecule will dispense with the antigen.
2. There are three unlinked families of genes responsible for immunoglobulin molecule structure. Two families are responsible for the chains (λ and κ chains) and one family for heavy chains.
3. Each light chain is encoded by three distinct segments, the variable (VL), the joining (JL), and the constant (CL) elements. The mammalian haploid genome contains over 500 VL segments, five or six JL segments, and 10 or 20 CL segments.

4. A VL segment is brought from a distant site on the same chromosome to a position closer to the region of the genome containing the JL and CL segments during the differentiation of a lymphoid B cell. This *DNA rearrangement* then allows the VL, JL and CL segments to be transcribed as a single mRNA precursor and subsequently processed to generate the mRNA for a specific antibody light chain. The immunity system can generate a diverse library of antigen specific immunoglobulin molecules by rearrangement of the various VL, JL, and CL segments in the genome. The DNA rearrangement is referred to as *V-J joining* of the light chain.
5. The *heavy chain* is encoded by four gene segments : the VH, the D (Diversity), the JH, and the CH DNA segment. The variable region of the heavy chain is generated by joining the VH with a D and a JH segment. The resulting VH-D-JH-DNA region is in turn linked to a CH gene. These CH genes ($C\mu$, $C\delta$, $C\gamma^3$, $C\gamma^1$, $C\gamma^{2b}$, $C\gamma^{2\alpha}$, C_α and C_e) determine the immunoglobulin class or subclass-IgM, IgG, IgA, etc. – of the immunoglobulin molecule.
6. A *B cell* that secretes antibody to a *specific antigen* will secrete antibodies of different classes having the same antigen specificity but different biologic roles during the differentiation. The different classes of immunoglobulins contain the same light chains and VH regions but different CH regions. Thus, a single B cell and its derivatives can undergo *class switching* which is the result of second type of DNA *rearrangement* occurring during differentiation of the immunity system. The *V-J joining* for light chain expression and the *V-D-J joining* for heavy chain expression *precede the class switching DNA rearrangement*.

V-J joining

1. In the undifferentiated cell (eg, germ line cell), the K-J gene (J_k) is closely linked to the C_k gene, but the gene segment for the k-variable region (V_k) is quite distant on the same chromosome.
2. Both the V_k–J_k and the similar V_λ–J_λ gene rearrangements to involve two short conserved sequence that exist in the direction 3′ *to the V-segment* and 5′ *to the J-segment*, close to the point of recombination.
3. There are two striking features of these conserved sequences. *First, the conserved sequences of the JL segments are inverse complements of the conserved sequences in the VL segments*. Second, the length of the nonconserved sequence separating the heptamers and the monomers is highly conserved.
4. The variable region of the heavy chain involves three DNA segments, VH, D, and JH, which must be joined in a process involving two DNA rearrangements since all three segments are separated.

Class Switching

1. Since differentiation proceeds and immunoglobulin production switches from IgM to IgA the V-D-J region of the parent B cell must be rearranged with C_α gene to permit the transcription of an mRNA precursor for an α-chain containing the same antigen–specific variable region.
2. The temporal order of the class switching is unidirectional within the physical order from left to right. Rearrangement of the CH genes seems to involve deletion of those CH genes 5′ to the CH gene joined to the V-D-J region.

The switch sites are different for different class switches but involve conserved sequences occurring in the appropriate regions of the genes to be rearranged.

Transcription Control

1. Glucocorticoids regulate gene expression. Once they enter the mammalian cell, they

bind to a steroid specific receptor molecule that undergoes a conformational change in the cytoplasm and enters the nucleus. The *gluco- corticoid receptor complex* in the nucleus binds to a specific *receptor-recognition site* on DNA, a few hundred base pairs 5′ upstream from the transcription start site, for steroid-responsive genes. The receptor recognition site on DNA influences the efficiency of utilization of the promoter by *RNA polymerase* and thereby influences the expression of the steroid responsive gene.

2. When an organism or its cultured cells are exposed to metal ions, such as zinc or cadmium, there is an accelerated rate of transcription of the metallothionein gene and a subsequent increase in the metallothionein protein to bind the potentially toxic heavy metal. Another structural gene, such as that for thymidine kinase, can be ligated to this '*metallothionein promoter region*' and the synthetic construct introduced to cultured cells, a small number of which will integrate the DNA into its own genome. When those cells are exposed to heavy metals, the metallothionein promoter region effects an induction of thymidine kinase.
3. There are many more primary tanscripts in the nucleus than are represented as messenger RNA molecules in the cytoplasm. There must exist regulatory decisions as to which transcripts will ultimately be expressed and which will be discarded. There is no information available regarding the mechanism of this process.

Regulation of messenger RNA stability provides another control mechanism

1. Most mRNAs in mammalian cells are very stable. Changes in the stability of a specific mRNA have the major effects in biologic processes.
2. Messenger RNAs exist in the cytoplasm as ribonucleoprotein particles (RNPs). Some of these proteins protect the mRNA from digestion by nucleases, while others under certain conditions, promote nuclease attack. Certain effectors, such as hormones, may regulate mRNA stability by increasing or decreasing the amount of the proteins.
3. Then ends of mRNA molecules are involved in mRNA stability. The 5′ cap structure in eukaryotic mRNA prevents attack by 5′ exonucleases, and the poly (A) tail prohibits the action of 3′ exonucleases. In mRNA molecules, a single endonucleolytic cut allows exonucleases to attack and digest the entire molecule. Other structures in the 5′ noncoding sequence (5′ NCS) the coding region, and the 3′NCS are thought to promote or prevent this initial endonucleolytic action.
4. Structures at the 3′ end including the poly (A) tail enhance or diminish the stability of specific mRNAs. The absence of poly (A) tail is associated with rapid degradation of mRNA and the removal of poly (A) from some RNAs results in their destabilization.

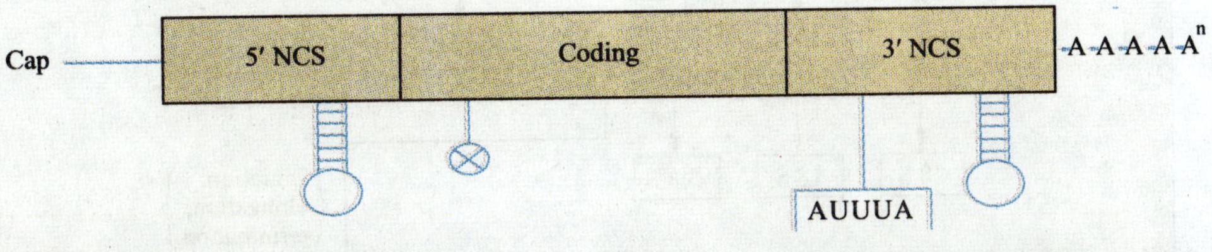

Fig. 26.13 : Structure of a typical eukaryotic mRNA. It has a 5′ noncoding sequence (5′NCS), a coding region, and a 3′NCS. All are capped at the 5′-end and most have a polyadenylate sequence at the 3′-end. The 5′ cap and 3′ poly(A) tail protect the mRNA against exonuclease attack

5. In AU-rich regions, many of which contain the sequence AUUUA which appears in mRNAs that have a very short half-life including some encoding oncogene proteins and cytokines.

CLINICAL ORIENTATION

- The change in the base sequence of DNA results in mutation. Mutations result from the faulty replication, repair of DNA and occur with a frequency of about one in every 10^6 cell division. A mutation in a germ cell is transmitted to offspring (a vertical transmission of hereditary disease).
- The rate of mutation is increased by viruses, chemicals, ultra-violet light and ionizing radiation.
- Mutations often affect somatic cells and are passed on to successive generations of cells within an organism.
- A number of diseases, mostly cancer, are due to horizontal transmission of induced mutations.
- The processing and splicing of mRNA transcripts are the causes of disease, eg, certain types of thalassemia.

CLINICAL ORIENTATION

- The genetic code is the foundation for explaining the way in which protein defects may cause genetic disease and for the diagnosis and treatment of these disorders.
- The pathophysiology of many viral infections is related to these agents to disrupt host cell protein synthesis.
- The mechanisms that control gene expression are used to respond to hormones and therapeutic agents. These processes may lead to the development of agents that inhibit the function of pathogenic organisms.

EXERCISE

1. What is codon? Describe the biosynthesis of proteins. (Mith. 75A)
2. Write notes on : codon, (M.U. 72A)
   Lac operon (B.H.U. '89A)
3. Discuss regulation of Gene Expression. (P.U. 88 A)

SCHEMATIC COMPENDIUM

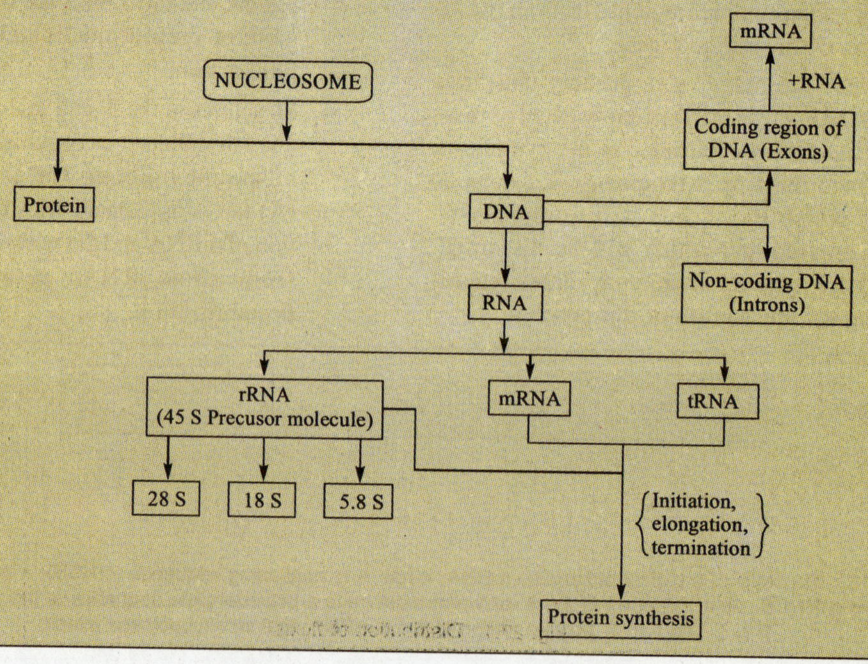

CHAPTER 27

WATER AND MINERAL METABOLISM

Distribution of fluids in the body

1. Water makes up 50 to 70 per cent of the weight of the adult human body and varies inversely as the fat content.
2. This mentioned amount of water is distributed throughout the body as the major component of the intracellular and extracellular fluids.
3. The intracellular fluids amount to about 50 per cent of the body weight in a lean individual and much less in an obese person.
4. The extracellular fluids represent about 20 per cent of the body weight.
5. Of the extracellular fluids, interstitial fluids amount to some 15 per cent and blood plasma about 5 per cent of the body weight.
6. Relatively small volumes are represented

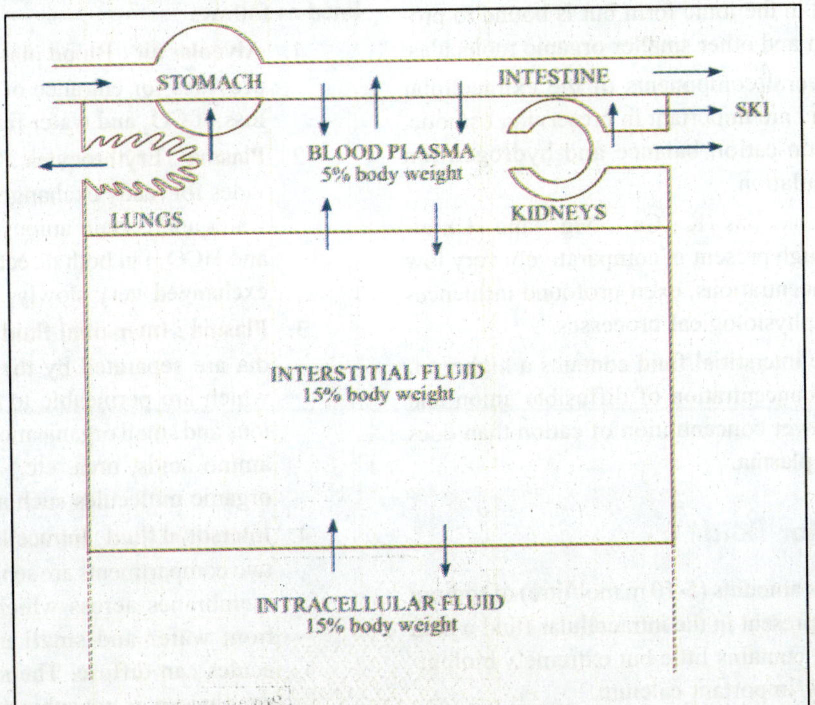

Fig. 27.1 : Distribution of fluids

by specialized fluids such as cerebrospinal fluid, ocular fluid, lymph, and synovial fluids, etc.

Extracellular fluid

1. All body cells exist in an environmemt of fluid collectively designated extracellular fluid. This includes the blood plasma, interstitial fluid and lymph.
2. 7 per cent protein is present in plasma, slightly less in hepatic lymph and 0.1 per cent protein in subcutaneous interstitial fluid.
3. They are solutions mainly of Nacl and $NaHCO_3$, with small amounts of Ca, Mg, K, H, phosphate, sulphate and organic acid ions, some nonelectrolytes (glucose, urea, lipids, etc.) and with pH values ranging from 7.35 to 7.45 under normal conditions.
4. The total concentration of the ionic constituents is about 310 m mol per litre of plasma.
5. Much of the intracellular magnesium is not in the ionic form but is bound to protein and other smaller organic molecules.
6. Several components of the extracellular fluid are important in preserving osmotic, anion-cation balance and hydrogen ion regulation.
7. The cations (K^+, Ca^{++}, Mg^{++} and H^+), although present in comparatively very low concentrations, exert profound influences on physiological processes.
8. The interstitial fluid contains a higher total concentration of diffusible anion and a lower concentration of cation than does the plasma.

Intracellular fluid

1. Less amounts (5-10 m mol/litre) of sodium are present in the intracellular fluid which also contains little but extremely biologically important calcium.
2. The chief cations of this fluid are potassium (about 160 m mol/litre) in muscle and magnesium (about 26 m mol/litre) in muscle.
3. The intracellular fluids contain much more phosphate and sulphate ions as well as proteins than the extracellular fluid. Chloride ion is practically absent from this fluid except in the case of erythrocytes, and cells of the kidney tubules, stomach and intestines. Both Na and K ions are able to cross the membrane more freely under certain physiological and pathological conditions.
4. Much of the magnesium is present as undissociated compounds of protein and organic phosphate and, therefore, is not in ionic form.

Exchanges between fluid compartments

Many of the substances entered into the body and produced by the cells are distributed to other tissues or excreted. The exchange systems are outlined as follows :

1. **Alveolar air : Blood plasma :** This system provides for entrance of oxygen into and loss of CO_2 and water from the body.
2. **Plasma : Erythrocytes :** This system provides for ready exchange of oxygen, CO_2, water and certain anions (particularly Cl^- and HCO_3^-) in both directions. Cations are exchanged very slowly.
3. **Plasma : Interstitial fluid :** These two media are separated by the capillary walls, which are permeable to water, inorganic ions and small organic molecules (glucose, amino acids, urea, etc.) but not to large organic molecules such as proteins.
4. **Interstitial fluid : Intracellular fluid :** These two compartments are separated by the cell membranes across which gases in solution, water and small unchanged molecules can diffuse. The small molecules, eg, glucose, is not subjected to simple diffusion but is carried across cell membrane

by active transport processes. The permeability of electrolytes follows biological pump mechanisms. These membranes are also relatively impermeable to large molecules such as proteins, except in special situations, namely the liver.

WATER METABOLISM

Distribution of water in the body

Water is the largest constituent of the body. The average body water is 60 to 70 per cent of the total body weight. Females contain a little less amount of water than males. Body water is distributed as follows :

1. *Intracellular fluid* (the fluid within the cells) contains 50 per cent of the total body weight.
2. *Extracellular fluid* (all of the fluid outside the cell) contains 20 per cent of the total body weight. This 20 per cent can be further divided into the following compartments:
   (a) Plasma (the extracellular fluid of the blood) : 4.5 per cent.
   (b) Interstitial and lymph fluids : 8 per cent.
   (c) Dense connective tissue, cartilage and bone : 6 per cent.
   (d) Transcellular fluid (found in salivary glands, pancreas, liver, thyroid gland, gonads, skin, mucous membranes of the respiratory and gastrointestinal tracts and the kidneys as well as the fluids in spaces within the eye, CSF, etc.) :1.5 per cent of the body weight.

Factors influencing the distribution of body water

The distribution of water is continuously changing. Osmotic forces are the principal factors for controlling the amount of fluid in the various compartments of the body. These are maintained by the solutes in the body water. The solutes are divided into three categories :

1. Organic compounds of small molecular size (glucose, urea, and amino acids etc.) : Since these substances diffuse freely across the cell membrane, they are not important in the distribution of water. But if they are present in large amounts, they can help in retaining water.
2. Organic substances of large molecular size (proteins) : These substances can throw effect in the transfer of fluids from one compartment to the other.
3. The inorganic electrolytes : These inorganic electrolytes are the most important both in the distribution and in the retention of body water.

Intake and loss of body water

A. **Water intake** : Water is supplied to the body by the following processes :
   1. Water taken orally.
   2. Along with food.
   3. Oxidation of foodstuffs : Fats, proteins and carbohydrates yield water after combustion. Fats produce 107 ml/100 gm, proteins 41 ml/100 gm and carbohydrates 56 ml/100 gm.

B. **Water losses** : Water is lost from the body by 4 routes :
   1. Evaporation from skin and lungs.
   2. Kidneys, as urine.
   3. The intestines, in the the feces.
   4. Perspiration.

C. **Additional water losses in disease** :
   1. Water loss is more in diarrhea and vomiting and these losses can be fatal in infants.
   2. In kidney disease, renal water loss is more.
   3. In fever, insensible losses may rise much higher than normal.
   4. Patients in high environmental temperatures also sustain extremely high extrarenal water losses.

Balance sheet of water [per day]			
Water intake		Water loss	
Drinks	1,350 ml	Lungs	500 ml
Solid food	900 ml	Skin	700 ml
Oxidation of food	450 ml	Urine	1,400 ml
		Feces	100 ml
	2,700 ml		2,700 ml

Water balance : An equilibrium persists between the intake and output of water in the body. In addition to other factors, certain hormones, such as ADH, vasopressin, oxytocin and aldosterone influence the regulatory mechanism of body water. The balance sheet of water is given above.

There is a continuous excretion of water in the form of digestive juices from the body into the alimentary canal and that water is reabsorbed along with the water of food and drink. The amount of this internal secretion is 7 to 10 litres per day. All of this amount is reabsorbed except about 100 ml which is excreted in feces. The secretion of saliva experiences the fact when we have a *sore throat*.

Physiological functions of water

1. *Specific heat* : More heat is required to raise the temperature of 1 gram of water through 1°C than almost any other solid or liquid. This clearly explains that heat produced as a result of chemical reactions in the cell can cause the minimum rise in body temperature.
2. *Latent heat of evaporation* : Water has the highest latent heat of evaporation than any other liquid. A certain amount of water can cause maximum cooling by evaporation so that the body temperature does not rise.
3. *Solvent power* : Water can form true solutions as well as colloidal solutions. Even the water insoluble substances are made water soluble by the hydrotropic action. Therefore, it is the most suitable solvent for cellular components,water thus brings various substances in contact for chemical reactions to proceed.
4. *Dielectric constant* : Oppositely charged particles can coexist in water. Therefore, it is a good ionizing medium. This increases the chemical reactions.
5. *Catalytic action* : A large number of chemical reactions in the body are accelerated by water due to its ionizing power. All chemical reactions in the body proceed in presence of water only.
6. *Lubricating action* : Water acts as a lubricant in the body to prevent friction in joints, pleura, conjunctiva and peritoneum.

Regulation of passage of water

1. If the capillary pressure is increased, more water will flow into the tissues.
2. A fall in blood pressure helps the passage of water from the tissues to the blood.
3. If the plasma proteins are decreased, water will flow into the tissues.
4. Dilution of blood by excessive ingestion of water can lower the osmotic pressure of the plasma proteins and thus may increase capillary pressure.

DEHYDRATION

Definition

When the losses of water exceeds the intake, the body's water content is reduced. That is, the body is in negative water balance and the condition is known as *dehydration*.

Causes

1. Primary dehydration :
   (a) Simple deprivation of water from desert travel, extreme weakness and mental patients refusing to drink causes dehydration. It occurs more quickly in fever or in high temperature of the environment.

(b) Excessive water loss due to vomiting, prolonged diarrhea, excretion of large quantities of urine or sweat.

In water depletion, the concentration of extracellular fluid increases. Water is drawn from the cells and both extracellular and intracellular compartments shrink. Extreme thirst results, the individual complains of hot and dry body, dry tongue.

2. **Secondary dehydration:** The concentration of electrolytes of the body fluids is maintained constant through the elimination or retention of water. The reduction or increase in the total electrolytes, which affects chiefly the basic radicals Na (extracellular) and K (intracellular) and the acid radicals HCO_3 and Cl_2 is accompanied by a corresponding increase or decrease in the volume of body water. This causes intracellular edema. There is slowing of circulation and impairment of urinal function. The individual becomes weaker.
3. **Dehydration due to injection of hypertonic solution :** When a highly concentrated sugar or salt solution is injected into the body, the osmotic pressure of blood will increase. This results in the flow of fluid from the tissues into the blood until equilibrium sets in. Consequently, the blood volume increases. This increased blood volume soon returns to normal by the loss of excess material through excretion. This causes a net loss of body water producing dehydration.

Effects of dehydration

1. Loss of weight due to the reduction in tissue water.
2. Disturbances in acid-base balance.
3. Rise in the nonprotein nitrogen of blood.
4. Rise in the plasma protein concentration and of chloride.
5. Rise in body temperature due to reduction in circulating fluid.
6. Increased pulse rate and reduced cardiac output.
7. Dryness, wrinkling and looseness of skin.
8. Exhaustion and collapse.

Correction of dehydration

1. Ordinarily sodium chloride solution may be given parenterally to repair the losses.
2. In the case of the removal of a fluid high in sodium and bicarbonate during fluid and electrolyte losses originating from the intestinal tract (as in prolonged diarrhea, pancreatic or biliary fistulas, etc.) a mixture of two-third isotonic saline solution and one-third sodium lactate solution (M/6) should be administered intravenously.
3. But dehydration is a problem in diabetes mellitus, Addison's disease, uremia, extensive burns and shock. This type of dehydration cannot be corrected by the application of above two processes.

Water intoxication

This condition is caused by the excess of water retention in the body and can occur due to renal failure, hypersecretion of ADH and excessive administration of fluids parenterally. Symptoms of water intoxication are headache, nausea and muscular weakness.

This condition also exhibits the decreased levels of hemoglobin and plasma proteins, lowered plasma electrolytes, increased urinary volume with low sp. gravity.

The condition is treated by withholding fluids by mouth and administering 3 to 5 per cent hypertonic saline.

MINERAL METABOLISM

The mineral elements present in the animal body is supplied by the diet. In poor diets consumed by a large majority in India and other developing countries, there occurs commonly the deficiencies of calcium and iron. Iodine deficiency occurs in people living in certain hilly tracts in India and in some other countries, where the soil and water are deficient in iodine. In tropical countries, addition of sodium chloride in the diet is of great importance, because of the loss of NaCl in sweat. The deficiencies of other minerals do not occur normally in average diets.

The mineral elements can be classified as *principal elements* (*macronutrients*) and *trace elements.*

Principal mineral elements (Macronutrients)

Seven essential elements : Calcium, magnesium, sodium, potassium, phosphorus, sulphur and chlorine.

1. Sodium, potassium and chlorine are involved mainly in the maintenance of acid-base balance and osmotic control of water metabolism.
2. Calcium, phosphorus and magnesium as constituents of bones and teeth.
3. Phosphorus as constituent of body cells of soft tissues, such as muscles, liver, etc.
4. Sulphur in cysteine and methionine, thiamine, biotin, lipoic acid and CoA.

Trace elements

These elements are present in living tissues in small amounts. They are subdivided into three groups– Essential, possibly essential and nonessential.

1. *Essential trace elements* (*micronutrients*) : Iron, iodine, copper, zinc, manganese, cobalt, molybdenum, selenium, chromium and fluorine.
2. *Possibly essential trace elements* : Nickel, tin, vanadium and silicon.
3. *Nonessential trace elements* : Aluminium, boron, germanium, cadmium, arsenic, lead and mercury.

Physiological importance

1. Iodine is required for thyroxine formation.
2. Iron and copper are required for hemoglobin formation.
3. Zinc is a constituent of carbonic anhydrase and insulin.
4. Cobalt is a constituent of vitamin B_{12} .

The ratio of one element to the other in the tissues is of physiological importance eg, normal ossification demands a proper ratio of calcium to phosphorus.

CALCIUM

Physiological functions

1. Calcium along with phosphorus is essential for the formation and development of bones and teeth.
2. Ionized calcium is required in blood coagulation process.
3. It regulates the excitability of nerve fibres and nerve centres.
4. It is essential for nerve impulses and muscular contraction.
5. It regulates the permeability of membranes.
6. It is essential for maintaining the integrity of intracellular material.
7. Calcium in the normal ratio with potassium maintains the normal activity of muscles.
8. It is required for the activation of several enzymes such as succinate dehydrogenase, ATPase and certain proteolytic enzymes.

Sources

Richest sources	: Milk and cheese.
Good sources	: Egg yolk, nuts, figs, beans, cabbage, cauliflowers, turnip greens and asparagus.

Daily requirements

Adult males and females	800 mg
Women during pregnancy and lactation	1.2 grams
Infants under 1 year	360-540 mg
Children 1-18 years	0.8-1.2 grams

Absorption

Calcium is taken in diet as calcium phosphate, carbonate, tartarate and oxalate. Calcium is actively absorbed in the upper small intestine. The active transport process is regulated by 1, 25-dihydroxycholecalciferol, a metabolite of vitamin D which is produced in the kidney in response to low plasma Ca^{++} concentrations. The absorption of it is influenced by the following factors :

1. Vitamin D promotes absorption of calcium.
2. Acidic pH favours calcium absorption; because calcium salts, particularly phosphates and carbonates, are quite soluble in acid solutions and are relatively insoluble in alkaline solutions. Hence, alkali decreases its absorption.
3. Organic acids, lactose and basic amino acids in the diet favour clacium absorption.
4. Higher levels of proteins in the diet help to increase the absorption of calcium.
5. Calcium: Phosphorus ratio of 1 : 1 (2:1 or 1:2) is the most convenient for the absorption of both.
6. High concentration of magnesium in the diet decreases absorption.
7. Phytic acid (inositol hexaphosphate), which occurs in cereal grains, forms insoluble salts (phytin) with calcium and magnesium resulting in the impaired absorption of calcium.
8. Fatty acids interfere with calcium absorption because of the formation of insoluble calcium salts of fatty acids which are excreted in the feces.
9. Presence of excess fibres in the diet interferes with the absorption of calcium.
10. Oxalic acid in the diet forms insoluble calcium oxalate which is excreted in the feces decreasing calcium absorption.
11. The percentage of calcium absorption decreases as its intake increases.
12. Parathyroid hormone increases the intestinal absorption of calcium.
13. After the age of 55 to 60 years, there is gradual diminution of intestinal transport of calcium.
14. Adrenal glucocorticosteroids diminish intestinal transport of calcium.

Distribution of calcium

Serum	9-11 mg/100 ml
C.S.F.	4.5-5 mg/ 100 ml
Muscle	70 mg/100 gm
Nerve	15 mg/100 gm

Blood calcium

There is virtually no calcium in erythrocytes. The calcium content of plasma (usually determined in serum) is 8.4 to 10.2 mg/100 ml. During infancy and early childhood, the average values approach the upper limit of this range and fall with advancing age. Calcium exists in the plasma in three fractions—ionized or diffusible calcium, protein bound (nondiffusible) and complexed (citrate and phosphate). In the usual determination of calcium, all these fractions are measured together.

About 2 mg of the total calcium occurs in ionized form, about 5 mg occurs in nonionized form and about 2 mg in the complex form.

Factors influencing blood calcium level

1. ***Parathyroid hormone*** : In the fasting state (i.e. no absorption from the intestine), the normal plasma calcium concentration is maintained primarily by its rate of excretion and its mobilization from the bones through the action of the parathyroid hormone.
2. ***Vitamin D*** : It enhances absorption of calcium from the intestine and thus maintain the normal plasma calcium concentration.
3. ***Plasma proteins*** : Half of the blood calcium (nondiffusible fraction) is bound to plasma protein (chiefly albumin) and thus any decrease in these proteins will be ac-

companied by a decrease in the total calcium level.

4. ***Plasma phosphate :*** A reciprocal relationship exists between the concentration of calcium and phosphate ions in plasma. The marked increase in serum phosphate causes a fall in serum calcium concentration.

5. ***Calcitonin :*** An increase in the ionized calcium levels in the plasma is the stimulus for the production of calcitonin which then causes a deposition of calcium in bone.

Excretion

Calcium is excreted in the urine, bile, and digestive secretions. Much of that is excreted in the feces which has escaped absorption. Under optimal conditions, 75 per cent of dietary calcium is absorbed. The remainder is the fecal calcium which is unabsorbed.

In man, about 10 grams of calcium are filtered in 24 hours by the renal glomeruli. Only about 200 mg appear in the urine which is in the ionic state as well as in the complexes with citrate and other organic anions.

70-90 per cent of the calcium eliminated from the body is excreted in the feces.

A very small amount is excreted into the intestine after absorption.

The daily loss of calcium in sweat is about 15 mg. Vigorous physical exercise increases the loss of calcium by the way of sweat.

Disease state

A. Effects of Parathyroid :

1. In *hyperparathyroidism*, the following changes occur:
   (a) Hypercalcemia (serum calcium 12-22 mg /100 ml).
   (b) Decrease in serum phosphate.
   (c) Diminished renal tubular reabsorption of phosphate.
   (d) Increased phosphatase activity.
   (e) Raised urinary calcium and phosphorus from bone decalcification and dehydration.
   (f) The extra calcium and phosphorus is lost from soft tissues and bones by increased bone destroying activity.
2. In *hypoparathyroidism*, the following changes occur :
   (a) The concentration of serum calcium may drop below 7 mg/100 ml.
   (b) Increased serum phosphate and decreased urinary excretion of calcium and phosphorus.
   (c) Normal or occasionally raised serum phosphatase activity.
   (d) Normal acid-base equilibrium.
   (e) Probably increased bone density.

B. Tetany :

Decreased ionized fraction of serum calcium causes tetany. This may be due to :

(a) An increase in the pH of blood.
(b) Poor absorption of calcium from the intestine.
(c) Decreased dietary intake of calcium.
(d) Increased renal excretion of calcium as in nephritis.
(e) Parathyroid deficiency.
(f) Increased retention of phosphorus as in renal tubular disease.

Symptoms :

(a) Muscles lose tone and become flabby.
(b) Affects the face, hands and feet.

C. Rickets :

This disease is characterized by faulty calcification of bones in children showing serum phosphate values 1 to 2 mg/ 100 ml. This may be due to :

(a) Vitamin D deficiency.
(b) A deficiency of calcium and phosphorus in the diet or a combination of both.

(c) Poor absorption of calcium from the intestine.
(d) Parathyroid deficiency.
(e) Increased serum alkaline phosphatase activity.

D. Osteoporosis :

The disease occurs in adults due to the following causes:

(a) Decalcification of bones as a result of calcium deficiency in the diet.
(b) Hypoparathyroidism.
(c) Low vitamin D content of the body.

Symptoms :

(a) Fractures of the brittle bones occur even after minor accident.
(b) Pain due to fractures of vertebrae which may radiate round the trunk, to the buttocks or down the legs.

E. Renal Rickets :

Renal Rickets are more accurately designated as *familial hypophosphatemic rickets* which is an inherited disease. Affected males show *hypophosphatemia* and severe rickets. Hypophosphatemic rickets are caused by the following reasons :

(a) Defective transport of phosphate by the intestine and by the renal tubules.
(b) Lowered serum phosphorus and *hyperphosphaturia.*
(c) Reduced intestinal absorption of calcium and phosphorus.

Vitamin D in ordinary doses does not relieve the disease. Hence, it is sometimes referred to as *vitamin D-resistant rickets.*

PHOSPHORUS

Physiological functions

1. It is essential for the formation and development of bones and teeth along with calcium.
2. It is required for the formation of phospholipids, nucleic acids, and phosphoproteins.
3. It is involved in the formation of organic phosphates, such as hexose phosphate, triose phosphate and creatine phosphate, etc.
4. It is required for the formation of energy rich compounds, such as ATP.
5. It forms coenzymes, such as NADP, ADP, AMP & B_6-PO_4 etc.
6. It functions in the buffering system in cells.
7. It is required in the absorption of glucose by phosphorylation.

Sources

Milk, cheese, egg-yolk, meat, fish and nuts, etc.

Daily requirements

Infants	240-400 mg
Children	800-1,200 mg
Adults	800 mg
Women during pregnancy and lactation	1.2 gms

Distribution

Serum (inorganic phosphorus) of	
(a) Children	4-7 mg/100 ml
(b) Adults	3-4.5 mg/100 ml
Muscle	170-250 mg/100 gm
Nerve	360 mg / 100 gm
Bones and teeth	22,000 mg / 100 gm

Absorption

1. Moderate amounts of fat or acid favour absorption of phosphorus.
2. High calcium diet and phytic acid (present in cereals) decrease phosphorus absorption.
3. The absorption is enhanced when the calcium and phosphorus ratio is 1:1 (2:1 or 1:2).

Blood phosphorus

The normal inorganic phosphate of plasma is 3 to 5.0 mg/100 ml in adults and is 3.5 to 5.0 mg/100 ml in children. It is somewhat higher in summer than in winter. It decreases during increased carbohydrate metabolism due to increased utilization for phosphorylation.

Phosphorus occurs in the blood in the following forms:

1. Inorganic phosphorus	2-5 mg/100 ml
2. Organic phosphorus	14-29 mg/100 ml
3. Phospholipids	8-18 mg/100 ml

Excretion

Inorganic phosphorus is excreted in the urine and feces. The total urine excretion is 0.8 to 2.0 gm/24 hr. The sources of urinary inorganic phosphorus is mainly that of plasma. On a balanced diet, urine phosphate constitutes about 60 per cent of the total excretion. The rest is excreted in the feces. The *renal threshold* for phosphate excretion is about 2 mg/100 ml of plasma. The reabsorption of phosphorus is inhibited by the parathyroid hormone.

Disease state

1. In rickets, serum phosphate level is as low as 1-2 mg/100 ml.
2. There is a temporary decrease in serum phosphate during absorption of carbohydrate and some fats.
3. A lower concentration of organic phosphorus but a higher concentration of inorganic phosphorus in the serum has been estimated in *diabetes mellitus*.
4. Phosphate retention causes the acidosis in severe renal disease. The resultant is the increase in serum phosphorus level.
5. Serum phosphorus levels are increased in hypoparathyroidism.
6. Blood phosphorus levels are decreased in hyperparathyroidism and in celiac disease.
7. In renal rickets, there is low blood phosphorus level with an increased alkaline phosphatase activity.
8. The deficiency of vitamin D is the cause of the low serum phosphorus and the defects in the calcification of bones.

MAGNESIUM

Physiological functions

1. 70 per cent of the total magnesium content (21 gm) of the body is combined with calcium and phosphorus in the complex salts of bone. The remainder is in the soft tissues and body fluids. It is the principal cation of the soft tissue.
2. Magnesium ions act as activators for many of the phosphate group transfer enzymes.
3. It is found in certain enzymes, such as *cocarboxylase*.
4. It functions as a cofactor for oxidative phosphorylation.

Sources

It is present in milk, eggs, cabbage, cauliflowers and fruits, etc.

Distribution

Whole blood	2-4 mg/100 ml
C.S.F.	3 mg/100 ml
Muscle	21 mg/100 gm

Daily requirement

Infants	100-150 mg
Children	150-200 mg
Adults	200-300 mg

Blood magnesium

The normal level of magnesium in blood is 1.7-2.6 mg/100 ml.

Absorption

1. A greater part (40 to 50 per cent) of the daily ingested magnesium (200 to 300 mg) is not absorbed.
2. Very high intake of fat, phosphate, calcium and alkalis diminish its absorption.
3. Parathyroid hormone increases its absorption.

Excretion

Two-third of the total excreted magnesium is excreted in the feces and the remaining one-third into the urine.

Disease state

1. Magnesium deficiency causes depression, muscular weakness, and liability to convulsions. The serum magnesium level is below 1 mg/100 ml.
2. Its deficiency has also been observed in chronic alcoholics with the low serum magnesium and muscular weakness.
3. In cases of kwashiorkor, the serum magnesium level is low causing weakness.
4. Low values for serum magnesium have been reported in uremia, normal and abnormal pregnancy, rickets, growth hormone treatment, hypercalcemia and the recovery phase of diabetic coma.

SODIUM

Physiological functions

1. It is the major component of the cations of the extracellular fluid and exists in the body in association with the anions chloride, bicarbonate, phosphate and lactate.
2. It is largely associated with chloride and bicarbonate in regulation of acid-base equilibrium.
3. It maintains the osmotic pressure of the body fluid and thus protects the body against excessive fluid loss.
4. Sodium ion plays an important role in the absorption of glucose and galactose as well as amino acids from the small intestine.
5. It maintains the normal water balance and distribution.
6. Sodium ion is involved in initiating and maintaining the heart beat.
7. It maintains the normal neuromuscular function.
8. It functions in the permeability of the cells.

Sources

Principal sources	:	The main source of sodium is the *sodium chloride* used in cooking and seasoning.
Rich sources	:	Bread, cheese, wheat germ, whole grains and oysters, etc.
Good sources	:	Carrots, cauliflowers, eggs, milk, nuts, spinach and turnips, etc.

Distribution

About one-third of the total sodium content of the body is present in the inorganic portion of the skeleton. Most of the sodium is found in the extracellular fluid.

Plasma	330 mg/100 ml
Cells	85 mg/100 gm
Muscle	60-160 mg/100 gm
Nerve	312 mg/100 gm

Daily requirement

For adults the daily requirement is 5 to 15 gm. In temperate region, the sodium chloride intake is less but in tropical countries the intake is more. A person suffering from hypertension should not take more than 1 gm of sodium.

Absorption

Normally, sodium is practically completely absorbed from the gastrointestinal tract. Less than 2 per cent of ingested sodium is eliminated in the feces. In subjects with diarrhea, large amounts are lost in the feces.

Blood sodium

The normal level of sodium is 136-146 mEq/L. In man, erythrocytes contain little or no sodium. Aldosterone increases plasma sodium level.

Excretion

The daily losses of sodium are as follows :

Urine	10-150 mEq
Stool	10-125 mg
Skin (not sweating)	25-25 mg

About 95 per cent of the sodium leaving the body is excreted in the urine since sodium is readily absorbed in the intestine. Therefore, feces contain very little except in diarrhea. There is the variable loss of sodium by way of the sweat. Heavy exercise, environmental heat and high fever can cause excessive losses of sweat which leads to more sodium losses. Loss of sodium by excessive sweating causes *heat cramps* with the intense and painful contractions of *skeletal muscle* of men working hard in hot humid climates.

Disease state

1. Adrenocortical steroids regulate the metabolism of sodium. In the insufficiency of adrenocortical steroids, the serum sodium level is decreased with an increase in sodium excretion.
2. In *chronic renal disease* when acidosis exists, sodium depletion occurs due to poor tubular reabsorption of sodium as well as to the loss of sodium in the buffering of acids.
3. In case, a person is not adapted to a high environmental temperature, large amount of sodium is lost in the sweat due to extreme sweating developing *muscular cramps* of the extremities and abdomen, headaches, nausea and diarrhea.
4. In *hyponatremia*, the serum sodium level is below normal. This causes severe dehydration, reduced blood pressure, decreased blood volume and circulatory failure. The following clinical conditions will develop :
   (a) Prolonged vomiting and diarrhea resulting in excessive loss of digestive juices rich in sodium ion.
   (b) Chronic renal disease with acidosis due to poor reabsorption of sodium in the tubules.
   (c) Adrenocortical deficiency leading to *Addison's disease*.
   (d) Loss of weight due to loss of water also.
5. In *hypernatremia*, the serum sodium level is higher than normal. This occurs in (i) hyperactivity of adrenal cortex as in *Cushing's syndrome*. (ii) Prolonged treatment of cortisone and ACTH as well as sex hormones. The symptoms of hypernatremia are :
   (a) Increased retention of water in the body.
   (b) Increase in blood volume.
   (c) Increase in blood pressure.
6. In certain stages of *pregnancy*, the steroid hormones cause the retention of sodium as well as water which results in *gain in weight*.

POTASSIUM

Physiological functions

1. Potassium is largely present in the intracellular fluid and it is also present in small amounts in the extracellular fluid because it influences cardiac muscle activity.
2. It plays an important role in the regulation of acid-base balance in the cell.
3. It maintains osmotic pressure.
4. It functions in water retention.

5. It is essential for protein biosynthesis by ribosomes.
6. The glycolytic enzyme *pyruvate kinase* requires K^+ for maximal activity.

Sources

The high content of potassium is found in chicken, beef liver, bananas, the juices of oranges, pineapples, jams, winter squash and potatoes, etc.

Distribution

Plasma	20 mg/100 ml
Cells	440 mg/100 gm
Muscles	250-400 mg/100 gm
Nerves	530 mg/100 gm

Daily requirement

The normal intake of potassium in food is about 4 gm. It is so widely distributed in food that its deficiency is rare except under pathological conditions.

Blood potassium

The normal level of potassium in serum is 3.5 to 4.5 mEq/L. In man, erythrocytes contain large amounts of potassium. Therefore, care must be taken to avoid hemolysis for the determination of serum potassium. The serum potassium decreases during increased *carbohydrate utilization* following administration of glucose or insulin. *Aldosterone* decreases serum potassium level.

Absorption

Normally, potassium is practically completely absorbed from the gastrointestinal tract and less than 10 per cent of potassium is eliminated in the feces. In subjects with *diarrhea*, large amounts of potassium are lost in the feces.

Excretion

Potassium is normally eliminated almost entirely in the urine and a small amount in the feces. *Aldosterone* exerts an influence on potassium excretion. In the presence of normal kidney function, potassium is very promptly and efficiently removed from the blood. Normally, 25-100 mEq of potassium is excreted in urine in 24 hours.

Disease state

1. Potassium is not only filtered by the kidney but is also secreted by the renal tubules. The excretion of potassium is greatly influenced by changes in acid-base balance and also by the activity of the adrenal cortex. The capacity of the kidney to excrete potassium is very great and, therefore, hyperkalemia does not occur even after the ingestion of potassium if the kidney function is unimpaired. Potassium should not be given intravenously unless circulatory collapse and dehydration are corrected.
2. *Hyperkalemia* occurs in patients in the following conditions : (i) Renal failure, (ii) Severe dehydration, (iii) Addison's disease due to decreased excretion of potassium by the kidney, (iv) Intravenous administration of excessive amounts of potassium salts, (v) Shock.

   The symptoms of *hyperkalemia* are cardiac and central nervous system depression. The heart signs include *bradycardia* and low heart sounds followed by peripheral vascular collapse leading to cardiac arrest. The other symptoms are mental confusion, numbness, weakness of respiratory muscles and flaccid paralysis of the extremities.

   The symptoms are corrected by *administration of deoxycorticosterone* which helps the excretion of potassium.
3. Prolonged *hypokalemia* causes injury to myocardium and kidneys. Potassium deficits occur in chronic wasting diseases with malnutrition, prolonged negative nitrogen balance, gastrointestinal losses and in metabolic alkalosis.

The clinical conditions exhibiting hypokalemia are :

(a) Prolonged diarrhea and vomiting with the loss of digestive juices.

(b) Intravenous administration of potassium-free fluid to replace digestive juices lost by prolonged vomiting.

(c) Overactivity of adrenal cortex (*Cushing's syndrome*) which causes increased excretion of potassium in urine.

(d) Prolonged use of diuretics.

(e) Heart failure treatment with digitalis.

(f) Diabetic coma treatment with insulin.

(g) In *familial periodic paralysis*, a rare disease, potassium is withdrawn from extracellular fluid and retained in the cells.

The *symptoms of hypokalemia* are muscular weakness, irritability, paralysis, tachycardia and dilatation of the heart with gallop rhythm and changes in the electrocardiogram (ECG).

CHLORINE

Physiological functions

1. As a component of sodium chloride, chloride ion is essential in acid-base equilibrium.
2. As chloride ion, it is also essential in water balance and osmotic pressure regulation.
3. It is also important in the production of hydrochloric acid in the gastric juice.
4. Chloride ion is important as an activator of amylase.

Sources

It is mainly available as sodium chloride.

Distribution

Plasma	365 mg/100 ml
Cells	190 mg/100 gm
C.S.F.	440 mg/100 ml
Muscle	40 mg/100 gm
Nerve	171 mg/100 gm

Daily requirement

The requirements of NaCl depend on the climate and occupation and on the salt content of the diet. Foods of animal origin contain more NaCl than those of vegetable origin. The daily requirement in tropical countries are given below :

Adults	10-20 gms
Children	5-10 gms
Women during pregnancy and lactation	10-15 gms

Excessive consumption of NaCl causes edema in protein deficiency and increases blood pressure in hypertension patients.

Blood Chlorine

The normal level of Cl in serum is 98-110 mEq/L. In man, erythrocytes contain smaller amounts of Cl. The distribution of Cl^- between plasma and erythrocytes is related to that of HCO_3^-. The Cl content of whole blood is relatively high in anemia and relatively low in *polycythemia*. The serum Cl may fall during active gastric secretion of HCl.

Absorption

Normally, Cl is practically completely absorbed from the gastrointestinal tract.

Excretion

Cl is chiefly eliminated in the urine. It is also excreted in the sweat. It is lost more during excessive sweating in hot climates under hard work. Its concentration in sweat is decreased by *aldosterone*.

Disease state

1. Chloride deficit also occurs when losses of sodium are excessive in diarrhea, sweating and certain endocrine disturbances.
2. There is a loss of chloride in the loss of gastric juice by vomiting or in pyloric or *duodenal obstruction*.
3. Hypochloremic alkalosis may develop in

Cushing's disease or after the administration of ACTH or cortisone.

SULFUR

Physiological functions

1. Sulfur is present primarily in the cell protein in the form of cysteine and methionine.
2. The *cysteine* is important in protein structure and in enzymic activity.
3. *Methionine* is the principal methyl group donor in the body. The activated form of methionine, *S-adenosylmethionine*, is the precursor in the synthesis of large number of methylated compounds which are involved in intermediary metabolism and detoxification mechanism.
4. Sulfur is a constituent of *coenzyme A* and *lipoic acid* which are utilized for the synthesis of acetyl-COA and S-acetyl lipoate, respectively.
5. Sulfur is a component of other organic compounds, such as heparin, glutathione, thiamine, biotin, ergothioneine, taurocholic acid, sulfocyanides, indoxyl sulfate, chondroitin sulfate, insulin, penicillin, anterior pituitary hormones and melanin.

Sources

Sulfur intake is mainly in the form of cystine and methionine present in proteins. Other compounds present in the diet contribute small amounts of sulfur.

Sulfur in blood

The normal concentration of sulfur in the serum is as follows :

Inorganic sulfur	0.5-1.5 mg/100 ml
Ethereal sulfate	0.1-1.0 mg/100 ml
Neutral sulfur	1.7-3.5 mg/100 ml

Absorption

Inorganic sulfate is absorbed as such from the intestine into the portal circulation. A small amount of sulfide may be formed in the bowel by the action of bacteria, but, if absorbed into the blood stream, this is rapidly oxidized to sulfate.

Excretion

Sulfur is excreted in the urine in three forms. The total sulfate excretion may be diminished in the presence of renal functional impairment and is increased in conditions accompanied by excessive tissue protein breakdown, such as high fever and increased metabolism.

Disease state

1. The serum sulfate concentration is increased in the presence of renal functional impairment, pyloric and intestinal obstruction and leukemia.
2. Marked sulfate retention in advanced *glomerulonephritis* cause the development of acidosis.
3. An increase in the blood *indican* concentration (indoxyl potassium sulfate) may occur in *uremia*.

TRACE ELEMENTS

Essential Trace Elements

IRON

The total iron content of the normal adult is about 4 to 5 gm. About 60 to 70 per cent of the total iron is present in hemoglobin; about 51 per cent is in storage as ferritin; 3 per cent as myoglobin and only about 0.1 per cent is carried in the plasma in combination with the β-globulin transport protein *transferrin*. The hemoprotein and flavoprotein enzymes together make up less than 1.0 per cent of the total iron. Large amounts are present as *hemosiderin*.

Physiological functions

1. Iron functions mainly in the transport of oxygen to the tissues (hemoglobin).
2. It is also involved in the processes of cellular respiration.
3. It is an essential component of hemoglobin, myoglobin, cytochromes and the respiratory enzyme systems (cytochrome oxidase, catalase and peroxidase).
4. The nonheme iron is completely protein-bound which exists in the form of storage and transport.
5. The nonheme iron is also utilized in the structure of xanthine dehydrogenase and succinate dehydrogenase and also in the iron-sulfur proteins of the respiratory chain.

Sources

Rich sources	: Liver, heart, kidney, spleen.
Good sources	: Egg-yolk, fish, nuts, figs, dates, beans, spinach, molasses, apples, bananas, etc.
Poor sources	: Milk, wheat flour, polished rice and potatoes etc. Human milk contains 0.3 to 0.6 μg iron/ml.

Distribution

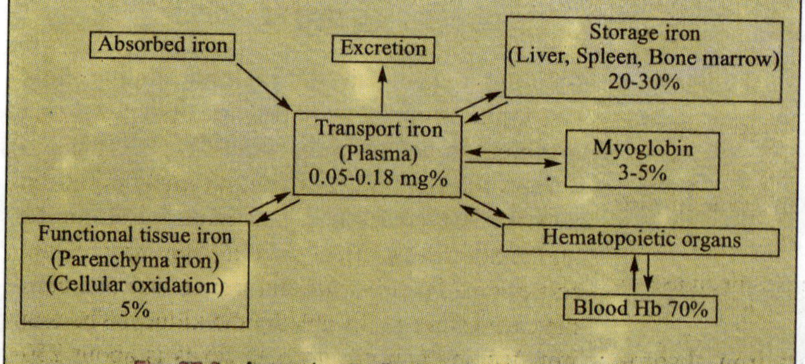

Fig. 27.2 : Approximate distribution of iron in the body

Daily requirement

About 10 per cent of the ingested iron is only absorbed.

Infants :	
Children (1 to 3 years of age)	15 mg
(4 to 10 years of age)	10 mg
Older children and adults (Males)	
(11 to 18 years of age)	18 mg
(above 19 years of age)	10 mg
Females :	
11-50 years of age and during pregnancy or lactation	18 mg
above 51 years of age	10 mg

In *adult women*, the average loss of blood during a menstrual period, which is a monthly loss, of 16-32 mg of iron or an additional average loss of 0.5-1.0 mg per day. This amount is easily obtained from the diet.

In *excessive menstrual blood* loss and in chronic iron-deficiency anemia, a supplement of 100 mg of iron per day is sufficient to respond. So during growth, pregnancy and lactation, iron demand is more.

In healthy adult male or in healthy women after menopause, the dietary requirement is negligible unless any deficiency or loss of iron occurs.

The iron deficiency occurs as a result of malabsorption from the gastrointestinal tract. A defect in hemoglobin synthesis in anemia is commonly found in *copper deficiency*.

Iron in the blood : The normal concentration of iron in blood is 65-170 μg/100 ml.

Absorption

Under normal conditions, very little dietary iron is absorbed (less than 10 per cent), the amount excreted in the urine are minimal. Infants and children absorb a higher percentage of iron from foods than adults. Iron deficiency in infants is due to a dietary deficiency. Iron-deficient children absorb twice as much as normal children.

Most of the iron in food occurs in the ferric (Fe^{+++}) state either as $Fe(OH)_3$ or as ferric organic compounds. These compounds are broken down into free ferric ions or loosely bound organic iron. The gastric hydrochloric acid and the organic acids of the foods are both important for this purpose. Reducing substances in foods, such as cysteine and ascorbic acid convert ferric ion into the ferrous (Fe^{++}) state. In this form it is more soluble and, therefore, readily absorbed. Iron absorption is enhanced by proteins of low molecular weight digestive products (peptides, amino acids) forming iron chelate. Heme enters the mucosal cells without being released from the porphyrin ring. In humans, dogs and rats, heme is broken down in the mucosa and iron appears in the plasma *transferrin*.

Factors effecting iron absorption

1. Absorption of iron occurs mainly in the stomach and the duodenum. Impaired absorption takes place in patients who have total removal of stomach or a removal of the considerable amount of the intestine.
2. A diet high in phosphate causes decreased absorption due to the formation of insoluble ferric phosphate ($FePO_4$). Very low phosphate favours increased absorption of iron.
3. Phytic acid (present in cereals) and oxalates interfere absorption.
4. Vitamin C increases absorption.
5. Gastric acidity increases absorption by converting $Fe(OH)_3$ to Fe^{+++}. Achlorhydria and administration of alkali decrease absorption.
6. Proteins of low molecular weight favour absorption.
7. Copper deficiency also causes decrease in absorption.
8. Alcohol ingestion favours iron absorption.

Mechanism of absorption

Previously, "*Mucosal Block*" theory was considered to be the controlling of iron absorption. The iron-binding protein, apoferritin, in the mucosal cells, was the controlling factor.Ferrous ion being oxidized to ferric ion combines with *apoferritin* to form iron containing protein.

Ferritin : It was believed that the absorption depended on the formation of ferritin. When apoferritin was saturated with iron, no further uptake of iron could take place.

More recently, evidences show that ferritin is involved in the major regulation of iron absorption. Iron taken into the mucosal cell is bound to specific carriers which regulate its passage across the cell to the blood. Intestinal ferritin, therefore, acts as a storage compound rather than the controlling of absorption.

Transport in the plasma

All the iron released from the mucosal cell enters the portal blood in the ferrous state. In the plasma, *ferrous* is oxidized to ferric state by *ceruloplasmin* (a copper binding plasma protein) exerts a catalytic activity (serum ferroxidase) in plasma. Human serum also contains a yellow cuproprotein (ferroxidase 11) which catalyzes the oxidation of ferrous ions. Ferric ion is then incorporated into a specific iron binding protein, transferrin or siderophilin, which is a glycoprotein of molecular weight 76,000 containing 5.3 per cent carbohydrate. Transferrin can bind 2 atoms of ferric ions per molecule of protein to form a red ferric-protein complex. Iron release from the mucosal cell is facilitated by a low degree of transferrin saturation by iron.

Under normal circumstances, almost all of the iron bound to transferrin is taken up readily by *bone-marrow*. Only the reticulocytes can utilize the ferric ion bound to transferrin, although reticulocytes and the mature erythrocytes can take up unbound ferric ion. The iron with transferrin makes a complex which is not filtrable by the kidney. The total iron binding capacity in both sexes is about 250-450 µg/dL.

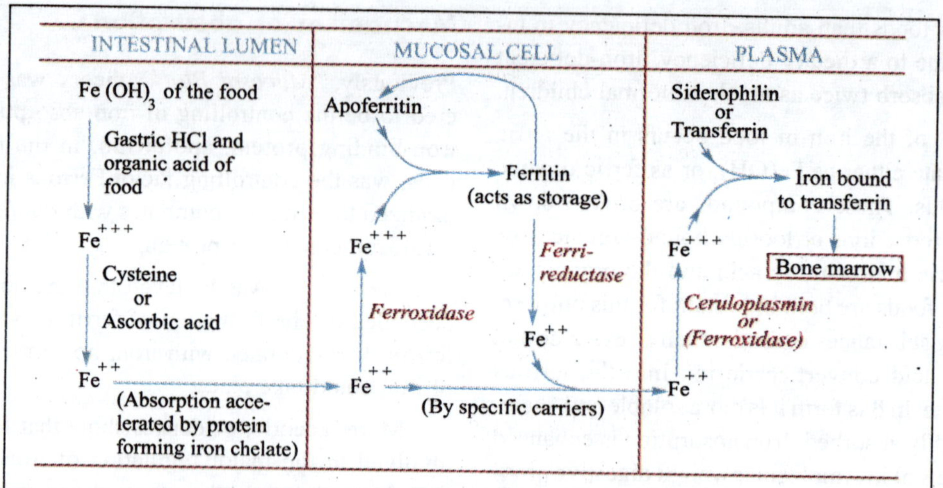

Fig. 27.3 : Iron absorption and transport mechanism

Losses of iron into the urine occur in proteinuria. In *nephrosis*, iron (1.5 mg/day) with protein may be excreted in the urine. In *hepatic disease*, both the bound iron and the total iron-binding capacity of the plasma is low.

Excretion

The body stores of iron are conserved very efficiently. Only minute amounts are excreted in the urine, feces and sweats. Relatively large amounts are lost in the menstrual flow. The bulk of the iron of the feces is unabsorbed food iron. In the tropics, iron loss is often much greater. During pregnancy, iron is lost to the fetus. Iron is also lost from the skin by means of sweat, hair loss and nail clippings.

The daily excretion of iron is as follows :

Adult male	0.5-1.5 mg in the urine.
Adult women during menstruation	Double the above in the urine.
Adult (Both Male & Female)	0.5-1.0 mg in the sweat.
Adult (Both Male & Female)	0.3-0.75 mg in the feces.

Abnormal iron metabolism

Ferritin and *hemosiderin*, the storage forms of iron, act as an internal iron reserve to protect against sudden losses of iron by bleeding. Ferritin is present not only in the intestine but also in liver (about 700 mg), spleen and bone-marrow. If more iron is administered parenterally exceeding the capacity of the body to store as ferritin, it accumulates in the liver as *hemosiderin*, a form of colloidal iron oxide in association with protein. The iron content of hemosiderin is 35 per cent by weight.

Iron metabolism is disturbed mainly by the following causes :

1. Decreased formation of hemoglobin .
2. Decrease in circulating hemoglobin.
3. Abnormalities in the serum iron concentration.
4. Abnormal deposition of iron-containing pigment in the tissues.

The disorders of iron metabolism are : 1. Siderosis; 2. Nutritional siderosis; 3. Hemochromatosis.

1. **Siderosis :** When excessive amounts of iron are released in or introduced into the body beyond the capacity for its utilization, the excess is deposited in the various tissues, mainly in the liver. This may occur due to repeated blood transfusion, excessive breakdown of erythrocytes in hemolytic types of anemia and inadequate synthesis of hemoglobin as in pernicious anemia.

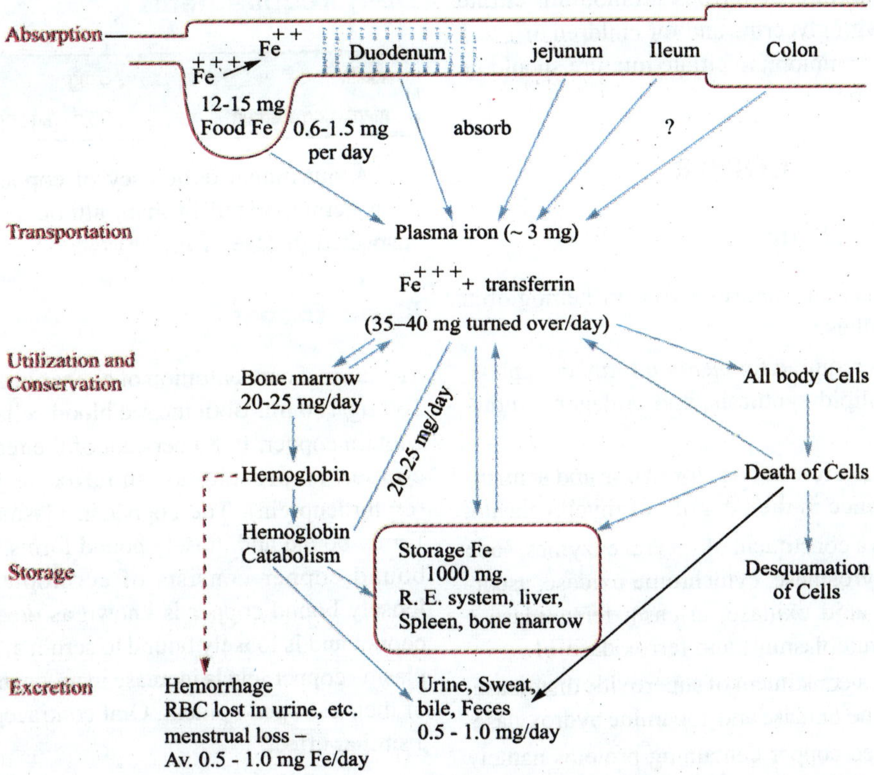

Fig. 27.4 : Schematic representation of iron metabolism in man, RBC = Red blood cells. RE system = Reticuloendothelial system

2. **Nutritional siderosis** : The disorder is found among Bantus in South Africa. Bantus cook their food in large iron pots and consume iron-rich food. The absorption of iron appears to be high leading to the development of nutritional siderosis. Livers of the Bantus contain large amounts of iron.
3. **Hemochromatosis** : Hemochromatosis is a rare disease in which large amounts of iron are deposited in the tissues, especially the liver, pancreas, spleen and skin producing various disorders. Accumulation of iron in the liver, pancreas and skin produces hepatic cirrhosis, bronze diabetes and bronze-state pigmentation, respectively.

Iron deficiency anemia

Iron deficiency anemia is widely prevalent among children, adolescent girls and nursing mothers. The hemoglobin content of the blood is 5 to 9 gm/100 ml.

Women of child bearing age : The clinical symptoms are breathlessness on exertion, giddiness and pallor of skin. In severe cases, there may be edema of the ankles.

Weaned infants and young children : The hemoglobin level is 5 to 9 gm/100 ml blood. The children are dull, and inactive and show pallor of the skin. The appetite is poor and growth and development are retarded.

Treatment : Anemic women should take ferrous sulphate tablets. For a child of below 12 months, a mixture of ferrous ammonium citrate sweetened with glycerine and for children of 1 to 5 years ferrous ammonium citrate mixture should be given in curing.

COPPER

Physiological functions

1. It has an important role in hemoglobin synthesis.
2. It is required for *melanin* formation, phospholipid synthesis, and collagen synthesis.
3. It has a role in bone formation and in maintenance in the integrity of myelin sheath.
4. It is a constituent of several enzymes, such as tyrosinase, cytochrome oxidase, ascorbic acid oxidase, uricase, ferroxidase 1 (ceruloplasmin) and ferroxidase 11.
5. It is a constituent of superoxide dismutase, amine oxidase and dopamine hydroxylase.
6. Three copper containing proteins namely cerebrocuprein, erythrocuprein and hepatocuprein are present in brain, RBC and liver, respectively.

Sources

Richest sources	: Liver, kidney, other meats, shellfish, nuts, and dried legumes.
Poor sources	: Milk and its products.
Cow milk	0.015 to 0.18 mg/L.
Human milk	1.05 mg/L (at the beginning of lactation).
Human milk	0.15 mg/L (at the end of lactation).

Distribution

The adult human body contains 100-150 mg of copper.

Muscles	64 mg of the total
Bones	23 mg of the total
Liver	18 mg of the total

The concentration of copper in the fetal liver is 5-10 times higher than that in liver of an adult.

Daily requirements

Adults	2.5 mg
Infants and children	0.05 mg/kg body weight

A nutritional deficiency of copper has never been demonstrated in man, although it has been suspected in case of *nephrosis*.

Blood copper

The normal concentration of copper in serum is 75-160 μg/100 ml. Both the red blood cells and serum contain copper. In 80 per cent of the red blood cell copper is present as superoxide dismutase (erythrocuprein). The copper in plasma occurs in firmly bound and loosely bound forms. The firmly bound copper consists of ceruloplasmin. The loosely bound copper is known as *direct reacting* copper and is loosely bound to serum albumin. The plasma copper levels increase in pregnancy because of their estrogen content. Oral contraceptives have a similar effect.

Absorption

Absorption of copper occurs in the human duodenum. 30 per cent of the normal daily diet of copper is absorbed in the duodenum.

Excretion

Only 0 to 32 μg of copper is excreted in normal urine in 24 hours. The daily biliary excretion of copper is 0.5 to 1.3 mg and 0.1 to 0.3 mg is excreted across the intestinal mucosa into the bowel lumen.

Effects of copper deficiency

1. Although iron absorption is not disturbed, the release of iron into the plasma is prevented due to the decreased synthesis of *ceruloplasmin*. As a result, hypoferrimia occurs which leads to the depressed synthesis of heme developing anemia in severe deficiency of copper.

2. The experimental animals on a copper-deficient diet lose weight and die.
3. In copper-deficient lambs, low cytochrome oxidase activity results in *neonatal ataxia*.
4. Copper deficiency produces marked skeletal changes, osteoporosis and spontaneous fractures.
5. Elastin formation is impaired in the deficiency of copper. Because a copper- containing enzyme plays an important role in the connective tissue metabolism, especially in the oxidation of lysine into aldehyde group which is necessary for cross-linkage of the polypeptide chains of elastin and collagen.
6. Copper deficiency results in the myocardial fibrosis in cows. It is suggested that reduction in cytochrome oxidase activity may lead to cardiac hypertrophy.

Disorders of copper metabolism

Wilson's Disease (*Hepato-lenticular degeneration*) : *Wilson's disease* is a rare hereditary disorder of copper metabolism. The following disorders have been observed in this disease :

1. The absorption of copper from the intestine is very high (about 50 per cent), whereas 2 to 5 per cent copper is absorbed in normal subjects.
2. Ceruloplasmin formation is very less. Hence, a greater part of serum copper remains loosely bound to serum protein—notably albumin and, therefore, copper can be transferred to the tissues, such as brain and liver or to the urine.
3. Excessive deposition of copper in the liver and the kidney causes hepatic cirrhosis and renal tubular damage, respectively. The renal tubular damage results in the increased urinary excretion of amino acids, peptides and glucose.

Clinical symptoms

1. Progressive hepatic cirrhosis of a coarse nodular type gradually leads to portal hypertension and finally to hepatic failure.
2. There is dysfunction of the lenticular region of the brain necrosis and sclerosis of the corpus structures cause basal ganglion syndromes in adolescence.
3. Defects in renal tubular reabsorption and glomerular filtration.
4. Copper deposition in Descemet's membrane of the eye causes a golden brown, yellow or green ring round the cornea. This lesion is called *Kayser-Fleischer ring*.
5. There is occasional pigmentation of the nails and skin.

Treatment : Improvement can be achieved by removing the excess of tissue copper by administering the copper chelating agent *penicillamine*. This brings about marked increase in the urinary secretion of the metal. Increased values of serum ceruloplasmin occur in many acute and chronic infectious diseases, hepatic and biliary tract disease, leukemia and other forms of malignancy, iron deficiency anemia, hyperthyroidism, myocardial infarction, and certain neurological diseases.

Menkes' Disease

It is an X-linked disorder of intestinal copper absorption. The absorption of copper and its uptake in the mucosal cells and its intracellular transport within the mucosal cells are normal in patients with this disease. But the transport of copper across the serosal aspect of the mucosal cell membrane is defective.

The clinical features show that unless therapy is started promptly at birth, many severe symptoms like mental retardation, temperature instability, abnormal bone formation and susceptibility to infection are not to be cured.

Ceruloplasmin

1. It is a copper-binding plasma protein.

Table 27.1 : Comparison of Wilson's disease and Menkes' disease

	Wilson's Disease	*Menkes' Disease*
Inheritance	Autosomal recessive	X-linked recessive
Gene Product	Cu^{2+}-binding P-type ATPase	Cu^{2+}-binding P-type ATPase
Mutations	Variety	Variety
Expression	Liver, Kidney, Placenta.	All tissues except liver.
Onset	Late childhood	At birth
Clinical Findings	Liver disease, neurologic signs, can survive to late adulthood.	Cerebral degeneration, abnormal hair, other signs, early death.
Cultured Cells	Normal in most patients.	↑ Cu^{2+} accumulation ↓ Cu^{2+} release
Laboratory Findings	↓ Serum Cu^{2+} ↓ Serum ceruloplasmin ↑ Urinary Cu^{2+} ↑ Liver Cu^{2+}	↓ Serum Cu^{2+} ↓ Serum ceruloplasmin ↑ Intestinal and Kidney Cu^{2+} ↓ Liver Cu^{2+}
Defect	Biliary Cu^{2+} excretion, Cu^{2+} incorporation into ceruloplasmin.	Intestinal Cu^{2+} absorption, deficiency of Cu^{2+}-dependent enzymes.
Treatment	Penicillamine.	No effective treatment.

2. Its molecular weight is about 151,000.
3. It contains about 8 atoms of copper per mole.
4. Normal plasma contains 30 mg of this protein per dl.
5. It functions as a ferroxidase enzyme during iron transport.

Superoxide dismutase

1. It is an enzyme which catalytically scavenges the toxic free radical superoxide ion (O_2^-) formed during aerobic metabolism.
2. Its molecular weight is about 32,000 and consists of two identical subunits.
3. It contains one Cu^{++} and one Zn^{++} per subunit.
4. Recent studies have shown that the copper proteins erythrocuprein, hepatocuprein and cerebrocuprein present in RBC, liver and brain, respectively, are identical with this enzyme.

IODINE

Physiological functions

Iodine is required for the formation of thyroxine and triiodothyronine hormones of the thyroid gland. These thyroid hormones are involved in cellular oxidation, growth, reproduction, and the activity of the central and autonomic nervous systems. Triiodothyronine is more active than thyroxine in many respects.

Sources

Rich sources are sea water, marine vegetation, sea foods, and vegetables as well as fruits grown on the seaboard. Plants (and animal tissues) grown at high altitudes are *deficient* in iodine because of its low concentration in the water. In such regions, iodide is commonly added to the drinking water or table salt in concentration of 1 : 5000–1 : 200,000.

Daily requirements

Adults	100-150 mg
In adolescence and in pregnancy	200 mg

Distribution

The body normally contains about 10 to 20 mg of iodine. 70 to 80 per cent of this is present in the thyroid gland. Muscles contain large amount of iodine. The concentration of iodine in the salivary glands, ovaries, pituitary gland, hair and bile is greater than that in muscle.

All the iodine in saliva is inorganic, but most tissues contain less amount of iodine in the inorganic form and most of the iodine is present in the organic form.

Blood iodine

Practically all of the iodine in the blood is in the plasma. The normal *concentration* of iodine in the plasma or serum is 3 to 6 μg/100 ml. Iodine protein bound is 3.5-7.5 μg/dL.

0.08 to 0.60 μg/100 ml is in the inorganic form and 4 to 8 μg/100 ml is in the organic form. The organic form is bound to protein and precipitated by protein-precipitating agents. 90 per cent of the organic form consists of thyroxine and the remainder is tri- and di-iodothyronine. About 0.05 per cent of the thyroxine is in the free state. *Erythrocytes* contain no organic iodine.

Absorption

Iodine and iodides are absorbed most readily from the small intestine. Organic iodine compounds (diiodotyrosine and thyroxine) are absorbed as such and a part is broken down in the stomach and intestine with the formation of iodides. Absorption also takes place from other mucous membranes and the skin.

Storage

90 per cent of the iodine of the thyroid gland is in organic combination and stored in the follicular colloid as "**thyroglobulin**" a glycoprotein of molecular weight 6,50,000 containing thyroxine, diiodotyrosine and smaller amounts of triiodothyronine.

On demand, these substances are mobilized and thyroxine as well as triiodothyronine are passed into the systemic circulation. They undergo metabolic degradation in the liver.

Excretion

1. Inorganic iodine is mostly excreted by the kidneys, liver, skin, lungs and intestine and in milk.
2. About 10 percent of circulating organic iodine is excreted in feces. This is entirely unabsorbed food iodine.
3. 40 to 80 per cent is usually excreted in the urine ; of which 20 to 70 μg daily in adults and 20 to 35 μg in children. The urinary elimination is largest when the intake is lowest.
4. Urine iodine is increased by exercise and other metabolic factors.

Iodine deficiency in human beings

1. ***In adults***, the thyroid gland is enlarged, producing the disease *goitre*. If treatment is started very early, the thyroid becomes normal. If treatment is delayed, the enlargement of the gland persists.
2. ***In children***, severe iodine deficiency results in the extreme retardation of growth which is known as ***cretinism***.

Prevention of goitre : Goitre can be prevented by the regular use of iodide salt or iodide added to the drinking water in the concentration of 1 : 5000 to 1 : 2,00,000.

Goitrogenic substances in foods

Foods such as cabbage, cauliflower and radish contain substances 1 to 5, vinyl-2-thio oxazolidone which react with the iodine present in the food and make it unavailable to the body. These substances are known as "*Goitrogenic*" substances.

FLUORINE

Physiological functions

1. Fluoride, in trace quantities, is essential for the development of teeth and bones.
2. It is, in combination with vitamin D, required for the treatment of *osteoporosis*.
3. Sodium fluoride is a powerful inhibitor of the glycolytic enzyme *enolase*.
4. Fluoroacetate acts as a powerful inhibitor of aconitase activity, responsible for the conversion of citrate to *cis*-aconitate, of the citric acid cycle.
5. Fluoride ions inhibit the metabolism of oral bacterial enzymes and diminish the local production of acids which are important in the production of *dental caries*.
6. Fluorine forms a protective layer of acid-resistant fluoroapatite with hydroxyapatite crystals of the enamel.

Sources

For humans, drinking water is the main sources of fluoride.

Daily requirement

Fluoride is present in small amounts in normal bones and teeth. Drinking water containing 1 to 2 ppm meet up the requirement of the body and prevent dental caries without producing any ill effect.

Distribution

It occurs in many tissues, notably the bones, teeth and kidneys. The amounts of fluoride in the soft tissues are very low and do not increase with age. It remains mostly in the extracellular water.

Absorption

Soluble fluorides are rapidly absorbed from the small intestine.

Excretion

It is excreted in the urine, in the sweat, and by the intestinal mucosa. Most of the fluoride that escapes retention by the bones and teeth is excreted rapidly into the urine.

Abnormalities

1. Intake of excessive amounts of fluoride (3 to 5 parts per million) in childhood causes "*dental fluorosis*" (*molted enamel*). The enamel of the teeth loses its lustre and becomes rough. Chalky white patches with yellow or brown staining are found over the surface of the teeth. The enamel becomes weak and in severe cases there occurs a profound loss of enamel with 'pitting' which gives both the surfaces a corroded appearance.
2. Highly excessive intake of fluorine (over 10 parts per million) results in increased density and hypercalcification of the bone of spine, pelvis and limbs. In addition, the ligaments of the spine become calcified and the collagen in the bone is also calcified. Neurological disturbances are common. Such individuals are crippled and cannot exhibit simple daily tasks, such as bending, squatting, etc. as the joints become stiff.
3. Drinking water containing less than 0.5 ppm fluorine causes dental caries in children.

Prevention of fluorosis

Fluorosis can be prevented by removing fluorides from the water by treatment with activated carbon or by some other suitable absorbents.

ZINC

Physiological functions

1. Zinc is an essential constituent of many enzymes, such as carbonic anhydrase, alkaline phosphatase, pancreatic carboxy-

peptidases, and crytosolic superoxide dismutase.

2. The retina contains a zinc metalloenzyme, *retinene reductase* which is required for the formation of retinene.
3. It maintains normal concentration of vitamin A in plasma.
4. It is required for the mobilization of vitamin A from the liver.
5. It is required for the preparation of insulin and increases the duration of insulin action when given by injection. Zinc is used in the β-cells of the pancreas to store and release insulin as required.
6. It is concerned with the healing of wounds.
7. It is essential for the normal growth and reproduction of animals.

Sources

Rich sources	:	Oysters and herrings.
Good sources	:	Meat, eggs, liver, and milk.
Fair sources	:	Cereals, pulses, nuts, oilseeds, vegetables and fruits.

Distribution

It is widely distributed in the tissues of the body. The whole body (70 kg weight) contains 1.4 to 2.3 gm zinc. 20 per cent of the total is present in skin. A certain amount is also present in the bones and teeth. High concentrations of zinc are present in spermatozoa, prostate and epididymis. The highest concentration occurs in the choroid of the eye.

Blood zinc

1. Zinc is present in higher concentration in erythrocytes than in plasma.
2. Normal plasma contains about 20 per cent of the zinc present in whole blood.
3. The concentration of zinc of human blood, plasma and erythrocytes are 0.8 mg 0.12 mg and 1.44 mg /100 ml, respectively.
4. About 3 per cent of zinc ion is contained in leukocytes. In certain types of chronic leukemia, there is a marked fall in the zinc content of peripheral leukocytes.
5. Most of zinc in erythrocyte is present in carbonic anhydrase.
6. The plasma concentration of zinc of human falls to 10 per cent of the normal level during later part of pregnancy and among those taking oral contraceptives.

Daily requirement

Breast-fed newborn babies	0.7-5 mg
Infants	3-5 mg
Children	10 mg
Adolescents	13 mg
Adults	15 mg
Pregnant women	30 mg
Women during lactation	25 mg

Absorption

1. Zinc present in animal foods are well absorbed in the small intestine, especially from the duodenum.
2. Zinc present in cereals, pulses, nuts and oilseeds are poorly absorbed due to the presence of phytic acid which interferes its absorption.

Excretion

1. Zinc given orally or by injection is mostly excreted in the feces.
2. Endogenous zinc is secreted into the small intestine in the pancreatic juice.
3. 90 per cent of zinc intake by healthy adult human is lost in the feces, about 5 per cent is excreted in the urine and 5 per cent is retained in the body.
4. Excessive sweating in the hot climate causes excessive loss of the metal.

Deficiency of Zinc

1. Zinc deficiency in man results in dwarf-

ism and hypogonadism (retarded genital development).

2. There is loss of taste acuity.
3. There is also poor growth, loss of appetite and hypogeusia in young malnourished children with subnormal hair zinc levels.
4. The deficiency of zinc causes hepatosplenomegaly, delayed closure of the epiphyses of the long bones, and anemia.

CLINICAL ASPECTS

- Deficiency of Zinc obstructs the storage and secretion of insulin causing *diabetes mellitus*.
- The leucocyte content of Zinc is reduced in *leukemias*.
- Zinc therapy is essential in some cases of *atherosclerosis*.
- Zinc level in plasma is decreased in acute *myocardial infarction*.
- Liver Zinc concentration is highly increased in *malignancies*.
- Serum zinc level is decreased in *cirrhosis of liver*.
- Zinc deficiency produces skin lesions.

COBALT

1. Cobalt is an essential component of vitamin B_{12}, which is necessary for normal red blood cell formation.
2. Certain enzymes, such as methylmalonyl-CoA mutase, methyltetrahydrofolate oxidoreductase, homocysteine methyltransferase, and ribonucleotide reductase require vitamin B_{12} for activity.

Sources

It is highly available in food.

Distribution

It is present in all tissues in small amounts. The total body content of cobalt is about 1.1 mg. The highest concentration occurs in liver, kidneys and bones. Most of the cobalt is present in vitamin B_{12}.

Daily requirements

Its requirement for man is very little. It is, required as vitamin B_{12}. As little as 1 to 2 μg of B_{12}, containing 0.045 to 0.09 μg of cobalt, is sufficient to maintain normal bone marrow function in pernicious anemia.

Absorption

Cobalt is readily absorbed from the small intestine (70 to 80 per cent). Only minute amounts are present in the tissues. Cobalt administered as a soluble salt is poorly absorbed and, therefore, largely eliminated in the feces.

Excretion

About 65 per cent of the amount ingested is excreted in the urine, the remainder in the feces. Injected isotopic cobalt is eliminated rapidly and almost completely by the kidneys into the urine.

Cobalt in ruminant nutrition

Nutritional anemia in cattle and sheep living in cobalt poor soil areas can be treated successfully with cobalt. Microorganisms in the rumens of these animals use cobalt to synthesize vitamin B_{12}.

Cobalt toxicity

Cobalt administered in large amounts to man or animals becomes toxic. It develops a condition known as *polycythemia* (increased number of erythrocytes in blood).

Cobalt deficiency

In ruminants only, cobalt deficiency causes anorexia, fatty liver, macrocytic anemia, wasting and hemosiderosis of spleen.

MANGANESE

1. Manganese is essential for normal bone structure, reproduction, and the normal functioning of the central nervous system.

2. Manganese ions activate glycosyltransferase which is concerned with the synthesis of the mucopolysaccharides of cartilage and also associated with the synthesis of glycoproteins (eg, prothrombin).
3. Pyruvate carboxylase and superoxide dismutase contain tightly bound manganese.
4. *Arginase* is activated by manganese ions.
5. It activates isocitrate dehydrogenase and phosphotransferases.
6. Manganese ions act as cofactor along with *glucose-6-phosphate dehydrogenase*.
7. Manganese ions inhibit lipid peroxidation reaction.

Sources

Rich sources	: Nuts and whole grains.
Good sources	: Vegetables and fruits.
Poor sources	: Meats, poultry, seafoods, and fish.

Distribution

The body of a normal adult (70 kg. weight) contains 12-20 mg. manganese. It is present in all tissues of the body. The kidney and the liver are the main storage organs for manganese. Mitochondria are the principal intracellular sites of manganese uptake.

Daily requirements

In humans, deficiency of manganese is not known. The average dietary intake of 2.5-7.0 mg is quite sufficient.

Manganese in blood

Normal blood contains 4-20 μg./100 ml. In human serum, manganese is bound to a specific β-globulin.

Absorption

Manganese is readily absorbed in the small intestine. Only 3 to 4 per cent of manganese present in the diet is absorbed.

Excretion

95 to 96 per cent dietary manganese is excreted in the feces. Only traces of manganese is excreted in the urine.

Manganese deficiency in animals

1. In manganese deficiency, the animals give birth to young ones which develop ***ataxia***. In more severe deficiency, sterility results. In poultry, egg production and hatchability are decreased, even in mild deficiency of the metal.
2. The livers of manganese-deficient rats contain large amount of fat. This fat accumulation is prevented by manganese or chlorine.
3. Liver ***arginase activity*** and **blood phosphatase activity** are reduced in manganese deficiency.
4. Bone deformities also occur in all animals in its deficiency.

Manganese toxicity

Miners who inhale large amount of manganese suffer from chronic manganese toxicity. There is the development of hepatolenticular degeneration resembling ***Parkinson's disease***.

MOLYBDENUM

Physiological functions

1. Molybdenum is an essential component of xanthine oxidase, aldehyde oxidase, and sulfite oxidase.
2. It is also present in nitrate reductase in plants, and nitrogenase, which functions in nitrogen fixation by microorganisms.
3. Traces of molybdenum are required for the maintenance of normal levels of xanthine oxidase in animal tissues.

Daily requirements

Adequate amounts of molybdenum are present in average diets. Therefore, exact requirement is unknown.

Absorption and excretion

About 50 to 70 per cent of the intake is readily absorbed in the small intestine. Half of the absorbed molybdenum is excreted in urine.

Toxicity

1. Molybednum-rich diet consumption causes severe diarrhea and ill-health in cattle.
2. Rats, on high molybdenum diet, lose body weight with marked anorexia.

SELENIUM

Physiological functions

1. Selenium is essential for normal growth, fertility and for the prevention of a wide variety of diseases in animals, although not known as essential for humans.
2. Glutathione peroxidase, a selenoprotein, catalyzes the peroxidation of glutathione. This enzyme is the protective agent against accumulation of H_2O_2, and organic peroxides within cells.
3. It is involved in immune mechanisms, ubiquinone synthesis, and mitochondrial ATP biosynthesis.
4. Selenoprotein also functions in the reductive deamination of glycine.

Sources

Selenium is largely available in different foodstuffs. The variation depends on the differences in soil selenium content.

Distribution

It is widely distributed in the animal body and highest concentration is present in renal cortex, pancreas, pituitary, and liver.

Daily requirement

Since average diet contains adequate amounts of selenium the requirement of it is not known.

Selenium deficiency

1. Selenium deficiency produces ***necrosis*** of the liver of rats.
2. Calves and lambs suffer from ***muscular dystrophy*** in selenium deficiency.
3. Chicks, on selenium-deficient diet, fail to grow and develop a diseased condition known as ***exudative diathesis***.

Deficiency in humans

A. Keshan Disease

1. This disease is reported from Keshan country of north eastern China.
2. The disease affects mainly children and younger women.
3. It is manifested by an acute or chronic cardiac enlargement, arrhythmia and E.C.G. changes.
4. Prophylaxis with sodium selenite is highly effective.

B. Kaschinbeck Disease

1. The disease is prevalent in several parts of eastern Asia and is characterized by ***degenerative osteoarthrosis*** affecting children between 5 and 13 years of age.
2. It ***retards growth*** by shortening fingers and bones with severe enlargement and dysfunction of the joints.
3. The disease becomes ***endemic*** in low selenium zones.

Relationship of selenium to Vitamin E

Both selenium and vitamin E are essential for curing certain diseases in experimental animals. When animals are given adequate amounts of vitamin E, selenium deficiency causes the following signs and symptoms :

1. Retardation of growth and muscular wasting in *rats*.
2. Retardation of growth and fertility in *chicks*.

These symptoms may be cured by the administration of both selenium and vitamin E because of their close metabolic relationship.

Toxicity

1. Chronic selenium poisoning develops "*alkali disease*". The symptoms of alkali disease are dullness, lack of vitality, roughness of coat, loss of hair from the body and tail, stiffness and lameness, cirrhosis of liver and anemia.
2. Acute selenium poisoning produces in animals salivation, grating of teeth, paralysis and blindness. Death results due to respiratory failure.

CLINICAL ASPECTS

- Kwashiorkor children exhibit decreased blood selenium level and selenium supplementation stimulates growth.
- The diminished levels of selenium are found in patients with *gastrointestinal cancers*, specially oesophageal cancers and oropharyngeal cancers.
- Selenium is recently involved in *cardiovascular disorders*.
- *Thrombosis* is caused by the low intake of selenium.

CHROMIUM

Physiological functions

1. Chromium potentiates the action of *insulin* in accelerating utilization of glucose in animal and humans.
2. It is effective in improving glucose tolerance in some patients suffering fron *diabetes mellitus*.
3. It maintains the normal *cholesterol level* in blood of rats.
4. It regulates the incorporation of certain amino acids in heart muscle in rats.

Sources

It is highly available in dietary foods.

Distribution

The chromium content of adult human body is estimated to be 6 mg. It is widely distributed in tissues.

Chromium in blood

Normal blood contains about 0.009 to 0.055 parts per million.

Requirements

Since average diet meets up the requirements, the exact necessity is unknown.

Absorption and excretion

It is readily absorbed in the small intestine. It is mobilized from the tissues in response to glucose administration.

Chromium is mainly excreted in urine, a small amount is lost in bile and feces.

Deficiency

Its deficiency is characterized by impaired growth, disturbances in glucose, lipid and protein metabolism.

Toxicity

Excessive amounts of chromium produces growth depression, liver and kidney damage in some experimental animals.

LEAD

Lead is not an essential component of our body but it is always present in our body due to extensive use and easy absorption in the body in various forms.

Physiological functions

1. Lead is the inhibitor of enzymes and acts as a toxic substance against enzymes.
2. Prolonged use of lead may lead to impotence or sterility.
3. Lead may not cause any toxic effect when enters the body over a long period even in low dose but may cause harm when its concentration becomes high in different tissues due to its cumulative property.
4. Abortion pills are made up of lead.
5. Chronic poisoning with lead may cause sterility but not impotence.
6. Lead has got local irritating action on the tissues.
7. Chronic lead poisoning causes blue lining in the gum.
8. Lead is extensively used in industries, agriculture, commerce, and for domestic purposes.
9. It is also used in the tin food container, batteries, paints, hair dyes, petrol, glass blowing and cosmetics as vermilion.

Sources

It is available as lead acetate (sugar of lead), lead carbonate (used in painting), lead tetraoxide (used as vermilion), tetra-ethyl lead (used in petrol and gasolene), lead chromate, toxic compounds such as lead sulphide, lead chloride, lead sulphate and lead iodide, etc.

Distribution

In chronic state, it is deposited in tissues, mostly in bones and also in liver and kidneys.

Absorption

1. Most of the lead compounds are soluble in gastric juice and are absorbed through the gastrointestinal tract.
2. Lead dust and fume are well absorbed through the respiratory tract.
3. Tetra-ethyl lead, lead tetraoxide (vermilion), some other dyes and cosmetics are absorbed through the skin.
4. Lead acetate, although more injurious locally, is soluble in water and easily absorbed when swallowed.

Excretion

1. It is mostly excreted through the urine although the excretion rate is very low 0 to 120 μg/24 hours.
2. It is also slightly excreted through bile and to a small extent through nails.
3. Much of the lead excreted through stool may be the unabsorbed ingested lead.

Lead in blood

The normal concentration of lead is 96-106 mEq/L of blood. The blood range between 0.1 mg% - 0.6 mg% usually shows clinical symptoms.

Toxicity

1. Ingestion of lead acetate causes burning pain in the throat, abdominal pain and vomiting. There are cramps in the abdomen, loose motion and the stool becomes dark due to lead sulphide. There may be thirst, dehydration, signs of collapse and death due to circulatory failure.
2. Chronic lead poisoning appears when the level of the accumulated lead exceeds the threshold level. This may occur in the industrial environment due to inhalation of lead dust or lead vapour producing from burning of paints, battery, glass blowing and polishing, enamel factories, dye, cosmetic and colour factories. Chronic lead poisoning in the early stage is manifested by facial pallor, anemia, blue line in the gum, basophilic stippling of red cells, retinal stippling and in the later stage there is constipation, palsy, encephalopathy, disturbance of genito-urinary and cardio-vascular system.

3. In acute poisoning with lead, the victims are usually children who chew substances painted with lead paints.
4. Chronic lead poisoning causes arteriosclerosis with resultant hypertension and hypertensive cardiopathy. There may also be chronic interstitial nephritis and alopecia in this poisoning.
5. Chronic lead poisoning too causes degeneration of anterior horn cells and demyelination leading to peripheral neuritis.

EXERCISE

1. Give an account of the distribution of fluids in the human body. Describe how water metabolism is regulated. (R.U. 70A; Luck. 69S)
2. State the distribution of the water in the body. Mention how extracellular fluid differs from intracellular fluid in composition. Describe the mechanism of exchange of fluids in the body. (P.U. 75S)
3. Describe the electrolyte composition of intracellular and extracellular compartments of body. Describe the importance of potassium ions and give the mechanism of its regulation. (P.U. 76A)
4. Describe the metabolism of calcium in the body. (P.U. 68S, 71A)
5. Describe the sources, requirements and physiological functions of calcium. (Mith. 71A)
6. What part does calcium play in bodily functions? How is the level of calcium in the blood regulated ? (R.U. 71S)
7. Mention the sources of iron in our diet. Describe the mechanism of iron absorption from G.I.T. Mention how iron is transported and stored in our body. (M.U. 75A)
8. State briefly the sources of iron in our diet. Mention the functions of iron in the body. Discuss the metabolism of iron. (R.U. 71S)
9. Write the metabolism of iron in the body. Describe the physiological functions of iron. Give the sources and daily requirements of iron. (M.U. 72A)
10. Discuss how iron is absorbed from the intestine and is utilized in the body. (P.U. 73S)
11. Discuss the metabolism of sodium in the body. (R.U. 72S)
12. State how copper is metabolized in the body. (Mith. 76A)
13. Discuss the metabolism of sulphur in the body. (R.U. 74S)
14. Write notes on :
    (a) Composition of extracellular fluid. (R.U. 66A)
    (b) Functions of fluorine in the body. (Mith. 72S)
    (c) Water intoxication. (Bh. U. 75A)
    (d) Functions of manganese in the body. (Mith. 77S)

CHAPTER 28

INBORN ERRORS OF METABOLISM

Inborn errors of metabolism are a group of metabolic 'diseases' which can be formed as a result of the lack of a single enzyme in a single metabolic pathway. These disorders are hereditary and tend to occur only in closely-related family groups.

Sir Archibald Garrod (1902) first identified only a small number of such metabolic disorders. A large number of such conditions which can be termed inborn errors have been recognized with the increase in knowledge of metabolic pathways and the improvement of more sensitive and searching methods of biochemical investigation.

HEREDITARY ANEMIA

Sickle cell anemia and thalassemia. See page 131, 129 (in hemoglobin chapter). Wilson's disease. See page 519.

CARBOHYDRATE METABOLISM

Galactosemia. See page 312.

Glycogen storage diseases (Glycogenoses). See page 313.

Pentosuria. See page 315.

Fructosuria. See page 315.

Essential pentosuria See page 310.

Hereditary fructose intolerance. See page 311.

Hurler's syndrome (Mucopolysaccharidosis-I)

This is a hereditary disorder in which acid mucopolysaccharides are excessively deposited in tissues. This disorder is due to a defect in catabolism rather than to excessive synthesis. The most important symptoms of this disorder are the followings:

1. Skeletal deformities and severe mental retardation.
2. Deafness of a conductive type and corneal opacity.
3. The voice is harsh and behaviour is apelike.
4. Dermatan sulfate and heparan sulfate are present in urine and tissues.
5. Dilatation of the ventricles.
6. Cardiovascular findings are prominent.

Hunter's syndrome (Mucopolysaccharidosis-II)

This is also a hereditary disorder in which acid mucopolysaccharides are excessively deposited in tissues. This disorder is also due to a defect in catabolism rather than to excessive synthesis. The most important symptoms are :

1. Severe skeletal deformities and moderate mental deterioration.
2. Deafness occurs early but there is no corneal opacity.

3. Dermatan sulfate and heparin sulfate are present in urine and tissues.
4. Dwarfing and abnormal facies are present.
5. Cardiovascular changes, obstruction and restriction of chest movement.
6. Death frequently occurs from cardiac troubles at an early age.

LIPID METABOLISM

Gaucher's disease

In this disease, the cerebroside content of the reticuloendothelial cells (eg, spleen) is very high. In the cerebroside molecule, the kerasin is characterized by glucose replacing galactose. The concentrations of cerebrosides are much higher in medullated than nonmedullated nerve fibers. The symptoms of this disease are mentioned below. The disease is caused by the deficiency of glucosyl ceremide hydrolase or glucocerebrosidase.

1. The spleen is significantly increased and there are signs of leukopenia and thrombocytopenia.
2. The liver is enlarged and the marrow cavity is widened.
3. Eyes show a yellow-brown wedge-shaped elevation.
4. In early infantile form, the disease starts early and death occurs before attaining the age of 2 years. Neurological symptoms are usually present.

Niemann-Pick disease

In this disease, excessive amounts of sphingomyelin are deposited in spleen, brain and liver. This results in the deficiency of sphingomyelinase. The hereditary disease is found in infancy and death occurs within the first two years of life. The clinical findings are :

1. Enlarged liver and spleen.
2. Mental retardation and fatal in early life.
3. Anemia and leukocytosis.
4. The nervous system is affected.
5. A cherry red spot may be seen over the retina.
6. Cholesterol deposits in the tissues are increased.

Tay-Sach's disease

This disease is characterized by the increased accumulation of Gangliosides (GM_2) in brain and spleen. The characteristic clinical symptoms are:

1. Mental retardation, blindness and muscular weakness.
2. A cherry red spot appears in the muscular region of the eye within the first year of life.
3. At about two years of age, the circumference of the head becomes 50 per cent greater than normal.
4. There is repeated respiratory tract infections of the patient and the patient expires at the third or fourth year.

Fabry's disease

In this disease, large amounts of ceramide trihexoside are accumulated in the kidney. The deficiency of the enzyme ceramide trihexosidase causes this disease. The clinical findings are:

1. Skin rash and kidney failure.
2. Male patients generally die due to progressive renal failure in the fourth or fifth decade of life.
3. Cardiac enlargement and edema of the extremities.
4. Some patients suffer from excessive pain in the joints.
5. Corneal opacities and vascular dilatation are frequent.

Refsum's disease

This disease occurs due to the accumulation of large amounts of phytanic acid (3, 7, 11, 15-tetramethyl hexadecanoic acid). The deficiency of the enzyme

phytanic acid oxidase causes the disease. The clinical symptoms are :

1. The early symptoms are signs of chronic polyneuropathy with distal muscular atrophy.
2. Severe pain in the knees.
3. The deep tendon reflexes are weak or absent.
4. Night blindness and narrowing of the visual fields.
5. Deafness and anosmia.
6. Cardiac involvement may lead to tachycardia.
7. The cerebrospinal fluid protein is always increased while the cell count is normal.

Krabbe's disease

This disease results in the deficiency of galactocerebrosidase which catalyzes the hydrolysis of galactocerebroside to form ceramide and galactose. Galactocerebroside is the important component of myelin.

The clinical manifestation of this disorder is severe mental retardation in infant. There is nearly total absence of myelin in the central nervous system which is replaced by gliosis and '**globoid bodies**' appear in the white matter. Diagnosis of the patients depends on the determination of the galactocerebrosidase activity in leukocytes.

PROTEIN METABOLISM

Albinism

1. This condition appears in the total absence of tyrosinase inside the melanocytes in the skin.
2. The black pigment melanin is not formed in the skin, eyes and hair.
3. This inherited condition occurs to a greater or less extent in all types of organism.
4. The diagnostic advice is for the prevention of exposure of sunlight and protection of the eyes by wearing dark glasses.

Tyrosinosis

1. This syndrome is due to the absence either of hepatic **p-hydroxyphenylpyruvate hydroxylase** or of **tyrosine transaminase** activities.
2. The patient excretes large quantities of tyrosine in the urine.

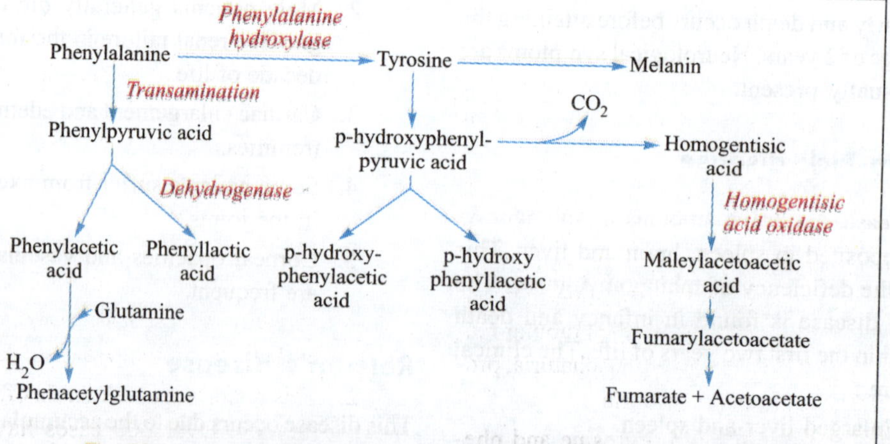

Fig. 28.1 : Overall representation of tyrosinosis, albinism, tyrosinemia, phenyl ketonuria and alkaptonuria

3. Diet rich in tyrosine causes the excretion of other p-hydroxyphenyl acids including 3, 4-dihydroxyphenylalanine (dopa) and p-hydroxyphenyl lactic acid.

Tyrosinemia

A. Neonatal tyrosinemia :

1. Neonatal tyrosinemia occurs in the new born.
2. Tyrosine, p-hydroxphenylpyruvic acid, p-hydroxyphenyl lactic acid and p-hydroxyphenyl acetic acid appear in the urine.
3. A transient deficiency of p-hydroxyphenylpyruvic acid oxidase causes this condition and the condition lasts for a few weeks.
4. Administration of ascorbic acid and reduction of protein intake brings about the normal condition.

B. Hereditary tyrosinemia :

1. This disorder is similar to neonatal tyrosinemia, but the amount of p-hydroxyphenyl lactic acid in the urine is greater.
2. This disorder is not controlled by the administration of ascorbic acid.
3. It is due to the inherited deficiency of **p-hydroxyphenyl pyruvic acid oxidase**.
4. Liver failure and death can occur six months after birth and the infant has a characteristic odour.
5. Other patients develop the clinical findings including hepatosplenomegaly, a nodular cirrhosis of the liver, abnormalities of tyrosine and methionine metabolism, multiple defects in renal tubular reabsorption, rickets, hyperphosphaturia, proteinuria and amino aciduria.
6. Diet containing low in tyrosine and phenylalanine improves renal function and also retards degenerative liver changes.

Phenylketonuria

1. This inherited disorder appears in the absence of phenylalanine hydroxylase which is responsible for the conversion of phenylalanine to tyrosine. As a result, alternative catabolites of phenylalanine are produced; these include phenylpyruvic acid, deamination product of phenylalanine ; phenyl acetic acid, the decarboxylation and oxidation product of phenylpyruvic acid. Much of the phenylacetylglutamine is excreted in the urine.
2. Mental retardation develops among infants and children.
3. Patients with phenylketonuria tend to have a deficiency of serotonin. This may be connected with the defect of myelin synthesis.
4. The accumulation of phenylalanine also impairs melanin synthesis and children with this defect tend to have fair skin and fair hair.
5. Excess of phenylalanine in the blood leads to excretion of the amino acid into the intestine. Here, it competes with the tryptophan for absorption and tryptophan is subjected to the action of intestinal bacteria.
6. Early diagnosis (shortly after birth) and extreme restriction of phenylalanine intake is effective in preventing this disorder.

Alkaptonuria

1. This condition is characterized by the excretion of homogentisic acid (dihydroxyphenyl acetic acid) in the urine owing to the lack of **homogentisic acid oxidase**.
2. This abnormal condition is often found in infancy. Over 600 cases have been reported. The incidence of alkaptonuria is 2-5 per million live births.

3. The most clinical manifestation is the dark urine due to the oxidation of homogentisic acid in air.
4. In this disorder, several grams of homogentisic acid are excreted daily.
5. This condition is present at birth and persists throughout life.
6. In later life, accumulation of dark pigment in cartilages and tendons gives rise to the condition known as ochronosis, which is accompanied by arthritic changes.

Maple syrup urine disease

1. This syndrome is characterized by the absence of the enzymes required for the oxidative decarboxylation of the keto acids derived from the branched chain amino acids–valine, leucine and isoleucine. As a result, these keto acids are accumulated in the blood and excreted in urine. The urinary excretion of these keto acids produce an odour like that of maple syrup or of burnt sugar.
2. This familial disorder is recognized by central nervous system manifestations of convulsions and attacks of flaccidity and apnea.
3. The patients may be well treated with amino acid mixtures low in the above three amino acids.
4. The infant is difficult to feed and may vomit. The patient suffers from a significant degree of lethargy.
5. Death may occur by the end of the first year of life without treatment.

Hartnup's disease

1. It is a hereditary disease characterized by a pellagra-like rash and mental deterioration in the abnormal metabolism of tryptophan.
2. The urine of the patients contain significantly increased amounts of indole acetic acid as well as tryptophan.
3. The urinary excretion comes to normal after administration of broad spectrum antibiotics.

Glycinuria

1. This condition is associated with the excess urinary excretion of glycine with a tendency to form oxalate renal stones in spite of the normal amount of excretion of oxalate in the urine.
2. The plasma content of glycine is normal in glycinuric patients while the urinary excretion of glycine ranges form 600-1,000 mg./d.
3. It is supposed that glycinuria is due to a defect in renal tubular transport of glycine and the decreased reabsorption of glycine by the renal tubule permits the amino acid to pass into the urine in high concentration.

Primary Hyperoxaluria

1. This is an inherited disorder characterized by increased urinary excretion of oxalate which is not related to dietary intake of oxalate.
2. The deficiency of the enzyme glycine transaminase together with some impairment of glycine oxidation to formate is the cause of excess oxalate in the form of glyoxalate formed from glycine.
3. The clinical features exhibit progressive bilateral calcium oxalate urolithiasis, nephrocalcinosis and recurrent infection of the urinary tract which may be due to renal failure or hypertension.
4. Death occurs in childhood or early adult life from renal failure or hypertension.

Cystinuria

1. It is an inherited metabolic disease in which lysine, arginine, ornithine and cystine are excreted in the urine in large amounts.

2. Cystinuria is due to renal transport defect.
3. Cystinuria is a misnomer, so that cystine-lysinuria may be preferred as the descriptive term for this disease.
4. Cystine is an insoluble amino acid which may precipitate in the kidney tubules to form cystine calculi in cystinuric patients. This is a major complication of this disease.
5. The mixed disulfide of L-cysteine and L-homocysteine present in the urine of cystinuric patients is more soluble than cysteine and hence reduces formation of cystine crystals and calculi.

Cystinosis (Cystine Storage Disease)

1. It is a rare lysosomal disorder characterized by defective carrier-mediated transport of cystine resulting cystine crystals.
2. These cystine crystals are accumulated in liver, spleen, bone-marrow, peripheral leukocytes, lymph nodes, kidney and cornea. These crystals are particularly accumulated in reticuloendothelial system.
3. Cystinosis is usually accompanied by aminoaciduria, glycosuria, polyuria, chronic acidosis leading to uremia and death.
4. This disorder may be due to the impaired conversion of cystine to cysteine in the involved tissues as a result of the deficiency of the enzyme cystine reductase.
5. The renal functions are seriously impaired for which patients die in the young stage from acute failure.
6. In adults, cystine gets deposited in cornea.
7. Cystine crystals can be detected easily in cornea by slit lamp microscopy and confirmed by chemical determination of cystine crystals of peripheral leukocytes or cultured fibroblasts.

Homocystinuria

1. In this abnormal condition, homocystine (300 mg/d) together with S-adenosylmethionine in some cases is excreted in the urine and plasma methionine levels are elevated.
2. The clinical findings of this disease are the occurrence of thrombosis, osteoporosis, dislocated lenses in the eyes and frequently mental retardation.
3. This condition appears due to the lack of cystathionine synthetase in the liver due to which both homocystine and methionine are accumulated in blood and urine.
4. A low methionine and a high cystine diet effectively prevent this condition if treated earlier.

Histidinemia

1. It is an inherited disorder of histidine metabolism in which the amounts of histidine in the blood and urine are increased. There is also increased excretion of imidazole pyruvic acid.
2. The metabolic block of histidine is due to the insufficient activity of liver histidase which impairs the conversion of histidine to urocanic acid.
3. Development of speech in this condition is retarded. Mental development is also retarded.
4. Histidine excretion is also increased during normal pregnancy but not in the toxaemia of pregnancy. This increase is not due to metabolic defect.
5. These histidinemic patients are treated well with a diet containing protein hydrolysate free from histidine instead of intact protein.

Hypervalinemia

1. In this abnormal condition, the infants suffer from stunted growth, muscle wasting and vomiting.

2. The valine content of blood and urine is very high.
3. A diet containing protein hydrolysate low in valine prevents this condition effectively.

NUCLEIC ACID METABOLISM

Lesch-Nyhan syndrome

1. This condition is characterized by the complete deficiency of phosphoribosyl transferase, which causes hypoxanthine or guanine to form a nucleotide with PRPP (5′- phosphoribosyl - 1′- pyrophosphate). These purines are thus available for the formation of uric acid.
2. This disorder is X-linked in its inheritance.
3. This appears in childhood as a severe neurological syndrome, which is sometimes accompanied by gout. The urinary uric acid amount is five to six times the normal.
4. Hypothyroidism, hypo- and hyperparathyroidism are accompanied by hyperuricemia.
5. Hypertension is accompanied by increased plasma uric acid. Such patients show an increased tendency to myocardial infarction, which is also a cause of hyperuricemia.
6. It can be prevented or diminished by the administration of allopurinol, an analogue of hypoxanthine. Allopurinol inhibits xanthine oxidase due to which uric acid cannot be formed.

Hereditary xanthinuria

1. In this rare genetic disorder, there is the deficiency of xanthine oxidase which leads to the diminished level of blood uric acid (1 mg/100 ml or less).
2. The urinary excretion contains large amounts of xanthine with lesser amounts of hypoxanthine.
3. Urinary calculi composed of xanthine may be produced.

Orotic aciduria

1. This is an inherited disorder which causes the excessive production of orotic acid. This occurs by the deficiency of orotate phosphoribosyl transferase.
2. The urinary excertion consists of large amounts of pyrimidine nucleotide precursor.
3. The urine becomes cloudy on cooling with the deposition of needle-shaped crystals of orotic acid.
4. Children affected by this condition develop a severe megaloblastic anemia with physical and mental retardation.
5. Administration of uridine improves this condition significantly.

Hypouricemia

1. This is due to a genetic defect or to severe liver damage.
2. The deficiency of the enzyme xanthine oxidase causes hypouricemia and increased excretion of hypoxanthine and xanthine.
3. Patients may exhibit Xanthinuria and Xanthine lithiasis in severe xanthine oxidase deficiency.

Von Gierke's Disease

1. This disease is associated with the purine over-production and hyperuricemia.
2. There is also enhanced generation of the PRPP.
3. The associated lactic acidosis elevates the renal threshold for urate, increasing total body urates. (The rest is already mentioned in Glycogen storage disease. Add this to it).

Reye's Syndrome

1. The orotic aciduria accompanies this syndrome.
2. This syndrome exhibits inability of severely damaged mitochondria to use carbamoyl phosphate which is available for cystosotic over- production of orotic acid.

EXERCISE

Write notes on :

1. Alkaptonuria. (R.U. 72S; P. U. 71S; Mith. 62S)
2. Phenylketonuria. (Muz. 75A; Mith. 61A; R. U. 70A)
3. Niemann-Pick disease. (R. U. 70A; Mith 75S)
4. Hunter's syndrome. (M. U. 76S)
5. Glycogen storage disease. (P. U. 72A; R. U. 65S)
6. Fructosuria. (M. U. 74S; Mith. 71A)
7. Galactosemia. (R. U. 64A)
8. Gaucher's disease. (P. U. 68A; R. U. 64S)
9. Refsum's disease. (Mith. 72S)
10. Tyrosinosis. (R. U. 72A; Bh. U. 74S)
11. Maple syrup urine disease. (P. U. 69S; R. U. 73S)
12. Homocystinuria. (Bh. U. 73S)
13. Lesch-Nyhan syndrome. (P. U. 74S)
14. Glycinuria. (M. U. 76A)

CHAPTER 29

DETOXIFICATION

Xenobiotics are compounds that are foreign to the body. The principal classes are drugs and carcinogen.

The metabolism of xenobiotics has two phases :

1. The major reaction involved is hydroxylation catalyzed by monoxygenase.
2. The hydroxylated compounds which are produced from phase I is converted by specific enzymes to various polar metabolites by conjugation with glucuronic acid, sulphate or by methylation.

Detoxification occurring in the body converts toxic substances introduced into or formed in the body into less toxic or non-toxic substances which are easily excreted out by the excretory routes.

The substances introduced into the body are the drugs used for treatment of diseases and the chemicals used for diagnostic purposes. The toxic substances are formed in the body by the bacterial action in the large intestine eg, indole, skatole, phenol, histamine, etc. The other toxic substance bilirubin is formed in the body by the breakdown of hemoglobin.

Liver is the principal organ for the detoxification processes to take place. The detoxification processes can be grouped as follows :

1. Oxidation.
2. Reduction,
3. Hydrolysis,
4. Conjugation.

1. Oxidation

Alcohols, aldehydes, amines and amine derivatives, hydrocarbons and sulphur compounds are detoxified by the process of oxidation.

Alcohols : Alcohols are oxidized to the corresponding acids via aldehydes :

Methyl alcohol ⟶ Formic acid

Ethyl alcohol ⟶ Acetic acid.

Benzyl alcohol ⟶ Benzoic acid.

Aldehydes : Aldehydes are oxidized to the corresponding acids :

Benzaldehyde ⟶ Benzoic acid

Chloral ⟶ Trichloracetic acid.

Amines and amine derivatives : Aliphatic amines are oxidized to the corresponding acids:

Aliphatic amine ⟶ Aliphatic acid + Urea.

Aromatic amines are oxidized to p-Phenols.

Aniline ⟶ p-amino phenol.

Acetanilide ⟶ p-Acetylaminophenol.

Hydrocarbons : Aromatic hydrocarbons are oxidized to phenols :

Benzene ⟶ Phenols.

Sulfur compounds : Sulfur compounds are oxidized to sulfuric acid :

Organic sulfur ⟶ Sulfuric acid.

2. Reduction

Less commonly reduction takes place in the body. Nitro or aldehyde groups are reduced.

Chloral ⟶ Trichlorethylalcohol

p-Nitrobenzaladehyde ⟶ p-Aminobenzaldehyde.

Picric acid ⟶ Picramic acid .

3. Hydrolysis

The therapeutic acitvity of some therapeutic agents becomes less in the body due to their hydrolysis. Some examples are mentioned below :

Atropine ⟶ Tropic acid + Tropine.

Digitalis ⟶ Aglycones + Sugars.

Aspirin ⟶ Salicylic acid + Acetic acid.

Procain ⟶ p-Aminobenzoic acid + Diethylaminoethanol.

4. Conjugation

It is the process in which the chemical to be detoxified combines with another chemical formed or supplied by the body. Eight types of chemical substances are used by the body for conjugation reactions. These are : glucuronic acid, sulphuric acid, glycine, glutamine, cysteine, acetic acid, active methyl groups and thiosulphate.

Glucuronic acid : This important conjugating agent is formed in the body from glucose. It produces the following products by conjugation :

Phenol + Glucuronic acid ⟶ Phenyl glucuronide.

Bilirubin + Glucuronic acid ⟶ Bilirubin diglucuronide.

Benzoic acid + Glucuronic acid ⟶ Benzoyl glucuronide.

Sulphuric acid : It detoxifies phenol, indoxyl, etc.

Phenol + Sulphuric acid ⟶ Phenol sulphate.

Indoxyl + Sulphuric acid ⟶ Indoxyl sulphate.

Glycine : The amino group of glycine reacts with the carboxyl goup of the compound to be detoxified.

Benzoic acid + Glycine ⟶ Hippuric acid.

Nicotinic acid + Glycine ⟶ Nicotinuric acid

Cholic acid + Glycine ⟶ Glycocholic acid.

Glutamine :

Phenylacetic acid + Glutamine ⟶ Phenylacetyl glutamic acid.

Dogs use glycine for detoxifying phenylacetic acid.

Cysteine :

Bromobenzene + Cysteine ⟶ p-Bromophenyl mercapturic acid.

Acetic acid :

Sulphanilamide + Acetic acid ⟶ Acetyl sulphanilamide

Active methyl groups :

Pyridine + Active methionine ⟶ n-Methyl pyridine.

Nicotinic acid + Active methionine ⟶ N-methyl nicotinic acid.

Thiosulphate : Inorganic cyanides are detoxified by thiosulphates converting the highly toxic cyanides into less toxic thiocyanate.

Cyanide + Sodium thiosulphate ⟶ Thiocyanate + Sodium sulphate.

EXERCISE

1. Discuss the detoxification mechanism of the body. (P.U. 72A)
2. Write Short note on :
   Detoxification. (Punj. 73A)

CHAPTER 30

IMMUNOCHEMISTRY

The lymphocyte is the primary cell for immune system. The T lymphocytes (T cells) are thymus derived. These are responsible for cellular immunity (i.e. delayed skin reactivity, antitumor immunity, cellular defense against fungi). The B lymphocytes (B cells) are derived from bone marrow in mammals. These are responsible for humoral immunity which is expressed by the production of specific circulating plasma proteins termed antibodies or immunoglobulins. B lymphocytes are mainly discussed in this chapter.

STRUCTURE OF IMMUNOGLOBULINS

1. The basic unit of all immunoglobulin molecules consists of 4 polypeptide chains linked by disulfide bonds shown in Fig. 30.1
2. There are two identical heavy (H) chains (MW 53,000-75,000) and two identical light (L) chains (MW 23,000).
3. Immunoglobulins composed of more than one basic monomeric unit are termed polymers.

   Important examples are IgA dimers (2 units), IgA trimers (3 units) and IgM pentamers (5 units).
4. The polypeptide chains are not straight sequences of amino acids but are folded 3-dimensionally with disulphide bonds to form areas called domains.
5. The part of the antibody molecule which combines with antigens is formed by a few amino acids in the F region of H and L chains.

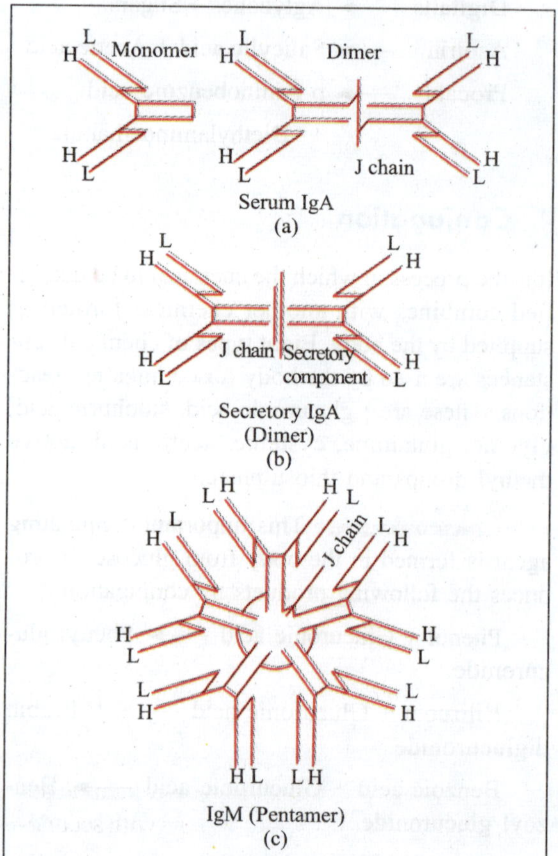

Fig. 30.1 : Polypeptide chains are represented by thick lines; disulfide bond linkings different polypeptide chains are represented by thin lines.

6. Papain cleaves IgG in the presence of cysteine into 3 fragments :

   2 Fab fragments (MW 52,000) and an FC fragment (MW 48,000). The Fab fragments bear the entire antibody combining site for the antigen.
7. There are two major types of L chains in man, the Kappa (κ) and lambda (λ) chains. About 70 per cent of the human immunoglobulin molecules carry κ light chains and 30 per cent carry λ light chains. In IgG, the H chain is termed a gamma (γ) chain; in IgA, an alpha (α) chain; in IgM, a mu (μ) chain; in IgD, a delta (δ) chain; and in IgE, an epsilon (ε) chain.
8. Carbohydrate residues are attached to the polypeptide chains. 2 carbohydrate units per molecule of IgG and 3 per molecule of IgM. The carbohydrate residues include D-mannose, D-galactose, L-fructose, D-acetylneuraminic acid and glucosamine.
9. Each chain is divided into specific domains or regions that have structural and functional significance. The half of the light chain (L) towards the carboxy terminus is termed as the constant region (CL), while the amino terminal half is the variable region (VL) of the light chain. About one quarter of the heavy (H) chain at the amino terminus is termed as variable region (VH), and the other three quarters of the heavy chain are referred to as the constant regions (CH^1, CH^2, CH^3) of that H chain.
10. The portion of the immunoglobulin molecule which binds the specific antigen is formed by the aminoterminal portions (variable regions) of both the H and L chains, i.e., the VH and VL domains. The domains of the Protein Chains do not simply exist as linear sequences of amino acids but form globular regions with secondary and tertiary structure.

Classification of immunoglobulins

On the basis of electrophoretic, immunologic and ultracentrifugal studies, the immunoglobulins have been divided into 5 groups :

1. IgG

(a) This is the major antibody-containing fraction which comprises 80 per cent of the gamma globulins.

(b) It is a single basic immunoglobulin unit with γ heavy chains.

(c) Its molecular weight is 150,000-160,000 and it contains 2-4 per cent carbohydrates.

(d) It has the slowest electrophoretic mobility and is distributed in the extracellular fluid and is capable of crossing the placenta.

(e) Its normal level in blood is 570-1,920 mg/dl

2. IgA

(a) It has a molecular weight of about 180,000-400,000 and its S rate is 6.6-13.

(b) It has a higher content of carbohydrate (5-10 per cent).

(c) It is present in high concentrations in the blood, in saliva and tears and in the secretions of gastrointestinal tract.

(d) It is single basic immunoglobulin unit with heavy α chains. Secretory IgA is made up of 2 basic units connected by a J chain.

(e) A 60,000 MW molecule called ***transport piece*** (t piece) is attached to the Fc portion. This is necessary for the transport of IgA molecules into the lumens of exocrine glands. Secretory IgA plays an important role in host defense mechanisms against viral and bacterial infections. IgA does not cross the placenta.

(f) Its normal level in blood is 70-400 mg/dl.

3. IgM

(a) It contains 576 amino acids and has a mass of 950,000 daltons.

(b) It is the first antibody to be formed in a new born animal or human.

(c) IgM (with IgD) is the major immunoglobulin expressed on the surface of B cells.

(d) It's carbohydrate content is 10-12 per cent. It is dissociated into subunits designated IgMs. Each monomer is composed of two L chains and two H chains (μ) with 2 combining sites, so that the intact molecule has 10 combining sites.

(e) IgM does not cross the placenta.

(f) The basic units of it are connected by disulfide bond bridges and small polypeptide J chain.

(g) Its normal level in blood in males is 30-250 mg/dl.

Its normal level in blood in females is 30-300 mg/dl.

4. IgD

(a) No antibody acitivity is associated with IgD.

(b) Its normal level in blood is 0-8 mg/dl.

5. IgE

(a) It is present in the serum in very low concentrations as a single basic unit with heavy E chains.

(b) Its molecular weight is about 190,000(8S). Half of patients with allergic diseases have increased serum IgE levels.

(c) The specific interaction between antigen and IgE bound to the surface of mast cells results in the release of inflammatory mast cell products such as serotonin and histamine.

(d) Its normal level in blood is < 0.025 mg/dl.

ELECTROPHORETIC DETERMINATION OF IMMUNOGLOBULINS

In 1937, Tiselius performed the separation of proteins in electrical fields. Owing to the relative complexity of his method, zone electrophoresis has replaced electrophoresis in a free solution.

Zone electrophoresis

1. Serum or other biologic fluid samples are placed on the cellulose acetate and separated by electrophoresis for 90 minutes using alkaline buffer solutions.
2. The strips are then stained and scanned.
3. This procedure separates normal serum proteins into 5 major electrophoretic bands (albumin, α_1-globulin, α_2-globulin, β-globulin, and γ-globulin).
4. This electrophoresis is very useful for the diagnosis of human paraprotein disorders such as multiple myeloma and hypogammaglobulinemia. In hypogammaglobulinemia, the decrease in serum γ-globulin is easily detected.

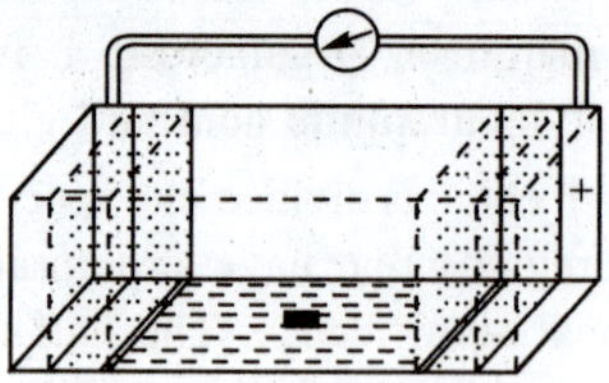

Fig. 30.2a : Electrophoresis of sample in electrolyte buffer is performed

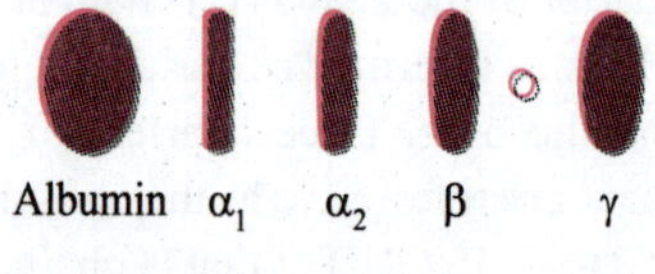

Fig. 30.2b : Separated protein bands are seen after being stained

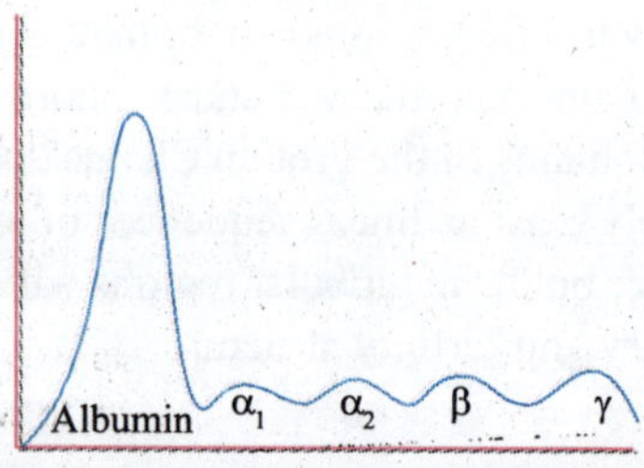

Fig. 30.2c : Graphical representation of albumin, α_1-globulin, α_2-globulins, β-globulins and γ-globulins

Immunoelectrophoresis (IEP)

1. It consists of both electrophoretic separation and immune precipitation of proteins.
2. A glass slide covered with buffered molten agar (pH 8.2) is required to perform the test.

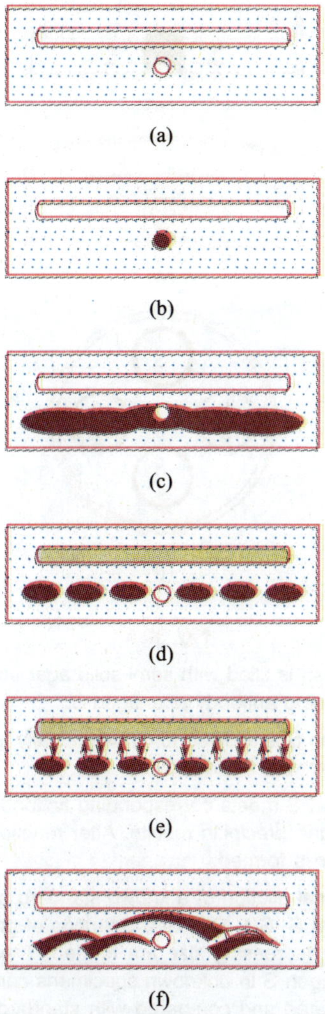

Fig. 30.3 : Immuno-globulin determination
(a) Semi-solid agar poured into glass slide and antigen well and antiserum trough cut out of agar.
(b) Antigen well filled with human serum.
(c) Serum separated by electrophoresis.
(d) Antiserum trough filled with antiserum to whole human serum.
(e) Serum and antiserum diffuse into agar.
(f) Precipitin lines form for individual serum proteins.

3. An antigen (serum) well and an antibody trough are cut in the agar after the agar is cooled.
4. The serum sample is placed in the well and the various proteins are then separated in an electrical field.
5. Antiserum is next placed in the trough and allowed to diffuse for 18-24 hours in the direction of the separated proteins.
6. When the various protein fractions meet the migrating antibody, precipitin lines are formed. These lines are stained or photographed to make a permanent record.
7. The absence of immunoglobulin classes or the presence of abnormal immunoglobulin molecules are detected by this process even in very low concentrations that are not apparent by zone electrophoresis.

QUANTITATIVE DETERMINATION OF IMMUNOGLOBULINS (Fig. 30.3)

1. The radial diffusion technique is utilized for the quantitative determination of immunoglobulins.
2. Wells are cut in an agar plate impregnated with a specific antiserum directed against a single human immunoglobulin class.
3. A circular precipitin ring will form after the human serum proteins placed in the well diffuse through the agar.
4. The diameter of the precipitin ring is proportionate to the concentration of serum immunoglobulin.
5. The level is determined by comparing the diameter of the unknown serum to that of a standard containing known levels of immunoglobulins.
6. This test does not differentiate between normal and abnormal immunoglobulin molecules as does the immunoelectrophoresis procedure.

7. The normal concentrations of the 3 major immunoglobulin classes are as follows :
   IgG : 570-1920 mg/dl.
   IgA : 70-400 mg/dl.
   IgM : Males : 30-250 mg/dl
   Females : 30-300 mg/dl

ANTIGENS

Definition

Substances which can give an immune response when introduced into an animal are called immunogens or antigens.

Although most antigens are macromolecular proteins, polysaccharides, synthetic polypeptides and other synthetic polymers may also be immunogenic. Although the characteristics of antigens are complex, certain conditions must be satisfied in order that a molecule be immunogenic, as follows :

1. The molecules that are foreign to the host are immunogenic.
2. Molecules smaller than MW 10,000 are only weakly immunogenic. Macromolecular proteins with molecular weights greater than 100,000 are the most potent immunogens.
3. Immunogenicity increases with structural complexity. Aromatic amino acids are more immunogenic than nonaromatic amino acids.
4. The ability to respond to a particular antigen varies with the genetic constitution of the animal.

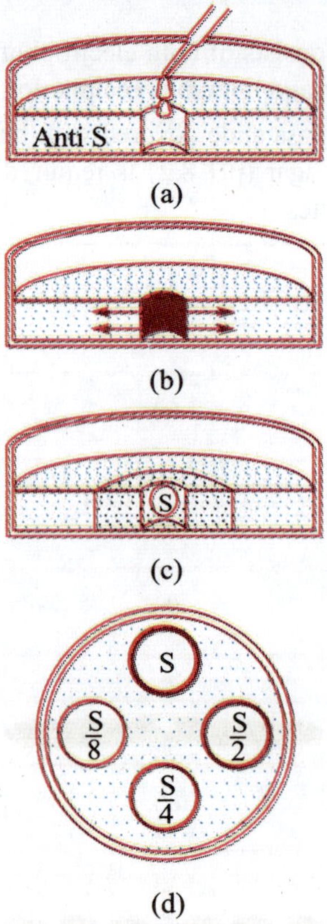

Fig. 30.4 :

(a) Petridish is filled with semi-solid agar solution containing antibody to antigen S.

(b) Antigen S is allowed to diffuse radially from the counter well 24-48 hours.

(c) Antigen S meets corresponding antibody to S in the agar, precipitin results. After reaction, a sharp border is formed.

(d) By serial dilution of a known standard quantity of antigen S- S/1, S/2, S/4, S/8-rings of progressively decreasing size are formed. The amount of antigen S in unknown specimens can be calculated and compared with standard.

ANTIGENIC DETERMINANTS

The initiation of immunoglobulin production requires binding of the antigen to the lymphocyte surface. The combining sites on the surface of the **lymphocytes** are antibody-like molecules called **antigen receptors**.

The portions of antigenic molecules which are involved in actual binding with antibody combining sites are termed **antigenic determinants**.

Haptens

1. Karl Landsteiner prepared haptens by covalently coupling diazonium deriva-

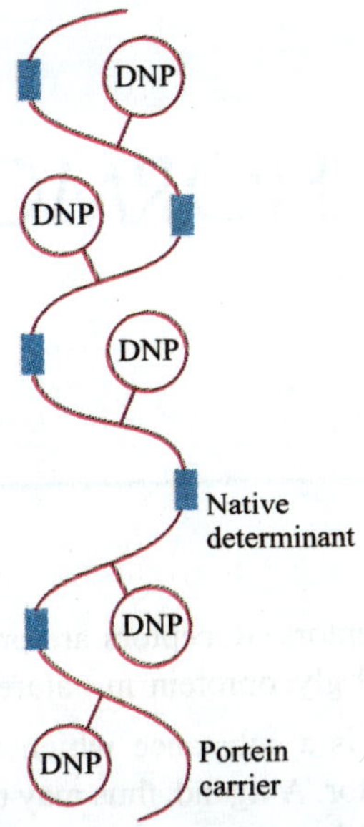

Fig. 30.5 : Hapten-protein conjugate. The conjugated dinitrophenyl (DNP) hapten introduces new antigenic determinants. The protein has several native or integral determinants denoted by thickened areas

tives of aromatic amines to lysine, tyrosine and histidine residues of immunogenic proteins. These proteins are called carriers.

2. These are small, chemically defined substances which, although not immunogenic, react with antibodies.
3. The protein-hapten conjugates form antihapten antibody. The conjugated haptens thus behave as the complete antigenic determinant of the molecules.

Size of antigenic determinants

1. Antigenic determinants and antibody combining sites are similar to a lock and key arrangement. The binding affinity between the antigen and antibody site is directly proportionate to the closeness of fit.
2. The antigenic determinant size was performed using single sugar (glucose), polysaccharides (dextran). These single-chain polysaccharides with few branch points were used to produce antibodies.
3. The hexasaccharide was the best inhibitor of the dextran-antidextran precipitant reaction.

IMMUNOPOTENCY

The ability of the region of the antigen molecule to act as an antigenic determinant and to induce the formation of specific antibodies is called immunopotency. The following factors influence immunopotency :

1. Exposure to the aqueous environment is important in immunopotency.
2. Charge residues contribute significantly to the specificity of antigens, because charged groups are hydrophilic and are in close contact with the environment.
3. Genetic factors play an important role in the ability of different animals to produce antibodies of different specificities against the same antigen. The dominant component of the antigenic determinant is termed immunodominant.

EXERCISE

1. Classify immunoglobulins. (M.U. '84A)
2. Write the structure of immunoglobulins. (C. U. '88A)
3. Write short notes on :
   (a) Antigens. (C.U. '87A)
   (b) IgM. (C.U. '88S)
   (c) Immunopotency. (P. U. '89A)

CHAPTER 31

HORMONES

Definition

The chemical substance or messenger is produced from one part of body by endocrine gland, enters into the circulation, carried to distal target organ or cell to modify their structure and function is called hormone.

General characteristics of hormones

The endocrine or ductless glands secrete hormones which catalyze and control metabolic processes. The hormones differ from enzymes in the following ways :

1. The hormones are produced in an organ in which they ultimately perform their function.
2. They are secreted into the blood prior to use.
3. Structurally, they are not always proteins.

The action of a hormone at a large organ is regulated by 5 factors :

(a) Rate of synthesis and secretion of the stored hormone from the endocrine gland.

(b) Specified transport systems in the plasma.

(c) Hormone-specific receptors in target cell cytosol or membranes.

(e) Ultimate degradation of the hormone, usually by the liver or kidneys.

Receptor : Chemical signals like hormone molecules reach the target cells where they become bound by receptors. Receptors are protein macromolecules and glycoprotein in nature.

A ligand is a substance which can combine with the receptor. A ligand, thus may be a hormone or drug molecule.

Agonist : Each type of receptor is specific for corresponding type of ligand. So the post receptor event will occur and this type of ligand is called agonist.

Antagonist : They are the blocks which have close chemical structure with the agonist molecule, so the antagonist molecule can and do occupy the receptor site but there is some structural difference also so that post receptor events do not occur and because of the fact that the receptor sites are now occupied, the agonist molecules cannot occupy the P sites.

Receptors which bind the peptide hormones are present on the cell membrane, that is they partly protrude in the ECF and partly remain within the membrane, such receptors are called *membrane bound receptors*.

Receptors for thyroid and steroid hormones are called *cytosol receptors* because they occur either in the cytosol or on the nucleus.

The receptor has two domains :

1. Recognition domain—binds hormone.
2. Coupling domain—generates signal.

Based on their chemical structures, the hormones have been classified.

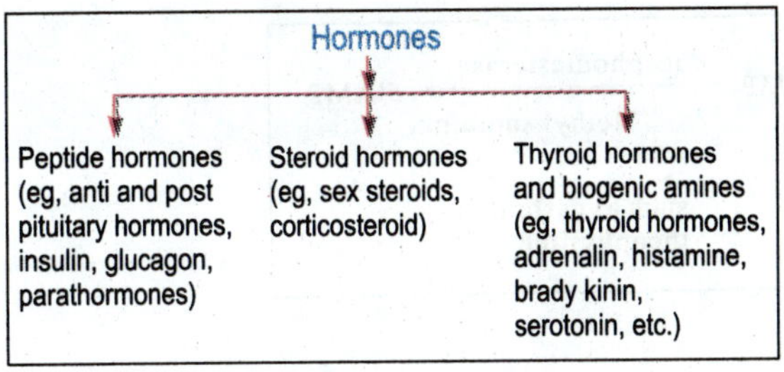

Classification of hormone

According to location of receptor :

1. Membrane bound receptor.
2. Intracellular receptor.

   Steroid hormone (Oestrogen, progesterone, androgen, glucocorticoid).

   Thyroid hormone—T3, T4.
3. Calcitriol.

Mechanism of action of hormones

Hormones that bind to intracellular receptor

1. These types of hormones are lipophilic in nature, so they can diffuse through the plasma membrane of all cells. The steroid hormone has a specific soluble, oligomeric receptor protein (mobile receptor) either in the cytosol and / or inside the nucleus.
2. The hormone-receptor complex next undergoes a temperature and salt dependent activation reaction that results in size, conformation and surface charge changes to favour its binding to the nuclear chromatin attached to nuclear matrix.
3. The hormone receptor complex binds to a specific region of DNA (called the hor-

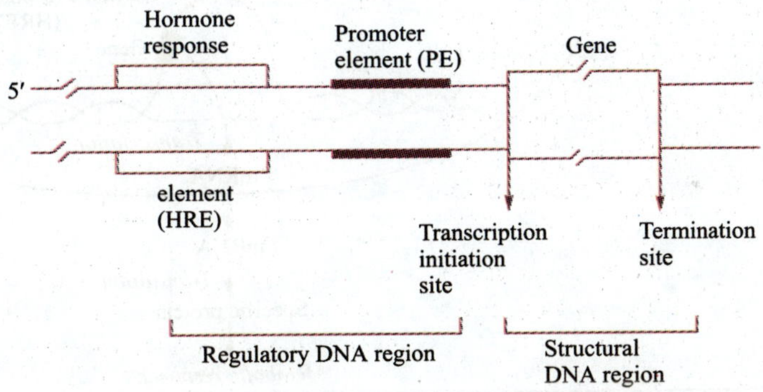

Fig. 31.1 : Structural requirements for hormonal regulation of gene transcription

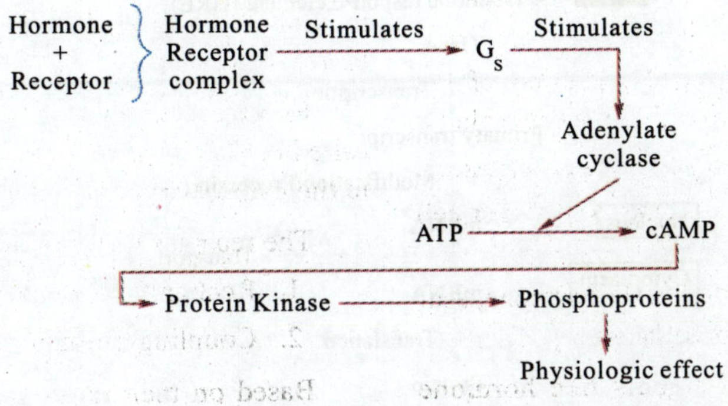

Fig. 31.2 : Hormones that bind to cell membrane receptor

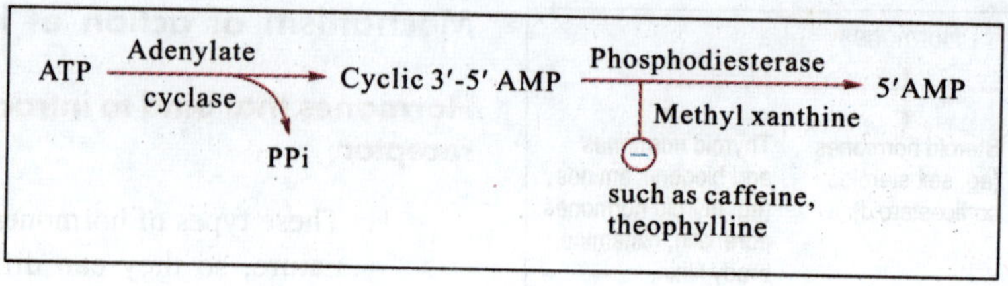

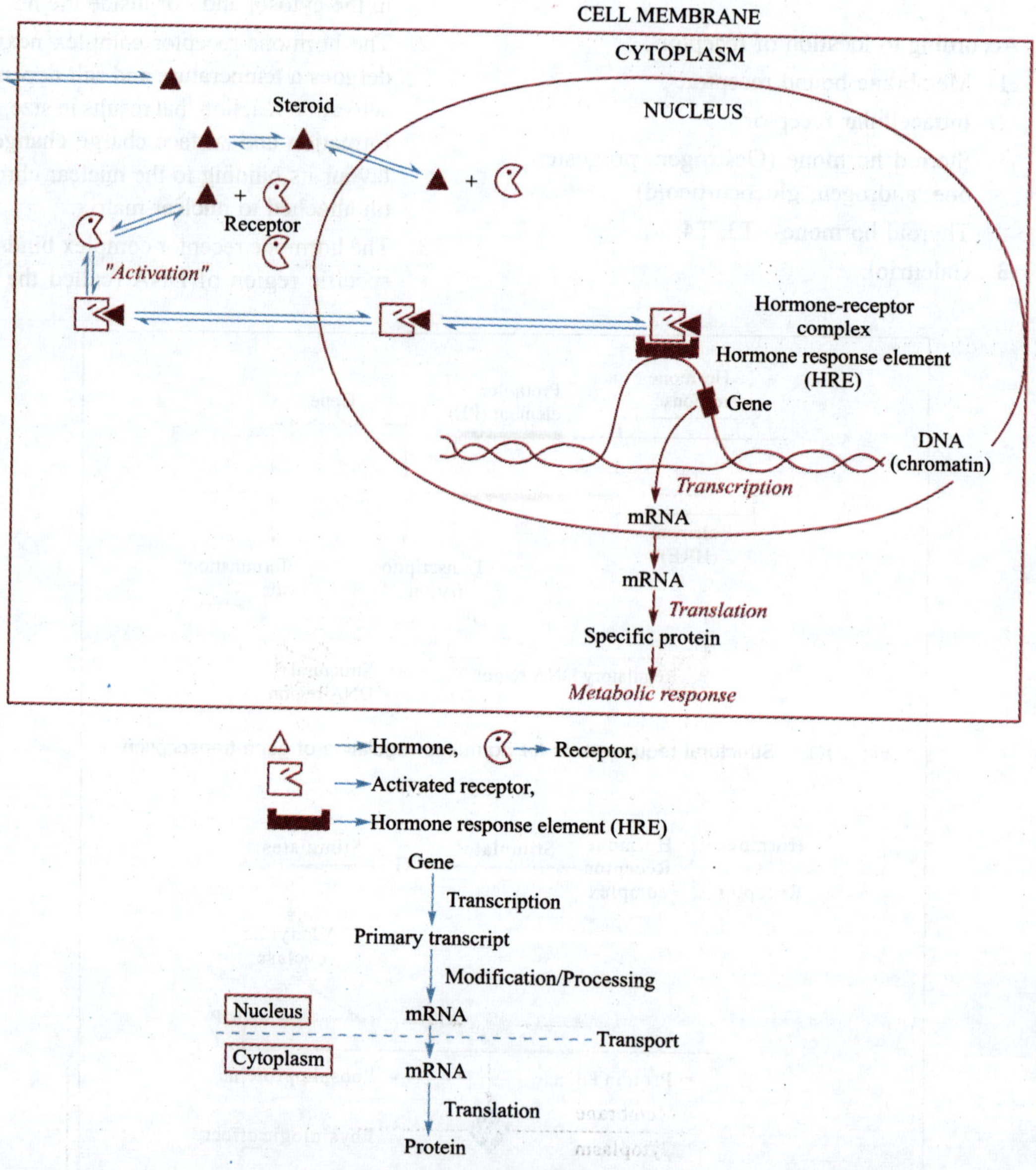

Fig. 31.3 : The information pathway where hormones can affect any of these steps

mone response element) and activates or inactivates specific genes.

4. Promoter element is generic since it is present in some form or other in all genes. This element specifies the site of RNA polymerase II attachment of DNA and so the accuracy of transcript initiation.
5. A second element, the hormone response element has been identified in many genes regulated by steroid hormones. The element modulates the frequency of transcript initiation and is less dependent on position and orientation, so it resembles the transcription enhancer elements found in other genes.

Hormones that bind to cell membrane receptor

A. cAMP as the second messenger

1. For the formation of cAMP from ATP needs : Receptor, GS protein, Adenylate cyclase.
2. cAMP (Cyclic adenosine 3′-5′ monophosphate) is formed from ATP by the action of the enzyme adenylate cyclase and converted to physiologically inactivated 5′-AMP by the action of enzyme phosphodiesterase.
3. Hormone receptor complex combines with G_s or G_i (s = stimulatory, i = inhibitory) type of GTP dependent trimeric nucleotide regulatory complex of the cell membrane.
4. Both G_s or G_i are made up of 3 subunits. G_s contains $\alpha_s\beta\gamma$ and G_i contains $\alpha_i\beta\gamma$.
5. The α subunit (either G_s or G_i) is bound to GDP. When binding of hormone to R_s or R_i results in a receptor-mediated activation of G, then GDP is exchanged for GTP and the α subunit separates from the combined β and γ subunits.

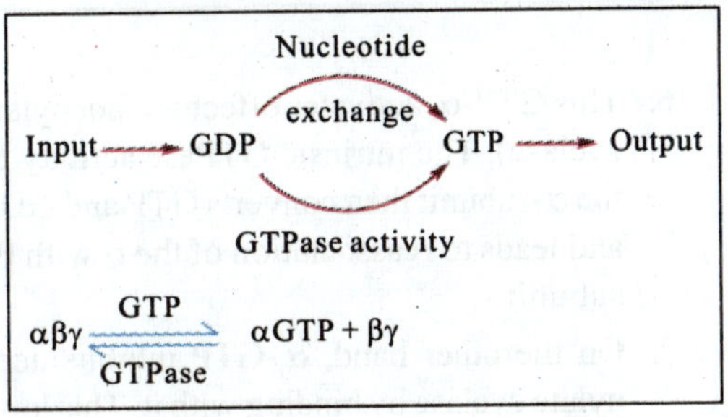

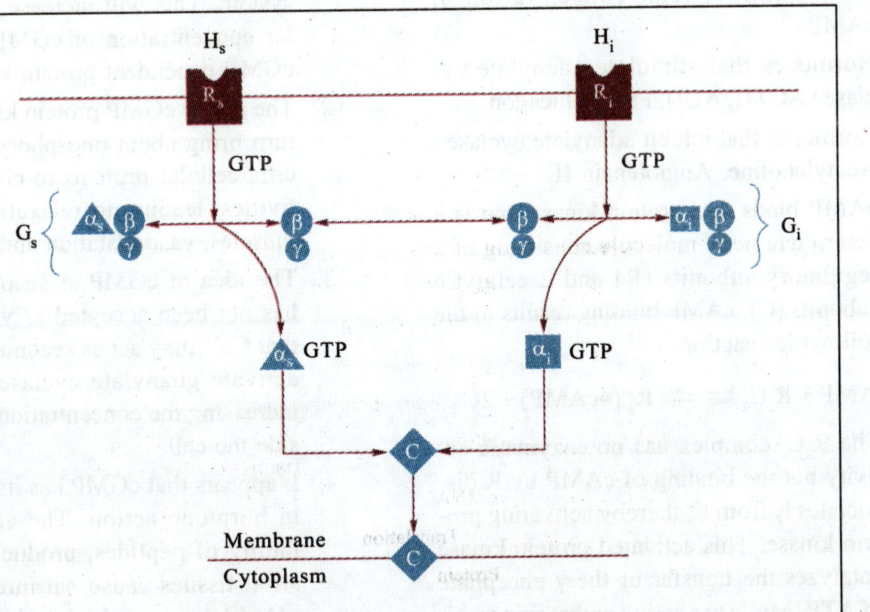

Fig. 31.4 : Effect on adenylate cyclase

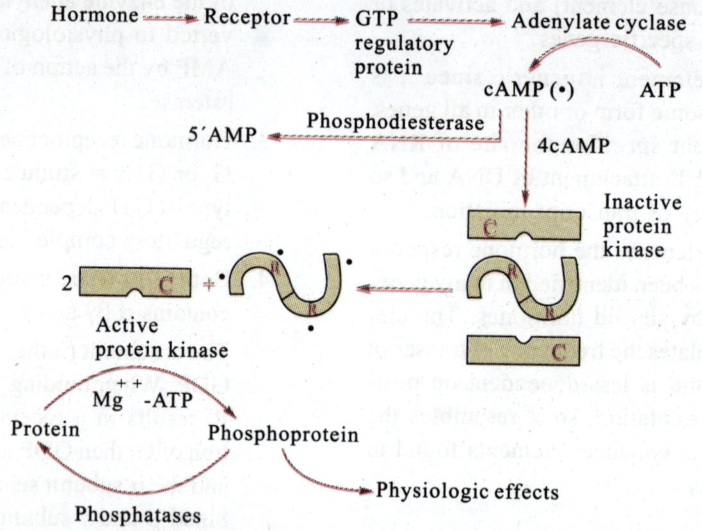

Fig. 31.5 : Hormonal regulation of cellular processes through cAMP dependent protein kinase

6. This GTP-α_s activates effectors (adenylate cyclase). The intrinsic GTPase activity of the α-subunit then converts GTP and GDP and leads to reassociation of the α with $\beta\gamma$ subunit.
7. On the other hand, α_i-GTP inhibits adenylate cyclase by binding with it. This lowers the intracellular concentration of cAMP

   Hormones that stimulate adenylate cyclase : ACTH, ADH, FSH, Glucagon.

   Hormones that inhibit adenylate cyclase : Acetylcholine, Angiotensin II.
8. cAMP binds to a protein kinase that is a hetero terameric molecule consisting of 2 regulatory subunits (R) and 2 catalytic subunits (C). cAMP binding results in the following reaction.

$$4cAMP + R_2C_2 \rightleftharpoons R_2\,(4cAMP) + 2C$$

9. The R_2C_2 complex has no enzymatic activity but the binding of cAMP by R dissociates R from C, thereby activating protein kinase. This activated protein kinase catalyzes the transfer of the γ phosphate of ATP (Mg^{++}) to a serine or threonine residue in a variety of proteins. Thus they regulate the conformational changes of phosphoprotein and physiologic effect occurs.

B. Role of cGMP in hormone action

1. Hormones such as insulin and growth hormone, affect the guanylate cyclase cGMP system. This will increase the intracellular concentration of cGMP and activate cGMP dependent protein kinases.
2. The active cGMP protein kinase would in turn bring about phosphorylation of specific cellular proteins to change their activities, leading to relaxation of smooth muscles, vasodilatation and other effects.
3. The idea of cGMP as second messenger has not been accepted as yet. It is likely that Ca^{++} may act as second messenger to activate guanylate cyclase and thereby increasing the concentration of cGMP inside the cell.
4. It appears that cGMP has its unique place in hormone action. The atriopeptins, a family of peptides, produced in cardiac atrial tissues cause natriuresis, diuresis, vasodilatation and inhibition of aldosterone secretion.

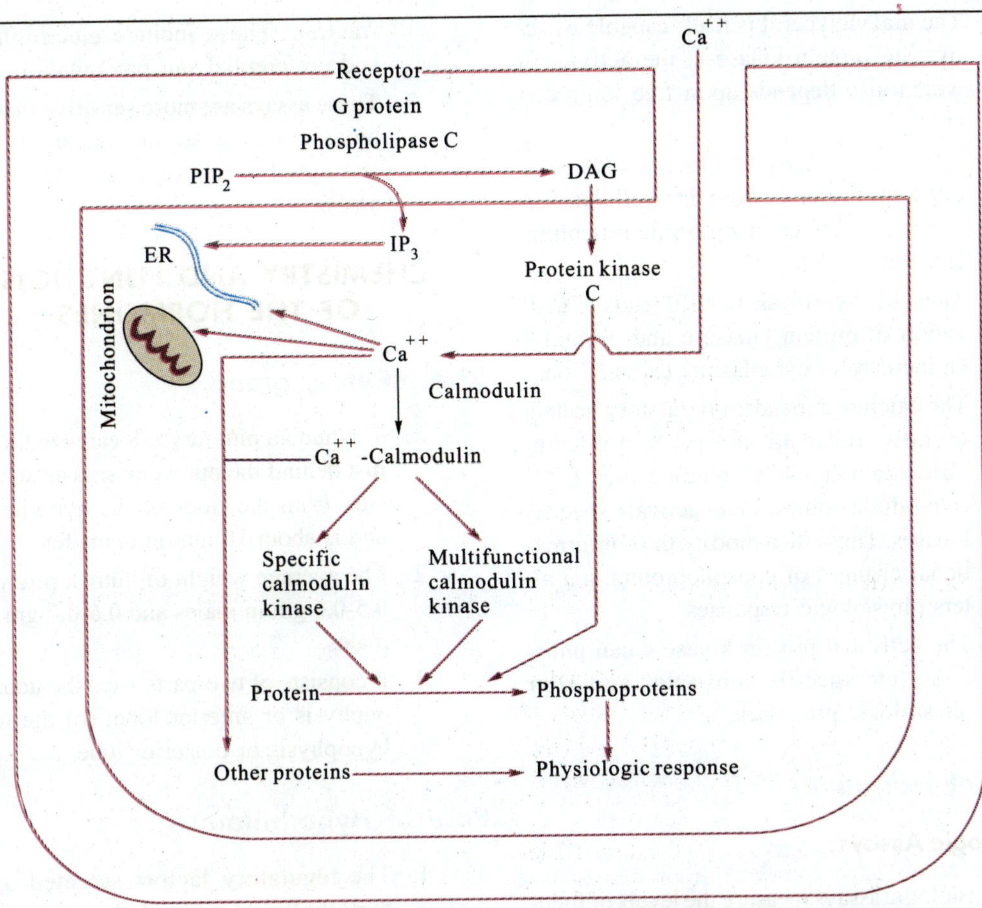

Fig. 31.6 : Role of calcium in hormone action

5. These peptides (eg, atrial natriuretic factor) bind to and activate the membrane bound form of guanylate cyclase. This results in an increase of cGMP.

C. Role of calcium in hormone action

1. It is suggested that ionized calcium of the cytosol is the important signal for hormone action than cAMP.
2. The extracellular calcium (Ca^{++}) concentration is about 5 mmol/L, the intracellular concentration of this free ion is much lower 0.1 - 10 μmol/L.
3. The hormones that bind cell membrane receptor enhance membrane permeability to Ca^{++} and thereby increase Ca^{++} influx. This is probably accomplished by an Na^+ / Ca^{++} exchange mechanism that has a high capacity but a low affinity for Ca^{++}. There is a $Ca^{2+}/2H^+$-ATPase dependent pump that extrudes Ca^{2+} in exchange for H^+. This has a high affinity for Ca^{2+} but a low capacity.
4. Cell surface receptors such as those for acetylcholine, ADH, when occupied by their respective ligands, potent activators of phospholipase c.
5. Receptor binding and activation of phospholipase c are coupled by a unique G protein.
6. Phospholipase c catalyzes the hydrolysis of phosphatidylinositol 4, 5-bisphosphate to inositol triphosphate and 1, 2 diacylglycerol.

7. The diacylglycerol is itself capable of activating protein kinase c, the activity of which also depends upon free ionic calcium.
8. Inositol triphosphate is an effective releaser of calcium from intracellular storage sites such as endoplasmic reticulum, and mitochondria.
9. Thus, the hydrolysis of PIP_2 leads to activation of protein kinase c and promotes an increase of cytoplasmic calcium ion.
10. The calcium dependent regulatory protein is now referred to as calmodulin. Calmodulin has $4Ca^{++}$ binding sites. Ca^{++}-calmodulin complex can activate specific kinases. These then modify the conformational changes of phosphoprotein and alters physiologic responses.
11. The activated protein kinase c can phosphorylate specific substrates and alter physiologic processes.

Assay of Hormones

A. Biologic Assays

1. Biologic assays measure the levels of functional activity of hormones.
2. These assays are not usually specific due to lack of sensitivity.

B. Chemical Assays

1. These assays measure the absolute quantity of a given hormone.
2. These are not generally applicable in case of protein hormones.

C. Radiodisplacement Chemical Assays

1. These assays are largely used nowadays for protein and nonprotein hormones.
2. The binding protein is a specific antibody, membrane receptor or serum transport protein. The unlabelled hormone, present as standard or unknown, displaces the labelled hormone and results in the increase in radioactivity in the unbound fraction. These include electrophoresis and preferential salt precipitation.
3. These assays are more sensitive than most bioassays because the concentrations of less than 1ng/ml can be detected by these assays.

CHEMISTRY AND FUNCTIONS OF THE HORMONES

The pituitary gland

1. The human pituitary is located in the brain just behind the optic chiasma as an extension from the floor of the hypothalamus and is about 10 mm in diameter.
2. The average weight of human pituitary is 0.5-0.6 gm in males and 0.6-0.7 gm in females.
3. It consists of two parts – (a) the adenohypophysis or anterior lobe; (b) the neurohypophysis or posterior lobe.

Role of Hypothalamus

1. The regulatory factors secreted by the hypothalamus control the secretion of hormones by the pituitary gland. The regulatory factors are considered to be hormones.
2. In addition to the secretion of regulatory hormones, neurophysial hormones are actually synthesized in the hypothalamus and migrate as granules down the nerve fibres and accumulate at the nerve endings from which they are secreted.

ADENOHYPOPHYSIS OR THE ANTERIOR PITUITARY GLAND

1. The anterior lobe is the largest and the most important portion of the pituitary.
2. In man, this lobe is about 70 per cent of the total weight of the gland.
3. This gland releases a number of hormones which are divided into two groups :

(a) Tropic hormones : Thyrotropic hormone or Thyroid stimulating hormone (TSH), Adrenocorticotropic hormone (ACTH), Luteinizing hormone (LH), Follicle stimulating hormone (FSH); (b) Growth hormone (GH, somatotrophin, prolactin and lipotrophins).

Thyrotropic Hormone; Thyroid Stimulating Hormone (TSH)

Chemistry

1. It is a glycoprotein of molecular weight about 30,000.
2. It consists of two polypeptide chains termed TSH-α and TSH-β.
3. It is very rich in sulphur containing amino acids consisting of 11 disulphide residues and contains fucose, mannose, galactose, glucosamine and galactosamine.

Functions

1. It binds to specific membrane receptors and activates thyroidal adenylate cyclase causing an increased cellular cAMP.
2. It increases the rates of the removal of inorganic iodide from blood by thyroid and incorporation of iodide in the thyroid hormone.
3. It increases the release of thyroxine from the thyroid gland to circulation.
4. It is used clinically to differentiate primary hypothyroidism (myxedema) from secondary hypothyroidism (pituitary insufficiency).

Control of release

The hypothalamus secretes thyrotropic releasing hormone (TRH), a regulatory factor. It is a tripeptide containing pyroglutamic acid, histidine, and prolinamide. This factor is abundant in the pineal gland and frog skin. Inhibitors of protein synthesis have no effect on the synthesis of this factor from the hypothalamus which indicates that the process is nonribosomal. Its action is calcium dependent. This factor specifically acts on the thyrotropin secreting cells. TRH also stimulates prolactin secretion.

ACTH (Adrenocorticotropic Hormone)

Chemistry

1. It is a straight chain polypeptide of molecular weight 4,500 containing 39 amino acids.
2. Only the first 23 amino acids are required for activity. These 23 amino acids in the peptide chain is the same in all species; whereas the remaining biologically inactive 16 amino acid chain varies according to the animal source.

Functions

1. It stimulates the synthesis of corticosteroids by the adrenal gland and also enhances their release from that gland.
2. It increases the total protein synthesis.
3. It causes the synthesis of steroid hormone from cholesterol.
4. It has a mild stimulating effect in dispersing melanin granules on the skin.
5. The adimnistration of ACTH causes (a) the increased excretion of nitrogen, potassium and phosphorus, (b) retention of sodium, chloride and secondary retention of water, (c) Elevation of fasting blood sugar, (d) increase in circulating free fatty acids, (e) increased excretion of uric acid, (f) decline in circulating eosinophils and lymphocytes.
6. Since it activates adenylate cyclase and increases intracellular levels of cAMP, it can increase lipolysis in adipose tissue and stimulate insulin secretion from the pancreas.
7. **Puromycin**, the blocking agent of protein synthesis, inhibits the effect of ACTH on the adrenal gland.

Control of ACTH Secretion

1. It is controlled by corticotropin-releasing hormone (CRH) of the hypothalamus.
2. Activation of the hypothalamic centres takes place via neurotransmitters in the central nervous system (eg, acetylcholine and serotonin). Therefore, stresses such as cold, insulin hypoglycemia, epinephrine, estrogens, psychic stimuli cause an increased production of ACTH leading to increased adrenal cortical activity. High levels of ACTH inhibits the further synthesis of ACTH.

The normal level of ACTH in serum is 15-17 pg/ml.

Lutenizing Hormone (LH)

Functions

1. In the female, it stimulates final maturation of the graafian follicle, ovulation, and the development of the corpora lutea. The secretion of estrogen and progesterone is also stimulated.
2. In the male, it stimulates testosterone production by the testis which maintains spermatogenesis and causes the developement of accessory sex organs such as the vas deferens, prostate, and seminal vesicles.
3. It increases cAMP. Both cAMP and LH are blocked by puromycin.

In serum the normal level of LH in males 1-15 mIU/ml & in females <1-30 mIU/ml.

Follicle Stimulating Hormone (FSH)

Functions

1. In the female, it stimulates the growth and maturation of graafian follicles and prepares them for ovulation and for the action of LH and enhances the release of estrogens. This hormone is very active during menstrual cycle.
2. In the male, it stimulates seminal tubule and testicular growth and also the early stages of spermatogenesis.
3. The secretion of this hormone is inhibited by the administration of testosterone, progesterone and high concentration of FSH.

Growth Hormone (GH, Somatotrophin)

Chemistry

1. It is a single polypeptide with a molecular weight of about 21,500.
2. The human growth hormone consists of 191 amino acids.
3. The activity of hormone resides only in a portion of the molecule.
4. Partial hydrolysis of the hormone does not stop its activity.
5. The growth hormone preparations from monkey and human are active in both humans and rats.

Functions

1. It increases total growth and causes gigantism in children. Its deficiency in children causes dwarfism.
2. It stimulates production of Somatomedins (Sulfation factors) from liver and kidney. somatomedins is similar to serum insulin-like activity.
3. It stimulates protein synthesis causing an increase in nitrogen and phosphorus retention. Blood amino acids and urea are decreased. Growth hormone increases synthesis of DNA and RNA in all tissues. It stimulates erythropoiesis.
4. It accelerates the mobilization of fat from the adipose tissues and the free fatty acids in the blood. It enhances oxidation of fats in liver and muscle. Increased ketogenesis occurs in the deficiency of insulin.
5. In muscle, it functions against the action of insulin causing the decrease in the utilization of glucose and stimulates the secretion of glucagon resulting in the increase in the blood glucose level. In liver, it inhibits hexokinase reaction and thus opposes the action of insulin.

6. It increases intestinal absorption of calcium as well as its excretion. As it stimulates the growth of the long bones at the epiphyses as well as the growth of soft tissues, retention of other minerals such as potassium, phosphorus, sodium, magnesium and chloride occurs. It causes the formation of sulphate esters for incorporation into cartilage by increasing somatomedin from the liver.

Control of secretion of growth hormone

1. The control of growth hormone is exerted by a specific growth hormone releasing factor (GHRF) also termed growth hormone releasing hormone (GRH) which is extracted from the hypothalamus.
2. Growth hormone release-inhibiting hormones (GH-RIH, GIH) or Somatostatin release-inhibiting hormone (SRIH) is negative modulator of growth hormone.
3. Secretion of growth hormone is influenced by exercise, hypoglycemia, starvation, excitement and exposure to cold.

Prolactin (PL)

Chemistry

1. It is a protein with a molecular weight of about 23,000.
2. It is produced by the pituitary acidophil cells.

Functions

1. It activates the corpus luteum and stimulates progesterone production by the developed corpus luteum.
2. It also stimulates enlargement of crop gland and formation of "crop milk" in pigeons.
3. It increases during pregnacy and stimulates mammary development and growth hormone like metabolic changes.
4. It is inhibited by a hypothalamic factor, prolactin inhibiting factor.

Neurohypophysis (The Posterior Lobe of the Pituitary) Oxytocin

Chemistry

1. It is a cyclic polypeptide containing 8 amino acids.
2. Its molecular weight is about 1000.
3. Its structure is quite similar to that of vasopressin. The differences are isoleucine of oxytocin which is replaced by phenylalanine of vasopressin and leucine of oxytocin is replaced by lysine of vasopressin.

Functions

1. It causes contraction of the smooth muscles in the mammary gland resulting in milk excretion. Its level is increased by suckling.
2. It is increased during labour. It causes uterine contraction and is used in obstetrics when induction of uterine contraction is required.
3. It stimulates the contraction of gall bladder, intestines and urinary bladder.

Vasopressin

Chemistry

1. It is a cyclic polypeptide containing 8 amino acids.
2. Its structure is quite similar to that of oxytocin except isoleucine which is replaced by phenylalanine and leucine which is replaced by lysine.

Functions

1. It has marked effect on the kidney tubules accelerating the rate of water reabsorption from the distal tubules and thus produces a marked antidiuretic effect. Hence, it is termed as antidiuretic hormone (ADH).
2. It is an effective inhibitor of the gonadotrophins, particularly LH.

3. ADH formation is prevented by the tumor in the hypothalamus or injury to hypothalamus. Diabetes insipidus occurs in the absence of ADH which is characterized by the large volume of excretion of urine—up to 30 L of urine per day. Alcohol also inhibits ADH secretion.
4. ADH secretion is increased by emotional and physical stress, electrical stimulation, acetylcholine, nicotine and morphine. These stimulations are associated with an increase in RNA synthesis in the neuron indicating an increased protein synthetic activity.

The Middle Lobe of the Pituitary

Intermedin or melanocyte-stimulating hormone (MSH)

Chemistry

1. There are two peptides (α-MSH and β-MSH). β-MSH is 50 times more than α-MSH.
2. α-MSH is smaller containing only 13 amino acids. Both have structural similarity with that of ACTH.
3. α-MSH is identical to the first 13 amino-acids of ACTH.
4. Amino acids 11-17 of β-MSH are common to both α-MSH and ACTH.
5. α-MSH has some corticotropic activity but not of β-MSH.

Functions

1. This hormone increases the deposition of melanin by the melanocytes of the human skin.
2. In Addison's disease when the production of the corticosteroids is inadequate, melanocyte stimulating hormone (MSH) is secreted causing the increased synthesis of melanin accompanied by brown pigmentation.
3. Both cortisone and hydrocortisone, epinephrine and norepinephrine inhibit the action of MSH.

Abnormalities of Pituitary Functions

A. Hyperpituitarism

Gigantism

1. It occurs from the hyperactivity of the gland during childhood or adolescence.
2. The long bones increase in length so that the individual attains an unusual height.
3. The growth of the legs and hands are relatively greater than the trunk.
4. The individuals are mentally subnormal.

Acromegaly

1. This condition occurs in adults after the epiphyses have closed and growth has ceased.
2. The individual exhibits enlarged nose, growth and enlargement of the hands, feet and thickening of the skin.
3. Sexual function is increased. Raised BMR, hyperglycemia and glycosuria result.

   Excess production of ACTH produces Cushing's disease.

B. Hypopituitarism

Dwarfism

1. It occurs as a result of hypoactivity of the gland in childhood.
2. Cessation of growth and sexual retardation.

Frohlich's syndrome

1. This occurs in pituitary destruction in childhood.
2. The children become stunted and quite stupid.
3. There are deposits of fats all over the body.
4. Obesity and diabetes insipidus occur.

5. Disturbances in sleep mechanism and body temperature occur. These children spend most of their time asleep.
6. The obesity is due to increased appetite and reduced energy output.

Simmond's Disease (Panhypopituitarism)

1. It occurs due to the deficiency of the function of the hypophysis in the adult.
2. BMR is reduced, subnormal body temperature and the heart rate is low.
3. Carbohydrate metabolism is entirely disturbed and the individual develops severe and sometimes fatal hypoglycemic coma.

Myxedema is due to lack of thyrotropin.

THE ADRENALS

The Adrenal Medulla

Epinephrine and norepinephrine

Chemistry

1. These hormones are structurally related to a group of organic compounds known as catechols. The aromatic nucleus of these hormones is that of catechol (1, 2-dihydroxybenzene) but the amino group is attaced to an aliphatic side-chain. Hence, the term **catecholamine**.
2. 80 per cent of the catecholamine activity is attributed to epinephrine. Naturally occurring epinephrine is the L-isomer. The unnatural D-form is only one-fifteenth active.
3. Epinephrine differs from norepinephrine only in that epinephrine is methylated on the primary amino group of the aliphatic side-chain shown in the figure 31.7.

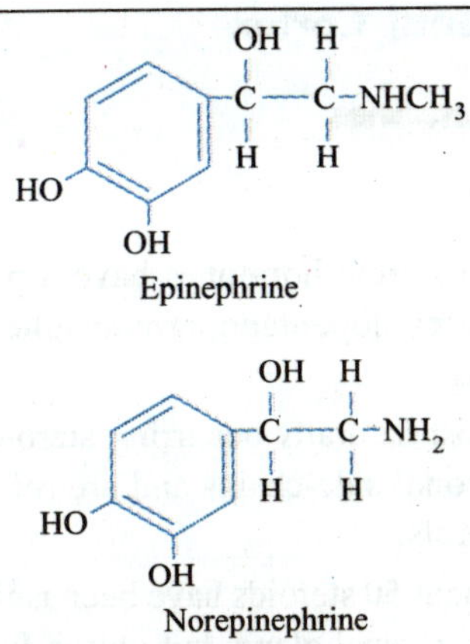

Fig. 31.7

Functions

1. Epinephrine gives a rapid physiologic response to emergencies such as cold, fatigue, shock etc. In this sense, it mobilizes the terms "**fight or flight**" mechanism.
2. These hormones are effective on parenteral administration. They are destroyed by oxidation when given orally.
3. They increase blood pressure.
4. Epinephrine increases cAMP by activating **adenylate cyclase**. cAMP stimulates phosphorylase which causes the breakdown of glycogen in the liver resulting in increased blood sugar. Norepinephrine is one-fifth potent in this respect.
5. Epinephrine causes increased breakdown of muscle glycogen to lactic acid, thereby increasing blood lactate.
6. Epinephrine acts on adipose tissue and releases free fatty acids into the circulation.
7. Epinephrine also increases BMR.
8. Epinephrine is an effective stimulant of heart action. It increases the irritability and the rate and strength of contraction of cardiac muscle and increases cardiac output. It causes vasodilatation of the arterioles of the skin and mucous membranes. Norepinephrine has less effect on cardiac output.
9. Epinephrine causes relaxation of the smooth muscles of the stomach, intestine, bronchioles and urinary bladder. This hormone is valuable in the treatment of asthmatic attacks.

In plasma the normal level of epinephrine is <0.1 ng/ml and norepinephrine is <0.5 µg/L.

The Adrenal Cortex

Steroid hormones

Chemistry

1. All steroid hormones have a parent ring, i.e. cyclopentanoperhydrophenanthrene ring.
2. Most naturally occurring steroids contain alcohol side-chains and are referred to as sterols.
3. About 50 steroids have been isolated from the adrenal gland, but only a few possess physiologic acitivity. The most important ones are cortisone, hydrocortisone (cortisol, 17- hydroxycorticosterone), aldosterone and the two androgens androstenedione and dehydroepiandrosterone. Cortisol is the major free circulating adernocortical hormone in human plasma.
4. A and B rings of the nucleus are joned in a *trans* or *cis* configuration. Estrogens do not show such isomerism since their A ring is aromatic.
5. In natural steroids, both the chains attached at C_{17} and various substitutions at C_{11} are in the β-configuration.

Demerization of glucocorticoid receptor

1. The amino terminal half contains most of the antigenic sites and has a region that modulates promoter function (*trans*-activation).
2. The carboxy terminal half contains the DNA and hormone binding domains.
3. The DNA binding domain is closer to the centre of the molecule, while the hormone binding domain is near the carboxyl terminus. Both of these domains are required for *trans*-activation of gene transcription.
4. A sequence of amino acids in the carboxy terminal regions is required for dimerization of 2 receptor molecules.
5. Two separate regions appear to be necessary for entry of the receptor into the nucleus.

Functions

A. The glucocorticoids

1. These hormones increase glucose, fatty acids and amino acids in the blood.
2. In the peripheral tissues (muscle and adipose tissues), they cause depletion of protein stores.
3. In adipose tissue, they increase lipolysis and in muscle, they cause depletion of protein stores.
4. They increase alanine-α-ketoglutarate and tyrosine transaminases as well as tryptophan pyrrolase.

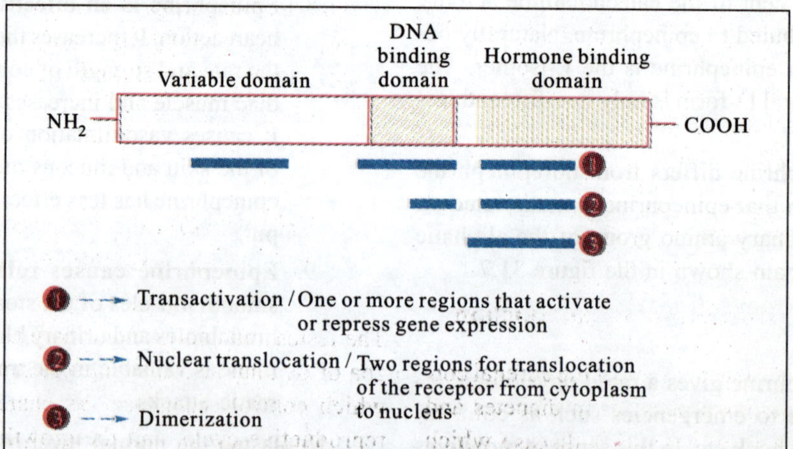

Fig. 31.8 : Glucocorticoid receptor

5. They increase the key enzymes in the regulation of gluconeogenesis (Pyruvate carboxylase, Phosphoenolpyruvate carboxykinase, fructose-1 6-diphosphatase and glucose-6- phosphatase).
6. In the liver, they act on the fixation of CO_2 at the level of pyruvate carboxylase which is the key enzyme in gluconeogenesis.
7. They are inactive on heart, brain and red cells.
8. They have anti-inflammatory effects and also immunosuppressive effects.
9. They have effects on bone, exocrine secretion, cyclic AMP and stress.

B. The Mineralocorticoids

1. They increase the absorption of sodium and chloride by the renal tubules and decrease their absorption by the sweat glands, salivary glands, and the gastrointestinal tract.
2. They cause, on administration, increased extracellular fluid volume, increased circulating blood volume and urinary output.
3. Aldosterone also increases the renal clearance of magnesium.

C. Sex Hormones (C-19 Corticosteroids)

1. The adrenocorticosteroids cause retention of nitrogen (a protein anabolic effect), phosphorus, potassium, sodium and chloride. If they are present in excessive amounts, they lead to masculinization in the female.
2. Estrogens and progesterones are produced in small amounts.

Abnormalities of adrenorcortical function

A. In humans, degeneration of the adrenal cortex due to tuberculous process or in diabetes and hypothyroidism causes Addison's disease which shows the following signs.

1. Decreased 17-hydroxycorticoid and aldosterone excretion.
2. Excessive loss of sodium chloride in the urine.
3. Elevated levels of potassium in the serum.
4. Low blood pressure and low body temperature.
5. Muscular weakness, gastrointestinal disturbances, hypoglycemia, and a progressive brownish pigmentation.

B. Adrenocortical hyperfunction is caused by malignant tumors of the cortex. This hyperfunction of the cortex causes Cushing's disease which exhibits the following signs.

1. Hyperglycemia and glycosuria (Diabetogenic effect).
2. Retention of sodium and water followed by edema, increased blood volume and hypertension.
3. Negative nitrogen balance (Protein antianabolic effect and gluconeogenesis).
4. Potassium depletion and hypokalemic alkalosis.

Orphan receptors

A comparison of several different steroid receptors with thyroid hormone receptors revealed a remarkable conservation of the amino acid sequence in regions, particularly in DNA binding domain. Receptors of the steroid / thyroid type constitute a large superfamily. Many related members of this family have no known ligand, so called orphan receptors.

THE ORGANS PRODUCING SEX HORMONES

The testes and ovaries not only provide spermatozoa or ova but also manufacture steroid hormones which control secondary sex characteristics, the reproductive cycle, and the growth and development of the accessory reproductive organs.

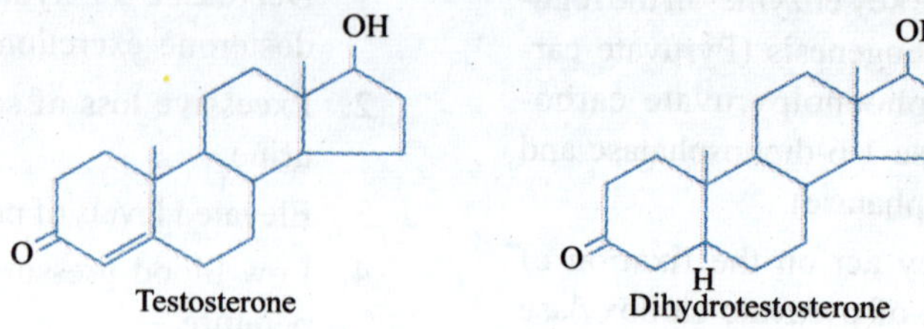

Fig. 31.9 : Structure of Testosterone and Dihydrotestosterone

Male Hormones

Testosterone

Chemistry

1. This is formed from cholesterol through pregnenolone, progesterone and hydroxyprogesterone.
2. The active hormone is dihydrotestosterone.

Functions

1. Testosterone promotes the growth and function of the epididymis, vas deferens, prostate, seminal vesicles, and penis.
2. It enhances and maintains the motility and fertilizing power of the sperms.
3. It stimulates the secretion of sebaceous glands.
4. It is largely responsible for the emotional make-up of the male.
5. It depresses the estrogenic over-activity in women with symptoms of dysmenorrhoea, painful breasts and stops lactation and menstruation.
6. It promotes protein synthesis in the body.
7. It increses the activity of glycolytic enzymes and decreases the activity of glutamic dehydrogenase and arginine synthetase.
8. It increases the rate of synthesis of fatty acids.

The normal level of testosterone in plasma of adult males is 30-100 mg/dL.

Female Hormones

The ovary secretes two hormones :

1. Estrogens formed by the follicular tissue.
2. Progesterone formed by the corpus luteum.

Estrogens

Chemistry

1. The estrogens are C-18 steroids and differ from androgens in lacking the methyl group C_{10}. The ring A is aromatic.
2. The androgens, testosterone, and androstenedione are precursors for the synthesis of the estrogens in testis, ovaries, adrenals and placenta. The chemical structure of estrogen is given below :

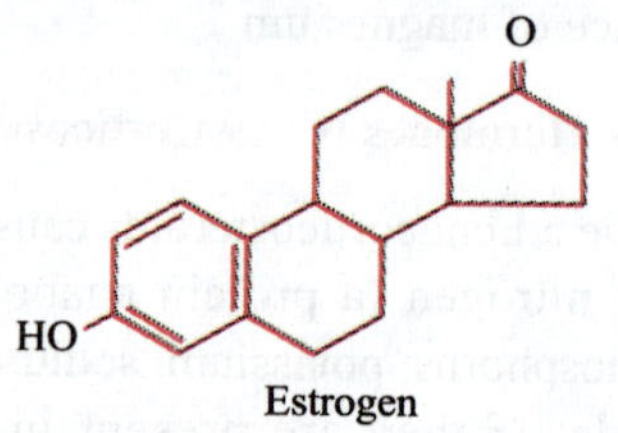

Fig. 31.10 : Structure of estrogen

Functions

1. Estrogens are responsible for the growth of the uterus, vagina, pelvis, breasts, pubic and axillary hair.
2. They influence the menstrual cycle and are essential for breast development.
3. On administration, they promote mitotic activity in the uterine muscle and endometrium.

4. The vaginal epithelium is sensitive to the action of estrogen.
5. They influence the secretion of the gonadotrophic hormones in the anterior pituitary.
6. They increase the plasma levels of thyroxine and cortisol binding globulins.
7. They cause rapid increase in RNA synthesis in uterine tissue.
8. They prevent lipid accumulation in liver when administered to animals having diets deficient in lipotropic factors i.e. methonine and choline.
9. They cause decrease in cholesterol level and other lipids in plasma. This is why the incidence of atherosclerosis is low in women as compared to men.
10. They regulate normal bone metabolism. Women after menopause develop osteoporosis.

Normally the females excrete estrogens 5-100 μg/24 hours in urine.

Progesterone

This hormone is formed in the corpus luteum of the ovary. It is also formed in the placenta during the latter part of pregnancy. It is secreted 1 or 2 days before ovulation takes place.

Chemistry

1. It is synthesized from cholesterol.
2. It is also formed in the adrenal cortex as a precursor of both C-19 and C-21 corticosteroids. The chemical structure is given below.

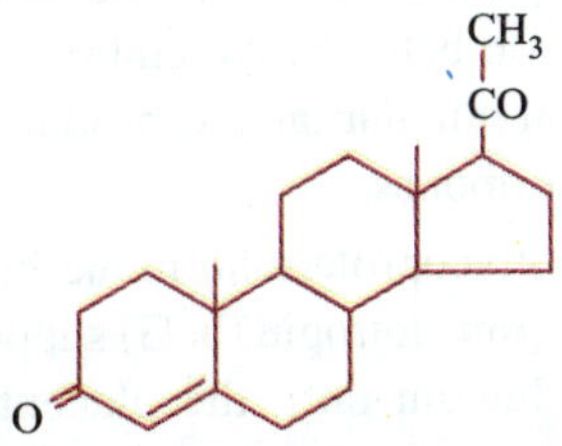

Progesterone

Fig. 31.11 : Structure of Progesterone

Functions

1. Progesterone causes the development of the endometrium preparing it for the implantation of fertilized ovum for conception.
2. It suppresses oestrus, ovulation and the production of luteinizing hormone. When pregnancy occurs, the ovulation and menstruation are suspended by the action of this hormone.
3. It stimulates the mammary glands.
4. It increases BMR during the luteal phase of normal menstrual cycle.
5. In the normal menstrual cycle, the antiovulatory effect of progesterone is the basis for the use of certain synthetic progestins as oral contraceptive agents.

The normal level of progesterone in serum is 0.2-1.5 mg/ml.

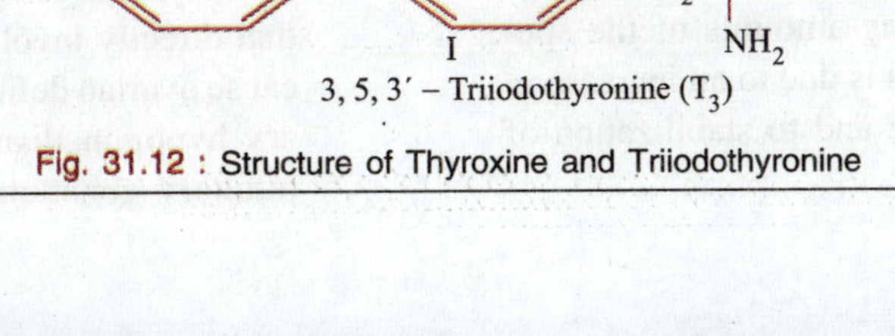

Fig. 31.12 : Structure of Thyroxine and Triiodothyronine

Placental hormones maintain pregnancy

1. The placenta provides the nutritional connection between the embryo and the maternal circulation and produces a number of hormones.
2. The glycoprotein hormone human chorionic gonadotropin (hcG) supports the corpus luteum until the placenta produces amounts of progesterone sufficient to support the pregnancy.
3. The corpus luteum is the major source of progesterone for the first 6-8 weeks of the pregnancy and then the placenta assumes this function. The placenta does not systhesize cholesterol and depends upon a maternal supply.
4. The level of plasma estradiol, estrone, and estriol is gradually increased throughout pregnancy. Estriol is produced in the largest amount which reflects a number of feto placental functions.
5. Estrogens and progestins are the candidates which affect uterine contractility. Catecholamines are involved in induction of labour. Oxytocin stimulates uterine contractility and is used to facilitate delivery.

Mammary gland development is stimulated by estradiol and progesterone and lactation by prolactin

1. Some growth of glandular tissue occurs during puberty with the deposition of adipose tissue; but extensive development occurs during pregnancy when glandular tissue is exposed to high concentration of estradiol and progesterone.
2. Prolactin, glucocorticoids and insulin increase the rate of synthesis of milk proteins by increasing amounts of the specific mRNAs. Ths is due to an increase in gene transcription and to stabilization of the mRNA.
3. Progesterone inhibits milk production and secretion in late pregnancy. Lactation starts when the level of this hormone decreases abruptly after delivery.
4. Lactation gradually decreases if suckling is not allowed. Suckling results in the release of oxytocin from the posterior pituitary. Oxytocin stimulates contraction of the myoepithelial cells that surround the alveolar ducts, thus expelling milk from the gland.

Menopause is complete with the loss of ovarian estrogen production

1. Menopause occurs with the loss of all follicles and ovarian estrogen production. No progesterone is produced but substantial amounts of a weak estrogen, estrone are produced by the peripheral aromatase from androstenedione. The levels of LH and FSH are highly increased since the levels of estrone are not sufficient to suppress pituitary gonadotropin.
2. GnRH is also involved at the beginning of menopause. This may be due to changes in the absolute level of the hormone.
3. Postmenopausal women face two problems associated with tissue catabolism. Estrone is not always able to prevent the atrophy of secondary sex tissues, particularly the epithelium of the lower urinary tract and vagina. Osteoporosis is a major health problem in older individuals. Women with the most severe decrease in bone mass have lower estrone levels than normal.

Hormones are involved in some pathophysiology of the female reproductive system

1. Primary hypogonadism is due to processes that directly involve the ovaries and thus cause ovarian deficiency, whereas secondary hypogonadism is due to the loss of pituitary gonadotropin function.

2. Gonadal dysgenesis (Turner's syndrome) is a frequent genetic disorder in which there is female internal and external genitalia, several developmental abnormalities, and delayed puberty.
3. Several syndromes are related to abnormal amounts of hormones. The most frequent is polycystic ovary syndrome in which there is obesity, irregular menses, and impaired fertility. The rare Leydig cell tumors and arrhenoblastoma produce testosterone; Granulosa-theca cell tumors produce estrogens; and intraovarian adrenal rests produce cortisol.
4. The radioimmunoassay of hcG is a diagnostic test for these dangerous conditions and is used to monitor efficacy of therapy.

THE THYROID

The thyroid gland weighs about 25-30 gm and consists of closely packed sacs (follicles) filled with proteinaceous colloid. The gland secretes the hormones thyroxine and tri-iodothyronine. Of a total of 50 mg of iodine in the body, about 10-15 mg are in thyroid.

Thyroxine and Triiodothyronine

Chemistry

1. Iodination of tyrosines in thyroglobulin occurs first in position 3 of the aromatic nucleus and then at position 5 forming mono-iodotyrosine and diiodotyrosine, respectively.
2. Coupling of two molecules of diiodotyrosine (I_2 Tyr.) then occurs within the thyroglobulin molecule to form tetraiodothyronine (Thyroxine).
3. Coupling of monoiodotyrosine (I Tys.) with diiodotyrosine (I_2 Tyr.) also occurs to form triiodothyronine (T_3).

Protein Bound Iodine (PBI) in Blood

The term PBI in blood represents iodine present in thyroid hormones. The PBI values for normal adults are 3.5-7.5 mg/100 ml of plasma. PBI is a reliable measure of thyroxine content of plasma. The vlaues for PBI in hypo- and hyperthyroidism are given below.

Myxoedema (Hypothyroidism)	0.2-2.5 mg/100 ml
Grave's disease (Hyperthyroidism)	8-18 mg/100 ml

Functions

1. In hypothyroid animals, the tissues show a low rate of oxygen consumption, the patient has a slow pluse, lowered vigor, obesity, blood cholesterol levels are increased, lipolysis and fatty acid liberation are decreased. In hyperthyroid states, reverse occurs.
2. Thyroxine causes increased intestinal glucose absorption. It increases glycogenolysis in liver and muscle. It promotes neoglucogenesis.
3. It increases RNA, amino acid transport and protein synthesis.
4. In hypothyroidism, there is marked increase in serum cholesterol and triglyceride, and phospholipid contents in blood are also increased. In hyperthyroidism, the serum cholesterol level is decreased.
5. In high concentration of thyroid hormones, calcium is mobilized from bone, there are losses of potassium and nitrogen in urine, and calcium and phosphorus increase in the urine.
6. High concentration of thyroid hormone causes uncoupled oxidative phosphorylation and increased swelling in the mitochondria. Such action results in the production of heat rather than storage of energy as ATP.
7. Thyroxine is essential for the conversion of β-carotene to vitamin A.

Abnormalities of thyroid function

Hypothyroidism causes cretinism in children and myxoedema in adults.

Cretinism

1. Growth in children is retarded.
2. The child is mentally defective. He has coarse scanty hair and thick yellowish scaly skin.
3. Cretinism occurs in areas where goitre is prevalent.

Myxoedema

1. Puffiness of the face and hands in adults.
2. Retention of water and NaCl in the body.
3. BMR is low.
4. Body temperature and pulse rate are subnormal.
5. Body weight is increased due to deposition of fat and retention of water.
6. Mental faculties are retarded.
7. Hypochlorhydria or achlorhydria is present.
8. Blood cholesterol and lipid levels are increased.

Hashimoto's disease

1. Thyroglobulin escapes from the cells of the gland and excites the production of antibodies which produce reactions with thyroid.
2. Fibrosis of the thyroid tissue develops leading to complete loss of thyroid function.

Severe hyperthyroidism leads to toxic goiter. This occurs mostly in women.

Toxic goitre

1. The patient complains of nervousness, restlessness, tiredness, undue sweating, breathlessness on exertion, tachycardia and palpitations.
2. The subject cannot tolerate warm climate but can tolerate severe cold climate.

Grave's disease

1. This disease is the result of hyperthyroidism and is caused by the production of **thyroid-stimulating IgG (TSI)** that activates the TSH receptor.
2. This causes a diffuse enlargement of the thyroid and excessive, uncontrolled production of T_3 and T_4.
3. Findings are multisystemic and include rapid heart rate, widened pulse pressure, nervousness, inability to sleep, weight loss in spite of increased appetite, weakness, excessive sweating, sensitivity to heat, and moist skin.
4. This disease is treated by blocking hormone production with an antithyroid drug, by ablating the gland with a radioactive isotope of iodide (such as ^{131}I). Occasionally, the gland is removed surgically.

THE PARATHYROIDS

The parathyroid gland consists of 4 small glands closely associated with the thyroid. In humans, the parathyroids are reddish or yellowish brown egg-shaped bodies. The total weight of 4 glands is 0.05-0.3 gm.

Parathormone

Chemistry

1. It is a polypeptide consisting of 84 amino acids.
2. Its molecular weight is 9,500.
3. This hormone from different species differ only slightly.
4. It is initially synthesized in the chief cells as preprohormone.
5. The secreted hormone is degraded rapidly; it has a half-life of about 18 minutes.

The normal level of parathyroid hormone in serum is 150-300 Pg/ml.

Functions

1. The parathormone stimulates the membrane bound adenylate cyclase causing increased synthesis of cAMP.
2. It mobilizes calcium and phosphorus from bones.

3. It increases serum calcium and lowers the serum phosphorus.
4. It increases urinary excretion of phosphate but decreases excretion of calcium.
5. It elevates serum alkaline phosphatase activity.
6. It increases the absorption of calcium and phosphorus from the intestine.
7. It activates vitamin D in renal tissue by increasing the rate of conversion of 25-hydroxycholecalciferol to 1, 25-dihydroxycholecalciferol.
8. It increases bone resorption by releasing calcium as well as collagenase, lyosomal enzymes, and hydroxyproline.
9. In the kidney, it affects renal tubular reabsorption of calcium and phosphorus.
10. It may stimulate protein synthesis in the osteoclasts which effect resorption of bone.

Control of release of parathormone

1. The concentrations of parathormone are decreased abruptly by administration of calcium ion and rise when circulating ionized calcium is lowered by the administration of the chelating agent.
2. Calcium loss associated with uremia results in an increase in circulating parathormone.
3. Vitamin A administration decreases parathormone by increasing calcium uptake into the parathyroid gland.

Calcitonin (Thyrocalcitonin)

Chemistry

1. Calcitonin is a polypeptide consisting of 32 amino acids.
2. Its molecular weight is 3,600.
3. There are large variations in the amino acid composition among different species.
4. It is required for biologic activity.
5. The half-life of the secreted calcitonin is about 4-12 minutes.

Functions

1. It is calcium-lowering hormone originated from the C cells of the thyroid gland.
2. It is directly effective on bone. It inhibits bone resorption and mobilization of calcium and phosphorus from bone.
3. In the kidney, it increases calcium excretion and inhibits synthesis of 1, 25-dihydroxycholecalciferol without affecting cAMP levels.
4. It decreases the excretion of urinary hydroxyproline by inhibiting the resorption of the organic phase of bone.

Abnormalities of parathyroid function

Hypoparathyroidism

1. The symptoms of hypoparathyroidism are muscular weakness, tetany and irritability.
2. In case, hypothyroidism begins early in childhood, there may be stunting of growth, defective tooth development, and mental retardation.
3. Serum calcium is low, serum phosphate is increased, urinary calcium is low, serum magnesium and hydroxyproline levels are decreased.
4. Calcium, parathyroid hormone, and vitamin D precursors are used in the treatment of hypoparathyroidism.

Hyperparathyroidism

1. Hyperparathyroidism occurs due to a tumor of the gland. The symptoms of hyperparathyroidism are hypercalcemia, decalcification of bones causing pain and deformities, renal stones.
2. Serum phosphorus is low, serum alkaline phosphatase activity is increased, and excretion of calcium in urine is highly increased.
3. Removal of tumor brings about prompt relief and injection of calcitonin helps to reduce serum calcium level.

1, 25(OH)$_2$–D$_3$ stimulates intestinal absorption of calcium and phosphate

1. 1, 25(OH)$_2$–D$_3$ is the only hormone that can promote the translocation of calcium against the concentration gradient which exists across the intestinal cell membrane. If controls the extracellular fluid Ca^{2+} in spite of marked fluctuations of the calcium content of food.
2. It maintains a proper concentration of calcium and phosphate for deposition on the collagen fibrils in bone.
3. In 1, 25(OH)$_2$–D$_3$ deficiency, new bone formation slows and bone remodelling is also impaired.
4. It also enhances the actions of parathyroid hormones (PTH) on renal calcium reabsorption.

1, 25(OH)$_2$-D$_3$ acts at the cellular level just like steroid hormones

1. There is nuclear accumulation of this hormone in pancreatic islet cells; some brain cells; some cells in the pituitary, ovary, testis, placenta, uterus, mammary gland, and thymus.
2. The 1, 25(OH)$_2$–D$_3$ receptor is a member of the steroid receptor family. The receptor has a DNA-binding domain that appears to contain the zinc finger motif characteristic of other steroid receptors.
3. The binding of the 1, 25(OH)$_2$–D$_3$ receptor to chromatin in the nucleus suggests that 1, 25 (OH)$_2$–D$_3$ stimulates gene transcription and the formation of specific mRNAs.
4. Type 1 vitamin D-dependent rickets is an inherited autosomal recessive trait characterized by a defect in the conversion of 25(OH)-D$_3$ to 1, 25(OH)$_2$–D$_3$.
5. The formation of 1, 25(OH)$_2$–D$_3$ is reduced by the loss of substantial renal parenchyma and calcium absorption decreases. Parathyroid hormone is increased due to hypocalcemia.

THE PANCREAS

Two hormones are secreted by this gland-insulin by beta cells of the islet of langerhans and glucagon by the alpha cells. Delta cells store and secrete somatostatin.

Insulin

In 1962, Abel isolated crystalline insulin from pancreas and the amino acid sequence was established by Sannger in 1953.

Chemistry

1. Insulin is a protein hormone secreted from the β-cells of the islet of langerhans.
2. Traces of zinc are required for the crystallization of insulin.
3. It consists of 51 amino acids and contains two polypeptide chains (A and B) linked together by two disulphide bridges [one is 7-7 and another in 20-19 of A & B chains, respectively]. A third intradisulphide bridge between 6 and 11 amino acids of A chain also occurs.
4. It has a molecular weight of 5,734.
5. Alkali or reducing agents can inactivate insulin by breaking the disulphide bonds.
6. Proteolytic enzymes digest the insulin protein and inactivates insulin. Hence, it cannot be given orally.

   The structure of insulin is given (Fig. 31.13).

Insulin receptor

1. Insulin acts on target tissues by binding to specific insulin recceptors which are glycoproteins.
2. The human insulin receptor gene is found on chromosome 19. The insulin receptors are being constantly synthesized and degraded. Their half life is 6 to 12 hours only.
3. It is synthesized as a single chain polypeptide, pro-receptor in the rough endoplas-

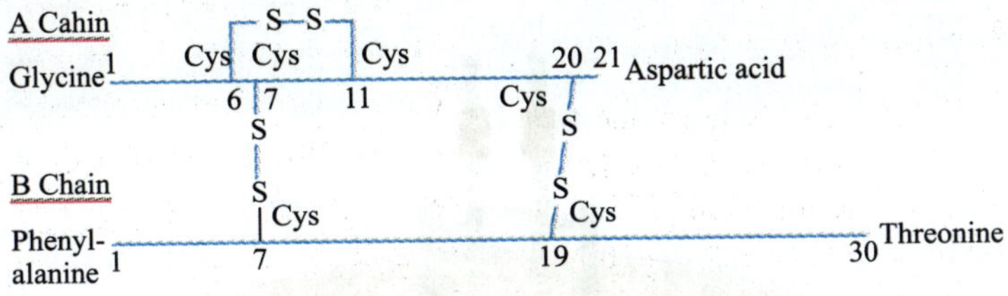

Fig. 31.13 : Structure of human insulin

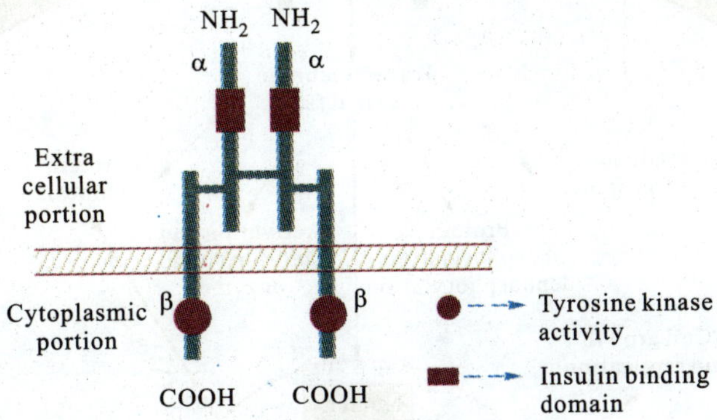

Fig. 31.14 : Insulin receptor

mic reticulum and is rapidly glycosylated in golgi region.

4. The pro-receptor is cleaved to form mature α and β subunits ($\alpha_2\beta_2$) which is heterodimer, linked by S–S bonds.
5. Both subunits are extensively glycosylated and removal of sialic acid and galactose decreases insulin binding and insulin action.
6. Insulin receptors are found in target cell membrane.
7. Though insulin receptor is a heterodimer consisting of 2 subunits, designated α and β ($\alpha_2\beta_2$) linked by disulphide bonds.
8. The α subunit is entirely extracellular and it binds insulin, probably via a cystine-rich domain.
9. The β subunit is a transmembrane protein that performs the second major function of a receptor, i.e. signal transduction and insulin action.
10. Binding of insulin to the receptor stimulates its tyrosine kinase activity, Tyrosine kinase enzyme phosphorylates the phenolic –OH group of tyrosine residues in specific protein including that of a tyrosine in the chain of insulin receptor itself to modulate their activities, ATP + tyrosine protein → ADP + phosphotyrosine protein.

    The cytoplasmic portion of β subunit has tyrosine kinase acitivity and an autophosphorylation site.

Insulin secrection

About 50 units of insulin are required per day. The human pancreas stores about 250 units.

Normal concentration of insulin (fasting) in plasma : 6-126 μU/ml.

Factors stimulating insulin secretion

1. Increased blood glucose level causes an increase in insulin secretion and decreased blood glucose level depresses insulin secretion.

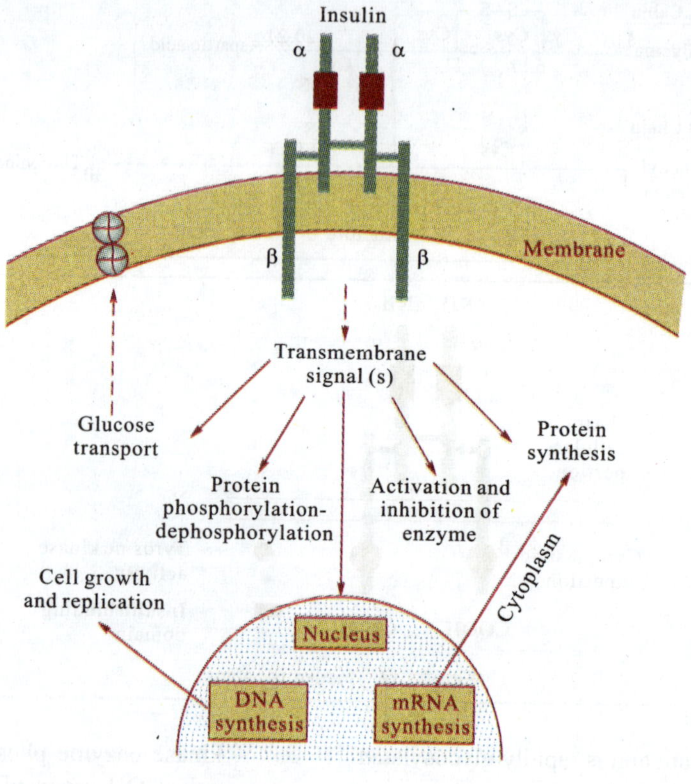

Fig. 31.15 : Relationship of the insulin receptor to insulin action

2. The hyperglycemia produced by glucagon enhances insulin production.
3. Since the growth hormone and glucocorticoids cause hyperglycemia they also stimulate insulin secretion.
4. Sugars which are readily metabolized—eg, mannose and fructose—can stimulate insulin release. But non-metabolised sugars such as glactose, L-arabinose and xylose do not stimulate.
5. Many agents, such as amino acids, fatty acids and some gastro—intestinal products can stimulate insulin release only in presence of glucose.
6. Insulin secretion is enhanced by cAMP, ACTH and thyrotropin.
7. Amino acids particularly leucine and arginine can stimulate pancreas to produce insulin in both vivo and vitro. Proteins like casein also increases secretion of insulin.
8. Central nervous system indirectly influences the release of insulin. Vagal stimulation causes an increase in insulin secretion.
9. Sulfonylureas, the hypoglycemic agent, may act on insulin secretion by a different mechanism than that of glucose.

Factors inhibiting insulin secrection

1. Epinephrine is the highly effective inhibitor of insulin secretion.
2. Starvation reduces insulin secretion.
3. Magnesium also inhibits insulin secretion.
4. Vagotomy reduces insulin secretion.

Metabolism of insulin

1. Insulin is degraded in liver and kidney by the enzyme glutathinone insulin transhydrogenase which brings about reductive cleavage of the S-S bonds that connect A and B chains of the insulin molecule. Reduced glutathione acts as a coenzyme.
2. The A and B chains are further degraded by proteolysis. But when insulin is bound to antibody, it is much less sensitive to enzymic degradation.

Functions

1. Insulin is firmly bound to the highly specific receptor site present in the cell membrane. The receptor may probably be a glycoprotein. The biologic activities of insulins are proportionate to their binding affinities. Insulin, thus, may carry out most of its function without entering the cell. The number of receptors declines where insulin levels are high.
2. Insulin exhibits transport at the membrane site, RNA synthesis at the nuclear site, translation at the ribosome for protein synthesis, and influence on tissue levels of cAMP. It is active in skeletal and heart muscle, adipose tissue, liver, the lens of the eye and leukocytes. It is inactive in renal tissue, red blood cells and gastrointestinal tract. The most metabolic function is centered in the muscle, adipose tissue and liver.
3. It facilitates the transport of glucose and related monosaccharides, amino acids, potassium ion, nucleosides, inorganic phosphate, and calcium ion in muscle and adipose tissue.
4. In muscle for adipose tissue, insulin increases the entry of glucose and thus leads to increased glycogen deposition, stimulation of HMP shunt resulting in increased production of NADPH, increased glycolysis, increased oxidation (Increase in oxygen uptake and CO_2 production), and increased fatty acid synthesis.
5. In adipose tissue, it increases lipid synthesis by means of fatty acid synthesis and glycerophosphate for triacylglycerol synthesis.
6. Insulin increases intracellular concentration of nonmetabolized sugars such as galactose, L-arabinose, and xylose. The hormone facilitates the entry of those sugars having the same configuration at carbons, 1, 2, and 3 as D-glucose. Since fructose having a ketone group at position 2 is not transported by insulin. Intracellular transport of glucose is enhanced by anoxia indicating that glucose transport requires energy.
7. It also increases the uptake of non-metabolizable amino acids such as alpha-aminoisobutyrate. It maintains muscle protein by decreasing protein degradation.
8. In adipose tissue, it quickly depresses the liberation of fatty acids caused by epinephrine or glucagon.
9. Insulin directly increases protein synthesis as the hormone promotes the incorporation of labelled intracellular amino acids into protein. At the ribosomal level, it increases the capacity of this organelle to translate information from messenger RNA to the protein-synthesizing machinery.
10. In the liver, it stimulates glycolysis by increasing the synthesis of glucokinase, phosphofructokinase, and pyruvate kinase. It also depresses the enzymes controlling gluconeogenesis such as pyruvate carboxylase, phosphoenolpyruvate carboxykinase, fructose-1, 6-diphosphatase, and glucose-6-phosphatase. Enzymes which are unimportant in the control of gluconeogenesis as well as glycolysis are not affected by insulin.

Pathophysiology of Insulin

1. About 90% of persons with diabetes have non-insulin dependent (Type II) diabetes mellitus (NIDDM). Such patients are usually obese, have elevated plasma insulin levels.

2. The other 10% have insulin dependent (Type 1) diabetes mellitus (IDDM).
3. A few individuals produce antibodies directed against their insulin receptors. These antibodies prevent insulin from binding to the receptor so that such persons develop a syndrome of severe insulin resistance.
4. Tumors of β-cell origin cause hyperinsulinism thereby hypoglycemia. Leprechaunism is caused by the role of insulin in organogenesis. The syndrome is characterized by low birth weight, decreased muscle mass, decreased subcutaneous fat, and early death.

Abnormal metabolism in diabetic states

1. In diabetes, hyperglycemia occurs due to the impaired transport and uptake of glucose into muscle and adipose tissue. Transport and uptake of amino acids are also depressed causing the raised level of amino acids into the blood, particularly, alanine, which supply fuel for gluconeogenesis in the liver. The amino acid breakdown during gluconeogenesis increases the production of urea nitrogen.
2. Lipid and fatty acid synthesis is decreased due to the decrease in acetyl-CoA, ATP, NADPH and glycerophosphate in all tissues. Stored lipids are hydrolysed by increased lipolysis and the liberated fatty acids interfere the carbohydrate phosphorylation in muscle and liver developing hyperglycemia.
3. Fatty acids in high concentration reaching the liver inhibit fatty acid synthesis by a feedback inhibition at the acetyl-CoA carboxylase step. Increased acetyl-CoA from fatty acids activates pyruvate carboxylase, stimulating gluconeogenic pathway for the conversion of amino acid carbon skeletons to glucose. Fatty acids

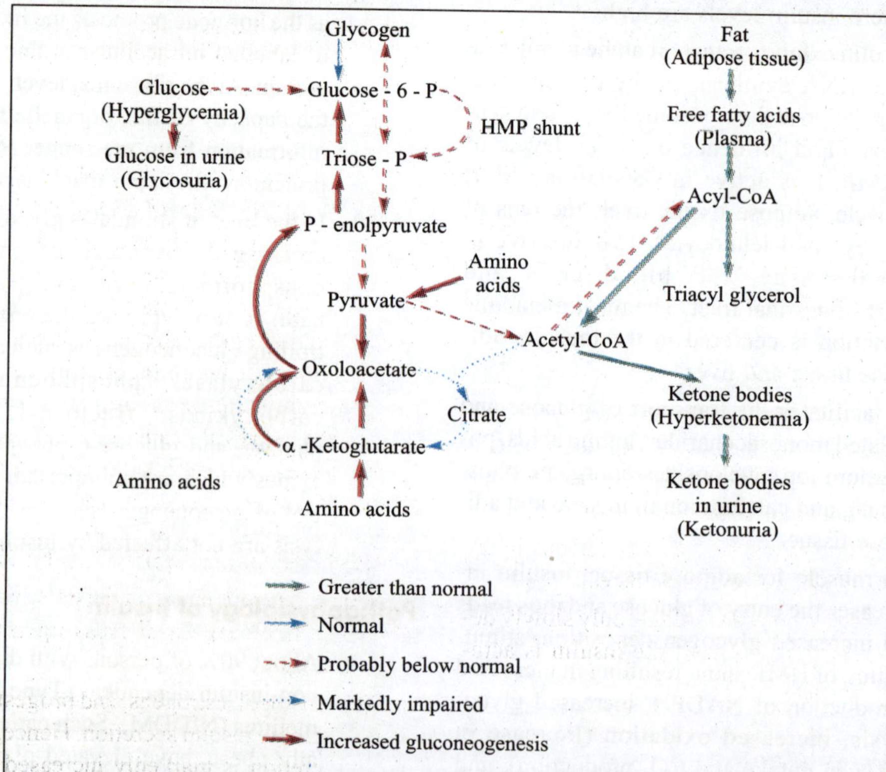

Fig. 31.16 : Abnormal metabolism in the liver during uncontrolled diabetes

also stimulate gluconeogenesis by entering the citric acid cycle and increasing production of citrate which is an inhibitor of glycolysis (at phosphofructokinase). Thus, the fatty acid cycle at the level of citrate synthetase and pyruvate and isocitrate dehydrogenases. The acetyl CoA, which cannot enter the citric acid cycle or cannot be used for fatty acids synthesis, is utilized in the synthesis of cholesterol or ketones or both. The rise in ketone bodies concentration in body fluids and tissues leads to acidosis.

4. Glycogen synthesis is diminished due to decreased glycogen synthetase activity, increased phosphorylase activity and increased ADP : ATP ratio. The phosphorylase activity is stimulated by epinephrine or glucagon.
5. The insulin deficiency causes hormonal imbalance and favours the action of corticosteroids, growth hormone and glucagon which enhance gluconeogenesis, lipolysis, and decreased intracellular metabolism of glucose. The excess glucose in the urine requires water to be excreted out causing dehydration.
6. In the degradation in insulin, both liver and kidney are required. Therefore, in renal or hepatic disease, insulin requirement is decreased. This is observed in some diabetics with associated kidney or liver disease.

Antibodies

1. The repeated injection of insulin produces low levels of an antibody to insulin in all subjects after 2 or 3 months of treatment.
2. The antibodies can produce lesions in the islet cells and severe diabetes.
3. Antibody-bound insulin is only slowly degraded; thus much of the insulin is actually wasted.

Experimental Diabetes

1. Experimental diabetes can be produced by total pancreatectomy or by a single injection of alloxan, a substance related to the pyrimidines or with streptozocin, an N-nitroso derivative of glucosamine.
2. Diabetes can also be produced by injection of diazoxide, a sulfonamide derivative which inhibits insulin secretion.
3. The injection of large amounts of antibodies to insulin is also considered to produce experimental diabetes.
4. **Phlorhizin diabetes** can be produced by the injection of the drug phlorhizin. This is actually a renal diabetes in which glycosuria is only produced by the failure of the reabsorption of glucose by the renal tubules.

Regulation of the Secretion of Insulin

40 to 50 units of insulin is daily secreted from human pancreas. This represents about 15 to 20 per cent of the hormone stored in the gland. Insulin secretion is an energy-requiring process. Different factors are involved in insulin release.

1. Glucose

(a) The increased concentration of glucose is the best regulator of insulin secretion.

(b) Among two ideas, one idea suggests that glucose combines with a receptor which is located on the B cell membrane that activates the release mechanism. The second idea suggests that intracellular metabolites pass through a pathway such as the HMP shunt, the TCA cycle, etc.

2. Hormonal factors

(a) Epinephrine inhibits insulin release.

(b) Beta adrenergic agonists stimulate insulin release by increasing intracellular cAMP.

(c) Cortisol, estrogens, and progestins also increase insulin secretion. Hence, insulin secretion is markedly increased during the later stages of pregnancy.

3. Pharmacologic agents

(a) Many drugs stimulate insulin secretion, but the **sulfonyl urea compounds** are used for therapy in humans.

(b) Drugs such as **tolbutamide** stimulate insulin release and effectively used in the treatment of type 11 (non-insulin-dependent) diabetes mellitus. This class of drug is binded by a receptor which has been derived from the pancreatic β cells.

Affect of Insulin on Gene Expression

(a) The actions of insulin are found to occur at the plasma membrane level or in the cytoplasm.

(b) The synthesis of phosphoenolpyruvate carboxykinase (PEPCK) which catalyzes a rate-limiting step in gluconeogenesis is decreased by insulin and hence gluconeogenesis is decreased.

Insulin deficiency

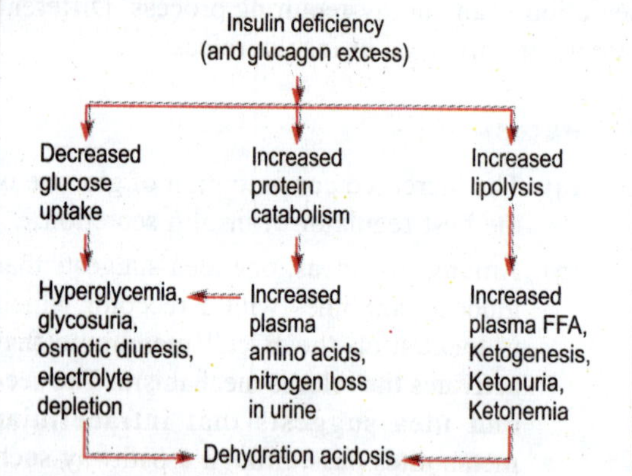

(c) Transcription is decreased due to the decreased amount of the primary transcript and of mature mRNA PEPCK which in turn is directly related to the decreased rate of PEPCK synthesis.

(d) More than 100 specific mRNAs are affeccted by insulin, and a number of mRNAs in liver, adipose tissue, skeletal muscle, and cardiac muscle. Some examples of the affect of insulin on gene transcription is noted (Table 31.1).

Table 31.1 : Messenger RNAs regulated by insulin
Intracellular enzymes
Phosphoenolpyruvate carboxykinase
Fatty acid synthase
Pyruvate kinase
Glucokinase
Glycerol-3-phosphate dehydrogenase
Secreted proteins and enzymes
Albumin
Amylase
Growth hormone
Structural proteins
δ-crystalline
Proteins involved in reproduction
Ovalbumin
Casein
Other proteins
Liver
Adipose tissue
Cardiac muscle
Skeletal muscle

Comparison between Insulin and Insulin-like Growth Factors (IGFs)

	Insulin	IGF-I	IGF-II
Number of amino acids	51	70	67
Source	Pancreatic β cells.	Liver & other tissues	Diverse tissues
Major physiologic role	Control of metabolism	Skeletal & cartilage growth	Perheps in embryonic development
Level regulated by	Glucose	Growth hormone	Unknown

GLUCAGON

Chemistry

1. Glucagon is a polypeptide with a molecular weight of 3,485.
2. It consists of 29 amino acid having 15 different amino acids. The amino acids are arranged in a straight chain.
3. It contains no cystine, proline or isoleu-

cine but contains sufficient amounts of methionine and tryptophan.

4. It can be crystallized in the absence of zinc or other metals.
5. A proglucagon precursor of about 9,000 daltons is also identified.
6. It originates primarily in the alpha-cells of the pancreas, although a considerable amount comes from extrahepatic alpha-cells in the stomach and other portions of the gastrointestinal tract.
7. A glucagon-like immunoreactive factor (GLI) is also identified in gastric and duodenal mucosa but less active than pancreatic glucagon.

Factors influencing secretions of glucagon

1. Low blood glucose increases secretion of pancreatic glucagon.
2. Most amino acids, particularly arginine, cause a rapid secretion of glucagon from the pancreas.
3. Fatty acids inhibit glucagon release.
4. Exercise stimulates the secretion of glucagon.
5. During mixed meals, both insulin and glucagon are secreted, but the carbohydrate meal causes insulin release. High protein meal favours glucagon secretion.
6. In stress, insulin secretion is inhibited but glucagon secretion is stimulated.

Functions

1. Glucagon is sensitive to the adenylate cyclase receptor sites in the liver and increases cAMP level. cAMP activates the enzyme protein kinase which further activates phosphorylase. Phosphorylase causes glycogenolysis making available of glucose in blood. cAMP suppresses glycogen synthetase. Thus, glucagon increases glycogen breakdown and inhibits synthesis of glycogen.
2. Glucagon can also activate various phosphoenzymes by activating portein kinase and inhibit dephosphoenzymes. Thus, glucagon can directly stimulate gluconeogenesis by activation of pyruvate carboxylase and fructose-1, 6-diphosphatase.
3. It also inhibits glucose oxidation by inhibiting pyruvate kinase and pyruvate dehydrogenase.
4. Elevated glucagon concentration also increases ketogenesis.
5. It increases potassium release from the liver.
6. In adipose tissue as well as in liver, it increases the breakdown of lipids to fatty acids and glycerol.
7. Crystalline glucagon polypeptide is used for the treatment of hypoglycemic persons. Glucagon is also used as a diagnostic test for glycogen storage disease.

Normal level of glucagon (fasting) in serum : 20-100 Pg/ml.

PINEAL BODY

Melatonin

Melatonin is formed from serotonin.

Functions

1. It is antagonist to melanocyte stimulating hormone and lightens skin colour.
2. Pinealectomy in female rats stimulates estrus and causes ovarian hypertrophy.

KIDNEY

Kidney secretes a hormone called erythropoietin.

Erythropoietin

Chemistry

1. It is a glycoprotein with a molecular weight of about 46,000.
2. It contains 30 per cent carbohydrate derivative.

Function

It is essential for the formation of red blood cells.

Hormones of gastrointestinal tract

Described in digestion and absorption chapter.

Somatostatin

1. Somatostatin is a cyclic peptide synthesized as a large somatostatin prohormone in the delta cells of the pancreatic islets.
2. It has a moelcular weight of 1,640 and contains 14 amino acids.
3. It is also found in gastrointestinal tissues and is a neurotransmitter.
4. It inhibits the release of other islet cell hormones through a paracrine action.
5. It also decreases the delivery of nutrients from the gastrointestinal tract into the circulation, because it prolongs gastric emptying, decreases gastric secretion as well as digestive secretion and slows sugar absorption.

Pancreatic polypeptide (pp)

1. It is a peptide of 36 amino acids.
2. It is the recently discovered product of the pancreatic F cells.
3. Its secretion in humans is increased by a protein meal, fasting, exercise, acute hypoglycemia and is decreased by somatostatin and intravenous glucose.
4. Its function is unknown but it has effect on hepatic glycogen levels and gastrointesti-nal secretion.

CLINICAL ORIENTATION

- Diseases of the endocrine system due to excessive or deficient production of hormones are the guide for the basic principles to clinical medicine. The physiologic and biochemical effects of the individual hormone enable one to know endocrine disease syndrome that results from hormone imbalance.

CLINICAL ORIENTATION

- The atrophy of the thyroid, adrenal cortex and gonads are due to the loss of anterior pituitary.
- The effects due to the absence of the hormones secreted by the target glands affect most body organs and tissue and many general processes like protein, fat, carbohydrate and fluid and electrolyte metabolism.
- The loss of posterior pituitary function results in diabetes insipidus.
- Radioactive iodine is widely used in the diagnosis and treatment of thyroid disorders.
- Excessive exposure of radioiodine is a major risk factor for thyroid cancer. This is true in infants and adolescents.
- Many glucocorticoid analogs are anti-inflammatory agents. Excessive or deficient plasma levels of these hormones result in serious, sometimes life-threatening, complications. A series of inherited enzyme deficiencies are involved in steroidogenesis.
- The hormones of the sympathoadrenal system are needed for adaptation to acute and chronic stress. Epinephrine and norepinephrine are the principal hormones in the response to severe stress. This response is involved in many complex processes in the organs like brain, muscles, and liver at the expense of other organs like skin, gastrointestinal system and lymphoid tissue which are less immediately involved.
- Catecholamines do not facilitate the stress response alone but are helped by the glucocorticoids, growth hormone, vasopressin, glucagon and angiotensin 11.
- The gonadal hormones are anabolic and are required for maintenance of metabolism of skin, bone and muscles.
- Insufficient action of insulin leads to diabetes mellitus and glucagon aggravates this condition. Many organ systems are involved by signs and symptoms and accurate diagnosis is very difficult unless the physicians are not quite aware of the syndrome.

EXERCISE

1. Discuss the influence of different hormones on carbohydrate metabolism. (R.U. 72A; M.U. 73A)
2. Give an account of the chemistry, secretion and functions of Insulin.(Punj. University 68A; P.U. 70A)
3. Discuss the mechanism of action of homones. (C.U. 92A)

CHAPTER 32

THE ENERGY REQUIREMENTS OF THE BODY

Atwater of U.S.A., during 1892-1902, established clearly the relations between food consumed, heat output, O_2 consumed and CO_2 output by conducting experiments with human subjects using an improved type of human respiration calorimeter. The other scientists H. Armsby, F. Benedict, and E. F. Du Bois did extensive study on energy metabolism.

The energy of the body is derived from the oxidation of the food. A small part of the daily energy output results in the breakdown of the animal's own tissues. The potential energy of the food provides the kinetic energy of the body which is mainly in the form of heat and work. The potential energy of the food is the heat given out when the food is completely burnt to CO_2, water and nitrogen. This is greater than the energy which the animal derives from the food because of the considerable potential energy contained in urea and other nitrogenous compounds being excreted out.

The unit energy contained in foods is expressed as a unit of heat, the calorie.

Calorie : A calorie is defined as the amount of heat required to raise the temperature of 1 gm of water by 1°C.

Kilocalorie (C) : It is 1,000 times the energy for the small calorie. [Since the calorie is very small the kilo calorie (C) is used in nutrition.]

The international unit of energy is the joule (10^7 ergs).

$\therefore$ 1 kilo calorie = 4.2 kilo joules (kJ).

Measurement of energy value of foods

The foodstuffs (carbohydrates, fats and proteins) on combustion by oxygen produces heat. This amount of heat can be measured in a **bomb calorimeter**.

Carbohydrate and fat are completely oxidized in the body to CO_2 and water like that of bomb calorimeter. But proteins are not completely burned because urea, the end product of protein metabolism, still contains some energy which is not available to the body. Therefore, the energy value of protein in the body (4.1 kcal/gm) is less than that obtained in the bomb calorimeter. The energy value of foods in the body is customary to express in round figures. The table below shows the energy value of foodstuffs.

	Kilocalories per gram (C)	
	In bomb calorimeter	In the body
Carbohydrate	4.1	4
Fat	9.4	9
Protein	5.6	4

THE ENERGY REQUIREMENT OF MAN

The energy requirement of man can be determined either directly by measuring his output in a calorimeter or indirectly by measuring his oxygen consumption.

A. Direct method

1. This is the most accurate method but is time-consuming, laborious, and demanding very expensive apparatus.
2. The subject is placed in an insulated chamber. His heat production is measured directly by recording the total amount of heat transferred to a weighed quantity of water circulating through the calorimeter.
3. The oxygen intake, the CO_2 output, and the nitrogen excretion in the urine and feces are also measured during the whole period of observation.

B. Indirect method

1. This method consists of measuring gas exchange and determining the respiratory quotient.
2. This method is simple and applicable to field studies and clinical analysis.

Respiratory Quotients (RQ) of foodstuffs : It is the ratio of the volume of CO_2 eliminated to the volume of oxygen utilized in the oxidation.

$$\text{i.e., R.Q.} = \frac{\text{Volume of } CO_2 \text{ eliminated}}{\text{Volume of } O_2 \text{ used}}$$

1. **Carbohydrates :** The complete oxidation of glucose is as follows :

$$C_6H_{12}O_6 + 6O_2 \longrightarrow 6CO_2 + 6H_2O$$

$$\therefore \text{R.Q.} = \frac{6CO_2}{6O_2} = 1.$$

2. **Fats :** The oxidation of tristearin is as follows :

$$2C_{57}H_{110}O_6 + 163O_2 \longrightarrow 114CO_2 + 110H_2O$$

$$\therefore \text{R.Q.} = \frac{114CO_2}{163O_2} = 0.70.$$

Fats have a lower R.Q. value because the oxygen content of their molecule in relation to carbon content is very low.

3. **Proteins :** Since the chemical structure of proteins is variable their oxidation cannot be so readily expressed. The R.Q. of proteins by indirect method has been calculated to be about 0.8.
4. **Mixed diets :** The R.Q. of mixed diets is about 0.85. The R.Q. is lowered if the carbohydrate metabolism is impaired.

Total heat production

To obtain the heat production, the grams of each foodstuff oxidized are multiplied by the caloric value of that food. The sum of these caloric values is equal to the total heat production of the diet. Therefore,

Carbohydrate	210.7 gm × 4 kcal	=	842.8 kcal
Fat	84.4 gm × 9 kcal	=	759.6 kcal
Protein	80.0 gm × 4 kcal	=	320.0 kcal
Total heat production		=	1,922.4 kcal

CLINICAL ORIENTATION

- R.Q. increases in acidosis since CO_2 output is greater than O_2 consumption during acidosis.
- In alkalosis, respiration is depressed and less CO_2 will be produced. Hence, R.Q. will fall.
- In febrile conditions, the body temperature rises due to high fever and causes increased breathing for which CO_2 production increases. Hence, R.Q. increases.
- Carbohydrate oxidation is less in diabetes mellitus with the increase in fat oxidation. Hence, R.Q. falls. But on insulin administration, carbohydrate oxidation increases with more CO_2 output. Hence, R.Q. rises.
- R.Q. falls in starvation due to oxidation of stored glycogen and fat.

SIGNIFICANCE OF R.Q.

1. R.Q. helps in the determination of metabolic rate.
2. It is the guide for assessing the type of food burning or the nature of synthesis taking place in the whole body or in any particular organ.

3. The determination of R.Q. aids in the diagnosis of acidosis, alkalosis and diabetes mellitus, etc.
4. Non-protein R.Q. is used for calculating the total energy output and the proportions of various foodstuffs being burnt.

BASAL METABOLISM

Definition

The total heat produced or the energy spent by the body under conditions to perform minimum possible work is known as basal metabolism. The lowest level of energy production consonant with life is the basal metabolic rate (BMR).

Conditions for measurement of BMR

1. Post-absorptive state : The patient should not take anything by mouth for the past 12 hours.
2. Complete mental and physical relaxation.
3. Patient should be awake.
4. Recumbent position during the test.
5. Enviromental temperature should be between 20°C-25°C.

Factors influencing BMR

1. *Surface area* : The BMR is directly related to the surface area of the subject. Larger the surface area greater would be the BMR.
2. *Age* : The BMR is inversely proportional to age. Children have larger BMR than adults.
3. *Sex* : Males have higher BMR than females. The BMR of females declines more rapidly between the ages of 5 and 17 than that of males.
4. *Climate* : The BMR is lower in warm climates.
5. *Racial variations* : The BMR of oriental women living in the USA is 10 per cent below the standard BMR of American women of the same age.
6. *Habit* : Persons accustomed to heavy exercise or hard physical work have a higher BMR than those involved in sedentary work.
7. *State of Nutrition* : The BMR is decreased in starvation and undernourishment.
8. *Pregnancy* : The BMR is not changed during pregnancy. The higher values of BMR in late pregnancy is due to the BMR of the foetus.
9. *Disease* : The BMR is increased in infectious and febrile diseases. The increase is usually proportional to the rise of the temperature. The BMR is also increased in increased activity of cells and, therefore, it increases in leukemia, cardiac failure, hypertension, polycythemia, dyspnea, and some types of anemia.
10. *Effects of Hormones* : The BMR is increased in hyperthyroidism and decreased in hypothyroidism. In adrenal insufficiency (Addison's disease), the BMR is subnormal.
11. *Drugs* : Caffeine and benzidine increase the BMR.
12. *Emotion* : The BMR is increased in emotional stress.
13. *Diet* : The BMR of strict vegetarians is 11% lower than that of meat eaters.

Measurement of BMR

1. The BMR can easily be measured either by the apparatus of Benedict and Roth (closed circuit device) or by Douglas bag method (open circuit device). Benedict-Roth's apparatus is abundantly used in hospitals and laboratories.
2. This method provides sufficient accuracy only by measuring the oxygen consumption of the patient for two 6-minute periods under basal conditions. The average of two such readings is mulitplied by 10 to obtain the hourly consumption of oxy-

gen and then multiplied by 4.825 kcal, the heat produced by one litre of oxygen. This gives the total heat produced in kcal/hour.

3. This above value is divided by the surface area of the subject to get the BMR in C/hr/sq.m.

The simple classical formula of Du Bois is applied which is as follows :

$$A = H^{0.725} \times W^{0.425} \times 71.84,$$

where A = surface area in cm^2.

H = height in cm.

W = weight in kg.

The above value obtained is in sq. cm. This should be divided by 10,000 to get the surface area in sq. metre.

A BMR between – 15% and + 20% is considered normal. The BMR may exceed + 50% to + 75% in hyperthyroidism. In hypothyroidism, the BMR may be – 30% to – 60%. The normal BMR is 39.5 C/m^2/hr. The BMR increases as much as 600%-800% over basal during exercise.

Significance of BMR

1. The determination of BMR is the principal guide for diagnosis and treatment of thyroid disorders.
2. The BMR – 10% of the normal develops moderate hyperthyroidism. In severe hypothyroidism, the BMR may be decreased to 40 to 50 per cent below normal.
3. It aids to know the BMR to have an idea of the total amount of food or calorie required to maintain body weight or calorie.
4. The BMR is below normal in starvation, under nutrition, hypothalamic disorders, Addison's disease, and lipoid nephrosis.
5. The BMR is above normal in fever, diabetes insipidus, leukemia and polycythemia.

SPECIFIC DYNAMIC ACTION (SDA)

Definition

It is defined as the extra heat production over and above the caloric value of a given amount of food when used by the body.

Example : An amount of protein which contains 100 kcal (25 gm) when metabolized in the body, the heat production is not 100 kcal but 130 kcal. This extra 30 kcal is the product of the SDA of protein. Similarly, in the body, a 100 kcal portion of fat produces 113 kcal, and a 100 kcal portion of carbohydrate produces 105 kcal. This extra heat is due to the activity of tissues metabolizing these foodstuffs.

When these above foods are taken in a mixed diet, the SDA is not the total of the SDA of each foodstuff when fed separately. According to Forbes, when the glucose and protein are combined, the SDA is 12.5% less than the sum of their individual effects. The SDA is 22% less when lard, glucose, and protein are combined, the SDA is 35% less in case of glucose-lard combination, and 54% less in case of protein-lard mixture than the sum of their individual SDA.

The high SDA of protein can be reduced depending on the quantities of other foodstuffs in the diet. It has been found in Forbes's data that fat (lard) has much influence on SDA than does any other nutrient, i.e., fat decreases the SDA more than any other nutrient.

When different amino acids are fed, alanine, glycine, and phenylalanine are found to produce high SDA. According to Krebs, the high SDA is due to two main factors :

1. The energy required for deamination of amino acids which is again derived by the oxidation of other metabolites.
2. The energy required for the synthesis of urea which is obtained by the oxidation of other metabolites present in tissues.

There is evidence also that the increment is

greater during the porcess of lipogenesis from glucose than during its oxidation to CO_2 and H_2O.

The SDA of carbohydrate is to represent the energy liberated in excess of that required for the conversion of glucose to glycogen.

The SDA of fat is due to the increased concentration of fat in the tissue fluids and to its more rapid oxidation.

The glands of internal secretion have no direct influence upon the SDA of protein. Thyroidectomy in animals reduces the SDA of cabohydrate and fat.

A diet very rich in protein is unfavourable to heavy muscular work. The SDA of protein is an important factor in the regulation of body temperature. With fat and carbohydrate, the extra heat is accelerated in the performance of work. When new tissue is formed, protein does not exert its usual SDA.

CALORIC REQUIREMENTS

The food requirements of a man depend upon his basal metabolism, S.D.A., the work to be done and the climatic conditions. Extra calorie is also required for a starved man to build up his tissues, a pregnant woman for the foetus and a child for his growth.

A man of 70 kg weight and 180 cm height with a surface area of 1.8 square metre, aged 30 and a BMR of 40 will have a basal metabolism of 1,800 calories per day. The allowance for S.D.A. will be 200 calories. Extra calories are required to be added to this on the basis of the work done by the individual. Total daily energy requirements of different categories of work done by individuals are mentioned below.

Very heavy work (lumbermen) entails the expenditure of 8,000 calories.

In long bicycle races, 10,000 calories are required in 24 hours. Calories expended per hour at various occupations are given below :

* This below table is according to the recommendations of the Food and Nutrition Board of the National Research Council.

One's food intake varies from day to day depending on his own habit. Sunday is made a high

Recommended Daily Caloric Requirements

		Male	Female
Adults * :			
Sedentary (Tailors, Shoemakers)		2,400	2,000
Moderate activity (Farmers, Soldiers)		3,000	2,400
Heavy work (Navies, Lumbermen, Stonemasons)		4,500	3,000
Light work (Carpenters, Painters)		2,500	2,000
Pregnancy (Latter half)		....	2,400
Lactation		....	3,000
Children :	16-20 years	3,800	2,400
	13-15 years	3,200	2,600
	10-12 years	2,500	2,500
	7-9 years	2,000	2,000
	4-6 years	1,600	1,600
	1-3 years	1,200	1,200
	under 1 year	110/kg	110/kg

* Men 70 kg, Women 56 kg

Approximate increments in caloric requirements (above basal) per hour for different occupations

Occupation or Activity	Increased requirement Cal./hour.
Sitting quietly	35
Standing quietly	40
Reading loudly	40
Tailor	70
Typing	75
Housework	110
Painter	145
Carpenter	150
Walking (Moderate)	235
Sawing wood	380
Walking fast	550
Walking upstairs	1,000

calorie day. Sunday's consumption is 36% above the average for the six days.

Women have a basal metabolic rate 7%-10% lower than that of men, and they are smaller in size, so their average caloric requirements are less. In a well-to-do home, the adults are not engaged in manual labour and may require less food than adolescents.

The average supplement for muscular work is 600 calories. The nervous and endocrine "make-up" or habits of an individual may alter his requirements. No certain evidence is yet known that mental work exhibits any significant increase in metabolism.

EXERCISE

1. Define B.M.R. and describe a method for measuring it. Discuss its significance. (R.U. 67, 70A ; Luck. 70A)
2. What is B.M.R. ? Discuss the factors affecting the B.M.R. (M.U. 73A)
3. What are the methods of determining energy requirement of man ? (Luck. 68A)
4. Write short notes on :
   (a) B.M.R. (P.U. 73A, 75A, 76A ; M.U. 72A ; Mith. 71A ; R.U. 76A)
   (b) R.Q. (R.U. 72A)
   (c) S.D.A. (P.U. 71A, 74S ; Luck. 76S)
   (d) Daily caloric requirements. (Luck. 72A ; Punj. 73S)

CHAPTER 33

PRINCIPLES OF NUTRITION

A correct diet must provide for the maintenance of the body as well as energy requirements, for growth and reproduction. The essential elements lost by the body by excretion must be replaced. The more important factors are :

1. Energy value.
2. Quality and quantity of primary foods, minerals and vitamins.
3. Variation in the diet.
4. Digestibility.
5. Cooking.
6. Psychological factors.
7. Cost.

1. Energy Value

The average caloric requirement of the adult male is 3,000 daily.

The figures of the daily caloric requirement of York in 1900 are given below :

Workhouse	3,702 Calories.
Prison (Class B)	3,038 Calories.
Prison (hard labour)	4,159 Calories.

2. Quality and Quantity of the Constituents of Food

(A) Primary foods (Protein, fat and carbohydrate) :

1. Protein, fat and carbohydrate are consumed in the ratio of 1 : 1 : 4 in this country.
2. 3,000 calories are provided by 1,00 grams of proteins, 100 grams of fat and 400 grams of carbohydrate.
3. Although the 1:1:4 ratio is prevalent in this country still wide variations of carbohydrate and fat may occur without harm.
4. It is advisable that 10%-15% of the total calories should be obtained from protein, 20%-35% from fat and 50%-66% from carbohydrate.

Proteins

Although the expected amount of protein is 100 gm per day, less amount can also maintain health and nitrogen equilibrium. Even 1 gm per kg body weight may be ample. Larger amounts are necessary for growth, pregnancy and lactation. Increased amounts may be required in wasting diseases. The protein requirements vary with the net protein utilization (NPU) of the dietary proteins. If the NPU is low, the requirements are high and if the NPU is high, the requirements are low.

The requirement of protein in the diet is not only quantitative, the qualitative aspect is also important since the metabolism of protein is connected with that of its constituent amino acids. Certain amino acids are called 'essential' because they must be obtained preformed and cannot be synthesized by the animal organism. The remainder, the so-called non-essential, are also required by the organism since they are found in the protein of the tissues, but they can be synthesized from α-keto acids by amination. Histidine in the diet is also necessary to maintain growth during childhood. Such amino acids are said to be relatively essential. The nutritive value of a protein depends on its content of essential amino acids.

Some proteins, if they are the sole source of nitrogen, will not support life, whereas others are sufficient. Hence, proteins are of two types—first class or good protein and second class or poor protein. First class protein can support life in the absence of other forms of protein. Proteins of animal origin (meat, milk and eggs) are almost completely utilized if taken alone whereas, 10% to 40% of vegetable proteins (peas, beans, potatoes) may remain unabsorbed and excreted in the feces. The absorption of vegetable proteins is improved if taken with other foodstuffs. Animal proteins are associated with fat, but very little carbohydrate (except milk), whereas vegetable proteins are almost associated with a large amount of carbohydrate.

It has been observed that the dietary requirement for protein is influenced markedly by the level in the diet of fat and carbohydrate. These latter foodstuffs have a "Protein-sparing" effect. Fat, primarily, functions as a fuel (9 cal/gm), carbohydrate also serves as a fuel (4 cal/gm), but is required for the synthesis of certain catalytic compounds of metabolic cycles (eg, oxaloacetate in the TCA cycle) and provides the carbon skeletons for the synthesis of the nonessential amino acids. At least 5 gm of carbohydrate per 100 calories must be supplied if nitrogen equilibrium is to be maintained. Protein has a more catalytic function in the form of enzymes. In the absence of fat and carbohydrate from the diet, protein is degraded to provide fuel (4 cal/gm) and catalytic compounds of metabolic cycles. Essential amino acids may be broken down to supply the materials for the synthesis of nonessential amino acids. For these increased burdens, the protein intake is ultimately increased. The protein requirement is decreased when fats and carbohydrates are taken alongwith.

Proteins differ in biologic value depending on their content of essential amino acids. The proteins of eggs, dairy products, kidney and liver have high biologic value since they contain all of the essential amino acids. Good quality proteins include soya beans, peanuts, potatoes and the muscle tissue of meats, poultry and fish. Fair proteins are cereals and most root vegetables. Poor biologic value includes nuts and legumes. The mixture of poor and fair can provide good biologic value.

Proteins are also the important sources of nitrogen, sulphur, and phosphorus for the body. Many amino acids have specific functions in metabolism such as methionine as a methyl donor, cystine as a source of – SH groups, the dicarboxylic acids (aspartic acid, glutamic acid) in transamination, tryptophan as precursor of niacin, arginine in the urea cycle, etc.

The protective substances in a diet are those which are essential for specific purposes other than energy production, i.e., first class proteins, mineral elements, and vitamins.

Fats

Fat has high fuel value. It increases the palatability of foods. It has got high capacity to be stored as energy in the body. Fat provides the essential fatty acids such as linoleic, linolenic and arachidonic acids. Fats are essential for the absorption of fat-soluble vitamins. Excess of saturated fat and cholesterol consumption is associated with hypercholesterolemia and atherosclerosis. A lack of fat causes a feeling of hunger shortly after a meal due to rapid digestion in its absence. Fat reduces the bulk of the food.

The consumption of fat usually is over 80 gm per day. It may increase up to 150 gm with an increase of family income.

As a source of energy there is little difference between animal and vegetable fats. Animal fats have a better biologic value because they contain vitamins A and D in particular. Animal fats are most used when the incidence of sunlight is low.

Fats provide fat-soluble vitamins and the essential fatty acid linoleic acid which is required for the synthesis of arachidonic acid from which prostaglandins are synthesized in the body. The major symptoms of essential fatty acid deficiency are a scaly dermatitis, hair loss, and poor wound healing. Linoleic acid is widely distributed in the lipid portion of both plant and animal foods; vegetable seed oils are especially rich sources.

In societies where fat is the principal food for energy intake, the population tends to develop coronary heart disease, obesity, and cancer of the bowel and breast.

Blood cholesterol level can be reduced significantly by changing the diet containing less saturated fatty acids or cholesterol. The serum cholesterol level is the major factor for the development of atherosclerosis.

Carbohydrates

Carbohydrate is the cheapest source of food. It can be readily digested, absorbed and utilized for producing energy. It is most efficient source of energy for vital processes. It can furnish 60%-80% of the total caloric intake in the scarcity of proteins and fats. It is bulky and liable to undergo fermentation producing acid (eg, lactic acid) if digestion is delayed. Hence, it is not wise entirely to replace fat by carbohydrate.

In gluconeogenesis, carbohydrate can be supplied from most amino acids as well as from the glycerol moiety of fats. A minimum of 5 gm of carbohydrate per 100 kcal of total diet is necessary to prevent the development of ketosis. Sufficient carbohydrate should be included in the diet to have complete oxidation of fat. The amount is near about 400 gm per day.

The chief carbohydrates are starch and sugars (from milk, fruits and vegetables as well as sucrose). Glycogen is only consumed in living tissues such as oysters. Carbohydrates are almost entirely derived from vegetable sources. Cellulose stimulates peristalsis. Some cereals, eg, oatmeal contains much phytin as to interfere with the absorption of calcium.

Galactose is synthesized in the body and lactose is also synthesized from galactose in the body. Hence, no need of giving lactose to lactating women. Pentoses are absorbed and excreted unchanged except ribose or deoxyribose.

Sugar is valuable for muscular work in minimizing fatigue.

Sucrose (common table sugar) is one of the major etiologic factors in dental caries. This sugar is not the cause of diabetes, heart disease, or obesity.

Within this century, decrease in fibre consumption in industrialized countries has been accompained by a high incidence of diverticulosis, colon cancer, cardiovascular disease, and diabetes. A large amount of dietary fibre decreases bowel transit time, alters the composition on intestinal bacteria, reduces enterohepatic circulation of cholesterol, delays intestinal sugar absorption, and reduces the absorption of certain minerals.

(B) Minerals

Some minerals are always required since small amounts of inorganic salts are always excreted. Fruits, vegetables and cereals are the chief sources of the mineral elements in the diet. Milk products supply the majority of the calcium and phosphorus in the diet.

Calcium

1. Increased amounts of calcium are required by children, pregnant and lactating women.
2. Children, whose normal intake is low, can absorb a higher percentage of calcium ingested than those with a high intake.
3. In many eastern countries, where little milk is taken, the average consumption by adults is about 400 mg daily but this low intake does not show any signs of calcium deficiency.
4. The average intake should be about 800 mg per day.

5. The best sources are cheese and milk. Other sources are eggs, green vegetables, oranges, nuts, buttermilk, beans, and carrots. Meat, fish and fruits are poor sources of calcium.

   Hard water is a source of calcium.

Phosphorus

1. This is taken in organic and inorganic form.
2. Children should intake 1 gm per day, adults 1.3 gm and a pregnant or lactating woman 1.9 gm.
3. The best sources of phosphorus are animal foods such as meat, fish, milk, cheese and eggs. Considerable proportions of the total phosphorus are provided by the vegetable foods such as beans, rye, oatmeal and lentils. Cereals provide phosphorus in the form of phytic acid which interferes with the absorption of calcium. Hence, animal sources of phosphorus are better.

Iron

1. It is an important constituent of the diet and its deficiency leads to anemia.
2. The more concentration of iron is from bread, meat and potatoes. Good sources are liver, kidney, egg yolk, green peas, cabbage, carrots and cereals.
3. Only less than 10% of iron is absorbed. In iron-deficient subjects about 20% may be absorbed.
4. Milk is very poor in iron. The infant is born with a store of iron which maintains for six months. After this period, iron is given along with food. Human milk contains 1-2 mg iron per litre, cow milk less.
5. Children up to 12 years require 15 mg to 20 mg iron daily, adults require 40 mg daily, and the daily requirement of iron during pregnancy and lactation is 50 mg.

Iodine

1. The daily requirement of iodine is 0.05 mg and iodine is obtained from water, vegetables and fish.
2. In certain localities, where the soil and water lack this element, simple goitre results due to dietary deficiency.
3. Fish, cod liver oil and vegetables can supplement iodine deficient diet.

Magnesium

1. The average daily consumption of about 0.2 gm is sufficient.
2. Good sources are meat, green vegetables and bread.

Copper

1. The daily adult requirement is 2 mg.
2. Good sources are liver, oysters, cocoa, nuts etc.
3. Human and cow milks only contain about 0.6 mg per litre.

Sodium Chloride

1. Ordinarily 20 gm of the salt is consumed per day.
2. The intake of the salt is only required if abnormal quantities are lost in sweat during severe muscular exercise at high temperature.
3. Heat stroke is prevented by ingestion of salt solutions instead of water.

Vitamins

1. The chemistry and physiologic functions of vitamin are already discussed in Chapter 15. Normal persons get sufficient vitamins in their diets and hence supplementation with vitamin is not required. Only in diseased state in which digestion and absorption are impaired, the supplementation with vitamin is most essential.

2. Many of the vitamins are destroyed by cooking and storage. Hence, fresh fruits and vegetables should be taken daily.
3. Cereals, when refined, lose B vitamins. Therefore, food is improved by the addition of vitamins.
4. The daily requirements of the vitamins are also discussed in Chapter 15.
5. Deficiency symptoms of water-soluble vitamins consist of dermatitis, anemia, digestive difficulties, and neurologic disorders.
6. Most of the water-soluble vitamins act as coenzymes in metabolism of substrates.
7. Vitamin B_{12} is synthesized by microorganisms. It is present only in meat and dairy foods. Therefore, strict vegetarians may be at risk for Vitamin B_{12} deficiency.
8. The fat-soluble vitamins are mainly present in fatty meats, liver, dairy fats, egg-yolks, vegetable seed oil and leafy green vegetables.
9. Fat-soluble hypervitaminosis produces symptom of toxicity.
10. Deficiencies of fat-soluble vitamins occur primarily in young children who lack adequate body stores. Deficiencies are rare in adults unless there is malabsorption, biliary obstruction, or other conditions that affect fat metabolism.

Water

Although water is not a food, it is ordinarily consumed in the diet. Hence, it is one of the components of food. The water requirements and functions are discussed in Chapter 27.

3. Variation in the Diet

1. There is a risk of missing some essential element or vitamin in a varied diet.
2. Eskimos live mainly on fish and meat but the poor Orientals chiefly live on rice with small amounts of fish.

4. Digestibility of the Food

1. The food is of no use if it is not digested in the alimentary canal.
2. Digestibility is more concerned with absorbability.
3. When fats and starch are largely used, vegetables and animal proteins are not absorbed.
4. Absorption is more enhanced with a mixed diet than the substance taken alone.

5. Cooking

1. In cooking, food is considerably changed. Harmful organisms are destroyed.
2. Cooking breaks down the connective fibres of meat and makes meat easier to masticate and helps digestion. Overcooking shrinks the coagulated protein and decreases the digestibilty.
3. Cooking increases the water content and digestibility of vegetables, cellulose framework is loosened and starch from starch grains is liberated.
4. Fats are little changed in cooking. Cooking enhances the flavour of the food.
5. When vegetables are cooked, Vitamins B_1 and C are destroyed.

6. Psychological Factors

1. Appetite is reduced by worry and anxiety and digestion is also upset due to imperfect mastication and secretion of digestive juices.
2. Consumption of food is increased while taken in pleasant surroundings and good company with different items.

7. Cost

1. Dietary food is much influenced by family income.
2. When the income is good, consumption

is high with all the protective foods. A poor income has poor protective foods.

3. The lowest income groups having the low protective foods suffer from rickets and nutritional anemia. The lower groups are less resistant to infectious diseases such as whooping cough, measles, diphtheria and tuberculosis.

BIOLOGIC VALUE OF PROTEINS

The biologic value of a protein is an expression of a number of the nutritional characteristics of the substance. Among these are :

1. The digestibility.
2. The availability of the digested products.
3. The presence and amounts of the various essential amino acids.

1. **The digestibility of proteins :** It is expressed as the percentage of the food nitrogen which has been absorbed.

Absorbed N = Food N – (fecal N- metabolic N).

Proteins of animal origin have the highest digestibility (95 to 100 per cent) with the wastage in digestion only 5% or less. The less is the digestibility, the less the biologic value.

2. **The availability of the digested products:** The greater the proportion of amino acids in the dietary protein which can serve for the construction of tissue protein, the greater will be its potential nutritive value. A smaller proportion of the amino acids will then be discarded.

3. **The presence and amounts of the various essential amino acids :**

An essential amino acid is defined as one that cannot be synthesized by the animal organism at a speed commensurate with the demands for normal growth and which must, therefore, be supplied in the diet. There are nine essential amino acids for man. These are : methionine, threonine, tryptophan, valine, leucine, isoleucine, phenylalanine, histidine and lysine.

Zein is almost completely free from tryptophan and lysine. Growth is continued if zein is given as diet along with tryptophan and lysine.

Gelatin is also lacking tryptophan. Casein lacks methionine. Normal growth will continue if these amino acids are given along with these diets.

A biologically good protein is one which contains all the essential amino acids in proportions not far different from those needed by the body. A much smaller intake of such a protein is required to produce nitrogen equilibrium than that of a protein which has a limited amount of one or more of the essential amino acids. The most suitable proteins for growth are those of animal origin and especially those which nature has provided for the nourishment of the growing animals, namely, Lactalbumin (of milk), ovalbumin (of hen's egg), ovovitellin (of hen's egg).

These support growth when given at a level of about 9 or 10 per cent in the diet.

Proteins in the order of biologic value

Proteins of meat, glutenin (wheat), casein (milk), glutelin (maize), glycinin (soyabean).

These support growth when given in a higher concentration.

The following vegetable proteins are incapable of supporting growth but are suitable for maintenance :

Gliadin (wheat), legumin (pea), legumelin (soya bean), hordein (barley).

The following incomplete proteins are unsuitable for either growth or maintenance.

Zein, gelatin.

BALANCED DIET

Definition

A balanced diet is one which contains all the food constituents in proper proportions to meet the energy and nutritional requirements of the individual.

The proportions of protein, fat and carbohydrate should approximately be 1:1:4 respectively. The balanced diet for an adult male of 70 kg requiring 3,000 calories per day should contain the following given in the right-hand column :

BALANCED DIET FOR ADULT MALE (WT. 70 kg)	
Protein	70 gm
Fats	50 gm
Carbohydrate	440 gm
Calcium	0.8 gm
Phosphorus	1.4 gm
Iron	40 mg
Vitamin A	7,300 I.U.
Vitamin B_1	1.8 mg
Vitamin C	200 mg

BALANCED DIET FOR ADOLESCENT BOYS AND GIRLS*

Dietary articles	BOYS				GIRLS	
	13-15 years		16-18 years		13-18 years	
	Vegetarian (gm)	Non-vegetarian (gm)	Vegetarian (gm)	Non-vegetarian (gm)	Vegetarian (gm)	Non-vegetarian (gm)
Cereals	430	430	450	450	350	350
Pulses	70	50	70	50	70	50
Green leafy vegetables	100	100	100	100	150	150
Other vegetables	75	75	75	75	75	75
Roots and tubers	75	75	100	100	75	75
Fruits	30	30	30	30	30	30
Milk	500	200	400	150	400	150
Fats and oils	35	40	45	50	35	40
Meat and fish	—	60	—	70	—	80
Egg	—	30	—	30	—	30
Sugar and jaggery	30	30	40	40	30	30
Groundnuts	—	—	50	50	—	—

* Adapted from the Report of the Nutrition Expert Group, ICMR (1968)

BALANCED DIET FOR CHILDREN*

Dietary articles	PRE-SCHOOL CHILDREN				SCHOOL CHILDREN			
	1-3 years		4-6 years		7-9 years		10-12 years	
	Vegetarian (gm)	Non-vegetarian (gm)	Vegetarian (gm)	Non-vegetarian (gm)	Vegetarian (gm)	Non-vegetarian (gm)	Vegetarian (gm)	Non-vegetarian (gm)
Cereals	150	150	200	200	250	250	320	320
Pulses	50	40	60	50	70	60	70	60
Green leafy vegetables	50	50	75	75	75	75	100	100
Other vegetables, roots and tubers	30	30	50	50	50	50	75	75
Fruits	50	50	50	50	60	60	60	60
Milk	500	200	400	200	500	200	400	200
Fats and oils	20	20	25	25	30	30	35	35
Meat, fish and egg	–	30	–	30	–	40	–	50
Sugar and jaggery	30	30	40	40	50	50	50	50

* Adapted from the Report of the Nutrition Expert Group, ICMR (1968)

BALANCED DIET FOR ADULT WOMAN *

Dietary articles	Sedentary work		Moderate work		Heavy work		Additional allowing during	
	Vegetarian (gm)	Non-vegetarian (gm)	Vegetarian (gm)	Non-vegetarian (gm)	Vegetarian (gm)	Non-vegetarian (gm)	Pregnancy (gm)	Lactation (gm)
Cereals	300	300	350	350	475	475	50	100
Pulses	60	45	70	55	70	55	–	10
Green leafy vegetables	125	125	125	125	125	125	25	25
Other vegetables	75	75	75	75	100	100	–	–
Roots and tubers	50	50	75	75	100	100	–	–
Fruits	30	30	30	30	30	30	–	–
Milk	200	100	200	100	200	100	125	125
Fats and oils	35	30	40	35	50	45	–	15
Meat and fish	–	30	–	30	–	50	–	–
Egg	–	30	–	40	–	30	–	–
Sugar and jaggery	30	30	30	30	40	40	10	20
Groundnuts	–	–	–	–	40	40	–	–

Adapted from the Report of the Nutrition Expert Group, ICMR (1968)

BALANCED DIET FOR ADULT MAN*

Dietary articles	Sedentary work		Moderate work		Heavy work	
	Vegetarian (gm)	Non-vegetarian (gm)	Vegetarian (gm)	Non-vegetarian (gm)	Vegetarian (gm)	Non-vegetarian (gm)
Cereals	400	400	475	475	650	650
Pulses	70	55	80	65	80	65
Green leafy vegetables	100	100	125	125	125	125
Other vegetables	75	75	75	75	100	100
Roots and tubers	75	75	100	100	100	100
Fruits	30	30	30	30	30	30
Milk	300	100	300	100	300	100
Fats and oils	40	35	45	40	50	45
Meat and fish	–	30	–	50	–	60
Egg	–	30	–	30	–	30
Sugar and jaggery	30	30	40	40	55	55
Groundnuts	–	–	–	–	50	50

Adapted from the Report of the Nutrition Expert Group, ICMR (1968)

RECOMMENDED DAILY DIETARY ALLOWANCES FOR DIFFERENT CATEGORIES

Sex	Age in years	Weight in kg(lbs)	Height in cm(in)	Calories	Protein (gm)	Calcium (gm)	Iron (mg)	Vit. A (I.U.)	Vit. B_1 (mg)	Vit B_2 (mg)	Niacin (mg)	Vit. D (I.U.)	Vit. C (mg)
Infants	Below 1.	8(18)	—	(kg × 11.5) ±15	(kg × 2.5) ±0.5	0.7	(kg×1)	1500	0.4	0.6	6.0	400	30
Children	1–3	13(29)	87(34)	1300	32	0.8	8	2000	0.5	0.8	9	400	40
	4–6	18(40)	107(42)	1600	40	0.8	10	2500	0.6	1.0	11	400	50
	7–9	24(53)	124(49)	2100	52	0.8	12	3500	0.8	1.3	14	400	60

Contd.

RECOMMENDED DAILY DIETARY ALLOWANCES FOR DIFFERENT CATEGORIES													
Sex	Age in years	Weight in kg(lbs)	Height in cm(in)	Calories	Protein (gm)	Calcium (gm)	Iron (mg)	Vit. A (I.U.)	Vit. B_1 (mg)	Vit B_2 (mg)	Niacin (mg)	Vit. D (I.U.)	Vit. C (mg)
Boys	10–12	33(72)	140(55)	2400	60	1.1	15	4500	1.0	1.4	16	400	70
	13–15	45(98)	156(61)	3000	75	1.4	15	5000	1.2	1.8	20	400	80
	16–18	61(134)	172(68)	3400	85	1.4	15	5000	1.4	2.0	22	400	80
Girls	10–12	33(72)	140(55)	2200	55	1.1	15	4500	0.9	1.3	15	400	80
	13–15	47(103)	158(62)	2500	62	1.3	15	5000	1.0	1.5	17	400	80
	16–18	53(117)	163(64)	2300	58	1.3	15	5000	0.9	1.3	15	400	70
Men	19–35	70(154)	175(69)	2900	70	0.8	10	5000	1.2	1.7	19	—	70
	36–55	70(154)	175(69)	2600	70	0.8	10	5000	1.0	1.6	17	—	70
	56–75	70(154)	175(69)	2200	70	0.8	10	5000	0.9	1.3	15	—	70
Women	19–35	58(128)	163(64)	2100	58	0.8	15	5000	0.8	1.3	14	—	70
	36–55	58(128)	163(64)	1900	58	0.8	15	5000	0.8	1.2	13	—	70
	56–75	58(128)	163(64)	1600	58	0.8	10	5000	0.8	1.2	13	—	70
Pregnancy	—	—	—	+200	+20	+0.5	+5	+1000	+0.2	+0.3	+3	400	+30
Lactation	—	—	—	+1000	+40	+0.5	+5	+3000	+0.4	+0.6	+7	400	+30

VALUES IN CALORIES OF SOME COMMON FOODS	
Foods	Calories
Banana	132/one full
Desi ghee	830–890/100 gm
Butter	720/100 gm
Egg	165/100 gm
Buffalo milk	109/100 ml
Meats	100–450/100 gm
Potato	83/one medium size
Cow milk	69/100 ml
Human milk	67/100 ml
Apple	66/one
Peas (cooked)	56/half cup
Fishes	50–150/100 gm
Tomato (raw)	20/one medium size

ENERGY CONTENT AND COMPOSITION OF DIFFERENT COMMON FOODSTUFFS							
Foodstuffs	Calories per 100 gm	Per 100 gms of edible portion					
		Protein (gm)	Fat (gm)	Carbohydrate (gm)	Calcium (mg)	Phosphorus (mg)	Water %
Butter	720	0.6	81	0.4	20	16	15.5
Eggs	165	12.8	11.5	0.7	54	210	74
Rice	360	7.5	1.7	77.0	39	303	12
Wheat	240	9.3	2.6	49.0	96	263	36.6
Barley	349	8.2	1.0	78.8	16	189	11.1
Cow milk	69	3.5	3.9	4.9	118	93	87
Cheese	398	25.0	32.2	2.1	725	495	37
Banana	132	1.2	0.2	23.0	8	28	74.8
Apple	66	0.3	0.4	14.9	6	10	84.1
Orange	52	0.9	0.2	11.2	33	23	87.2
Carrots	50	1.2	0.3	9.3	39	37	88.2
Potatoes	90	2.0	0.1	19.1	11	56	77.8

PROTEIN AND AMINO ACID REQUIREMENTS AND INTAKES IN HUMANS				
	Requirements (mg/kg body weight/day)			Intake (gm/day)
	Infant (4–6 months)	Child (10–12 years)	Adult	Adult (70 kg) Allowance
Protein :	1100	1000	800	56
Animal	....	....	....	....
Vegetable	....	....	....	....
Essential amino acids :	(3–4 months)			
Histidine	28	12	10	0.70
Isoleucine	70	28	10	0.70
Leucine	161	42	14	0.98
Lysine	103	44	12	0.84
Methionine (and cysteine)	58	22	13	0.91
Phenylalanine (and tyrosine)	125	22	14	0.98
Threonine	87	28	07	0.49
Tryptophan	17	03.3	03.5	0.25
Valine	93	25	10	0.70

Data from Recommended Dietary Allowances 10th edition. National Academy of Science, 1989

NUTRITIONAL VALUE OF SOME COMMON FOODSTUFFS											
Foodstuffs	Proteins (gram %)	Fats (gram %)	Carbohydrates (gram %)	Fe (mg %)	Ca (mg %)	NaCl (mg %)	Vit B_1 (mg %)	Vit B_2 (mg %)	Niacin (mg %)	Vit C (mg %)	Vit A I.U.
Cow milk	3.5	3.9	4.9	0	118	123	46	200	80	0	180
Buffalo milk	4.2	8.8	5.3	0	21.2	0	0	0	0	0	162
Hen's egg	12.8	11.5	5.7	25	54	141	trace	353	0	0	882
Mutton (Goat)	14.8	10.9	0	1.8	123	141	49	219	5.3	0	28
Butter	0.6	81	0.4	0	20	1764	0	0	0	0	2716
Rice	7.5	1.7	77.0	2.8	39	88	198	53	1.8	0	0
Atta	9.3	2.6	49.0	7.4	56	116	451	120	4.9	0	60
Musoor dal	26.1	0.7	60.1	21	134	88	444	0	1.4	0	250
Beans	1.1	0	2.5	0.7	46	0	71	46	0.4	11	120
Onion	1.1	0	11.6	0.7	169	0	35	11	0.4	11	0
Apple	0.3	0.4	14.9	1.4	6	0	116	28	0.4	4	trace
Banana	1.2	0.2	23.0	0.4	8	282	78	32	0.4	4	trace

COMPOSITION OF COCOA AND CHOCOLATE				
	Protein (gm %)	Fat (gm %)	Carbohydrate (gm %)	Calories per 100 gm
Cocoa	18.1	26.8	40.3	489
Chocolate	4.8	31.1	59.9	555

CHOLESTEROL CONTENT OF SOME IMPORTANT HUMAN FOODS

Foods	Cholesterol (mg per 100 gm edible portion)
Cow milk	11
Fish	70
Chicken	60
Egg yolk	1,500
Butter	250
Liver	300
Brain	2,000
Vegetable oils	Nil

PERCENTAGE OF LINOLEIC ACID IN DIFFERENT EDIBLE OILS

Safflower oil	80 per cent
Sunflower oil	70 per cent
Peanut oil	30 per cent
Coconut oil	5 per cent

DIFFERENCE BETWEEN THE COMPOSITION OF HUMAN MILK AND COW MILK

	Solids (%)	Water (%)	Protein (gm %)	Fats (gm %)	Carbo-hydrates (gm %)	Calcium (mg %)	Phos-phorus (mg %)	Magne-sium (mg %)	Sodium (mg %)	Potas-sium (mg %)	Chloride (mg %)
Human milk	12.5	87.5	1 to 2.5	1.0 to 8.0	4.5 to 8.0	40	30	5	15	60	40
Cow milk	13.5	87.0	2.0 to 6.0	1.5 to 6.5	2.0 to 6.0	120	90	20	50	140	110

	Vit. A (μg %)	Vit. D (I.U.)	Vit. C (mg %)	Vit. B_1 (μg %)	Vit. B_2 (μg %)	Niacin (μg %)	Vit. B_6 (μg %)	Pantothenic acid (μg %)	Calories (per 100 gm)
Human milk	50	5.0	4.5	15	45	180	10	200	67
Cow milk	35	2.5	2.0	45	200	80	50	350	69

NUTRITIVE VALUE OF FOODS

Milk

1. Milk is nearest to a perfect and complete food.
2. It contains not only protein, fat and carbohydrate but also minerals and vitamins.
3. The chief proteins are caseinogen and lactalbumin. Caseinogen is a phosphoprotein. Caseinogen is associated with calcium as calcium caseinogenate. Small amounts of other proteins are also present.
4. Milk fat is in the form of a very fine emulsion. It contains all saturated fatty acids as well as unsaturated fatty acids.
5. The carbohydrate of milk is lactose which is less sweet than sucrose.
6. It is rich in calcium, potassium, sodium, chlorine and phosphorus. The iron content is very low which is insufficient for baby.
7. It is rich in vitamin A and vitamin B_2 and good in vitamin B_1 and nicotinic acid but poor in C and D.
8. Milk kept on unsterilised becomes sour due to formation of lactic acid from lactose by bacteria present in milk and this lactic acid is sufficient to precipitate (curdle) the caseinogen.
9. When milk comes in contact with rennin or proteolytic enzymes, it becomes clotted.
10. The protein of milk neutralises gastric HCl to allow clotting.
11. The percentage composition of milk of different species is given below :

Species	Protein	Fat	Carbohydrate	Ash
Woman	1.0	2.9	6.7	0.2
Cow	3.0	3.6	4.8	0.7
Buffalo	5.0	7.0	2.4	0.6
Goat	4.3	4.8	4.7	0.8
Since iron is less in the milk it is added to the milk				

12. Condensed milk is prepared by the removal of water by evaporation in vacuo. Vitamins A and B are preserved. Dried milk is prepared by passing a film of milk over hot rollers or spraying milk into hot air to evaporate water.

Egg

1. The hen's egg contains about 30% yolk, 59% white and 11% shell. The chemical composition of duck's egg is similar to that of hen's egg.
2. The white contains proteins and salts. The greater part of the protein is ovalbumin. The other proteins are conalbumin, ovoglobulin and ovomucoid (a glycoprotein). The pale yellow colour is due to riboflavin. The white contains 10.7% protein, 0.1% fat and 1.4% carbohydrate.
3. The yolk contains 51% water, 15% protein, 33% fat of which 4% cholesterol, and 1% mineral. The proteins are vitellin (a phosphoprotein) and livetin (a globulin). The minerals are calcium, iron and phosphate. The yolk is rich in vitamins A, B_1, B_2 and D but not C. The other B group vitamins and E are also present.

Meat

1. Meat contains about 22% protein and B group vitamins but does not contain vitamins A, C or D.
2. Nearly all the carbohydrates disappear during hanging, and lactic acid as well as acid phosphates are formed. These acids con-

vert collagen to gelatin when the meat is cooked. In boiling, there is the loss of salts, gelatin and extractives. During roasting this loss is reduced.

3. The flavour of meat is due to the organic substances extracted from meat by boiling water.

Fish

1. Fish is free from carbohydrate just like meat. The fat content ranges from a trace to 5%. It is fair source of B group vitamins. Fatty fish contains some vitamins A and D.
2. Large fish is rich in phosphorus but deficient in calcium. Small fish eaten with bones are good sources of calcium.
3. It contains 22% protein, 1.4% fat, 75% water. The dried fish may contain about 5% carbohydrate because of the use of flour.

VEGETABLE FOODS

A. Cereals

1. The crude cereals (wheat, rice, barley, rye, maize, oats) contain 11% protein, 70% carbohydrate, 0.5% to 8% fat, 11% water and 2% minerals. Oatmeal is the richest in protein and fat—rice is the poorest.
2. The main proteins of cereals are glutelins and gliadins. Small amounts of albumins and globulins are generally present.
3. The fats contain considerable amounts of olein to make them liquid at ordinary temperatures.
4. The carbohydrate is almost completely starch in the form of grains covered by a thin membrane of cellulose. Small amounts of sugar are also often present in cereals.
5. The abundant mineral elements are calcium and phosphate. The phosphate is partly in the form of phytic acid (inositol hexaphosphate) which interferes the absorption of calcium. Some cereals (wheat and rye) contain the enzyme phytase which hydrolyzes phytic acid and reduces the anticalcifying effect.
6. There is a great loss of vitamins B and E, and mineral elements, especially Ca, P and Fe in the roller milled white flour. Carotene and riboflavin are lost by chemical blenching.

B. Pulses

1. The dried pulses (peas, beans, lentils etc.), contain large amounts of protein. The main protein is a globulin called legumin. The dried pulses contain 20-25% protein, 11% water, less than 2% fat.
2. The soyabean contains high protein, high fat and low carbohydrate. Soyabean oil is used for frying "fish and chips".
3. The dried pulses are the good sources of many B group vitamins and minerals but deficient in vitamins A, D, B_{12} and C.

C. Nuts

1. Nuts contain high protein and fat but low carbohydrate. The protein content is slightly lower than that of dried pulses.
2. They can be used for the preparation of milk substances which can be fed to infants over 6 months of age and to young children of some areas where milk is not available in sufficient amounts.

D. Roots and Tubers

1. The most important tuber used in this country is the potato, sweet potato. These are the chief sources of starch. These can substitute sugar and cereals and are the valuable sources of iron and vitamin C. New potatoes contain about 28 mg vit. C. per 100 gm but on storage 5 mg vit. C per 100 gm is reduced.

2. The common roots (carrots, beetroots, turnips) are almost free from starch. Their caloric value is completely due to sugars (sucrose, fructose and glucose). Carrots, the richest in sugars, are the sources of carotene. Roots contain valuable salts but negligible protein and fat. Arrowroot contains traces of protein and salts.

E. Green Leafy Vegetable

1. Green leafy vegetables are rich in carotenes.
2. They are good sources of calcium, riboflavin, folic acid and vitamin C.
3. They are the cheapest among the protective foods.
4. Yellow pumpkin is a fair source of carotene.

F. Fruits

1. Fresh fruits are the protective foods. Their energy value is due to sugars and starch. The protein and fat are usually less than 0.5%. Many fruits contain pentoses and pectins.
2. Tartaric acid, malic acid and citric acid are present in fruits.
3. Cooking destroys vitamin C and makes fruits more digestible by softening the cellulose.
4. Some fruits, notably the banana, contains starch as well as sugar. It has high protein content.
5. The carbohydrate of dates is all sugar, not starch. The protein content is low (2%).

G. Tea, Coffee, Cocoa

1. The black tea sold consists of the leaves of young shoots of the tea plant which are fermented and dried by heat. It has negligible caloric value. It has a stimulant and a diuretic effect. Strong tea disturbs gastric digestion due to tannic acid.
2. Coffee is the roasted seed of coffea arabica. The odour is due to an oil, caffeol, formed when the beans are roasted. It has little caloric value. It contains caffeine and tannic acid. It cannot be regarded as food unless taken with milk and sugar.
3. The seeds are obatined from the pods of the cocoa tree after fermentation and roasting. They contain 50% fat. Cocoa as beverage is of little importance. It becomes nourishing when taken with milk and sugar. Chocolate consists of ground cocoa- nibs mixed with sugar. Starch and flavouring are frequently added.

ALCOHOL

1. Alcohol has an energy value of 7 calories per gram. Hence, it has been admitted as food. Its consumption is restricted due to its pharmacological action. It is consumed in the form of wines, spirits or liquors.
2. Beers are the fermented product of malt and contain 4-8% alcohol. It has small amount of protein and some sugar.
3. Wines are the products of fermentation of fruit juices (usually grapes). The alcohol content is 10-20% and sugar 0.1-4%. They contain large amount of organic acids (tartaric, malic, succinic, etc.).
4. Spirits are the distillation product of many fermented products. Whisky is a distilled beer and brandy is a distilled wine. They are practically free from sugar. The alcoholic content is 30-50%.
5. Liquors are alcohol sweetened with cane sugar and flavoured with essences. The sugar content is 30% and alcohol is 35-55%.

Principles in planning a balanced diet

In planning a balanced diet it is to be aimed that the diet must contain various groups of foodstuffs such as energy yielding foods, body building foods and protective foods in the correct proportions. The constituents of balanced diet differ according to age, sex, physical activity, economic status and the physiological condition.

A. Energy yielding foods

1. This group contains high carbohydrates and also pure fats and carbohydrates. They are divided into two groups : (a) cereals, roots and tubers and (b) pure carbohydrates and fats.
2. Cereals provide proteins, certain minerals and vitamins in addition to energy in the diets of the low-income groups.
3. Roots and tubers provide some amounts of proteins, minerals and vitamins.
4. Pure carbohydrates and fats provide only energy.

B. Body building foods

These contain high protein and are divided into two groups : (a) Milk, egg, meat and fish. (b) Pulses, oilseeds and nuts.

C. Protective foods

The protective foods are rich in proteins, vitamins and minerals. These are classified into two groups :

(a) Foods rich in vitamins, minerals and proteins of high biologic value eg, milk, egg, fish and liver.

(b) Foods rich in certain vitamins and minerals only eg, green leafy vegetables and some fruits.

Balanced diets at high cost : Large amounts of costly foods such as milk, eggs, meat, fish and fruits and moderate quantities of cereals, pulses, nuts and fats.

Balanced diets at moderate cost : Moderate amounts of milk, egg, meat, fish, fruits, fat and large amounts of cereals, pulses, nuts and green leafy vegetables.

Balanced diets at low cost : Small amounts of milk, egg, meat, fish, fat and large amounts of cereals, pulses, nuts and green leafy vegetables.

Food Toxins and Additives

1. A good number of potentially harmful compounds are present in foods. Some occur naturally. They include harmful substances such as the neurotoxins from shellfish or mushrooms, goitrogens from plants of the cabbage family, bean compounds that interfere with collagen formation.
2. Pesticides and packaging materials are added to food through inadvertent contamination.
3. Many compounds are added to foods in order to preserve them or to add colour, flavour or texture. Some of the additives may be harmful.
4. Consumption of fresh and unprocessed foods are advised in order to minimize intake of natural toxins and additives.

CLINICAL ASPECTS

- The nutritional deficiency is present among the poor or among the growing children, pregnant or lactating women, ill and convalescing patients, alcoholics, or individuals on restricted diets from necessity.
- The deficiencies are more widespread in more deprived populations eg, Kwashiorkor in deficiency of protein, Xerophthalmia in deficiency of Vitamin A, anemia in deficiency of iron.
- Malabsorption leads to deficiency of nutrients causing pathologic conditions.
- Obesity is associated with dietary excess.
- The excess intake of particular nutrients develops atherosclerosis, coronary heart disease, cancer of the breast and colon, cardiovascular disease, cirrhosis of liver, diabetes.

APPLIED NUTRITION

Primary Nutritional Diseases

1. Starvation and Anorexia Nervosa

Starvation

(a) Hormone production

1. There is impaired secretion of pituitary gonadotrophins and testosterone as well as 17-oxysteroid concentration in plasma falls.
2. Plasma Insulin level is reduced because the stimulating substances, glucose and amino acids, are not absorbed during fasting.
3. The secretion of pituitary growth hormone tends to be increased and this favours fat mobilization.
4. In prolonged starvation, the concentration of 3, 5, 3′- triiodo-thyronine falls which is responsible for the reduced metabolism.

(b) Clinical Features

1. The patients are thin having loss of weight.
2. The hair is dry and lustreless.
3. The eyes are dull and sunken.
4. The skin is thin, dry, and inelastic.
5. Dirty brown splotches of pigmentation may appear over the face and trunk.
6. Polyuria at night is a frequent troublesome symptom.
7. Oedema first starts in the face when lying down and the ankle oedema is found when the patient gets up and walks about.
8. The blood pressure is low ; the diastolic pressure may be impossible to estimate, while the systolic pressure may be as low as 70 mm of Hg. In severe cases, the pulse rate is often below 40/min.

(c) Psychological disturbances

1. The mind is never fixed for long on single subject except the desire for food. Existence of mental restlessness.
2. The patient becomes self-centred and indifferent to the troubles of others.
3. The patient is worried and sensitive to noise and other petty irritations which may make him quarrelsome.
4. In the last stages of starvation, the personality may completely disintegrate. A mother may even steal from her child.

(d) Infections

1. Starvation patients often suffer from infections—malaria, cholera, typhus, pneumonia, gastroenteritis.
2. Cancrum oris, an infective gangrene of the mouth eroding the lip and cheeks, is a dreadful catastrophe which occasionally occurs in famines both among children and adults.
3. The atrophied intestinal glands and the paper-thin walls of the digestive tract are unable to digest and absorb properly even a bland diet.
4. In almost all famines, there is outbreak of diarrhoea without bacteriological organisms.

(e) Treatment

1. Most famine victims, owing to alimentary dysfunction, cannot consume large quantities of food. The patient's desire for food is immense and no guide to his digestive capacities. Food intake should necessarily be limited.
2. The choice of food is vital. Many starving people in the prison camp of World War II died from diarrhoea and collapsed after they had been given bullybeef, baked beans which they could not readily digest. Only bland foods can be tolerated by the thin-walled intestines lacking essential digestive enzymes.
3. Frequent small feeds of skimmed milk, 100

ml or so at a time–as often as the patient is willing and able to take them– is a good way to avert death from starvation. This requires constant personal attention and nursing care.

4. A variety of mild flavouring essences may be useful to stimulate the appetite. Slightly sour foods are usually acceptable.
5. Starving patients may tolerate moderate amounts of fat or edible oils, which provide a larger energy intake. There may be a temporary increase in oedema with refeeding, so the intake of salt should be restricted.
6. A time may come in severe starvation when the patient refuses all food. The outlook is then very grave. Nasogastric or parenteral feeding provides the only hope.

(f) Prognosis

1. Most people with primary undernutrition recover rapidly, once they have a free access to food. Appetite may be sufficient. Over 5,000 k cal/day may be consumed with a weekly gain in weight of 1.5 to 2 kg.
2. In some patients, even after careful nursing and a good diet, low blood pressure and diarrhoea may persist. If, after one or two weeks, they show little improvement, this suggests strongly that irreversible changes in the myocardium or small intestine have developed and that the prognosis is poor. After any severe famine there are some who may linger on in this condition for many months if supported by good medical and nursing skill, death rather than recovery is the usual end.

Anorexia Nervosa

This is a psychiatric disease arising from a refusal to eat which often leads to severe emaciation. Patients are usually middle class and often above average intelligence. The disease affects mainly the adolescent girls and young women, but cases are seen in older women and also in men. The patients come from homes where plenty of food is available and often some of the family are obese.

(a) Clinical Features

1. The weight may be reduced to 35 kg or less. Patients often deny all normal sense of fatigue.
2. The pulse is slow; the blood pressure, peripheral blood flow, and skin temperatures are all low.
3. There is usually no anemia, there may be fine downy hair. Amenorrhoea is a characteristic feature but secondary sexual characteristics are present.
4. Urinary excretion of gonadotrophins and oestrogens is diminished.
5. Menstruation starts at puberty when weight rises above 47 kg and stops when it falls below 47 kg. The plasma potassium may be abnormally low.
6. Anorexia is common in patients with an anxiety state, a depressive illness.

(b) Treatment

1. The primary aim is to get the patient to eat. Severe cases should be treated by a psychiatrist and under close supervision in hospital until a satisfactory weight has been achieved.
2. Chlorpromazine in doses up to 150 to 200 mg three times daily makes the patient more amenable, counteracts vomiting and is claimed to increase appetite.
3. Disturbed relationships within the family need attenion. Regular interview with relatives conducted by a psychiatric social worker over many months should reduce guilt, misunderstanding and intolerance.

(c) Prognosis

1. Restoration of weight and menstruation may occur within several months in favourable cases.

2. 50 per cent may make a partial recovery continuing to restrict their diet and remaining abnormally thin. There is a mortality of about 5 per cent from suicide.

2. Obesity

Is is the condition in which an excess of fat has accumulated. In most cases it can be detected by visual inspection.

(a) Clinical features

1. Some of the patients appear to be in good health and leading normal lives, they are likely to have a reduced exercise tolerance with shortness of breath on exertion and to be unduly fatigued by continuing physical activity. This is due to the burden of the increased weight that they carry always and to reduced capacity of the circulatory and respiratory systems that work under handicaps imposed by masses of internal fat and fatty infiltration of muscle.
2. They are also at increased risk when under anaesthesia ; surgical operations are more difficult and post-operative complications are more likely.
3. In obesity grade III, the everyday activities of the patients are seriously restricted by their enormous mass and they are likely to be suffering from diabetes, hypertension, gall bladder diseases, fatty liver, gout, osteoarthritis, hernias, etc. Life expectation is low. They have serious psychological disturbances.

(b) Treatment

1. If a patient eats a diet providing 500-1,000 kcal less than which is needed for the activities of daily life, then and only then the excess reserves of energy in adipose tissue be drawn upon and there will be loss of weight at the rate of 0.5-1 kg each week.
2. Any curtailment of food intake is liable to reduce intake of essential nutrients. All reducing regimes should include ample fruits and vegetables and preferably whole meal bread. These should supply ample vitamins and minerals. They also supply dietary fibre and this may help to make the inevitably small meals more satisfying. It also prevents constipation and there is no need to take supplements of minerals and vitamins. To prevent depletion a protein supplement should be given.
3. Patients should appreciate that there are no slimming foods although advised in magazines and daily newspapers and by radio and television. All foods are fattening if taken in excess. Foods legitimately advertised as aiding in a slimming diet are modified forms of conventional foods and beverages usually with a lower energy density.
4. Even if the exercise contributes only a little to a negative energy balance, it benefits by promoting physical fitness. Exercise prevents atrophy of muscle. Insufficient use of the muscle of the trunk and limbs of the respiratory system and the myocardium increases the risk of patients to orthopaedic, respiratory and circulatory disorders.
5. Obese patients may be advised to walk, to climb stairs, to swim and to gardenings. They are positive and enjoyable.
6. Amphetamine and its derivatives have been much used as 'slimming agents'. They are psychomotor stimulants and also have an anorectic action. They may also cause insomnia, irritability, increases heart rate, raised blood pressure and severe psychotic reaction. Serious withdrawal symptoms may occur on discontinuing the drug. So amphetamine should not be prescribed.

   Diethylpropion, phenmetrazine and fenfluramine are the durgs which have some chemical resemblance to amphetamine. They are also psychomotor drugs and they may be prescribed for cases refractory obesity for periods of up to six weeks. After this period their appetite-suppressing effect usually wears off. Patients with a history of depression or other psychological disturbance should not be treated by these drugs.

Thyroxine stimulates metabolism and for this reason it has had an extensive trial in the treatment of obesity. In euthyroid people thyroxine produces no increase in metabolism unless given in doses which cause tremor, diarrhoea, palpitation and tachycardia. Hence, the administration of thyroxine to obese euthyroid patients is not only useless but potentially dangerous.

Methyl cellulose is indigestible and adds bulk to the diet. It has little effect in promoting weight loss. However, it is quite harmless.

Sedatives and tranquilisers can play no part in the treatment of obesity but they may be useful for some obese patients who suffer from an anxiety state.

Diuretics are potentially dangerous and are of no value in promoting weight loss unless the patient has oedema.

7. Diets providing only 500 kcal daily or a period of total starvation are justifiable in some patients who are not responding well to loss of vigorous restrictions. In selected patients with no orthopaedic or cardiovascular complication, it is possible to increase physical activities. Patients have been kept for up to six weeks on diets providing only 400 kcal, whilst they walk 10 miles daily. Negative energy balances of up to 3000 kcal/day, and weight losses of up to 3kg a week followed. If dietary intake is below 1,000 kcal per day, multivitamin and mineral supplements may be necessary. Excessive losses of potassium and nitrogen may occur and urinary output of these should be checked regularly.

8. Surgical treatment to reduce food intake may be considered in patients in whom medical treatment has failed and whose life has been made miserable by severe obesity. Various forms of jejunoileostomy, which, by creating a bypass of the small intestine, lead to malabsorption—have been carried out. The operation is unphysiological. Complications with features of the malabsorption syndrome and the blind loop syndrome frequently arise.

Gastroplasty is an operation in which the stomach is reduced to a small reservoir, about 60 ml. in capacity, in the fundus which drains through a narrow channel, about 12 mm in diameter, along the greater curvature and into the duodenum. This is much more safe.

Surgical removal of large masses of fat from the abdomen, thighs or arms is contraindicated. There may be appearance of irregular ugly lumps of fat at the operation sites.

MARASMUS

Marasmus is due to a continued restriction of both dietary energy and protein as well as other nutrients. The marasmic form of the syndrome occurs in infants under one year and more frequently in towns. During the nineteeth century in the industrial towns in Europe and North America, marasmus, resulting from the poor diets and numerous infections, took a toll of infant lives like many Asian, African and South American towns today. It is due to rapid succession of pregnancies and early and abrupt weaning followed by dirty and unsound artificial feeding of the infants with very dilute milk products given in inadequate amounts to avoid expense. Thus the diet is low in both energy and proteins. Poor houses due to lack of equipments make the preparation of clean food almost impossible. Repeated infections develop, especially of the gastrointestinal tract. They are deprived of mother's breast milk.

Clinical Features

1. Diarrhoea is frequent. Many infants are hungry, but some are anorexic.
2. The child is wizened and shrunken and there is little or no subcutaneous fat.
3. There is often dehydration. The weight is much below the standard for age.
4. The temperature may be subnormal.
5. If the disease is of long duration, the length of the child is also below the standard, but less so than the weight.
6. There is usually watery diarrhoea with acid stools.
7. The abdomen may be shrunken or distended with gas.
8. Peristalsis may be easily visible because of the thinness of the abdominal wall.
9. The muscles are weak and atrophic and this together with the lack of subcutaneous fat makes the limbs appear as skin and bone.
10. The skin and mucous membranes may be dry and atrophic.
11. Psychological disturbances, resulting from a lack of mother's love and care, can depress the appetite.

KWASHIORKOR

Kwashiorkor is due to a quantitative and qualitative deficiency of protein, but in which energy intake may be adequate. It is mainly a disease of rural areas occurring in the second year of life. This disease occurs when the child is weaned into the traditional family diet ; this may be low in protein because of poverty, insufficient land and poor agricultural practice. There is no supplement of milk. The disease is frequently precipitated by outbreaks of febrile illnesses, such as malaria, measles or gastroenteritis. It arises as a result of poverty and ignorance. Many mothers have received no satisfactory instruction in infant feeding although there is money to buy food. In Africa, the disease is commoner.

(a) Clinical Features

1. There are oedema, anorexia, diarrhoea and a generalised unhappiness or apathy. An infection often precipitates at the onset for which the child is brought to the doctor.
2. Failure of growth is an early sign. Oedema is more marked in the lower limbs.
3. The characteristic dermatosis consists of areas of both hypo- and hyperpigmentation. The skin first becomes thickened as if varnished. This then peels and appears like "flaky paint" leaving cracks or denuded areas of shallow ulceration. In moderate cases the dermatosis resembles crazy paving ; when severe, the desquamated part looks as if there has been a burn. The lower limbs, buttocks and perineum are usually most affected but ulcers can occur over pressure points and deep cracks in skinfolds.
4. The hair is sparse, soft and thin. Negro children lose their characteristic curl. There may be changes in pigmentation with diffuse patches or streaks which may be red or grey in colour.
5. Angular stomatitis, cheilosis and a smooth atrophic tongue are commonly seen, as in ulceration around the anus.
6. Watery diarrhoea or large semisolid, acid stools are usual. The liver can generally be palpated and is firm, not tender.
7. The muscles are always wasted and as a result many children may no longer be able to walk or crawl.
8. Some degree of anemia is always present and may be severe.
9. Apathy is a characteristic feature and the child appears constantly unhappy. Neurological features are unusual, but some children during recovery have tremors resembling parkinsonism.
10. Many patients show a mixture of some of the features of both marasmus and kwashiorkor. These children are said to have marasmic kwashiorkor.

11. Some children adapt to prolonged insufficiency of food by a marked retardation of growth. They resemble children a year or more younger.

(b) Biochemical and metabolic disorders

1. A high body water content, loss of the fat stores, and loss of protein from the wasted muscles and other tissues greatly alter the chemical composition of the child.
2. Plasma concentrations of essential amino acids, especially branched chain amino acids and tyrosine are low, but those of some non-essential amino acids may be higher than normal. Sooner the treatment starts with protein, the concentration of amino acids in the plasma rises and there may be an overflow aminoaciduria.

   The plasma albumin level is low owing to a failure of synthesis in the liver. In severe cases, it is usually below 20 and sometimes below 10 g/I. Plasma IgG is often raised if infections are present but other immunoglobulins are usually normal. Plasma transferrin is lowered, especially in severe cases, and may be a better guide to prognosis than plasma albumin.

   The plasma concentrations of some enzymes such as cholinesterase, alkaline phosphatase, amylase, and lipase are lowered.

   The blood urea is usually low and may fall to 6 mg/100 ml. This indicates a reduced protein intake rather than a lowered rate of protein catabolism. Urinary creatinine is also reduced reflecting decreased muscle mass.
3. There is fatty liver and the excess of fat in the liver is triglyceride. Plasma triglyceride and cholesterol are low due to a decreased ability of the liver cells to mobilize lipid in the form of lipoproteins.
4. Blood glucose is usually normal. Hypoglycemia may occur and is a complicated matter to be kept in mind.
5. Plasma potassium level is below normal due to diarrhoea. Plasma magnesium level is also low due to increased losses in the stool.
6. Plasma [H^+] may be either raised or lowered. Acidosis is probably due to poor circulation and consequent tissue hypoxia. Alkalosis may be associated with potassium depletion and a failure of the kidneys to excrete bicarbonate.
7. The total body water may increase from 60 per cent to 80 per cent. The severity of clinical oedema is not closely associated with the size of the increase in total body water or with the level of plasma albumin. The cause and nature of the oedema still remains in part a mystery. The oedema may shift from one part of the body to another. Clinical dehydration and shock may be found in a child, associated with gross oedema in parts of the body.
8. Malnourished children are likely to have infections and may have complications requiring drug treatment. Some antibiotics and antimalarials act by interfering with nutrition more in the micro-organism than in the human host, and they need to be used with care in Protein Energy Malnutrition (PEM). The antibiotics streptomycin, chloramphenicol and the tetracyclines inhibit protein synthesis by interfering with the action of messenger or transfer RNA. The antimalarial trimethoprim is folate antagonist.

   Some drugs are carried in the circulation bound to plasma proteins. Kwashiorkor plasma with a low albumin content has a reduced binding capacity with the drugs. Higher concentrations of the free form of the drug increase the risk of toxic effects. Many drugs are detoxicated in the liver by the microsomal enzyme-oxidising system and its function may be impaired in PEM. Therefore, all drugs should be used with caution.

(c) Changes in the organs and systems of the body

1. The atrophy of the cells of the pancreas and intestinal mucosa cannot produce digestive enzymes in normal amounts. Duodenal contents contain reduced amounts of amylase, trypsin and lipase. The activity of lactase, sucrase, and maltase are greatly reduced in the atrophic mucosa which is also associated with impaired absorption of nutrients.
2. The fat first accumulates in small droplets within liver cells, situated at the periphery of the lobules. The droplets increase in size and extend from the periphery to the centre of the lobules. In severe cases, all the liver cells may be filled with big fat droplets, pushing aside the cell nucleus and reducing the cytoplasm to a narrow rim. Yet the liver function is well maintained and severe liver failure is unusual. Plasma bilirubin is normal ; prothrombin concentrations are often reduced, but return to normal on treatment with vitamin K. In case the plasma concentrations of alanine amino-transferase and isocitrate dehydrogenase are found raised, they suggest the presence of damage from a bacterial or viral infection. With proper treatment the lipid accumulated in the liver cells is all cleared with return to a normal structure.
3. Plasma concentration of growth hormone may be raised as the pituitary responds effectively to the stimulus of protein depletion. Plasma concentrations of cortisol and other adrenocortico- steroids are normal or raised. Plasma thyroxine is often low, but free T_4 is usually normal or raised.
4. Atrophy of the heart leads to a reduced cardiac output and a poor circulation. In many severe cases the extremities are cold and cyanosed and the pulse small or impalpable. The electrocardiogram (ECG) shows low-voltage changes in the QRS complex and the T-wave may be depressed or inverted. Some of the changes are rapidly reversed by potassium therapy.
5. There is no specific structural or functional abnormality of kidneys. The glomerular filtration rate may be low, but this is probably due to dehydration or reduced cardiac output. The concentrating power of the kidneys is often poor, but this may be due to the depression of tubular function by electrolyte deficiencies. They are all reversible by treatment.
6. The immune responses of the body are produced by cells arising in the thymus, lymph nodes and spleen, the lymphoreticular organs. These are very immature at birth and develop rapidly in the first two years of life. In kwashiorkor, the thymus, tonsils, spleen, and other lymphoid tissues are atrophied. These changes are accompained by a delayed or absent tuberculin response and other skin hypersensitivity reactions; reduced complement activity in the serum, especially the C_3 component; reduced numbers of thymus-dependent lymphocytes (T cells) in the blood; reduced lymphocytes transformation in response to phytohemagglutinin. The bactericidal action of neutrophil leucocytes is also impaired. These signs of reduced cell-mediated immunity are in contrast to a usually unimpaired humoral immunity; the response to injected antigens is often normal and plasma IgG may be higher than normal.

   The depression of cell-mediated immunity is mainly due to protein deficiency. Once the immunological system has reached maturity it is much less susceptible to malnutrition.

(d) Treatment

1. Children who are seriously ill require treatment in hospital as their recovery depends on high standards of clinical skill and nursing care with some laboratory support.

When the acute phase of the illness is over, they need several weeks of special feeding and some medical supervision before recovery is complete.

2. Most children admitted to hospital have had repeated attacks of diarrhoea and often of vomiting and as a result are dehydrated. Some of these respond to oral rehydration therapy but if the response is not prompt or the dehydration is severe, intravenous fluid should be given at once.
3. A small infusion of plasma is beneficial when there is severe peripheral circulatory failure and of whole blood or red cells when there is severe anemia. Under these conditions the myocardium may be damaged by hypoxia and there may be acute heart failure, infusion should be low and a fast diuretic (frusemide) given at the start.
4. Many patients have malaria, pneumonia, dysentery and other infections and infections do not cause fever in the malnourished. It is better to give a short course of procaine benzylpenicillin and ampicillin. If ampicillin is not available, chloramphenicol and tetracycline may also be given.
5. Hypothermia is often present and needs urgent treatment. It is, therefore, wise to let their mothers sleep with them in the hospital.
6. The child should be given a dilute milk feed with added sugar from the first or second day. Thus the strength can be increased and a vegetable oil added to give extra energy.
7. Infants who are seriously ill will improve when given 1 gm protein per kg body weight daily. Good recovery can be obtained with 2 gm and recovery is not accelerated by giving more than 3.5 gm.
8. A high energy diet is required and this can only be made by including large amounts of fat. Some infants can tolerate at first the amount of fat in whole cow milk. Feeds, therefore, be made up from skimmed milk powder.
9. As soon as children are able to take normal food and infection or other complications are controlled, they should be discharged to a centre where their nutritional rehabilitation can be supervised.

(e) Prognosis

1. A child may suffer for a short period from one of the forms of PEM and make a complete recovery. If growth is retarded slightly, the child may reach the normal size for its age quickly, provided the dietary supply is satisfactory. If growth is retarded for a long period, the child may be stunted and develop into a small size, but healthy adult.
2. If the disease is so severe as to demand treatment in hospital, the prognosis is uncertain and often bad.
3. Children who survive severe PEM in early childhood perform less well in intelligence tests than controls. A poor performance can be due to their growing up in an unfavourable psychological environment rather than to a short period of malnutrition in early life.
4. It is proper to request government that failure to provide adequate nutritional services for mothers and young children may well lead to a school population with a diminished capacity for learning. On the other hand, a mother whose child has suffered a period of severe malnutrition and made a good recovery may be reassured that subsequent mental development is not likely to be impaired seriously, if at all.

GOITRE

The term "Goitre" is used to denote enlargement of the thyroid gland of whatever kind. Simple goitre is said to be present when the gland is visible and palpable, but the subject has no symptoms either of hypothyroidism or hyperthyroidism. Such goitres do not usually affect health, but sometimes they may be complicated with serious consequences. Environmental factors determine the prevalence of most simple goitres—especially dietary factors, of which iodine deficiency is the major one.

(a) Clinical symptoms

The classification of goitre is recommended as :

Grade Oa	Thyroid not palpable or if palpable not larger than normal.
Grade Ob	Thyroid distinctly palpable but usually not visible with the head in a normal or raised position; considered to be definitely larger than normal.
Grade I	Thyroid easily palpable and visible with the head in either a normal or a raised position.
Grade II	Thyroid easily visible with the head in a normal position.
Grade III	Goitre visible at a distance.
Grade IV	Monstrous Goitres.

(b) Clinical effects and complications

In the majority of cases of simple goitre there are not clinical manifestations due to hypofunction or hyperfunction of the thyroid gland. Simple colloid goitre may require surgical treatment because of pressure effects on the adjacent structures. The following complications occur rarely : 1. hypothyroidism, (ii) hyperthyroidism, (iii) cretinism, (iv) deaf-mutism. Such complications are more likely encountered in regions where endemic goitre is prevalent.

(c) Treatment

1. A simple goitre in a non-endemic area rarely requires treatment. It happens so that in time it gets smaller without causing any harm. If this does not occur and the goitre becomes disfiguring, iodine therapy is seldom effective, thyroxine 0.2 to 0.3 mg/day may be given. This inhibits productions of TSH by the pituitary gland and so reduces the size of the thyroid gland. If there is no response to thyroxine and the goitre continues to be disfiguring, thyroidectomy should be considered, and this is indicated if the size of the goitre leads to obstruction of the trachea.
2. Cretinism may affect up to 5 per cent of the population in places where endemic goitre is severe. There are two types of endemic cretinism. In nervous cretinism there is mental deficiency, deaf mutism, spasticity and ataxia but features of hypothyroidism are rare. In myxoedematous cretinism there is dwarfism, signs of myxoedema and no goitre. Nervous cretinism can be prevented by giving a single injection of iodized poppy-seed oil to the women of child-bearing age but it must be given before pregnancy starts. It thus appears that iodine is required for the early development of the nervous system before the fetal thyroid appears in the third month of gestation. In myxoedematous cretins nervous system develops normally in the critical early months but the thyroid gland fails to adapt adequately by hypertrophy to severe iodine deficiency. The thyroid glands are small and uptake of radioiodine very low. Consequently, such cretins have low plasma T3 and T4 with clinical signs of hypothyroidism, including dwarfism. Both types of cretinism are seen in endemic areas but the proportion of the two varies from region to region.

(d) Prevention

1. Iodization of table salt greatly reduces the prevalance of goitre in many countries. The iodized salt must contain 76 kg of iodine/gm salt and daily consumption of salt is betwen 2 and 6 gm. Lower levels of iodization are used in most other countries and are effective. Potassium iodide is used for the iodization of the crude moist salt consumed in many countries.

2. Where goitre is endemic daily intakes of dietary iodine are likely to be less than 50 µg. This should be increased to the normal range of 100 to 300 µg. The amount of iodine added to the salt should be related to the usual consumption of salt in the community so that individual iodine intakes fall within this range. Iodine given for medical purposes may cause allergic skin rashes and induce hyperthyroidism, but the risk of these adverse effects from iodized salt is minimal.
3. Iodized oil injections provide an alternative means of prevention in parts of the world where the use of iodized salt is impossible and where endemic goitre is most severe and accompanied by cretinism. Goitres are not merely a cosmetic problem, but impair the vitality of many of the people by causing hypothyroidism, and cretinism retards the intellectual development of children. A single injection has been found to correct the deficiency for a period of two to three years. Iodized oil has now been used successfully in many countries.
4. Iodization of the water supply in a remote village, where distribution of iodized salt and injection of iodized oil are impractical, reduces the prevalence of goitre. Iodinators consisting of cannisters containing iodine crystals are connected to main water pipes and a fraction of the water diverted through them.

XEROPHTHALMIA

Xerophthalmia (Greek xeros, dry ; ophthalmos, eye) is a condition caused by vitamin A deficiency. In its milk form it is confined to the conjunctiva which is very common in many countries. There is danger of corneal ulceration when it spreads to the cornea and a permanent defect in vision. In severe cases, there is softening of the cornea, keratomalacia, which, if not immediately treated, soon leads to permanent blindness. Keratomalacia is associated in young children with protein-energy malnutrition.

Xerophthalmia arises when the diet contains practically no whole milk and butter and very limited amounts of fresh vegetables and fruit and so lacks both retinol and carotenes. Xerophthalmia and keratomalacia both occur in the first year of life amongst artificially fed infants but are rare amongst the breast fed. Children in poorly nourished communities are born to mothers who have had small intakes of vitamin A and consequently their liver stores are small at birth.

(a) Clinical Features

1. The bulbar conjunctiva is dry, thickened, wrinkled and pigmented due to a failure to shed the epithelial cells, and consequent keratinization. The pigmentation gives the conjuctiva a peculiar 'smoky' appearance. The pigment is diffused and especially marked in the interpalpebral fissure. Dryness, thickening and pigmentation characteristic of the condition are also caused by long periods of exposure to glare, dust and infections. This is common in older children and adults in the tropics.
2. When dryness spreads to the cornea, this takes on a dull, hazy, lack-lustre appearance. This is due to the keratinisation which is the result of Vitamin A deficiency on all epithelial surface. The cornea often becomes insensitive to touch with a wisp of cotton wool. Corneal ulceration may occur from many causes and be unrelated to Vitamin A deficiency. The characteristic feature is a loss of substance (erosion) of a part or the whole of the corneal thickness. Unless there is secondary infection, there are no signs of inflammation. The lesion only heals by scarring. Corneal xerosis may progress suddenly and rapidly to keratomalacia.
3. Softening and dissolution of the cornea follow and are known as colliquative

necrosis. This presents a grave emergency. When the process involves only part of the cornea, there is ulceration but the inflammatory reaction is mild. If the process is not stopped by treatment, perforation of the cornea leads to prolapse of the iris, extrusion of the lens and infection of the whole eyeball. The chances of saving any useful vision are slight. Healing results in scarring of the whole eye and frequently in total blindness. The retinol content of the plasma is below 200 μg/I, the lower limit of the range.

4. Night blindness is an early symptom of vitamin A deficiency and is often present without any signs of xerophthalmia.
5. The prolonged Vitamin A deficiency in school children or young adults may show lesions appearing as spots, either white or yellow, scattered along the sides of the blood vessels. The spots may fuse and the lesions are most numerous on the periphery of the fundus and never appear on the macula.
6. Corneal scars are white, opaque patches on the cornea and the result of healing of an older ulcer. Vision may be seriously affected, depending on the size of the scars. There are other causes of corneal scars but Vitamin A deficiency is the most effective.

(b) Treatment

1. Vitamin A in a dose of 30 mg of retinol (100,000 I.U.) should be administered daily for three days immediately the diagnosis is made or strongly suspected. It is recommended that half the dose should be given orally in the form of halibut oil and half intramuscularly as water-miscible retinol palmitate. An oil solution should not be injected as the retinol is then absorbed very slowly from the injection site.
2. Cod liver should not be instilled directly into the eye. During convalescence 9 mg of retinol in the form of a fish liver oil orally is adequate. It is also most essential that the diet is satisfactory in regard to other nutrients.
3. Secondary bacterial infection should be treated by the use of antibiotics which are of great value. Local treatment of the eye will only be required provided the diagnosis is present.

(c) Prevention

1. Most cases occur in urban poor and rural peasants in under doctored areas, and keratomalacia develops without the children being seen by a doctor. Therefore, to prevent this, doctors, nurses and midwives, and other paramedical staffs associated with maternal and child health clinics should be trained.
2. Pregnant women should be advised to eat dark supplements rich in Vitamin A in prophylactic doses. This helps to build up stores of retinol in the fetal liver and should be continued during lactation. Mothers should be advised to include in the weaning foods dark green leafy vegetables or yellow and orange fruits, which are locally available, cheap and known to be good sources of β-carotene.
3. In blindness from keratomalacia single large prophylactic doses of retinol in oily solution are recommended. This is given as a capsule to be taken orally. The dose is safe and adverse effects are rare. All prophylactic programmes should be evaluated by periodic field surveys of the prevalence of xerophthalmia.
4. Nutritional disorders of the eye are only one of several causes of blindness. There are more than ten million blind people in the world. Most of them are blind before the age of 5. Trachoma, a virus infection, is the most important eye disease in the world. It is common in children and, if untreated, often causes progressive loss of

vision and blindness. Small pox often affect the eyes and is responsible for 1/5th of the blindness in India. Onchocercosis is caused by a filarial worm which is transferred from man to man by flies. It causes nodules in the skin and thence microfilaria invade the eye which leads to blindness. The disease commonly affects people who live besides river. In some villages, all the adult population are blind. Venereal diseases, congenital syphilis and gonococcal ophthalmia neonatorum are important causes of blindness in parts of the world where the maternity services are inadequate or totally lacking. Accidents in the home and at work are common causes of blindness. Many young children lose their sight in this way. Diabetes, cataract and glaucoma are important causes of blindness in elderly people.

Vitamin A deficiency is thus one of the important causes of blindness and the most easily preventable.

RICKETS AND OSTEOMALACIA

Rickets is a disease of children in which the bones are softened and deformed. It arises as a result of deficiency of Vitamin D and a failure to absorb calcium from the small intestine. Osteomalacia, which means softening of bone, arises when there is Vitamin D deficiency in adults. The resultant calcium deficiency leads to demineralisation of the bones. The severe forms of both diseases with gross deformities of the skeleton are now rare but cases may be seen occasionally in hospitals in most countries. Only a few foods are good sources of Vitamin D and about 90 per cent of the vitamin in our bodies comes from photosynthesis in the skin.

(a) Risk factors

1. In northern latitudes during long winters with only a few hours of daylight greatly reduce exposure to ultraviolet radiation. Vitamin D deficiency is then a risk for all children and adolescents, as they have greater need for the vitamin than adults, and also for all elderly people and others with disabilities restricting outdoor activities.
2. Puppies developed rickets when white bread in their diet was replaced by unrefined oatmeal. Asian communities whose stable food is chapati made from high extraction wheat flour appear to be at increased risk of both rickets and osteomalacia.
3. Strict Hindus and others who eat no animal food provide an increased proportion of cases of osteomalacia seen in Britain. This complete exclusion of Vitamin D from the diet does not normally lead to rickets or osteomalacia but increases the risk.
4. When an infant is fed exclusively on milk from a Vitamin D deficient mother for more than three months, the risk of infantile rickets rises.
5. Heavy pigmentation reduces synthesis of Vitamin D in the skin. This can only be a minor risk factor since in Britain rickets is much less common in the darker West Indian than in the lighter Asian communities.

(b) Secondary osteomalacia and rickets

1. Osteomalacia may arise in patients with the malabsorption syndrome after many months. This is due to impaired calcium absorption.
2. Osteomalacia and other bone disorders arise in patients with chronic renal failure. Impaired formation of 1,25 $(OH)_2$ D in the kidney may be responsible.
3. Osteomalacia is sometimes found in patients with cirrhosis of the liver due to failure to form 25 (OH) D.
4. Five different genetic defects that lead to rickets are known. All are uncommon.

(c) Clinical Features

Rickets

1. The child is restless, fretful and pale, with flabby and toneless muscles which allow the limbs to assume unnatural postures (acrobatic rickets).
2. Excessive sweating on the head is common.
3. The abdomen is distended as a result of the weak abdominal muscles, the atony of the intestinal musculature and the intestinal fermentation that may arise from excessive carbohydrate in the diet.
4. Gastrointestinal upsets with diarrhoea are common.
5. The infant or child is prone to respiratory infections.
6. Development is delayed, the teeth often erupt late and there is failure to sit up, stand, crawl and walk at the normal ages.
7. There is extension and widening of the epiphyses at the growing points, where cartilage meets bone. The earliest bony lesions are usually enlargement of the epiphyses at the lower end of the radius and at the costochondral junctions of the ribs or '**rickety rosary**', an early and important diagnostic feature. Later features are '**bossing**' of the frontal and parietal bones and delayed closure of the anterior fontanelle.
8. There may be deformities of the chest such as undue prominence of the sternum (pigeon chest) and a transverse depression, passing from the costal cartilages towards the axillae which deepens with inspiration. In unusually severe cases, respiratory functions can be seriously impaired by the combination of respiratory infection and a rachitic chest.
9. If rickets continues into the second and third year of life, these signs may persist or be magnified. Deformities such as kyphosis of the spine develop as a result of the new gravitational and muscular strains, caused by sitting up and crawling. At the same time there may be enlargement of the lower ends of the femur, tibia and fibula. When the rachitic child begins to walk, deformities of the shafts of the leg bones develop, so that '**knock knees**' or '**bow legs**' are added to the clinical picture. Anterolateral bowing of the tibiae at the junction of the middle and lower third is frequently noted in young children with rickets.
10. The spinal kyphosis is often replaced by lordosis. Pelvic deformities may follow and lead years later to serious difficulties at child birth.
11. When ionized calcium in the plasma is reduced, infantile tetany may result, with spasm of the hands and feet and of the vocal cords. The latter causes a high-pitched, distressing cry and great difficulty in breathing.

(d) Diagnosis

A flabby baby towards the end of its first year, unable to pull itself up, fretful and easily irritated, with too few teeth showing and liable to profuse sweats, should always be suspected of having rickets. Early evidence of rickets may be overlooked in a child ill with broncho-pneumonia or diarrhoea, especially in the first year of life. If there is any doubt, a radiograph of the wrist may show characteristic changes at the epiphyses; the outline of the joint is blurred and hazy, and the epiphyseal line becomes broadened. Later, in older children, as a result of decalcification of the metaphysis and the effects of movements and stresses the classical metaphysis and the effects of movements and stresses the classical concave '**saucer**' deformity is clearly shown radiographically. The opinion of an experienced radiologist may be needed to distinguish the picture from that of scurvy. The diagnosis is supported by a raised plasma alkaline phosphatase and confirmed if plasma 25 (OH) D is low.

It is sometimes necessary to distinguish rickets from other rare disorders involving the bones, such as congential syphilis, achondroplasia and osteogenesis imperfecta. Radiographs of the bone are helpful in differentiating these disorders.

Osteomalacia

(a) Clinical Features

1. Deformities of the spine, pelvis and legs are now rarely seen.
2. The common features are pain and muscular weakness. Pain ranges from a dull ache to severe pain. The affected sites are the ribs, sacrum, lower lumbar vertebrae, pelvis and legs.
3. Bone tenderness on pressure is common.
4. Muscular weakness is often present and the patient may find difficulty in climbing stairs or getting out of a chair.
5. Spontaneous fractures may occur.
6. Tetany may be manifested by carpopedal spasm and facial twitcing.

(b) Diagnosis

The early symptoms may resemble those present in osteoporosis and rheumatic disorders. The measurement of plasma 25 (OH) D should clear the diagnosis. The distinction between osteomalacia and osteoporosis.

(c) Treatment

1. A daily oral dose of 25-125 mg (1,000-5,000 I.U.) of Vitamin D cures rickets and osteomalacia. This should be reduced to 10 μg, the prophylactic dose, when plasma alkaline phosphatase has returned to normal and radiographs show that healing is established.
2. Children can be given halibut liver oil in a very small dose (1 ml) since it contains 30 to 40 times the concentration of Vitamin D of cod-liver oil. For severe cases, synthetic calciferol is useful.
3. When an infant or young child may be seen once by an emergency medical service and perhaps not again for months, a single massive dose of Vitamin D (150,000 I.U.) (three strong calciferol tablets) can be given by mouth with reasonable safety and curative effects. The single dose can be given by injection but this has no proved advantage over the oral route. A daily small dose is recommended to avoid danger of overdosage.
4. If there is evidence of malabsorption in osteomalacia, the dose of Vitamin D should be up to 1.25 mg (50,000 I.U.) daily and it may have to be given intramuscularly at weekly or monthly intervals. If the disease is secondary to kidney or liver disease large doses and either 1,25 $(OH)_2D$ or 25 (OH) D are indicated.
5. An adequate intake of calcium is essential. The best source is milk and at least 500 ml should be drunk daily. When this is not practical and in severe cases, calcium lactate, taken by mouth, should be prescribed.
6. An egg and butter daily is required to increase the dietary intake of Vitamin D.
7. Mothers of young children require tactful education in feeding and general care, as do elderly patients.
8. Unnecessary clothing should be removed and there should be every opportunity to go out to enjoy the sunshine.

(d) Prognosis

Rickets is not a fatal disease but the untreated rachitic child is a weakling with an increased risk of infections, notably broncho-pneumonia. The skeletal changes usually tend to heal spontaneously as the child gets older. The bony deformities, if mild, usually right themselves as growth proceeds, but in severe cases pigeon chest, contracted pelvis, knock knees or bow legs may persist. With early and sufficient treatment these changes are entirely avoided.

In osteomalacia, Vitamin D quickly relieves the pain and muscular weakness but it takes many weeks or months to restore the bones to their normal strength.

	Osteomalacia	Osteoporosis
Clinical features :		
Skeletal pain	A major complaint usually persistent.	Episodic and usually associated with a fracture.
Muscle weakness	Usually present and producing disability and a characteristic gait when severe.	Absent.
Fractures	Relatively uncommon; healing delayed.	The usual presenting features; heals normally.
Skeletal deformity	Common, especially kyphosis.	Only occurs where there is a fracture.
Radiographic features :		
Loss of density of bones	Widespread.	Irregular and often most marked in the spine.
Loss of bone detail	Characteristic.	Not a features.
Looser's zones	Diagnostic.	Absent.
Biopsy :		
Histological changes	Excess osteoid tissue with bone present in normal quantity.	Bone reduced in quantity but fully mineralized.
Biochemical changes :		
Plasma Ca and P	Often low.	Normal.
Plasma alkaline phosphatase	Often high.	Normal.
Urinary calcium	Often low.	Normal or high.
Response to treatment :		
Vitamin D	Dramatic.	None.

(e) Prevention

1. All people can be protected from rickets by a supplement of 10 μg of Vitamin D daily. The dose is effective and safe. Supplements are necesary in all countries with long dark winters for all children up to 5 years of age, and for their mothers during pregnancy and lactation. All children on anticonvulsive drugs should continue to receive a supplement. In some communities there is now evidence that large numbers of the elderly require extra Vitamin D to protect them against osteomalacia.
2. Rickets occurs predominantly in families which are poor and where the mother lacks education. An intensive programme of health education is needed to ensure that the supplements are taken regularly ; this should also provide advice on diet, clothing and general hygiene.
3. Children in countries with abundant sunlight should not normally need a Vitamin D supplement. Where rickets is present, teaching of mothercraft, emphasising that young children must not be excessively protected from the sun, is necessary, protein-energy malnutrition is sometimes associated with rickets and routine dietary supplement may be needed.
4. In tropical countries, fish-liver oils are more easily obtained. They contain useful amounts of Vitamin A. Sometimes it may be advantageous to give children a single massive dose (1-2 mg) of Vitamin D. This is stored in the liver, liberated slowly and protects a child for several months.
5. Smoke abatement, slum clearances, and provision of open-air playgrounds were mainly responsible for the marked fall in the prevalence of rickets in the early part of the present century. The great improvement in these matters have to be maintained and extended where necessary. These are public health measures of major importance.

BERIBERI AND THE WERNICKE KORSAKOFF SYNDROME

Beriberi

Beriberi is a nutritional disorder formerly widespread in the rice-eating people of the East. `Beri' means weak. The main features of beriberi are due to deficiency of thiamine. Three forms of the disease occur.

1. Wet beriberi, characterised by oedema often associated with high-output cardiac failure.
2. Dry beriberi, a polyneuropathy.
3. The infantile form.

Raw fermented fish which contains thiaminase and fermented tea leaves both reduce the availability of the small amount of thiamine in the diet.

(a) Clinical Features

1. At first there is anorexia and ill-defined malaise, associated with heaviness and weakness of the legs. This may cause some difficulty in walking.
2. There may be a little oedema of the legs or face and the patient may complain of precordial pain and palpitation.
3. The pulse is usually full and moderately increased in rate.
4. There may be tenderness of the calf muscles on pressure and complains of `pins and needles' and numbness in the legs.
5. The tendon jerks are usually sluggish, but occasionally slightly exaggerate.
6. Anaesthesia of the skin, especially over the tibiae, are common.

Such conditions may persist for months or even years with only minor alterations in the symptoms. Patients can work to earn their living with low efficiency. At any time this chronic malady may develop into severe forms.

Wet beriberi

1. Oedema is the most prominent feature and may develop rapidly in the legs as well as in the face, in the trunk and also in the serous cavities.
2. Palpitations are marked and there may be breathlessness.
3. Anorexia and dyspepsia are commonly present.
4. There may be pains in the legs after walking.
5. The calf muscles are frequently tense, slightly swollen and tender on pressure.
6. The neck veins become distended and show visible pulsations.
7. The apex beat of the heart is displaced outwards.
8. In the arteries there is often a lowered diastolic pressure and systolic pressure is disproportionately higher ; hence on auscultation over the femoral and other large arteries, a curious 'pistol shot' sound may be heard.
9. The pulse is generally fast.
10. If the circulation is well maintained, the skin is warm to the touch owing to the associated vasodilatation.
11. When the heart begins to fail, the skin becomes cold and cyanotic, particularly on the face.
12. ECG often shows no changes but in some cases there are low voltage of the QRS complex, inverted T waves or evidence of disturbed conduction.
13. The urinary volume is diminished, but there is no albuminuria.
14. The mind is usually clear.
15. The patient is in danger of sudden increase in the oedema, acute circulatory failure, extreme dyspnoea and death.

Dry beriberi

1. The essential feature is a polyneuropathy.
2. The muscles become progressively more wasted and weak, and walking becomes increasingly difficult.

3. The thin patients need at first one stick, then two and finally become bedridden.
4. The disease is a chronic malady, which may be arrested at any stage by improving the diet.
5. Bedridden patients are very susceptible to infections.
6. Dysentery or tuberculosis are often fatal unless prompt and efficient treatment is given.

Infantile beriberi

1. This occurs in breast-fed infants, usually between the second and fifth months.
2. The mother may have no clinical signs of beriberi although they must have been eating a diet and secreting milk with a low thiamine content.
3. Frank beriberi can develop in late pregnancy and the puerperium.
4. It exists in an acute and chronic form.
5. In acute form cardiac failure may develop abruptly ; the mother may have noticed that the infant is restless, cries a lot, is passing less urine than normal and shows signs of puffiness. The infant then may suddenly become cyanosed with dyspnoea and tachycardia and die within 24 to 48 hours. Other serious signs are convulsions and coma.
6. In the chronic form, which is much less common, the main symptoms are due to gastrointestinal disturbances. There is constipation and vomiting. The child is fretful and sleeps poorly. The muscles are soft and toneless, but not markedly wasted. There is often intense pallor of the skin with cyanosis round the mouth. Cardiac failure and sudden death are common.
7. Infantile beriberi is the chief cause of death between the ages of 2 and 5 months in rice-eating rural areas.

(b) Differential diagnosis

1. In mild and chronic cases of beriberi, there may be few or no physical sings and the diagnosis fully depends on the interpretation of symptoms and the dietary history.
2. The oedema of wet beriberi has to be distinguished from that associated with hepatic and renal disease and heart failure. The warm extremities in cardiac beriberi and the absence of protein in the urine are useful diagnostic points. Famine oedema should seldom be a diagnostic difficulty if a proper dietary history is taken.
3. Cardiovascular beriberi has to be distinguished from other causes of high output cardiac failure, notably hyperthyroidism and severe anemia.
4. In all doubtful cases of wet beriberi the therapeutic response to thiamine usually settles the diagnosis.
5. The diagnosis of dry beriberi is mainly based on the dietary history. In endemic areas the disease may be confused with neuritic leprosy, but this is characterized by palpable, cord-like superficial nerves and areas of skin anaesthesia. These two diseases do not infrequently occur together and when they do, dry beriberi, if mild, may be overlooked.
6. The diagnosis of infantile beriberi may be difficult. Neither oedema nor paralysis is an early sign and sudden death may occur before either is present. A history of sudden death of a previous child between the ages of 2 and 5 months is suggestive. Infantile beriberi may be confused with PEM and the two may occur together.

(c) Treatment

Wet beriberi

1. Treatment must be started as soon as the diagnosis is made because of the sudden fatal heart failure.
2. Complete rest is essential and thiamine should be given at once intramuscularly 25 mg twice daily for three days. Thereafter an oral dose of 10 mg two or three times

a day should be continued until convalescence is established.

3. Thiamine treatment to a patient with cardiovascular beriberi is the most effective in medical science. Within a few hours the breathing is easier, the pulse rate slower, the extremities cooler and a rapid diuresis begins to dispose of the oedema. Within a few days the size of the heart is restored to normal. Muscular pain and tenderness are also dramatically improved.
4. During convalescence and rehabilitation, a good mixed diet with less rice is needed. Another cereal should be substituted for part of the rice in the diet. Pulses have a well-deserved reputation for curing and preventing beriberi.

Dry beriberi

1. Patients are generally undernourished and, if they take sufficient of a good mixed diet to enable them to gain weight, slow improvement may be expected.
2. If the dietary intake is adequate, there is no need to continue with supplementary thiamine.
3. Infections and intercurrent disease should be treated and appropriate physiotherapy given.

Infantile beriberi

1. The mother's milk is the simplest way of treatment. The mother should receive 10 mg thiamine twice daily–in severe cases this should be administered by injection.
2. The infant should be given thiamine in doses of up to 10 to 20 mg intramuscularly once a day for three days. This should be followed by 5 to 10 mg orally twice a day.
3. In severe heart failure of convulsions and coma the initial dose may be increased to 25 to 50 mg given intravenously very slowly.

(d) Prevention

1. Beriberi can be prevented by the use of undermilled, home pounded or parboiled rice or by increased use of pulses and other foods containing thiamine. Medicinal preparations of thiamine are also available and cheap.
2. Changed milling practices are probably the most important, and in many areas rice is not so highly polished as to remove all the bran.
3. Changed milling practice of rice is the most important without removing all the bran.
4. Improvement in social and economic conditions and the consumption of a better diet with more thiamine-containing foods are essential.
5. The establishment of maternal and child Health Centres has led to many pregnant and lactating women getting good advice on diet as well as vitamin supplements ; infants may also receive them or extracts of rice bran.

Thiamine deficiency in those whose staple diet is not rice

Thiamine deficiency, in countries where rice is not the staple food, arises in the great majority of cases in persons whose diet has been greatly restricted usually as a result of chronic alcoholism. It also arises, but rarely, secondary to carcinoma of the stomach and other conditions associated with prolonged partial starvation. Hence there is always a lack of other essential nutrients.

(a) Alcoholic Neuropathy

Alcoholics who have restricted their food intake for many weeks often develop a disorder of peripheral nerves, sometimes indistinguishable from dry beriberi. The administration of thiamine leads to no dramatic improvement, but if the patient takes a good diet and gives up alcohol completely a slow diminution of the symptoms may be expected. The nerves of the lower limbs are affected more severely than those in the upper limbs. There is dysfunction of both sensory and motor fibres. The effects on sensory nerves may be paraesthesiae (pins and needles) or sometimes severe nerve pains, as in the burning feet syndrome ; there may be loss of sensation, either numbness of the extremities or loss of position sense. Signs of motor nerve involvement are foot drop, muscle wasting and impaired knee and ankle jerks.

Other causes of polyneuropathy are :

1. Deficiency diseases : Pellagra, subacute combined degeneration, burning feet syndrome and pyridoxine deficiency.
2. Metabolic diseases : Diabetes Mellitus, Uremia, Porphyria etc.
3. Chemical poisoning : Heavy metals (lead, arsenic, mercury) and some drugs.
4. Infective : Diphtheria, Leprosy etc.
5. In association with carcinoma.
6. Rare genetic types : Refsum's disease.

(b) Occidental beriberi heart disease

Cardiac failure with generalised oedema, pulmonary congestion and dyspnoea sometimes develop in chronic alcoholics. Damage to the myocardium may be due to direct action of ethanol or thiamine deficiency.

Less frequently there is a sudden circulatory collapse with lactic acidosis. It is wise to give all alcoholics with evidence of heart failure and sufficient thiamine to replenish their tissues ; in most cases no benefit is likely but it will do no harm and in few may be life-saving.

Wernicke-Korsakoff Syndrome

Wernicke in 1881 described a neurological disorder in three patients, two of them alcoholics and the third a seamstress who had persistent vomiting after ingestion of sulphuric acid. It is characterised by weakness of eye muscles, so that the patient cannot look upwards or sideways and a state of disorientation and apathy. Sometimes there is jerky, rhythmical movements of the eyes (nystagmus) and if the patient can stand he is unsteady (ataxia). It was found that many of the cases recovered dramatically after large doses.

Korsakoff in 1887 described a psychosis, also occurring in alcoholics, characterised by a severe defect in memory and learning. Confabulation is a characteristic feature, though not always present. The patient can remember past events with verifiable accuracy. He cannot remember what he did earlier in the same day but tends to provide a superficially convincing tale rather than say he has forgotten.

Wernicke's disease and Korsakoff's psychosis are manifestations of thiamine deficiency. After carrying out 82 postmortem cases it had been found out that there were symmetrical lesions in various parts of the brain stem, diencephalon and cerebellum, the areas commonly affected being the mamillary bodies, the nuclei of the thalamus and the periaqueductal grey matter. In the most advanced lesions there was virtually complete tissue necrosis; the less severe lesions were characterised by destruction of myelin with less damage to neurones. Small haemorrhages are characteristic but not always present.

The same histological features were seen in cases of Wernicke's disease and of Korsakoff's psychosis. Most of the patients who recovered from the acute confusional state subsequently developed some memory defect.

Treatment with thiamine is effective in these cases when given promptly. Large doses are also needed.

Diagnosis is confirmed by the RBC transketolase test. Although the syndromes occur in alcoholics, it may arise secondary to any disorder which seriously impairs nutrition.

(e) Comment

Thiamine deficiency can thus lead to an encephalopathy or a cardiomyopathy or a peripheral neuropathy. Two or rarely all three diseases can occur together in a patient but it is surprising how often they do not. We cannot yet explain why the brain is affected in one person, the heart in another and the peripheral nerves in a third. Alcoholism is widespread and body stores of thiamine in alcoholics are likely to be small, but thiamine deficiency should be registered. There are still many unsolved problems in the relation between ethanol, thiamine and brain damage.

Pellagra

Pellagra is nutritional disease endemic among poor peasants who subsist chiefly on maize. It has been called the disease of the three Ds: dermatitis, diarrhoea and dementia. But diarrhoea and mental changes are not always present in mild and early cases and the mental symptom is usually depression and not dementia. The clinical features are loss of weight, increasing debility, an erythematous dermatitis characteristically affecting parts of the skin exposed to sunlight, gastro-intestinal disturbance especially diarrhoea and glossitis, and mental changes.

(a) Clinical Features

Skin

1. There is an erythema resembling severe sunburn, especially the backs of the hands, the wrists, the forearms, face and neck.
2. Exposure to trauma or mechanical irritation of the skin, especially over bony prominences, may also determine the site of the lesion.
3. The skin in the affected areas is at first red and slightly swollen ; it itches and burns.
4. In acute cases the skin lesions may progress to vesiculation, cracking, exudation and crusting with ulceration and sometimes secondary infection ; but in chronic cases the dermatitis occurs as a roughening and thickening of the skin with dryness, scaling and brown pigmentation.

Digestive system

1. There is usual digestive upset; and diarrhoea is not always present.
2. There may be nausea, a burning sensation in the epigastrium, and sometimes constipation in chronic cases.
3. The digestive symptoms may be aggravated by the presence of intestinal parasites.
4. The mouth is sore and often shows angular stomatitis and cheilosis.
5. The tongue characteristically has a `raw beef' appearance – red, swollen and painful, though usually without loss of papillae.
6. Secondary infection of the mouth with Vincent's organisms is common.
6. A non-infective inflammation followed by mucosal atrophy may involve the gastrointestinal tract and account for the diarrhoea which is characteristically profuse and watery, sometimes with blood and mucus in the stools.
8. The rectum and anus are frequently affected and chronic gastritis with reduction or absence of acid secretion is a common finding.
9. Vaginitis and amenorrhoea may occur.

Nervous System

1. In mild cases the symptoms consist of weakness, tremor, anxiety, depression and irritability ; in severe acute cases delirium is common and dementia occurs in the chronic form.
2. In chronic cases there may be decreased sensation in the feet to touch and loss of vibration and position sense. The loss of position sense may give rise to ataxia.
3. Spasticity and exaggerated tendon reflexes give evidence of involvement of the pyramidal tracts. The features are those of subacute combined degeneration of the cord and may be due to associated Vitamin B_{12} deficiency.

(b) Diagnosis

1. The skin lesions are of diagnostic importance since they are only found in pellagra, whereas the gastrointestinal and mental features may be present in many other diseases. A variety of erythemas and exfoliative skin lesions may mimic pellagra. The two characteristic features of cutaneous pellagra are its symmetrical distribution, determined by the clothes of the patient and exposure to sunlight, and the therapeutic response to nicotinic acid.

2. A nutritional glossitis identical with the tongue changes seen in pellagra may occur without the other signs of the disease in people who have been all the times indoors, out of sunlight.
3. Pellagra is a disease affecting poor people on bad diets. Hence it is often accompained by signs of protein energy malnutrition by anaemia and by deficiencies of thiamine and other vitamins. These together with chronic infections may complicate the clinical picture.

(c) Laboratory Findings

The fasting plasma tryptophan ranges from 1 to 4.8 mg/l in pellagrins and from 6.5 to 8.8 mg/l in healthy adults. Plasma tryptophan may prove to be a convenient test for confirming a diagnosis of pellagra.

(d) Prognosis

Mental symptoms, especially dementia, are the most serious features and may be permanent. Occasionally a fulminating form develops,with fever and severe prostration which can be fatal. In the past many deaths were due to secondary infections (notably tuberculosis and dysentery) or to emaciation due to general dietary failure, intensified by the diarrhoea.

(e) Treatment

1. Nicotinic acid or nicotinamide are the standard treatment for quick relief of symptoms. Nicotinamide is to be preferred because it does not cause the unpleasant flushing and burning sensations that often result from taking nicotinic acid. These are transitory and harmless, but may alarm the patient. The vitamin is rapidly absorbed from the stomach, despite severe digestive disorders. There is no need to give intravenous or intramuscular injections. The immediate response to nicotinamide is usually dramatic; within 24 hours the erythema diminishes, the tongue becomes paler and less painful and the diarrhoea ceases. Often there is striking improvement in the patients's behavior and mental attitude. But nicotinamide alone is usually insufficient to restore health due to other associated deficencies, notably of protein and other components of Vitamin B complex. Therefore, B complex should be given as a routine and if there are signs of peripheral neuropathy or subacute combined degeneration of the cord larger doses of thiamine or Vitamin B_{12} are indicated.
2. To restore the patient to normal weight, the diet should provide ample energy and good quality protein, as is present in milk, eggs, meat or fish. In severely ill patients it is necessary to climb the dietetic ladder cautiously. The food should be low in bulk to avoid further diarrhoea. The diet may be poorly tolerated because of the mental state of the patient and the sore mouth which may make eating difficult. Alcohol should be forbidden.
3. Rest in bed and sedation are necessary for severely ill pellagrins, especially those with marked mental symptoms. If the dermatitis is associated with much crusting or secondary infection, gentle washing with a bland solution is indicated.

(f) Prevention

1. Enrichment of maize meal with vitamins is technically simple and inexpensive but is difficult to implement for subsistence farmers who grow their own maize. It is wise to avoid dependence on a single cereal crop, such as maize.
2. Animal husbandry should be encouraged in all areas where pellagra is endemic so that the production of milk and milk products, and meat is increased. Encouraging the planting of opaque-2 maize may help. It contains about three times as much tryptophan and twice as much lysine as conventional maize.

Scurvy

Scurvy is a nutritional disease which results from prolonged subsistence on diets practically devoid of fresh fruit and vegetables. Lack of ascorbic acid causes a disturbance in the structure of connective tissue, leading to swollen, bleeding gums, and haemorrhages into the skin and elsewhere.

(a) Clinical Features

1. The gums are swollen, particularly in the region of the papillae between the teeth, sometimes producing the appearance of 'scurvy buds'. These may be so extensive that they project beyond the biting surface of the teeth and almost completely conceal them. The spongy gums are livid in colour and bleed on the slightest touch. There is always some infection.
2. The first sign of cutaneous bleeding is often to be found on the lower thighs, just above the knees. These haemorrhages are perifollicular – tiny points of bleeding around the orifice of a hair follicle. The condition in scurvy can be distinguished by its appearance from the follicular keratosis sometimes associated with Vitamin A deficiency. In the latter condition there is usually a horny plug of keratin projecting from the orifice of the hair follicle. In scurvy there is a heaping up of keratin-like material on the surface around the mouth of the follicle, through which a deformed 'cockscrew' hair characteristically projects. Perifollicular haemorrhages may appear on the buttocks, abdomen, legs and arms. African patients often present with pain in a leg due to haemorrhage into intermuscular septa in the thigh or calf.
3. Anaemia is present in most patients. In patients the bone narrow may be normoblastic or megaloblastic and associated deficiency of iron or folate is often responsible, but there may be other unidentified factors. Destruction of erythrocytes in muscle haematomata may lead to bilirubinaemia and mild jaundice.
4. Osteoporosis may occur in scurvy, since ascorbic acid is necessary for the synthesis of collagen in all parts of the body, including the bones.
5. Haemorrhages into any of the internal organs may occur and a patient dies suddenly and without warning, apparently from cardiac failure.
6. There is scurvy in infants. Until the teeth have errupted, scorbutic infants do not develop gingivitis. When this occurs, the gums have the classical appearance of 'scurvy buds'. The first sign of bleeding is usually a large subperiosteal haemorrhage immediately overlying one of the long bones–frequently the femur–producing the characteristic 'frog legs' position. This gives rise to intense pain, especially on movement. The infant may cry continuously and agonisingly, and scream even louder when lifted.

(b) Diagnosis

1. The inflamed rim of the gums is bright red in colour, in contrast to the cyanotic appearance in scurvy, and there is usually much less swelling. In Vincent's angina the gums are acutely inflamed, ulcerated and painful, but the bright red appearance of the lesions is distinctive. Poisoning with heavy metals, particularly lead and mercury, produces a gingivitis in which the gum margin is stained blue; but there is usually little swelling and the appearance is easily distinguished from scurvy. Phenytoin, a drug used in epilepsy, may cause marked swelling of the gums, but they preserve their normal colour and do not bleed.
2. Scurvy in infants and children may sometimes be mistaken for rheumatic fever or osteomyelitis, because of the pain caused by a subperiosteal haemorrhage. The refusal of the child to use one leg may cause the disease to be mistaken for poliomyelitis.
3. Blood ascorbic acid can easily be determined. If the concentration of ascorbic acid in blood is low, it is a sure case of scurvy.

(c) Treatment

1. Adequate amounts of synthetic ascorbic acid should be given at once because of the danger of sudden death. The vitamin is very soluble and quickly absorbed from the digestive tract. It can be given intravenously, but a large part of it is immediately lost in the urine. It is the aim to saturate the body with ascorbic acid. The fully saturated body contains about 5 grams of the vitamin. Therefore, a dose of 250 mg by mouth four times daily should achieve this within a week, despite some loss in the urine.
2. Scurvy arises among people far removed from supplies of synthetic ascorbic acid (eg, among prisoners of war) ; in such conditions valuable therapeutic effects are obtained by the use of natural sources of the vitamin, such as fresh fruit and vegetables.
3. In case of anemic patient ferrous sulfate and folate tablets should be given by mouth.
4. With adequate and immediate treatment no patient dies of scurvy, but if it is delayed he may die.

(d) Prevention

1. Scurvy generally occurs at the two extremes of age. The prevention of scurvy in infants has been accomplished by the better education of mothers and helped by the distribution of cheap, concentrated orange juice of standard ascorbic acid content. For old people living alone, the provision of proper meals is the best means of preventing scurvy. This should be the responsibility of their family. If there are no relations, responsibility falls on the social services. In cases where an old person is unwilling to eat foods containing the vitamin, he should be given ascorbic acid tablets.
2. In cases of persons who travel in barren lands or make long sea voyages, they should be given ascorbic acid tablets to carry along with them.
3. In times of drought and famine, when fresh vegetables are not available, ascorbic acid can be obtained by the germination of pulses or cereals. 30 grams of dried pulse on germination yields 9 to 15 mg ascorbic acid which is sufficient to prevent scurvy.

OTHER NUTRITIONAL DISORDERS OF THE NERVOUS SYSTEM

A. Burning feet syndrome

1. The earliest symptom is aching, burning or throbbing in the feet. This becomes more intense and is followed by sharp, stabbing, shooting pains, which may spread up as far as the knee like an electric shock, causing excruciating agony. They come on in paroxysms and are usually worse at night.
2. Most patients get some relief by walking about, and sufferers may spend the night limping up and down outside their quarters.
3. Some patients manage to get relief by wrapping their feet in cold wet cloth or sitting with their feet in a pail of cold water. Continuous pain and loss of sleep produce a thin, exhausted, irritable patient.
4. The tendon jerks may be normal but may be exaggerated.
5. The syndrome has been associated with the prolonged consumption of a diet deficient in protein and the B group of vitamins.
6. Patients who suffer from it may also develop the orogenital syndrome or nutritional amblyopia, but rarely beriberi.
7. The syndrome can be seen sometimes in chronic alcoholics and patients with diabetic and other neuropathies, and rarely in other disorders.

B. Spinal ataxia

1. Patients living on unbalanced diets for long periods develop neurological signs which indicate that the principal lesion is

in the dorsal columns of the spinal cord involving particularly proprioceptive sensation. The gait is unsteady and the patient is unable to stand upright without swaying when the eyes are closed. Vibration sense in the legs is often lost.

2. In tropical ataxic neuropathy, Vitamin B_{12} plays only a secondary role. In the fully developed syndrome there is sensory spinal ataxia, retrobulbar neuropathy or optic atrophy and sometimes bilateral nerves deafness. Signs indicating mild involvement of the lateral or pyramidal tracts may be found. Epidemiologically the condition is found in people who regularly consume large amounts of cassava. Cassava contains a cyanogenic glycoside, Linamarin, which can be broken down to yield free hydrogen cyanide by enzymes in the plant tissue if it is crushed or left standing in water.
3. The cyanide is detoxified by sulphur containing amino acids, which convert it to thiocyanate, and by hydroxycobalamin which forms cyanocobalamin. Patients with tropical ataxic neuropathy have increased plasma concentration and urinary excretion of thiocyanate with increased Vitamin B_{12} and reduced cystine in the plasma. The condition can be prevented in part by cooking methods which wash out the glycoside or boil off the HNC.

C. Cerebellar cortical degeneration :

This condition in which the characteristic clinical feature is ataxia of the legs arises from degenerative changes limited to the anterior superior part of the vermis of the cerebellum. It is associated with alcoholism and poor nutrition but the response to vitamin thereby and nutritional rehabilitation is less consistent that in the Wernicke-Korsakoff syndrome.

D. Vitamin B_{12} neuropathy

1. Early symptoms are tingling, coldness and numbness in the extremities due to peripheral neuropathy.
2. Motor weakness and ataxia appear later and become increasingly severe as the cord is involved.
3. The physical signs depend on the relative involvement of the peripheral nerves and the dorsal and lateral columns of the cord.
4. In severe cases, ataxia is the outstanding feature with loss of reflexes especially in the lower limbs. Sometimes the pyramidal tracts are involved, and spasticity, increased reflexes and an extensor plantar response are present.
5. If the brain is affected, there may be an organic psychosis and this may be the first evidence of Vitamin B_{12} deficiency.

E. Spastic paraplegia

1. If khesari dal (pulse) eaten in excessive amount and for a long period, it gives rise to lathyrism.
2. The onset of lathyrism is sudden and is often preceded by exertion or exposure to cold. A patient may go to bed well and wake up paralysed; or he may fall down at the plough. Sometimes backache and stiffness of the legs precede the onset of the paralysis by a few days. The condition is a spastic paralysis of the lower limbs, due presumably to precisely localised lesion of the lower parts of the pyramidal tracts. The motor nerves to the muscles of the trunk, upper limbs and sphincters are spared. The sensory nervous system is not involved. In mild cases there is only stiffness and weakenss of the legs and exaggerated knee and ankle jerks. In more severe cases, the patients walk with bent knee on tiptoe. The legs are often crossed ; a 'scissors gait' develops and walking is only possible with the aid of sticks. In severe cases, paraplegia in flexion follows, and walking becomes impossible. The patient can only move about by pushing himself along, supporting his body on his hands, buttocks and heels. The paraple-

gia is typically spastic with greatly increased ankle and knee jerks and with clonus. The final stage of the disease is completely incapacitating and the sufferers may move to the cities where they are easily recognised amongst the beggars.

3. Epidemic lathyrism is mainly a disease of famine. When the price of wheat rises, many of the poor increase thier consumption of the pulse and cases of lathyrism arise. Prevention requires change in agricultural and economic policies. The toxin can be extracted from khesari dal (pulse) by heating it in four volumes of water for an hour.
4. There is no specific treatment. All patients need a good diet.

LESS NUTRITIONAL DISORDERS

Disorders of the Skin

A. Follicular hyperkeratosis

1. The follicles become blocked with plugs of keratin derived from their epithelial lining which has undergone squamous metaplasia. This is due to Vitamin A deficiency. Halibut-liver oil, red palm oil or other oils rich in Vitamin A or carotene may produce a striking clinical improvement, vegetable oils are also likely to be rich in essential fatty acids and Vitamin E, and the condition has responded to Vitamin E therapy. Other factors may contribute to its development, such as exposure to sunlight and lack of cleanliness.
2. Slight follicular keratosis may be found in people who are adequately nourished in respect of Vitamin A. Thus, the condition is not a specific or constant feature of Vitamin A deficiency.
3. The typical distribution is over the backs of the upper arms and the fronts of the thighs, but it may extend over the buttocks and indeed over the whole trunk. Only the feet, hands, and face may be spared. Some degree of xeroderma is commonly associated. The horny plugs that project from the follicular orifices can often be pulled out with a pair of forceps ; they give the skin a characteristic feeling of roughness. Because of its appearance, the condition has been called 'toad-skin' or phrynoderma.
4. Folliculosis is sometimes mistaken for follicular hyperkeratosis. The follicles are raised above the surface, but no horny plug projects from the follicular orifices.

B. Xeroderma

It means dryness of the skin. The skin feels dry and often rough. On uncovering the legs, a cloud of fine, brawny dandruff is often seen. Xeroderma is commonly but not constantly associated with follicular keratosis and 'cracked skin'.

C. Crazy-paving skin

In this condition the appearance indicates a layer of lacquer painted on the surface, which on drying has broken up into individual islands of varying size. There is often some desquamation from the borders of each island, while the intervening gaps may become fissured. The commonest site for this lesion is the skins, and it seems probably that exposure to dirt and alternate heat and moisture is often responsible.

D. Pachyderma (elephant skin)

1. The affected skin areas are thick, rough and thrown into folds like the skin of an elephant.
2. It starts as a roughness of skin on the back of the hands and feet, and the skin of the whole body may be affected. The changes are most marked at the back of the elbows and front of the knees.
3. Fissures may occur round the heels. The condition is seen most often in boys and in the dry season.

E. Pigmentary changes and colour

Nutritional failure can affect the colour of the skin in many different ways. In pellagra, there is typically an erythema with subsequent desquamation and pigmentation. The areas of skin especially involved are those exposed to sunlight or affected by friction. In anaemia the skin may be unduly pale. The hands of underfed children are often cyanosed, even in warm weather; while in cold and damp climates, they may be affected by chilblains.

F. Tropical Ulcer

1. They are chronic ulcers, affecting chiefly the lower limbs, occurring in hot, damp climates among people whose tissues are vitiated by malnutrition.
2. They are often caused by minor injuries in people living in poor hygienic surroundings, infested by disease, such as dysentery and malaria.
3. They are not attributable to lack of single nutrient, but their presence in any community or labour force is an indication that the diet and hygienic conditions are unsatisfactory.

G. Angular stomatitis

1. This is an affection of the skin at the angles of the mouth, characterised by heaping-up of greyish-white sudden epithelium into ridges, giving the appearance of fissures radiating outwards from the mouth.
2. Secondary infection and staining by food may give the lesion a yellowish colour. It may extend across the mucocutaneous boundary and produce whitish patches on the mucous membrane lining the cheeks.
3. It often responds rapidly to large doses of riboflavin and sometimes to pyridoxine.
4. It occurs in association with iron deficiency anemia and other diseases.

H. Cheilosis

1. This is a zone of red, denuded epithelium at the line of closure of the lips.
2. It is frequently seen in pellagrins and is often associated with angular stomatitis.
3. It only appears during periods of drought and lack of fresh foods; but it is unlikely that lack of any one specific vitamin or nutrient is the sole cause. The condition overlaps with chapped lips, seen in healthy people who have been exposed to cold winds or excessive sunlight.

I. Orogenital syndrome

1. There is angular stomatitis, but in addition there are changes in the epithelium of the mouth, tongue and lips, and other mucocutaneous junctions are affected.
2. The earliest sign is oedema and milky opacity of the buccal mucosa which goes onto patchy or diffuse desquamation of the lips, tongue and sometimes soft palate. These areas are red and sensitive. Secondary infection with superficial ulceration may occur.
3. Soggy, whitish patches at the outer angles of the eyes, within the ears, at the vulva or prepuce of the penis, and around the anus are often present. Along with these changes there is often corneal vascularisation and a scaly, greasy eczema at the angles of the nose, on the lips, chin and behind the ears.
4. A dry, intensely, itching, erythematous dermatitis, with a well-defined edge, may appear on the genitalia–the scrotum or mons pubis, over the perineum and down the inner sides of the thighs. There is often secondary infection.
5. In health, the hair is sleek and glossy, often with a natural wave or curl. In malnourished or undernourished people, the hair frequently becomes dull and lustreless; it is not easily brushed and tends to stand up straight (staring hair). At the same time the colour of the hair may change. In fair people, it may turn to a dirty brown, while in black-haired people, there may be loss of pigment, with a change of colour ranging from brown, rusty red to almost white. This occurs in kwashiorkor.
6. Dietary factors such as deficiency of pan-

tothenic acid or biotin can change the colour of the hair of black rats to grey, but the white or grey hair of human middle age has no nutritional significance. Nor is baldness a manifestation of nutritional failure.

7. In chronic iron deficiency anaemia the fingernails may be spoon-shaped (Koilonychia).
8. In other forms of malnutrition, the nails may be brittle or thickened or lined on the surface, either transversely or longitudinally ; but these changes may also be seen in well-nourished people.
9. Severe protein deficiency may result in transverse white bands in the nails, occurring symmetrically on both hands.

Disorder of the Eye

A. Night blindness

1. Night blindness is a frequent complaint in underdeveloped communities who have no night lights and where the diet is grossly lacking in retinol and β-carotene. Children who stray from home after dark may get lost, or fall down a well or injure themselves in other ways.
2. Many factors besides retinol deficiency may contribute to complaints of night blindness. These include fatigue, emotional disturbances associated with acute danger and also chronic anxiety states.
3. There are organic causes such as retinitis pigmentosa.
4. Night blindness arising from Vitamin A deficiency always respond to suitable vitamin therapy and it is unwise to make the diagnosis before adequate therapeutic trials have been carried out.

B. Bitot's spot

1. Greyish or glistening white plaques formed of desquamated thickened conjunctival epithelium, usually triangular in shape and firmly adherent to the underlying conjunctiva.
2. Sometimes the spots are covered with material resembling dried foam which can be scraped away but forms again. It consists of epithelial debris, fatty globules and often masses of xerosis bacilli. The spots are generally bilateral, on the temporal sides of the cornea, and in coloured races are often surrounded by dense brown pigmentation.
3. Pigmentation of the conjunctiva is frequently associated with exophthalmia. Pigment may be deposited round the cornea (pigmented ring), in the lower eyelid (pigmented gutter), and over the sclera equatorially in the area commonly occupied by Bitot's spots. Various forms of irritation appear to play a major role in its causations.

C. Corneal vascularisation

1. The essential lesion in this condition is an invasion of the normally avascular cornea by capillary blood vessels. These vessels cannot be seen with the naked eye, nor with an ordinary hand lens.
2. Small greyish-white opacities may also be seen on the surface of the cornea.
3. The patient often complains of a burning sensation in the eyes, misty vision, lachrymation and photophobia–the latter symptom may make slit-lamp examination difficult.
4. There is often injection of the conjunctiva with diluted blood vessel which are easily visible on simple inspection.
5. The presence of an injected conjunctiva should not allow the assumption that a vascular cornea is also present.
6. Corneal vascularisation may be associated with the orogenital syndrome, with keratomalacia and with ariboflavinosis.
7. Nutritional amblyopia is a major nutritional disorder of the eyes.

Disorders of the Mouth

A. Nutritional glossitis

1. Deficiencies of nicotinic acid, riboflavin,

Vitamin B_{12}, folic acid and iron may all give rise to glossitis.

2. It is a feature of pellagra, sprue and the various types of nutritional anemia.
3. The tongue seems to be particularly susceptible to metabolic disorder to all kinds.
4. If the deficiency is partial and extends over months or years, chronic atrophic glossitis is more often seen.
5. In acute glossitis the tongue is swollen, sometimes to such an extent that it is continually pressed against the lower jaw and well-marked dental impressions are visible.
6. The papillae are usually very prominent.
7. The colour of the tongue is characteristically red, but in some cases it may have a purplish blue.
8. The mucous membrane sometimes desquamates in pitches leaving areas of red raw surface.
9. Deep irregular fissuring is common and shallow ulcers may occur, especially on the sides or tips.
10. The tongue may be extremely painful, so much so that fear of pain may prevent the patient from eating.
11. In chronic atrophic glossitis, the tongue is small, with an atrophic mucous membrane and small or absent papillae so that its surface appears smooth, moist and abnormally clean. Fine fissuring may be present. It is usually not painful.

B. Parotid gland enlargement

1. The condition may be confused with mumps. Histological examination of the swollen gland shows hypertrophy of the acini. In the final state, fibrosis develops with cystic dilatation of the ducts—a parotid cirrhosis.
2. The parotid galnds are sometimes enlarged temporarily during the refeeding of people who have been severely undernourished.

DIET AND OTHER DISEASES

Inborn Errors of Metabolism

Phenylketonuria (PKU)

PKU occurs due to a defect in the metabolism of amino acid phenylalanine. The affected children show normal physical development but impaired mental development to a varying degree. The untreated children become mentally defective adults.

1. The biochemical and genetic defects

1. PKU had been named by the finding of a ketone, phenylpyruvic acid in the urine. The primary defect is in the phenylalanine hydroxylating system which converts the amino acid into tyrosine.
2. This system when cannot convert all of the phenylalanine derived from the protein in a mother's milk, the level in the blood rises and this causes the impaired development of the nervous system. Some of the excess phenylalanine is deaminated to phenylpyruvic acid which is excreted in the urine.
3. Hydroxylation is effected by phenylalanine hydroxylase (PH) and a coenzyme 5, 6, 7, 8-tetrahydrobiopterin (BH_4). The coenzyme is oxidized to BH_2 from which it is reformed by another enzyme, dihydrobiopterin reductase (DHPR). Hydroxylation may be impaired by genetic defects in production of pH of DHPR and of an enzyme responsible for the formation of BH_2.
4. The genes responsible for the control of the hydroxylation are not closely linked and may be on different chromosomes. The defects are transmitted by autosomal recessive inheritance. Unaffected heterozygous individuals act as carriers.

2. Classical phenylketonuria

1. When the baby is at the age of 8 to 10 months, the parents may become anxious

because their child is slow in learning to sit and handle things and is generally unresponsive.

2. About 25 p.c. of the affected children develop eczema.
3. The retarded development becomes obvious and there may be signs of severe birth damage, such as myoclonic epilepsy and marked hyperactivity.
4. Most affected children grow up to become physically sound but are mentally defective.

3. Variant forms

Besides classical PKU, other forms of hyperphenylalanemia are known and at least nine types have been recorded.

4. Dietary Management

1. Clinical symptoms do not arise in case the affected infant is put on a low phenylalanine diet soon after birth and kept on it for a long period.
2. A severe emotional strain is imposed on a young child on an entirely artificial diet.
3. As soon as the diagnosis is made, breast feeding should be stopped and the infant is bottle-fed with a low phenylalanine milk substitute.
4. Greater difficulty arises when the baby has to be weaned. A mother then has to prepare a low phenylalanine diet for her child from five lists of foods. Therefore, she needs continuing help from a dietician. The lists are :
   (a) Basic foods containing negligible phenylalanine which can be used freely (these include sugar, sweets, jams, solid vegetable oils and cooking oils).
   (b) Fruits and vegetables which can be taken freely since they provide negligible phenylalanine and protein in a normal helping.
   (c) A basic list of 59 mg phenylalanine exchanges of foods.
   (d) Manufacturer's foods of negligible phenylalanine content.
   (e) Exchanges of foods containing 50 mg of phenylalanine (by calculation taking one gram of protein as 50 mg phenylalanine).

Galactosemia

This hereditary defect due to reduced activity of glucose-1-phosphate uridyl transferase impairs the metabolism of galactose raising its concentration in the blood. Toxic signs appears soon after birth when an infant begins to take milk and are due to accumulation of galactose-1-phosphate within the cells. This defect is less common than phenylketonuria.

1. Clinical Features

1. The disorder is found in infants in its severe form within two or three weeks after birth with the manifestation of vomiting, difficulty in feeding, loss of weight and the onset of jaundice.
2. The spleen may be palpable and the liver greatly enlarged and very firm, and ascites may be present.
3. Sugar (galactose) and protein are found in urine.
4. If not treated properly, cataracts may develop which may lead to blindness.
5. Mental and physical retardation are likely to occur.
6. Without immediate dietetic treatment of such severe cases death rapidly occurs.
7. Lack of galactokinase also causes rarer form of this disorder in which mental development is normal and the liver is not damaged but severe cataracts leading to blindness occur early in childhood.

2. Diagnosis

The raised blood galactose and the reduced glucose-l-phosphate uridyl transferase activity in erythrocytes confirm the disorder.

3. Treatment

1. Breast feeding should be stopped immediately and the infant should be given a milk powder in which lactose has been replaced by dextrin, dextrose and maltose.
2. Milk, milk products and food preparations containing these should be excluded from the diet. It is necessary to continue these restrictions throughout life.
3. The intake of galactosides, present in small quantities in most foods, and widely used in the food industry as a filler or flavouring agent, should be reduced.
4. Lactose is used in the pharmaceutical industry in the formulation of many drug products. Such preparations can normally be taken with safety but when a patient requires large doses for a long time it may be wise to use an alternative drug.

Refsum's Disease

This is due to a defect in the enzyme systems responsible for the metabolism of phytanic acid (3, 7, 1, 15-tetramethyl-hexadecanoic acid) which accumulates in the plasma and tissues. Phytanic acid is derived from phytol, a product of the hydrolysis of chlorophyll.

1. Clinical Features

1. The main features are peripheral neuropathy, cerebellar ataxia, nerve deafness and retinitis pigmentosa.
2. Symptoms usually first appear in childhood and progress slowly, patients become severely disabled between the age of 20 to 30 years.

2. Treatment

1. Great improvement is possible by removing phytanic acid by plasma exhange and by a diet low in chlorophyll.
2. There should be restriction of many fruits and vegetables, butter and ruminant fat.

Maple-Syrup Urine Disease

A defect in the oxidative deamination of the branched-chain amino acids, leucine, isoleucine and valine, leads to accumulation of their oxyacids in the blood and these are excreted in the urine imparting an odour of maple syrup.

1. Clinical Features

1. Soon after birth the infant has difficulty in feeding; loss of reflexes, convulsions and coma follow and in severe cases, death within a month.
2. Long survivors are mentally defective unless fed with a formula diet low in leucine, isoleucine and valine.
3. A few children have been reared successfully and when they are older, require a low protein diet based on gelatin, gluten-free flour, butter, sugar and fruits.

Fructose Intolerance

The lack of the enzyme aldolase which converts fructose-l-phosphate to dihydroxyacetone phosphate, and glyceraldehyde cause this disorder. When fructose is ingested, fructose-l-phosphate accumulates in the liver. This inteferes with release of glucose from the liver and leads to severe hypoglycemia.

1. Clinical Features

1. There may be vomiting and hypoglycemic fits.
2. A series of episodes may lead to jaundice and enlargement of the liver.
3. The teeth of the patient do not show caries.

2. Treatment

Sucrose and fruit should be excluded from the diet.

Von Gierke's Disease (Glycogen Storage Disease, Type 1)

This is due to a defect in the low activity of glucose-6-phosphatase for which glycogen is not mobilised and large amounts accumulate in the liver. Most patients survive into adult life in this rare disorder.

1. Clinical Features

1. Growth is retarded and there is marked enlargement of the liver so that the abdomen is protruded.
2. Hypoglycemia and ketoacidosis may occur in the new born and attacks may continue throughout life, often brought on by an infection or temporary starvation.
3. Mental development is retarded only if episodes of hypoglycemia have been frequent and severe.
4. As the child grows up attacks of hypoglycemia become less severe.

2. Treatment

1. A diet high in protein accelerates gluconeogenesis from amino acids and so helps to maintain blood sugar.
2. In severe cases, frequent feeds every 3 to 4 hours may be required.
3. A moderate amount of carbohydrate is necessary, but this should be in the form of glucose or starch.
4. Both sucrose and lactose should be avoided, because fructose and galactose are readily converted to glycogen in the liver.

Hyperlipidemia

Hyperlipidemia may arise from an increased concentration of either cholesterol or triglycerides and frequently both are raised.

There is slow rise of cholesterol with age starting at 20 years but ceasing at about 60 years when it may be followed by a slight fall. The rise is less steep in women until the menopause when it may increase abruptly. After 60 years there is little or no difference between the sexes.

A high fasting plasma triglyceride is often associated with a high plasma cholesterol, but is not by itself a risk factor for CHD.

Dietary cholesterol has little effect on plasma cholesterol over the range of intake since increasing intake inhibits endogenous production. Very high intakes, only obtainable by eating two or more eggs daily, raise plasma concentration and this can be lowered by eliminating the diets—eggs, meat, etc. Dietary cholesterol is much less important than dietary fat in determining plasma cholesterol.

People who subsist on diets high in starchy foods, like rice or maize, do not have high plasma triglycerides unless they are obese. Sucrose in large amounts may have a greater effect than starch in raising plasma triglycerides. Plasma cholesterol usually falls with increasing carbohydrate in the diet.

Alcohol favours hepatic lipogenesis and this stimulates the synthesis of VLDL as well as leading to a fatty liver.

Familial hypercholesterolaemia

This disorder consists of an increase in low density lipoproteins (LDL), the main carrier of cholesterol in the blood, and is due to genetic failure of LDL receptors.

1. Clinical Features

1. There is a greatly increased risk of coronary heart disease. A heart attack due to myocardial infarction before the age of 30 is not uncommon and 50 p.c. of patients present evidence of CHD before they are 50 years old. Thus, it greatly reduces life expectancy.
2. Xanthomata, swellings or small tumours containing cholesterol and sometimes triglycerides, commonly occur. Common sites are tendons, especially the Achilles tendon and tendons on the back of the

hands and skin, and around the eye where they are known as Xanthelasmata.

3. Corneal arcus occurs early and when seen in someone under 40 years is probably due to the disorder.

2. Risk Factors of Hyperlipidemia

1. Smoking.
2. BMI > 25 kg/m^2
3. Apple shape obesity-waist circumference

Men	Women
≥ 102 cm	≥ 88 cm

4. Triglycerides ≥ 150 mg/dl.
5. HDL cholesterol

Men	Women
< 40 mg/dL	< 50 mg/dl

6. Blood pressure ≥ 130 ≥ 85 mm of Hg.
7. Fasting glucose ≥ 110 mg/dl.

3. Management

1. The coronary heart disease (CHD) may be prevented by lowering plasma LDL cholesterol level throughout life and beginning in early childhood. This needs life long dietary restriction and the use of drugs.
2. Since clinical symptoms rarely arise before the patient is an adult, early diagnosis depends on the examination of the blood. Whenever the diagnosis is made in a new adult patient, a biochemical examination of the blood should be made on as many of the near relatives as possible, especially the children. All affected members of the family should then be advised to begin life-long dietary restrictions to reduce plasma cholesterol and often prolonged periods on a drug.
3. The diet should be low in saturated fats and cholesterol, but polyunsaturated fats from vegetable oils and in other foods are permitted.
4. Two drugs lower plasma cholesterol effectively. Cholestyramine is an anion exchange resin that absorbs bile salts and increases their excretion in the stools. Although this leads to increased synthesis of cholesterol, plasma cholesterol falls, gastrointestinal side effect may be severe. Nicotinic acid in pharmacological doses (3-6 g/day) inhibits lipolysis and mobilization of free fatty acids from adipose tissue and also reduces plasma VLDL. Both drugs have been shown to be very safe in the short term. Nicotinic acid has vasodilator effects which cause hot flushings in some individuals but these are not serious. Reduction of plasma cholesterol for long periods increases the risk of stone formation in the bile and the incidence of gall bladder disease. A third drug, clofibrate, that lowers plasma cholesterol is no longer prescribed since its continued use has been associated with an increased number of deaths from a variety of causes.

DISEASES OF THE CARDIOVASCULAR SYSTEM

Atherosclerosis

Atherosclerosis is defined by the variable combination of focal accumulation of lipids, complex carbohydrates, blood and constituents, fibrous tissue and calcium deposits combined with its changes of the media (as per WHO). So it is a patchy, nodular type of arteriosclerosis.

The three major clinical forms arise from narrowing of the coronary arteries, the cerebral arteries and the femoral artery and its branches, but other arteries can be affected, such as the renal or mesenteric.

Coronary atherosclerosis is almost invariably associated with aortic atherosclerosis.

Types of lesion

It is customary to separate the lesions of atherosclerosis into fatty streaks, plaques and complicated lesions.

Fatty streaks are short, thin, slightly raised yellow lines running longitudinally along the internal surface of arteries and consist of an intracellular accumulation of lipids within the intima.

Plaques are the lesions of atherosclerosis. They are raised, focal, circumscribed lesions up to 1 cm in diameter, consisting of various amounts of fibrous tissue and lipid. The lipid accumulates mostly in extracellular amorphous masses; plaques in which this process is prominent are called soft or **atheromatous plaques**. In others, fibrous tissue is prominent and lipid is widely scattered or localized to the deeper portions of the lesion; these are called **hard** or **fibrous plaques**.

Four other processes may complicate the lesions : (1) The endothelium may be lost so that the surface ulcerates and the fatty contents may be exposed to the blood stream. (2) Fibrin is commonly deposited, and thrombosis occurs on the plaque surface. (3) Free blood can be found in plaque. (4) Calcification may occur.

Conventional Risk Factor for (CAD) (Coronary Artery Disease)

1. Dyslipidemia – elevated LDL-C and low HDL-C.
2. Smoking.
3. Diabetes.
4. Hypertension.
5. Obesity.
6. Sedentary life style.
7. Positive family history.
8. Advancing age.
9. Postmenopausal state in women.

New and Emerging State in women

1. Hyperhomocysteinaemia.
2. Small dense LDL phenotype.
3. Inflammation and infectious agents.
4. Hyperfibrinogenaemia.
5. Elevated LP (a) [Lipoprotein (a)].

Development with ageing

A few small lesions are seen in most adolescents and they increase in number and size throughout life. It does not usually lead to clinical disease until middle age.

Thrombogenesis

An arterial thrombus almost always forms at the site of an atheromatous plaque, these are present in most of us from an early age. Small thrombi form frequently in the circulation. These microthrombi are then attached to the arterial wall where they are rapidly lysed. A failure of the fibrolytic mechanism could in this way allow a thrombus to grow. Hence one of the primary faults responsible for atherosclerosis might lie in an increased tendency to thrombosis or in an efficient fibrinolytic mechanism.

Coronary Heart Disease (CHD)

Coronary heart disease (CHD) or ischaemic heart disease (IHD) are synonymous terms for a group of syndromes arising from failure of the coronary arteries to supply sufficient blood to the myocardium. These syndromes are in most cases associated with atherosclerosis of the coronary arteries. They include myocardial infarction, angina pectoris and sudden death without infarction.

Myocardial infarction

1. This is necrosis or destruction of part of the heart muscle due to failure of the blood supply (ischaemia).
2. It may lead to sudden death or heal leaving a scar.
3. Patients with healed lesions may be severely disabled or may be able to return to their normal life with little or no restriction of their physical activities, but they carry an increased risk of a second infarct.
4. The infarction is usually due to a thrombus forming in an atherosclerotic coronary artery and blocking the lumen.

5. Sometimes there is no thrombus and the infarct arises because the lumen of a coronary artery has been so narrowed by atherosclerosis that the blood flow is insufficient to supply the oxygen needed to maintain the cardiac muscle.
6. Occasional cases of myocardial infarction are seen in which neither thrombosis nor significant narrowing of the lumen can be recognised.

Angina pectoris (Pain in the chest)

1. In this condition exercise or excitement causes severe chest pain.
2. Patients may live for many years and remain free of further disability, so long as they keep within the limits of their exercise tolerance.
3. The patients carry an increased risk of sudden death or myocardial infarction, especially if they undertake any unusual exertion.
4. Emotional stress may also bring on angina.

Sudden death

1. A proportion of sudden death results from angina pectoris or myocardial infarction.
2. The death is presumed to be due to CHD.
3. The most death is unexpected but autopsy shows evidence of old myocardial infarction or extensive atheroma of the coronary arteries.

1. Clinical Features

1. The prominent symptom is a severe, pressing or constricting pain, poorly localized deep in the centre of the chest and radiating down the left or both arms.
2. In angina pectoris this pain comes on with exertion or excitement; it forces the patient to stop and when he does so, that pain passes off in a few minutes.
3. Patients with angina can suffer several attacks of pain in a day and they can go on like this for long periods of time.
4. In myocardial infarction the pain often comes on at rest; it is very severe, lasts for hours and is only relieved by strong opiates like morphine. It is accompained by general symptoms such as weakness, collapse, cardiac arrhythmias and circulatory shock.
5. Some patients have a mild myocardial infarction without noticing pain (silent coronary).
6. There are also several other causes of pain in and around the chest which can mimic CHD, such as Pericarditis, Pulmonary embolism, Oesophageal and respiratory disease.
7. The E.C.G. often changes temporarily during an attack of angina, it is usually nomal between attacks. But soon after the onset of myocardial infarction, the ECG may undergo a permanent change and show the infarct pattern.
8. Patients with angina or who have had a myocardial infarct are liable to develop cardiac failure or arrhythmias.

2. Risk factors

A. Sex and age

1. Men are more prone to CHD than women and incidence rises with age in both sexes.
2. Greatly increased incidence are found in women after the menopause, and by the age of 70 there is no difference between the sexes.
3. The secretion of the ovarian hormones during the reproductive life of women aids in the relative immunity against the rise in cholesterol level.
4. The concentration of plasma total cholesterol is lower in women aged 20 to 45 years than men of the same age group but HDL cholesterol is higher.

5. If both ovaries of women are removed before the age of 35, the plasma total cholesterol will increase with the increase in CHD.

B. Family history

1. Some families are more susceptible to CHD than others.
2. Family clustering of any disease may be due to either inherited susceptibility or family sharing of environmental experience eg, an atherogenic diet. More than 50 per cent of plasma cholesterol is genetically determined.
3. In families susceptible to CHD, there is often a high incidence of other diseases which have multiple aetiology, eg, hypertension, diabetes mellitus, gout and hyperlipidemia.

C. Behaviour patterns and personality traits

People with behaviour pattern A above an increased incidence and prevalence of CHD. Such people show an excessive sense of time urgency, a preoccupation with vocational deadlines and enhanced aggressiveness and competitive drive. This behaviour pattern is probably constitutional.

D. Hyperlipidemia

A patient with a very high plasma cholesterol is at high risk of developing the disease at an early age but for the majority with values not greatly raised the increased risk is small.

E. Hypertension

The incidence of CHD in men aged 45 to 65 years with blood pressures exceeding 160/95 was more than five times that in normotensive men (blood presure 140/90 or less). Elevations in both diastolic and systolic pressures correlate positively with CHD, the diastolic pressure perhaps being more important in younger people.

F. Obesity

1. Obesity is associated with an increased risk of death from heart disease.
2. A moderately increased plasma cholesterol or mild hypertension or Grade I obesity (plumpness), each by itself, carries only a small risk of CHD, a combination of any two of them increases it much more, and in a patient with all three the prognosis is very uncertain unless they are reduced by appropriate treatment.

G. Diabetes Mellitus

This is an important risk factor both in its clinically recognised and latent forms. This association is not due solely to blood lipid disturbances. Diabetics are particularly prone to hypertension and proliferative lesions of the small blood vessels.

H. Hyperuricaemia and gout

There is a positive association between these abnormalities and CHD, although not all hyperuricaemia represents pre-gout.

I. Electrocardiographic abnormalities

1. Persons with an abnormal ECG or with ECG changes or left ventricular hypertrophy have an increased risk of developing CHD.
2. If the changes are present when the patient is at rest, they may represent early CHD.
3. Abnormalities which appear on exercise usually indicate relatively advanced ischaemia.

J. Cigarette smoking

1. Peripheral vascular disease and CHD, particularly myocardial infarction, occur much more frequently in heavy cigarette smokers than in those who do not smoke.
2. There is a clear relationship between CHD and the amount of smoking.
3. The mechanism by which nicotine or some other constituent of tobacco causes this adverse effect is not clear. However, it may be due to the vasoconstrictor action of nicotine, to inhalation of carbon monoxide, or to some undesirable effect on the coagulability of the blood.

4. Heavy smoking may be a manifestion of a susceptible personality and a reaction to stress and strain.
5. Patients with CHD should give up smoking.

K. Diet

1. A high intake of saturated fatty acids may raise plasma cholesterol and so promote the formation of atheroma.
2. A low intake of polyunsaturated fatty acids may modify platelet function and so promote the formation of thrombi.
3. An intake of energy above requirements leads to obesity and high intake of salt may lead to hypertension in susceptible individuals.
4. It has been found that some individuals suffer a heart attack early in life. In that case, diet is less important than genetic hypercholesterolaemia or cigarette smoking in the aetiology of coronary artery disease.
5. Excessive intake of alcohol should be avoided.
6. A higher proportion of energy should come from bread, potatoes, and other vegetables and lower proportion from separated fats and fatty meats.

L. Lack of exercise

1. Physical activity protects aganist CHD. Bus conductors have a lower incidence than bus drivers, and postmen who deliver letters have a lower incidence than telephonists and post office clerks.
2. Active recreations are also important. In middle-aged Harvard alumni, the incidence of heart attacks was lower in those who have been physically active.

M. Emotional stress and tension

1. The modern prosperous communities suffer from more stress and strain than their less wealthy predecessors. Increased risk of myocardial infarction could be due to acute or chronic adrenergic overdrive caused by stress. Possible mediating factors include plasma noradrenaline and FFA.
2. The stress of racing driving raise plasma triglycerides and FFA but not plasma cholesterol. The same reaction probably occurs in some urban drivers.

N. Drinking water

The harder the drinking water the lower the death rate from cardiovascular disease. Calcium and magnesium in hard waters may have a protective action and there are several trace elements in hard water that may be beneficial.

Soft waters, being more acidic, are more likely to dissolve potentially toxic trace elements like lead and cadmium from pipes or rocks.

Since there is no concrete evidence to be based on it will be premature to advise modification of water supplies in the hope of preventing cardiovascular disease.

O. Coffee and alcohol

1. Myocardial infarction is associated with high consumption of coffee though has not yet been confirmed.
2. Moderate alcohol consumption is not a risk factor, and in patients with disorders directly attributable to alcohol the incidence of CHD is no higher than normal.

3. Prevention and management

1. Smoking must be stopped.
2. The diet should be moderate so as to prevent being overweight.
3. Physical exercise should be taken as much as possible without causing undue breathlessness or fatigue.

CEREBROVASCULAR DISEASE

Cerebrovascular disease is second only to heart disease as a cause of premature death. Atherosclerosis is common to both conditions, the overwhelming risk factor is a high blood pressure. Cerebrovascular disease has little association with hypercholesterolaemia, but much more with obesity than does CHD.

Hypertension

Hypertension is an important cause of several major disease and it is amenable to treatment. Hence, measurement of blood pressure is a part of all routine clinical examinations of adults. Hypertension is a major risk factor for the development of cardiovascular disease. It is defined as systolic blood pressure of 140 mm of Hg or greater and/or diastolic blood pressure of 90 mm of Hg or greater or any level of blood pressure in patients taking antihypertensive medications.

The pathological changes accompained by hypertension are a thickening of the arterioles with hyaline material and, later, hypertrophy of the myocardium of the left ventricle. If untreated, moderate hypertension eventually leads to cardiac failure, with dilatation of the left ventricle and congestion of the pulmonary or systemic veins. It also causes ischaemic changes in the kidney, nephrosclerosis.

The blood pressure if very high is said to be malignant hypertension. The heart, kidneys, retinal and other arteries are quickly affected and it shows immediate risk of a dangerous vascular accident. Therefore, vigorous treatment is most essential.

80 per cent of cases suffer from essential hypertension. Treatment has to be symptomatic. In the remaining cases the hypertension is usually secondary to renal disease, eg, glomerulonephritis or pyelonephritis, and less commonly, to an endocrine disorder, eg, Cushing's syndrome, acromegaly, pregnancy also show hypertension.

1. Classification of Blood Pressure

Category	Systolic (mm of Hg)		Diastolic (mm of Hg)
Optimal	< 120	and	< 80
Normal	< 130	and	< 85
High normal	130 - 139	or	85 - 89
Hypertension			
Stage - 1	40 - 159	or	90 - 99
Stage-2	160 - 179	or	100 - 109
Stage-3	≥ 180	or	≥ 110
Isolated Systolic Hypertension			
Stage-1	140 - 159		< 90
Stage-2	≥ 160		< 90

A. Heredity

1. Hypertension frequently affects several members of a family and there is a high correlation between the blood pressures of identical twins.
2. Individual susceptible to hypertension is related to decreased activity of (Na^+, K^+) ATPase, the enzyme that through the sodium pump maintains intracelluar concentrations of Na^+ and K^+. The activity of the enzyme is genetically determined.

B. Obesity

1. Hypertension commonly accompanies obesity though it may occur in thin people.
2. Obese patients with hypertension, if reduce their weight to normal value, their blood pressue frequently returns to normal without any other treatment. This indicates that in susceptible individuals obesity is a direct cause of hypertension.

C. High salt intake

High salt intake in some countries causes hypertension; but in some other countries high salt intake shows no effect. It is, therefore, assumed that there is a threshold for salt intake, as yet undefined, above which hypertension commonly arises in susceptible individuals and that susceptibility is genetically determined.

D. Other possible risk factors

1. Stress, either mental or physical, raises the blood pressure immediately but this is only temporarily.
2. Environmental stresses may contribute to hypertension and of the theory that individuals with a poor capacity for relaxation are more prone to hypertension indicate that stress is not a major risk factor.
3. Inactivity may be a risk factor but is only a minor one.
4. Alcohol intake (> 3 units/day for men; > 2 units/day for women).
5. Physical activity.
6. Smoking.
7. Drugs (eg, NSAID, Oral contraceptives).
8. Renal diseases.
9. Pheochromocytoma.
10. Conn's syndrome (tetany, muscle weakness, polyuria, hypokalemia).

Factors involved in pathogenesis of hypertension

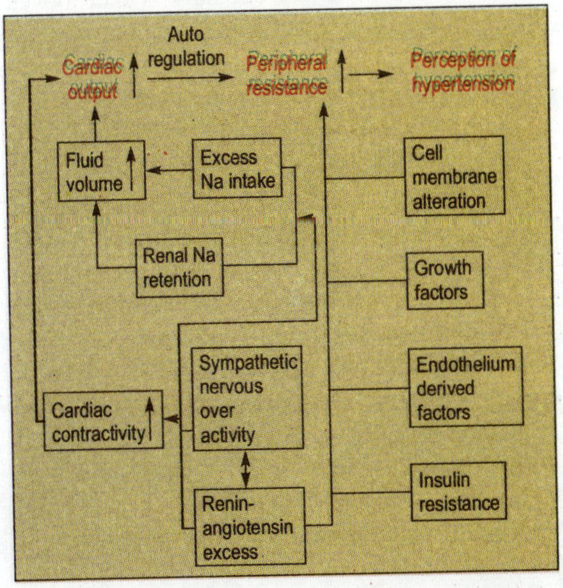

Laboratory investigations

Urine for protein, glucose and microscopic (RBCs/other segments) haemoglobin, fasting blood glucose, serum creatinine, potassium and total cholesterol.

ECG

Additional investigations

Lipid profile, Uric acid

Chest X-ray

Echocardiogram.

2. Treatment

Non-pharmacological measures

Life style measures

1. Maintain normal weight for adults (body mass index 20-25 kg/m^2).
2. Reduce salt intake to < 100 mmol/day (< 6 gm NaCl or < 2.4 gm Na^+/d).
3. Limit alcohol consumption to ≤ 3 units for men and ≤ 2 units/day for women.
4. Aerobic exercise (eg, walking, cycling, swimming) ≥ 3 min/day (at least three days of the week).
5. Consume at least five portions/day of fresh fruits and vegetables. Fish consumption rather than meat should be recommended. High intake of fibres, grains, vegetables and fruits improve BP control.
6. Reduce the intake of total and saturated fat.

Pharmacological measures

1. Achieve gradual reduction of blood pressure.
2. Use low doses of antihypertensive drugs to initial therapy.
3. Choice of an antihypertensive agent is influenced by age, concominant risk factors, pressure of target organ damage, other co-existance diseases and socio-economic condition.
4. Use of long acting drugs providing 24 hr efficacy ensures smooth and sustained control of blood pressure. Once daily application also improves patient compliance.

Guidelines for selecting drug treatment of hypertension		
Class of drugs	**Definite indications**	**Definite contraindication**
Diuretics	Heart failure Elderly patients Systolic hypertension	Gout
Beta-blockers	Angina After myocardial infarction Tachy arrhythmias	Asthma and chronic Obstructive pulmonary disease Heart block
ACE inhibitors	Heart failure Left venticular dysfunction Post myocardial infarct Diabetes	Pregnancy Hyperkalemia Bilateral renal artery Stenosis
Calcium channel blocker	Angina Elderly systolic hypertension	Heart block
Alpha blocker	Prostatic hypertrophy	—
Angiotensin II antagonists	ACE inhibitor Cough	Pregnancy and lactation Bilateral renal artery stenosis. Hyperkalemia

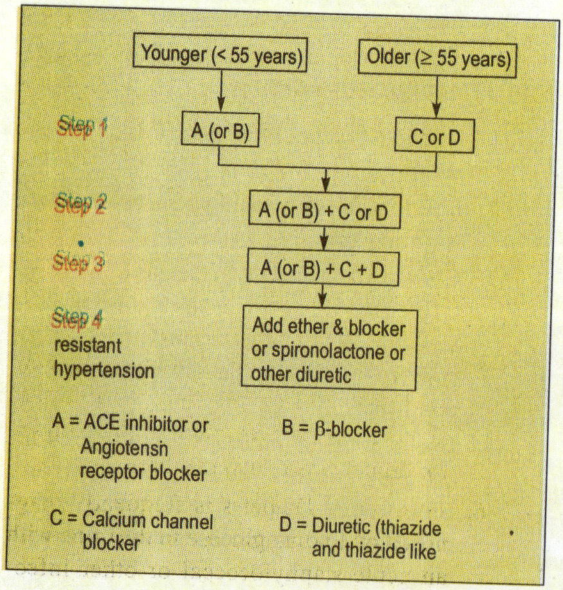

3. Prevention

1. The dietary restriction is the reduced intake of salt.
2. It is advised widely not to use table salt and to use a minimum salt in home cooking. It is more important for the public to know that prepared meats contain about 20 times as much salt as is present in the original fresh meat. Chips and crisp potatoes are very rich sources of salts. Salted and canned fish are fine for an occasion but if eaten frequently may increase salt intake greatly. Many snack foods are rich in salt.
3. As a preservative, salt is added to manufactured foods. But it should be such as to be ineffective. Manufacturers are not unresponsive to medical opinion and have greatly reduced the previously high salt content of infant foods when it was shown that this was a cause of hyperosmolar dehydration, a serious disorder in infants. A significant amount of salt is added to bread, butter and cheese. It is to be looked upon as medicinal products that low salt bread, butter and cheese are obtained easily. If there is a popular demand for low salt foods of all sorts, manufacturers will produce and advertise them.
4. Legislation to restrict the addition of salt to foods will be difficult to enforce. A more practical approach will be to provide more money and support for appropriate health education.

Heart Failure

In severe CHD the heart fails because its muscle is weakened; in hypertension it fails because the muscle is working against an increased load. In mild degrees of heart failure the only manifesation is an inability to increase the cardiac output in response to strenuous exertion. In moderate grades there is venous congestion. Acute and severe cardiac fallure causes pulmonary oedema and cardiogenic shock.

1. Treatment

1. Complete rest.
2. The administration of diuretics.
3. A diet low in sodium, and also in energy if the patient is obese.
4. In special circumstances a digitalis preparation.

   Heart failure is usually precipitated by a respiratory infecion, arrhythmia or some other complications. If these are diagnosed and effectively treated, the patient may well get over the episode and be able to lead a normal life.

2. Dietary management

1. Since there is oedema developed, sodium intake should be reduced. Because oedema is always associated with sodium retention.
2. Sufficient fluid must be allowed to mitigate thirst and to make the patients comfortable.
3. Owing to the congestion of the digestive organs, each feed must be small in quantity and easily digestible. In the initial treatment, fluid or semifluid food is advisable for patients who are seriously ill.
4. In patients who are seriously ill after a severe myocardial infarction, it is important to prevent ketosis arising and for this purpose they should be given ample glucose or other carbohydrate.

Diabetes Mellitus

Diabetes mellitus is a syndrome due to different diseases characterized by a raised glucose concentration in the blood, due to diminished effectiveness of insulin. The disorder is chronic and also affects the metabolism of fat and protein. It has an increased risk of atherosclerotic diseases and of certain obstetrical difficulties.

It is the commonest endocrine disorder. There are two major types –Type 1 insulin dependent diabetes which was formerly known as juvenile onset diabetes occurring between 10 and 12 years of age. Type 2 non-insulin diabetes occur in middle age or later.

Genetic and dietary factors, infections and possibly stress may increase the risk of developing diabetes.

Primary diabetes

1. **Genetic factors :** Many separate genetic mechanisms increase the risk of diabetes and its various manifestations, and these differ in type 1 and type 2 diabetes.
2. **Obesity :** Although most type 2 diabetics are obsese, only a minority of obese patients develop diabetes. In simple obesity there is insulin resistance, particularly in muscle, and hyperinsulinemia. There is impaired insulin uptake by receptors in target tissues. In general, the more carbohydrate tolerance is impaired in obese diabetics, the more deficient the insulin secretory response to various stimuli. Obese people in general are less physically active than those whose weight is normal. It is possible that physical exercise may reduce the risk of diabetes in susceptible individuals.
3. **Dietary restrictions :** Restrictions on the food supply of community, affect diabetes. Rationing is beneficial to individuals susceptible to diabetes.
4. **Sugar intake :** A high intake of sugar is definitely associated with a high prevalence of obesity. Sucrose has a specific diabetogenic effect, though the very high intake may contribute to the high prevalence of diabetes.
5. **Dietary fibre :** The high fibre content of the diet causes the reduced prevalence of diabetes. Most diets now recommended for diabetics are high in fibre..
6. **Infections :** Diabetes is frequently diagnosed by finding glucose in the urine with an acute staphylococcal or other infec-

tions. Infections cause a non-specific outpouring of catabolic hormones which antagonise insulin action and this may trigger the onset of the disorder. More evidences show that type 1 diabetes especially in younger patients is caused by virus infection. The virus may trigger an autoimmune reaction in the pancreatic islets and this impairs insulin secretion and ultimately destroys the beta cells.

7. **Stress :** Stress causes a sudden increase in secretion of catabolic hormones which may precipitate the disorder. It probably does not cause diabetes in people who would never have developed it. The impaired secretion by pancreatic islet cells may cause the disorder. Many environmental factors may lead to such impairment. Genetic factors appear as the main determinant of susceptibility to such environmental factors, those leading to overweight and obesity being the most important in type 2 diabetes and viral infections in type 1.

Secondary diabetes

A minor cases of diabetes occurs as a result of pancreatitis, haemochromatosis, carcinoma of the pancreas and pancreatectomy.

Diabetes may also accompany endocrine disorders which increase concentrations of catabolic hormones or modify the regulation of insulin receptors.

Clinical Features

Type 1 diabetes : This usually appears in early age (between 10 to 12 years of age) in patients of normal or less than normal weight. Symptoms are usually severe and develop rapidly. Severe Ketoacidosis occurs and is often fatal without insulin treatment. Since insulin is required for their survival an alternative name for this patient is **insulin dependent**.

Type 2 diabetes : This usually appears in middle age or later in patients who are often obese and their hyperglycemia is controlled by dietary means alone or by an oral hypoglycemic drug. The patients are less prone to develop Ketosis. Therefore, type 2 is less severe disease than type 1.

Symptoms

(a) Some patients complain of some or all of the classical symptoms which are thirst, polyuria, nocturia, tiredness, loss of weight, reduced visual activity, white marks on clothing, polydipsia.

(b) Many patients are first found to have glycosuria in the course of some routine examination eg, for insurance, for employment purposes. They may have few or no symptoms.

(c) Ketoacidosis may occur in diabetics in infection. Epigastric pain and vomiting may be the main complaints. These are usually type 1 diabetes.

(d) Patients may complain one of the symptoms like failing vision ; parasthesia in the limbs or pain in the legs; impotence; infection of the skin, lungs or urinary tract.

Physical signs : There is dehydration, loose dry skin, dry furred tongue with cracked lips. The pulse is rapid and the blood pressure is low. Breathing may be deep and rapid. The sweet smell of acteone may be noticeable in the breath. There may be coma.

Early signs of diabetic neuropathy are depression of the ankle jerks and impaired vibration sense in the legs. The presence of neuropathy may be indicated by proteinuria.

Diagnosis

Urine testing : (a) Glycosuria (b) Ketonuria.

Random blood sugar : The oral glucose tolerance test.

Management

Diabetic patients no longer die in Ketoacidosis in any number as they once did. The increased death rate of treated diabetic patients is due to coronary heart disease. Many of those whose duration of life has been extended are chronic invalids. They may live for many years with cerebral, coronary, or peripheral vascular disease, or with renal disease or serious visual impairment.

Treatment

1. Diet alone.
2. Diet and oral hypoglycemic drugs.
3. Diet and insulin.

About 40 per cent of new case of diabetes can be controlled adequately by diet alone, about 30 per cent require insulin and another 30 per cent need an oral hypoglycemic drug. Insulin is needed for juvenile diabetes; older patients do not require insulin except when control of their diabetes by an illness, infection or operation.

1. Diet

(a) In all diabetics the amount and time of food intake, especially the carbohydrate, should be controlled to prevent the fluctuations of blood glucose beyond the normal range.

(b) Intake of refined sugars should be low because their consumptions is followed by absorption and a high peak of blood glucose.

(c) Patients should avoid fasting or feasting; their intake from day to day should be maintained with adjustments for exercise and appetite; they should not miss a meal or over-indulge.

(d) Type 1 patients require insulin and their food, particularly carbohydrate, should be adjusted to match the time of action of their insulin. This depends on the type of insulin being used and whether the patient is having a single injection or more than one each day. The balance between insulin and meals has to be adjusted from time to time. They usually want to take moderate and sometimes strenuous exercise. They, therefore, require a generous amount of dietary energy.

(e) Type 2 patients are usually obese. Being middle aged or elderly they may not take much exercise. For these reasons the daily energy intake should be restricted to about 1,000 k cal. The flying career of airline pilots depends on avoiding insulin or drugs.

(f) The nature of carbohydrate is important. Sucrose should be eliminated or greatly reduced. Starchy foods rich in dietary fibre are beneficial to diabetics. There are two essential points about a diabetic diet. First, energy intake should be adjusted to maintain ideal body weight. Secondly, all patients taking insulin should follow a regular pattern of meals, matched to the injected insulin.

(I) Carbohydrate

(a) A minimum of 100 grams is needed to prevent ketonuria.

(b) Foods rich in sucrose and other sugars should be kept to a minimum.

(c) 60 grams of fructose should be taken a day; obese diabetics should not use fructose and sorbitol as they have the same energy value as other sugars. Although fructose may not raise blood glucose as much as sucrose or glucose, it may raise plasma triglycerides more.

(II) Protein

Amino acids stimulate insulin secretion in both normal subjects and in those with type 2 diabetes. A smaller rise in blood glucose also occurs when carbohydrate is consumed along with protein. A minimum amount of about 50 grams of protein should be specified in all diabetic diets unless the patient is obese.

(III) Fat

As diabetic patients have an increased risk of death from coronary heart disease and as this may be related to the amount of saturated fat in the diet, the total amount of fat should be restricted even in those who are not obese.

2. Types of diet

There are two types of diet :

(a) **Measured diet :** The amount of food to be eaten at each time of the day is specified.

(b) **Unmeasured diet :** The patient is supplied with a list of foods grouped in three categories – foods with a high concentrated carbohydrate content which are to be avoided altogether, foods with a relatively stable unconcentrated carbohydrate content which are to be eaten in moderation only, and non-carbohydrate foods which may be eaten as desired.

Alcohol : Patients may take alcohol if they need to have energy value and carbohydrate content. Beer may contain 10 to 30 grams of carbohydrate per half litre and this provides 150 to 400 kcal depending on the strength of the beer.

Sweetening agents : Saccharin and aspartate may be used but have no energy value.

Drugs : A good number of compounds reduce hyperglycemia in patients who would require insulin. The sulphonylurea compounds, tolbutamide, chlorpropamide, glibenclamide, glipizide, gliquidone, metformin have a place in the management of 30 per cent of diabetic patients. It is dangerous to attempt to control juvenile-onset diabetes with these compounds.

Insulin : Most diabetics now manage with two injections daily, one before breakfast and one before the evening meal, each containing soluble and depot insulin.

(a) **Soluble insulin :** This is a clear solution whereas depot insulins are cloudy. It begins to lower the blood glucose in 30 minutes : the effect is maximal in 4 to 6 hours and ends after 6 to 10 hours. Soluble insulin is essential in the following circumstances :

1. For new cases with severe dehydration or Ketoacidosis.
2. For emergencies associated with ketosis, such as acute infection, gastroenteritis or some surgical operations.
3. For the treatment of nearly all young patients.

(b) **Depot insulins :** Depot insulins do not lower blood sugar befor 5 to 6 hours; the effect is maximal at 8 to 14 hours and ends only after 20 to 30 hours.

Choice of therapeutics

(a) All young patients who develop diabetes before the age of 40 years require treatment with insulin. The majority are best controlled by a combination of short-acting and intermediate-acting (depot) insulin injected twice daily, before breakfast and before the evening meal.

(b) Most patients developing the disease over the age of 40 years can be controlled by diet alone. Obese patients should be treated by dietary restriction and weight reduction but others may do well on dietary therapy alone. Insulin and the sulphonylureas increase the appetite and thus may increase weight and intensify disability.

(c) Those over the age of 40 who are not controlled by dietary measures alone usually respond well to sulphonylureas. If adequate control is not achieved by one drug, a combination of sulphonylurea and biguanide may be tried. If this fails, insulin is needed.

(d) Elderly patients who require insulin often do well with a small dose (20 units) of a depot insulin alone. But those who require more than 40 units a day should be given soluble insulin in addition.

Clinical features of Ketoacidosis

The most common cause of Ketoacidosis is neglect of treatment due to carelessness, misunderstanding or illness.

There is intense thirst and polyuria. Constipation, muscle cramps and altered visions are common. Sometimes, there is abdominal pain with or without vomiting. Weakness and drowsiness are commonly present.

The signs include dry tongue, soft eyeballs due to dehydration, hyperventilation indicated by rapid, deep, sighing respirations and rapid, weak pulse, with low blood pressure and acetone may be smelt in the breath. Sometimes there is abdominal rigidity and tenderness. Ultimately, coma appears.

Laboratory tests show heavy glycosuria and ketonuria, blood glucose usually between (360 and 720 mg/100 ml), and low plasma bicarbonate and blood pH.

Treatment

This condition should be treated with the utmost urgency in hospital. Intravenous therapy is required although the patient is able to swallow. Extracellular fluid is repleted first with sodium chloride infusions. It is better to give low dose insulin starting with 6 to 9 units per hour and halving the dose when the blood glucose has returned to normal. In the majority of cases potassium therapy should be started from the outset. Intracellular fluid is replaced once the blood glucose has fallen below 250 mg/100 ml by infusing glucose solution. Intensive medical care is needed and the blood glucose, pH, electrolytes and ketones have to be monitored, hourly at first.

Vascular disorders

Atherosclerosis occurs commonly and extensively in diabetics. Diabetics are more prone to myocardial infarction and gangrene of the toes and feet at an earlier age than other people.

The peripheral pulses in the legs are often diminished, ischaemic changes in the feet are frequently apparent. Defective circulation in the legs results in the dangerous complication of gangrene. Diabetic gangrene usually starts in one foot. Toxic absorption from necrotic tissue and secondary infection may kill the patient unless the limb is amputated. Amputation of a toe, a foot or even a whole leg is sometimes necessary to save life.

Cataract

Cataract is more prevalent in old people having diabetes. Rarely a specific type of opacity of the lens occurs in diabetic children whose disease has not been adequately controlled.

Infections

Poor control of diabetes has lowered resistance to infection. The following forms are especially important.

(a) **Carbuncle :** The development of a carbuncle may unmask diabetes and may even precipitate ketosis and coma. Cleanliness is very important in the prevention of skin infection in diabetes. Once infection has occurred a suitable antibiotic is needed.

(b) **Urinary tract infections :** The presence of glucose in the urine provides a favourable medium for the growth of bacteria. Intractable infections of the urinary tract frequently occur and for this reason catheterisation should be avoided. Once infection has been started treatment consists of controlling the glycosuria and the administration of suitable antibiotics.

(c) **Pulmonary tuberculosis :** In countries where this is prevalent all new diabetic patients should have a chest radiograph.

(d) **Vulvitis :** Candida albicans is nearly always present in diabetic women. In the majority, the treatment is abolition of glycosuria which brings rapid relief.

Problems in Management

1. Children

(a) Diabetes is not common in childhood, but when it occurs it is relatively severe and always requires treatment with insulin. The problem of matching the dose of insulin to the food intake raises practical difficulties.

(b) As the children should be growing, their energy requirements are large and difficulties arise in meeting them. It is important to make sure that the child does not become too fat because too much insulin can lead to excessive appetite and hence to obesity. It is better to encourage them to take responsibility of their own care. They should be trained to swim under supervised pools.

(c) Children are not expected to lead a steady life and their activities fluctuate unexpectedly. Excessive activity may result in hypoglycemia, and lethargy in hyperglycemia. Hyperglycemia may be caused by infectious disease. A combination of one of the depot insulins and soluble insulin before breakfast and a second dose of soluble insulin before supper.

2. Pregnancy

(a) Pregnancy in a diabetic woman carries certain risks. In the later stages, she may develop an excessive accumulation of amniotic fluid; in addition the fetus is soemtimes unusually large leading to difficulty in labour. There is also an increased risk of her baby having a neural tube defect or other error in development. A planned pregnancy reduces the risk.

(b) A pregnant diabetic patient requires close supervision by a team consisting of physician, obstetrician, anesthetist, nurse and dietician. After the diagnosis of pregnancy, the patient should be seen at first fortnightly and later at weekly intervals. In the later stages of pregnancy, lactosuria occurs and may lead to confusion. Therefore, blood glucose estimation should be done by finger-prick.

(c) Nowadays, many pregnancies are allowed to go on to full term with improved glycemic control and caesarian section is less used.

3. Diabetes and surgery

Any surgical operation causes a metabolic stress which the diabetic is less able to meet. The position is worse if there is tissue wasting with much breakdown of fat and protein. It is to be kept in mind that there is the need to provide an adequate supply of energy for the tissues and the need to be on the alert for acidosis. Diabetes is to be first diagnosed before operation.

Prognosis

The prognosis of diabetes has improved steadily since the introduction of insulin. It is difficult to estimate the prognosis of an individual patient because so many variable factors have to be considered. The incidence of the complications of diabetes is mainly related to the duration of the disease but probably also to the precision with which it has been controlled.

Prevention

Diabetes is a disease of the prosperous and in wealthy countries. It is one of the major health problems. Sufficient exercise and avoidence of excess diet have repeatedly been stated. Diabetes, like obesity and atherosclerosis, is likely to occur in persons who eat too much and exercise too little.

Dysglycemia

Dysglycemia is a term used to describe blood glucose levels which are higher than normal but not involved in the diagnosis of diabetes. In addition to the prediabetic states namely IFG and IGT, it also includes blood glucose levels just above the normal levels. It is commonly associated with the metabolic syndrome characterized by the clustering of CHD risk factors like obesity, hypertension and dyslipidemia in a single individual. These risk factors increase the risk of development of type 2 diabetes and also have a prominent effect on the risk for development of CHD. Recognition and treatment of the metabolic syndrome will help in the prevention of type 2 diabetes and CHD. Lifestyle modification plays a very important role in preventing the development of type 2 diabetes and CHD. The measurement of fasting glucose is the principal means of diagnosis of diabetes.

Type 2 diabetes is a strong risk factor for CHD in both men and women. The risk for CHD is high even at the time of diagnosis of type 2 diabetes since the prediabetic states are also associated with an increased risk of CHD. Dysglycemia includes, in addition to the prediabetic states (IFG and IGT), blood glucose values over this threshold.

Table 1: Categories of glucose tolerance

Category	Fasting plasma Glucose (mg/dl)	2 hr post glucose plasma glucose (mg/dl)
Normal	< 100	< 140
IFG	100 - 125	< 140
IGT	< 100	140 - 199
Diabetes	≥ 126	≥ 200

The term prediabetes is a practical term for impaired fasting glucose and impaired glucose tolerance which places individuals at risk of developing diabetes and its complications. Both IGT and IFG appear well before type 2 diabetes is diagnosed. IFG and IGT are associated with the metabolic syndrome which includes obesity, dyslipidemia of the high triglyceride and low HDL type and hypertension. IGT is more strongly associated with CHD outcomes. Individuals identified as having both IFG and IGT are at higher risk for diabetes as well as CHD. Dysglycemia is, therefore, a qualitative term used to describe blood glucose that is abnormal without defining a threshold.

The metabolic syndrome is characterized by insulin resistance and is also known as the insulin resistance syndrome. The pathogenesis of this syndrome has multiple origins. Obesity and sedentary lifestyle coupled with diet and largely unknown genetic factors clearly interact to produce the syndrome. These risk factors increase the risk of developing diabetes and have vast effect on the development of cardiovascular disease. At least three components such as upper body obesity, increased triglycerides, and decreased HDL cholesterol are required to confirm the diagnosis of metabolic syndrome.

The metabolic syndrome is a powerful determinant of diabetes and cardiovascular disease. Non-diabetic glycemic persons with metabolic syndrome have an increased mortality from all causes including cardiovascular disease. The increased risk of progression from IGT to type 2 diabetes and the strong association of both prediabetic dysglycemia and type 2 diabetes with increased cardiovascular risk emphasize the need for diabetes prevention. There is increasing evidence that intensive lifestyle changes or pharmacological interventions can reduce progression from IGT to type 2 diabetes. Weight loss and increased physical activity may decrease insulin resistance and related metabolic disorders. The role of drugs like acarbose, metformin and the glitazones has also been set up.

The oral antidiabetic agents are acarbose and metformin. Acarbose has the most beneficial effect on age and sex. Acarbose treatment shows the significant reduction in cardiovascular events and myocardial infarction. Metformin can reduce the cardiovascular morbidity and mortality in obese patients with type 2 diabetes. The role of metformin is not as good as that of lifestyle medications.

Life-style interventions and weight loss are important components for preventing or delaying the progression to type 2 diabetes in prediabetic individuals. However, they are not effective in all subjects. The conditions like arthritis and other cultural issues may present barriers to successful intervention. On the other hand, pharmacological interventions may be implemented more easily in such individuals.

Prediabetic dysglycemia is an important entity associated with the metabolic syndrome carrying a significant risk of type 2 diabetes as well as an unacceptable burden of cardiovascular disease. Recognition of individuals with the metabolic syndrome based on clinical criteria is important in identifying individuals at high risk of type 2 diabetes and cardiovascular disease. Life-style changes and appropriate medication in these individuals are helpful in reducing the epidemic of type 2 diabetes and cardiovascular disease.

Gout and Hyperuricaemia

Gout is a characteristic arthritis which affects single joints, often the big toe giving pains that last only a few days but are liable to recur. Middle aged men are chiefly affected. It is caused by the deposition of urate crystals in the joint. The increased concentration of urate in the plasma causes hyperuricaemia.

Hyperuricaemia may be asymptomatic. Such individuals carry a greatly increased chance of the clinical complications, gouty arthritis or uric acid stones in the urinary tract. Those who have recurrent gout and hyperuricemia over long time are liable to develop **tophi**, accumulations of urate in tendons or cartilage.

Clinical Features

(a) The patient has suddenly stabbing pain in one joint. The big toe is easily stubbed or injured by an ill-fitting shoe. The wrists, ankles and knees are much less common sites. The spine is practically immune.

(b) The joint is swollen and exquisitely tender; pain is aggravated by the least movement.

(c) The overlying skin is tense, red and shiny with distended veins and may later show oedema.

(d) Fever, malaise, loss of appetite, gastrointestinal upset and scanty highly coloured urine are common.

(e) The patient may be very irritable.

(f) Blood E.S.R. rate is raised.

(g) As the inflammation subsides the skin over the joint becomes scaly and itches. The joints recover completely.

(h) If the hyperuricaemia is high and not regulated, attacks tend to occur with increasing frequency and to last longer. Eventually a stage may be reached where there is chronic persistent though generally less painful arthritis of several joints.

Complications

People with hyperuricaemia have a greatly increased liability to form **uric acid stone** in the urinary tract. These can present with renal colic and are one cause of the chronic renal disease that is an important late complication in gout. In gout, the kidneys may receive decreased amounts of glutamine because more has been diverted into purine synthesis.

Degenerative renal disease occurs in patients with chronic gout, and hypertension itself appears to reduce uric acid clearance. Raised plasma urates have also been found in patients with cerebral vascular disease.

Diagnosis

(a) When the big toe is affected, clinical diagnosis is confirmed by a raised plasma urate.

(b) There is another type of arthritis caused by deposition of crystals of calcium pyrophosphate in joints, sometimes called **pseudogout**.

(c) Large joints are usually involved, especially the knees, which show calcification of articular cartilage. It does not respond to colchicine and the plasma urate is normal.

Treatment

1. Drug

(a) There are drugs which relieve the acute arthritis, but do not affect plasma urate concentrations. The oldest effective remedy for acute gout is **colchicine**. It acts by inhibiting the polymorph response to urate crystals and thus breaks the vicious cycle. The dose is 1 mg two hourly until relief is obtained or diarrhoea supervenes. If the response with colchicine is too slow or diarrhoea too troublesome, it may be replaced by a short course of **phenylbuta-**

zone or of corticotrophin or one of the adrenocorticosteroids. During the acute attack the patient gets little rest from pain which is always worse at night. He should be in bed and the joint made as comfortable as possible by supporting it on pillows and protecting it from pressure and knocks.

(b) The second group of drug is Probenecid which impairs the reabsorption of uric acid by the renal tubules and so increases uric acid excretion in the urine by 30 to 50 per cent. As its administration is continued, plasma urate falls and later the frequency of arthritis decreases and tophi regress.

(c) When there is gross overproduction of uric acid, the newer drug allopurinol is the best for treatment. It inhibits xanthine oxidase so that uric acid production is reduced and the patient excretes instead xanthine and hypoxanthine which are more soluble. Allopurinol has no direct effect on gouty arthritis and prophylactic doses of colchicine 0.5 mg three times a day should be given for the first two or three months. Other adverse effects are unusual with these drugs.

2. Diet

(a) If the patient is overweight, he should be advised to bring his weight down by a gentle dietary regimen.

(b) Fasting should be avoided. Heavy, rich meals high in purines or fat are likely to raise the plasma urate and may be followd by an acute attack.

(c) Excessive alcohol precipitates gout. The patient should be advised not to drink alcohol at all.

(d) It is always better to avoid foods rich in purine, eg, liver, kidneys, sweet breads, fish and meat extracts.

(e) The patient should be advised to drink water before going to bed. Coffee and tea—although contain methyl xanthines such as caffeine can be drunk because caffeine is not converted into uric acid in the body.

(f) Lastly, it is wise to check fasting plasma lipids and treat hyperlipemia if present and to give some advise to gouty patients on diets which may reduce the risk of coronary heart disease.

DISEASES OF THE KIDNEYS AND URINARY TRACT

Principles of dietetics in renal disorders

(a) The major task of the kidneys is the elimination of surplus nutrients from the body. When renal function is greatly reduced, homeostasis may be maintained by reducing oral consumption of water, sodium, potassium and magnesium. Urea production is reduced by minimizing the intake of protein.

(b) In other renal disorders the capacity of the kidney to excrete sodium is impaired. This can be treated by reducing the dietary intake of sodium.

(c) The nephrotic syndrome damages the glomerular capillaries and result in massive loss of albumin and other proteins from the plasma into the urine. Increased dietary intake of protein allows increased hepatic synthesis of albumin which compensates the urinary loss.

(d) Diseases involved in the renal tubules can result in excessive urinary losses of water, sodium, potassium, phosphate and other substances. These losses can easily be replaced by increasing the oral intake.

1. Acute glomerulonephritis

This is characterized by acute inflammation of the glomeruli with congestion. Renal blood flow and glomerular filtration rate are reduced by 50 per cent. The urine volume falls and sodium excretion is greatly reduced. The urine contains moderate amounts of protein, red and white cells in abundance and casts of the renal tubules formed by precipitation of protein and red cells in the tubular system.

When the patient continues to ingest normal quantities of sodium and water, oedema develops and the blood pressure rises leading to headaches and swelling of the face and hands in the morning and of the ankles at night.

Treatment

Fluid intake should be restricted to 500 ml daily. During the first few days of treatment, the fluid given should be less. Sodium intake can be relaxed when oedema resolves and the blood pressure falls. Protein is restricted only when the blood urea is raised.

2. Nephrotic syndrome

This is characterised by heavy proteinuria, hypoalbuminaemia and peripheral oedema. It occurs when glomerular capillaries are damaged resulting in the increased losses of plasma proteins from the body into the urine. This syndrome can arise in diabetes mellitus, amyloidosis, multiple myeloma.

Treatment

Salt should not be added at table and only small quantities should be added during cooking. Fresh meat and fish can be used to supplement the protein intake. Egg can be used freely but cheese should be reserved for special occasions.

3. Acute renal failure

This is a catastrophic event. The causes are :

(a) Loss of blood from any cause including complications of pregnancy, trauma or gastrointestinal bleeding.

(b) Loss of plasma as in burns and crush injuries.

(c) Loss of fluid from severe vomiting, diarrhoea, acute intestinal obstruction.

(d) Serious infections especially septicaemia.

(e) Acute haemolytic disorders.

Treatment

Protein intake should be reduced. A daily intake of 100 grams of sugar has a marked protein-sparing effect. If the patient is vomiting, dextrose in water has to be given intravenously. The diet should contain potassium chloride by mouth.

4. Chronic renal failure (Uremia)

Uremia is a term used to describe general renal failure from any cause. As a result, many complex biochemical changes occur which are more responsible for the clincial features than the elevation of blood urea. These changes include disturbances in hydrogen ion concentration and abnormalities in water and electrolyte balance.

Clinical features

(a) The failing kidney is unable to compensate for large fluctuations in salt intake and for other increased metabolic demands.

(b) Uncompensated losses of water and sodium result in dehydration and salt depletion, with a fall in plasma volume, arterial blood pressure, renal blood flow and glomerular filtration rate.

(c) Renal function is lost and mild renal failure progresses to severe uremia.

(d) Tiredness, breathlessness on exertion may arise from anemia. A tendency to bleed due to abnormal platelet function.

(e) Anorexia, nausea and vomiting may result from the accumulation of urea, creatinine or an unknown uremic toxin. When GFR falls below 5 ml/min the kidneys may be unable to excrete even normal quantities of sodium and water. Many patients at this stage develop hypertension, oedema and features of water intoxication.

(f) The execution of hydrogen ions is impaired, the plasma bicarbonate concentration falls and an observer may notice compensatory hyperventilation of which the patient is often unaware.

(g) In the final stages death can result from hypertension, uremic coma, pulmonary oedema, gastrointestinal haemorrhage, hyperkalemia or severe infection.

Treatment

(a) In mild cases, active steps should be taken to control hypertension, to correct salt and water imbalance and to treat active urinary tract infection. Fluids and electrolytes should be given intravenously. A high protein diet is usually indicated in nephrotic patients with excessive losses of protein in the urine. Sodium restriction may be necessary in some patients, but others may need extra salt to compensate for urinary losses. Sodium bicarbonate may be needed for treatment of acidosis.

(b) Vitamin D can be used for the treatment and prevention of metabolic bone disease. Increase in plasma calcium may damage the kidney and accelerate loss of renal function.

(c) As renal failure progresses and the patient develops symptoms of uremia, more active measures are necessary to compensate for the loss of renal function. Patients may be treated by dietetic measures alone, by regular hemodialysis or by renal transplantation.

Stones in the urinary tract

Stones may form in the bladder (vesical calculi) or the kidney (renal calculi) 95 per cent of stones are made up of calcium salts. About 3 per cent are uric acid salts and about 1 per cent are cystine. Most stones are a mixture of calcium oxalate, calcium phosphate and magnesium ammonium phosphate, but about one-third are pure calcium oxalate. Stones form more readily infected urine in which bacteria have converted urea into ammonia so making the urine more alkaline.

Vesical calculus

Blood stones usually occur either in boys, in young men or in old men (in whom it is generally associated with prostatic obstruction or other cause of urinary stagnation).

Renal calculus

Renal colic, the severe pain caused by the passage of stone down the urinary tract. It stops when the stone is passed naturally or removed by a surgeon. Most stones remain in the kidney and they produce no symptoms (silent stones). They may grow there sometimes to a very large size. Infection may lead to pyelonephrosis, the commonest cause of chronic renal failure.

Prevention

(a) **Fluid intake :** A good flow of urine washes out particles of gravel. So water should be drunk before going to bed as urine flow is lowest at night. All patients who have suffered from stones should drink sufficient water to produce 2.5 litres of urine daily.

(b) **Diet :** The patient should drink only moderate amounts of milk and tea and eat only moderate amounts of milk products, meat and fish. Vegetables and fruits may be of benefit by increasing fibre intake.

(c) **Drugs : Bendrofluazide** reduces urinary calcium by about 30 per cent. **Pyridoxine** may be given to patients who form oxalate stones in the hope of diverting glycine metabolism towards serine and away from oxalate. **Penicillamine** given by mouth is useful in cystinuria, as it combines with cystine which is then excreted in a more soluble form.

DISEASE OF THE GASTROINTESTINAL TRACT

Diseases of the mouth

Lesions primarily nutritional in origin are angular stomatitis, cancrum oris, nutritional parotitis and nutritional glossitis. Glossitis is often a presenting feature in pellagra, the sprue syndrome, pernicious anemia and iron deficiency anemia of long standing.

1. The tongue

(a) The tongue may be dry in mouth breathers and coated with whitish yellow fur in those persons who smoke excessively.
(b) A clean red tongue which is inflamed and painful (acute glossitis) suggests an acute primary deficiency of some members of the Vitamin B complex.
(c) A clean pale and smooth tongue (chronic atrophic glossitis) suggests pernicious anemia or a long-standing iron-deficiency anemia.
(d) A local ulcer may be due to an ill-fitting denture or malignant disease, but rarely nowadays syphilis or tuberculosis.

2. The teeth, gums and mouth

(a) A bad taste in the mouth may be due to pyorrhoea.
(b) Inflammatory and hemorrhagic lesions in the mouth and gums can result from many causes, eg, infections, haemolytic streptococci, drug reactions and blood diseases (acute leukemia, aplastic anemia).
(c) Any inflammatory condition of the mouth may contribute to a nutritional disorder.

Treatment

If lesions are causing pain on chewing or swallowing, a fluid or semiliquid diet must be given until the condition is brought under control.

Diseases of the oesophagus

(a) Difficulty in swallowing (dysphagia) is the main feature and may lead to choking and even inhalation of food causing pneumonia or death. Dysphagia results from a functional defect with failure of onward movement of the peristaltic waves; alternatively the wave may be adequate but a block caused by spasm, inflammation or malignant disease prevents the food from getting through the affected area.
(b) Appropriate relaxation and contraction of the smooth muscle at the cardia allow food to enter the stomach and prevent regurgitation of stomach contents into the oesophagus. If this neuromuscular mechanism is disturbed, dysphagia or heartburn may ensue.
(c) Dysphagia may be produced by a neurological disorder which damages the motor pathway between the cerebral cortex and peripheral muscle. Common causes are stroke and achalasia.

Dyspepsia

(a) It means indigestion or difficulty in digestion. Any gastrointestinal symptom associated with the taking of food is called dyspepsia, eg, nausea, heartburn, epigastric pain, discomfort or distension.
(b) Dyspepsia may be a symptom of any organic disorder of the alimentary canal. It may also be caused by disease or disorder of structure outside the alimentary tract, eg, the gall-bladder, pancreas, etc.
(c) It may be a symptom of a general disease, eg, chronic nephritis and cardiac failure.

Dyspepsia and acid secretion

(a) When dyspepsia occurs with achlorhydria, it is probably due to chronic gastritis, cancer of the stomach or disease of the gall-balder or to emotional states.
(b) Hyperchlorhydria may be found in peo-

ple who have never suffered from dyspepsia. It is frequently found in patients with duodenal ulcer.

(c) Hydrochloric acid may be partially responsible for the pain and dyspepsia of the acute stage of peptic ulcer.

Management of dyspepsia

(a) Dyspepsia in young people may be due to overworking or overworrying or eating meals when excessively tired or has been smoking excessively or taking too much alcohol. The patient should be advised to give up such habits. His symptoms will clear up rapidly.

(b) Dyspepsia occurring in middle age for the first time accompanied by weight loss should be carefully investigated without delay.

(c) Patients with functional dyspepsia need dietary advice. Patient may find that certain foods bring on their symptoms whereas other can be taken with impurity. Hence bland diets have been prescribed.

Peptic ucler

The term peptic ulcer is used because it appears to develop from a loss of ability of the mucosa to withstand the digestive action of pepsin and HCl. A balance exists between acid pepsin secretion and mucosal resistance. In patients with gastric ulcer, the secretion of acid is often within normal limits, but patients with duodenal ulcer nearly always have a high output of acid. The great importance of gastric hypersecretion is supported by the intractable peptic ulceration of the **Zollinger-Ellison syndrome** in which gastrin produced by a tumour of the non-β-islet cells of the pancreas stimulates excessive gastric secretion by day and by night. Mucin is a protective agent. It adheres to the stomach wall as a thin but resistant coating. It is secreted in response to local, nervous and hormonal influences.

Caffeine, ethanol, aspirin, and nicotine promote peptic ulcers. Peptic ulcers occur more frequently in persons with blood group "O" than in those in other groups and possibly with those with HLA-B_5 antigens.

1. Clinical features and diagonsis

(a) The commonest symptom is pain or discomfort in the upper central abdomen. It is usually described as burning or gnawing in character. The pain comes and goes and is related to meals.

(b) In duodenal ulcer it usually occurs when the stomach is empty and is relieved by meals. The pain of gastric ulcer often comes shortly after eating.

(c) Other symptoms are loss of weight, heartburn or vomiting. In some patients an ulcer causes no symptoms until a complication such as haemorrhage occurs.

(d) An ulcer bleeds slowly and there is melaena (black stools) and anemia. With a larger haemorrhage there is usually haematemesis; the blood which is vomited is changed to a dark brown colour.

(e) The spasm of the pyloric canal can give rise to a characteristic feeling of sickness and distension; this prevents some patients from taking food which would relieve their symptoms.

(f) Acid output is usually above the normal range in patients with duodenal ulcer, and low or absent in patients with carcinoma of the stomach.

2. Medical treatment

Principles of treatment

1. Rest, both physical and psychological.
2. A bland diet, given in small amounts at frequent intervals.
3. Drugs—antacids and secretory inhibitors.
4. Giving up smoking.

Drugs

(a) **Cimetidine** and **allied drugs** by blocking the H_2 receptors in the gastric mucosa reduce acid secretion. This relieves symptoms and may promote healing. The drug is given throughout the day for one to three months. These drugs have no adverse effects. Some physicians keep patients on the drugs for longer in the hope of preventing recurrence.

(b) Insoluble antacid powders (aluminium hydroxide, mangesium oxide or trisilicate) usually bring immediate relief of pain.

Giving up smoking : The early deaths are not due to surgery or disease of the stomach but mainly to disease known to be associated with smoking (carcinoma of the lung, chronic bronchitis and coronary heart disease).

Diet

(a) Patients should avoid large meals by taking small amounts at a time for a number of times. This will reduce the risk of exposing the gastric and duodenal mucosa to excessive amounts of acid.

(b) When a patient has severe symptoms with pylorospasm or has had haemorrhage he should be given milk, eggs and fruit juice as the main ingredients.

(c) It is important to take Vitamin C since some patients have low reserves as a result of previous self-imposed dietary restrictions.

Rest

(a) When a patient curtails his business and social activities, symptoms are often relieved.

(b) Both physical and mental rest promote healing of an ulcer.

(c) Because of anxiety and emotional difficulties, they often need simple psychological support and hence they are admitted to hospital. Otherwise not required to be admitted to hospital.

Complications of medical treatment

(a) Scurvy may result from the intake of milk diets for a long period by adults. An excess of soluble alkalis can lead to alkalosis with tetany. A condition known as the milk alkali syndrome may occur in patients who have taken large amounts of milk (more than 1 litre daily) and soluble alkali over long periods. Weakness, anorexia and lethargy are the characteristic features and there may be psychological disturbances.

(b) Hypercalcaemia may give rise to calcification in the kidneys and elsewhere.

Gastritis

(a) Ingestion of alcohol, drugs or other chemical irritants may be responsible for gastritis.

(b) The commonest drug causing gastritis is aspirin, often taken for headaches and menstrual pain.

(c) Atrophic gastritis may be due to an autoimmune reaction and this is responsible for the failure to secrete intrinsic factor and HCl in pernicious anemia. It is also present in patients with severe iron deficiency anemia.

(d) It may also be present in metabolic disorders, eg, uremia, carcinoma of the stomach.

(e) Clinical features are mild anorexia, vague discomfort, nausea and heartburn to severe and repeated vomiting accompained by diarrhoea if there is associated enteritis. Sometimes the clinical picture may simulate acute peptic ulcer, and massive gastric haemorrhage may occur. Nausea, abdominal fullness, heartburn and pain occur before breakfast and improve as the day goes on.

Acute gastritis

(a) The symptoms are nausea, pain and vomiting and commonly follow an excess of alcohol, aspirin or other drugs.

(b) Treatment consists of stopping alcohol or the drug, sometimes washing out the stomach and giving alkalis.

(c) Water and electrolyte losses can be replaced by an oral rehydration fluid. With improvement of the condition the patient is given small feeds of milk and gradually returns to a normal diet within 1 to 2 days.

Disorders of the intestine

The small intestine is the main site of absorption of nutrients. Normally absorption of all nutrients begins in the jejunum and is completed in the ileum except that of water and electrolytes which is completed in the colon.

Diarrhoea leads to depletion of water and electrolytes and may be due to disorders of the small or large intestine. The malabsorption syndrome arises when there is failure of digestion and absorption due to disorders of the small intestine. Failure to absorb fat is the main feature leading to steatorrhoea. Disorders of the colon leads to constipation.

Acute diarrhoeal diseases

These diseases are caused by infections of the small and large intestine by pathogenic viruses, bacteria or protozoa. The term covers illnesses such as acute gastroenteritis, bacterial food poisoning, traveller's diarrhoea, infantile diarrhoea and weanling diarrhoea, as well as the specific infections, bacillary dysentery and cholera.

The losses of water and electrolytes in the diarrhoea lead to dehydration of the body. If the diarrhoea is severe, death from dehydration may occur within one or two days, especially in young children, in the very old and in the undernourished.

The serious dehydration can be prevented by giving an oral dehydration solution (ORS) early in the disease. The solutions used consist mainly of common salt, sodium bicarbonate and glucose or other source of carbohydrate. They are cheap and can be administered safely by any mother in a primitive home.

1. Clinical features

(a) The diarrhoea is usually accompained by abdominal discomfort and nausea and often by vomiting.

(b) Fever, if present, is seldom high.

(c) When there is severe dehydration, there is circulatory collapse with a marked fall in blood pressure. This may be fatal or lead to death later from acute renal failure.

(d) Infections not only interfere with intestinal absorption but the cholera toxin causes active secretion of chloride by the intestinal mucosa. Then with the chloride, sodium and water flow out from the tissues into the lumen of the gut. In severe cholera as much as 1 litre of fluid may be lost in an hour. By this way death from cholera may arise within a few hours of the onset of the disease.

(e) Vomiting also increases the fluid loss and the accompanying loss of acid leads to an alkalosis which is a cause of drowsiness.

(f) Sweating may lead to a loss of a litre or more during the course of a hot day in the tropics by an adult, even when at rest. Fever also leads to sweating.

(g) An attack of diarrhoea reduces food intake. Repeated attacks in a young child cause protein-energy malnutrition. In this way infections are responsible for retarded growth and development of young children in poor communities and also for a large part of the deaths from malnutrition.

2. Treatment

(a) WHO has recommended an oral rehydration solution which comprises of the followings :

Sodium chloride (table salt)	3.5 gm	Dissolve in 1 litre of potable water
Sodium bicarbonate (baking soda)	2.5 gm	
or trisodium citrate, dihydrate	2.9 gm	
Potassium chloride	1.5 gm	
Glucose	20 gm	

The solution was first developed in Bangladesh for the treatment of cholera and weanling diarrhoea. It is there firmly established.

(b) Absorption of oral fluids is fully dependent on a good blood supply to the alimentary canal. Circulatory collapse, detected by a fall in blood pressure, indicates a need to replace fluid intravenously. This is also necessary when vomiting is severe.

(c) In young children, a lower concentration of NaCl may be desirable to avoid any risk of hypernatremia.

(d) Intestinal loss of potassium should be replaced and sodium bicarbonate ($NaHCO_3$) may be needed to correct acidosis caused by starvation; rarely NH_4Cl may be required to correct alkalosis arising from vomiting.

(e) Most physicians do not prescribe any antibiotic drug for an attack of acute diarrhoea. They select the appropriate drug after microbiological diagnosis.

Malabsorption Syndrome

Lack of digestive secretions and injury to the epithelial surface of the small intestine impair absorption of nutrients. The clincial features of undernutrition and malnutrition for a long time is known as the malabsorption syndrome.

1. Clinical features

(a) Loss of weight and edema in cases of long standing.

(b) Chronic diarrhea with abdominal discomfort and distension.

(c) Steatorrhea, increased fat in the feces, is frequently present, its severity depends on the amount of fat in the diet.

(d) There is usually diarrhea with a bulky stool that has an offensive smell and floats on water.

(e) Anemia is usually present due to impaired absorption of iron and folic acid.

(f) Nutritional glossitis, angular stomatitis and peripheral neuropathy arise from the deficiency of the B group vitamins.

(g) Prolonged failure of calcium absorption may lead to evidence of osteomalacia and to tetany.

(h) Haemorrhages may occur due to Vitamin K deficiency.

Tropical Sprue

Sprue is the name given to a tropical disease in which the presenting features are sore mouth, fatty diarrhoea and associated secondary manifestations of undernutrition and malnutrition.

Although there is defective absorption of fat, the absorption of water, electrolytes, glucose, vitamins and minerals is also impaired. These defects are associated with atrophy of the jejunal villi.

Sprue is a serious disease and may prove fatal without proper medical and dietary care. Full recovery is possible with proper treatment and if they leave the tropics.

Intestinal obstruction

(a) The cause of intestinal obstruction is mechanical or due to paralysis of the intestinal muscle (paralytic ileus).

(b) The common causes of mechanical obstruction are external hernias, volvulus, tumours of the colon, adhesions due to previous inflammatory disease or operation.

(c) Paralytic ileus is usually a consequence of peritonitis, resulting from any cause, eg, a gastric or intestinal perforation or an abdominal operation.

1. Features

(a) The chief features are vomiting, complete constipation and colicky pain which may be absent or slight in paralytic ileus.

(b) A serious loss of water and electrolytes results from the vomiting and from the stagnation of intestinal secretions in the dilated paralysed loops.

(c) The loss of fluid from the circulation from the latter source may be several litres in 24 hours and this may lead to prerenal uremia.

(d) The loss of potassium causes apathy, mental confusion and muscular weakness.

2. Treatment

(a) Since intestinal obstruction is always serious it should be treated only in a hospital where surgical and biochemical help are available.

(b) Immediate operation is required for the relief of mechanical obstruction while it is strongly contraindicated in paralytic ileus.

(c) In paralytic ileus the distension of the paralysed gut must be treated by continuous suction through a tube passed into the stomach or jejunum, and continued until the bowel recovers from its paralysed state.

(d) In both types of obstruction the loss of fluid and electrolytes must be made good by appropriate infusions, and intravenous feeding is often needed.

Constipation

Constipation is delay in passage of the feces. The presence of food in the stomach is a stimulus to a gastrocolic reflex which causes movements of the colon and these may lead to feces entering the rectum. The reflex usually occurs after the first meal of the day. In some people the presence of liquid in the stomach initiates the reflex and a drink on rising may be sufficient to stimulate defecation. Some healthy people do not defecate every day and a few do so once or twice a week. They should not be considered constipated, and constipation should only be diagnosed when delay in defecation causes discomfort and indigestion.

The two common causes of constipation are a small fecal bulk and persistent neglect of the call to defecate. Many diseases are associated with constipation.

The bulk of the feces is mainly water and the amount of water depends on the amount of dietary fibre present and the capacity of the fibre to bind water. Low fibre diets cause constipation.

If the call to defecate is persistently neglected, the reflex mechanism becomes less sensitive and constipation results. Going to the toilet should become a habit early in life.

Gastrointestinal diseases give rise to constipation and the irritable bowel syndrome. Carcinoma of the colon and rectum sometimes present as constipation. Constipation is common in psychiatric disorders which cause depression. Any neurological disease causing lesions in the lumbar cord may affect the reflex centres responsible for defecation and lead to constipation.

Pregnant women and old people are often constipated. The pressure of the gravid uterus on the colon may delay movements of the contents. In old people the sensitivity of the neuromuscular reflexes in the colon may be impaired or they may become less aware of the presence of feces in the rectum.

When constipation occurs due to low intake of dietary fibres, there is often pain in the left side of the abdomen along the line of descending colon, and the feces may be passed as hard pellets. Passage of feces relieves the pain. When the call to defecate is repeatedly ignored, a mass of inspissated fecal matter may accumulate in the descending colon. The fluid contents of the colon may run down the side of the mass and cause a watery diarrhoea.

Treatment

(a) The intake of dietary fibre should be increased by the consumption of fruits and vegetables. The fibre has got water-holding capacity. The fruits and vegetables are oranges, apples, carrots and cabbages.

(b) A young child should be advised to go to the toilet at a regular time each day. An adult should not neglect to go to the toilet due to laziness, hurry or a lack of suitable accommodation. Once the reflex has been lost, it cannot be regained to move the bowels at the same time each day.

(c) Many laxatives are available. Their continued use may lead to excessive losses of potassium, sodium and water in the feces and is not recommended. Two mild laxatives that are recommended for short periods are lactulose and senna. Lactulose is a sugar which is not absorbed in the small intestine but passes to the colon where it is partially broken down by bacteria. It reduces the absorption of water from the colon and increases the bulk of the feces. Senna is a glycoside which is broken down in the small intestine to emodin ; this is absorbed into the blood stream and stimulates the muscles of the colon for 6 to 12 hours after administration of the senna.

DIEASES OF THE LIVER, BILIARY TRACT AND PANCREAS

Many disorders and diseases damage liver function. Moderate damage causes jaundice. Serious damage is caused by many infections and toxic chemicals. Persistent infection causes liver failure. Dietary therapy is an important part of the treatment of liver failure and of its serious complications.

Liver damage

1. Hepatitis

(a) The liver becomes tender and enlarged following continued excessive intake of alcohol. The patient may become jaundiced.

(b) If alcohol intake is continued, acute liver failure may arise with deep jaundice and several cerebral symptoms.

(c) The patients may complain of general malaise, anorexia and a variety of digestive disturbances and become progressively undernourished.

(d) During the course of chronic hepatitis there is slow progressive fibrosis leading to major disturbances of liver structure and function.

2. Cirrhosis of the liver

(a) It is now referred to any diffuse chronic disease of the liver.

(b) Venous congestion leads to enlargement of the spleen which is detected by palpation. It leads to varicosities which may bleed.

(c) Oesophageal bleeding is a common and serious consequence of cirrhosis of the liver.

(d) Damage to liver cells leads to jaundice that may be severe. The clinical features and disordered sleep rhythm, restlessness or drowsiness, impaired intellectutal function with confusion and, in severe cases, stupor and coma. The onset may be sudden. Sometimes these features are the first manifestations of liver failure to appear and then they may be mistaken for a psychiatric disorder. The condition is due to unidentified toxic substances reaching the brain in the systemic circulation. These may arise from disturbances of intermediary metabolism or to substances absorbed from the gut that the damaged hepatocytes are unable to detoxify.

Jaundice

(a) This is caused by the increase in the bilirubin in the blood. The skin and sclera of the eyes appear yellow and the urine is usually dark yellow or brown. It is not detectable until the plasma bilirubin rises above 8 mg/I.

(b) There are three types of jaundice. Hepatocellular jaundice is due to damage to hepatocytes by toxic or infectious agents interfering with the uptake and conjugation of bilirubin by the cells or to blocking of the bile canaliculi (cholestatic jaundice).

Prehepatic (hemolytic) jaundice is due to increased bilirubin from excessive destruction of R.B.C. It may arise from congential defects causing the erythrocytes to be unduly fragile (spherocytosis, sickle cell anemia, thalassemia or from some drugs, incompatible blood transfusions).

Posthepatic jaundice is due to an obstruction of bile flow between the liver and duodenum. Common causes are impacted gallstones and cancer of the head of the pancreas. The excess of bilirubin in the plasma is then in the conjugated form, whereas in prehepatic jaundice it is unconjugated.

(c) Jaundice is sometimes accompained by itching (pruritus) and this may be severe. The cause is retention of bile salts in the blood and it may be relieved by cholestyramine. This is an anion exchange resin which binds bile salts in the gut and so interrupts the enterohepatic circulation and increases fecal excretion of bile salts.

Agent damaging the bile salts

1. Dietary deficiency

Low protein intake causes the fatty changes in the liver in kwashiorkor and there is evidence of reduced capacity to secrete β-lipoproteins but not of choline deficiency in man. Choline deficiency is not at all responsible for liver damage in man.

Fatty changes in the liver are common whenever there is a high proportion of fat in the metabolic mixture, eg, in uncontrolled diabetes, in starvation, in some cases of obesity.

2. Infective agents

Hepatitis A virus is excreted in the stools and spreads by the fecal oral route. Children or adolescents are affected by this infection where hygienic condition is low. The initial symptoms are loss of appetite, nausea and malaise and usually mild fever. Jaundice appears after 4 to 7 days and the liver is enlarged and tender. The symptoms usually last from 2 to 6 weeks but complete recovery may take many months.

Very rarely there is fulminant hepatitis with symptoms of acute liver failure and the death usually occurs within 10 days.

Some healthy people, who have no clinical history of the attack of the disease, excrete the virus in their feces. Such carriers may cause epidemic. Some of the cases of cirrhosis in which no cause can be found may be due to long continuing exposure of the liver to the virus.

Hepatitis B virus produces a disease known as homologous serum jaundice because it arises after transfusion of blood. It can also be spread by the use of one needle to give injections to a series of patients without effective sterilisation. The virus is never spread by the fecal-oral route and the infection arises from close personal contact, as between sexual partners. They cause similar type of illness as that in Hepatitis A virus. A carrier can be detected by the presence of the antigen of the virus (HBeAG) in the serum.

Yellow fever virus has a specific affinity for hepatocytes. It is much more virulent. More than 20 per cent of patients died of acute liver failure. Yellow fever is now very rare, as the mosquito responsible for its transmission is a domestic species and easily controlled.

3. Other infective agents

In three diseases the liver is especially involved. In Weil's disease leptospira invade all organs and tissues but markedly the liver and jaundice is the common feature.

Amoebiasis is primarily a disease of the large intestine but protozoa may migrate up the portal vein and cause amoebic hepatitis and liver abscesses.

In schistosomiasis with infection of the large intestine by Schistosoma mansoni, ova frequently move up the portal vein and cause multiple granulomata in the liver; these become fibrosed and are a common cause of cirrhosis of the liver.

4. Toxic agents

(a) Alcohol, other undetected toxic chemicals and infective agents may act synergistically to produce progressive liver damage.

(b) Foods may be contaminated by moulds that produce toxin. Aflatoxins are known to cause liver disease.

(c) Many drugs can damage the liver in susceptible individuals and produce a clinical and biochemical picture similar to acute viral hepatitis. Examples are paracetamol, a much used analgesic, sometimes used in the treatment of depression. Cholestatic jaundice may follow the use of methyltestosterone and very rarely oral contraceptives.

(d) Workers in the chemical industry may be exposed to many chemicals that damage the liver and produce jaundice. Examples are hydrocarbons, carbon tetrachloride, arsenic, trinitrotoluene, etc.

(e) Excess stores of iron, copper, galactose and glycogen may accumulate in the liver and in time lead to cirrhosis. These conditions are usually due to hereditary defects. Hereditary defects in the transport and conjugation of bilirubin may lead to jaundice. People suffering from Gilbert's disease have no symptoms but juandice, anorexia, nausea and abdominal pain may arise during a period of stress, such as an infection, fasting or strenuous exercise.

Dietetic management and treatment

1. Acute hepatitis

(a) There is no antiviral agent for use against the hepatitis virus but metronidazole is very effective in amoebic hepatitis. Patients with viral hepatitis do well with rest.

(b) No specific dietary treatment is needed but as patients lack appetite and suffer from nausea, meals should be well-cooked and attractively served. Several small meals may be better tolerated than three large ones.

(c) Fat intake should be restricted. Carbohydrate and protein diet should be taken. Vitamins and minerals should be given through fruits and vegetables.

2. Hepatic failure

(a) Hemorrhages as a result of lack of clotting factors synthesized by the liver lead to shock and renal failure. Profound hypoglycemia and potassium depletion occur.

(b) Intravenous infusion of hypertonic glucose and electrolytes is required with repeated monitoring of blood level.

(c) Sedatives may be needed for delirium but are potentially dangerous due to the state of the liver.

(d) Despite all treatment majority of the patients die.

3. Chronic hepatitis

This is common in alcoholics and occasionally follows viral hepatitis. When it is due to alcoholics, most patients do well when alcohol intake is stopped.

Some chronic alcoholics are malnourished because they eat irregularly and seldom take proper meal. They need to change dietary habits and require professional advice.

4. Cirrhosis

(a) Many patients with cirrhosis are seriously malnourished, especially those who are alcoholic and require nutritional rehabilitation with a high energy, high protein diet. This may be difficult to achieve on account of poor appetite, and admission to hospital is necessary.

(b) The fat-soluble vitamins are poorly absorbed because of the poor absorption of fat owing to the absence of bile salts. Therefore, supplements may be needed. A monthly intramuscular injection of Vitamin K (10 mg), of Vitamin A (30 mg) and of Vitamin D (25 mg) provides for maintenance. As bleeding is common and often severe, Vitamin K should be given at once and continued.

(c) **Osteomalacia** is a well-known complication of chronic liver disease and also reserves of Vitamin A in the liver is low. Vitamins A and D should be given in proper doses and should not be given more which may cause toxic effect.

5. Ascites

(a) Fluid accumulates in the peritoneal cavity because of increased capillary pressure in the portal vessels and decreased oncotic pressure owing to a failure of the liver to synthesize plasma albumin. Sodium is retained in the fluid.

(b) The **ascitic fluid** often amounts to 10 litres and contains protein in concentrations of 10 to 20 g/I.

(c) It depletes plasma proteins and carries a risk of introducing infection.

(d) Treatment consists of a diet rich in protein and low in sodium, and also diuretics to increase urinary output of sodium and fluid. With all of them there is a danger of electrolyte imbalance. Overdosage may lead to sodium deficiency and also to potassium depletion.

(e) The patient should avoid salt-rich foods.

6. Portal hypertension

Bleeding from oesophageal and gastric varices is common and often serious and a cause of death. The anemia is treated with iron and repeated blood transfusion are often necessary. Pressure in the veins can be relieved by a surgical operation but such operations are dangerous and increase the risk of death from liver failure.

7. Biliary cirrhosis

Stagnation of bile in the canaliculi of the liver may lead to hepatitis and subsequently to cirrhosis. This may be a consequence of unrelieved obstruction of the bile duct and then secondary bacterial infection is often an additional factor. This disease occurs most commonly in middle-aged women.

Prompt surgical treatment prevents any damage to the liver from progressing jaundice fluctuates in intensity as does pruritis which is often serious and then may be relieved by cholestyramine.

Diseases of the gall bladder and bile ducts

The function of the gall-bladder and bile ducts is to concentrate, store and deliver bile into the duodenum at suitable times to assist digestion; hormonal and nervous factors play a part in this process. The stimulus for this activity is the entry of food into the small intestine; this causes the mucosa of the duodenum and jejunum to secrete a hormone, **cholecystokinin**; which is carried in the blood to the gall-bladder and causes it to contract. Fats and foods rich in fats are especially effective for this purpose. Vagal stimulation causes contraction of the gall-bladder and relaxation of the sphincter, while stimulation of the sympathetic nerves produces the reverse effects.

Gall-stones (Cholelithiasis)

The bile is concentrated in the gall-bladder and when it is supersaturated gall-stones are likely to form. Supersaturation arises when there is an insufficiency of solubilising agents such as bile acids and lecithin to keep cholesterol and bile pigments in solution. The most common gall-stones are composed of cholesterol, bile pigment and various calcium salts including calcium palmitate. In the centre is a protein nidus. This suggests an infective origin. The stones are usually multiple and their surfaces faceted. Pigment gall-stones are more prevalent in the eastern world and may be due to increased hemolysis arising from the sickle cell trait and thalassemia.

Gall-stones are more common in women than in men. Advancing age, repeated pregnancies and sedentary life and the use of oral contraceptives are contributing factors.

The formation of cholesterol gall-stones depends on the concentration in the bile of cholesterol relative to that of the conjugated bile acids and lecithin. High cholesterol concentration may arise from either excess of cholesterol or lack of bile acids. Excess cholesterol may be due to increased activity of HMG-CoA reductase. Deficiency of bile acid may be due to deficiency of the enzyme that controls their formation (cholesterol-7α-hydroxylase).

The symptoms due to gall-stones are removed surgically often a stone causes no symptoms and then there is no strong case for its removal.

Some gall-stones can be dissolved in the gall-bladder by administration of the bile acid, chenodeoxycholic acid. In a large trial complete dissolution occurred in 13.5 per cent of cases after two years with some adverse effects—diarrhoea and a slight rise of plasma cholesterol. So it is only for treatment of elderly, frail or who are losing weight before operation.

Acute cholecystitis

It is rarely due to infection of the gall-bladder but almost always occurs in association with obstruction to the cystic duct or neck of the gall-bladder, upon which infection is usually superimposed. In most cases gall-stones are the cause of obstruction.

1. Treatment

(a) Cholelithiasis with accompanying cholecystitis is the choice of the surgeon to postpone operation until the acute infection has subsided.

(b) The patient should be in bed and given suitable analgesics and antibiotics.

(c) Heat should be applied to the gall-bladder region and ample fluid is given intravenously if the patient is vomiting.

(d) So long as the gall-bladder is acutely inflamed, it is advisable to keep the organ at rest as far as possible.

(e) For acute cases an entirely fluid diet of at least 2 to 3 litres daily, given in small feeds at hourly or two-hourly intervals, is advisable for a few days.

(f) When the condition is improved within two or three days, clear soups, milk, fruit jellies, and cereals may be added, and the diet is rapidly built up to normal.

(g) After recovery if there are gall-stones or non-functioning gall-bladder, cholecystectomy should be advised.

Chronic cholecystitis

If it is decided not to remove the gall-bladder or if for any reason operation is suspended, dietetic treatment should be given.

(a) Foods which aggravate the symptoms should be avoided.

(b) The cooked meats rich in fat and fried foods be given.

(c) Milk, butter and cream cheese is permitted, and may promote drainage of the gall-bladder.

(d) Eggs may be permitted in moderation, if they do not cause symptoms.

(e) The diet should be bland and contain adequate protein.

(f) Care should be taken to avoid large meals and indigestible articles of food, and ample quantities of fluids should be taken in the morning and between meals.

Diseases of the pancreas

Acute pancreatitis

This is a serious disorder which may lead to hemorrhagic necrosis of the pancreas, peritonitis and death. It usually occurs in middle-aged and elderly persons. The main symptom is the sudden onset of agonising pain in the epigastrium which may radiate to the back. It may follow a heavy meal or an excess of alcohol. Nausea and vomiting are frequently present. Moderate fever occurs and jaundice may develop. In addition there are signs of peritonitis and profound shock.

1. Treatment

(a) Medical management consists of the relief of pain and the control of shock.

(b) Continuous gastrointestinal suction is essential to reduce vomiting and distension.

(c) It also removes acid gastric juice, a stimulus to pancreatic secretion.

(d) Production of gastric jucie is inhibited by cimetidine and ranitidine.

(e) Antibiotics should be administered to prevent secondary infection.

The Anemias

There are three main causes of anemia :

1. Loss of blood from the circulation, i.e. external or internal hemorrhage.
2. Hemolysis, i.e. increased destruction of R.B.C.
3. Reduced production of erythrocytes and hemoglobin-dyshaemopoiesis.

For the production of R.B.C. many nutrients are needed. The most important are iron, folic acid, and Vitamin B_{12}, but others are protein, pyridoxine, Vit. C, copper and Vit. E. The anemia occurs in a healthy person as a result of poor diet. The diet often contains insufficient one or more of the essential nutrients to meet increased needs caused by chronic hemorrhage, infection and genetic defects affecting the R.B.C. disorders of the alimentary tract often leads to impaired absorption of the essential nutrients and so to anemia. So the extra needs are to be met by increased nutrients.

1. Clinical features of anemia

(a) Symptoms of anemia arise when the transport of oxygen by the blood is insufficient to meet the needs of the body. The need for oxygen is related to physical activity. A person leading a sedentary life may have a moderate degree of anemia but entirely free of symptoms.

(b) The severity of the clinical features does not depend on the degree of anemia but depends on the rapidity of its development.

(c) Common symptoms are general fatigue and lassitude, breathlessness on exertion, giddiness, dimness of vision, headache, insomnia, palor of the skin, palpitation, anorexia and dyspepsia, tingling and 'pins and needles' in the fingers and toes.

(d) Angina pectoris (due to myocardial hypoxia) is sometimes present.

(e) Physical signs include palor of mucous membranes and finger-nails, tachycardia, functional systolic murmurs, evidence of cardiac dilatation and in severe cases, oedema of the ankles and crepitations at the bases of the lungs.

(f) There are signs of nutritional deficiency, particularly angular stomatitis, koilonychia and glossitis.

(g) Atrophy of the papillae and mucous membranes gives the tongue a smooth glazed appearance. The atrophy begins at the edges and later affects the whole tongue. As a result the tongue appears moist and exceptionally clean.

2. Diet

(a) The most valuable dietary sources are meats and liver; they should perferably be eaten once a day.

(b) Eggs also have a high iron content but this is poorly absorbed because of phospholipid inhibitors in the yolk.

(c) Less expensive sources of iron are beans, especially soyabeans and nuts.

(d) Fresh fruits and vegetables are of greater value because of their ascorbic acid content which facilitates iron absorption.

(e) Milk is a poor source of dietary iron.

3. Genetic defects of R.B.C.

There are hereditary defects of R.B.C. that make them more susceptible to hemolysis and persons who carry these genes are liable to become anemic. Heterozygotes are at increased risk of becoming severely anemic.

(i) Thalassemia (Greek Thalassa means sea):

(a) It is defined by a defect in the synthesis of part of the polypeptide chain of hemoglobin A which is partially compensated by persisting synthesis of fetal hemoglobin (Hb-F).

(b) There are also abnormalities of the red cell membrane.

(c) Homozygotes have thalassemia major, a severe hemolytic anemia and rarely survive into adult life.

(d) Heterozygotes may or may not have a mild anemia, thalassemia minor.

(ii) Sickle cell trait :

(a) This is due to an abnormal hemoglobin, hemoglobin S, differing only in having a single molecule of valine instead of one of glutamic acid in one of the polypeptide chains.

(b) The configuration of the molecules of Hb-S distort the R.B.C. into a characteristic sickle shape. Such corpuscles are abnormally sensitive to hypoxia.

(c) Persons with the sickle cell trait are at risk of hemolytic crisis when flying and cabin pressure is reduced.

(d) Heterozygotes have the sickle cell trait and are only occasionally anemic.

(e) Homozygotes have a severe anemia, sickle cell disease.

(iii) Spherocytosis :

(a) This is the commonest congenital defect of R.B.C.

(b) The abnormality lies in the cell membranes which are more than normally permeable to sodium ions.

(c) The cells assume the shape of spheres which are more easily trapped in the microcirculation of the spleen, where hemolysis takes place.

(d) In most cases the increased loss of cells by hemolysis is slight and can be made good by increased production in the bone marrow.

(e) When anemia develops, removal of the spleen reduces hemolysis and cures the anemia but does not change the defect.

(iv) Glucose-6-phosphate dehydrogenase deficiency : Persons with the defect are liable to develop anemia when treated with oxidant drugs ; these include antimalarials, sulphonamides, antipyretics and analgesics.

Anaemia in tropical countries

The demands for iron may be greatly increased by the loss of hemoglobin in the feces and urine resulting from hemorrhage due to parasitic diseases, such as hookworm. In addition, significiant amounts of iron may be lost in the sweat during muscular work in hot climates. The protozoal, helminthic infections, and bacterial infections which are common in the tropics may be responsible for anemia. The nutritional megaloblastic anemias of tropical climates are always associated with diets poor in animal protein and fresh vegetables, and hence low in Vitamin B_{12} and especially folate.

1. Malaria

(a) An attack of malarial fever due to plasmodium falciparum is always accompanied by hemolysis and in a severe or prolonged attack severe anemia may ensure.

(b) After the parasites have been removed from the blood, hemolysis may continue due to a complement-mediated immune response, and there may also be a mild depression of erythropoiesis.

(c) Malarial infection, if associated with pregnancy and malnutrition, the anemia may present as megaloblastic anemia due to folate deficiency, but is more usually hypochromic and microcytic because of iron deficiency.

(d) Since the blood destruction is intravascular, most of the iron liberated from the destroyed red cells is retained in the body and can be used again for synthesis of hemoglobin.

(e) A vicious circle develops in communities suffering from chronic malaria-sickness, weakness and anemia, economic inefficiency, poverty, malnutrition, bad housing and social conditions, reinfection.

2. Hookworm infection

(a) Infection with hookworm is a common cause of anemia where there is 'wet' cultivation of the land.

(b) Hemorrhages occur at the site of the attachment of the worms to the intestinal mucous membrane. These are certainly in part responsible for the anemia.

(c) A patient with a heavy infection, namely about 1,000 worms, can sustain a heavy loss of blood and anemia may quickly develop.

(d) Heavy hookworm infection usually occurs in populations whose dietary intake of iron is unsatisfactory.

(e) The hypochromic anemia is in part due to a poor diet and in part to the worms. This combination causes much ill-health; it may reduce greatly the working capacity of all and in these directly responsible for the poverty of many families in the tropics.

(f) Heavy infections may cause severe anemia with hemoglobin levels below 4 gm/100 ml. In such seriously ill patient, before administering a vermifuge, blood transfusion should be given and the general condition of the patient improves by bed rest, diet and iron.

3. Other causes of anemia

(a) A mild or moderate anemia may develop due to chronic infection, particularly if fever is present.

(b) Many drugs can cause anemia by impairing red cell formation, by causing hemolysis or by loading to bleeding. Small repeated hemorrhages in the stomach due to aspirin is a common example.

(c) In chronic renal disease anemia is common but the bone marrow remains cellular until renal damage is marked. Deficiency of erythropoietin is a possible cause.

(d) If megaloblastic anemia occurs this is probably due to primary malnutrition and is found in chronic alcoholics with cirrhosis.

(e) Malignant disease causes anemia. They include impaired appetite, malabsorption or blood loss from the alimentary tract, multiple deposit in the bone marrow and increased hemolysis.

(f) Abnormal utilization of iron by the marrow may cause anemia. Some cases respond to pyridoxine therapy.

DISEASES OF THE NERVOUS, RESPIRATORY SYSTEMS AND OF THE SKIN

Epilepsy

Epilepsy is a common disorder liable to periodic attacks associated with a disturbance of consciousness and commonly manifest as fits. In a small minority of cases a local lesion in the brain or a generalised metabolic disturbance is responsible. In the great majority of patients the cause is unknown and they are said to have idiopathic epilepsy.

Antiepileptic drugs are very effective in preventing the seizures. A period of starvation was effective to reduce their frequency. Such benefits may be due to a depression of the brain by the accompanying ketoacidosis. Ketogenic diets are given when the fits are not controlled by drugs.

Childhood myoclonic epilepsy are resistant to drugs and associated with brain damage and patients have many attacks throughout the day. Growing child should not be ketoacidotic. A diet low in carbohydrate and high in fat may help to control the fits.

Antiepileptic drugs cause changes in liver metbolism which may increase requirements of Vitamin D and folic acid. Patients taking doses over the years should be checked regularly for signs of anemia and rickets or osteomalacia.

Patients with epilepsy should take normal well-balanced meals at regular intervals. Epileptic children should not be allowed to take very large meals as these may predispose them to fits.

Psychiatric disorders

Deficiencies of thiamine, niacin and Vitamin B_{12} may each be associated with psychological disorders. The psychological symptoms are occasionally so marked and other symptoms and signs of the deficiency so slight that the correct diagnosis is missed and the patient is treated as if he is a neurotic or has a psychotic disorder.

Some of the patients have requirements for a vitamin in excess of the recommended intake. Psychiatric disorders may lead to abnormal feeding behaviour and so to anorexia nervosa and obesity.

Chronic Rheumatic Diseases

The main members of this groups are rheumatoid arthritis, osteoarthrosis, and non-articular rheumatism. The rheumatic group of disease is second only to bronchitis in men and first in women.

1. Rheumatoid arthritis

(a) It occurs in people of all ages. The peak incidence is at the age of 40. It occurs in women at least three times as in men.

(b) It is mainly a disease of temperate climates which are associated with cold and damp.

(c) The principal tissue affected is the synovial membrane of joints which becomes inflamed and thickened.

(d) Muscular stiffness develops first and is followed later by pain and swelling of many joints, starting frequently with the small joints of the hands and feet.

(e) During the active stage of the disease the patient suffers from general malaise and fatigue; fever is sometimes present and the appetite is poor.

(f) There is loss of weight and some degree of anemia.

Treatment

(a) Anemia and loss of weight by diet should be corrected.

(b) Foods rich in protein, iron and ascorbic acid should be tried first; additional vitamins and minerals should be prescribed.

2. Osteoarthrosis

(a) Osteoarthrosis is characterised by degeneration of the articular cartilage and the formation of bony outgrowths at the edges of the joints. A generalised form of the disease occurs normally in middle-aged women in whom the small joints of the fingers, the carpometacarpal joint of the thumb and the interfacetal joints of the spine are particualrly affected.

(b) When one joint is particularly affected, there is frequently a history of an injury to that joint some years before.

(c) Following fractures of the long bones gives rise to osteoarthrosis in adjacent joints.

(d) Symptoms are proved to develop in the weight-bearing joints or those joints subjected to excessive strain at work.

(e) Obesity imposes to 'osteoarthrosis' of the weight-bearing joints in the lower half of the body.

(f) The joints invloved are those of the spine,

the hips, knees, elbows, and the terminal joints of the fingers.

(g) The symptoms are gradual in onset. Pain is at first intermittent and of the aching character, appearing after the joint has been used, and relieved by rest.

(h) As the disease progresses, movement in the affected joints become increasingly limited, at first by muscular spasm and later by the loss of joint cartilage and the formation of osteophytes. General health is usually excellent.

Treatment

(a) Rest, graduated physical exercises and physiotherapy.

(b) Analgesic should be advised according to the patient's needs.

(c) Prevention and correction of obesity is needed.

3. Non-articular Rheumatism

(a) This is characterised by pain and stiffness, often of sudden onset, affecting mainly the neck, shoulders, back and gluteal regions.

(b) Exposure to cold and damp, excessive or unaccustomed muscular activity, injury to muscles and tendons and poor posture may each be held responsible.

(c) Muscular pain and stiffness arise from strain or injury to ligamentous or articular structures.

(d) Most cases of brachial neuralgia, lumbago and sciatica may result from degenerative changes in the intervertebral discs.

(e) In the acute stage the patient may be severely incapacitated by pain and stiffness.

(f) In the more chronic stage pain is felt most often after rest and improve with moderate activity.

(g) Muscular spasm may be marked and movement limited.

Treatment

(a) Rest in bed is essential, together with heat to the affected part and analgesics in ample amounts.

(b) At a later stage physiotherapy (heat, massage and graduated exercise) is essential.

(c) In cases with lesions of the intervertebral disc causing pressure, operation should be considered.

(d) Diet has no role in curative treatment but if the patient is obese, reduction in weight is the relief of symptoms.

Respiratory System

The main causes of disease in the respiratory system are : (1) infection, (2) inhaled irritants (3) allergy, (4) vascular accidents, (5) malignant disease.

1. **Infection :** Infection may be caused by a virus as in the case of influenza, and virus pneumonia, or by bacteria, as in bronchitis, bronchopneumonia, lobar pneumonia and tuberculosis.
2. **Irritants :** Exposure to tobacco smoke, industrial pollution or other inhaled irritants over many years leads to chronic bronchitis. When severe and prolonged and complicated by heart failure, this may lead to severe undernutrition and a state of cachexia. This is probably due to anorexia caused by chronic hypoxemia.
3. **Allergy :** Allergy plays an important role in bronchial asthma. Besides inhaled dust or pollen, ingested foods such as eggs, milk, wheat, etc. can be responsible for asthma. The detection of a food allergen and its exclusion from the diet may be an important factor in the treatment of asthma.
4. **Vascular accidents :** The vascular accident in the lung is embolism from a thrombosis formed usually in the veins of the lower limb or pelvis. If this dislodges, it travels through the right side of the heart and

obstructs one of the pulmonary arteries. It occurs especially in elderly people confined to bed, eg, after an operation.

5. Malignant disease : Malignant disease of the bronchial tree and lungs has greatly increased in recent years. Tobacco smoke and pollution of the atmosphere with smoke and fumes are of great importance. There is little evidence that nutritional factors play any part in the causation of bronchial carcinoma.

EXERCISE

1. Describe the important principles of balanced diet and the role of each component contained in it. (Mith. 62S)
2. Define balanced diet. Described briefly the basic principles followed for the preparation of balanced diet for a woman in late pregnancy. (Mith. 71S)
3. What factors are to be considered in planning a balanced diet ? Give the composition of such a diet for an average Indian man of sedentary habits. (M.U. 75A).
4. Describe the principle to be followed to prepare the balanced diet (i) for an active adult of 25 years age (ii) in late pregnancy. (Bh. U.75S)
5. What are the essential nutritive principles of human diet ? Describe briefly the role and effects of inadequate intake of protein in the diet. (R.U. 64A).
6. Discuss the role of protective substances in nutrition. (R.U. 68A).
7. Discuss the role of protein in nutrition. (Bh. U. 76A ; R.U. 73A).
8. Discuss the role of fat in human nutrition. (R. U. 69A).
9. Describe a diet for a normal adult in terms of total calorie, protein, fat and carbohydrate requirements. Give reasons which determine the proportion of these food substances. (Pun. 65A)

CHAPTER 34

TOXICANTS IN FOOD

Toxic compounds are naturally occurring in some plants and animal foods or foods may be contaminated by toxicants during storage. These toxicants cause severe illness in man and in some cases lead to death even. The toxicants are mainly divided into three classes : (A) Naturally occurring toxicants. (B) Toxicants from pathogenic microorganisms. (C) Contamination of food with toxic chemicals and pesticides.

A. Naturally occurring toxicants

Lathyrism

1. This is a crippling disease accompanied by paralysis of the leg muscles occurring mostly in adults who consume large quantities of the seeds of L-sativus or other lathyrus species for a long period.
2. The disease is found to occur in Bihar, Uttar Pradesh and Madhya Pradesh as well as Spain, Algeria, France and Italy.

Symptoms

1. First of all weakness in the lower limbs with spasticity of leg muscles. As a result, the movement of the ankle and knee joints are restricted and painful.
2. Flexion of the knee is prominent in the second stage and there is inversion of foot with a tendency to walk on toes.
3. In the third stage, the above symptoms become more prominent and the individual can walk only with the help of sticks.
4. In the fourth stage, the knee becomes completely flexed and walking becomes quite impossible. The thigh and leg muscles become atrophy.

Ackee fruit poisoning

1. The ackee fruit is cultivated in Nigeria. This fruit is consumed after boiling for 15 to 20 minutes.
2. The poisonous properties of the fruit are due to unusual amino acid, hypoglycin A, hypoglycin B. Both hypoglycin A and B have strong hypoglycemic action resulting in coma and death.
3. The signs and symptoms of this fruit poisoning in young children are found that the children sometimes vomit and they show drowsiness, convulsions or coma. These symptoms are due to severe hypoglycemia. They are treated by intravenous glucose for recovery.

Goitrogens

1. Many food stuffs contain organic compounds which have goitrogenic properties.

2. The active goitrogenic principle present in brassicae family in 1, 5-vinyl-2-thio-oxazolidone which is present in cabbage and turnip.
3. Certain oilseeds namely rapeseed, mustard etc. contain thioglycosides which act as goitrogens.
4. The red skin of groundnut contains phenolic glycosides which possess goitrogenic properties.

Pressure Amines

1. A number of amines namely histamine, tyramine, serotonin and norepinephrine found in some foods have profound physiological activity. Most of them are inactivated by mono-amino oxidase in the intestinal tract. The poisoning effect of pressure amines due to consumption of aged cheese has been reported in patients receiving mono-amino oxidase inhibiting drugs.
2. In Africa, the serotonin intake from plantains as a staple food may reach 100 to 200 mg per day. The endomyocardial fibrosis may occur as a result of large amount of serotonin.
3. The pressure amine foods are mainly plantains (green and ripe), juices of pineapple and tomato, banana, lemon etc.

Argemone seed oil poisoning

1. During the harvest of rapeseed, argemone seeds are mixed up with rapeseed which grow as weeds. The rapeseed oil obtained from a mixture of rapeseed and argemone seed causes epidemic dropsy in man.
2. The toxic substance in argemone seed is sanguinarine.

B. Toxicants from microorganisms

1. The harmful microorganisms contaminate the raw foods such as meat, fish, milk, etc. purchased from the market.
2. These microorganisms are destroyed during cooking or processing but some of the microorganisms survive due to insufficient heat.
3. Pathogenic fungus infects foodgrains and oilseeds when stored in humid atmosphere causing serious illness.

The pathogenic microorganisms responsible for contaminating foods are :

Salmonella : This group frequently causes food poisoning developing the symptoms like vomiting, diarrhoea, rise in body temperature. Infection is transmitted mainly to the processed animal foods.

Staphylococci : The contaminated food is poisoned owing to the presence of exotoxin produced by this organism. Precooked animal foods cause poisoning developing the symptoms such as vomiting, diarrhoea, and abdominal pain.

Streptococci : This group is the agent for sore throat producing the symptoms like vomiting, diarrhoea, and abdominal pain.

Shigella : Bacillary dysentery is caused by this group. Loose motions with mucus and blood and abdominal pain are the main symptoms. Flies are the carriers for the bacteria from infected material to food when cooked foods are kept exposed. The consumption of raw milk to which unclean water is added also causes the disease.

Fungal contamination

Fungi, eg, Aspergillus flavus, Penicillium islandicum, Fusariums, Claviceps purpurea (ergot) produce a good number of toxic compounds (mycotoxins).

Aspergillus flavus : This fungus has been found to be grown in cotton seed, cereals, moist groundnut, and soyabean. It produces a toxic substance named as aflatoxins which can develop cancer and cirrhosis of the liver in experimental animals. The aflatoxin poisoning has recently been occurred in Rajasthan and Gujarat owing to consumption of maize highly contaminated with Aspergillus flavus.

Penicillium islandicum : The yellow discolouration in rice has been reported in Japan by the contamination of rice by Penicillium islandicum during storage and develops toxic symptoms in man.

Claviceps purpurea (Ergot) : This parasitic fungus infects foodgrains such as rye and pear millet during cultivation. The disease 'Ergotism' occurs as a result of the consumption of the contaminated grain. The symptoms of this disease are nausea, vomiting, diarrhea, giddiness, severe burning sensation in the extremities, painful cramps in limbs, gangrene in the fingers and toes, depression, weakness and convulsions. The fungus produces the toxic alkaloids causing the disease.

Fusarium and cladosporium : The millet is infected by this fungus when left unharvested in the field during winter. The toxic compound formed by this fungus causes local inflammatory response, acute gastro-enteritis, nausea, and vomiting within 1 to 3 days after the consumption of the infected grain. The progessive pathological changes in the bone-marrow due to the toxin leads to leukopenia, agranulocytosis and anemia. There is also petechial hemorrhages followed by the development of necrotic ulcers of the skin. In less advanced cases, the recovery of the disease takes place, in case, the patient is put on a good diet with proper treatment.

Parasitic infection

The contaminated foods transmit some parasitic diseases, eg, amoebiasis, ascariasis and hookworm, when raw vegetables grown on sewage are consumed.

Contamination of foods with toxic chemicals, pesticides and insecticides

Lead, mercury, arsenic, antimony, D.D.T., and B.H.C., etc. can contaminate foods due to the following reasons :

1. The toxic chemicals such as barium carbonate, arsenic oxides, lead arsenate, etc. used as rat poisoning are accidentally mixed with food.
2. Accidental contamination of food with pesticides and insecticides.
3. Some toxic chemicals or minerals are also present in certain marine foods.

Table 34.1 : Toxic effects of some metals and chemicals

Metals and Chemicals	Foods	Toxic effect
Arsenic	Fruits sprayed by lead arsenate	Chills, Cramps, Paralysis
Barium	Foods contaminated by rat poison	Muscular twitching and convulsions
Cadmium	Fruits juices, soft drinks	Excessive salivation, liver and kidney damage, prostate cancer
Cobalt	Water, beer	Cardiac failure
Lead	Some processed foods	Paralysis, brain damage
Mercury	Mercury fungicide treated seed grains	Paralysis, brain damage
Tin	Canned foods	Colic, vomiting
Zinc	Foods stored in galvanised ironware	Dizziness, vomiting
Copper	Acid foods in contact with tarnished copperware	Vomiting, diarrhoea, abdominal pain
Pesticides	All types of food	Damage to liver, kidney, brain, and nerves leading to death
Diethyl stilbesterol	Present in meat of stilbesterol fed animals and birds	Carcinogenesis
Antibiotics	Meat from animals fed antibiotics	Drug resistance, hardening of arteries, heart disease

4. The presence of large amounts of certain food additives.

Toxic metals : The toxic element lead causes toxic symptoms after contamination with food, the pathological changes in the kidneys, liver, and arteries are brought about by lead. The common signs of lead poisoning are nausea, abdominal pain, anemia, muscular paralysis, and brain damage. The toxic effects of methyl mercury are neurological. When the brain is affected, the subject becomes blind, deaf, and paralysis of the various muscles makes him cripple. Cadmium, arsenic, antimony, cobalt, etc. are toxic in small doses.

Pesticides : The organic pesticides are D.D.T., B.H.C., and malathion, etc. and these are the toxic compounds. The presence of large amounts of pesticides causes toxic effects.

Additives : In U.S.A. and other western countries, the additives such as diethyl stilbesterol and antibiotics are added to animal and poultry feeds. These are present in the meat of animals fed on feeds containing these chemicals. **Stilbesterol** can cause leukemia and cancer even in small doses. Antibiotics can cause drug resistance and hardening of arteries.

EXERCISE

1. Describe how foods are toxicated. (P.U. 89A)

CHAPTER 35

BLOOD, LYMPH AND CEREBROSPINAL FLUID

Blood is a tissue consisting of different types of cells—the red blood cells (RBC), the white blood cells (WBC) and the platelets suspended in a liquid medium called plasma. It circulates in a closed system of blood vessels. The red colour of blood is due to hemoglobin present in the RBC.

Functions of blood

1. Blood transports oxygen from the lungs to the tissues and CO_2 from the tissues to the lungs.
2. It transports absorbed food materials to the tissues.
3. It transports metabolic waste products to the kidneys, lungs, skin and intestines for removal.
4. In association with the kidneys and lungs it maintains the acid-base equilibrium of the body by its efficient buffering action.
5. It maintains the steady osmotic pressure in the tissues and fluids of the body being assisted by the kidneys and the skin.
6. The plasma proteins assist in the exchange of water in the body from the tissue to blood and vice versa.
7. It maintains the body temperature at a constant level during its circulation.
8. It transports hormones from the site of production to different tissues.
9. By clotting, it protects body from hemorrhage.
10. The WBC form a defence against micro-organisms.
11. It transports metabolites from one tissue to another, eg, lactic acid formed in muscle is transported to liver and so on.
12. Other substances in blood combat toxic agents; they are antitoxins, agglutinins, precipitins.

Normal blood

Blood volume :

Males : 69 ml/kg body wt.

Females : 65 ml/kg body wt.

Plasma volume :

Males : 39 ml/kg body wt.

Females : 40 ml/kg body wt.

Red blood cells (RBC) : It is circular, non-nucleated bi-concave disc. The count of RBC in adult male is about 4.5-6.2 millions and in the adult female is 4.0-5.5 millions per cu. mm. The life span is about 120 days. The formation and destruction of RBC are going on continuously.

Packed cell volume and hematocrit ratio : The ratio is 45/55.

Hemoglobin : In normal males and females, the hemoglobin contents of blood are 14.5-16.5 gm and 12-14.5 gm per 100 ml of blood respectively. One gram of hemoglobin can carry 1.34 ml oxygen under optimal conditions.

Table 1 : Components of blood

I. Cellular fraction (Volume : 45 per cent) :

1. Red blood corpuscles (RBC) 5,000,000 per cu. mm.
2. White blood corpuscles (WBC) 6,000 per cu. mm.
3. Platelets 2,50,000 per cu. mm.

II. Plasma fraction (Volume : 55 per cent) :

A. Non-diffusible constituents :

1. Albumin. 2. Globulin. 3. Fibrinogen.
4. Enzymes, lipids, etc.

B. Diffusible constituents :

1. Hormones, vitamins, etc.
2. Anabolic constituents : Glucose, amino acids, creatine, etc.
3. Electrolytes : Na^+, K^+, Ca^{++}, Mg^{++}, Cl^-, HCO_3^-, HPO_4^-, etc.
4. Catabolic products : Urea, creatinine, uric acid, etc.

Chemistry

Table 2 : Constituents present in normal blood

Constituents	Values per 100 ml or dl
Hemoglobin :	
Men	14.5 - 16.5 gm
Women	12.0 - 14.5 gm
Plasma proteins :	
Total protein	6.3 - 7.9 gm
Albumin	3.5 - 5.3 gm
Globulin	1.8 - 3.6 gm
Fibrinogen	0.2 - 0.6 gm
Nutrients :	
Glucose (true) Fasting	70 - 115 mg
Reducing sugars Fasting	80 - 120 mg
Glucose (true) 2 hr after meals	115 - 160 mg
Free amino acid	3.0 - 5.5 mg
Lipids :	
Total lipids	550 - 800 mg
Cholesterol (total)	140 - 220 mg
Cholesterol esters	50 to 60 per cent of total cholesterol
Phospholipid	
Male	125 to 275 mg
Female	35 to 135 mg
Free fatty acids	8 - 25 mg
Minerals in serum :	
Calcium	8.4 - 10.2 mg
Chloride	98 - 110 mEq/L
Bicarbonate	24 - 28 mEq/L
Iron	65 - 170 μg
Iron binding capacity	250 - 450 μg
Iodine (total)	3 - 6.0 μg
Iodine protein bound	3.5 - 7.5 μg
Magnesium	1.7 - 2.6 mg
Phosphorus	
Adult	3 - 5.0 mg
Child	3.5 - 5.0 mg
Potassium	3.5 - 4.5 mEq/L
Sodium	136 - 146 mEq/L
Copper	75 - 160 μg
Vitamins :	
Vitamin A	0.15 - 0.60 μg
Ascorbic acid	0.4 - 1.5 mg
Folic acid free	0.6 - 2.0 μg
Vitamin B_{12}	200 - 850 μg
Waste products :	
Urea	15 - 45 mg
Uric acid	2.5 - 7.0 mg
Bilirubin	0.3 - 1.1 mg
Creatinine	0.6 - 1.2 mg
Enzymes :	
Serum amylase	25 - 125 U/L
Lactic dehydrogenase	85 - 190 I.U./L
Acid phosphatase	< 4 U/L
Alkaline phosphatase	
Adult	20 - 70 U/L
Child	20 - 150 U/L
Lipase	0 - 160 U/L
Transaminases :	
SGOT	8 - 40 Karmen units
SGPT	0 - 40 Karmen units
Lactic acid	4.5 - 19.8 mg
Non-protein nitrogen	12 - 30 mg
Pyruvic acid	0.6 - 1.0 mg

Properties of blood

Specific gravity : The specific gravity of normal blood usually lies between 1.056 and 1.06. The specific gravity of plasma lies between 1.024 and 1.038 and roughly proportional to the protein content.

Viscosity : The viscosity of blood is important in the sense that it determines the blood pressure. Human blood is 5 times as thick as water. The high viscosity is due to the cells, plasma has a very much lower viscosity. The viscosity of blood is affected by the change in the numbers or size of red cells or white cells. Abnormal values are found in leukemias, severe hemorrhage and pernicious anemia.

pH : Blood is slightly alkaline and its pH lies between 7.35 and 7.45. In the resting individuals, the arterial blood is very slightly more alkaline (about 0.02) than venous blood. This difference is increased by muscular exercise due to more formation of lactic acid. Under normal conditions the pH of blood of an individual is maintained in the region of 7.4. The pH of an individual is below 7.35 is considered in a condition of acidosis and over pH 7.45 is under the condition of alkalosis.

Osmotic pressure : Normally the osmotic pressure of blood is constant which is equal to 0.945% NaCl. The osmotic pressure of blood is slightly reduced on the ingestion of large amounts of water and increased on strenuous exercise.

Clotting of blood : Blood is clotted or coagulated within 5 or 10 minutes after shedding if it is left undisturbed. Different factors are involved in blood clotting which is discussed elaborately in physiology.

Composition : Sugar and urea are fairly equally distributed between plasma and corpuscles but others are almost entirely confined to one or the other, eg, Na and Ca in the plasma and K in corpuscles. Bromide, iodide, indican, enzymes, anti-enzymes, hormones, vitamins and various antibodies are circulating in blood. Plasma contains 8% to 9% solids, composed largely of proteins. The constituents of plasma are as follows :

Proteins	7 %
Lipids	0.7%
Inorganic substances	0.75%
Other organic substances	0.15%

In normal individuals, the plasma proteins vary from 6.0% to 8.5%. The proteins are as follows :

Serum albumin	4.5% of plasma
Serum globulin	2.2% of plasma
Fibrinogen	0.3% of plasma

Other proteins such as glycoproteins, lipo- proteins, enzymes and hormones are present in small amounts.

Plasma Proteins

Chemistry

1. The human plasma proteins are a mixture of simple proteins, glycoproteins, lipoproteins and other conjugated proteins.
2. These proteins are separated by salt precipitation, electrophoresis and immunoelectrophoresis.
3. The three major plasma proteins (albumin, globulin and fibrinogen) are separated by the use of varying concentrations of sodium or ammonium sulfates. These fractions are separated further by electrophoresis. These are as follows :

Albumin	52-56%	of total protein
Globulin	29.5-54%	of total protein
α_1-Globulin	2.5-5%	of total protein
α_2-Globulin	7-13%	of total protein
β-Globulin	8-14%	of total protein
γ-Globulin	12-22%	of total protein
Fibrinogen	6.5%	of total protein

A/G ratio 1.2 : 1

The electrophoretic representation of normal human plasma is shown on next page.

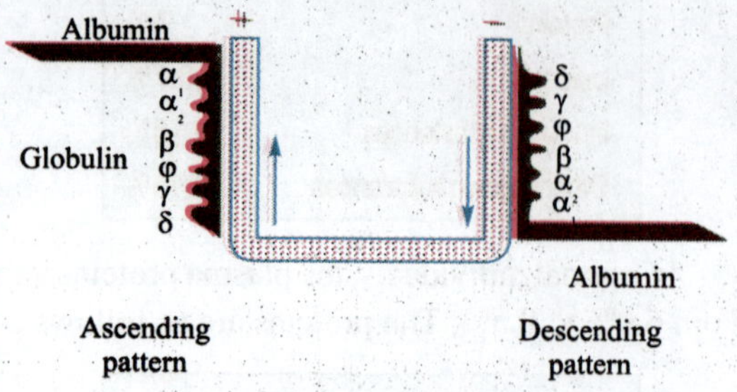

Fig. 35.1 : Electrophoresis of normal human plasma

4. The characteristics of individual plasma proteins are as follows :

I. Albumin

1. About half of the total plasma proteins is albumin.
2. It consists of 610 amino acids arranged in a single peptide chain.
3. It has a molecular weight of about 69,000 and is synthesized in the liver.
4. It is precipitated by full-saturation with ammonium sulphate.
5. The normal concentration of serum albumin is 6.3 - 7.9 gm/100 ml by precipitation method.
6. It exerts 80 per cent of the colloid osmotic pressure of plasma.
7. It plays an important role in the exchange of water between tissue fluid and blood.
8. Its concentration decreases in severe protein deficiency, liver diseases and nephritis leading to the development of edema.

II. Globulins

1. Globulins are separated by half-saturation with ammonium sulphate.
2. The molecular weight ranges from 90,000 to 1,300,0000.
3. Globulins are separated into α_1-, α_2-, β- and γ-globulin fractions. α- and β- globulins are synthesized in liver but γ- globulins are formed in the cells of reticulo-endothelial system.
4. The nature of different globulin fractions are discussed below :

α_1-Globulins : This fraction is of several complex proteins containing carbohydrates and lipids. These are orosomucoid, α_1- glycoprotein and α_1-lipoproteins. Its normal concentration in serum is 0.2-0.4 gm/100 ml.

α_2-Globulins : This fraction also includes complex proteins such as α_2-glycoprotein, plasminogen, prothrombin, haptoglobulins, ceruloplasmin and α_2-macroglobulin. The normal concentration of it in serum is 0.5-0.9 gm/100 ml.

β-Globulins : This fraction consists of lipid containing three different β-lipoproteins. It also contains siderophilin. Its normal concentration in serum is 0.6-1.2 gm/ 100 ml.

γ-Globulins : These are immunoglobulins having antibody activity. They are classified as IgG, IgA ad IgM on the basis of their electrophoretic mobility.

Mucoproteins and glycoproteins : In addition to globulins they contain carbohydrates such as galactose, mannose, fructose and hexosamine. They are found in α_1- and α_2-globulin fractions. Those which contain more than 4% hexosamine are designated as mucoproteins and those which contain less than 4% hexosamine are glycoproteins. Some glycoproteins have specific binding function for thyroxine and cortisol.

Lipoproteins : Two lipoproteins (α and β) are found in the α- and β-globulin fractions. Human serum lipoproteins are separated into different fractions. Fraction A contains the β lipoproteins with densities less than 1.063; fraction B contains the α_2-lipoproteins with densities of 1.063-1.107; and fraction C contains the α-lipoproteins with densities of 1.107-1.220. The lipoproteins are large molecules having molecular weights of 1,300,000. They are the major carriers of the lipids of the plasma.

Metalloproteins : The two important metalloproteins are transferrin (containing iron) and ceruloplasmin (containing copper). Transferrin transports iron and ceruloplasmin transports copper into different tissues. Transferrin is found in the β-globulin fraction while ceruloplasmin in α_2-globulin fraction. Metalloprotein concentration is reduced in pernicious anemia, chronic infections or liver diseases and its concentration is increased in iron deficiency or pregnancy.

III. Fibrinogen

1. Its molecular weight is between 3,50,000 and 4,50,000.
2. It is the precursor of fibrin (the substance of the blood clot).
3. Like globulin, it is precipitated by half-saturation with ammonium sulphate.
4. It normally constitutes 4-6% of the total proteins of the plasma.
5. It is formed in the liver and its concentration in blood falls rapidly in the excessive destruction of liver tissue.
6. It is a large asymmetrical molecule which is highly elongated having an axial ratio of about 20 : 1.

Functions

1. **Nutritive :** Albumin is largely involved in the nutritive functions of the plasma proteins owing to its high concentration. It is effective as a source of protein in hypoproteinemic patients.
2. **Water distribution :** The colloid osmotic pressure of plasma proteins plays an important role in the distribution of water between the blood and the tissues. Plasma albumin is responsible for this function due to its low molecular weight and quantitative dominance over other proteins. In kidney diseases where protein loss from the body is more, large amount of water moves to tissues producing edema.
3. **Buffering actions :** The serum proteins can combine with acids or bases to maintain the pH of blood. They act as acids and combine with cations (mainly sodium) at the normal pH of the blood.
4. **Transport :** The plasma proteins transport lipids and the fat-soluble vitamins (eg, A, D & E). Bilirubin is associated with albumin and also with fractions of the α-globulins. Thus, bilirubin is transported along with them. β_1-metal combining globulins (siderophilin) is responsible for the transport of iron in the plasma. Thyroxine is transported in association with an α-globulin (thyroxine-binding protein, TBP) and cortisol by a mucoprotein (transcorrin). Many drugs and dyes are transported in the plasma in combination with albumin. Half of the calcium of plasma is bound to protein for transport. Hemoglobin liberated intravascularly is carried to the reticuloendothelial system by complexing with the hepatoglobins.
5. **Viscosity :** Because of the presence of protein in plasma, it is a viscous fluid. The viscocity of blood provides resistance to flow of blood in the blood vessels to maintain blood pressure at normal level.
6. **Coagulation :** Plasma contains prothrombin, fibrinogen and other factors involved in caogulation of blood.
7. **Immunity :** γ-globulins are present in plasma and these γ-globulins protect body against bacterial infections.
8. **Enzymes :** Plasma contains several enzymes of diagnostic importances in some diseases.

Leukocytes

1. They contain proteins, nucleoproteins, fats, lecithin, cholesterol, purines, enzymes and inorganic salts.
2. They are highly supplied with proteases.

Blood platelets

1. They are unnucleated and so they contain no DNA, but RNA is present.
2. They contain glycogen, ADP, ATP, and the enzymes of glycolysis.

3. Large amounts of catecholamines, 5- hydroxytryptamine (serotonin), and histamine are present in them.
4. Phospholipids are also present in the form of lipoproteins which activates prothrombin in blood coagulation.
5. They also contain contractile protein (thrombosthenin) which is involved in the process of clot retraction.

Blood groups

Human blood is classified into 4 main groups and several sub-groups. Agglutinogens are neutral nitrogenous mucopolysaccharides with molecular weights ranging from 200,000 to 300,000. The four main groups are A, B, AB and O. The minor groups are M, N, P and Rh.

The plasma contains antibodies called *agglutinin*. The distribution of agglutinogens in RBC and agglutinins in plasma in the four groups are noted.

Blood group	Agglutinogen in RBC	Agglutinin in Plasma	Description of the group
A	A	β (anti-B)	Aβ
B	B	α (anti-A)	Bα
AB	A and B	Nil	AB
O	Nil	α + β (Anti-A + Anti-B)	Oαβ

To determine the blood group of an individual, an isotonic saline suspension of RBC is mixed with a test serum containing agglutinin α or agglutinin β on a slide. When no agglutination occurs, the cells separate and are evenly distributed. When agglutination occurs, the cells clump together. The results are interpreted as follows :

1. Blood of group A is agglutinated by plasma of group B containing α-agglutinin.
2. Blood of group B is agglutinated by plasma of group A containing β-agglutinin.
3. Blood of group AB is agglutinated by the plasma of blood group A and B containing agglutinins β and α, respectively.
4. Blood of group O is not agglutinated by the plasma of group A, B or AB. Therefore, persons of group O are called *universal donors*. Anybody can receive their blood.

Rh blood groups

It is an antigen of the Rhesus monkey and is present in the blood of 85% of white people and may be transmitted from father to child. The commonest Rh antigen is D and its antibody is anti D. If the mother is Rh negative, she develops antibodies to it and these antibodies pass through the placenta to the fetus and cause severe destruction of red blood cells in the new-born child.

Hemolysis

Red blood cells (RBC) are destroyed by hemolysis. Hemolysis can take place in the following ways :

1. Osmotically

(i) The RBC membrane is permeable to water. When the blood is placed in hypotonic solution, red cells swell owing to water passing in. If the solution is more hypotonic, the red cells swell up more and hemoglobin comes out of the red cell due to the rupture of its membrane.

(ii) If the cells are placed in a hypertonic solution, water passes out of the cells and the cells shrink due to reduction in volume. The process is called crenation.

2. By hemotoxins

(i) Snake venom and certain other bacteria can produce hemolysis in vivo. The hemolytic effect is due to the presence of the enzyme phospholipase A which hydrolyzes phospholipids and thus disrupts the cell membrane.

(ii) The W.R. (Wassermann reaction) for syphilis employs a hemolytic system, produced by the injection of foreign red cells into an animal such as a rabbit.

3. By certain drugs

Quinine, nitrites and chlorates cause hemolysis.

4. By lipid solvents

Alcohol, chloroform, ether, bile salts and saponin cause hemolysis due to their solvent action on the lipid of the plasma membrane.

5. By mechanical means

Grinding, vigorous stirring or shaking causes destruction of erythrocytes.

6. Other hemolytic agents

Heat, pH changes, ultraviolet light, etc. can cause hemolysis.

Stop of bleeding is said to be hemostasis. There are four phases of hemostasis :

(i) The first phase is the constriction of the injured vessel to reduce blood flow.

(ii) The second phase consists of formation of a loose platelet plug or white thrombus at the site of injury. Collagen is exposed at the site of injury and acts as a binding site for platelets.

(iii) The third phase is the formation of red thrombus (blood clot).

(iv) The fourth phase is the partial or complete dissolution of the clot.

Clots are of three types : The white thrombus is composed of platelets and fibrin and is poor in erythrocytes. It is formed at the site of injury, particularly in areas of rapid blood flow (arteries). The second type of thrombus is deposited in small vessels (capillaries). The third type is the red thrombus and consists of red cells and fibrin.

Initiation of clot formation to tissue injury is carried out by extrinsic pathway and the initiation of the pure red thrombus in an area of restricted blood flow without tissue injury is carried out by intrinsic pathway. The intrinsic and the extrinsic pathway merge into a final common pathway – the activation of prothrombin to thrombin and thrombin catalyzed fibrinogen to fibrin clot.

The blood clotting factors are mentioned below :

Factor	Name
I	Fibrinogen.
II	Prothrombin.
IV	Calcium.
V	Labile factor, proaccelerin, accelerator (AC)-globulin.
VII	Proconvertin, serum prothrombin conversion accelerator (SPCA), cothromboplastin, autoprothrombin 1.
VIII	Antihemophilic factor, antihemophilic globulin (AHG).
IX	Plasma thromboplastin component (PTC) (Christmas factor).
X	Stuart-Power factor.
XI	Plasma thromboplastin antecedent (PTA)
XII	Hageman factor
XIII	Laki-Lorand factor (LLF).

The Conversion of Fibrinogen to Fibrin by Thrombin

1. Fibrinogen is a soluble plasma glycoprotein whose length is 46 nm. Its molecular weight is 340,000 and it consists of 6 polypeptide chains synthesized in liver. The 6 chains are two Aα chains, two Bβ chains, and two γ chains making the structure Aα_2 Bβ_2 γ_2. All three genes (Aα, Bβ and γ) are genetically linked.

2. The ends of the fibre-shaped fibrinnogen molecule are highly negatively charged. These negatively charged termini of the fibrinogen molecules contribute to its water solubility and also repulse the termini of other fibrinogen molecule, for which aggregation is prevented.

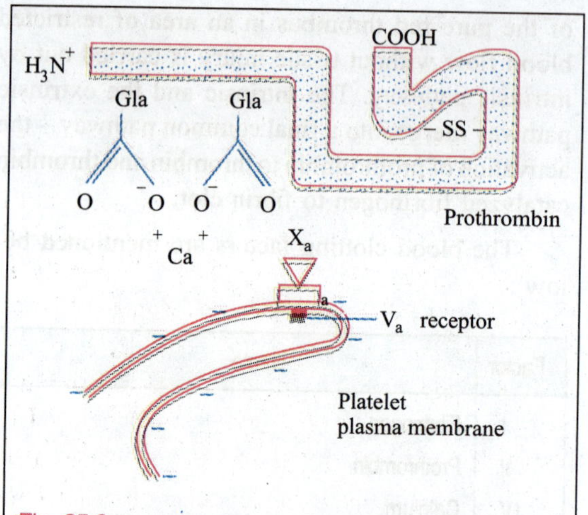

Fig. 35.2 : Diagrammatic representation of the binding of factors Va, Xa, Ca^{++}, and prothrombin to the platelet plasma membrane. [Gla = γ-carboxyglutamic acid.]

3. The molecular weight of **Thrombin** is 34,000 and it consists of two polypeptide chains and hydrolyzes four peptide bonds in fibrinogen. Removal of A and B portions of the fibrinogen molecule releases the negatively charged **fibrinopeptides** and generates the **fibrin** monomer. The long insoluble fibrin monomers form the insoluble **fibrin polymer** clot–which traps red cells, platelets, and other components to form the red thrombus or the white thrombus (platelet plug). The initial fibrin clot is weak and held together only by **fibrin monomers**.
4. **Thrombin** also converts factors XIII to active factor XIII (XIIIa) which is a **transglutaminase**. The **transglutaminase** covalently cross-links fibrin monomers. Persons with an inherited **deficiency of factor** XIII have a bleeding tendency due to unstable fibrin clot.
5. Prothrombin is a single chain glycoprotein with a molecular weight of 72,000. It is activated on the platelet and its activation requires platelet anionic phospholipid, Ca^{++}, factor Va, and factor Xa. The phospholipids bind Ca^{++} and prothrombin. The platelets also contain factor V and is activated as Va which binds to specific receptors in the platelet membrane. Factor Va also acts as a receptor for factor Xa which in turn binds prothrombin. Factor Xa is a serine protease and cleaves the catalytically inactive prothrombin and the amino portion of prothrombin is released.
6. Factor Va, generated by thrombin, is subsequently inactivated by thrombin providing a means of limiting the activation of prothrombin to thrombin.

Activation of factor Xa

1. Activation of factor Xa takes place at the conceptual site where the intrinsic and extrinsic pathways join to form the **final common pathway**.
2. Factor X is a zymogen having molecular weight of 55,000 of a serine protease and contains Gla residues.
3. The Gla residues of factor X are responsible for the calcium mediated binding of factor X to the acidic phospholipids of platelet membranes.

The extrinsic pathway for generating factor Xa

1. The extrinsic pathway is very rapid in response to tissue injury.
2. Factor VII is the precursor of factor VIIa and it is the Gla-containing glycoprotein synthesized in the liver. It can also be cleaved by thrombin or factor Xa. It is a zymogen and has high endogenous activity.
3. The tissue factor necessary to accelerate the attack of factor VII or VIIa on factor X is abundant in placenta, lung, and brain.
4. There is only 0.01 mg of factor X per ml of plasma. This requires the amplification provided by the clotting system, conversion of factor X to Xa is an autocatalytic process and, therefore, an amplification system.

The intrinsic pathway for generating factor Xa

1. The intrinsic pathway fo the generation of Xa starts with the exposure of prekallikrein, high molecular-weight kininogen, factor XII, and factor XI to an activating surface.
2. Factor XIIa is generated by kallikrein and attacks prekallikrein to generate more kallikrein, setting up a reciprocal activation. It releases bradykinin from high molecular-weight kininogen and activated factor XI to XIa.
3. Factor IX, a Gla-containing zymogen, is activated in a two step reaction by factor XIa.
4. Factor IXa slowly activates factor X by clearing the same Arg-IIe bond that factor VIIa of the extrinsic system hydrolyzes in the presence of calcium and acid phospholipids. The factor IXa-catalyzed activation of factor X is accelerated about 500-fold by the presence of factor VIII or VIIIa.
5. Factor VIII requires activation by minute quantities of thrombin to form factor VIIIa.
6. The intrinsic pathway is slow becaue it involves many factors operating in a cascade mechanism to generate factor Xa.

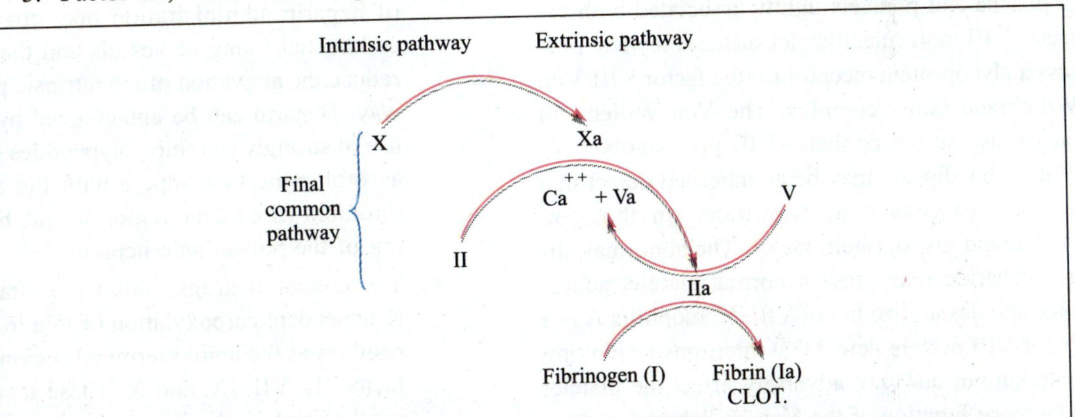

Fig. 35.3 : Relationship between the intrinsic, extrinsic and final common pathways of blood clotting

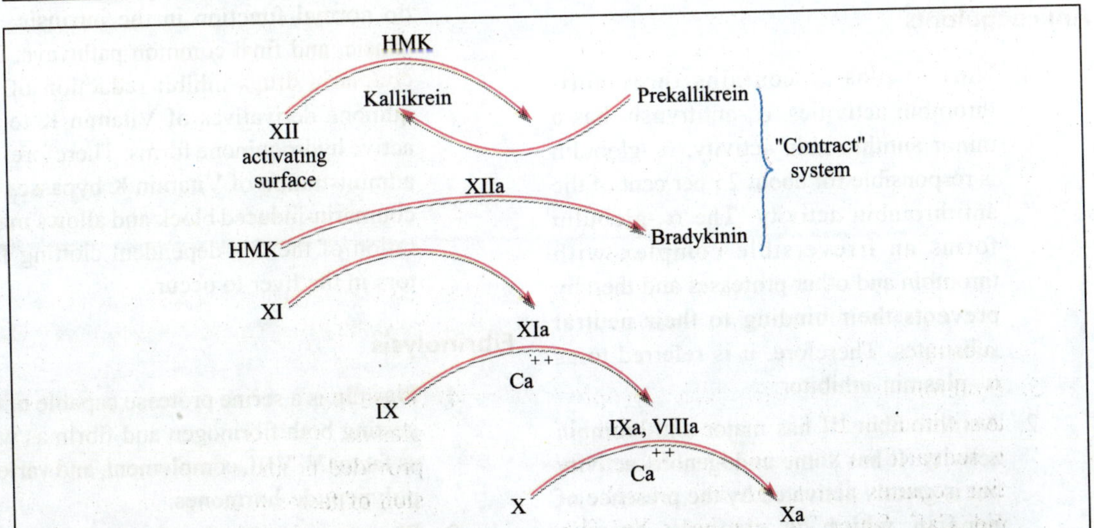

Fig. 35.4 : The intrinsic pathway for activating factor X to Xa. HMK is high molecular weight Kininogen

Hemorrhagic disorders and their abnormalities

Hemophilia A : The common deficiency of factor VIII produces a disease known as hemophilia A. The X chromosome-linked deficiency of factor VIII has played a major role in the history of the royal families of Europe.

Von Willebrand's disease : This disease is caused by the defect in platelet adherence and a deficiency of factor VIII clotting activity (an antigenic material). The platelet adherence factor (Von Willebrand factor) is a large glycoprotein synthesized in vascular endothelial cells and megakaryocytes (platelet precursor cells). It exists in plasma and platelets tightly associated with the factor VIII molecule. Platelet surfaces seem to possess a glycoprotein receptor for the factor VIII/Von Willebrand factor complex. The Von Willebrand factor also stabilizes factor VIII procoagulant activity. This disease may be an inherited defect in a specific oligosaccharide moiety on the Von Willebrand glycoprotein factor. The abnormal oligosaccharide may prevent normal platelet adherence and destabilize factor VIII. Hemophilia A is a factor VIII protein defect that interrupts its clotting function but does not adversely affect the platelet adherence function of the Von Willebrand factor.

Anticoagulants

1. Normal plasma contains three antithrombin activities. α_1-antitrypsin has a minor antithrombin activity. α_2-globulin is responsible for about 25 per cent of the antithrombin activity. The α_2-globulin forms an irreversible complex with thrombin and other proteases and thereby prevents their binding to their neutral substrates. Therefore, it is referred to as α_2-plasmin inhibitor.
2. Antithrombin III has major antithrombin activity. It has some endogenous activity but is greatly activated by the presence of heparin which is strongly anionic proteoglycan. Heparin binds to a specific cationic site of antithrombin III inducing a conformational change that accelerates the binding of antithrombin III to all serine proteases including trypsin, chymotrypsin, and plasmin. Antithrombin III inhibits the activity of thrombin, IXa, Xa, XIa, and XIIa. Antithrombin III deficiencies develop frequent and severe widespread clots.
3. Heparin is frequently used in clinical medicine to inhibit clotting. The high-dose heparin therapy may cause the prothrombin time prolonged but usually is not prolonged by continuous intravenous administration of heparin. The low-doses of heparin administration may coat the endothelial lining of vessels and thereby reduce the activation of the intrinsic pathway. Heparin can be antagonized by the use of strongly cationic polypeptides such as protamine to compete with the antithrombin III cationic region for the binding of the polyanionic heparin.
4. The coumarin drugs inhibit the vitamin K-dependent carboxylation of Glu to Gla residues at the amino-terminal regions of factors II, VII, IX, and X. These factors, synthesized in the liver, are dependent upon the Gla residues for maturation, and do normal function in the intrinsic, extrinsic, and final common pathways. The coumarin drugs inhibit reduction of the quinone derivatives of Vitamin K to the active hydroquinone forms. Therefore, the administration of Vitamin K bypasses the coumarin-induced block and allows maturation of the Gla-dependent clotting factors in the liver to occur.

Fibrinolysis

1. Plasmin is a serine protease capable of digesting both fibrinogen and fibrin as well as factor V, VIII, complement, and various polypeptide hormones.
2. Plasmin normally exists in plasma in the inactive form plasminogen. Upon forma-

tion of fibrin, plasminogen activator cleaves plasminogen to generate plasmin. When plasmin digests fibrin, the plasminogen activator no longer remains active and proteolysis stops providing a well-regulated fibrinolytic process.

3. **Urokinase**, the proteolytic enzyme of urine, is also a serine protease and can cleave plasminogen at two sites activating the protease activity of plasmin.
4. **Plasminogen** normally coprecipitates with fibrin and is incorporated into fibrin deposits. When it is activated, it digests the fibrin in clots to soluble fragments dissolving the clot. Cross-linked fibrin clots are less sensititve to dissolution by plasmin.
5. The plasminogen activators increase in cancers and shock. In cirrhosis of the liver, the antiplasmin activities are also impaired. Some bacterial products can activate plasminogen without cleavage and are responsible for the diffuse hemorrhage sometimes observed in patients with disseminated bacterial infections.

Clotting Tests

1. **Prothrombin time :** This is a test of the extrinsic pathway and the final common pathway of clotting. It is performed by adding a tissue factor to the unknown plasma and determining the time necessary for the formation of visible clot. The tissue factor along with factor VII activates factor X to Xa in the presence of factor Va, Ca^{++}, and platelet phospholipids. The complex converts prothrombin to thrombin which in turn catalyzes the formation of fibrin from fibrinogen. The deficiency of factor II, V, VII, or X or a severe deficiency of fibrinogen prolongs the prothrombin time.
2. **The clotting time :** This is determined by introducing freshly drawn whole blood into small glass tubes. Care must be taken that no tissue factor is introduced from the site of venipuncture. The blood is gently agitated at regular intervals to determine the time required for coagulation. The intrinsic pathway is activated by exposure of whole blood to the glass surface of the tube. The clotting time is dependent upon all clotting factors except factors VII and XIII. This test is not sensitive to mild deficiencies of factors VIII, IX or XI.
3. **The activated partial thromboplastin time :** This involves the addition of acid phospholipids and kaolin to plasma to make certain that platelets are not rate-limiting and that factors XI and XII are fully activated. The normal response time is 35-45 seconds. This is more useful than the clotting time determination.
4. **The thromboplastin generation test :** This test activates the intrinsic system to form a prothrombinase or "plasma thromboplastin". The prothrombinase does not itself clot the fibrinogen but it is capable of activating prothrombin to thrombin which in turn forms fibrin clot from fibrinogen. Factor VII is not involved in this test. This test involves mixing the patient's plasma after adsorption (containing factors I, V, VIII, XI, XII) with the patient's serum (containing factors VII, IX, X, XI, XII) and normal platelets and Ca^{++}. If no clot forms and the addition of normal serum or normal plasma corrects the defect, there is sure of the missing factor or factors. If plasma alone and serum alone both correct the defect, factor XI or XII must be missing from both the patient's serum and plasma. If only normal serum corrects the defect, factor IX or X must be defective in the patient's serum. If only normal plasma corrects the defect, factor V or VIII must be defective in the patient's plasma. Antisera specific to the various factors are also available for determining which specific component of the clotting system is missing or defective.

Lymph

1. The term "lymph" denotes a fluid not only present in the lymphatic vessels but also the fluid which bathes the cells, the tissue or interstitial fluid.
2. It resembles plasma in its content of substances which can permeate the capillary wall but there are some differences in electrolyte concentrations.
3. About 2 litres of lymph are drained into the bloodstream per day. Lymph flow is very slow.
4. The protein content of lymph varies widely. The fluid from the leg contains 2-3% protein; whereas that from the intestines contain 4-6% and that from the liver 6-8%.
5. The lacteals absorb the majority of fat from the intestine.
6. The ratio between albumin and globulin in lymph is the same as in plasma.
7. The protein present in entire plasma in an adult human is about 210 grams. About one-third of this leaks out into the lymph through the interstitial fluid and is returned to the blood at the thoracic duct. This process goes on continuously. If the thoracic lymph is drained off, the concentration of plasma proteins falls and the blood volume decreases.

Cerebrospinal Fluid

Chemistry

1. The cerebrospinal fluid is formed by the ultrafiltration of the plasma by the choroid plexuses of the brain.
2. The normal fluid is water-clear with a specific gravity of 1.003-1.008 and pH 7.34-7.43.
3. Normally, the protein content is low, about 20-45 mg/100 ml with an albumin-globulin ratio of 3 : 1.
4. It contains only a small number (less than 5) of lymphocytes per cu. mm.
5. The total volume of C.S.F. in an adult human is about 130 ml and this is renewed about 6 to 8 times a day.
6. The glucose in C.S.F. is somewhat less than in blood, 40-70 mg/100 ml in the fasting adult.
7. The chloride concentration is normally 118-132 mEq/L of the fluid (expresed as NaCl).
8. The concentration of calcium in normal human C.S.F. is 2.43 ± 0.05 mEq/L; that of magnesium 2.40 ± 0.14 mEq/L. The ratio of calcium to magnesium is 1.01 ± 0.06.

Functions

1. The C.S.F. acts as a buffer.
2. It acts as a reservoir to regulate the contents of the cranium. If the quantity of blood supply to brain increases, the C.S.F. drains away.
3. It serves to a limited extent as a medium for nutrient exchange in the nervous system, even though the brain derives its nutrients mainly from blood.

CLINICAL ASPECTS

- In the inflammatory lesions of the central nervous system, the protein content of the fluid is increased as high as 125 mg to over lgm/100 ml of the fluid.
- In uremia, pneumonia and typhoid fever, the protein content is raised in the fluid. In neurosyphilis, encephalitis, abscess and tumor of the brain, the concentration of protein in C.S.F. is elevated above normal of 20-300 mg/100 ml.
- In the various forms of suppurative meningitis the total protein content of C.S.F. is high (125 to 300 mg per 100 ml) consisting of globulin, albumin, small amounts of fibrinogen and many other proteins. High values are also found in tuberculous meningitis (200 to 2,000 mg) and in acute luetic meningitis (180-540 mg). In some cases of early meningitis the fluid protein may be normal.
- In leukemic and carcinomatous infiltration of the meningitis, the sugar content can be very low and the protein level is high.

CLINICAL ASPECTS

- The protein content of C.S.F. is increased in myxedema. Values as high as 100-200 mg/100 ml are observed.
- The sugar content of C.S.F. is raised in encephalitis, central nervous system, syphilis, abscesses, and tumors. It is decreased in purulent meningitis.
- In most forms of meningitis, the chloride is decreased, the reduction is most marked in tuberculous meningitis. The chloride content is unchanged is syphilis, encephalitis, poliomyelitis and other diseases of the central nervous system.
- Increased level of calcium is observed in all cases of meningitis and epidemic encephalitis.
- Aminotransferases, lactate dehydrogenase and glucose phosphate isomerase also occur in C.S.F. Isomerase activity is increased in malignant brain tumors. Elevations also occur in meningitis and cerebral thrombosis. Lactate dehydrogenase and aspartate aminotransferase are also increased in cerebral vascular accidents. In convulsive disorders, increased levels of creatine kinase, aspartate aminotransferase and lactate dehydrogenase are found in C.S.F.
- In many diseases, the amounts of various plasma proteins are changed which can be identified by electrophoresis.
- The activities of certain enzymes of plasma are altered signifying diagnostic importance in a number of pathologic conditions.
- Hemorrhagic and thrombotic states pose terrible medical emergencies.
- The major causes of death are the thrombosis in the coronary and cerebral arteries.
- Anemias are caused by the reduction of the number of RBCs and of their content of hemoglobin.

CLINICAL ASPECTS

- Some of the blood group systems present on the membranes of RBCs and other blood cells are extremely important in relation to blood transfusion and tissue transplantation.
- Neutrophils play a central role in acute inflammation and other WBCs, such as lymphocytes, play important roles in chronic inflammation.
- Leukemias affect precursor cells of any of the major classes of WBCs; common types are acute and chronic myelocytic affecting precursors of the neutrophils and acute and chronic lymphocytic leukemias.
- The increased levels of glucose in C.S.F. are found in encephalitis, poliomyelitis, cerebral abscess.
- The decreased levels of glucose in C.S.F. are also found in coccal meningitis, tuberculous meningitis, viral meningitis. The chloride levels are also decreased in these meningitis.
- C.S.F. protein levels are increased in uremia, multiple myelomatosis, multiple sclerosis, polyneuritis, acute meningitis.

EXERCISE

1. Discuss the chemistry, compostion and functions of blood. (Punb. 61S)
2. Discribe the chemistry and function of plasma proteins. (Mith. 64A; P.U. 62S)
3. Write short notes on :
   (a) Blood groups. (R.U. 66a)
   (b) Hemolysis. (M.U. 73S)
   (c) Lymph. (Bhag. U. 74A)
   (d) C.S.F. (P.U. 68A)

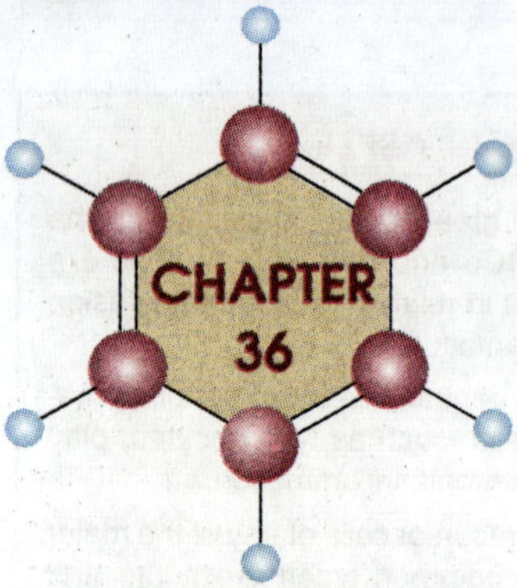

EXCRETION

The Kidneys

A large number of waste products are produced in the body as a result of metabolic activities. The main waste products are carbon dioxide, water, and nitrogenous compounds. The retention of these products produces a harmful effect on the normal health. Therefore, the removal of these products from the body is a must. Carbon dioxide is removed mainly through lungs and water as well as nitrogenous compounds are removed through urogenital system. The kidneys are the most important component of this system.

The kidneys are two in number, usually bean shaped, and exist behind the peritoneum on either side of the vertebral column extending from the 12th thoracic to the 3rd lumbar vertebra. Each kidney weighs about 120-170 grams and is about 11-13 cms. long, the left being larger than the right one.

Each kidney is found to consist of two main parts by section. The outer part is called cortex and the inner one is medulla. The cortex consists of a large number of glomeruli and convoluted tubules. The medulla is composed of renal tubules projecting into a cavity towards the inner region of the kidney called the *pelvis*, the region where the renal artery and vein enters and leaves the kidney respectively.

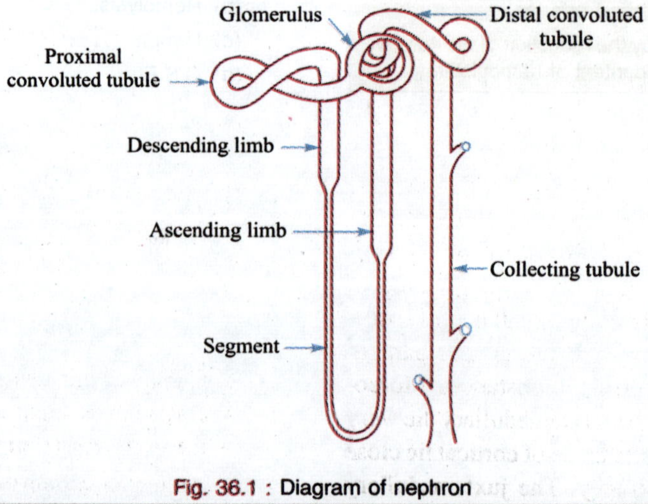

Fig. 36.1 : Diagram of nephron

Nephron : It is a functional basic unit of kidney. Each kidney is provided with about one million nephrons containing the glomerulus and the tubule. The glomerulus is a network of afferent and efferent capillaries. Each glomerulus is surrounded by a double-walled epithelial sac known as *Bowman's Capsule* which leads to the tubule which is divided into three parts—proximal convoluted tubule, loop of Henle, and the the distal convoluted tubule.

The Proximal Convoluted Tubule (PCT) is about 45 mm long and 50 mm in diameter. This lies in the cortex along with glomerulus. Its lumen is continuous with that of the Bowman's Capsule. It consists of cells with scalloped outline and brush border. The brush border is formed by numerous microvilli which increases the surface enormously for absorption.

The loop of Henle consists of three parts—the descending limb, a thin segment, and an ascending limb. The proximal convoluted tubule opens into the descending limb which is continued into the thin segment from where the ascending limb arises. The whole loop of Henle is lined by a single layer of flattened epithelial cells.

The ascending limb of the loop of Henle continues into the distal convoluted tubule (DCT) which finally opens into a collecting tubule or duct which carries the urine to the renal pelvis from where it is carried to the bladder by the ureter. The distal convoluted tubule commences near the pole of the glomerulus and establishes a close proximity to the afferent arteriole of its parent glomerulus. The DCT contains cuboidal epithelium.

Nephrons are mainly of two types—cortical and juxtamedullary. The loop of Henle of the juxtamedullary is long and dips deep into the substance of the medulla. But the loop of Henle of cortical is short and only a very small part of it dips into the medullary tissue and the greater part remains embedded in the cortical substances. Moreover, the glomeruli of the juxtamedullary lie very close to the medulla while those of cortical lie close to the surface of the kidney. The juxtamedullary nephrons constitute 20 per cent of nephrons, while the cortical nephrons constitute 80 per cent of the total nephrons. These two types of nephrons have the same common function.

Blood supply of the kidneys : The short renal artery arising from the abdominal aorta supplies the blood to the kidney. The renal artery after entering the kidney divides into a number of arterioles—the afferent arterioles which further branch into capillaries and enter into each glomerulus. The capillaries then join to form another arteriole—the efferent arteriole which opens into another set of capillaries called *peritubular capillaries* surrounding the proximal tubule, the loop of Henle, and the distal tubule. Ultimately, the capillary set opens into a venule which joins with other venules to form the renal vein. The renal vein then opens into the inferior vena cava.

Blood flow through the nephron : The blood flows through both the kidneys of an adult weighing 70 kg at the rate of about 1200 ml/mt. The portion of the total cardiac output (about 560 ml/mt) which passes through the kidneys is called the **renal fraction**. This is about 560/1200 ml per minute, i.e., about 21 per cent.

There are two sets of capillaries—the glomerulus and the peritubular. These two capillaries are separated from each other by the efferent arteriole which contributes sufficient resistance to blood flow. The glomerular capillary bed provides a high pressure of about 70 mm Hg, while the peritubular bed provides a low pressure about 13 mm Hg. The pressures in the artery and vein are 100 mm of Hg. and 8 mm of Hg respectively. The high pressure in the glomerulus exerts the filtering of fluids continually into the Bowman's Capsule. The low pressure in the peritubular capillary system, on the other hand, functions in the same way as the usual venous ends of the tissue capillaries with the fluid being absorbed continually into the capillaries.

Functions of kidney

1. Kidney eliminates excess of certain nutrients such as sugar and amino acids when their concentration increases in the blood.
2. It removes certain non-volatile waste products such as urea, uric acid, creatinine, and sulphates, etc. from the body.

3. It eliminates certain foreign or toxic substances such as iodides, pigments, drugs, and bacteria, etc. from the blood.
4. It regulates hydrogen ion concentration of the blood by removing excess of non-volatile acids and bases.
5. It maintains the osmotic pressure of the blood by regulating the excretion of water and inorganic salts and thus preserves the constant volume of the circulating blood.
6. It regulates the arterial blood pressure by causing the secretion of the hormone **renin**.
7. It maintains the erythrocyte production by excreting the secretion of the hormone **erythropoietin**.

Urine formation

The regulatory activities of kidneys form urine as a by-product. Urine formation involves three main steps—the glomerular filtration, the tubular reabsorption, and the tubular secretion.

1. **Glomerular filtration (Ultrafiltration) :** Glomerulus filters out substances of low molecular weight from the blood with the retention of substances of high molecular weight, especially the proteins. Therefore, proteins are retained in the glomeruli and are not normally found in urine. If protein is detected in the urine, it indicates the kidney damage or other disease which effect the glomerular membrane. In normal adult, two million nephrons filter one litre of blood each minute to give about 1200 ml of glomerular filtrate (primary urine) at Bowman's Capsule. Therefore, the Glomerular Filtration Rate (GFR) in adult is about 120 ml per minute.

   The hydrostatic pressure of the blood in the glomerular capillaires (P_g) is the main force for driving the fluid (water and solute) out of the glomerulus. The pressure is opposed by two forces :

   (i) The hydrostatic pressure of the Bowman's Capsule fluid (P_{BC}).

   (ii) The osmotic pressure of the plasma proteins (P_{pp}).

   Therefore, the effective filtration pressure (P_{ef}) is calculated by the following relation :

$$P_{ef} = P_g - (P_{PP} + P_{BC})$$

$$\therefore \; P_{ef} = 74 - (30 + 20) \text{ mm of Hg}$$

$$\therefore \; P_{ef} = 24 \text{ mm of Hg.}$$

   Thus, by substituting the normal values of the various forces, it has been found that the calculated effective (net) filtration pressure (P_{ef}) is 24 mm Hg.

   A fall in blood pressure may reduce the P_{ef} which results in less amount of urine. When the aortic systolic pressure is reduced to 70 mm Hg, the hydrostatic pressure of the blood in glomerular capillaries is reduced to 50 mm. Hg. This reduces the P_{ef} to Zero [50 – 50] and thus filtration will be ceased. Under such circumstances, urine will not be formed (anuria) until the blood pressure is maintained.

2. **Tubular reabsorption :** The rate of formation of the primary urine is 120 ml/minute, while the rate of urine passing to the bladder under the same condition is 1-2 ml/minute. Therefore, it indicates that about 99 per cent of the glomerular filtrate is reabsorbed during its passage through the different segments of the renal tubule. Although, the glomerular filtrate contains nearly the same concentration of glucose as in plasma, the urine contains nil or very little glucose. Hence, glucose is also practically completely reabsorbed in the tubules when the blood sugar level is normal. The capacity of reabsorption depends on the renal threshold of that substance.

   The reabsorption of different solids takes place at different sites in the renal tubules. Amino acids, glucose, and small amounts of protein that pass through the glomerulus are reabsorbed in the first part of the proximal tubule. Sodium, chloride, and bicarbonate are reabsorbed uniformly along the entire length of the proximal tubule and also in the distal tubule. Potassium is reabsorbed in the proximal and secreted in the distal tubule.

The glomerular filtrate produces about 170 litres in a day; whereas the tubules reabsorb about 168.5 litres of water, 170 gm of glucose, 100 gm of NaCl, 360 gm of $NaHCO_3$, and small amounts of phosphate, sulphate, amino acids, urea, uric acid, etc. and excrete about 60 gm of NaCl, urea and other waste products in about 1.5 litres of urine. Most of these solids are reabsorbed by active transport mechanism, while some (eg, urea) are reabsorbed by passive transport mechanism.

In diseases, the reabsorption mechanism is altered developing glycosuria, phosphaturia, and amino aciduria.

3. **Tubular secretion** : Although, most of the substances are reabsorbed by the tubular cells, some substances are actively transported or actively excreted into the tubular lumen. The secreted substance by the tubular epithelium in man are **creatinine** and **potassium**. The tubular epithelium also removes a number of foreign substances that are introduced into the body for therapeutic and diagnostic purposes. These foreign substances are penicillin, p-Aminosalicylic acid, phenosulphonphthalein (PSP), p-Aminohippuric acid, and diodrast. The hydrogen ions and ammonia formed in the distal tubular cells are also actively excreted into tubular lumen and thus pass to urine.

   Hormonal regulation : The function of kidney is regulated by three important hormones. These hormones are aldosterone (from adrenal cortex), parathormone (from parathyroid), and vasopressin (from hypophyseal posterior lobe).

   Aldosterone restricts the excretion of Na^+ and stimulates the excretion of K^+. **Parathormone** stimulates excretion of phosphate. **Vasopressin**, the antidiuretic hormone, is held responsible mainly for the reabsorption of water. In the absence of this hormone, a large amount of very dilute urine is excreted.

Composition of urine

A. Volume

1. A normal adult excretes daily from 1000 ml to 1800 ml of urine. The average is 1500 ml containing 60 gm of solids.
2. The quantity depends on the water intake, external temperature, the diet and the individual's mental and physical condition.
3. A high protein diet increases excretion because the urea formed as a result of catabolism of protein has a diuretic action.
4. The diuretic action of tea, coffee and coca is due to caffeine.
5. The decreased volume of urine in hot weather is due to an increased loss of water by perspiration.
6. Nervousness or excitement causes increased urinary volume.
7. Increased urine volumes are observed in diabetes insipidus, diabetes mellitus and certain types of kidney diseases and decreased volumes are found in acute nephritis, fevers, diseases of the heart, diarrhea and vomiting.

B. Specific Gravity

1. The specific gravity of urine in 24 hours lies between 1.003 and 1.030 and varies according to concentration of solutes in the urine. The figures in the second and third decimal places, multiplied by 2.66 (Long's coefficient) give roughly the total solids in the urine in gm/L. 50 gm of solids are the average normal for the day.
2. The specific gravity of the urine varies with the food, water intake, and the activity of the individual.

3. In chronic interstitial nephritis, the specific gravity is lowered.
4. The specific gravity is increased in the excretion of abnormal substances such as albumin or glucose (eg, diabetes mellitus).

C. Colour

1. Normal urine is pale yellow or amber. The colour is roughly proportional to the specific gravity. Very dilute urine is colourless.
2. **Terochrome**, composed of a polypeptide and urobilin, is the chief pigment of urine. Traces of **coproporphyrin**, **urobilinogen** and **uroerythrin** are also found in urine.
3. Reddish urine is due to the ingestion of naturally coloured foods (eg, beetroot, blackberries). In fever, the urine may be dark yellow or brownish because of concentration. In liver disease, the urine may be green, brown, or deep yellow due to bile pigments. Blood or hemoglobin develops smoky to red colour. The urine is dark brown due to **methemoglobin** and **homogentisic acid**. Methylene blue gives the urine a green appearance.
4. The urine is transparent. A turbidity is developed in alkaline urine by precipitation of calcium phosphate. Strongly acid urine is pink due to the precipitation of uric acid salts.

D. Odour

1. Fresh urine is normally aromatic.
2. The odour is modified by the ingestion of certain foods or drugs. This is noticed after eating asparagus; the odour is due to methyl mercaptan.
3. The ammoniacal smell of urine is due to the action of bacteria on urea.
4. In ketosis, the odour of excreted acetone is detected.

E. pH

1. The mixed sample of normal urine in 24 hours has a pH 6.0. Individual samples vary from 4.6 to 8.0.
2. The urine is acid in high protein intake because excess phosphate and sulfate are formed in the catabolism of protein. Acidity is also increased in acidosis and in fever.
3. The urine becomes alkaline on standing due to the conversion of urea to ammonia and loss of CO_2 to air. It may be alkaline in alkalosis such as after excessive vomiting and after meals due to H^+ secretion in the stomach (the "**alkaline tide**").
4. The acidity of urine is increased after strenuous muscular exercise (elimination of lactic acid), by ingestion of ammonium salts of strong acids. An alkaline urine may be produced by ingestion of sufficient $NaHCO_3$. Ammonium carbonate does not produce an alkaline urine because ammonia is rapidly converted into urea.

Normal constituents of urine

A. Urea : Nitrogenous constituents

1. Urea is the main end product of catabolism of protein in mammals. Its excretion is directly proportional to the protein intake. It consists of 80-90% of the total urinary nitrogen.
2. In fever, diabetes, or excess adrenocortical activity, urea excretion is increased due to increased protein catabolism.
3. Decreased urea excretion is due to decreased urea production in the last stages of fatal liver disease.
4. In acidosis, there is decreased urea excretion.

B. Ammonia

1. Ammonia is formed by the kidney from glutamine or amino acids in acidosis.
2. There is a high ammonia output in the urine in uncontrolled diabetes mellitus in which renal function is unimpaired.

C. Creatinine and creatine

1. Creatine is excreted by children and pregnant women and much smaller amounts in men. The excretion in men is 6% of the total excretion of creatinine.
2. Creatinine is formed from creatine. It is excreted in relatively constant amounts regardless of diet.
3. The creatinine coefficient is the ratio between the amount of creatinine excreted in 24 hours and the body weight in kg. It is usually 20-26 mg/kg/day in normal men and 14-22 mg/kg/day in normal women.
4. Creatinine excretion is decreased in many pathological conditions.
5. Creatine excretion is also found in pathological states such as starvation, hyperthyroidism, impaired carbohydrate metabolism and infections.
6. Creatine excretion is decreased in hypothyroidism.

D. Uric acid

1. It is the end product of the oxidation of purines in the body. It is not only formed from dietary nucleoprotein but also from the breakdown of cellular nucleoprotein in the body.
2. It is slightly soluble in water and precipitates readily from acid urine on standing.
3. Uric acid excretion is increased in leukemia, severe liver disease and various stages of gout.
4. The concentrated urine on cooling forms a brick-red deposit which is mainly acid urate.
5. Pure uric acid is colourless. Deposits of uric acid and urates are coloured by absorbed urinary pigments, particularly the red uroerythrin.
6. The specificity of the analysis of uric acid is increased by treatment with uricase, the enzyme (from hog kidney) which converts uric acid to allantoin.

E. Amino acids

1. About 150-200 mg of amino acid nitrogen is excreted in the urine of adults in 24 hours.
2. The infant at birth excretes about 3 mg amino acid nitrogen per pound of body weight, and up to the age of 6 months the value reaches to 1 mg/pound which is maintained throughout childhood. Premature infants excrete 10 times amino acid nitrogen than that of full-term infant.
3. The low excretion of amino acid nitrogen is due to its high renal threshold value.
4. Increased amounts of amino acids are excreted in liver disease and in certain types of poisoning.
5. In cystinuria, 4 amino acids-arginine, cystine, lysine and ornithine are excreted in urine.

F. Allantoin

1. It is the partial oxidative products of uric acid. Small quantities of the allantoin are excreted in human urine.
2. In other subprimate mammals, allantoin, the principal end product of purine metabolism, is excreted.

G. Sulphates

The urine sulphur is derived from sulphur containing amino acids such as methionine and cystine and therefore, its output varies with protein intake. The urine sulphur exists in 3 forms :

I. Inorganic (sulfate) sulfur

1. This is the completely oxidized sulfur precipitated from urine.
2. It is proportionate to the ingested protein with a ratio of 5 : 1 between urine nitrogen and inorganic sulfate.

II. Ethereal sulfur (conjugated sulfates)

1. It is about 10% of the total excreted sulfur.

This includes the organic combination of sulfur excreted in the urine.

2. It consists of the sulphuric esters of certain phenols.
3. It forms no precipitate on addition of acidified $BaCl_2$. Some of the phenols are derived from putrefaction of protein in the large intestine.
4. Clinically, the ethereal sulfate is that of indoxyl indican which is formed from bacterial decomposition of tryptophan in the large intestine.
5. Normally, 5-20 mg of indican are excreted and the amount increases in constipation. In cholera, typhus gangrene of lung, sufficient indican is excreted. Indoxyl liberated from indican is oxidized to indigo blue on exposure to air.

III. Neutral sulfur

1. These are unoxidized sulfur and contained in cystine, taurine, thiocyanate or sulfides.
2. They do not vary with the diet.
3. They are mainly the products of endogenous metabolism.

Other organic compounds

H. Chlorides

These are excreted as NaCl and output varies with intake.

I. Phosphates

1. The urine phosphates consist of sodium and potassium phosphates as well as calcium and magnesium phosphates.
2. The greater part of the excreted phosphates is derived from ingested food which contains organic phosphates, eg, nucleoprotein, phosphoprotein and phospholipids. Phosphates of food are not completely absorbed. Some phosphate is also derived from cellular breakdown.
3. Phosphate excretion is increased in certain bone diseases such as osteomalacia, wasting diseases of the nervous system and in renal tubular rickets.
4. Marked increase of phosphate excretion is also observed in hyperparathyroidism and decrease in hypoparathyroidism and in infectious diseases.

J. Oxalates

1. The amount of oxalate in the urine is low (20 mg/day) and found as calcium oxalate crystals in urinary deposits.
2. The excretion of oxalate is increased by ingestion of fruits and vegetables containing high oxalates (spinach).
3. Large quantities of oxalate are excreted in urine in inherited metabolic diseases.
4. The oxalates present in urine are composed of partly unchanged ingested acid and partly oxidative products of other compounds.

K. Minerals

1. The 4 cations of the extracellular fluid—sodium, potassium, calcium and magnesium—are present in the urine.
2. Sodium content varies with intake. Urine potassium increases when the intake is increased or in excessive tissue catabolism. The excretion of potassium is affected by alkalosis. Sodium and potassium excretion are also controlled by the activity of the adrenal cortex.
3. Calcium and magnesium are not completely absorbed and their presence in the urine is low. But their presence in the urine varies in certain pathological states, particularly those involving bone metabolism.

L. Enzymes

1. Traces of many enzymes are excreted in urine including pancreatic amylase, pepsin, trypsin and lipase.

2. The pancreatic amylase excretion is increased in pancreatic disease.

M. Hormones and vitamins

1. Certain hormones (sex hormones) and vitamins (eg, B_1 and C) are found in urine.
2. The vitamin needs are assessed by studying the urinary output after test doses. The pregnancy test is also performed by the urinary sex hormones.

Abnormal constituents of the urine

A. Proteins

Proteinuria (albuminuria) is the presence of albumin and globulin in the urine in abnormal concentrations. The traces of protein (10-150 mg) present in normal urine cannot be detected by the ordinary simple tests. Pathologically, several proteins, such as serum albumin, serum globulin, hemoglobin, mucus, protease, Bence-Jones proteins are found in urine.

1. **Physiologic proteinuria** : In this condition, less than 0.5% protein is present in urine which occurs after severe exercise, after a high protein meal or as a result of some temporary impairment in renal circulation when a person stands errect. In 30-35% of pregnancy, there is proteinuria.
2. **Pathologic proteinuria** : Proteinuria is marked in glomerulonephritis. In nephrotic syndrome, a marked proteinuria occurs. The proteinuria increases with the increasing severity of the renal lesion. Proteinuria also results in poisoning of the renal tubules by heavy metals like mercury, arsenic or bismuth.
3. Hemoglobin is also present as a result of hematuria due to hemorrhage from the kidneys or urinary tract, clotting may occur due to sufficient fibrinogen on passing of much blood.
4. **Mucus** is the term for an unidentified protein precipitated by acetic acid in the cold. It is mucin. The mucus is increased in infection of the bladder.
5. Proteose may be found which is of little clinical significance.
6. Bence-Jones proteins found in the urine are the peculiar proteins which are light chain fragments of globulins. Most commonly they occur in multiple myeloma and rarely in leukemia. They are precipitated when the urine warmed to 50-60°C and redissolved almost completely at 100°C and precipitated again on cooling.

B. Glucose

1. Normal individuals excrete not more than 16-300 mg of sugar per day which is difficult to detect by simple test. It is said to be glycosuria when more than this quantity is found in urine.
2. There are different causes of glycosuira. Transient glycosuria is observed after emotional stress such as exciting atheletic contest. 15% of cases of glycosuria are not due to diabetes glycosuria suggesting diabetes must be confirmed by blood glucose studies to eliminate the probability of renal glycosuria.
3. The presence of glucose must be tested by Benedict's test. But in case of pregnant women and lactating mother, the Osazone test must be performed for urine glucose to eliminate the lactose present in urine.

C. Other sugars

1. **Fructosuria** : Fructosuria is due to the disturbance in fructose metabolism but not other carbohydrates.
2. **Galactosuria and lactosuria** : These may occur occasionally in infants, pregnant women and lactating mother. Galactosuria may occur in inherited diseases due to the nonconversion of galactose to glucose.
3. **Pentosuria** : This may occur transiently after intake of food containing large quantities of pentoses, such as grapes, cherries

and plums. It may take place in inherited diseases in which pentoses are not metabolized.

To detect these other sugars in urine it is wise to perform Osazone test.

D. Ketone bodies

1. Only less than 1 mg of ketone bodies are excreted in urine normally in 24 hours.
2. Increased amount of ketone bodies are excreted in urine in starvation, diabetes mellitus, pregnancy, ether anesthesia, and some types of alkalosis.
3. Excess fat metabolism may induce a ketonuria in many animals.
4. Increased amount of ammonia is excreted in acidosis accompanying ketosis.

E. Bilirubin and Bile salts

1. Bilirubin is found in the urine in cases of obstructive or hepatic jaundice.
2. Bilirubinuria is accompanied by the excretion of bile salts.
3. Bile salts may be excreted in urine without bile pigment in certain stages in liver disease.
4. In excessive hemolysis, traces of bilirubin without bile salts are excreted in urine.

F. Blood

1. In the lesion of the kidney or urinary tract blood is excreted in the urine in addition to its presence in nephritis.
2. Free hemoglobin is also found in urine after quick hemolysis, eg, in black water fever (a complication of malaria) or after severe burns.

G. Urobilinogen

1. In excessive hemolysis, eg, hemolytic jaundice or pernicious anemia, part of the bile pigment formed by breakdown of hemoglobin is excreted in urine as urobilinogen.
2. Urobilin is formed from colourless urobilinogen when the urine is exposed to air. This gives the urine an orange colour.
3. In liver disease or temporarily in constipation, large amounts of urobilin are found in urine.
4. Urobilinogen excreted in urine normally are 1-3.5 mg/24 hrs.

H. Porphyrins

1. Coprophyrins excreted in urine normally are 50-250 μ gm/day.
2. Coproporphyrins are excreted more in certain liver diseases.
3. The increased amount of coproporphyrins in the urine is a characteristic of the urine of patients suffering from porphyria.

Mechanism of action of diuretics

1. Diuretics, the drugs, enhance losses of water and salt via the urine through interference with normal reabsorptive mechanisms.
2. Osmotic diuretics are nonreabsorbable substances which increase tubular osmolarity. The osmotic substances which limit the amount of water. Osmotic diuresis is responsible for the serious dehydration which accompanies diabetic ketoacidosis.
3. Diamox is the inhibitor of carbonic anhydrase. It blocks both HCO_3^- reabsorption in the proximal tubule and regeneration in the distal tubule.
4. Thiazide diuretics, furosemide, ethacrynic acid and mercurials all inhibit chloride reabsorption in the ascending limb.

Renal function tests

Clearance is measured to assess quantitatively the rate of excretion of a given substance by the kidney. This is a volume of blood or plasma which contains the amount of the substance which is excreted in the urine in one minute.

A. Inulin clearance

1. Inulin is a polysaccharide which is filtered at the glomerulus but not secreted or reabsorbed by the tubule. Therefore, it is a measure of glomerular filtration rate. Mannitol can also be used for the same purpose.
2. These clearances vary with the body size. The clearance is calculated on the basis of ml/1.73 m^2.
3. To measure inulin clearance it is wise to maintain a constant plasma level of the test substance during the period of urine collections. The clearance is measured according to the following formula :

$$C_{in} = \frac{U \times V}{P}$$

where C_{in} = Clearance of inulin (ml/min)
U = Urinary inulin (mg/100 ml)
V = Volume of urine (ml/min)
P = Plasma inulin (mg/100 ml)

B. Endogenous creatinine clearance

1. Creatinine is filtered at the glomerulus but not secreted or reabsorbed by the tubule. Its clearance is measured to get the GFR.
2. This method is convenient for the estimation of the GFR because it does not require the intravenous administration of a test substance.
3. Normal values for creatinine clearance are in males : 130 ± 20 ml/mt and females : 120 ± 15 ml/mt.

C. The phenolsulphonephthalein (PSP) Test

1. The dye is almost completely eliminated within 2 hours.
2. If less than 25 per cent of the dye is not excreted in 15 minutes, it is an indication of impairment of renal function.

D. Other functional tests

1. Dilution test (water excretion test)
2. Urine concentration test (specific gravity test)
3. Vasopressin (ADH) test
4. Urine acidification test
5. Blood NPN, urea and creatinine
6. Urea clearance test.

1. Dilution test

(a) After emptying the bladder of the individual after overnight fast, he is advised to drink 1200 ml water in 30 minutes.

(b) During four hours after drinking, the urine is collected at hourly intervals.

(c) In normal individuals in cold climates, 1200 ml of urine is excreted in four hours.

(d) This test is not applicable to warm climates since the greater part of the ingested water is lost in perspiration during summer.

(e) In case of impaired renal function, the amount of water eliminated in four hours will be less than 1200 ml depending on the degree of impairment and specific gravity of urine is often 1.010 or higher in conditions of oliguria.

2. Urine concentration test (specific gravity test)

(a) The bladder is emptied on the day of the test at 7 a.m. and the urine is discarded.

(b) The urine is collected at 8 a.m. and the specific gravity is measured. If the sp. gr. is 1.022, the test may be rejected.

(c) If the sp. gr. is below 1.022, another urine specimen should be collected at 9 a.m. and the sp. gr. is determined.

(d) In case, the urine does not have a sp. gr. of 1.022, it is sure that the renal concentrating power is impaired either due to tubular defects or decreased secretion of ADH (diabetes insipidus). If the urine volume is large and the sp. gr. is below 1.022, the ADH test must be carried out.

3. Vasopressin (ADH) test

(a) The individual is not allowed any food or water after 6 p.m. on the night before the test. Vasopressin (5 units) is injected intramuscularly at 7 p.m. in the night.

(b) The urine is collected at 7 a.m. and 8 a.m. and the sp. gr. is determined. If the sp. gr. is 1.022, it is quite confident that the individual suffers from diabetes insipidus and ADH injection is effective in controlling it.

4. Urine acidification test

(a) This test should not be done on individuals who have acidosis or poor liver function.

(b) No dietary or other restrictions are involved in carrying out this test. The bladder is emptied at 8 a.m. Thereafter, hourly specimens of urine are collected until 6 p.m. At 10 a.m., ammonium chloride in a dose of 0.1 gram/kg body weight is given. A portion of each specimen is transferred to stoppered bottles and sent immediately to the laboratory for pH determination.

(c) In normal individuals, all urine specimens collected after 2 hours from the time of administration of ammonium chloride should have a pH between 4.6 and 5.0 but in patients with renal tubular acidosis, the pH does not fall below 5.3.

5. Blood non-protein nitrogen

(a) In acute nephritis, the NPN values are increased and range from a slight increase (NPN-45 mg, urea N-25 mg, creatinine-2 mg per 100 ml) to very high values (NPN-200 mg, urea N-160 mg creatinine-25 mg per 100 ml).

(b) NPN increase and retention are due to impaired renal function and excessive protein catabolism.

Congenital tubular function defects

1. Diabetes insipidus

(a) This disease is developed due to the non-production of ADH. The individual passes large volume of urine (5-20 litres in 24 hours). The individual has to drink large amount of water to make up the loss.

(b) The reabsorption of water in the distal tubules does not take place in the absence of ADH.

2. Vitamin D resistant rickets

(a) The tubular reabsorption of phosphate does not take place under this condition.

(b) Excessive loss of phosphate in urine leads to the development of a type of rickets which does not respond to usual doses of Vitamin D.

3. Renal glycosuria

In this condition, the tubular reabsorption of glucose is affected. Although the blood sugar is within normal level but glucose is excreted in urine due to defective reabsorption by the tubules.

4. Idiopathic hypercalcinuria

Calcium is not reabsorbed by the renal tubules in this condition. Hence, large amounts of calcium are excreted in the urine. Renal calculi may be developed owing to the presence of large amounts of calcium in urine.

5. Salt losing nephritis

(a) Large amounts of sodium and chloride ions are excreted in urine in this condition due to the defect in the tubular reabsorption of these ions resulting in severe dehydration, hyponatremia and hypochloremia.

(b) Blood urea is increased due to the reduced glomerular filtration rate.

(c) This condition does not respond to aldosterone adminstration but responds to parenteral administration of sodium chloride solution.

6. Renal tubular acidosis

(a) In this condition, the urine becomes alkaline or neutral due to the defect in the sodium and hydrogen ion exchange mechanism in the distal tubules. There is a loss of sodium in the urine.

(b) The acidosis is accompanied by excessive mobilization and urinary excretion of calcium and potassium.

(c) These abnormalities led to clinical manifestation of dehydration, hypokalemia, defective mineralisation of bones and nephrocalcinosis.

7. Fanconi Syndrome

(a) A number of defects in tubular reabsorption exist in this condition. The defects are renal amino acid in renal glycosuria, hyperphosphaturia, metabolic aciduria, with increased urinary excretion of Na, Ca and K.

(b) In some individuals, cystinosis prevails due to the abnormality of cystine metabolism in which cystine crystals are deposited in macrophages in the liver, kidney, spleen, bone marrow, lymph nodes and cornea.

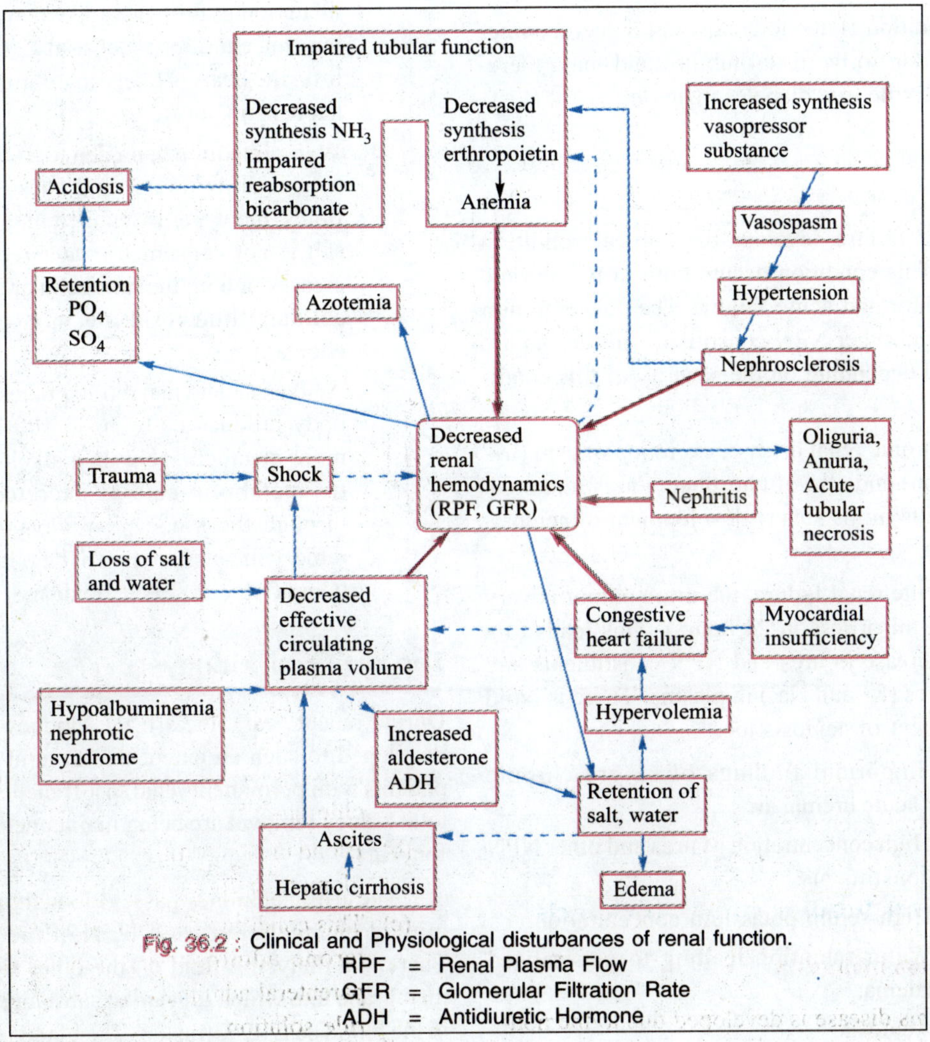

Fig. 36.2 : Clinical and Physiological disturbances of renal function.
RPF = Renal Plasma Flow
GFR = Glomerular Filtration Rate
ADH = Antidiuretic Hormone

8. Hartnup syndrome (Hard Syndrome)

(a) In this condition, a number of amino acids are not reabsorbed owing to the defect in tubular reabsorption mechanism.

(b) Disturbances in tryptophan metabolism is suggested by the presence of increased amounts of tryptophan, indican and indole acetic acid in urine.

(c) The clinical symptoms are of niacin deficiency—a pellagra like skin lesions and mental deficiency.

9. Nephrogenic diabetes insipidus (water-losing nephritis)

This condition is due to congenital defect in water reabsorption in the distal tubules and may, therefore, resemble true diabetes insipidus.

Uremia

The renal failure develops the clinical condition uremia. This condition occurs both in the chronic renal failure and acute failure. The concentration of urea and other NPN constituents in plasma are increased depending on the severity of this condition.

In chronic renal disease, excretion of acid (hydrogen ion) and also of phosphate ion is impaired. This results in the steady development of acidosis in uremia.

In acute renal failure, the urine output is very low (300 ml or less in 24 hours). This leads to a steady increase in urea and NPN constituents and electrolytes (K^+ and Na^+) in plasma. There is rapid development of acidosis too.

The important findings of severe chronic uremia or acute uremia are :

1. High concentration of urea and other NPN constituents.
2. High serum potassium concentration.
3. Water retention leading to generalised edema.
4. Acidosis.

Uremic coma occurs in serious cases.

1. **Urea and NPN :** The concentration of urea and other NPN constituents of blood are very much increased (i.e., 10 times the normal level) in severe renal failure.
2. **Potassium :** The potassium ion level may be slightly increased in chronic uremia. But in acute uremia, the concentration in serum is very much increased. Potassium is released from the cells due to the breakdown of cellular proteins. This released potassium passes into the blood and interstitial fluid. When the concentration of potassium ion increases to 8 m. Eq/litre, it exerts a cardiotoxic effect resulting in the dilatation of the heart and when potassium ion concentration reaches at 12 to 15 mEq/litre, the heart is likely to be stopped. This happens in severe uremia.
3. **Water retention and edema :** If the uremic patient drinks water and consumes other fluids, the water is retained in the body. If salt is not consumed, water retention increases in both the intracellular and extracellular fluid resulting in extracellular edema.
4. **Acidosis :** The metabolic processes in the body produce daily 50 to 100 m mol of more metabolic acid than alkali. This extra metabolic acid is excreted mainly through the kidneys. Acidosis develops rapidly in acute uremia. The patient faces 'Coma' due to severe acidosis.

The artificial kidney

During recent years, the artificial kidney has been developed to such an extent that several thousand patients with permanent renal insufficiency or even total kidney removal are being maintained in health for years.

The artificial kidney passes blood through very minute channels bounded by thin membranes. There is a dialyzing fluid on the other side of the membrane into which unwanted substances present in the blood pass by diffusion. The blood is pumped continually between two thin sheets of cellophane; the dialyzing fluid is on the outside of the sheets. The cellophane is porous enough to allow all constituents of the plasma except the plasma proteins to diffuse freely in both directions—from plasma into the dialyzing fluid and from the dialyzing fluid into the plasma.

The rate of flow of blood through the artificial kidney is several hundred ml per minute. Heparin is infused into the blood as it enters the kidney to prevent clotting of blood. To prevent bleeding as a result of heparin, an anti-heparin substance, such as protamine, is infused into the blood as it is returned to the pateint.

The dialyzing fluid : Sodium, potassium and chloride concentrations in the dialyzing fluid and in normal plasma are identical; but in uremic plasma, the potassium and chloride concentrations are considerably greater. These two ions diffuse through the dialyzing membrane so rapidly that their concentrations fall to equal those in the dialyzing fluid within three to four hours, exposure to the dialyzing fluid. On the other hand, there is no phosphate, urea, urate or creatinine in the dialyzing fluid. When the uremic patient is dialyzed, these substances are lost in large quantities into the dialyzing fluid, thereby removing major proportions of them from the plasma. Thus, the constituents of the dialyzing fluid are such that those substances in excess in the extracellular fluid in uremia be removed at rapid rates, while the essential electrolytes remain quite normal.

Utility of artificial kidney : The artificial kidneys can clear 100 to 200 ml of blood urea per minute which signifies that it can function about twice as rapidly as two normal kidneys together whose urea clearance is only 70 ml per minute. However, the artificial kidney can be used for not more than 12 hours once in three to four days because of danger from excess heparin and infection to the subject.

Hormones of the kidney

1. Not only the kidney performs excretory functions but it acts as an endocrine organ. It liberates many hormones which affect other organs and tissues and some hormones which locally act within the kidney itself. It also destroys several hormones which are liberated from other endocrine organs.
2. The juxtaglomerular cells of the renal cortex produce the proteolytic enzyme **rennin** and secrete it into the blood. Rennin acts on α_2-globulin which is normally present in blood plasma, although it is produced in the liver. Rennin splits off a polypeptide fragment called angiotensin I which is decapeptide containing 10 amino acids. Another enzyme of the lung acts on angiotensin I to split off 2 amino acids and thus form the octapeptide angiotensin II.

   Angiotensin increases the force of the heartbeat and constricts the arterioles. It raises blood pressure and causes contraction of smooth muscle. It is destroyed by the enzyme **angiotensinases** present in normal kidneys, plasma and other tissues. Recent studies suggest that **rennin angiotensin** system is important in the maintenance of normal blood pressure.
3. **Prostaglandins** are the other hormones of the kidney. They cause relaxation of smooth muscles. They cause vasodilatation and a decrease in blood pressure. They also increase renal blood flow. **Kininogen** which is produced by the kidney has an antihypertensive effect.
4. The two hormones **erythropoietin** and **erythrogenin** have an effect on bone marrow to stimulate production of red cells. Kidney plays an important role in the release of erythropoietin and thus in control of red cell production. **Hypoxia** stimulates production of erythropoietin.

Urinary deposits

The commonest deposits are phosphates, oxalates and urates and are frequently seen in normal urine.

Phosphates

1. They are usually found in alkaline urines. The commonest is ammonium magnesium phosphate which forms a characteristic crystal.
2. A less common form is calcium hydrogen phosphate which forms long prisms.
3. Amorphous calcium and magnesium phosphates may be deposited from alkaline urines.

The deposition of phosphates is due to a change in pH after the urine has been passed.

Calcium oxalate : This is found in acid urine but may be found in alkaline urine. The crystals are of two types—octahedra, dumb-bells. Calcium oxalate is insoluble in acetic acid.

Urates

1. They are usually found in acid urines.
2. Uric acid separates into different forms including prisms, barrels, hexagons and needles which are always pigmented.
3. Urates are redissolved on warming the urine.
4. The cause of deposition of urates is the cooling of urine after it has been passed.

Feces

Amount

1. The quantity of feces varies from day to day and with the diet.
2. Vegetable food increases the bulk of feces but meat diet which is largely absorbed diminishes the bulk.
3. An adult taking a mixed diet passes from 60 to 250 gm of moist feces containing 25-45 gm of solids per day.
4. The young children release high bulk.

Composition

1. Normal adult feces have a water content of 65-80%.
2. One-third of the dry matter of feces is represented by bacteria. The rest are the remainings of intestinal secretions, substances excreted by the large intestine (eg, Ca, Fe) and small amounts of food residues. The food residues are cellulose, fruit skins and seeds.
3. The only substances in feces are fat, nitrogen and mineral elements. Dry feces contain the following :

Nitrogen	5-10%
Fat	10-20%
Ash	10-20%

Colour

1. The normal colour of feces is brown which is due to stercobilin (urobilin).
2. Milk, cereals and meat, coffee, cocoa, black berries, etc. form darker stools.
3. Greenish colour due to excessive consumption of green vegetables.
4. The colour of the feces of newly-born infants is dark or blackish-green due to biliverdin and porphyrin.
5. The stools of young infants on a milk diet is yellow due to bilirubin because of nondevelopment of intestinal bacterial flora.
6. The presence of blood in the stool gives the feces a red colour. Excessive hemolysis will give a dark-brown feces.
7. Iron or bismuth gives a black stool.

Odour

1. The normal odour is due to indole and skatole.
2. Mercaptans and H_2S may produce odour.
3. A meat diet produces a more intense odour than a vegetable one and a milk diet least of all.

pH

1. Feces are normally slightly alkaline, pH 7.0-7.5.
2. They may be slightly acid with a large proportion of carbohydrate or fat.

Fat

1. The total fat of feces is divided into two : (a) Split fat (fatty acids), (b) Unsplit fat (neutral fats, phospholipids, sterols, pigments).
2. The amount of fat is independent of the diet. The split fat diminishes in amount on low-fat diets.
3. In diseases, fecal fat is increased when digestion or absorption of fat is impaired.

Nitrogen

1. Fecal nitrogen is very little affected by the amount of protein ingested if the protein is well masticated and well assimilated.
2. An adult of a mixed diet usually excretes about lgm of fecal nitrogen per day.
3. In diseases, fecal nitrogen is greatly increased by failure of digestion or absorption of protein.

Salts

1. Moist feces contain about 2-3% of salts. Most abundant are calcium and phosphate.
2. Small amounts of magnesium, iron, sodium, potassium, chloride and sulphate are also present in feces. The proportion of calcium is higher on a milk diet and that of magnesium on a meat diet.

Sweat

1. Sweat is a more dilute fluid and is always hypotonic.
2. The pH is about 4.5. But if the skin is washed and dried, sweat secreted is slightly alkaline (pH 7.0-7.4).
3. It contains most of the diffusible constituents of plasma. The most abundant constituent is NaCl.
4. The lactic acid content is more than normally found in blood.

CLINICAL ORIENTATION

- The specific gravity of urine is high in febrile conditions due to smaller volume executed and low in diabetes insipidus owing to large volume excreted, but it is high in case of diabetes mellitus due to the presence of sugar in urine.
- The urinary acidity increases due to the ingestion of large amounts of proteins, vegetables and fruits containing organic acids, other fruits producing hippuric acid, and due to starvation also.
- The urine exhibits fruity odour due to acetone excretion in diabetic ketoacidosis, mousy odour due to phenyl-acetyl glutamine excretion in phenylketonuria and unpleasant odour due to methyl mercaptan excretion in abnormal metabolism of methionine in case of hepatic failure.
- Mucus appears in the urine in greater amounts in inflammatory conditions of urinary tracts, after prostatic massage of semen content of vaginal discharge.
- The amount of urinary chloride is decreased in oedema, diabetes insipidus, infectious diseases, etc.
- Decreased urinary phosphate concentration is associated with diarrhoea, nephritis, acute infections, pregnancy, etc.
- Increased urinary sulphur is found in cystinuria, melanuria, obstructive jaundice, chloroform anaesthesia.
- The excretion of potassium increases in primary aldosteronism (Cohn's syndrome).
- Excessive sodium excretion takes place in excessive sweating and diarrhoea, etc.
- Increased urinary excretion of calcium is associated with hyperparathyroidism, hypervitaminosis D, multiple myeloma, renal stones, certain drugs, etc.
- The high urea level in urine is found in fever, diabetes mellitus, increased adrenocortical activity but the decreased level is found in cirrhosis of liver, severe acidosis, and nephritis.

CLINICAL ORIENTATION

- The urinary uric acid level is elevated in gout, leukemia, Wilson's disease.
- The oxalic acid content of urine is increased in diabetes mellitus, certain liver disease.
- Increased aminoaciduria is observed in Fanconi syndrome and Wilson's disease, muscular dystrophies.
- Proteinuria is associated with degenerative lesions of the pelvis of the kidney, ureter, bladder, prostate, bleeding in genito-urinary tract.

EXERCISE

1. Discuss the normal constituents of urine. (B.U. 76A)
2. Discuss the abnormal constituents of urine. (R.U. 71S)
3. Explain fully the composition of urine. (P.U. 68S)
4. Write notes on :
   (a) Hormones of the kidney. (Pun. U. 67A)
   (b) Renal function test. (Mith U. 72A)
   (c) Feces. (Bhag. U. 74S)

CHAPTER 37

LIVER FUNCTION TESTS

Various important metabolic functions are carried out by liver. These include the following :

1. Destruction of red blood cells and formation and excretion of bile pigments.
2. Secretion of bile.
3. Synthesis of plasma proteins, fibrinogen and prothrombin.
4. Detoxifying functions.
5. Carbohydrate metabolism.
6. Fat metabolism.
7. Protein metabolism.
8. Storage of glycogen, vitamins and iron.
9. Removal or excretion of drugs, hormones and other substances.

The undermentioned tests are required to be done for assessing the functions of liver.

1. Tests for bile pigment metabolism in jaundice

(a) Serum bilirubin estimation and Van den Bergh reaction.
(b) Biochemical tests for jaundice.
(c) Icteric Index.
(d) Urine bilirubin.
(e) Urine and fecal stercobilinogen and urobilinogen.
(f) Congenital hyperbilirubinemia.

2. Tests for carbohydrate metobolism

(a) Galactose tolerance test.
(b) Fructose tolerance test.

3. Tests for plasma protein concentration

(a) Plasma albumin, globulin and fibrinogen concentration.
(b) Flocculation tests.
(c) Amino acids in blood and urine.

4. Tests for lipid metabolism

(a) Serum total, free and ester cholesterol.
(b) Fecal fat.

5. Tests for detoxifying functions—excretion of foreign substances

Hippuric acid synthesis.

6. Tests for excretion of foreign substances

Bromsulphthalein test

7. Tests for blood coagulation

Prothrombin time determination.

8. Tests for serum enzymes derived from liver

Alkaline phosphatase, transaminases, isocitrate dehydrogenases, pesudocholine esterase and 5-nucleotidase.

9. Tests for conversion of ammonia to urea

Blood ammonia determination.

10. Determination of glutamine in cerebrospinal fluid.

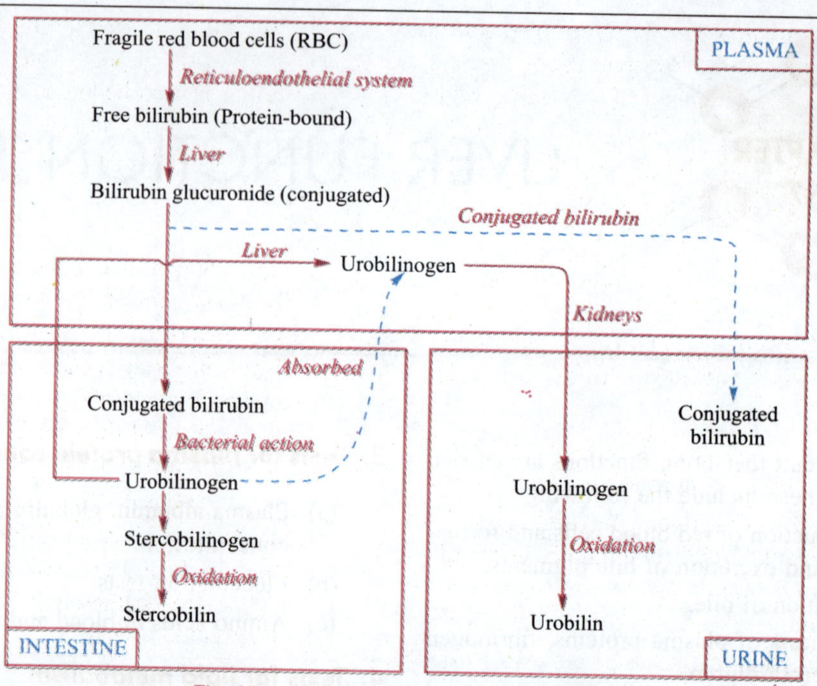

Fig. 37.1 : Bilirubin formation and excretion

1. Bile pigment metabolism in health and in jaundice

When the red blood cells have lived out their life span (averaging 120 days), their cell membranes rupture and the released hemoglobin is phagocytozed by the reticuloendothelial cells throughout the body. The hemoglobin is first split to heme and globin, and the heme ring is opened to give a straight chain of four pyrrole nuclei from which bile pigments are formed. The first pigment formed is **biliverdin** which is rapidly reduced to **free bilirubin**. This free bilirubin immediately combines very strongly with the plasma albumin and is transported throughout the blood and interstitial fluids. Even when bound with the plasma protein, this bilirubin is still called "**free bilirubin**". 80 per cent of the free bilirubin then conjugates in the liver with glucuronic acid to form **bilirubin glucuronide**; 10 per cent conjugates with sulphate to form **bilirubin sulphate** and the rest 10 per cent conjugates with a multiple of other substances. This conjugated bilirubin is excreted by an active transport process into the bile canaliculi. A small portion of the conjugated bilirubin formed by the hepatic cells returns to the plasma either directly or indirectly by absorption. Therefore, a small portion of the conjugated bilirubin is always available in the extracellular fluid.

Bilirubin is then converted by bacterial action into **mesobilirubin** which is reduced to mesobilirubinogen in the intestine. This compound is further reduced to **stercobilinogen** (**Urobilinogen**) which is highly soluble. Some of the urobilinogen is reabsorbed through the intestinal mucosa into the blood. Most of this is reexcreted by the liver back into the gut; but about 5 per cent of it is excreted by the kidneys into the urine. After exposure to air in the urine, the urobilinogen becomes oxidized to urobilin. The stercobilinogen in the feces becomes oxidized to **stercobilin**. This explanation is represented by the figure 37.1.

Jaundice

"**Jaundice**" means a yellowish tint to the body tissues, including yellowness of the skin and also of the deep tissues. The usual cause of jaundice is the increased bilirubin content in the extracellular fluids, either free or conjugated bilirubin. The normal plasma concentration of bilirubin (both free and conjugated forms) averages 0.5 mg per 100 ml of plasma. But in certain abnormal conditions this can rise to as high as 40 mg per 100 ml. The skin becomes yellow when the concentration rises to about three times the normal.

Causes

1. Increased destruction of red blood cells with rapid release of bilirubin into the blood.
2. Excessive production of bilirubin beyond the capacity of normal liver to excrete it.
3. Dysfunction of liver cells resulting in failure to convert bilirubin into bilirubin glucuronide form in the liver and excrete it in the bile.
4. Obstruction of the bile ducts to make usual flow of the bile to the duodenum.

Classification

Rolleston and McNee classified jaundice as follows :

1. Hemolytic jaundice.
2. Obstructive jaundice.
3. Hepatic jaundice.

Rich classified jaundice as follows :

1. Retention jaundice (overproduction of unconjugated bilirubin).
2. Regurgitation jaundice (due to necrosis of liver cells and obstruction of bile ducts).

A. Hemolytic jaundice

1. The excretory function of the liver is not at all impaired.
2. The red blood cells are hemolyzed rapidly by hemolytic anemias by the action of some drugs, by malarial or viral infections, and by incompatible blood transfusion. The hepatic cells simply cannot excrete the bilirubin as rapidly as it is formed. Therefore, the plasma concentration of free bilirubin rises markedly.
3. The rate of formation of urobilinogen in the intestine is greatly increased and much of this is absorbed into the blood and later excreted in the urine. The colour of the feces becomes dark brown.

B. Obstructive Jaundice

1. This happens due to the obstruction in the bile duct preventing the flow of bile into the intestine. The obstruction is caused by the blocking of the bile passage by gallstones, by enlarged glands due to tumour of the head of the pancreas, and by stricture of narrowing of the bile duct as a result of surgery.
2. The rate of bilirubin formation is normal; but the bilirubin cannot pass from the blood into the intestines. The free bilirubin then enters the liver cells and becomes conjugated in the usual way.
3. The conjugated bilirubin is then returned to the blood probably by rupture of the congested bile canaliculi and direct emptying of the bile into the lymph leaving the liver.
4. The excess bilirubin is excreted in the urine producing a deep yellow or brownish colour. The stools become clay coloured and bulky containing excessive amount of fat. The serum alkaline phosphatase concentration is usually high.

C. Hepatic jaundice (Toxic and infective jaundice)

1. Hepatic jaundice is caused by infection, toxins and liver poisons. Infection with virus is the most common cause. The infective organisms cause damage to the liver parenchymal cells.
2. The conjugation of bilirubin in the liver is thereby affected. Hence, both free and conjugated bilirubin concentration is increased in the serum.
3. The urine becomes highly coloured due to the presence of conjugated bilirubin and urobilinogen.

(i) **Van den Berg reaction to differentiate between hemolytic and obstructive jaundice diagnostically :**

If a freshly preapared diazotised sulphanilic acid reagent is added to serum, conjugated bilirubin gives a reddish violet colour within a minute known as the "**direct**" **Van den Berg reaction**.

The free bilirubin (unconjugated bilirubin) of the serum does not develop any colour within one minute but the colour is formed if alcohol is added to the mixture. Alcohol precipitates the protein and makes free the free bilirubin from its protein complex so that it can then combine with the **Van den Berg reagent**. This result is called the "**indirect**" Van den Berg reaction.

Therefore, in hemolytic jaundice an indirect Van den Berg reaction occurs (increased free bilirubin) and in obstructive jaundice a direct Van den Berg reaction takes place (increased conjugated bilirubin).

If a faint pink colour is formed after one minute and deepening of the colour results in 2 or 3 minutes in some cases, that indicates clearly that both conjugated and free bilirubins are present in the serum. This reaction is said to be "**biphasic**" reaction.

In the total obstruction of bile flow, no bilirubin can reach the intestines to be converted into urobilinogen by bacteria. Therefore, urobilinogen is not reabsorbed into the blood and is not excreted by the kidneys into the urine. So in total obstructive jaundice, tests for urobilinogen in the urine are completely negative. The stool becomes clay coloured for lack of **stercobilin**, but not free bilirubin. Therefore, in severe obstructive jaundice, large quantities of conjugated bilirubin appear in the urine. This can be known by shaking the urine and observing the foam, which becomes intense yellow in colour.

(ii) **Biochemical tests for jaundice :** The biochemical tests are the following :

1. Serum bilirubin concentration and nature of Van den Berg reaction.
2. Serum alkaline phosphatase activity and SGPT activity.
3. Urine test for urobilinogen and bilirubin.
4. Thymol flocculation test and colloidal gold test.
5. Feces colour.

The biochemical findings in three types of jaundice are given in table 37.2.

Table 37.1 : Biochemical changes in jaundice

No.	Tests	Hemolytic	Obstructive	Hepatic
1.	Serum bilirubin	2 to 3 mg/100 ml (free bilirubin)	Up to 50 mg/100 ml (conjugated bilirubin)	Up to 20 mg/100 ml (both types of bilirubin)
2.	Van den Berg reaction	Indirect	Direct	Biphasic
3.	Test for bile pigments in urine	Urobilinogen (Schlesinger's test positive)	Conjugated bilirubin (Fouchet's test and Gmelin's test positive)	Urobilinogen and conjugated bilirubin
4.	Feces colour	Dark brown	Clay colour	Varies
5.	SGPT activity	5 to 30 units (normal)	35 to 100 units (Raised slightly)	100 to 300 units (Raised markedly)
6.	Serum alkaline phosphatase activity	3 to 12 K.A. units and up to 25 in children (normal)	30 to 100 K.A. units (Raised markedly)	15 to 30 K.A. units (Raised slightly)
7.	Serum 5-nucleotidase activity	Negative	Raised markedly	Raised slightly
8.	Thymol flocculation test	Negative	Negative	Positive
9.	Colloidal gold test	Negative	Negative	Positive

(iii) **Icteric Index :** The icteric index shows the degree of jaundice by measuring the intensity of yellow colour of the serum. Serum is diluted with normal saline until it matches the colour of 1 in 10,000 solution of potassium dichromate. The dilution factor is termed **Icteric Index**.

The Icteric Index in normal person is 4 to 6 units. In latent jaundice, it is 3 to 14 units. In clinical jaundice, it is higher than 15 units. Carotene present in serum interferes its determination.

(iv) **Urine bilirubin :** The urobilinogen in urine can be detected by Schlesinger's test. The concentration of urobilinogen is more in hemolytic and hepatic jaundice. Fouchet's and Gmelin's test are both positive in obstructive jaundice. These tests indicate the presence of conjugated bilirubin in urine in obstructive jaundice.

(v) **Bile pigments in feces :** The quantity of stercobilinogen depends on the quantity of bilirubin entering into the intestine. Stercobilinogen content of feces of patients with stone obstruction is nearer the normal level (10 to 150 mg/ day in adults) while it is low or absent (0 to 5 mg/day) in those with malignant obstruction.

(vi) **Congenital hyperbilirubinemia :**

(a) **Gilbert's syndrome :** In this condition, there is defective intracellular transport of bilirubin. In some cases, there is defective hepatic conversion of bilirubin to bilirubin diglucuronide. Therefore, the serum-free bilirubin concentration is high ranging from 4 to 6 mg/100 ml, sometimes it may rise to 12 mg 100 ml; The serum shows indirect Van den Berg reaction.

(b) **Lucey-Driscoll syndrome :** This condition appears in newborn infants. The serum bilirubin concentration is very high (up to 60 mg/100 ml). This syndrome occurs due to the presence of a substance (probably steroid) which inhibits the conversion of bilirubin into bilirubin diglucuronide in the liver. This inhibitor disappears one month after the birth of the infants having this syndrome.

(c) **Dubin-Johnson syndrome :** In this condition, there is defective excretion of conjugated bilirubin by liver cells into bile. Conjugated bilirubin is found in urine. Alkaline phosphatase level of serum is normal.

(d) **Crigler-Najjar syndrome :** This syndrome appears in new born infants. The serum bilirubin level rises to 20 mg/100 ml or more within a few days after birth. This condition is a familial incidence. Bilirubin is not converted to bilirubin diglucuronide due to the deficiency of the enzyme glucuronyl transferase in the liver. The bile does not contain conjugated bilirubin. The serum bilirubin level comes down to normal level in those infants who survive.

2. Tests for carbohydrate metabolism

(i) **Galactose tolerance test :** This test is done in the morning after overnight fast. Fasting blood sample is taken. The individual is then given to ingest a galactose solution containing 40 grams of galactose in 300 ml of water. Blood is drawn at half an hour interval for 2 hours. Galactose content of the blood samples are then determined after removing the glucose by fermenting with yeast.

The normal blood galactose level is 0 to 160 mg/100 ml. In infective and toxic hepatitis, values may go up to 500 mg/ 100 ml of blood. In **cirrhosis of the liver**, values up to 500 mg/100 ml of blood are also found depending on the severity of the disease.

(ii) **Fructose tolerance test :** This test is also performed in the morning after overnight fast. The individual is administered 50 grams of fructose dissolved in 300 ml of water. Fasting blood sample is taken. Blood is also taken at half an hour intervals for 2 hours after fructose ingestion. The total blood sugar (glucose + fructose) is estimated.

In normal subjects, the highest blood sugar level does not exceed the fasting level by more than 30 mg/100 ml. Blood sugar levels up to 150 mg/100 ml are found in patients with infective hepatitis. This test is less sensitive than the galactose tolerance test.

3. Tests for plasma protein concentration

Albumin, fibrinogen, and some of the α- and β-globulins are synthesized in the liver. In advanced liver diseases, the albumin content is decreased and globulin content increased. Edema may develop when the plasma albumin level falls below 2.5 per cent. The globulin content may increase up to 5 per cent in some cases. Fibrinogen values in normal persons range from 0.2 to 1.03 per cent and it may fall to 0.1 gm in severe liver disorder, such as acute hepatic necrosis.

(i) **Electrophoretic separation of plasma proteins :** The percentage of different proteins determined by paper electrophoresis in normal subjects are as follows :

Albumin	55.2%
Globulin	44.8%
Albumin/Globulin ratio	1.23%
α_1-Globulin	5.3%
α_2-Globulin	8.7%
β-Globulin	13.4%
γ-Globulin	11.0%
Fibrinogen	6.5%

The flowing results are obtained in liver diseases :

Chronic infective hepatitis	....	γ-globulin increased
Cirrhosis of the liver	....	Albumin content decreased
Necrosis of the liver	....	Fibrinogen content decreased

(ii) **Flocculation tests.**

(a) **Thymol turbidity test :** The degree of trubidity is measured against standards containing 10, 20, 30, 40, . . . 100 mg per 100 ml of protein when serum is mixed with a buffered solution of thymol. A turbidity equal to that of the 10 mg protein standard is taken as 1 unit by McLagan.

In normal subjects, the thymol units range from 0 to 4 units. In infective hepatitis, the values range from 5 to 20 units. In obstructive jaundice, only 8 per cent give positive result. The thymol flocculation test will be positive in all cases in which the turbidity is positive.

(b) **Serum colloidal gold test :** The results obtained in this test in subjects suffering from liver diseases are similar to those obtained with thymol turbidity test.

(c) **Zinc sulphate test :** This test is positive in all cases of infective hepatitis and cirrhosis. In normal persons, serum γ-globulin content is 2 to 8 units but the values rise from 15 to 80 units in infective hepatitis and cirrhosis.

4. Tests for detoxifying functions

(a) **Hippuric acid synthesis test :** The liver detoxicates benzoic acid by reacting it with glycine to form hippuric acid which is excreted in urine. The liver is able to synthesize sufficient glycine to conjugate with benzoic acid to form hippuric acid.

The test should begin at least 3 hours after a light breakfast. The patient empties the bladder and drinks sodium benzoate in about 200 ml of water. Urine is collected for a period of 4 hours from the time of ingestion of sodium benzoate. The amount of hippuric acid excreted is determined.

In normal persons, 60 per cent of the benzoic acid taken should be excreted as hippuric acid. The excreted should be 4.5 grams. Smaller quantities are excreted in acute or chronic liver damage.

5. Test for excretion of foreign substances

Bromsulphthalein (BSP) test : When bromsulphthalein dye is injected, it circulates in the blood in combination with albumin. Normal subjects after injection of 5 mg BSP per kg body weight retain less than 10 per cent of the dye in 30 minutes, and 7 per cent in 45 minutes. At 60 minutes, no dye is retained. This is the most sensitive and dependable liver function test. It is particularly useful for evaluating suspicious or slightly positive results obtained by flocculation tests in the absence of hyperbilirubinemia.

If the liver function is impaired, the dye is excreted slowly and up to 50 per cent of the dye will be retained in the body at the end of 45 minutes after injection. The test is more useful in the diagnosis of liver cell damage without clinical jaundice, in chronic hepatitis and in cirrhosis of the liver.

6. Tests for blood coagulation

Prothrombin time test : Prothrombin (Factor II) and factors VII, IX and X involved in the coagulation of blood are synthesized in the liver in the presence of Vitamin K. Deficiencies of these can occur for two reasons : (i) in the presence of parenchymal cell damage, synthesis is impaired despite adequate supplies of Vitamin K. (ii) In the absence of bile as in cholestasis and obstructive jaundice, the Vitamin K is not absorbed from the intestines and their synthesis is affected.

Shortening of prothrombin time after parental Vitamin K therapy suggests cholestasis; while lack of response to Vitamin K indicates liver damage.

7. Tests for serum enzymes

Certain enzymes are released from liver into the blood due to the damage to liver cells. The levels of SGOT, SGPT, LDH and isocitrate dehydrogenase are increased. The levels of these enzymes are increased in viral hepatitis and reach their maximum soon after the onset of jaundice, and then decrease slowly. Very high levels occur in toxic hepatic necrosis. Levels in cirrhosis may be moderately raised but only if the process is active. Choline esterase levels are decreased in liver cell dysfunction.

8. Test for the conversion of ammonia to urea

The normal range of blood ammonia is 40 to 75 mg ammonia nitrogen per 100 ml. In cirrhosis of the liver, blood ammonia may be increased to over 250 mg/100 ml. High values are found in hepatic coma.

9. Glutamine content of cerebrospinal fluid

The normal range of glutamine in cerebrospinal fluid is from 6 to 14 mg per 100 ml. In cirrhosis of the liver, higher values ranging from 16 to 31 mg/100 ml have been reported. In hepatic coma, still higher values ranging from 30 to 54 mg/100 ml have been reported.

MISCELLANEOUS METABOLIC FUNCTIONS OF THE LIVER

1. **Storage of vitamins :** The liver has a particular propensity for storing vitamins. Vitamin A is stored in the liver to the greatest extent but large quantities of vitamin D and Vitamin B_{12} are normally stored as

well. Sufficient amount of Vitamin A can be stored to prevent Vitamin A deficiency for one to two years, sufficient Vitamin D and Vitamin B_{12} can be stored to prevent their deficiency for one to four months.

2. **Storage of iron** : Iron is stored in the liver in the form of ferritin. When the iron in the circulating body fluids reaches a low level, the ferritin releases the iron. Therefore, the apoferritin-ferritin system of the liver acts as a blood iron buffer and also as an iron storage medium.

3. **Removal or excretion of drugs, hormones and other substances** : The liver excretes into the bile many different drugs including sulfonamides, penicillin, ampicillin and erythromycin. In the same manner, different hormones are either chemically altered or excreted including thyroxine and essentially all of the steroid hormones such as estrogens, cortisol, aldosterone and so forth. Finally, one of the major routes for excreting calcium from the blood is into the bile and then into the feces.

EXERCISE

1. Write short notes on liver function test. (M.U. 82A)

CHAPTER 38

MEMBRANES

Membranes are viscous solutions which exist around all living cells. They form closed compartments around the protoplasm of the cells and separate one cell from another. They act as barriers with selective permeabilities to materials and thus maintain the difference between inside and outside. The selective permeabilities are regulated by gates and pumps as well as by specific receptors for enzymes, substrates, and hormones.

Membranes are mostly composed of lipids and proteins but also contain carbohydrates. Membranes anchor protein molecules where they carry out specific function of the organelle, the cell, and the organism.

LIPIDS

The lipid composition of membranes are phospholipids, glycolipids, and sterols-cholesterol in mammalian membranes.

Phospholipids

1. The commonest one is phosphoglycerides which consist of glycerol, two fatty acids, and a phosphorylated alcohol. The fatty acids contain even number carbon atoms and they are unbranched and can be saturated or unsaturated.
2. The other phospholipids are the sphingomyelins which contain sphingosine rather than glycerol. A fatty acid is attached to the amino group of sphingosine. The primary alcoholic group of sphingosine is esterified to phosphorylcholine.

Glycolipids

1. The glycolipids are the cerebrosides and gangliosides which are also derived from sphingosine.
2. A cerebroside contains a single glucose or galactose, but a ganglioside contains a branched chain of seven glucose or galactose.

Sterols

1. Cholesterol, the most common sterol in membranes, exists in the plasma membranes of mammalian cells but also found in mitochondria, golgi complex and nuclear membranes in less quantity.
2. It is more abundant in the outside of the plasma membrane.
3. The major lipids in membranes are amphipathic which contains both hydrophobic and hydrophilic regions. If the hydrophobic region is separated from the rest of the molecule, it will be insoluble in water but soluble in oil. The reverse happens in case of hydrophilic region. Detergents are amphipathic molecules which have much biochemical importance.

PROTEINS

1. The phospholipids of the membrane are the solvent for the membrane proteins. The functional groups attached to the alpha carbon of twenty amino acids involved in the primary structure of protein are strongly hydrophobic in 6, weakly hydrophobic in a few, and hydrophilic in the remainder.
2. The alpha helical structure of protein minimizes the hydrophilic character of the peptide bonds. Proteins are, therefore, amphipathic. Different proteins perform different functions in membranes.
3. The outside location of the carbohydrates attached to membrane proteins provides an inside outside asymmetry.

Integral and peripheral membrane proteins

1. The membrane proteins are mostly integral components of the membrane. The integral proteins are usually globular in shape and are amphipathic. They contain two hydrophilic ends separated by a hydrophobic region.
2. The integral proteins are asymmetrically distributed across the membrane bilayer. If they are dissolved in detergent and the detergent is then slowly removed, the phospholipids and the integral proteins will assemble.
3. Peripheral proteins do not interact directly with the phospholipids in the bilayer but are weakly bound to the hydrophilic regions of specific integral proteins.
4. The immunoglobulin molecules on the plasma membranes of lymphocytes are integral membrane proteins and are released by the shedding of small fragments of the membrane.
5. Many hormone receptor molecules are integral proteins and the specific polypeptide hormones which bind to the receptor molecules are considered peripheral proteins.

MEMBRANE ASSEMBLY

1. The enzymes responsible for the synthesis of phospholipids reside on the cytoplasmic aspects of the vesicles of endoplasmic reticulum. The lipid vesicles migrate to the golgi apparatus which in turn fuses with the plasma membrane. Both the golgi complex and the endoplasmic reticulum vesicles exhibit transverse asymmetries of both lipid and protein and these asymmetries are maintained during fusion with the plasma membrane. The inside of the vesicle after fusion becomes the outside of the plasma membrane and the cytoplasmic side of the vesicles remains the cytoplasmic side of the membrane.
2. The integration of proteins into membrane can be explained by two hypotheses. The signal hypothesis supports that the protein is inserted into the membrane simultancously with the translation of its mRNA on polyribosomes. As the leader sequence of the protein emerges from the ribosome, it is recognized by a signal recognition particle (SRP) that blocks further translation after about 70 amino acids have been polymerized—40 buried in the large ribosomal complex and 30 exposed. The SRP contains 6 proteins and has associated with it a 7sRNA.
3. Integral membrane proteins do not completely cross the membrane and are prevented by a hydrophilic C-terminal anchor region. Secreted proteins completely traverse the membrane bilayer and are discharged into the lumen of the endoplasmic reticulum. Carbohydrate moieties are already attached before they reach the interior of that vesicle. The secretory proteins are also found in the lumen of the golgi apparatus.

4. The membrane trigger hypothesis minimizes the role of catalysis in membrane assembly. This hypothesis does not require specific ribosome membrane interaction but this does not mean that protein synthesis on membranes cannot occur.
5. Both the signal and trigger mechanisms exist even in the same cell. Some membrane proteins and secreted proteins are synthesized on membrane-bound polysomes, while others are formed on free cytoplasmic polysomes.

INTERCELLULAR CONTACT AND COMMUNICATION

Cells have developed specialized regions on their membranes for intercellular communication. Gap junctions mediate and regulate the passage of ions and small molecules through a narrow hydrophilic pore connecting the cytoplasms of adjacent cells. These pores are composed of subunits called "connexions". Connexions consist of 6 protein subunits that span the membrane and connect with the analogous structures on the adjacent cells. In response to specific chemical stimuli, the subunits rearrange themselves relative to one another to provide a tangential central opening about 2 nm in diameter and through this central opening, ions and small molecules can pass from one cytoplasm to another in a regulated manner.

Signal transmission

1. Neurotransmitters, hormones, and immunoglobulins bind to specific receptors (integral proteins) exposed to the outside of cellular membranes and transmit information through these membranes to the cytoplasm.
2. The β-adrenergic receptor, which binds catecholamines stereospecifically, is asymmetrically located on the outer aspect of plasma membranes of target cells such as the erythrocytes. The binding of the catecholamine on the outside stimulates the catalytic activity of adenylate cyclase which is asymmetrically located on the inside of the membrane. Adenylate cyclase generates cAMP from ATP. Thus, the information on the outside of the second messenger cAMP.
3. The catecholamine when binds to the beta receptor, the latter by conformational change activates phospholipid methyltransferase I (an integral enzyme). This enzyme generates phosphatidyl-N-monomethylethanolamine from phosphatidylethanolamine. The phosphatidyl-N-monomethyl-ethanolamine flip-flops toward the outside of the membrane and is further methylated to phosphatidyl-N-dimethylethanolamine and subsequently to phosphatidylcholine by the integral enzyme phospholipid methyltransferase II, located toward the outside of the membrane. The increased local concentration of phosphatidylcholine enhances the local membrane fluidity. The increased fluidity exposes more pre-existing betareceptor molecules to the outside membrane surface.
4. Phospholipid methylation serves as an initial common pathway for the transmission of many receptor-mediated biologic signals through membranes. In some systems, the increased fluidity due to phospholipid methylation allows the rapid influx of calcium which can bind to calmodulin and activate specific enzymes including phospholipases. The phospholipases from phosphatidylcholine can in turn generate free fatty acids including arachidonic acid which is an immediate precursor of prostaglandins. The prostaglandins are involved in signal transmission in several transmembranes, hormonally responsive systems.

Endocytosis

1. Endocytosis is a transport process which allows cells to internalize extracellular material and forms endocytotic vesicles

during that process. Endocytotic vesicles are formed when segments of the plasma membrane invaginate enclosing a volume of extracellular fluid and pinch off as the fusion of plasma membranes seals the neck of the vesicle on the original site of invagination. Subsequent fusion of the endocytotic vesicle with other membrane structures performs the transport of its contents to other cellular compartments or even back to the cells exterior.

2. Endocytosis are of two types-Phagocytosis and Pinocytosis. Phagocytosis occurs only in specialized phagocytic cells present in blood. Pinocytosis leads to all cellular uptake of fluid and fluid contents. The adsorptive pinocytosis in a receptor- mediated, selective process primarily responsible for the uptake of macromolecules for which there is finite number of binding sites on the plasma membrane. These high affinity receptors permit pinocytosis to concentrate ligands from the medium and to minimize the uptake of fluid or soluble unbound macromolecules.

3. Several hormones, the other macromolecules, are subject to adsorptive pinocytosis and form receptosomes, vesicles which avoid lysosomes and deliver their contents to the Golgi system.

4. Adsorptive pinocytosis of extracellular glycoproteins requires glycoproteins that carry specific carbohydrate recognition signals. These recognition signals are bound by membrane receptor molecules which play a role analogous to that of the LDL receptor. Acid hydrolases taken up by adsorptive pinocytosis in fibroblasts are recognized by their mannose-6-phosphate moieties. The mannose-6-phosphate moiety also plays an important role in the intracellular segregation of the acid hydrolases to the lysosomes of the cells in which they are sythesized.

5. The other type of pinocytosis is a fluid phase pinocytosis which is a nonselective fluid phase process in which the uptake of a solute is simply proportionate to its concentration in the surrounding medium. This type of pinocytosis forms small vesicles and it is an extraordinarily active process utilizing up to 50 per cent of the plasma membrane per hour in some cells. The components of the membrane must be recycled to maintain cellular integrity.

Plasma membranes are involved in facilitated diffusion, active transport, and other processes

(a) Transport system is described on the basis of function according to the number of molecules moved and the direction of movement.

(b) A uniport system moves one type of molecules bidirectionally.

(c) The cotransport systems cause the transfer of one solute depending upon the sequential transfer of another solute.

(d) A symport system moves these solutes in the same direction (eg, Na^+-amino acid transporters in mammalian cells).

(e) The antiport systems move two molecules in opposite directions (eg, Na^+ in and Ca^{2+} out).

(f) The molecules that cannot pass freely through the lipid bilayer membrane by themselves can pass with the help of carrier proteins. Facilitated diffusion, active transport, and highly specific transport systems are involved in this mechanism.

(g) Facilitated diffusion and active transport show specificity for ions, sugars, and amino acids.

1. Facilitated diffusion

(a) Some specific solutes diffuse across cell membranes under electrochemical gradients more rapidly.

(b) Many facilitated diffusion systems are stereospecific, but require no metabolic energy.

(c) In the "Ping Pong" mechanism, the carrier protein exists in two principal ways. The high concentration of solute and the binding of solutes to specific sites on the carrier protein are found in the "Pong" state. The transport occurs when the carrier protein is exposed to a lower concentration ("Ping" state).

(d) The facilitated diffusion is determined by the following factors :

  (i) The concentration gradient across the membrane.

  (ii) The amount of carrier available.

  (iii) The rapidity of the solute-carrier interaction.

  (iv) The rapidity of the conformational change for both the loaded and unloaded carrier.

(e) Hormones regulate facilitated diffusion by changing the number of transporters available. Insulin enhances amino acid transport in liver and other tissues. Growth hormone increases amino acid transport in all cells, and estrogens do so in the uterus.

2. Active transport

(a) This process requires energy and the energy is derived from the hydrolysis of ATP from electron movement or from light.

(b) Cells maintain a low intracellular Na^+ concentration and a high intracellular K^+ concentration along with a net negative electrical potential inside. The pump that maintains this gradient is an ATPase which is activated by Na^+ and K^+. The ATPase requires phospholipids for activity. The ATPase has catalytic centres for both ATP and Na^+ on the cytoplasmic side of the membrane but the K^+ binding site is located on the extracellular side of the membrane. Quabain (or digitalis) inhibits the ATPase.

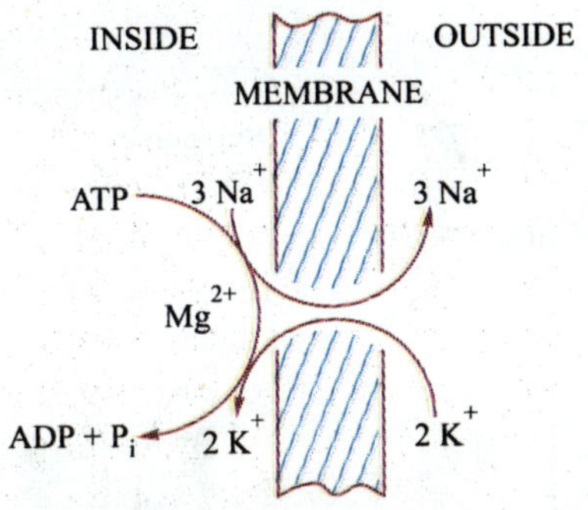

Fig. 38.1 : Stoichiometry of the Na^+-K^+ ATPase pump

3. Nerve impulses

(a) The membrane forming the surface of neuronal cells maintains electrical potential and is electrically excitable.

(b) The gates in the membrane are opened when suitably stimulated by a chemical signal to allow the rapid influx of Na^+ or Ca^{2+}, so that the voltage difference rapidly collapses and that segment of the membrane is depolarized. But the gradient is quickly restored by the action of the ion pumps in the membrane.

(c) When the large areas of the membrane are depolarized, the electrochemical disturbance propagates in a wave-like form down the membrane generating a nerve impulse.

(d) The myelin sheets wrap around nerve fibers and provide an electrical insulator that surrounds most of the nerve and greatly speeds up the propagation of the wave (signal) by allowing ions to flow in and out of the membrane only where the membrane is free of the insulation.

(e) The myelin membrane is composed of phospholipids, cholesterol, proteins, and glycosphingolipids.

(f) Demyelination and impaired nerve conduction cause certain diseases like multiple sclerosis and Guillain-Barre syndrome.

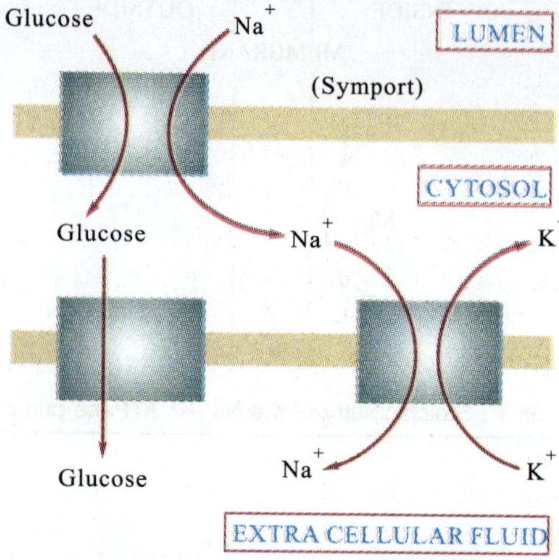

Fig. 38.2 : The transcellular movement of glucose in an intestinal cell. Glucose at high concentration within the cell moves downhill into the extracellular fluid by facilitated diffusion (a uniport mechanism)

4. Glucose transport

(a) Glucose enters in adipocytes and muscle by a specific transport system that is enhanced by insulin.

(b) The changes in transport of glucose are mainly due to alterations of V_{max} but changes in K_m is also noticed.

(c) Glucose and Na^+ bind to different sites on the glucose transporter. Na^+ moves into the cell down its electrochemical gradient and drags glucose with it. The greater the Na^+ gradient, the more glucose enters. Glucose transport stops if Na^+ in extracellular fluid is low. This Na^+-glucose symport is dependent on gradients generated by an Na^+-K^+ pump that maintains a low intracellular Na^+ concentration. Similar mechanisms are also used to transport others sugars and amino acids.

(d) The transcellular movement of sugars requires a uniport which allows the glucose to move across a different surface toward a new equilibrium that occurs in intestinal and renal cells.

DISORDERS OF MEMBRANES

Mutations affecting the protein constituents of membranes can cause many diseases. Proteins in membranes are classified as receptors, transporters, ion channels, enzymes, and structural components. All these classes are often glycosylated, so that mutations affecting this process may change their function. Almost 30 genetic disorders have been attributed to mutations affecting various proteins involved in the transport of amino acids, sugars, lipids, urate, anions, cations, water, and vitamins across the plasma membrane. Mutations in genes encoding mitochondrial membrane proteins involved in oxidative phosphorylation can cause neurological and other problems. Membrane proteins are also affected by conditions other than mutations.

Myasthenia gravis is caused by the formation of autoantibodies to the acetylcholine receptor in skeletal muscle.

Ischemia can promptly affect the integrity of various ion channel in membranes.

Abnormalities of membrane constituents other than protein are also harmful. Excess of cholesterol (eg, in familial hypercholestrolemia), excess of lysophospholipid (eg, the venom after biting of certain snakes contains phospholipases), or excess of glycosphingolipids (eg, in a sphingolipidosis) can affect membrane function.

Multiple sclerosis : It is a somewhat confusing neurologic disorder in which nervous tissue is demyelinated by an unknown process. The demyelinization removes neuronal insulation and profoundly decreases the velocity of nerve impulse transmission.

Pseudohypoparathyroidism : It is an inherited disorder which signifies defective signalling across the plasma membranes of several types of target cells. In that affected cell, the parathormone signal is normally mediated by activation of adenylate cyclase. There is a defective coupling protein in the membrane that prevents transmission of the extracellular hormone signal to the adenylate cyclase.

I cell disease : It is a recessive disorder due to inability of the patient's fibroblasts to attach the mannose-6-phosphate signal to the acid hydrolases synthesized in various cell types. The acid hydrolases do not segregate normally to the primary lysomes but are secreted into the extracellular space leaving the intracellular lysosomes deficient in these important hydrolases. Thus, the lysosomes accumulate excessive debris and thereby generate intracellular cytoplasmic inclusions (I). These abnormal secreted hydrolases are not subject to adsorptive pinocytosis by the patient's or even by normal fibroblasts, since they lack the necessary mannose-6-phosphate recognition signal.

Each chain is divided into specific domains or regions that have structural and functional significance. The half of the light chain (L) toward the carboxy terminus is termed as the constant region (CL), while the amino terminal half is the variable region (VL) of the light chain. About one quarter of the heavy (H) chain at the amino terminus is termed as variable region (VH), and the other three quarters of the heavy chain are referred to as the constant regions (CH^1, CH^2, CH^3) of the H chain.

The portion of the immunoglobulin molecule which binds the specific antigen is formed by the aminoterminal portions (variable regions) of both the H and L chains, i.e., the VH and VL domains. The domains of the protein chains do not simply exist as linear sequences of amino acids but form globular regions with secondary and tertiary structure.

Table 38.1 : Some diseases resulting from abnormalities of membranes

Disease	Abnormality
Wilson's disease	Mutations in the gene encoding a copper-dependent ATPase.
Familial hypercholesterolemia	Mutations in the gene encoding the LDL receptor.
Cystic fibrosis	Mutations in the gene encoding the CFTR protein, a Cl^- transporter.
Metastasis	Abnormalities in the oligosaccharide chains of membrane glycoproteins and glycolipids are also involved.

Table 38.2 : Some diseases resulting from abnormalities of membranes

Disease	Abnormality
Congenital long QT syndrome	Mutations in genes encoding ion channels in the heart.
Paroxysmal nocturnal hemoglobinuria	Mutation resulting in deficient attachment of the GP1 anchor to certain proteins of the red cell membrane.
Achondroplasia	Mutations in the gene encoding the fibroblast growth factor receptor 3.
Hereditary spherocytosis	Mutations in the gene encoding spectrin, a structural protein in the red cell membrane.

CLINICAL ASPECTS

- Water balance and ion flux and even every process within the cell are affected by the gross change of membrane structure.
- The change in the components of membrane or particular deficiencies result in a variety of diseases, eg, the lysosomal absence of acid maltase causes type II glycogen storage disease; the lack of an iodide transporter causes congenital goiter; and defective endocytosis of low-density lipoproteins (LDL) results in enhanced hypercholesterolemia and coronary artery disease.
- Normal membranes cause normal cellular function.
- Peroxisome is an important organelle where there is the metabolism of many molecules including fatty acids and other lipids, amino acids, purines, and hydrogen peroxide. Each peroxisome contains about 50 enzymes of which catalase and urate oxidase are the marker enzymes. Zellweger syndrome and other related diseases have been observed due to abnormalities of peroximal proteins or enzymes. Zellweger syndrome is characterized by profound neurologic impairment and victims often die within a year. Peroxisomes are virtually absent in patients with this syndrome. Biochemical findings include an accumulation of very long chain fatty acids, abnormalities of the synthesis of bile acids, and a marked production of plasmalogens. The condition is due to mutations in genes encoding certain proteins (so called peroxins) involved in many steps of peroxisomes biogenesis. Two closely related conditions are neonatal adrenoleukodystrophy and infantile Refsum disease.

CLINICAL ASPECTS

- Mutations affecting the protein constituents of membrane result in many diseases or disorders. Proteins in membranes can be classified as receptors, transporters, ion channels, enzymes, and structural components. Members of all of these classes are often glycosylated for which mutations affecting this process may alter their function. About 30 genetic diseases or disorders have been ascertained to mutations affecting various proteins involved in the transport of amino acids, sugars, lipids, urate, anions, cations, water, and vitamins across the plasma membrane. Mutations in genes encoding mitochondrial membrane proteins can cause neurologic and other problems. Formation of autoantibodies to the acetylcholine receptor in skeletal muscle causes myasthenia gravis. Ischemia can quickly affect the integrity of various ion channels in membranes. Abnormalities of membrane constituents other than proteins can also be harmful. Excess of cholesterol, of lysophospholipid or of glycosphingolipids can all affect membrane function.

EXERCISE

1. Discuss in details the membrane.
2. Write short notes on : (B.H.U. '90A)
   (a) Membrane assembly. (B.U. '91A)
   (b) Disorders of membranes. (P.U. '92A)

CHAPTER 39

GLYCOPROTEINS, PROTEOGLYCANS, AND GLYCOSAMINOGLYCANS

These three types of molecules have the common structure, synthesis, degradation, and even function.

Glycoproteins differ from other proteins with the oligosaccharide chains covalently attached to their polypeptide backbones.

Proteoglycans are also proteins to which oligosaccharide chains are covalently attached to the polypeptide backbone, but the oligosaccharides chemically differ from those of glycoproteins. The oligosaccharides consist of repeating units of (i) glucosamine or galactosamine, (ii) a uronic acid (except for keratan sulphate) and (iii) covalently attached sulphate groups (except for hyaluronic acid).

Glycosaminoglycans are oligosaccharide structures that have been removed from the protein backbone of their proteoglycan precursor.

All these three classes of molecules are mostly present in the extracellular space but are synthesized intracellularly.

Glycoprotein	:	Protein–oligosaccharide
Proteoglycan	:	Protein–oligosaccharide (uronic acid/or SO_4) – Protein
Glycosaminoglycan	:	Oligosaccharide (uronic acid/or SO_4)

GLYCOPROTEINS

1. Their molecular weight ranges from 15,000 to over 1 million containing 15 or fewer sugar units per chain.
2. Their carbohydrate contents range from 1 to 85 per cent by weights.
3. They are present in plants, bacteria, fungi, viruses and animals. The most membrane proteins and secreted proteins are glycoproteins.
4. They act as structural molecules in cell walls, collagen, elastin, fibrins, and bone matrix; as lubricants and protective agents in mucins, mucus secretions.
5. They are utilized as transport molecules for vitamins, lipids, minerals and trace elements; as immunologic molecules for immunoglobins, histocompatibility antigens, complement and interferon; as hormones in chorionic gonadotropin, thyrotropin (TSH); as enzymes in proteases, nucleases, glycosidases, hydrolases, and clotting factors; as recognition sites in cell-cell, virus-cell, bacterium-cell, and hormone receptors.

Oligosaccharides of Glycoproteins

1. The oligosaccharide chains contain nine different sugar residues. Glucose (Glc) is found only in collagen, but galactose (Gal)

and mannose (Man) are more common and widely distributed. The hexoses are N-acetylgalactosamine (Gal NAC) and N-acetylglucosamine (Glc NAC). Fucose (Fuc) is a common constituent. Two pentoses-arabinose (Ara) and Xylose (Xyl) are found and the ninth are the sialic acids (Sial) of which N- acetylneuraminic acid (Nana) is an example. The fucose and Nana residues are more distal in the chain, frequently at terminal sites.

2. The oligosaccharide chains are attached to the polypeptide backbone at one of five amino acid residues–asparagine (Asn), serine (Ser), threonine (Thr), hydroxylysine (Hyl), or hydroproline (Hyp). Two types of chemical bonds that provide the attachment sites are (a) O-glycosidic links and (b) N-glycosidic links.

(a) O-glycosidic Links

(i) The O-glycosidic links occur through the free alcoholic groups of Ser or Thr residues of the polypeptide in a tripeptide sequence of Asn-Y-Ser (Thr), where Y is an amino acid other than aspartic acid.

(ii) Gal NAC is the most common sugar residue attached directly to the Ser or Thr residue. Six different types of oligosaccharide are attached to this Gal NAC-Ser (Thr) linkage.

(iii) The initiation and extension of different types of oligosaccharide chains of glycoproteins occur by the stepwise donation of sugar residues from pyrimidine or purine nucleotide sugars.

(iv) Oligosaccharides may be linked to proteins via O-glycosidic bonds to Hyl or Hyp which are amino acid residues found in collagens and some fibrous proteins of plants.

(b) N-glycosidic linkage

(i) The N-linked oligosaccharide consists of a core region with the structure Man-β-1, 4-Glc NAC-β-1, 4-Glc NAC-Asn. This core region is of two types—the high mannose (simple) type and the complex type. A single protein can contain oligosaccharide chains of both high mannose and complex types.

(ii) Although all high-mannose oligosaccharides are synthesized from nucleotide sugars, there exists an important lipid-linked precursor oligosaccharide that is transferred en bloc from a lipid carrier to the Asn of the protein.

(iii) The complex N-linked oligosaccharides also contain the β-man-di-N-acetylchitobiose core structure but consist also of a variable number of outer chains containing Sial, Gal, and Fuc residues linked to the core.

(iv) Complex N-linked oligosaccharide structures are found only in higher animals; whereas the high mannose types are common in primitive organisms.

The synthesis of Complex Carbohydrates of Glycoproteins

(i) The nucleotide of sialic acid, CMP-Sial, is formed from CTP by **sialyltransferases** located in the **Golgi complex** and in the nucleoplasm.

(ii) In animal cells, the sugars are linked to the nucleotides by the alpha-linkage with the exception of the beta-linkage of L-fucose to GDP. The alpha-bridges are converted to the beta-bridges and vice-versa during the transfer of the sugar moiety to the oligosaccharide.

(iii) A number of specific glycosyl transferase enzymes catalyze the transfer of the sugar moieties to generate the complex glycoproteins. These enzymes require Mn^{++}.

(iv) A golgi-localized enzyme, **UDP Glc NAC transferase 1**, can then denote a Glc NAC to a linear or branched alpha-Man moiety to form Glc NAC-β-1, 2-Man linkages. A second transferase, **UDP Glc NAC trans-**

ferase 11, will denote its Glc NAC moiety only to a branched structure by the transferase 1 enzyme.

(v) Fucosyltransferases can then act on the products of GLc NAC transferase 1 or transferase 11. The galactosyltansferase enzymes are also located on the Golgi complex and attach a galactosyl residue usually to the end of a chain. The galactosyl residues are linked to Glc NAC by beta-1, 4-linkages but occasionally by beta-1, 6-linkages.

(vi) Four different sialyltransferase enzymes are found in the Golgi complex and use CMP-sialic acid as donor for the sialation of protein-linked oligosaccharides.

(vii) The elongation process generating the complex type oligosaccharides of glycoproteins occurs exclusively in the Golgi complex. Each linkage is carried out by a specific glycosyl-transferase; thus, there seems to be a "one linkage, one glycosyltransferase" synthetic arrangement.

Lectins can be used to purify glycoproteins

1. Lectins, the sugar-binding protein, that precipitate glycoconjugates. Immunoglobulins that react with sugars are not lectins. Lectins contain at least two sugar-binding sites; proteins with a single sugar-binding site will not precipitate glycoconjugates.
2. Enzymes, toxins, and transport proteins can be classed as lectins if they bind carbohydrate.
3. Lectins such as concanavalin A (con A) can be attached covalently to inert supporting media such as sepharose. The resulting sepharose–con A may be used for the purification of glycoproteins.
4. Smaller amounts of certain lectins are required to cause agglutination of tumor cells than of normal cells.
5. When mammalian cells in tissue culture are exposed to appropriate concentrations of certain lectins (eg, Con A), most are killed, but a few resistant cells survive. Such cells are found to lack certain enzymes involved in oligosaccharide synthesis. The cells are resistant because they do not produce oligosaccharide chains that interact with the lectin used.

BLOOD GROUP ANTIGENS

In 1900, Landsteiner described the ABO blood groups. Today, there are more than twenty blood group systems expressing more than 160 distinct antigens. These erythrocyte antigens are linked to specific membrane proteins by O-glycosidic bonds in which Gal NAC is the most proximal sugar residue. The specific oligosaccharides exist in three forms : (i) as glycosphingolipids and glycoproteins on the surfaces of erythrocytes and other cells, (ii) as oligosaccharides in milk and urine, and (iii) as oligosaccharides attached to mucins secreted in the gastrointestinal, genitourinary, and respiratory tracts.

Four independent gene systems are related to the expressions of these oligosaccharide antigens.

Genetic Locus		Alleles
H	....	H, h.
Secretor	....	Se, se.
ABO	....	A, B, O.
Lewis	....	Le, le.

The H Locus : This codes for a fucosyltransferase that attaches a fucose residue in alpha-1, 2-linkage to a Gal residue, itself attached in beta-1, 4-linkage to an oligosaccharide. Fuc-α-1, 2- Gal-β-R is a precursor for the formation of both the A and B oligosaccharide antigens. The h allele of the H locus codes for an inactive fucosyltransferase. Therefore, individuals with the hh genotype cannot generate this necessary precursor of the A and B antigens. Hence hh genotypic persons will be type O.

The secretor locus : This controls the appearance of the **H-specific Fuctransferase** in some secretory organs, such as the exocrine glands, but not in the erythrocytes. Accordingly, individuals with the Hh or HH genotype and an Se allele will generate the A and B antigen precursor in the exocrine glands that form saliva. The individuals who are SeSe or Sese and possess an H allele will be secretors of the A or B antigens (or both), when the A- or B-specific transferase are present. Individuals who are sese genotype will not secrete A or B antigens; but if they possess an H allele and A or B allele, their erythrocytes will express the A, B or both antigens.

The ABO locus : These codes for two specific transferases that act to transfer specific Gal moieties to the Fuc-α-1, 2-Gal-β-R precursor oligosaccharide formed by the action of the H allele-coded fucosyltransferase. Persons possessing an A allele will attach a Gal NAC moiety to the precursor generated by the H allele transferase and an individual possessing a B allele will transfer a Gal moiety to the same precursor. Individuals possessing both A and B allele will generate both A and B alleles (OO homozygotes) will not attach either Gal NAC or Gal to the precursor. When neither Gal NAC nor Gal is at the reducing terminus of this oligosaccharide, it will not be recognized by either anti-A or anti-B antisera, and the blood group antigen is said to be **type O**. Individuals with the hh genotype incapable of attaching the Fuc moiety to the appropriate Gal-β-R oligosaccharide is incapable of expressing the A or the B antigen determinant and thus is considered to be of the O type blood group.

The Lewis Locus : The Lewis-dependent **fucosyltransferase** is not specific about what is not the Gal-1, 3-β group. When no H allele is present (hh), the product of the Lewis α-1, 4-fucosyl- transferase is referred to as the Le^a antigen which cannot have A or B antigenicity even when the A or B transferases are also present. When both the H allele and the Le allele fucosyltransferases have acted on the Gal-1, 3-R oligosaccharide, the product is referred to as the Le^b antigen. The Le^b antigen may also exist without A antigenicity or B antigenicity on the same molecule. The le allele codes for an inactive Lewis transferase, and thus neither Le^a nor Le^b antigens will be formed in a person with lele genotype.

Role of glycoproteins in fertilization

(a) A sperm has to traverse the zona pellucida (ZP) which contains three glycoproteins ZPI-3, particularly ZP3, (an O-linked glycoprotein that functions as a receptor for the sperm) to reach the plasma membrane of an oocyte.

(b) A protein on the sperm surface interacts with oligosaccharide chains of ZP3. This interaction induces the acrosomal reaction in which proteases and hyaluronidase, and other contents of the acrosome of the sperm are released. The liberation of these enzymes helps the sperm to pass through the zona pellucida and reach the plasma membrane of the oocyte.

(c) Another glycoprotein pH-30 is important in binding of the PM of the sperm to the PM of oocyte. These interactions enable the sperm to enter and thus fertilize the oocyte.

(d) It is also possible to inhibit fertilization by developing drugs or antibodies that interfere with the normal functions of ZP3 and PH-30 and which thus act as contraceptive agents.

Certain diseases in the abnormalities in the synthesis of glycoproteins

(a) Many cancer cells exhibit different profiles of oligosaccharide chains on their surfaces and some of which contribute to metastasis.

(b) **Carbohydrate-deficient glycoprotein syndrome** (**CDGS**) is caused due to defects in the biosynthesis of N-glycan chains (eg, due to deficiency of GIc NAC transferase (II) particularly affecting the central nervous system.

(c) **Paroxysmal nocturnal hemoglobinuria** is an acquired mild anemia characterized by the presence of the hemoglobin in urine due to hemolysis of red cells, particularly during sleep. This disease is caused by the acquired defect in the biosynthesis of glucophosphatidylinositol (GPI) structures of decay accelerating factor (DAF) and another protein CD59.

(d). I-cell disease is an uncommon condition characterized by severe progressive psychomotor retardation and a variety of physical signs with death even occurring in the first decade. The patients suffering from this disease lack almost all of the normal lysosomal enzymes. These enzymes ultimately lack Man 6-P and are secreted from cells (eg, into the plasma) rather than targeted to lysosomes. These lysosomes are thus deficient in certain hydrolases, do not function properly and accumulate partly digested cellular material manifesting as inclusion bodies.

(e) Pseudo-Hurler polydystrophy is another genetic disease closely related to I-cell disease. It is a milder condition and patients may survive to adulthood. In this disease, the Glc NAC phosphotransferase interacts with lysosomal enzymes and the retention of some catalytic activity results in a milder condition.

(f) The degraded glycoproteins can lead to various diseases. The best recognized of these diseases are mannosidosis, fucosidosis, sialidosis, aspartylglycosaminuria, and Schindler's disease due to respectively deficiencies of α-mannosidase, α-fucosidase, α-neuraminidase, aspartylglucosaminidase, and N-acetylgalactosaminidase. These diseases usually exhibit mental retardation or other neurologic abnormalities. Vacuolization of cells are also observed in some disorders. The abnormal degradation products eg, oligosaccharides that accumulate due to enzyme deficiency are present in urine in these diseases. No confirmative treatment has yet been established for these diseases.

(g) Glycoproteins have also direct or indirect roles in a number of other diseases such as

(i) The influenza virus has a neuraminidase that eludes the newly synthesized progeny from infected cells.

(ii) HIV-I, the causative agent of AIDS, attaches to cells via one of its surface glycoproteins, gp 120.

(iii) Rheumatoid arthritis is associated with an alteration in the glycosylation of circulating immunoglobulin-γ (IgG) molecules, such that they lack galactose in their Fc regions and terminate in Glc NAC. Mannose-binding protein, a C-lectin synthesized by liver cells and secreted into the circulation, binds mannose, Glc NAC, and certain other sugars. It can thus bind the galactosyl IgG molecules contributing to chronic inflammation in the synovial membranes of joints. This protein can also bind the above mentioned sugars when they are present on the surfaces of certain bacteria, fungi, and viruses preparing the pathogens for destruction by the complement system. Deficiency of this protein in young infants, due to mutation, renders them very susceptible to recurrent infections. The effective treatments for diseases in which these molecules are involved fully depend on the studies of glycoproteins and other glycoconjugates.

PROTEOGLYCANS

Each polysaccharide of proteoglycans consists of repeating disaccharide units in which D-glucosamine or D-galactosamine is always present. Each disaccharide contains a uronic acid, glucuronic acid (GlC UA), L-iduronic acid (ldUA). All polysaccharides contain sulphate groups with the exception of hyaluronic acid.

The linkage of the polysaccharides to their polypeptide chain is one of three types :

(i) An O-glycosidic bond between Xyl and Ser, a bond that is unique to proteoglycans.

(ii) An O-glycosidic bond between Gal NAC and Ser (Thr), present in keratan sulphate II.

(iii) An N-glycosylamine bond between GlC NAC and the amide nitrogen of Asn.

The elongation process of chain involves the nucleotidyl sugars acting as donors. The reactions are performed by the substrate specificities of the specific glycosyltransferases. Thus, "one enzyme, one linkage" relationship holds. The specificity of these reactions is dependent upon the nucleotide sugar donor, the acceptor oligosaccharide.

The polysaccharide chain growth termination results from (i) capping effects of isolation by the specific sialyl transferases, (ii) sulfation, particularly at the 4-positions of the sugars, and (iii) the progression of the particular polysaccharide away from the site in the membrane where the catalysis occurs. After formation of the polysaccharide chain, numerous chemical modifications take place.

Inherited defects in the degradation of the polysaccharide chains lead to the group of diseases known as **mucopolysaccharidoses** and **mucolipidoses**.

Seven types of polysaccharides are covalently attached to the proteins of proteoglycans. Six of them contain alternating uronic acid and hexosamine residues. Except hyaluronic acid all contain sulphated sugars. These seven types of polysaccharides are distinguished by their monomer composition, their glycosidic linkage, and the amount and location of their sulphate substituents.

Functions of Glycosaminoglycans and Proteoglycans

1. **Glycosaminoglycans** can interact with **extracellular macromolecules**, plasma proteins, cell surface components, and **intracellular macromolecules** :
2. Because of their polyanionic nature the binding of this is generally electrostatic.
3. These with IdUA bind proteins with greater affinities than those containing GlcUA as their only uronic acid constituent.
4. The binding between these and other extracellular macromolecules contributes to the structural organization of connective tissue matrix.

A. Interactions with extracellular macromolecules

(i) All glycosaminoglycans except those that lack sulphate groups or carboxyl groups bind to collagen electrostatically at neutral pH. Tighter binding is promoted by the presence of IdUA and the proteoglycans interact more strongly than glycosaminoglycans.

(ii) The chondroitin sulphate and keratan sulphate chains of proteoglycans aggregate with hyaluronic acid.

B. Interactions with plasma proteins

(i) **Dermatan sulphate** binds plasma lipoproteins and appears to be the major glycosaminoglycan synthesized by arterial smooth muscle cells. This dermatan sulphate may play an important role in the development of **atherosclerosis**.

(ii) **Heparin** with its high negative charge density (due to the IdUA and sulphate residues) interacts strongly with several plasma components. It interacts with antithrombin 111. Heparin sulphate is also capable of accelerating the action of antithrombin 111, but is much less potent than heparin. Heparin can bind to **lipoprotein lipase** present in capillary walls and causes a release of that triglyceride-degrading enzyme into the circulation. Hepatic lipase also binds heparin but with lower affinity.

C. Cell surface molecules

(i) Heparin associates with blood platelets, arterial endothelial cells, and liver cells.

Chondroitin sulphate, dermatan sulphate, and heparan sulphate bind to independent sites on surface of cells such as fibroblasts. At those sites, the glycosaminoglycans and proteoglycans are taken up by fibroblasts and degraded.

(ii) Some proteoglycans serve as receptors and carriers for macromolecules. These proteoglycans are involved in the regulation of cell growth.

D. Intracellular macromolecules

(i) Proteoglycans and their glycosaminoglycan components have effects on protein synthesis and intranuclear functions. Glycosamino-glycans are found in significant quantities in nuclei from different cell types.

(ii) The acid hydrolases in lysosomes may be naturally complexed with glycosaminoglycans to provide a protected and inactive form. Chondroitin sulphates, dermatan sulphates, and heparin can affect the activities of various lysosomal acid hydrolase in negative or positive ways.

(iii) Many storage or secretory granules such as the chromaffin granules in adrenal medulla, the prolactin secretory granules in the pituitary gland, and the basophilic granules in mast cells contain sulphated glycosaminoglycans. The glycosaminoglycan-peptide complexes that occur in these granules play a role in the release of biogenic amines.

Tunicamycin inhibits N- but not O-Glycosylation

1. Many compounds are involved in inhibiting various reactions of glycoproteins. Tunicamycin, deoxynojirimycin, and swainsonine are the agents which can be used experimentally to inhibit various stages of glycoprotein biosynthesis. If cells are grown in the presence of tunicamycin, no glycosylation of their normally N-linked glycoproteins will occur. In certain cases, lack of glycosylation increases the susceptibility of these proteins to proteolysis.
2. Inhibition of glycosylation does not have a consistent effect upon the secretion of glycoproteins that are normally secreted.
3. The inhibitors of glycoprotein processing do not affect the biosynthesis of O-linked glycoproteins. The extension of O-linked chains can be prevented by GalNAC-benzyle. This compound competes with natural glycoprotein substrates.

Glycoproteins are involved in many biologic processes and in many diseases

In addition to the different functions of glycoproteins, it is also involved in two specific processes — fertilization and inflammation.

1. Importance of glycoproteins in fertilization

(a) A sperm has to traverse the Zona Pellucida (ZP), a thick, transparent, noncellular envelope that surrounds the oocyte to reach the plasma membrane of an oocyte.

(b) The Zona Pellucida contains three glycoproteins, ZPI-3.ZP3 is an O-linked glycoprotein that functions as a receptor for the sperm. A protein on the sperm surface, galactosyl transferase, interacts with oligosaccharide chains of ZP3. This interaction induces the acrosomal reaction in which enzymes (proteases and hyaluronidase) of the sperm are released and this helps the sperm to pass through the Zona Pellucida and reach the plasma membrane of the oocyte.

(c) Another glycoprotein, pH-30, is important in binding of the PM of the sperm of the oocyte and also in the fusion of the two membranes. These interactions enable the sperm to enter and thus fertilize the oocyte.

(d) It is also possible to inhibit fertilization by developing drugs or antibodies that interfere with the normal functions of ZP3 and PH-30 and thus act as contraceptive agents.

2. Selectins play key roles in inflammation

(a) Leukocytes have important roles in many inflammatory and immunologic phenomena. Leukocytes and endothelial cells contain specific **lectins**, called selectins, on their surfaces.

(b) Selectins are single-chain Ca^{2+}-binding transmembrane proteins that contain a number of domains. Their amino terminal-ends contain the lectin domain which is involved in binding to specific carbohydrate ligands.

(c) The adhesion of neutrophils to endothelial cells of postcapillary venules occurs in several stages. The initial baseline stage occurs by slowing or rolling of the neutrophils being mediated by selectins. During this stage, activation of the neutrophils by various chemical mediators occurs resulting in a change of shape of the neutrophils and firm adhesion of these cells to the endothelium. An additional set of adhesion molecules is involved in firm adhesion, namely, LFA-1 and Mac-1 on the neutrophils and ICAM-1 and ICAM-2 on endothelial cells.

(d) The selectins bind **sialylated** and **fucosylated oligosaccharides**, and in particular all bind **sialyl-Lewis**, a structure present on both glycoproteins and glycolipids. Sulfatides may be ligands in certain instances. It is tried to synthesize compounds that block selectin-ligand interactions and thus may inhibit the inflammatory response.

(e) Cancer cells exhibit sialyl-Lewis and other selectin ligands on their surfaces. These ligands play a role in the invasion and metastasis of cancer cells.

CLINICAL OREINTATION

- A good number of blood group substances are glycoproteins and others are glycosphingolipids.
- Certain hormones (eg, chronic gonadotropin) are glycoproteins.
- Abnormal genetic regulation results in **cancer**. The remarkable problem in cancer is **metastasis** in which cancer cells leave their tissue of origin, migrate through the blood stream to some distant site in the body, and grow there in an unregulated manner with catastrophic results for the affected individual.
- It is the idea of many researches that change in the structure of glycoproteins and other glycoconjugates on the surface of cancer cells are responsible for the phenomenon of metastasis.

EXERCISE

1. Write about glycoproteins. (Delhi University '89A)
2. What do you know about proteoglycans? (P.U. '91A)
3. Discuss blood group antigens. (B.H.U. '92A)

CHAPTER 40

BIOCHEMICAL ANALYSIS OF NERVE TISSUE

Introduction

Two main systems namely the nervous system and the endocrine system regulate and control the functions of the body. The endocrine system mainly regulates the metabolic functions of the body which are comparatively sluggish. The nervous system controls the rapid activities of the body namely muscular contractions, respiration, and the rates of secretion of some endocrine glands such as adrenal medulla and neurohypophysis.

The nervous system includes brain, spinal cord, cranial and peripheral nerves, ganglia and plex-

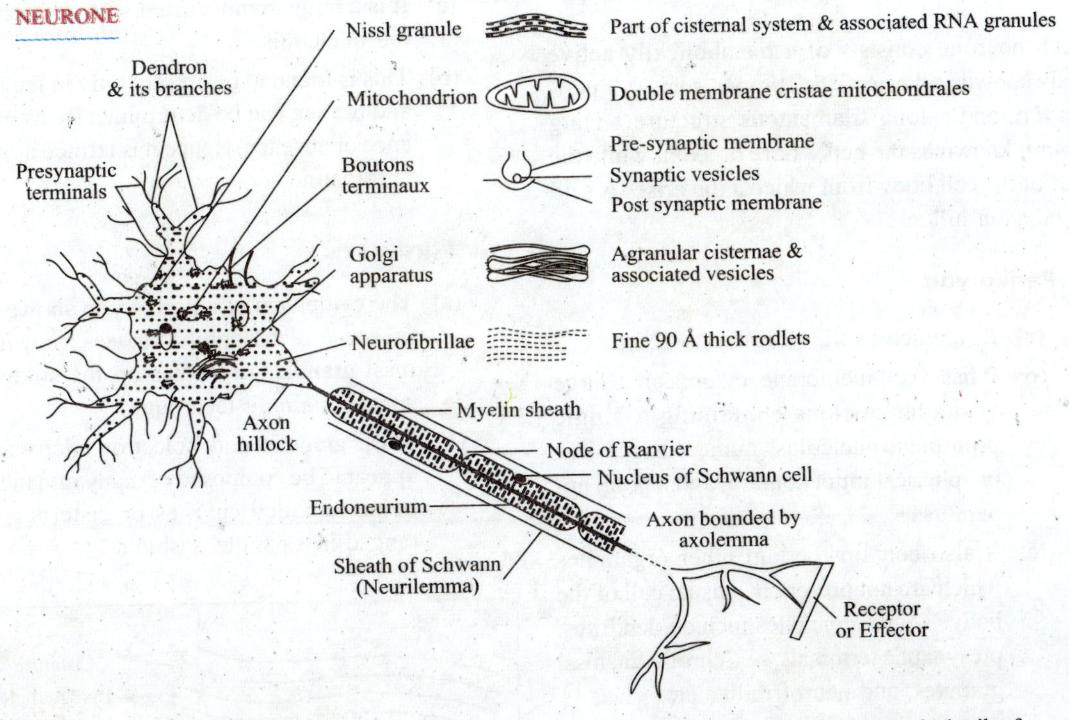

Left : Diagrammatic representation of the structure found in the nerve cell under the light microscope

Right : The electron microscopic details of these structures

Fig. 40.1 : Structure of nerve tissue

uses. This system has the capacity to transmit rapidly a number of informations at a time from one part of the body to another. Although the detailed informations regarding the functions of nervous system are not yet known, still some of the biochemical informations as regards to the transmission of the nervous system have been discussed.

Structural and functional units

Nevous system comprises about 2.5 per cent of the total body weight. This consists of three sub-systems namely central, peripheral, and autonomic nervous system. The central nervous system includes the brain and the spinal cord. The peripheral includes various cranial and peripheral nerves. The autonomic nervous system is divided into sympathetic and parasympathetic systems. The nervous tissue consists of a group of highly specialised cells known as neurones.

Structure of a neurone

Each neurone consists of a metabolically active body known as the nerve cell proper, soma-, or perikaryon, and a long filamentous structure arising from it known as the nerve fibre or axon; while the part of the cell body from which axon arises is said to be axon hillock.

1. Perikaryon

(a) Its diameter varies from 5μ to 125μ.

(b) It has a cell membrane, cytoplasm, a large vesicular nucleus containing a single prominent nucleolus, numerous rod-like or spherical mitochondria, and a golgi apparatus.

(c) It also contains certain other organelles which are not present in a usual cell of the body. These organelles include dendrites, presynaptic terminals, sex-chromatin, nissl granules, and neurofribillae etc.

Dendrites

(a) The cell membrane is projected in the form of a number of filamentous structures known as dendrites.

(b) They may vary from almost none to one metre in length (peripheral sensory nerve fibre).

(c) Each dendron contains nissl granules, mitochondria, and neurofibrillae.

(d) Each dendron on leaving the cells goes on branching repeatedly.

Presynaptic terminals

(a) These are small knob like structures which exist on the surfaces of the dendrites and perikaryons.

(b) They are the ends of the nerve fibrils that originate in other neurones next to on which they come to exist.

Sex-chromatin

(a) It is a large granule often seen adjacent to the nucleolus.

(b) This is found to be present only in females and the sex can be determined by its presence or absence. Hence it is termed as sex-chromatin.

Nissl granules

(a) The cytoplasm of a perikaryon shows the presence of basophilic masses known as nissl granules or bodies on the study of special staining techniques.

(b) These granules under electron microscope appear to be composed of many thin membrane bounded cavities or cisternae arranged in a parallel fashion.

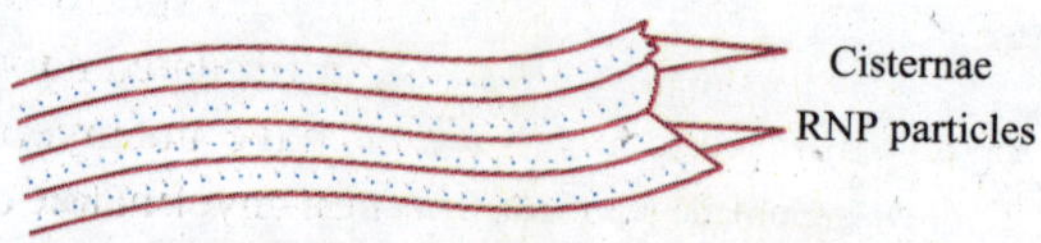

Fig. 40.2 : Electron microscopic structure of nissl granules

(c) Their sizes and number vary with the physiological conditions of the cell, eg, fatigue, the action of certain poisons. The granular disintegrate into a fine dust which ultimately disappears on the section of the axon.

(d) These granules are similar to the (less numerous) rough surfaced membranes found in the cytoplasm of most cells which collectively make-up the endoplasmic reticulum.

(e) The minute particles consist chiefly of ribose nucleoprotein (RNP) which stain with basic dyes.

Neurofibrillae

These are aggregation of thin, thread-like structures of variable length and 60-100Å in diameter forming a loose network of fibrils in the cytoplasm.

2. Axon

(a) The axon or nerve fibre arises from the axon-hillock of the perikaryon which is rich in neurofibrillae but has no nissl granules.

(b) It consists of a central core of semifluid axoplasm which flows from the cell body to the axon and not in the reverse direction.

(c) Axoplasm also contains mitochondria, nissl granules, neurofibrillae running parallel to the long axis of the axon. Axoplasm is bounded by a cell membrane known as axolemma which is continuous with the cell membrane of parent cell body.

(d) The axon is further surrounded by a sheath known as myelin sheath or medullary sheath which is not present in all the fibres. The fibres having myelin sheath are known as myelinated or medullated nerve fibres and other which have no myelin sheath are known as non-myelinated or non-medullated nerve fibres. The myelinated nerve fibres are also said to be white nerve fibres on account of the presence of lipids in the myelin sheath. The non-myelinated nerve fibres appear grey for which they are known as grey nerve fibres. All the post-ganglionic fibres of the autonomic nervous system are non-medullated; while all the pre–ganglionic fibres are medullated.

(e) A thin layer of fine reticular fibres surrounding the myelin sheath forms the endoneurium or sheath of Henle.

(f) The axolemma at the terminal end of the axon is also thrown into similar projections—dendrites which come in contact either with the dendrites of the perikaryon of next neuron forming synapses or receptor or effector (eg, myoneuronal junction).

3. Myelin sheath and Myelinogenesis

(a) A specialized set of Schwann cells arranged in a linear manner all around the axolemma forms myelin sheath by a process known as myelinogenesis. In this process, the schwann cells grow round the axon and envelop it completely along its entire length so that cell membrane of the Schwann cell can be divided into two parts—one surrounding the axon, i.e., inner part and other outer part which are connected to one another by the double mesaxon.

(b) The Schwann cells start to rotate and go on rotating for several times. As a result, many closely packed, helically arranged layers of double membrane are wrapped around the axon as shown in figure 40.3. This set of membranes is known as myelin sheath.

(c) Each membrane is composed of two layers of lipid substances sandwiched between layers of protein and they form the myelin sheath of the nerve fibre.

(d) The thickness of the myelin sheath is ascertained by the number of membrane layer wrapped round the axon.

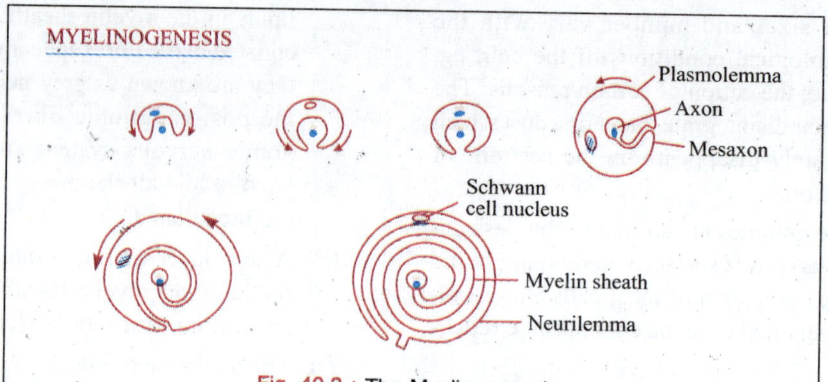

Fig. 40.3 : The Myelinogenesis

(e) The phospholipid sphingomyelin is an excellent insulator that prevents almost all flow of ions outside the nerve fibre.

(f) A small uninsulated area at the junction between each two successive Schwann cells along the axon remains where ions can flow with the case between the extracellular fluid and the axon. The area is said to be the node of Ranvier.

(g) The higher permeability at the nodes of Ranvier causes the impulses to be conducted from node to node by the myelinated nerve fibre rather than continuously along the entire fibre as happen in the unmyelinated nerve fibre. This process is called as saltatory conduction and is helpful in the higher velocity of nerve transmission in myelinated fibres than that in unmyelinated fibres.

(h) The nerve fibres in the central nervous system are myelinated by a different types of cells known as oligodendroglia and by a different process.

(i) The Schwann cells although surround the axon of unmyelinated nerve fibres, they do not spin a myelin sheath around them.

(j) Neurones in different parts of the body differ markedly from one another in the size and shape of the cell body; length, size and number of dendrites; the length and size of the axon; the number of presynaptic terminals which may range from hundred to several thousand.

Structure of a typical nerve

A nerve consists of groups of nerve fibres. Bundles of nerve fibres are enclosed in a connective tissue sheath known as perineurium. A number of such bundles are bounded together by another sheath of connective tissue known as the epineurium to form a nerve.

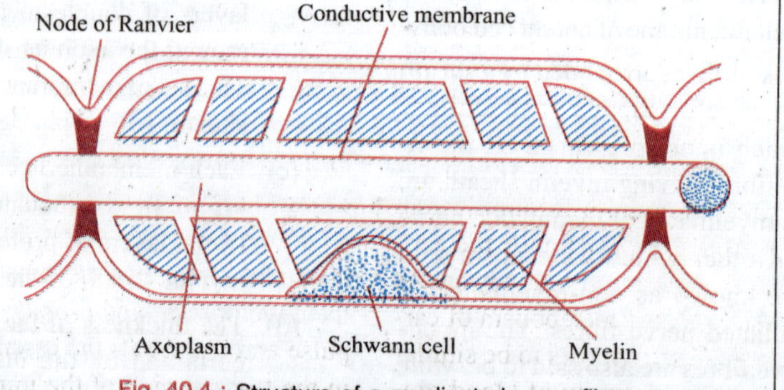

Fig. 40.4 : Structure of a myelinated nerve fibre

Chemical composition of the nerve tissue

Nerve tissue is composed of :

Water .... 80 per cent.

Solids .... 20 per cent.

The solids are mainly composed of proteins, lipids, small amounts of organic extracts and inorganic salts. Proteins are about 38 to 40 per cent of the total solids. They include different globulins, nucleoproteins, and a characteristic albuminoid called *neurokeratin*. The lipid contents are 50 to 54 per cent of the total solids. The important lipids are phospholipids, cholesterol, cerebrosides, aminolipids, and sulphur containing lipids. The principal inorganic salts are potassium phosphate and chloride, with smaller amounts of sodium and other alkaline elements. Potassium is highly significant in the nerve impulse.

The water content of brain is little more than that of spinal cord. In brain too the grey matter which represents a concentration of nerve cell bodies contains more water than the whole matter where the nerve fibres are mainly found. A fraction of the brain proteins remains combined with copper forming ceruloplasmin. An increased deposition of copper is found in the brain tissue in Wilson's disease. White matter and the peripheral nerves contain a little more cerebrosides, free cholesterol, and sphingomyelin than the brain grey matter. The considerable increase in the concentration of sphingomyelin is found in Niemann-Pick disease.

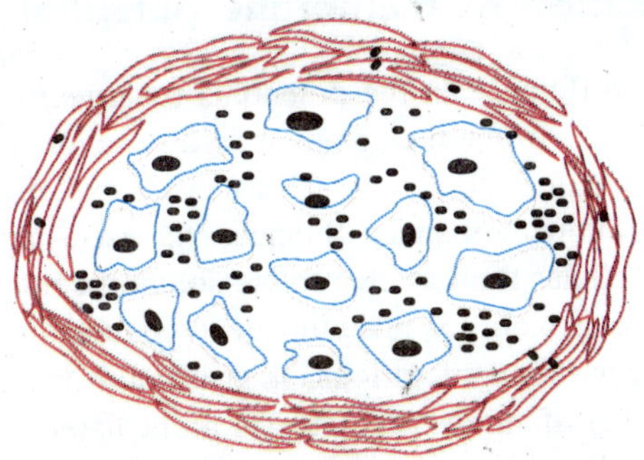

Fig. 40.5 : Cross-section of a small nerve trunk containing myelinated (largely few comprising most of area) and unmyelinated (many more, small and lying between large ones) fibres

Metabolism of the nerve tissue

Since the respiratory quotient (R.Q.) of the metabolic nerve is near about 1 this indicates that the nerve tissues utilize carbohydrates almost exclusively for the energy purposes. Lactic acid and pyruvic acid are formed under anaerobic conditions in the nerve tissue as a result of carbohydrate metabolism and these acids disappear very slowly in the presence of oxygen. Hence, the metabolism of carbohydrates in the nerve tissue appears to be similar to that of muscles. A minute supply of blood glucose is specially important to the nervous system because of the less glycogen storage in the nerve tissue. This may be the vital cause for the prominence of nervous symptoms such as mental confusion, dizziness, delirium, etc. in hypoglycemia.

As regards to protein metabolism, glutamic acid is the only amino acid metabolized by brain tissue. This acid serves as a precursor of γ-aminobutyric acid (GABA), one of the chemical transmitters, and as a major acceptor of ammonia produced either by brain itself or delivered to brain when the arterial blood ammonia is increased and thus protects the brain tissue against its toxic effects. Different workers have also demonstrated the synthesis and exchange of lipids in nerve tissue.

Nerve impulse

Nerve impulse may be defined as an electro-chemical change which is transmitted by nerve fibres. It must not be confused with the stimulus which is the external force (eg, chemical, physical, biological) which sets up the impulse. The chemical changes in the nerve fibres are concerned with the recovery processes which follow the activity. The electrical change in the most certain indicator of the development and propagation of the nerve impulse and represents the essential process involved in the transmission of the impulse along the nerve fibre as discussed in the following pages.

Electrical or membrane potential

The difference in the potentials (charges) on the two sides of the membrane is said to be membrane or electrical potentials. This can be explained in the following manner. Suppose, the extra- and intra-cellular fluid as electrolyte solution containing approximately 155 mEq/litre of anions and the same concentration of cations, then usually an excess number of anions accumulates immediately inside the cell membrane separating the two fluids along its inner surface and an equal number of cations outside the membrane on its outer surface. This results in the development of membrane potential or electrical potential.

Membrane potential is of two types – resting membrane potential and action potential. Resting membrane potential is the potential which exists during the resting state of membrane. Action potential exists during the active state of membrane.

Electrical potential exists across the membranes of essentially all cells of the body.

Resting membrane potential

The resting membrane potential may be defined as the net difference in the charge between inside and outside membrane requiring little or no expenditure of energy. In a neurone, this is about –85mV so long as this neurone is in a resting stage and is maintained almost constant. It is believed to arise (1) from the unequal distribution of ionic substances on the two sides of the membrane due to sodium and potassium pumps. (2) Due to semi- permeability of the membranes.

In the resting condition, the concentration of Na^+ is about 142 mEq/litre outside the cell and 10 mEq/litre inside the cell; while that of K^+ (main diffusible cation) is 5 mEq/litre outside the cell and 140 meq/litre inside the cell due to sodium and potassium pump respectively. There is also a very high concentration of non-diffusible anions, 150 mEq/litre inside the cell in contrast to a very low concentration of only 5 mEq/litre outside the cell. Simultaneously, K^+ is easily diffusible through the membrane; while Na^+ and non-diffusible anions diffuse very poorly due to selective permeability of the membrane. Therefore, K^+ diffuses from inside to outside of the membrane removing positive (+) charge from the inner side of the membrane while building up positive (+) charge on outer surface of the membrane. This results in overall excess of negativity on the inner side as compared to the positivity on the outerside. This comes out to be in between –75 and –95 millivolts with an average of –85 millivolts. This means that the interior of the

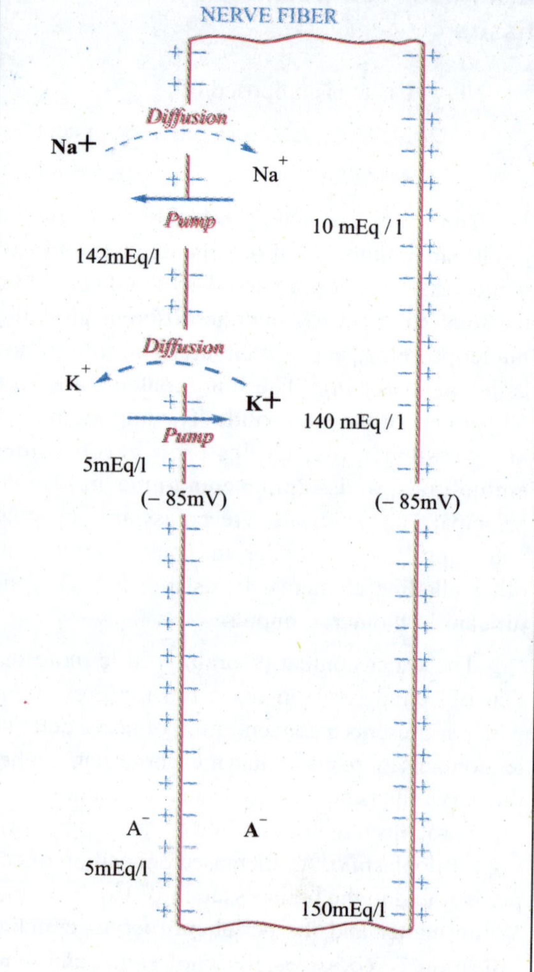

Fig. 40.6 : Establishment of a membrane potential of -85 mV in the normal resting nerve fibre and development of cencentration difference of Na^+, K^+ and A^- (the non-diffusable protein ions) between the two sides of the membrane. The dashed arrows represent diffusion and the solid arrows represent active transport (pumps)

axon (or nerve cell body) is about 85 mv negative with respect to the exterior in the resting stage of the nerve.

As this potential is dependent on diffusion of K^+, it is also called as *diffusion electrical potential*. There can be another type of electrical potential due to active transport of ions.

Calculation of the membrane potential : The resting membrane potential can be calculated by **Nernst equation**. When a concentration difference of ions across a membrane causes diffusion of ions through the membrane creating a membrane potential, the magnitude of the potential is determined by the difference in tendency of the ions to diffuse in one direction versus in the other direction. This is determined by the following formula :

$$E(mv) = \frac{RT}{\eta F} \times 2.3 \log_{10} \frac{[C_t]}{[C_o]}$$

where, E(mv) = Membrane potential in millivolts.

R = Gas constant = 8.3 joules/°C

T = Absolute temperature = 273°C

n = Valency of the diffusible ions

F = Faraday = 96,500 coulombs

C_t = Concentration of diffusible ions inside the cell.

C_0 = Concentration of diffusible ions outside the cell.

At a temperature of 38°C (body temperature), this equation results in :

$$E(mV) = \frac{8.3 \times (273+38)}{\eta \times 96{,}500} \times 2.3 \log_{10} \frac{[C_t]}{[C_0]}$$

$$= \frac{61.5}{\eta} \times \log_{10} \frac{[C_t]}{[C_0]}$$

As K^+ ions are the main diffusible ions and their concentration inside the cell is about 30 times as that outside the cell, so the electrical potential created by this concentration difference of K^+ ions will be :

$$E = \frac{61.5}{1} \log_{10} \frac{[30]}{[1]} = -90 \text{ mV (approximately)}$$

This shows that the calculated value for the K^+ diffusion potential (–90 mV) is normally slightly more than the actual measured value of the resting membrane potential (–85 mV). This is due to non-consideration of Na^+ and non-diffusible protein anions which also have a slight tendency to diffuse to the interior and exterior of the cell respectively. Obviously, either of these effects would reduce the potential to slightly below that calculated for potassium diffusion alone.

Action potential

So long the membrane of the nerve fibre remains completely undisturbed, the membrane potential remains about –85 mV. This is said to be the resting membrane potential. But when this membrane is disturbed by a stimulus (eg, electrical, chemical, mechanical) which can suddenly increase the permeability of this membrane to Na^+ then a sequence of rapid changes in the resting membrane potential takes place. This lasts a minute fraction of a second and is followed immediately by return of the membrane potential to its resting value. This sequence of rapid changes in the electrical potential is said to be the **action potential** which is discussed below :

It has been found that in the resting condition, the external surface of the membrane is positive with respect to its interior. When this membrane is suddenly stimulated, the permeability of the membrane to Na^+ increases transiently lowering the relative permeability to K^+ and, therefore, many of the Na^+ rush to the interior of the fibre carrying sufficient positive charge to the interior to cause complete disappearance of the normal resting potential and usually enough charge to develop a positive (+) state inside the fibre instead of the normal negative (–) state. This positive state inside the nerve fibre is called as the reversal (from negative to positive) potential and this process is said to be the **depolarisation of membrane.**

As the positive potential inside the nerve reaches a peak which is known as the peak potential or spike potential, the process of depolarisation stops and the membrane again becomes al-

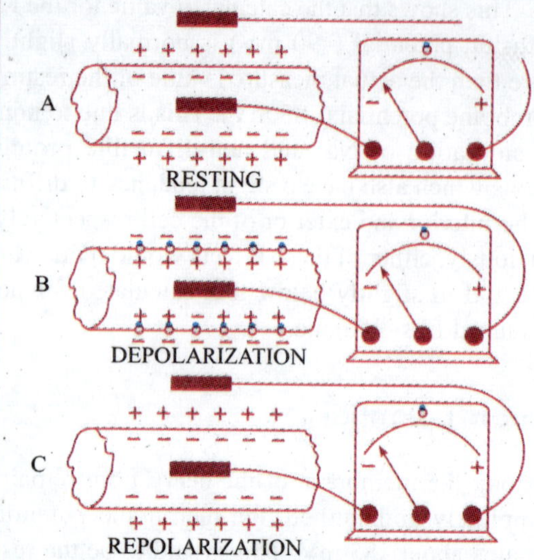

Fig. 40.7 : Sequential events during the action potential

A = The normal resting potential

B = Development of a reversal potential during depolarisation

C = Re-establishment of the normal resting potential during repolarisation

most completely impermeable to Na^+ and relatively more permeable to K^+ so that the K^+ ion leaves the fibre faster than the Na^+ enters the fibre. Hence, the reversal potential inside the fibre disappears and the normal resting membrane potential returns to the site of stimulation and the fibre once more becomes responsive to further stimulation. This process of restoration of the resting membrane potential from the reversal potential is said to be the repolarisation of the membrane.

Mechanism of action potential

The exact mechanism responsible for action potential is not yet known. However, one of the theories to explain it is as follows :

It is postulated that under normal circumstances the sodium "channels" in the membrane are lined with Ca^{++} repelling the Na^+ and other cations. Hence, Ca^{++} resists the passage of Na^+ through these sodium channels. Further, when the membrane is stimulated, these Ca^{++} are dislodged from their binding sites and some Na^+ begin to move inwards. The inward movement of the in–rushing Na^+ theoretically dislodge more and more Ca^{++} from their ending site for which more Na^+ rush inward, and the process is continued till no Ca^{++} remain to resist the passage of Na^+ through Na^+ channels. The diffusion of the Na^+ inside the membrane causes reversal potential with the development of positivity inside the membrane and negativity outside. This positivity inside the membrane reaching the peak of spike potential opposes further inflow of the positively charged Na^+ allowing the Ca^{++} to begin rebinding with the binding sites along the Na^+ channel and the highly concentrated K^+ begin moving outwards through the so-called potassium "channels" – probably separate from sodium channels. As the Ca^{++} bind, Na^+ conductance decreases which allows still more Ca^{++} to bind and thus another vicious cycle sets in which operates in the opposite direction and continues until the membrane becomes almost totally impermeable to Na^+ once again and acquires the similar resting membrane potential.

Propagation of action potential

The electrical stimulation of a nerve fibre initiates at the point of stimulation action potential which in turn excites the adjacent portion of the membrane resulting in the propagation of nerve impulse. Fig. 40.8 shows that when a nerve fibre is stimulated at any point, a sort of current flows inwards through this point and outwards through a part of resting membrane ahead of this, thus completing a circuit. In some unknown way, the current flowing through resting membrane now increases the membrane's permeability to sodium which immediately allows Na^+ to diffuse inward through the membrane and thus the action potential is developed in this area also by the same mechanism as discussed earlier. The newly activated area causes local circuits of current flow still further along the membrane causing progressively more and more depolarisation. Thus, the depolarisation process propagates in both directions from the point of stimulation along the entire length of the nerve fibre. This transmission of the depolarisation process along a nerve

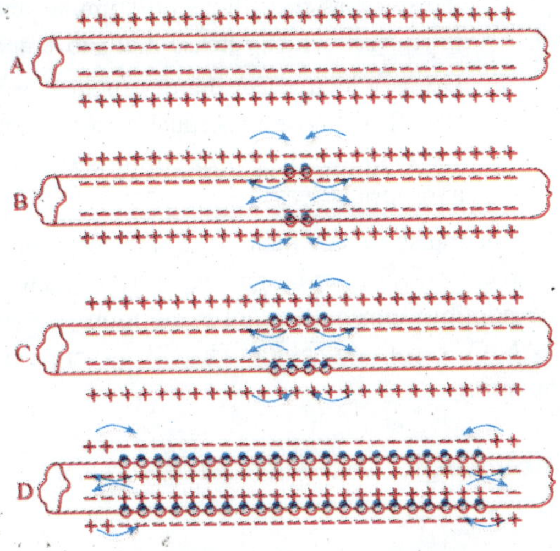

Fig. 40.8 : Propagation of action potential in both directions along a conductive fibre
A = Resting nerve fibre
B = Same nerve fibre has been excited in its mid portion
C and D = Depolarisation of successive portion of the same nerve fibre membrane

or muscle fibre is called as the nerve or muscle impulse.

Within few moments the depolarisation is followed by repolarisation which also propagates in the same direction in that depolarisation had previously propagated as shown figure 40.9 :

Transmission of nerve impulse from one neurone to other

The successive neurones are separated from one another by a narrow slit called as synaptic cleft; while the junction between one neurone and the next is termed as a synapse. Synapse consists of a presynaptic terminal and a part of perikaryon which

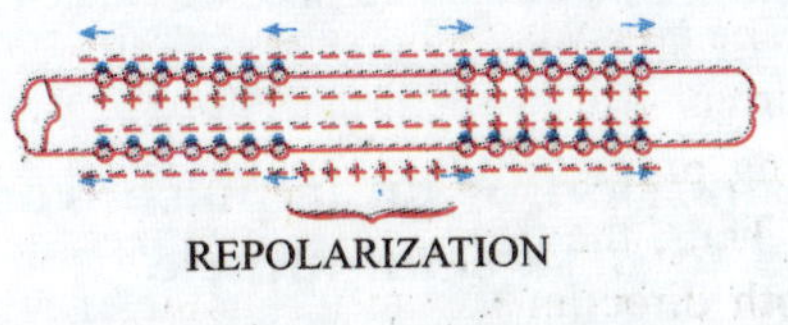

Fig. 40.9 : Propagation of repolarisation in both directions along a conductive fibre

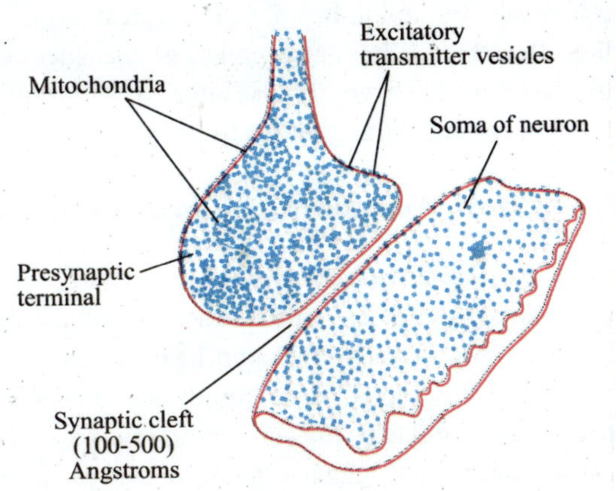

Fig. 40.10 : Physiologic anatomy of the synapse

is separated by a synaptic cleft having an average width of about 200Å.

The presynaptic terminal contains mitochondria and synaptic vesicles. The synaptic vesicles contain chemical transmitter which may be excitatory and then the presynaptic terminal is termed as excitatory presynaptic terminal and the parent neurone as excitatory neurone or inhibitory to the next neurone and then the presynaptic terminal is called as inhibitory presynaptic terminal and the parent neurone as inhibitory neurone. The action potential reaching the presynaptic terminal causes these vesicles to release transmitter substance in the synaptic cleft. The transmitter immediately increases the permeability of sub-synaptic somal membrane either to the Na^+ when it is excitatory in nature or to K^+ when it is inhibitory in nature. When the permeability of Na^+ increases, Na^+ enters the interior of the perikaryon of the next neurone. Thus, it brings about a sequence of changes resulting in the action potential called as excitatory post synaptic potential (EPSP) and thus this neurone gets excited.

But if the permeability increases to K^+ due to release of inhibitory transmitter, K^+ moves out rapidly through the membrane because of excess of K^+ inside the cell leaving a greater degree of negativity inside the neurone. This state is known as hyperpolarisation and the potential so developed as Inhibitory Postsynaptic Potential (IPSP) because this results in the inhibition of the next neurone or the effector cell. It is also called as the chemical transmission of nerve impulses due to the involvement of chemical transmitters in the above process.

Nature of the chemical transmitters

Loewi in 1921 first discovered the process of chemical transmission. He applied the fluid from a perfused frog heart to a second heart and showed that when the vagus nerve supplying to the first heart was stimulated, the fluid from this heart also caused the typical vagal effects of depression of the second heart. Similarly, the stimulation of sympathetic nerve supply to the first heart also caused the stimulation of second heart. Loewi called the substance so liberated by vagus stimulation as 'vagusstoff' and that liberated by sympathetic stimulation as ''acceleranstoff''. Later on these substances had been shown to be acetylcholine and noradrenaline respectively.

A substance to act as a transmitter must satisfy the following needs :

1. Nerve stimulation must release the substance (the transmitter) from the nerve terminals.
2. Drugs which antagonise the action of that substance at the post-synaptic membrane should also block the effects of the nerve stimulation transmitted by that very substance.
3. Substances (drugs) which block the synthesis or storage of this substance should block the effects of nerve stimulation transmitted by this substance.
4. There must be mechanism for the destruction or removal of the proposed transmitter, eg, the enzyme acetyl cholinesterase is needed for the destruction of acetylcholine; so this enzyme must be present in the neurons whose transmitter is acetyl choline.
5. It must be synthesized and stored within the neurones.
6. The action of the substance on the innervated organ if added from outside must be the same as the effects on stimulation of the nerves which release that substance.
7. Drugs which affect the destruction or removal of transmitter must also influence the response to nerve stimulation in an appropriate way.

The transmitters which satisfy the above needs are acetylcholine, epinephrine, norepinephrine. The possible transmitters are γ-aminobutyric acid (GABA), serotonin, glutamic acid, glycine, and dopamine.

Cholinergic and Adrenergic nerve fibres

Dale expressed that the nerve fibres which form and release acetylcholine as transmitter are termed as cholinergic fibres; while those which form and release noradrenaline are termed as adrenergic fibres. Cholinergic nerve fibres are as follows :

1. All the pre-ganglionic fibres of both parasympathetic as well as sympathetic systems.
2. All the post-ganglionic fibres of the parasympathetic systems.
3. Certain post-ganglionic fibres of the sympathetic system, eg, nerve fibres to the sweat glands.
4. The motor nerve fibres to skeletal muscles.
5. Some of the neurones of the central nervous system.
6. Efferent fibres supplying to adrenal medulla.

The adrenergic fibres are as follows :

1. Most of the post-ganglionic fibres of the sympathetic system.
2. Some of the neurones of the central nervous system.

Transmission of an impulse from the nerve to a skeletal muscle

Figure 40.11.A. shows a junction between a motor nerve fibre and a skeletal muscle fibre which is said

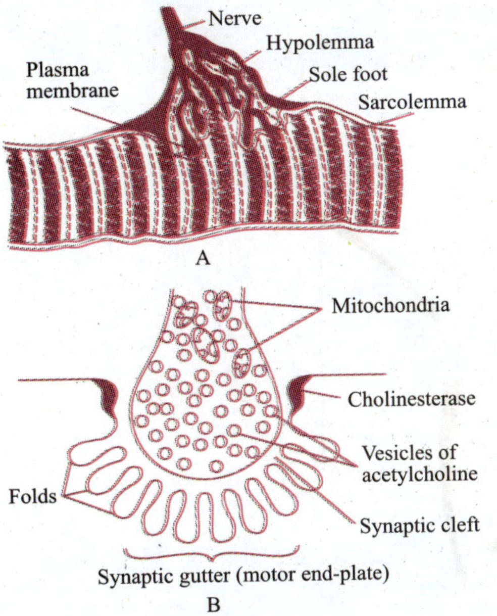

Fig. 40.11 : A. The neruromuscular junction
B. Invagination of a sole foot into the membrane of the muscle fibre

to be neuromuscular junction. A motor neurone together with the group of muscle fibres which it innervates is called a motor unit.

As the myelinated nerve fibre approaches to a skeletal muscle fibre, it loses its medullary sheath and branches at its terminal end into terminal ramifications. At the tips of these terminal ramifications of the nerve fibre are still other structure called as sole feet which make an extensive contact with a specialised part of the muscle fibre membrane to form the actual neuromuscular junction. The part of muscle fibre membrane lying in contact with a sole foot is called as motor end plate (Fig. 40.11B). It is thrown into several folds which increase the surface area at which the synaptic transmitter can act.

The large aggregates of the enzyme cholinesterase which is capable of destroying the synaptic transmitter (acetylcholine) exist around the rim of the motor end plate. The sole foot is separated from the synaptic gutter (motor end plate) by a gap known as synaptic cleft which remains filled with a gelatinous "ground substance" through which the extracellular fluid diffuses. The sole foot contains a number of small vesicles which store the synaptic transmitter synthesized probably by mitochondria which also supply the energy for synthesis.

As a result of the discussion of the structural details of neuromuscular junction, the sequence of events that occur at the neuromuscular junction during the transmission of a nerve impulse from motor nerve fibre to a skeletal muscle fibre can be discussed as follows :

1. Arrival of an impulse at motor nerve terminal.
2. Rupture of synaptic vesicles containing acetylcholine by these calcium ion (Ca^{++}) through the membrane of the sole foot and thus release of acetylcholine into the synaptic cleft and its attachment to the end plate receptor (cholinergic receptors). Acetylcholine release is inhibited by magnesium ion (Mg^{++}).
3. Movements of Ca^{++} (hypothetical) from the extracellular fluid into membrane of the sole foot.
4. Propagation of the so produced end-plate potential when it reaches a certain critical magnitude (threshold).
5. Increase in permeability of muscle membrane to Na^+ allowing rapid influx of Na^+ into the muscle fibre and thus production of a potential called the local end-plate potential which is entirely analogous to the excitatory post-synaptic potential.
6. Removal of acetylcholine from its receptor site within about two milli seconds partly by diffusion back into sole foot and partly by destruction by cholinesterase. The sequence of these events is represented in the figure 40.12.

When the nerve fibre is stimulated at a rate greater than 150 times per second continuously for many minutes, the quantity of acetylcholine release with each impulse diminishes; so that impulses even fail to pass into the muscle fibre. This is referred to

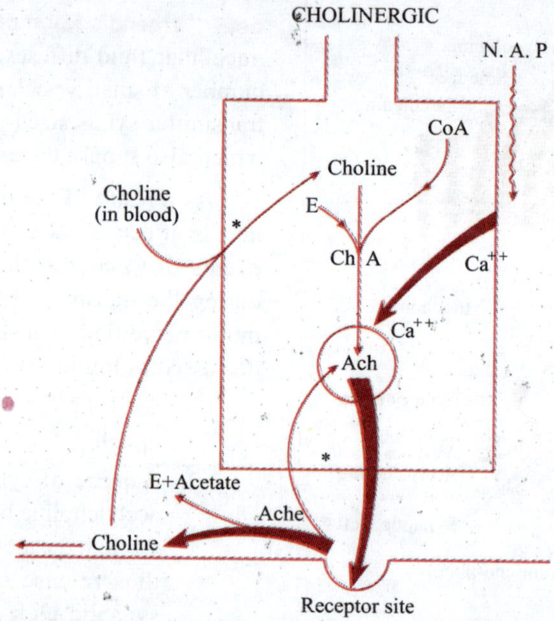

Fig. 40.12 : Diagrammatic representation of the processes involved in synthesis, release, and disposal of acetylcholine at cholinergic nerve terminal and receptor site. Ach = Acetylcholine; chA = choline acetylase. Ache = Acetylcholinesterase; CoA = Coenzyme A. E = Energy → Sites of active transport; NAP = Nerve action potential

as the fatigue of the neuromuscular junction and is analogous to fatigue of a synapse. However, under normal conditions, fatigue of the neuromuscular junction does never occur since it is rate that more than 150 impulses per second reach even the most active neuromuscular junction.

CHAPTER 41

METABOLISM OF XENOBIOTICS.

Introduction

Humans are subjected to exposure to various foreign chemicals (Xenobiotics), may be drug, food additives, pollutants, etc. Chemical pollution and other abuses of the natural resources of the earth are the causes of threat to life. A knowledge of handling xenobiotics at the cellular level is the most essential.

Xenobiotics must be metabolized to be Excreted

The principal classes of xenobiotics of medical branch are drugs, chemical carcinogens and various compounds which are found in the environment by one or another way, such as polychlorinated biphenyls (PCBs) and certain insecticides. Most of these compounds are metabolized in the liver of human body; sometimes xenobiotics may be excreted unchanged. The metabolism of xenobiotics can be divided into two phases:

Phase I : The major reaction is hydroxylation catalyzed by a class of enzymes known as monooxygenases or cytochrome P-450 species, other types of reactions in phase I are reduction and hydrolysis.

Phase 2 : The compounds produced in phase I are converted by specific enzymes to various metabolites by **conjugation** with glucuronic acid, sulfate, acetate, glutathione, or certain amino acids, or by methylation.

The purpose of the two phases to increase their water solubility is to facilitate their excretion from the body. Highly hydrophobic xenobiotics persist in adipose tissue almost indefinitely if they are not converted to more polar forms. Phase I metabolic reactions, in certain cases, convert xenobiotics from inactive to active compounds. In this case, the original xenobiotics are referred to as prodrugs or procarcinogens. In less number of cases conjugation reactions increase the biologic activity of a xenobiotic. Detoxification is not always an appropriate term because in some cases the reactions to which xenobiotics are involved actually increase their biologic activity and toxicity.

Cytochrome P-450 Hydoxylates xenobiotics in phase I of their metabolism

The responsible enzymes in hydroxylation in phase I are called monooxygenases or cytochrome P-450 species. The reaction catalyzed by a monooxygenase is

$$RH + O_2 + NADPH + H^+ \rightarrow R\text{-}OH + H_2O + NADP^+$$

RH represents a very wide variety of drugs, carcinogens, pollutants, and certain endogenous compounds such as steroids. The reaction catalyzed by cytochrome P-450 can also be represented as:

$$\text{Reduced cyt. P-450} \longrightarrow \text{oxidized cyt. p-450}$$

$$RH + O_2 \longrightarrow R\text{-}OH + H_2O$$

The major monooxygenases in the ER are cytochrome P-450 species. Cytochrome P-450 is notable because about 50% of the drugs ingested by patients are metabolized by species of cytochrome P- 450. The same enzyme also acts on various carcinogens and pollutants.

Cytochrome P-450 species

1. They are hemoproteins.
2. They are present in high concentration in the membranes of the endoplasmic reticulum (ER) of *liver*. They are also present in the mitochondria as well as in the ER of the *adrenal*.
3. Six closely related species of cytochrome P-450 present in liver ER that acts on a wide variety of drugs, carcinogens, and other xenobiotics. In recent years, the genes for these species of cytochrome P-450 have been isolated and studied in detail.
4. The enzyme that uses NADPH in the reaction mechanism of cytochrome P-450 is called *NADPH-cytochrome P-450 reductase*.
5. The suitable lipid *phosphatidyl choline*, the major lipid found in the membranes of ER, is the component of the cytochrome P-450 system.
6. Most species of cytochrome P-450 are inducible, eg, the administration of phenobarbital(PB) causes a hypertrophy of the smooth ER and a 3- to 4-fold increase of the amount of cytochrome P-450 within 4-5 days. The mechanism of induction involves increased transcription of mRNA for cytochrome P-450. This can be explained by an example : suppose that a patient is taking the anticoagulant Dicoumarol to prevent blood clotting. This drug is metabolized by the cytochrome P-450 system. It has also been found that the patient also needs PB to treat certain type of epilepsy. The administration of PB is started but the dose of dicoumarol is unchanged. After 5 days, the level of cytochrome P-450 in the patients liver will be increased 3- to 4-fold. This means that the dicoumarol will be metabolized much more quickly than before, and its dosage will have become inadequate. Therefore, the dose must be increased. The patient will be at risk of bleeding since the high dose of dicoumarol will be even more active than it was before, because the level of cytochrome P-450 will decline once PB has been stopped.
7. Another species named as cytochrome P-448 is specific for the metabolism of polycyclic aromatic hydrocarbons (PAHs) and related molecules; hence it is said to be aromatic hydrocarbon hydroxylase (AHH). This enzyme is very important in the metabolism of PAHs and in **carcinogenesis** produced by these agents. Smokers have higher levels of this enzyme in some of their cells and tissues than do nonsmokers. It is reported that the activity of this enzyme may be increased in the placentae of women who smoke, thus altering the quantities of metabolites of PAHs to which the fetus is exposed.

Phase 2 reactions

Glucuronidation : Molecules as 2- acetylaminofluorene (a carcinogen), aniline, benzoic acid, phenol, and many steroids are excreted as glucuronides. The glucuronide may be attached to oxygen, nitrogen or sulfur groups of the substrates.

Sulfation : Some alcohols, arylamines, and phenols are sulfated. The sulfate donor is adenosine 3′-phosphate -5′-phosphosulfate.

Conjugation with glutathione : A number of potentially toxic electrophilic, xenobiotics (such as certain carcinogens) are conjugated to the nucleophilic GSH.

$$R + GSH \longrightarrow R—S—G$$

where R = an electrophilic xenobiotic. The enzymes catalyzing these reactions are called *glutathione S-transferases* and are present in high concentration in liver cytosol and lower concentration in other tissues. GSH is an important defense mechanism against certain toxic compounds. Glutathione–conjugates are further metabolized prior to excretion. The glutamyl and glycinyl group belonging to glutathione are removed by specific enzymes and an acetyl group is added to the amino group of the remaining cysteinyl moiety. The resulting compound is a mercapturic acid which is then excreted in urine. Glutathione participates in the *decomposition of partially toxic hydrogen peroxide*. It is an important intracellular reductant, helping to maintain essential SH groups of enzyme in their reduced state. When GSH acts as a reducing agent, its SG becomes oxidized and forms a disulfide link with another molecule of glutathione.

$$GSH + GSH \rightleftharpoons G{-}S{-}S{-}G$$

G—S—S—G can be reduced to GSH by the action of *glutathione reductase* using NADPH.

$$G{-}S{-}S{-}G + NADPH + H^{+} \rightarrow 2GSH + NADP$$

GSH as a carrier is involved in the *transport of certain amino acids across membranes in* the kidney.

–Amino acid + GsH → γ–Glutamyl amino acid+ Cysteinylglycine. The enzyme catalyzing the above reaction is γ–Glutamyltransferase (GGT). The enzyme has diagnostic value since it is released into the blood from hepatic cells in various hepatobiliary diseases.

Other reactions :

(a) Acetylation : it is represented by :

$$X + Acetyl - CoA \rightarrow Acetyl - X + CoA$$

where X represents xenobiotics. This reaction is catalyzed by acetyltransferase present in the cytosol of various tissues, particularly liver.The drug *isoniazid*, used in the treatment of tuberculosis, is subject to acetylation. The acetyltransferases are classified as *slow* and *fast acetylators* and help in the clearance of drugs from blood.

(b) Methylation : Some xenobiotics are subject to *methylation* by methyltransferases using S-adenosylmethionine as the methyl donor.

Affects of age, sex, and other factors on the activities of xenobiotic metabolizing enzyme

1. The activity of the xenobiotic metabolizing enzyme may differ among species. The possible toxicity or carcinogenicity of xenobiotics cannot be extrapolated freely from one species to another.
2. Significant difference in enzyme activities among individuals may exist due to genetic factors.
3. The activities of some of these enzymes vary according to age and sex.
4. Intake of PCBs or certain hydrocarbons can cause enzyme induction. It is to know whether an individual is exposed to inducing agent or not.
5. Metabolites of certain xenobiotics can inhibit the activities of xenobiotic-metabolizing enzymes.

Responses to xenobiotics

When the xenobiotic is a drug, Phase I reactions may produce its active form or terminate its action if it pharmacologically active in the body without prior metabolism. Here it is important to mention that drugs act through biochemical mechanisms.

Certain xenobiotics are very *toxic*(eg, cyanide) and some xenobiotics are not so much toxic even sufficient amounts of these are administered. The toxic effects of xenobiotics cover the wide spectrum. Three general types of effects are mentioned here.

1. The first one is *cell injury*, which can result in cell death. The macro-molecular targets include DNA, RNA, and protein.
2. The reactive species of a xenobiotic may bind to a protein. The xenobiotic is said to act as a hapten. The resulting antibody can damage the cell by several immunologic mechanisms.
3. Reactions of activated species of chemical carcinogens with DNA are of great importance in *chemical carcinogenesis*.

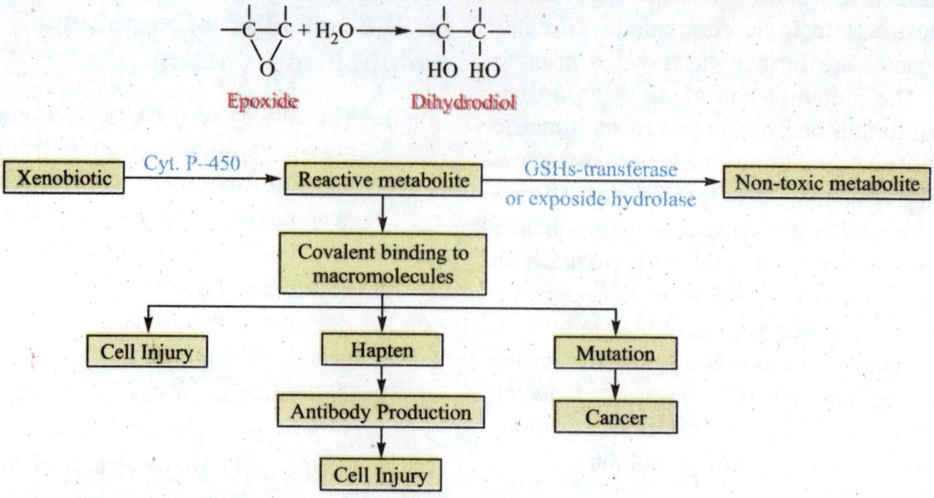

Fig. 41.1 : Simplified scheme showing how metabolism of a xenobiotic can result in cell injury, immunologic damage, or cancer

Some chemicals require activation by monooxygenases in the ER to become carcinogenic. The activities of the monooxygenases and of other xenobiotic–metabolizing enzymes present in the ER help to determine whether such compounds are carcinogenic or are detoxified. Other chemicals can react directly with DNA without undergoing intracellular chemical activation.

The enzymes *epoxide hydrolase* can exert a protective effect against certain carcinogens. The action of certain monooxygenases on some procarcinogen substrates produces epoxides which are highly reactive and mutagenic or carcinogenic. Epoxide hydrolase acts on epoxide forming dihydrodiols which are much less reactive.

CLINICAL ASPECTS

- The metabolism of xenobiotics provides a basic knowledge to the understanding of pharmacology, toxicology, cancer research, drug addiction. All these areas involve exposure to xenobiotics.

SCHEMATIC COMPENDIUM

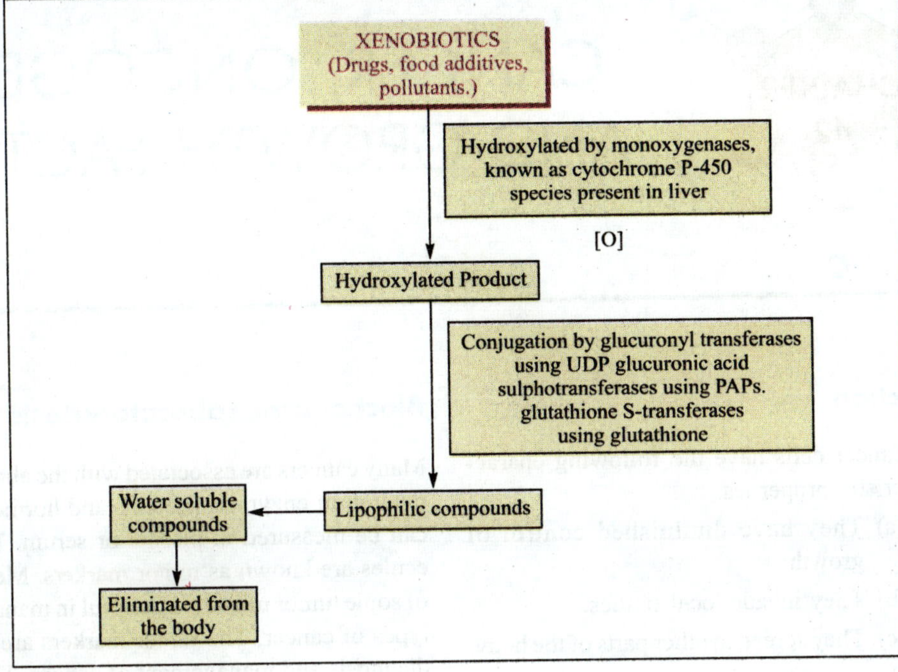

CHAPTER 42

CANCER, ONCOGENES AND GROWTH FACTORS

Introduction

1. Cancer cells have the following characteristic properties.
   (a) They have diminished control of growth.
   (b) They invade local tissues.
   (c) They spread to other parts of the body.
2. The main object is to explain in biochemical terms the uncontrolled growth of cancer cells and their capacity to invade and metastasize.
3. The genes controlling growth and interactions with other normal cells are apparently abnormal in structure or regulation in cancer cells.
4. Information regarding the control of cell growth (both normal and pathologic) is limited and knowledge of specific genes involved in growth regulation is even very scanty. Little is also known as yet about the biochemical basis of metastasis.
5. Some types of cancer (eg, certain leukemias) can be regarded as examples of abnormal differentiation. But little is known about the molecular basis of differentiation. Many workers in this area think that further research on oncogenes and growth factors will provide information on the nature of the disturbed control of growth, of differentiation, and of cell-cell interaction exhibited by cancer cells.

Biochemical Laboratory tests

Many cancers are associated with the abnormal production of enzymes, proteins and hormones which can be measured in plasma or serum. These molecules are known as tumor markers. Measurement of some tumor markers is helpful in managing some types of cancer. The tumor markers are applied in diagnosis and management of cancer. Three major conclusions are derived from the study of tumor markers.

1. No single marker is useful for all types of cancer or for all patients with a given type of cancer.
2. Markers are most often detected in advanced stages of cancer rather than early stages.
3. The most successful use of markers is to monitor the responses to therapy and the detection of early recurrence.

Table 42.1. Clinically useful tumor markers

Marker	Associated Cancer
Carcinoembryonic antigen (CEA)	Colon, Lung, Breast, Pancreas
Alpha-fetoprotein (AFP)	Liver, Germ Cell
Human chorionic gonadotropin (HCG)	Trophoblast, Germ Cell
Calcitonin (CT)	Thyroid (medullary carcinoma)
Prostatic acid phosphatase (PAP)	Prostate

Causes of Cancer

The following are the groups causing cancer :

1. Radiant energy.
2. Chemical componds.
3. Viruses.

1. Radiant energy

(a) Ultraviolet rays, X-rays, and γ-rays damage DNA in several ways.

(b) Ultraviolet radiation may cause pyrimidine dimers to form.

(c) Apurinic or apyrimidinic sites may form by elimination of corresponding bases. Single and double-strand breaks or cross-linking of strands may occur.

(d) The basic mechanism of carcinogenicity with radiant energy is to cause damage to DNA.

(e) **Free radicals** are formed in tissues by X-rays and γ-rays. The resultant OH, superoxide, and other radicals can interact with DNA and other macromolecules, leading to molecular damage and thereby probably contributing to carcinogenic effects of radiant energy.

2. Chemical Compounds

(a) It has been estimated that environmental factors, principally chemicals, can cause up to 80 per cent of human cancers.

(b) Exposure to such compounds can occur because of a person's occupation (eg, benzene, asbestos); diet (eg, aflatoxin B_1 which is produced by the mold Aspergillus flavus and sometimes found as a contaminant of peanuts and other foodstuffs); Life style (eg, cigarette smoking); or in other ways (eg, certain therapeutic drugs can be carcinogenic).

(c) The carcinogenic substances may be both organic and inorganic molecules. The following table shows some chemical compounds which may cause cancer.

Table 42.2. : Some Chemical Carcinogens

Class	Compound
Polycyclic aromatic hydrocarbons	Benzo [a] Pyrene.
Aromatic amines	2-Acetylaminofluorene, N-methyl-4-aminoazobenzene (MAB)
Nitrosamines	Dimethylnitrosamine, diethynitrosamine
Various drugs	Alkylating agents (eg, Cyclophosphamide), diethylstilbesterol
Naturally occurring compounds	Dactinomycin, aflatoxin B_1
Inorganic compounds	Arsenic, asbestos, beryllium, cadmium, chromium

Action

(a) The organic carcinogens, such as nitrogen mustard and β-propiolactone, interact directly with target molecules (direct carcinogens), but others require prior metabolism to become carcinogenic (Procarcino-

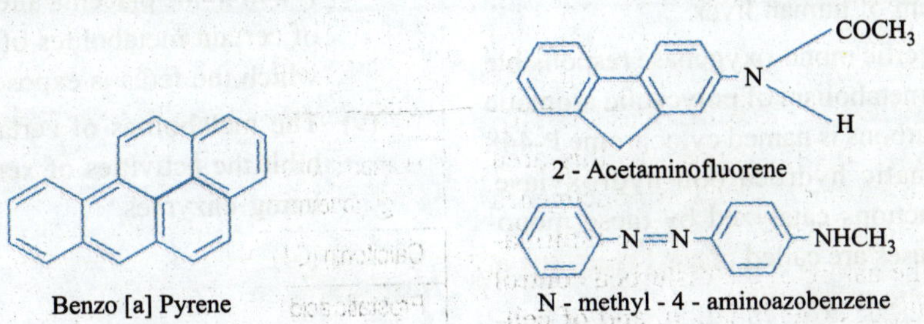

Fig. 42.1 : Structures of three important experimentally used chemical carcinogens

gens). The process by which one or more enzyme-catalyzed reactions convert procarcinogens to active carcinogens is called metabolic activation. Any intermediate compound formed is proximate carcinogens, and the final compound that reacts with cellular components (eg, DNA) is the ultimate carcinogen.

Procarcinogen → Proximate carcinogen A → Proximate carcinogen B → Ultimate carcinogen.

(b) The ultimate carcinogen is highly reactive and is usually electrophile (molecule deficient in electron). This readily attacks nucleophilic (electron-rich) group in DNA, RNA, and protein.

Mono-oxygenases and Transferases

(a) Mono oxygenases and transferases cause the metabolism of procarcinogens and other xenobiotics.

(b) The heme-containing mono-oxygenases of the cytochrome P-450 type located in the endoplasmic reticulum are mainly responsible for the metabolic activation of procarcinogens.

(c) The mono-oxygenases catalyze the hydroxylation of various procarcinogens and other xenobiotics using molecular oxygen as the source of oxygen and NADPH as a reducing source.

$$R-H + O_2 + NADPH + H^+ \rightarrow R-OH + H_2O + NADP^+$$

(d) At least 6 or many more such mono-xygenases are present in the endoplasmic reticulum of human liver.

(e) The specific mono-oxygenase responsible for the metabolism of polycyclic aromatic hydrocarbons is named cytochrome P-448 or aromatic hydrocarbon hydroxylase. The reactions catalyzed by these mono-oxygenases are called phase I reactions of xenobiotic metabolism.

(f) In phase 2 reactions of xenobiotic metabolism, the hydroxylated xenobiotics are conjugated with various moieties (eg, Glucuronate, sulfate, acetate, glutathione). These reactions usually detoxify the reactivity of the compounds involved and make them ready for excretion, mainly in the urine.

(g) In some cases, conjugation actually increases the biologic activity or chemical reactivity of a molecule. The enzymes catalyzing the above conjugation reactions are usually cytosolic in location, although some are also present in the endoplasmic reticulum. The various glutathione transferases use glutathione transferases utilizing glutathione itself as the donor.

Factors affecting enzymes metabolizing xenobiotics

The following factors affect the activities of the enzymes metabolizing xenobiotics :

(i) The activities of these enzymes may differ substantially among species.

(ii) Significant differences are found in enzyme activities among individuals, many of which are due to genetic factors.

(iii) The activities of some of the enzymes vary according to age and sex.

(iv) Intake of phenobarbital, PCBS, or certain hydrocarbons can also increase the activities of many enzymes by a process known as enzyme induction. Hydrocarbon inhalation from cigarette smoking during pregnancy induces the activity of cytochrome P-448 in the placenta altering the amounts of certain metabolites of hydrocarbons to which the fetus is exposed.

(v) The metabolites of certain drugs can inhibit the activities of xenobiotic-metabolizing enzymes.

Mutagens

(a) Most of the chemical carcinogens are mutagens.

(b) The use of bacteria is mutagenicity tests creates a problem that they do not contain the spectrum of mono-oxygenases found in higher animals.

Initiation and promotion

(a) The stage of carcinogenesis of the skin of experimental mice caused by the application of benzo [a] pyrene is called initiation and this stage is rapid and irreversible. It is supposed to involve an irreversible modification of DNA resulting in one or more mutations. Benzo [a] pyrene is thus called an initiating agent.

(b) The second stage of carcinogenesis, resulting from the application of croton oil, is called promotion and croton oil is, therefore, a promoter. Promoters can cause initiation.

(c) Most carcinogens can act as both initiating and promoting agents.

(d) A good number of compounds including phenobarbital and saccharin can act as promoters in different organs.

(e) The active agent of croton oil is a mixture of phorbol esters. The most active phorbol ester is 12-0-tetradecanoylphorbol-13-acetate (TPA) which has numerous effects.

(f) Protein Kinase C can act as a receptor for TPA. The enzyme being stimulated by interaction with TPA may result in the phosphorylation of a number of membrane proteins resulting in the effects on transport and other functions.

Role of DNA

DNA is the premier target molecule in carcinogenesis which is being established by the following facts :

(i) Cancer cells beget cancer cells, i.e., the required changes responsible for cancer are transmitted from mother to daughter cells. This is consistent with the behaviour of DNA.

(ii) DNA is damaged by both irradiation and chemical carcinogens which are capable of causing mutations in DNA.

(iii) Many tumor cells exhibit abnormal chromosomes.

(iv) Transfection experiments show that purified DNA (oncogenes) from cancer cells can transform normal cells into cancer cells. Epigenetic factors may also play a role in carcinogenesis.

3. Oncogenic viruses

(i) Oncogenic viruses contain either DNA or RNA as their genome.

(ii) Polyomavirus and SV 40 viruses have played an important role in the development of current ideas about viral oncogenesis. Both of them are small and their circular genomes code for only about 5-6 proteins. Under certain circumstances, appropriate cells being infected with these viruses can result in malignant transformation. Specific viral proteins are involved too.

(iii) In case of SV 40, these proteins (often called antigens) are known as T and t, and in case of polyomavirus, they are known as T, mid–T, and t (T refers to the first of these proteins detected in a tumor).

(iv) The T antigens are to bind tightly to DNA and cause alteration in gene expression. These proteins show cooperative effects, suggesting that alteration of more than one reaction or process is required for transformation.

(v) Transformation of certain animal cells are caused by some types of adenovirus.

(vi) Epstein-Barr virus is associated with Burkitt's Lymphoma and nasopharyngeal carcinoma in humans.

(vii) Herpes Simplex virus is associated with cancer of the cervix, and hepatitis B virus is also associated with some cases of liver cancer in humans.

Transformation

The cultured cells undergo malignant transformation when they are infected with certain oncogenic viruses. These changes affect cell shape, motility, growth, and a number of biochemical processes. They reflect the conversion from the normal to the malignant state. Acquisition by cells of the changes collectively known as transformation does not mean that such cells will display the same biologic properties as tumor cells in vivo; cells must yield tumors when injected into a suitable host animal.

Oncogenes

Oncogenes are genes capable of causing cancer. These were first recognised as unique genes of tumor-causing viruses that are responsible for the process of transformation (viral oncogenes).

Oncogenes of Rous Sarcoma Virus

(i) The genome of this retrovirus contains four genes named gag, pol, env, and src.

(ii) The gag gene codes for group-specific antigens of the virus, pol for the revese transcriptase that characterizes retroviruses, and env for certain glycoproteins of the viral envelope. A protein-tyrosine kinase was shown to be the product of src (i.e., the sarcoma-causing gene) that is responsible for transformation.

(iii) Certain glycolytic enzymes become target proteins for the src protein-tyrosine kinase. This shows that transformed cells often show increased rates of glycolysis. The product of src may also catalyze phosphorylation of phosphatidylinositol to phosphatidylinositol mono- and biphosphate.

(iv) When phosphatidylinositol 4, 5-biphosphate is hydrolyzed by the action of phospholipase C, 2 second messengers are released : inositol triphosphate and diacylglycerol. The first compound mediates release of Ca^{++} from intracellular sites of storage (eg, the endoplasmic reticulum).

(v) Diacylglycerol stimulates the activity of the plasma membrane-bound proteins kinase C which in turn phosphorylase a number of proteins, some of which may be components of iron pumps.

(vi) Mild alkalinization of the cell brought about by activation of an Na^+/H^+ antiport system can play a role in stimulating mitosis.

(vii) The product of src may, therefore, affect a large number of cellular processes by its ability to phosphorylate various target proteins and enzymes and by stimulating the pathway of synthesis of the polyphosphoinositides.

Oncogenes of other Retroviruses

(i) About 20 oncogenes of other retroviruses have been identified. Almost half of the products are protein kinases, mostly of the tyrosine type.

(ii) Some of these encode protein kinases, the remainder encode various other proteins with interesting biologic activities.

(iii) The product of the ras oncogene of murine sarcoma viruses binds GTP, has GTPase activity, and is related to the proteins that regulate the activity of the important plasma membrane enzyme, adenylate cyclase.

Mechanism by which proto-oncogenes become Oncogenes

A. Promoter insertion

(i) When the particular viruses infect cells, a DNA copy (cDNA) of their RNA genome is synthesized by reverse transcriptase, and the cDNA is integrated into the host genome. The integrated double-stranded cDNA is called a Provirus.

(ii) Following infection of chicken B lymphocytes by certain avian leukemia viruses, their proviruses become integrated near the myc gene. The myc gene is activated by an upstream, adjacent viral long terminal repeat acting as a promoter, resulting in transcription of both the corresponding myc mRNA and translation of its product in such cells.

B. Enhancer Insertion

(i) In certain cases, the provirus is inserted downstream from the myc gene or upstream from it but oriented in the reverse direction, the myc gene never become activitated. Such activation cannot be due to promoter insertion.

(ii) Enhancer sequences present in the long terminal repeat sequences of the retroviruses.

(iii) The above two mechanisms—promoter and enhancer insertion—commonly operate in viral oncogenesis.

C. Chromosomal Translocations

(i) Many tumor cells show chromosomal abnormalities. Translocation is a type of chromosomal change seen in cancer cells.

(ii) A piece of one chromosome being split off joins to another chromosome and if the second chromosome donates material to the first, the translocation is said to be reciprocal.

(iii) A number of tumor cells show characteristic translocations. One important translocation is the philadelphia chromosome occurring in chronic granulocytic leukemia.

(iv) Burkitt's Lymphoma is a fast-growing cancer of human B Lymphocytes.

(v) Synthesis of greatly increased amounts of the DNA-binding protein coded for by the myc gene acts to ''drive'' or ''force'' the cell towards becoming malignant by an effect on the regulation of mitosis.

D. Gene Amplification

(i) One method is shown in respect of gene amplification in tumors by administration of the anticancer drug methotrexate, an inhibitor of the enzyme dihydrofolate reductase. Tumor cells can become resistant to the action of this drug.

(ii) Certain cellular oncogenes can also be amplified and are thus activated.

(iii) Increased amounts of the products of certain oncogenes produced by gene amplification may play a role in the progression of tumor cells to a more malignant state.

E. Single-point mutation

(i) The product of murine retroviruses, a protein of MW 21000 is related to the G proteins that modulate the activity of adenylate cyclase and thus play a key role in cellular responses to many hormones and drugs.

(ii) The lower activity of GTPase can result in chronic stimulation of the activity of adenylate cyclase which normally is diminished when GDP is formed from GTP. The resulting stimulation of the activity of adenylate cyclase can result in a number of effects on cellular metabolism exerted by the increased amount of cAMP affecting the activities of various cAMP-dependent protein kinases.

General comments on activation of oncogenes

(i) Increased amounts of the product of an oncogene may be sufficient to push a cell towards becoming malignant.

(ii) The presence of a structurally abnormal key regulatory protein in a cell may also be sufficient to tip the scales towards cancer.

(iii) Oncogenes have been isolated from only about 15 per cent of human tumors.

(iv) Recent work has shown that activation of C-ras in rat mammary cancers induced by nitrosomethylurea was apparently due to a specific G $\rightarrow$ A transition type of mutation, demonstrating that oncogenes are probably involved in chemical carcinogenesis.

(v) Further research is essential to examine the possible involvement of oncogenes in the phenomena of initiation, promotion, tumor progression, and metastasis.

Mechanisms of action of oncogenes

(i) They may act on key intracellular pathways involved in growth control uncoupling them from the need for an exogenous stimulus.

(ii) The products of oncogenes may also imitate the action of a polypeptide growth factor.

(iii) The products may also imitate an occupied receptor for a growth factor.

Polypeptide growth factors

(i) The growth factors affect many different types of cells, eg, cells from the blood, nervous system, mesenchymal tissues, and epithelial tissues.

(ii) They exert a mitogenic response on their target cells.

(iii) Platelet-derived growth factor (PDGF) released from the alpha granules of platelets plays a role in normal wound healing. Various growth factors play key roles in regulating differentiation of stem cells to form various types of mature hematopoietic cells. Growth inhibitory factors also exist. Thus, chronic exposure to increased amounts of a growth inhibitory factor can alter the balance of cellular growth.

Endocrine, paracrine, and autocrine actions of growth factors

Growth factors may act in the following ways :

(i) Their effects may be endocrine, like hormones, they may be synthesized elsewhere in the body and may pass in the circulation to their target cells.

(ii) They may be synthesized in certain cells and secreted from them to affect neighbouring cells. The cells that synthesize the growth factor are not themselves affected because they lack suitable receptors. This mode of action is called paracrine.

(iii) Some growth factors can affect the cells that synthesize them. This third mode of action is called autocrine.

BIOCHEMICAL MECHANISM OF ACTION OF GROWTH FACTORS

(i) The growth factors like hormones must transmit a message across the plasma membrane to the interior of the cells. The message ultimately affects one or more processes involved in mitosis.

(ii) Most growth factors have high affinity protein receptors on the plasma membrane of target cells. The genes for the receptors for epidermal growth factor (EGF) and insulin have been cloned.

(iii) A number of receptors have been found to exhibit protein-tyrosine kinase activities. The kinase activity located in the cytoplasmic domains causes autophosphorylation of the receptor protein and also phosphorylates other target proteins. The receptor-ligand complexes are subjected to endocytosis in coated vesicles.

(iv) Phospholipase C is stimulated following exposure of cells to PDGF resulting in hydrolysis of phosphatidylinositol 4, 5-biphosphate to form inositol triphosphate and diacylglycerol. These two second messengers can effect intracellular release of Ca^{++} and stimulation of the activity of protein kinase C respectively, thus affecting a larger number of cellular reactions.

(v) The hydrolysis of diacylglycerol of phospholipase A_2, liberating arachidonic acid, can also result in the production of prostaglandins and leukotrienes which themselves may exert many biologic activities.

(vi) Exposure of cells to PDGF can cause rapid activation of certain cellular proto-oncogenes.

(vii) Gene activation is involved in the action of most growth factors.

Progression of tumors

(i) A cell once becomes a tumor cell, tendency for malignancy is increased. This is signified by increasing rates of growth, and increasing tendency to invade and metastasize.

(ii) Cells having faster rates of growth will have a selective advantage.

(iii) The biochemical affairs of highly malignant cells may be very different from that of normal cells. Some changes are secondary to rapid growth rates and others are due to chromosomal instability.

(iv) The fast-growing cells tend to increse the anabolic processes involved in growth (eg, DNA and RNA synthesis), cut down on catabolic functions (eg, catabolism of pyrimidines), and dispense with the differentiated functions shown by their normal ancestors. In fact, they are mostly concentrated upon growth. They also show biochemical changes that reflect altered gene regulation, such as the synthesis of certain fetal proteins.

Drugs used in cancer chemotherapy

1. The patients with cancer are well treated by surgery, radiotherapy, and chemotherapeutic agents. It is always advisable to see that the drugs (natural products or synthetic) that kill cancer cells effectively must not toxic to normal cells.
2. Many of the agents are used because they inhibit DNA synthesis. For this reason, they cause damage to normal tissues whose cells divide continuously (eg, the gut and bone marrow).
3. The growth fraction (the cells of tumor that are constantly in cycle) is an important factor in cancer chemotherapy; tumors with a high growth fraction are usually more responsive to cancer chemotherapy. It is wise to classify anticancer drugs as cell cycle specific (CCS) or cell cycle non-specific (CCNS).
4. CCS agents are further subdivided into phase-specific agents (eg, agents acting on the S phase of the cycle [eg, cytarabine] or on the G_2-M phase [eg, bleomycin].
5. In all cancer therapy, it is vital to diagnose and treat the tumor early; otherwise, the tumor burden may be too great for therapy to be successful. During therapy, it is wise to eliminate all clonogenic cells ("tumor stem cells").

Table 42.3 : Some drugs used in cancer chemotherapy

Class of compound	Example	Site of action	Treatment use
Alkylating agents	Melphalan	Alkylates DNA & other molecules	Myeloma
Antitumor antibiotics	Doxorubicin	Intercalates in DNA	Hodgkin's disease
Sex hormones	Estrogens in prostatic	Block effect of androgens tumors	Cancer of the prostate
Corticosteroids	Prednisone	Inhibits proliferation of lymphocytes	Myeloma
Purine antagonists	Mercaptopurine	Inhibits purine synthesis	Acute myelocytic leukemia
Pyrimidine antagonists	Fluorouracil	Inhibits thymidylate synthetase	Colorectal cancer
Plant compounds	Vinblastine	Binds tubulin and inhibits microtubule formation	Kaposi's sarcoma
Other agents	Cisplatin	Causes strand breakage in DNA	Carcinoma of the lung

6. Combination chemotherapy (combining three or four agents) is successful in some cancers because the drugs may act synergistically, the onset of resistance may be delayed, and toxicity is often less.

P-Glycoprotein plays an important role in multidrug resistance in cancer chemotherapy

1. A particular drug is never effective against the target tumor cells.
2. The acquired resistance of cancer cells to chemotherapeutic agents is to reflect their high mutation rates. Resistance is not only to the drug but also to other structurally unrelated anticancer agents, eg, if resistance to methotrexate develops, the tumor cells may also be found to be resistant to antitumor antibiotics such as doxorubicin and to plant compounds such as vincristine. This is known as **multi drug resistance** which is important as its development is frequently the cause of failure of cancer chemotherapy, a common event during the treatment of cancer.
3. The protein, **P-glycoprotein** (P = permeability) acts as an energy-dependent pump, ejecting a wide variety of anticancer drugs from cancer cells and thus diminishing their efficacy.
4. P-glycoprotein has been the subject of much study because of its therapeutic importance. In cultured cells, gene amplification and point mutations have been found in resistant cancer cells showing elevated amounts of P-glycoprotein.
5. P-glycoprotein has physiologic functions since it is present in normal organs, such as kidney and gut; it may play a role in the excretion of potentially toxic compounds from cells of these organs.
6. P-glycoprotein can be inhibited by chemosensitizers such as verapamil or cyclosporine.

Metastasis

Metastasis is a complex process to analyze in humans. It is the spread of cancer cells from a primary site of origin to other tissues where they grow as secondary tumors, it is the major problem presented by the disease. Biochemical knowledge regarding this is quite restricted.

Much attention has been put to the comparisons of the biochemistry of the surfaces of normal and malignant cells.

At present, metastasis has been studied by considerable research work on developing suitable animal model systems. Much studies are being carried out to identify the possible roles of certain proteases (eg, type 4 collagenase) and of certain glycoproteins and glycosphingolipids of the cell surface is the process of metastasis. For instance, it may be possible that changes in the oligosaccharide chains of cell glycoproteins may be effective in permitting metastasis to occur. The biochemical mechanisms involved in metastasis may provide a basis for the rational development of more effective anticancer therapies.

Table 42.4 : Some new and developing cancer therapies

Therapy	*Example or underlying principle*
Cell cycle inhibitors	Develops agents capable of selectively inhibiting key steps of the cell cycle of cancer cells
Glycosylation inhibitors	Alters patterns of sugars on cell surface glycoproteins in order to inhibit metastasis
Metastasis inhibitors	Inhibitors of various proteases may inhibit metastasis
Protein kinase inhibitors	Many tumor cells have elevated amounts of various protein kinases involved in signalling, etc.
Ribozymes	Can degrade RNA molecules that are relatively unique to tumor cells
Telomerase inhibitors	May act on tumor cells in which telomerase is active
Vaccines	Inject peptides from relatively "tumor-specific" proteins to boost the T cell immune response

Table 42.5 : Some genes that increase susceptibility to cancer

	Gene	Chromosome	Clinical Expression
Familial breast and ovarian cancer	BRCA1	17q21	Development of cancer before age 30.
Li-Fraument syndrome	P^{53}	17P13	Cancers at different sites, developing at an early age, and exhibiting multiple primary tumors.
Wilm's tumor	WT1	11P13	A specific type of kidney tumor developing in childhood
Neurofibromatosis type 2	NF2	22q12	Development of bilateral acoustic neuromas
Familial adenomatous polyposis	APC	5q21	Development of many early onset adenomatous polyps which are immediate precursors of colorectal cancers

The isolation of a number of genes involved in the increased susceptibility to cancer

(a) Much progress has been made in isolating genes that increase susceptibility to cancer. Recently, BRCA1, a gene that influences susceptibility to early-onset breast and ovarian cancer has been isolated.

(b) BRCA1 is a protein of 1863 amino acids containing a zinc finger domain in its amino terminal region and is a transcription factor.

(c) About 12 other genes associated with familial types of cancer have been identified to specific chromosomal locations.

(d) The second gene of these 12 BRCA2 is associated with increased susceptibility to breast cancer only and mapped to chromosome 13q12-13.

(e) The further study of these genes will facilitate early diagnosis of different types of familial cancer and will provide to know the mechanisms of development of cancer in various tissues.

P^{53} tumor suppressor gene acts as a guardian of the genome

(a) P^{53} is nuclear phosphoprotein of molecular mass 53KDA.

(b) It is located on short arm of chromosome 17.

(c) It binds specific sequences in dsDNA.

(d) It acts as a transcriptional regulator regulating genes that participate in cell division.

(e) It binds various viral proteins forming inactive oligomeric complexes.

(f) It acts as G_1 checkpoint control for DNA damage. If DNA is excessively damaged, the activity of P^{53} is increased resulting in inhibition of cell division and allowing time for repairing.

(g) It participates in the initiation of apoptosis which is defined as programmed cell death occurring in many physiologic and pathologic conditions.

(h) It hastens the death of potentially dangerous cells which have the potential to become malignant. In such affair, P53 has been referred to as a "guardian of the tissues".

Mutations in P^{53} occur in many human tumors.

(a) Mutations in the P^{53} gene are the most common genetic change in human cancer and are often in colon, breast, and lung cancers.

(b) At least 100 different mutations have been detected in this gene.

(c) The mutations are found at highly conserved codons.

(d) The mutational spectrum differs in different cancers.

(e) Transitions occurring at CpG dinucleotides (a hot spot) are frequent in neoplasms of the colon, brain, and lymphoid tissue.

(f) Transversions are frequent in lung cancer and liver cancer and are found mainly at one specific site in certain hepatocellular carcinomas occurring in areas where exposure to aflatoxin B_1 and hepatitis B virus.

The malignancy of tumor cells tends to progress

1. When a cell becomes a tumor cell, the composition and behaviour of the cell do not remain static. Rather, tendency for malignancy is increased. This is manifested by increasing rates of growth, and an increasing tendency to invade and metastasize.
2. The mutations in DNA repair genes are found in this phenomenon creating a mutator phenotype. The activation of additional oncogenes occurs. Cells are found with faster rates of growth.
3. The biochemical profile of highly malignant cells may be very different from that of normal cells. Many changes in enzyme profile also occur.
4. The fast growing cells tend to maximize the anabolic processes involved in growth, cut down on catabolic functions. Briefly speaking, they are concentrating mostly upon growth.
5. They also show biochemical changes that reflect altered gene regulation, such as the synthesis of certain fetal proteins and the manufacture of hormones.

Telomerase is active in many cancer cells

1. Telomeres are compound of DNA and proteins that are located at the ends of chomosomes of lower and higher eukaryotes. They confer stability on the ends of chromosomes.
2. In humans, telomeres consist of 1000 or more arrays of TTAGGG repeats at the terminals of the 3′ ending strands. These repeats are maintained in germ-line cells by the action of **telomerase**.
3. Telomeres are observed to shorten during sequence of primary cultures of cells and cultured primary cells are found to lack telomerase activity.
4. Telomers as well as telomerase are proved to be useful targets for anticancer therapy. A good number of normal cells contain telomerase activity. The enzyme telomerase is active in normal or malignant cells with a high proliferation rate. There is evidence that telomerase is active in some tumors and not others. Measurement of its activity may help in the staging of certain tumors.

Table 42.6 : Application of tumor markers

Detection	:	Screening in asymptomatic persons
Diagnosis	:	Differentiating malignant from benign conditions
Monitoring	:	Predicting effect of therapy and detecting recurrent cancer
Classification	:	Choosing theraphy and predicting tumor behaviour
Staging	:	Defining extent of disease
Localization	:	Nuclear scanning of injected radioactive antibodies
Therapy	:	Cytotoxic agents directed to marker-containing cells

CLINICAL ORIENTATION

- After cardiovascular disease, cancer is the second most common cause of death in the world.
- Humans of all ages are victimised by cancer and different organs are affected by it.
- Many cancers are related to the increase of age ; the development of this disease is directly associated with the longer life.
- Proper treatment by surgery, radiotherapy and chemotherapeutic agents in early stages of the disease can control the spreading of it. Otherwise, metastasis can develop rapidly with acute and uncurable symptoms causing death in the long run.

SCHEMATIC COMPENDIUM

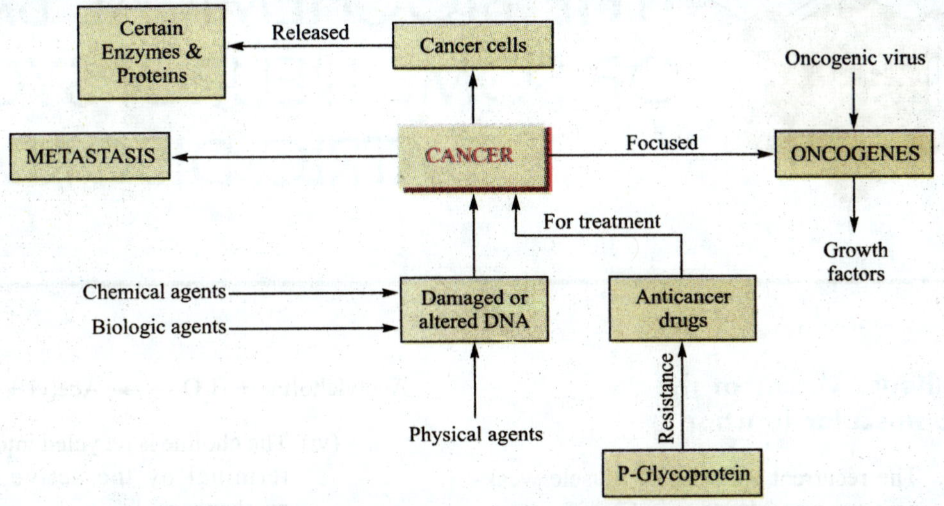

CHAPTER 43

THE BIOCHEMICAL BASIS OF SOME NEUROPSYCHIATRIC DISORDERS

Myasthenia Gravis at the neuromuscular junction

1. The recurrent episodes of muscle weakness produce Myasthenia Gravis.

   This is improved by administration of drugs that inhibit acetylcholinesterase.
2. The body forms autoantibodies against the acute acetylcholine receptor (ACHR) of the neuromuscular junction. These antibodies destruct the receptors.
3. The neuromuscular junction consists of a single nerve terminal separated from the postsynaptic region by the synaptic clefts. The overall process at the junction takes place in six steps :

   (i) In the cytosol of the nerve terminal, acetylcholine is synthesized by the enzyme choline acetyltransferase.

$$\text{Acetyl–CoA} + \text{choline} \longrightarrow \text{Acetylcholine} + \text{CoA}$$

   (ii) Acetylcholine is then incorporated into synaptic vesicles and stored therein.

   (iii) Acetylcholine is released from these vesicles into the synaptic cleft.

   (iv) The released one diffuses rapidly across the synaptic cleft to its receptors in the junctional folds.

   (v) The acetylcholine is then hydrolysed by acetylcholinesterase in the following manner.

$$\text{Acetylcholine} + H_2O \longrightarrow \text{Acetyl} + \text{Choline}$$

   (vi) The choline is recycled into the nerve terminal by the active transport mechanism.

Inhibitors of cholinesterase for the treatment of myasthenia gravis

1. There are two types of inhibitors of cholinesterase reversible and irreversible. Both are short-acting and long-acting reversible inhibitors. The former is mostly used in diagnosis. The diagnosis depends on history, clinical findings, and laboratory tests.
2. Pyridostigmine and nesotigmine are the effective agents for the treatment of this disease.
3. Disopropylflurophosphate (DFP) binds to a serine residue in the active centre of acetylcholinesterase. It is the toxic nerve gas and is also used as an insecticide.
4. Severe cases may require administration of azathioprine.

Huntington's Disease

1. This is an inherited disease in an autosomal dominant manner. Fifty percent of the children of an affected parent are at risk. It usually begins between 35 and 50 years of age and the duration is 15 years after onset.

2. This disease affects the neurons in the corpus striatum. Many die and are partly replaced by glial cells.
3. The levels of some neurotransmitters (eg, GABA) and acetylcholine of some enzymes involved in their synthesis are decreased.
4. The levels of dopamine and norepinephrine are increased.

Neuronal death in Huntington's disease.

1. This disease is characterized by the death of certain neuron due to certain chemicals (excitotoxins) which may be exogenous or endogenous.
2. Glutamate and aspartate are excitatory amino acids in the brain. Glutamate receptors have been subdivided into five classes.
3. The NMDA (N-methyl-D-aspartate) receptor plays a central role in long-term potentiation which results in an increase of synaptic efficacy in the hippocampus and other areas and may be involved in memory and learning.

Excitotoxins & other biochemical mechanisms are involved in brain damage due to store

1. In a stroke,brain damage occurs due to diminished blood flow. Strokes are the third leading cause of death.
2. Most strokes are caused by thromboses in the cerebral arteries. This reduces the supply of oxygen and glucose to the brain.
3. Ischemia causes depolarization of the neuronal membrane, leading to release of glutamate which stimulates the AMPA.
4. Increased intracellular Na^+ activates Na^+ / Ca^{++} transporters. Voltage gated Ca^{++} channels are activated by depolarization. Inositol triphosphate releases Ca^{++} which releases additional glutamate for exciting neighbour neurons.
5. The high level of intracellular Ca^{++} activates Ca^{++}-dependent nucleases, Proteases and phospholipases. Degradation of phospholipids can also lead to formation of platelet -activating factor (PAF) and release of arachidonic acid which forms eicosanoids.
6. The disease may be prevented by treatment of risk factors such as hypertension, diabetes mellitus, smoking and hyperlipidemia. Low dose aspirin therapy is effective in lowering the risk of stroke.

Parkinson's disease reflecting a deficiency of dopamine

1. Parkinson disease is characterized by tremor, bradykinesia, rigidity and postural instability. It U-rarely occurs before 40 years of age.
2. The pathologic characteristic of the disease is degeneration of the pigmented cells in the substantia nigra. When normal, these cells synthesize and use dopamine as a neurotransmitter and thus are said to be dopaminergic.
3. There is no evidence for a significant genetic component in this disease.

   The cell damage may reflect the process of ageing.
4. Viral infections can cause this disease. Drugs are important causes of Parkinsonism which can be reversible when therapy is stopped.
5. Mazindol inhibits the uptake of dopamine increasing the concentration of dopamine at the postsynaptic junction.
6. Surgical treatment includes the injection of minced adrenal autografts.
7. The antiviral agent amantadine induces release of dopamine from the presynaptic terminals of dopaminergic cells.

Alzheimer's Disease

1. It is an uncurable neuropsychiatric condition and its prevalence increases as more people live longer.
2. This disease is the commonest cause of dementia which ensures the decline in intellectual functions due to an organic cause. It imposes a heavy burden on families and on health care system and most patients cannot look after themselves.
3. The usual age in the starting of this disease is over 65 years, but it can occur at the age of 40 also. The survival period ranges from 2 to 20 years. The first symptom is the loss of memory.
4. The pathologic finding is characterized by the loss of cells in certain areas of the brain. Deposits of amyloid are common in small blood vessels.
5. Amyloid β protein, the major constituent of the plaques is present in this disease. Amyloid is a group of extracellular protein deposits found in many different diseases. Amyloid proteins stain blue with iodine. AβP is derived from a larger precursor protein named amyloid precursor protein (APP) whose gene is located on chromosome 21 close to the area affected in Down's syndrome. Patients with Down's syndrome who survive till age 50 often suffer from Alzheimer's disease.
6. APP is a protein of 770 amino acids and AβP is a peptide of 39-42 amino acids derived from APP by proteolytic cleavage.
7. One of the genes of Alzheimer's disease is encoded by APP and the mutations in this gene can lead to the increased deposition of AβP in neurons.
8. AβP-containing fragments is directly or indirectly neurotoxic. Some protein kinases are regulated by levels of Ca^{2+}.

Table 43.1 : Genes involved in familial types of Alzheimer's disease

Gene	Type of Alzheimer's disease	Chromosome	Protein
APP	Familial with amyloid precursor protein mutation	21	APP[2]
S182	Early onset, aggressive	14	Integral membrane protein
E5-1	Early onset	1	Integral membrane protein'
APOE4	Adult onset	19	APOE4

9. Recently, two genes have been isolated in connection with Alzheimer's disease which encode transmembrane proteins and show homology to the SPE-4 protein of caenorhalditis elegans. The protein participate in the fibrous body membrane organelle (FBMO) complex during spermatogenesis. The FBMO plays a role in the transport of soluble membrane bound proteins.

Role of other factors in Alzheimer's disease

1. Infection with a slow virus is the cause of the disease although no such virus has been isolated.
2. Another finding shows that plaques from patients with this disease often contain large amounts of aluminium.
3. The brains of patients who have died from Alzheimer's disease often show a marked reduction in choline acetyltransferase and acetylcholine. The levels of other neurotransmitters are also decreased.
4. The levels of APP become lower in the cerebrospinal fluid of patients with Alzheimer's disease.
5. It has become possible to change the deposition of AβP in tissues by drugs although

no specific drug therapy for Alzheimer's disease is yet available.

Causes of schizophrenia

1. The schizophrenic disorders are mental disorders characterized by psychotic symptoms having disturbances of thinking, feeling, and behaviour. There are a number of subtypes, eg, disorganized, catatonic, and paranoid.
2. This disorder starts befoe the age of 45 and turns to a chronic course. It is a major medical problem with devastating consequences for the victim and family members.
3. The causes of schizophrenia are unknown. Various psychologic, social, developmental, environmental, anatomic, genetic, biochemical and other factors are involved.

Schizophrenia due to lack of Replication

1. A child born to parents both of whom are schizophrenic has a 39 per cent chance of developing the condition.
2. A disease occurs in a family does not mean that it has a genetic origin. Poor psychologic interactions among family members can lead to abnormal behaviour patterns.
3. It has multiple genetic causes.

Dopamine has been implicated in schizophrenia

1. Acetylcholine, GABA, norepinephrine, peptides, and other molecules are involved in the causation of this disorder.
2. Neuroleptic drugs are acting to lower levels of dopamine. Neuroleptics act primarily by reducing dopamine activity in mesolimbic dopamine neurons.
3. There is the increase in the amounts of dopamine in brain tissue samples of patients with schizophrenia as per the reports of a number of observers.
4. Prolonged treatment with neuroleptics leads to a fall in the levels of homovanillic acid in cerebrospinal fluid.
5. The antipsychotic potencies of most neuroleptics correlate with their binding to D2 receptors.
6. All of the five classes of receptors are transmembrane proteins, some are glycoproteins, and most appear to be coupled to G proteins. D4 is the first receptor in the catecholamine family.
7. On the basis of analysis of postmortem specimens, it has been found that the density of D2 receptor is increased in the brains of schizophrenics.
8. The low dopamine activity in the prefrontal cortex shows negative symptoms of schizophrenia.

Techniques to establish the molecular basis of neuropsychiatric disorders

Mutations in the genes encoding rhodopsin, peripherin, and the β-subunit of rod phosphodiesterase are responsible for various forms of retinitis pigmentosa, a common cause of blindness.

Gene therapy for neurologic disorders is started. The technique of MRI-guided stereotaxis is used to inject mouse fibroblasts carrying a retroviral vector containing the gene for the thymidine kinase of the herpes simplex virus into inoperable human brain tumors. The retrovirus then infect neighbouring tumor cells, introducing the gene for thymidine kinase and thus rendering the recipient tumor cells susceptible to the drug ganciclovir. Normal cells are not affected by this compound. This approach has been called "molecular surgery".

CLINICAL ASPECTS

- Clinically, neuropsychiatric disorders are important among other reasons because they are devastating since they can destroy intellectual functions.
- Neurosciences provide rational bases for effective therapy of many extremely disabling conditions.

CHAPTER 44

BIOCHEMISTRY OF DISEASE

Biochemical Bases of Health and Disease

At present, the causes of many important diseases, such as Alzheimer's disease, atherosclerosis, rheumatoid arthritis, and schizophrenia are still unknown. So the treatment of such conditions are generally symptomatic and not very effective. The economic and human costs of such ignorance are much.

The molecular nature of a disease cannot be treated because of limitations—technological or other. In case of Sickle Cell anemia, it is not yet possible to correct the underlying abnormality in DNA. The space of research on gene therapy suggests, this statement may not apply for very long to this disease or to many other genetic disorders.

The major causes of disease are given in table 44.1.

Major Classes of Genetic Disease

Genetic diseases can be divided into three classes:

1. Chromosomal disorders.
2. Monogenic (classic Mendelian) disorders.
3. Multifactorial disorders

1. Chromosomal disorder

These are the conditions in which there is an excess or loss of chromosomes, deletion of part of a chromosome, or a translocation. The best known condition is Down syndrome. They can be known by analysis of the karyotype (chromosomal pattern) of an individual. Chromosomal translocations are important in a activating oncogenes.

Table 44.1 : Various biomolecules in diseases

Biomolecule	Property affected	Disease	Fundamental cause
DNA	Structure	Hbs	Mutation.
RNA	Structure	Thalassemia	Mutation leading to faulty splicing of mRNA.
Protein	Structure	Hbs	Mutation
Lipid	Amount (increase)	Tay-sachs	Mutation resulting in defective hexosaminidase A.
Glycogen	Amount (increase)	Glycogen storage disease	Mutation in gene for enzyme degrading glycogen (eg, phosphorylase).
Electrolyte (Cl^-)	Amount in sweat (increase)	Cystic fibrosis	A mutation in a membrane protein affecting transport of Cl^-.
Water	Amount(decrease)	Cholera	Infection of small intestine by vibrio cholerae leading to massive loss of water and electrolytes.

Table 44.2 : Examples of three major classes of genetics		
Class	**Example**	**Comment**
Chromosomal	Down syndrome	Prevalence increases with maternal age.
Chromosomal	Chronic myelogenous leukemia	Presence of Philadelphia chromosome.
Monogenic	Familial hypercholesterolemia	Autosomal dominant.
Monogenic	Cystic fibrosis	Autosomal recessive.
Monogenic	PKU	Autosomal recessive.
Monogenic	Duchenne muscular dystrophy.	X-linked.
Monogenic	Hemophilia.	X-linked.
Multifactorial	Ischemic heart disease.	Complex genetics.
Multifactorial	Essential hypertension	A single-gene theory also has its proponents.

2. Monogenic disorders

They are classified as (a) autosomal dominant (b) autosomal recessive (c) X–linked.

Dominant is used to mean that the mutation will be clinically evident even if only one chromosome is affected. Recessive denotes that both chromosomes must be affected (homozygous) X– linked indicates that the mutation is present on the X- chromosome.

As females have two chromosomes, they may be either homozygous or heterozygous for the affected gene. Thus X-linked inheritance in females can be dominant or recessive. Males have only one X- chromosome, so that they will be affected if they inherit the mutant gene.

3. Multifactorial disorders

Multifactorial disorders involve the action of a number of genes. The pattern of inheritance of these conditions does not follow classical Mendelian genetics. Less is known about this class, but importance is more because common adult diseases such as ischemic heart disease and hypertension are members of this group.

Pathologic Consequences by Genetic Diseases

In case, an enzyme is affected by mutation, an inborn error of metabolism may result. An inborn error of metabolism is a genetic disorder in which a specific enzyme is affected, producing a metabolic block, which leads to pathologic consequences as shown below :

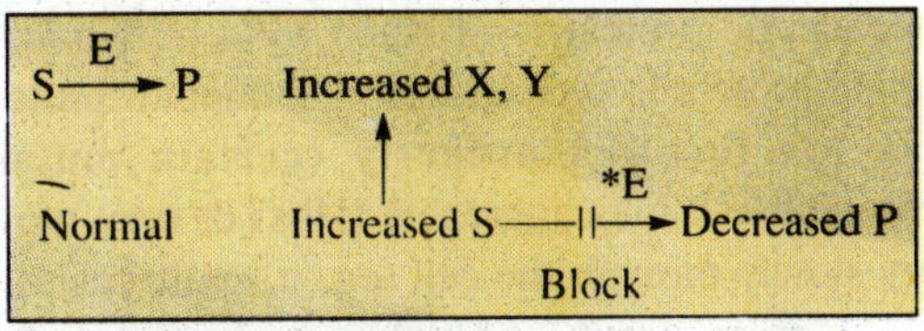

where*E = mutant enzyme and X, Y = alternative products of the metabolism of S. A block has three results.

1. Decreased formation of the product P.
2. Accumulation of the substrate S behind the block.
3. Increased formation of metabolites (X,Y) of the substrate S, resulting from its accumulation.

Any one of these three results may have pathologic effects.

In the case of *phenylketonuria* (*PKU*), the mutant enzyme is phenylalanine hydroxylase, resulting in the following condition :

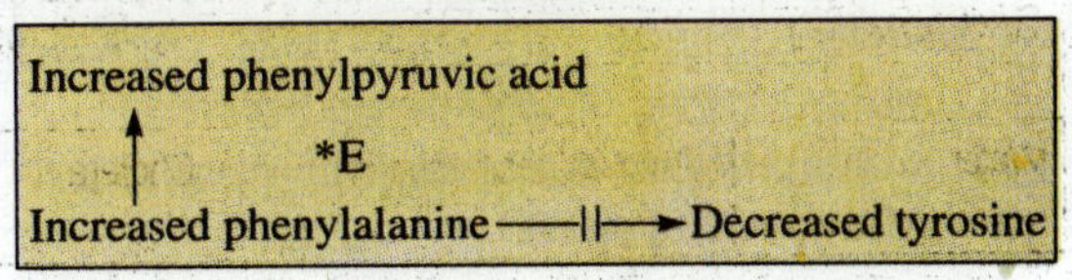

So, patients with PKU synthesize less tyrosine, have increased phenylalanine levels in plasma, and also show increased amounts of phenylpyruvate and other metabolites of phenylalanine in their body fluids and urine. This causes decreased availability of tyrosine for protein and neurotransmitter synthesis in the brain and also the inhibitory effects of high level of phenylalanine on the transport of other aminoacids into the brain. Both decreased formation of product and accumulation of substrate or other metabolites behind a block as well as change of feedback regulation alter the flux through metabolic pathway leading to pathologic effects.

Some inborn errors of metabolism are harmless. These are usually blocked in peripheral areas of metabolism, where neither low formation of product nor accumulation of its precursor perturbs the cell.(eg, pentosuria).

In many cases, if a structural gene for a noncatalytic protein is affected by a mutation, a mutant protein is synthesized. The mutant protein may not function properly (certain mutant hemoglobins), may aggregate (Hbs) or may move very slowly through the cell (eg, α_1-antitrypsin).

Early Diagnosis of Certain Inborn Errors to Avoid Permanent Damage

Treatment must be commenced immediately in certain inborn errors, otherwise the victimized will be permanently damaged (eg, PKU and galactosemia). A number of useful clues are given below :

Table 44.3 : Useful clues for the diagnosis of an inborn error of metabolism
(a) Family history of a genetic disease.
(b) Positive screening test (eg, for PKU).
(c) Sick newborn with other conditions (infections, cardiovascular, etc.) excluded.
(d) Physical findings, such as enlarged organs (eg, hepatomegaly in Gaucher's and Niemann-Pick disease).
(e) Unusual odour from breath or urine (eg, maple syrup urine disease).
(f) Low blood sugar (eg, Glycogen storage disease).
(g) Low blood pH (eg, due to accumulation of organic acids in maple syrup urine disease).

The sources of material that can be analyzed and the major tests used in investigating patients or fetuses suspected of having genetic diseases are listed below. Chorionic villus sampling and amniocentesis apply only to investigation of the fetus.

Material	:	Plasma red cells, white cells, fibroblasts, urine, organ biopsy, chorionic villus sample, amniotic cells.
General tests	:	Blood and urine glucose, blood pH, blood ammonia, amino acids in plasma and urine, detection of organic acids in urine.
Specific tests	:	Measurement of low amount of product or increased amount of precursor (eg, phenylalanine in PKU).

Measurement of activity of enzyme in red cells, white cells, or tissue biopsy.

Electrophoresis (eg, for Hbs).

Analyses by Southern blotting of restriction fragment length polymorphisms (RFLPs) and other features of DNA structure linked to or causing specific diseases (eg, Hbs, Huntington's chorea, DNA, etc.)

Identification of novel metabolite in urine of plasma by GLC-Ms.

Successful treatment for some genetic diseases

One of the four following applications is employed in general for the treatment of genetic diseases.

1. Attempts to correct the metabolic consequences of the disease by administration of the missing product or limiting the availability of substrate.
2. Attempts to replace the absent enzyme or protein or to increase its activity.
3. Attempts to remove excess of a stored compound.
4. Attempts to correct the basic genetic abnormality.

Enzyme replacement therapy has so far obtained mostly limited success. Problems arise in

Table 44.4 : Major classes of treatments available for genetic diseases			
Class of Treatment	**Principle**	**Disease**	**Treatment or comment**
1	Replacement of missing product	Familiar goitre	Administration of L-thyroxine.
1	Limitation of substrate	PKU	Diet low in phenylalanine.
2	Replacement of mutant enzyme	Gaucher's disease.	Injections of β-glucosidase.
2	Replacement of missing Protein	Hemophilia.	Injections of factor VIII (AHG).
3	To increase activity of mutant enzyme by supplying large amount of cofactor	Methylmaonic aciduria	Injections of Vitamin B_{12}
3	To increase activity of mutant enzyme by induction	Criggler-Najjar syndrome	Administration of phenobarbital.
4	Replacement of diseased organ carrying defective gene by normal organ	Galactosemia.	Liver transplant.

obtaining good sources of human enzymes (placentae have been useful in some instances), targeting enzymes in sufficient amounts to the suitable organ, and maintaining their activities in tissues if they are rapidly degraded. This is specially difficult in the case of brain, where administered enzymes must be made to cross the blood-brain barrier. Liposomes have been tried to use to deliver enzymes to target organs, but success has not been obtained.

Gene therapy could involve (i) gene replacement (ii) correction, or (iii) augmentation. In (i) the mutant gene would be removed and replaced with a normal gene. In (ii) only the mutated area of the affected gene would be corrected, the remainder being left unchanged. Augmentation involves introduction of foreign genetic material into a cell to compensate for the defective product of the mutant gene. It is already possible to introduce a normal gene into affected cells (eg, by transfusion). Such cells contain both the mutant and the foreign gene. The foreign gene is introduced at random sites on chromosomes, may interrupt the expression of certain host genes and is not subject to normal regulatory mechanisms. Gene therapy is a key area in medical research.

BIOCHEMICAL CASE HISTORIES

Case No. 1 : Xeroderma pigmentosa

Classification : Physical exposure to UV irradiation

History and Physical Examination

1. A 12-year old boy presented at Dermatology clinic with a skin tumor on his right cheek.
2. He had always avoided exposure to sunlight since it made his skin blister.
3. His skin had scattered areas of hyperpigmentation.
4. The dermatologist made a tentative diagnosis of xeroderma pigmentosa (pigmented dry skin).

Laboratory investigations

1. Histologic examination of the excised tumor showed that it was a squamous carcinoma (a common type of skin cancer in older people, but not in a boy of this age).
2. A sample of skin was taken for the preparation of fibroblasts.

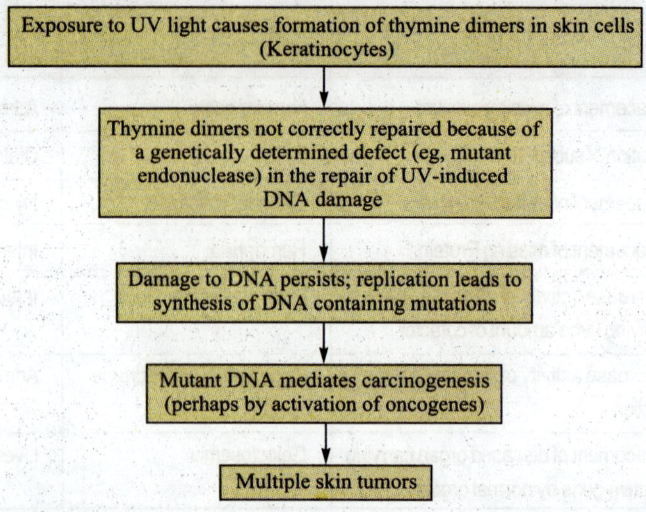

Fig. 44.1 : Summary of mechanisms involved in the causation of Xeroderma pigmentosa (XP)

3. A research lab in radiobiology was set up to measure the amount of thymine dimers formed by exposure to UV light.
4. The patient's fibroblasts and control fibroblasts were exposed to UV light and cell samples were taken at 8-hourly intervals for a total of 32 hours post-irradiation.
5. Extracts of DNA were prepared and the numbers of dimers remaining at each time point indicated were determined.
6. Only 24% of the dimers formed persisted in DNA extracted from the normal cells at 32 hours, about 95% were found in the extract from the patient's cells at 32 hours.

This confirmed the diagnosis of xeroderma pigmentosa.

Comments

1. Xeroderma pigmentosa (XP) is a relatively rare autosomal recessive condition in which the mechanisms for repair of DNA subsequent to damage inflicted by UV irradiation are impaired.
2. Thymine dimers are formed by the damage of DNA by UV irradiation. The covalent bonds are formed between carbons 5 and 5 and carbons 6 and 6 of adjacent intrachain thymine residues. These can be removed by an endonuclease. There is defect in this enzyme in XP.
3. In case, UV damage is not repaired, mutations in DNA will result and may cause cancer.
4. Patients with XP often suffer from variety of skin cancers from an early age.
5. The parents of this boy were advised that he would have to be watched throughout life for the development of new skin cancers.
6. The boy was advised to continue to avoid sunlight and to use an appropriate sunscreen ointment.
7. It has been estimated that patients with XP have a 1000-fold greater chance of developing skin cancer than do normal individuals.

Case No. 2 : Myocardial infarction

Classification : Lack of oxygen
History and Physical Examination

1. A 46-year old businessman was admitted to the emergency department of his local hospital complaining of severe restrosternal pain for 2 hours.

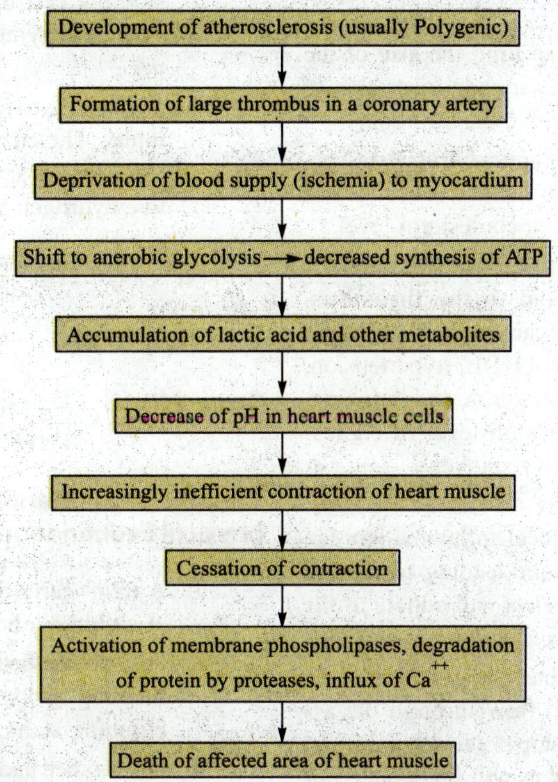

Fig. 44.2 : Summary of mechanisms involved in acute myocardial infarction

2. He has previously been admitted to hospital once for treatment of a small myocardial infarction (MI), still he continued to smoke heavily.
3. His blood pressure was 150/90. Pulse 60/min, and he was sweating continuously.
4. He was given morphine to relieve his pain and immediately transferred to a cardiac care unit, where ECG was started at once.

Laboratory tests

1. The initial ECG showed S-T segment elevation and other changes in certain leads, indicative of an acute anterior transmural left ventricular infarction.
2. Blood was taken at 4 hours for measurement of the MB isozyme of CK. At 4 hours the activity of this isozyme was slightly increased and at 12 hours the elevation was 4-fold.
3. The plasma cholesterol was moderately elevated and triglyceride levels were normal.

Treatment

1. The attending cardiologist decided to administer streptokinase (SK) by cardiac catheterization, because of the diagnosis of another transmural MI seen within 4 hours of the onset of symptoms.
2. Chest pain began to disappear after 12 hours and the patient felt comfortable.
3. He was discharged from hospital ten days later under the care of his family doctor with the advice to start a cholesterol lowering regimen and to stop smoking.

Discussion

1. The decision was to limit the size of the infract by intracoronary administration of SK which can dissolve the thrombus.
2. For long term therapy, a drug was prescribed that inhibits HMG-CoA reductase to reduce the plasma cholesterol level.
3. The causes of *atherosclerotic lesion* in the coronary artery led to the thrombus. Atherosclerosis is caused by high levels of LDL, low levels of HDL, hypertension, high levels of cholesterol, and smoking. This patient had a high level of cholesterol and was a heavy smoker.
4. The lesion is now an intimal plaque. Hemorrhage and local inflammation can occur into the plaque leading to rupture of its surface. Platelets will adhere to the exposed collagen and a thrombus results.
5. In case, the thrombus persists 90% of the vessel wall, blood flow through the affected vessel may stop (total ischemia) and capillary hemoglobin will be depleted of oxygen very rapidly.
6. The affected myocardium shows the reduction of ATP, an accumulation of lactic acid, the development of severe acidosis, and a marked reduction in contractile force.
7. The accumulation of lactate and H^+ inhibits glycolysis and thus stops ATP formation. Different studies have interpreted the depletion of ATP, activation of intracellular phospholipases (which can damage to cell membranes), activation of proteases, and accumulation of intracellular Ca^{++}.
8. Irreversible damage of some cells had been made by one hour of complete ischemia. So administration of SK is a must.
9. Reperfusion injury can occur after the administration of SK. This can damage to the plama membrane resulting in the entry of calcium ion from the plasma. High levels of intracellular Ca^{++} can pass inside the cell by activating or inhibiting various enzymes in an unregulated manner.
10. *Superoxide* may play a role in reperfusion injury. The enzyme *superoxide dismutase* can convert superoxide to the less reactive hydrogen peroxide as shown below :

$$O_2^- + O_2^- + 2H^+ \xrightarrow[\text{dismutase}]{\text{Supexoide}} H_2O_2 + O_2$$

Case No. 3 : Duchenne muscular dystrophy

Classification : Genetic History and physical Examination

1. A four-year old boy was brought to clinic at children's hospital.
2. His mother had noticed that her son was walking awkwardly, and had difficulty climbing stairs.
3. His mother had a brother who died at the age of 19 of muscular dystrophy.
4. The physician noted muscle weakness in both the pelvic and shoulder girdle.
5. There was the modest enlargement of the calf muscle.
6. The pediatrician made a tentative diagnosis of Duchenne muscular dystrophy (DMD) due to the patients' muscle weakness and its distribution.

Lab and other tests

1. Serum creatine phosphokinase activity was markedly increased.
2. Electromyographic findings showed muscular dystrophy.
3. Nerve conduction studies were normal.
4. A biopsy of calf muscle showed areas of muscle necrosis and some variation in the size of muscle fibres.
5. Muscle biopsy by electrophoresis showed an absence of the protein dystrophin.

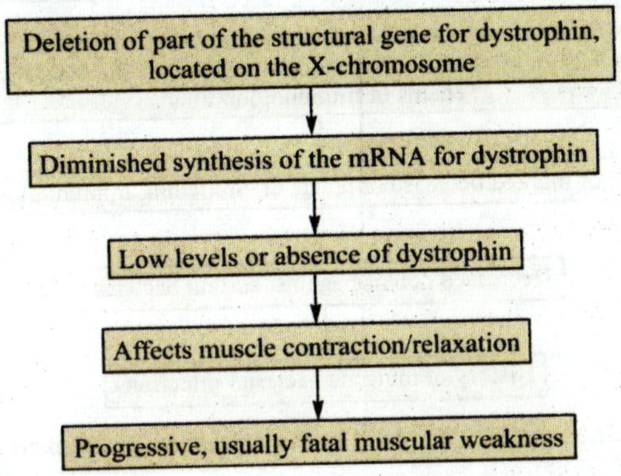

Fig. 44.3 : Summary of mechanisms in the causation of Duchenne-muscular dystrophy

Discussion

1. The lab findings confirmed the pediatrician's provisional diagnosis of DMD.
2. This is severe X-chromosome-linked degenerative disease of muscle.
3. This happens 1 in 3500 live male births.
4. It affects young boys who show loss of strength in their proximal muscles leading to difficulty in standing up and to very severe weakness.
5. There had been the defect to the middle of the short arm of the X-chromosome and to subsequent identification of a segment of DNA that was deleted in patients with DMD.
6. Antibodies against dystrophin were located in the sarcolemma (plasma membrane) of normal muscle and was absent in the sarcolemma of patients with DMD.
7. Further studies have shown that the gene coding for dystrophin is one of the largest human genes recognized to date.
8. Dystrophins are being produced by recombinant DNA technology. The availability of cDNA probes for dystrophin facilitates prenatal diagnosis of DMD by chorionic villus sampling.
9. The demonstration of dystrophin as a cause of DMD is one of the major accomplishments of the application of the new molecular biology to human disease.

Treatment

1. The boy was encouraged to exercise, and regular attendance at a specialized muscular dystrophy clinic was started for the treatment of the complications if arose.
2. The mother was advised to take genetic instruction.

Case No. 4 : X-Linked Agammaglobulinemia

Classification : Immunologic, genetic History and Physical Examination

1. A-3-year old boy was referred to children's hospital with a high temperature, pain in the chest, and difficulty in breathing.
2. His family doctor diagnosed pneumonia.
3. Since his older brother had previosly been diagnosed as having X-linked agammaglobulinemia, the attack of pneumonia raised the suspicion that the younger brother might also lack the potential to synthesize gammaglobulins.

Lab Results

1. Electrophoretic analysis showed the levels of albumin and α- and β-globulins normal, but only a trace of γ-globulins was detected.

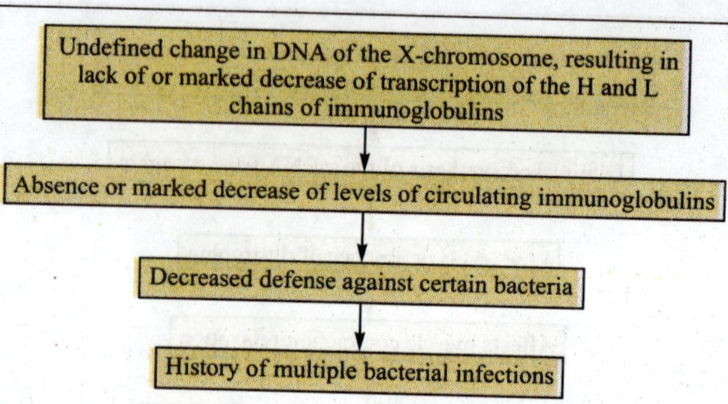

Fig. 44.4 : Summary of mechanisms involved in the causation of X-linked agammaglobulinemia

2. B lymphocytes were markedly low in the blood but the T-cell count was normal.
3. A biospy of lymph node showed an absence of plasma cells.
4. Streptococcus pneumoniae was demonstrated in samples of sputum.

Discussion

1. Disorders of the immune system can affect either the B-cell system(circulating immunoglobulins) or the T-cell system (cell- mediated immune responses).
2. This is an immunologic condition, although the underlying cause is genetic.
3. X-linked agammaglobulinemia affects young boys and is a relatively uncommon condition.
4. The actual nature of the defect in this condition has not been established but involves the X-chromosomes.
5. Although only a few circulating B cells are present, normal numbers of pre-B cells are present in bone narrow.

Case No. 5 : Diabetes Mellitus with Ketoacidosis

Classification : Endocrine deficiency, Lack of Insulin History and physical examination

1. A 14-year old girl was admitted to children's hospital in coma.
2. She developed a sore throat and fever two weeks before according to her mother's statement.
3. Some days before admission, she began to complain of thirst and strarted to get up several times during the night for urination.
4. On the day of admission the girl had started to vomit.
5. On examination she was dehydrated, her skin was cold, she was deeply breathing, and her breath had a fruity odour.
6. Her blood pressure was 90/60 mm of Hg. and her pulse rate was 115/min.
7. A diagnosis of severe type I insulin-dependent diabetes mellitus with resulting ketoacidosis and coma was made by the intern on duty.

Lab results

(A) **Plasma** — **Normal Values**

Glucose	:	35 mmol/L(3.6-6.1)
Acetoacetate	:	2.8 mmol/L (< 0.2)
Bicarbonate	:	5 mmol/L(24-28)
Urea	:	12 mmol/L(2.9-8.9)
Potassium	:	5.8 mmol/L(3.5-5.0)
Creatinine	:	160 μmol/L (60-132)

(B) **Urine**

Glucose	:	(++++)
Ketone bodies	:	(++++)

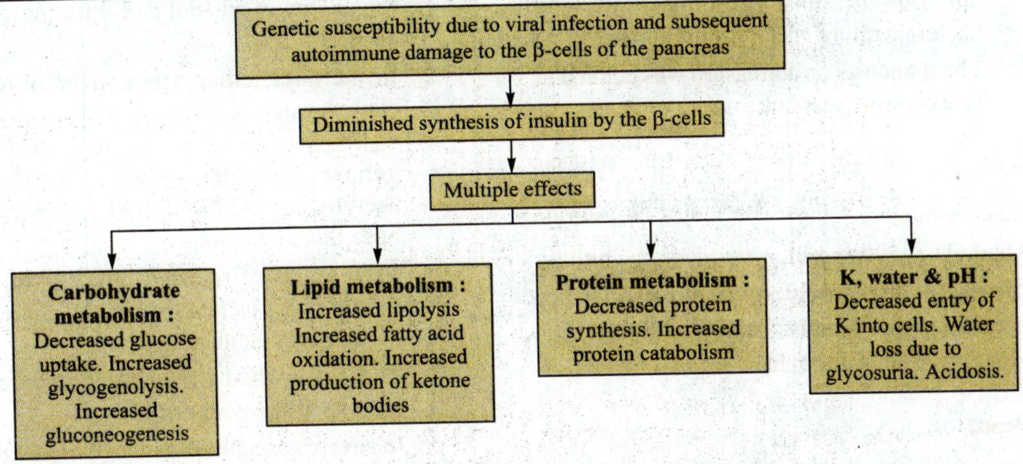

Fig. 44.5 : Summary of mechanisms involved in the causation of the Ketoacidosis of type I diabetes mellitus

Treatment

1. In the treatment of diabetic Ketoacidosis (DKA), Insulin administration and saline fluids are most essential. The patient was given insulin (10 μ/h) added to 0.9% NaCl. Glucose was withheld until the level of plasma glucose fell below 250 mg/dl.
2. KCl was also administered very cautiously, continuous checking of K levels is very important in the management of DKA, because insuffcient management of K balance is the major cause of death in this condition.
3. Bicarbonate may be necessary if acidosis is very severe.

Discussion

1. This type of diabetic patients have a genetic susceptibility, which may be due to a viral infection.
2. The infection and consequent inflammatory reaction may alter the antigenicity of the surface of the pancreatic β-cells and establish an autoimmune reaction involving cytotoxic antibodies and T lymphocytes. This causes destruction of β-cells resulting in type I diabetes mellitus.
3. The sore throat that the patient had several weeks before reflected the initiating viral infection.
4. The low pH indicated a severe acidosis due to the increased production of acetoacetic acid and β-hydroxybutyric acid.
5. The levels of urea and creatinine indicated some renal impairment, dehydration, and increased catabolism of protein.
6. A high plasma potassium level is often found in DKA due to the lowered uptake of potassium by cells in the absence of insulin.
7. The increased *osmolality* of plasma due to hyperglycemia causes the development of coma in DKA.

Case No. 6 : Acute intoxication with ethanol

Classification : Chemical History and Physical Examination

1. A 52-year old male was admitted to the emergency department in coma.
2. He had been a moderate drinker before the death of his wife. After her death he had become depressed and started increasing alcohol consumption. He had been eating poorly.
3. On sunday morning his married daughter found him unconscious on the living room couch.
4. His breathing was deep and noisy, alco-

hol could be smelled on his breath, and his temperature was 35.5°C (normal 37.4)

5. The diagnosis on admission was coma due to excessive alcohol intake.

Lab tests

Blood results

Alcohol : 500 mg/dL
Glucose : 2.7 mmol/L (normal 3.6-6.1)
Lactate : 8.0 mmol/L (normal 0.5-2.22)
pH : 7.21 (normal 7.4)

Treatment

1. Hemodialysis was immediately started to eliminate the toxic ethanol from the body due to high level of blood alcohol and the coma.
2. The level of blood alcohol fell rapidly and the patient became conscious.
3. Intravenous glucose (5%) was administered after dialysis was stopped to counteract the hypoglycemia.
4. The patient made a good recovery and was referred to for psychiatric advice.

Discussion

1. The remarkable problem is the development of liver cirrhosis in individuals who are accustomed to high intake of ethanol (eg, 80 gm of ethanol daily) for more than 10 years.
2. In the liver, ethanol is converted to acetate through acetaldehyde by alcohol dehydrogenase and acetaldehyde dehydrogenase. Acetate is then converted to acetyl-CoA. Reduced NAD is formed in both these reactions.
3. High levels of reduced NAD favour the formation of lactate from pyruvate resulting in lactic acidosis.
4. The diminished pyruvate concentration inhibits gluconeogenesis.
5. In severe cases, liver glycogen level is diminished and hypoglycemia results due to nonavailability of glycogenolysis.
6. The microsomal cytochrome P-450 also produces acetaldehyde which is a highly reactive molecule and can form adducts with proteins, nucleic acids, and other molecules. The ability of acetaldehyde to react with various molecules is involved in the causation of the toxic effects of ethanol.
7. The biological membranes are affected by ethanol and are excitable by that affect. This results in alterations of their action potentials, and also affects neurotransmitter release.
8. All of these depress cerebral function and, if severe enough, can produce coma and death from respiratory paralysis.

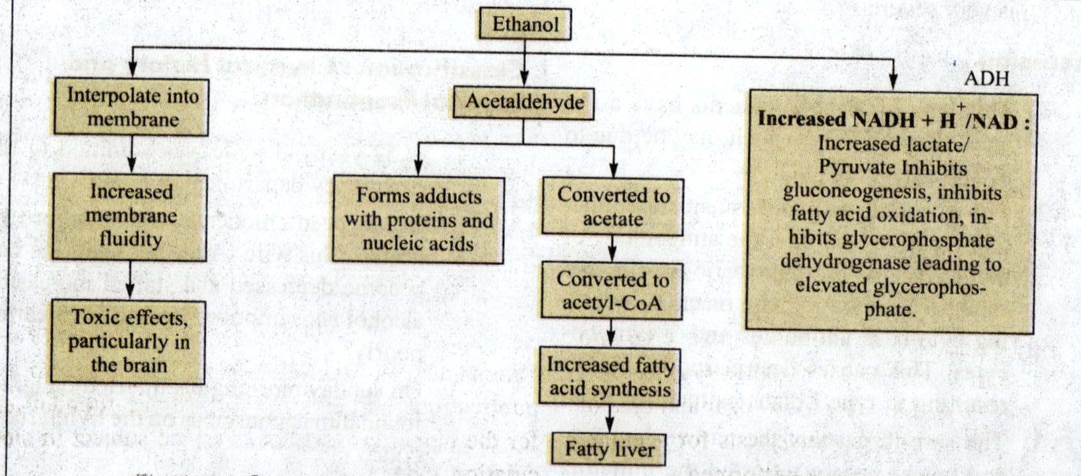

Fig. 44.6 : Summary of mechanisms involved in the causation of toxicity by ethanol

CHAPTER 45

HISTORY AND ADVANCEMENT OF CLINICAL BIOCHEMISTRY

From the time of Hippocrates (C.460-C.375 B.C.) the crude examination of body fluids including urine had been done. Thomas Willis, in the second half of the 17th century wrote a dissertation on urine and did the sweet taste to differentiate between the two types of diabetes.

Boerhaave of Leyden (1688-1738), who had much influence on medicine, taught quite confidently that medical phenomena could be interpreted in terms of chemistry. Physicians had long been interested in urinary calculi. Richard Bright in 1836 showed the relation of albuminous urine with Kidney disease. He did this by heating urine in a spoon and showed that cloudiness developed just before boiling.

In nineteenth century, Henry Bence Jones whose name is associated with "protein" and J. W. L. Thudichum became lecturer in chemical pathology at St. Thomas' Hospital, London. Alfred Garrod, just over 100 years ago, first did quantitative analytical methods applied to blood and at about this time, a number of well-known qualitative tests for urine had been done. By the end of the nineteenth century much of modern urine testing had already been done.

At the commencement of the 20th century "Clinical Chemistry" really began to develop into the discipline which we know today. Quantitative tests were carried out in regard to sugar and urea of urine and a series of qualitative tests for acetone, albumin, bile, and sugar were performed.

Different techniques now began to appear, mainly in the United States, for quantitative estimations applied to blood. In 1912, Folin and Denis described the phosphotungstic acid reagent for uric acid; in 1913, Bang described his micromethod for blood sugar, and Van den Bergh used Ehrlich's diazo-reagent for the determination of bilirubin. In 1915, Maclean and Van Slyke published their iodometric method for chloride. In the same year, the Liebermann- Burchard reaction was applied to the determination of cholesterol in blood. The newly introduced clinical application of insulin to the treatment of diabetes greatly stimulated urine and blood sugar estimations as well as those of electrolytes.

The range of quantitative tests performed in the clinical chemistry laboratory increased remarkably during the years between the two world wars. The B.M.R. was also estimated as well as urea concentration tests, the phenol red excretion test, the fractional test meal, and the test for occult blood. Pancreatic disease was studied by means of urinary diastase, fecal fat, and muscle fibres in the feces. In relation to neurological disease, cerebrospinal fluid was fairly well investigated. In 1930, there appeared Harrison's Chemical Methods in Clinical Medicine and Stewart and Dunlop's clinical chemistry in Practical Medicine. A year later, the two volumes of Peters and Van Slyke were published. These three publications undoubtedly formed the foundation for the numerous text books on the subject in circulation today.

In the late thirties in the United States and in the early forties in the United Kingdom, various types of photoelectric colorimeters had appeared resulting in the speeding up of routine clinical chemical work with accuracy. Another great progress, in the middle and late forties, there was the appearance of the flame photometer which enabled the estimation of sodium and potassium in the body fluids to be done quite quickly and very accurately. This had helped the clinicians to treat the diseases such as diabetic coma.

In 1938, Callow published the use of the Zimmerman reaction to determine 17- Ketosteroids. This was the first of the steroid determinations applied for routine purposes. Satisfactory methods had been developed for the estimation of catecholamines and their metabolites, particularly in urine. The more recent development of radioimmunological assay and protein-binding techniques has led to the estimation of the hormones themselves, both in blood and other biological fluids.

Bodansky as well as King and Armstrong, in the early thirties, introduced their techniques for phosphates estimation. Specific inhibition techniques then led to the estimation of prostatic acid phosphatase which played a great part in the diagnosis and control of therapy of prostatic carcinoma. The estimation of aminotransferases, lactate dehydrogenase and a number of other intracellular enzymes had been done. Isoenzyme determination had also been performed. Techniques also became available for the estimation of trace elements by absorption spectrophotometer which gave accurate results for routine purposes.

Grabar and his associates in 1953 made a firm basis on Immunoelectrophoresis which gave detailed study on Para- Proteinemias. Individual serum proteins could be investigated also by other immunological techniques, including radio- immunoassay.

Tswett in 1906 described absorption chromatography, Adams and Holmes in 1936 ion-exchange chromatography, Martin and Synge in 1941 partition chromatography, Consden, Gordon and Martin in 1944 paper chromatography, Mottier, James and Martin in 1952 gas-liquid chromatography. All these techniques are in current use by clinical chemists and have greatly increased the range of substances which can be investigated in relation to disease.

In the fifties and sixties, the improvement in the control of respiratory disorders, open-heart surgery, and renal dialysis had been made by the use of the Astrup technique. Hereditary metabolic disorders had been detected by microbiological methods.

Since clinical chemistry had achieved the status of a recognized discipline in its own right national societies devoted to the subject were formed in various countries. First formed in the Netherlands in 1947 and was soon followed by those in the United States of America and in the United Kingdom. Societies of clinical chemistry have now grown up all over the world. The first journal, Clinical Chemistry, was published in 1955 by the United States. This was soon followed by **Clinica Chemica Acta**, and now numerous journals have been published in many parts of the world. In addition, there are a good number of excellent books dealing with details of analytical methodology.

Since the national societies had been formed, it was realized that problems related to clinical chemistry should be considered at an international level. This has led to the formation of the clinical chemical section of the International Union of Pure and Applied Chemistry (IUPAC) and the International Federation of Clinical Chemistry (IFCC). These organizations have a number of joint committees. Much of the account of the history of clinical biochemistry given in this chapter was originally produced for the joint commission on Education of these two bodies.

This discipline is, of course, advancing rapidly. High resolution analysis is progressing speedily. This includes pressure chromatography and techniques such as mass spectroscopy and electron spin resonance. Sophisticated techniques of this kind will enable the clinical chemist to estimate relatively large numbers of metabolites in relatively to all amounts of blood, tissue fluid, tissue culture, and biopsy samples. Such development must certainly lead to greater understanding of disease processes as well as to more accurate diagnostic procedures and therapeutic control. The normal range of these estimations should also be known for detecting diseases. It is now becoming almost mandatory to abolish the old fashioned concept of "normal range" and replace it by "reference values".

Normal Range

The so-called normal ranges were at first determined on specimens obtained from healthy medical students or laboratory staff. Factors such as bed-rest could effect certain important biochemical values.

Whatever method is used to determine ranges, it soon becomes obvious that these values are dependent on a whole variety of factors including method of collection and handling of the sample for analysis, time of collection, seasonal changes, laboratory analytical method employed, laboratory accuracy, patient's age, sex, ethnic group, social class, diet, physiological factors such as pregnancy or environment, and so on. There is also the effect of therapeutic agents on the analytical method employed as well as on the actual blood levels of certain constituents. Many of these variables can be standardized, and then it is possible to obtain so-called reference populations. These must be defined in terms of the other factors, such as sex, age, ethnic group, social class, etc. It should be pointed out here that in this regard biochemical values resemble many other physical signs obtained at the bedside or by ancillary methods of investigation.

It may also be mentioned that there is the effect of disease itself on biochemical values, which in themselves are not diagnostic of that disease. For example, in renal disease, in addition to an increase of blood urea there is frequently an increase in blood urate. There is a rough clinical correlation. The clinician learns to associate certain blood urate levels with appropriate increase in blood urea. If the level of the former is too high in relation to the latter, then the possibility of Gout is considered. The latter disease can also be associated with an increase in the blood urea as well as of blood urate and a similar reasoning process must be employed.

Under the above discussion, it indicates that, in relation to diagnosis, biochemical values are merely additional and frequently very important for physical signs. Biochemical findings, as with most physical signs, have significance only in relation to the history of the patient's illness and the physician's findings, at the bedside as well as from ancillary methods of investigation.

Maximum difficulty arises to obtain normal reference values in the elderly because of the presence of frequent diseases in them. Therefore, it is difficult to define the "normal" state. The meaningful values are just begun to obtain in pediatrics possibly because of difficulties of sample collection and emotional obstacles.

It will be essential to standardize analytical methodology to obtain meaningful "reference intervals" in all groups of patients. Necessity will come into force to set up reliable reference methods, by which other methods can be standardized. This is particularly the case in relation to enzyme determination. It is also vital to report laboratory findings in terms of standard units. For this purpose, it has already been decided to adopt SI Units.

SI Units

The conference Generele des Poids et Mesures, in 1960, confirmed the Systeme International d' Unites (SI), originally adopted in 1954. This system is now based on eight base units relation to eight basic kinds of quantity.

The katal has been included in the following table 45.1. It should be pointed out that a number of authorities believe that the concept catalytic amount is not really viable and its unit would really be derived rather than base. One could introduce the concept catalytic activity per litre (derived coherent unit) which would give a figure which would be the same as if the katal were employed.

The mole and the katal have already been defined. In order to define the Kelvin, it is necessary to state that 273.16 K represents the temperature interval between the absolute zero and the triple point of water. Temperature is more commonly expressed in terms of celsius temperature, which is the thermodynamic temperature minus 273.16 K, since the celsius temperature at the triple point water is 0.01 degree celsius (0.01°C). The former term "centigrade" should no longer be used. The steam point celsius temperature is 100°C above the celsius triple point of water. Other units can be derived from the base units.

Table 45.1 : Base Units of SI system

Kind of Quantity	SI Base Unit	Symbol for Unit
Length	meter	m
Mass	kilogram	kg
Time	second	S
Electric current	ampere	A
Thermodynamic Temperature	kelvin	K
Luminous intensity	candela	cd
Amount of substance	mole	mol
Katalytic amount	katal	kat

Derived Coherent Units

These are the units constructed exclusively from base units. In this category, unit area (unit length × unit length) or unit velocity (unit length divided by unit time) can also be included. The coherent unit of volume is the cubic metre. For the purpose of clinical biochemistry, the litre has been retained as the unit and redefined as exactly one-thousandth part of a cubic metre.

Derived Noncoherent Unit

These are the units constructed from base units and numerical factors, for example, milligram per decilitre. In other words, the numerical factors are multiples or submultiples, which can be denoted by prefixes as shown in table 45.2.

Table 45.2 : Prefixes denoting decimal factors

Factor	Name	Symbol
10^{12}	tera	T
10^{9}	giga	G
10^{6}	mega	M
10^{3}	kilo	k
10^{2}	hecto	h
10^{1}	deca	da
10^{-1}	deci	d
10^{-2}	centi	c
10^{-3}	milli	m
10^{-6}	micro	μ
10^{-9}	nano	n
10^{-12}	pico	p
10^{-15}	femto	f
10^{-18}	atto	a

Certain derived coherent units are used in clinical biochemistry. The unit of force is the Newton (N) and 1N = 1 kg m/s². the unit of energy is the joule (J) and 1J= 1 Nm : 4.2 kg corresponds to 1 kcal (kilocalorie), and the latter unit, which is the medical calorie, will be placed out. The unit of power is watt (W) and 1W = 1J/s. The electrical units are volt (V), farad (F) and ohm (Ω) :

$$1V = 1\ W/A;\ 1F = 1\ A/s/V;\ 1\Omega = 1V/A.$$

Concentration can be expressed in two types of unit. Mass concentration is the mass component divided by the volume of the mixture (kg/l, gm/l, mg/l, etc). Substance concentration is the amount of substance of the component divided by the volume of the mixture (mol/l, m mol/l, etc.). The latter form of expression is preferable if the molecular weight of the substance is known. It can also be used for formula units.

Substance concentration has not been expressed in SI units in a number of chapter of this book, since expressions of concentration in common clinical usage have been employed (mg/100 ml, mEq/litre, etc.). To convert these values into substance concentration in SI units (m mol/l, mmol/ I, etc.), it is necessary to multiply by certain factors, some of which are shown in table 45.3.

The multiplication factor is easily derived from the former ways of expressing mass concentration, since it equals $\frac{\eta}{V \times MW}$, where η is the numerical

Table 45.3 : Multiplication factor for conversion substance concentration in Plasma				
System	Component	Kind or Quantity	Numerical Value	Unit
Serum	Calcium (Total)	Substance concentration	2.5	m mol/l

Table 45.4 : Multiplication factors for conversion to substance concentration in Plasma

Substance	Current concentration expression	Substance concentration	Multiplication factor for conversion to substance
Aminoacid N	mg/100 ml	m mol/l	0.714
Ammonium	μg/100 ml	μ mol/l	0.587
Ascorbic acid	mg/100 ml	μ mol/l	56.8
Bilirubin	mg/100 ml	μ mol/l	17.1
Calcium ion	mg/100 ml	m mol/l	0.250
	mEq/1	m mol/l	0.5
Cholesterol	mg/100 ml	m mol/l	0.0259
Copper	μg/100 ml	μ mol/l	0.157
Creatinine	mg/100 ml	μ mol/l	88.4
Glucose	mg/100 ml	m mol/l	0.0555
Iron	μg/100 ml	μ mol/l	0.179
Lead	μg/100 ml	μ mol/l	0.0483
Magnesium	mg/100 ml	m mol/l	0.411
	m Eq/1	m mol/l	0.5
Phosphate (Inorganic P)	mg/100 ml	m mol/l	0.323
Protein-bound iodine	μg/100 ml	m mol/l	78.8
Sodium ion	m Eq/l	m mol/l	1.0
Urate	mg/100 ml	m mol/l	0.0595
Urea	mg/100 ml	m mol/l	0.166

value in a stated volume V (measured in litres), MW is the molecular weight of the substance concerned. When concentration has been expressed as mEq/l, it is merely necessary to divide by the valency of the component being determined.

Certain standard abbreviations are also recommended as shown below :

Whole blood	B
Arterial blood	aB
Venous blood	vB
Day	d (preferably 24 hr)
Fasting measurement	fpt
Feces	F
Night	n
Plasma	P
Serum	S
Spinal fluid	Sp
Urine	U

A laboratory report of a measurement should ideally always indicate :

1. the system.
2. the component.
3. the kind of quantity.
4. the numerical value.
5. the unit.

It is also doubtful whether actually the words "substance concentration" will be included in the average laboratory report.

References

Caraway, W.T. : The Scientific Development of Clinical Chemistry to 1948, clin. chem. 19:373, 1973.

Henry, R.J. : Clinical Chemistry—Principles and Techniques, Ed. 2. New Work, Harper and Row, 1973.

Robinson, R. : Clinical Chemistry and Automation, London, Chas. Griffin and Co. Ltd. 1971.

Tietz. N. W (ed.) : Fundamentals of Clinical Chemistry, Philadelphia, W.B. Saunders Co., 1970

IUPAC & IFCC Information Bulletin : Quantities and Units in Clinical Chemistry, 1972.

Young, D. S., Thomas, D. W. Friedman, and Pestaner, L.C. : Bibliography : Drug Interferences with Clinical Laboratory Tests, Clin. Chem. 18 : 1041, 1972.

CHAPTER 46

ADVANCED TECHNIQUES FOR BIOCHEMICAL INVESTIGATION

Ion-Exchange Chromatography

(i) Ion-exchange resins are nothing but cross-linked polymers. The polymers must have negligible solubility but be porous enough for the ions to diffuse freely through it.

(ii) Ion-exchange resins are of cation and anion exchangers. Strong cation-exchange resins contain sulphuric acid groups ($-SO_3^-$), weak ones carboxylic acid groups ($-COO^-$), whereas strong anion-exchange resins have $-N\ (R_1\ R_2\ R_3)$ and weak ones $—N\ (R_1\ R_2)$.

(iii) The most important resins are polystyrene resins formed by condensation of styrene (vinyl benzene) and divinyl benzene. Acidic or basic groups are introduced before or after polymerising.

(iv) Strong alkaline cellulose treated with chloroacetic acid introduces the carboxy methyl group to give the weak cation-exchange resin carboxymethyl-cellulose (CM-cellulose) while condensation with 2-chlorotriethylamine gives the weak anion-exchange diethylaminoethyl-cellulose (DEAE-cellulose). Cellulose ion-exchange materials are specially suitable for protein separations.

(v) The resins can be looked on as insoluble acids or bases which form insoluble salts shown below.

$$H^+ - Resin^- + Na^+ \rightarrow Na^+ - Resin^- + H^+ \quad \text{(Cation-exchanger)}$$

$$Resin^+ - OH^- + Cl^- \rightarrow Resin^+ - Cl^- + OH^- \quad \text{(Anion-exchanger)}$$

The more strongly acidic the ion-change resin, the greater is the ionisation of the acidic group and the lower the pH at which it will exchange.

(vi) Ion-exchange resins have been widely used for the separation of amino acids and peptides. These have also been used to separate organic weak cations or anions from inorganic salts – ion-exchange resulting.

(vii) Mixed beds of anion and cation exchange resins have the property of replacing cations and anions of any salt in water by equivalent amounts of H^+ and OH^- respectively. This process is used in the preparation of "deionised" water in the laboratory. Non-ionic contaminants are not removed.

Partition Chromatography

This term covers liquid-liquid partition chromatography (Paper chromatography and thin layer chromatography). Gel chromatography, Gas liquid chromatography.

Paper chromatography of amino acid mixture

Introduction

In biology and chemistry, it is most often necessary to separate components of a mixture which are very similar and are difficult to separate by chemical or physical methods. Chromatography and electrophoresis are two powerful modern methods utilised for such purpose.

Tswett (1906) the Russian biologist, first appreciated the possibilities of chromatography and he put the term "Chromatography". Consden, Gorden, and Martin (1944) described paper chromatography in which separations were done mainly by partition.

Principle

The separation of components of a mixture by a chromatographic system depends on multiple partition process. Small differences in partitioning of each component of a mixture are multiplied many fold. The greater such differences, the greater is the ease of separation.

A small drop of solution containing the mixture of compounds is put on a strip of filter paper and allowed to dry. A suitable solvent (mixture of two solvents) is allowed to flow along the filter paper over this spot. The substances in the initial spot are extracted by the flowing solvent and carried forward along the filter paper to a distance which appears related to their partition coefficient between the free and bound solvent phases of the filter paper. After the solvent has run for a suitable distance along the paper, the paper is removed, dried and subjected to suitable tests to locate the various compounds.

R_f value is defined as the ratio of the distance travelled by the component to the distance covered by the solvent. R_f value depends on the nature of the solvent, the temperature, and the presence of other substances.

Procedure

Whatmann filter paper is cut into 35 × 15 cm sheet.

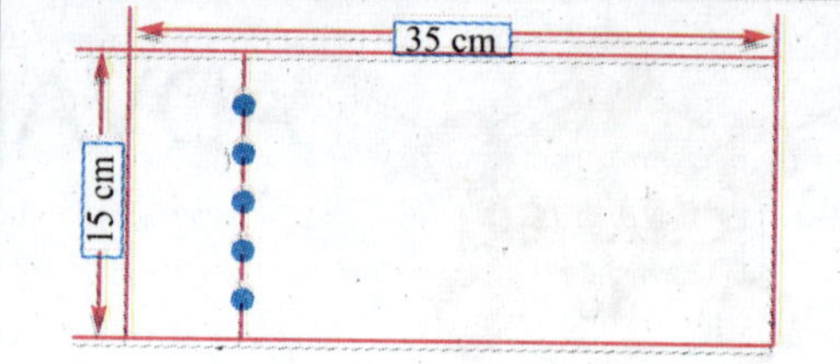

Fig. 46.1 : Serum Protein electrophoretogram

A pencil line is drawn about 3 cm above the shorter edge of the paper and 5 points are marked at equal spacing leaving 2.5 cm from the two edges. On the middle point, a mixture of four amino acids is applied with the help of a fine capillary. It is dried with the help of hot air blown by a hair dryer. Again, another small quantity of the mixture is spotted at the same place and dried. This process is repeated 2 to 3 times more. On the other points, the individual amino acids of the mixture are similarly applied. The positions of these amino acids are marked with a pencil.

The solvent for developing the chromatogram is a mixture of *n*- butanol : acetic acid: water (12 : 3 : 5) respectively. This mixtue is freshly prepared.

If a chromatographic tank is not available, specimen jar (about 40 × 15 cm) with fitting lid can very well serve the purpose. In such a jar, put about 100 to 150 ml of the solvent mixture and replace the lid so that the lid is airtight. In case of doubts, apply vaseline to the lid to avoid leakage. In an hour, the inside atmosphere will be saturated with the solvent vapour. Now fold the paper in which sample has been applied in the shape of a

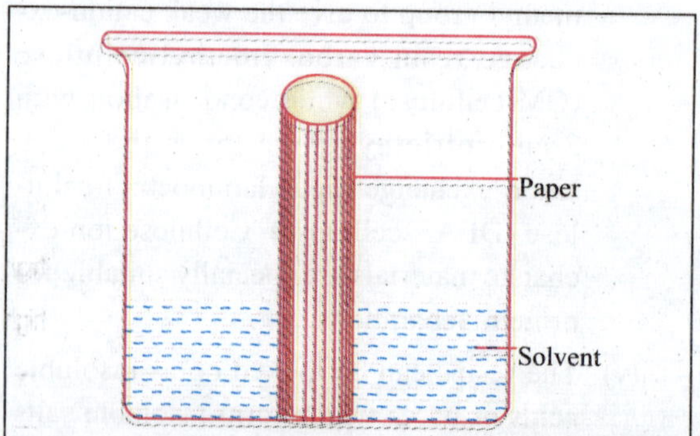

Fig. 46.2 : The chromatographic jar containing the solvent and the paper

cylinder and tie the opposing ends of the paper with staples or thread. Open the lid and place this folded paper in upright position in the jar, the pencil line lower most and about a centimeter above the solvent. Replace the lid. The paper should stand absolutely vertically. Leave the chromatogram to develop for 10 to 15 hours or earlier if the solvent has ascended quite near the upper margin of the paper.

Take out the paper at the desired time, cut the stitches and let it dry completely in the air.

After the paper has dried thoroughly, the location reagent (0.2 per cent ninhydrin in acetone) is sprayed uniformly on the paper with the help of an all glass sprayer. The paper is then allowed to dry first in the air and then in a hot air oven at 105°C for 3 minutes. Purple coloured amino acid spots are seen on the chromatogram.

Identify the amino acids in the mixture with the help of spots produced by known amino acids. The R_f values of the amino acids can be calculated as :

$$R_f = \frac{\text{Distance travelled by particular amino acid}}{\text{Distance covered by the solvent from the point of origin}}$$

Reagents

1. N-butanol (Chromatographic grade).
2. Glacial acetic acid.
3. Solvent mixture of butanol, acetic acid, and distilled water in the proportion of (12 : 3 : 5)
4. Ninhydrin : 0.2 per cent solution of ninhydrin in acetone. It is prepared just before use.

Thin Layer Chromatography (TLC)

Introduction

In recent years, thin layer chromatography has been developed. This technique consists of a thin layer of absorbents (silica gel, alumina, Kiselguhr or cellulose) on a glass plate or plastic sheet. Since absorbent is used, it is also termed as absorption chromatography. This technique also provides superior results than that of paper chromatography. The spots are more compact with better resolution and the run is comparatively of shorter duration. Therefore, quicker run is possible.

Preparation of plates

Chromatographic plates (20 × 20 cm) of 200μ thickness are prepared by using a suspension of 30 grams of silica gel G in 63 ml of 0.1 M Na_2CO_3 solution by shaking vigorously for 90 seconds. Only these plates are used which appear to be uniform in both

Table 46.1 : Electrophoretic pattern changes in some conditions. '—' mark indicate no change

Disease	Albumin	α_1-globulin	α_2-globulin	β-globulin	γ-globulin
Normal (gm/100ml)	3.5 to 5.5	0.1 to 0.4	0.4 to 0.8	0.5 to 1.0	0.7 to 1.5
Infancy	—	high	high	—	—
Pregnancy	low	high	high	high	—
Acute infection	—	high	high	—	—
Chronic infection	low	high	high	—	high
Cirrhosis	low	high	—	—	high
Nephrotic syndrome	low	high	very high	—	low
Myelomatosis	—	—	—	—	high
Hypogammaglobinemia	—	—	—	—	very low
		Separate band occassionally seen in α and β position			

transmitted and reflected light. These plates are activated at 110°C for 30 minutes immediately prior to use.

Procedure

Samples (5-100 μL) are applied as a spot of less than 5 mm diameter on the lower right corner of the plates under a stream of warm air. Plates are first developed in a standard Brinkmann developing chamber previously staturated with the vapour of the solvent mixture with chloroform : methanol : acetic acid: water (250 : 74 : 19 : 3 ; V/V). When the solvent front migrates about 15 cm, plates are dried in air for 15 minutes and develop in the second dimension (90° rotation clockwise) with chloroform: methanol : 7 M. ammonium hydroxide (230 : 90 : 15 ; V/V). The solvent front is again allowed to move about 15 cm.

Developed plates are then dried in air for 5 minutes and exposed to iodine vapour in a sealed chamber for 30 to 60 seconds. The pale yellow areas are quickly outlined using a dental probe and the plates are exposed to air until the iodine has evaporated from the spots. When a permanent record of developed plates is desired, plates are sprayed lightly with 10 N. H_2SO_4 and then heated at 110°C for 15 minutes.

The silica gel in each spot is scraped with the aid of a sharp edged polyethylene blade on paper. The weighing papers are then transferred to a 12 ml conical centrifuge tube and eluted by different solvents for estimation by photoelectric colorimeter.

Discussion

The constituents of the mixture of amino acids, and the constituents of neutral lipids and phospholipids are separated and estimated in a short time.

Gel Chromatography

(i) This type of Chromatography uses a porous gel. The dry gel particles are first allowed to take up the chosen solvent. This is accompained by swelling; the liquid taken up constitutes the stationary phase.

(ii) These swollen particles are when made into a column with the same solvent, the spaces between them are filled by solvents. The mobile phase is the "void volume" (Vo); the gel particles are sponge like and the channels within them are of similar diameter (pore size) for a particular grade of gel.

(iii) Molecules can enter the stationary phase only within the gel if their diameter is less than this pore-size. Thus small molecules have the whole fluid volume (bed volume), but molecules above this limit are confined to the void volume and are rapidly washed through the column as mobile phase.

(a) This type of separation is usually performed using the pressure of a small head of the liquid phase to force this through the column. By perfusing the mobile phase at much higher pressures the separation can be carried out quickly with high resolution.

(b) Dense column packing gives high resolution and high pressure metering systems maintain extremely reproducible solvent flows.

(c) This equipment consists of a column, a solvent delivery system, a sample injecter and a detection. The column is of stainless steel of 10-50 cm long and 2-5 mm internal diameter. The solvent delivery is effected by solvent pumps which generate pressure at several thousand Psi and direct a pulse-free delivery of the solvent. Constant pressure and constant volume pumps are used for the purpose. The sample injection is done with a syringe. The detectors may be ultraviolet with fixed wavelength Hg lamp 254 nm, refractive index detectors of deflection type, fluorescence detectors.

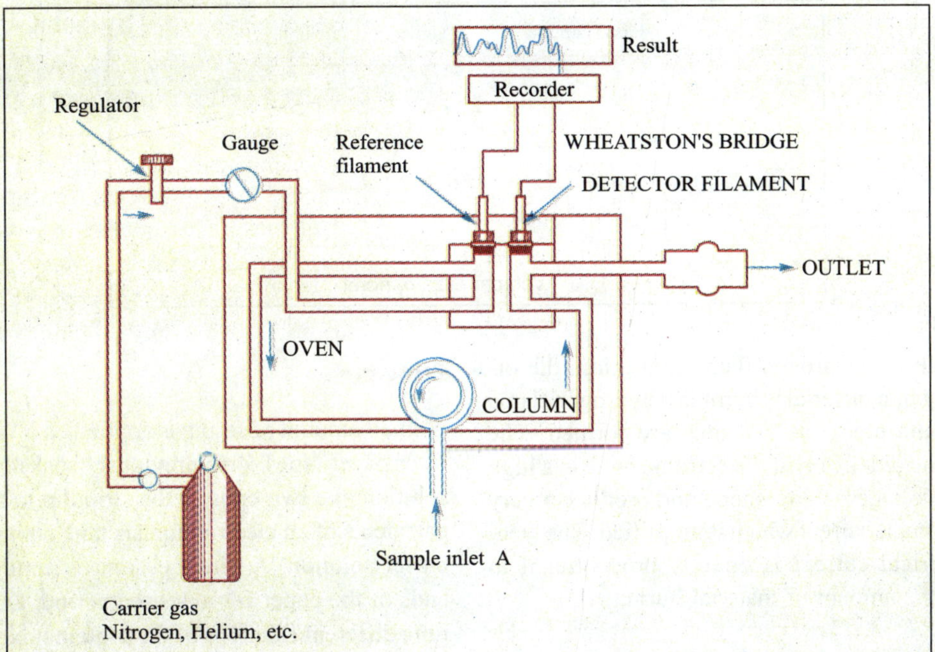

Fig. 46.3 : Gas Liquid Chromatography (GLC)

(d) This method is highly applicable for the separation of carbohydrates, proteins, peptides, aminoacids, vitamins, steroids, neuropeptides, hormones and drugs. Separations are achieved very quickly. Hence this method is indispensable in laboratories where advance research and sophisticated analyses are performed.

Gas Liquid Chromatography (GLC)

(i) The substances to be separated are carried as vapours in an inert gas like nitrogen, argon or helium over liquids when there is partitioning of the substances between the gas and the liquid. The liquids used are silicone, oils, lubricating greases, etc. held in inert solids like diomaceous earth or ground firebrick.

(ii) Glass or metal tube 1–2 metres long and of 0.2 – 2 cm diameter can be used.

(iii) If fatty acids are to be separated, they are first converted to methyl esters which are easily evaporated. The vapours are swept constantly by nitrogen through the tube containing the liquid phase at a temperature of 170-225°C, so that the vapours of the esters may remain as vapours. Separation takes places owing to the partitioning of the esters between the gas and the liquid. The ingredients can be identified by physical, chemical means or flame ionisation.

Identification by thermal conductivity cell or hydrogen flame ionisation

(i) Two fine coils of wire with a high temperature coefficient of resistance are placed in parts of metal C_1 and C_2. Electrical resisters are inserted in the circuit of C_1 and C_2 to form Wheatston's bridge.

(ii) Electrical resistance will change and the bridge remains unbalanced. Different samples cause different resistances. The extent of imbalance is measured by potentiometer.

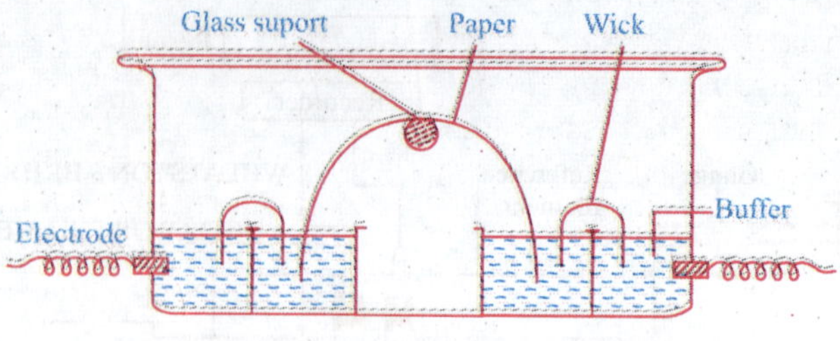

Fig. 46.4 : Vertical electrophoretic tank

(iii) In the hydrogen flame ionisation, the organic material is burnt in a hydrogen flame and electrons and ions are formed. The negative ions and electrons move in a high voltage field to anode and produce a very small current which is amplified. The electrical current is directly proportional to the amount of material burnt.

Paper electrophoresis of serum protein

Introduction

Electrophoresis is a method by which electrically charged particles of a mixture of substances are separated under an electric field. The positively charged particles migrate towards cathode, and the negatively charged particles towards anode. The rate of migration depends on the number of charges of each particle carries. As a result of different rates of migration, a mixture of proteins (such as plasma) can be separated into a number of fractions having similar mobility.

Tiselius (1937) first performed electrophoresis in a fluid electrolyte. In this method, the differently charged particles moved inside the buffer. This method was called "moving boundary electrophoresis". Few workers only used this method because of the costly and complicated apparatus.

Consden, Gordon, and Martin (1946) introduced filter paper to hold the electrolyte in the electrophoretic technique. Filter paper was the first anticonvection medium to be widely used. Later on, other media such as cellulose acetate, agar gel, starch gel, starch block and acrylamide gel have been used in place of paper with better results.

Principle

A small amount of serum is applied on a Whatmann filter paper No. 1 strip previously soaked in buffer solution, the two ends of the strip dip in two compartments of an electrophoresis tank containing the buffer solution. A direct current is applied to the ends of the paper for a desired period. During this time different fractions of the protein mixture separate into bands depending upon the amount of charge, the shape and size of the protein particles. After that electric current is disconnected, paper is removed from the tank, dried and stained with a protein stain to demonstrate different protein bands. The fastest band is albumin, behind it are α_1-globulin, α_2-globulin, β-globulin and γ-globulin (shown in fig. 46.5). The stained fractions are eluted in suitable solvent and the eluted colour is measured in a colorimeter.

Two types of electrophoretic tanks are available

1. Vertical and
2. Horizontal.

1. In the vertical type, there are five compartments. The electrodes are present in two outermost compartments, which is connected to adjoining compartments by wicks. By this arrangement the pH change occurring in the electrode compartment is prevented from affecting the adjoining compartments in which the ends of the paper dip. In the middle, there is a central compartment. The arrangement is shown in Fig. 46.4. All compartments excepting the central one are filled by buffer.

2. In the horizontal type, paper lies in horizontal position supported by strings. The paper ends dip in two compartments on the two sides of the tank. These compartments are connected by means of wicks to adjoining compartments containing electrodes. Buffer is kept in all compartments to equal levels.

Procedure

Pour sufficient buffer into the four compartments up to the same level, leaving the central compartment empty, cut Whatmann No. 1. paper into strips of 35 cm long and 5 cm wide. Mark the centre of the strip with a pencil and mark the point of application of sample by another line 2.5 cm in length parallel to and 5 cm from the mid-line (for the vertical tank) or 7.5 cm away from the mid-line for the horizonal tank. Now apply the paper in such a way that the two ends of the paper dip in the two buffer compartments adjoining the electrode compartments. Replace the lid of the tank. In about 2 hours time the whole paper is soaked with the buffer. Now apply 10μl. of the serum (or plasma) by a micropipette on the line marked for sample application. Replace the paper in its place and cover the tank. At a time, several such strips can be set up for different samples. Now connect one electrode with the cathode and the other with anode of direct current with constant voltage. The electrode nearer the site of application is connected to the cathode. Switch on the current and allow the electrophoresis to run for 16 to 18 hours. A voltage gradient of 3.2 volts per cm length of the paper, or a current of 1.5 to 2.0 milli amps for each 5 cm wide strip is suitable. At the end of the desired length of time, first disconnect the instrument from the mains, remove the strip from the tank, and dry it at 100°C to 110°C for 30 minutes.

Staining

Pour sufficient stain in a large size glass or enamel tray and put the dried paper strips in it for at least 6 hours (overnight staining is always preferable). After removing from staining tray, wash for 6 minutes each time in two changes of wash solution, then in the fixative solution for a further 6 minutes. Dry the paper in air or in an oven at 100°C – 110°C.

After staining 5 bands (Fig. 46.5) corresponding to albumin, α_1–, α_2–, β, and γ-globulins are noticed. It is possible to guess from the electrophoretogram if some protein fraction is absent, present in decreased or increased amount, or some extra band such as myeloma protein is present.

However, the relative amount of different fractions can be accurately estimated by scanning the paper under photoelectric scanner or different bands are cut and eluted in 6 ml of 0.01 N. NaOH for 30 minutes. These eluted coloured solutions are read in a photoelectric colorimeter at 540 mμ.

Reagents

1. **Buffer (pH 8.6, ionic strength 0.083 M) :** Dissolve 3.12 grams of diethylbarbiturate in water and make up to 1 litre.
2. **Dye solution :** Add 25 ml of 95 per cent ethanol to 0.1 gram of bromphenol blue and 50 grams of $ZnSO_4$. Mix well, add 5 per cent acetic acid (V/V), mix to dissolve and make up to a litre with further 5 per cent acetic acid.
3. **Wash solution :** Acetic acid (5 per cent, V/V).
4. **Fixation solution :** Acetic acid (5 per cent) containing 0.3 per cent of sodium acetate ($CH_3COONa, 3H_2O$).

Fig. 46.5 : Serum protein electrophoretogram

Gel Electrophoresis

(i) Microscopic glass slides coated with gels of agar, starch and polyacryl amide can be used.

(ii) Gel electrophoresis is done in about 30 minutes by which a lot of time is saved.

(iii) Molecular sieving factor helps in better resolution causing in adsorption. Problems and optical measurements are done with better accuracy.

(iv) Different fractions like prealbumin, post albumin, haptoglobins, etc. are separated by starch column electrophoresis.

(v) Many additional fractions of serum proteins are separated by means of disc electrophoresis using polyacrylamide gel. In this disc electrophoresis, the polyacrylamide gel is used as a vertical cylindrical column about 10 cm long and the serum is applied on the top. The process is conducted with the use of suitable buffers. When the gel is taken on glass plates, it is called an open block type of electrophoresis. Disc electrophoresis is the column zone type, whereas agar slide electrophoresis is the open block type.

(vi) In the immunoelectrophoresis, the proteins of serum are first separated by ordinary electrophoresis on agar slides. Then these proteins are allowed to be reacted upon by specific anti-serum from rabbit, horse, etc. by applying the same in the same agar slide parallel to the path of proteins. The antibodies of the immune serum diffuse and meet the antigenic proteins resulting in the antigen-antibody reaction with the formation of opaque lenses in the form of precipitins. About 30 fractions of serum proteins have been detected by immunoelectrophoresis.

Isoelectric focussing

(i) This is a very recent improved type of electrophoresis which shows the heterogeneity and micro-heterogeneity of proteins and enzymes. Proteins move under the influence of applied electric current as primary gradient in the presence of the secondary gradient pH resulting in the molecular species to resolve further at equilibrium position. The pH of buffer in the usual electrophoresis does not vary.

(ii) Proteins or enzymes are put into a system of suitable carrier ampholytes (eg, Polyaminopoly- carboxylic acid) and electrolysis is carried out. The more acid ampholytes with a lower iso-electric point move to the anode and the more alkaline ones with a higher isoelectric point move to the cathode. Thus the pH shows increasing values from anode to the cathode. Hence each protein or enzyme focuses quickly at its respective iso-electric pH in the resulting pH gradient.

(iii) In this technique, the micro-heterogeneity has been evaluated for gamma globulins, ovalbumins, globin chains of hemoglobin, myeloma proteins, lipoproteins, α-fetoproteins, etc. Aldolase is separated into five components and L-amino acid dehydrogenase is separated into 18 components with iso-electric pH values ranging from 5.2 to 8.4. Enzyme, Enzyme-substrate complex, enzyme-NAD^+ combination, etc. have been shown by this process.

(iv) The immuno-isoelectric focussing involves the coupling of isoelectric focussing with immuno-diffusion in which the micro- heterogeneity of protein antigens can be analysed easily.

Determination of pH

The electrometric method is the most accurate method for the determination of pH of a solution. This method measures the electromotive force set up due to the existence of H^+ and OH^- in the solution. The instrument which measures pH in this manner is known as pH-meter.

The other method is the colorimetric method which is also fairly accurate and can be used in the absence of a pH-meter. The basis of it is that certain organic dyes (indicators) change colour with change in pH. Each indicator in its ionised state gives one colour and in unionised state gives another colour.

Each indicator has its useful range of pH within which it can indicate change in pH by change of colour and beyond which there is no further change in colour.

It is possible to prepare an indicator whose pH range covers almost the entire pH scale. Such an indicator is known as 'Universal indicator'. The universal indicator enables us to determine the approximate pH of a solution and thus helps us to choose an appropriate indicator and buffers for subsequent more exact determination of pH of the solution.

Principle

When equal volumes of the unknown solution and a series of standard buffer solutions with closely varying pH values are treated with a suitable indicator, the colour of the unknown solution will be found to match with one of the standard buffer solutions. The pH of the unknown is, therefore, the same as the pH of the buffer with which its colour matches.

Permanent standard made from glass discs of the appropriate colour and shade for each indicator may also be used in place of the standard buffer and indicator. Hellige's comparator with its permanent standard discs made of coloured glass is a very convenient instrument for this purpose.

pH determination of a colourless solution

First determine the approximate pH of the solution with the help of a universal indicator and colour chart. Take a small piece of universal indicator paper and just dip its one end in the given solution. After about 30 seconds compare the colour of the paper with the standard colour chart provided and know the approximate pH. If universal indicator solution is to be used in place of paper, put a drop of it in about 2 ml of the solution and compare the colour of the solution with that of the colour chart. After having known the approximate pH of the solution, choose from the set of indicators an indicator which more closely covers the pH range of the solution under examination. Then add a drop of this indicator to approximately 2 ml of the solution, and match the shade of the colour produced against the series of coloured standard buffers. The standard colour solutions are prepared by adding one drop of the same indicator to 2 ml of each of the buffer solutions (chosen in the pH range of the unknown). The pH of the unknown solution is that pH of the buffer with which the colour matches. In case, the colour of the unknown is intermediate between two colours produced by two consecutive buffers, the pH of the unknown lies between the two known standards. The accuracy of this method can be increased by employing set of buffer solutions with closely differing pH values.

pH determination of coloured solution (or Urine)

Place a test tube containing plain urine of coloured solution in front of the coloured buffer standards. View the two test tubes together and then match this against the urine or coloured solution containing the same indicator in front of which a test tube containing water is to be placed. The use of a comparator makes the matching easier, otherwise, a double rowed test tube rack is also convenient.

Reagents

1. Set of indicators.
2. Standard buffer solutions.

Electrometric method of pH determination

The switch of the instrument is put on and after a few minutes when the instrument is warm enough, the pointer is adjusted to 0 mV or 7.0 pH position by set zero control. A standard buffer solution is taken in a clean glass or polythene beaker. Electrodes are lowered so that they are immersed in the solution to a depth of about one inch. The temperature of the buffer solution is measured by a thermometer and the instrument is again adjusted by temperature compensate control to this value. The pointer is again set at 7.0 pH by means of set control. The Selector switch is now turned to proper range, i.e. 0-7 or 7-14. Pointer will now move to show the pH of the buffer and by the help set buffer control, the pointer is set to the exact pH value of the buffer. Now the pH meter is standardized and is ready for determination of pH of any given solution.

To measure the pH of a given solution, the selector is brought back to Zero position and the electrodes are cleaned by distilled water. The given solution is taken in a clean beaker and the electrodes are immersed in it. Selector switch is now put on expected pH range and the reading of the pointer on the pH scale is noted. This value is the pH of the given solution.

Photometry

Photometry is the technique used to study the concentration of substances by adopting the property of absorption of light of a definite wavelength by molecules or particles. Small quantities of substances are used for simple and rapid estimation to maintain high order of accuracy. Photometry cannot help in separating substances but help in estimating substances.

When monochromatic light is passed through the solution of a substance, it comes out with diminished intensity since a part of it is absorbed. The absorption of light by the substance obeys Beer-Lambert's laws.

Beer's laws state that the intensity of a ray of monochromatic light decreases exponentially as the concentration of absorbing material increases.

$$\log\frac{I_0}{I} = K_1C$$

where I_0 is the intensity of light entering the solution and I, the intensity of light emerging from the solution. C is the concentration of the solution in gm/litre or mg/100 ml and K_1 is a constant.

Lambert's laws state that the proportion of light absorbed by an absorbing substance is independent of the intensity of the incident light.

$$\log\frac{I_0}{I} = K_2 l$$

where l is the length of the solution traversed, combining the two laws :

$$\log_{10}\frac{I_0}{I} = KCl$$

The logarithmic ratio of I_0 and I is called the optical density (OD) or the extinction E or Absorbance. Substituting E in the above equation, E = KCl. Therefore, when light passes into a solution of an absorbing substance in a non-absorbing substance, the extinction E is direcly proportional to the depth of the solution (l) and the concentration (C) of the absorbing substance.

Transmission is the ratio of the transmitted light to that of the incident light,

$$T = \frac{I}{I_0}$$

Percentage transmission is the transmission when I_0 is equal to 100. Relation between E and percentage transmission,

$$E = KCl = \log_{10}\frac{I_0}{I}$$

$$\therefore \quad E = \log_{10}\frac{100}{I} (\text{where } I_0 = 100)$$

$$\therefore \quad E = \log_{10}100 - \log_{10} I$$

$$\therefore \quad E = 2 - \log_{10} I$$

$$\therefore \quad E = 2 - \log_{10} T.$$

[I is the percentage transmission T]

The above principle is employed in estimating the substances in all the instruments used in photometry. If the product of the reaction is coloured, the wavelength of light used is in the region of the spectrum (400-800 mμ). Even colourless products can be estimated by photometry for which a special instrument called spectrophotometer is required (the colourless purine bases absorb light at about 260 mμ.)

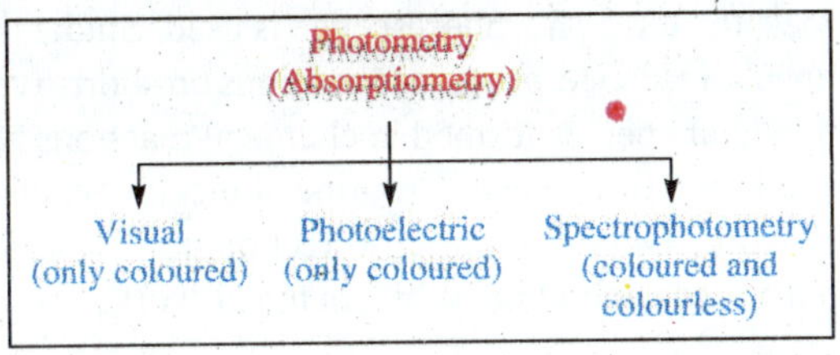

Colorimetry

Coloured solution is estimated by colorimetry. The principal colorimeters are (1) the Visual Colorimeters (2) the Photoelectric Colorimeters. Standard solutions of the substances are treated in the same way as with the test samples.

In the visual colorimeter, the length of the column through which light passes is changed with the help of plungers so that the intensity of transmitted light of both the standard and the test as seen through the eyepiece is equal. The intensity of incident light (I_0) is also same for both.

If C_1 the concentration of the standard, I_1 the length of the solution column corresponding to standard and I_2 the length corresponding to the test solution are known, the only other unknown, i.e. C_2, the concentration of the test sample is calculated.

$$E_{std.} = KC_1I_1$$

$$E_{test} = KC_2I_2$$

We know, $E = \log_{10} \frac{I_0}{I}$

Since I_0 and I are the same, E_{std} and E_{test} are also the same. So

$$KC_1I_1 = KC_2I_2$$

$$\therefore \frac{C_1}{C_2} = \frac{I_2}{I_1}$$

But in photoelectric colorimeters, the principle is different from that of visual colorimeter. No colour matching is done in this colorimeter. A photocell and a galvanometer are used for measuring the actual intensity of the transmitted light. The electrical energy formed from the light energy by the photocell is measured in a galvanometer. The percentage transmission or the optical density is read from the galvanometer. The light of desired wavelength is set up by means of suitable filters.

$$E_{std} = KC_1I_1$$

$$E_{test} = KC_2I_2$$

$I_1 = I_2$, since the lengths of solutions are constant.

$$\therefore \frac{E_{std}}{E_{test}} = \frac{C_1}{C_2}$$

If the extinctions due to the known concentration (C_1) of the substance and that of the test are known, the concentration of the unknown (C_2) can be easily calculated.

A graph is desired to be drawn to have the actual concentration and extinction of the standard by plotting the different concentrations and extinctions of the standard substance.

The extinction, if any, produced by reagent blank should be subtracted from that due to the test and the standard.

The colorimeters are of single photocell type and double photocell type. The photocells are of two types (1) the barrier type, (2) the high vacuum cell type. The barrier type consists of the photosensitive surface of selenium on a plate of copper or iron covered by a light transmitting layer of metal. The high vacuum cell type consists of a highly evacuated cell with a composite caesium cathode and requires application of a potential.

Spectrophotometer

The spectrophotometer has got the development over the photoelectric colorimeter since it functions at various range of wavelength. Ultraviolet spectrophotometers work at wavelengths below 400 mμ, whereas infra-red spectrophotometers work at the wavelengths above 800 mμ. The former is more useful in biochemical laboratory in the estimation of many dehydrogenase enzymes which needs the extinction to be taken at about 340 mμ, and at the extinction at about 260 mμ the purines and the Pyrimidines of nucleic acids are estimated. A good number of spectrophotometers work over a wide range of wavelength eg, Beckmann model DK, works from 185 to 3500 mμ, and model DU works from 210 to 1000 mμ.

Superiority of the Spectrophotometer over the photoelectric colorimeter

1. The Spectrophotometer uses prisms as monochromator and the wavelength of incident light can be adjusted precisely.
2. It maintains greater accuracy.
3. It provides for the use of smaller volumes of liquids and even samples of solids and gases which can never be used in photoelectric colorimeter.

Fluorometry

The fluorometric technique allows measurement of concentrations as low as 1 ng/ I in suitable cases. The emission of electromagnetic radiation is shared by the processes of fluorescence, phosphorescence and luminescence. If the excited state persists for less than 10 nanoseconds, the process is called fluorescence but for excited states with longer survival times, the term phosphorescence is used. Studies of phosphorescence gives informations on short-lived molecular species formed in chemical reactions but have found little place in routine clinical biochemical work. Only fluorometric technique is considered for routine clinical biochemical work.

The fluorescent light produced as a result of the irradiation on substances like thiochrome, riboflavin, etc. is used in the fluorometer and it is emitted in all directions. The photocell is placed at right angles to the path of the incident beam.

The suitable wavelength for the exciting or primary radiation is selected from the light source and then falls on to a solution of substance under study in a cuvette. The fluorescent light is emitted in all directions but it is usual to select that part emerging at right angles to the incident beam, the unabsorbed part of which passes directly through the cuvette and enters a light trap. The appropriate wavelength is selected from the emitted or secondary light and the beam then falls on to the detector. The output from this is amplified and displayed. Fluorometers are usually divided into filter fluorometers and spectrofluorometers depending on the method of wavelength selection. In the

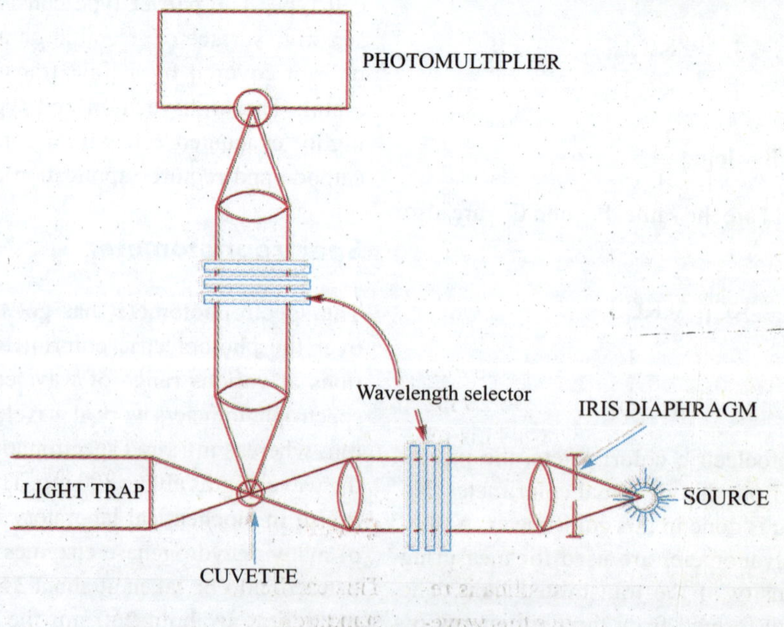

Fig. 46.6 : Fluorometer

former group the selection of the primary wavelength is made by one or more individual filters. In spectrofluorometers, the wavelength selection is by monochromators which usually employ diffraction gratings but some prism instruments are available.

Light sources

1. High intensity of radiation is achieved using an arc lamp.
2. The mercury vapour lamp has its energy concentrated into a relatively small number of sharply defined lines.
3. Special lamps using other elements are available for particular analyses, eg, Zinc emitting at 472 nm for cortisol and thallium for oestrogens at 538 nm.
4. A high intensity source with a continuous spectrum extending down to 220 nm is the Xenon arc originally developed for war-time searchlights.

Wavelength selection

1. The relatively monochromatic nature of the exciting beam gives satisfactory performance.
2. Usually the grating on the primary monochromator is blazed at about 300 nm and that of the secondary one at about 500 nm to have maximum energy in the most useful ranges.
3. Both monochromator slits are kept the same for maximal passage of radiant energy. The energy is proportional to the square of the slit width and the bandwidth depends directly on the slit width.
4. Under good conditions the bandwidth may be as low as 0.5 nm increasing to 20 nm when maximal energy transmission is required.

Cuvettes

1. Quartz or preferably synthetic silica is required below 320 nm, but above this wavelength, glass cells can be employed.
2. All faces of the cuvette should be clear since light has to pass through the cuvette in all directions.
3. The Cuvette should be large enough to avoid exciting radiation.
4. The cuvette temperature should be kept constant.

Detector

1. A high sensitivity detector is needed to maintain the overall sensitivity of the instrument. In almost all cases the detector is a photomultiplier and this is essential for the spectrofluorometer for which it is sometimes described as **spectrophotofluorometer**.
2. The output from the photomultiplier is passed through a linear amplifier.
3. Many instruments incorporate a switch to allow amplification to be adjusted precisely thereby covering different ranges of sensitivity.

Use of fluorometers

1. It is essential to prepare a blank solution. The reading of this is subtracted from the sample reading and should be relatively low. At high sensitivity it may be necessary to purify the reagents including solvents.
2. If blanks are high, it needs to modify the exciting and emission wavelengths to improve.
3. The presence of buffer ions can also affect the energy state of excited molecules and may reduce it sufficiently to abolish fluorescence. Thus chloride and phosphate ions quench quinine and indole of fluorescence respectively.

Checking of performance

This can be done by using a set of fluorescent glass standards, or by testing a known concentration of quinine sulphate. This gives information on the accuracy of wavelength calibration in both beams and allows the sensitivity and linearity of response to be checked.

Flame Photometry

In flame photometry a solution containing the substance to be determined is passed under carefully controlled conditions as a very fine spray into the air supply of a burner. In the flame the solution evaporates and the substance is first converted to the atomic state in which the electrons in the outermost shell are in their lowest energy state closest to the nucleus – the ground state. As the temperature rises, the thermal energy of the flame excites these electrons. If the energy supplied increases it can become great enough for the electrons to escape from the atom to form an ion. The electrons in the higher energy orbits are in a metastable state and are prone to lower energy orbits including the ground state. In doing so the energy previously absorbed is released as quanta of light, the wavelengths of which are characteristics of the substance, thus giving rise to the emission spectrum. The more energy supplied, the more complex the spectrum.

In the emission flame photometry, the relatively low energy source only excites a few elements mainly the alkali metals. A portion of the light which is emitted in all directions is collected by a reflector and falls on a detector. The detector output is directly proportional to the concentration of the substance in the flame. Under the usual conditons only a small proportion of the atoms present in the flame are excited into emission 1 to 5 percent in the case of alkali metals and less for the other elements. Most atoms, therefore, remain in the ground state. This technique has been mostly used to determine sodium and potassium, and recently also lithium.

Burner and flame

1. The burner produces a steady flame when supplied with fuel and air or oxygen at constant pressure.
2. Most burners for low temperature flames have a series of holes where the gas burns in the presence of an adequate mixture of air producing a series of clear blue cones without yellow streaks. This appearance is used to set the flame correctly.
3. Most fuel-air mixtures burn at about 2000°C, whereas the same fuel mixed with oxygen burns at 2700°C to 3000°C
4. The propane air flame is the most commonly employed in instruments used in clinical biochemistry where they are concerned mainly with the determination of sodium, potassium and lithium.

Wavelength selection

The principal wavelengths are Na – 589 nm ; K – 766 nm ; Li – 671 nm ; Ca – 554 nm.

Detectors

1. The less expensive instruments usually use barrier layer photocells; selenium cells sensitive in the near infrared are available for thc 766 nm line of potassium.
2. The more expensive instruments employ phototubes or photomultipliers.

Output Devices

1. Most instrument for clinical biochemistry are at least dual channel, i.e. they are able to give an output for sodium and potassium concentrations simultaneously. There are then two independent light paths, detectors and displays which allow both determinations to be made on a single specimen.
2. As the dilution needed for sodium is greater than is necessary for potassium, which is present in much lower concentration, the detector response in the potassium channel needs amplification as the simple barrier layer cell is usually insufficiently sensitive.

Autoanalysers

These instruments are involved in the automation to the routine biochemical analysis of hospital specimens. About 60 to 300 samples of blood or serum per hour for as many as 18 constituents can be analysed by the autoanalysers.

The principal parts of an autoanalyser are the followings :

1. Sampler.
2. Proportionating pump.
3. Dialyser.
4. Constant temperature.
5. Flow through colorimeter.
6. Recorder.

1. The simpler

(a) This module holds the batch of samples awaiting analysis in separate cups on a circular tray which is rotated at intervals.

(b) A probe connected by plastic tubing to the proportionating pump enters each sample serially.

(c) The volume of sample aspirated is determined by the pumping rate and the adjustable dwell time of the probe in the sample.

(d) As the probe movement into the sample cup was slow, the volume aspirated varied somewhat with the depth to which this was filled.

(e) The transit time between reservoir and sample is short.

2. Proportionating Pump

(a) This module determines the relative flow rates of sample and all reagents and replaces the use of different sizes of pipettes in manual methods.

(b) The pumping technique involves the peristaltic action produced by a series of rollers passing along an array of parallel plastic "Pump tubes".

(c) Each roller compresses all tubes so that the rate flow in each tube is proportional to the square of the pump tube diameter.

(d) Colour-coded tubes with a range of nominal diameters and pumping rates are available in three materials.

(e) The normal tygon tubing is suitable for most reagents but Acidflex tubing is preferred for reagents containing corrosive acid and Salvaflex tubing for certain organic reagents.

3. The Dialyser

(a) This module achieves the separation of small and large molecules by allowing the former to pass through a semipermeable membrane from the donor (sample) stream of liquid and air bubbles to a recipient stream of liquid again segmented by air bubbles.

(b) The dialysis rate depends on the temperature but complete passage of small molecules into the recipient stream is rarely achieved and may be only a few percent of the total.

(c) The analytical process then requires that a constant fraction should dialyse and this is not always the case when simple aqueous and protein-containing solutions are compared.

(d) It is important to ensure that the two streams flow in the same direction.

(e) Care must be taken to ensure that the output from the recipient stream is the one which enters the remainder of the analytical system. If the sample stream is greatly diluted, the difference may not be seen easily.

4. The constant temperature

(a) It is to maintain the reaction mixture at a constant temperature for a defined time to bring about the required chemical change under controlled conditions.

(b) The incubator bath consists of a glass delay coil mounted in a thermostatically controlled oil bath. This is sealed and stirred constantly.

(c) Most baths are set at 37°C or 95°C but some have adjustable thermo-regulators which allow operation up to 120°C or even higher.

5. Flow through colorimeter

(a) The colorimeter is to measure the intensity of colour produced in the reaction and to provide a graphical display of change in colour with time.

(b) The use of double beam spectrophotometer is costly and is rarely justified by analytical requirements. The single beam colorimeters have insufficient stability to operate reliably over the long period required.

(c) The Autoanalyser MK I colorimeter combines double beam operation with interference filters to select the wavelength.

6. The recorder

The servo-potentiometer recorder is used to record the ratio of the responses from the two detectors and these responses are proportional to the intensity of light reaching the detectors.

Functioning of the instruments

1. The flow of the analytical stream is directed through plastic tubing from one module to another.
2. The samples are then loaded into the cup of the sampler and the channels of the proportionating pump aspirate them as well as dilute them.
3. The diluted serum samples are led through one of the dialyser unit.
4. The pump introduces suitable reagents through the other side of the dialyser.
5. The two streams run side by side being separated only by dialysing membrane.
6. A portion of dialysable constituents of serum passes across the membrane to the reagent stream.
7. Further treatment like incubation at suitable temperature is given in the constant temperature bath.
8. The intensity of the colour developed is measured in a colorimeter and recorded in the recorder.
9. Suitable standards are treated in the same way.

Lately, multiple analysers have been introduced with multiple channels to estimate as many as 18 selected constituents including the nondialysable proteins, enzymes like alkaline phosphatase, triacylglycerols and electrolytes. About 60 to 300 samples can be investigated per hour simultaneously for 18 parameters.

Autoanalysers have gained much popularity for routine biochemical analysis in clinical laboratories for their speed, flexibility of the methodology, increased quantum of clinical biochemical analysis, easy operation, and accuracy of results.

Disadvantages

1. Prohibited cost.
2. "Carry-over" errors.
3. The usual hazards of any complicated equipment.

Enzyme Immuno Assay

Hormones, Proteins, etc. present in blood in minute concentration can be assayed by the recent advanced technique of "Enzyme Immuno Assay" without involving any disadvantage.

The basic reaction is the interaction between an antibody and an antigen.

Enzyme immuno assay (EIA) is classified into two ways :

1. Homogeneous enzyme immuno-assay (EHIA)

Fig. 46.7 : Antigen-Antibody reaction

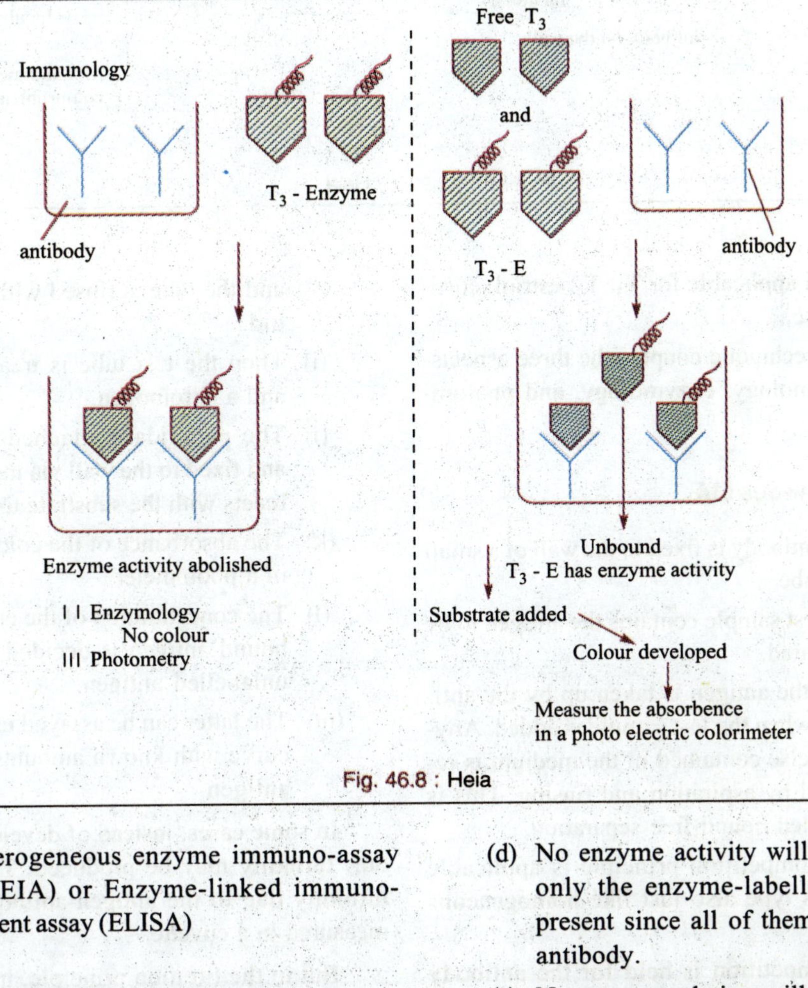

Fig. 46.8 : Heia

2. Heterogeneous enzyme immuno-assay (HTEIA) or Enzyme-linked immuno-sorbent assay (ELISA)

1. HEIA

(a) It has got only liquid phase.

(b) Haptens like hormones, drugs, etc. are labelled with enzyme like lysozyme, malate dehydrogenase, etc. with no loss of enzyme activity. They will be made to react with a limited number of antibody molecules.

(c) There is competition between the enzyme-labelled hapten and the free hapten for the limited number of antibody molecules.

(d) No enzyme activity will be available if only the enzyme-labelled haptens are present since all of them unite with the antibody.

(e) No accommodation will be attained for some enzyme-labelled haptens depending on the concentration of free haptens in a mixture of free haptens and enzyme-labelled haptens. A colour will be produced after the reaction of this with the substrate.

(f) The intensity of colour is measured in a photometer.

(g) It is called homogeneous EIA since the reaction takes place in liquid phase only

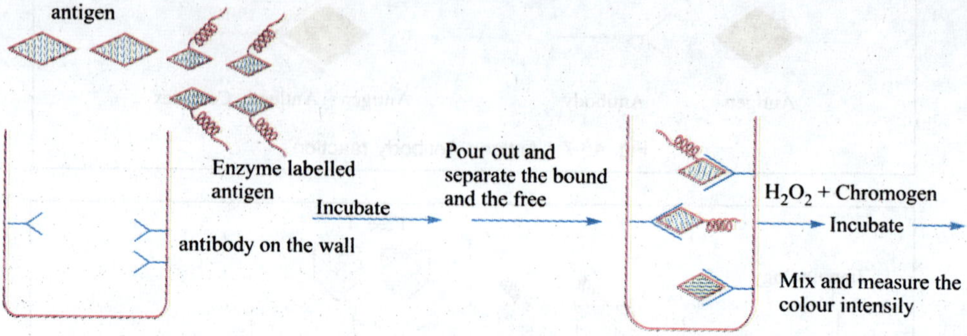

Fig. 46.9 : Heia

and is applicable for T_3, T_4, estriol, cortisol, etc.

(h) This technique couples the three aspects- immunology, enzymology, and photometry.

2. Heterogeneous EIA

(a) The antibody is fixed to the wall of a small test tube.

(b) The test sample contains the antigen to be estimated.

(c) Only the antigen is taken up by the antibody when the test sample is added. Anything else contained in the medium is removed by aspiration and rinsing. This is so-called bound-free separation.

(d) The competition principle is applicable in this type also just like homogeneous EIA.

(e) A competition is held for the antibody sites between the unlabelled antigen and labelled antigen.

(f) Some antigens which are bound may be labelled while some others may be unlabelled.

(g) The extent of the labelled antigen binding to the antibodies is governed by the concentration of unlabelled antigen in the specimen.

(h) The specimen liquid is then aspirated out and the tube is rinsed with a suitable liquid.

(i) Then the test tube is treated with H_2O_2 and a chromogen.

(j) The peroxidase attached to the antigen and fixed to the wall via the antibody then reacts with the substrate to form a colour.

(k) The absorbance of the colour is measured in a photometer.

(l) The concentration of the enzyme-labelled bound antigen is decided by that of the unlabelled antigen.

(m) The latter can be assayed using a standard curve with known amounts of unlabelled antigen.

In some cases, instead of development of colour, turbidity may be produced. The increase in turbidity due to the antigen-antibody reaction is measured in a cuvette.

But in the titration principle, insulin labelled with enzyme is made to bind to antibodies in competition with unlabelled insulin. The sites are filled with enzyme-labelled insulin in titrable quantities.

The antigen can bind to more than one site of the antibody when applied to α–fetoprotein on the basis of Sandwich principle. This states as follows.

1. It is involved in filling the antibody sites fixed to the wall with α–fetoproteins in the serum of a pregnant woman.

2. The antigen has multiple sites for antibody binding.
3. The antibody labelled with the enzyme is now added. This labelled antibody attaches to the site still free on the antigen.
4. The contents are aspirated.
5. The greater the amount of antigen, the greater the amounts of labelled antibody remaining on the wall.
6. Colour reaction is performed for estimating the enzyme.

Important points to be noted

1. Except glass tube, polystyrene micro-titre plate, cellulose, isothiocyanate and polyacrylamide can be used for fixing the antibodies.
2. The enzyme label for the antibody may be alkaline phosphatase, glucose oxidase, β-Galactosidase, glucose-6-phosphate dehydrogenase, and penicillinase.
3. Except hormones, immunoglobulins, α-fetoproteins, etc. the antigens can be purified virus and virus components, bacterial antigens, tetanus or diphtheria toxoids, fungal antigens etc.

Radioimmunoassay

1. It has significant role in the diagnosis of diseases.
2. The gravimetric and the colorimetric methods fail when the substances to be analysed is in very low concentrations.
3. Even the nanogram amounts of radio-labelled substance give suitable counts detectable by electronic counters. These counts are related to the concentration of the substance assayed.
4. Radioimmunoassay is employed for the estimation of Vitamins like B_2, and folic acid ; hormones like insulin, thyroxine (T_4), triiodothyronine (T_3), cortisol, testosterone, dihydrotestosterone, estrogens; trophic hormones like ACTH, FSH, LH; drugs like digoxin, digitoxin ; antigens like the Australia antigen.

The principle of the method

1. A radio-labelled antigen (eg, insulin labelled with I^{125}) is made to compete with an unstable antigen, eg, insulin to be estimated in the serum of a patient) for a limited number of binding sites of a specific antibody raised against insulin.
2. The antigen binds to the antibody.
3. Owing to inadequate binding sites, some of the antigens will be free and will include radio-labelled antigen also.

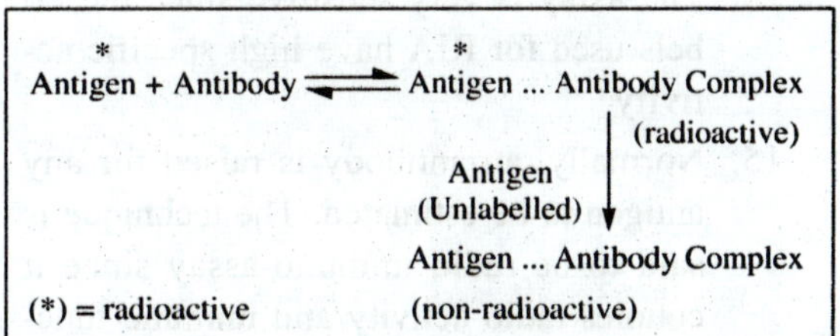

Fig. 46.10 : Radio-labelled antigen reaction

4. After equilibrium, the antigen-antibody complex is precipitated by using suitable reagents.
5. The supernatant is separated from the precipitate by centrifugation.
6. Both the precipitate (the bound antigen, B-form) and the supernatant (the free antigen, F) will have radioactivity since they have I^{125} - insulin.
7. Dextran - coated charcoal if is used, it removes the free antigen. The supernatant will have the bound form.
8. The extent of radioactivity of the two forms is measured in gamma ray well type scintillation counter.
9. The magnitude of radioactivity of the free form may be related to the concentration of the unlabelled antigen.
10. Alternately, the radioactivity of the sound

form or the ratio of B/F is also related to the concentration of the unlabelled antigen.

11. Different concentrations of the unlabelled insulin standard are used separately with the same concentration of the labelled insulin.
12. A graph is prepared with the radioactivity of the free form and the concentrations of the standard insulin. The serum whose insulin content is to be determined is treated in the same way.
13. From the radioactivity of the free form and extrapolation, the concentration of the insulin in serum is determined.
14. The assay is very sensitive since the labels used for RIA have high specific activity.
15. Normally, an antibody is raised for any antigen to be estimated. The technique is said to be radio immuno-assay since it couples radio activity and immune function (antigen binding to antibody).

Importance of RIA

1. The receptor sites can even be assayed in addition to the estimation of hormones proper.
2. The patient does not require to take the radioisotope as the label is used only in vitro.
3. RIA has significance in diagnostic biochemistry. RIA comes in handy for the estimation of T_3 and T_4 separately.
4. In some thyroid disorders, T_4 may be normal but T_3 is elevated as in subclinical hyperthyroidism, ophthalmic Graves disease, autonomous thyroid nodules, T_3 toxicosis, etc.
5. An apparent increase of the thyroid hormones is observed in pregnancy or in persons taking oral contraceptives due to the increased level of the serum thyroxine-binding globulin.
6. RIA can help to differentiate the basic biochemical lesion in endocrinology whether the increased level of a hormone is due to the production of the hormone as such or the tropic hormone.
7. This technique offers safety to the patient in the use of drugs if there is only a narrow margin between the therapeutic and toxic dosage. This is applied during digitalisation in the management of congestive heart failure.
8. Low levels of serum digoxin mean under digitalisation whereas high levels are toxic. The patient is given the optimum digitalisation by RIA of digoxin in serum.
9. This technique is also useful in diagnosing insulinomas, sex hormone sensitive tumors, etc. and this facilitates proper treatment of the diseases.

Respirometry (Manometry)

In many instances, the study of oxygen uptake by tissue slices, homogenates, or subcellular particles has been made. The level of enzyme is expressed in terms of oxygen uptake and for this Warburg's manometer (Respirometer) is used. It actually measures the change in pressure at constant volume and at constant temperature. It consists of the following :

1. A special flask with a side arm and a centre well is used.
2. In the side arm, the homogenate or enzyme peparation or serum is taken while the medium or the substrate is taken in the flask.
3. In the centre well, a strong alkali with a filter paper strip is taken to absorb the CO_2 produced during reaction.
4. The flask is attached by means of a glass-tight joint to a capillary manometer containing a device for returning the manometric liquid to a fixed volume mark.
5. The homogenate and the medium of the enzyme and the substrate are equilibrated at 37°C in an atmosphere of oxygen or air.

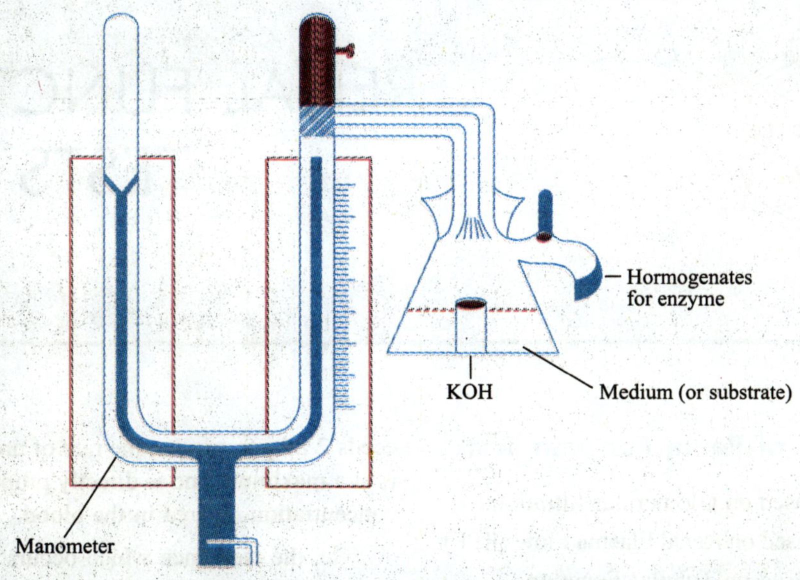

Fig. 46.11 : Manometer

6. The substances are then mixed by tilting and the Warburg's flask is agitated at 37°C in airtight condition.
7. As a result of respiration, oxygen will be consumed while the carbon dioxide formed is absorbed by potassium hydroxide in the centre well.
8. The contraction in volume due to carbon dioxide is measured at different time intervals and is equivalent to the oxygen consumed.
9. It is the pressure difference which is noted in the manometer.
10. But the volume is calculated by using the flask constant. From the volumes of oxygen consumed for a definite time interval and the volume of the homogenate, weight of the tissue, etc., the amount of oxygen respired by the tissue can be calculated for unit time.
11. In enzymology, the quantity of enzyme in a sample can be expressed as micromoles of oxygen consumed by unit weight of the tissue in unit time, when the enzyme of the tissue reacts with the required substrate in the flask.

CHAPTER 47

RENAL FUNCTION TESTS (RFT)

Classification of Renal Function Tests

A. Tests based on Glomerular filtration.

B. Tests based on Renal Plasma Flow (RPF).

C. Tests based on Tubular Function.

D. Certain Miscellaneous Tests.

A. Glomerular Filtration Tests

1. Urea clearance test.
2. Endogenous creatinine clearance test.
3. Inulin clearance test.
4. Recently, Cr^{51}-EDTA clearance test.

The clearance of a substance is defined as the volume of blood/plasma which contains the amount of the substance which is excreted in the urine in one minute.

1. Urea clearance test

Blood urea clearance is expressed as the number of millilitre (ml) of blood/plasma which are completely cleared of urea by the kidney per minute.

The plasma is not completely cleared of urea. Only about 10 per cent of the urea is removed while 750 ml of plasma passes through the kidney per minute. Hence, 10 per cent of urea is removed which is equivalent to completely clearing 75 ml of plasma per minute. If a larger volume than normal is cleared per minute, renal function is satisfactory. But the renal function is impaired when a smaller volume is cleared. In case, the urine volume exceeds 2 ml per minute, the rate of urea elimination is at a maximum and is directly proportional to the concentration of urea in the blood.

So, the clearance which occurs when the urinary volume exceeds 2 ml per minute is termed as maximum urea clearance (C_m).

$$\therefore\ C_m = 75\ [75\ (\text{normal range}) \pm 10]$$

$$\text{Normal range} = \frac{U \times V}{B} = \frac{1000 \times 2.1}{28} = 75.$$

[U = Concentration of urea, V = Volume of urine in ml/minute, B = Concentration of urea in blood.]

The rate of urea elimination is reduced when the urinary volume is less than 2 ml per minute. Such clearance is termed as standard urea clearance (C_s).

$$\therefore\ C_s = \frac{U \times \sqrt{V}}{B} = 54\ [54\ (\text{normal range}) \pm 10]$$

Sometimes, the result of urea clearance test is expressed as the percentage of the normal maximum or of the normal standard urea clearance depending on whether urinary output is greater or lesser than 2 ml per minute.

So, normal percentage is expressed as

$$C_m = \frac{U \times V}{B} \times \frac{100}{75} = 1.33 \text{ per cent.}$$

$$C_s = \frac{U \times \sqrt{V}}{B} \times \frac{100}{54} = 1.85 \text{ per cent.}$$

The urea clearance is proportional to the surface area of the body, and if the result is to be expressed as a percentage of normal, a correction must be made in the case of children and those of abnormal stature.

Procedure of the test

The test must be performed between breakfast and lunch since excretion is more uniform during this period.

(a) The patient at rest throughout the test is given a light breakfast and 2 to 3 glasses of water.

(b) The bladder is made emptied and the exact time of urination is noted.

(c) The urine is collected after one hour and the urea content of blood is also noted.

(d) At the end of another hour, a second specimen of urine is collected.

(e) The volume of each specimen of urine is measured accurately and the concentration of urea in the specimen of blood and urine is determined. The average value of two specimens of urine is the urea content of urine.

Interpretation

(a) Urea clearance of 70 per cent indicates the satisfactory excretion of the kidney.

(b) In mild impairment of the renal function, values show between 40 to 70 per cent; in moderate impairment, values show between 20 to 40 per cent; in severe impairment, values show below 20 per cent.

(c) In acute renal failure, C_m or C_s is lowered usually less than half the normal values.

(d) In chronic nephritis, the urea clearance falls rapidly and shows half or less of the normal values.

(e) In nephrotic syndrome, the urea clearance is usually normal until the onset of renal insufficiency sets in.

(f) Terminal uremia appears when the urea clearance falls to about 50 per cent of the normal values.

(g) In benign hypertension, a normal urea clearance is usually maintained indefinitely except in few cases.

(h) Low urea clearance value is found in normal persons with a very low protein diet.

2. Endogenous creatinine clearance test

The metabolite creatinine is filtered at the glomerulus but neither secreted nor reabsorbed by the tubules. Hence, its clearance gives the GFR. This method is convenient for estimation of GFR for the following reasons :

(a) It is a normal metabolite in the body.

(b) It does not require the intravenous administration of any test material.

(c) Creatinine estimation is also simple.

(d) The longer collection period of creatinine minimizes the timing error.

Procedure

(a) The total volume of the collected 24-hour urine specimen ending at 7 AM is measured accurately.

(b) A blood sample is also collected for the determination of serum creatinine.

(c) The urinary creatinine concentration is also estimated.

$$\therefore\ C_{cr} = \frac{U \times V}{P}$$

where U = urine creatinine concentration in mg/dl

P = serum creatinine in mg/dl

V = volume of the urine in ml/minute

Normal values for creatinine clearance varies from 95 to 105 ml/minute.

3. Inulin clearance test

Inulin (fructosan, a homopolysaccharide) is not metabolised in the body, is excreted entirely through glomerular filtration, and is not secreted or reabsorbed by the renal tubules.

Procedure

(a) The test is done preferably in the morning. The patient should be hospitalised overnight and kept reclining during the test.

(b) A light breakfast consisting of half a glass of milk and one sliced toast is given at 7.30 a.m.

(c) 10 grams of inulin dissolved in 100 ml. of saline at body temperature is injected intravenously at the rate of 10 ml per minute at 8 a.m.

(d) At 9 a.m. the bladder is emptied and the urine is discarded.

(e) After one and two hours, the volume of the collected urine is measured and analysed for inulin content.

(f) 8 to 10 ml of blood is withdrawn in oxalated bottle at the mid-point of each collection of urine. Plasma is separated and analysed for inulin concentration. The average of the values of two samples of blood is noted.

Calculation

$$C_{Inu} = \frac{U \times V}{P}$$

where, U = mg of inulin/100 ml of urine.

V = ml of urine/minute

P = mg of inulin/dl of plasma (the average of two values)

Normal value : Inulin clearance in an adult of 1.73 sq. m = 125 ml of plasma cleared of inulin/minute.

Range = 110 to 150 ml.

Remark

(a) One glass of water is given at 6.30 AM and is repeated every half an hour until the test is completed. This step should be eliminated during the administration of fluid.

(b) This test is undoubtedly superior for the determination of GFR but requires tolerable tedious chemical procedure for determination.

4. Determination of Cr^{51}-EDTA clearance

Cr^{51}-EDTA plasma clearance is widely used for routine assessment of glomerular filtration rate (GFR) in adults as well as in children. It is especially convenient in children where it is not easy to collect 24 hour urine sample. It is used for children younger than one year old.

4.5 μci/kg body weight of Cr^{51}-EDTA is injected IV. Capillary blood samples are drawn at 5, 15, 60, 90 and 120 minutes after the injection and simultaneously the hematocrit (hct) is determined.

The radioactivity is calculated as measured activity in 0.2 ml. Capillary blood/1-hct. Cr^{51}-EDTA plasma clearance is determined as the ratio between the injected amount of the 'tracer' (Q_0) and the total area under the plasma activity curve c(t).

The plasma clearance (Cl) is then calculated as

$$Cl = \frac{Q_0}{C_1/b_1 + C_2/b_2}$$

B. Tests for Renal Blood Flow

Measurement of Renal plasma flow (RPF)

Para-amino hippuric acid (PAH) is filtered at the glomeruli and secreted by the tubules.

PAH is removed completely during a single circulation of the blood through the kidneys. Tubular capacity for excreting PAH is great.

Therefore, the amount of PAH in the urine becomes a measure for the value of plasma cleared of PAH in a unit time, i.e., PAH clearance at low blood levels measures renal plasma flow (RPF).

∴ RPF (for a surface area of 1.73 sq. m) = 574 ml./minute.

Filtration Fraction (FF)

Filtration Fraction (FF) is the fraction of plasma passing through the kidney which is filtered at the glomerulus. This is derived by dividing the inulin clearance by the PAH clearance.

$$\therefore \text{ FF} = \frac{C_{In}}{C_{PAH}} = \frac{GFR}{RPF}$$

If GFR = 125 and RPF = 574, then FF = $\frac{125}{574}$ = 0.217 (21.7%).

Normal range = 0.16 to 0.21 in an adult.

Interpretations

(a) The filtration fraction becomes normal in early essential hypertension. FF is increased as the disease progresses.

(b) The filtration fraction (FF) is highly increased in malignant phase of hypertension.

(c) FF is also increased in congestive cardiac failure.

C. Tests Based on Tubular Function

The sufficient renal tubular function depends on the adequate renal blood flow. Arteriolar nephrosclerosis and other diseases diminish blood flow and hence are unable to concentrate or dilute the urine resulting in "isosthenuria" ("fixation" of specific gravity at 1.010).

Concentration Tests

The tests are conducted either (a) under conditions of restricted fluid intake, or (b) by inhibiting diuresis by injection of ADH.

(i) Fishberg concentration test

This test is performed in a shorter period of time with less intake of fluid.

Procedure

(a) The patient is given an evening meal at 7.0 p.m. containing high protein and a fluid containing less than 200 ml.

(b) The urine if passed in the night is discarded.

(c) The patient is kept without any fluid after 7.0 p.m. to 10.0 a.m. next morning.

(d) The urine specimens must be collected in the next morning at 8.0 a.m., 9.0 a.m. and 10.0 a.m. to determine the specific gravity of each sample.

Interpretation

(a) In case of normal tubular function, the specific gravity of at least one of the specimens should be greater than 1.025 after suitable correction for temperature, albumin and glucose.

(b) At least one of the urine specimens shows a specific gravity of 1.020 or less in case of impaired tubular function and the specific gravity becomes fixed at 1.010 in severe renal damage.

(c) A false result is also expected in case of congestive cardiac failure and diabetes insipidus owing to the inability of the elimination to be concentrated.

(ii) Water dilution / Elimination test

The ability of the kidneys to eliminate water is tested by ingesting a large volume of water after measuring the urinary output.

Procedure

(a) The patient should take an evening meal and should not take anything after 8.0 p.m.

(b) On the next morning, he should empty his bladder at 8.0 a.m. which is discarded and then within half an hour he must be allowed to drink 1200 ml of water.

(c) His bladder is emptied at 9.0 a.m., 10.0 a.m., 11.0 a.m., and 12.0 noon. The volume and the specific gravity of these four specimens should be measured.

(d) The patient must be in bed during the test, since elimination of water is maximum in the horizontal position.

Interpretation

(a) In case of normal renal function, the excretion of water is more than 80 per cent in four hours and the specific gravity of at least one specimen must be 1.003 or less.

(b) But in case of impaired renal function, less than 80 per cent of water is excreted in four hours and the specific gravity of at least one specimen does not fall to 1.003. The specific gravity of one specimen rather becomes fixed at 1.010 in severe renal damage.

(iii) Concentration test with posterior pituitary extract

(a) The subcutaneous injection of 0.5 ml of vasopressin to a normal person inhibits the diuresis.

(b) This test is performed within a short period of time and does not require any preparation of the patient.

(c) In case of congestive heart failure and diabetes insipidus, the posterior pituitary extract inhibits the diuresis causing sufficient concentration to determine degree of tubular function.

(d) In case of person with normal kidney function, the specific gravity of excreted urine becomes 1.020 or more and failure to this degree indicates renal damage.

(iv) Phenol sulfthalein (PSP) excretion test

This test is reliable and sensitive if the amount of dye in the first 15 minutes is taken as the criterion of renal function.

Procedure

1.0 ml of PSP (6 mg) is injected intravenously to a patient. Observation should be done after 15 minutes.

Interpretation

(a) Normal kidneys excrete 30 to 50 per cent of the dye during the first 15 minutes.

(b) Impaired renal function is confirmed if less than 23 per cent of the dye is excreted during this period.

D. Certain Miscellaneous Test

(i) Test of renal ability to excrete acid

Procedure

(a) Half a gram of gelatin coated capsules should be given over a period of an hour.

(b) The bladder should be emptied an hour later and the specimen must be discarded.

(c) All passed urine specimens must be collected during the next 6 hours and the bladder should be emptied at the end of that period.

(d) The urine must be collected in cleaned vessels containing a crystal of thymol.

(e) The pH of the urine specimens should be measured and the ammonia content of the combined urine specimens must also be determined.

Interpretation

(a) In case of normal persons, the pH is 5.3 of the passed urine during the 6-hour period and the value of ammonia excretion is between 30 and 90 micro equivalents/minute.

(b) The pH of the passed urine falls but ammonia excretion is low in case of renal failure.

(c) The pH of the urine is between 5.7 and 7.0 and ammonia excretion is also low in renal tubular acidosis.

(ii) Intravenous Pyelography

(a) After the injection of radio-opaque organic compounds intravenously, certain compounds of iodine are excreted by the kidneys in adequate concentrations to form a shadow of the renal calyces, renal pelvis, ureters, and the baldder on an X-ray film and provides sufficient information regarding size, shape, and functioning of the kidneys.

(b) The most recently used substance is diodone (30 per cent) which gives better results and available as pyelosil (Glaxo).

The excretion of diodone is used as a rough indication of renal function.

(c) Intravenous pyelography is used in the investigation of urinary tract diseases and it becomes a routine procedure for the investigation of patients suffering from renal calculi, repeated urinary infections, renal pain, hematuria, prostatic enlargement and suspected tumors, etc.

(d) This pyelography should not be done in case of patients suffering from acute nephritis, severely impaired liver function, congestive cardiac failure, hypersensitivity, etc.

(iii) Radio-active Renogram

(a) The radioactivity of each kidney is recorded graphically after the administration of I^{131}-labelled hippuran intravenously. This hippuran-I^{131} is actively secreted by the kidney tubules and is not concentrated in the liver.

(b) 15 to 60 μ_{ci} of hippuran-I^{131} is given slowly intravenously by a single dose.

(c) This investigation gives the information regarding the major asymmetry in function between the two kidneys, the assessment of overall renal function, and the presence of obstruction to urine flow in renal pelvis or ureters.

(iv) Radio-active scanning

(a) This is the recent developed technique and in this technique, Hg^{197}-labelled chlormerodrin is injected intravenously. A renal scan can be obtained by a scintillation counter over the lumbar regions.

(b) This technique is useful for the detection of abnormalities in size, shape, and position of the kidneys.

(c) This technique also shows renal tumors and renal infarcts which are not detected in pyelography.

CHAPTER 48

GASTRIC FUNCTION TESTS

Gastric analysis is performed for the following purposes :

1. Diagnosis of gastric ulcer.
2. Diagnosis of pulmonary tuberculosis.
3. Diagnosis of Zollinger-Ellison syndrome.
4. Confirmation of the completeness of surgical vagotomy.

Tests for determining gastric functions

1. Examination of resting contents.
2. Fractional gastric analysis using a test meal.
3. Examination of the contents after stimulation.
4. Tubeless gastric analysis.

Stomach tubes

(a) The stomach tube is of rubber or plastic with an external diameter of 4 mm.

(b) The Ryle's tube which is commonly used has a covered end having a small weight of lead which acts as a radio opaque. The holes in the tube are at a short distance from the end.

(c) The Rehfuss tube has an uncovered metal end with openings like the bore of the tube.

(d) Both the tubes are marked in the form of rings for the facility of swallowing the tubes by the patient. When the single ring reaches the lips, the tip reaches the cardiac end. When the double ring reaches the lips, the tube is almost to pylorus.

(e) The tube should be boiled in water and should be lubricated with glycerol before use.

(f) Ryle's tube is easier to swallow and does not cause trauma; but it tends to block easily.

Collection of contents of stomach

1. The resting gastric contents arc completely removed for examination by introducing a stomach tube through nasogastric route into the stomach after removing the contents by aspiration.
2. After a test meal, small samples of 5 to 6 ml of the gastric contents are removed at an interval of 15 minutes and the samples are collected in small sterile clean bottle.
3. The common errors are :
   (a) Sometimes, the tube may be blocked by mucus or food residues showing wrong assumption that the stomach is empty.
   (b) In case, the tube is not properly placed in the stomach for which no specimen can be obtained or if saliva is swallowed, saliva is only sent for analysis which shows the wrong diagnosis of achlorhydria.

(c) If the tube is too much swallowed, the aspiration will give heavily bile stained duodenal contents.

1. Examination of resting contents

After a night's fast the stomach contents are completely removed by passing the tube.The following characteristics are important in the diagnosis of diseases of stomach :

1. Volume

(a) Only 20 to 50 ml of resting contents are obtained in normal cases.

(b) An increase in volume may be due to hypersecretion of gastric juice, retention of gastric contents owing to delayed emptying of the stomach, and regurgitation of the duodenal contents.

2. Consistency

(a) The normal gastric juice is fluid in consistency and does not contain any food residue and may contain small amounts of mucus.

(b) Food residues are present in case of carcinoma of the stomach.

3. Colour

(a) In case of normal person, the gastric residue is clear or it may slightly yellow or green due to regurgitation of bile from duodenum.

(b) A dark red or brown colour may be observed due to the presence of altered blood or fresh-blood.

4. Bile

Increased quantities of bile shows abnormality which is a result of intestinal obstruction or ileal stasis.

5. Blood

(a) Blood is not present in normal cases.

(b) Presence of small amount of fresh blood may be traumatic.

(c) Brown or reddish-brown blood may occur in gastric ulcer and sometimes in gastric carcinoma due to the formation of dark brown acid hematin as a result of the hemolysis of red blood cells by HCl.

(d) Bleeding may also occur from gastritis.

6. Mucus

(a) A small amount of mucus may be present in normal cases.

(b) Increased amount of mucus is present in gastritis and in gastric carcinoma. Presence of mucus is inversely proportional to the amount of HCl present.

(c) Swallowed saliva may contain excess of mucus.

7. Free and total acidity

(a) The acidity is determined by titration with a standard solutionof NaOH using methyl orange or Topfer's reagent which indicates end point by the change of red to yellow colour or using phenolphthalein indicator which shows end point by the change of yellow to red colour.

(b) The presence of the amount of free HCl is free acidity; the complete titration shows the total acidity which is composed of protein hydrochloride and any organic acid; the difference between two titrations gives the combined acid.

(c) The result is expressed as ml of 0.1 N HCl per 100 ml of gastric contents. This is same as mEq/litre. This figure is obtained by multiplying the above titration by 10.

(d) The normal values of free acid is 0 to 30 mEq/L and that of total acid is 10 to 40 mEq/L.

8. Organic acids

(a) The presence of large amounts of lactic acid and butyric acid in achlorhydria and hypochlorhydria indicates the remaining of residual foods in the stomach. In absence of HCl, the microorganisms ferment the food residues producing lactic acid and butyric acid.

(b) Achlorhydria is associated with retention of food residues and is found in carcinoma stomach.

2. Fractional gastric analysis using test meals

This consists of :

1. Introduction of Ryle's tube in stomach of a fasting individual.
2. Analysis of residual gastric contents after collection.

(These two points are already discussed)

3. Ingestion of test meal.
4. Collection of 5 to 6 ml gastric contents after meal by aspiraton using a syringe and analysis of the samples.

(i) Test meals

(a) Oatmeal is prepared by adding 2 tablespoonfuls of oat meal to one quart of boiling water.

(b) "Ewald" test meal consists of two pieces (35 gm) of toast and 250 ml light tea.

(c) Either of the meal is consumed by the patient and either of the tubes in introduced after one hour.

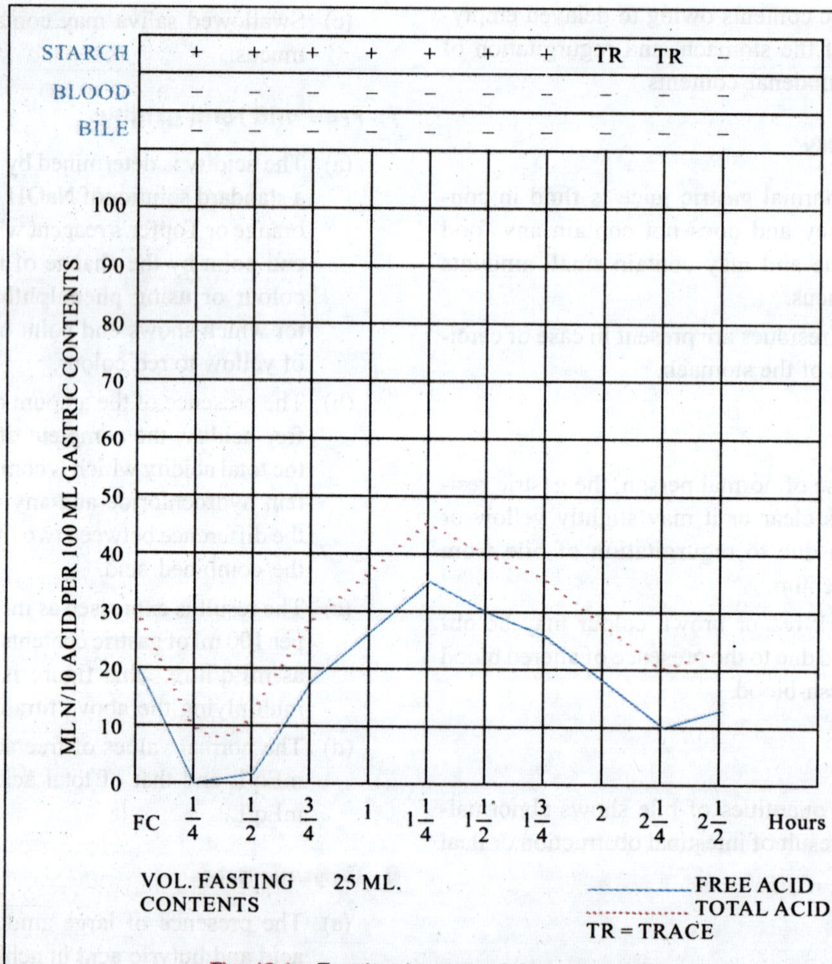

Fig. 48.1 : Fractional test meal. Normal result

(ii) Collection of samples

(a) About10 ml gastric contents are collected at an interval of exactly 15 minutes by means of syringe attached to the tube.

(b) If the stomach is not empty at the end of 3 hours, the remaining stomach contents are removed and the volume is also noted.

(c) Each sample is strained through a fine cloth.

(d) The strained samples are analysed for free and total acidity and the residue on the cloth is examined for mucus, bile, blood, and starch.

(iii) Results and Interpretation

(a) In normal health, after taking the meal, the free acid is found after 15 to 45 minutes (See figure below). The free acid then steadily rises to reach the maximum at about 15 minutes to 1/2 hour, after which the concentration of free acid begins to fall. The free acid ranges from 15 to 45 mEq/litre at the maximum with total acid at about 10 units higher. Blood is not present and appreciable amount of bile is also not present.

(b) In **hyperchlorhydria**, free acidity exceeds 45 mEq/litre but the combined acidity remains the same as in nromal persons. Hyperacidity is found in duodenal ulcer in which a climbing type of curve is formed in gastric ulcer in which 50 per cent cases give normal results, and blood may be present, in gastric carcinoma in which small percentage show hyperacidity and blood, in jejunal and gastrojejunal ulcers in which there may be hyperacidity after operation.

(c) In **hypochlorhydria**, low acidities are found in carcinoma of stomach and in atonic dyspepsia. Free HCl is absent in gastric secretion in pernicious anemia.

(d) In **achlorhydria**, no HCl secretion but pepsin is present.

(e) In **achylia gastrica**, gastric secretion is completely absent due to advanced cases of cancer of stomach, advanced cases of gastritis, and acute pernicious anemia.

3. Examination of contents after stimulation

A. Alcohol Stimulation

(i) The Ryle's tube is passed into the stomach after overnight fast and resting contents are collected for analysis.

(ii) 100 ml of 7 per cent ethyl alcohol is administered. Samples of gastric contents are collected at an interval of 15 minutes and all the samples are analysed for free and total acidity, peptic activity, presence of bile, blood and mucus.

(iii) The advantages of alcohol test meal are the followings :

(a) More easily administered and prepared.

(b) Consumed better.

(c) The gastric response is more rapid and more intense.

(d) Quick emptying of the stomach.

(e) Specimens are clear and easily analysed.

(iv) The disadvantages of this test are :

(a) Stimulus with alcohol is more vigorous.

(b) Stimulus is not so strictly physiological.

(c) Free acidity levels are higher and normal limits are wider.

B. Caffeine Stimulation

(i) The Ryle's tube is introduced into the stomach after overnight fast and the resting gastric contents are collected and analysed.

(ii) Caffeine sodium benzoate (500 mg dissolved in 200 ml of water) is administered orally. Samples of stomach contents are collected at an interval of 15 minutes and analysed for free and total acidity, peptic activity, presence of bile, blood and mucus.

(iii) Advantages of this stimulation is similar to that of alcohol stimulation.

C. Histamine Stimulation test

Histamine is a powerful stimulant for the secretion of HCl in the normal stomach. It increases the cAMP level which causes the increased secretion of highly acidic gastric juice with low pepsin content.

(i) Standard histamine test

(a) The Ryle's tube is passed into the stomach after overnight fast and the stomach contents are collected for analysis.

(b) A subcutaneous injection of histamine (0.01 mg/kg body weight) is inserted. 10 ml stomach contents are collected at an interval of 10 minutes for one hour and samples are analysed for free and total acidity, peptic activity, presence of bile, blood and mucus.

(c) Achylia gastrica ("true" achlorhydria) is indicated by the absence of free HCl in the secretion after histamine administration. More juice may be secreted in duodenal ulcer.

(ii) Augmented histamine test : It is a more powerful stimulus test and it shows an inability to secrete acid. Larger doses of histamine sometimes causes an unwanted severe reactions.

(a) The Ryle's tube is introduced after an overnight fast and the gastric contents are collected for analysis. The resting çontents are collected at an interval of 20 minutes for an hour. Halfway of this period, 4 ml anthisan is given intramuscularly.

(b) At the end of the hour, histamine (0.04 mg histamine acid phosphate per kg body weight) is given subcutaneously and gastric contents are collected at an interval of 15 minutes for one hour for analysis.

(c) In pernicious anemia, no free HCl is secreted after histamine stimulation. In duodenal ulcers, higher values of acid are obtained.

(d) Recently, histalog is used in place of histamine. No side effects like histamine are observed by its use. The recommended dose of histamine is 10 to 50 mg. This histalog is highly effective in stimulating gastric secretion.

D. Insulin stimulation test

Hypoglycemia due to insulin administration is an active stimulus of gastric acid secretion. The blood sugar level below 45 mg per cent is essential for a reliable test.

(i) The Ryle's tube is passed into the stomach after an overnight fast and the stomach is made empty.

(ii) 15 units of soluble insulin is injectied intravenously and about 10 ml gastric contents are collected at an interval of 15 minutes for $2\frac{1}{2}$ hours. The samples are analysed for free and total acidity, peptic activity, and presence of blood, bile, starch. Starch should not be present.

(iii) In duodenal ulcer, acid level is more in response to insulin. The concentration of free acid may be over 100 mEq/litre. After vagotomy no response of insulin is found and the gastric acidity remains at 15 to 20 mEq/litre before and after insulin injection.

E. Pentagastrin test

Pentagastrin is a synthetic peptide and it is butyloxycarbonyl β-alanine. It is an active stimulator.

(i) The Ryle's tube is passed into the stomach after an overnight fast and the resting contents are completely removed. After emptying the stomach two 15 minute specimens are collected to have the "basal secretion".

(ii) Pentagastrin (6μg/kg body weight) is injected subcutaneously and specimens are collected at an interval of 15 minutes for analysis.

(iii) The normal basal secretion rate is 1 to 2.5 mEq/hour. The maximum secretion in normal person after pentagastrin stimulus varies from 20 to 40 mEq/hour.

(iv) In **duodenal ulcer**, the range is 15 to 83 mEq/hour. This test is of little value in **gastric ulcer**. The "true" achlorhydria is found in **cancer of the stomach**. The reduced acid level is observed in **acute gastritis**. The "true" achlorhydria is also noted in **Pernicious anemia**. The **Zollinger-Ellison syndrome** is characterised by a high basal secretion usually above 10 mEq/hour and no further rise is found after giving Pentagastrin.This syndrome is characterised by peptic ulcer, gastric hypersecretion and diarrhoea in patient with "gastrin". This syndrome is also accompanied by parathyroid adenomas with hyperparathyroidism. The secretion of pepsin occurs after stimulation with pentagastrin.

4. Tubeless gastric analysis

The modified test is done with the introduction of "**DiagnexBlue**" prepared by reacting carbacrylic cation exchange resin with "Azure A" an indicator. The hydrogen ions of the resin is exchanged with "Azur A" ions. The reaction is reversed in the stomach when acid is in a concentration having pH less than 3.0. The indicator "Azur A" is released by the action of acid. The released one is absorbed in the small intestine and excreted in the urine, the colour of which is matched with known standards.

This test is valuable if it is used as "**screening test**" only. A positive result indicates the secretion of acid by the stomach. A negative result is an unreliable indicator of "true" achlorhydria. This test is not reliable in patients suffering from renal diseases, urinary retention, malabsorption, pyloric obstruction. Vitamin preparation should not be taken on the day preceeding the test which may contain substances decolourised by vitamin C.

CHAPTER 49

ACID-BASE REGULATION

Acid-base Balance in Normal Health

An acid may be defined as substance that produces H^+ (protons) in solution, and a base is a particle that combines with hydrogen ions (H^+).

Example : Carbonic acid is an acid which is dissociated into H^+ and HCO_3^- and its anionic component is a base. $NaHCO_3$ acts as a base since it produces HCO_3^- which can combine with hydrogen ion (H^+).

Other examples :

Acid		Base
CH_3COOH	$\rightleftharpoons$	$H^+ + CH_3COO^-$
H_2PO_4	$\rightleftharpoons$	$H^+ + HPO_4^{--}$
HSO_4^-	$\rightleftharpoons$	$H^+ + SO_4^{--}$

Water (H_2O) can act either as acid or base.

$$\underset{\text{(Acid)}}{H_2O} \rightleftharpoons H^+ + OH^-$$

$$\underset{\text{(Base)}}{H_2O + H^+} \rightleftharpoons \underset{\text{(Strong acid)}}{H_3O^+} \text{ (Hydroxonium ion)}$$

Hydrochloric acid (HCl) is a strong acid since it is extensively dissociated into H^+ and Cl^-. Chloride ion (Cl^-) is a very weak base since it has got very little capacity in combining firmly with H^+. But HCO_3^-, HPO_4^{--}, and protein$^-$ are comparatively strong bases since they have got strong affinity for H^+ forming weak acids.

The stronger the acid, the weaker is the base. Such pairs are termed as conjugate acid base pairs.

Chloride ion (Cl^-) is the conjugate base of the acid HCl.

The cations such as Na^+, K^+, Ca^{++}, Mg^{++}, cannot donate or accept protons. So, they are neither acids nor bases. Such cations are termed as aprotes.

Buffers

Definition

Buffers are the mixtures of weak acids and their salts of strong bases.

Example : ($CH_3COOH + CH_3COONa$).

Principles of buffers

$$HAC + NaAC \rightarrow Na^+ + H^+ + 2AC^-$$

[where HAC = Acetic acid; NaAC = Sodium acetate]

If alkali (NaOH) is added to this system, it will form salt and no free H^+ or OH will be available.

$$HAC + NaAC + NaOH \rightarrow 2NaAC + H_2O$$

If acid (HCl) is added to this system, it will also form salt and no free H^+ or OH^- will be available.

$$HAC + NaAC + HCl \rightarrow NaCl + 2HAC$$

In either cases, there is no change in hydrogen ion concentration. The buffer acts almost as if it were "absorbing" the added free hydrogen or hydroxyl ions.

Major sources of acids produced in the body

1. Carbonic acid (H_2CO_3) : This acid is produced in the body during oxidation in the cells. The C-compounds on oxidation produces CO_2. About 300 litres of CO_2 are produced and eliminated from the body of an adult daily.
2. Sulphuric acid (H_2SO_4) : This acid is formed by the oxidation of sulphur-containing amino acids eg, methionine and cystine or cysteine.
3. Phosphoric acid : The dietary phosphoproteins, nucleoproteins, and phosphatides are metabolized to produce this acid.
4. Abnormal quantities of pyruvic acid, lactic acid, acetoacetic acid, and β-hydroxybutyric acid are formed from the oxidation of carbohydrates, fats and proteins.
5. When some medicines like mandelic acid, ammonium chloride are administered in excess, they may increase hydrogen ion concentration of blood.

Mechanisms of regulation of pH

The following factors are involved in the regulation of blood pH.

1. Buffer systems in the blood.
2. Respiratory mechanisms.
3. Renal mechanisms.
4. Dilution factor.

Physiological buffer systems

E.C. fluids are capable of transporting acids from the cells to the lungs and kidneys without the change in pH depending mainly on the efficient buffer systems in these fluids and in the erythrocytes.

The most important blood buffer systems are as follows :

(a) Plasma buffers :

$$\frac{NaHCO_3}{H_2CO_3};\ \frac{Na_2HPO_4}{NaH_2PO_4};\ \frac{Na\text{-}Pr}{H\,Pr};$$

$$\frac{\text{K organic acid}}{\text{H organic acid}}$$

(b) Buffers of red blood cells :

$$\frac{K\cdot Hb}{H\cdot Hb};\ \frac{KHCO_3}{H_2CO_3};\ \frac{KHbO_2}{H\cdot HbO_2};\ \frac{K_2HPO_4}{KH_2PO_4}$$

$$\frac{\text{K}\cdot\text{organic acid}}{\text{H}\cdot\text{organic acid}}$$

The buffer systems in the interstitial fluids and lymph are almost the same as in the blood plasma except the proteins which are present in muscle in smaller quantities.

The followings are the most important buffers :

(a) In plasma : The bicarbonate buffer and plasma proteins.

(b) In erythrocytes : The bicarbonate buffer and Hb-system.

Role of different buffer systems

1. Bicarbonate buffer system

This consists of carbonic acid and sodium bicarbonate. The normal ratio of it in blood is 20 : 1. These chief buffers constitute the alkali reserve. The bicarbonate buffers neutralize the strong and non-volatile acids entering the ECF. Thus, lactic acid is buffered as follows :

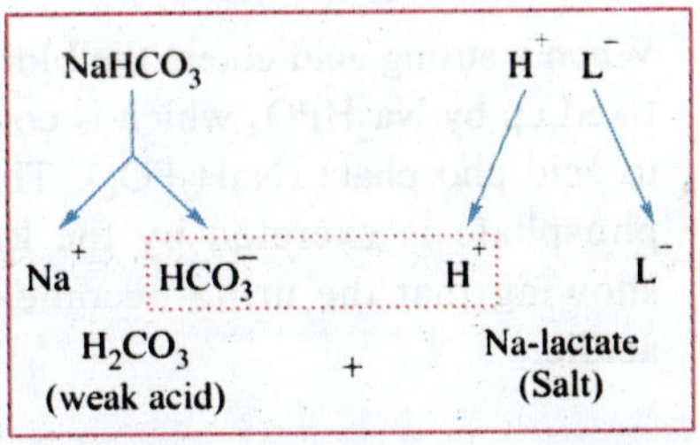

A strong and non-volatile acid (lactic acid) is converted into weak and volatile acid at the expense of $NaHCO_3$ (the salt component of the buffer). H_2CO_3 thus formed is eliminated through diffusion of CO_2 through alveoli of lungs.

$$H_2CO_3 \xrightarrow{\text{Carbonic anhydrase}} H_2O + CO_2\uparrow$$

Therefore, bicarbonate buffer system is directly linked up with respiration.

When NaOH enters the ECF, it reacts with H_2CO_3 of the buffer system :

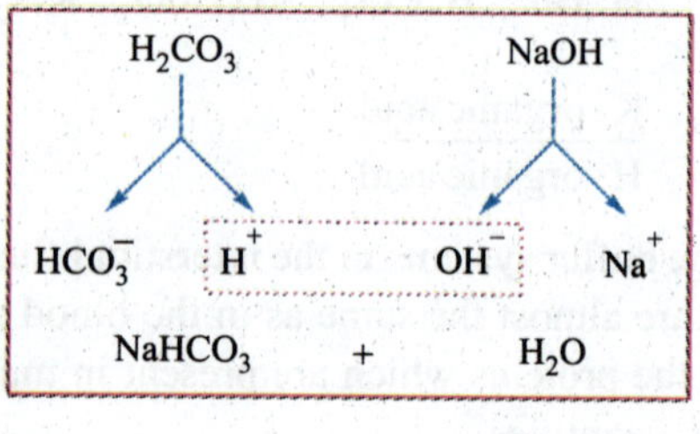

$NaHCO_3$ concentration in the blood is represented as the alkali reserve since it does not combine with strong and non-volatile acid.

Advantage : Bicarbonate buffer system is quite efficient as compared to other buffer systems since it is present in very high concentration and it produces H_2CO_3 from which CO_2 is exhaled out.

Disadvantage : Bicarbonate buffer is very weak as a chemical buffer and hence PKa is far away from the physiological pH.

2. Phosphate buffer system

The normal ratio of Na_2HPO_4/NaH_2PO_4 in plasma is 4 : 1 and this ratio is kept constant by the help of the kidneys for which phosphate buffer system is directly related to the kidneys.

(i) When a strong acid enters the blood, it is fixed up by Na_2HPO_4 which is converted to acid phosphate (NaH_2PO_4). This acid phosphate is excreted by the kidneys showing that the urine becomes more acidic.

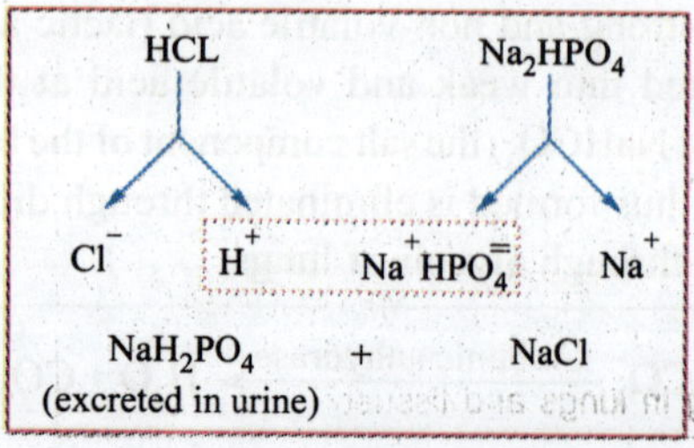

(ii) When an alkali enters, it is buffered by the acid phosphate resulting the alkaline phosphate which is excreted in urine showing the increased alkalinity of urine.

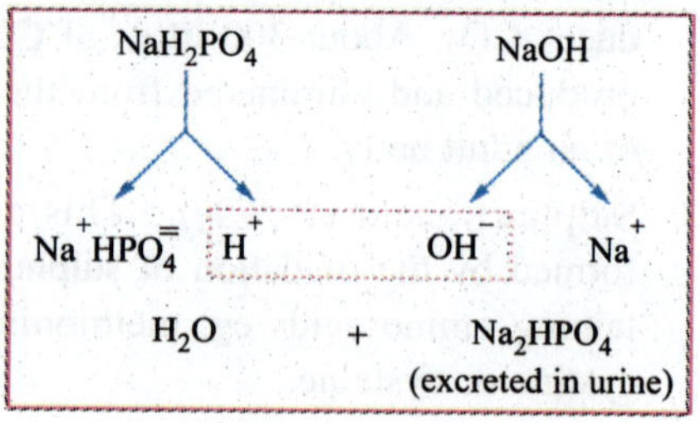

So, phosphate buffer system works in conjugation with the kidneys.

Advantage : It is very effective as a chemical buffer and PKa approaches physiological pH better.

Disadvantage : Phosphate buffer is low in concentration in blood and less efficient as a physiological buffer.

3. Protein buffer system

The capacity of plasma protein buffer (Na^+Pr^- / H^+Pr^-) is less than that of Hb which is only involved in erythrocytes.

Buffering action of proteins :

(i) In acidic medium, protein acts as a base and NH_2 group takes up H^+ from the medium forming NH_3^+. Protein becomes positively charged.

(ii) In alkaline medium, protein acts as an acid and acid group (COOH) dissociates into COO^- and H^+. This H^+ combines with OH^- producing a molecule of H_2O. Proteins become negatively charged.

(iii) The salt component (Na^+ proteinate) combines with strong acid producing weak acid (H^+Pr^-).

$$\underset{}{Na^+Pr^-} + \underset{\text{(Strong acid)}}{H^+L^-} \longrightarrow \underset{\text{(Salt)}}{NaL} + \underset{\text{(Weak acid)}}{H^+Pr^-}$$

(iv) CO_2 is removed by other factor with the formation of carbamino compounds. The binding of CO_2 is performed without passing through the carbonic acid stage.

$$PrNH_2 + CO_2 \rightleftharpoons PrNHCOOH$$

(Carbamino compound)

4. Hemoglobin as a buffering agent

The capacity of Hb as a buffer fully depends on the number of dissociable buffering groups, namely –COOH group, $-NH_2$ group, guanido group and the important imidazole group.

The physiological buffering action of Hb to maintain the pH range of blood from 7.0 to 7.8 is due to imidazole group of histidine.

Imidazole contains two groups :

(a) Fe^{++} containing group which is concerned with carriage of O_2.

(b) Imidazole N_2 group which can give up H^+ and accept H^+ depending on the pH of the medium.

Therefore, the buffering capacity of Hb entirely depends on the presence of imidazole nitrogen group which remains dissociated in acidic medium and conjugated base forms.

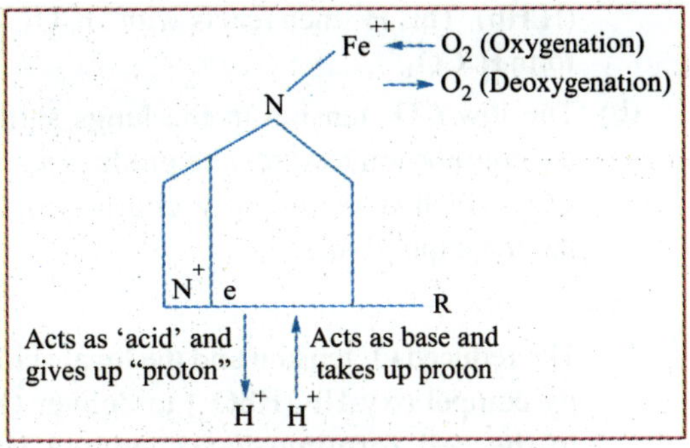

(a) The imidazole N_2 group being oxygenated acts as an acid and donates protons in the medium. **Oxygenated Hb is a stronger acid than de-oxygenated Hb.**

(b) The deoxygenated Hb is less acidic and the imidazole N_2 group acts as a base and takes up protons from the medium.

(c) The delivery of O_2 is favoured by the acidity of the medium.

(d) The oxygenation of Hb is favoured by the alkalinity of the medium.

In the lungs :

(a) H^+ is released during the formation of OXy-Hb (HbO_2) from deoxygenated Hb

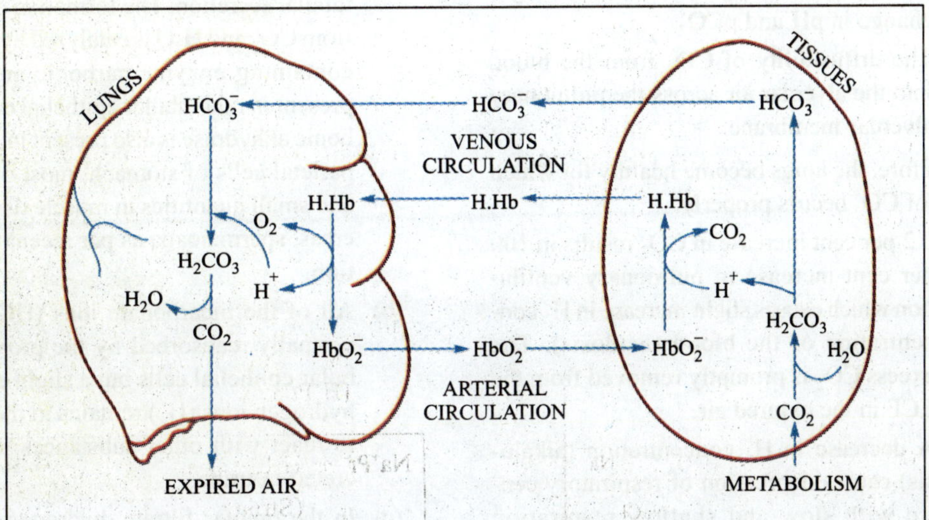

Fig. 49.1 : Sequence of events occurring in lungs and tissues

(H.Hb). This H^+ then reacts with HCO_3^- to form H_2CO_3.

(b) The low CO_2 tension in the lungs shifts the equilibrium towards the production of CO_2 which is continually eliminated in the the expired air.

In the tissues :

(a) The reduced O_2 tension and the local acidity compel oxy-Hb (HbO_2) to deliver O_2 to the cell forming deoxygenated Hb (H.Hb).

(b) The metabolic CO_2 in the cells is hydrated to form H_2CO_3 which is dissociated to form H^+ and HCO_3^-.

(c) The deoxygenated Hb (H.Hb) is formed from oxygenated Hb(HbO_2) by the effect of H+ with the little change in pH.

Isohydric transport of CO_2 : The newly formed H^+ in the tissues from H_2CO_3 does not bring about a change in pH. This circumstance in the role of Hb buffers is sometimes referred to as isohydric transport of CO_2.

Role of Respiration in Acid-base Regulation

The respiratory mechanism depends on :

(a) The response of the respiratory centre (RC) in medulla oblongata to a very slight change in pH and pCO_2.

(b) The diffusibility of CO_2 from the blood into the alveolar air across the pulmonary alveolar membrane.

Therefore, the lungs become healthy for which diffusion of CO_2 occurs properly.

(i) 0.2 per cent increase in CO_2 results in 100 per cent increase in pulmonary ventilation which causes slight increase in H^+ concentration of the blood (acidosis). The excess CO_2 is promptly removed from the ECF in the expired air.

(ii) A decrease in H^+ concentration (alkalosis) causes depression of respiratory centre with slow and shallow respiration (hypoventilation) resulting in the retention of CO_2 in the blood until the normal pCO_2 and pH are restored.

Hence, the respiratory mechanism tends to maintain the normal $BHCO_3/H_2CO_3$ ratio in the EC fluids.

Renal Mechanisms for Regulation of Acid-base Balance

Kidneys are also involved in acid-base equilibrium.

(a) By eliminating the non-volatile acids, eg, lactic acid, sulphuric acid, acetoacetic acid, β-hydroxy butyric acid, etc. after being buffered with Na^+ through the glomerular filtration.

(b) Na^+ exists as $NaHCO_3$ (alkali reserve) in the renal tubules by reabsorption by the renal tubules in exchange of H^+ which are secreted.

Three mechanisms are involved in the above mechanism :

1. Bicarbonate mechanism.
2. Phosphate mechanism.
3. Ammonia mechanism.

1. Bicarbonate mechanism

(a) Carbonic acid (H_2CO_3) is ionized as H^+ and HCO_3^- and this H^+ is mobilized for tubular secretion. The formation of H_2CO_3 from CO_2 and H_2O is catalyzed by the zinc-containing enzyme carbonic anhydrase, present in real tubular epithelial cells. Carbonic anhydrase is also present in RB cells, parietal cells of stomach, most of tissues, and small quantities in muscle tissue, pancreas, spermatozoa as per recent information.

(b) All of the bicarbonate ions (HCO_3^-) are normally reabsorbed by the proximal tubular epithelial cells but a slight excess of hydrogen ions (H^+) remains in the tubules to react with other substances to be excreted in urine.

(c) In the tubular filtrate, hydrogen ions are exchanged against sodium ions.

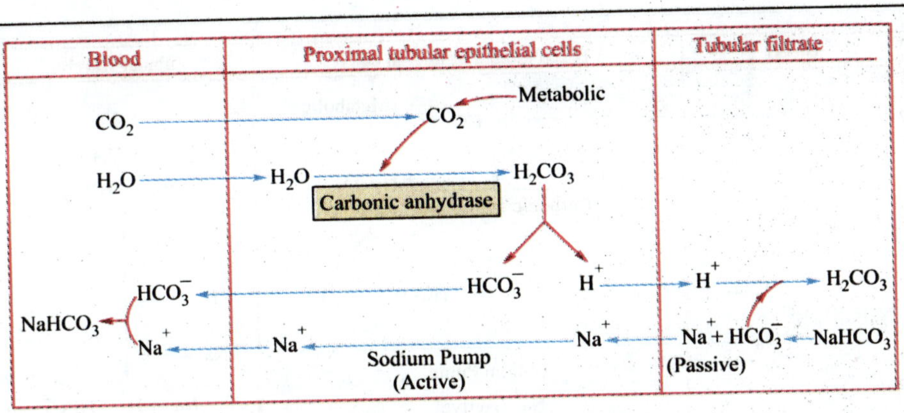

Fig. 49.2 : Bicarbonate mechanism

(d) Increase in pCO_2 accelerates formation of H_2CO_3 and thus increases H^+ secretion by the renal tubular epithelial cells which help in the reabsorption of HCO_3^-.

(e) In the excess administration of K^+, K^+ quickly enters into the cell in exchange of H^+ and hydrogen ions are buffered by bicarbonate of ECF for which plasma bicarbonate content is diminished. As a result, the renal tubular epithelial cells secrete less hydrogen ions (H^+), and less bicarbonate ions (HCO_3^-) are absorbed. The excretion of bicarbonate in urine becomes more and the urine becomes alkaline although ECF is acidic.

(f) The clinical importance shows that potassium ions (K^+) leave the cell in K^+ deficiency and hydrogen ions (H^+) enter the cells producing intracellular acidosis. ECF becomes alkaline. Increased excretion of H^+, NaH_2PO_4, and NH_4Cl increase the titratable acidity. ECF is though alkaline, highly acidic urine is excreted. This condition is referred to as paradoxic aciduria. This occurs in patients treated for long with cortisone, Cushing's syndrome, and in patients with potassium free fluids.

(g) Increase in Cl^- causes decrease in HCO_3^-.

(h) The enzyme carbonic anhydrase is inhibited by diamox. The urine becomes alkaline when the drug is administered and increased amount of $NaHCO_3$ appears in urine. There is reduction in titratable acidity and ammonia excretion in the urine. Drug is used clinically as a diuretic to induce a loss of Na^+ and water in congestive cardiac failure and in hypertensive heart diseases.

2. Phosphate mechanism

(a) The ratio of plasma Na_2HPO_4 and NaH_2PO_4 determines the pH of the urine. The ratio of these two in plasma is maintained to 4 : 1 if the concentration of Na_2HPO_4 exceeds that of NaH_2PO_4. But the ratio becomes 9 : 1 if the concentration of $NaHPO_4$ in urine exceeds that of Na_2HPO_4.

(b) The pH 7.4 of glomerular filtrate is changed to pH 6.0 or even as low as 4.8 in urine.

(c) Na^+ is exchanged for secreted H^+ which changes Na_2HPO_4 to NaH_2PO_4 resulting the increase in acidity of urine.

(d) This mechanism takes place in the distal tubule of kidney.

3. Ammonia mechanism

Na^+ is conserved by the elimination of H^+ for the production of NH_3 by the renal distal tubular epithelial cells.

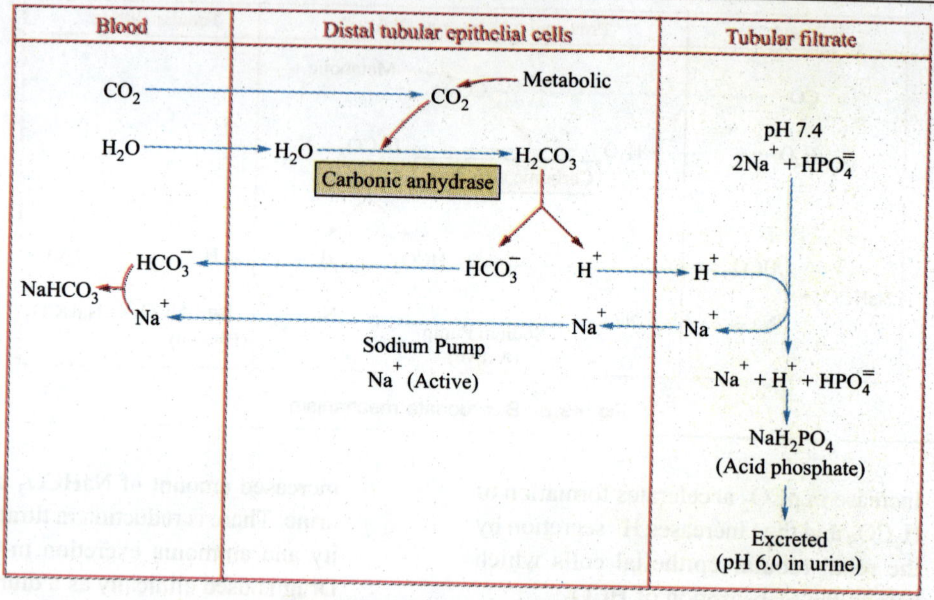

Fig. 49.3 : Phosphate mechanism

Sources of NH_3

(a) Glutamine is hydrolysed to produce NH_3 by the enzyme **glutaminase** present in these cells.

$$\text{Glutamine} + H_2O \xrightarrow{\text{Glutaminase}} \text{Glutamic acid} + NH_3$$

(b) It can be formed from other amino acid by oxidative deamination.

(c) It is also formed from glycine by **glycine oxidase**.

Formation of NH_4^+

NH_3 is diffused to the tubular filtrate where it combines with H^+ to form NH_4^+.

The renal glutaminase acitivity is enhanced by acidosis. NH_3 production is highly increased in metabolic acidosis and negligible in alkalosis.

ACID-BASE IMBALANCE

Acidosis

1. Metabolic acidosis;
2. Respiratory acidosis.

1. Metabolic acidosis

(a) Metabolic acidosis occurs when a reduction in plasma HCO_3^- ($BHCO_3$) takes place. In case, there is the deficit HCO_3^-, the ratio of $[HCO_3^-]/[H_2CO_3] = 20/1$ is decreased, i.e., pH is decreased causing metabolic acidosis.

(b) Acidosis stimulates the respiratory centre causing deep and rapid breathing. This increased ventilation results in CO_2 loss and reduction in $[H_2CO_3]$. Increased ventilation causes reduction in PCO_2 which in turn depresses the respiratory centre.

(c) Renal mechanisms can correct the disturbances by conserving cations and by increasing NH_3 formation, H^+ excretion in distal tubules, and reabsorption.

Causes

(i) Abnormal increase in "anions" other than HCO_3^- caused from :

(a) Endogenous excessive production of acid ions occur in diabetic acidosis,

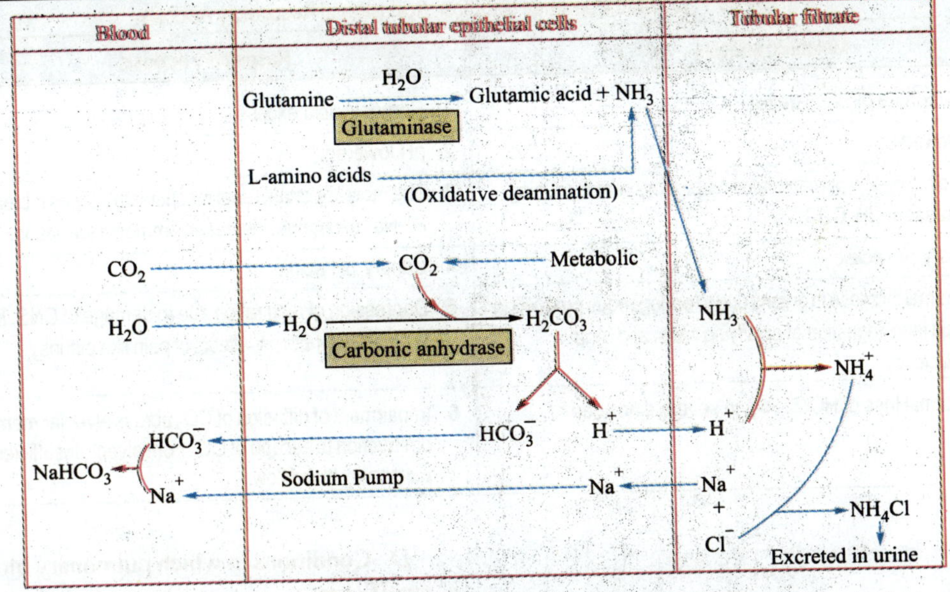

Fig. 49.4 : Formation of NH_4^+

lactic acidosis, starvation, high fever, violent exercise, hemorrhage, shock, and anoxia.

(b) Ingestion of excessive quantities of acids like phosphoric acid, hydrochloric acid, salicylic acid, and mandelic acid, etc.

(c) Renal insufficiency causing retention of acids commonly observed in the terminal stages of nephritis, and destructive renal lesions such as pyelonephritis, pyonephrosis, renal T.B., etc.

The main factors of these causes are :

(a) Decreased H^+ - Na exchange.

(b) Decreased NH_3 formation.

(c) Decreased glomerular filtration with retention of acid radicals.

(d) Nephritic acidosis due to accumulation of certain organic acids.

(ii) Abnormal loss of HCO_3^- due to loss of base occurring in the loss of excessive intestinal secretions as in severe diarrhoeas, small bowel fistulas, and severe biliary fistulas.

2. Respiratory acidosis

This is caused due to the increase in H_2CO_3 in the blood following eliminaton of $CO_2(PCO_2)$ in the pulmonary alveoli. This results from breathing air containing abnormally high percentage of CO_2, and conditions in which elimination of CO_2 through lungs is retarded.

Mechanism

(a) In the depression of respiratory centre, the excretion of CO_2 through lungs is impaired causing more accumulation of CO_2 in blood for which there is excess formation of H_2CO_3. So the ratio of $[HCO_3^-]/[H_2CO_3]$ is lowered resulting the decrease in pH.

(b) The increased CO_2 tension (PCO_2) causes the elevated stimulation to respiratory centre resulting increased ventilation. This mechanism is less effective.

(c) In renal mechanism, more HCO_3^- are reabsorbed from tubules responding to raised PCO_2 in blood and the ratio of $[HCO_3^-]/[H_2CO_3]$ is restored towards 20 : 1. This mechanism is an important one.

Table. 49.1 : Difference between Metabolic Acidosis and Respiratory Acidosis

Metabolic acidosis	Respiratory acidosis
1. Bicarbonate (HCO_3^-) deficit.	1. Carbonic acid excess.
2. pH lowered.	2. pH lowered.
3. Respiratory-low pH stimulates respiratory centre causing decrease in $[H_2CO_3]$.	3. The most important renal mechanism causes increase in H^+-Na^+ exchange, increased ammonia formation.
4. Urinary pH-acidic.	4. Urinary pH-acidic.
5. Abnormal increase in anions due to endogenous production of acid ions, ingestion of acidifying salts, and renal insufficiency.	5. Depression of respiration due to damage to CNS, loss of ventilatory functions, effects of pain like pleurisy.
6. Abnormal loss of HCO_3^-, eg, in severe diarrhoea.	6. Impairment of diffusion of CO_2 across alveolar membrane, obstruction to escape of CO_2 from alveoli, insufficient pulmonary blood flow.

Causes

A. Conditions Causing depression of respiration :

(a) Damage to CNS :

(i) Brain damage—trauma, inflammation, and convulsive disorders.

(ii) Drug Poisoning—Morphine and barbiturates.

(iii) Excessive anaesthesia.

(iv) Bulbar Polio.

(b) Loss of ventilatory functions :

(i) Tension.

(ii) Pulmonary and mediastinal tumors.

(iii) Emphysema.

(c) Effect of pain, eg, pleurisy.

B. Conditions causing impairment of diffusion of CO_2 across alveolar membrane.

(a) Emphysema.

(b) Pulmonary edema and congenital alveolar dysplasia.

C. Conditions in which there is an obstacle to the escape of CO_2 from

The alveoli :

(a) Obstruction to respiratory tract-Laryngeal obstruction and asthma.

(b) Rebreathing from a closed space.

D. Conditions in which pulmonary blood flow is insufficient :

(a) Certain congenital heart diseases.

(b) Ayerza's disease.

Alkalosis

1. Metabolic Alkalosis

(a) The accumulation of excess of HCO_3^- causes an increase in the ratio of $[HCO_3^-]/[H_2CO_3]$, i.e., pH is elevated.

(b) Alkalosis inhibits the respiratory centre (RC) developing irregular breathing. This reduced ventilation causes CO_2 retention and carbonic acid level $[H_2CO_3]$ elevation.

(c) Since the levels of both in blood are increased the ratio of $[HCO_3^-]/[H_2CO_3]$ are restored towards 20 : 1.

(d) Decreased ventilation raises PCO_2 which stimulates the respiratory centre.

(e) In renal mechanisms, there is the increased excretion of cations, bicarbonates, K^+ but reduced excretion of NH_3.

(f) There is the decreased urinary acidity and titratable acidity.

Causes

(a) The excessive loss of HCl from stomach is

associated with high intestinal obstruction, pyloric obstruction, pylorospasm in infants.

(b) Na^+ and K^+ are mainly retained in the body in the form of bicarbonate due to the loss of Cl^- from the blood. Thus, a neutral salt (NaCl)is replaced by an alkaline salt ($NaHCO_3$).

(c) Excessive intake of bases like $NaHCO_3$, and acetates, lactates, citrates of sodium and potassium. Lactates and citrates are converted into HCO_3^-.

(d) Potassium deficiency.

(e) Deep X-ray therapy, radium therapy, and prolonged exposure to ultra violet rays cause a decrease in the H^+ concentration in the blood plasma.

(f) Excessive vomiting.

Clinical manifestations

Alkalosis is observed in :

(a) Decreased ionic calcium in blood producing tetany.

(b) Hypokalemia due to the increased excretion of K^+ from the distal tubules.

(c) Kidney damage caused by the degenerative changes in the tubules (nephrosis) occurring with oliguria and nitrogen retention.

(d) Ketosis and ketonuria due to excessive vomiting for the insufficient carbohydrate intake.

2. Respiratory Alkalosis

Respiratory alkalosis occurs when there is a decrease in $[H_2CO_3]$ with no change in HCO_3^- in plasma. As a result, pH is increased.

Mechanism

(a) $[H_2CO_3]$ is reduced as a result of increasd loss of CO_2 and the ratio of $[HCO_3^-]/[H_2CO_3]$ is elevated for which pH is increased.

(b) Owing to increased loss of CO_2 less bicarbonate is reabsorbed by the renal tubules

Table 49.2 : Difference between metabolic alkalosis and respiratory alkalosis

Metabolic alkalosis	Respiratory alkalosis
1. HCO_3^- excess.	1. H_2CO_3 deficit.
2. pH increased.	2. pH increased.
3. Disproportionate increase in $[HCO_3^-]$, $[H_2CO_3]$, PCO_2, pH.	3. Disproportionate decrease in $[HCO_3^-]$, $[H_2CO_3]$, PCO_2, pH.
4. Depression of respiratory centre and hyperventilation leading to retention of CO_2.	4. In renal mechanism, there is decreased H^+-Na^+ exchange, decreased excretion of acid and ammonia, increased excretion of HCO_3^- and K^+, retention of Cl^-.
5. In renal mechanism, there is increased NH_3 formation and H^+- Na^+ exchange, increased K^+ excretion, decreased reabsorption, retention of Cl^-.	5. Hypoventilation due to respiratory high pH and low PCO_2 and increase in H_2CO_3.
6. Urine shows alkaline, decreased NH_3, and decreased titratable acidity.	6. Urine shows alkaline, decreased NH_3 and decreased titratable acidity.
7. Low Ca^{++} leading to tetany, hypokalemia, ketosis and ketonuria, degenerative changes in tubules leading to nitrogen retention.	7. Low Ca^{++} leading to tetany, hypokalemia, ketosis and ketonuria, kidney damage leading to nitrogen retention.
8. Causes are excessive loss of HCl, high intestinal obstruction, pyloric obstruction, alkali ingestion, excessive loss of K^+, X-ray therapy, ultra violet radiation.	8. Causes are CNS diseases like meningitis and encephalitis, salicylate poisoning, hyperpyrexia, hysteria, high altitude ascending, apprehensive blood donors, injudicious use of respirator, some cases of hepatic coma.

for which the ratio of $[HCO_3^-]/[H_2CO_3]$ returns towards nomal (20 : 1).

(c) The respiratory centre is depressed by alkalosis and PCO_2 and CO_2 excretion is reduced.

(d) In renal mechanism, there are excretion of alkali in the form of bicarbonate, decreased excretion of acid, decreased excretion of ammonia, and retention of chloride ion (Cl^-) in the blood.

(e) Other features are the same as that of metabolic alkalosis.

Causes

(a) The **respiratory centre (RC)** is stimulated by CNS diseases like meningitis, encephalitis due to hyperventilation in some cases of these diseases manifesting hyperpnea over long periods.

(b) **Large doses of salicylate** cause stimulation of respiratory centre (RC) with hyperventilation showing alkalosis.

(c) **Hyperventilation** is caused by the increased respiratory rate associated with increased body temperature.

(d) Hysterical attacks cause hyperventilation and hyperventilation tetany with alkalosis is also observed in hyperexcitable donors.

(e) Hyperpnoea occurring in untrained individuals ascending to high altitudes results in H_2CO_3 deficit and alkalosis.

(f) Hepatic coma is also another cause of respiratory alkalosis.

(g) Injudicious use of respirator.

CHAPTER 50

CARDIAC PROFILE

In industrial societies as well as in developing countries, the cause of death in adult life is the cardiovascular disease.

Two-thirds of the mass of the heart extends to the left of the mediastinum. The significant structure situated anterior to the heart is the thymus gland. Posteriorly, the oesophagus lies behind the left atrinum. The left main and right coronary arteries arise from the left and right coronary sinuses. The sinoatrial (SA) node is supplied by the right coronary artery in about 60 per cent of individuals and the atrioventricular (AV) node in 90 per cent. The heart is supplied by both sympathetic and parasympathetic supply. Adrenergic nerves supply muscle fibres in the atria and ventricles and the electrical conducting system. β_2-adrenoceptors predominate in vascular smooth muscle. Parasympathetic pre-ganglionic fibres and sensory fibres reach the heart through vagus nerve. Cholinergic nerves supply the AV and SA nodes vit. muscarinic (M2) receptors. The parasympathetic and sympathetic fibres may exert profound influences on control of heart rate and arrhythmia generation. Myocardial cells (myocytes) are about 100 μm long and each cell branches and interdigitates with adjacent cells. An intercalated disc permits electrical conduction and force conduction to adjacent cells. The basic unit of contraction is the sarcomere. The shortening of sarcomere results from the actin and myosin molecules. Contraction is initiated in the presence of calcium. Cardiac output is determined by the product of stroke volume and heart rate. Small pulse volume occurs in cardiac failure and vasoconstriction. Large pulse volume occurs in vasodilatation, anemia. The first heart sound results from closure of the mitral and is best heard at the apex. The second heart sound is produced by closure of the aortic and pulmonary valves and is best heard at the base of the heart. Additional sounds are produced as a result of ventricular filling. They are low pitched and occur early (third sound) or late (fourth sound) in diastole. Cardiac murmurs are associated with turbulent blood flow. When cardiac output is greatly increased by exercise, pregnancy or severe anemia, blood flows through a normal aortic valve. Diastolic murmurs are indicative of abnormal pathology. Continuous murmurs result from a combination of systolic and diastolic flow. Electrocardiography (ECG) is used to find out cardiac arrhythmias and conduction defects, and to diagnose and localise myocardial hypertrophy, ischaemia or infarction. It also gives informtion regarding electrolyte imbalance and the toxicity of certain drugs. The basic concept of electrocardiography is that the electrical activation of a heart muscle cell causes a depolarisation of its membrane. The depolarisation is propagated along the length of a cell or fibre, and transmitted to adjoining cells. The moving wavefront depolarisation passes through the heart and sets up electric currents which are detected by surface electrodes, and amplified and displayed as the electrocardiograms.

Echo cardiography (ECHO) uses reflected ultrasound to study blood flow, the structure of the heart and the movement of valves, and cardiac muscle.

Major manifestations of heart disease

A. Chest pain

1. Chest pain is a common presentation of cardiac disease, but it can also signify disease of the lungs and the gastrointestinal system.
2. Cardiac pain is centrally located in the chest on account of the derivation of the nerve supply to the heart and mediastinum. Pain is only experienced at a peripheral site in the chest.
3. This pain is dull, constricting or heavy. It may not be appreciated as pain but as discomfort, and the constricting sensation can be described as breathlessness.
4. The pain is very sudden in onset (seconds). Myocardial infarction pain takes several minutes or even longer to develop.
5. The severe pain of myocardial infarction is accompained by sweating, nausea and vomiting.
6. Angina is a choking or constricting chest pain which comes with exertion and is relieved by rest. The pain becomes worse by large meals or a cold wind. It is relieved by nitrates.
7. Myocardial infarction is more severe and does not respond to nitrates, vomiting is common.
8. Oesophageal pain is precipitated by exercise and may be relieved by nitrates.

B. Heart failure

1. Heart failure may be diagnosed whenever a patient with significant heart disease develops the symptoms of low cardiac output, pulmonary congestion or systemic venous congestion.
2. Heart failure is frequently due to coronary artery disease and tends to affect elderly subjects.
3. Heart failure may develop suddenly, as in myocardial infarction.
4. A low cardiac output causes fatigue. The peripheries are cold and the blood pressure is low, poor renal perfusion leads to uremia.
5. Chronic heart failure is sometimes associated with marked weight loss caused by impaired absorption due to gastrointestinal congestion.
6. In advanced heart failure, there occurs hypokalemia, hyponatremia, impaired liver function and thromboembolism. Patients with heart failure should be advised to avoid a high dietary salt intake, excess alcohol, salt or fluid retaining drugs.
7. Diuretics are usually the first line of treatment in this case.

Other manifestations of cardiovascular disease

A. Oedema

1. This is a symptom of chronic heart failure and is due to excessive salt and water retention.
2. It usually affects the ankles, legs, thighs and lower abdomen.
3. Oedema of nephrotic syndrome tends to be more severe and often affects the face and arms.

B. Palpitations

1. This is an abnormal subjective awareness of the heart beat.
2. Palpitation with a regular rhythm and a normal heart rate may be due to sudden vasodilatation.

C. Syncope

1. This is a loss of consciousness resulting from an inadequate blood supply to the

brain. This is due to sudden vasodilatation.

2. Exertional syncope may occur under conditions of aortic stenosis.

Ischaemic (coronary) heart disease

It is the commonest form of heart disease and the most important cause of premature death in the developed world. The following factors are responsible for coronary heart disease;

1. Age and male sex : These are the risk factors which cannot be corrected. Hormone replacement therapy reduces the disease.
2. Smoking : Tobacco is the most important cause of coronary disease. The risk is highest in young people and becomes lower within six months of quitting.
3. Hypertension : The coronary artery disease increases as blood pressure rises. Antihypertensive drugs can reduce the disease.
4. Hypercholesterolemia : Patients with hyperlipidemia have a high incidence of premature coronary disease. The exces risk is closely related to the plasma concentration of LDL : cholesterol.
5. Diabetes mellitus : Insulin resistance is associated with obesity and physical inactivity and is also a potent risk factor for coronary heart disease.
6. Hemostatic factors : High levels of fibrinogen and factor VII are associated with an increased risk of myocardial infarction (coronary thrombosis).
7. Physical activity : Regular exercise (walking, cycling or swimming) has a protective effect which has the ability to increase HDL : cholesterol, lower blood pressure, reduce blood clotting.
8. Alcohol : A moderate intake of alcohol (2 to 4 units a day) offers some protection from coronary disease. Heavy drinking is associated with hypertension and an excess of cardiac events.
9. Mental stress : Stress can produce the symptoms of heart disease.
10. Other dietary factors : Diets deficient in polyunsaturated fatty acids are associated with an increased risk of coronary disease. Decreased levels of Vitamin C, Vitamin E and other antioxidants may promote the production of oxidized LDL.

Prevention for coronary disease

1. Smoking is prohibited.
2. Regular exercise is a must.
3. Ideal body weight must be maintained.
4. Mixed diet with fruits and vegetables as well as meat and dairy products must be taken as food.
5. Fats containing polyunsaturated fatty acids must be taken.

Angina Pectoris

This is described as discomfort due to transient myocardial ischaemia and forms a clinical syndrome rather than a disease. It may occur due to an imbalance between myocardial oxygen supply and demand.

Oxygen demand	Oxygen supply
Cardiac work	Coronary blood flow
• Heart rate	• Duration of diastole
• Blood pressure	• Coronary perfusion
• Myocardial contractility.	• Coronary vasomotor tone : Oxygenation
	• Hemoglobin
	• Oxygen saturation

Clinical Symptoms

1. Stable angina is characterised by left-sided or central chest pain.
2. Most patients describe a sense of oppression or tightness in the chest — "like a band round the chest".
3. Many patients report a "choking" sensation. Breathlessness is sometimes a prominent feature.

4. The pain may radiate to the neck and is often accompanied by discomfort in the arms.
5. Symptoms become worse after a heavy meal, in the cold, in walking uphill, into a strong wind and in lying flat.
6. Care must be taken for important risk factors like nicotine stains, hypertension, hyperlipidemia, diabetes mellitus, myxoedema, contributory disease like obesity, anemia, thyrotoxicosis.

Advice to patients with angina

1. Smoking is strictly prohibited.
2. Ideal body weight must be maintained.
3. Regular exercise should be taken.
4. Severe unaccustomed exertion, vigorous exercise after a heavy meal or in very cold weather must be avoided.
5. The anti-anginal drugs, eg, nitrates, beta-adrenoceptor antagonists (beta-blockers) and calcium antagonists must be taken.

Myocardial infarction

It occurs due to the formation of thrombus at the site of rupture of an atheromatus plaque in a coronary artery. The thrombus often undergoes spontaneous lysis by the next few days.

Clinical Symptoms

1. Pain is the principal symptom; but breathlessness, vomiting and collapse or syncope are common features.
2. The pain is more severe and lasts longer. It is often descibed as a tightness, heaviness or constriction in the chest.
3. Most patients are breathless. Syncope occurs due to profound hypotension.
4. Vomiting and sinus bradycardia often occur due to vagal stimulation and are particularly common in patients with this disease.
5. Some myocardial infarcts are unrecognised. These painless myocardial infarcts are common in elderly and diabetic patients.
6. Sudden death from ventricular fibrillation may occur immediately and many deaths occur within the first hour. If the patient survives this most critical stage, the dangerous state remains, but diminishes as each hour goes by.
7. The development of cardiac failure shows the extent of myocardial damage and is the major cause of death in those who survive the first few hours of infarction.
8. There are signs of tissue damage causing fever, leucocytosis, high ESR.

Investigations

1. The ECG is usually a sensitive and specific way of confirming the diagnosis.
2. The plasma enzymes creatine kinase (CK), aspartate aminotransferase (AST) and lactate dehydrogenase (LDH) should be estimated. CK starts to rise at 4-6 hours, peaks at about 12 hours and falls to normal within 48-72 hours.

   AST starts to rise about 12 hours after infarction and reaches a peak on the first or second day, returning to normal within 3 or 4 days.

   LDH starts to rise after 12 hours, reaches a peak after 2 or 3 days and may remain elevated for a week or more. Measurements of LDH are appropriate when a patient presents several days after a possible infarct. Unfortunately, LDH is highly concentrated in red cells and abnormal results may be obtained due to very mild hemolysis.
3. The ESR becomes raised and may remain so for several days.

Early management of acute myocardial infarction

1. Bed rest is most essential.
2. Oral aspirin is needed.

3. High flow oxygen should be administered.
4. Beta adrenoceptor antagonist should be considered.
5. I.V. analgesia with opiates should be arranged.

Risk factor modification

1. Giving up smoking is the most effective decision for the future not to face coronary attack.
2. Lipid lowering therapy is ideal to reduce the risk of attack.
3. Maintaining an ideal body weight, taking regular exercise, having good control of hypertension and diabetes may improve the long-term outlook.

Prognosis

1. Death occurs within a few minutes without medical care in about one fourth of all cases of myocardial infarction.
2. Half the deaths occur within two hours and three quarters within 24 hours of the onset of symptoms of myocardial infarction.
3. The unfavourable features in this case include poor left ventricular functions and heart block.
4. Old age, stress and social isolation are the main cause for the enhancement of this disease.
5. Among the patients who survive an acute attack, more than 80 per cent live for a further year, about 75 per cent for 5 years, 50 per cent for 10 years and 25 per cent for 20 years.

VASCULAR DISEASE

Hypertension

Hypertension is defined as the levels above generally accepted 'normal', eg, 140/90 at the age of 20, 160/95 at the age of 50.

The risks associated with a particular blood pressure are dependent upon the combination of risk factors which include age (risk increases with age), gender (males > females), race (Blacks > whites), diet (high salt), smoking and coronary artery disease.

Anxiety, discomfort and unfamiliar surroundings can lead to a transient rise in blood pressure.

In most cases, a specific underlying cause of hypertension is not found. Such patients are said to have essential hypertension.

Factors influencing essential hypertension

1. Genetics and familial.
2. Socio economic (related to social deprivation).
3. Dietary factors (high salt intake, high alcohol, caffeine).
4. Hormonal factors (high renin).
5. Neurotransmitters (acetylcholine, norepinephrine, substance p, serotonin, dopamine).

Causes of secondary hypertension

1. Renal disease (parenchymal renal disease, polycystic kidney disease, renal artery stenosis).
2. Endocrine disorders (Cushing's syndrome, hyperparathyroidism, acromegaly, primary hypothyroidism).
3. Drugs (oral contraceptives containing oestrogens, corticosteroids, anti-inflammatory drugs).
4. Pregnancy (± pre-eclampsia).

Clinical symptoms

1. Headache or polyuria.
2. Angina, left ventricular failure and cerebrovascular disease.
3. Palpitations and sweating.

Complications

The adverse effects of hypertension involve the central nervous system, the retina, the heart and the kidneys.

Central nervous system

1. Stroke may result from cerebral hemorrhage.
2. Transient cerebral ischemic attacks are more common.
3. Transient disturbances of speech or vision, fits and loss of consciousness.
4. Papilledema is common.

Retina

1. Arteriolar damage.
2. "Cotton wool" exudates are associated with infarction.
3. Small white dense deposits of lipid persist for years.

Heart

1. A pressure load on the heart and left ventricular hypertrophy and ultimately left ventricular failure.
2. Higher incidence of coronary artery disease.

Kidneys

1. Damage to the renal vessels.
2. Proteinuria and progressive renal failure.

Investigations

1. Urine analysis (protein, glucose, hematuria).
2. Plasma (urea, creatinine).
3. Plasma electrolytes (sodium, potassium).
4. Plasma (cholesterol, triglycerides).
5. ECG.
6. Chest radiography.

Management

1. Weight reduction diet, reduction of heavy alcohol consumption, avoidence of excess salt consumption.
2. Regular exercise programme is encouraged.
3. Stopping of smoking.
4. Testing of cholesterol including HDL, LDL and triglyceride.

Anti-hypertensive drug therapy

1. Beta-adrenoceptor antagonists—especially relevant in the presence of angina.
2. Thiazide diuretic—for the elderly.
3. Calcium antagonists—when beta-adrenoceptor not tolerated, can be used in combination with them.
4. Aspirin—indicated to reduce cardiac and stroke complications.
5. Angiotensin—converting enzyme inhibitors—may be used as an alternative to beta-adrenoceptor antagonist or calcium antagonist.

CHAPTER 51

AIDS

The full term for the word AIDS is Acquired Immunodeficiency Syndrome. There is a relation between infection with the human immunodeficiency virus (HIV) and the development of AIDS. HIV infection is caused by the exchange of semen, vaginal secretions, milk or blood which is infected by virus. The transmission for HIV occurs through anal or vaginal intercourse. The infection with HIV is higher with anal than with vaginal intercourse. Intravenous drug abusers (IVDA) are affected by HIV infection due to the practice of "needle sharing" which allows transmission of HIV infected blood from one individual to another. Transmission of the virus is also caused by transfusion of infected blood or blood products (coagulation factors for hemophilia). The disease specially occurs among homosexuals, contaminated bisexuals, promiscuous heterosexuals, drug abusers, babies from affected mother's milk, people receiving blood transfusion. AIDS is a viral disease due to retrovirus, now called HIV.

HIV

(a) HIV belongs to the subfamily of retroviruses. Retroviruses have an RNA genome and transcribe a DNA copy of the RNA genome after penetration of the host cell. The DNA is used as a template to transcribe new RNA viral copies—thus the term "retrovirus".

(b) HIV has a core consisting of the RNA genome and core protein surrounded by an envelope with high lipid content rendering it sensitive to organic compounds.

(c) The virus gains entry to host cells by binding to the CD4 receptor using the viral surface membrane glycoprotein 120. The CD4 receptor is present abundantly on T-helper lymphocytes. After the penetration of the host cell, the viral RNA is transcribed by the viral enzyme reverse transcriptase into a DNA copy which becomes incorporated into the host cell genomic DNA. The viral DNA may then lie dormant within the cell or undergo replication resulting in transcription of RNA copies and translation to viral proteins causing new virus formation and assembly. Viruses then bud from the cell surface. New virus is then available to infect other cells and repeat the process.

(d) HIV is of two types—HIV-1 and HIV-2. HIV-1 is extremely antigenically variable and many different strains are found within a single patient. HIV-1 is responsible for disease in Western Europe, North America and Central Africa. HIV-2 causes clinical syndrome similar to HIV-1 infection and it is chiefly found in West Africa.

(e) HIV also infects cells in the central nervous system (CNS) by the migration of HIV infected monocytes to the brain.

(f) The number of CD4 cells circulating in peripheral blood is gradually decreased during the course of HIV disease.

(g) HIV principally affects CD4 helper T-lymphocytes (TH cells) and even can infect and impair the function of many other cell types, mainly cells of the monocyte, dendritic cells and B-lymphocytes.

(h) HIV infection is confirmed by detecting the presence of antibodies to HIV in the patient's serum. Such antibodies are detected by the current ELISA test which is simple and cheap and has the advantage of a very low false negative rate. All positive results are normally confirmed by the more precise Western blot test. Following infection, the detection of antibodies to HIV may not occur for three months and sometimes much longer. This indicates that serial testing may need to be performed on a patient following a high risk of exposure to HIV in order to definitely exclude infection. However, ELISA becomes the routine screening test for the "same day" testing service.

(i) The drugs for regular clinical use are reverse transcriptase inhibitors that inhibit the spread of infectious virus into uninfected cells but do not affect replication of the HIV genome once it is integrated into the host cell. The drugs in clinical use at the moment are zidovudine (AZT), didanosine (dideoxyinosine, DDI) and zalcitaline (dedeoxy- cytidine, DDC). Zidovudine is although generally recommended still it has got side-effects including nausea, vomiting, headaches, anemia, neutropenia, and occasionally leucopenia and thrombocytopenia. Macrocytosis also develops in most patients.

Clinical features

(a) HIV infection is followed by seroconversion which coincides with a brief clinical illness for about two weeks after which HIV antibodies appear in the blood. Symptoms of this illness include fever, headache, malaise, macular papular rash, tender lymphadenopathy, encephalitis. All individuals infected with HIV proceed to AIDS which is fatal.

(b) Some HIV infected patients develop persistent generalised lymphadenopathy (PGL) which expresses the presence of enlarged lymph nodes greater than 1 cm in diameter in two anatomically distinct sites for more than three months in the absence of other detectable causes of lymphadenopathy. Most of the patients are asymptomatic, although a few have fever and weight loss.

Disseminated disease in AIDS

1. Cytomegalovirus (CMV)

(a) CMV is a herpes virus, and infection is very common. 90 per cent of homosexual men are victimised.

(b) CD4 count is normally below 50/mm^3 when clinical disease appears.

(c) Many organs are involved including the eyes, CNS, liver, gut, adrenals, mouth and lung.

(d) CMV can cause colitis presenting with diarrhea, weight loss, anorexia and fever.

2. Bacterial infections in HIV

(a) Bacterial infections are very prevalent in AIDS causing tuberculosis. The diagnosis in this case is very difficult.

(b) TB may accelerate HIV disease and the prognosis is poor.

(c) There is the threat of multi-drug resistant TB.

3. Fungal infections

(a) Oral candidosis is almost common at some stage of HIV infection. Lesions in the mouth may initially respond to topical anti-fungal agents such as nystatin or amphotericin but in advanced stages fluconazole is required.

(b) When candida spreads to oesophagus, painful dysphagia is frequent and barium swallow may reveal a very ragged looking mucosal surface.

(c) The candida infection if is firmly established, anti-fungal therapy is to be continuous and liver function tests should be monitored.

(d) The initial symptoms of cryptococcal meningitis are fatigue, fever and weight loss followed by headache, nausea, vomiting. Diagnosis is made by Indian ink staining of the CSF to identify the organism. CT scan of the brain is usually normal and CSF examination may show a monocytosis and raised protein. Amphotericin is the standard therapy.

(e) Protozoal infection is very common in man. The brain is the most common site for lesions which is usually present with focal neurological symptoms, convulsions, confusion, lethargy or coma. Retina is also involved. Diagnosis is usually made by cranial CT scan which shows a characteristic ring enhancing lesion surrounded by cerebral oedema.

4. Secondary neoplasms of AIDS

(a) DNA has been identified in AIDS—associated Kaposi's sarcoma which is most common in homosexual male AIDS patients and sexually transmitted HIV infection. This suggests that a cofactor transmitted by the sexual route is required for its development. The most common site of involvement is the skin. The mouth, the hard palate, tip of nose, penis and lower legs are also favoured sites.

(b) Lymph nodes are the second most common site. Stomach or rectal involvement may present with pain, bleeding or obstruction.

(c) Treatment with doxarubicin has produced good responses.

(d) There is an increased incidence of cervical dysplasia and neoplasia in HIV infected women and anal carcinoma in HIV infected homosexual men. Women with HIV infection should have regular cervical smears. The association between these carcinomas and HIV is due to a greater incidence of infection by the Human Papilloma virus in HIV—infected patients.

(e) Skin disease is very common in HIV infected patients. The common skin diseases are seborrhoeic dermatitis, secondary syphilis, fungal infections, drug eruptions. Skin disease consists of weight loss, diarrhoea and dermatitis.

(f) Oral disease is prominent in HIV infection and maintenane of good dental and oral hygiene is very important in these patients. Careful inspection of the mouth is essential.

(g) Nausea and vomiting are common in AIDS and are the result of drug therapy, particularly in high dose of co-trimoxazole in the treatment of pneumonia. But this usually responds to metoclopramide.

(h) Pulmonary disease is very common to AIDS. Pneumocystis infection is largely confined to the lung. There is often background of fatigue, weight loss and fever and other signs of HIV infection. The respiratory rate is increased with cyanosis indicating severity of the disease. The treatment for pneumocystis pneumonia is high dose of co-trimoxazole, initially given intravenously.

(i) Meningitis may occur during the period of a seroconversion following HIV infection. The other causes of meningitis include bacteria, tuberculosis, syphilis, other fungal infections and HIV itself.

(j) A positive HIV test shows a variety of reactions ranging from anger, guilt, anxiety to depression. Patients with HIV infection may present with symptoms of organic brain syndromes.

(k) Nephropathy is usually seen in AIDS patients who are intravenous drug users or

due to heroin use than HIV infection. A large number of nephrotoxic drugs like amphotericin B, foscarnet and pentamidine are commonly used in AIDS patients.

Diagnosis

(a) More attention is applied on the suggestive features in the presence of persistent fever, continuous loss of weight, oral candidiasis, diarrhoea, cough with thick expectoration, leucopenia, anemia, lymphadenopathy and thrombocytopenia.

(b) These above cases should be selected for ELISA test for HIV infection. Positive ELISA test should be confirmed by other tests like Western blot (WB) and Immunofluorescence test. Virus can also be confirmed by electron microscopy.

Treatment

(a) The main theory today depends on the following :

1. Direct effect on the virus or its enzymes.
2. Immunodulation. It means boosting up of immune response of the patient by which the virus may be eliminated by interleukin-2 therapy and alpha and gamma interferon therapy.

(b) Several antiviral agents have also been tried without satisfactory results. No success has also been obtained with the trial of autologous bone marrow transplant, palliative radiotherapy, chemotherapy and vaccine.

(c) Drug therapy is also not effective for prevention, progression or spread of the disease.

(d) Counselling is only the most important and valuable.

(e) Different Antiviral Treatment.

	Aggressive	Conservative
When to start	Time of diagnosis	CD4 + T cell count
What to start with	AZT plus lamivudine plus protease inhibitor	Didanosine AZT plus zalcitabine AZT plus didanosine AZT plus lamivudine
When to change	10% decrease in CD4 + T cell count	CD4 + T cell count
What to change to	Two new nucleosides plus a new protease inhibitor	Two new nucleosides plus a protease inhibitor

Terminal care

(a) Good support from a partner, family and friends can be of great help during this potentially distressing time.

(b) Effective attention is given to the control of anorexia, nausea, vomiting, dry mouth and diarrhoea.

(c) Dysphagia may be a problem and many medications can be given rectally.

(d) Intramuscular injections should be avoided.

(e) Syringe drivers for subcutaneous or intravenous delivery of antiemetics and opiates can be extremely useful.

CHAPTER 52

BIOCHEMICAL AND GENETIC BASES OF DISEASE

All diseases have a biochemical basis

1. Life depends upon biochemical reactions.
2. Health fully depends on the regulated and harmonious functioning of many biochemical reactions and processes that occur in normal cells.
3. Diseases are caused by the alternations of structures; or the amounts of certain biomolecules; or disturbances of important biochemical reactions or processes.
4. Diseases are found as abnormalities in the structure and function of cells, organs, and systems generated by biochemical mechanisms.
5. Patients experience illnesses which reflect Psychologic, cultural, and other factors.
6. Physicians must treat the patients considering social, psychological, cultural, economic, and other factors, but always depending on a sound knowledge of biochemical, physiologic, and pathologic mechanisms.

Diseases are considered from a biochemical standpoint

1. **Many diseases are determined genetically.**
2. **All classes of biomolecules are affected in structure and function.**

Biomolecules can be affected in genetic diseases. The primary defect resides in DNA, and the structures or functions of other biomolecules are affected secondarily.

Table 52.1 : Examples of various biomolecules in diseases

Biomolecule	Property affected	Disease	Fundamental cause
DNA	Structure	Sickle cell anemia	Mutation
RNA	Structure	Thalassemia	Mutation leading faulty splicing of mRNA
Protein	Structure	Sickle cell anemia	Mutation
Lipid (GM_2)	Amount ↑	Tay—Sachc	Mutation roulting in defective hexosaminidase A
Glycogen	Amount ↑	Glycogen storage disease	Mutation in gene for phosphorylase

3. **Biochemical changes that cause disease may occur rapidly or slowly.**
   (a) Some diseases progress rapidly causing death within minutes after coronary thrombosis on the fact that brain and heart particularly are very sensitive to lack of oxygen and fuel.

(b) Life is threatened within hours at the excessive loss of water and electrolytes in cholera.

(c) Rapid and large changes of distribution in the body of certain electrolytes (eg, K^+) become hazardous very quickly because of the sensitivity of myocardial muscles to such changes.

(d) Extensive change of pH can be tolerated for a short time. Several years may be required for the build-up of a biomolecule to affect organ function.

4. **Diseases are caused by deficiency or excess of certain biomolecules.**

(a) Deficiency of Vitamin A causes night blindness; whereas excessive intake of Vitamin A results in chronic states of toxicity.

(b) Deficiency of Vitamin D results in rickets, but excess causes severe hypercalcemia.

(c) The general causes of nutritional secondary deficiency are the insufficient absorption, increased requirement, inadequate utilization, and increased excretion. Each of these causes occur by a number of diseases or conditions.

5. **Almost every cell organelle is involved in the genesis of various diseases.**

Table 52.2 : Involvements of the major intracellular organelles in various diseases

Organelles	Diseases	Mechanism
Nucleus	Most genetic diseases	Mutations in DNA.
Mitochondrion	Several diseases including mitochondrial myopathies.	Mutations in mitochondrial DNA affecting the structures of proteins.
Endoplasmic reticulum.	Chemical toxicities, e.g., following intake of CCl_4.	Enzymes in the endoplasmic reticulum such as cytochrome P 450 activate various chemicals to potentially toxic species.
Plasma membrane.	Metastasis of cancer cells.	Changes in the oligosaccharides of glycoproteins in this organelle are important in permitting metastasis.
Lysosome	Lysosomal storage diseases	Decreased activities of the various hydrolases present in lysosomes
Peroxisomes	Zellweger syndrome and others.	Decreased biogenesis of peroxisomes and decreased activity of certain peroxisomal enzymes, such as dihydroxyacetone phosphate acyltransferase

6. **Different biochemical mechanisms can produce similar Pathologic, clinical, and laboratory findings.**

(a) Many different bacteria and viruses can cause acute or chronic inflammation.

(b) Hepatomegaly occurs from accumulation of glucosylceramide, but more commonly it is due to heart failure or metastases.

(c) Fibrosis of the liver (cirrhosis) is caused by chronic intake of ethanol, excess of copper causes Wilson disease, and excess of iron causes primary hemochromatosis.

(d) A variety of inborn errors of metabolism can lead to mental retardation, and many conditions can result in ketosis.

(e) Calcium oxalate, magnesium ammonium phosphate, uric acid, and cystine—all form renal calculi (stones), but they accumulate for different biochemical reasons.

(f) Distinct biochemical causes can produce the same pathologic finding (eg, cirrhosis), clinical finding (eg, mental retardation), or laboratory finding (eg, ketosis).

(g) It is possible to distinguish among diseases by the history, physical examination, and suitable laboratory tests.

Table 52.3 : The major Pathologic Processes that occur in the human body

Pathological Process	Main Cause	Example of disease	Example of biomolecules involved
Chronic inflammation	Infections, bacterial or viral	Pneumonia	Mediators of inflammation
Degenerations	Various chemicals	Fatty liver	Ethanol
Enlargement of Liver	Accumulation of a compound	Gaucher's disease	Glucosylceramide
Atrophy (Decrease in size) of an organ	Diminished blood supply	Atrophy of a kidney	Decrease of various nutrients supplied by the blood
Anemia	Lack of mineral	Iron deficiency anemia	Iron
Fibrosis	Often follows cell death	Cirrhosis	Accumulation of collagen
Cell death	Diminished blood supply	Myocardial infarction	Lack of oxygen

The major classes of genetic disease

1. Genetic diseases may be arranged into three major classes : chromosome disorders, monogenic disorders, and multifactorial disorders. Two additional classes are somatic disorders and mitochondrial disorders.
2. In chromosomal disorders, there is an excess or loss of chromosomes. The best-known such condition is Down syndrome. Chromosomal translocations are important in activating oncogenes.
3. The monogenic disorders involve single mutant genes. They are classified as autosomal dominant, autosomal recessive, or X-linked. The term "dominant" signifies that the mutation will be clinically evident if only one chromosome is affected (heterozygous); the term "recessive" denotes that both the chromoses must be affected (homozygous). "X-linked" disorders are those in which the mutation is present on the X-chromosome. Females have two X-chromosomes, they may be either heterozygous or homozygous for the affected gene. So, X-linked inheritance in females can be dominant or recessive. Males have only one X-chromosome, so that they will be affected if they inherit the mutant gene.
4. Multifactorial disorders are associated with a number of genes and environmental factors. Less is known about this category of diseases, but it is gaining importance because of the common adult diseases such as ischemic heart disease and hypertension that are members of this group.
5. Mutations are highly involved in genetic disorders. Mutations of the mitochondrial genome have been demonstrated in a number of other neurologic diseases.

Genetic diseases produce their Pathologic Consequences

1. If an *enzyme* (protein) is affected by mutation, an *inborn error of metabolism* may result. This produces a metabolic block which may have pathologic consequences as explained in the next page.

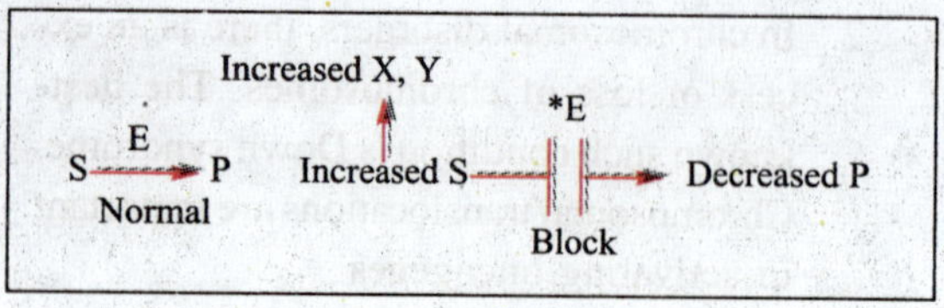

where *E = mutant enzyme, and X, Y = alternative products of the metabolism of S. A block has got three results : (1) decreased formation of the product P, (2) accumulation of the substrate S, and (3) increased formation of metabolites (X, Y) of the substrate S, resulting from its accumulation. Any one of these three results may have pathologic effects. In the case of *phenylKetonuria*, the mutant enzyme is phenylalanine hydroxylase causing the following situation :

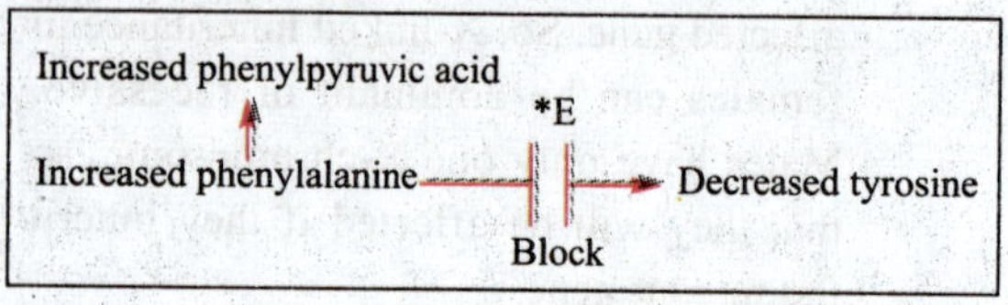

Therefore, patients with PKU have less tyrosine which is the main source for melanin formation. Hence, they are fair skinned. The plasma high levels of phenylalanine produces the inhibitory effects on the transport of other amino acids into brain. So the accumulation of substrate leads to pathologic effect.

2. Some inborn errors of metabolism are essentially harmless. These are usually due to blocks in peripheral areas of metabolism, where diminished formation of product or accumulation of its precursor perturbs the cell (eg, pentosuria).
3. A mutant protein is synthesized if a structural gene for a non-catalytic protein is affected. Even a change of one amino acid can have serious pathologic consequences. The mutant protein does not function properly or move very slowly through the cell.

Early diagnosis of certain inborn errors is the most essential to avoid permanent damage

In some cases, treatment must be started immediately for saving the infant from permanent damage (eg, phenylketonuria, galactosemia). A number of cues are mentioned in Table 52.4.

Table 52.4 : Cues to the diagnosis of an inborn error of metabolism

- Family history of a genetic disease.
- Positive screening test (eg, for PKU).
- History of poor physical and mental development and of failure to thrive.
- Physical findings such as enlarged organs.
- Unusual odour from breath or urine (eg, maple syrup urine disease).
- Low blood sugar (eg, Glycogen storage disease).
- Low blood pH (eg, due to accumulation of organic acids in maple syrup urine disease).

The sources of material that can be analyzed and the main tests used in investigating patients suspected of having genetic diseases are mentioned in table 52.5.

Table 52.5 : Major tests used in the diagnosis of genetic diseases

Material :	Plasma, red cells, white cells, urine, organ biopsy, amniotic cells.
General tests :	Blood and urine glucose, blood pH, blood ammonia, amino acids in plasma and urine, Organic acids in urine.
Specific tests :	Electrophoresis (eg, for HbS); measurement of low amount of product or elevated amount of precursor (eg, phenylalanine in PKU); measurement of activity of candidate enzyme in red cells, white cells, or tissue biopsy; analysis by Southern blotting of restriction fragment length polymorphisms (RFLPs) and other features of DNA structure linked to or causing specific diseases (eg, HbS, Huntington disease, Duchenne muscular dystrophy).

Successful treatment is available for some genetic diseases

1. One of the following points is taken into accounts for the treatment of genetic disease :
   (a) Attempts to correct the metabolic consequences of a disease by administration of the missing product or by arranging the availability of substrate.
   (b) Attempts to replace the absent enzyme or protein or to increase its activity.
   (c) Attempts to remove excess of a stored compound.
   (d) Attempts to correct the basic genetic abnormality by gene therapy.
2. Some of these plannings are undoubtedly effective for certain disorders, eg, the dietary treatments of phenylketonuria and galactosemia, replacement therapy for hemophilia and agammaglobulinemia, and removal of iron by periodic bleeding in hemochromatosis. But attempts at enzyme therapy have limited success.
3. Enzyme therapy is difficult in the case of the brain, because administered enzymes must be made to cross the blood-brain barrier.
4. Because of the tremendous progress in the field of recombinant DNA technology, sufficient efforts are being devoted to **gene therapy**.

Table 52.6 : Four major classes of treatment for genetic diseases

Class of treatment	Principle	Disease	Treatment or Comment
(a)	Replace missing product	Familial goiter	Administration of levothyroxine
	Limit substrate	Phenylketonuria	Diet low in Phenylalanine
(b)	Replace mutant enzyme	Gaucher disease	Injections of β-glucosidase
	Replace missing protein	Hemophilia	Injections of factor VIII (AHG)
(c)	Increase activity of mutant enzyme by supplying large amounts of cofactor.	Methylmalonic aciduria	Injections of B_{12}
	Increase activity of mutant enzyme by induction	Crigler-Najjar Syndrome	Administration of Phenobarbital
(d)	Replace diseased organ carrying defective gene by normal organ	Galactosemia	Liver transplantation
	Introduce Protein into somatic cells by gene therapy	Many possible candidates	Clinical trials already conducted on individuals with adenosine deaminase deficiency, cystic fibrosis, familial hypercholesterolemia, etc.

Gene therapy in clinical trials is feasible, but its efficacy must be greatly improved

1. Firstly, the major criteria should be satisfied to permit the use of gene therapy in humans (shown in table 52.7). Somatic gene therapy is only permissible in humans at present, because germ cell gene therapy transmits genetic alternations to offspring.

2. Gene therapy can involve gene replacement, correction, or augmentation. In *replacement*, the mutant gene should be removed and replaced with a normal gene. In *correction*, only the mutated area of the affected gene should be corrected and the remainder remains unchanged. In augmentation there is the introduction of foreign genetic material into a cell to compensate for the defective product of the mutant gene and is the sole type of gene therapy available at present.
3. The foreign material can be introduced into affected cells by any of the methods mentioned in table 52.8 for gene therapy in humans, the gene of interest is usually administered via a viral vector or via plasmid liposome complexes. The cells which take up the gene will contain both the mutant and the exogenously derived gene.
4. Various methods are being developed to target genes to specific sites in order to increase uptake and thus efficacy. Target cells have included bone marrow cells, fibroblasts, epithelial cells of the respiratory tract, hepatocytes and various tumor cells.
5. The three major routes by which genes have been introduced into humans are via retroviruses, adenoviruses, and plasmid—liposome complexes. In the case of retroviruses, the target cells must be actively growing, since cell division is required if the gene is to be integrated into the genome. For adenoviruses, their genome does not integrate with the host cell genome.
6. Attempts are created to develop *in utero gene therapy* for disorders such as α-thalassemia and adenosine deaminase deficiency. This can be accomplished by mixing fetal blood cells with a suitable retroviral vector carrying the gene requiring correction and retransfusing the cells back into the fetus.

Table 52.7 : Some criteria to be satisfied prior to initiation of gene therapy
• The prognosis of the disease should be accurately predictable and should be adversed.
• The gene should be isolated and its regulatory regions defined.
• The target cells should be identified and safe methods of introducing the gene into them should be available.
• Evidence should be available that the gene functions adequately and produces no deleterious effects.

Table 52.8 : Some methods of introducing genes into cells for gene therapy
• Intranuclear injection.
• Transfection (eg, with calcium phosphate).
• Electroporation.
• Retroviruses (Specially constructed).
• Adenovirus (used for introduction of the cystic fibrosis gene into cells of the respiratory tract).
• Plasmid—liposome complexes.
• Site—directed recombination.

Antisense or Triplex therapy is useful in the treatment of certain diseases

1. A triplex molecule is formed by the bending of oligonucleotides (about 20 bases long) through base-pairing to a specific mRNA or to double-stranded DNA.
2. Specific oligonucleotides is most essential to be synthesized. Such oligonucleotides can act as drugs (genetic medicine) inhibiting the synthesis of the protein products of specific genes involved in various diseases or inhibiting the genes themselves. One such oligonucleotide blocks the replication of HIV-I virus by combining with its gag gene and can thus prove useful in the treatment of AIDS. Other compounds are being investigated as treatments for autoimmune disorders or some type of cancer.
3. Some problems have been noted in the development of clinically useful antisense

compounds. Many oligonucleotides do not enter cells easily, and they may also be degraded inside cells by nucleuses. Therefore, chemists have synthesized modified oligonucleotides in which a critical oxygen atom in each nucleotide is replaced by a sulphur atom. They have also an undesirable capacity to bind to various proteins producing unwanted side effects. In certain cases, control oligonucleotides have been found to produce effects similar to those observed with antisense compounds. It is hoped that antisense molecules will prove of therapeutic use in a variety of diseases.

Molecular medicine is advancing rapidly

Some of the major advances (shown in table 52.9) have been made in the area of molecular medicine; all have been dependent upon the development of recombinant DNA technology.

Table 52.9 : Some recent major achievements in or relating to molecular medicine
• Increased knowledge of homeobox genes and other genes relating to embryonic development and differentiation.
• Development of gene and antisense therapy.
• Production of biotechnology-derived products based on recombinant DNA technology.
• Development of various DNA-based methods for diagnosis and screening of disease.
• Elucidation of the molecular bases of some major Mendelian disorders (eg, cystic fibrosis, Duchenne muscular dystrophy).
• Isolation of oncogenes, tumor suppressor genes, and cancer susceptibility genes and proposal of specific model for the development of colorectal cancer.
• Discovery that mutations in mitochondrial genes can cause diseases.

SCHEMATIC COMPENDIUM

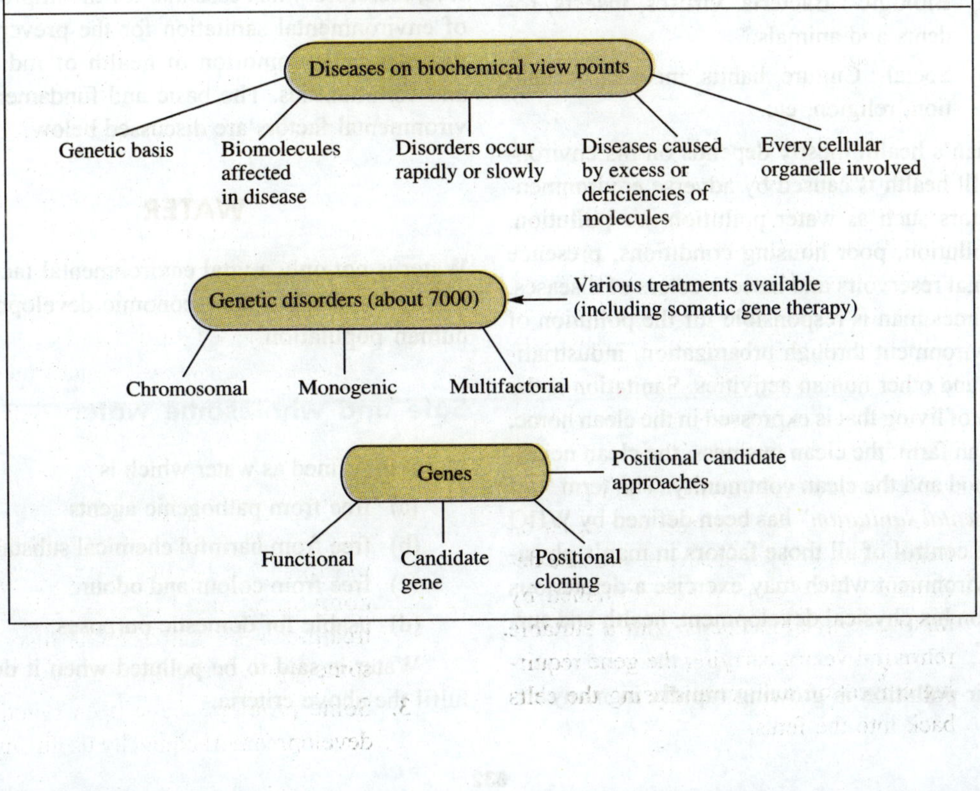

CHAPTER 53

ENVIRONMENT AND POLLUTION

Environment includes all the external factors (living and non-living, material and non-material) which surround man. According to modern concept, it includes not only the water, air and soil but also the social and economic conditions under which we live.

Environment has been divided into three components.

(i) Physical : Water, air, soil, housing, wastes, radiation, etc.

(ii) Biologic : Bacteria, viruses, insects, rodents and animals.

(iii) Social : Culture, habits, income, occupation, religion, etc.

Man's health mostly depends on his environment. Ill health is caused by adverse environmental factors such as water pollution, air pollution, soil pollution, poor housing conditions, presence of animal reservoirs and insect vectors of diseases. Sometimes man is responsible for the pollution of his environment through urbanization, industrialization and other human activities. Sanitation is the quality of living that is expressed in the clean home, the clean farm, the clean business, the clean neighbourhood and the clean community. The term "*environmental sanitation*" has been defined by WHO as "the control of all those factors in man's physical environment which may exercise a deleterious effect on his physical development, health and survival."

Air pollution is growing rapidly in many urban centres. Industrial growth has created environmental pollution by industrial wastes. Radio-active pollution of the environment has been formed by the development of nuclear technology.

The poor environmental sanitation, i.e., unsafe water, polluted soil, unhygienic disposal of human excreta, poor housing, insects and rodents is responsible for most of the ill health in India. The defective environmental sanitation causes the high death rate, infant mortality rate, sickness rate, etc. It is, therefore, most essential for the improvement of environmental sanitation for the prevention of diseases and promotion of health of individuals and communities. The basic and fundamental environmental factors are discussed below.

WATER

Water is not only a vital environmental factor, but it plays a role in socio-economic development of human population.

Safe and wholesome water

It is defined as water which is

(a) free from pathogenic agents.

(b) free from harmful chemical substances.

(c) free from colour and odour.

(d) usable for domestic purposes.

Water is said to be polluted when it does not fulfil the above criteria.

Uses of water

1. Domestic use : Water is required for drinking, cooking, washing, bathing, flushing of toilets, gardening, etc.
2. Public purposes : Water is used for cleaning streets, swimming pools, public fountains, fire protection and public parks.
3. Industrial purposes : Water is used for processing and cooling.
4. Agricultural purposes : It is used for irrigation.
5. Power purposes : It is used for power production from hydropower and steam power.
6. Waste removal purpose : It is also used for the removal of waste from all manner of establishments and institutions.

Sources of water supply

Three main sources of water are :

1. Rain.
2. Surface water : Rivers and streams, impounding reservoirs, tanks, ponds and lakes.
3. Ground water : Shallow wells, deep wells, springs.

1. Rain

(a) A part of the rain water sinks into the ground, some forms streams and rivers which flow ultimately into the sea.

(b) Some of the water in the soil is taken up by the plants and is evaporated by the leaves.

(c) Physically, it is clear, bright and sparkling.

(d) Chemically, it is very soft water containing traces of dissolved solids.

(e) Bacteriologically, it is free from pathogenic agents.

(f) It becomes impure as it picks up suspended impurities from the atmosphere such as dust, micro organisms and gases such as carbon dioxide, nitrogen, oxygen and ammonia.

(g) Gaseous sulphur and nitrogen oxides are eliminated from power plants. These gases react with atmospheric water forming dilute solutions of sulphuric acid and nitric acid. The acid rain causes serious affects on plants and on surface water quality.

2. Surface water

It originates from rain water. It is the main source of water supply in many areas. It is contaminated from human and animal sources and hence it is never safe for human consumption unless purified before use

(i) Impounding reservoirs

(a) These are artificial lakes constructed of earthwork or masonry in which large quantities of surface water is stored.

(b) Dams and mountain streams also provide large reserves of surface water.

(c) Algae and other microscopic organisms are grown in storing water in reservoirs imparting bad odours and tastes to water.

(d) These reservoirs are having good quality of water. The water is usually clear and palatable if the surrounding hills are covered with peat. The water is soft and free of pathogenic organisms.

(ii) Rivers

(a) The river water is on the whole polluted and is quite unfit for drinking without treatment.

(b) River water contains dissolved and suspended impurities of all kinds. The bacterial count in it is very high.

(c) The impurities of river water are derived from sewage, industrial and trade wastes, and drainage from agricultural areas. The water is polluted by bathing, animal washing and disposal of the dead.

(d) Self-purification of river water accurs by dilution, sedimentation, aeration, oxidation, sunlight. But this is not sufficient to render the water potable. The water needs purification before drinking purposes.

(iii) Tanks

(a) Surface water is stored in tanks and tanks are having contaminated water of all sorts.

(b) Tank water contains colloidal matters immediately after the rains. The older tanks are full of aquatic vegetation.

(c) Tanks are used for washing of clothes, cattle, humans, cooking pots and are also used for swimming purposes. Hence, the water becomes fully contaminated and the water cannot be drunk without boiling.

(d) Sea water contains 3.5 percent of salts in solution, 19,000 mg/litre of chloride, 10,600 mg/litre of sodium, and 1270 mg/litre of magnesium. Heavy expenditure is involved in desalting and demineralization.

3. Ground water

Ground water is superior to surface water, because it has an effective filtering medium. It is mostly free from pathogenic agents and does not require any treatment. It is less contaminated than surface water. It contains salts of calcium and magnesium which make the water hard. The ground water sources are wells and springs. Wells are classified into shallow and deep wells, dug and tube wells.

(i) Wells

(a) Shallow wells produce limited quantities of water and the water is liable to pollution unless well constructed.

(b) Deep wells are usually machine-dug and are several hundred meters deep. These wells are satisfactory sources of water supply.

(c) Shallow wells are polluted from neighbouring sources of contamination such as latrines, urinals, drains and collections of manure.

(d) According to construction, wells may be classified into dug wells and tube wells. Dug wells are of two types—*katcha* well and *pucca* well. In these wells, there is personal contact between the user and the water. Some people may even wash their faces, hands and feet which is a common Indian custom.

(e) Tube wells are beneficial as a source of drinking water in many parts of India. They produce water which is bacteriologically safe and also cheap.

(ii) Springs

(a) Ground water when comes to the surface and flows freely under natural pressure is called a "spring".

(b) Springs are of two types—Shallow springs and deep springs. Shallow springs dry up quickly during summer months, whereas deep springs do not show seasonal effect in the flow of water.

(c) Springs are exposed to contamination. Well protective structures are essential to safeguard water supply.

WATER POLLUTION

1. Pure water does not occur in nature. It contains natural and man-made impurities.
2. The natural impurities are dissolved gases (carbon dioxide, nitrogen, hydrogen sulphide, etc.), suspended impurities (clay, sand, mud, etc.), microscopic organisms and dissolved minerals (salts of calcium, magnesium, sodium etc.).
3. Water pollution is seriously caused by urbanization and industrialization. Sewage contains decomposable organic matter and pathogenic agents. Industrial wastes contain toxic agents. Agricultural pollutants contain fertilizers and pesticides. Physical pollutants contain heat and radioactive substances.

4. Even after treatment properly, water pollution still occurs due to corrosion of pipelines, leaky joints, and cross-connections between water supply pipes and sewage drainage pipes.

Water-related diseases

A. Biological (water-borne diseases)

1. Those caused by the presence of an infective agent :
   (a) Protozoal : amoebiasis, giardiasis.
   (b) Helminthic : roundworm, threadworm.
   (c) Viral : viral hepatitis A, hepatitis E, poliomyelitis.
   (d) Bacterial : typhoid and paratyphoid fever, bacillary dysentery, cholera, E. Coli diarrhoea.
   (e) Leptospiral : Weil's disease.
2. Those due to the presence of an adequate host :
   (a) Snail : Schistosomiasis.
   (b) Cyclops : Fish tape warm, Guinea worm.

B. Chemical

1. Chemical pollutants include detergent solvents, cyanides, minerals, heavy metals, organic acids, bleaching agents, dyes, pigments, sulphides, ammonia, and toxic substances.
2. Chemical pollutants may affect man's health by accumulating fish used as human food.
3. Some new pollutants are not easily removed by water treatment or purification processes.
4. High levels of fluoride cause mottling of the dental enamel.
5. Methemoglobinemia may occur when surface water from farmland, treated with a fertilizer, gain access to the water supply.
6. Hardness of water causes beneficial effect against cardiovascular disease.
7. Some diseases are transmitted due to inadequate use of water like trachoma, conjunctivitis, ascariasis.
8. Some diseases are related to the disease carrying insects breeding in or near water like malaria, filaria, etc.

Water Pollution Law

In India, water pollution is a serious problem. In 1947, parliament passed the Water (prevention and control of pollution) Act to protect water from being contaminated. The Act seeks to provide legal action against the spread of water pollution.

Purification of water

Purification of water is done :

1. On a large scale.
2. On a small scale.

1. Purification of water on a large scale : Ground water (wells and springs) does require no treatment except disinfection. Surface water (river water) requires extensive treatment since it is turbid and polluted. Purification system depends on one or more of the following methods.
   (i) Storage.
   (ii) Filtration.
   (iii) Disinfection.

(i) Storage

(a) As a result of storage, 90 per cent of the suspended impurities settle down in 24 hours by gravity.

(b) The aerobic bacteria oxidize the organic matter present in the water by the help of dissolved oxygen. The content of free ammonia is reduced.

(c) The pathogenic organisms gradually die out. The total bacterial count drops.

(d) If the water is stored for long periods, there is development of vegetable growths such as algae which cause a bad smell.

(ii) Filtration

(a) 99 per cent of the bacteria are removed by filtration in addition to other impurities.

(b) There are two types of filters—the "biological" or "slow sand" filters and the "mechanical" or "rapid sand" filters.

(c) The cost of construction of slow sand filters is cheaper than that of rapid sand filters. The physical, chemical and bacteriological quality of filtered water is very high.

(d) Rapid sand filters are of two types—the gravity vpe (Paterson's filter) and the pressure type (Candy's filter). The filter beds occupy less space. The filtration is rapid and the washing of the filter is easy.

Comparison of rapid and slow sand filters

	Rapid sand filter	Slow sand filter
1. Space	Occupies very little space	Occupies large area.
2. Preliminary treatment	Chemical coagulation and sedimentation	Plain sedimentation
3. Operation :	Highly skilled	Less skilled
4. Washing	By back-washing	By scrapping the sand bed
5. Removal of turbidity	Good	Good
6. Removal of colour	Good	Fair
7. Removal of bacteria	98 – 99 percent	99.9 – 99.99 per cent
8. Effective size of sand	0.4 – 0.7 mm	0.2 – 0.3 mm

(iii) Disinfection

(a) Disinfection is capable of destroying the pathogenic organisms present.

(b) Chlorine kills pathogenic bacteria but it has no effect on spores and certain viruses except in high doses. It oxidizes iron, manganese and hydrogen sulphide. It destroys some taste and odour producing constituents. It controls algae and slime organisms.

(c) Ozone eliminates undesirable odour, taste and colour, and removes all chlorine from the water. It has a strong virucidal effect. After the work, ozone decomposes and disappears and it does not have germicidal effect. Therefore, ozone is used as a pretreatment of water to destroy not only viruses and bacteria but also organic compounds.

AIR

1. All forms of life fully depend on air. Human beings need a continuous supply of air to exist on earth. Disease agents are conveyed by air.
2. Air is a mixture of gases like nitrogen (78.1 per cent), oxygen (20.93 per cent), carbondioxide (0.03 per cent) and other gas in traces, eg, argon, neon, krypton, xenon, and helium. Air also contains water vapour, traces of ammonia, dust, bacteria, spores and vegetable debris.
3. Air becomes impure by respiration of men and animals, combustion of coal, gas, oil, etc; decomposition of organic matter; trade, traffic and manufacturing processes.
4. Certain self-cleansing mechanisms may operate in nature to remove the impurities of air, eg. Wind sweeps away the impurities by its movement, sunlight oxidizes impurities and kills bacteria, rain removes the suspended and gaseous impurities, the green plants utilize the carbon dioxide and generate oxygen and this becomes reverse during the night time.
5. Discomfort is experienced in ill-ventilated and crowded rooms due to increased carbon dioxide and decreased oxygen resulting from respiration. The causes of discomfort are not due to chemical changes but physical changes.

AIR POLLUTION

"Air pollution" signifies the presence of gases, mixtures of gases and particulated matter in the surrounding atmosphere. Air pollution is one of the present-day health problems throughout the world.

Sources of air pollution

1. Automobiles : Motor vehicles emit hydrocarbons, Carbon monoxide, lead, nitrogen oxides and particulate matter. Certain of these hydrocarbons and oxides of nitrogen are converted into "photochemical" pollutants of oxidizing nature by the strong sunlight.
2. Industries : Combustion of fuel to generate heat and power produces smoke, sulphur dioxide, nitrogen oxides and fly ash. Petrochemical industries generate hydrogen fluoride, hydrochloric acid and organic halides. Many industries discharge carbon monoxide, carbon dioxide, ozone, hydrogen sulphide and sulphur dioxide.
3. Domestic sources : Domestic combustion of coal, wood or oil produces smoke dust, sulphur dioxide and nitrogen oxides.
4. Tobacco smoke : It is the most direct source of air pollution affecting the health of many people.
5. Miscellaneous : Pesticide spraying, fungi, molds and bacteria contribute to air pollution.

Air pollutants

1. Carbon monoxide : It is a product of incomplete combustion of carbon containing materials. Concentrations in urban areas depend on weather and traffic density.
2. Sulfur dioxide : It results from the combustion of sulphur containing fossil fuel, the smelting of sulfur containing ores, and other industrial processes. Domestic fires can also produce it.
3. Lead : The mining and smelting of lead ores create pollution problem. Children up to 6 years of age are a population at increased risk for lead exposures. The neurological effects of lead occur at lower threshold in children than in adults.
4. Carbon dioxide : This is not commonly regarded as air pollutant. It is a natural constituent of air. It does not take part in any significant chemical reactions with other substances in the air. However, its global concentration is rising above the natural level by an amount that can increase global temperature enough to affect climate markedly.
5. Hydrocarbons : Hydrocarbons formed by combustion of coal, wood, processing and use of petroleum take part in the chemical reactions to exert their pollutant action.
6. Ozone : It is the strongest oxidizing agent. It is formed by chemical reactions that occur in the air. It acts as the pollutant in the air.
7. Cadmium : Tobacco contains cadmium, and smoking may contribute significantly to the uptake of cadmium.
8. Polynuclear aromatic hydrocarbons (PAH) : These are formed by the incomplete combustion of organic materials. Its role is to induce lung cancer.

Indoor air pollution

It is the most environmental problem. It contributes to acute respiratory infections in young children, chronic lung disease and cancer in adults, and adverse pregnancy outcome. Pneumonia is the chief killer of young children.

Sources of indoor air pollutants

Pollutant	Sources
Respirable particles	Tobacco smoke, stove
Carbon monoxide	Combustion equipment, stove, gas heaters
Nitrogen dioxide	Gas cookers, Cigarettes
Sulfur dioxide	Coal combustion
Carbon dioxide	Combustion, respiration
Ozone	UV light sources

Sources of indoor air pollutants	
Pollutant	**Sources**
Formaldehyde	Carpet adhesives, insulation
Asbestos	Insulation, fire proofing
Mineral fibres	Appliances
Radon	Building material

Effects of air pollution

1. Health aspects

(a) The immediate effects of air pollution are on the respiratory systems and the resulting state of it is acute bronchitis. If the air pollution is intense, it may result in immediate death.

(b) Air pollution damages the cardiorespiratory system in various ways.

(c) Lead poisoning is dangerous to children developing brain and nervous system.

(d) Elevated lead levels in children are associated with neuropsychologic development as measured by loss of IQ, poor school performances and behavioural difficulties.

Major air pollutants, their sources and adverse effects on health		
Noxious agents	**Sources**	**Adverse effects**
Oxides of nitrogen	Automobile exhaust, gas stoves and heaters	Respiratory tract irritation, impaired lung defences, bronchial hyperactivity.
Hydrocarbons	Automobile exhaust, cigarette smoking	Lung cancer.
Ozone	Automobile exhaust, high altitude aircraft cabins	Cough, substernal discomfort, respiratory tract irritation.
Sulphur dioxide	Power plants, oil refineries, kerosene space heaters	Respiratory tract irritation, death may occur in severe exposure.
Lead	Automobile exhaust	Impaired neuropsychological development in children.

2. Social and economic aspects

(a) Air pollution causes destruction of plant and animal life, corrosion of metals, damage of buildings, cost of cleaning and maintenance and repairs.

(b) It also reduces visibility in towns and it can soil and damage clothings.

Prevention and control of air pollution

(a) Containment is the prevention of escape of toxic substances into the ambient air. It is done by the engineering methods such as enclosure, ventilation and air cleaning. "Arrestors" are the development for the removal of contaminants.

(b) Smoke reduction is done by the increased use of electricity, natural gas and central heating in place of coal. In Delhi and Bombay deleaded petrol is being used.

(c) Some air pollutants are readily removed by vegetation. The establishment of "Green belts" between industrial and residential areas is an attempt at dilution. The capacity for dilution is limited and trouble occurs when the atmosphere is overburdened with pollutants.

(d) Air pollution is controlled in many countries by suitable legislation, eg, clean Air Acts. The Govt. of India have enacted "The Air Act" in 1981 to decrease the nuisance of air pollution.

(e) The WHO has established an international network of laboratories for the monitoring and study of air pollution. The international centres for this network issue warnings of air pollution where and when necessary.

Disinfection of air

1. Mechanical ventilation reduces vitiated air and bacterial density.
2. Ultraviolet radiation is effective in operation theatres and infectious disease wards.

Ultraviolet rays are also effective for general use in public assembly and school rooms.

3. Dust control is made by the application of oil to floors of hospital wards. This oil reduces the bacterial content of the air.

NOISE

Definition

Noise is "wrong sound in the wrong place at the wrong time".

Noise pollution

It signifies the vast sounds that are being produced in the modern life leading to health hazards.

Sources

1. Noise levels are acute near railway junctions, bus terminuses, airports and traffic round-abouts.
2. The other sources are pressure horns, loudspeakers with full volume during festivities particularly at night.
3. The domestic noises in daily life generates from radios, tape recorders, T.V. sets.

Effects

1. Noise has side effects such as whistling and buzzing in the ears.
2. The pathological effect is deafness or hearing loss. Repeated or continuous exposure to noise around 100 decibels may result in a permanent hearing loss. When the complete destruction of the organs of corti occurs as a result of occupation in industries, it is called "occupational hearing loss". Noise above 160 decibels may rupture the tympanic membrane and cause permanent loss of hearing.
3. Neurotic people are more sensitive to noise than balanced people.
4. Workmen under high intensities of noise become irritated, short tempered and impatient. They disrupt production and become more agitated.
5. Reduction in noise is found to increase work output.
6. The direct result of noise shows a rise in blood pressure, an increase in heart rate and breathing and an increase in sweating. The symptoms such as giddiness, nausea and fatigue may also occur.
7. Noise causes a narrowing of pupils, affects colour perception and reduces night vision.

Control of Noise

1. **Careful planning of cities :** Residential areas should be separated from the main streets by means of wide green bolts. House fronts should lie not less than 15 meters from the road. The main streets should be widened to reduce the level of noise penetration into dwellings.
2. **Control of vehicles :** Heavy vehicles should not be routed into narrow streets. Vehicular traffic on residential streets should be reduced. The use of pressure horn should be prohibited.
3. **Improved insulation of building :** It is wise to construct detached buildings rather than a single large building. Buildings should be sound-proof.
4. **Industries and railways :** Noise should be controlled in industries. Special areas must be earmarked outside residential areas for industries and for railways.
5. **Legislation :** Many states have adopted legislation for controlling noise. Workers have the right to claim compensation if they have suffered a loss of ability to understand speech.
6. **Education :** Education is needed to highlight the importance of noise as a community hazard.

EXCRETA DISPOSAL

Public health importance

1. Human excreta is an important cause of environmental pollution.
2. Improper excreta disposal causes soil pollution, water pollution, contamination of foods and propagation of flies. The resulting diseases are typhoid and paratyphoid fever, dysenteries, diarrheas, cholera, hookworm disease, viral hepatitis and similar other intestinal infections.
3. Proper disposal of human excreta is a fundamental environmental health service.

Carrying of disease from excreta

The fecal-borne diseases are transmitted to a new host through various channels such as water, fingers, flies, soil and food.

Methods of excreta disposal

Some methods are applicable to unsewered areas and some to sewered areas.

1. Unsewered areas

(i) Service type : Night soil is collected from pail by human agency and later disposed of by burying or composting.

(ii) Non-service type : Bore hole latrine, Dugwell latrine, water-seal type of latrine, septic tank, aqua privy.

(iii) Latrines for camps : Shallow trench latrine, deep trench latrine, pit latrine, bore hole latrine.

2. Sewered areas

Sewage treatment

(a) Primary treatment : Screening, removal of grit, plain sedimentation.

(b) Secondary treatment : Trickling filters, activated sludge process.

(c) Other methods : Sea outfall, river outfall, sewage farming, oxidation ponds.

1. Service type

(a) The collection and removal of night soil from bucket by human agency is called the service type, and the latrines are called service latrines.

(b) Service latrines are a source of insanitation. They have the drawbacks and faults which tend to carry the disease cycle of fecal-borne diseases in the community.

(c) The night soil is exposed to flies and there is always the possibility of water and soil pollution.

(d) The emptying occupation of buckets is not always satisfactory.

(e) It is also difficult to recruit adequate staff for the collection of night soil. If the sweepers go on strike, the entire machinery collapses with bad results to public health.

(f) The human labour for the collection of night soil is unfavourable with human dignity.

(g) The environmental hygiene committee has, therefore, recommended that the service latrine should be replaced by sanitary latrine.

2. Sanitary latrines

A Sanitary latrine fulfils the following criteria :

(a) Excreta should not contaminate the ground or surface water.

(b) Excreta should not pollute the soil.

(c) Excreta should not be available to flies, rodents, animals.

(d) Excreta should not create a nuisance due to odour.

CHAPTER 54

FAMILY PLANNING

Definition

A way of thinking and living that is adopted voluntarily upon the basis of knowledge, attitudes and responsible decisions by individuals and couples in order to promote the health and welfare of the family group and thus contribute effectively to the social development of a country.

Objectives

The objectives of the family planning are :

1. To avoid unwanted births
2. To bring about wanted births
3. To regulate the intervals between pregnancies.

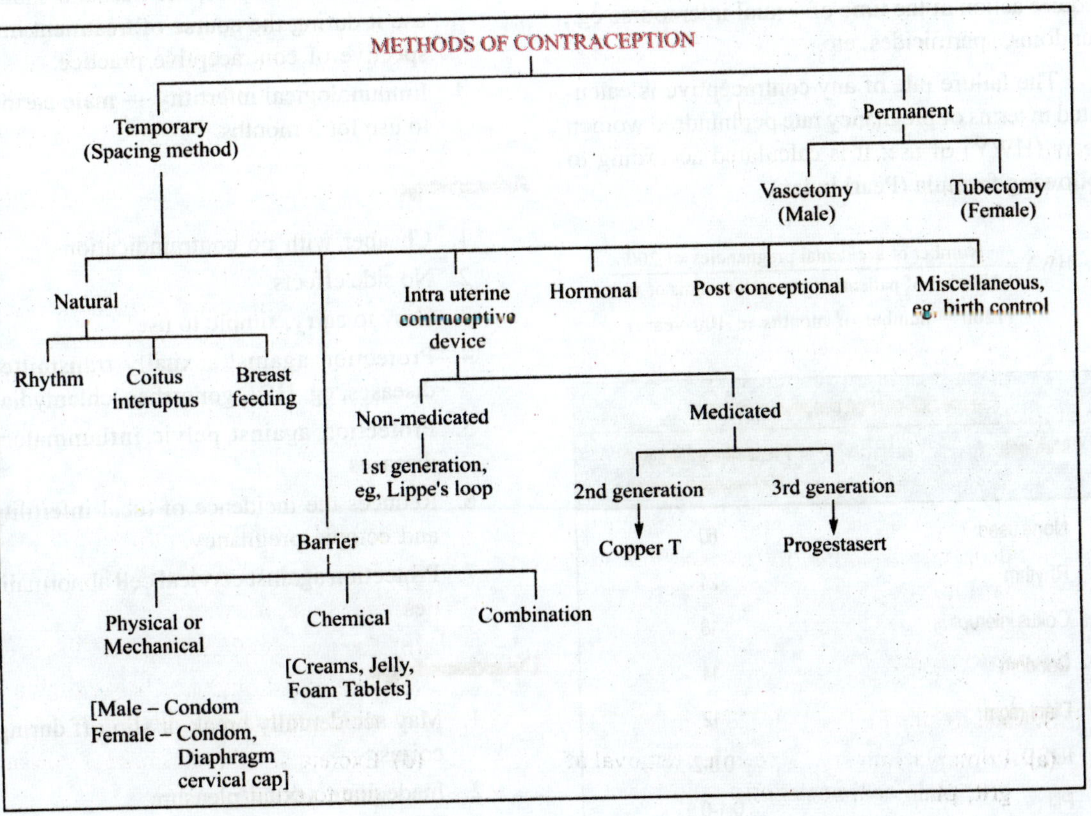

4. To control the time at which births occur in relation to the ages of the parent.
5. To determine the number of children in the family.

Cafeteria choice

It offers all the methods from which an individual can choose according to his needs and wishes and to promote family planning as a way of life.

Ideal contraceptive methods should fulfil the following criteria :

1. Widely acceptable
2. Inexpensive
3. Simple to use
4. Safe
5. Highly effective
6. Requiring minimal motivation, maintenance and supervision.

Conventional contraceptive : The methods that require action at the time of sexual intercourse eg., condoms, spermicides, etc.

The failure rate of any contraceptive is calculated in terms of pregnancy rate per hundred women years (HWY) of use. It is calculated according to following formula (Pearl index).

$$HWY = \frac{\text{Number of accidental pregnancies} \times 1{,}200}{\text{Number of patients observed} \times \text{months of use}}$$

(1200 = number of months in 100 years)

Failure rate of contraceptive methods	
Methods	**Pregnancy rate per 100 women years**
None used	80
Rhythm	24
Coitus interuptus	18
Condom	14
Diaphragm	12
IUCD	0.5-2
Pill	0.1-0.5

Hormonal

Oral pills :	Depot formulation :
1. Combined pill	1. Injectable DMPA
2. Progestogen only pill	2. Subcutaneous, eg,. – norplant
3. Post coital pill	3. Vaginal ring contaning levonorgestrel
4. Once a month pill	
5. Male pill	

BARRIER METHODS

CONDOM

Use

1. Elective contraceptive method.
2. Interim form of contraception during pill use, following vasectomy operation, and if an IUCD is thought lost until a new IUCd can be fitted.
3. During the treatment of trichomonal vaginitis of the wife, the husband should use it during the course of treatment irrespective of contraceptive practice.
4. Immunological infertility — male partner to use for 3 months.

Advantage

1. Cheaper with no contraindications
2. No side effects.
3. Easy to carry, simple to use.
4. Protection against sexually transmitted diseases, eg, HIV, gonorrhea, chlamydia.
5. Protection against pelvic inflammatory diseases.
6. Reduces the incidence of tubal infertility and ectopic pregnancy.
7. Protection against cervical cell abnormalities.

Disadvantage

1. May accidentally break or slip off during coitus.
2. Inadequate sexual pleasure.
3. Disposable after one coital act.

Diaphragm

It is an intravaginal device made of rubber with flexible metal or spring ring at margin.

The device is introduced up to 3 hr before intercourse and is to be kept for at least 6 hr after the last coital act.

Female condom

It is a pouch made of polyurethane which lines the vagina and also the external genitalia. Inner ring at the closed end is smaller compared to the outer ring.

[There are 3 brands of condoms used for male : Dry Nirodh, Delux Nirodh, etc.

The condom is fitted on the erect penis before intercourse. The air must be expelled from the teat end to make room for the ejaculation. The condom must be held carefully when withdrawing it from the vagina to avoid spilling seminal fluid into the vagina after intercourse. A new condom should be used for each sexual act.]

INTRAUTERINE CONTRACEPTIVE DEVICE (IUCD)

Mode of action of IUCD

A. IUCD

↓

Foreign body reaction in uterus.

↓ result in

Biochemical & histological changes

↓ Impair

Vitality of gamete and reduces chance of fertilization

B. Lysosomal disintegration from macrophages attached to device

↓

Liberation of prostaglandin

↓

Antifertility effect

C. Cu^{++} ion affects :

(a) enzymes in the uterus, eg, carbonic anhydrase.

(b) It enhances cellular response in endometrium.

Contraindications for insertion

1. Presence of pelvic inflammatory disease.
2. Dysfunctional uterine bleeding.
3. Suspected pregnancy.
4. Prolapse uterus.
5. Severe dysmenorrhoea.
6. Within 6 weeks following caesarean section.
7. Distortion of shape of uterine cavity, eg, fibroid.

Time of insertion

1. 2-3 days after period is over.
2. Post abortal — following termination of pregnancy.
3. Post partum.
4. Post placental delivery.

Complication

1. Cramp like pain.
2. Syncopal attack.
3. Perforation.
4. Abnormal menstrual bleeding.
5. Pelvic inflammatory disease.
6. Spontaneous expulsion.

Indications for removal

1. Persistent excessive regular or irregular uterine bleeding.
2. Flaring up salpingitis.
3. Perforation of uterus.
4. Downward displacement of device into cervical canal.
5. Missing thread.
6. One year after menopause.

Examples of IUCD

• Cu T 200 • Multiload 250 • Progestasert

• Cu T 380A • Multiload 375 • LNG 20

COMBINED ORAL CONTRACEPTIVES PILLS

Mechanism of action

1. It prevents ovulation or makes the menstrual cycle anovulatory :
   (a) By supressing LH surge.
   (b) By supressing the pulsatile sensation of GnRH from hypothalamus.
   (c) By supressing FSH, the proper development of Graafian follicle fail to occur.
2. By making the endometrium altered that proper implementation of fertilised ovum does not occur.
3. By making the cervical mucous a thick, teneceous so that penetration of sperm is inhibited.
4. It also alter fallopian tube motility.

   Example : Mala N, Mala D, Ovral L, Femilon, etc.

Non-controceptive benefit of OCP

1. Decreased incidence of endometrium and ovarian carcinoma.
2. Dimimised premenstrual tension.
3. Relief of menorrhagia, dysmenorrhoea.
4. Relief of iron deficiency anemia.
5. Positive health benefits are reduction in risk of
   - pelvic inflammatory disease,
   - ectopic pregnancy,
   - fibroid uterus.

Contraindications to use of OCP

Absolute	Relative
1. Thrombophlebitis	1. Migraine
2. Coronary artery disease	2. Hypertension.
3. Known or suspected	3. Depression.
4. Breast carcinoma	4. Varicose vein.
5. Pregnancy	5. Asthma.
	6. Hepatitis.

Complication of OCP

1. Nausea, vomiting, headache, leg cramps
2. Mastalgia
3. Weight gain
4. Depression, mood change, libido, sleep
5. Disturbances
6. Hypertension
7. Thromboembolic phenomenon.

How to prescribe pill

Packing of 28 tabs, there should no break between packs.

Seven of pills are dummies and contain either iron or vitamin preparations.

However, a women can start the pill up to day 5 of the bleeding. In that case she is advised to use a condom additionally for the next 7 days.

The pill should be started on the day after abortion.

Following child birth in non-lactating woman, it is started after 3 weeks and in lactating woman it is to be withheld for 6 months.

Follow up—should be examined after 3 months, then after 6 months and then yearly.

The patient above the age of 35 should be checked more frequently.

Indications for withdrawal

1. Severe migraine
2. Visual disturbances

3. Sudden chest pain
4. Severe cramp and pain in leg
5. Excessive weight gain
6. Severe depression
7. Patient wanting pregnancy.

STERILISATION

It is a surgical method whereby the reproductive function of an individual male or female is purposefully and permanently destroyed.

The operation done in male is called vasectomy and in female called tubectomy.

VASECTOMY

This operation is done in male where a segment of vas deferens of both the sides are resected and the cut ends are ligated.

Advantage

1. It can be done in outdoor procedure or in a mass camp even in remote villages.
2. It is simple and can be performed by one with minimal training.
3. Failure rate is minimal.

Drawbacks

Additional contraceptive protection is needed for about 2-3 months following operations, i.e., till the semen becomes free of sperm.

Selection of candidate

Sexually active and psychologically adjusted husband having the desired number of children is an ideal one.

Complications

1. Wound sepsis
2. Scrotal hematoma
3. Impotency
4. Sperm granuloma
5. Spontaneous recanalisation.

TUBECTOMY

It is an operation where resection of a segment of both the fallopian tubes is done to achieve permanent sterilisation.

The approach may be abdominal and vaginal.

Indications

1. Family planning purpose.
2. Socio-economic.
3. Medico-surgical indications.

Time of operation

1. During puerperium — 24-48 hours following delivery.
2. Interval — following menstrual period in the proliferative phase.
3. Concurrent with MTP.
4. Complete her family.

Different technique

1. Pomeroy
2. Park land
3. Irving
4. Madlener
5. Minilap.

CHAPTER 55

THE HUMAN GENOME PROJECT

The determination of the sequences of the human genome and those of other organisms will alter biology and medicine for all time. As regards to human genome, new information on our origins, on disease genes, on diagnosis, and approaches to therapy are already being established. Progress is being accelerated in the fields of genomics, proteomics, bioinformatics, biotechnology, and the pharmacogenomics.

The Human Genome Project (HGP) is an international effort being started in 1990 whose principal goals were to sequence the whole human genome and the genome of several other organisms for the basic study of genetics. Initially, a number of short-term goals were established for the United States effort eg, producing a human genetic map with markers 2-5 centimorgans (cM) apart and constructing a physical map of all 24 human chromosomes with markers with spaced at about 100,000 base pairs (bP). The following figure (Fig. 55.1) summarises the differences between a genetic map, a cytogenetic map, and a physical map of a chromosome.

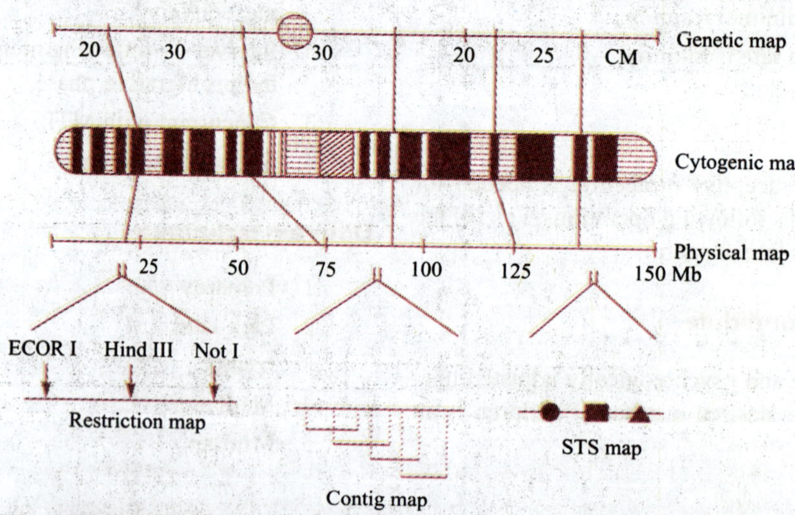

Fig. 55.1 : Principal methods are used to identify and isolate normal and disease genes. For the genetic map, the positions of several hypothetical genetic markers are shown along with the genetic distances in centimorgans between them. The circle shows the position of the centromere. For the cytogenic map, the classic banding pattern of a hypothetical chromosome is shown. For the physical map, the approximate physical positions of the above genetic markers are shown along with the relative physical distances in megabase pairs. Examples of a restriction map, a contig map, and an STS map are shown

Procedure	Comments
Use of rodent or human somatic cell hybrids	Permits assignment of a gene to one specific chromosome but not to a subregion.
Fluorescence in situ hybridization	Permits localization of a gene to one chromosomal band.
Use of probes to define marker loci	Probes identify STSs, RFLPs, SNPs, etc. Thousands, covering all the chromosomes, are now available. It is desirable to flank the gene on both sides, clearly delineating it.
Extensive linkage studies	Large families with defined pedigrees are desirable. Dominant genes are easier to recognize than recessives.
Radiation hybrid mapping	Now the most rapid method of localizing a gene or DNA fragment to a subregion of a human chromosome and constructing a physical map.
Detection of specific cytogenetic abnormalities	For instance, a small deletion of band Xp 21.2 was important in cloning the gene involved in Duchenne muscular dystrophy.
PCR	Can be used to amplify fragment of the gene; also many other applications.
Chromosome jumping	By cutting DNA into relatively large fragments and circularizing it, one can move more quickly and cover greater lengths of DNA than with chromosome walking.
DNA sequencing	Establishes the highest resolution physical map. Identifies open reading frame. Facilities with many high throughout instruments could sequence millions of base pairs per day.
Cloning via YACs, BACs, cosmids, phages, and plasmids	Permits isolation of fragments of varying lengths.
Use of pulse-field gel electrophoresis (PFGE) to separate large DNA fragments	Permits isolation of large DNA fragments obtained by use of restriction endonucleases (rare cutters) that result in very limited cutting of DNA
Detection of expression of mRNAs in tissues by Northern blotting using one or more fragments of the gene as a probe	The mRNA should be expressed in affected tissues.
Databases	Comparison of DNA and protein sequences obtained from unknown gene with unknown sequences in databases can facilitate gene identification.
Chromosome walking	Involves repeated cloning of overlapping DNA segments : the procedure is laborious and can usually cover only 100-200 Kb.

Table 55.1 : Principal methods adopted to identify and isolate normal and disease genes

Abbreviation : STS = Sequence tagged site.

RFLP = Restriction fragment linked polymorphism.

SNP = Single nucleotide polymorphism.

YAC = Yeast artificial chromosome.

BAC = Bacterial artificial chromosome.

PCR = Polymerase chain reaction.

SNPs = Many single nucleotide polymorphisms are being detected and catalogued. These are stable and frequent, and their detection can be automated. It is anticipated that they will be particularly useful for mapping complex traits, such as diabetes mellitus.

An STS marker is typed against a radiation hybrid panel by using its two oligonucleotide primers to perform a PCR assay against the DNA from each hybrid cell line of the panel. If enough markers are typed on one panel, continuous linkage can be established along each arm of a chromosome and the markers can be assembled into the map as a single linkage group.

Table 55.2 : Major findings reported in the rough drafts of the human genome

- Over 90 per cent of the human genome has been sequenced. Gaps still exist, small and large, and the quality of the some of the sequencing data will be refined since some of the findings are probably not exactly right.
- Estimated number of protein-coding genes ranges from 30,000 to 40,000.
- Only 1.1-1.5 per cent of the genome consists of exons. About 24 per cent consists of introns, and 75 per cent of sequences lying between genes.
- Marked differences are being observed among individual chromosomes in many features, such as gene number per megabase, density of single nucleotide polymorphisms (SNPs), GC content, number of transposable elements and CpG islands, and recombination rate.
- Human genes do more work than those of simpler organisms.
- The human proteome is more complex than that of invertebrates.
- Repeat sequences probably constitute more than 50 per cent of the human genome.
- About 100 coding regions are estimated to have been copied and moved by RNA-based transposons (retrotransposons).
- Over 200 genes may be derived from bacteria by lateral transfer.
- More than 3 million SNPs have been identified. It is hoped that they will prove invaluable for certain aspects of gene mapping.

Table 55.3 : Major classes of proteins encoded by human genes

Class of protein	Number (%)[2]
Unknown	12,809 (41%)
Nucleic acid enzymes	2,308 (7.5%)
Transcription factors	1,850 (6%)
Receptors	1,543 (5%)
Hydrolases	1,227 (4%)
Select regulatory molecules (eg, G proteins, cell cycle regulators)	988 (3.2%)
Protooncogenes	902 (2.9%)
Cytoskeletal structural proteins	876 (2.8%)
Kinases	868 (2.8%)

The percentages are derived from a total of 26,383 genes reported in the rough draft by celera genomics. Classes containing more than 2.5% of the total proteins encoded by the genes identified when this rough draft was written arbitrarily listed as major.

Further work is planned on the human and other genomes

The principal task is required for filling in the gaps and identifying new genes, their locations, and functions. Regulatory regions will be identified, and the sequences of other large genomes will be obtained. Additional SNPs will be identified; a complete catalog of these variants is expected to be of great value in mapping genes associated with complex traits and for other uses too. Existing databases will be added to as new information flows in, and new databases will probably be established to serve specific purposes. Different studies in functional genomics will also be undertaken.

Implications for proteomics, biotechnology, and bioinformatics

The knowledge of the human genome will influence many regions without any doubt. Only a few are briefly discussed here.

(a) Proteomics is the study of all proteins encoded in an organism (i.e., the proteome), including their structures, modifications, functions, and interactions. In a short sense, it indicates the identification and study of multiple proteins linked through cellular actions but not the entire proteome.

(b) As regards to humans, many individual proteins will be identified and characterized; their interactions and levels will be determined in physiologic and pathologic states, and the resulting information will be entered into appropriate databases.

(c) Two-dimensional electrophoresis, mass spectrometry, and antibody arrays will be central for the expansion of this rapidly growing field. In the long run, proteomics will highly advance our knowledge of proteins at the basic level and will also nourish biotechnology as new proteins that are likely to have diagnostic, therapeutic, and other uses are discovered and methods for their economic production are improved.

(d) Specialists in bioinformatics will be in great demand, as this field becomes strong enough to manage, analyse and utilize the flood of data from genomic and proteomic studies.

Implications for medicine

(a) Every area of medicine will be practically affected by the new information obtained from knowledge of the human genome. The tracking of disease genes will be sufficiently facilitated.

(b) SNP maps will greatly help in the determination of genes involved in complex diseases.

(c) Probes for any gene will be available if required, leading to improved diagnostic testing for disease genes and for genes directly involved in the causation of specific diseases.

(d) The field of pharmacogenomics is already expanding greatly and it is assumed that in the future, drugs will be tailored to accommodate the variations in enzymes and other proteins involved in drug action and metabolism found among individuals.

(e) Studies of genes involved in behaviour may lead to new insights into the causation and possible treatment of psychiatric disorders.

(f) Many ethical issues eg, privacy concerns and the use of genomic information for commercial purposes will have to be noticed.

(g) In Third world countries it will also be vital that medical and economic benefits as regards to individuals from the effects on health services and the diagnosis and treatment of disease.

CHAPTER 56

NUTRIGENOMICS

Nutrigenomics may be defined as the application of genomic tools to study the integrated effects of nutrients on gene regulation. However, it holds great promise in increasing the understanding of how nutrients affect molecular events in an organism for development and progression of various diseases.

The working definition of nutrigenomic is that it provides a genetic and molecular understanding for how common dietary chemicals (i.e., nutrients) affect the balance between health and disease by altering the expression and/or structure of an individual's genetic make-up. The new branch of genomic and nutritional research can finely be summarized with the following five points :

1. Common dietary chemicals and nutrients directly or indirectly act on the human genome to alter gene expression or structure.
2. Under certain circumstances and in some individuals, diet can be a serious risk factor for a number of diseases.
3. Some diet-regulated genes are susceptible genes and likely to play a role in the onset, incidence, progression, and/or severity of chronic diseases.
4. The degree to which diet influences the balance between healthy and disease states may depend on an individual's genetic make-up (eg, efficient genetic polymorphism and nutrient metabolism).
5. Dietary process based on knowledge of nutritional requirements, nutrition states, and genotype (i.e., "individualized nutrition") can be used to prevent, mitigate, or cure chronic disease.

Many chemicals in foods are nutrients, i.e., these are metabolized to energy or involved in key metabolic reactions (eg, vitamins). But some naturally occurring chemicals in foods are ligands for transcription factors and directly alter gene expression. Other dietary chemicals alter signal transduction pathways and chromatin structure to indirectly affect gene expression. Studies have shown that intake of different diets are associated with the incidence and severity of chronic diseases. Overconsumption of proteins, fat or carbohydrates, or lack of key micronutrients are associated with obesity, cardiovascular diseases, certain cancers, developmental defects, and neurological diseases such as alzheimer's.

At the cellular level nutrient may :

1. act as a ligand for transcription factor receptor.
2. be metabolized by primary, secondary pathways, thereby altering concentration of substrate or intermediates.
3. positively or negatively affects signal pathway.

Dietary chemical may interact with one or more variants to increase or decrease disease risk.

Table 56.1 : Nutrients deficiency and DNA damage		
Nutrients	DNA damage	Health effect
Folic acid	Chromosome breaks and hampers DNA repair	Colon cancer, heart disease, brain dysfunction
Vitamin B_{12}	Unknown	Same as folic acid, memory loss.
Vitamin B_6	Unknown	Same as folic acid
Niacin	Hampers DNA repair	Nerve problem, memory loss
Vitamin C	Mimics radiation damage	Cataract, cancer
Vitamin E	Mimics radiation damage	Colon cancer, heart disease, immune dysfunction.
Vitamin D	Prevent gene variation	Colon, breast, prostate cancer.
Zinc	Chromosome breaks	Brain and immune dysfunction.

Micronutrients, Macronutrients : Effects on Gene

Human diets almost require 40 micronutrients. The specific micronutrients are associated with CVD (B Vitamin, carotinoids, Vitamin E), cancer (folate, carotinoids), neural tube defect (folate) and bone mass. The deficiency of vitamin B_{12}/B_6/folic acid/niacin/, vitamin C and E or iron and zinc appear to mimic radiation in damaging DNA by causing single and double strand breaks, oxidative lesion or both. A good number of other degenerative diseases of aging are also being associated with low fruits and vegetables intake. The specific mechanisms are being determined for the role of certain minerals (calcium, magnesium, manganese, copper, and selenium) and vitamins in heart disease from work in humans in cell culture systems. Unbalanced intake of any of the three major macronutrients, carbohydates, fats, proteins causes the initiation, progression, and severity of chronic disease.

The single macronutrient or micronutrient is unable to prevent chronic diseases. Dietary imbalances and dietary supplements from micronutrients deficiencies to overconsumption of macronutrients of dietary supplements are the modifiers of metabolism and potentiates of chronic diseases.

Specific subgroups are already targeted with "subgroup nutrition" (eg, cholesterol-lowering margarines) without stressing the genetic background of variation of nutritional response. A big challenge is in front of us in validating the combined action of these minor-impact polymorphisms and their practical effect on the relation between nutrition and health on the basis of scientific point of view (Fig. 56.1).

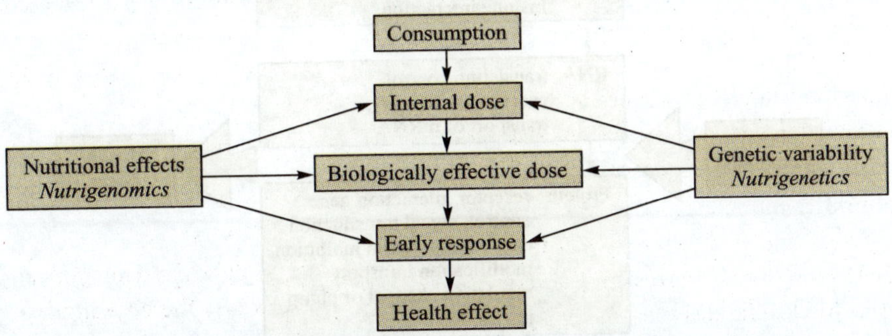

Fig. 56.1 : Nutritional efficacy is subject to external and internal variability. The human genome affects the relation between nutrition and health in two ways :

1. Genetic variation resulting in inter-individual differences in response with implications toward susceptible subgroups in the population (nutrigenetics).
2. The effect of the many bioactive compounds in our nutrition on gene repression and the resulting changes in physiology (nutrigenomics).

Genotype and Environment Interactions

It is defined as a different effect of an environmental exposure on disease risk in persons with different genotypes or a different effect of a genotype on disease risk in persons with different environmental exposures.

In humans, many diseases are related to deficits of essential nutrients, imbalance of macronutrients, micronutrients, even toxic concentrations of certain food compounds. It is finally realised that the nutrition and health relationship is perfectly anchored in interactions on the levels of DNA, RNA, protein, and metabolites (Fig. 56.2).

Our diet is a complex mixture of many possible bioactive chemical compounds, chronically administered in different compositions, and with a multitude of biological effects. The majority of these biological responses are mediated through effector genes, effects on enzyme concentration or activity, and changes in metabolite concentration.

Each of our genes contains ten deviations in its code from the "standard gene". But not all of these polymorphisms have a functional impact. A small number of these polymorphism have serious health implications and may even be lethal. This is the domain of clinical genetics. Many polymorphisms have only a mild effect on the functionality of the resulting protein. A large variety in response to nutrition have been observed within certain limits of "health". Many examples have been set up in which nutritional compounds directly cause DNA damage or modulate susceptibility against DNA damage through regulation of specific pathways in many processes involved in these events.

Role of folic acid in nutrigenomics

Up to this date 1,000 human diseases gene are being identified. 97 per cent of which cause monogenic diseases. Most of the chronic diseases (obesity, diabetes, cardiovascular diseases, cancer) are due to complex interaction between several genes and environmental factors. More complete single nucleotide polymorphism (SNP) and halo type maps are helpful in identifying the genes involved in the disease.

Deficiency of folic acid and other macro and micronutrients appear to mimic radiation in damaging DNA by causing single and double strand breaks, oxidative lesion or both. Nutrient deficiencies are orders of magnitude more important than radiation because of constancy of exposure to mi-

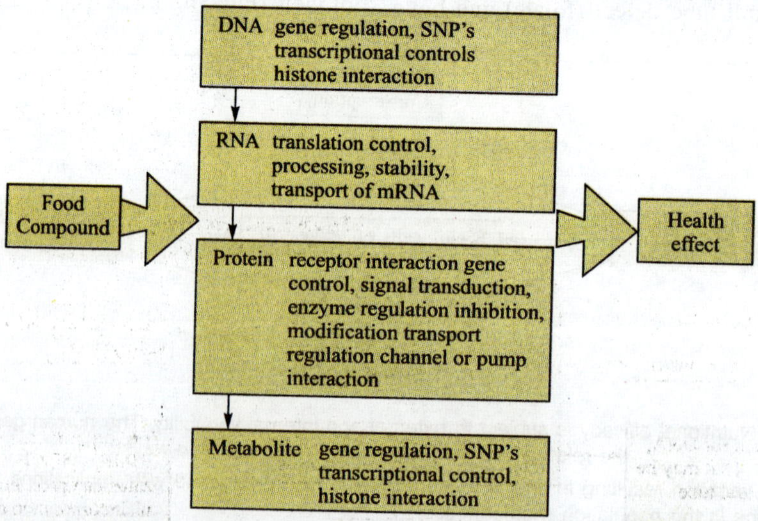

Fig. 56.2 : Health effects of food compounds are related mostly to specific interactions on a molecular level. SNP (single nucleotide polymorphism)

lieu promoting DNA damage. Folate deficiency breaks chromosome due to substantial incorporation of uracil in human DNA. Single strand break in DNA are formed during base excision repair, with two nearby single-strand breaks on opposite DNA strands leading to chromosomal fragmentation.

In humans, folate level and variation of different genes that code the folate-dependent enzymes are linked to many diseases like cancer, vascular diseases, birth defects and complications of pregnancy. The genomic machinery is very much sensitive to folate and vitamin B status and responsible to interaction between folate nutrition and folate-dependent enzyme polymorphism (folate nutrigenomics). Mechanisms that may affect :

1. Maintenance of genomic CpG methylation pattern (which regulate gene expression).
2. Synthesis of nucleotide to prevent DNA damage.
3. Influence plasma homocysteine status, thus risk of vascular diseases.

This complex relationship is shown in Fig. 56.3.

Currently, worldwide interest on folate research is due to discovery of several single nucleotide polymorphisms (SNP) which modulate risk of several diseases (Table 56.2).

Dietary folate interacts with proteins that are encoded by various genes and reduces the risk to development of various diseases and gives protection over the diseases.

Direct biochemical effects

Folate stabilizes the polymorphic enzyme encoded by $C_{677}T$ variant gene by preventing it from relinquishing its flavin cofactors.

Since 5, 10 methylenetetrahydrofolate reductase is a flavin protein people with TT recessive genotype may respond more rapidly to riboflavin supplements as well as folate to lower homocysteine.

Nucleotide biosynthesis

dTMP synthesized by the thymidylate synthetase from dUMP and requires the one carbon unit of 5, 10 methylenetetrahydrofolate. dTMP is used by DNA. If there is low level of folate, uracil misincorporation occurs leading to breakage of DNA strand which predisposes to cancer.

The polymorphic enzyme coded by $C_{677}T$ variant genes can enhance the synthesis of dTMP nucleotide if folate status is good, and this is thought to afford protection against colon cancer and leukaemia.

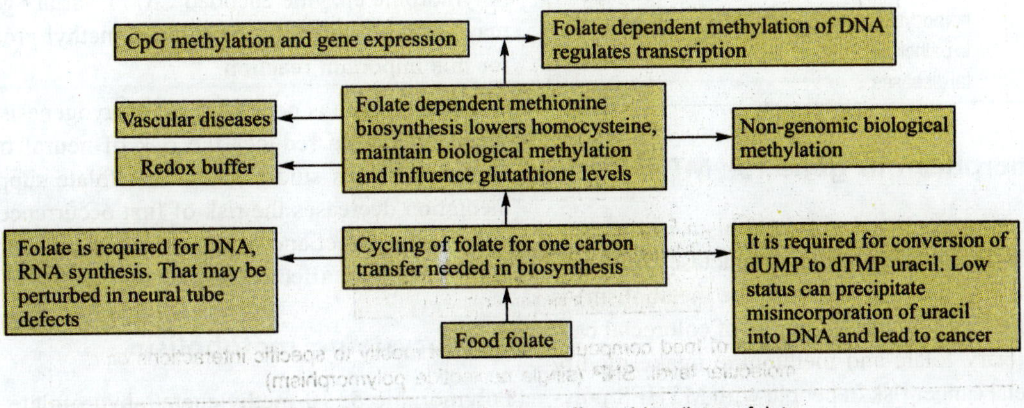

Fig. 56.3 : Molecular mechanism affected by dietary folate

Table 56.2 : Some important single nucleotide polymorphism (SNP) of B-vitamins genes associated with various clinical conditions	
C677T variant of 5, 10 methylenetetrahydrofolate reductase gene	• Colon cancer. • Spina bifida. • Down syndrome. • Leukaemia. • Oral cleft. • Risk of vascular diseases due to elevated homocysteine. • Complication of pregnancy (pre-eclampsia, recurrent pregnancy loss, fetal growth retardation).
A1298C variant of 5, 10 methylenetetrahydrofolate reductase gene	• Spina bifida. • Leukaemia.
A2756G variant of methionine synthase gene	• Thromboembolic diseases.
A66G variant of methionine synthase gene	• Spina bifida. • Down syndrome
C1420T variant of serine hydroxymethylenetransferase gene	• Leukaemia
2R3R variant thymidylate synthetase gene	• Leukaemia
C1561T variant of glutamate carboxy peptidase gene	• May affect cardiovascular diseases.

Table 56.2 : Shows the consequence of SNP of 5, 10 methylenetetrahydrofolate reductase in terms of dTMP nucleotide biosynthesis, DNA methylation, homocysteine metabolism. All of these are related to pathology of cancer, vascular and developmental diseases.

Polymorphism in gene for MTHFR

A common functional polymorphism in the gene of methylenetetrahydrofolate reductase (MTHFR, a major enzyme involved in folate metabolism) is associated with an increased risk of colorectal cancer. Dietary folate and methionine intake modify colorectal cancer risk in people with MTHFR polymorphism.

Nurses health study showed that folate in women who used alcohol had a 25 per cent reduction in breast cancer risk.

Recently a team of American and Chinese researchers showed that folic acid have protective effect against breast cancer. The effect of it had been pronounced when taken with other vitamins especially B_6, B_{12}, and methionine. Researchers believed that folic acid exerts its protective effect by preventing errors in DNA replication and by helping to regenerate methionine, a vital component of DNA synthesis, vitamin B_6, B_{12} and act as cofactors required for folic acid to "do its job".

If folate status is poor, single nucleotide polymorphism may confer risk rather than protection.

Biological methylation

Since dietary methionine cannot provide all methyl groups for cellular methylation reaction there is requirement of de novo synthesis of methionine from folate one carbon pool. S-adenosylmethionine regulates protein, biogenic amine, lipid, and DNA methylation. The S-adenosylmethionine dependent DNA methylation of specific CpG site regulates gene expression and play a critical role in the developmental process.

Methylation of cluster of CpG sites associated with promoter regions tends to silence gene expression.

A deficiency of methyl group may alter the normal control of proto-oncogene expression. The polymorphic enzyme encoded C677T variant gene may reduce availability of de novo methyl groups for this important reaction.

Since folate is necessary in embryogenesis its supplementation reduces the risk of neural tube defects. Various studies show that folate supplementation decreases the risk of first occurrence of neural tube defect and recurrent defects in women with previously affected pregnancy.

Homocysteine metabolism

Polymorphic 5, 10 methylenetetrahydrofolate reductase reduces one carbon flux to methylfolate, the donor molecule for conversion of homocysteine into methionine. This single nucleotide polymorphism may thus elevate homocysteine which is the independent risk factor for the cardiovascular diseases.

Homocysteine is atherogenic and undergoes redox cycling in the presence of transition metal ions, forming radical that causes oxidative damage to low density lipoprotein. It also reacts with cysteine SH groups and modifies apolipoprotein. It is also a hypertensive compound. It also reacts with endothelium—derived relaxation factor to form S-nitrosohomocysteine and superoxide. This leads to loss of vasodilatation action. Since homocysteines promote atherosclerosis through oxidative stress hyperhomocysteinemia is being associated with coronary artery diseases. Recent researchers have studied that folic acid supplementations reduce the risk of CAD (coronary artery diseases).

It also inhibits and downregulates anticoagulants including prostacycline synthesis, activation Protein C, thrombomodulin expression, heparin sulphate expression and fibrinolysis. People with inflammatory bowel diseases (ulcerative colitis, Crohn's disease) have high risk of thromboembolic events, such as stroke and peripheral venous thrombosis. It is also observed that patients with Crohn's disease may be benefited from supplementation of folic acid. It also activates factor V and tissue clotting factor. The final effects of it are to chelate copper and inhibit lysyl oxidase which impairs crosslinking of collagen and elastin and leads to connective tissue abnormalities.

Conclusion

The new science of nutritional systems biology is emerging, taking up the challenge of exploiting all available data generated by genomics technology in a complete description of a biological system. This new method is ideally fit for the evaluation of many vital changes in biological activity as propagated by nutrition. A good number of bioactive compounds act simultaneously and chronically in constantly changing combinations.

The recent unrevealing of human genomic and the coinciding technological developments, genotyping, transcriptomics, proteomics, and metabolomics are now available to nutritional research. In future we are likely to see new screening tools for the selection of bioactive nutrients, new biomarkers for the in vivo efficacy of nutrients, and better insight into the influence of genetic polymorphisms on nutrient metabolism.

CHAPTER 57

EFFECT OF WHOLE GRAINS IN HEALTH

Grains are the most important food source of the Indian population. Asian Indians in India consume a high carbohydrate diet about 60-67% as compared to the Asian Indians settled in UK and USA. Foods made by the whole grains are the import source of fibres, trace minerals, and vitamins. Diets containing whole grains are most essential in reducing the risk of coronary heart disease (CHD), type 2 diabetes mellitus (T2DM), certain types of cancers and obesity. In India the commonly used grains are wheat, rice, bajra, jowar etc. In USA, the most commonly consumed grains are wheat, rice, maize, oats, and rye.

Structure and definition of whole grain

All grains have a protective hull, beneath which are the endosperm, bran, and germ. The germ contains the plant embryo. The endosperm supplies food for the growing seedling. Surrounding the germ and the endosperm is the bran which protects the grain from the weather, insects, moulds, and bacteria. About 50-75 per cent of the endosperm is starch which supplies major energy to the embryo during germination of the kernel. The endosperm also contains storage proteins about 8-18 per cent, few vitamins, minerals, and fibre. The germ is a minor contributor to the dry weight of most grains.

The whole grain is the entire grain seed of a plant. This seed is known as the kernel which is made up of the bran, the germ, and the endosperm. The whole grains may be eaten whole, cracked, or flaked. These are, most often, milled into flour and used to make breads, cereals, pastas, and crackers etc.

Whole grains and health

1. Obesity

(a) Grains rich in viscous soluble fibre tend to increase intraluminal viscosity, prolong gastric emptying time, and slow nutrient absorption in the small intestine.

(b) A healthy eating pattern, including the consumption of whole grains, is associated with smaller gains in body mass index (BMI).

(c) As per the data of UK researchers it was found that a higher intake of whole grains (about 3 servings/day) was associated with lower BMI and less central adiposity (abdominal fat).

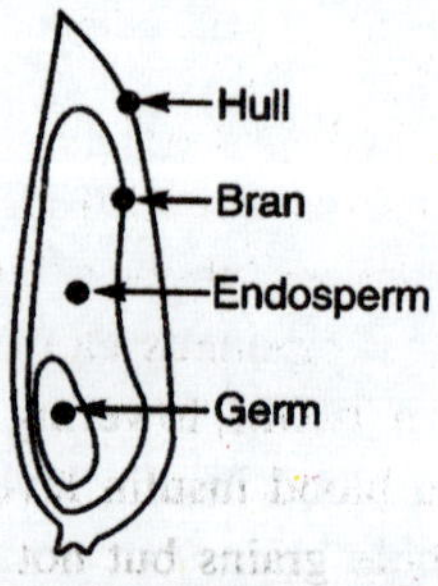

Fig. 57.1: Structure of whole grain

2. Insulin resistance

(a) People who eat three or more servings of whole grains a day, mainly from high fibre cereals, are less likely to develop insulin resistance and the metabolic syndrome, common precursors of both T2DM and CHD.

(b) The dietary product (WG) favourably influences metabolic risk factors (like insulin resistance) for T2DM independent from the amount of body weight lost during a hypo-energetic diet.

3. Dyslipidemia

(a) The most common symptoms of dyslipidemia in people with the metabolic syndrome and diabetes include elevated serum triglyceride (TG), low HDL cholesterol (HDL-C) and predominance of small dense LDL cholesterol (SDLDL-C).

(b) The healthy diet of consumption of barley containing increased soluble fiber can reduce cardiovascular risk factors.

(c) Diets having abundant whole grains tend to decrease LDL cholesterol, triglycerides, blood pressure, and increase HDL cholesterol.

4. Diabetes

(a) German researchers had observed that those who ate cereal fiber the most had a 27 per cent lower risk of developing T2DM than those who ate the least.

(b) Whole grains containing magnesium, fiber, vitamin E, phytic acids, lectins, and phenolic compounds can contribute reduction of T2DM, lowering of blood glucose and blood insulin levels. The fiber from whole grains but not from fruit or vegetable sources can only exert the protective effect in reducing risk for developing T2DM.

(c) Some of the whole grain products can control blood glucose for upto ten hours. Barley has the strongest effect in this regard.

5. Coronary Heart Disease

(a) It has been strongly suggested that the regular consumption of whole grains significantly reduces the risk of CHD.

(b) It was also found that those eating two to six servings of whole grain cereal a week reduced their risk of heart failure by 22 per cent, while those eating whole grains daily reduced the risk by 28 per cent.

(c) It had been found that those with the highest whole grain intake (about 40 g per day) cut CHD risk by almost 20 per cent, but even those eating just 25 g cut their CHD risk by 15 per cent.

6. Hypertension

(a) Higher whole grain intake is associated with a reduced risk of hypertension in middle-aged and older women. This suggests a potential role for increasing whole grain intake in the primary prevention of hypertension and its cardiovascular complications.

(b) The DASH diet consisting of whole grain products and other healthy foods can lower blood pressure.

7. Polycystic ovarian syndrome (PCOS)

(a) It has been suggested that a diet low in saturated fat and high in fiber is recommended.

(b) Diet and exercise are the instructions for the overweight/obese women to lose weight. This diet includes an increase in fiber and a decrease in refined carbohydrates.

8. Non-alcoholic fatty liver disease (NAFLD)

(a) Carbohydrates that are high in indigestible and fermentable fiber and low in glycemic index can be helpful in maintaining glucose, insulin, and free fatty acid concentrations in individuals with insulin resistance and non-alcoholic steatohepatitis (NASH).

(b) Fiber and certain starches found in whole grains ferment in the colon to help reduce transit time and improve gastrointestinal health.

(c) Whole grains also contain antioxidants that may help protect against oxidative damage which may play a role in cancer development.

(d) Other bioactive components in whole grains may affect hormone levels and lower the risk of hormone dependent cancers. But it is not yet clear between whole grain intake and breast cancer.

Definition of whole grain serving

A serving of whole grain is defined as $\frac{1}{2}$ cup cooked brown rice or other cooked grain, $\frac{1}{2}$ cup cooked whole grain pasta, $\frac{1}{2}$ cup cooked hot cereal such as oatmeal, 1 ounce uncooked whole grain pasta, brown rice or other grain, 1 slice 100 per cent whole grain bread, 1 very small 100 per cent whole grain muffin, and 1 cup 100 per cent whole grain ready to eat cereal.

One can count a serving of whole grain as 16 grams of whole grain ingredients.

Health effects of some whole grains

(a) **Barley**: The fiber in barley is definitely healthy. It may lower cholesterol more effectively than oat fiber.

(b) **Oats**: Oats contain a special kind of fiber called beta glucan which is effective in lowering cholesterol. Oats have an antioxidant that protects blood vessels from the damaging effects of LDL cholesterol.

(c) **Jowar**: It is a gluten-free grain and is very popular among those with celiac disease.

(d) **Ragi**:
1. Ragi products are consumed in various states of the southern region.
2. It is used to make flat breads (named bhakari), dosa, and rotis.
3. Ragi grain is malted too and mixed with milk, water or yoguri and eaten at breakfast time.
4. The sprouted ragi is used to make baby food. It is also useful for elderly people since it is easy to digest.
5. It contains high fiber, proteins, calcium, B complex vitamins and also vitamin E.
6. It is rich in minerals and helps in lowering cholesterol.

(e) **Bajra (whole)**: It is used in the preparation of khichri in many parts of Rajasthan.

(f) **Rice bran oil**: This oil is rich in vitamin E and it has also cholesterol lowering effects.

(g) **Corn (Maize) oil**: It is a highly desired vegetable oil owing to its high level of linolenic fatty acid and its excellent flavour.

(h) **Fermented cereal products**: The term fermentation means the breakdown of carbohydrate like matter under aerobic or anaerobic conditions. The organisms involved may be bacteria or moulds. The gas during fermentation produces flavouring substances. The vitamin B and C contents are also increased after fermentation. The commonly consumed fermented products in different parts of India are idli, dosa, dhokla, bread etc.

APPENDIX - I

MULTIPLE CHOICE QUESTIONS & ANSWERS

Appendix 1

BIOPHYSICS

Write the correct answer number of the followings

1. The pH of gastric juice of infants

   (a) 2.0, (b) 4.0, (c) 4.5, (d) 5.0. **Ans.** d.

2. The pH of water is 7.0 at the temperature in centigrade

   (a) 22, (b) 30, (c) 37, (d) 40. **Ans.** a.

3. The pH of blood is 7.4 when the ratio between [$NaHCO_3$] and [H_2CO_3] is

   (a) 10 : 1, (b) 20 : 1, (c) 25 : 1, (d) 30 : 1. **Ans.** b.

4. The difference in pH between arterial and venous blood is rarely more than

   (a) 0.02, (b) 0.03, (c) 0.04, (d) 0.06. **Ans.** c.

5. The pH of buffer is determined by $pH = P_{Ka} + \log \frac{[Salt]}{[acid]}$ which is also known as the equation of

   (a) Henderson-Joules, (b) Henderson-Smith, (c) Henderson-Harris, (d) Henderson-Hasselbalch. **Ans.** d.

6. The chief buffering system in the blood

   (a) K_2HPO_4 and KH_2PO_4, (b) B. Protein and H. Protein, (c) $NaHCO_3$ and H_2CO_3, (d) B. Hemoglobin and H. Hemoglobin. **Ans.** c.

7. During severe muscular exercise when the blood lactic acid content rises over 100 mg per 100 ml the pH of blood

   (a) Slightly increases, (b) Highly increases, (c) Slightly decreases, (d) Markedly decreases. **Ans.** d.

8. In case of non-homogeneous solution, diffusion is expressed as $\frac{ds}{dt} = DA\frac{dc}{dx}$ by

   (a) Fick's Law, (b) Peterson's Law, (c) Faraday's Law, (d) Dick's Law. **Ans.** a.

9. The average pH of urine is

   (a) 5.6, (b) 6.0, (c) 6.4, (d) 6.8, **Ans.** b.

10. The osmotic pressure of a solution increases with the rise in

    (a) Temperature, (b) Cold, (c) Humidity, (d) Rancidity. **Ans.** a.

11. The osmotic pressure of a solution relating to solute molecules depends on the

    (a) Size, (b) Shape, (c) Number, (d) Volume. **Ans.** c.

12. If a cell is immersed in a concentrated solution, it follows the phenomenon

    (a) Turgor, (b) Plasmolysis, (c) Hemolysis, (d) Paralysis. **Ans.** b.

13. Osmosis is opposite to

    (a) Effusion, (b) Affusion, (c) Confusion, (d) Diffusion. **Ans.** d.

14. The intracellular fluid of red cells and the red cell membrane in 0.92% NaCl solution maintains a relation

    (a) Hypertonic, (b) Hypotonic, (c) Isotonic, (d) None of the above. **Ans.** c.

15. The purgative action of Epsom ($MgSO_4$) salt follows

    (a) Osmosis, (b) Diffusion, (c) Adsorption, (d) Absorption. **Ans.** a.

16. Hemolysis is caused by the dilution of RBC by

(a) Diffusion, (b) Osmosis, (c) Effusion, (d) Imbibition. Ans. b.

17. The surface tension of a solution is decreased by

(a) Calcium sulphate, (b) Barium sulphate, (c) Magnesium phosphate, (d) Potasssium permanganate. Ans. d.

18. The surface tension of a solution is lowered by

(a) Ammonia, (b) Sodium hydroxide, (c) Potassium hydroxide, (d) Aluminium hydroxide. Ans. a.

19. The surface tension of a solution is increased by

(a) Bile salts, (b) Bile acids, (c) Concentrated sulphuric acid, (d) Acetic acid. Ans. c.

20. Lipids and proteins which are both effective in lowering surface tension are found concentrated in the cell wall following the principle of

(a) Johnson-John, (b) Gibbs-Thomson, (c) Peterson-Pollen, (d) None of the above. Ans. b.

21. Adsorption decreases with the rise in

(a) Cold, (b) Humidity, (c) Dryness, (d) Temperature. Ans. d.

22. The process of adsorption is applied in the purification of

(a) Enzymes, (b) Hormones, (c) Vitamins, (d) Coenzymes. Ans. a.

23. The following is the hydrotropic substance

(a) Hydrochloric acid, (b) Nitric acid, (c) Hippuric acid, (d) Salicylic acid. Ans. c.

24. Bile salts make emulsification with fat for the action of

(a) Amylase, (b) Lipase, (c) Pepsin, (d) Trypsin.

Ans. b.

25. The viscosity of a liquid increases due to the presence of

(a) Suspended particles, (b) Soluble particles, (c) Small sized particles, (d) None of the above.

Ans. a.

26. The size of each colloidal particle in nm

(a) 4 to 40, (b) 6 to 60, (c) 8 to 80, (d) 10 to 100.

Ans. d.

27. The serum colloidal protein particles may be precipitated by the addition of large amount of

(a) Calcium sulphate, (b) Ammonium sulphate, (c) Barium sulphate, (d) Magnesium sulphate.

Ans. b.

28. Some of the calcium phosphate of blood is held in colloidal suspension by the protective action of

(a) Lipids, (b) Carbohydrate, (c) Proteins, (d) Minerals. Ans. c.

29. Water is not expelled by squeezing in

(a) Imbibition, (b) Precipitation, (c) Combination, (d) Dilution. Ans. a.

30. In simple diffusion $\frac{ds}{dt}(C_0 - C_1)$ can be expressed by the modification of

(a) Rongent's Law, (b) Fabry's Law, (c) Pollinger's Law, (d) Fick's Law. Ans. d.

31. Fatty acids can be transported into and out of mitochondria through

(a) Active transport, (b) Facilitated transfer, (c) Non-facilitated transfer, (d) None of the above.

Ans. b.

32. The absorption of intact protein from the gut in the foetal and newborn animals takes place by

(a) Pinocytosis, (b) Passive diffusion, (c) Simple diffusion, (d) Active transport. Ans. a.

33. Thyroid function is determined by the use of the isotopes

(a) Na^{24}, (b) K^{42}, (c) Ca^{45}, (d) I^{131}. Ans. d.

34. Pernicious anemia is diagnosed by the radioactive substance

(a) Cl^{36}, (b) P^{32}, (c) CO^{60}, (d) Fe^{59}. Ans. c.

35. The protein biosynthesis has been studied by using

(a) N^{15}, (b) Ca^{45}, (c) Na^{24}, (d) Au^{198}. Ans. a.

Fill in the blanks of the followings

1. Solution of mixed indicators having a number of colour changes over a wide range of pH are called ________ indicators. Ans. Universal.
2. The logarithm of the reciprocal of hydrogen ion concentration is said to be ________. Ans. pH.
3. Temperature cannot alter the pH of a solution of ________. Ans. 0.1 N HCl.
4. The mixture of weak acids and their salts of strong bases are said to be ________.

Ans. Buffers.

5. The spreading of solute molecules throughout the water molecules is called _______ .

Ans. Diffusion.

6. Diffusion takes place against ________ .

Ans. Gravity.

7. The most selective artificial membrane is ______ .

Ans. $Cu_2Fe(CN)_6$.

8. Soaps have _______ osmotic pressure.

Ans. Lower.

9. If a cell is kept in a hypotonic solution, the cell is called ________ . Ans. Turgid.

10. The force with which the surface molecules are held together is called _______ .

Ans. Surface tension.

11. Surface tension is involved in the process of _____ .

Ans. Digestion.

12. The process by which water-insoluble substances are made water-soluble by hydrotropic substances is called _______ . Ans. Hydrotropy.

13. The resistance experienced by one layer of a liquid in moving over another is called ________ .

Ans. Viscosity.

14. The sparingly soluble cholesterol and the calcium salt of bilirubin is kept in colloidal solution by the protective colloids _______ .

Ans. Bile Salts.

15. The osmotic pressure of serum protein is about ______ . Ans. 30 mm Hg.

16. The process of separation of crystalloids from colloids by diffusion through a membrane by osmotic force is called _______ . Ans. Dialysis.

17. Dialysis is applied in medicine in the _______ .

Ans. Artificial Kidney.

18. An emulsoid may be changed into a suspensoid by ______ . Ans. Dehydration.

19. The cell membrane consists of small water-filled pores of radius about _______ nm. Ans. 0.4.

20. The atoms having the same atomic number but different atomic weights are said to be _______ .

Ans. Isotopes.

21. Absorption of fat is studied by using _______ .

Ans. I^{131}-Oleic acid.

22. The life span of RBC has been determined by tagging _______ to RBC. Ans. Cr^{51}.

23. The absorption, mobilization and transport of iron have been studied by the use of _______ .

Ans. Fe^{59}.

24. Plasma volume is measured by using ______ labelled serum albumin. Ans. I^{131}.

25. The formation of antibody in the reticuloendothelial system has been studied by ________ .

Ans. P^{32}.

26. ________ is the end product of Au^{198} after emitting β- and γ- rays. Ans. Hg^{198}.

27. Some of the carrier proteins are called ______ and others are called ________ .

Ans. Uniports; Symports.

28. Na^+-K^+ ATPase catalyzes the hydrolysis of _____ to _____ . Ans. ATP; ADP.

29. Na^+-K^+ ATPase is a ______ made up of ______ and ________ subunits. Ans. Heterodimer; α; β.

30. Na^+ and K^+ transport occur through _____ subunit.

Ans. α.

31. β subunit is a ________ . Ans. Glycoprotein.

32. Lyophobic colloids are mostly _______ , whereas lyophilic colloids are mostly ________ .

Ans. Inorganic; Organic.

33. Lyophobic colloids undergo _______ coagulation and lyophilic colloids undergo ________ coagulation. Ans. Irreversible; Reversible.

34. Lyophobic colloids do not __________ readily, whereas lyophilic colloids __________ readily.

Ans. Getatinise; Gelatinise.

35. Lyophilic colloids are precipitated by addition of _______ of the _______ charge.

Ans. Colloids; Opposite.

36. Acidosis is two types, _______ and _______ .

Ans. Metabolic; Respiratory.

37. The pH of blood is maintained 7.4 when the ratio of $[B_2HPO_4] : [BH_2PO_4]$ is _______ . Ans. 4 : 1.

Indicate "True" or "False" of the followings

1. The pH range of solutions is 0 to 14 only.

Ans. True.

2. The buffer acts almost as if it is "absorbing" the added free hydrogen or hydroxyl ions.

Ans. True.

3. The buffering system of lymph, cerebrospinal fluid, etc. are similar to that of blood. Ans. True.
4. In cases of gas, diffusion is slow. Ans. False.
5. Absorption from gastrointestinal tract does not follow the principles of osmosis. Ans. False.
6. Adsorption is a surface phenomenon. Ans. True.
7. Heat is not given off in all adsorption. Ans. False.
8. Adsorption decreases with the rise in temperature. Ans. True.
9. Surface adsorption helps to combine enzymes with substrates to give reaction product. Ans. True.
10. Phospholipids are hydrotropic substances. Ans. False.
11. The unit of Viscosity is the "Poise". Ans. True.
12. Brownian movement is quite regular. Ans. False.
13. In the body equilibrium may be established owing to the removal of "excreted" ions in other reactions. Ans. False.
14. Emulsoids are easily precipitated by salts. Ans. False.
15. Formaldehyde can remove water from the gel. Ans. True.
16. The mitochondrial membrane is permeable to citrate. Ans. True.
17. The transport of most ions occur more slowly than the non-electrolytes. Ans. True.
18. Active transport is involved in the absorption of lactose from the small intestine. Ans. False.
19. Mineral metabolism is studied by S^{35}. Ans. False.
20. The half-life of I^{131} is 8.04 days. Ans. True.

Match the following

	Solution		pH	Answer
(a)	Saliva	(i)	7.1	(a)...... (ii)
(b)	Milk	(ii)	6.8	(b)...... (i)
(c)	Tears	(iii)	12.0	(c)...... (iv)
(d)	Pancreatic Juice	(iv)	7.2	(d)...... (v)
(e)	Gastric Juice (adult)	(v)	8.0	(e)...... (vi)
(f)	N/100 NaOH	(vi)	1.4	(f)...... (iii)

Appendix 2

CARBOHYDRATES

Write the correct answer number of the following

1. The heptose ketose sugar formed as a result of chemical reaction in HMP shunt.

   (a) Glucoheptose, (b) Galactoheptose, (c) Sedoheptulose, (d) Mannoheptose. Ans. c.

2. The general formula for polysaccharide.

   (a) $(C_6H_{10}O_5)_n$, (b) $(C_6H_{12}O_6)_n$, (c) $(C_6H_{12}O_5)_n$, (d) $(C_6H_{10}O_6)_n$. Ans. a.

3. The number of isomers of glucose.

   (a) 4, (b) 8, (c) 12, (d) 16. Ans. d.

4. The epimers of glucose.

   (a) Fructose, (b) Galactose, (c) Ribose, (d) Deoxyribose. Ans. b.

5. Human heart muscle contains

   (a) D-Arabinose, (b) D-ribose, (c) D-Lyxose, (d) D-Xylose. Ans. c.

6. The intermediate in hexose monophosphate shunt.

   (a) D-ribulose, (b) D-Arabinose, (c) D-Xylose, (d) D-Lyxose. Ans. a.

7. Honey contains the hydrolytic product of

   (a) Lactose, (b) Maltose, (c) Inulin, (d) Starch. Ans. c.

8. On boiling Benedict's solution is not reduced by

   (a) Sucrose, (b) Lactose, (c) Maltose, (d) Fructose. Ans. a.

9. Glycosides are found in many

   (a) Vitamins, (b) Drugs, (c) Minerals, (d) Nucleoproteins. Ans. b.

10. Erythromycin contains

    (a) Diethyl amino sugars, (b) Trimethyl amino sugars, (c) Dimethylamino sugars, (d) Triethyl amino sugars. Ans. a.

11. Galactose on oxidation with concentrated HNO_3, produces

    (a) Gluconic acid, (b) Saccharic acid, (c) Saccharolactone, (d) Mucic acid. Ans. d.

12. The distinguishing test between monosaccharides and disaccharides.

    (a) Bial's test, (b) Seliwanoff's test, (c) Barfoed's test, (d) Hydrolysis test. Ans. c.

13. Barfoed's solution is not reduced by

    (a) Glucose, (b) Mannose, (c) Sucrose, (d) Ribose. Ans. c.

14. Cellulose is made up of the molecules of

    (a) α-Glucose, (b) β-Glucose, (c) Both of the above, (d) None of the above. Ans. b.

15. Iodine solution produces no colour with

    (a) Cellulose, (b) Starch, (c) Dextrin, (d) Glycogen. Ans. a.

16. Glycogen structure includes a branch in between α-glucose units

    (a) 4 – 101, (b) 6 – 12, (c) 8 – 14, (d) 12 – 18 Ans. d.

17. Amylose contains glucose units

    (a) 100 – 200, (b) 200 – 300, (c) 300 – 400, (d) 500 – 600 Ans. c.

18. Each branch of amylopectin is at an interval of glucose units

    (a) 14 – 20, (b) 24 – 30, (c) 34 – 40, (d) 44 – 50 Ans. b.

19. N-acetylneuraminic acid is an example of

(a) Sialic acid, (b) Mucic acid, (c) Glucuronic acid, (d) Hippuric acid. Ans. a.

20. The molecular weight of hyaluronic acid in millions ranges from

(a) 1 – 2, (b) 1 – 4, (c) 1 – 6, (d) 1 – 8 Ans. b.

21. In place of glucuronic acid chondroitin sulphate B contains

(a) Gluconic acid, (b) Gulonic acid, (c) Iduronic acid, (d) Sulphonic acid. Ans. c.

22. Heparin has a molecular weight of about

(a) 14,000, (b) 15,000, (c)16,000, (d) 17,000. Ans. d.

23. Blood group substances consist of

(a) Lactose, (b) Maltose, (c) Fucose, (d) Mucose. Ans. c.

24. The component of cartilage and cornea is

(a) Kerato sulphate, (b) Chondroitin sulphate, (c) Cadmium sulphate, (d) Antimony sulphate. Ans. a.

25. Benedict's test is less likely to give weakly positive results with concentrated urine due to the action of

(a) Urea, (b) Uric acid, (c) Ammonium salts, (d) Phosphates. Ans. b.

26. Salivary amylase is activated by

(a) Na^+, (b) K^+, (c) HCO_3^-, (d) Cl^- Ans. d.

27. Active transport of sugar is depressed by the agent

(a) Oxaloacetate, (b) Fumarate, (c) Malonate, (d) Succinate. Ans. c.

28. The absorption of glucose is interfered by the deficiency of

(a) Vitamin A, (b) Thiamine, (c) Magnesium sulphate, (d) Ferrous sulphate. Ans. b.

29. Glucose absorption may be decreased in

(a) Oedema, (b)Nephritis, (c) Rickets, (d) Osteomyelitis. Ans. a.

30. Glycogen synthetase activity is depressed by

(a) Glucose, (b) Insulin, (c) Cyclic AMP, (d) Fructokinase. Ans. c.

31. Glucose is removed from the blood following a meal by

(a) Hexokinase, (b) Glucokinase, (c) Both of the above, (d) None of the above. Ans. b.

32. The branching enzyme acts on the glycogen when the glycogen chain has been lengthened to between glucose units.

(a) 1 and 6, (b) 2 and 7, (c) 3 and 9, (d) 6 and 11. Ans. d.

33. Cyclic AMP is formed from ATP by the enzyme adenylate cyclase which is activated by the hormone.

(a) Insulin, (b) Epinephrine, (c) Testosterone, (d) Progesterone. Ans. b.

34. The synthesis of adenylate cyclase is increased by

(a) Thyroid hormones, (b) Growth hormones, (c) ACTH, (d) FSH. Ans. a.

35. UDPG is essential for the synthesis of

(a) Lactose, (b) Maltose, (c) Sucrose, (d) Starch. Ans. a.

36. Phosphorylase a in muscle is a tetramer containing 4 mol of

(a) ATP, (b) NAD^+, (c) Pyridoxal phosphate, (d) CoA Ans. c.

37. Inactive muscle phosphorylase b is activated by active protein kinase which is being stimulated by

(a) ATP, (b) NAD^+, (c) Cyclic GMP, (d) Cyclic AMP. Ans. d.

38. Hexokinase has a high affinity for glucose than

(a) Fructokinase, (b) Galactokinase, (c) Glucokinase, (d) All of the above. Ans. c.

39. Fructose-1.6-diphosphate is converted to Fructose-6-phosphate by the enzyme Fructose-1, 6-diphosphatase which is stimulated by

(a) Glucagon, (b) Insulin, (c) ACTH, (d) None of the above. Ans. a.

40. Dihydroxyacetone phosphate and Glyceraldehyde–3–phosphate are interconverted by

(a) Triose isomerase, (b) Phosphotriose isomerase, (c) Diphosphotriose isomerase, (d) Dihydroxyacetone phosphorylase. Ans. b.

41. In the liver Glyceraldehyde–3–phosphate is converted to

(a)Glycol, (b) Formaldehyde, (c) Formic acid, (d) Glycerol. Ans. d.

42. Phosphoglycerate kinase responsible for the conversion of 1, 3–diphosphoglycerate to 3–phosphoglycerate is inhibited by

(a) Arsenate, (b) Fumarate, (c) Citrate, (d) Cyanate.

Ans. a.

43. The reduced lipoate is reoxidized by

(a) NAD^+, (b) $NADP^+$, (c) FAD^+, (d) FMN. Ans. c.

44. The approximate number of mols. of Pyruvate dehydrogenase in pyruvate dehydrogenase complex.

(a) 19, (b) 29, (c) 39, (d) 49. Ans. b.

45. Pyruvate is accumulated by the dietary deficiency of vitamin.

(a) B_6, (b) Folic acid, (c) B_{12}, (d) Thiamine.

Ans. d.

46. Citrate is converted to isocitrate by aconitase which contains

(a) Ca^{++}, (b) Fe^{++}, (c) Zn^{++}, (d) Mg^{++}. Ans. b.

47. The reaction involving to succinyl-CoA to succinate requires

(a) CDP, (b) ADP, (c) GDP, (d) $NADP^+$. Ans. c.

48. The carrier of the citric acid cycle

(a)Succinate, (b) Fumarate, (c) Malate, (d) Oxaloacetate. Ans. d.

49. Glucose–6–phosphatase is absent from

(a) Adipose tissue, (b) Intestine, (c) Kidney, (d) Heart.

Ans. a.

50. 6–Phosphogluconate is converted to 3-keto-6-phospho-gluconate in presence of coenzyme.

(a) FAD^+, (b) NAD^+, (c) $NADP^+$, (d) ATP.

Ans. c.

51. UDPG is oxidized to UDP glucuronic acid by UDP dehydrogenase in presence of.

(a) FAD^+, (b) NAD^+, (c) $NADP^+$, (d) ADP.

Ans. b.

52. Fructokinase is present in

(a) Intestine, (b) Adipose tissue, (c) Heart, (d) Brain.

Ans. a.

53. Fructose–1–phosphate is splitted into glyceraldehyde and dihydroxyacetone phosphate by the enzyme

(a) Enolase, (b) Aldolase A, (c) Aldolase A & B, (d) Aldolase B. Ans. d.

54. Galactose is phosphorylated by galactokinase to form.

(a) Galactose–6–phosphate, (b) Galactose–1, 6–diphosphate, (c) Galactose–1–phosphate, (d) All of the above. Ans. c.

55. In galactosemic individual UDP Galactose is formed by epimerization from.

(a) Glucose, (b) UDP Glucose, (c) CDP Glucose, (d) ITP Glucose. Ans. b.

Fill in the blanks of the followings

1. 2-6 monosaccharide units are produced on hydrolysis of _______ . Ans. Oligosaccharides.

2. Mucopolysaccharides are _______ .

Ans. Heteropolysaccharides.

3. Cis-trans isomerism occurs in compounds with _______ bonds. Ans. Double.

4. When equal amounts of dextrorotatory and levorotatory isomers are present in a mixture, the mixture is said to be _______ .

Ans. Racemic.

5. In the body epimerization takes place by the enzyme _______ . Ans. Epimerase.

6. Sugars forming five-membered rings are called _______ . Ans. Furanoses.

7. A method for synthesis of monosaccharides was first proposed by _______ . Ans. Kiliani.

8. Galactose is not _______ by yeast.

Ans. Fermented.

9. The compounds formed by the condensation reaction between a sugar and the hydroxyl group of an aglycone which may or may not be another sugar are called _______ .

Ans. Glycosides.

10. The inhibitor of the Na^+–K^+–ATPase of cell membranes is _______ . Ans. Ouabain.

11. Sugars containing an amino group are called _______ . Ans. Aminosugars.

12. 3-amino-D-ribose is contained in _______ .

Ans. Carbomycin.

13. Glucose when heated wtih conc. HI is converted into _______ . Ans. Iodohexane.

14. Galactose on reduction yields _______ . Ans. Dulcitol.

15. Fructose on reduction produces _____ and _____.
Ans. Mannitol; Sorbitol.

16. Glucose on treatment with strong mineral acids produces _____. Ans. Levulinic acid.

17. Galactose forms a different osazone owing to _____ in the structure. Ans. Carbon 4.

18. Maltose can exist in _____ or _____ forms.
Ans. α ; β.

19. Sucrose solution is _____ rotatory but after hydrolysis with dil. HCl it shows _____ rotation.
Ans. Dextro; Laevo.

20. Sucrose is used to be referred to as _____.
Ans. Invert sugar.

21. The glycogen content of _____ is more than that of _____. Ans. Liver; Muscle.

22. The molecular weight of glycogen ranges from _____ to _____. Ans. 10^5 ; 10^8.

23. Starch is formed by _____ chain.
Ans. Alpha-glucosidic.

24. Each turn of the helix of amylose contains _____ units. Ans. Six glucose.

25. The molecular weight of amylopectin is about _____. Ans. 5,00,000.

26. The amylopectin chains have at least _____ branches. Ans. 80.

27. Dextrins are formed by the partial hydrolysis of _____ by an enzyme _____.
Ans. Starch; Salivary amylase.

28. Achrodextrin gives no colour with _____.
Ans. Iodine.

29. Inulin is a _____. Ans. Fructosan.

30. Mucosubstances have _____ functions.
Ans. Protective.

31. The hyaluronidase present in hyaluronic acid increases the rate of _____. Ans. Diffusion.

32. The richest source of the hyaluronidase in mammals is the _____. Ans. Testis.

33. Chondroitin sulphates may regulate the flow and _____ of _____ round the cells.
Ans. Concentration; Cations.

34. Heparin is used in _____ as an _____
Ans. Medicine; anticoagulant.

35. The strong _____ of _____ solution can polymerize small quantities of glucose when vigorously boiled. Ans. Alkali, Fehling's.

36. In the stomach some of the sucrose present in the food are hydrolysed by the action of _____.
Ans. HCl.

37. The action of _____ ceases due to high _____ in the stomach. Ans. Amylase; acidity.

38. The pancreatic amylase converts starch and glycogen into a mixture of _____ and _____.
Ans. Maltose; Isomaltose.

39. Cellulose is not digested in human G.I.T. due to the absence of _____. Ans. Cellulase.

40. _____ and _____ are absorbed at a faster rate than fructose. Ans. Galactose; Glucose.

41. The free energy required for the active transport is obtained from the hydrolysis of _____ linked to a sodium pump which expels _____ from the cell. Ans. ATP ; Na^+.

42. The hydrolysis of lactose proceeds at half the rate for _____. Ans. Sucrose.

43. The patient with _____ damage excrete more disaccharides. Ans. Intestinal.

44. Carbohydrate after absorption into the portal blood passes to the _____ circulation through the _____. Ans. Systemic; Liver.

45. There is synthesis of glucose by the _____ and _____ from non-carbohydrate sources.
Ans. Liver; Kidney.

46. Glucose is converted to _____ when _____ storage is exceeded.
Ans. Fatty acids; Glycogen.

47. The process of synthesis of glycogen from _____ in the liver and muscle is called _____.
Ans. Glucose; glycogenesis.

48. The process of breakdown of _____ to glucose in the liver and pyruvate and lactate in the muscle is said to be _____.
Ans. Glycogen; Glycogenolysis.

49. The process of breakdown of glucose and _____ to pyruvate and lactate is called _____.
Ans. Glycogen; Glycolysis.

50. The method of formation of glucose and glycogen from _____ sources is said to be _____.
Ans. Non-carbohydrate; Gluconeogenesis.

51. Glucose-1-phosphate reacts with UTP to form ______ catalyzed by the enzyme ______ .

Ans. UDPG; UDPG Pyrophosphorylase.

52. Glucose is converted to Glucose-6-phosphate by the enzyme glucokinase being stimulated by ______ with the help of the coenzyme ______ .

Ans. Insulin; ATP.

53. In muscle, glycogen synthetase exists in two forms ______ and ______ .

Ans. Synthetase D; Synthetase I.

54. Glycogen synthetase is stimulated by ______ and ______ . Ans. Insulin; glucose.

55. Cyclic AMP is destroyed by ______ .

Ans. Phosphodiesterase.

56. In liver ______ increases the activity of phosphodiesterase. Ans. Insulin.

57. The debranching enzyme causes the hydrolytic splitting of ______ linkages of ______ .

Ans. 1, 6; Glycogen.

58. Glucose-6-phosphate is converted to glucose in the liver by the enzyme ______ with the release of ______ .

Ans. Glucose-6-phosphatase ; Phosphate.

59. In liver, phosphorylase exists in both ______ and ______ form. Ans. Active; Inactive.

60. In muscle, phosphorylase is present in two forms ______ and ______ .

Ans. Phosphorylase a; Phosphorylase b.

61. Conversion of Phosphorylase b to ______ signifies the mechanism for increasing ______ .

Ans. Phosphorylase a; Glycogenolysis.

62. A muscle when contracts under anaerobic condition ______ and ______ become the principal end product. Ans. Pyruvate; Lactate.

63. Pyruvate and Lactate disappear if the muscle contracts under ______ . Ans. Aerobic condition.

64. When ______ is in short supply, ______ is reoxidized by being coupled to the reduction of pyruvate to lactate. Ans. Oxygen; NADH.

65. The enzyme ______ splits fructose-1, 6-diphosphate into two ______ phosphates.

Ans. Aldolase; Triose.

66. 2-phosphoglycerate is catalyzed by ______ to produce phosphoenolpyruvate in presence of ______ . Ans. Enolase; Mg^{++}.

67. Pyruvate is reduced to lactate by ______ in presence of ______ .

Ans. Lactate dehydrogenase; Reduced NAD.

68. Glyceraldehyde-3-phosphate dehydrogenase responsible for the conversion of ______ to 1, 3-diphosphoglycerate is inhibited by ______ .

Ans. Glyceraldehyde-3-phosphate; Iodoacetate.

69. ______ inhibits enolase involved in the conversion of 2- phosphoglycerate to ______ .

Ans. Fluoride; Phosphoenolpyruvate.

70. The reduced NAD enters the ______ chain producing ______ ATP. Ans. Respiratory; 3.

71. Pyruvate dehydrogenase is inhibited by ______ .

Ans. Arsenite.

72. The pyruvate dehydrogenase complex consists of about ______ mol of pyruvate dehydrogenase. Ans. 29.

73. Kreb's cycle consists of a series of reactions in mitochondria which catabolyze the oxidation of ______ to ______ and H_2O in aerobic condition. Ans. Acetyl-CoA; CO_2.

74. α-Ketoglutarate dehydrogenase is inhibited by ______ and causes ______ to accumulate.

Ans. Arsenite; α-ketoglutarate.

75. Gluconeogenesis is a reversal of ______ .

Ans. Glycolysis.

76. Propionyl-CoA undergoes ______ fixation to form D- methylmalonyl-CoA with the help of the coenzyme ______ . Ans. CO_2; Biotin.

77. The conversion of muscle ______ to glucose in liver and its re-entry into muscle is called ______ .

Ans. Lactate; Cori cycle.

78. Transketolase takes the help of the coenzyme ______ and the cofactor ______ .

Ans. TPP; Mg^{++}.

79. The deficiency of ______ in erythrocytes is associated with a tendency to ______ by primaquine.

Ans. Glucose-6-phosphate dehydrogenase; Haemolysis.

80. Ascorbic acid cannot be synthesized in ______ .

Ans. Man.

81. The K_m for fructokinase in ______ is very ______ . Ans. Liver; Low.

82. Galactosemia is an ______ metabolic ______ .
Ans. Inherited; Disease.

83. Glycogenosis Type I occurs due to the deficiency of ______ for which ______ cannot be broken down. Ans. Glucose-6-phosphatase; Glycogen.

84. Type VII glycogenosis is a disease due to the deficiency of ______ .
Ans. Phosphofructokinase.

85. Lactosuria occurs in women during the period of ______ and appears more frequently in the ______ . Ans. Lactation; Afternoon.

86. Insulin is destroyed by ______ and also by ______ . Ans. Insulinase; Peptidase.

87. Thyroxine accelerates hepatic ______ with the rise in blood ______ .
Ans. Glycogenolysis; Sugar.

88. Glucagon stimulates gluconeogenesis in ______ .
Ans. Liver.

89. The renal tubular reabsorption capacity is at the rate of ______ mg/ ______ . Ans. 350; minute.

90. The venous blood sugar level when exceeds 160-180 mg% is called ______ for glucose.
Ans. Renal threshold.

91. Glucocorticoids are the ______ to insulin.
Ans. Antagonists.

92. In juvenile diabetes, the ______ of pancreas are exhausted. Ans. β-cells.

93. Normal urine contains ______ of glucose which cannot be detected by ______ test.
Ans. Traces; Benedict's.

94. Glycosuria results in the use of ether due to hypersecretion of ______ . Ans. Epinephrine.

95. In renal glycosuria carbohydrate utilization is ______ . Ans. Normal.

96. GTT gives the assessment of the degree of severity of ______ . Ans. Diabetes Mellitus.

97. HMP shunt is closely associated with ______ due to the supply of reduced ______ .
Ans. Lipogenesis; NADP.

98. In mild diabetes the blood sugar returns to the fasting level after ______ hours. Ans. 4.

99. ______ and ______ are present in the urine of a person suffering from diabetic coma.
Ans. Ketone bodies; Glucose.

100. In diabetes mellitus ______ occurs due to increased blood ______ level.
Ans. Glycosuria; glucose.

101. The glycogen primer is formed on a ______ primer known as ______ which is a ______ .
Ans. Protein; Glycogenin; Protein.

102. Glycogen exerts an ______ on its own ______ .
Ans. Inhibition; Formation.

103. The inhibition of ______ enhances net ______ .
Ans. Glycogenesis; Glycogenolysis.

104. ______ stimulates the oxidation of glucose in the ______ whereas ______ stimulates the oxidation of glucose in the ______ .
Ans. Hexokinase; Muscle; Glucokinase; Liver.

105. Glucose and ______ binds to different sites on the glucose ______ . Ans. Na^+; transporter.

106. The key enzyme ______ regulates the process of glycogenesis. Ans. Glycogen synthase.

107. The key enzymes of glycolysis are ______ , ______ , and ______ .
Ans. Glucokinase; Phosphofructokinase-1; Pyruvate kinase.

108. The hormone ______ stimulates the key enzymes responsible for glycolysis and antagonizes the key enzymes responsible for ______ .
Ans. Insulin; Gluconeogenesis.

109. The hormones ______ and ______ activate cAMP-dependent protein kinase.
Ans. Epinephrine; Glucagon.

110. Phosphofructokinase-1 is activated by ______ and inhibited by ______ and ______ .
Ans. AMP; ATP; Citrate.

111. In hypoxia, the concentration of ______ in the cells is decreased with the increase in the concentration of ______ . Ans. ATP; AMP.

112. cAMP activates ______ which stimulates the key enzyme ______ of glycogenolysis.
Ans. Protein kinase; Phosphorylase.

113. Just after the onset of muscle contraction ______ is highly increased in ______ by the rapid activation of phosphorylase.
Ans. Glycogenolysis; muscle.

114. Phosphorylase kinase is activated by ______ .
Ans. Ca^{2+}

115. The further activation of phosphorylase kinase is caused by ______ for ______ .

Ans. Calmodulin; Glycogenolysis.

116. The increased pyruvic acid from carbohydrate diet inhibits ______ for which ______ pyruvate dehydrogenase is formed.

Ans. Pyruvate dehydrogenase kinase; active.

117. Acetyl-CoA and ______ formed during diabetes mellitus activate ______ .

Ans. NADH; Pyruvate dehydrogenase kinase.

118. Glycolysis is inhibited due to the less catabolism of ______ and decreased formation of active ______ .

Ans. Pyruvic acid; pyruvate dehydrogenase.

119. The main function of TCA cycle is to provide ______ through ______ and ______ .

Ans. Energy; respiratory chain; oxidative phosphorylation.

120. The activity of TCA cycle depends on ______ and the availability of ______ .

Ans. NAD^+; ADP.

121. The activity of the TCA cycle is determined by the adequate ______ and the rate of doing ______ .

Ans. O_2; Work.

122. Fructose 2, 6-bisphosphate is the positive allosteric effector of ______ and inhibitor of ______ in liver.

Ans. Phosphofructokinase-1; fructose-1, 6-bisphosphatase.

123. Fructose 2, 6-bisphosphate is formed by the ______ of fructose 6-phosphate by ______ .

Ans. Phosphorylation; phosphofructokinase-2.

124. Fructose 2, 6-bisphosphatase is under the ______ control of ______ .

Ans. Allosteric; fructose 6-phosphate.

125. When glucose is short, cAMP inactivates ______ and activates ______ by phosphorylation.

Ans. Phosphofructokinase-2; fructose-2, 6-bisphosphatase.

126. In glucose shortage, gluconeogenesis is stimulated by a decreased concentration of fructose 2, 6-bisphosphate which deactivates ______ and deinhibits ______ .

Ans. Phosphofructokinase-1; fructose-1, 6-bisphosphatase.

127. Glucagon stimulation of glycogenolysis in liver results in ______ release rather than ______ .

Ans. Glucose; Glycolysis.

128. The control point in glycogen metabolism involves a cycle of phosphorylation and dephosphorylation catalyzed by the enzymes ______ and ______ .

Ans. Glycogen synthase; phosphorylase.

129. The key enzymes of TCA cycle are ______ , ______ , and ______ .

Ans. Citrate synthase; Isocitrate dehydrogenase; α-ketoglutarate dehydrogenase.

130. Citrate synthase is allosterically inhibited by ______ and ______ .

Ans. ATP; Long chain fatty acyl-CoA.

131. Succinate dehydrogenase is inhibited by ______ whose availability is controlled by ______ .

Ans. Oxaloacetate; malate dehydrogenase.

132. The key enzymes of gluconeogenesis are ______ , ______ , ______ , and ______ .

Ans. Pyruvate carboxylase; phosphoenolpyruvate carboxykinase; fructose-1, 6-bisphosphatase; glucose-6-phosphatase.

133. High carbohydrate diets increase the ratio of ______ and thus reduce the activities of ______ of gluconeogenesis.

Ans. Insulin/Glucagon; Key enzymes.

134. Fructose-1, 6-bisphosphatase is activated by ______ and ______ .

Ans. ATP; Citrate.

135. The high level of ______ in liver cells inhibits fructose-1, 6-bisphosphatase and thus reduces ______ .

Ans. AMP; Gluconeogenesis.

136. Pyruvate carboxylase is ______ activated by ______ .

Ans. Allosterically; acetyl-CoA.

137. Increased ______ allosterically inhibits ______ .

Ans. ADP; Pyruvate carboxylase.

138. The synthesis of pyruvate carboxylase is stimulated by the hormones ______ , ______ , ______ .

Ans. Glucagon; epinephrine; glucocorticoids.

139. The first reaction of HMP shunt pathway catalyzed by ______ is the ______ step.

Ans. Glucose-6-phosphate dehydrogenase; "rate limiting".

140. The rate limiting step is mainly regulated by the cytoplasmic levels of ______ and ______ .

Ans. $NADP^+$; NADPH.

141. The increased ______ ratio of ______ enables to enhance the shunt pathway.

Ans. Cytoplasmic; $NADP^+$ / NADPH.

142. The hormone insulin ______ the synthesis of both the dehydrogenases and thus ______ the activity of HMP shunt pathway.

Ans. Stimulates; enhances.

143. Thyroid hormone also stimulates the activity of ______ of HMP shunt pathway.

Ans. Glucose-6-phosphate dehydrogenase.

144. The transketolation reaction is affected since the enzyme ______ cannot bind ______ .

Ans. Transketolase; TPP.

145. The subjects suffering from essential pentosuria excrete an increased amount of ______ on administration of ______ .

Ans. Xylulose; Glucuronic acid.

146. Lactate formed by fructolysis is completely oxidized to ______ and ______ .

Ans. CO_2 ; H_2O.

147. In diabetic individuals, sorbitol level is highly ______ in the ______ of the eye.

Ans. Increased; Lens.

148. Fructose by parenteral use can cause depletion of ______ in ______ .

Ans. Adenine nucleotides; liver.

149. Sorbitol by intravenous administration is converted to ______ which is fermented in the ______ producing ______ and ______ level.

Ans. Fructose; Colon; acetate; hydrogen.

150. Hypoglycemia is caused due to the inhibition of ______ by ______ .

Ans. Phosphoglucomutase; fructose 1-phosphate.

Indicate "True" or "False" of the following

1. Cis-trans isomerism occurs in compounds with double bonds. Ans. True.
2. When the OH group around the carbon atom adjacent to the terminal primary alcohol carbon is on the right, the sugar is a member of L-series. Ans. False.
3. The separation of optically active isomers from a racemic mixture is called resolution. Ans. True.
4. Isomers formed as a result of interchange of the –OH and –H on carbon atoms 2, 3 and 4 of glucose are known as epimers. Ans. True.
5. Ribose and Deoxyribose are important constituents of many coenzymes. Ans. True.
6. The aglycone may be –COOH containing substances. Ans. False.
7. The amino sugars are formed by the replacement of hydroxyl group attached to carbon atom 3 of the sugar by an amino group. Ans. False.
8. Mannosamine is an important constituent of mucoprotein. Ans. True.
9. Glucose on reduction yields mannitol. Ans. False.
10. Gluconic acid on heating produces lactones. Ans. True.
11. The osazone is formed by phenylhydrazine only. Ans. False.
12. In maltose molecule there is one β-1, 2 glucosidic linkage. Ans. False.
13. Sucrose is also said to be "invert sugar". Ans. True.
14. Each glycogen molecule contains 5000 to 10000 glucose molecules. Ans. True.
15. Starch is a fructosan. Ans. False.
16. In amylopectin each branch is at an interval of 24-30 glucose units. Ans. True.
17. Inulin produces colour with iodine. Ans. False.
18. Hyaluronic acid does not occur in the skin. Ans. False.
19. Chondroitin sulphates have a marked capacity to hold water. Ans. False.
20. Galactose and fucose are also present in blood group substances. Ans. True.
21. Disaccharides are absorbed and used in the body. Ans. False.

22. Fructose is actively transported. Ans. False.

23. The absorption of glucose is retarded in hypothyroidism. Ans. True.

24. Disacchariduria results in the absence of disaccharidase. Ans. True.

25. Glucuronic acid is not involved in detoxication reactions. Ans. False.

26. Cyclic AMP depresses phosphorylase activity. Ans. False.

27. Thyroid hormones may decrease the synthesis of adenylate cyclase. Ans. False.

28. PGE_5 at low concentration stimulates cGMP accumulation. Ans. False.

29. Conversion of phosphorylase b to phosphorylase a signifies the mechanism for increasing glycogenolysis. Ans. True.

30. Muscle phosphorylase is affected by glucagon. Ans. False.

31. Glucose-6-phosphate is not an important compound in the metabolic pathway. Ans. False.

32. Aldolase requires a coenzyme for its activity. Ans. False.

33. Glycolysis in erythrocytes even under aerobic condition forms lactate. Ans. True.

34. Insulin stimulates phosphofructokinase. Ans. True.

35. S-acetyl lipoate reacts with coenzyme A in presence of dihydrolipoyl transacetylase. Ans. True.

36. Fluoride inhibits pyruvate dehydrogenase. Ans. False.

37. Citrate synthetase forms citrate in the presence of the coenzyme CoA. Ans. False.

38. Under anaerobic conditions, one molecule of glucose produces 8 ATP. Ans. False.

39. One molecule of acetyl-CoA in TCA cycle for complete oxidation forms 12 ATP. Ans. True.

40. Succinate dehydrogenase takes the help of the coenzyme NAD^+. Ans. False.

41. Pyruvate is converted to oxaloacetate by pyruvate decarboxylase. Ans. False.

42. The citric acid cycle is amphibolic in nature. Ans. True.

43. The enzymes of the TCA cycle are located at the cytosol. Ans. False.

44. Oxaloacetate inhibits succinate dehydrogenase. Ans. True.

45. In mammals, the liver and the muscles are the principal organs for gluconeogenesis. Ans. False.

46. The glycolytic activity is low when there is active gluconeogenesis. Ans. True.

47. Dihydroxy acetone phosphate is also converted to glucose in the liver. Ans. True.

48. Acetyl-CoA is permeable to pass through the mitochondrial membrane. Ans. False.

49. The glucogenic amino acids are converted to the intermediates of TCA cycle. Ans. True.

50. CO_2 formed in HMP shunt is utilized for the synthesis of fatty acids and purine bases, etc. Ans. True.

51. The reduced NADP produced in HMP shunt is required for ketone bodies formation. Ans. False.

52. In essential pentosuria large quantities of L-xylulose appear in the urine. Ans. True.

Match the following :

		Answer
(i) Fructokinase	(a) occurs due to the deficiency of branching enzyme in the liver.	(i) .... (j)
(ii) Hexokinase	(b) are the stimulators of insulin release.	(ii) .... (i)
(iii) Galactose	(c) occurs due to the deficiency of lysosomal enzyme.	(iii) .... (h)
(iv) Galactosemia	(d) occurs due to the lack of galactose-l-phosphate uridyl transferase in the liver and red blood cells.	(iv) .... (d)
(v) Von Gierke's disease	(e) has no clinical significance except the wrong idea for glycosuria.	(v) .... (g)
(vi) Pompe's disease	(f) accelerates the conversion of glucose to fat.	(vi) .... (c)
(vii) Amylopectinosis	(g) children tend to develop hypoglycemia.	(vii) .... (a)
(viii) Pentosuria	(h) is also required as a constituent of glycolipids and mucoproteins.	(viii) .... (e)
(ix) Insulin	(i) has less affinity for fructose.	(ix) .... (f)
(x) Sulfonylureas	(j) the activity is not affected by fasting or by insulin.	(x) .... (b)
(xi) Adrenocortical hormones	(k) muscle glycogen does not serve directly as a source of glucose.	(xi) .... (r)
(xii) Epinephrine	(l) the symptoms are sweating, headache, semiconsciousness.	(xii) .... (p)
(xiii) In the liver	(m) normal blood sugar but sugar is present in urine.	(xiii) .... (n)
(xiv) During hypoglycemia	(n) galactose and fructose are converted to glucose.	(xiv) .... (k)
(xv) Administration of diabenese	(o) the symptoms are dehydration, abdominal pain and vomiting.	(xv) .... (q)
(xvi) In hypoglycemic coma	(p) accelerates glycogenolysis in the liver and muscles.	(xvi) .... (l)
(xvii) In diabetic coma	(q) in moderate diabetes mellitus.	(xvii) .... (o)
(xviii) Emotional glycosuria	(r) stimulates gluconeogenesis.	(xviii) .... (s)
(xix) In renal glycosuria	(s) occurs in worried athletes and candidates for competitive examinations.	(xix) .... (m)
(xx) Phlorrhigin	(t) is incapable of synthesizing ribose.	(xx) .... (v)
(xxi) Skeletal muscle	(u) is said to be the common metabolic pathway.	(xxi) .... (t)
(xxii) Malate	(v) poisons the kidney and prevents reabsorption of glucose from the tubules.	(xxii) .... (w)
(xxiii) TCA cycle	(w) is readily diffused from mitochondria.	(xxiii) .... (u)
(xxiv) Adenylate cyclase	(x) depresses active transport of glucose.	(xxiv) .... (z)
(xxv) A mobile carrier	(y) is needed for transport of glucose.	(xxv) .... (y)
(xxvi) Cyanide	(z) occurs in cell membrane.	(xxvi) .... (x)

Appendix 3

LIPIDS

Write the correct answer number of the followings

1. Fats are solids at

   (a) 10°C, (b) 20°C, (c) 30°C, (d) 40°C. Ans. b.

2. Esters of fatty acids with higher alcohols other than glycerol are said to be.

   (a) Waxes, (b) Fats, (c) Both of the above, (d) None of the above, Ans. a.

3. The combination of an amino-alcohol, fatty acid and sialic acid form

   (a) Phospholipids, (b) Sulpholipids, (c) Glycolipids, (d) Aminolipids. Ans. c.

4. Hydrolysis of fat by alkali is called

   (a) Saponification number, (b) Saponification, (c) Both of the above, (d) None of the above. Ans. b.

5. The number of millilitres of 0.1 N. KOH required to neutralize the insoluble fatty acids from 5 grams of fat is called

   (a) Acid number, (b) Acetyl number, (c) Halogenation, (d) Polenske number. Ans. d.

6. The rate of fatty acid oxidation is increased by

   (a) Phospholipids, (b) Glycolipids, (c) Amino lipids, (d) All of the above. Ans. a.

7. Cardiolipin found in mitochondria is formed from

   (a) Lipositol, (b) Phosphatidyl ethanolamine, (c) Phosphatidyl glycerol, (d) None of the above. Ans. c.

8. Lecithin contains a nitrogenous base named as

   (a) Ethanolamine, (b) Choline, (c) Inositol, (d) All of the above. Ans. b.

9. Lecithins contain an unsaturated fatty acid at position

   (a) α–, (b) α– and β–, (c) β–, (d) None of the above. Ans. c.

10. Lecithins are soluble in ordinary fat solvents except

    (a) Benzene, (b) Ethyl alcohol, (c) Methyl alcohol, (d) Acetone. Ans. d.

11. When lecithins exposed to air become

    (a) Black, (b) Brown, (c) Red, (d) Yellow. Ans. b.

12. Lecithins combine with protein to form

    (a) Phosphoprotein, (b) Mucoprotein, (c) Lipoprotein, (d) Glycoprotein. Ans. c.

13. Phosphatidyl inositol is found in

    (a) Cabbages, (b) Soyabeans, (c) Cauliflowers, (d) Apples. Ans. b.

14. Instead of ester link plasmalogens possess an ether link in position

    (a) α–, (b) β–, (c) γ–, (d) None of the above. Ans. a.

15. The alkyl radical in plasmalogen is an alcohol

    (a) Saturated, (b) Unsaturated, (c) Both of the above, (d) None of the above. Ans. b.

16. The concentrations of sphingomyelins are increased in

    (a) Gaucher's disease, (b) Fabry's disease, (c) Febrile disease, (d) Niemann-Pick disease. Ans. d.

17. Sphingomyelins contain a complex amino alcohol named as

(a) Serine, (b) Lysolecithin, (c) Sphingosine, (d) Glycol. Ans. c.

18. The types of sphingomyelins are

(a) 2, (b) 3, (c) 4, (d) 5. Ans. b.

19. Glycolipids contain an amino alcohol

(a) Sphingosine, (b) Iso-sphingosine, (c) Both of the above, (d) None of the above. Ans. c.

20. Cerebrosides may also be classified as

(a) Sphingolipids, (b) Sulpholipids, (c) Aminolipids, (d) Glycolipids. Ans. a.

21. Kerasin contains

(a) Nervonic acid, (b) Hydroxynervonic acid, (c) Cerebronic acid, (d) Lignoceric acid. Ans. d.

22. Oxynervon contains

(a) Nervonic acid, (b) Hydroxynervonic acid, (c) Lignoceric acid, (d) Hydroxylignoceric acid. Ans. b.

23. Gaucher's disease is characterized specially by the increase in

(a) Lignoceric acid, (b) Nervonic acid, (c) Cerebronic acid, (d) Hydroxynervonic acid. Ans. c.

24. Gangliosides are the glycolipids occurring in

(a) Liver, (b) Brain, (c) Kidney, (d) Muscle. Ans. b.

25. Some gangliosides also contain

(a) Sphingosine, (b) Dihydrosphingosine, (c) Both of the above, (d) None of the above. Ans. b.

26. Most of the gangliosides contain sialic acid upto the molecule number

(a) 1, (b) 2, (c) 3, (d) 4. Ans. c.

27. The percentage of triacylglycerol of lipoproteins is

(a) 30, (b) 35, (c) 40, (d) 45. Ans. d.

28. Cholesterol and cholesteryl esters in per cent present in lipoproteins is

(a) 15, (b) 20, (c) 25, (d) 28. Ans. a.

29. Lipoprotein present in cell membrane is by nature

(a) Hydropholic, (b) Hydrophilic, (c) Both of the above, (d) None of the above. Ans. b.

30. The density of lipoproteins increases as the protein content

(a) Rises, (b) Decreases, (c) Highly decreases, (d) Slightly and promptly decreases. Ans. a.

31. Lipoproteins may be identified more accurately by means of

(a) Electrophoresis, (b) Centrifugation, (c) Immunoelectrophoresis, (d) Ultra centrifugation. Ans. c.

32. Very low density lipoproteins are also known as

(a) β-Lipoproteins, (b) Pre-β-lipoproteins, (c) α-lipoproteins, (d) None of the above. Ans. b.

33. The protein moiety of lipoprotein is known as

(a) Apoprotein, (b) Pre-protein, (c) Post-protein, (d) Pseudoprotein. Ans. a.

34. Apoprotein constitutes chylomicrons in percentage is

(a) 0.4, (b) 0.6, (c) 0.8, (d) 1.0. Ans. d.

35. The β-lipoprotein fraction increases in severe

(a) Diabetes Mellitus, (b) Uremia, (c) Nephritis, (d) Muscular dystrophy. Ans. a.

36. Sulpholipids have been isolated from

(a) Heart, (b) Liver, (c) Brain, (d) Intestine. Ans. c.

37. Δ^9 indicates a double bond between carbon atoms of the fatty acids

(a) 8 and 9, (b) 9 and 10, (c) 9 and 11, (d) 9 and 12. Ans. b.

38. The number of carbon atoms in decanoic acid present in butter

(a) 6, (b) 8, (c) 10, (d) 12. Ans. c.

39. Lignoceric acid present in peanut oil contains carbon atoms

(a) 18, (b) 20, (c) 22, (d) 24. Ans. d.

40. Arachidonic acid contains the number of double bonds

(a) 2, (b) 3, (c) 4, (d) 5. Ans. c.

41. The prostaglandins are synthesized from

(a) Arachidonic acid, (b) Oleic acid, (c) Linoleic acid, (d) Linolenic acid. Ans. a.

42. Chaulmoogric acid was used many years ago in the treatment of

(a) Bronchitis, (b) Leprosy, (c) Nephritis, (d) Oedema. Ans. b.

43. The iodine number of essential fatty acids of vegetable oils

(a) High, (b) Very high, (c) Very low, (d) Low. Ans. d.

44. The essential fatty acids retard

(a) Atherosclerosis, (b) Diabetes Mellitus, (c) Nephritis, (d) Oedema. Ans. a.

45. The shape of arachidonic acid.

(a) L, (b) M, (c) U, (d) V. Ans. c.

46. Waxes contain higher alcohols named as

(a) Methyl, (b) Ethyl, (c) Phytyl, (d) Cetyl. Ans. d.

47. The example of cardiac glycosides

(a) Digitonin, (b) Strobanthin, (c) Lycopyll, (d) Digitalis. Ans. b.

48. The reduction product of cholesterol by bacteria in the intestine occurs in feces.

(a) Ergosterol, (b) Demosterol, (c) Coprosterol, (d) Lanosterol. Ans. c.

49. Lieberman-Burchard reaction is performed to detect

(a) Cholesterol, (b) Glycerol, (c) Fatty acid, (d) Vitamin D. Ans. a.

50. Lipase present in the stomach cannot hydrolyze fats owing to

(a) Alkalinity, (b) Acidity, (c) High acidity, (d) Neutrality. Ans. c.

51. Before the action of lipase the fat is emulsified by

(a) Lipoproteins, (b) Phospholipids, (c) Ergosterols, (d) Digitonin. Ans. b.

52. The free glycerol of the total amount of triacylglycerol in the intestinal lumen is present in per cent about

(a) 16, (b) 18, (c) 20, (d) 22. Ans. d.

53. The great majority of absorbed fat appears in the form of

(a) HDL, (b) Chylomicrons, (c) VLDL, (d) LDL. Ans. b.

54. More metabolic water is available on oxidation of

(a) Fatty acids, (b) Glycerol, (c) Both of the above, (d) None of the above. Ans. a.

55. Fatty acids are oxidized by

(a) α–, (b) β–, (c) ω–, (d) All of the above. Ans. d.

56. The fatty acids containing even number and odd number of carbon atoms as well as the unsaturated fatty acids are oxidized by

(a) α-oxidation, (b) β-oxidation, (c) ω-oxidation, (d) All of the above. Ans. b.

57. Long chain fatty acids are first activated to acyl-CoA in the

(a) Cytosol, (b) Mitochondria, (c) Microsomes, (d) Lysosomes. Ans. a.

58. Activation of lower fatty acids occurs within the

(a) Cytosol,(b) Mitochondria, (c) Ribosomes, (d) Microsome. Ans. b.

59. Long chain acyl-CoA penetrates mitochondria in the presence of

(a) Palmitate, (b) Carnitine, (c) Sorbitol, (d) DNP. Ans. b.

60. Acyl-CoA dehydrogenase converts Acyl-CoA to α, β-unsaturated acyl-CoA in presence of the coenzyme

(a) NAD^+, (b) $NADP^+$, (c) ATP, (d) FAD^+. Ans. d.

61. For the activation of long chain fatty acids the enzyme thiokinase requires the cofactor

(a) Mg^{++}, (b) Ca^{++}, (c) Mn^{++}, (d) K^+. Ans. a.

62. ω-oxidation takes place by the hydroxylase in microsomes involving.

(a) Cytochrome b, (b) Cytochrome c, (c) Cytochrome P–450, (d) Cytochrome a_3. Ans. c.

63. Carboxylation of acetyl-CoA to malonyl-coA takes place in presence of

(a) FAD^+, (b) Biotin, (c) NAD^+, (d) $NADP^+$. Ans. b.

64. Malonyl-CoA reacts with the central

(a) – SH group, (b) – NH_2 group, (c) – COOH group, (d) – CH_2OH group. Ans. a.

65. Fatty acid synthesis takes place in presence of the coenzyme

(a) NAD^+, (b) Reduced NAD, (c) $NADP^+$, (d) Reduced NADP. Ans. d.

66. Fatty acids are activated to acyl-CoA by the enzyme thiokinase using.

(a) NAD^+, (b) $NADP^+$, (c) CoA, (d) FAD^+. Ans. c.

67. Phospholipase A_1 attacks the ester bond of phospholipids in position

(a)1, (b) 2, (c) 3, (d) All of the above. Ans. a.

68. Phospholipase C release, 1, 2-diacylglycerol and a phosphoryl base attacking the ester bond in position

(a)1, (b) 2, (c) 3, (d) 4. Ans. c.

69. Phospholipids help the oxidation of
(a) Glycerol, (b) Fatty acids, (c) Glycerophosphates, (d) None of the above. Ans. b.

70. The desaturation and chain elongation system of polyunsaturated fatty acids are greatly diminished in the absence of
(a) Insulin, (b) Glucagon, (c) Epinephrine, (d) Thyroxine. Ans. a.

71. Prostaglandins are liberated in the circulation by the stimulation of
(a) Anterior pituitary glands, (b) Posterior pituitary gland, (c) Adrenal gland, (d) Thyroid gland. Ans. c.

72. Prostaglandins have a common structure based on prostanoic acid which contains carbon atoms
(a) 12, (b) 16, (c) 18, (d) 20. Ans. d.

73. The carbon chains of prostanoic acid are bonded at the middle of the chain by a.
(a) 5-membered ring, (b) 6-membered ring, (c) 8-membered ring, (d) None of the above. Ans. a.

74. All active prostaglandins have at least one double bond between positions.
(a) 7 and 8, (b) 9 and 10, (c) 11 and 12, (d) 13 and 14. Ans. d.

75. The synthesis of prostaglandins is inhibited by
(a) Aspirin, (b) Arsenite, (c) Fluoride, (d) Cyanide. Ans. a.

76. The synthesis of prostaglandins requires the consumption of two molecules of oxygen and two molecules of reduced
(a) NAD, (b) NADP, (c) Glutathione, (d) Liopoate. Ans. c.

77. Prostaglandins lower cyclic AMP in
(a) Thyroid, (b) Adipose tissue, (c) Platelets, (d) Lung. Ans. b.

78. The inhibitory activity of prostaglandins in liver is effective on
(a) Glycogen synthetase, (b) Hexokinase, (c) Phosphorylase, (d) Glucose-6-phosphatase. Ans. a.

79. In adipose tissue prostaglandins decrease.
(a) Lipogenesis, (b) Ketogenesis, (c) Lipolysis, (d) Ketolysis. Ans. a.

80. Prostaglandins increase intestinal motility and cause
(a) Constipation, (b) Loose motion, (c) Diarrhoea, (d) Dysentery. Ans. b.

81. The enzyme responsible for the metabolism of Prostaglandins is blocked by the introduction of a methyl group at the
(a) C_9 position, (b) C_{11} position, (c) C_{13} position, (d) C_{15} position. Ans. d.

82. In adipose tissue less glycerol-3-phosphate is formed in
(a) Diabetes Mellitus, (b) Nephritis, (c) Coronary thrombosis, (d) Heart failure. Ans. a.

83. LDL contains the apoprotein
(a) C – I, (b) C – II, (c) C – III, (d) B. Ans. d.

84. The normal concentration of β-lipoproteins in mg. per cent in plasma
(a) 100, (b) 200, (c) 300, (d) 400. Ans. c.

85. The concentration of triacylglycerol in Pre-β-lipoproteins in per cent
(a) 40, (b) 45, (c) 50, (d) 55. Ans. c.

86. Chylomicrons and VLDL both are released from the intestine or hepatic cell by reverse
(a) Pinocytosis, (b) Diffusion, (c) Osmosis, (d) Passive diffusion. Ans. a.

87. HDL is synthesized and secreted from
(a) Pancreas, (b) Liver, (c) Kidney, (d) Muscle. Ans. b.

88. Serum LDL has been found to be increased in
(a) Obstructive jaundice, (b) Hepatic jaundice, (c) Hemolytic jaundice, (d) Septicemia. Ans. a.

89. The enzyme systems for lengthening and shortening, for saturating and desaturating of fatty acids occur in
(a) Intestine, (b) Muscle, (c) Kidney, (d) Liver. Ans. d.

90. The lowered glucokinase leading to diminished fatty acid synthesis in the liver is caused by the effect of
(a) Feeding, (b) Overfeeding, (c) Starvation, (d) Diarrhoea. Ans. c.

91. The increased levels of plasma free fatty acids resulting from mobilization of fat from

(a) Muscle, (b) Adipose tissue, (c) Kidney, (d) Intestine. Ans. b.

92. Fatty appearance and enlargement of the liver are caused by

(a) Pregnancy, (b) Nephritis, (c) Abortion, (d) Toxaemia of pregnancy. Ans. d.

93. Triacylglycerol is accumulated even in normal rate of fatty acid synthesis by the deficiency of

(a) Lipotropic factor, (b) Vitamin A, (c) Calcium, (d) Vitamin C. Ans. a.

94. The lipotropic activity is possessed by

(a) Casein, (b) Cellulose, (c) Phospholipids, (d) Glycolipids. Ans. a.

95. Fatty liver is caused by

(a) CH_3Cl, (b) CCl_4, (c) $MgSO_4$, (d) CH_3COOH. Ans. b.

96. Fatty liver results in the deficiency of

(a) Stearic acid, (b) Caproic acid, (c) Vitamin A, (d) Pantothenic acid. Ans. d.

97. The normal concentration of ketone bodies in mg. per cent in blood not exceeding.

(a) 1.0, (b) 1.5, (c) 2.0, (d) 2.5. Ans. a.

98. Ketosis generally occurs in

(a) Nephritis, (b) Oedema, (c) Infective hepatic disease, (d) Coronary thrombosis. Ans. c.

99. Ketone bodies are utilized in

(a) Mitochondria, (b) Extrahepatic tissues, (c) Nuclei, (d) Chromosomes. Ans. b.

100. The excretion of ketone body in the urine involves the deficiency of

(a) Na^+, (b) Fe^{++}, (c) Ca^{++}, (d) Mg^{++}. Ans. a.

101. The total amount of cholesterol in grams in the body of man weighing 70 kg is about

(a) 200, (b) 180, (c) 160, (d) 140. Ans. d.

102. The blood cholesterol level is increased in the deficiency of

(a) Vitamin D, (b) Vitamin B_2, (c) Pyridoxine, (d) Inositol. Ans. c.

103. Eicosanoids are formed from

(a) Arachidonate, (b) Palmitate, (c) Stearate, (d) Butyrate. Ans. a.

104. Three groups of eicosanoids synthesized from eicosanoic acid are

(a) PG, TX, LT, (b) PG, TT, LT, (c) TG, TX, LT, (d) TG, TX, DT. Ans. a.

105. Eicosanoic acids are derived from

(a) Linoleic acid, (b) Linolenic acid, (c) Arachidonic acid, (d) All of the above. Ans. d.

106. Prostaglandin endoperoxide synthase catalyze

(a) 4 Mols of O_2, (b) 3 Mols of O_2, (c) 2 Mols of O_2, (d) 1 Mole of O_2. Ans. c.

107. Cyclooxygenase is termed as

(a) Inhibiting enzyme, (b) Suicide enzyme, (c) Oxdizing enzyme, (d) Reducing enzyme. Ans. b.

108. PGH (endoperoxide) is converted to Prostaglandins

(a) D, (b) E, (c) F, (d) All of the above. Ans. d.

109. Leukotrienes are formed in

(a) Leukocytes, (b) Mastocytoma, (c) Platelets, (d) All of the above. Ans. d.

110. Leukotrienes are formed by only

(a) 5-Lipoxygenase, (b) Lipase, (c) Phospholipase, (d) None of the above. Ans. a.

111. Leukotriene A_4 is metabolized to leukotriene

(a) C_2, (b) C_3, (c) C_4, (d) D_4. Ans. c.

112. Leukotriene C_4 is formed by the addition of

(a) Ascorbic acid, (b) Glutathione, (c) Glutamate, (d) Aspartate. Ans. b.

113. PG_3 and TX_3 inhibit the release of

(a) Oleic acid, (b) Palmitoleic acid, (c) Palmitic acid, (d) Arachidonic acid. Ans. d.

114. TXA_3 is a weaker aggregator than

(a) TXA_2, (b) TXA_4, (c) TXA_5, (d) TXA_6. Ans. a.

115. Prostaglandins increase cAMP in

(a) Platelets, (b) Thyroid, (c) Corpus luteum, (d) All of the above. Ans. d.

116. The Eskimos have low plasma concentrations of

(a) Cholesterol, (b) Triacylglycerol, (c) LDL, (d) All of the above. Ans. d.

117. The contraction of smooth muscle is caused by prostaglandins with the minimum concentration of

(a) 1 mg/ml, (b) 10 mg/ml, (c) 100 mg/ml, (d) 1 ng/ml. Ans. d.

118. Leukotrienes are important in

(a) Allergic reaction, (b) Oxidation reaction, (c) Reduction reaction, (d) Inhibitory reaction. Ans. a.

Fill in the blanks of the followings

1. The lipids are a _______ group of compounds.
   Ans. Heterogeneous.
2. The lipids produce metabolites through _______ in the tissues.
   Ans. Oxidation.
3. Glycerides, cholesterol and cholesteryl esters are _______ lipids because they are _______.
   Ans. Neutral; Unchanged.
4. Fats are esters of _______ with _______.
   Ans. Fatty acids; Glycerol.
5. The spreading effect of fat is to lower _______.
   Ans. Surface tension.
6. The number of _______ of KOH required to neutralize 1gm. of fat is called _______.
   Ans. Milligrams; Saponification number.
7. The number of _______ of iodine absorbed by _______ gm of fat is called iodine number.
   Ans. Grams; 100.
8. The number of milligrams of KOH required to neutralize the _______ obtained by saponification of 1gm. of fat after it has been _______ is said to be acetyl number.
   Ans. Acetic acid; acetylated.
9. Waxes are esters of _______ with _______.
   Ans. Fatty acids; higher alcohols.
10. In the human body, the commonest waxes are esters of _______. Ans. Cholesterol.
11. Phospholipids act as carriers of _______ across the membranes. Ans. Inorganic ions.
12. Lecithin can exist in _______ or _______ forms.
    Ans. α ; β.
13. Acetylcholine formed from choline has an important role in the _______ of nerve impulses across _______. Ans. Transmission; synapses.
14. Choline is the most important _______ agent.
    Ans. Lipotropic.
15. Phosphatidyl inositol are more _______ than the other phospholipids. Ans. Acidic.
16. Plasmalogens possess an _______ link in α-position instead of ester link. Ans. Ether.
17. _______ are in much higher concentration in medullated nerve fibres. Ans. Cerebrosides.
18. In chylomicrons the predominant lipid is _______. Ans. Triacylglycerol.
19. In HDL the predominant lipid is _______.
    Ans. Phospholipid.
20. Aminolipids are _______ and _______.
    Ans. Phosphatidyl ethanolamine; Serine.
21. The end methyl carbon of fatty acids is known as the _______ carbon. Ans. ω–.
22. In the structure of linoleic acid the unsaturated carbon atoms are at _______ & _______.
    Ans. Δ^9 ; Δ^{12}.
23. Linolenic and arachidonic acids are formed from _______. Ans. Linoleic acids.
24. The poly unsaturated fatty acids which are not synthesized in the body but are taken from natural sources are called _______.
    Ans. Essential fatty acids.
25. The essential fatty acids of vegetable oils have low _______ and _______.
    Ans. Melting points; iodine number.
26. The example of saponins is _______.
    Ans. Digitonin.
27. Cholesterol in ester form is often referred to as _______ cholesterol esters which are normally rich in _______. Ans. Bound; Linoleic acid.
28. _______ is essential for cholesterol absorption.
    Ans. Bile.
29. Blood cholesterol level increases with _______ and also during _______.
    Ans. Ages; Pregnancy.
30. Cholesterol controls the red cells from being easily _______. Ans. Hemolysed.
31. Cholesterol acts as an antagonist to _______.
    Ans. Phospholipids.
32. Plasma cholesterol level is not changed by the high intake of _______. Ans. Protein.
33. _______ in large amounts causes lowering of plasma cholesterol. Ans. Nicotinic acid.
34. The conversion of acetate to cholesterol is depressed by _______ salts and increased by _______ salt. Ans. Iron; Manganese.

35. Physical exercise _______ the serum cholesterol level. Ans. Lowers.

36. The principal organ for cholesterol synthesis is _______. Ans. Liver.

37. The HMG-CoA reductase activity is not _______ in diabetes mellitus. Ans. Reduced.

38. The activity of HMG-CoA reductase is _______ by cholesterol feeding. Ans. Inhibited.

39. HMG-CoA reductase activity is increased by the administration of the hormone _______ or _______. Ans. Insulin; Thyroxin.

40. Cholesterol synthesis is _______ by cyclic AMP. Ans. Inhibited.

41. Atherosclerosis is characterised by the deposition of _______ and _______ in the connective tissue of the arterial walls. Ans. Cholesteryl ester; Other lipids.

42. The severe atherosclerosis is accompanied by the prolonged elevated levels of _______ and _______ in the blood. Ans. LDL; VLDL.

43. Polyunsaturated fatty acids can _______ serum cholesterol level. Ans. Decrease.

44. Drugs like _______ and _______ can cause the increased fecal excretion of cholesterol and bile acids. Ans. Choloxin; Neomycin.

45. The total carbon numbers in linolenic acid and arachidonic acids are _______ and _______ respectively. Ans. 18; 20.

46. Acetoacetyl-CoA is splitted to _______ by the enzyme _______. Ans. Acetyl-CoA; thiolase.

47. Acetoacetate and β-hydroxyburate are readily oxidized by _______ tissues; but acetone utilization is very _______. Ans. Extrahepatic; Slow.

48. Ketonemia is due to increased production of _______ by the _______. Ans. Ketone bodies; Liver.

49. Acetoacetate reacts with _______ to form acetoacetyl-CoA by the enzyme _______. Ans. Succinyl-CoA; thiophorase.

50. The enzyme responsible for the activation of acetoacetate to _______ is absent from _______ for which these are utilized in extrahepatic tissues. Ans. Acetoacetyl- CoA; Liver.

51. HMG-CoA is splitted into acetoacetic acid and _______ by _______ present in mitochondria. Ans. Acetyl-CoA; HMG-CoA Lyase.

52. The condition in which there is a high concentration of ketone bodies in tissues and _______ is called _______. Ans. Blood; Ketosis.

53. The substances that prevent the accumulation of fat in the _______ are known as _______. Ans. Liver; Lipotropic factor.

54. Vitamin _______ and _______ which are important in hematopoiesis are also able to produce lipotropic effect. Ans. B_{12}; folic acid.

55. Inositol exerts a limited _______ effect in _______ free diet. Ans. Lipotropic; fat.

56. Choline is synthesized using labile _______ groups donated by _______ in the process of transmethylation. Ans. Methyl; Methionine.

57. Fatty liver causes the metabolic block in the synthesis of lipoproteins from _______ and _______. Ans. Lipid; Apoprotein.

58. The hypoglycemia stimulates _______ hormone production which stimulates _______. Ans. Growth; Lipolysis.

59. LDL is formed from _______ and chylomicrons. Ans. VLDL.

60. HDL is synthesized and secreted from _______ and _______. Ans. Liver; Intestine.

61. A mixture of leukotrienes is the _______ reacting substance of _______. Ans. Slow; Anaphylaxis.

62. Leukotrienes are _______ and 5-Lipoxygenase is found in _______ walls. Ans. Vasoactive; Arterial;

63. The mixture of leukotrienes together with leukotriene _______ causes vascular _______. Ans. B_4; Permeability.

64. The prostaglandins and thromboxanes are local _______ and act near the site of _______. Ans. Hormones; Synthesis.

65. Thromboxanes cause _______ and _______ aggregation. Ans. Vasoconstriction; Platelets.

66. PG_3 and TX_3 inhibit the formation of _______ and _______. Ans. PG_2; TX_2.

67. Thromboxanes and ______ are ______. Ans. Prostacyclins; Antagonistic.

68. The symptoms of deficiency of ______ are not relieved by ______. Ans. EFA; Prostaglandins.

69. The subsequent removal of glutamate and glutamate generates Leukotriene ______ and Leukotriene ______. Ans. D_4; E_4.

70. The Leukotrienes are a family of ______ trienes formed from ______.

Ans. Conjugated; Eicosanoic acids.

71. HPETE is formed from ______ by three different ______.

Ans. Arachidonic acids; Lipoxygenases.

72. The half-life of Prostaglandins can be prolonged by blocking the enzyme action with ______ or ______. Ans. Indomethacin; sulphasalazine.

73. ______ and ______ inhibit cycloxygenase.

Ans. Aspirin; Indomethacin.

74. Prostanoid is synthesized by ______ and ______. Ans. Cycloxygenase; Peroxidase.

75. Eicosanoic acids are derived from ______ and directly derived from ______ and ______.

Ans. Linoleic acid; Linolenic acid; Arachidonic acid.

76. Phospholipase A_2 activity is responsible for the synthesis of PG_2, ______ and ______ compounds. Ans. TX_2; LT_4.

77. Eicosanoids are formed from ______ and some other ______ fatty acids.

Ans. Arachidonate; C_{20}

78. PG, LTTX are ______ and ______ active compounds.

Ans. Physiologically; Pharmacologically.

79. The inactivation of Prostaglandin is ______ in most ______ tissues. Ans. Rapid; Mammalian.

80. Oxygen is inserted into 5, ______ and ______ positions of arachindonate to have HPETE.

Ans. 12, 15

81. Niemann-Pick disease is ______ and caused by the deficiency of ______.

Ans. Lipidoses; Sphingomyelinase

82. Gaucher's disease is caused by the deficiency of the enzyme ______. Ans. Glucocerebrosidase.

83. Tay Sach's disease is characterized by increased accumulation of ______ in ______ and ______. Ans. GM_2 ganglioside, brain; spleen.

84. GM_1 gangliosidosis is due to deficiency of the enzyme ______. Ans. β-galactosidase.

85. Eicosanoic acid is formed by ______ and ______. Ans. Linolenic acid; Arachidonic acid.

86. Eicosanoid is composed of ______ and ______.

Ans. Prostanoid; Leukotriene.

87. The modified form of -oxidation in peroxisomes forms ______ and ______ which is broken down by ______.

Ans. Acetyl-CoA; H_2O_2; Catalase.

88. The enzymes in peroxisomes do not attack ______ fatty acids. Ans. Shorter chain.

89. Peroxisomal β-oxidation shortens the side chain of ______ in ______ formation.

Ans. Cholesterol; Bile acid.

90. Carnitine palmitoyl transferase is not present in ______. Ans. Peroxisomes.

91. Fatty acid synthase is a ______.

Ans. Multienzyme complex.

92. Acetyl-CoA carboxylase is a ______ and requires ______. Ans. Multienzyme protein; Biotin.

93. The rate limiting enzyme of fatty acid synthesis is ______ which is an ______ enzyme with ______ as its activator and ______ as its inhibitor. Ans. Acetyl-CoA carboxylase; Allosteric; Citrate; Palmityl-CoA.

94. Acetyl-CoA combines with –SH group of ______ being catalyzed by ______.

Ans. Cysteine; Acetyl transacylase.

95. Malonyl-CoA combines with –SH group on 4'-phosphopantotheine of ______ being catalyzed by ______. Ans. ACP; Malonyl transacylase.

96. Butyryl-CoA is the ______ molecule in ______.

Ans. Primer; Mammary gland.

97. Acetyl-CoA combines with carnitine by ______ to form acyl carnitine which is cleaved into the matrix by ______ to form acetyl-CoA.

Ans. Carnitine palmitoyl transferase 1; Carnitine palmitoyl transferase 11.

98. Acyl-CoA inhibits the mitochondrial ______ preventing activation of the enzyme by ______.

Ans. Tricarboxylate transporter; Citrate.

99. The rate of lipogenesis is ______ in well fed animal whose diet contains a high proportion of ______ . Ans. Higher; Carbohydrate.

100. Lipogenesis is higher when ______ is fed instead of ______ . Ans. Sucrose; Glucose.

101. ______ activates acetyl-CoA carboxylase and inhibits ______ . Ans. Insulin; Lipolysis.

102. Acyl-CoA ______ pyruvate dehydrogenase by inhibiting the ______ exchange transporter of the inner mitochondrial membrane.

Ans. Inhibits; ATP-ADP.

103. Krabbe's disease is due to the deficiency of the enzyme ______ . Ans. β-galactosidase.

104. In uncontrolled diabetes mellitus ______ appearance and ______ of liver occurs.

Ans. Fatty; Enlargement.

105. Alcoholism leads to ______ in liver causing ______ . Ans. Triacylglycerol; Cirrhosis.

106. The antibiotic ______ causes ______ liver.

Ans. Puromycin; fatty.

107. The catabolism of VLDL occurs by ______ with ______ acting as cofactor.

Ans. Lipoprotein lipase; Phospholipid.

108. VLDL acts as a vehicle for the transport of ______ from the liver to ______ .

Ans. Triacylglycerol; Extrahepatic tissue.

109. IDL is formed when ______ is largely removed by the action of ______ .

Ans. Triglyceride; Lipoprotein lipase.

110. The receptors recognize the ______ component of the LDL and bind ______ .

Ans. Apo B-100, Apo E.

111. The human LDL receptor is made up of a ______ region of ______ amino acid residues.

Ans. Cysteine-rich; 292.

112. Cholesterol ______ the synthesis of new ______ receptors. Ans. Inhibits; LDL.

113. Sulfonylureas ______ fatty acid oxidation by inhibiting ______ .

Ans. Reduce; Carnitine Palmitoyl transferase.

114. Jamaican vomiting sickness occurs from the eating of ______ fruit of the ______ .

Ans. Unripe; Akee tree.

115. Zellweger's syndrome occurs in the inherited absence of ______ in all ______ .

Ans. Peroxisomes; Tissues.

116. Refsum's disease has an inherited defect in ______ that allows accumulation of ______ .

Ans. α-oxidation; phytanic acid.

117. Niacin in the dosage of ______ thrice daily ______ cholesterol and triacylglycerol.

Ans. $\frac{1}{2}$ to 1 gm.; Lowers.

118. Sunflower oil contains a ______ proportion of ______ fatty acids and olive oil contains a high concentration of ______ fatty acids.

Ans. High; Polyunsaturated; Monounsaturated.

119. Diets rich in ______ inhibits the conversion of ______ to bile acids.

Ans. Palmitate; Cholesterol.

120. Saturated fatty acids cause the formation of smaller ______ particles that contain more ______ . Ans. VLDL; Cholesterol.

121. Moderate alcohol consumptions have a lower incidence of ______ disease because of elevation of ______ concentration.

Ans. Coronary heart; HDL.

122. Red wine is particularly ______ because of its content of ______ .

Ans. Beneficial; antioxidants.

123. Increased levels of very long chain ______ have been found in the ______ of patients with Zellweger's syndrome.

Ans. Polyenoic acids; brains.

124. Garlic in dietary preparations lowers ______ and ______ . Ans. Cholesterol; triglyceride.

125. The intermediates from squalence to cholesterol are attached to a special ______ known as the ______ . Ans. Carrier protein; squalene and sterol carrier protein.

126. Farnesyl pyrophosphate is the ______ for the ______ of dolichol.

Ans. Branch point; synthesis.

127. Cholesterol level is ______ due to efflux of cholesterol from the ______ to lipoproteins.

Ans. Decreased; membrane.

128. In the form of cholesterol-sterol carrier protein cholesterol is converted to ______ and ______.
Ans. Steroid hormones; bile acids.

129. Squalene is converted to ______ by ______.
Ans. Squalene epoxide; Squalene epoxidase.

130. Probucol causes to increase ______ catabolism but its ______ properties are more important.
Ans. LDL; antioxidant.

Indicate "True" or "False" of the followings

1. Some lipoproteins are also glycoproteins. Ans. True.
2. The apoproteins present in chylomicrons are B, C–I, C–II, C–III. Ans. True.
3. α-lipoproteins occupy the β-globulin region after electrophoresis. Ans. False.
4. Free fatty acids are classified with the other plasma lipoproteins. Ans. False.
5. Apoprotein B is synthesized by ribosomes in the rough endoplasmic reticulum. Ans. True.
6. Lipolysis is controlled by the amount of cAMP present in the tissue. Ans. True.
7. Nicotinic acid and Prostaglandin E_1 stimulate the synthesis of cAMP. Ans. False.
8. Insulin inhibits the release of free fatty acids from adipose tissue and enhances lipogenesis. Ans. True.
9. Glucocorticoids and Thyroid hormones increase lipolysis. Ans. False.
10. Administration of caffeine causes no change in plasma-free fatty acids in humans. Ans. False.
11. When glucose utilization in adipose tissue is reduced, the minimum portion of it is utilized to form acylglycerol. Ans. False.
12. Triacylglycerol is hydrolyzed by a hormone-sensitive lipase to form free fatty acids and glycerol. Ans. True.
13. Prostaglandins are used to control inflammation. Ans. True.
14. Prostaglandins are also used as contraceptives to prevent conception. Ans. True.
15. PGE_2 does not irritate the mucosa lining of the throat. Ans. False.
16. The direct inhibitory effect is produced on glycogen synthetase in liver by prostaglandins. Ans. True.
17. Since the prostaglandins are synthesized from essential fatty acids they relieve symptoms of essential fatty acid deficiency. Ans. False.
18. Fourteen prostaglandins have been isolated from male reproductive tract. Ans. True.
19. Phospholipase A_2 catalyzes the hydrolysis of the ester bond in position 2 of glycerophospholipids to form sphingosine. Ans. False.
20. Phospholipids do not facilitate lipid transport between tissues. Ans. False.
21. Cardiolipin is formed from phosphatidyl glycerol. Ans. True.
22. Choline is directly activated to form phosphatidyl choline. Ans. False.
23. Butyryl-CoA is the primer molecule in mammalian liver and mammary gland. Ans. True.
24. In mammalian systems, free palmitate is liberated from the enzyme complex by hydrolysis. Ans. True.
25. If biotin is bound to the protein avidin of egg-white, acetyl-CoA carboxylase is inhibited. Ans. True.
26. Fatty acid oxidation captures high-energy phosphate as 48 per cent. Ans. False.
27. Unsaturated fatty acids are not oxidized by β-oxidation. Ans. False.
28. The reduced NAD is required for the conversion of β-hydroxyacyl-CoA to β-keto acyl-CoA. Ans. False.
29. Fatty acids with odd number of carbon atoms are not oxidized by β-oxidation. Ans. False.
30. "Fatty acid oxidase" are found in the mitochondrial matrix adjacent to the respiratory chain. Ans. True.
31. Thiokinases are found both inside and outside the mitochondria. Ans. True.
32. The lipids occur in animal kingdom only. Ans. False.

33. Fat serves as an efficient source of energy when stored in liver only. Ans. False.

34. The phosphatides of blood platelets are involved in the production of thromboplastin actvity in the early stages of blood clotting. Ans. True.

35. Lipids produce metabolites through oxidation in tissues. Ans. True.

36. Oils are liquids at 30°C. Ans. False.

37. Phospholipids are esters of fatty acids and nitrogen containing base. Ans. False.

38. Fats act as insulator for the gain of body heat. Ans. False.

39. The number of milligrams of KOH requird to neutralize the free fatty acids of one gram of fat is called acid number. Ans. True.

40. Peroxides are formed at the double bonds of unsaturated fatty acids by oxidation when exposed to air. Ans. True.

41. Waxes resemble the fat and are usually solid. Ans. True.

42. Phospholipids act as prosthetic group to enzymes. Ans. False.

43. Phosphatidic acid is important in the synthesis of prostaglandins. Ans. False.

44. Lecithins constitute valuable agents for the emulsification of fats and oils. Ans. True.

45. Lecithins form complexes with many different substances especially with other lipids, proteins, carbohydrates. Ans. True.

46. Choline is not the lipotropic agent. Ans. False.

47. Phosphatidyl inositol contains no base but has inositol in its place. Ans. True.

48. Plasmalogens give a negative reaction when tested for aldehydes with Schiff's reagent. Ans. False.

49. Cerebrosides are chief constituents of skeletal muscle. Ans. False.

50. Gangliosides contain ceramide and glucose only. Ans. False.

51. Lipoproteins may be identified more accurately by means of immunoelectrophoresis. Ans. True.

52. The structural integrity of cell surface and subcellular particles are maintained by lipoproteins. Ans. True.

53. Stearic acid contains 16 carbon atoms. Ans. False.

54. Palmitoleic acid is a saturated fatty acid. Ans. False.

55. Essential fatty acids effect the prolongation of clotting time and increase fibrinolytic acitivity. Ans. True.

56. The lipids of gonads also contain a high concentration of polyunsaturated fatty acids. Ans. True.

57. Cholesterol occurs in animal fats as well as in plant fats. Ans. False.

58. Ergosterol is the precursor of Vitamin D. Ans. True.

Match the following :		Answer
1. Glycerol	(a) allows long chain acyl groups to penetrate the mitochondria.	1. ...... (e)
2. Long chain fatty acids	(b) formed from the β-oxidation of fatty acids with odd number of carbon atoms is converted to Succinyl-CoA.	2. ...... (c)
3. Carnitine-Palmityl acyl transferase	(c) are oxidized in liver, heart, kidney, testis, brain and adipose tissue.	3. ...... (a)
4. Crotonase	(d) yields 129 ATP by β-oxidation.	4. ...... (f)
5. Propionyl-CoA	(e) is utilized by the tissues where glycerokinase is present abundantly.	5. ...... (b)
6. α-oxidation	(f) catalyzes the formation β-hydroxy acyl-CoA from α, β-unsaturated acyl-CoA by the addition of water.	6. ...... (g)
7. Palmitic acid	(g) does not require CoA intermediates.	7. ...... (d)
8. Stearic acid	(h) undergoes 8 times β-oxidation and produces 9 molecules of acetyl-CoA.	8. ...... (h)
9. Reduced NADP	(i) is required in the initial carboxylation of acetyl-CoA to malonyl-CoA.	9. ...... (k)
10. Carbon dioxide	(j) is present in maternal venous blood immediately before uterine contractions in normal spontaneous labor.	10. ...... (i)
11. , ATP and CoA	(k) is required for the synthesis of fatty acids in the extra mitochondria.	11. ...... (m)
12. Reduced NAD	(l) is an effective nasal decongestant.	12. ...... (o)
13. Prostaglandins	(m) activate fatty acids to acyl-CoA by thiokinase.	13. ...... (n)
14. Prostaglandin F	(n) are a group of naturally occurring substances synthesized primarily in the prostate.	14. ...... (p)
15. PGF 2α	(o) is required for the catalytic conversion of stearyl-CoA to oleonyl-CoA by an enzyme system.	15. ...... (j)
16. PGE_1	(p) has a hydroxyl group at position 9.	16. ...... (l)
17. 15-hydroxy prostaglandin dehydrogenase	(q) performs important functions in the metabolism of lipids and it is under nervous control.	17. ...... (t)
18. Adipose tissue	(r) does not contain apoprotein C but only apoprotein A and it is formed by the liver.	18. ...... (q)
19. Theophylline	(s) in the daily excretion of normal person are less than one milligram.	19. ...... (z)
20. Thyroid hormone	(t) metabolizes prostaglandin quickly and is present in most mammalian tissues.	20. ...... (y)
21. Chylomicrons	(u) is converted to lanosterol by ring closures.	21. ...... (x)
22. Nascent HDL	(v) is increased by the consumption of excessive amounts of sucrose.	22. ...... (r)
23. Liver	(w) is the site for the synthesis of bile acids from cholesterol.	23. ...... (w)
24. Ketone bodies	(x) derived from intestinal absorption of triacylglycerol.	24. ...... (s)
25. Serum cholesterol level	(y) inhibits phosphodiesterase activity.	25. ...... (v)
26. Squalene	(z) inhibits cyclic 3′, 5′–nucleotide phosphodiesterase.	26. ...... (u)

Appendix 4

PROTEINS

Write the correct number of the followings

1. Out of 200 different amino acids found in nature the number of amino acids present in protein.
(a) 20, (b) 25, (c) 30, (d) 35. **Ans.** a.

2. Enzyme-catalyzed hydrolysis of proteins produces amino acids of the form.
(a) D–, (b) DL–, (c) L–, (d) All of the above. **Ans.** c.

3. The ionizable groups of amino acids are at least
(a) 1, (b) 2, (c) 3, (d) 4. **Ans.** b.

4. The carboxyl groups of amino acids exist almost entirely as the conjugated base at pH
(a) 6.6, (b) 6.8, (c) 7.2, (d) 7.4. **Ans.** d.

5. The pH of Arginine is
(a) 10.5, (b) 10.6, (c) 10.8, (d) 11.0. **Ans.** c.

6. The neutral amino acid is
(a) Leucine, (b) Lysine, (c) Proline, (d) Histidine. **Ans.** a.

7. The amino acid containing hydroxyl group
(a) Alanine, (b) Isoleucine, (c) Arginine, (d) Threonine. **Ans.** d.

8. The sulphur containing amino acid
(a) DOPA, (b) Homoserine, (c) Methionine, (d) Valine. **Ans.** c.

9. The basic amino aicd is
(a) Glycine, (b) Histidine, (c) Proline, (d) Serine. **Ans.** b.

10. All amino acids are optically active except
(a) Glycine, (b) Serine, (c) Threonine, (d) Tryptophan. **Ans.** a.

11. The amino acid which synthesizes many hormones
(a) Valine, (b) Phenylalanine, (c) Alanine, (d) Histidine. **Ans.** b.

12. Amino acids are insoluble in
(a) Acetic acid, (b) Chloroform, (c) Ethanol, (d) Benzene. **Ans.** d.

13. The melting point of amino acid is above.
(a) 100°C, (b) 180°C, (c) 200°C, (d) 220°C. **Ans.** c.

14. From two amino acids peptide bond formation involves removal of one mole of
(a) Water, (b) Ammonia, (c) Carbon dioxide, (d) Carboxylic acid. **Ans.** a.

15. Polymers of more than 100 amino acids are termed
(a) Proteins, (b) Polypeptides, (c) Both of the above, (d) None of the above. **Ans.** a.

16. Insulin degradation or disulfide bond formation is effected by
(a) Pyruvate dehydrogenase, (b) Xylitol reductase, (c) Glutathione reductase, (d) Xanthine oxidase. **Ans.** c.

17. The example of globulins
(a) Leucosin, (b) Tuberin, (c) Oryzenin, (d) Legunelin. **Ans.** b.

18. The example of scleroproteins
(a) Glutenin, (b) Gliadin, (c) Salmine, (d) Elastin. **Ans.** d.

19. The example of phosphoprotein

(a) Mucin, (b) Ovovitellin, (c) Ovomucoid, (d) Tendomucoid. Ans. b.

20. The example of metalloprotein

(a) Siderophilin, (b) Osseomucoid, (c) Elastin, (d) All of the above. Ans. a.

21. The example of chromoprotein

(a) Salmine, (b) Catalase, (c) Zein, (d) Gliadin. Ans. b.

22. Protein structures are confirmed by weak bonds, the example of which is

(a) Hydrophobic, (b) Disulphide, (c) Peptide, (d) All of the above. Ans. a.

23. Proteases produce polypeptides from proteins by

(a) Oxidizing, (b) Reducing, (c) Hydrolyzing, (d) None of the above. Ans. c.

24. Proteins react with biuret reagent which is suggestive of 2 or more

(a) Hydrogen bonds, (b) Peptide bonds, (c) Disulphide bonds, (d) Hydrophobic bonds. Ans. b.

25. Insulin is oxidized to separate the protein molecule into its constituent polypeptide chains without affecting the other part of the molecule by the use of

(a) Performic acid, (b) Oxalic acid, (c) Citric acid, (d) Malic acid. Ans. a.

26. The disulphide bond is not broken under the usual conditions of

(a) Filtration, (b) Reduction, (c) Oxidation, (d) Denaturation. Ans. d.

27. Each hydrogen bond is quite.

(a) Strong, (b) Weak, (c) Both of the above, (d) None of the above. Ans. b.

28. A coiled structure in which peptide bonds are folded in regular manner by

(a) Globular proteins, (b) Fibrous proteins, (c) Both of the above, (d) None of the above. Ans. a.

29. In many proteins, the hydrogen bonding produces a regular coiled arrangement called

(a) β-helix, (b) α-helix, (c) Both of the above, (d) None of the above. Ans. b.

30. Many globular proteins are stable in solution although they lack in

(a) Hydrogen bonds, (b) Salt bonds, (c) Non-polar bonds, (d) Disulphide bonds. Ans. d.

31. Each turn of α-helix contains the number of amino acids

(a) 2.8, (b) 3.2, (c) 3.4, (d) 3.6. Ans. d.

32. The distance travelled per turn of α-helix in nm is

(a) 0.34, (b) 0.44, (c) 0.54, (d) 0.64. Ans. c.

33. The space covered by each amino acid residue of α-helix in nm is

(a) 0.09, (b) 0.12, (c) 0.15, (d) 0.18. Ans. c.

34. α-helix is disrupted by certain amino acid like

(a) Proline, (b) Arginine, (c) Histidine, (d) Lysine. Ans. a.

35. α-helix is stabilized by

(a) Hydrogen bonds, (b) Disulphide bonds, (c) Salt bonds, (d) Non-polar bonds. Ans. a.

36. Glutamic dehydrogenase is a

(a) Monomer, (b) Tetramer, (c) Dimer, (d) None of the above. Ans. b.

37. Aldolase molecule is a

(a) Dimer, (b) Trimer, (c) Both of the above, (d) None of the above. Ans. b.

38. Foetal haemoglobin contains.

(a) Two α and two γ chains, (b) Two α and two β chains, (c) Both of the above, (d) None of the above. Ans. a.

39. When haemoglobin takes up oxygen there is a change in the structure due to the moving closer together of

(a) β chains, (b) α chains, (c) γ chains, (d) α and γ chains. Ans. a.

40. The hydrogen bonds between peptide linkages are interfered by

(a) Guanidine, (b) Uric acid, (c) Salicylic acid, (d) Oxalic acid. Ans. a.

41. The hydrogen bonds in the secondary and tertiary structure of proteins are directly attacked by

(a) Salts, (b) Alkalis, (c) Detergents, (d) All of the above. Ans. b.

42. The digestibility of certain denatured proteins by proteolytic enzymes is

(a) Decreased, (b) Increased, (c) Normal, (d) None of the above. Ans. b.

43. The antigenic or antibody functions of proteins by denaturation are frequently.

(a) Not changed, (b) Changed, (c) Both of the above, (d) None of the above. Ans. b.

44. In case of severe denaturation of protein, there is

(a) Reversible denaturation, (b) Moderate reversible denaturation, (c) Irreversible denaturation, (d) None of the above. Ans. c.

45. Bovine ribonuclease of single polypeptide chain of 124 amino acid residues with small molecular weight contains the number of disulphide bonds.

(a) 2, (b) 3, (c) 4, (d) 6. Ans. c.

46. When egg albumin is heated till it is coagulated, the secondary and tertiary structures of the proteins are completely lost resulting in a mixture of randomly arranged.

(a) Dipeptide chains, (b) Tripeptide chains, (c) Polypeptide chains, (d) All of the above.

Ans. c.

47. In glycoproteins the carbohydrate is in the form of disaccharide units, the number of units are

(a) 50 – 100, (b)200 – 300, (c) 400 – 500, (d) 600 – 700.

Ans. d.

48. The disaccharide units of glycoproteins are attached to the peptide chain, one per

(a) 3.4 amino acid residues, (b) 4.4 amino acid residues, (c) 5.4 amino acid residues, (d) 6.4 amino acid residues. Ans. d.

49. The milk protein in the stomach of the infants is digested by

(a) Pepsin, (b) Trypsin, (c) Chymotrypsin, (d) Rennin. Ans. d.

50. Achylia gastrica is said to be when absence of

(a) Pepsin only, (b) Both pepsin and HCl, (c) HCl only, (d) All of the above. Ans. b.

51. The pH of gastric juice become low in

(a) Hemolytic anemia, (b) Pernicious anemia, (c) Both of the above, (d) None of the above.

Ans. b.

52. In the small intestine, trypsin hydrolyzes peptide linkages containing

(a) Arginine, (b) Histidine, (c) Serine, (d) Aspartate.

Ans. a.

53. Chymotrypsin in the small intestine hydrolyzes peptide linkage containing

(a) Phenylalanine, (b) Alanine, (c) Methionine, (d) Valine. Ans. a.

54. Carboxypeptidase B in the small intestine hydrolyzes peptides containing.

(a) Leucine, (b) Isoleucine, (c) Arginine, (d) Cysteine.

Ans. c.

55. The transport of amino acids is regulated by active processes of different numbers –

(a) 1, (b) 2, (c) 3, (d) 4. Ans. c.

56. The third active process for amino acids transport involves

(a) Basic amino acids, (b) Neutral amino acids, (c) Acidic amino acids, (d) Sulphur containing amino acids. Ans. b.

57. The neutral amino acids for absorption need

(a) TPP, (b) B_6-PO_4, (c) NAD^+, (d) $NADP^+$. Ans. b.

58. If one amino acid is fed in excess, the absorption of another is

(a) Slightly accelerated, (b) Moderately accelerated, (c) Highly accelerated, (d) Retarded.

Ans. d.

59. Under normal conditions, food proteins are generally readily digested upto the per cent.

(a) 67 to 73, (b) 74 to 81, (c) 82 to 89, (d) 90 to 97.

Ans. d.

60. By overheating the nutritional value of cereal proteins is

(a) Increased, (b) Lowered, (c) Unchanged, (d) None of the above. Ans. b.

61. More than half of the protein of the liver and intestinal mucosa are broken down and resynthesized in

(a) 10 days, (b) 12 days, (c) 15 days, (d) 18 days.

Ans. a.

62. The half-life of antibody protein is about

(a) 4 weeks, (b) 3 weeks, (c) 2 weeks, (d) 1 week.

Ans. c.

63. Protein anabolism is stimulated by

(a)ACTH, (b) Testosterone, (c) Glucagon, (d) Epinephrine. Ans. b.

64. The metabolism of protein is integrated with that of carbohydrate and fat through

(a) Oxaloacetate, (b) Citrate, (c) Isocitrate, (d) Malate.

Ans. a.

65. The building up and breaking down of protoplasm are concerned with the metabolism of

(a) Carbohydrate, (b) Fat, (c) Protein, (d) Minerals. **Ans.** c.

66. The amino acids abstracted from the liver are not utilized for repair or special synthesis but are broken down to

(a) Keto acids, (b) Sulphur dioxide, (c) Water, (d) Ammonia. **Ans.** d.

67. The unwanted amino acids abstracted from the tissues are either used up by the tissue or in the liver converted into

(a) Ammonia, (b) Urea, (c) Ammonium salts, (d) Uric acid. **Ans.** b.

68. Amino acids provide the nitrogen for the synthesis of

(a) The bases of the phospholipids, (b) Uric acid, (c) Glycolipids, (d) Chondroitin sulphates. **Ans.** a.

69. The metabolism of all proteins ingested over and above the essential requirements is called

(a) Endogenous metabolism, (b) Exogenous metabolism, (c) Both of the above, (d) None of the above. **Ans.** b.

70. Sulphur containing amino acids after catabolism produces a substance which is excreted

(a) SO_2, (b) HNO_3, (c) H_2SO_4, (d) H_3PO_4. **Ans.** c.

71. Ethereal sulphate is synthesized from the amino acid

(a) Neutral, (b) Acidic, (c) Basic, (d) Sulphur containing. **Ans.** d.

72. Keratin, the protein of hair, is synthesized from the amino acid

(a) Glycine, (b) Serine, (c) Proline, (d) Methionine. **Ans.** d.

73. In human and other ureotelic organisms, the end product of amino acid nitrogen metabolism is

(a) Bile acids, (b) Ketone bodies, (c) Urea, (d) Barium sulphate. **Ans.** c.

74. The end product of amino acid nitrogen metabolism in uricotelic organisms (eg. reptiles and birds) is

(a) Bilirubin, (b) Urea, (c) Uric acid, (d) Biliverdin. **Ans.** c.

75. The transaminase activity needs the coenzyme

(a) ATP, (b) $B_6 - PO_4$, (c) FAD^+, (d) NAD^+. **Ans.** b.

76. Transamination is a

(a) Irreversible process, (b) Reversible process, (c) Both of the above, (d) None of the above. **Ans.** b.

77. Most amino acids are substrates for transamination except

(a) Alamine, (b) Threonine, (c) Serine, (d) Valine. **Ans.** b.

78. Oxidative conversion of many amino acids to their corresponding α-keto acids occurs in mammalian

(a) Liver and kidney, (b) Adipose tissue, (c) Pancreas, (d) Intestine. **Ans.** a.

79. The α-keto acid is decarboxylated by H_2O_2 forming a carboxylic acid with one carbon atom less in absence of the enzyme

(a) Catalase, (b) Decarboxylase, (c) Deaminase, (d) Phosphatase. **Ans.** a.

80. The activity of mammalian L-amino acid oxidase, an FMN-flavoprotein, is quite

(a) Slow, (b) Rapid, (c) Both of the above, (d) None of the above. **Ans.** a.

81. From dietary protein as well as from the urea present in fluids secreted into the gastrointestinal tract intestinal bacteria produce

(a) Carbon dioxide, (b) Ammonia, (c) Ammonium sulphate, (d) Creatine. **Ans.** b.

82. The symptom of ammonia intoxication includes

(a) Blurring of vision, (b) Constipation, (c) Mental confusion, (d) Diarrhea. **Ans.** a.

83. Ammonia intoxication symptoms occur when brain ammonia levels are

(a) Slightly diminished, (b) Highly diminished, (c) Increased, (d) All of the above. **Ans.** c.

84. Ammonia production by the kidney is depressed in

(a) Acidosis, (b) Alkalosis, (c) Both of the above, (d) None of the above. **Ans.** b.

85. Ammonia is excreted as ammonium salts during metabolic acidosis but the majority is excreted as

(a) Phosphates, (b) Creatine, (c) Uric acid, (d) Urea. **Ans.** d.

86. Synthesis of glutamine is accompanied by the hydrolysis of

(a) ATP, (b) ADP, (c) TPP, (d) Creatine phosphate.

Ans. a.

87. In brain, the major mechanism for removal of ammonia is the formation of

(a) Glutamate, (b) Aspartate, (c) Asparagine, (d) Glutamine. **Ans.** d.

88. Carbamoyl phosphate synthetase structure is markedly changed in the presence of

(a) N-Acetyl glutamate, (b) N-Acetyl aspartate, (c) Neuraminic acid, (d) Oxalate. **Ans.** a.

89. In bacteria, the synthesis of carbamoyl phosphate takes place from the substance

(a)Ammonium salts, (b) Ammonia, (c) Glutamine, (d) Aspartate. **Ans.** c.

90. The competitive inhibitor of arginine is

(a) Citrulline, (b) Malate, (c) Lysine, (d) Serine.

Ans. c.

91. The biosynthesis of urea occurs mainly in the liver

(a) Cytosol, (b) Mitochondria, (c) Microsomes, (d) Nuclei. **Ans.** b.

92. One mol of urea is synthesized at the expense of the mols of ATP

(a) 2, (b) 3, (c) 4, (d) 5. **Ans.** b.

93. Urea biosynthesis occurs mainly in the liver involving the number of amino acids

(a) 3, (b) 4, (c) 5, (d) 6. **Ans.** d.

94. The normal daily output of urea through urine in grams

(a) 10 to 20, (b) 15 to 25, (c)20 to 30, (d) 25 to 35.

Ans. c.

95. In severe acidosis, the output of urea is

(a) Decreased, (b) Slightly increased, (c) Highly increased, (d) Moderately increased. **Ans.** a.

96. Uremia occurs in

(a)Cirrhosis of the liver, (b) Nephritis, (c) Diabetes Mellitus, (d) Coronary thrombosis. **Ans.** b.

97. Clinical symptom in urea cycle disorder is

(a) Mental retardation, (b) Drowsiness, (c) Diarrhoea, (d) Oedema. **Ans.** a.

98. The sparing action of methionine is

(a) Tyrosine, (b) Cystine, (c) Arginine, (d) Tryptophan.

Ans. b.

99. NH_4^+ aminates glutamate to form glutamine requires ATP and

(a) Mg^{++}, (b) Ca^{++}, (c) Na^+, (d) K^+. **Ans.** a.

100. Glutathione is a

(a) Dipeptide, (b) Tripeptide, (c) Polypeptide, (d) None of the above. **Ans.** b.

Fill in the blanks of the followings

1. The amino acids which cannot be synthesized by the organism from substances ordinarily present in the diet according to physiological requirements are called _______ .

Ans. Essential amino acids.

2. Two amino acids _______ and _______ which are required for animals, are _______ for humans.

Ans. Arginine; Histidine; Nutritionally semi-essential.

3. The omission of an essential amino acid from the diet results in the _______ .

Ans. Negative nitrogen balance.

4. Alanine is formed from pyruvate by _______ in presence of the coenzyme _______ .

Ans. Transamination; B_6-PO_4.

5. Glutamate is synthesized in animals by _______ in presence of the coenzyme _______ .

Ans. Glutamate dehydrogenase; Reduced NADP.

6. Asparagine is synthesized from aspartate by _______ with the help of coenzyme _______ and Mg^{++}. **Ans.** Asparagine synthetase; ATP.

7. Glycine is synthesized from _______ as well as _______ . **Ans.** Serine; Choline.

8. Cysteine is formed from _______ .

Ans. Methionine.

9. Phenylalanine is converted to tyrosine by _______ with the incorporation of one atom of oxygen the para position of phenylalanine and the other atom is reduced forming water, the reducing power supplied by _______ is immediately provided as tetrahydrobiopterine.

Ans. Phenylalanine hydroxylase complex; Reduced NADP.

10. The only ketogenic amino acid ______ .
Ans. Leucine.

11. Arginine is converted to ______ by arginase wtih the removal of ______ . **Ans.** Ornithine; Urea.

12. Histidine on deamination produces ______ which is converted to 4-imidazolone-5-propionate by ______ . **Ans.** Urocanic acid; Urocanase.

13. Serine forms ______ by ______ rich in liver of rats. **Ans.** Pyruvate; Serine dehydratase.

14. Cystine is converted to cysteine by an ______ dependent ______ .
Ans. NADH; Cystine reductase.

15. Threonine aldolase cleaves threonine to ______ and ______ . **Ans.** Acetaldehyde; Glycine.

16. Hydroxyproline is converted to ______ and ______ . **Ans.** Pyruvate; Glyoxalate.

17. P-hydroxyphenylpyruvate hydroxylase converts P-hydroxyphenyl-pyruvate to ______ with cofactor ______ .
Ans. Homogentisic acid; Ascorbic acid.

18. Homogentisate oxidase, an ______ metalloprotein of mammalian liver, forms maleylacetoacetate; the reaction is inhibited by a ______ agent that binds iron. **Ans.** Iron; Chelating.

19. ______ , the black pigment present in the skin, hair and retina of the eye, is formed from phenylalanine in the ______ present in the skin.
Ans. Melanin; Melanoblasts.

20. DOPA is converted to dopamine by a ______ with ______ as cofactor.
Ans. Decarboxylase; B_6-PO_4.

21. The hydroxylation of dopamine forms norepinephrine in the presence of ______ and molecular ______ . **Ans.** Ascorbic acid; Oxygen.

22. Epinephrine is formed from norepinephrine by ______ from ______ .
Ans. Methylation; Methionine.

23. Two peptides are liberated during the conversion of fibrinogen into ______ . One of the peptides contains ______ .
Ans. Fibrin ; Tyrosine-o-sulphate.

24. Tryptophan oxygenase enzyme is an ______ porphyrin metalloprotein which is present in the ______ of mammals. **Ans.** Iron; Liver.

25. The chief inducing agent ______ of tryptophan oxygenase is blocked by ______ .
Ans. Adrenal corticosteroids; Puromycin.

26. Kynurenine is hydroxylated with molecular ______ in presence of ______ .
Ans. Oxygen; Reduced NADP.

27. In the deficiency of vitamin ______ kynurenine derivatives reach the extrahepatic tissues where they are converted to ______ .
Ans. B_6; Xanthurenic acid.

28. In the deficiency of vitamin B_6, the synthesis of ______ and ______ in the tissues is impaired owing to the non-conversion of tryptophan to niacin. **Ans.** NAD^+; $NADP^+$.

29. ______ mg of tryptophan produces ______ mg. of niacin. **Ans.** 60; 1.

30. The vitamins ______ and ______ are required for the formation of niacin from tryptophan in the body. **Ans.** B_6; B_2.

31. 5-hydroxytryptophan on ______ produces ______ , a vasoconstrictor and is stored in platelets. **Ans.** Decarboxylation; Serotonin.

32. Serotonin on acetylation and ______ produces the hormone ______ .
Ans. Methylation; Melatonin.

33. Methionine is first activated by ______ forming S-adenosyl methionine which is also termed as ______ . **Ans.** ATP; Active methionine.

34. L-methylmalonyl-CoA is converted to ______ by methylmalonyl-CoA isomerase with ______ as coenzyme. **Ans.** Succinyl-CoA; B_{12}.

35. Leucine on catabolism yields ______ and ______ . **Ans.** Acetoacetate; Acetyl-CoA.

36. Both valine and isoleucine on catabolism produce ______ . **Ans.** Succinyl-CoA.

37. Imidazolone-3-propionic acid is converted into ______ by ______ . **Ans.** FIGLU; Hydrolase.

38. In human beings and animals, deficiency of ______ or ______ causes an excessive excretion of FIGLU. **Ans.** Folic acid; Vitamin B_{12}.

39. FIGLU is converted into ______ resulting in the transfer of C_1 carbon atom to ______ .
Ans. Glutamate; Terahydrofolic acid.

40. Histidine is converted into ______ and ______ by minor pathways. **Ans.** Carnosine; Aniserine.

41. Carnosine is a peptide of _______ and _______ .
Ans. Histidine; β-alanine.

42. Ergothionine is present in high concentration in human _______ . Ans. Erythrocytes.

43. _______ acts as a coenzyme in the formation of prostaglandin, PGE_2 from _______ .
Ans. Glutathione; Arachidonic acid.

44. Skeletal muscle contains about _______ per cent creatine and heart muscle about _______ that amount. Ans. 0.5; Half.

45. Creatinine is formed largely in the _______ by the irreversible and nonenzymic removal of water from _______ .
Ans. Muscle; Creatine phosphate.

46. The synthesis of creatine is completed in the liver by the methylation of _______ in which _______ methionine is the methyl donor.
Ans. Glycocyamine; Active.

47. In the methylation of creatine _______ and _______ are required. Ans. ATP; Oxygen.

48. The activation of methionine occurs requiring ATP, _______ and _______ as well as methionine activating enzyme. Ans. Mg^{++}; Glutathione.

49. In the methylation of glycocyamin, guanidoacetate methylferase found in the _______ of mammals requires _______ for its optimal activity. Ans. Liver; glutathione.

50. The rate of creatine biosynthesis is dependent on kidney _______ activity.
Ans. Transamidinase.

51. Creatinine is the _______ of creatine and is formed by the _______ means in muscle.
Ans. Anhydride; Nonenzymatic.

52. Creatine is reabsorbed in the _______ at _______ concentrations.
Ans. Renal tubules; Lower.

53. The number of milligrams of creatinine (plus creatine) nitrogen excreted per kg body wt. in 24 hours is called _______ .
Ans. Creatinine coefficient.

54. The creatinine coefficient is an _______ of the amount _______ muscle tissue in the body.
Ans. Index; Active.

55. Positive nitrogen balance exists when _______ exceeds _______ . Ans. Intake; Output.

56. All body proteins continually undergo _______ and _______ . Ans. Degradation; Synthesis.

57. Amino acids have at least two ionizable groups _______ and _______ . Ans. –COOH; NH_4^+.

58. The pH at which the amino acid has no net charge and hence does not move in an electric field is called _______ . Ans. Isoelectric pH (pI).

59. The ion at the isoelectric point which carries + and – charges internally neutralized is called _______ . Ans. Zwitterion.

60. Owing to the opposite reactions depending on the acidity or alkalinity of the solution the amino acids are called _______ . Ans. Ampholytes.

61. Albumins are also widely distributed in plants especially in the _______ and _______ .
Ans. Seeds; Fruits.

62. Scleroproteins are similar to _______ and _______ . Ans. Albumins; Globulins.

63. Degradation of protein in one tissue is accompanied by _______ in others. Ans. Synthesis.

64. Antibody protein induced by immunization also undergo continual _______ and _______ .
Ans. Breakdown; Synthesis.

65. The metabolic pool is contributed by substances derived from catabolism of protein in the tissues by stimulations of the excessive amounts of _______ hormones and _______ hormones.
Ans. Thyroid; 11-oxygenated adreno-cortical.

66. The protein contains nitrogen _______ per cent.
Ans. 6.25.

67. The globular proteins have _______ and _______ polypeptide chains. Ans. Folded; Coiled.

68. The axial ratio of globular protein is not greater than _______ to _______ . Ans. 3; 4.

69. Most fibrous proteins fulfil structural roles in _______ and _______ .
Ans. Skin; connective tissue.

70. Silk fibroin contains small quantities of bulky _______ ; these bulky residues interrupt the _______ regions. Ans. Amino acids; β-sheet.

71. Tropocollagen consists of _______ polypeptide chains and each polypeptide forms a _______ helix. Ans. 3; left-handed.

72. Glutamate dehydrogenase is _______ and aldolase is _______ . Ans. Tetamer; trimer.

73. The amino acids obtained after hydrolysis of protein are separated and identified by ______ or ______ chromatography. **Ans.** HPLC; Ion exchange.
74. ______ and ______ agents disrupt disulphide bridges. **Ans.** Oxidizing; reducing.
75. Automated sequencing instruments operate most efficiently on polypeptides of ______ to ______ residues. **Ans.** 20;60.
76. When pH_2 is applied to the column, the amino acids bind via ______ exchange with ______. **Ans.** Cations; Na^+.
77. The columns are eluted with sodium citrate under the conditions of ______ and ______. **Ans.** pH; temperature.

Indicate "True" or "False" of the followings

1. The free amino acids after hydrolysis of peptides are separated and identified by chromatography or by electrophoresis. **Ans.** True.
2. Taurine is an amino acid but not a constituent of protein. Therefore, it is not present in the classification of amino acids. **Ans.** True.
3. Ninhydrin is a powerful oxidizing agent. **Ans.** True.
4. The neutral amino acids absorb ultraviolet light. **Ans.** False.
5. D-amino acids are also present in proteins. **Ans.** False.
6. Substitution of a single amino acid for another in a sequence of 100 or more amino acids may abolish biologic activity with serious consequences. **Ans.** True.
7. All hormones are proteins. **Ans.** False.
8. The proteins are involved in blood clotting. **Ans.** True.
9. The proteins act as the defence against infections by means of protein antibodies. **Ans.** True.
10. Proteins may be defined as the high molecular weight mixed polymers of α-amino acids joined together with peptide linkage. **Ans.** True.
11. The denatured protein is highly increased in solubility as its isoelectric point. **Ans.** False.
12. The native proteins are said to be occurring in plants only. **Ans.** False.
13. The native proteins possess no sedimentation rate. **Ans.** False.
14. Denaturation is used to know the enzyme catalyzed reaction of an extract at the loss of the enzyme activity when boiled or acidified. **Ans.** True.
15. Denatured ribonuclease when freed from urea and β-mercaptoethanol by dialysis slowly regains enzymic activity. **Ans.** True.
16. In severe malnutrition, blood albumin level is unchanged. **Ans.** False.
17. Albumin has no role in fatty acid transport. **Ans.** False.
18. The cause of hypoalbuminemia is impaired synthesis. **Ans.** True.
19. Pathological proteinuria may occur as a result of increased glomerular permeability. **Ans.** True.
20. Glycoproteins are secreted by the sublingual glands of some animals. **Ans.** False.
21. The glycine content of many albumins is high. **Ans.** False.
22. Some glycoproteins have many identical disaccharide units attached to the polypeptide chain. **Ans.** True.
23. Visual purple of the retina, a chromoprotein in which the prosthetic group is a carotenoid pigment. **Ans.** True.
24. Chromoproteins of certain animal fibres in which the prosthetic group is melatonin. **Ans.** False.
25. Scleroproteins are readily attacked and dissolved by alkali. **Ans.** False.
26. Scleroproteins cannot form the supporting structures of animals. **Ans.** False.
27. Nitroprusside test is positive for the amino acid cysteine. **Ans.** True.
28. Millon's test is negative for the amino acid tyrosine. **Ans.** False.
29. Arginine is an aromatic amino acid. **Ans.** False.
30. Tryptophan is a basic amino acid. **Ans.** False.

31. The intestinal mucosa also contains tripeptidase, dipeptidase, etc. Ans. True.

32. The polypeptides formed in the stomach are digested there by trypsin. Ans. False.

33. If one amino acid is fed in excess, it favours the absorption of another. Ans. False.

34. D-amino acids are absorbed by simple diffusion. Ans. True.

35. The amino acid associates with the carrier and Na^+ in the microville. Ans. True.

36. Sometimes the whole protein is absorbed into the blood. Ans. True.

37. The insoluble fibrous protein is also hydrolysed by enzymes of the human digestive tract. Ans. False.

38. Endogenous metabolism is involved in the formation of tissue and the other nitrogenous subtances and also the normal breakdown of tissue proteins. Ans. True.

39. Aromatic amino acids form glutathione. Ans. False.

40. In ammonotelic organisms, uric acid is the end product of amino acid nitrogen metabolism. Ans. False.

41. δ-amino group of ornithine undergoes transamination. Ans. True.

42. Both L- and D-amino acid oxidase activities occur in mammalian brain also. Ans. False.

43. The reduced FMN or FAD is reoxidized directly by molecular oxygen forming H_2O without the help of electron carriers. Ans. False.

44. If catalase is present, the α-keto acid is decarboxylated by H_2O forming a carboxylic acid with one carbon atom less. Ans. False.

45. Certain hormones also depress glutamate dehydrogenase activity. Ans. False.

46. Ammonia is produced in the kidney from glutamine catalyzed by renal glutaminase. Ans. True.

47. Glutamate dehydrogenase uses reduced NAD or NADP as cosubstrate. Ans. False.

48. The normal blood concentration of ammonia is 10-20 mg/100 ml. Ans. False.

49. Formation of glutamine is catalyzed by glutamine oxidase, a mitochondrial enzyme. Ans. False.

50. Asparaginase and glutaminase are employed as antitumor agents. Ans. True.

51. The amino acids ornithine and citrulline occur in protein. Ans. False.

52. Small quantities of arginase also occur in brain, mammary gland, testicular tissue and skin. Ans. True.

53. High erythrocyte level of arginase causes hyperargininemia. Ans. False.

54. Creatine cannot be synthesized in the pancreas of mammals. Ans. False.

55. Dietary creatine or high blood creatine has no effect on the rate of synthesis of creatine in the liver. Ans. True.

56. Hypothyroidism is associated with reduced kidney transamidinase activity. Ans. False.

57. Excessive creatinuria has been observed in hyperthyroidism during pregnancy and the menstrual period. Ans. True.

58. Normal values of creatinine coefficient range from 30 to 36 mg creatinine for men. Ans. False.

59. Less creatinuria may occur in starvation. Ans. False.

60. Glutamate, aspartate and ammonia are utilized for purine and pyrimidine synthesis. Ans. True.

Match the following :		Answer
1. Edema	(a) are the active agents in the processes of deamination and amination through the participation in transamination reactions.	1. ...... (f)
2. Cardiac edema	(b) produces a prompt excretion of water, sodium and chloride.	2. ...... (i)
3. Edema of the brain	(c) may occur as a result of increased glomerular permeability.	3. ...... (g)
4. Furosemide	(d) may occur in fever, after intense physical exertion and after minor emotional stimuli.	4. ...... (b)
5. SGOT activity is significantly increased	(e) is defined as the quantitative difference between the nitrogen intake and the nitrogen output.	5. ...... (j)
6. Acidic amino acids	(f) is recognized by the increase in the volume of interstitial fluid.	6. ...... (a)
7. Methionine	(g) occurs in acute glomerulonephritis.	7. ...... (m)
8. In nephrotic syndrome	(h) urinary albumin : globulin ratios are high.	8. ...... (h)
9. In amyloid diseases of the kidneys	(i) is produced by increased venous pressure.	9. ...... (o)
10 Transitory proteinuria	(j) in acute myocardial infarction.	10. ...... (d)
11. Nitrogen balance	(k) splits urea into ammonia and carbon dioxide at the optimum temperature and pressure.	11. ...... (e)
12. Creatine synthesis	(l) is participated by glycine, arginine and methionine.	12. ...... (l)
13. Serum albumin level is decreased	(m) is involved in the detoxification of nicotinic acid for the formation of the excretory product trigonelline.	13. ...... (n)
14. Pathological proteinuria	(n) in sudden haemodilution.	14. ...... (c)
15. Plant urease	(o) urinary albumin : globulin ratios are low.	15. ...... (k)
16. Ammonia intoxication	(p) is a stimulant of the central nervous system and also promotes peristalsis.	16. ...... (t)
17. Uses of antibiotics	(q) if introduced into the blood stream, appear in the urine.	17. ...... (w)
18. In hyperammonemia type 11	(r) takes part as a coenzyme in the prevention of peroxidation of polyunsaturated fatty acids in tissues.	18. ...... (y)
19. 5-hydroxytryptamine	(s) is found in daily excretion of creatine in urine of a normal male on a balanced diet.	19. ...... (p)
20. Aniserine	(t) is caused by all disorders of urea synthesis.	20. ...... (v)
21. Glutathione	(u) is found in the daily excretion of creatinine in urine of a normal adult female.	21. ...... (r)
22. Blood normal level of creatine :	(v) is a peptide of methylhistidine and β-alanine.	22. ...... (z)
23. Nil	(w) are to decrease the absorption of ammonia formed by bacterial decomposition in the intestine.	23. ...... (s)
24. 0.8 to 1.8 grams	(x) may occur in febrile and wasting diseases, diabetes mellitus, congenital muscle dystrophies.	24. ...... (u)
25. Excessive creatinuria	(y) patients suffer from a deficiency of ornithine transcarbamoylase.	25. ...... (x)
26. Egg albumin and other foreign proteins,	(z) 2.6 - 7.6 mg/100 ml.	26. ...... (q)

Appendix 5

NUCLEIC ACIDS

Write the correct answer number of the followings

1. The best role of purine and pyrimidine nucleotides is to serve as the monomeric precursors of

   (a) RNA, (b) DNA, (c) Both of the above, (d) None of the above. Ans. c.

2. The Purine nucleotides act as the components of

   (a) FAD^+, (b) NAD^+, (c) $NADP^+$, (d) All of the above. Ans. d.

3. The pyrimidine nucleotides act as the high energy intermediates

   (a) UDPG, (b) ATP, (c) ADP, (d) AMP. Ans. a.

4. The chemical name of thymine

   (a) 2-oxy-4-aminopyrimidine, (b) 2, 4-dioxy-5-methylpyrimidine, (c) 2, 4-dioxypyrimidine, (d) None of the above. Ans. b.

5. The chemical name 2-amino-6-oxypurine is said to be

   (a) Adenine, (b) Xanthine, (c) Guanine, (d) Hypoxanthine. Ans. c.

6. The lactam form is the predominant tautomer of

   (a) Uracil, (b) Cytosine, (c) Adenine, (d) Xanthine. Ans. a.

7. N_7-methylguanine has been found more recently in the nucleic acids of the cells of

   (a) Bacteria, (b) Yeast, (c) Plant, (d) Mammals. Ans. d.

8. Hypoxanthine and ribose constitute

   (a) Adenosine, (b) Inosine, (c) Guanosine, (d) Cytidine. Ans. b.

9. Thymine and deoxyribose form

   (a) Deoxycytidine, (b) Deoxyadenosine, (c) Deoxythymidine, (d) Deoxyuridine. Ans. c.

10. The most abundant intracellular free nucleotide

    (a) ATP, (b) FAD^+, (c) NAD^+, (d) $NADP^+$. Ans. a.

11. The concentration of ATP in living mammalian cells in mM is near

    (a) 0.2, (b) 0.4, (c) 0.6, (d) 1.0. Ans. d.

12. The intracellular cAMP concentrations in μM are near

    (a) 3.0, (b) 2.0, (c) 1.0, (d) 0.5. Ans. c.

13. The epimerization of galactose to glucose and vice versa takes place by

    (a) UTP, (b) CTP, (c) GTP, (d) ATP. Ans. a.

14. The biosynthesis of phosphoglycerides in animal tissue requires

    (a) ATP, (b) CTP, (c) GTP, (d) TPP. Ans. b.

15. The chemical name 4-hydroxypyrazole pyrimidine is used for

    (a) Thioguanine, (b) Mercaptopurine, (c) Azathioprine, (d) Allopurinol. Ans. d.

16. "Transforming factor" is used for the name of

    (a) RNA, (b) DNA, (c) tRNA, (d) None of the above. Ans. b.

17. Guanosine nucleotide is held by the cytosine nucleotide by the number of hydrogen bonds

    (a) 1, (b) 2, (c) 3, (d) 4. Ans. c.

18. Within the single turn of DNA the number of base pair exists

(a) 4, (b) 6, (c) 8, (d) 10. Ans. d.

19. Each turn of DNA structure has a pitch in nm of

(a) 1.4, (b) 2.4, (c) 3.4, (d) 4.4. Ans. c.

20. The double-stranded DNA molecule loses its viscosity upon

(a) Denaturation, (b) Filtration, (c) Sedimentation, (d) Concentration. Ans. a.

21. DNA molecule contains the number of nucleotides

(a) 800 to 4000, (b) 1000 to 6000, (c) 1200 to 8000, (d) 1600 to 9000. Ans. d.

22. DNA is denatured by

(a) Heat, (b) Acid, (c) Alkali, (d) All of the above. Ans. d.

23. Chromatin contains the number of repeating units in nm

(a) 10, (b) 15, (c) 20, (d) 25. Ans. a.

24. In RNA structure, its guanine content does not necessarily equal its cytosine content, and its adenine content does not necessarily equal its uracil content since it is a

(a) Double stranded molecule, (b) Single-stranded molecule, (c) Stable molecule, (d) Unstable molecule. Ans. b.

25. Each RNA molecule contains the number of nucleotides

(a) 40 to 4000, (b) 50 to 5000, (c) 60 to 6000, (d) 70 to 7000. Ans. c.

26. Messenger RNA has a molecular weight of

(a) 15000 to 30000, (b) 20000 to 35000, (c) 25000 to 40000, (d) 30000 to 50000. Ans. d.

27. Of the total cellular RNA molecules the Pc. of transfer RNA molecules amount to

(a) 5 to 10, (b) 8 to 16, (c) 10 to 20, (d) 15 to 30. Ans. c.

28. Each transfer RNA molecule contains the number of nucleotides

(a) 70, (b) 75, (c) 80, (d) 85. Ans. b.

29. In every cell the number of tRNA molecules are at least

(a) 10, (b) 15, (c) 20, (d) 25. Ans. c.

30. All tRNA molecules have a common CCA sequence at the

(a) 3′ termini, (b) 5′ termini, (c) 3′, 5′ termini, (d) All of the above. Ans. a.

31. In nearly all tRNA molecules, there is a loop containing the nucleotides of

(a) Uridine, (b) Thymidine, (c) Ribothymine and Pseudouridine, (d) Cytidine. Ans. c.

32. In tRNA molecules, there is another loop containing the minor base

(a) Uracil, (b) Dihydrouracil, (c) Cytosine, (d) Dihydrocytosine. Ans. b.

33. The PC of the ribosomal RNA of the RNA within the cell

(a) 60, (b) 70, (c) 80, (d) 90. Ans. c.

34. The mammalian ribosme contains the number of major nucleoprotein subunits

(a) 2, (b) 3, (c) 4, (d) 5. Ans. a.

35. Gene is a segment of the DNA molecule containing base pairs about

(a) 300, (b) 400, (c) 500, (d) 600. Ans. d.

36. The fragments of DNA attached to an RNA initiator component were discovered by

(a) Watson and Crick, (b) Okazaki, (c) Peterson, (d) Nelson. Ans. b.

37. Synthesis of RNA molecule is terminated by a signal which is recognized by

(a) ρ (Rho) factor, (b) δ-factor, (c) α-factor, (d) None of the above. Ans. a.

38. The promoter site at which the synthesis of new RNA molecule begins with the help of another factor.

(a) σ, (b) γ, (c) β, (d) α. Ans. d.

39. The binding of Prokaryotic DNA-dependent RNA polymerase to promoter sites of genes is inhibited by the antibiotic.

(a) Septran, (b) Streptomycin, (c) Aureomycin, (d) Rifampin. Ans. d.

40. The N atoms at positions 3 and 9 of Purine base are derived from the amide nitrogen of

(a) Glutamate, (b) Glutamine, (c) Asparagine, (d) Aspartate. Ans. b.

41. The carbon atoms at positions 4 and 5 and the N

atom at position 7 of purine base are supplied from

(a) Valine, (b) Alanine, (c) Glycine, (d) Serine.

Ans. c.

42. 5-phosphoribosylamine reacts with glycine to produce glycinamide ribosylphosphate by glycinamide kinosynthetase in presence of

(a) ATP, (b) GTP, (c) UTP, (d) CTP. Ans. a.

43. The carbon atoms at position 4, 5 and 6 and the N atom at position 3 of pyrimidine base are derived from

(a) Glutamic acid, (b) Aspartic acid, (c) Glycine, (d) Serine. Ans. b.

44. The end product of purine catabolism in other mammals except man.

(a) Uric acid, (b) Allantoin, (c) Ammonia, (d) Creatinine. Ans. b.

45. The net excretion of total uric acid in normal men in mg in 24 hours.

(a) 100 - 300, (b) 200 - 400, (c) 300 - 500, (d) 600 - 700.

Ans. d.

46. A portion of uric acid is converted to urea and ammonia by intestinal.

(a) Urogenolysis, (b) Ureolysis, (c) Uricolysis, (d) Ureotolysis. Ans. c.

Fill in the blanks of the followings

1. Xanthine is oxidized to uric acid by _______ .

Ans. Xanthine oxidase.

2. Xanthine oxidase is inhibited by _______ for which uric acid cannot be formed.

Ans. Allopurinol.

3. The chief end product of purine catabolism in man is _______ . Ans. Uric acid.

4. Urinary Uric acid is both _______ and _______ in origin. Ans. Endogenous; Exogenous.

5. Birds and reptiles do not possess _______ activity. Ans. Uricase.

6. _______ and _______ also increase the urinary excretion of uric acid by inhibiting renal tubular reabsorption.

Ans. ACTH; Adrenocortical oxysteroids.

7. Blood uric acid level also _______ in hemolytic anemia and thalassemia. Ans. Increases.

8. Sodium urate crystals are deposited in soft tissues and these urate deposits are referred to as _______ . Ans. Tophi.

9. The synthesis of pyrimidine starts with the formation of carbamoyl phosphate from _______ , _______ and CO_2 being catalyzed by carbamoyl phosphate synthetase present in the cytosol of the cell. Ans. Glutamine; ATP.

10. Dihydroorotic acid on dehydrogenation by dihydroorotate dehydrogenase utilizing _______ as coenzyme is converted into _______ .

Ans. NAD^+; Orotic acid.

11. The main site for the catabolism of pyrimidines is _______ . Ans. Liver.

12. The major end products of cytosine, uracil and thymine are _______ and _______ respectively.

Ans. β-alanine; β-aminoisobutyric acid.

13. For each gene in the DNA molecule, there is a _______ strand and its complimentary _______ strand. Ans. Sense; Antisense.

14. The strand which is transcribed into an RNA molecule is referred to as the _______ of the DNA.

Ans. Sense strand.

15. RNA and DNA chain elongation is inhibited by _______ . Ans. Actinomycin D.

16. DNA and RNA synthesis is inhibited by _______ .

Ans. Anthramycin.

17. Tea contains the purine base containing substituent _______ . Ans. Theophylline.

18. Cocoa contains the purine base containing substituent _______ . Ans. Theobromine.

19. The base thymine of DNA is replaced by the base _______ in _______ . Ans. Uracil; RNA.

20. Cyclic AMP is destroyed in tissues by its conversion to _______ by the enzyme _______ .

Ans. AMP; cAMP phosphodiesterase.

21. S-adenosylmethionine serves as a form of _______ methionine. Ans. Active.

22. cAMP is formed from _______ by the enzyme _______ . Ans. ATP; Adenylate cyclase.

23. cGMP is catabolized by _______ to produce _______ .

Ans. Phosphodiesterase; 5′-monophosphate.

24. cGMP acts antagonistically to _______ .
Ans. cAMP.

25. Uridine diphosphoglucuronic acid serves as the _______ glucuronide for the formation of _______ in the liver.
Ans. Active; Bilirubin glucuronide.

26. CTP acts as the precursor for the polymerization of _______ into _______.
Ans. CMP; Nucleic acids.

27. Ceramide and CDP - choline are responsible for the formation of _______ and other substituted _______. Ans. Sphingomyelin; Sphingosine.

28. _______ and _______ are used in the chemotherapy of cancer and viral infections.
Ans. Cytarabine; Vidarabine.

29. Azathioprine is useful in organ _______.
Ans. Transplantation.

30. Both _______ and _______ are used clinically to inhibit the catabolism of intracellular cAMP.
Ans. Aminophylline; Theophylline.

31. DNA is referred to as _______ .
Ans. Transforming factor.

32. Two strands of double helical molecules are _______ . Ans. Anti-parallel.

33. DNA molecules are long and its length is _______ times greater than its breadth. Ans. 250.

34. Chromatin consists of a long double- stranded _______ molecule. Ans. DNA.

35. The messenger RNA is the most _______ in size and stability. Ans. Heterogeneous.

36. The heterogeneous nuclear RNA (HnRNA) molecules are processed to generate the _______ molecules which then enter the cytoplasm to serve as templates for _______ synthesis.
Ans. mRNA; Protein.

37. One tRNA corresponds to each of the _______ amino acids required for _______ synthesis.
Ans. 20; Protein.

38. The _______ loop at the end of a base-paired stem of tRNA recognizes the codon of the template _______.
Ans. Anticodon; mRNA.

39. Ribosomes are _______ particles and reticular granules of _______ in diameter.
Ans. Nucleoprotein; 100 - 150 Å.

40. The 60S subunits contain a 5S rRNA, a _______ rRNA and a _______ rRNA. Ans. 5.8S; 28S.

41. The 40S subunit contains _______ rRNA and about _______ polypeptide chains.
Ans. 18S; 30.

42. The unit of genetic information is the _______ or _______ . Ans. Gene; Cistron.

43. RNA can be synthesized by _______ which is dependent on the presence of _______ acting as a template. Ans. RNA-polymerase; DNA.

44. DNA produces a _______ which helps in placing _______ in the code for protein synthesis.
Ans. mRNA; Amino acids.

45. The formation of RNA template is directed by _______ . Ans. Nuclear DNA.

46. Nucleoproteins are conjugated proteins containing _______ and a basic protein like _______ or _______ .
Ans. Nucleic acids; Protamine; Histones.

47. The initiation of DNA synthesis requires _______ by a short length of _______
Ans. Priming; RNA.

48. The selection of proper deoxyribonucleotide is dependent upon proper _______ with the other strand of _______ molecule. Ans. Pairing; DNA.

49. The newly synthesized DNA strand attached to _______ primer is called _______ fragments.
Ans. RNA; Okazaki.

50. Each strand of DNA acts as a _______ for the formation of a new strand. This process is called _______ . Ans. Template; replication.

51. Replication begins at a specific _______ point and this is a unique sequence of bases called _______ . Ans. Initiation; Ori.

52. In the initiation process of DNA about _______ DnaA protein molecule each with a bound _______ . Ans. 20; ATP.

53. In the elongation process of replication of DNA, two operations occur such as _______ and _______ . Ans. Leading strand synthesis; Lagging strand synthesis.

54. Deoxyribonucleotides are added to this ________ by ________ . *Ans.* Primer; DNA polymerase III.

55. ________ acts to relieve the stress generated by ________ . *Ans.* DNA gyrase; helicases.

56. When the new okazaki fragment is complete, the ______ primer is removed by ______ .

Ans. RNA; DNA polymerase 1.

57. Each ________ is extended by ________ .

Ans. Primer; DNA polymerase III.

58. Transcription is the process by which the synthesis of ______ molecule is ______ and ______ .

Ans. RNA; initiated; terminated.

59. the sequences recognized by ________ polymerase are called ________ . *Ans.* RNA; Promoter.

60. RNA polymerase consists of ________ enzyme and ________ factor. *Ans.* Core; sigma (σ).

61. The first incoming ________ binds ______ at the start point of initiation.

Ans. NTP; RNA polymerase.

Appendix 6

PROTEIN SYNTHESIS AND GENETIC CODE

Fill in the blanks of the followings

1. In the nucleotide sequence of the mRNA molecule, code words exist for each amino acid. This is referred to as the __________ .

   Ans. Genetic code.

2. The adaptor molecules translating the code words into the amino acid sequence are the __________ molecules.

   Ans. Transfer RNA (-tRNA).

3. The process by which the cell translate information by the machinery from the nucleotide sequence of an mRNA into the sequence of amino acids of the corresponding specific ________ is said to be __________ .

   Ans. Protein; Translation.

4. The rough endoplasmic reticulum is the factory of __________. **Ans.** Polyribosomes.

5. Code words consisting of 2 nucleotides each can provide for only ________ specific code words.

   Ans. 16.

6. Each code word termed as ________ consists of a sequence of ________ nucleotides.

   Ans. Codon; 3.

7. 64 triplet codes in mRNA with 4 nucleotides are to provide for ________ amino acids and three codons do not code for specific amino acids and these have been termed ________.

   Ans. 20; Nonsense codon.

8. Only a single amino acid is indicated for any specific codon; the genetic code is ________.

   Ans. Unambiguous.

9. Once the reading is started at a specific codon, there is no ________ between codons.

   Ans. Punctuation.

10. Amino acids are activated by the enzyme ________ in the presence of the coenzyme ________ .

    Ans. Aminoacyl-tRNA synthetase; ATP.

11. The binding of the mRNA to the 40S ribosomal subunit requires the presence of __________ .

    Ans. IF–3.

12. The initiation factor is a ________ factor.

    Ans. Protein.

13. The complex formed by anticodon of tRNA and IF-I attaches the 60S ribosomal subunits with the release of ________, ________, ________.

    Ans. IF–I, IF–2, IF–3.

14. The complete ribosome contains two sites ________ site and ________ site on the mRNA.

    Ans. P; A.

15. During the process of initiation, the complete ________ ribosome is formed. **Ans.** 80S.

16. ________ forms a complex with ________ and the entering aminoacyl-tRNA. **Ans.** EF–1; GTP.

17. The alpha amino group of the new aminoacyl-tRNA in the A site combines with the carboxyl group of the peptidyl - tRNA occupying the P site in presence of ________ of the ________ ribosomal subunits. **Ans.** Peptidyl transferase; 60S.

18. The newly formed peptidyl-tRNA at the A site is translocated into the vacated P site by ________ and ________. **Ans.** EF–2; GTP.

19. After multiple cycles of elongation the _______ codon of mRNA appears in the _______ site.
Ans. Nonsense; A.

20. The releasing factor hydrolyzes the bond between the peptide and the _______ occupying the _______ site. Ans. tRNA;P.

21. On hydrolysis and release the 80S ribosome dissociates into its _______ and _______ subunits which are then recycled. Ans. 40S;60S.

22. The releasing factors are _______.
Ans. Proteins.

23. The antibiotic _______ has structural similarity with that of tyrosinyl-tRNA is incorporated via the A site on the ribosome. Ans. Puromycin.

24. Since EF–2 is inactivated by _______ and thereby inhibits mammalian protein synthesis.
Ans. Diphtheria toxin.

25. A single ribosome is capable of translating about _______ codons in _______ seconds into a protein with a molecular weight of 40,000.
Ans. 400; 10.

26. There are only two types of gene regulation _______ regulation and _______ regulation.
Ans. Positive; Negative.

27. A double negative has the effect of acting as a _______. Ans. Positive.

28. The gene is the smallest segment of the _______ molecule containing about _______ base pairs.
Ans. DNA;600.

29. When a pair carries genes with the same characteristics, say tallness, then the individual is said to be _______. Ans. Homozygous.

30. When one of the pair tallness and the other gene shortness, the individual is _______.
Ans. Heterozygous.

31. Lac operon is nothing but _______ and _______.
Ans. Structural gene; operator gene.

32. The regulator gene induces the synthesis of protein macromolecules called _______.
Ans. Repressors.

33. The operon becomes active because the repressor system is itself inactivated. The phenomenon is said to be _______. Ans. Derepression.

34. The operator locus is a region of double-stranded DNA of _______ base pairs long with a 2-fold rotational symmetry in a region that is _______ base pairs long. Ans. 27;21.

35. The operator locus lies between the _______ at which the DNA- dependent RNA polymerase attaches to commence transcription and the beginning of the _______ gene.
Ans. Promoter site : Z.

36. Normally _______ repressor molecules are present and one or two operator loci per cell are also present. Ans. 20-40.

37. The binding of the inductor to the repressor molecule involves the amino acid residues in positions _______ and _______. Ans. 74;75.

38. The larger hnRNA contains _______ that will form the mature _______. Ans. Exons; mRNA.

39. Exons are spliced to form _______.
Ans. mRNA.

40. 61 codons code for _______ amino acids. So there must be _______ in the genetic code.
Ans. 20; degeneracy.

41. Every amino acid except _______ is represented by several _______. Ans. Methionine; codons.

42. Codons that represent same _______ are called as _______. Ans. Amino acid; synonyms.

43. For any specific codon, only a single _______ is indicated. So the genetic code is _______.
Ans. Amino acid; Unambiguous.

44. Mitochondria require only _______ tRNA molecules to read their genetic code, whereas cytoplasmic translation system passes a full complement of _______ tRNA species. Ans. 22;31.

45. When the first base in anticodon is _______ or _______ the pairing with the third base in codon is _______ or _______. Ans. C;A;G;A.

46. When the first base in the anticodon is 1, the third base may be _______, _______, or _______.
Ans. A;C;U.

47. Three codons for glycine _______, _______ and _______ can form a base pair for one anticode _______. Ans. GGU; GGC; GGA; CCI.

48. The chain terminating codons are _______, _______, _______. Ans. UAA;UAG;UGA.

49. An alteration in the sequence of ______ in a gene resulting in an altered gene product is called ______. **Ans.** Bases; mutation.

50. In transitions, a given pyrimidine is changed to the other ______ or a given purine is changed to other ______. **Ans.** Pyrimidine; Purine.

51. In transitions, a purine is change to either of the 2 ______ or the change of a pyrimidine into either of the 2 ______.

Ans. Pyrimidines; Purines.

52. A nonsense codon may appear that would result in the ______ termination of ______ incorporation into a peptide chain.

Ans. Premature; amino acid.

53. The process of initiation needs ______, ______, ______ and at least ______ eukaryotic initiation factors. **Ans.** tRNA; rRNA; mRNA; 10.

54. The binding of two eukaryotic initation factors ______ and ______ to the 40S subunit favours dissociation of the 80S ribosome into its ______ and ______ subunits.

Ans. eIF-3; eIF-1A; 40S; 60S.

55. Reassociation is also prevented by the binding of ______ to the ______ subunits.

Ans. eIF-3A; 60S.

56. The eIF-2 consists of ______ and ______ subunits.

Ans. α; β.

57. The eIF-2 is phosphorylated by a ______ which is activated when a cell is under ______.

Ans. cAMP-independent protein kinase; stress.

58. The ______ helps the binding of mRNA to the ______ preinitiation complex.

Ans. Methyl-guanosyl triphosphate cap; 40S.

59. ______ hydrolysis is required in the association of mRNA with the ______ preinitiation complex.

Ans. ATP; 40S.

60. The eIF-3 is a ______ since it binds ______ to and the ______ ribosomal subunits.

Ans. Key protein; mRNA; 40S.

61. eIF-4A is an ______ that helps unwind the ______ and ______.

Ans. RNA-dependent ATPase; RNA; p220.

62. 4E is important since ______ binding is a ______ step in translation.

Ans. Cap; rate-limiting.

63. The phosphorylation of 4E and activation of the protein are caused by ______ and ______.

Ans. Insulin; mitogenic growth factors.

64. A family of proteins 4E-BPI and ______ binds to and inactivates ______. **Ans.** 4E-BP2; 4E.

65. The hydrolysis of the GTP bound to ______ by ______ is caused by the binding of the ______ ribosomal subunits to the ______ initiation complex. **Ans.** eIF-2; eIF-5; 60S; 40S.

66. The association of ______ and ______ subunits forms the 80S ribosome. **Ans.** 40S; 60S.

67. Elongation is a ______ process consisting of several steps catalyzed by ______.

Ans. Cyclic; eFF.

68. The aminoacyl-tRNA binds to the A site of the complete ______ ribosome requiring proper ______ recognition. **Ans.** 80S; codon.

69. The amino acyl-tRNA enters the A site with the release of ______ and ______.

Ans. eEF-1α-GDP; Phosphate.

70. The alpha amino group of the new aminoacyl-tRNA in the A site has a ______ attack on the esterified carboxyl group of the ______ occupying the P site.

Ans. Nucleophilic; peptidyl-tRNA.

71. ______ has a direct role in protein synthesis.

Ans. RNA.

72. The amino acid on the amino acyl-tRNA is already ______ for which no further ______ is required for this reaction.

Ans. Activated; energy.

73. The ______ and ______ are responsible for the translocation of the newly formed peptidyl-tRNA at the A site into the empty ______ site.

Ans. eEF-2; GTP; P.

74. ATP is hydrolyzed to ______ which is equivalent to the ______ of two ATPs to two ADPS and phosphates in the charging of the tRNA molecule with the amino acyl moiety.

Ans. AMP; Hydrolysis.

75. The hydrolysis of GTP is caused by the translocation of the newly formed _______ in the A site into the P site by _______.

Ans. Peptidyl-tRNA; eEF-2.

76. The formation of one peptide bond causes the hydrolysis of two _______ molecules and two _______ molecules. Ans. ATP;GTP.

77. _______ amino acids are incorporated per second in case of _______ ribosome and _______ amino acids are incorporated per second in case of _______ ribosome.

Ans. 6; eukaryotic; 18; prokaryotic.

78. The termination signal in the A site is recognized by the _______ which are _______ that hydrolyze the _______ bond when a _______ codon occupies the A site.

Ans. eRF; Proteins; Peptidyl-tRNA; nonsense.

79. The 80S ribosome dissociates into its _______ and _______ subunits upon _______ and _______.

Ans. 40S; 60S; hydrolysis; release.

80. Ferritin synthesis is stimulated by _______ by the release of a cytoplasmic _______ that binds to 5′ nontranslated region of _______.

Ans. Elemental iron; Protein; ferritin mRNA.

81. Other viruses inhibit host cell protein synthesis by preventing the association of _______ with the _______ ribosome. Ans. mRNA;40S.

82. In animal cells, insulin is synthesized as a _______ that folds to allow the _______ bridges to form. Ans. Prohormone; disulfide.

83. Many effective antibiotics _______ with the proteins of prokaryotic ribosomes and thus _______ protein synthesis. Ans. Interact; inhibit.

84. Protein synthesis on all ribosomes are inhibited by _______ which is structurally similar to _______. Ans. Puromycin; tyrosinyl-tRNA.

85. Cycloheximide inhibits _______ in the _______ ribosomal subunit in _______ by binding to an _______ component.

Ans. Peptidyltransferase; 60S; eukaryotes; rRNA.

86. Diphtheria toxin catalyzes the _______ of _______ in mammalian cells.

Ans. ADP-ribosylation; eFF-2.

87. Many ribosomes on the same _______ molecule form a _______. Ans. mRNA; Polysome.

88. A single mammalian ribosome can synthesize about _______ peptide bonds _______.

Ans. 100; Per minute.

89. Polysomes actively synthesizing _______ can exist as _______ particles in the cellular cytoplasm.

Ans. Protein; Free.

90. Some of the proteins synthesized by the attached _______ are packaged by the _______.

Ans. Polysomes; golgi apparatus.

91. More than _______ specific _______ are effected by insulin. Ans. 100; mRNAs.

92. Transcription is decreased due to the _______ amount of the _______ transcript and of mature _______ PEPCK which in turn is directly related to the decreased rate of _______ synthesis.

Ans. Decreased; Primary; mRNA; PEPCK.

93. The synthesis of _______ which catalyzes a _______ step in gluconeogenesis is decreased by _______ and hence gluconeogenesis is _______.

Ans. Phosphoenolpyruvate carboxykinase (PEPCK); rate-limiting; insulin; decreased.

Appendix 7

HEMOGLOBIN

Write the correct answer number of the followings

1. Hemoglobin contains the number of gram atoms of iron per mole in the ferrous state—
(a) 1, (b) 2, (c) 3, (d) 4. **Ans.** d.

2. The molecular weight of hemoglobin
(a) 44, 450, (b) 54, 450, (c) 64, 450, (d) 74, 450. **Ans.** c.

3. The porphyrins are cyclic compounds formed through methylene bridges by the linkages of pyrrole rings number
(a) 4, (b) 3, (c) 2, (d) 1. **Ans.** a.

4. The porphyrins are found in nature in which the various side chains are substituted for the hydrogen atoms number
(a) 2, (b) 4, (c) 6, (d) 8. **Ans.** d.

5. A porphyrin with a completely symmetrical arrangement of the substituents is classified as a porphyrin of
(a) Type I, (b) Type II, (c) Type III, (d) All of the above. **Ans.** a.

6. The more abundance found in nature is
(a) Type I series, (b) Type II series, (c) Type III series, (d) None of the above. **Ans.** c.

7. In the biosynthesis of porphyrins the activation of glycine needs the coenzyme
(a) NAD^+, (b) $B_6 - PO_4$, (c) FAD^+, (d) ATP. **Ans.** b.

8. The anemia has been observed in the deficiency of vitamin
(a) Biotin, (b) Inositol, (c) Niacin, (d) Pantothenic acid. **Ans.** d.

9. The dipyrryl compounds are of two types
(a) A and B, (b) B and C, (c) A and C, (d) B and D. **Ans.** a.

10. A type III porphyrin results with the condensation of the components
(a) Two of the (A), (b) One (A) and one (B), (c) Two of the (B), (d) One (A) and one (C). **Ans.** b.

11. The enzyme coproporphyrinogen oxidase is able to act on coproporphyrinogen
(a) Type I, (b) Type II, (c) Type III, (d) All of the above. **Ans.** c.

12. In mammalian liver the reaction of conversion of corporporphyrinogen to protoporphyrin requires
(a) Molecular oxygen, (b) Water, (c) ATP, (d) $B_6 - PO_4$. **Ans.** a.

13. Heme is synthesized by the incorporation of ferrous ion (Fe^{++}) into protoporphyrin III being catalyzed by the enzyme
(a) Ferroxidase, (b) Ferroreductase, (c) Ferrochelatase, (d) None of the above. **Ans.** c.

14. The globin of the hemoglobin is a protein composed of closely packed polypeptide chains of
(a) 6 parallel layers, (b) 4 parallel layers, (c) 3 parallel layers, (d) 2 parallel layers. **Ans.** b.

15. Two α-chains of globin have identical amino acid composition of
(a) 111, (b) 121, (c) 131, (d) 141. **Ans.** d.

16. The total number of amino acid in globin

(a) 544, (b) 554, (c) 564, (d) 574. Ans. d.

17. Hemoglobin takes up the number of molecules of oxygen

(a) 1, (b) 2, (c) 4, (d) 6. Ans. c.

18. Carboxyhemoglobin is formed by

(a) CO, (b) CO_2, (c) HCO_3, (d) HCN. Ans. a.

19. One mol of hemoglobin contains histidine residues

(a) 5, (b) 15, (c) 25, (d) 35. Ans. d.

20. Methemoglobin is formed as a result of the oxidation of hemoglobin by oxidizing agent

(a) Oxygen of air, (b) Hydrogen peroxide, (c) Potassium Ferricyanide, (d) Potassium permanganate. Ans. c.

21. Methemoglobin can be reduced to hemoglobin by

(a) Removal of hydrogen, (b) Vitamin C, (c) Glutathione, (d) Creatinine. Ans. b.

22. The colour of cyanomethemoglobin

(a) Yellow, (b) Pink, (c) Brown, (d) Bright red. Ans. d.

23. The iron of heme is coordinated in β-chains at positions

(a) 43 and 72, (b) 53 and 82, (c) 63 and 92, (d) 73 and 102. Ans. c.

24. Myoglobin contains the number of gram atom of iron per mole

(a) 1, (b) 2, (c) 3, (d) 4. Ans. a.

25. Catalases contain the number of gram atoms of iron per mole

(a) 1, (b) 2, (c) 3, (d) 4. Ans. d.

Fill in the blanks of the followings

1. Myoglobins are the ______ pigments occurring in the ______ cells of vertebrates and invertebrates. Ans. Respiratory; Muscle.
2. The molecular weight of catalases is about ______. Ans. 225000.
3. Tryptophan pyrrolase catalyzes the oxidation of ______ to ______.

   Ans. Tryptophan; Formyl Kynurenine.
4. The fundamental role of cytochromes is in ______. Ans. Cellular respiration.
5. Cytochrome C has a molecular weight of about ______ and contains ______ iron.

   Ans. 13000; 0.43 per cent.
6. The peptide chain of human heart cytochrome C contains ______ amino acids. Ans. 104.
7. Cytochrome oxidases can carry three general types of reactions eg, oxygen transfer ______, ______.

   Ans. Mixed function oxidation; Electron transfer.
8. Hemoglobin A has a molecular weight of ______ and contains ______ pairs of peptide chains.

   Ans. 64,456; Two.
9. The α-chains of hemoglobin A contains ______ and β-chain contains ______ amino acids.

   Ans. 141; 146.
10. Fetal hemoglobin A contains ______ and β-chain contains ______ amino acids.

    Ans. 141; 146.
11. Fetal hemoglobin comprises ______ per cent of the total hemoglobin in the new born.

    Ans. 50 to 90.
12. Fetal hemoglobin takes up ______ more readily at low oxygen tension and releases ______ more readily than adult hemoglobin (A).

    Ans. O_2; CO_2.
13. Hemoglobin F is gradually replaced by ______ during the first 6 months of extrauterine life.

    Ans. Hemoglobin A.
14. Fetal hemoglobin is more resistant to denaturation by ______ and is more susceptible to conversion to methemoglobin by ______.

    Ans. Alkali; Nitrites.
15. Some of the abnormal hemoglobins are easily differentiated by their electrophoretic mobilities and have given rise to the concept of ______.

    Ans. Molecular disease.
16. Acidic amino acid is replaced by a ______ or ______ amino acid for the formation of abnormal hemoglobin from normal hemoglobin.

    Ans. Basic; Neutral.

17. The abnormality of hemoglobin C is found in the β-chain at position ______ , the amino acid glutamic acid is replaced by ______ .
Ans. 6; Lysine.

18. Hemoglobin C is characterized by the mild ______ with a tendency to ______ .
Ans. Anemia; Infarction.

19. In hemoglobin S ______ anemia develops and the ______ becomes long and boat-shaped.
Ans. Sickle cell; RBC.

20. The abnormality in hemoglobin S occurs in ______ chain, glutamic acid at position 6 is replaced by ______ . **Ans.** β; Valine.

21. If HbF is present in large amounts in the blood of adults, it gets ______ rapidly producing ______ . **Ans.** Hemolysed; Thalassemia major.

22. The abnormality of hemoglobin M is found in the α-chain, the histidine residues in ______ and ______ positions are replaced by ______ .
Ans. 58; 87; Tyrosine.

23. Methemoglobin is not reduced to ______ by ______ agents. **Ans.** Hemoglobin; Reducing.

24. Trypsin splits the peptides in hemoglobin at points where only ______ and ______ occur.
Ans. Lysine; Arginine.

25. Methemoglobin is an ______ hemoglobin ______ is in the firm combination.
Ans. Oxidized; Oxygen.

26. When the concentration of methemoglobin reaches ______ per 1000 ml of blood, ______ usually develops. **Ans.** 3 grams; Cyanosis.

27. Methemoglobin is present in normal ______ about ______ per cent of the total hemoglobin.
Ans. Erythrocytes; 0.4.

28. ______ hemoglobin combines with ______ to form sulfhemoglobin. **Ans.** Reduced; H_2S.

29. Sulfhemoglobin is also observed in subjects with marked ______ in the presence of ______ producing bacteria in the intestine.
Ans. Constipation; Nitrite.

30. Carboxyhemoglobin is formed by the combination of ______ with the ______ in the hemoglobin molecule. **Ans.** CO; Iron.

31. Carboxyhemoglobin is produced by the excessive exposure to artifical illuminating ______ and to automobile ______ in closed or poorly ventilated rooms. **Ans.** Gas; Exhaust gases.

32. The condition by which the excretion of both coproporphyrin and uroporphyrin increases is said to be ______ . **Ans.** Porphyria.

33. Erythropoietic prophyria has got tendency to ______ and defective ______ .
Ans. Hemolysis; Erythropoiesis.

34. In erythropoietic protoporphyria, there is increased ______ and ______ in the circulating erythrocytes, plasma and the feces.
Ans. Protoporphyrin; Uroporphyrin.

35. Acute intermittent porphyria is due to a marked increase of hepatic ______ .
Ans. ALA synthase.

36. Increased serum PBI and hypercholesterolemia occur in acute ______ porphyria.
Ans. Intermittent.

37. In porphyria variegata there is increased hepatic ______ . **Ans.** ALA synthase.

38. Hereditary coproporphyria causes increased urinary output of ______ and ______ during acute attacks. **Ans.** Porphobilinogen; ALA.

39. Acquired porphyria is caused by severe ______ diseases and ingestion of certain ______ .
Ans. Liver; Toxins.

40. In acquired porphyria, there is increased excretion of ______ in urine. **Ans.** Uroporphyrin.

41. Porphyria cutanea tarda shows frequent rise in serum ______ . **Ans.** Iron.

42. The porphyrins are ______ compounds with ______ structure. **Ans.** Cyclic; tetra pyrrole.

43. Iron in the ______ state is bound to the ______ atom of the pyrrole rings.
Ans. Ferrous; nitrogen.

44. Iron is also internally linked to the nitrogen of the ______ ring of ______ of the polypeptide chains. **Ans.** Imidazole; histidine.

45. The propionic acid of ______ and ______ position of heme of III and IV pyrroles are also linked to the amino acids ______ and ______ of the polypeptide chain respectively.
Ans. Sixth; Seventh; ariginine; lysine.

46. The globin of hemoglobin is a ______ which is composed of ______ parallel layers of closely packed ______ chains.

Ans. Protein; 4; Polypeptide.

47. Each of two alpha chains of globin has ______ amino acids and each of the rest two chains has ______ amino acids. Therefore, the total number of amino acids in globin is ______ .

Ans. 141; 146; 574.

48. Hemoglobin molecule and its subunits contain ______ amino acids internally and ______ amino acids on their surfaces. So they form ______ .

Ans. Hydrophobic; hydrophilic; "heme pockets".

49. In "heme pockets" subunits allow the entry of ______ molecule but the same entry in subunit is blocked by ______ residue.

Ans. Oxygen; Valine.

50. The amount of hemoglobin broken each day in normal adult is about ______ which contain ______ of iron. Ans. 8 gm; 27 mg.

51. The protoporphyrin yields about ______ of bilirubin. Ans. 300 mg.

52. When hemoglobin is catabolized in the body, globin is reutilized ______ or in the form of its constituent ______ and iron enters ______ for reuse. Ans. As such; amino acids; ferritin "pool".

53. Choleglobin is formed from hemoglobin by ______ in presence of ______ .

Ans. Oxygen; ascorbic acid.

54. Bilirubin is formed from biliverdin by ______ in presence of ______ or ______ .

Ans. Bilirubin reductase; $NADP^+$; NAD^+.

55. Bilirubin conjugates with UDP-glucuronic acid by the enzyme ______ to form ______ which can pass through ______ .

Ans. UDP-glucuronyltransferase; bilirubin diglucuronide; glomerular filter.

56. Glucuronyltransferase activity is inhibited by ______ . Ans. Novobiocin.

57. Urobilinogen on exposure to oxygen is oxidized to ______ which gives the ______ of stools.

Ans. Urobilin; darkening.

58. In sickle cell anemia, hemoglobin ______ appears due to abnormality of ______ chain where ______ is replaced by ______ at position ______ . Ans. δ; β; glutamic acid; valine; 6.

59. Sickle cells are more ______ and causes ______ and ______ .

Ans. Fragile; hemolytic anemia; jaundice.

60. Persons suffering from sickle cell anemia show an increased resistance to ______ and ______ infection is found more in this disease.

Ans. Malaria; Salmonella.

Indicate "True" or "False" of the followings

1. Chlorophyll is magnesium-containing porphyrin and the photosynthetic pigment of plants.

Ans. True.

2. Chlorophyll and heme of hemoglobin are synthesized in living cells by different pathways.

Ans. False.

3. The condensation of active succinate and glycine is catalyzed by the enzyme Amlev synthetase.

Ans. True.

4. Two mols of Amlev condense to form porphobilinogen catalyzed by the enzyme Amlev dehydrase. Ans. True.

5. Tripyrrylmethane is formed by the condensation of 2 mols of porphobilinogen. Ans. False.

6. In low oxygen tension, oxyhemoglobin takes up more oxygen. Ans. False.

7. Histidine contained in hemoglobin exerts its buffering action through its basic imidazole ring for which hemoglobin plays an important role in regulating the acid-base balance of blood.

Ans. True.

8. Methemoglobin can carry oxygen in blood.

Ans. False.

9. Sulphemoglobin formed by the administration of certain drugs continues to remain in the blood and can be converted into hemoglobin.

Ans. False.

10. Reduced hemoglobin in the absorption spectra shows one single broad band in the green region. Ans. True.

11. Oxyhemoglobin in the absorption spectra shows three bands. **Ans.** False.
12. The biosynthesis of hemoglobin takes place in the α-cells of pancreas. **Ans.** False.
13. Iron in the ferrous state is incorporated into protoporphyrin to form heme. **Ans.** True.
14. The iron of heme is coordinated to 2 imidazole nitrogen of histidine at positions 36 and 85. **Ans.** False.
15. The function of erythrocruorins is dissimilar to hemoglobin. **Ans.** False.
16. Catalases are zinc porphyrin enzymes. **Ans.** False.
17. In plants, the activities of catalases are significant. **Ans.** False.
18. The cytochromes are iron porphyrins and act as electron transfer agent in oxidation - reduction reactions. **Ans.** True.
19. The reduced form of cytochrome C is autooxidizable. **Ans.** False.
20. Cytochrome Q_3 is found in kidney. **Ans.** False.
21. Methemoglobin also functions as oxygen carrier. **Ans.** False.
22. Porphyrias are of two types - congenital and acquired. **Ans.** True.
23. Erythropoietic porphyria causes highly increased excretion of uroporphyrin I, and to a lesser extent coproporphyrin I in both urine and feces. **Ans.** True.
24. Erythropoietic coproporphyria shows large amounts of coproporphyrin II in the erythrocytes. **Ans.** False.
25. Acute intermittent porphyria causes periodic attacks of abdominal pain which is associated with fever and leukocytosis. **Ans.** True.

Match the following :		Answer
1. Catabolism of hemoglobin means	(a) in the reticuloendothelial cells of the liver, spleen and bone-marrow.	1. ...... (g)
2. About 8 grams of hemoglobin	(b) yields about 300 mg of bilirubin.	2. ...... (i)
3. The protoporphyrin	(c) formed by the oxidation of hemoglobin by oxygen in presence of ascorbic acid is said to be choleglobin.	3. ...... (b)
4. When hemoglobin is catabolized in the body,	(d) is formed by the reduction of biliverdin by the enzyme bilirubin reductase requiring NAD^+ or $NADP^+$	4. ...... (f)
5. The porphyrin portion is broken down mainly	(e) 35 mg of bilirubin which is carried in loose association with plasma protein.	5. ...... (a)
6. The green conjugated protein	(f) the iron enters ferritin "pool" for reuse.	6. ...... (c)
7. Biliverdin, the chief pigment of the bile in birds	(g) conversion of hemoglobin to bile pigments and metabolism of bile pigments.	7. ...... (k)
8. Bilirubin, the chief pigment in human bile,	(h) 0.3 to 1.1 mg per cent.	8. ...... (d)
9. Biliverdin jaundice is largely limited to	(i) contains 27 mg of iron.	9. ...... (j)
10. One gram of hemoglobin yields	(j) biliary obstruction as a result of carcinoma and severe parenchymal liver disease.	10. ...... (e)
11. Normal serum bilirubin level	(k) is formed after removal of iron and cleavage of the porphyrin ring of heme.	11. ...... (h)
12. The feces acquires a green tinge	(l) appear in the urine in the concentration of 1-3.5 mg per day.	12. ...... (p)
13. L-Stercobilinogen	(m) when the concentration of the bile pigment in blood is more and diffuses into the tissues. producing a yellow pigmentation.	13. ...... (q)
14. L-Stercobilin	(n) is caused by the milder defect in the bilirubin conjugated system.	14. ...... (o)
15. Some of the urobilinogen	(o) is an orange yellow pigment which gives the normal colour of the feces and is strongly levorotatory.	15. ...... (l)
16. The unabsorbed urobilinogen	(p) when the intestinal flora is diminished by administration of antibiotics, the bilirubin is auto-oxidized to biliverdin in contact with air.	16. ...... (r)
17. Jaundice occurs	(q) is the reduction product of bilirubin.	17. ...... (m)
18. Crigler-Najjar syndrome, Type II	(r) is excreted in the stool in the concentration of 40-280 mg per day.	18. ...... (n)
19. Gilbert's disease	(s) urobilinogen increased and bilirubin present in urine and urobilinogen decreased in feces.	19. ...... (u)
20. In hepatic jaundice	(t) urobilinogen is absent in urine and urobilinogen trace is absent in feces.	20. ...... (s)
21. In hemolytic jaundice,	(u) is caused by the defect in the hepatic clearance of bilirubin.	21. ...... (w)
22. In obstructive jaundice,	(v) is owing to a primary metabolic defect in the conjugation of bilirubin.	22. ...... (t)
23. Crigler-Najjar syndrome, Type I	(w) urobilinogen increases and bilirubin is absent in urine and urobilinogen increases in feces.	23. ...... (v)
24. Unconjugated hyperbilirubinemia	(x) results from liver dysfunction caused by chloroform, CCl_4, hepatitis virus and cirrhosis.	24. ...... (x)

Appendix 8

ENZYMES AND COENZYMES

Write the correct answer number of the followings

1. The example of extracellular enzymes

(a) Glucose-6-phosphatase, (b) Hexokinase, (c) Glucokinase, (d) Pancreatic amylase. Ans. d.

2. Enzymes are usually named by adding the suffix-ase to the main part of the name of the substrate on which they act except

(a) Erepsin, (b) Maltase, (c) Lactase, (d) Sucrase.

Ans. a.

3. Some enzymes are named by their functions only, eg,

(a) Ptyalin, (b) Pepsin, (c) Reductases, (d) Trypsin.

Ans. c.

4. Some enzymes acting on the substrates are freely described by the adjectives, eg,

(a) Lactate dehydrogenase, (b) Lipolytic, (c) Cytochrome oxidase, (d) Phosphatase. Ans. b.

5. The enzyme which uses oxygen as hydrogen acceptor eg,

(a) Tyrosinase, (b) Succinate dehydrogenase, (c) Aconitase, (d) Carboxylase. Ans. a.

6. The enzyme which uses H_2O_2 as substrate eg,

(a) Catalase, (b) Malate dehydrogenase, (c) Ascorbic oxidase, (d) Phosphorylase. Ans. a.

7. The enzyme which acts on single hydrogen donor with incorporation of oxygen eg,

(a) Succinate thiokinase, (b) Glycogen synthetase, (c) Tryptophan oxygenase, (d) Uricase. Ans. c.

8. The enzyme which acts on paired donors with incorporation of oxygen into one donor eg,

(a) Pyruvate kinase, (b) Enolase, (c) Steroid hydroxylase, (d) Glycerokinase. Ans. c.

9. The example of transglycosidase

(a) Hexokinase, (b) Phosphorylase, (c) GPT, (d) Amino acid transacetylase. Ans. b.

10. Enzyme acting on peptide bonds eg,

(a) Hexokinase, (b) Chymotrypsin, (c) GOT, (d) Glucose-6-phosphatase. Ans. b.

11. The peroxidase has the coenzyme,

(a) FAD^+, (b) NAD^+, (c) $NADP^+$, (d) None of the above.

Ans. d.

12. Enzyme exergonic reaction means the system undergoes a loss of free energy, eg,

(a) Synthetase, (b) Phosphatase, (c) Hexokinase, (d) Urease. Ans. d.

13. The optimum temperature of plant urease in centigrade

(a) 40, (b) 37, (c) 45, (d) 60. Ans. d.

14. The maximum activity of the most of the enzymes is at the optimum pH

(a) Between 2 and 3, (b) Between 4 and 5, (c) Between 5 and 9, (d) Between 7 and 12. Ans. c.

15. Oxidases are generally inhibited by

(a) Cyanides, (b) Fluorides, (c) Salts of mercury, (d) Salts of silver. Ans. a.

16. The group transferring coenzyme

(a) CoA, (b) NAD^+, (c) $NADP^+$, (d) FAD^+. Ans. a.

17. The coenzyme containing an aromatic hetero-ring in the structure
(a) Biotin, (b) TPP, (c) Sugar phosphate, (d) Coenzyme Q. **Ans.** b.

18. The example of hydrogen transferring coenzyme
(a) $B_6 - PO_4$, (b) $NADP^+$, (c) TPP, (d) ATP. **Ans.** b.

Fill in the blanks of the followings

1. Enzymes which are used in the cells which make them are said to be _______ enzymes.
Ans. Intracellular.

2. An extracellular enzyme which is secreted ready for action is called a _______ secretion.
Ans. Zymase.

3. Trypsinogen is activated by _______ to give active trypsin. **Ans.** Enterokinase.

4. Prothrombin is activated by _______ to give active thrombin. **Ans.** Thromboplastin.

5. A dipeptide can be attacked by _______ enzymes.
Ans. Three.

6. Enzymes catalyzing hydrolysis of ester, peptide by the addition of water are called _______ .
Ans. Hydrolases.

7. Enzymes catalyzing the linking together of two compounds couple to the breaking of a pyrophosphate bond in _______ or a similar compound are called _______ . **Ans.** ATP; Ligases.

8. Enzymes that catalyze removal of groups from substrates without addition or removal of water are called _______ . **Ans.** Lyases.

9. The imidazole ring and the alcoholic group of serine are responsible for the union between substrate and enzyme. These groups are termed _______ and the region of the protein surface at which they are located is termed _______ .
Ans. Active groups; Active centre.

10. Many enzymes are _______ proteins.
Ans. Conjugated.

11. Enzymes, the organic catalysts, differ from inorganic catalysts in their extraordinary _______ .
Ans. Specificity.

12. Many substrates form _______ with enzymes.
Ans. 3 bonds.

13. Oxidoreductases function in biosynthetic processes in mammalian system tend to use _______ as reductant but those function in degradative processes tend to use _______ as oxidant.
Ans. Reduced NADP; NAD^+.

14. In liver, about _______ of the _______ enzyme occurs extramitochondrially.
Ans. 90%; NADP-specific.

15. The _______ enzyme of mitochondria is specifically activated by _______ .
Ans. NAD-specific; ADP.

16. Trypsin of pancreatic juice is a mixture of _______, _______ and _______ .
Ans. Trypsin; Chymotrypsin A ; Chymotrypsin B.

17. Most active enzyme preparations are made by various processes of _______ involving _______ to remove inorganic impurities.
Ans. Purification; Dialysis.

18. The insoluble substrates must be made soluble by the help of _______ substances whenever required for good mixing with _______ .
Ans. Hydrotropic; Enzymes.

19. The substrate concentration that produces _______ velocity is called _______ or _______ .
Ans. Half-maximal; Michaelis constant; km value.

20. At the temperature O°C, most enzymes are practically _______ . **Ans.** Inactive.

21. Pepsin works only in _______ medium and is inactivated by making the medium _______ .
Ans. Acid; Alkaline.

22. At low pH values E will be _______ and lose its _______ charge. **Ans.** Portonated; Negative.

23. The loss of enzyme activity by oxidation may be regained by _______ or _______ .
Ans. Cysteine; Glutathione.

24. High energy radiation forms _______ which causes _______ of the enzyme resulting loss in enzyme activity. **Ans.** Peroxides; Oxidation.

25. Some enzymes which are activators are called _______ . **Ans.** Kinases.

26. The best preservative for enzyme solution is _______ . **Ans.** toluene.

27. Enzymes are destroyed by _______ .
Ans. Formaldehyde.

28. The enzymes depending for their activity on the presence of free-SH groups can be inactivated by ______ which reacts with the free-SH groups.
Ans. Mercuric chloride.

29. Sulphanilamide competes ______ .
Ans. P-aminobenzoic acid.

30. Enzymes which show "allosteric" inhibition have two sites namely ______ and ______ .
Ans. Isosteric site; Allosteric site.

31. Inhibition of enzymes may also be reversed by removal of the inhibitor by the treatment with ______ .
Ans. H_2S.

32. It is believed that the mucous membrane contains suitable ______ for which the alimentary canal is not digested by its own secretions.
Ans. Anti-enzymes.

33. The non-protein organic molecules and the prosthetic groups of enzymes are called ______ .
Ans. Coenzymes.

34. The chief function of CoA is to carry ______ groups.
Ans. Acyl.

35. The coenzyme TPP is also called ______ which carries ______ group.
Ans. Co-carboxylase; Active aldehyde.

36. The chief function of tetrahydrofolic acid as a carrier of ______ and it is used in the synthesis of ______ and ______ .
Ans. Formate; Purines; Pyrimidines.

37. In dehydrogenation reactions ______ and ______ coenzymes function as hydrogen acceptors.
Ans. NAD^+; $NADP^+$.

38. Lysosomes can be regarded as a type of ______ .
Ans. Ground substance.

39. Lysosomes are responsible for post-mortem ______ .
Ans. Autolysis.

40. Lysozyme has a molecular weight of about ______ and consists of a single polypeptide chain of ______ amino acid residues having no ______ or ______ cofactors.
Ans. 15000; 129; Coenzymes; Metal ion.

41. Lysozyme has a central ______ site with ______ subunits.
Ans. Catalytic; 6.

42. The enzymes that occur in a number of different forms and differ each other chemically, electrophoretically and immunologically are called ______ .
Ans. Isoenzymes.

43. Lactate dehydrogenase exists in ______ isozymic forms.
Ans. Five.

44. In the presence of acetyl-CoA, the enzyme ______ acts on oxaloacetate to form ______ .
Ans. Citrate synthase; Citrate.

45. When only oxaloacetate is present, the specific enzyme ______ acts on oxaloacetate to form ______ . Ans. Malate dehydrogenase; Malate.

46. Maltase catalyzes the hydrolysis of ______ glycosides but not ______ glycosides.
Ans. α; β.

47. Oxidoreductases function in ______ processes and tend to use ______ as reductant, but those which function in degradative processes tend to use ______ as oxidant.
Ans. Biosynthetic; NADPH; NAD^+.

48. In liver, about 90 per cent of the NADP-specific enzymes occurs ______ , but NAD-specific enzyme of mitochondria is activated by ______ .
Ans. Extramitochondrially; ADP.

49. Potassium containing enzyme is ______ and nickel containing enzyme is ______ .
Ans. Pyruvate kinase; Urease.

50. Selenium containing enzyme is ______ and molybdenum containing enzyme is ______ .
Ans. Glutathione peroxidase; xanthine oxidase.

Indicate "True" or "False" of the followings

1. Lysosomes can be regarded as a type of "Ground substance". Ans. True.
2. The enzyme systems present in lysosomes are for oxidation – reduction processes. Ans. False.
3. Lysozyme destroys the cell-wall of many airborne gram-positive bacteria in tears and nasal mucus. Ans. True.
4. Isozymes are present in the serum and tissues of mammals, birds, insects, plants and unicellular organisms. Ans. True.
5. The isozymic forms of lactate dehydrogenase are trimers. Ans. False.

6. Splitting and reconstitution of lactate dehydrogenase I_5 produces new isozymes. Ans. False.
7. LD_5 is usually predominant in the tumors. Ans. True.
8. The kinetic or collision theory states that the molecules to react must collide and must possess sufficient energy to overcome the energy barrier for reaction. Ans. True.
9. If some molecules have insufficient energy to react, increased temperature which increases their kinetic energy will decrease the rate of the reaction. Ans. False.
10. Hydrophobic groups and charged groups both are not involved in substrate binding. Ans. False.
11. In the absence of substrate, the catalytic and the substrate-binding groups are at several bond distances removed from one another. Ans. True.
12. The catalytic activity of certain key enzymes can be irreversibly decreased or increased by small molecules. Ans. False.
13. Many enzymes catalyze a reaction between two or more substrates producing one product. Ans. False.
14. Reactions whose rates vary as regards to changes in H^+ or H_3O^+ concentration but are independent on the concentration of other acids or bases present in the solution are said to be specific acid or specific base catalysis. Ans. True.
15. More than 25 per cent of the enzymes contain tightly-bound metal ions for their activity. Ans. True.
16. Most kinases form substrate-bridge complexes of the type Enz-nucleotide - M. Ans. True.
17. The metal ion in pyruvate kinase also holds one substrate (ATP) to activate it. Ans. True.
18. Enzymes whose concentration in a cell is dependent of an added inducer are termed constitutive enzymes. Ans. False.
19. Permeases have many properties in common with enzymes and perform functions like cytochromes in electron transport. Ans. True.
20. Protein synthesis takes place in the nuclei. Ans. False.
21. Increase in the enzyme activities in C.S.F. is reflected in the blood. Ans. False.
22. Lactate dehydrogenase activity in C.S.F. is increased in meningitis. Ans. True.
23. The serum acid phosphatase concentration is increased in metastatic prostatic carcinoma. Ans. True.
24. Serum amylase concentration is decreased in liver disease. Ans. True.
25. Creatine phosphokinase level in serum is increased in muscular dystrophy. Ans. True.

Match the following :

		Answer
1. Serum Aldolase level	(a) in normal pregnancy from fourth month.	1. ...... (g)
2. The normal level of serum lipase is	(b) 20-70 U/L.	2. ...... (d)
3. CoA	(c) is seen in untreated pernicious anemia.	3. ...... (i)
4. Serum oxytocinase level is increased	(d) 0-160 U/L.	4. ...... (a)
5. The normal level of SGOT is	(e) in Wilson's disease.	5. ...... (j)
6. Serum ceruloplasmin (Ferro-xidase activity) level is decreased	(f) in cirrhosis and hemolytic jaundice.	6. ...... (e)
7. The normal level of alkaline phosphatase is	(g) is increased in hemolytic anemia.	7. ...... (b)
8. A great rise in serum LDH	(h) 85-190 I.U/L.	8. ...... (c)
9. SGOT is higher than SGPT	(i) is involved in ketone body formation.	9. ...... (f)
10. SGPT is always higher than SGOT	(j) 8-40 Karmen units/dL.	10. ...... (n)
11. Normal level of serum LDH is	(k) in nephrotic syndrome.	11. ...... (h)
12. Serum lipase level is decreased	(l) in muscle trauma, polymyositis, MC Ardle's syndrome, etc.	12. ...... (q)
13. Serum cholinesterase level in increased	(m) occur in parietal cells of the stomach.	13. ...... (k)
14. Serum Glucose-6-phosphate dehydrogenase level is decreased	(n) in viral hepatitis.	14. ...... (x)
15. Serum isocitrate dehydroge-nase level is increased	(o) 25-125 U/L.	15. ...... (y)
16. Normal serum amylase level is	(p) the enzymes for hydroxylation are present.	16. ...... (o)
17. Normal Serum CPK level is	(q) in Vitamin A deficiency and diabetes mellitus.	17. ...... (w)
18. Serum CPK level is increased	(r) in liver and kidney as well as in erythrocytes. Hence serum levels are not affected by hemolysis.	18. ...... (l)
19. Much larger quantities of carbonic anhydrase	(s) zinc-protein complex.	19. ...... (m)
20. CPK is apparently not present	(t) during myocardial injury.	20. ...... (r)
21. Transaminases exist	(u) in acute diseases of the pancreas.	21. ...... (v)
22. Carbonic anhydrase is a	(v) in isozyme forms.	22. ...... (s)
23. Carbonic anhydrase is involved	(w) 170 U/L.	23. ...... (z)
24. LDH is liberated to the blood stream	(x) congenital deficiency causes hemolytic anemia.	24. ...... (t)
25. Serum trypsin level is increased	(y) in liver diseases.	25. ...... (u)
26. In microsomes	(z) in hydrogen ion secretion.	26. ...... (p)

Appendix 9

BIOENERGETICS

Write the correct answer number of the followings

1. The equation $\Delta G = \Delta H - T\Delta S$ depends on

   (a) Constant temperature and pressure, (b) Constant temperature and volume, (c) Constant pressure and volume, (d) All of the above. Ans. a.

2. Some autotropic bacteria utilize the following reaction

   (a) $Fe^{+++} \rightarrow Fe^{++}$, (b) $Fe^{++} \rightarrow Fe^{+++}$, (c) $Cu \rightarrow Cu^{++}$, (d) $Mg^{++} \rightarrow Mg$. Ans. b.

3. Lipmann introduced the symbol ~Ⓟ indicating

   (a) Phosphate bond, (b) Low energy phosphate bond, (c) High energy phosphate bond, (d) Energy phosphate bond. Ans. c.

4. The number of high energy phosphate contained by ATP is

   (a) 1, (b) 2, (c) 3, (d) None of the above. Ans. b.

5. AMP contains

   (a) High energy phosphate bond, (b) phosphate bond, (c) Energy phosphate bond, (d) Low energy phosphate bond. Ans. d.

6. The free energy is derived from respiratory chain oxidation within

   (a) Cytosol, (b) Microsomes, (c) Mitochondria, (d) Lysosomes. Ans. c.

7. In TCA Cycle at the succinyl thiokinase step there is the direct generation of

   (a) One ~Ⓟ, (b) Two ~Ⓟ, (c) Three ~Ⓟ, (d) All of the above. Ans. a.

8. Phosphagens act as storage forms of

   (a) Low energy phosphate, (b) High energy phosphate, (c) Energy phosphate, (d) No energy phosphate. Ans. b.

9. Phosphagens maintain ATP concentration in

   (a) Muscle, (b) Liver, (c) Kidney, (d) Heart. Ans. a.

10. When ATP becomes abundant, its concentration can cause

    (a) Direct reaction, (b) Irreversible reaction, (c) Reverse reaction, (d) All of the above. Ans. c.

11. The enzyme adenylate kinase is said to be

    (a) Glucokinase, (b) Pyruvate kinase, (c) Phosphoglycerokinase, (d) Myokinase. Ans. d.

12. The activation of long chain fatty acids produces

    (a) Inorganic phosphate, (b) Organic phosphate, (c) Adenosine diphosphate, (d) No phosphate. Ans. a.

13. The activation via the pyrophosphate pathway results in the loss of

    (a) One ~Ⓟ, (b) Two ~Ⓟ, (c) Three ~Ⓟ, (d) None of the above. Ans. b.

14. The followings take part in phosphorylations in the cell

    (a) UTP, (b) GTP, (c) CTP, (d) All of the above. Ans. d.

15. Adenylate Kinase is a specialized

    (a) Monophosphate Kinase, (b) Diphosphate Kinase, (c) Triphosphate Kinase, (d) All of the above. Ans. a.

Fill in the blanks of the followings

1. The rate of ______ release is controlled by ______ hormones. **Ans.** Energy; Thyroid.
2. Obesity is caused by the ______ of ______ energy. **Ans.** Storage;Surplus.
3. Death results from ______ when energy reserves are ______ . **Ans.** Starvation; Depleted.
4. Energy production from food is the role of normal ______ and ______ .

   Ans. Nutrition; Metabolism.
5. Nucleoside triphosphates are similar to ______ but contain alternative base to ______ .

   Ans. ATP; Adenine.
6. Nucleoside triphosphates can be synthesized from their ______ by means of the enzyme ______ . **Ans.** Diphosphates; Nucleoside diphosphate kinase.
7. Inorganic ______ is formed during the ______ of long chain fatty acids.

   Ans. Pyrophosphate; Activation.
8. Myokinase catalyzes the interconversion of ATP and ______ on the one hand and ______ on the other. **Ans.** AMP;ADP.
9. Under physiologic conditions, ATP is rapidly used as a source of ______ for ______ contraction.

   Ans. Energy; Muscle.
10. Creatine phosphate is present in vertebrate ______ and ______ but arginine phosphate is present in ______ ______ .

    Ans. Muscle; Brain; Invertebrate; Muscle.
11. When ATP acts as a phosphate ______ , the phosphate group is converted to one of ______ .

    Ans. Donor; Low energy.
12. A net formation of 2~Ⓟ results from the formation of ______ from one mole of ______ in glycolysis. **Ans.** Lactate; Glucose.
13. Three major sources of ~Ⓟ take part in ______ or ______ .

    Ans. Energy Conservation ; Energy capture.
14. The term group transfer potential is referred to ______ . **Ans.** High energy bond.
15. ATP transfers free energy from the ______ to the ______ processes.

    Ans. Exergonic; Endergonic.
16. Heterotrophic organisms obtain free ______ by copuling their ______ .

    Ans. Energy; Metabolism.
17. ATP transfers ______ radicals in the process of ______ . **Ans.** Phosphate; Phosphorylation.
18. The reaction is ______ if ΔG is negative in sign and the reaction ______ if ΔG is positive in sign.

    Ans. Exergonic; Endergenic.
19. The total energy of a system plus its ______ remain ______ .

    Ans. Surroundings; Constant.
20. The total entropy of a system must ______ if a process is to occur ______ .

    Ans. Increase; Spontaneously.
21. Biological systems are ______ and utilize chemical ______ for the living processes.

    Ans. Isothermic; Energy.

Appendix 10

BIOLOGIC OXIDATION

Write the correct answer number of the followings

1. Oxidases are conjugated proteins having the prosthetic group
(a) Magnesium, (b) Manganese, (c) Copper, (d) Iron.
Ans. c.

2. Cytochrome oxidase is
(a) a_3, (b) aa_3, (c) a, (d) None. **Ans.** b.

3. Cytochrome oxidase is poisoned by
(a) Cyanide, (b) Sulfide, (c) Sulfite, (d) Sulfate.
Ans. a.

4. Phenolase is an enzyme containing
(a) Cobalt, (b) Iron, (c) Zinc, (d) Copper. **Ans.** d.

5. Monoamine oxidase oxidizes
(a) Epinephrine, (b) Norepinephrine, (c) Glucagon, (d) Glutathione. **Ans.** a.

6. Aerobic dehydrogenases have the prosthetic group
(a) ATP, (b) NAD^+, (c) FAD^+, (d) $NADP^+$. **Ans.** c.

7. Uricase catalyzes the oxidation of uric acid to
(a) Carbon dioxide, (b) Ammonia, (c) Glyoxal, (d) Allantoin. **Ans.** d.

8. Xanthine dehydrogenase converts purine bases to
(a) Uric acid, (b) Hypoxanthine, (c) Xanthine, (d) Urea. **Ans.** a.

9. Alcohol dehydrogenase from liver contains
(a) Copper, (b) Zinc, (c) Sodium, (d) Potassium.
Ans. b.

10. NADP-linked dehydrogenases in the extramitochondria are found to synthesize
(a) Urea, (b) Steroid, (c) Ascorbic acid, (d) Niacin.
Ans. b.

11. The endoplasmic reticulum cytochrome
(a) b, (b) c, (c) a_1, (d) P-450. **Ans.** d.

12. Mono-oxygenases are found in the
(a) Mitochondria, (b) Microsomes, (c) Nuclei, (d) Cytosol. **Ans.** b.

13. The mitochondrial superoxide dismutase contains
(a) Mg^{++}, (b) Zn^{++}, (c) Mn^{++}, (d) CO^{++}. **Ans.** c.

14. When substrates are oxidized through an NAD-linked dehydrogenase, the P : O ratio is
(a) 1, (b) 2, (c) 3, (d) 4. **Ans.** c.

15. The antibiotic piericidin A inhibits the site of the respiratory chain
(a) I, (b) II, (c) III, (d) All of the above. **Ans.** a.

16. The uncoupling agent of oxidative phosphorylation
(a) Antimycin A, (b) Dicoumarol, (c) Barbiturates, (d) Penicillin. **Ans.** b.

17. The oxidation and phosphorylation in intact mitochondria is completely blocked by
(a) Streptomycin, (b) Gentamycin, (c) Puromycin, (d) Oligomycin. **Ans.** d.

18. The high-energy compound is
(a) UDPG, (b) ATP, (c) ADP, (d) Arginine phosphate.
Ans. a.

Fill in the blanks of the followings

1. Mitochondria are generally impermeable to ________ and other ________. **Ans.** Protons; ions.
2. Uncouplers like DNA increase the permeability of mitochondria to ________ reducing the electrochemical potential for the generation of ________. **Ans.** Protons ; ATP.
3. The P/H$^+$ (transported out) quotient of the ATP ________ is ________. **Ans.** Synthetase; ½.
4. The electrochemical potential difference is utilized to drive membrane-located ________ which in the presence of ________ forms ATP. **Ans.** ATP synthetase; ADP + Pi.
5. The respiratory chain is folded into ________ oxidation/reduction loops in the membrane. **Ans.** Three.
6. A single loop consists of a ________ carrier and an ________ carrier. **Ans.** Hydrogen; Electron.
7. The phosphorylating subunits consist of several proteins collectively known as an ________ subunit which projects into the matrix containing the ________. **Ans.** Fi; ATP synthetase.
8. For every proton pair passing through the ________ complex, one ________ molecule is formed from ADP + Pi. **Ans.** Fo - Fi; ATP.
9. A proton pair attacks one oxygen of Pi to form ________. **Ans.** H_2O.
10. Actively respiring mitochondria accumulate ________. **Ans.** Cations.
11. Coenzyme Q has a structure very similar to vitamin ________ and Vitamin ________. **Ans.** K; E.
12. Coenzyme Q is the constituent of ________ lipids. **Ans.** Mitochondrial.
13. Thyroxine is an uncoupling agent and causes ________ of the mitochondria. **Ans.** Swelling.
14. The process by which ADP is phosphorylated by Pi to ________ in the respiratory chain is called ________. **Ans.** ATP; Oxidative phosphorylation.
15. Various oxidations occur in other parts of the cell except mitochondria and liberate ________ instead of ________. **Ans.** Heat; Energy.
16. Site III of respiratory chain is inhibited by ________ and ________. **Ans.** Dimercaprol; Antimycin A.
17. There must be a redox potential of about ________ volts between components of the respiratory chain if that particular site is to support the coupled formation of 1 mole of ________. **Ans.** 0.2; ATP.
18. The toxicity of oxygen is due to its conversion to ________. **Ans.** Superoxide.
19. Superoxide is formed when ________ flavins are reoxidized by molecular ________ in the respiratory chain. **Ans.** Reduced; Oxygen.
20. Superoxide can be removed by the specific enzyme ________ in the presence of ________. **Ans.** Superoxide dismutase; protons.
21. The sequence of enzymes and carriers responsible for the transport of reducing equivalents from substrate to molecular oxygen is known as ________. **Ans.** Respiratory chain.
22. In oxidation and reduction reactions, the free energy exchange is proportionate to the tendency of reactants to donate or accept electrons is called ________. **Ans.** Redox potential.
23. Many drugs are metabolized in the liver by a system that utilizes a specific ________, cytochrome P450 and the utilization of ________ by this cytochrome is highly ________. **Ans.** Hemoprotein; heme; increased.
24. Hydroxylation reactions take place by the enzymes ________ or ________. **Ans.** Monooxygenases ; Cytochrome P450.
25. Cytochrome P450 is considered the most versatile ________. **Ans.** Biocatalyst.
26. The dual fate of oxygen is responsible for naming of monooxygenases as ________. **Ans.** Mixed-function oxidases.
27. Monooxygenases are termed cytochrome P450 because the enzyme was prepared from ________ that exhibited a distinct pack at ________ on exposure to ________. **Ans.** Microsomes; 450 nm; CO.
28. Cytochrome P450 act on various ________ and ________. **Ans.** Carcinogens; pollutants.

29. The isoforms of cytochrome P450 are about ________ and the symbol CYP denotes ______ .

Ans. 150; Cytochrome P450.

30. CYPA1 denotes a cytochrome P450 that is a member of family ______ and subfamily ______ .

Ans. 1; A.

31. In the adrenal, cytochrome P450 are found in ________ as well as in the _______ .

Ans. Mitochondria; endoplasmic reticulum.

32. The various hydroxylases present in the adrenal play an important role in _______ and _______ biosynthesis. Ans. Cholesterol; steroid.

33. There are at least ______ isoforms of cytochrome P450 present in the _______ of human _______ , each is acting on _______ and _______ .

Ans. Six; endoplasmic reticulum; liver; xenobiotics; endogenous compounds.

34. The enzyme that uses ______ to form the reduced cytochrome P450 is called _______ .

Ans. NADPH; NADPH-cytochrome P450 reductase.

35. The suitable lipid _________ found in the membranes of the _________ is the component of the cytochrome P450.

Ans. Phosphatidyl choline; endoplasmic reticulum.

36. The example of enzyme induction is ______ which is induced by consumption of ________ .

Ans. CYP2EI; ethanol.

37. Certain isoforms of cytochrome P450 are involved in the metabolism of ____________ and related ________ .

Ans. Polycyclic aromatic hydrocarbons (PAHs); molecules.

38. CYP2D6 is involved in the metabolism of ________ and shows slow metabolism of certain drugs resulting in their ________ .

Ans. Debrisoquin; accumulation.

39. Cytochrome oxidase contains ________ and has a high affinity for ________ .

Ans. Copper; Oxygen.

40. Cytochrome oxidase is the only _______ in the chain which signifies the ________ reaction.

Ans. One; irreversible.

41. Site 1 of respiratory chain is inhibited by _____ , _____ , _____ . Ans. Amobarbital; Piericidin A; Fish poison rotenone.

42. Respiration is totally blocked by _________ , _________ , _________ . Ans. H_2S; CO; Cyanide.

43. Cytochrome C is quite stable to ________ and ________ .

Ans. Heat; acids.

44. The peptide chain of human heart cytochrome C contains ________ amino acids and ______ is the N-terminal amino acid and _______ is the C-terminal amino acid.

Ans. 104; acetylglycine; glutamic acid.

45. Two cysteine residues of cytochrome C are located at positions ________ and ______ .

Ans. 14; 17.

46. The linkage of iron in heme occurs through the ________ residue at position _______ in the peptide chain.

Ans. Imidazole nitrogen of histidine; 18.

47. The iron porphyrin group of cytochrome C is attached to _______ more firmly than ______ .

Ans. Protein; hemoglobin.

48. Cytochrome C has a molecular wt. of _______ .

Ans. 13,000.

49. The reduced form of cytochrome C is not ________ .

Ans. Autooxidizable.

50. Carboxin and _______ inhibit transfer of reducing equivalents from succinate dehydrogenase to ______ .

Ans. TTFA; Q.

Appendix 11

INBORN ERRORS OF METABOLISM

Fill in the blanks of the followings

1. Inborn errors of metabolism are a group of metabolic ______ which can be formed as a result of the lack of a single ______ in a single metabolic pathway. **Ans.** Diseases; Enzyme.
2. The hereditary disorder in which acid mucopolysaccharides are excessively deposited in tissues is called ______ in which there is corneal opacity. **Ans.** Hurler's syndrome.
3. Gaucher's disease is caused by the deficiency of ______. **Ans.** Glucocerebrosidase.
4. In Gaucher's disease, the spleen is significantly ______ and there are signs of leukopenia and ______.

   Ans. Increased; Thrombocytopenia.
5. The deficiency of sphingomyelinase causes ______ which occurs in ______.

   Ans. Niemann-Pick disease; Infancy.
6. In Niemann-Pick disease, the liver and spleen are ______. **Ans.** Enlarged.
7. Tay-sach's disease caused by the deficiency of the enzyme ______ in tissues is characterized by the increased accumulation of ______ in brain and spleen.

   Ans. Hexosaminidase A; Gangliosides.
8. In Tay-sach's disease, there is repeated respiratory tract ______ of the patient and the patient expires at the third and fourth ______.

   Ans. Infections; Years.
9. The deficiency of the enzyme ceramide trihexosidase causes ______ in which large amounts of ______ are accumulated in the kidney. **Ans.** Fabry's disease; Ceramide trihexoside.
10. The deficiency of phytanic acid oxidase causes ______ in which large amount of ______ is accumulated.

    Ans. Refsum's disease; Phytanic acid.
11. In Refsum's disease, there is night blindness and narrowing of the ______ fields. **Ans.** Visual.
12. Krabbe's disease results in the deficiency of the enzyme ______ which catalyzes the hydrolysis of ______ to form ceramide and galactose.

    Ans. Galactocerebrosidase; Galactocerebroside.
13. Galactocerebroside is the important component of ______. **Ans.** Myelin.
14. Albinism appears in the total absence of ______ inside the ______ in the skin.

    Ans. Tyrosinase; Melanocytes.
15. Tyrosinosis is due to the absence either of hepatic ______ or of ______ activities.

    Ans. P-hydroxyphenyl pyruvate hydroxylase; Tyrosine transaminase.
16. Hereditary tyrosinemia is caused by the deficiency of ______ and this disorder is not controlled by the administration of ______.

    Ans. P-hydroxyphenylpyruvic acid oxidase; Ascorbic acid.
17. Neonatal tyrosinemia occurs in the ______ and the urine contains ______, ______, P-hydroxyphenyllactic acid and P-hydroxyphenylacetic acid. **Ans.** Newborn; Tyrosine; P-hydroxyphenylpyruvic acid.

18. Hereditary tyrosinemia is similar to neonatal tyrosinemia, except the large amount of ________ is present in the urine.

Ans. P-hydroxyphenyl lactic acid.

19. Phenylketonuria is caused by the absence of the enzyme ________ for which phenylpyruvic acid, phenyllactic acid, phenylacetic acid are formed; the later is converted to ________ which is excreted in the urine.

Ans. Phenylalanine hydroxylase; Phenyl acetylglutamine.

20. Patients with phenylketonuria tend to have a deficiency of ________ . Ans. Serotonin.

21. Alkaptonuria is characterized by the excretion of ________ in the urine owing to the lack of ________ . Ans. Homogentisic acid; Homogentisic acid oxidase.

22. In alkaptonuria, the urine is ________ due to the oxidation of ________ in air.

Ans. Dark; Homogentisic acid.

23. In later life, the accumulation of dark pigment in cartilages and tendons gives rise to the condition known as ________ which is accompanied by ________ changes.

Ans. Ochronosis; Arthritic.

24. Maple syrup urine disease is characterized by the absence of the enzymes required for the ________ of the ________ acids derived from the branched chain amino acid-valine, leucine and isoleucine.

Ans. Oxidative decarboxylation; Keto.

25. The urinary excretion of the keto acids derived from branched chain amino acids produces an odour like that of ________ or ________.

Ans. Maple syrup; Burnt sugar.

26. The infant suffering from maple syrup urine disease is difficult to ________ and may ________.

Ans. Feed; Vomit.

27. Hartnup's disease is characterized by a ________ like skin rash and mental deterioration in the abnormal metabolism of ________.

Ans. Pellagra; Tryptophan.

28. The urine of the patients suffering from Hartnup's disease contains highly increased amounts of ________ as well as ________ .

Ans. Indole acetic acid; Tryptophan.

29. Glycinuria is associated with the excess urinary excretion of ________ with a tendency to form ________ renal stones. Ans. Glycine; Oxalate.

30. The clinical findings ________ are the occurrence of thrombosis, osteoporosis, dislocated lenses in the eyes and frequently mental retardation.

Ans. Homocystinuria.

31. In homocystinuria, a low ________ and a high ________ diet effectively prevent this condition if treated earlier. Ans. Methionine; Cystine.

32. The metabolic block of histidine is due to the insufficient activity of liver ________ which impairs the conversion of histidine to ________.

Ans. Histidase; Urocanic acid.

33. Histidine excretion is also increased during normal ________ but not in ________ of pregnancy.

Ans. Pregnancy; Toxaemia.

34. In hypervalinemia, the infants suffer from ________ growth, ________ wasting and vomiting. Ans. Stunted; Muscle.

35. Lesch-Nyhan syndrome is characterized by the complete deficiency of ________ which causes hypoxanthine or guanine to form a nucleotide with ________.

Ans. Phosphoribosyl transfers; PRPP.

36. Lesch - Nyhan syndrome appears in childhood as a severe neurological syndrome, which is sometimes accompanied by ________. Such patients show an increased tendency to ________.

Ans. Gout; Myocardial infarction.

37. In hereditary xanthinuria, there is the deficiency of ________. The urinary excretion contains large amounts of ________ with lesser amounts of ________ . Ans. Xanthine oxidase; Xanthine; Hypoxanthine.

38. Orotic aciduria causes the excessive production of ________ due to the deficiency of ________. The urinary excretion consists of large amounts of ________ nucleotide precursor.

Ans. Orotic acid; Orotate phosphoribosyl transferase; Pyrimidine.

39. In orotic aciduria, the urine of the patients becomes cloudy on cooling with the deposition of needle-shaped crystals of ________. Children affected by this condition develop a severe

__________ anemia with physical and mental __________. **Ans.** Orotic acid; Megaloblastic; Retardation.

40. Homocystinuria appears due to the lack of __________ in the liver due to which both __________ and __________ are accumulated in blood and urine. **Ans.** Cystathionine synthetase; Homocystine; Methionine.

41. Cystine is an __________ amino acid which may precipitate in the kidney tubules to form __________ in cystinuric patients.
Ans. Insoluble; Cystine Calculi.

42. Cystinuria is a misnomer, so that __________ may be preferred as the descriptive term for this disease.
Ans. Cystine-Lysinuria.

43. The plasma content of glycine is normal in glycinuric patients while the urinary excretion of glycine ranges from __________ /d.
Ans. 600 - 1000 mg.

44. Maple syrup urine disease is recognized by central nervous system manifestations of __________ and attacks of flaccidity and __________.
Ans. Convulsions; Apnea.

Appendix 12

WATER AND MINERAL METABOLISM

Write the correct answer number of the followings

1. Of the total body weight the average body water in per cent

   (a) 55 to 65, (b) 50 to 60, (c) 60 to 70, (d) 70 to 80.

   Ans. c.

2. Intracellular fluid contains the per cent of the total body weight

   (a) 45, (b) 50, (c) 55, (d) 60. **Ans.** b.

3. The per cent of plasma in the extracellular fluid

   (a) 3.5, (b) 3.8, (c) 4.2, (d) 4.5. **Ans.** d.

4. Transcellular fluid in per cent of extracellular fluid

   (a) 1.4, (b) 1.5, (c) 1.6, (d) 1.8. **Ans.** b.

5. Organic compounds of small molecular size

   (a) Urea, (b) Uric acid, (c) Creatinine, (d) Phosphates.

   Ans. a.

6. Organic substances of large molecular size

   (a) Starch, (b) Inulin, (c) Lipids, (d) Proteins.

   Ans. d.

7. 100 grams of fat on combustion produces water in ml

   (a) 107, (b) 105, (c) 102, (d) 98. **Ans.** a.

8. 100 grams of protein on combustion yields water in ml

   (a) 35, (b) 38, (c) 41, (d) 44. **Ans.** c.

9. The daily intake and output of water through different sources in ml

   (a) 2500, (b) 2700, (c) 2800, (d) 3000. **Ans.** b.

10. The amount of internal secretion per day in litres

    (a) 4 to 7, (b) 5 to 8, (c) 6 to 9, (d) 7 to 10. **Ans.** d.

11. The amount of water excreted per day in feces in ml

    (a) 90, (b) 95, (c) 100, (d) 105. **Ans.** c.

12. The regulatory mechanism of body water is influenced by the hormone

    (a) Oxytocin, (b) ACTH, (c) FSH, (d) Epinephrine.

    Ans. a.

13. All chemical reactions in the body proceed in presence of

    (a) Organic salts, (b) Inorganic salts, (c) Water, (d) Alkali. **Ans.** c.

14. The symptom of water intoxication is

    (a) Muscular weakness, (b) Anemia, (c) Paralysis, (d) Fever. **Ans.** a.

15. Dehydration may be ordinarily corrected by parenteral ingestion of the solution of

    (a) $ZnCl_2$, (b) $MgCl_2$, (c) $CaCl_2$, (d) NaCl. **Ans.** d.

16. A mixture of two-thirds isotonic saline solution and one-third lactate solution should be administered intravenously in case of

    (a) Constipation, (b) Prolonged diarrhea, (c) Malaria, (d) Jaundice. **Ans.** b.

17. Dehydration is a problem as regards to

    (a) Vomiting, (b) Typhoid, (c) Uremia, (d) Hot climates. **Ans.** c.

18. The principal mineral elements are also said to be

    (a) Macronutrients, (b) Micronutrients, (c) Semimicronutrients, (d) None of the above. **Ans.** a.

19. The trace elements are subdivided into

    (a) Two groups, (b) Three groups, (c) Four groups, (d) None of the above. Ans. b.

20. The essential trace elements include

    (a) Lead, (b) Nickel, (c) Boron, (d) Cobalt. Ans. d.

21. The number of principal mineral elements

    (a) Five, (b) Six, (c) Seven, (d) Eight. Ans. c.

22. Zinc is a constituent of

    (a) Carbonic anhydrase, (b) Malate dehydrogenase, (c) Aldolase, (d) Amylase. Ans. a.

23. Hemoglobin formation needs both

    (a) Iron and zinc, (b) Iron and calcium, (c) Iron and copper, (d) Iron and magnesium. Ans. c.

24. Cobalt is a constituent of

    (a) Folic acid, (b) Vitamin B_{12}, (c) Niacin, (d) Biotin. Ans. b.

25. Calcium is required for the activation of the enzyme

    (a) Isocitrate dehydrogenase, (b) Fumarase, (c) Succinate thiokinase, (d) ATPase. Ans. d.

26. The absorption of calcium is increased by the dietary higher levels of

    (a) Fats, (b) Proteins, (c) Cereals, (d) Vitamin A. Ans. b.

27. Calcium absorption is interfered by

    (a) Fatty acids, (b) Amino acids, (c) Vitamin D, (d) Vitamin B_{12}. Ans. a.

28. Calcium in the complex form is about

    (a) 8 mg%, (b) 6 mg%, (c) 4 mg%, (d) 2 mg%. Ans. d.

29. The percentage of calcium in milligram in non-ionized form is about

    (a) 3, (b) 4, (c) 5, (d) 6. Ans. c.

30. In man, the number of grams of calcium filtered in 24 hours by the renal glomeruli is about

    (a) 8, (b) 10, (c) 12, (d) 14. Ans. b.

31. The percentage of calcium excreted in the feces

    (a) 40 to 60, (b) 50 to 70, (c) 60 to 80, (d) 70 to 90. Ans. d.

32. The daily loss of calcium in mg in sweat is about

    (a) 12, (b) 13, (c) 14, (d) 15. Ans. d.

33. The concentration of serum calcium may drop below 7 mg/100 ml in

    (a) Hyperparathyroidism, (b) Hypoparathyroidism, (c) Tetany, (d) Rickets. Ans. b.

34. Serum phosphate concentration is 1 to 2 mg/100 ml in

    (a) Rickets, (b) Tetany, (c) Osteoporosis, (d) Hyperthyroidism. Ans. a.

Fill in the blanks of the followings

1. There is virtually no calcium in ________. Ans. Erythrocytes.
2. Calcium exists in the plasma in ________, ________, ________ forms. Ans. Diffusible; Nondiffusible; Complexed.
3. The active transport process of calcium is regulated by ________, a metabolite of Vitamin ________ which is produced in the kidney in response to low plasma ________ concentration. Ans. 1, 25-dihydroxycholecalciferol; D; Ca^{++}.
4. Ionized calcium is required in blood ________ process. Ans. Coagulation.
5. Under optimal conditions, the percentage of calcium absorbed is ________. Ans. 75.
6. Decreased ionized fraction of serum calcium causes ________. Ans. Tetany.
7. ________ which occurs in cereal grains forms insoluble salts ________ with calcium and magnesium resulting in the impaired absorption of calcium. Ans. Phytic acid; Phytin.
8. An increase in the ________ calcium levels in the plasma is the stimulus for the production of ________ which then causes a deposition of calcium in bone. Ans. Ionized; Calcitonin.
9. Succinate dehydrogenase requires ________ for its activation. Ans. Calcium.
10. In rickets, serum alkaline phosphatase activity is ________. Ans. Increased.
11. Renal rickets are more accurately designated ________ which is an inherited disease. Ans. Familial hypophosphatemic rickets.
12. Vitamin D in ordinary doses does not relieve renal rickets. Hence, it is sometimes referred to as ________. Ans. Vitamin D-resistant rickets.

13. Phosphorus is essential for the formation and development of bones and ________ along with ________. Ans. Teeth; Calcium.

14. Phosphorus is required in the absorption of glucose by ________. Ans. Phosphorylation.

15. High calcium diet and phytic acid ________ phosphorus absorption. Ans. Decrease.

16. The normal inorganic phosphate of plasma is ________ per 100 ml in adults.

Ans. 3 to 5.0 mg

17. The reabsorption of phosphorus is inhibited by ________. Ans. Parathyroid hormone.

18. Magnesium is the principal ________ of the soft tissue. Ans. Cation.

19. Co-carboxylase is provided with ________.

Ans. Magnesium.

20. Parathyroid hormone ________ the absorption of magnesium. Ans. Increases.

21. In case of Kwashiorkor, the serum magnesium level is ________ causing ________.

Ans. Low; Weakness.

22. In uremia, the serum magnesium level is ________. Ans. Low.

23. Sodium is largely associated with ________ and ________ in regulation of acid-base equilibrium.

Ans. Chloride; Bicarbonate.

24. In the absorption of glucose and galactose ________ ion plays an important role.

Ans. Sodium.

25. In man, erythrocytes contain ________ or ________ sodium. Ans. Little; No.

26. ________ increases plasma sodium level.

Ans. Aldosterone.

27. Loss of sodium by excessive sweating causes ________ with the intense and painful contractions of ________ of men working hard in hot humid climates.

Ans. Heat cramps; Skeletal muscle.

28. The metabolism of sodium is regulated by ________. Ans. Adrenocortical steroids.

29. Hypernatremia occurs in prolonged treatment of ________ and ________ as well as ________ hormones. Ans. Cortisone; ACTH; Sex.

30. In certain stages of pregnancy, the steroid hormones cause the retention of ________ as well as ________ which results in gain in weight.

Ans. Sodium; Water.

31. Pyruvate kinase requires ________ for maximal activity. Ans. K^+.

32. Protein biosynthesis by ribosomes requires ________. Ans. Potassium.

33. ________ decreases serum potassium level.

Ans. Aldosterone.

34. In man, ________ contain large amounts of potassium. Ans. Erythrocytes.

35. The serum potassium level ________ during increased carbohydrate utilization following administration of ________.

Ans. Decreases; Insulin.

36. The symptoms of hyperkalemia are corrected by the administration of ________ which helps the excretion of ________.

Ans. Deoxycorticosterone; Potassium.

37. Prolonged hypokalemia causes injury to ________ and ________.

Ans. Myocardium; Kidneys.

38. Hypokalemia is exhibted in heart failure treatment with ________. Ans. Digitalis.

39. Chloride ion is important as an activator of ________. Ans. Amylase.

40. Excessive consumption of ________ causes ________ in protein deficiency.

Ans. NaCl; Edema.

41. Sodium concentration in sweat is decreased by ________. Ans. Aldosterone.

42. Hypochloremic alkalosis may develop in ________ disease or after the administration of ________ or ________.

Ans. Cushing's; ACTH; cortisone.

43. Sulfur is present primarily in the cell protein in the form of ________ and ________.

Ans. Cysteine; Methionine.

44. A small amount of sulfide if absorbed into the blood stream is rapidly oxidized to ________.

Ans. Sulfate.

45. An increase in the blood indican concentration may occur in ________. Ans. Uremia.

46. Marked sulfate retention in advanced glomerulonephritis causes the development of ________. Ans. Acidosis.

47. The nonheme iron is completely ________ which exists in the form of ________ and ________.

Ans. Protein-bound; Storage; Transport.

48. Iron absorption is enhanced by proteins of ________ digestive products forming ________.

Ans. Low molecular weight; Iron chelate.

49. Most of the iron in food occurs in the ________ state. Ans. Ferric.

50. Phytic acid and oxalates ________ with iron absorption. Ans. Interfere.

Indicate "True" or "False" of the followings

1. The total iron content of the normal adult is about 4 to 5 grams. Ans. True.
2. About 60 to 70 per cent of the total iron is present in hemoglobin. Ans. True.
3. About 1.0 per cent of the total iron is carried in the plasma in combination with β globulin transport protein transferrin. Ans. False.
4. Iron is not involved in the cellular respiration. Ans. False.
5. The richest sources of iron are egg-yolk, fish, figs and apples. Ans. False.
6. A defect in hemoglobin synthesis in anemia is commonly found in copper deficiency. Ans. True.
7. In humans, heme is broken down in the mucosa and iron appears in the plasma transferrin. Ans. True.
8. Vitamin D and alkali increase iron absorption. Ans. False.
9. Apoferritin, in the mucosal cells, is the controlling factor for iron absorption. Ans. False.
10. Transferrin or siderophilin is a specific iron binding protein. Ans. True.
11. Only the reticulocytes can utilize the ferric ion bound to transferrin, although reticulocytes and mature erythrocytes can take up unbound ferric ion. Ans. True.
12. In hepatic diseases both the bound iron and the total iron- binding capacity of the plasma is high. Ans. False.
13. Ferritin is not present in the intestine and also in liver. Ans. False.
14. Iron deficiency anemia is not prevalent among children and adolescent girls. Ans. False.
15. Copper is required for melanin formation, phospholipid synthesis and collagen synthesis. Ans. True.
16. In 80 per cent of the red blood cell copper is present as superoxide dismutase (erythrocuprein). Ans. True.
17. The daily biliary excretion of copper is 0.5 to 1.3 mg. Ans. True.
18. Wilson's disease is not a rare hereditary disorder of copper metabolism. Ans. False.
19. The absorption of copper from the intestine is very low. Ans. False.
20. Excessive deposition of copper in the liver causes hepatic cirrhosis. Ans. True.
21. The excess of tissue copper can be removed by administering the copper chelating agent penicillamine. Ans. True.
22. Ceruloplasmin is a copper binding plasma protein. Ans. True.
23. Triiodothyronine is less active than thyroxine in many respects. Ans. False.
24. The body normally contains about 10 to 20 grams of iodine. Ans. False.
25. All the iodine in saliva is organic. Ans. False.
26. Muscles contain less amount of iodine. Ans. False.
27. Erythrocytes contain no inorganic iodine. Ans. False.
28. Most tissues contain less amount of iodine in the inorganic form and most of the iodine is present in the organic form. Ans. True.
29. 90 per cent of the iodine of the thyroid gland is in organic combination. Ans. True.
30. The urinary elimination of iodine is largest when the intake is high. Ans. False.
31. Urinary iodine is increased by exercise and other metabolic factors. Ans. True.

32. In children, severe iodine deficiency results in the extreme retardation of growth which is known as cretinism. **Ans.** True.

33. Fluorine in combination with Vitamin C is required for the treatment of osteoporosis. **Ans.** False.

34. Sodium fluoride is a powerful inhibitor of phosphoglycerate kinase. **Ans.** False.

35. Fluoride ions inhibit the metabolism of oral bacterial enzymes and diminish the local production of acids which cause the production of dental caries. **Ans.** True.

36. The retina contains a zinc metalloenzyme, retinene reductase which is required for the formation of retinene. **Ans.** True.

37. Zinc is not concerned with the healing of wounds. **Ans.** False.

38. High concentration of zinc are present in spermatozoa, prostate and epididymis. **Ans.** True.

39. The lowest concentration of copper occurs in the choroid of the eye. **Ans.** False.

40. Zinc present in cereals is highly absorbed owing to the presence of phytic acid. **Ans.** False.

41. The deficiency of zinc causes hepatosplenomegaly, delayed closure of the epiphyses of the long bones and anemia. **Ans.** True.

42. Methyltetrahydrofolate oxidoreductase and homocysteine methyltransferase require Vitamin B_{12} for activity. **Ans.** True.

43. Cobalt in large doses develops a condition known as polycythemia. **Ans.** True.

44. Manganese ions inhibit lipid peroxidation reactions. **Ans.** True.

45. The heart and the brain are the main storage organs for manganese. **Ans.** False.

46. Manganese toxicity develops hepatolenticular degeneration resembling Parkinson's disease. **Ans.** True.

47. Liver and kidney contain less amount of molybdenum than other tissues. **Ans.** False.

48. Selenium is essential for normal growth and fertility of human beings. **Ans.** False.

49. Glutathione peroxidase, a selenoprotein, catalyzes the peroxidation of glutathione. **Ans.** True.

50. Selenium is involved in immune mechanism and ubiquinone synthesis. **Ans.** True.

51. Chromium potentiates the action of insulin in accelerating utilization of glucose in humans. **Ans.** True.

52. Chromium deficiency is characterized by impaired growth, disturbances in glucose, lipid and protein metabolism. **Ans.** True.

Match the following :		**Answer**
1. Calcium is required for the activation of	(a) Prolonged vomiting and diarrhea resulting in execessive loss of digestive juices rich in sodium ion.	1. ..... (f)
2. The daily requirement of calcium of adults	(b) Overactivity of adrenal cortex (Cushing's syndrome) which causes increased excretion of potassium in urine.	2. ..... (i)
3. The Calcium distributed in C.S.F.	(c) Muscular weakness, irritability, paralysis, tachycardia and dilatation of the heart.	3. ..... (j)
4. The renal theshold for phosphate excretion	(d) For oxidative phosphorylation.	4. ..... (e)
5. The normal level of magnesium in blood	(e) About 2 mg/100 ml. of plasma.	5. ..... (o)
6. Magnesium functions as a cofactor	(f) Succinate dehydrogenase, ATPase and certain proteolytic enzymes.	6. ..... (d)
7. Magnesium distributed in muscle	(g) 136 – 146 mEq/L	7. ..... (r)

Contd.

Match the following :

		Answer
8. Sodium ion is involved in	(h) Initiating and maintaining the heart beat.	8. .... (h)
9. The normal level of sodium in blood	(i) 800 mg.	9. .... (g)
10. The daily requirement of sodium for adults	(j) 4.5 – 5.0 mg/100 ml .	10. .... (u)
11. The normal daily losses of sodium	(k) Severe dehydration, reduced blood pressure, decreased blood volume and circulatory failure.	11. .... (w)
12. The symptom of hyponatremia	(l) 98 – 110 mEq/L	12. .... (a)
13. The symptom of hypernatremia	(m) 65 –170 μg/dL	13. .... (p)
14. Hypernatremia occurs in	(n) Prolonged treatment of cortisone and ACTH as well as sex hormones.	14. .... (n)
15. Hyponatremia causes	(o) 1.7 –2.6 mg/100 ml.	15. .... (k)
16. Potassium is distributed in nerves	(p) Increased retention of water in the body.	16. .... (s)
17. The normal level of potassium in serum	(q) Injury to myocardium and kidneys.	17. .... (t)
18. The daily requirement of potassium	(r) 21 mg/100 grams.	18. .... (x)
19. The symptoms of hyperkalemia	(s) 530 mg/100 grams.	19. .... (v)
20. Prolonged hypokalemia causes	(t) 3.5 – 4.5 mEq/L	20. .... (q)
21. The clinical condition of hypokalemia	(u) 5 – 15 grams.	21. .... (b)
22. The symptoms of hypokalemia are	(v) Mental confusion, numbness, weakness of respiratory muscles and flaccid paralysis of the extremities.	22. .... (c)
23. The normal level of chloride in serum	(w) 10 – 150 mEq/24 hr	23. .... (l)
24. Sulfur is constituent of	(x) About 4 grams.	24. .... (z)
25. The total sulfate excretion is increased	(y) In conditions accompanied by excessive tissue protein breakdown as in high fever and increased metabolism.	25. .... (y)
26. The normal level of iron in blood	(z) Coenzyme A and lipoic acid	26. .... (m)
27. The normal level of copper in serum	(z_1) 3-6 μg/dL	27. .... (z_2)
28. The normal concentration of iodine in serum	(z_2) 75-160 μg/dL	28. .... (z_1)

Appendix 13

DIGESTION AND ABSORPTION FROM THE GASTROINTESTINAL TRACT

Write the correct answer number of the followings

1. The amount of saliva secreted each day in ml
(a) 600 – 800, (b) 700 – 900, (c) 800 – 1000, (d) 1000 – 1500. **Ans.** d.

2. The submaxillary secretion contains most of
(a) Glycoproteins, (b) Glycolipids, (c) Both of the above, (d) None of the above. **Ans.** a.

3. A saliva contains inorganic material in per cent
(a) 0.1, (b) 0.2, (c) 0.3, (d) 0.4. **Ans.** b.

4. The amylase is helped to be stabilised by
(a) Ca^{++}, (b) Mg^{++}, (c) Na^{+}, (d) K^{+}. **Ans.** a.

5. Salivary amylase is readily inactivated at pH
(a) 6.0, (b) 5.0, (c) 4.0, (d) None of the above. **Ans.** c.

6. The continued gastric secretion is regulated by the hormone
(a) Glucagon, (b) Gastrin, (c) Epinephrine, (d) ACTH. **Ans.** b.

7. The gastric juice is stimulated to be secreted by
(a) Protamine, (b) Histamine, (c) Prolamine, (d) Albumin. **Ans.** b.

8. Hydrochloric acid secreted by the oxyntic cells contains the pH
(a) 0.5, (b) 0.6, (c) 0.8, (d) 0.9. **Ans.** d.

9. The percentage of organic materials in gastric juice is about
(a)0.1, (b) 0.2, (c) 0.4, (d) 0.8. **Ans.** c.

10. On ordinary diet, the daily amount of secretion of gastric juice by an adult in litres
(a) 1 – 2, (b) 2 – 3, (c) 3 – 4, (d) 4 – 5. **Ans.** b.

11. The intrinsic factor (HCL and mucoproteins) present in the gastric juice helps in the absorption of
(a) Vitamin B_2, (b) Biotin, (c) Folic acid, (d) Vitamin B_{12}. **Ans.** d.

12. Hydrochloric acid stimulates duodenum to liberate
(a) Secretin, (b) Pepsin, (c) Trypsin, (d) Enterocrinin. **Ans.** a.

13. Pepsin contains large amount of amino acids
(a) Neutral, (b) Basic, (c) Acidic, (d) Sulfur containing. **Ans.** c.

14. Pepsin has a molecular weight of
(a)31.600, (b) 32.700, (c) 33.300, (d) 34.400. **Ans.** b.

15. Rennin changes the casein of milk to paracasein in presence of
(a) Mg^{++}, (b) Mn^{++}, (c) Ca^{++}, (d) Ba^{++}. **Ans.** c.

16. Lipase can act only at the pH
(a) 3 to 4.5, (b) 3.5 to 5.0, (c) 4 to 6, (d) 5 to 7. **Ans.** d.

17. Pancreatic juice contains percentage of solids
(a) 1.6, (b) 1.8, (c) 2.0, (d) 2.2. **Ans.** b.

18. The amount of pancreatic juice secreted each day in ml
(a) 300 to 500, (b) 400 to 600, (c) 500 to 700, (d) 600 to 800. **Ans.** d.

19. Trypsin attacks peptide linkages containing the amino acid residue

(a) Arginine, (b) Glycine, (c) Tryptophan, (d) Serine. Ans. a.

20. Chymotrypsin attacks peptide linkages containing the amino acid residue

(a) Leucine, (b) Valine, (c) Phenylalanine, (d) Aspartic acid. Ans. c.

21. Carboxypeptidase contains

(a) Zinc, (b) Copper, (c) Manganese, (d) Magnesium. Ans. a.

22. Carboxypeptidase B hydrolyses terminal peptide linkages containing

(a) Glycine and valine, (b) Serine and tryptophan, (c) Leucine and isoleucine, (d) Lysine and arginine. Ans. d.

23. Collagenase hydrolyses collagen present in

(a) Eggs, (b) Soyabeans, (c) Meat, (d) Milk. Ans. c.

24. RNAse and DNAse are capable of cleaving internal

(a) Phosphodiester bonds, (b) Diester bonds, (c) Ester bonds, (d) None. Ans. a.

25. The amount of intestinal juice secreted each day by an adult in litres

(a) 1 to 2, (b) 2 to 3, (c) 3 to 4, (d) 4 to 5. Ans. b.

26. The percentage of solids in intestinal juice is about

(a) 0.5, (b) 1.0, (c) 1.5, (d) 2.0. Ans. c.

27. The inorganic substances present in the solids of intestinal juice is nearly

(a) One fourth, (b) One third, (c) Half, (d) Two third. Ans. d.

28. Bile is produced by the

(a) Liver, (b) Gall-bladder, (c) Pancreas, (d) Intestine. Ans. a.

29. Bile formed daily in adult human beings in mol. of about

(a) 200 to 800, (b) 300 to 1200, (c) 700 to 1500, (d) 800 to 1800. Ans. b.

30. The pH of hepatic bile

(a) 6.9 – 7.7, (b) 7.1 – 7.3, (c) 7.4 – 7.8, (d) 7.6 – 8.1. Ans. b.

31. Bile acids are synthesized from cholesterol in the

(a) Duodenum, (b) Intestine, (c) Gall-bladder, (d) Liver. Ans. d.

32. Bile acids are derived from the parent acid called

(a) Taurocholic acid, (b) Prostanoic acid, (c) Cholanic acid, (d) Cholic acid. Ans. c.

33. In human bile, sodium glycocholate is greater than sodium taurocholate by

(a) Two times, (b) Three times, (c) Four times, (d) Five times. Ans. b.

34. Bile salts activate

(a) Pancreatic lipase, (b) Cholesterol esterase, (c) Both, (d) None. Ans. c.

35. In obstructive jaundice, bile salts in the blood are

(a) Decreased, (b) Highly decreased, (c) Increased, (d) Greatly increased. Ans. d.

36. The amount of bile salts in mg per day are not absorbed and is eliminated in the feces

(a) 500, (b) 700, (c) 900, (d) 1200. Ans. a.

37. Bile is an important source of

(a) Acid, (b) Alkali, (c) Salt, (d) None of the above. Ans. b.

38. Arginine is decarboxylated by the intestinal bacteria into

(a) Agmatine, (b) Arginamine, (c) Arginatine, (d) Argininemine. Ans. a.

39. By a series of reactions in the large intestine tryptophan forms

(a) Indole, (b) Methylindole, (c) Both of the above, (d) None of the above. Ans. c.

40. In the large intestine, cysteine by a series of transformations forms

(a) Ethyl mercaptan, (b) Methyl mercaptan, (c) H_2S, (d) All of the above. Ans. d.

41. The quantity of ammonia transported from the large intestine to the blood is reduced by the antibacterial action of

(a) Streptomycin, (b) Neomycin, (c) Penicillin, (d) Chloramphenicol. Ans. b.

42. The reduction of biliverdin to urobilinogen takes place in the

(a) Duodenum, (b) Small intestine, (c) Large intestine, (d) All of the above. Ans. c.

43. Cathepsins occur in the
(a) Mitochondria, (b) Cytosol, (c) Nuclei, (d) Lysosomes. **Ans.** d.

44. Enterocrinin is released from the
(a) Duodenal mucosa, (b) Pancreas, (c) Large intestine, (d) Small intestine. **Ans.** a.

45. The secretion of enterogastrone by the duodenal mucosa is influenced by the dietary
(a) Carbohydrate, (b) Fat, (c) Protein, (d) Minerals. **Ans.** b.

Fill in the blanks of the followings

1. Hepatocrinin liberated from the duodenum stimulates ________ to secrete more ________. **Ans.** Liver; Bile.
2. Enterogastrin inhibits the secretion of ________ and ________. **Ans.** HCl ; Pepsin.
3. Enterocrinin stimulates the secretion of ________. **Ans.** Intestinal juice.
4. CCK-PZ is a ________ containing ________ amino acids. **Ans.** Polypeptide; 33.
5. CCK-PZ represents as one of the ________ influencing ________ release during oral administration of glucose. **Ans.** Gut factors; Insulin.
6. CCK-PZ is the important stimulator for the ________ enzyme secretion and ________ contraction. **Ans.** Pancreatic; Gall-bladder.
7. CCK-PZ performs many of the actions of ________ and ________ on water, bicarbonate and acid changes. **Ans.** Gastrin; Secretin.
8. Gastrin stimulates the secretion of HCl, ________ and ________ from gastric mucosa. **Ans.** Pepsin; Intrinsic factor.
9. Gastrin contains ________ amino acids and in Gastrin 11 tyrosine is sulphated in position ________. **Ans.** 17; 12.
10. Secretin is present in ________ and ________ mucosa. **Ans.** Duodenal; Jejunal.
11. Secretin contains ________ amino acids, 14 of which are identical to those found in ________. **Ans.** 27; Glucagon.
12. Secretin stimulates ________ secretion in the stomach and inhibits ________ secretion as well as intestinal ________ activity. **Ans.** Pepsin; Gastric acid; Motor.
13. Secretin is administered to assess the pancreatic ________ function. **Ans.** Exocrine.
14. After death, the intracellular enzymes digest the tissues when kept under ________ condition; this self-digestion of tissues are called ________. **Ans.** Sterile; Autolysis.
15. The group of intracellular proteases present in all mammalian tissues is called ________. **Ans.** Cathepsin.
16. Atropy is a process of ________ in the living animal. **Ans.** Autolysis.
17. In man, intestinal bacteria synthesize certain vitamins, particularly ________ and possibly certain members of the ________. **Ans.** Vitamin K; B-complex.
18. In herbivora, the intestinal bacteria also synthesize ________ and ________. **Ans.** Essential amino acids; Vitamins.
19. Intestinal bacteria putrefy nitrogenous substances to form ________ which is absorbed into the portal ________. **Ans.** Ammonia; Circulation.
20. Intestinal bacteria cause the decarboxylation of lysine into ________. **Ans.** Cadaverine.
21. After reduction cholesterol is converted to ________. **Ans.** Coprosterol.
22. Lecithin is decomposed to ________ and ________. **Ans.** Choline; Neurine.
23. Urate stones are formed due to ________ and ________. **Ans.** Hyperuricemia; Gout.
24. Renal calculi formation is due to urine ________ and ________. **Ans.** Infection; Stagnation.
25. Mixed gall stones are composed of ________, ________. **Ans.** Cholesterol; Bile pigments.
26. Gall-stones are formed due to defects in the ________ circulation and with the diseases of the terminal ileum as well as in patients with ________ **Ans.** Entrohepatic; Cirrhosis.
27. Choleic acid formed by ________ assists the absorption of many important ________ compounds. **Ans.** Deoxycholic acid; insoluble.

28. Bile salts stimulate intestinal __________. Ans. Peristalsis.
29. Many substances such as fatty acids, phenols, naphthalene, etc. combine with deoxycholic acid to form __________. Ans. Choleic acid.
30. During digestion the gall-bladder contracts by the stimulation of the hormone __________ which is produced by the __________. Ans. Cholecystokinin; Small intestine.
31. The composition of the bile in the gall-bladder is modified by the addition of __________ and other substances and by removal of water, __________ and chloride by reabsorption by the bladder mucosa. Ans. Mucin; Bicarbonate.
32. Enterokinase converts inactive __________ to active __________. Ans. Trypsinogen; Trypsin.
33. Fats, proteins and carbohydrates entering the duodenum and upper jejunum stimulate the secretion of __________ which induces the flow of __________. Ans. Enterocrinin; Intestinal juice.
34. Pancreatic lipase is specific for the hydrolysis of primary __________ linkages which occurs in position __________ of triglycerol. Ans. Ester; 1 and 3.
35. Pancreatic amylase is an __________ and an __________. Ans. α-amylase; Endoamylase.
36. Carboxypeptidases are __________ and hydrolyze only the terminal __________ linkage. Ans. Exopeptidase; Peptide.
37. Carboxypeptidase is a __________ containing enzyme. Ans. Zinc.
38. Pancreozymin stimulates the pancreas to produce a viscous fluid low in __________ but high in __________ content. Ans. Bicarbonate; Enzyme.
39. The secretion of pancreatic juice is controlled by both __________ and __________ means. Ans. Nervous; Hormonal.
40. Paracasein is acted on by __________. Ans. Pepsin.
41. Pepsin converts native proteins into __________ and __________. Ans. Proteoses; Peptones.
42. Hydrochloric acid stimulates duodenum to liberate __________. Ans. Secretin.
43. Carbonic anhydrase catalyzes the formation of H_2CO_3 from __________ and __________. Ans. H_2O; CO_2.
44. Saliva facilitates swallowing by the lubricating action of the __________. Ans. Glycoprotein.
45. __________ or __________ mechanisms cause the initiation of gastric secretion. Ans. Nervous; Reflex.
46. Salivary amylase is activated by __________ ion. Ans. Chloride.
47. Production of no hydrochloric acid leads to the condition __________. Ans. Achlorhydria.
48. Renin causes the __________ of milk. Ans. Coagulation.

Appendix 14

DETOXIFICATION AND IMMUNOCHEMISTRY

Fill in the blanks of the followings

1. Detoxification converts ______ substances introduced or formed in the body into ______ substances which are easily excreted out by the excretory routes. **Ans.** Toxic; Non-toxic.

2. The substances introduced into the body are the ______ used for treatment of diseases and the ______ used for diagnostic purposes.

   Ans. Drugs; Chemicals.

3. The toxic substances are formed in the body by the ______ action in the large intestine.

   Ans. Bacterial.

4. The other toxic substance ______ is formed in the body by the ______ of hemoglobin.

   Ans. Bilirubin; Breakdown.

5. The detoxification processes take place in the ______. **Ans.** Liver.

6. Alcohols and aldehydes are detoxified by the process of ______. **Ans.** Oxidation.

7. Chloral is oxidized to ______.

   Ans. Trichloracetic acid.

8. Aniline is oxidized to ______.

   Ans. P-aminophenol.

9. Acetanilide is oxidized to ______.

   Ans. P-acetylamino phenol.

10. Benzene is oxidized to ______. **Ans.** Phenol.

11. Chloral on reduction produces ______.

    Ans. Trichlorethylalcohol.

12. Picric acid on reduction yields ______.

    Ans. Picramic acid.

13. Atropine on hydrolysis forms ______ and ______. **Ans.** Tropic acid; Tropine.

14. Aspirin is hydrolyzed to ______ and ______.

    Ans. Salicylic acid; Acetic acid.

15. Procaine on hydrolysis yields ______ and ______.

    Ans. P-aminobenzoic acid; Diethylaminoethanol.

16. The process by which the chemical to be detoxified combines with another chemical formed by the body is called ______. **Ans.** Conjugation.

17. Bilirubin diglucuronide is formed by the combination of ______ and ______.

    Ans. Bilirubin; glucuronic acid.

18. Indoxyl sulphate is the product from the combination of ______ and ______.

    Ans. Indoxyl; Sulphuric acid.

19. ______ and ______ form hippuric acid.

    Ans. Benzoic acid ; Glycine.

20. Nicotinuric acid is formed from ______ and ______. **Ans.** Nicotinic acid; Glycine.

21. Bromobenzene and cysteine form ______.

    Ans. P-bromophenyl mercapturic acid.

22. N-methyl pyridine is formed from ______ and ______. **Ans.** Pyridine; Active methionine.

23. Cyanide and sodium thiosulphate form ______ and ______.

    Ans. Thiocyanate; Sodium sulphate.

24. The lymphocyte is the primary cell for _______ system. **Ans.** Immune.

25. The T lymphocytes are _______ derived. **Ans.** Thymus.

26. The B lymphocytes are responsible for _______ immunity which is expressed by the production of specific circulating plasma proteins termed _______. **Ans.** Humoral; Antibodies.

27. The basic unit of all immunoglobulin molecules consists of _______ chains linked by _______ bonds. **Ans.** Four polypeptide; Disulfide.

28. Immunoglobulins composed of more than one basic _______ unit are termed _______. **Ans.** Monomeric; Polymers.

29. IgM are _______ and IgA are _______. **Ans.** Pentamers; Trimers.

30. The polypeptide chains are folded 3-dimensionally with _______ bonds to form areas called _______. **Ans.** Disulfide; domains.

31. The part of the antibody molecule which combines with antigens is formed by a few amino acids in the F region of _______ and _______ chains. **Ans.** H; L.

32. There are two major types of L chains in man, the _______ and _______ chains. **Ans.** Kappa; Lamda.

33. About 70 per cent of the human immunoglobulin molecules carry _______ light chains and 30 per cent carry _______ light chains. **Ans.** Kappa; Lamda.

34. In IgG, the H chain is termed as _______ chain; and in IgE, as _______ chain. **Ans.** Gamma; Epsilon.

35. The half of the light (L) chain towards the carboxy terminus is termed as the _______, while the amino terminal half is the _______ of the light chain. **Ans.** Constant region; Variable region.

36. The portion of the immunoglobulin molecule which binds the _______ is formed by the _______ portions of both the H and L chains. **Ans.** Specific antigen; Aminoterminal.

37. The domains of the protein chains form globular regions with _______ and _______ structure. **Ans.** Secondary; Tertiary.

38. IgG comprises _______ per cent of the gamma globulins and contains 2-4 per cent _______. **Ans.** 80; Carbohydrates.

39. IgA has _______ per cent carbohydrate and does not cross the _______. **Ans.** 5 to 10; Placenta.

40. IgM contains _______ amino acids and has a mass of _______ daltons. **Ans.** 576; 9,50,000.

41. IgM is the major immunoglobulin expressed on the surface of _______ cells and its carbohydrate content is _______ per cent. **Ans.** B; 10 to 12.

42. IgE is present in the serum in very _______ concentrations as a single basic unit with _______ E chains. **Ans.** Low; Heavy.

43. No antibody activity is associated with _______. **Ans.** IgD.

44. Half of the patients with allergic diseases have increased serum _______ levels. **Ans.** IgE.

45. Immunoelectrophoresis consists of both _______ separation and immune precipitation of _______. **Ans.** Electrophoretic; Proteins.

46. The normal concentration of IgM _______ / dl. **Ans.** 30-250 mg.

47. The normal concentration of IgG _______ / dl. **Ans.** 570-1920 mg.

48. Substances which can give an immune response when introduced into an animal are called _______. **Ans.** Antigens.

49. The combining sites on the surface of the lymphocytes are antibody-like molecules called _______. **Ans.** Antigen receptors.

50. The portions of antigenic molecules which are involved in actual binding with antibody combining sites are termed _______. **Ans.** Antigenic determinants.

51. The ability of the region of the antigen molecule to act as an antigenic determinant and to induce the formation of specific antibodies is called _______. **Ans.** Immunopotency.

Appendix 15

BLOOD, LYMPH AND CEREBROSPINAL FLUID

Write the correct answer number of the followings

1. Under optimal conditions one gram of hemoglobin can carry oxygen in ml
   (a) 1.0, (b) 1.14, (c) 1.24, (d) 1.34. Ans. d.
2. Cellular fraction of blood in volume per cent
   (a) 40, (b) 45, (c) 50, (d) 55. Ans. b.
3. Plasma fraction of blood in volume per cent
   (a) 40, (b) 45, (c) 50, (d) 55. Ans. d.
4. The diffusible constituent of plasma
   (a) Urea, (b) Vitamins, (c) Hormones, (d) All of the above. Ans. d.
5. The catabolic products of diffusible constituents of plasma
   (a) Uric acid, (b) Creatinine, (c) Both of the above, (d) None of the above. Ans. c.
6. The normal concentration of fibrinogen of blood per 100 ml
   (a) 0.2 – 0.6 grams, (b) 0.4 – 0.8 grams, (c) 0.6 – 1.0 grams, (d) 0.8 – 1.2 grams. Ans. a.
7. The normal level of free fatty acids in blood per 100 ml
   (a) 110 – 400 mg, (b) 125 – 450 mg, (c) 140 – 480 mg, (d) 8 – 25 mg. Ans. d.
8. The normal blood total lipids concentration in mg per 100 ml
   (a) 500 – 750, (b) 550 – 800, (c) 570 – 820, (d) 600 – 850. Ans. b.
9. The normal concentration of PBI of blood per 100 ml
   (a) 2 –6 µg, (b) 3.5–7.5 µg, (c) 6–10 µg, (d) 8–12 µg. Ans. b.
10. The normal level of blilirubin in serum in mg per 100 ml
    (a) 0.3 – 1.1, (b) 0.2 – 0.9, (c) 0.3 – 1.2, (d) 0.4 –1.5. Ans. a.
11. The normal concentration of serum lactate dehydrogenase per litre
    (a) 50 – 100 I.U., (b) 60 – 120 I. U., (c) 80 – 150 I. U., (d) 85 – 190 I.U. Ans. d.
12. The normal level of NPN of blood in mg per cent
    (a) 10 – 25, (b) 12 – 30, (c) 20 – 40, (d) 25 –45. Ans. b.
13. The normal concentration of pyruvic acid in blood in mg per cent
    (a) 0.6 – 1.0, (b) 0.8 – 2.2, (c) 1.0 – 2.5, (d) 1.5 – 3.0 Ans. a.
14. Human blood is thicker than water by the number of times.
    (a) 2, (b) 3, (c) 4, (d) 5. Ans. d.
15. The normal osmotic pressure of blood is equal to
    (a) 0.915% NaCl, (b) 0.925% NaCl, (c) 0.945% NaCl, (d) 0.955% NaCl. Ans. c.
16. The substance almost entirely confined to corpuscles is
    (a) Sodium, (b) Calcium, (c) Potassium, (d) Magnesium. Ans. c.

17. The substances circulating in blood

(a) Enzymes, (b) Vitamins, (c) Hormones, (d) All.

Ans. d.

18. Plasma contains solids in per cent

(a) 7 to 8, (b) 8 to 9, (c) 9 to 10, (d) 10 to 12.

Ans. b.

19. The human plasma proteins are a mixture of

(a) Simple proteins (b) Glycoproteins, (c) Lipoproteins, (d) All of the above. Ans. d.

20. The plasma proteins are separated by

(a) Salt precipitation, (b) Electrophoresis, (c) Immunoelectrophoresis, (d) All of the above. Ans. d.

21. The numbers of amino acids arranged in a single peptide chain constituting albumin are

(a) 580, (b) 600, (c) 610, (d) 620. Ans. c.

22. The normal concentration of serum albumin in gram per cent by precipitation method

(a) 3.5-5.3, (b) 5.0, (c) 5.2, (d) 5.4. Ans. a.

23. The serum albumin concentration decreases in

(a) Severe protein deficiency, (b) Liver diseases, (c) Nephritis, (d) All of the above. Ans. d.

24. The normal concentration of serum globulin in grams per 100 ml by precipitation method

(a) 2.0, (b) 1.8-3.6, (c) 2.9, (d) 3.2. Ans. b.

25. β-globulins contain

(a) Siderophilin, (b) Ceruloplasmin, (c) Both of the above, (d) None of the above. Ans. a.

26. Serum β-globulin's normal concentration in gram per cent

(a) 0.82, (b) 0.85, (c) 0.6-1.2, (d) 0.95. Ans. c.

27. The substances which contain more than 4% hexosamine are designated as

(a) Glycoproteins, (b) Mucoproteins, (c) Both of the above, (d) None of the above. Ans. b.

28. The normal concentration of serum α_1-globulins in gram per cent

(a) 0.32, (b) 0.35, (c) 0.38, (d) 0.2-0.4 Ans. d.

29. Metalloprotein concentration is reduced in

(a) Pernicious anemia, (b) Chronic infections, (c) Liver diseases, (d) All of the above. Ans. d.

30. Fibrinogen is precipitated by half-saturation with

(a) Ammonium sulphate, (b) Magnesium sulphate, (c) Barium sulphate, (d) Calcium sulphate.

Ans. a.

31. Leukocytes contain

(a) Proteins, (b) Nucleoproteins, (c) Fats, (d) All of the above. Ans. d.

32. Blood platelets contain large amounts of

(a) Catecholamines, (b) Serotonin, (c) Histamine, (d) All of the above. Ans. d.

33. The minor blood groups are

(a) M, (b) N, (c) Rh, (d) All of the above. Ans. d.

34. The commonest Rh antigen is

(a) A, (b) B, (c) D, (d) AB. Ans. c.

35. Hemolysis of blood is caused by

(a) Quinine, (b) Ether, (c) Both of the above, (d) None of the above. Ans. c.

36. Thrombin consists of the numbers of polypeptide chains

(a) 1, (b) 2, (c) 3, (d) 4. Ans. b.

37. The total volume of C.S.F. in an adult human is about

(a) 130 ml, (b) 140 ml, (c) 150 ml, (d) 160 ml.

Ans. a.

38. In the fasting adult, the sugar in C.S.F. in mg per 100 ml

(a) 40 – 60, (b) 40 – 70, (c) 0 – 110, (d) 90 – 130

Ans. b.

Fill in the blanks of the followings

1. The sugar content of C.S.F. is decreased in purulent ________. Ans. Meningitis.
2. The reduction in chloride of C.S.F is most marked in ________. Ans. Tuberculous meningitis.
3. Increased level of calcium of C.S.F. is observed in all cases of ________ and ________.

   Ans. Meningitis; Epidemic encephalitis.
4. Isomerase activity of C.S.F. is increased in malignant brain ________. Ans. Tumors.
5. The protein content of C.S.F. is increased in ________. Ans. Myxedema.
6. The term ________ denotes a fluid not only

present in the lymphatic vessels but also the fluid which bathes the cells. Ans. Lymph.

7. The ratio between albumin and globulin in lymph is the same as in ______. Ans. Plasma.

8. The C.S.F. is formed by the ultrafiltration of the ______ by the choroid plexuses of the ______.

Ans. Plasma ; Brain.

9. Plasmin normally exists in ______ in the inactive form ______. Ans. Plasma; Plasminogen.

10. Urokinase, the ______ enzyme for urine, is also a serine ______. Ans. Proteolytic ; Protease.

11. Antithrombin 111 has a major ______ activity and has some ______ activity.

Ans. Antithrombin ; Endogenous.

12. The common deficiency of factor ______ produces a disease known as hemophilia A.

Ans. VIII.

13. Von Willebrand's disease is caused by the defect in ______ and a deficiency of factor ______ clotting activity. Ans. Platelet adherence ; VIII

14. Factor VII is the GLa-containing ______ synthesized in the ______. Ans. Glycoprotein; Liver.

15. Thrombin converts factor XII to ______ which is a ______. Ans. XIIIa ; Transglutaminase.

16. Prothrombin is a single chain ______ with a molecular weight of ______.

Ans. Glycoprotein ; 72000.

17. The initial fibrin clot is ______ and held together only by fibrin ______.

Ans. Weak ; Monomers.

18 Fibrinogen is a soluble plasma ______ whose length is ______ .

Ans. Glycoprotein ; 46 mm.

19. The white thrombus is composed of ______ and ______ and is poor in ______.

Ans. Platelets ; Fibrin ; Erythrocytes.

20. Stop of bleeding is said to be ______.

Ans. Hemostasis.

21. The hemolytic effect is due to the presence of the enzyme ______ which hydrolyzes ______.

Ans. Phospholipase A ; phospholipids.

22. Persons of group O are called ______ donors.

Ans. Universal

23. Blood platelets contain a ______ protein which is involved in the process of clot ______.

Ans. Contractile ; Retraction.

24. Blood platelets are ______ and contain no ______.

Ans. Unnucleated . DNA.

25. Fibrinogen is highly elongated having a axial ratio of about ______. Ans. 20 : 1

26. Some glycoproteins have specific binding function for ______ and ______.

Ans. Thyroxine ; Cortisol.

27. Metalloprotein concentration is increased in ______ deficiency or ______

Ans. Iron ; Pregnancy.

Indicate "True" or "False" of the following

1. The viscosity of blood provides resistance to the flow of blood in the blood vessels to maintain blood pressure at normal level. Ans. True.

2. Albumin exerts 60 per cent of the colloid osmostic pressure of plasma. Ans. False.

3. γ-globulins are formed in the liver. Ans. False.

4. α- and β-globulins are synthesized in the kidney. Ans. False

5. γ-globulins are immunoglobulins having antibody activity. Ans. True.

6. α_1-globulins are of several complex proteins containing carbohydrates and lipids. Ans. True.

7. The normal concetration of α_2-globulins in serum is 0.42 gm/100ml. Ans. False.

8. About one third of the total plasma proteins is albumin. Ans. False.

9. Plasma contains 50% to 16% solids composed largely of proteins. Ans. False.

10. The pH of blood of an individual is below 7.3 is considered in a condition of alkalosis.

Ans. False.

11. Blood is clotted within 2 or 3 minutes after shedding if it is left undisturbed. Ans. False

12. The viscosity of blood is affected by the change in the number of red cells or white cells.

Ans. True.

13. The normal concentration of chloride in C.S.F. is 118-132 mEq/L. Ans. True

14. Plasma is about 2.83 litres in an adult weighing 70 kg. Ans. True
15. Enzymes and lipids are the diffusible constituent of plasma. Ans. False.
16. In the normal concentration of serum total iodine is 3-6.5 mg/100ml. Ans. False
17. Na^+, K^+, Ca^{++} are the non-diffusible constituents of plasma. Ans. False
18. The specific gravity of plasma lies between 1.05 and 1.06. Ans. False.
19. Plasma has a very much higher viscosity. Ans. False.
20. The osmotic pressure of blood increases in strenuous exercise. Ans. True.
21. In normal individuals, the plasma proteins vary from 6.3-7.9 g/dL Ans. True.
22. Albumin has a molecular weight of about 69,000 and is synthesized in liver. Ans. True.
23. Albumin plays an important role in the exchange of water between tissue fluid and blood. Ans. True.
24. The molecular weight of globulin ranges from 9000 to 130000. Ans. False.
25. Bilirubin is associated with γ-globulins. Ans. False.
26. Thyroxine is transported in association with β-globulins. Ans. False.
27. Many drugs and dyes are transported in the plasma in combination with globulins. Ans. False.
28. Albumin is largely involved in the nutritive functions of the plasma proteins owing to its high concentration. Ans. True.
29. Half of the calcium of plasma is bound to protein for transport. Ans. True.
30. In kidney diseases, large amount of water moves to the tissue producing edema. Ans. True.

Appendix 16

EXCRETION

Write the correct answer number of the followings

1. The weight of each kidney in grams is about
   (a) 90 –130, (b) 100 – 140, (c) 110 – 150, (d) 120 – 170. **Ans.** d.
2. The length of each kidney in cm is about.
   (a) 10 – 12, (b) 11 – 13, (c) 13 – 15, (d) 15 – 17. **Ans.** b.
3. Urine formation involves the main steps
   (a) The glomerular filtration, (b) The tubular reabsorption, (c) The tubular secretion, (d) All of the above. **Ans.** d.
4. The daily normal total amount of urine contains the number of grams of solids
   (a) 35, (b) 40, (c) 60, (d) 80. **Ans.** c.
5. The urinary excretion increases by the following substance having a diuretic action
   (a) Urea, (b) Cellulose, (c) Ammonium salts, (d) Glycerol. **Ans.** a.
6. The diuretic action of tea, coffee and cocoa is due to
   (a) Theophylline, (b) Caffeine, (c) Bromophylline, (d) None of the above. **Ans.** b.
7. Increased urine volumes are observed in
   (a) Diabetes insipidus, (b) Diabetes mellitus, (c) Both of the above, (c) None of the above. **Ans.** c.
8. Decreased urine volumes are found in
   (a) Acute nephritis, (b) Fevers, (c) Diseases of the heart, (d) All of the above. **Ans.** d.
9. The function of kidney is regulated by the hormone
   (a) Aldosterone, (b) Parathormone, (c) Vasopressin, (d) All of the above. **Ans.** d.
10. The chief pigment of urine
    (a) Coproporphyrin, (b) Terochrome, (c) Urobilinogen, (d) Uroerythrin. **Ans.** b.
11. In liver disease the colour of urine may be
    (a) Green, (b) Brown, (c) Deep yellow, (d) All of the above. **Ans.** d.
12. The colour of urine is drak brown due to
    (a) Methemoglobin, (b) Homogentisic acid, (c) Both of above, (d) None of the above. **Ans.** c.
13. A turbidity is developed in alkaline urine by precipitation of
    (a) Calcium phosphate, (b) Magnesium sulphate, (c) Both of the above, (d) None of the above **Ans.** a.
14. Strongly acid urine is pink due to the precipitation of
    (a) Chloride salts, (b) Ammonium salts, (c) Uric acid salts, (d) None of the above. **Ans.** c.
15. The urine is acid in high protein intake because of the formation of excess.
    (a) Phosphates, (b) Sulphates, (c) Both of the above, (d) None of the above. **Ans.** c.
16. Urea excretion is increased in
    (a) Fever, (b) Diabetes, (c) Excess adrenocortical activity, (d) All of the above. **Ans.** d.

17. Ammonia is formed by the kidney from

    (a) Glutamate, (b) Glutamine, (c) Aspartate, (d) Aneurine. Ans. b.

18. There is high ammonia output in the urine in

    (a) Uncontrolled diabetes mellitus, (b) Cirrhosis of the liver, (c) Uremia, (d) Nephritis. Ans. a.

19. Creatine excretion is also found in

    (a) Starvation, (b) Hyperthyroidism, (c) Infections, (d) All of the above. Ans. d.

20. Uric acid excretion is increased in

    (a) Leukemia, (b) Severe liver disease, (c) Various stages of gout, (d) All of the above. Ans. d.

21. Deposits of uric acid and urates are coloured by absorbed urinary pigments, particularly the red

    (a) Uroerythrin, (b) Uracil, (c) Urocateil, (d) Uropsin. Ans. a.

22. The number of mg of amino acid nitrogen excreted in the urine of adults daily is about

    (a) 125 – 180, (b) 150 – 200, (c) 160 – 240, (d) 180 – 280 Ans. b.

23. Premature infants excrete more amino acid nitrogen than that of full-term infant by the number of times

    (a) 6, (b) 8, (c) 10, (d) 12. Ans. c.

24. Increased amounts of amino acids are excreted in

    (a) Liver disease, (b) Certain types of poisoning, (c) Both of the above, (d) None of the above. Ans. c.

25. The amino acids excreted in urine in cystinuria

    (a) Arginine, (b) Cystine, (c) Lysine, (d) All of the above. Ans. d.

26. Out of the total excreted sulphur the percentage of ethereal sulfate is about.

    (a) 8, (b) 10, (c) 12, (d) 14. Ans. b.

27. The number of mg of indican excreted normally

    (a) 5 – 20, (b) 8 – 25, (c) 10 – 30, (d) 15 – 40. Ans. a.

28. Neutral sulfur is contained in

    (a) Cystine, (b) Taurine, (c) Thiocyanate, (d) All of the above. Ans. d.

29. The greater part of the excreted phosphates is derived from ingested food.

    (a) Nucleoprotein, (b) Phosphoprotein, (c) Phospholipids, (d) All of the above. Ans. d.

30. Phosphate excretion is increased in

    (a) Osteomalacia, (b) Wasting diseases of the nervous system, (c) Renal tubular rickets, (d) All of the above. Ans. d.

31. Decrease in phosphate excretion is observed in

    (a) Hypoparathyroidism, (b) Infectious diseases, (c) Both of the above, (d) None of the above. Ans. c.

32. The normal amount of oxalate in urine in mg per day

    (a) 20, (b) 25, (c) 30, (d) 35. Ans. a.

33. The excretion of oxalate is increased by ingestion of

    (a) Fruits, (b) Vegetables, (c) Both of the above, (d) None of the above. Ans. c.

34. Sodium and potassium excretion are controlled by the activity of

    (a) Adrenal medulla, (b) Adrenal cortex, (c) Thyroid, (d) None of the above. Ans. b.

35. The number of mg of protein present in normal urine cannot be detected by the ordinary simple tests

    (a) 30 – 200, (b) 50 – 250, (c) 80 – 380, (d) 100 –400. Ans. a.

36. Proteinuria also results in poisoning of the renal tubules by heavy metals

    (a) Mercury, (b) Arsenic, (c) Bismuth, (d) All of the above. Ans. d.

Fill in the blanks of the followings

1. Mucus is a _______ and is increased in _________ of the bladder. Ans. Mucin, Infection.
2. Bence-Jones proteins found in the urine are light chain fragments of ____________ and most commonly occur in _________ and rarely in ________. Ans. Globulins ; Multiple myeloma ; Leukemia.
3. Normal individuals excrete not more than _________ of sugar per day. Ans. 16-300 mg.

4. In case of pregnant women and lactating mother, the __________ must be performed for urine glucose to eliminate the __________ present in urine.
Ans. Osazone test ; Lactose.

5. Increased amount of ammonia is excteted in __________ accompanying __________ .
Ans. Acidosis ; Ketosis.

6. Bilirubinuria is accompanied by the excretion of __________.
Ans. Bile salts.

7. Bile salts may be excreted in urine without __________ in certain stages in __________ disease.
Ans. Bile pigment ; Liver.

8. In excessive hemolysis, part of the bile pigment formed by breakdown of __________ is excreted in urine as __________.
Ans. Hemoglobin ; Urobilinogen.

9. Urobilin is formed from colourless urine __________ when exposed to __________.
Ans. Urobilinogen; air.

10. Coproporphyrins are excreted more in certain __________ diseases.
Ans. Liver.

11. Diuretics, __________ , enhance losses of water and salt via the urine through interference with normal __________ mechanisms.
Ans. The drugs ; Reabsorptive.

12. Osmotic diuresis is responsible for the serious __________ which accompanies diabetic __________.
Ans. Dehydration ; Ketoacidosis.

13. Diamox blocks both __________ reabsorption in the __________ tubule.
Ans. HCO_3^- ; Proximal.

14. Furosemide and mercurials inhibit __________ resbsorption in the __________ limb.
Ans. Chloride ; Ascending.

15. Clearance is measured to assess quantitatively the rate of __________ of a given substance by the __________.
Ans. Excretion ; Kidney.

16. Diabetes insipidus is developed due to the non-production of __________ and the individual passes __________ litres of urine in 24 hours.
Ans. ADH ; 5 to 20.

17. In salt losing nephritis, there is severe __________, __________ and __________ .
Ans. Dehydration ; Hyponatremia ; Hypochloremia.

18. Renal tubular acidosis is accompanied by excessive mobilization and urinary excretion of __________ and __________ .
Ans. Calcium ; Potassium.

19. In Hartnup syndrome, the disturbances in tryptophan metabolism is suggested by the presence of increased amount of __________, __________ and __________ in urine.
Ans. Tryptophan ; Indican ; Indole acetic acid.

20. The clinical symptoms of Hartnup syndrome are of __________ deficiency.
Ans. Niacin.

Indicate "True" or "False" of the followings

1. In uremia, the concentration of urea and other NPN constituents in plasma are increased.
Ans. True.

2. Acidosis develops due to the impairment of acid and phosphate excretion in acute uremia.
Ans. True.

3. Not only the kidney performs excretory functions but it acts as an endocrine organ.
Ans. True.

4. Kidney also destroys several hormones which are liberated from other organs.
Ans. True.

5. The juxtraglomerular cells of the renal cortex produce the proteolytic enzyme renin.
Ans. True.

6. Renin acts on α_1-globulin which is normally present in plasma.
Ans. False.

7. Renin splits off a polypeptide fragment celled angiotensin I which is decapeptide containing 16 amino acids.
Ans. False.

8. Angiotensin decreases the force of the heart beat and relaxes the arteriols.
Ans. False.

9. Renin angiotensin system is important in the maintenance of normal blood pressure.
Ans. True.

10. Prostaglandins are the other hormones of kidney and increase renal blood flow.
Ans. True.

11. Kininogen is produced by the kidney as an antihypertensive effect.
Ans. True.

12. Erythropoietin and erythrogonin have an effect on bone marrow to stimulate production of red cells. **Ans.** True.
13. Hypoxia stimulates production of erythropoietin. **Ans.** True.
14. Amorphous calcium and mangesium phosphates may be deposited from acid urines. **Ans.** False.
15. Calcium oxalate is found in alkaline urine but may be found in acid urine. **Ans.** False
16. Urates are usually found in alkaline urines. **Ans.** False.
17. Meat diet increases the bulk of feces. **Ans.** False.
18. One-third of the dry matter of feces is represented by intestinal secretions. **Ans.** False
19. Greenish colour of feces is due to excessive consumption of green vegetables. **Ans.** True.
20. Iron or bismuth gives a green stool. **Ans.** False
21. Feces are normally slightly acidic. **Ans.** False.
22. In diseases, fecal fat is increased when digestion or absorption of fat is impaired. **Ans.** True.
23. An adult on a mixed diet usually excretes about 3 gm of fecal nitrogen per day. **Ans.** False.
24. Moist feces contain about 6 - 7% of salts. **Ans.** False.
25. The pH of sweat is about 4.5 ; but if the skin is washed and dried, sweat is slightly alkaline. **Ans.** True.
26. The lactic acid content of sweat is more than normally found in blood. **Ans.** True.
27. The tubular epithelium also removes a number of foreign substances. **Ans.** True.
28. Mercaptans and H_2S may produce odour of feces. **Ans.** True.

Appendix 17

THE ENERGY REQUIREMENTS AND THE PRINCIPLES OF NUTRITION

Write the correct answer number of the followings

1. The R.Q. of mixed diets is about
   (a) 0.72, (b) 0.75, (c) 0.80, (d) 0.85. Ans. d.
2. The BMR is influenced by the factor
   (a) Sex, (b) Climate, (c) Habit, (d) All of the above. Ans. d.
3. The BMR is subnormal in
   (a) Addison's disease, (b) Hyperthyroidism, (c) Hypothyroidism, (d) All of the above. Ans. a.
4. The BMR is below normal in
   (a) Starvation, (b) Lipoid nephrosis, (c) Both of the above, (d) None of the above. Ans. c.
5. The BMR is above normal in
   (a) Diabetes insipidus, (b) Leukemia, (c) Polycythemia, (d) All of the above. Ans. d.
6. When lard, glucose and protein are combined, the percentage of SDA is less than the sum of their individual SDA by
   (a) 20, (b) 22, (c) 24, (d) 25. Ans. b.
7. When protein and lard are combined, the percentage of SDA is less than the sum of their individual SDA by
   (a) 45, (b) 50, (c) 54, (d) 58. Ans. c.
8. The SDA is decreased more by
   (a) Carbohydrate, (b) Fat, (c) Protein, (d) All of the above. Ans. b.
9. When different amino acids are fed, high SDA is found to be produced by
   (a) Glycine, (b) Alanine, (c) Phenylalanine, (d) All of the above. Ans. d.
10. The glands of internal secretion have no direct influence upon the SDA of
    (a) Protein, (b) Fat, (c) Carbohydrate, (d) All of the above. Ans. a.
11. Thyroidectomy in animals reduces the SDA of
    (a) Carbohydrate, (b) Fat, (c) Both of the above, (d) None of the above. Ans. c.
12. It is unfavourable to heavy muscular work by a diet rich in
    (a) Carbohydrate, (b) Protein, (c) Fat, (d) All of the above. Ans. b.
13. The extra heat is accelerated in the performance of work with
    (a) Carbohydrate, (b) Fat, (c) Both of the above, (d) None of the above. Ans. c.
14. When new tissue is formed, the usual SDA is not exerted by
    (a) Carbohydrate, (b) Fat, (c) Vitamins, (d) Protein. Ans. d.
15. A man of 70 kg weight and 180 cm height with a surface area of 1.8 square metre, aged 30 and a B.M.R. of 40 will have a basal metabolism of
    (a) 1,750 calories, (b) 1,800 calories, (c) 1,875 calories, (d) 1,925 calories. Ans. b.
16. Lumberman entails the expenditure of
    (a) 8,000 calories, (b) 9,000 calories, (c) 10,000 calories, (d) 12,000 calories. Ans. a.

17. In long bicycle races in 24 hours, there is the expenditure of

(a) 8,000 calories, (b) 9,000 calories, (c) 10,000 calories, (d) 12,000 calories. **Ans.** c.

18. Above the average for the six days, the percentage of Sunday's consumption is

(a) 24, (b) 28, (c) 32, (d) 36. **Ans.** d.

19. Women have a basal metabolic rate lower than that of men by

(a) 7% – 10%, (b) 8% – 12%, (c) 10% – 15%, (d) 12% – 20%. **Ans.** a.

20. The average supplement for muscular work in calories

(a) 400, (b) 600, (c) 800, (d) 1000. **Ans.** b.

21. The daily average caloric requirement of the adult male.

(a) 2000, (b) 2200, (c) 2500, (d) 3000. **Ans.** d.

22. If the NPU (Net Protein Utilization) is low, the requirements for proteins are

(a) Low, (b) High, (c) Moderate, (d) None of the above. **Ans.** b.

23. The relatively essential amino acid.

(a) Glycine, (b) Serine, (e) Alanine, (d) Histidine. **Ans.** d.

24. Vegetable proteins are almost associated with a large amount of

(a) Minerals, (b) Vitamins, (c) Carbohydrate, (d) Fat. **Ans.** c.

25. Animal proteins are associated with fat but very little carbohydrate except.

(a) Milk, (b) Peas, (c) Beans, (d) Potatoes. **Ans.** c.

26. The foodstuffs having the "protein-sparing" effect

(a) Carbohydrate, (b) Fat, (c) Both of the above, (d) None of the above. **Ans.** a.

27. Protein has a more catalytic function in the form of

(a) Hormones, (b) Enzymes, (c) Antigens, (d) Antinogens. **Ans.** b.

28. Proteins differ in "biologic value" depending on their contents of

(a) Non-essential amino acids, (b) Semi-essential amino acids, (c) Essental amino acids, (d) All of the above. **Ans.** c.

29. Good quality proteins include

(a) Soyabeans, (b) Peanuts, (c) Potatoes, (d) All of the above. **Ans.** d.

30. High quality proteins include

(a) Eggs, (b) Dairy products, (c) Kidney, (d) All of the above. **Ans.** d.

31. Proteins are also important sources of

(a) Nitrogen, (b) Sulphur, (c) Phosphorus, (d) All of the above. **Ans.** d.

32. The protective substances in a diet are

(a) First class proteins, (b) Mineral elements, (c) Vitamins, (d) All of the above. **Ans.** d.

33. The bulk of the food is reduced by

(a) Carbohydrate, (b) Fat, (c) Protein, (d) Minerals. **Ans.** b.

34. The daily consumption of fat in grams is usually over

(a) 50, (b) 60, (c) 80, (d) 90. **Ans.** c.

35. Animal fats have a better biological value because thay contain

(a) Vitamin A, (b) Vitamin D, (c) Both of the above, (d) Vitamin B_{12}. **Ans.** c.

36. The major symptoms of essential fatty acid deficiency are

(a) A scaly dermatitis, (b) Hair loss, (c) Poor wound healing, (d) All of the above. **Ans.** d.

37. In societies where fat is the principal food for energy intake, the population tends to develop

(a) Coronary heart disease, (b) Obesity, (c) Cancer of the bowel, (d) All of the above. **Ans.** d.

38. Carbohydrate is the cheapest source of food and for producing energy it can be readily.

(a) Digested, (b) Absorbed, (c) Utilized, (d) All of the above. **Ans.** d.

Fill in the blanks of the followings

1. Blood cholesterol level can be reduced significantly by changing the diet containing less ________ fatty acids or ________.

Ans. Saturated ; Cholesterol.

2. Carbohydrate can furnish ________ of the total caloric intake in the scarcity of ________ and ________.
Ans. 60% – 80%; Proteins; fats.

3. In gluconeogenesis, carbohydrate can be supplied from most ________ as well as from ________ moiety of fats. **Ans.** Amino acids ; Glycerol.

4. Carbohydrate is bulky and liable to undergo ________ producing ________ if digestion is delayed. **Ans.** Fermentation ; Lactic acid.

5. A minimum of ________ of carbohydrate per 100 k cal of total diet is necessary to prevent the development of ________.
Ans. 5 grams ; Ketosis.

6. Some cereals contain much ________ as to interfere with the absorption of ________.
Ans. Phytin, Calcium.

7. Cellulose stimulates ________.
Ans. Peristalsis.

8. Pentoses are absorbed and excreted ________ except ________ or ________.
Ans. Unchanged ; Ribose ; Deoxyribose.

9. Sucrose is one of the major etiologic factors in ________ and it is not the cause of ________, ________, or ________,
Ans. Dental caries ; Diabetes ; Heart disease ; Obesity.

10. Large amounts of dietary fibre decrease bowel ________ time and alter the composition of ________ bacteria. **Ans.** Transit ; Intestinal.

11. Milk products supply the majority of the ________ and ________ in the diet.
Ans. Calcium ; Phosphorus.

12. The best sources of calcium are ________ and ________. **Ans.** Cheese ; Milk.

13. Human milk contains ________ iron per ________.
Ans. 1–2 mg ; litre.

14. Milk is nearest to a ________ and ________ food.
Ans. Perfect; Complete.

15. The chief proteins of milk are ________ and ________. **Ans.** Caseinogen ; Lactalbumin.

16. Milk kept on unsterilised becomes sour due to formation of ________ which is sufficient to precipitate the ________.
Ans. Lactic acid ; Caseinogen.

17. Vitamin B_{12} is synthesized by ________ and is present only in ________ and ________.
Ans. Microorganisms ; Meat ; Dairy products.

18. The extra heat production over and above the caloric value of a given amount of food when used by the body is said to be ________.
Ans. Specific dynamic action.

19. The total heat produced or the energy spent by the body under conditions to perform minimum possible work is known as ________.
Ans. Basal metabolism.

20. The ratio of the volume of CO_2 eliminated to the volume of oxygen utilized on oxidation is called ________. **Ans.** Respiratory Quotients.

21. Zein is almost completely free from ________ and ________. **Ans.** Tryptophan; Lysine.

22. Gelatin is also lacking ________ and casein lacks ________.
Ans. Tryptophan; Methionine.

23. The incomplete proteins are ________ and ________. **Ans.** Zein; Gelatin.

24. The most suitable proteins for growth are ________, ________, ________.
Ans. Lactalbumin; Ovalbumin; Ovovitellin.

25. Proteins of animal origin have the ________ digestibility with wastage in digestion only ________ or ________. **Ans.** Highest; 5%; less.

26. A balanced diet is one which contains all the food constituents in proper ________ to meet the enrgy and ________ requirements of the individual. **Ans.** Proportions ; Nutritional.

27. Milk fat contains all ________ fatty acids as well as ________ fatty acids.
Ans. Saturated; Unsaturated.

28. The white portion of egg contains ________ and ________. **Ans.** Proteins; Salts.

29. The egg-yolk contains ________ proteins, ________ cholesterol, and ________ mineral.
Ans. 15%; 4%; 1%.

30. The proteins of egg-yolk are ________ and ________. **Ans.** Vitellin; Livertin.

31. Meat contains about ______ protein and ______ vitamins. **Ans.** 22%; B group.
32. During hanging nearly all the carbohydrates of meat are converted to ______ as well as ______. **Ans.** Lactic acid; Acid phosphates.
33. Large fish is rich in ______ but deficient in ______. **Ans.** Phosphorus; Calcium.
34. The crude cereals contain ______ protein, ______ carbohydrate. **Ans.** 11%; 70%.
35. The main proteins of cereals are ______ and ______. **Ans.** Glutelins; Gliadins.
36. The abundant mineral elements in cereals are ______ and ______. **Ans.** Calcium; Phosphate.
37. The dried pulses contain ______ protein, less than ______ fat. **Ans.** 20%-25%; 2%.
38. Nuts contain ______ protein and fat but ______ carbohydrate. **Ans.** High ; Low.
39. The most important tuber used in this country is the ______ and ______. **Ans.** Potato ; Sweet potato.
40. The tubers are the valuable sources of ______ and ______. **Ans.** Iron ; vitamin C.
41. The common roots are almost free from ______. **Ans.** Starch.
42. Carrots, the richest in ______, are the sources of ______. **Ans.** Sugars ; Carotene.
43. Roots contain valuable salts but negligible ______ and ______. **Ans.** Protein ; Fat.
44. Arrowroot contains traces of ______ and ______. **Ans.** Protein ; Salts.
45. Yellow pumpkin is a fair source of ______. **Ans.** Carotene.
46. The energy value of fruits is due to ______ and ______. **Ans.** Sugars ; Starch.
47. Many fruits contain ______ and ______. **Ans.** Pentoses ; Peptone.
48. Banana contains ______ as well as ______. **Ans.** Starch ; Sugar.
49. Tea has a ______ and ______ effect. **Ans.** Stimulant ; diuretic.
50. Strong tea disturbs ______ digestion due to ______. **Ans.** Gastric ; Tannic acid.
51. Coffee contains ______ and ______. **Ans.** Caffeine ; Tannic acid.
52. Chocolate consists of ground ______ mixed with ______. **Ans.** Cocoa-nibs ; Sugar.

Indicate "True" or "False"

1. Tea has negligible caloric value. **Ans.** True.
2. Cereals provide proteins, certain minerals and vitamins in additions to energy in the diets of the low-income groups. **Ans.** True.
3. Pure carbohydrates and fats provide energy only. **Ans.** True.
4. The protective foods are rich in proteins, vitamins and minerals. **Ans.** True.
5. Foods rich in vitamins, minerals and proteins of high biological value, eg, milk, eggs, meat, fish. **Ans.** True.
6. Tartaric acid, citric acid and malic acid are present in cereals. **Ans.** False.
7. Green leafy vegetables are lack in calcium and Vitamin C. **Ans.** False.
8. Nuts contain high fat and carbohydrate but low proteins. **Ans.** False.
9. The main protein of pulses is a globulin called legumin. **Ans.** True.
10. The fats of cereals contain considerable amounts of olein to make them liquid at ordinary temperature. **Ans.** True.
11. All cereals contain the enzyme phytase which hydrolyzes phytic acid. **Ans.** False.
12. Carotene and riboflavin of cereals are lost by heating. **Ans.** False.
13. Small fish eaten with bones are good sources of calcium. **Ans.** True.
14. The dried fish may contain about 20% carbohydate. **Ans.** False.
15. The flavour of meat is due to the organic substances extracted from it by boiling water. **Ans.** True.
16. The minerals in egg-yolk are sodium, potassium and carbonate. **Ans.** False.
17. The egg-yolk is very rich in Vitamin C. **Ans.** False.

18. The other proteins of egg-white are conalbumin, ovoglobulin and ovomucoid. Ans. True.
19. The proteins of milk neutralize gastric HCl to allow clotting. Ans. True.
20. Milk is rich in Vitamin B_1, C and D. Ans. False.
21. The iron content of milk is very high which is quite sufficient. Ans. False.
22. Proteins of meat, glutenin (wheat), and glycinin (soyabean) support growth when given in a higher concentration. Ans. True.
23. The vegetable proteins legumin (pea) and hordein (barley) are only suitable for maintenance. Ans. True.
24. Zein and gelatin are neither suitable for growth nor maintenance. Ans. True.
25. The proportions of protein, fat, and carbohydrate should approximately be 1 : 1.2 : 3.8. Ans. False.
26. Vitamin B_{12} is fairly available in cereals and vegetables. Ans. False.
27. Human and cow milk contain copper only about 0.6 mg per litre. Ans. True.
28. The more concentration of iron is from bread, meat, and potatoes. Ans. True.
29. A diet lacking in fat causes a feeling of hunger shortly after a meal. Ans. True.
30. The biologic value of protein depends on the content of essential amino acids. Ans. True.
31. Hard water is a source of calcium. Ans. True.
32. In prolonged starvation, the concentration of triiodothyronine falls which is responsible for the reduced metabolism. Ans. True.
33. Starvation patients often suffer from infections —malaria, cholera, pneumonia and gastroenteritis. Ans. True.
34. In nervous cretinism there is mental deficiency, deaf mutism, spasticity and ataxia. Ans. True.
35. In myxoedematous cretinism there is dwarfism. Ans. True.
36. Where goitre is endemic daily intake of dietary iodine are likely to be less than 100 μg. Ans. False.
37. Xerophthalmia means dry eye is a condition caused by Vitamin A deficiency. Ans. True.
38. Xerophthalmia arises when the diet contains whole milk and butter. Ans. False.
39. Osteomalacia is sometimes found in patients with cirrhosis of the liver due to failure to form 25(OH)D. Ans. True.
40. In osteoporosis there is muscle weakness. Ans. False.
41. In osteomalacia the plasma alkaline phosphatase is normal. Ans. False.
42. All people can be protected from rickets by a supplement of 10 μg of Vitamin D daily. Ans. True.
43. In wet beriberi, anorexia and dyspepsia are not present. Ans. False.
44. In the treatment of infantile beriberi, the mother should receive 10 mg thiamine twice daily in severe cases. Ans. True.
45. Wernicke's disease and Korsakoff's psychosis are manifestations of thiamine deficiency. Ans. True.
46. In pellagra the tongue characteristically has a 'raw beef' appearance. Ans. True.
47. Pellagra is often accompanied by signs of protein energy malnutrittion. Ans. True.
48. Tropical ulcers are often caused by minor injuries in people living in poor hygienic conditions infested by dysentery and malaria. Ans. True.
49. Cheilosis is characterized by a zone of red, denuded epithelium at the line of closure of the lips. Ans. True.
50. Angular stomatitis occurs in association with iron deficiency, anemia and other diseases. Ans. True.
51. In corneal vascularisation small greyish-white opacities may also be seen on the surface of the cornea. Ans. True.
52. Nutritional glossitis is a feature of pellagra, sprue and the various types of nutritional anemias. Ans. True.
53. In chronic atrophic glossitis the tongue is small with an atrophic mucous membrane. Ans. True.
54. In Von Gierke's disease growth is retarded and marked enlargement of the liver. Ans. True.
55. Emotional stress may bring an angina. Ans. True.

Appendix 18

VITAMINS

Write the correct answer number of the followings

1. Retinal is reduced to retinol by retinene reductase in presence of the coenzyme

(a) NAD^+, (b) $NADP^+$, (c) $NADH + H^+$, (d) $NADPH + H^+$. **Ans.** c.

2. Retinol exists as an ester with higher fatty acids in

(a) Liver, (b) Kidney, (c) Lung, (d) All of the above. **Ans.** d.

3. Retinol is transported to the blood as retinol attached to

(a) α_1-Globulin, (b) α_2-Globulin, (c) β-Globulin, (d) γ-Globulin. **Ans.** a.

4. Carotenes are transported with the

(a) Proteins, (b) Lipids, (c) Lipoproteins, (d) Minerals. **Ans.** c.

5. The dietary Vitamin A which is the form of Vitamin A esters are hydrolyzed in the lumen of the intestine by the enzyme

(a) Amylase, (b) Lipase, (c) Peptidase, (d) Nuclease. **Ans.** b.

6. In the blood the vitamin esters are attached to

(a) α_1-Lipoproteins, (b) α_2-Lipoproteins, (c) β-Lipoproteins, (d) γ-Lipoproteins. **Ans.** c.

7. The percentage of Vitamin A in the form of esters is stored in the liver

(a) 80, (b) 85, (c) 90, (d) 95. **Ans.** d.

8. The preformed Vitamin A is supplied by foods such as

(a) Butter, (b) Egg, (c) Fish liver oil, (d) All of the above. **Ans.** d.

9. One I.U. is equal to the number of micrograms of β-carotene

(a) 0.3, (b) 0.4, (c) 0.6, (d) 0.8. **Ans.** c.

10. The non-protein part of rhodopsin is

(a) Retinal, (b) Retinol, (c) Carotene, (d) Repsin. **Ans.** a.

11. Lumirhodopsin is stable only at a temperature below

(a) –35°C, (b) –40°C, (c) –45°C, (d) –50°C. **Ans.** d.

12. Metarhodopsin is hydrolyzed to retinal and opsin at a temperature below

(a) –8°C, (b) –10°C, (c) –15°C, (d) –20°C. **Ans.** c.

13. The normal concentration of Vitamin A in blood in I.U./dl

(a) 20 - 55, (b) 24 - 60, (c) 30 - 65, (d) 35 - 70. **Ans.** b.

14. Continued intake of excessive amounts of Vitamin A especially in children produces

(a) Irritability, (b) Anorexia, (c) headache, (d) All of the above. **Ans.** d.

15. Vitamin D_2 is also said to be

(a) Activated ergosterol, (b) Ergocalciferol, (c) Viosterol, (d) All of the above. **Ans.** d.

16. The poor sources of Vitamin D

(a) Egg, (b) Butter, (c) Milk, (d) Liver. **Ans.** c.

17. The normal concentration of Vitamin D in blood in I.U./L

(a) 600 - 2,500, (b) 700 - 3,100, (c) 800 - 4,100, (d) 850 - 4,700. Ans. b.

18. The activity of tocopherols is destroyed by

(a) Oxidation, (b) Reduction, (c) Conjugation, (d) All of the above. Ans. a.

19. Some tocopherols are

(a) Terpenoid in structure, (b) Dionol in structure, (c) Isoprenoid in structure, (d) Farnesyl in structure. Ans. a.

20. The methyl groups in the aromatic nucleus of α-tocopherol are

(a) 2, (b) 3, (c) 4, (d) 5. Ans. b.

21. Vitamin E is stored in

(a) Mitochondria, (b) Microsomes, (c) Both of the above, (d) None of the above. Ans. c.

22. Vitamin E protects the polyunsaturated fatty acids from oxidation by molecular oxygen in the formation of

(a) Superoxide, (b) Peroxide, (c) Trioxide, (d) All of the above. Ans. b.

23. The tocopherols prevent the oxidation of

(a) Vitamin A, (b) Vitamin D, (c) Vitamin K, (d) Vitamin C. Ans. a.

24. Vitamin E protects enzymes from destruction in

(a) Muscles, (b) Nerves, (c) Gonads, (d) All of the above. Ans. d.

25. Vitamin K_2 was originally isolated from

(a) Soyabean, (b) Oysters, (c) Putrid fish meal, (d) Alfalfa. Ans. c.

26. The poor sources of Vitamin K_1 are

(a) Milk, (b) Meat, (c) Fish, (d) All of the above. Ans. d.

27. Vitamin K regulates the synthesis of blood clotting factors

(a) VII, (b) IX, (c) X, (d) All of the above. Ans. d.

28. Ascorbic acid can reduce

(a) 2, 6 - dichlorophenolindophenol, (b) 2, 4 - dinitrobenzene, (c) 2, 6 - dioxypyridine, (d) 2, 6 - dibromobenzene. Ans. a.

29. Sterilised milk is devoid of

(a) Vitamin A, (b) Vitamin D, (c) Vitamin C, (d) Vitamin B_1. Ans. c.

30. Vitamin C is required in the metabolism of

(a) Phenylalanine, (b) Tryptophan, (c) Both of the above, (d) None of the above. Ans. c.

31. The symptoms of scurvy are

(a) Poor healing of wounds, (b) Loosening of teeth, (c) Anaemia, (d) All of the above. Ans. d.

32. Thiamine is also said to be

(a) Antiberiberi substance, (b) Antineuritic vitamin, (c) Aneurine, (d) All of the above. Ans. d.

33. Lipoic acid is also termed as

(a) Thioctic acid, (b) Protogen, (c) Acetate replacement factor, (d) All of the above. Ans. d.

34. Folic acid is also termed as

(a) SLR factor, (b) Pteroyl-glutamic acid, (c) Liver lactobacillus casei factor, (d) All of the above. Ans. d.

35. Thiamine is oxidized to thiochrome in alkaline solution by

(a) Potassium permanganate, (b) Potassium ferricyanide, (c) Potassium dichromate, (d) Potassium chlorate. Ans. b.

36. The requirement of Vitamin B_1 is increased when metabolism is elevated as in

(a) Fever, (b) Hyperthyroidism, (c) Increased muscular activity, (d) All of the above. Ans. d.

37. Riboflavin in alkaline solution when exposed to ultra-violet light is converted into lumiflavin which in ultra-violet light has a

(a) Greenish yellow fluorescence, (b) Bluish yellow fluorescence, (c) Reddish yellow fluorescence, (d) Light yellow fluorescence. Ans. a.

38. The normal concentration of riboflavin in plasma in mg/100 ml

(a) 1.5 to 3.5, (b) 2.5 to 4.0, (c) 3.5 to 5.5, (d) 4.5 to 7.5. Ans. b.

39. Riboflavin is involved in the regulatory functions of some hormones connected with the metabolism of

(a) Carbohydrate, (b) Fat, (c) Protein, (d) Minerals. Ans. a.

40. FMN is a constituent of

(a) Warburg yellow enzyme, (b) Cytochrome C reductase, (c) L- amino acid dehydrogenase, (d) All of the above. Ans. d.

41. The majority of the excess nicotinic acid is excreted in the urine in the form of
(a) N-methyl nicotinamide, (b) 6-pyridone of N-methyl nicotinamide, (c) N-methyl nicotinic acid, (d) All of the above. Ans. d.

42. Niacin is present in the maize in the form of
(a) Niatin, (b) Niacytin, (c) Nicotin, (d) Nicyn.
Ans. b.

43. The normal concentration of niacin in blood in mg./100 ml
(a) 0.3 to 0.5, (b) 0.4 to 0.6, (c) 0.5 to 0.8, (d) 0.6 to 0.99. Ans. c.

44. Nicotinic acid is essential for the normal functioning of
(a) Skin, (b) Intestinal tract, (c) Nervous system, (d) All of the above. Ans. d.

45. Pyridoxine is a mixture of
(a) Pyridoxine, (b) Pyridoxal, (c) Pyridoxamine, (d) all of the above. Ans. d.

46. Pyridoxine produces a coloured compound with
(a) 2 : 6 dichloroquinone chlorimide, (b) 2 : 6 dichloroquinone, (c) 2 : 4 nitroquinone, (d) All of the above. Ans. a.

47. Pyridoxal phosphate is involved in the desulphuration of
(a) Cysteine, (b) Homocysteine, (c) Both of the above, (d) None of the above. Ans. c.

48. In some subjects pyridoxine deficiency causes
(a) Lymphopenia, (b) Peripheral neuropathy, (c) Both of the above, (d) None of the above.
Ans. c.

49. In the deficiency states of Vitamin B6, there are inborn errors of metabolism including
(a) Cystathioninuria, (b) Familial xanthurenic aciduria, (c) Some pyridoxal-responsive anemias, (d) All of the above. Ans. d.

50. Pantothenic acid exists in the tissues as
(a) β-mercaptoethylamine, (b) Coenzyme A, (c) Pantoic acid, (d) β-alanine. Ans. b.

51. Pantothenic acid deficiency causes
(a) Nausea, (b) Irritability, (c) Anemia, (d) All of the above. Ans. d.

52. Folacin is made up of
(a) Pteridine nucleus, (b) P-aminobenzoic acid, (c) Glutamic acid, (d) All of the above. Ans. d.

53. Folic acid coenzymes take part in the synthesis of
(a) Purines, (b) Thymine, (c) Both of the above, (d) None of the above. Ans. c.

Fill in the blanks of the followings

1. The folic acid inhibitors are ______ and ______. Ans. Aminopterine; Amethopterine.

2. Aminopterine has been used in the treatment of ________, particularly in children.
Ans. Leukemia.

3. Folic acid deficiency causes increased output of ________ in the urine after histidine loading.
Ans. Formiminoglutamic acid (FIGLU).

4. Folic acid is converted to tetrahydrofolate catalyzed by __________ which use ______.
Ans. Folic acid reductase; Reduced NADP.

5. Folic acid as a coenzyme is involved in the transfer and utilization of the single ________ moiety. Ans. Carbon.

6. The single (formyl) carbon present in the tetrahydrofolic acids is utilized as a source of carbons ________ and ________ in the purine nucleus. Ans. 2:8.

7. Folic acid is transported to the plasma as ________ bound to ________.
Ans. Methyltetrahydrofolate; Protein.

8. The normal level of folic acid in serum is ________ per dL of serum in healthy subjects.
Ans. 0.6-2.0 μg.

9. The central structure of Vitamin B_{12} is referred to as a __________ ring system. Ans. Corrin.

10. Vitamin B_{12} contains cobalt ________ per cent.
Ans. 4 to 5.

11. Vitamin B_{12} activity is slowly destroyed by the reducing action of ________.
Ans. Ascorbic acid.

12. The absorption of Vitamin B_{12} depends on the presence of __________ and a constituent of

normal gastric juice called the intrinsic factor (IF), a ________ secreted by the parietal cells of the gastric mucosa. Ans. Hydrochloric acid; Mucopolysaccharide.

13. Cobalamin is bound to intrinsic factor in the proportion of ____________ mol of cobalamin to ________ mol of IF. Ans. 2;1.

14. Intrinsic factor is found in the ________ and ________ of the stomach.

Ans. Cardia; Fundus.

15. Vitamin B_{12} also binds to the proteins of gastric juice, ________ and ________ .

Ans. Bile; Saliva.

16. The intrinsic factor possesses two receptor sites, one for ________ and the other for ideal intestinal ________ . Ans. Vitamin B_{12} ; Microvili.

17. Intrinsic factor protects ________ against bacterial attack. Ans. Vitamin B_{12}-peptide.

18. Transcorrin I is a ________ binder of cobalamin and Transcorrin 11 is a ________ binder of cobalamin. Ans. Strong; Weaker.

19. The total amount of cobalamin in the bodies of adults is ________ of which about ________ is in the liver. Ans. 2.5 mg; 1.5 mg.

20. The normal concentration of Vitamin B_{12} in serum is ____________ per dL.

Ans. 200-850 μg.

21. Vitamin B_{12} is present only in the foods of ____________ origin. Ans. Animal.

22. Vitamin B_{12} is along with ____________ is required for the development of red blood cells beyond ________ stage.

Ans. Folic acid ; Megaloblast.

23. Vitamin B_{12}, as the coenzyme, is involved with ________ in the synthesis of labile ________ groups.

Ans. Tetrahydrofolate; Methyl.

24. Vitamin B_{12} is also a coenzyme for the ____________ which converts methyl malonyl-CoA into ____________. Ans. Mutase; Succinyl-CoA.

25. The carboxyl group of biotin combines with the terminal nitrogen of ________ residue of enzyme protein forming ________ .

Ans. Lysine; Biocytin.

26. The conversion of pyruvate to oxaloacetate is also a ____________ dependent carboxylation reaction catalyzed by ________.

Ans. Biotin; Pyruvate carboxylase.

27. Biotin is involved in the fixation of ________ for the formation of carbon ________ in purine synthesis. Ans. CO_2;6.

28. The deficiency of biotin may result from the adequate intake of raw ____________ which contains the protein ____________.

Ans. Egg white; Avidin.

29. Inositol exists in the form of phospholipids, ________ and ____________ in plant tissues.

Ans. Phytic acid; Phytin.

30. Large amounts of inositol containing phosphatides are present in ______ and ______.

Ans. Mitochondria; Microsomes.

31. Along with ____________ inositol exerts a ________ effect. Ans. Choline; Lipotropic.

32. Inositol is oxidized to ____________ in the liver by ____________.

Ans. Glucuronic acid; Oxygenase.

33. Choline takes part in ____________ reactions in the formation of ____________ from homocystine.

Ans. Transmethylation; Methionine.

34. Choline acts as a ________ donor only after oxidation to ________ . Ans. Methyl; Betaine.

35. Choline is first oxidized to ____________ and then to ________ .

Ans. Betaine aldehyde; Betaine.

36. Osteomalacia can be prevented by the administration of ____________ and ____________ .

Ans. Vitamin D; Calcium.

37. In pernicious anemia the tongue is ____________ and ____________ . Ans. Sored ; Inflammed.

38. The antagonists of vitamin B6 are ____________ and ____________ .

Ans. Isonicotinic acid hydrazide; Hydralazine.

39. The fat-soluble vitamins are ____________ molecules and are generally stored in ________ .

Ans. Hydrophobic; liver.

40. Cones are specialized for ____________ and detail vision in ________ . Ans. Colour; bright light.

41. Rods are specialized for ________ acitivity in ________ . Ans. Visual; dim light.

42. The basic factor stimulating the rod is conformational changes in the ________ due to ________ . Ans. Rhodopsin; cGMP.

43. Rhodopsin is a ________ which contains opsin as its ________ and ________ as its prosthetic group. Ans. Conjugated protein; apoprotein; retinene.

44. Rhodopsin has a light ________ property due to ________ group of 11-cis retinal.

Ans. Absorbing; polyene.

45. Retinene or retinal present in rhodopsin is ________ . Ans. 11-cis retinal.

46. Vitamin A helps in the preservation of the ________ and the normal ________ of the cell membrane.

Ans. Structural integrity; permeability.

47. Retinoic acid is found to be more involved in ________ synthesis probably forming its ________ .

Ans. Glycoprotein; phosphate.

48. Vitamin A behaves similar to ________ hormone.

Ans. Steroid.

49. The keratinization of ________ tissue progresses to ________ . Ans. Occular; blindness.

50. Defective vision due to ________ of ________ of cornea. Ans. Keratinization; epithelium.

51. Retinol deficiency ________ the resynthesis of ________ and interferes with the function of ________ resulting in ________ .

Ans. Depresses; rhodopsin; rods; night blindness.

52. β-carotene is an ________ and effective at ________ oxygen concentration.

Ans. Antioxidant; low.

53. Vitamin D binds to the ________ of target tissue and expresses the genes for calcium binding ________ as well as ________ in intestinal cells.

Ans. Chromatin; Protein; Ca^{2+}-ATPase.

54. Vitamin D stimulates the transcription of ________ for calcium binding ________ .

Ans. mRNA; Protein.

55. Vitamin D increases the excretion of ________ and ________ the concentration of serum phosphate.

Ans. Phosphate; decreases.

56. Vitamin D stimulates the activity of ________ which catalyzes the hydrolysis of ________ in intestine.

Ans. Phytase; Phytic acid.

57. Type I ricket is caused due to the deficiency of ________ for which no conversion of ________ 1, $25(OH)_2D_3$ to 1, takes place.

Ans. 1α-hydroxylase; $25(OH)D_3$.

58. Type II ricket is an autosomal recessive disorder where there is a single amino acid change in one of ________ fingers of the ________ sites for receptors. Ans. Zinc; DNA binding.

59. Vitamin D deficiency occurs due to ________ defect and less ________ defect.

Ans. Absorption; synthesis.

60. When renal parenchyma is lost, it is unable to form ________ and calcium absorption is ________ .

Ans. Calcitriol; impaired.

61. Hypocalcemia leads to increase in ________ which acts on bone to increase ________ .

Ans. Parathyroid hormone ; Ca^{2+}.

62. William syndrome appears due to ________ and over dosage of ________ .

Ans. Hypoparathyroidism; vitamin D.

63. A free radical is an ________ that has one or more unpaired ________ .

Ans. Atom; electrons.

64. A free radical with an unpaired electron may take away ________ from a ________ group of ________ fatty acid.

Ans. Hydorgen; methylene; polyunsaturated.

65. Free radical like OH^- is formed during the action of some ________ , such as microsomal ________ .

Ans. Oxidoreductases; NADPH oxidase.

66. When oxygen is reduced to water by ________ , ________ electrons are required.

Ans. Cytochrome oxidase; 4.

67. The individual molecules in univalent reaction are ________ reactive and ________ damaging the tissues. Ans. Highly; Potentially.

68. Superoxide is produced during the metabolism of ________ by ________ .

Ans. Xenobiotics; Cytochrome P450.

69. Superoxide anion is produced by peroxidation or ________ of membrane ________ .

Ans. Antioxidation; lipids.

70. Vitamin E acts as a defense against ________ of ________ fatty acids.

Ans. Peroxidation; Polyunsaturated.

71. The phospholipids of mitochondria possess affinities for ________ and ________ .

Ans. α-tocopherol; Vitamin E.

72. The tocopherol acts as ________ breaking ________ chain reactions. **Ans.** Antioxidant; free radical.

73. The phenoxy free radical reacts with ________ to regenerate ________ .

Ans. Vitamin C; tocopherol.

74. Antioxidant action of ________ is effective at high ________ concentration.

Ans. Tocopherol; oxygen.

75. Vitamin E along with other factors prevent the ________ effect of NO_2 on respiratory ________ .

Ans. Peroxidative; membrane.

76. Vitamin E prevents the hepatic ________ produced by the lack of ________ containing amino acids in the diet.

Ans. Necrosis; sulfur.

77. Selenium is required for normal ________ function which is nesessary for the absorption of ________ and ________ .

Ans. Pancreatic; lipids; vitamin E.

78. Selenium is a component of ________ and helps to destroy ________ .

Ans. Glutathione peroxidase; peroxides.

79. The diminished peroxidation greatly reduces the ________ requirement for the maintenance of ________ integrity. **Ans.** Vitamin E; membrane.

80. Primaquine has a capacity to produce ________ and to damage the cell membane of ________ by increasing the rate of oxidation of hemoglobin to ________ . **Ans.** O_2^-; RBC; methemoglobin.

81. In the erythrocytes, ________ is required for the reduction of oxidized glutathione to reduced glutathione catalyzed by ________ .

Ans. NADPH; glutathione reductase.

82. Reduced glutathione removes ________ from the erythrocytes by ________ containing ________ .

Ans. H_2O_2; glutathione peroxidase; selenium.

83. Vitamin E deficiency causes an increased oxidation of polyunsaturated fatty acids in the muscle with the rise in ________ consumption and ________ production which favours an increase in ________ producing ________ .

Ans. O_2; Peroxide; hydrolase activity; muscular dystrophy.

84. Vitamin E deficiency causes ________ breast disease and ________ .

Ans. Fibrocystic; atherosclerosis.

85. Vitamin K deficiency is caused by surgery due to ________ circuiting of the bowel and ________ is alone useful in such cases.

Ans. Short; Vitamin K_3.

86. Vitamin K is converted to its hydroquinone in liver ________ by dehydrogenase using ________ .

Ans. Microsomes; NADPH.

87. K-hydroquinone acts as ________ for carboxylase and converts γc-glutamate to ________ with the help of ________ .

Ans. Coenzyme ; γ-carboxyglutamate; CO_2.

88. Hydroquinone may change to 2, 3 epoxide which is reduced to ________ by microsomal ________ which is inhibited by ________ .

Ans. Quinone; epoxride reductase; dicoumarol.

89. γ-carboxyglutamate provides ________ sites in ________ portion. This brings together ________ and ________ close to the phospholipid membrane of platelets enhancing blood clotting.

Ans. Calcium binding; N-terminal; activated factor; accelerin.

90. Water-soluble vitamins are not ________ in the body except ________ .

Ans. Stored; Vitamins B_{12}.

91. Water-soluble vitamins are synthesized from ________ except ________ .

Ans. Plant tissue; Cobalamine.

92. Vitamin contains a ________ and ________ ring and they are linked by ________ bridge.

Ans. Pyrimidine; thiazole; methylene.

93. Infantile beriberi occurs in ________ infants between ________ to ________ months.

Ans. Breast fed; 2nd; 5th.

94. In acute form of infantile beriberi, the infant may be ________ with ________ and ________ .

Ans. Cyanosed; dyspnoea; tachycardia.

95. Wernicke's encephalopathy is associated with ________ deficiency and is frequently found in ________ . Ans. Thiamine; Chronic alcoholics.

96. Flavo protein may unite with metals like ________ and ________ forming metalloflavoproteins.

Ans. Iron; molybdenum.

97. Riboflavin is converted into ________ by ________ during absorption in intestine.

Ans. FMN; flavokinase.

98. Free riboflavin undergoes ________ .

Ans. Phosphorylation.

Indicate "True" or "False" of the followings

1. Renal ricket is caused by defective transport of phosphate by the renal tubules. Ans. True.
2. The symptom of osteomalacia is the progressive calcification of bones, and bones become hard. Ans. False.
3. In infantile, scurvy purpura occurs in the sternum. Ans. False.
4. Palpitation and breathlessness appear in the wet beriberi. Ans. True.
5. Pellagra is not caused by chronic infection. Ans. False.
6. In pernicious anemia the RBC count is very high. Ans. False.
7. The best sources of Vitamin K are cauliflower and soyabean. Ans. False.
8. Vitamin K is involved in oxidative phosphorylation in animal tissues. Ans. True.
9. The daily requirement of Vitamin C by adult is 80 mg. Ans. False.
10. Vitamin C is involved in the hydroxylation of steroids in the adrenal cortex. Ans. True.
11. Vitamin C is required for the formation of growth hormone. Ans. False.
12. The deficiency of Vitamin C causes susceptibility to infection. Ans. True.
13. The daily requirement of thiamine by pregnant women is 0.5 to 0.8 mg. Ans. False.
14. In dry beriberi the muscles become wasted and weak. Ans. True.
15. No toxicity results in the intake of excessive amounts of thiamine. Ans. True.
16. Vitamin B_2 is destroyed when kept in cold. Ans. False.
17. The best sources of riboflavin are cereals and germinating wheat. Ans. False.
18. Niacin is chemically pyridine 3-carboxylic acid. Ans. True.
19. Nicotinamide when heated in a strong alkaline or acid solution, it is converted into nicotinic acid. Ans. True.
20. The best sources of niacin are fruits and vegetables. Ans. False.
21. The daily requirement of niacin by children is 5 to 8 mg. Ans. False.
22. The reduced NADP is also required in enzyme reaction in the formation of tetrahydrofolate. Ans. True.
23. In large doses, nicotinic acid causes burning sensation. Ans. True.
24. Pyridoxamine contains a primary amine side-chain in no. 4 of pyridine nucleus. Ans. True.
25. Pyridoxal phosphate acts as a cotransaminase in the transamination reactions. Ans. True.
26. Liver phosphorylase also contains pyridoxal phosphate as coenzyme. Ans. False.
27. Pyridoxal phosphate is concerned with immune response. Ans. True.
28. The fair source of pantothenic acid is jelly. Ans. False.
29. The daily requirement of pantothenic acid by infants is 5 to 10 mg. Ans. False.
30. Pantothenic acid is a constituent of coenzyme A. Ans. True

31. Coenzyme A is involved in the metabolism of propionate and of branch chain fatty acids. Ans. True.

32. Riboflavin diminishes the photo-oxidation of folic acid. Ans. False.

33. Dihydrofolate is reduced to tetrahydrofolate by folic acid reductase. Ans. True.

34. 2 to 5 mg of folic acid is excreted in the urine daily. Ans. False.

35. Folic acid is incorporated into the erythrocytes during erythropoiesis. Ans. True.

36. Pregnant women require 800 mg of folic acid daily. Ans. False.

37. Rhodopsin is a conjugated protein with a molecular weight of about 40,000. Ans. True.

38. Retinene reductase is very similar to alcohol dehydrogenase of liver. Ans. True.

39. Hypervitaminosis A causes drowsiness, severe headache, sluggishness and peeling of the skin about the mouth. Ans. True.

40. Vitamin D is stored largely in heart and pancreas. Ans. False.

41. Eggs and butters are the poor sources of Vitamin D. Ans. False.

42. Adults daily require 800 I.U. of Vitamin D. Ans. False.

43. The major function of Vitamin D is to stimulate transcription of the mRNA for a calcium transport protein. Ans. True.

44. The effect of Vitamin D is not much related to the activity of parathormone and calcitonin. Ans. False.

45. In fully growing bones in adults, there is a type of defective mineralization of osteoid tissue termed "Osteomalacia". Ans. True.

46. The symptoms of hypervitaminosis D are thirst, constipation and polyuria. Ans. True.

47. Vitamin E prevents the hepatic necrosis produced by the lack of sulphur containing amino acids in dietary proteins. Ans. True.

48. The tocopherols play a part in the cellular oxidation. Ans. False.

49. The normal concentration of Vitamin E in blood is 20 mg/l. Ans. False.

50. The tocopherols are largely methyl derivatives of the parent compound tocol. Ans. True.

51. Vitamin E refers to a group of compounds known as tocopherols. Ans. True.

52. Vitamin E is involved in heme synthesis. Ans. True.

53. Vitamin C is stable below pH 6.8 at room temperature and readily oxidized in an alkaline medium. Ans. True.

54. Human body cannot synthesize Vitamin C. Ans. True.

Appendix 19

METABOLISM OF XENOBIOTICS

Fill in the blanks of the followings

1. Xenobiotics are metabolized in the ________ and sometimes excreted ________.

   Ans. Liver; Unchanged.

2. The major reaction of xenobiotics is ________ catalyzed by the enzyme ________.

   Ans. Hydroxylation ; Monooxygenases.

3. Other types of xenobiotics reactions are ________ and ________.

   Ans. Reduction; Hydrolysis.

4. Highly ________ xenobiotics persists in ________ tissue. Ans. Hydrophobic; Adipose.

5. The original xenobiotics are referred to as ________ or ________.

   Ans. Prodrugs ; Procarcinogens.

6. ________ reactions increases the ________ activity of a xenobiotic.

   Ans. Conjugation; Biologic.

7. The major monooxygenases in the ________ are cytochrome ________ species.

   Ans. E.R.; P-450.

8. Cytochrome P-450 acts on various ________ and ________. Ans. Carcinogens ; Pollutants.

9. NADPH cytochrome P-450 reductase uses ________ in the reaction mechanism of cytochrome ________. Ans. NADPH ; P-450.

10. Cytochrome ________ is specific for the metabolism of ________. Ans. P-448; PAHs.

11. Cytochrome P-448 is said to be ________.

    Ans. AHH.

12. The activity of cytochrome P-448 is increased in the ________ of women who ________.

    Ans. Placentae; Smoke.

13. GSH is an important ________ mechanism against certain ________ compounds.

    Ans. Defense; Toxic.

14. GSH as a carrier involved in the transport of certain ________ across membranes in the ________.

    Ans. Amino acids ; Kidney.

15. ________ of certain xenobiotics can ________ the activities of the enzymes.

    Ans. Metabolites; Inhibit.

16. The ________ effects of Xenobiotics cover the wide ________. Ans. Toxic ; Spectrum.

17. The enzyme ________ has a protective effect against certain ________.

    Ans. Epoxide hydrolase; Carcinogens.

Appendix 20

CANCER

Write the correct answer number of the followings

1. Many cancers are associated with the abnormal production of

   (a) Carbohydrates, (b) Fats, (c) Proteins, (d) Minerals. **Ans.** c.

2. The tumor markers are

   (a) Enzymes, (b) Hormones, (c) Proteins, (d) All of the above. **Ans.** d.

3. Cancer is caused by the groups

   (a) Radiant energy, (b) Chemical compounds, (c) Viruses, (d) All of the above. **Ans.** d.

4. DNA of gene is damaged by

   (a) Ultraviolet rays, (b) X-rays, (c) γ-rays, (d) All of the above. **Ans.** d.

5. Ultraviolet radiation may cause the formation of

   (a) Purine dimers, (b) Pyrimidine dimers, (c) Nucleoside dimers, (d) Nucleotide dimers. **Ans.** b.

6. Carcinogenicity with radiant energy is to cause damage to

   (a) DNA, (b) RNA, (c) mRNA, (d) tRNA. **Ans.** a.

7. Chemicals can cause human cancers up to

   (a) 60 percent, (b) 70 percent, (c) 80 percent, (c) 90 percent. **Ans.** c.

8. The carcinogenic substances may be

   (a) Organic molecules, (b) Inorganic molecules, (c) Both organic and inorganic molecules, (d) All of the above. **Ans.** c.

9. The ultimate carcinogen attacks nucleophilic group in

   (a) DNA, (b) RNA, (c) Protein, (d) All of the above. **Ans.** d.

10. The specific mono-oxygenase responsible for the metabolism of polycyclic aromatic hydrocarbons is termed as

    (a) Cytochrome P-448, (b) Cytochrome P-450, (c) Cytochrome b, (d) Cytochrome C. **Ans.** a.

11. Promoters can cause.

    (a) Promotion, (b) Initiation, (c) Acceleration, (d) Demotion. **Ans.** b.

12. The premier target in carcinogenesis is

    (a) DNA, (b) RNA, (c) rRNA, (d) None of the above. **Ans.** a.

13. Many tumour cells exhibit abnormal

    (a) Golgi bodies, (b) RNA, (c) Chromosomes, (d) All of the above. **Ans.** c.

14. Burkitt's Lymphoma is a fast-growing cancer caused by

    (a) Monocytes, (b) B-lymphocytes, (c) Neutrophils, (d) Eosinophils. **Ans.** b.

Fill in the blanks of the followings

1. Oncogenic viruses contain either ________ or ________ as their genome. **Ans.** DNA ; RNA.

2. The _________ antigens are to bind tightly to _________. **Ans.** T; DNA.
3. The _______ of this retrovirus contains _________ genes. **Ans.** Genome; Four.
4. The integrated double-stranded _________ is called a _________. **Ans.** cDNA ; Provirus.
5. Translocation is a type of _________ change seen in _________ cells.

   Ans. Chromosomal; Cancer.
6. The lower activity of _________ can cause stimulation of the activity of _________.

   Ans. GTPase; Adenylate cyclase.
7. Metastasis is a _________ process to analyse in _________. **Ans.** Complex ; Humans.
8. Changes in the _______ chains of cell glycoproteins may be effective in _______ to occur.

   Ans. Oligosaccharide; Metastasis.
9. Markers are most often detected in _________ stages of cancer rather than _________ stages.

   Ans. Advanced; Early.
10. Apurinic sites may form by _________ of corresponding _________.

    Ans. Elimination; Bases.
11. The process by which procarcinogens are converted to _________ carcinogens is called _________.

    Ans. Active; Metabolic activation.
12. Monooxygenases and _________ cause the metabolism of _________ and other xenobiotics.

    Ans. Transferases; Procarcinogens.
13. Most of the chemical _______ are _______.

    Ans. Carcinogens; Mutagens.
14. The _______ fraction is an important concept in cancer _________.

    Ans. Growth ; Chemotherapy.
15. The basic aim is to use drugs that kill _________ cells but not _________ to the normal cells.

    Ans. Cancer; Toxic.
16. Cancer cells which exhibit MDR have _________ amounts of _________.

    Ans. Increased; Proteins.
17. Humans of all _________ are _________ by cancer. **Ans.** Ages; victimised.
18. Gene _________ is involved in the action of most _________ factors.

    Ans. Activation; Growth.
19. A cell once becomes a _________ cell, tendency for _________ is increased.

    Ans. Tumor ; Malignancy.
20. Some _________ factors can affect the cells that _________ them.

    Ans. Growth; Synthesize.
21. Exposure of cells to _________ can cause rapid activation of cellular _________.

    Ans. PDGF ; Proto-oncogens.
22. BRCA1, a gene that influences susceptibility to early-onset _________ and _________ cancer has been isolated. **Ans.** Breast; ovarian.
23. BRACA1 is a protein of _________ amino acids containing a _________ finger domain in its amino terminal region and is a _________ factor.

    Ans. 1863; Zinc; transcription.
24. BRACA2 is associated with increased susceptibility to _______ cancer only and mapped to chromosome _______. **Ans.** Breast; 13q12-13.
25. P^{53} is a nuclear _____ of molecular mass _____.

    Ans. Phosphoprotein; 53KDA.
26. P^{53} acts as a transcriptional _______ regulating _______ that participate in cell division.

    Ans. Regulator; genes.
27. P^{53} acts as _______ checkpoint control for _______ damage. If DNA is excessively damaged, the activity of P^{53} is _______ resulting in _______ of cell division. **Ans.** G_1; DNA; increased; inhibition.
28. P^{53} hastens the _________ of potentially dangerous cells which have the potential to become _________. **Ans.** Death; malignant.
29. At least _________ different mutations have been detected in this _________. **Ans.** 100; gene.

30. The mutational ________ differs in different ________ . Ans. Spectrum; Cancers.

31. Transversions are frequent in ________ cancer and ________ cancer. Ans. Lung; liver.

32. Estrogens block effects of ________ in ________ tumors. Ans. Androgens; Prostatic.

33. Melphalan alkylates ________ and other molecules in ________ . Ans. DNA; myeloma.

34. Tumor markers are detected in ________ stages rather than ________ stages of cancer. Ans. Advanced; early.

35. Single tumor marker is not used in ________ types of ________ . Ans. All ;Cancer.

36. The tumor marker calcitonin is used in ________ and ________ is used in the cancer of liver, germ cell. Ans. Medullary carcinoma; Alpha-fetoprotein.

Appendix 21

ACID-BASE REGULATION

Write the correct answer of the followings

1. The amount of CO_2 in litres are produced and eliminated from the body of an adult daily about

   (a) 300, (b) 250, (c) 350, (d) 400 **Ans.** a

2. The oxidation of carbohydrates, fats, and proteins forms abnormal quantities of

   (a) Oxalic acid; (b) Lactic acid; (c) Succinic acid; (d) Tartaric acid. **Ans.** b.

3. The hydrogen ion concentration of blood may be increased by the administration of excess amount of some medicines like.

   (a) Hipuric acid; (b) Mandelic acid; (c) Carbolic acid; (d) Neomycin. **Ans.** b.

4. The buffer systems in the interstitial fluids and lymph are almost the same as in the blood plasma except

   (a) Fats; (b) Vitamins; (c) Proteins; (d) Carbohydrates. **Ans.** c.

5. Respiration is directly linked up with the buffer systems of

   (a) Bicarbonate; (b) Phosphate; (c) Protein; (d) Hemoglobin. **Ans.** a.

6. In blood there is low concentration of buffer

   (a) Hemoglobin; (b) Protein; (c) Phosphate; (d) Oxyhemoglobin. **Ans.** c.

7. The lungs become healthy due to the proper diffusion of

   (a) SO_2; (b) NO; (c) H_2O_2; (d) CO_2. **Ans.** d.

8. ECF is though alkaline, highly acidic urine is excreted showing the condition referred to as.

   (a) Chronic aciduria; (b) Paradoxic aciduria; (c) Mild aciduria; (d) Symptomatic aciduria. **Ans.** b.

9. The enzyme carbonic anhydrase is inhibited by

   (a) Chloramphenicol; (b) Tetracycline; (c) Aspirin; (d) Diamox. **Ans.** d.

10. The renal glutaminase activity is enhanced by

    (a) Acidosis; (b) Alkalosis; (c) Phosphatases; (d) Oxidases. **Ans.** a.

11. Endogenous excessive production of acid ions occurs in

    (a) Citric acidosis; (b) Starvation; (c) Overfeeding; (d) High tension **Ans.** b.

12. Depression of respiration is caused by

    (a) Excessive anaesthesia; (b) Tartaric acid; (c) Guanine; (d) Cold. **Ans.** a.

Fill in the blanks of the followings

1. Decreased ventilation raises ______ which stimulates the ______ centre.

   Ans. PCO_2; Respiratory.

2. Hypokalemia is due to the increased excretion of ______ from the ______ tubules.

   Ans. K^+; Distal.

3. Respiratory alkalosis occurs when there is a decrease in _______ with no change in _______ in plasma.
Ans. $[H_2CO_3]$; HCO_3^-.

4. Large doses of _______ cause stimulation of respiratory centre with hyperventilation showing _______.
Ans. Salicylate; Alkalosis.

5. In acidic medium, protein acts as a _______ and NH_2 group takes up _______ from the medium.
Ans. Base; H^+.

6. Phosphate buffer is _______ in concentration in blood and less efficient as a physiological _______.
Ans. Low; Buffer.

Appendix 22

CARDIAC PROFILE

Write the correct answer number of the followings

1. Myocardial infarction does not respond to

   (a) Sulphates; (b) Nitrates; (c) Phosphate; (d) Oxalate. **Ans.** b.

2. Oesophageal pain is precipitated by

   (a) Exercise; (b) Complete bed rest; (c) Sleeping; (d) Reading. **Ans.** a.

3. Heart failure is frequently due to

   (a) Pancreatitis; (b) Stomatitis; (c) Coronary disease; (d) Tension. **Ans.** c.

4. A low cardiac output leads to

   (a) Uremia; (b) Natremia; (c) Uricemia; (d) Lipidemia. **Ans.** a.

5. An increased risk of myocardial infarction is associated with

   (a) Protein; (b) Fibrinogen and factor VII; (c) Lipids; (d) Calcium. **Ans.** b.

6. Coronary disease is protected from intake of alcohol daily amounting to

   (a) 6 to 8 units; (b) 10 to 12 units; (c) 0.5 to 1 unit; (d) 2 to 4 units. **Ans.** d.

7. Ischaemic (coronary) heart disease can be prevented by the diets

   (a) Vegetables and fruit; (b) oils; (c) Sweets; (d) Cold drinks. **Ans.** a.

8. The patients with angina pectoris are adviced to take

   (a) Oxalate; (b) Beta-blockers; (c) Citrate; (d) Butarate. **Ans.** b.

9. Early management of acute myocardial infarction is done by

   (a) Dicuomerol; (b) Caffeine; (c) Cocoa; (d) Aspirin. **Ans.** d.

10. Myocardial infarction is enhanced by

    (a) Idle life; (b) Social isolation; (c) Busy life; (d) Comfortable life. **Ans.** b.

Appendix 23

AIDS

Write the correct answer number of the followings

1. AIDS is a viral disease due to

   (a) HIV; (b) MID; (c) TDD; (d) NAD. Ans. a.

2. The durgs clinically used for the inhibition of the spread of infectious virus are

   (a) Didanosine; (b) Imidazole; (c) Bodyache; (d) Sweating Ans. a.

3. Symptoms developed by HIV infection is

   (a) Drowsiness; (b) Malaise; (c) Bodyache; (d) Sweating. Ans. b.

4. HIV disease is accelerated by

   (a) Diarrhoea; (b) TB; (c) Pneumonia; (d) Respiratory distress. Ans. b.

5. The initial symptoms of cryptococcal meningitis are

   (a) Weight gain; (b) Chest pain; (c) Weight loss; (d) Nephritis. Ans. c.

6. Treatment produces good response in case of secondary neoplasms of aids by

   (a) Doxarubicin; (b) Cytidine; (c) Septran; (d) Chloramphenical. Ans. a.

7. The commonest symptoms in AIDS are

   (a) Appetite; (b) Nausea and vomiting; (c) Chest pain; (d) Diarrhoea. Ans. b.

8. The treatment of pneumonia is performed by

   (a) high dose of co-trimoxazole; (b) Low dose of penicillin; (c) High dose of tetracycline; (d) Low dose of septran. Ans. a.

9. The commonest disease in AIDS is

   (a) Diarrhoea; (b) Pulmonary disease; (c) Nephritis; (d) Pancreatitis. Ans. b.

10. CT scan of the brain affected by AIDS is usually

    (a) Normal; (b) High; (c) Low; (d) Very high. Ans. a.

Appendix 24

ENVIRONMENT AND POLLUTION

Write the correct answer number of the following

1. The term "environmental sanitation" has been defined by

   (a) I.C.U.D.; (b) W.H.O.; (c) M.C.T.D.; (d) N.C.R.D.

   Ans. b.

2. Surface water originates mainly from

   (a) Rain water; (b) Sea water; (c) River water; (d) Waterfalls. Ans. a.

3. The number of milligrams per litre of magnesium in sea water is

   (a) 1200; (b) 1225; (c) 1270; (d) 1295. Ans. c.

4. Ground water is hard due to the presence of the salts of

   (a) Sodium and Potassium; (b) Manganese and Cobalt; (c) Iron and nickel; (d) Calcium and Magnesium. Ans. d.

5. Chemical pollutants of water include

   (a) Organic acids; (b) Inorganic acids; (c) Carbides; (d) Sulphides. Ans. a.

6. Hardness of water causes beneficial effect against

   (a) Cardiovascular disease; (b) Pancreatitis; (c) Nephritis; (d) Hepatitis. Ans. a.

7. The aerobic bacteria oxidize the organic matter present in the water by the help of dissolved

   (a) Nitrogen; (b) Sulphur dioxide; (c) Oxygen; (d) Carbon dioxide. Ans. c.

8. The pathogenic bacteria are killed by

   (a) Chlorine; (b) Florine; (c) Bromine; (d) Iodine.

   Ans. a.

9. The pretreatment of water in order to destroy not only viruses and bacteria but also organic compounds is done by

   (a) Nitrites; (b) Ozone; (c) Hydrogen peroxide; (d) Sulfur dioxide Ans. b.

10. If the water is stored for long periods, there is the production of bad smell due to the development of

    (a) Algae; (b) Fungi; (c) Spermatozoa; (d) Sulphangitis. Ans. a.

11. Domestic combustion of coal, wood or oil produces

    (a) Sulphur dioxide; (b) Phosphorus pentaoxide; (c) Calcium oxide; (d) Magnesium oxide. Ans. a.

12. Children up to 6 years of age are the population at increased risk for the exposures to

    (a) Silicon; (b) Antimony; (c) Lead; (d) Nickel.

    Ans. c.

13. Air pollutant is not commonly regarded by

    (a) Sulphur dioxide; (b) Carbon dioxide; (c) Carbon monoxide; (d) Nitrogen dioxide. Ans. b.

14. Smoking may contribute significantly to the uptake of

    (a) Cadmium; (b) Argon; (c) Neon; (d) Silicon.

    Ans. a.

15. Lung Cancer is formed by

(a) TSH; (b) DMCH; (c) PAH; (d) RSTH. Ans. c.

16. The neuropsychological development is observed in children due to the elevated levels of

(a) Manganese; (b) Magnesium; (c) Cobalt; (d) Lead. Ans. d.

17. The operation theatres and infectious disease wards are significantly affected by

(a) Sunlight; (b) Ultraviolet radiation; (c) Sulfur dioxide; (d) Nitrogen dioxide. Ans. b.

18. The tympanic membrane is ruptured by noise above

(a) 120 decibels; (b) 135 decibels; (c) 145 decibels; (d) 160 decibels. Ans. d.

19. Noise causes the reduction of

(a) Night vision; (b) Nearest vision; (c) Distant vision (d) Long distant vision. Ans. a.

20. The fecal-borne diseases are transmitted to a new host through various channels like

(a) Stools; (b) Flies; (c) Vomiting; (d) Urine. Ans. b.

21. House fronts should exist from the road not less than

(a) 7 metres; (b) 12 metres; (c) 15 metres; (d) 21 metres. Ans. c.

Appendix 25

BIOCHEMICAL AND GENETIC BASES OF DISEASES

Fill in the blanks of the followings

1. The mutation in gene for ______ causes ______.
   Ans. Phosphorylase; Glycogen storage disease.
2. Life is threatened within hours at the excessive loss of ______ and ______ in cholera.
   Ans. Water; Electrolytes.
3. Excessive intake of vitamin A results in ______ states of ______.
   Ans. Chronic; Toxicity.
4. Most genetic diseases are caused by mutations in ______ involving the organelles ______.
   Ans. DNA; Nucleas.
5. Metastasis of cancer cells is caused by the changes in the ______ of ______.
   Ans. Oligosaccharides; Glycoproteins.
6. Zellweger syndrome and others involves the organelles ______ causing decreased activity of ______.
   Ans. Peroxisomes; Peroxisomal enzymes.
7. Hepatomegaly occurs from ______ of ______.
   Ans. Accumulation; Glucosylceramide.
8. Fibrosis indicates the accumulation of ______ in the disease ______.
   Ans. Collagen; Cirrhosis.
9. Cell death occurs due to the lack of ______ involving the disease ______.
   Ans. Oxygen; Myocardial infarction.
10. Multifactorial disorders are associated with a number of ______ and ______ factors.
    Ans. Genes; Environmental.
11. Pateints with PKU have less ______ which is the main source for ______ formation.
    Ans. Tyrosine; Melanin.
12. The plasma high levels of phenylalanine produces the inhibitory effects on the transport of other ______ into ______.
    Ans. Amino acids; Brain.
13. Gaucher disease is treated by the injection of ______.
    Ans. β-glucosidase.
14. Crigler-Najjar syndrome is treated by the administration of ______.
    Ans. Phenobarbital.
15. A triplex molecule is formed by the binding of ______ through base-pairing to a specific ______.
    Ans. Oligonucleotides; mRNA
16. Oligonucleotides being combined with its ______ blocks the replication of HIV-1 virus in the treatment of ______.
    Ans. Gag gene; AIDs.

Appendix 26

THE HUMAN GENOME PROJECT

Write the correct answer number of the following

1. The human genome project (HGP) is an international effort being started in

   (a) 1990; (b) 1992; (c) 1994; (d) 1996. Ans. a.

2. The human genetic map constructs a physical map of human chromosomes number

   (a) 20; (b) 22; (c) 24; (d) 26. Ans. c.

3. The method which can be used to amplify fragments of the gene may be

   (a) TCR; (b) PCR; (c) UCR; (c) MCR. Ans. b.

4. The highest resolution physical map is being established by

   (a) RNA sequencing; (b) polymerization; (c) Deamination; (d) DNA sequencing. Ans. d.

5. The estimated number of protein-coding genes ranges from

   (a) 30,000 to 40,000; (b) 35,000 to 42,000; (c) 40,000 to 46,000; (d) 50,000 to 60,000. Ans. a.

6. The genome consists of exons by per cent

   (a) 0.5 – 1.0; (b) 0.8 – 1.4; (c) 1.0 – 1.2; (d) 1.1 – 1.5. Ans. d.

7. The number of genes may be derived from bacteria by lateral transfer is over

   (a) 200; (b) 220; (c) 240; (d) 260. Ans. a.

8. The number of SNPs has been identified which may be more than

   (a) 2 million; (b) 2.5 million; (c) 3 million; (d) 4 million. Ans. c.

9. The number of nucleic acid enzymes encoded by human genes is

   (a) 2,304; (b) 2,308; (c) 2,312; (d) 2,316 Ans. b.

10. In the long run, biotechnology will also be nourished by

    (a) Lipomics; (b) Neuromics; (c) Dentomics; (d) Proteomics. Ans. d.

11. The determination of genes involved in complex diseases will be greatly helped by

    (a) DTP maps; (b) SNP maps; (c) DNP maps; (d) MCP maps. Ans. b.

12. The number of kinases encoded by human genes is

    (a) 868; (b) 872; (c) 876; (d) 882. Ans. a.

13. The percentage of human genome which has been sequenced is over

    (a) 80; (b) 85; (c) 90; (d) 95. Ans. c.

14. The percentages of proteins encoded by human genes are derived from a total gene number

    (a) 26, 282; (b) 26, 292; (c) 26, 322; (d) 26, 383. Ans. d.

APPENDIX - II

NORMAL LABORATORY VALUES

Appendix II

NORMAL LABORATORY VALUES

HEMATOLOGY

Bleeding time	: 1–7 minutes (Ivy method).
Coagulation time	: 4–8 minutes (Lee and White)
Clot retraction	: 40–65% Starts in 1–3 hours Completes in 6–24 hours No clot lysis in 24 hours
Fibrinogen	: 200–400 mg/dl
Activated partial thromboplastin time	: 24–35 seconds
Thrombin time	: 9–10 seconds (control + 3 seconds).
Hemoglobin (in blood)	: Males = 14.5–16.5 gm/dl Females = 12.0–14.5 gm/dl
Hemoglobin (in serum)	: 2–3 mg/dl
Packed cell volume	: Males = 38–50% Females = 34–43%
Mean corpuscular volume	: 78–94 cμ
Erythrocyte Sedimentation Rate (ESR) [Westergren method]	: Males = 0–15 mm at the end of 1 hour Females = 0–20 mm at the end of 1 hour
Cellular measurement of red cells	: Average diameter = 7.3 μm Mean corpuscular volume (MCV) : Males = 80–94 fl Females = 81–99 fl Mean Corpuscular Hemoglobin (MCH) = 27–32 pgm Mean Corpuscular Hemoglobin Concentration (MCHC) = 32–36%
Fragility of red cells	: Starts at 0.5% NaCl Completes at 0.3% NaCl
Red Blood Count (RBC)	: Males = 4.5–6.2 million/μl Females = 4.0–5.5 million/μl
Average plasma volume	: 45 ≠ 5 ml/kg body weight. 1487 ≠ 112 ml/sq metre body surface
Platelets	: 1,50,000–4,00,000/cmm
Reticulocyte count	: 0.5–2% of erythrocytes
Total leucocyte count	: 4000–11,000/cmm

Differential count of WBC or Leucocyte

Myelocytes	0%
Juvenile neutrophils	0%
Band neutrophils	0.5%
Segmented neutrophils	60-70%
Lymphocytes	20-30%
Eosinophils	1-7%
Monocytes	1-6%
Basophils	0-1%
Lymphocytes (Total)	1500-4000/μl
B cell	5-25%
T cell	60-88%
Suppressor	10-43%
Helper	32-66%
H : S	> 1.

BODY FLUIDS

Body Fluids

Total volume : 50 per cent (in obese) to 70 per cent (lean) of body weight.

Intracellular : 0.3–0.4 of body weight.

Extracellular : 0.2–0.3 of body weight.

Blood

Total volume : Males = 69 ml/kg body weight

Females = 65 ml/kg body weight

Plasma volume : Males = 39 ml/kg body weight

Females = 40 ml/kg body weight

Red blood cell volume : Males = 30 ml/kg body weight, (1.15–1.21 L/m^2 body surface area)

Females = 25 ml/kg body weight, (0.95–1.00 l/m^2 body surface area)

Blood Acid-base

		Normal range	
		Arterial	Venous
Actual pH	:	7.35–7.45	7.32–7.42
PCO_2 (mm of Hg)	:	35–45	41–51
Total CO_2 (mmol/l plasma)	:	23–27	26–30
Actual bicarbonate (mmol/l plasma)	:	22–26	24–28
Oxygen saturation (%)	:	96–97	40–70
Oxygen tension PO_2 (mm of Hg)	:	80–100	25–40

SYNOVIAL FLUID

Cells	200 cells/mm^3
Polymorphonuclear cells	25%
Fibrin clot	None
Crystals	None
Glucose	Same as serum
Hyaluronic acid	2.45–3.97 gm/l
pH	7.32–7.64
Protein	2.5 gm/dl
Albumin	63%
α_1-globulins	7%
α_2-globulins	7%
β-globulins	9%
γ-globulins	14%
Uric acid	Same as serum

SEMEN

Liquefaction	Complete in 15 minutes
Morphology	60% normal forms
Motility	75% motile forms
pH	7.2–8.0
Volume	2.0–6.6 ml
Spermatocrit	10%
Spermatocyte count	50 million/dL

STOOL

Bulk	
Dry weight	< 66.4 gm/24 hours
Wet weight	< 197 gm/24 hours.
Water	Approximately 65%
Fat (on a diet containing 30 gm fat/day)	< 7.2 gm/24 hours or 30% dry weight
Nitrogen	< 2.2 gm/24 hours
Coproporphyrin	12–832 mg/24 hours
Urobilinogen	40–280 mg/24 hours

CEREBROSPINAL FLUID (CSF)

Appearance	Clear and colourless
Specific gravity	1.003–1.008
Cells	<5/mm^3, all lymphocytes
pH	7.34–7.43
Pressure	50–180 mm of H_2O
Glucose	40–70 mg/dl or 60% of serum glucose
Chloride	118–132 mEq/l
Protein, lumber	15–45 mg/dl
Albumin	58%
α_1-globulins	9%
α_2-globulins	8%
β-globulins	10%
γ-globulins	10%
Protein, cisternal	15–25 mg/dl
Protein, ventricular	5–15 mg/dl

CLINICAL CHEMISTRY

Constituents	Blood	Serum	Plasma	Urine
Glucose (fasting)			70 – 115 mg/dl	Qualitative — nothing detected
Glucose (2 hr after meal)			115 – 160 mg/dl	Quantitative — 16-300 mg/24 hours
pH	7.35 – 7.45			4.6 – 8.0
Specific gravity	1.056 (varies with Hb and protein concentration			1.003 – 1.030
Cholesterol (total)		140–220 mg/dl	140–220 mg/dl	
HDL cholesterol				
Male		3.5 – 5.5 mg/dl		
Female		4.5 – 6.5 mg/dl		
Triglycerides		40 – 160 mg/dl		
Phospholipids				
Male		125 – 275 mg/dl		
Female		35 – 135 mg/dl		
Free fatty acids		8 – 25 mg/dl		
Protein (Total)				
Albumin		6.3 – 7.9 g/dl		Qualitative — none detected
Globulin		3.5 – 5.3 g/dl		Quantitative — 10 – 150 mg/24 hours
Fibrinogen		1.8 – 3.6 g/dl	0.2 – 0.6 g/dl	

Constituents	Blood	Serum	Plasma	Urine
Protein electrophoresis				
Alpha – 1		0.2 – 0.4 gm/dl		
Alpha – 2		0.5 – 0.9 gm/dl		
Beta		0.6 – 1.2 gm/dl		
Gamma		0.7 – 1.7 gm/dl		
Chylomicrons		Absent		
Alpha lipoprotein		13 – 24%		
Beta lipoprotein		60 – 75%		
Pre-beta lipoprotein		8 – 20%		
Urea		15 – 45 mg/dl	15 – 45 mg/dl	
Urea nitrogen		7 – 21 mg/dl	7 – 21 mg/dl	
Uric acid		2.5 – 7.0 mg/dl	2.5 – 7.0 mg/dl	600 – 700 mg/24 hours
Bilirubin (total)				
Adult		0.3 – 1.1 mg/dl		none detected
New born		0.4 – 4.0 mg/dl		
At 3 days		1.0 – 10.0 mg/dl		
Direct (conjugated)		0.2 – 0.3 mg/dl		
Indirect		0.3 – 0.6 mg/dl		
Creatinine		0.6 – 1.2 mg/dl (Male/Female)	0.6 – 1.2 mg/dl (Male/Female)	0 – 200 mg/24 hours (Adult) 7 – 11 mg/24 hours (Children)
Ascorbic acid			0.4 – 1.5 mg/dl	
Sulphate		0.5 – 1.5 mg/dl		
Viscosity		1.4 – 1.8 mg/dl (Serum compared to H_2O)		
Lactate	4.5 – 19.8 mg/dl			
Pyruvate	0.6 – 1.0 mg/dl			
Chloride		98 – 110 mEq/l	98 – 110 mEq/l	120 – 240 mEq/24 hours (varies with dietary intake)
Bicarbonate		24 – 28 mEq/l		
Ammonia			19 – 80 μg/dl	20 – 50 mEq/24 hours
Vitamin A		0.15 – 0.60 μg/dl		
Vitamin D :				
25-Hydroxycholecalciferol		8 – 55 ng/ml		
1, 25-dihydroxycholecalciferol		26 – 65 pg/ml		
24, 25-dihydroxycholecalciferol		1 – 5 ng/ml		
Vitamin B_{12}		200 – 850 μg/dl		
Porphobilinogen				Qualitative : None detected Quantitative : 0 – 24 mg/24 hr
Urobilinogen				1 – 3.5 mg/24 hours

Constituents	Blood	Serum	Plasma	Urine
Coproporphyrin				50 – 250 µg/24 hours
Uroporphyrin				10 – 30 µg/24 hours
Acetone and Acetoacetate		0.2 – 2.0 mg/dl		Less than 1 mg/day
Sodium		136 – 146 mEq/l	136 – 146 mEq/l	10 – 150 mEq/24 hours (varies with dietary sodium intake)
Potassium		3.5 – 4.5 mEq/l	3.5 – 4.5 mEq/l	25 – 100 mEq/24 hours
Calcium		8.4 – 10.2 mg/dl		0.05 – 0.31 mEq/kg body wt./24 hours in normal diet
Calcium ionized		4.15 – 5.20 mg/dl		
Copper		75 – 160 µg/dl	75 – 160 µg/dl	0 - 32 µg/24 hours
Zinc		50 – 150 µg/dl		
Iron		65 – 170 µg/dl		
Iron binding capacity		250 – 450 µg/dl		
Lead		96 – 106 mEq/l	96 – 106 mEq/l	0 – 120 µg/24 hours
Magnesium		1.7 – 2.6 mg/dl	1.7 – 2.6 mg/dl	
Ferritin		Male : 30 – 300 ng/ml Female : 20 – 120 ng/ml Children : 7 – 140 ng/ml		
Hepatoglobin		40 – 170 mg of Hb binding capacity		
Phenylalanine		3 mg/dL		
Phosphorus				
Adult		3 – 5.0 mg/dl		0.8 – 2.0 g/24 hours
Child		3.5 – 5.0 mg/dl		
Ceruloplasmin				
Males		36 ± 5.6 mg/dl		
Females		40.9 ± 6.8 mg/dl		
Serolonin	5 – 20 µg/dl			
Transferrin		200 – 400 mg/dl		
Aldolase		Up to 7.6 U/l		
Amylase		25 – 125 U/l		25 – 260 somogyl unit/hr
Acid phosphatase		< 4U/l		
Acid phosphatase, prostatic		< 1.2 U/l		
Alkaline phosphatase				
Adult		20 – 70 U/l		
Child		20 – 150 U/l		
Glucose-6-phosphate dehydrogenase		12.1 ± 2.09 U/gm Hb		
Lipase		0 – 160 U/l		

Constituents	Blood	Serum	Plasma	Urine
Creatinine Phosphokinase (total)		170 U/l		
Creatinine phosphokinase isoenzymes				
MM fraction		94 – 95%		
MB fraction		0 – 5%		
BB fraction		0 – 2%		
γ-glutamyl transpeptidase				
Male		9 – 50 U/l		
Female		8 – 40 U/l		
SGOT		8 – 40 karmen units/dl		
SGPT		0 – 40 karmen units/dl		
LDH isoenzymes				
LDH_1		20 – 34%		
LDH_2		28 – 41%		
LDH_3		15 – 25%		
LDH_4		8 – 16%		
LDH_5		6 – 15%		
Gastrin		< 200 pg/ml		
Immunoglobulin, Quantitative (Adults)				
IgG		570 – 1920 mg/dl		
IgE		< 0.025 mg/dl		
IgA		70 – 400 mg/dl		
IgM				
Males		30 – 250 mg/dl		
Females		30 – 300 mg/dl		
IgD		0 – 8 mg/dl		

Hormones	Serum	Plasma	Urine
ACTH	15 – 70 pg/ml		
Insulin (Fasting)		6 – 126 μU/ml	
Hypoglycemia		< 5 μU/ml	
Glucagon (fasting)	20 – 100 pg/ml		
Cartisol	5 – 25 μg/dl		20 – 100 μg/24 hours
FSH	Males : 6 – 18 μU/dl Females : 5 – 20 μU/dl (Follicular phase) 5 – 15 μU/dL (Luteal phase) 50 μU/dL (Postmeno-pausal)		
Parathyroid hormone	150 – 300 pg/ml		
Calcitonin	Not detectable		
Catecholamines			100 μg/24 hours (total) Epinephrine : 10 μg/24 hours Norepinephrine : 100 μg/24 hours
Metanephrine			1.3 mg/24 hours
Thyroid stimulating hormone (TSH)	< 10 μU/ml		
Thyroxine (T_4)	Total : 4.5 – 11.5 μg/dl Free : 0.8 – 2.4 μg/dl		
Thyroxine index (free)	1 – 4 mg/dl		
Tri-iodothyronine (T_3)	70 – 190 mg/dl		
Thyroxin-binding globulin capacity	15 – 25 μg T_4/dl		
Radioactive iodine ^{131}I Uptake (RAIU)	5 – 25% at 24 hours (varies with iodine intake)		
Renin activity			
Normal diet		Supine : 1.1 ± 0.8 mg/dl Upright : 1.9 ± 1.7 mg/dl	
Low-Sodium diet		Supine : 2.7 ± 1.8 mg/dl Upright : 6.6 ± 2.5 mg/dl	
Diuretic and low-sodium diet		10.0 ± 3.7 mg/dL	
Testosterone			
Total plasma bound			
Adolescent males		100 mg/dl	
Adult males		30 – 100 mg/dl	
Adult females		25 – 90 mg/dl	
Unbound			
Adult males		3 – 24 mg/dl	
Adult females		0.09 – 1.30 mg/dl	
17-Ketogenic steroids			
under 8 years			0 – 2 mg/24 hours
Adolescent			0 – 18 mg/24 hours
Adult males			8 – 18 mg/24 hours
Adult females			5 – 15 mg/24 hours

Hormones		Serum	Plasma	Urine
Estrogens				
Total :	Males			4 – 25 µg/24 hours
	Females			5 – 100 µg/24 hours
Estriol :	Males			1 – 11 µg/24 hours
	Females			0 – 65 µg/24 hours
Estradiol :	Males			0 – 6 µg/24 hours
	Females			0 – 14 µg/24 hours
Estrone :	Males			3 – 8 µg/24 hours
	Females			4 – 31 µg/24 hours
Estriocholandone (Female)				1.2 µg/24 hours
Norepinephrine			< 0.5 µg/l	100 µg/24 hours
Epinephrine			< 0.1 ng/ml	
Leuteinizing hormone (LH)		Male : 1 – 15 mIU/ml Female : < 1 – 30 mIU/ml Castrate : > 30 mIU/ml		
Antidiuretic hormone (ADH, Vasopressin)			0 – 2 pg/ml	
Insulin – like growth factor-1 (IGF – 1)			400 – 2000 units/l	
Progesterone				
Follicular phase		0.2 – 1.5 mg/ml		
Luteal phase		6 – 32 ng/ml		
Pregnancy		> 24 ng/ml		
Vanillylmandelic acid (VMA)				1 – 8 µg/24 hours

RENAL FUNCTION TESTS

Specific gravity of urine	:	1.003–1.030
Maximal glucose reabsorptive capacity	:	Males : 300–450 mg/min.
		Females : 250–350 mg/min.
Inulin clearance (GFR)	:	Males : 110–150 ml/min.
		Females : 105–132 ml/min.
Creatinine clearance $\left(\frac{U_{Cr} \times V}{P_{Cr}}\right)$	:	Males : 130 ± 20 ml/min.
		Females : 120 ± 15 ml/min.
U_{Cr} = Urine Creatinine (mg/dl)		
P_{Cr} = Plasma Creatinine (mg/dl)		
V = Urine volume (ml/24 hr)		
P-Aminohippurate (PAH) clearance (RPF)	:	Males : 560–830 ml/min.
		Females : 490–700 ml/min
Maximal PAH excretory capacity (Tm PAH)	:	80–90 mg/min
Filtration fraction (FF)	:	Males : 17–21%
[FF = GFR/RPF]		Females : 17–23%
Anion gap	:	10 ± 2 mEq/dl
$Na^+ - (HCO_3^- + Cl^-)$		
Osmality	:	292 – 297 m osm/l

$$\text{Osmality (Serum)} = 2\ Na^+ \text{ (mEq/L)} + \frac{\text{BUN(mg/dl)}}{2.8} + \frac{\text{Glucose(mg/dl)}}{18}$$

Bicarbonate (HCO_3^-) deficit = Body weight (kg) × 0.4 (desired HCO_3^- – observed HCO_3^-)

$$\text{Renal plasma flow (RPF)} = \frac{U_{Pah} \times V}{PPah}$$

: Males : 700 ± 100 ml/min.

Females : 600 ± 100 ml/min.

where U_{Pah} = Urine para-aminohippuric acid

V = Urine volume/24 hr

PPah = Plasma para-aminohippuric acid

$$\text{Fractional excretion of sodium (FE } Na^+) = \frac{(U/P)Na^+}{(U/P)Cr} \times 100$$

[Adapted from Current Medical Diagnosis and Treatment, 2003]

$$\text{Milliequivalent per litre (mEq/l)} = \frac{\text{mg per 100 ml} \times 10}{\text{Equivalent weight}}$$

or,

mmol/l

INDEX

A

C

D

H

I

J

K

L

M

N

O

Q

R

S

T

U

V

W

X

Z